Instructor's Annotated Edition

P9-DFS-605

Opt. 3

Exploring Intermediate Algebra

A GRAPHING APPROACH

Instructor's Annotated Edition

Exploring Intermediate Algebra

A GRAPHING APPROACH

Richard N. Aufmann
Palomar College

Joanne S. Lockwood
Plymouth State College

Laurie Boswell
Plymouth State College

HOUGHTON MIFFLIN COMPANY   Boston   New York

Publisher: *Jack Shira*
Senior Sponsoring Editor: *Lynn Cox*
Senior Development Editor: *Dawn Nuttall*
Assistant Editor: *Lisa Pettinato*
Project Editor: *Merrill Peterson*
Senior Production/Design Coordinator: *Carol Merrigan*
Manufacturing Manager: *Florence Cadran*
Senior Marketing Manager: *Ben Rivera*

Cover image: 2003 © Sergio Spada/Images.com, Inc

PHOTOGRAPH CREDITS: *p. 1* Paul Conklin/PhotoEdit, Inc.; *p. 14* David Young-Wolff/PhotoEdit; *p. 91* David Frazier/The Image Works; *p. 108* Becky Luigart-Stayner /CORBIS; *p. 131* Craig Lovell/COR-BIS; *p. 145* David Lees/CORBIS; *p. 146* Cathy Charles/PhotoEdit; *p. 157* Bill Lai/The Image Works; *p. 177* Cathy Melleon Resources/PhotoEdit; *p. 178* Archivo Iconografico, S.A./ CORBIS; *p. 179* Morandi/Granata/The Image Works; *p. 233* Tim David/CORBIS; *p. 271* David Young-Wolff/PhotoEdit; *p. 272* Michael Johnson; *p. 274* Bettmann/CORBIS; *p. 307* David Grossman/The Image Works; *p. 358* AP/Wide World Photos; *p. 375* AP/Wide World Photos; *p. 463* Tom Sanders/CORBIS; *p. 544* Donald C. Johnson/CORBIS; *p. 547* Tom Wagner/ CORBIS SABA; *p. 551* James Marshall/CORBIS; *p. 563* ER Productions/CORBIS; *p. 586* Bill Bachman/PhotoEdit, Inc.; *p. 600* Charles Krebs/CORBIS; *p. 619* CORBIS; *p. 621* Roy Morsch/CORBIS; *p. 628* Richard Glover, Ecoscene/CORBIS; *p. 652* AFP/Corbis; *p. 655* Joseph Sohm; ChromoSohm Inc./CORBIS; *p. 701* Dallas and John Heaton/CORBIS; *p. 804* AP/Wide World Photos.

Copyright © 2004 by Houghton Mifflin Company. All rights reserved.

No part of this work may be reproduced or transmitted in any form or by any means, electronic or mechanical, including photocopying and recording, or by any information storage or retrieval system without the prior written permission of Houghton Mifflin Company unless such copying is expressly permitted by federal copyright law. Address inquiries to College Permissions, Houghton Mifflin Company, 222 Berkeley Street, Boston, MA 02116-3764.

Printed in the U.S.A.

Library of Congress Control Number: 2002109359

Student Text ISBN: 0-618-15696-8
Instructor's Annotated Edition ISBN: 0-618-15697-6

123456789-VH-07 06 05 04 03

CONTENTS

CHAPTER **4**

Linear Functions 233

CHAPTER **5**

Systems of Linear Equations and Inequalities 307

CHAPTER **7**

Rational Expressions and Equations   463

CHAPTER **10**

Exponential and Logarithmic Functions   701

Additional Topics in Algebra   803

APPENDIX **A**   # Keystroke Guide for the TI-83 and TI-83 Plus   845

P R E F A C E

Exploring Intermediate Algebra: A Graphing Approach is a new text designed to help students make connections between mathematics and its applications. Our goal is to develop a student's mathematical skills through appropriate use of applications and to use technology to establish links between abstract mathematical concepts and visual or concrete representations. In response to the more prevalent use of technology in the intermediate algebra curriculum, we have written this text as graphing calculator dependent. Although any calculator can be used, we have used the TI-83/TI-83 Plus in many of the examples.

Our hallmark *interactive approach*, which encourages students to practice a skill or concept as it is presented and get immediate feedback, is also highlighted in this text. Each section contains one or more sets of matched-pair Example/You Try It examples. The numbered example in each set is worked out; the second example, the *You Try It*, is for the student to work. By solving this problem, the student actively practices concepts as they are presented in the text. There are complete worked-out solutions to the You Try It problems in an appendix. Students can compare their solution to the solution in the appendix and thereby obtain immediate feedback on the concept.

Through the use of applications, we demonstrate to students that mathematics has a vast array of tools that can be used to solve meaningful problems. Modeling, analytic representation, and verbal representations of problems and their solutions are encouraged. We have also integrated numerous data analysis exercises throughout the text and encourage students to use technology to assist them in deriving meaningful conclusions about the data.

To promote and support problem solving, we offer students a systematic procedure to solve application problems. For each application example, we take students through a four-step process that asks the student to **State the goal**, **Devise a strategy**, **Solve the problem**, and **Check your work**. For the corresponding You Try It, we ask students to follow that procedure. To reinforce this process, the solution to the You Try It in the appendix demonstrates how the four-step approach could be used to solve the problem.

In some cases, we have incorporated into an exercise a writing component that asks the student to write a sentence explaining the meaning of an answer in the context of the problem. Additional writing exercises are integrated throughout every exercise set. These exercises ask students to make a conjecture based on some given facts, restate a concept in their own words, provide a written answer to a question, or research a topic and write a short report.

We have paid special attention to the standards suggested by AMATYC and have made a serious attempt to incorporate those standards in the text. Problem solving, critical analysis, the function concept, connecting mathematics to other disciplines through applications, multiple representations of concepts, and the appropriate use of technology are all integrated within this text. Our goal is to provide students with a variety of analytic tools that will make them more effective quantitative thinkers and problem solvers.

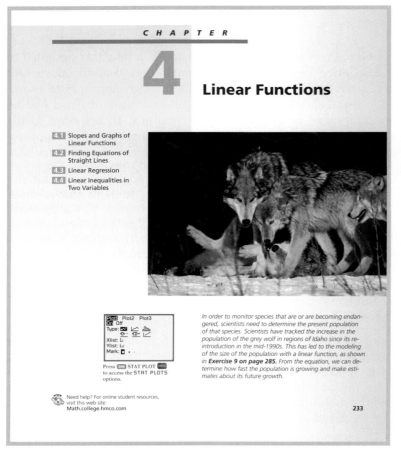

■ Chapter Opener

Each chapter begins with a **Chapter Opener** that illustrates a specific application of a concept from the chapter. There is a reference to a particular exercise in the chapter that asks the student to solve a problem related to the chapter opener topic.

The 🌐 at the bottom of the page lets students know of additional online resources at **math.college.hmco.com/ students**.

■ Prep Test and Go Figure

Prep Tests occur at the beginning of each chapter and test students on previously covered concepts that they must understand in order to succeed in the upcoming chapter. Answers are provided in the Answer Appendix. Section references are also provided for students who need to review specific concepts.

The **Go Figure** problem that follows the Prep Test is a playful puzzle problem designed to engage students in problem solving.

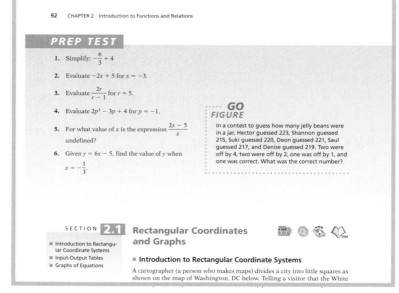

❓ QUESTION Why does it not make sense for the domain of $f(x) = -0.05x + 18$, discussed above, to exceed 360?

EXAMPLE 1

Suppose a 20-gallon gas tank contains 2 gallons when a motorist decides to fill up the tank. The gas pump fills the tank at a rate of 0.08 gallon per second. Find a linear function that models the amount of fuel in the tank x seconds after fueling begins.

Solution Because there are 2 gallons of gas in the tank when fueling begins (at $x = 0$), the y-intercept is $(0, 2)$.

The slope is the rate at which fuel is being added to the tank. Because the amount of fuel in the tank is increasing, the slope is positive and we have $m = 0.08$.

To find the linear function, replace m and b in $f(x) = mx + b$ by their values.

$f(x) = mx + b$

$f(x) = 0.08x + 2$ • Replace m by 0.08; replace b by 2.

The linear function is $f(x) = 0.08x + 2$, where $f(x)$ is the number of gallons of fuel in the tank x seconds after fueling begins.

YOU TRY IT 1

The boiling point of water at sea level is 100°C. The boiling point decreases 3.5°C for each 1-kilometer increase in altitude. Find a linear function that gives the boiling point of water as a function of altitude.

Solution See page S14.

■ Find Equations of Lines Using the Point–Slope Formula

For each of the previous examples, the known point on the graph of the linear function was the y-intercept. This information enabled us to determine b for the linear function $f(x) = mx + b$. In some instances, a point other than the y-intercept is given. In this case, the *point–slope formula* is used to find the equation of the line.

Point–Slope Formula of a Straight Line

Let $P_1(x_1, y_1)$ be a point on a line, and let m be the slope of the line. Then the equation of the line can be found using the point–slope formula

$$y - y_1 = m(x - x_1)$$

❓ ANSWER If $x > 360$, then $f(x) < 0$. This would mean that the tank contained negative gallons of gas. For instance, $f(400) = -2$.

page 257

■ An Interactive Approach

Exploring Intermediate Algebra: A Graphing Approach uses an interactive approach that provides the student with an opportunity to try a skill as it is presented. Each section contains one or more sets of matched-pair examples. The first example in each set is worked out; the second example, called *You Try It*, is for the student to work. By solving this problem, the student actively practices concepts as they are presented in the text.

There are <u>complete worked-out</u> solutions to these problems in an appendix. By comparing their solution to the solution in the appendix, students obtain immediate feedback on, and reinforcement of, the concept.

YOU TRY IT 1   Let x represent the number of kilometers above sea level and y represent the boiling point of water.

Since the boiling point of water at sea level is 100°C, $x = 0$ when $y = 100$. The y-intercept is $(0, 100)$.

The slope is the decrease in the boiling point per kilometer increase in altitude.

Since the boiling point decreases 3.5°C per 1-kilometer increase in altitude, the slope is negative; $m = -3.5$.

To find the linear function, replace m and b in $f(x) = mx + b$ by their values.

$f(x) = mx + b$

$f(x) = -3.5x + 100$

The linear function is $f(x) = -3.5x + 100$, where $f(x)$ is the boiling point of water x kilometers above sea level.

page S14

■ Question/Answer

At various places during a discussion, we ask the student to respond to a **Question** about the material being read. This question encourages the reader to pause and think about the current discussion and to answer the question. To make sure the student does not miss important information, the **Answer** to the question is provided as a footnote at the bottom of the page.

■ *AIM for Success* Student Preface

This "how to use this book" student preface explains what is required of a student to be successful in mathematics and how this text has been designed to foster student success through the Aufmann Interactive Method (AIM). *AIM for Success* can be used as a lesson on the first day of class or as a project for students to complete to strengthen their study skills. There are suggestions for teaching this lesson in the *Instructor's Resource Manual* and on the *Class Prep* CD.

AIM FOR SUCCESS

Welcome to *Exploring Intermediate Algebra: A Graphing Approach*. As you begin this course we know two important facts: (1) We want you to succeed. (2) You want to succeed. To do that requires an effort from each of us. For the next few pages, we are going to show you what is required of you to achieve that success and how you can use the features of this text to be successful.

Motivation   One of the most important keys to success is motivation. We can try to motivate you by offering interesting or important ways mathematics can benefit you. But in the end, the motivation must come from you. On the first day of class, it is easy to be motivated. Eight weeks into the term, it is harder to keep that motivation.

TAKE NOTE

Motivation alone will not lead to success. For instance, suppose a person who cannot swim is placed in a boat, taken out to the middle of a lake, and then thrown overboard. That person has a lot of moti-

To stay motivated, there must be outcomes from this course that are worth your time, money, and energy. List some reasons why you are taking this course. Do not make a mental list—actually write them out.

_____

_____

_____

page xxv

PROGRAM SOLVING

YOU TRY IT 3

Find the equation of the line that passes through $P(4, 3)$ and whose slope is undefined.

Solution   See page S14.

EXAMPLE 4

Judging on the basis of data from the Kelley Blue Book, the value of a certain car decreases approximately $250 per month. If the value of the car 2 years after it was purchased was $14,000, find a linear function that models the value of the car after x months of ownership. Use this function to find the value of the car after 3 years of ownership.

State the goal.   Find a linear model that gives the value of the car after x months of ownership. Then use the model to find the value of the car after 3 years.

Devise a strategy.   Because the function will predict the value of the car, let y represent the value of the car after x months.

Then $y = 14,000$ when $x = 24$ (2 years is 24 months).

The value of the car is decreasing $250 per month. Therefore, the slope is -250.

Use the point–slope formula to find the linear model.

To find the value of the car after 3 years (36 months), evaluate the function when $x = 36$.

Solve the problem.
$$y - y_1 = m(x - x_1)$$
$$y - 14,000 = -250(x - 24)$$
$$y - 14,000 = -250x + 6000$$
$$y = -250x + 20,000$$

A linear function that models the value of the car is $V(x) = -250x + 20,000$.

$$f(x) = -250x + 20,000$$
$$f(36) = -250(36) + 20,000 \quad \bullet \text{ Evaluate the function at } x = 36.$$
$$= -9000 + 20,000$$
$$= 11,000$$

The value of the car is $11,000 after 36 months of ownership.

Check your work.   An answer of $11,000 seems reasonable. This value is less than $14,000, the value of the car after 2 years.

The graph of the function is shown at the right. Pressing **TRACE** 24 shows that the or-

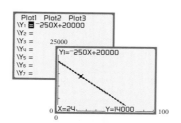

■ Problem-Solving Strategies

The text features a carefully developed approach to problem solving. Students are encouraged to develop their own strategies—drawing diagrams, for example, or writing out the solution steps in words—as part of their solution to a problem. In each case, model solutions consistently encourage students to

State the goal.
Devise a strategy.
Solve the problem.
Check your work.

Having students describe a strategy is a natural way to incorporate writing into the math curriculum.

---

page 259

EXAMPLE 5

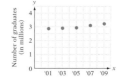 In 2002, the computer service America Online offered its customers the option of paying $23.90 per month for unlimited use. Another option was a rate of $4.95 per month with 3 free hours plus $2.50 per hour thereafter (*Source:* AOL web site, March 2002). How many hours per month can you use this second option if it is to cost you less than the first option? Round to the nearest whole number.

State the goal. The goal is to determine how many hours you can use the second option ($4.95 per month plus $2.50 per hour after the first 3 hours) if it is to cost you less than the first option ($23.90 per month).

Devise a strategy. Let x represent the number of hours per month you use the service. Then $x - 3$ represents the number of hours you would be paying $2.50 per hour for service under the second option.

Cost of first plan: 23.90
Cost of second plan: $4.95 + 2.50(x - 3)$

Write and solve an inequality that expresses that the second plan is less expensive (less than) the first plan.

Solve the problem.
$$4.95 + 2.50(x - 3) < 23.90$$
$$4.95 + 2.50x - 7.50 < 23.90$$
$$2.50x - 2.55 < 23.90$$
$$2.50x < 26.45$$
$$x < 10.58$$

The greatest whole number less than 10.58 is 10.

In order for the second option to cost you less than the first option, you can use the service for up to 10 hours per month.

Check your work. One way to check racy of our work. We can also make a p

page 203

Applications

One way to motivate an interest in mathematics is through applications. Wherever appropriate, the last portion of a section presents applications that require the student to use problem-solving strategies, along with the skills covered in that section, to solve practical problems. This carefully integrated applied approach generates student awareness of the value of algebra as a real-life tool.

Applications are taken from many disciplines, including agriculture, business, carpentry, chemistry, construction, Earth science, education, manufacturing, nutrition, real estate, and sociology.

Fit a Line to Data

8. 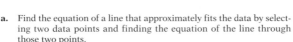 *Demography* The table and scatter diagram show the projected number of U.S. high school graduates, in millions (*Source:* National Center for Education Statistics).

Year, x	'01	'03	'05	'07	'09
Number of graduates, y (in millions)	2.90	2.98	2.99	3.13	3.25

a. Find the equation of a line that approximately fits the data by selecting two data points and finding the equation of the line through those two points.
b. What does the slope of your line mean in the context of the problem?
c. What does the y-intercept mean in the context of the problem?

9. 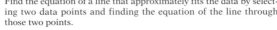 *Zoology* The table and scatter diagram show the increase in the grey wolf population in regions of Idaho after that species's re-introduction in the mid-1990s (*Source:* U.S. Fish and Wildlife Service).

Year, x	'95	'96	'97	'98	'99	'00
Number of wolves, y	14	42	71	114	141	185

a. Find the equation of a line that approximately fits the data by selecting two data points and finding the equation of the line through those two points.
b. What does the slope of your line mean in the context of the problem?

page 285

■ Real Data

Real-data examples and exercises, identified by , ask students to analyze and solve problems taken from actual situations. Students are often required to work with tables, graphs, and charts drawn from a variety of disciplines.

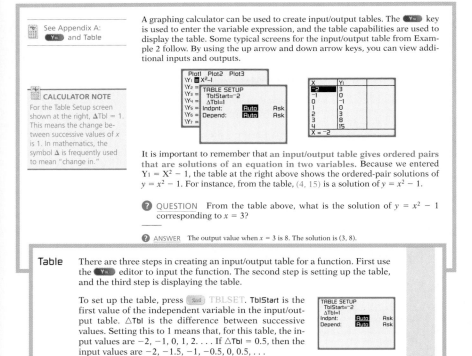

SECTION 3.4 Inequalities in One Variable **201**

EXAMPLE 2

Solve and graph the solution set of $7x < -21$. Write the solution set in set-builder notation and interval notation.

ALGEBRAIC SOLUTION

$7x < -21$

$\dfrac{7x}{7} < \dfrac{-21}{7}$ • Divide each side of the inequality by 7. Because 7 is a positive number, the inequality symbol is not reversed.

$x < -3$ • Simplify.

The solution set is $\{x \mid x < -3\}$ or $(-\infty, -3)$.

GRAPHICAL CHECK

YOU TRY IT 2

Solve and graph the solution set of $-3x \geq 6$. Write the solution set in set-builder notation and interval notation.

Solution See page S11.

page 201

See Appendix A:
Y= and Table

A graphing calculator can be used to create input/output tables. The Y= key is used to enter the variable expression, and the table capabilities are used to display the table. Some typical screens for the input/output table from Example 2 follow. By using the up arrow and down arrow keys, you can view additional inputs and outputs.

CALCULATOR NOTE
For the Table Setup screen shown at the right, ΔTbl = 1. This means the change between successive values of x is 1. In mathematics, the symbol Δ is frequently used to mean "change in."

It is important to remember that **an input/output table gives ordered pairs that are solutions of an equation in two variables.** Because we entered $Y_1 = X^2 - 1$, the table at the right above shows the ordered-pair solutions of $y = x^2 - 1$. For instance, from the table, $(4, 15)$ is a solution of $y = x^2 - 1$.

? QUESTION From the table above, what is the solution of $y = x^2 - 1$ corresponding to $x = 3$?

? ANSWER The output value when $x = 3$ is 8. The solution is $(3, 8)$.

Table There are three steps in creating an input/output table for a function. First use the Y= editor to input the function. The second step is setting up the table, and the third step is displaying the table.

To set up the table, press 2nd TBLSET. TblStart is the first value of the independent variable in the input/output table. ΔTbl is the difference between successive values. Setting this to 1 means that, for this table, the input values are $-2, -1, 0, 1, 2 \ldots$. If ΔTbl = 0.5, then the input values are $-2, -1.5, -1, -0.5, 0, 0.5, \ldots$

pages 96 and 855

■ Integration of Technology

We have used a TI-83/TI-83 Plus graphing calculator throughout the text to help students make connections between abstract mathematical concepts and a concrete representation provided by technology. This is one way in which students are encouraged to think about and use multiple representations of a concept.

For appropriate examples within the text, we have provided both an algebraic solution and a graphical representation of the solution. This enables the student to visualize the algebraic solution. For other graphing calculator examples, an algebraic verification of a graphing calculator solution is presented. This promotes the link between the algebraic and graphical components of a solution.

■ Calculator Note

These margin notes provide suggestions for using a calculator in certain situations.

■ Graphing Calculator Appendix A

A TI-83/TI-83 Plus graphing calculator appendix contains some of the common calculator keystrokes that are used in the text. Students are referred to this appendix by appropriately placed *See Appendix A* notes indicating which calculator feature is in use.

In addition, a convenient **calculator bookmark** containing a synopsis of major calculator functions is included in the front of the text. The calculator bookmark can be removed from the text and used to mark the student's current lesson.

STUDENT PEDAGOGY

This text was designed as a resource for students. Special emphasis was given to readability and effective pedagogical use of color to highlight important words and concepts.

■ Icons

The 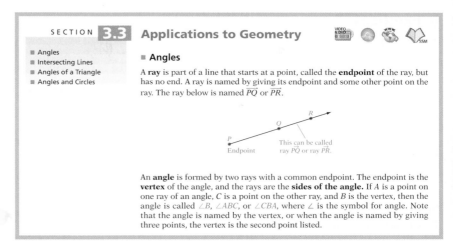, ○, 🌐, ⌂_SSM at each objective head remind students of the many and varied additional resources available for each objective.

■ Key Terms and Concepts

Key terms, in bold, emphasize important terms.

page 183

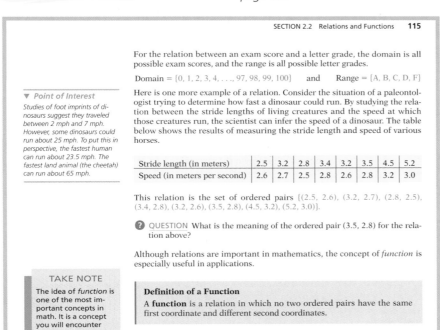

Key Concepts are presented in green boxes in order to highlight these important concepts and to provide for easy reference.

■ Point of Interest

These margin notes contain interesting sidelights about mathematics, its history, or its application.

■ Take Note

These margin notes either alert students to a point requiring special attention or amplify the concept under discussion.

page 115

■ Annotated Examples

Examples indicated by ➡ use annotations in blue to explain what is happening in key steps of the complete, worked-out solutions.

As shown in the example that follows, the Addition Property of Inequalities applies to variable terms as well as to constants.

➡ Solve $3x - 4 \leq 2x - 1$. Write the solution set in interval notation.

$$3x - 4 \leq 2x - 1$$
$$3x - 2x - 4 \leq 2x - 2x - 1 \qquad \text{• Subtract } 2x \text{ from each side of the inequality.}$$
$$x - 4 \leq -1 \qquad \text{• Simplify.}$$
$$x - 4 + 4 \leq -1 + 4 \qquad \text{• Add 4 to each side of the inequality.}$$
$$x \leq 3 \qquad \text{• Simplify.}$$

The solution set is $(-\infty, 3]$. ⬅

page 199

Exercises

The exercise sets of *Exploring Intermediate Algebra: A Graphing Approach* emphasize skill building, skill maintenance, and applications. Concept-based writing or developmental exercises have been integrated with the exercise sets.

Icons identify appropriate writing ✎ , group 🧑 , and data analysis 🥧 exercises.

Before each exercise set are **Topics for Discussion**, which ask students to discuss or write about a concept presented in the section. Used as oral exercises, these can lead to interesting classroom discussions.

page 284

4.3 EXERCISES

Topics for Discussion

1. What does it mean to "fit a line to data"?

2. What is a regression line?

3. What might be the purpose of determining the regression equation for a set of data?

4. Determine whether the statement is always true, sometimes true, or never true.
   a. For linear regression to be performed on two-variable data, the points corresponding to the data values must lie on a straight line.
   b. If the correlation coefficient is equal to 1, the data exactly fit the regression line. If the correlation coefficient is equal to -1, the data do not fit the regression line.
   c. A scatter diagram is a graph of ordered pairs.

58. **Sound** The distance sound travels through air when the temperature is 75°F can be approximated by $d(t) = 1125t$, where d is the distance, in feet, that sound travels in t seconds. What is the slope of this function? What is the meaning of the slope in the context of this problem?

59. **Biology** The distance that a homing pigeon can fly can be approximated by $d(t) = 50t$, where $d(t)$ is the distance, in miles, flown by the pigeon in t hours. What is the slope of this function? What is the meaning of the slope in the context of this problem?

60. **Construction** The American National Standards Institute (ANSI) states that the slope for a wheelchair ramp must not exceed $\frac{1}{12}$.
   a. Does a ramp that is 6 inches high and 5 feet long meet the requirements of ANSI?
   b. Does a ramp that is 12 inches high and 170 inches long meet the requirements of ANSI?

page 250

Applying Concepts

In Exercises 25 to 36, you were asked to find the number in the domain of a function for which the output was the given number. Frequently in mathematics, we express directions such as these by using a combination of words and symbols. For instance, in Exercise 25, we could have written "Find the value of a in the domain of $f(x) = 3x - 4$ for which $f(a) = 5$." Recall that $f(a)$ is the output of a function for a given input a. Thus $f(a) = 5$ says the output is 5 for an input of a. We will use this terminology in Exercises 70 to 75.

70. Find the value of a in the domain of $f(x) = 2x + 5$ for which $f(a) = -3$

71. Find the value of a in the domain of $g(x) = -2x - 1$ for which $g(a) =$

72. Find the value of a in the domain of $h(x) = \frac{3}{2}x - 2$ for which $h(a) = 4$

73. Find the value of a in the domain of $F(x) = -\frac{5}{4}x - 1$ for which $F(a) =$

74. For $f(x) = x^3 - 4x - 1$, how many different values, a, in the domain satisfy the condition that $f(a) = 1$? (You do not have to find the values, just determine number of possible values for a.)

75. For $f(x) = 0.01(x^4 - 49x^2 + 36x + 252)$, how many different values, a, in the domain of f satisfy the condition that $f(a) = 1$? (You do not have to find the values; just determine number of possible values for a.)

In Exercises 76 to 79, find two numbers in the domain of the functions for which the values of the functions are equal. *Hint:* Apply the method of finding the point of intersection of two graphs twice, once to each point of intersection.

76. $f(x) = x^2 + 4x - 1, g(x) = 3x + 5$

77. $f(x) = x^2 - x - 1, g(x) = -3x + 2$

78. $f(x) = 2x - 7, g(x) = x^2 - 4x - 2$

79. $f(x) = x + 4, g(x) = x^2 + 3x - 4$

EXPLORATION 🧑

1. *Calculator Viewing Windows* A graphing calculator screen consists of **pixels**,[2] which are small rectangles of light that can be turned on or off. When a calculator draws the graph of an equation, it is turning on the pixels that represent the ordered-pair solutions of the equation. The jagged appearance of the graph is a consequence of the solutions being approximations; the pixel nearest the ordered pair is turned on.

[2] Digital cameras are rated in pixels; some cameras have over 3 million pixels. A typical graphing calculator has just over 5800 pixels. Because of this, the graphs on these screens are not so sharp as an image in a digital camera.

Included in each exercise set are **Applying Concepts**, which present extensions of topics, require analysis, or offer challenge problems.

Explorations are extensions of a concept presented in the section. These Explorations can be used in cooperative learning situations or as extra-credit assignments.

CHAPTER **4** SUMMARY

Key Terms

coefficient of determination [p. 281]
constant function [p. 243]
correlation coefficient [p. 281]
half-plane [p. 294]
linear function [p. 235]
linear inequality in two variables
  [p. 294]
linear regression [p. 280]
line of best fit [p. 278]

negative reciprocals [p. 265]
parallel lines [p. 262]
perpendicular lines [p. 264]
regression line [p. 280]
scatter diagram [p. 277]
slope [p. 236]
solution set of a linear inequality in
  two variables [p. 294]

Essential Concepts

Slope of a Line
Let $P_1(x_1, y_1)$ and $P_2(x_2, y_2)$ be two points on a line. Then the slope m of the line
through the two points is the ratio of the change in the y-coordinates to the
change in the x-coordinates. [p. 237]

$$m = \frac{\text{change in } y}{\text{change in } x} = \frac{y_2 - y_1}{x_2 - x_1}, x_1 \neq x_2$$

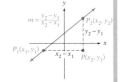

Slope–Intercept Form of a Straight Line
The equation $y = mx + b$ is called the slope–intercept form of a straight line.
The slope of the line is m, the coefficient of x. The y-intercept is $(0, b)$.

page 301

■ Chapter Summary

At the end of each chapter there is
a Chapter Summary that includes
Key Terms and **Essential Concepts** that
were covered in the chapter. These
chapter summaries provide a single
point of reference as the student pre-
pares for a test. Each concept is accom-
panied by the page number from the
lesson where the concept is introduced.

---

302 CHAPTER 4 Linear Functions

CHAPTER **4** REVIEW EXERCISES

1. Find the slope of the line that contains the
    points $(-1, 3)$ and $(-2, 4)$.

2. Find the slope of the line that passes through the
    points $(-6, 5)$ and $(-6, 4)$.

3. Graph the line that has slope $\frac{1}{2}$ and passes
    through the point $(-2, 4)$.

4. Graph $y = -\frac{2}{3}x + 4$ by using the slope and
    y-intercept.

5. Graph $x + 2y = -4$.

6. Graph $y = 3$.

■ Chapter Review Exercises

Review exercises are found at the end of
each chapter. These exercises are selected to
help the student integrate all of the topics
presented in the chapter.

■ Chapter Test

The Chapter Test exercises are designed to
simulate a possible test of the material in
the chapter.

CHAPTER **4** TEST

1. Find the slope of the line that contains the
    points $(-2, 6)$ and $(-1, 4)$.

2. Find the slope of the line that passes through the
    points $(-4, 3)$ and $(-8, 3)$.

■ Cumulative Review Exercises

Cumulative Review Exercises, which appear
at the end of each chapter (beginning with
Chapter 2), help students maintain skills
learned in previous chapters.

The answers to all Chapter Review Exercises,
all Chapter Test exercises, and all Cumulative
Review Exercises are given in the Answer
Section. Along with the answer, there is a
reference to the section that pertains to
each exercise.

◄ CUMULATIVE REVIEW EXERCISES

1. In how many different ways can a panel of four on–off switches be set if
    no two adjacent switches can be off?

2. Given the operation $a @ b = a + ab$, evaluate $(x @ y) @ z$ for $x = 2, y = 3$,
    $z = 4$.

3. Let $E = \{0, 5, 10, 15\}$ and $F = \{-10, -5, 0, 5, 10\}$. Find $E \cup F$ and $E \cap F$.

pages 302, 303, 305

The Instructor's Annotated Edition includes the following features:

Instructor Notes give suggestions for teaching concepts, warnings about common student errors, or historical notes. Next to each Example there is a special *Instructor Note* that refers the instructor to an even-numbered exercise in the exercise set that is similar to the Example and can be used as an additional in-class example.

Suggested Activities can be used in class to explore concepts that are presented. Some of these are alternative strategies for teaching the concept (such as using a discovery approach). Others are designed for cooperative learning activities. Some Suggested Activities refer to the **Student Activity Manual** that can accompany the student text. This manual contains all the worksheets that students need to complete such activities. This manual, including answers, is also found in the *Instructor's Resource Manual*.

---

page 186 (excerpt)

186   CHAPTER 3   First-Degree Equations and Inequalities

Suggested Activity
Suppose an obtuse angle of two intersecting lines is 135°. What are the measures of the three other angles?
[Answer: 45°, 135°, 45°]
Sketch a pair of nonperpendicular intersecting lines. Label one of the acute angles x. Label one of the obtuse angles $2x + 78°$. Find x.
[Answer: 34°]

Adjacent angles of intersecting lines are supplementary angles. This is summarized by the following equations:

$$m\angle x + m\angle y = 180°$$
$$m\angle y + m\angle z = 180°$$
$$m\angle z + m\angle w = 180°$$
$$m\angle w + m\angle x = 180°$$

The nonadjacent angles formed when two lines intersect are called **vertical angles.** For the intersecting lines m and n above, $\angle x$ and $\angle z$ are vertical angles; $\angle w$ and $\angle y$ are also vertical angles. Vertical angles have the same measure. Thus,

$$m\angle x = m\angle z$$
$$m\angle w = m\angle y$$

INSTRUCTOR NOTE
Exercise 28 can be used for a similar in-class example.

EXAMPLE 2

In the figure at the right, what is the value of x?

State the goal.   The goal is to find the value of x.

Devise a strategy.   The labeled angles are vertical angles of intersecting lines. Therefore, the angles are equal.

Solve the problem.
$$3x - 50 = x + 50$$
$$2x - 50 = 50$$
$$2x = 100$$
$$x = 50$$

The value of x is 50.

Check your work.   Replace x by 50 in the equation $3x - 50 = x + 50$ to ensure that the solution checks.

YOU TRY IT 2

The measures of two adjacent angles for a pair of intersecting lines are $2x + 20°$ and $3x + 50°$. Find the measure of the larger angle.

Solution   See page S10.   116°

INSTRUCTOR NOTE
The definitions which follow are not dependent upon the lines being parallel.

Parallel lines never meet; the distance between them is always the same. The symbol for parallel lines is $\parallel$.

A line that intersects two other lines at two distinct points is called a **transversal.** If the lines cut by a transversal are parallel lines and the transversal is perpendicular to the parallel lines, then all eight angles formed are right angles. For the diagram at the right, $p \parallel q$.

page 186

Next to many of the graphs or tables in the text, there is a that indicates that a Microsoft **PowerPoint®** **slide** of that figure is available. These slides (along with PowerPoint Viewer) are available on the *Class Prep CD* and can also be downloaded from our web site at **math.college.hmco.com/instructors.** These slides can also be printed as transparency masters.

A **Suggested Assignment** is provided for each section.

Answers for all exercises are provided.

page 251

---

page 245 / 251 (excerpt)

4.1 EXERCISES   Suggested Assignment: 9–95, odds

Topics for Discussion

1.   Give an example of a linear equation in two variables that **a.** is in slope–intercept form and **b.** is in standard form.
   Answers will vary. For example,   **a.** $y = 3x - 4$,   **b.** $2x - 5y = 10$.

2.   What is the formula for slope? Explain what each variable in the formula represents.
   $m = \dfrac{y_2 - y_1}{x_2 - x_1}$, $x_1 \neq x_2$, where m is the slope of a line, and (x_1, y_1) and (x_2, y_2) are two points on the line.

page 245

3.   Explain the difference between zero slope and no slope.
   The graph of a line with zero slope is horizontal. The graph of a line with no slope is vertical.

66.   *Sports*   Lois and Tanya start from the same place on a jogging course. Lois is jogging at 9 kilometers per hour, and Tanya is jogging at 6 kilometers per hour. The graphs below show the total distance traveled by each jogger and the total distance between Lois and Tanya. Which lines represent which distances?

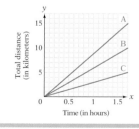

A - Lois
B - Tanya
C - Distance between

INSTRUCTOR RESOURCES

* *

Exploring Intermediate Algebra: A Graphing Approach has a complete set of teaching aids for the instructor.

Instructor's Annotated Edition   This edition contains a replica of the student text and additional items just for the instructor. These include *Instructor Notes, Suggested Activity* notes, *PowerPoint transparency icons,* and *Suggested Assignments.* Answers to all exercises are also provided.

Instructor's Resource Manual with Testing   The *Instructor's Resource Manual* includes a lesson plan for the *AIM for Success* student preface as well as the complete *Student Activity Manual,* with answers. The testing consists of a *Printed Test Bank* providing a printout of one example of each of the algorithmic items in *HM Testing* and four ready-to-use printed *Chapter Tests* per chapter.

Instructor's Solutions Manual   The *Instructor's Solutions Manual* contains worked-out solutions for all exercises in the text.

HM ClassPrep with HM Testing CD-ROM   *HM ClassPrep* contains a multitude of text-specific resources for instructors to use to enhance the classroom experience. These resources can be easily accessed by chapter or resource type and can also link you to the text's web site. *HM Testing* is our computerized test generator and contains a database of algorithmic test items as well as providing **on-line testing** and **gradebook** functions.

Instructor Text-specific website   The resources available on the *Class Prep CD* are also available on the instructor web site at math.college.hmco.com/instructors. Appropriate items are password protected. Instructors also have access to the student part of the text's web site.

STUDENT RESOURCES

* *

Student Activity Manual   This manual contains worksheets for the optional *Suggested Activities* referenced in the Instructor's Annotated Edition.

Student Solutions Manual   The *Student Solutions Manual* contains complete solutions to all odd-numbered exercises in the text.

Math Study Skills Workbook by Paul D. Nolting   This workbook is designed to reinforce skills and minimize frustration for students in any math class, lab, or study skills course. It offers a wealth of study tips and sound advice on note taking, time mangement, and reducing math anxiety. In addition, numerous opportunities for self-assessment enable students to track their own progress.

HM eduSpace® online learning environment   *eduSpace®* is a text-specific online learning environment that combines an algorithmic tutorial program with homework capabilities. Specific content is available 24 hours a day to enhance your understanding of your textbook.

HM mathSpace™ Tutorial CD-ROM   This tutorial CD ROM allows you to practice skills and review concepts as many times as necessary by providing algorithmically generating exercises and step-by-step solutions for practice.

SMARTHINKING™ live, online tutoring   Houghton Mifflin has partnered with SMARTHINKING to provide an easy-to-use and effective online tutorial service. **Whiteboard Simulations** and **Practice Area** promote real-time visual interaction.

Three levels of service are offered:

- **Text-specific Tutoring** provides real-time, one-on-one instruction with a specially qualified "e-structor."
- **Questions Any Time** allows students to submit questions to the tutor outside the scheduled hours and receive a reply within 24 hours.
- **Independent Study Resources** connect students with around-the-clock access to additional educational services, including interactive web sites, diagnostic tests and Frequently Asked Questions posed to SMARTHINKING e-structors.

Houghton Mifflin Instructional Videos and DVDs   This text offers text-specific videos and DVDs, hosted by Dana Mosely, covering all sections of the text and providing a valuable resource for further instruction and review. Next to every objective head, serves as a reminder that the objective is covered in a video/DVD lesson.

Student Text-specific web site   Online student resources can be found at this text's web site at **math.college.hmco.com/students.**

ACKNOWLEDGMENTS

• •

The authors would like to thank all the people who reviewed this manuscript and provided many valuable suggestions.

Dianne Adams, *Hazard Community College, KY*
Richard B. Basich, *Lakeland Community College, OH*
Laurette Blakey Foster, *Prairie View A&M University, TX*
Anne Haney
Sandeep H. Holay, *Southeast Community College-Lincoln, NE*
Glenn Hunt, *Riverside Community College, CA*
Jerry Kissick, *Portland Community College, OR*
Charyl Link, *Kansas Community College, KS*
Michelle Merriweather, *Southern Connecticut State University, CT*
Kim Nunn, *Northeast State Technical Community College, TN*
Scott Reed, *College of Lake County, IL*
Russ Reich, *Sierra Nevada College, NV*
Deana Richmond
Karl Zilm, *Lewis & Clark Community College, IL*

Special thanks to Christi Verity for her diligent preparation of the solutions manuals and for her contribution to the accuracy of the textbook.

AIM FOR SUCCESS

INSTRUCTOR NOTE
See the *Instructor's Resource Manual* or *Class Prep CD* for suggestions on how to teach this lesson.

Welcome to *Exploring Intermediate Algebra: A Graphing Approach.* As you begin this course we know two important facts: (1) We want you to succeed. (2) You want to succeed. To do that requires an effort from each of us. For the next few pages, we are going to show you what is required of you to achieve that success and how you can use the features of this text to be successful.

Motivation

One of the most important keys to success is motivation. We can try to motivate you by offering interesting or important ways in which mathematics can benefit you. But in the end, the motivation must come from you. On the first day of class, it is easy to be motivated. Eight weeks into the term, it is harder to keep that motivation.

For you to stay motivated, there must be outcomes from this course that are worth your time, money, and energy. List some reasons why you are taking this course. Do not make a mental list—actually write them out.

_____

_____

_____

TAKE NOTE

Motivation alone will not lead to success. For instance, suppose a person who cannot swim is placed in a boat, taken out to the middle of a lake, and then thrown overboard. That person has a lot of motivation to swim, but there is a high likelihood the person will drown without some help. Motivation gives us the desire to learn but is not the same as learning.

Although we hope that one of the reasons you listed was an interest in mathematics, we know that many of you are taking this course because it is required for graduation, because it is a prerequisite for a course you must take, or because it is required for your major. Although you may not agree that this course is necessary, it is! If you are motivated to graduate or complete the requirements for your major, then use that motivation to succeed in this course. Do not become distracted from your goal to complete your education!

Commitment

To be successful, you must make a commitment to succeed. This means devoting time to math so that you achieve a better understanding of the subject.

List some activities (sports, hobbies, talents such as dance, art, or music) that you enjoy and at which you would like to become better.

ACTIVITY	TIME SPENT	TIME WISHED SPENT

Thinking about these activities, put next to each activity the number of hours that you spend every week practicing that activity. Then indicate how many hours per week you would like to spend on each activity.

Whether you listed surfing or sailing, aerobics or restoring cars, or any other activity you enjoy, note how many hours a week you spend doing it. To succeed in math, you must be willing to commit the same amount of time. Success requires some sacrifice.

The "I Can't Do Math" Syndrome

There may be things you cannot do, such as lift a 2-ton boulder. You can, however, do math. It is much easier than lifting the 2-ton boulder. When you first learned the activities you listed above, you probably could not do them well. With practice, you got better. With practice, you will be better at math. Stay focused, motivated, and committed to success.

It is difficult for us to emphasize how important it is to overcome the "I Can't Do Math" Syndrome. If you listen to interviews of very successful athletes after a particularly bad performance, you will note that they focus on the positive aspect of what they did, not the negative. Sports psychologists encourage athletes to always be positive—to have a "Can Do" attitude. Develop this attitude toward math.

Strategies for Success

Textbook Reconnaissance Right now, do a 15-minute "textbook reconnaissance" of this book. Here's how:

First, read the table of contents. Do it in three minutes or less. Next, look through the entire book, page by page. Move quickly. Scan titles, look at pictures, notice diagrams.

A textbook reconnaissance shows you where a course is going. It gives you the big picture. That's useful because brains work best when going from the general to the specific. Getting the big picture before you start makes details easier to recall and understand later on.

Your textbook reconnaissance will work even better if, as you scan, you look for ideas or topics that are interesting to you. List three facts, topics, or problems that you found interesting during your textbook reconnaissance.

_____

_____

_____

The idea behind this technique is simple: It's easier to work at learning material if you know it's going to be useful to you.

Not all the topics in this book will be "interesting" to you. But that is true of any subject. Surfers find that on some days the waves are better than others; musicians find some music more appealing than other music; computer gamers find some computer games more interesting than others; car enthusiasts find some cars more exciting than others. Some car enthusiasts would rather have a completely restored 1957 Chevrolet than a new Ferrari.

Know the Course Requirements To do your best in this course, you must know exactly what your instructor requires. Course requirements may be stated in a *syllabus*, which is a printed outline of the main topics of the course, or they may be presented orally. When they are listed in a syllabus or on other printed pages, keep them in a safe place. When they are presented orally, make sure to take complete notes. In either case, it is important that you understand the requirements completely and follow them exactly. Be sure you can answer the following questions.

1. What is your instructor's name?
2. Where is your instructor's office?
3. At what times does your instructor hold office hours?
4. Besides the textbook, what other materials does your instructor require?
5. What is your instructor's attendance policy?
6. If you must be absent from a class meeting, what should you do before returning to class? What should you do when you return to class?
7. What is the instructor's policy regarding collection or grading of homework assignments?
8. What options are available if you are having difficulty with an assignment? Is there a math tutoring center?
9. Is there a math lab at your school? Where is it? What hours is it open?
10. What is the instructor's policy if you miss a quiz?
11. What is the instructor's policy if you miss an exam?
12. Where can you get help when studying for an exam?

Remember: Your instructor wants to see you succeed. If you need help, ask! Do not fall behind. If you are running a race and fall behind by 100 yards, you may be able to catch up, but it will require more effort than if you had not fallen behind.

Time Management We know that there are demands on your time. Family, work, friends, and entertainment all compete for your time. We do not want to see you receive poor job evaluations because you are studying math. However, it is also true that we do not want to see you receive poor math test scores because you devoted too much time to work. When several competing and important tasks require your time and energy, the only way to manage the stress of being successful at both is to manage your time efficiently.

Instructors often advise students to spend twice as much time outside of class studying as they spend in the classroom. Time management is important if you are to accomplish this goal and succeed in school. The following activity is intended to help you structure your time more efficiently.

List the name of each course you are taking this term, the number of class hours each course meets, and the number of hours you should spend studying each subject outside of class. Then fill in a weekly schedule like the one below. Begin by writing in the hours spent in your classes, the hours spent at work (if you have a job), and any other commitments that are not flexible with respect to the time that you do them. Then begin to write down commitments that are more flexible, including hours spent studying. Remember to reserve time for activities such as meals and exercise. You should also schedule free time.

We know that many of you must work. If that is the case, realize that working 10 hours a week at a part-time job is equivalent to taking a three-unit

TAKE NOTE

Besides time management, there must be realistic ideas of how much time is available. There are very few people who can *successfully* work full-time and go to school full-time. If you work 40 hours a week, take 15 units, spend the recommended study time given at the right, and sleep 8 hours a day, you will use over 80% of the hours in a week. That leaves less than 20% of the hours in a week for family, friends, eating, recreation, and other activities.

	Monday	Tuesday	Wednesday	Thursday	Friday	Saturday	Sunday
7–8 a.m.							
8–9 a.m.							
9–10 a.m.							
10–11 a.m.							
11–12 p.m.							
12–1 p.m.							
1–2 p.m.							
2–3 p.m.							
3–4 p.m.							
4–5 p.m.							
5–6 p.m.							
6–7 p.m.							
7–8 p.m.							
8–9 p.m.							
9–10 p.m.							
10–11 p.m.							
11–12 a.m.							

class. If you must work, consider letting your education progress at a slower rate to allow you to be successful at both work and school. There is no rule that says you must finish school in a certain time frame.

Schedule Study Time As we encouraged you to do by filling out the time management form above, schedule a certain time to study. You should think of this time the way you would the time for work or class—that is, reasons for missing study time should be as compelling as reasons for missing work or class. "I just didn't feel like it" is not a good reason to miss your scheduled study time.

Although this may seem obvious, list a few reasons why you might want to study.

_____

_____

_____

Of course we have no way of knowing what reasons you listed, but from our experience, one reason given quite frequently is "To pass the course." There is nothing wrong with that reason. If that is the most important reason for you to study, then use it to stay focused.

One method of keeping to a study schedule is to form a ***study group.*** Look for people who are committed to learning, who pay attention in class, and who are punctual. Ask them to join your group. Choose people with similar educational goals but different methods of learning. You can gain insight from seeing the material from a new perspective. Limit groups to four or five people; larger groups are unwieldy.

There are many ways to conduct a study group. Begin with the following suggestions and see what works best for your group.

1. Test each other by asking questions. Each group member might bring two or three sample test questions to each meeting.
2. Practice teaching each other. Many of us who are teachers learned a lot about our subject when we had to explain it to someone else.
3. Compare class notes. You might ask other students about material in your notes that is difficult for you to understand.
4. Brainstorm test questions.
5. Set an agenda for each meeting. Set approximate time limits for each agenda item and determine a quitting time.

And finally, probably the most important aspect of studying is that it should be done in relatively small chunks. If you can study only three hours a week for this course (probably not enough for most people), do it in blocks of one hour on three separate days, preferably after class. Three hours of studying on a Sunday is not as productive as three hours of paced study.

Text Features That Promote Success

Preparing for a Chapter Before you begin a new chapter, you should take some time to review previously learned skills. There are two ways to do this. The first is to complete the ***Cumulative Review,*** which occurs after every chapter (except Chapter 1). For instance, turn to page 305. The questions in this review are taken from the previous chapters. The answers for all these exercises can be found on page A15. Turn to that page now and locate the answers for the Chapter 4 Cumulative Review. After the answer to the first exercise, which is 8 you will see the section reference [1.1]. This means that this question was taken from Chapter 1, Section 1. If you missed this question, you should return to that section and restudy the material.

A second way of preparing for a new chapter is to complete the ***Prep Test.*** This test focuses on the particular skills that will be required for the new chapter. Turn to page 234 to see a Prep Test. The answers for the Prep Test are the first set of answers in the answer section for a chapter. Turn to page A11 to see the answers for the Chapter 4 Prep Test. Note that a section reference is given for each question. If you answer a question incorrectly, restudy the section from which the question was taken.

Before the class meeting in which your professor begins a new section, you should browse through the material, being sure to note each word in bold type. These words indicate important concepts that you must know in order to learn the material. Do not worry about trying to understand all the material. Your professor is there to assist you with that endeavor. The purpose of browsing through the material is so that your brain will be prepared to accept and organize the new information when it is presented to you.

Turn to page 2. Write down the title of Section 1.1. Under the title of the section, write down the words in the section that are in bold print. It is not necessary for you to understand the meaning of these worlds. You are in this class to learn their meaning.

_____   _____   _____   _____

_____   _____   _____   _____

_____   _____   _____   _____

_____   _____   _____   _____

_____   _____   _____   _____

See Appendix A:
Graphing Linear
Inequalities

page 295

Using Technology There are many places in the text where a graphing calculator is used to assist you in making connections between abstract mathematical concepts and a graphical representation provided by the calculator. To benefit from this feature, you must be able to use your calculator effectively. Whenever appropriate, there are *See Appendix A* margin notes. These refer you to a **graphing calculator appendix** at the end of the text that demonstrates many of the major keystrokes you will need in this course. An abbreviated **calculator bookmark** at the beginning of the text can be removed and used to mark your current lesson. The calculator bookmark is a quick keystroke reference for many of the calculator's functions.

Math Is Not a Spectator Sport To learn mathematics you must be an active participant. Listening and watching your professor do mathematics is not enough. Mathematics requires that you interact with the lesson you are studying. If you filled in the blanks above, you were being interactive. There are other ways this textbook has been designed to help you be an active learner.

Annotated Examples A green arrow indicates an example that has explanatory remarks next to solution steps. These examples are used for two purposes. The first is to provide additional examples of important concepts. Second, these examples illustrate important techniques or principles that are often used in the solution of other types of problems.

➡ Find the equation of the line that contains the point $(4, -1)$ and has slope $-\frac{3}{4}$.

$$y - y_1 = m(x - x_1)$$ • The slope and a point other than the y-intercept are given. Use the point–slope formula.

$$y - (-1) = -\frac{3}{4}(x - 4)$$ • $(x_1, y_1) = (4, -1)$ and $m = -\frac{3}{4}$.

$$y + 1 = -\frac{3}{4}x + 3$$ • Simplify the left side. Use the Distributive Property on the right side.

$$y = -\frac{3}{4}x + 2$$ • Subtract 1 from each side of the equation. The equation is now in the form $y = mx + b$.   ⬅

After you review the example, get a clean sheet of paper. Write down the example, and then try to complete the solution without referring to your notes or the book. When you can do that, move on to the next part of the section. Leaf through the book now, and write down the page numbers of two other occurrences of an "arrowed" example.

Example/You Try It Pairs One of the key instructional features of this text is Example/You Try It pairs. Note that each example is completely worked out and the You Try It following the example is not. Study the worked-out example carefully by working through each step. Then work the You Try It. If you get stuck, refer to the page number following the You Try It, which directs you to the page on which the You Try It is solved—a complete worked-out solution is provided. Try to use the given solution to get a hint for the step you are stuck on. Then try to complete the solution yourself.

YOU TRY IT 2

$$y - y_1 = m(x - x_1)$$

$$y - 2 = -\frac{1}{2}[x - (-2)]$$

$$y - 2 = -\frac{1}{2}x - 1$$

$$y = -\frac{1}{2}x + 1$$

page S14

EXAMPLE 2

Find the equation of the line that passes through $P(1, -3)$ and that has slope -2.

Solution $y - y_1 = m(x - x_1)$ • Use the point–slope formula.

$y - (-3) = -2(x - 1)$ • $m = -2$, $(x_1, y_1) = (1, -3)$

$y + 3 = -2x + 2$

$y = -2x - 1$

In this example, we wrote the equation of the line as $y = -2x - 1$. We could have written the equation in functional notation as $f(x) = -2x - 1$.

YOU TRY IT 2

Find the equation of the line that passes through $P(-2, 2)$ and has slope $-\frac{1}{2}$.

Solution   See page S14.

page 258

When you have completed your solution, check your work against the solution we provided. (Turn to page S14 to see the solution of You Try It 2.) Be aware that frequently there is more than one way to solve a problem. Your answer, however, should be the same as the given answer. If you have any question about whether your method will "always work," check with your instructor or with someone in the math center.

Browse through the textbook and write down the page numbers where two other Example/You Try It pairs occur.

Remember: Be an active participant in your learning process. When you are sitting in class watching and listening to an explanation, you may think that you understand. However, until you actually try to do it, you will have no confirmation of the new knowledge or skill. Most of us have had the experience of sitting in class thinking we knew how to do something, only to get home and realize that we didn't.

Word Problems Word problems are difficult because we must read the problem, determine the quantity we must find, think of a method to find it, actually solve the problem, and then check the answer. In short, you must *state the goal, devise a strategy, solve the problem,* and *check your work.*

TAKE NOTE

There is a strong connection between reading and being a successful student in math or in any other subject. If you have difficulty reading, consider taking a reading course. Reading is much like other skills. There are certain things you can learn that will make you a better reader.

Note in the example below that solving a word problem includes stating the goal, devising a strategy, solving an equation, and checking the answer. If you have difficulty with a word problem, write down the known information. Be very specific. Write out a phrase or sentence that states what you are trying to find. Ask yourself whether there are known formulas that relate the known and unknown quantities. Do not ignore the word problems. They are an important part of mathematics.

EXAMPLE 4

A doctor has prescribed 2 cc (cubic centimeters) of medication for a patient. The tolerance is 0.03 cc. Find the lower and upper limits of the amount of medication to be given.

State the goal. The goal is to find the lower and upper limits of the amount of medication to be given.

Devise a strategy. Let p represent the prescribed amount of medication, T the tolerance, and m the given amount of medication. Solve the absolute value inequality $|m - p| \leq T$ for m.

Solve the problem.
$$|m - p| \leq T$$
$$|m - 2| \leq 0.03$$
$$-0.03 \leq m - 2 \leq 0.03$$
$$1.97 \leq m \leq 2.03$$

The lower and upper limits of the amount of medication to be given to the patient are 1.97 cc and 2.03 cc.

Check your work. Be sure to check your work by doing a thorough check of your calculations. As an estimate, the answers appear reasonable in that the amounts of medication are close to 2 cc.

YOU TRY IT 4

A machinist must make a bushing that has a tolerance of 0.003 inch. The diameter of the bushing is 2.55 inch. Find the lower and upper limits of the diameter of the bushing.

Solution See page S13.

page 219

Rule Boxes Pay special attention to rules placed in boxes. These rules give you the reasons why certain types of problems are solved the way they are. When you see a rule, try to rewrite the rule in your own words.

Find and write down two page numbers on which there are examples of rule boxes.

TAKE NOTE

If a rule has more than one part, be sure to make a notation to that effect.

Absolute Value Inequalities of the Form $|ax + b| < c$

To solve an absolute value inequality of the form $|ax + b| < c$, $c > 0$, solve the equivalent compound inequality $-c < ax + b < c$.

page 216

Chapter Exercises When you have finished studying a section, do the exercises in the exercise set that correspond to that section. Math is a subject that needs to be learned in small sections and practiced continually in order to be mastered. Doing all of the exercises in each exercise set will help you master the problem-solving techniques necessary for success. As you work through the exercises for a section, check your answers to the odd-numbered exercises with those at the back of the book.

Preparing for a Test There are important features of this text that can be used to prepare for a test.

- Chapter Summary
- Chapter Review Exercises
- Chapter Test

After completing a chapter, read the Chapter Summary. This summary is divided into two sections: *Key Terms* and *Essential Concepts*. (See page 301 for the Chapter 4 Summary.) This summary highlights the important topics covered in the chapter. The page number following each topic refers you to the page in the text on which you can find more information about the concept.

Following the Chapter Summary are Chapter Review Exercises (see page 302) and a Chapter Test (see page 303). Doing the review exercises is an important way of testing your understanding of the chapter. The answer to each review exercise is given at the back of the book, along with its section reference. After checking your answers, restudy any section from which a question you missed was taken. It may be helpful to retry some of the exercises for that section to reinforce your problem-solving techniques.

The Chapter Test should be used to prepare for an exam. We suggest that you try the Chapter Test a few days before your actual exam. Take the test in a quiet place, and try to complete the test in the same amount of time you will be allowed for your exam. When taking the Chapter Test, practice the strategies of successful test takers: (1) Scan the entire test to get a feel for the questions; (2) Read the directions carefully; (3) Work the problems that are easiest for you first; And, perhaps most important, (4) try to stay calm.

When you have completed the Chapter Test, check your answers. If you missed a question, review the material in that section and rework some of the exercises from that section. This will strengthen your ability to perform the skills in that section.

Your career goal goes here. → Is it difficult to be successful? YES! Successful music groups, artists, professional athletes, chefs, and _____ have to work very hard to achieve their goals. They focus on their goals and ignore distractions. The things we ask you to do to achieve success take time and commitment. We are confident that if you follow our suggestions, you will succeed.

1

Fundamental Concepts

```
ERR:SYNTAX
1: Quit
2: Goto
```

Be sure to use the correct symbol for a minus sign ⊟ versus a negative sign (-) or your calculator will display this error message.

*What cargo might this freighter be carrying? Agricultural goods, cars, computers, tobacco products? These are only a few of the products exported from the United States each year. The more goods a country exports, the better the likelihood for a favorable balance of trade. In order to avoid a trade deficit, a country needs to export more than it imports. The **Exploration on page 52** shows how to calculate a country's balance of trade and determine whether it is favorable or unfavorable. These calculations involve operations with integers.*

Need help? For online student resources, visit this web site:
Math.college.hmco.com

PREP TEST

For Exercises 1 to 4, add, subtract, multiply, or divide.

1. $875 + 49$ 924

2. $1602 - 358$ 1244

3. $39(407)$ 15,873

4. $456 \div 19$ 24

5. What is 127.1649 rounded to the nearest hundredth?   127.16

6. Which of the following numbers are greater than -8?
   a. -6 **b.** -10 **c.** 0 **d.** 8 a, c, d

7. Match each fraction with its decimal equivalent.

   a. $\dfrac{1}{2}$ **A.** 0.75

   b. $\dfrac{7}{10}$ **B.** 0.89

   c. $\dfrac{3}{4}$ **C.** 0.5

   d. $\dfrac{89}{100}$ **D.** 0.7

   a and C; b and D; c and A; d and B

8. What is the least whole number that both 8 and 12 divide evenly into?   24

9. What is the greatest whole number that divides into both 16 and 20 evenly?   4

10. Without using 1, write 21 as a product of two whole numbers. $3 \cdot 7$

INSTRUCTOR NOTE
The Prep Test is a means to test your students' mastery of *prerequisite* material that is assumed in the coming chapter. All answers, along with the section to review (if necessary), are provided in the Answers to Selected Exercises appendix.

GO FIGURE

If $\boxed{5} = 4$ and $\enclose{circle}{5} = 6$ and $y = x - 1$, which of the following has the greatest value?

SECTION **1.1**   **Problem Solving**

- Problem Solving
- Inductive Reasoning
- Deductive Reasoning

■ Problem Solving

A group of students is standing, equally spaced, around a circle. The 43rd student is directly opposite the 89th student. How many students are there in the group?

Solving a problem like the one above requires problem-solving strategies. One way to organize these strategies was expressed by George Polya (1887–1985) as a four-step process.

1. **Understand the problem and state the goal.**
2. **Devise a strategy to solve the problem.**
3. **Solve the problem by executing the strategy, and state the answer.**
4. **Review the solution and check your work.**

▼ *Point of Interest*

George Polya was born in Hungary and moved to the United States in 1940. He lived in Providence, Rhode Island, where he taught at Brown University until 1942, when he moved to California. There he taught at Stanford University until his retirement. While at Stanford, he published 10 books and a number of articles for mathematics journals. Of the books Polya published, How To Solve It *(1945) is one of his best known. In this book, Polya outlines a strategy for solving problems. This strategy, although frequently applied to mathematics, can be used to solve problems from virtually any discipline.*

Each of these steps is described below.

Understand the Problem and State the Goal.

This part of problem solving is often overlooked. You must have a clear understanding of the problem.

- Read the problem carefully and try to determine the goal.
- Make sure you understand all the terms or words used in the problem.
- Make a list of known facts.
- Make a list of information that, if known, would help you solve the problem. Remember that it may be necessary to look up information you do not know in another book, in an encyclopedia, at the library, or perhaps on the Internet.

Devise a Strategy to Solve the Problem.

Successful problem solvers use a variety of techniques when they attempt to solve a problem.

- Draw a diagram.
- Work backward.
- Guess and check.
- Solve an easier problem.
- Look for a pattern.
- Make a table or chart.
- Write an equation.
- Produce a graph.

Solve the Problem.

- If necessary, define what each variable represents.
- Work carefully.
- Keep accurate and neat records of your attempts.
- When you have completed the solution, state the answer carefully.

Review the Solution and Check Your Work.

Once you have found a solution, check the solution against the known facts and check for possible errors. Be sure the solution is consistent with the facts of the problem. Another important part of this review process is to ask whether your solution can be used to solve other types of problems.

We will apply this process to the problem stated earlier: A group of students is standing, equally spaced, around a circle. The 43rd student is directly opposite the 89th student. How many students are there in the group?

State the goal.   We need to determine the number of students standing around the circle, given that the 43rd student is standing directly opposite the 89th student.

Devise a strategy.   One strategy for this problem is to draw a diagram of the situation. This approach might lead to a method by which to solve the problem.

Solve the problem.   First draw a diagram of the students standing around a circle.

Note that if the 43rd and 89th students are standing opposite each other, then these two students divide the group into two equal parts. The difference between 89 and 43 is one-half of the total number of students.

$$89 - 43 = 46$$

There are 46 students in half of the group.

Multiply 46 by 2 to find the total number of students in the group.

$$46(2) = 92$$

There are 92 students in the group.

Check your work.   The answer "92 students" makes sense in the context of this problem. For example, we know that there have to be more than 89 students because we are told the 89th student is in the group.

INSTRUCTOR NOTE
Exercise 12 can be used for a similar in-class example.

EXAMPLE 1

Find the sum of the first 10,000 natural numbers.

State the goal.   Do you understand the meaning of all terms in the problem? For instance, do you know the meaning of the term *natural number*? Do you know the meaning of the word *sum*?[1] If you do, you will know this problem is asking you to find $1 + 2 + 3 + \ldots + 9998 + 9999 + 10,000$.

Devise a strategy.   One strategy for this problem would be to use a calculator and just start adding the numbers. However, this plan may lead to mistakes because of all the numbers to enter and then add. Even if we

_____

[1] A *natural number* is one of the numbers 1, 2, 3, 4, 5, 6, . . . , where the . . . means that the list of natural numbers continues on and on and that there is no largest natural number. A *sum* is the result of adding numbers.

could enter numbers accurately and quickly—say, one number every two seconds—it would take over 5 hours to get the answer. Therefore, we will try a different strategy: try to solve an easier problem first. The idea is to see whether solving an easier problem will lead to a strategy for solving the original problem.

Solve the problem.   Our easier problem will be to find the sum of the first 10 natural numbers. Note that when the natural numbers are paired as shown below, each pair has the same sum.

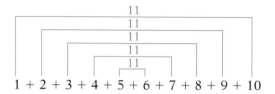

There are 5 pairs ($10 \div 2 = 5$) whose sum is 11.

Because there are 5 pairs whose sum is 11, the sum of the first 10 natural numbers is the product 11 times 5.

Sum of each pair
Number of pairs

$$1 + 2 + 3 + 4 + 5 + 6 + 7 + 8 + 9 + 10 = 11 \cdot 5 = 55$$

Using the easier problem as a model suggests that a strategy we can use to find the sum of the first 10,000 natural numbers is to pair the numbers, find the sum of one pair, and then multiply that sum by the number of pairs.

Extending the pattern for the easier problem to the original problem, we have

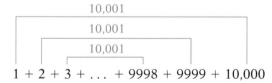

There are 5000 pairs ($10,000 \div 2 = 5000$) whose sum is 10,001. Thus,

$$1 + 2 + 3 + \ldots + 9998 + 9999 + 10,000 = 10,001 \cdot 5000 = 50,005,000$$

The sum of the first 10,000 natural numbers is 50,005,000.

Check your work.   By repeating the calculations, you can verify that the solution is correct and is consistent with the problem we were given to solve.

Suggested Activity

See the *Student Activity Manual,* Section 1.1, for an activity on problem-solving strategies.

Suggested Activity

Part of the review process is to ask whether the strategy can be used to solve other problems. Ask students whether the strategy will work for 47 + 48 + . . . + 85 + 86. What about 28 + 29 + . . . + 55 + 56? If not, is there a modification that will work?

YOU TRY IT 1

The product of the ages of three teenagers is 4590. How old is the oldest if none of the teens are the same age?

Solution   See page S1.   18 years old

■ Inductive Reasoning

Looking for patterns is one of the techniques used in *inductive reasoning.* Let's look at an example.

Suppose you take 6 credit hours each semester. The total number of credit hours you have taken at the end of each semester can be described in a list of numbers.

$$6, 12, 18, 24, 30, 36, \ldots$$

The list of numbers that indicates the total credit hours is an ordered list of numbers called a **sequence.** Each number in a sequence is called a **term** of the sequence. The list is ordered because the position of a number in the list indicates the semester at the end of which that number of credit hours has been completed. For example, the 5th term of the sequence is 30, and a total of 30 credit hours have been completed after the 5th semester.

? QUESTION What is the 3rd term of the sequence?*

Now consider another student who is taking courses each semester. The total number of credit hours taken by this student at the end of each semester is given by the sequence

$$9, 18, 27, 36, 45, 54, \ldots$$

? QUESTION Assuming that the pattern continues in the same manner, what will be the total number of credit hours completed after the 8th semester?†

The process you used to discover the next number in the above sequence is inductive reasoning. **Inductive reasoning** involves making generalizations from specific examples; in other words, we reach a conclusion by making observations about particular facts or cases.

INSTRUCTOR NOTE
Exercise 32 can be used for a similar in-class example.

EXAMPLE 2

Use inductive reasoning to find the three missing terms in the sequence.

A 2 3 4 B 6 7 8 C 10 11 12 D 14 15 16 ... __ __ __ 64

Solution The pattern of the sequence is that the numbers 1, 5, 9, 13, . . . are replaced by consecutive letters of the alphabet, beginning with the letter A.

Think of each four terms as a group. The groups end with 4, 8, 12, 16, . . . , which are the multiples of 4.

$64 \div 4 = 16$. Because 64 is the 16th multiple of 4, we are looking for the 16th group. The 16th letter of the alphabet is P.

The missing terms in __ __ __ 64 are P, 62, 63.

? ANSWERS * The 3rd term of the sequence is 18. † The 6th term is 54. The 7th term is 63. The 8th term is 72. The total number of credit hours taken after the 8th semester will be 72.

A portion of the beads on the string shown below is not visible. How many beads are not visible along the dashed portion of the string?

Solution   See page S1.   116 beads

INSTRUCTOR NOTE
Exercise 36 can be used for a similar in-class example.

TAKE NOTE

In a proper fraction, the numerator is greater than 0 but less than the denominator.

EXAMPLE 3

Using a calculator, determine the decimal representation of several proper fractions that have a denominator of 11. For instance, you may use $\frac{2}{11}$, $\frac{5}{11}$, and $\frac{9}{11}$. Then use inductive reasoning to explain the pattern, and use your reasoning to find the decimal representation of $\frac{8}{11}$ without a calculator.

Solution $\frac{2}{11} = 0.181818\ldots$; $\frac{5}{11} = 0.454545\ldots$; $\frac{9}{11} = 0.818181\ldots$

Note that $2(9) = 18$, $5(9) = 45$, and $9(9) = 81$. The repeating digits of the decimal representation of the fraction equal 9 times the numerator of the fraction.

The decimal representation of a proper fraction with a denominator of 11 is a repeating decimal in which the repeating digits are the product of the numerator and 9.

Using this reasoning, we know that $\frac{8}{11} = 0.727272\ldots$

 CALCULATOR NOTE

To find the decimal representation of $\frac{2}{11}$, press

2 ÷ 11 ENTER .

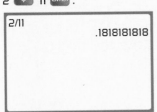

INSTRUCTOR NOTE
Have students check their solutions to You Try It 3 by using a calculator.

Using a calculator, determine the decimal representation of several proper fractions that have a denominator of 33. For instance, you may use $\frac{2}{33}$, $\frac{10}{33}$, and $\frac{25}{33}$. Then use inductive reasoning to explain the pattern, and use your reasoning to find the decimal representation of $\frac{19}{33}$ without a calculator.

Solution   See page S1.

$\frac{2}{33} = 0.060606\ldots$; $\frac{10}{33} = 0.303030\ldots$; $\frac{25}{33} = 0.757575\ldots$; $\frac{19}{33} = 0.575757\ldots$

The conclusion formed by using inductive reasoning is often called a **conjecture** because the conclusion may or may not be correct. For example, predict the next letter in the following list.

O, T, T, F, F, S, S, E, . . .

You might predict that the next letter is E because you see two T's, followed by two F's, followed by two S's. However, note that there is only one O at the beginning of the list.

Suggested Activity
Remind students that one of the characteristics of a good problem solver is to try to extend the solution of a problem. Ask students how the pattern in Example 3 continues for improper fractions with a denominator of 11—for example, $\frac{14}{11}$.

The next letter in the pattern is N, because the letters are chosen by using the first letter in the English words used to name the counting numbers.

One, **T**wo, **T**hree, **F**our, **F**ive, **S**ix, **S**even, **E**ight, **N**ine, . . .

■ Deductive Reasoning

Another type of reasoning that is used to reach conclusions is called deductive reasoning. **Deductive reasoning** is the process of reaching a conclusion by applying a general principle or rule to a specific example.

INSTRUCTOR NOTE
You may want to tell your students that this is an example of the law of syllogism, which states that if p implies q and q implies r are true conditions, then p implies r is true.

For example, suppose that during the last week of your math class, your instructor tells you that if you receive an 88 or better on the final exam, you will earn an A in the course. When grades for the final exam are posted, you learn that you received an 89 on the final exam. By using deductive reasoning, you can conclude that you will earn an A in the course.

Deductive reasoning is also used to reach a conclusion from a sequence of known facts. For example, consider the following:

If Gary completes his thesis, he will pass the course. If Gary passes the course, he will graduate.

From these statements, we can conclude that "If Gary completes his thesis, he will graduate." This is shown in the diagram below.

Gary completes his thesis. $\xrightarrow{\text{means}}$ Gary passes the course. $\xrightarrow{\text{means}}$ Gary will graduate.
Gary completes his thesis. $\xrightarrow{\hspace{3cm}\text{means}\hspace{3cm}}$ Gary will graduate.

Example 4 is another example of reaching a conclusion from a sequence of known facts.

INSTRUCTOR NOTE
Exercise 42 can be used for a similar in-class example.

EXAMPLE 4

If ◇◇◇◇◇ = ‡‡‡ and ‡‡‡ = ▽▽▽▽, how many ◇'s equal ▽▽▽▽▽▽▽▽?

Solution We are given that ◇◇◇◇◇ = ‡‡‡ and that ‡‡‡ = ▽▽▽▽.

These are the same.

◇◇◇◇◇ = ‡‡‡   and   ‡‡‡ = ▽▽▽▽

Therefore, these are equal.

◇◇◇◇◇ = ▽▽▽▽

Because 4 ▽'s = 5 ◇'s, 8 ▽'s = 10 ◇'s. That is,
▽▽▽▽▽▽▽▽ = ◇◇◇◇◇◇◇◇◇◇.

YOU TRY IT 4

Given that ¥¥¥ = △△△△ and that △△△△ = ΩΩ, how many Ω's equal ¥¥¥¥¥¥¥¥¥¥?

Solution See page S1.   6 Ω's

INSTRUCTOR NOTE
Exercise 46 can be used for
a similar in-class example.

EXAMPLE 5

Determine whether the following argument is an example of inductive reasoning or deductive reasoning:

During the past 10 years, this tree has produced fruit every other year. Last year this tree did not produce fruit, so this year the tree will produce fruit.

Solution   The conclusion is based on observation of a pattern. Therefore, it is an example of inductive reasoning.

YOU TRY IT 5

Determine whether the following argument is an example of inductive reasoning or deductive reasoning:

All kitchen remodeling jobs cost more than the contractor's estimate. The contractor estimated the cost of remodeling my kitchen at $35,000. Therefore, it will cost more than $35,000 to have my kitchen remodeled.

Solution   See page S1.   deductive reasoning

Deductive reasoning, along with a chart, is used to solve problems like the one in Example 6.

INSTRUCTOR NOTE
Exercise 52 can be used for
a similar in-class example.

EXAMPLE 6

Each of four neighbors, Chris, Dana, Leslie, and Pat, has a different occupation (accountant, banker, chef, or dentist). From the following statements, determine the occupation of each neighbor.

1. Dana usually gets home from work after the banker but before the dentist.
2. Leslie, who is usually the last to get home from work, is not the accountant.
3. The dentist and Leslie usually leave for work at about the same time.
4. The banker lives next door to Pat.

Solution   From statement 1, Dana is not the banker or the dentist. In the chart on the next page, write X1 (which stands for "ruled out by statement 1") for these conditions.

From statement 2, Leslie is not the accountant. We know from statement 1 that the banker is not the last to get home, and we know from statement 2 that Leslie usually is the last to get home; therefore, Leslie is not the banker. In the chart, write X2 for these conditions.

From statement 3, Leslie is not the dentist. Write X3 for this condition. There are now X's for three of the four occupations in Leslie's row; therefore, Leslie must be the chef. Place a √ in that box. Since Leslie is the chef, none of the other three people can be the chef. Write X3 for these conditions. There are now X's for three of the four occupations in Dana's row; therefore, Dana must be the accountant. Place a √ in that box. Since Dana is the accountant, neither Chris nor Pat is an accountant. Write X3 for these conditions.

Suggested Activity

Have students write problems similar to Example 6 and You Try It 6. This activity requires imagination in coming up with a scenario, as well as reasoning to determine statements to write that will generate the desired outcome. The problems can be solved by fellow students.

From statement 4, Pat is not the banker. Write X4 for this condition. Since there are three X's in the banker's column, Chris must be the banker. Place a √ in that box. Now Chris cannot be the dentist. Write X4 in that box. Since there are 3 X's in the dentist's column, Pat must be the dentist. Place a √ in that box.

	Accountant	Banker	Chef	Dentist
Chris	X3	√	X3	X4
Dana	√	X1	X3	X1
Leslie	X2	X2	√	X3
Pat	X3	X4	X3	√

Chris is a banker, Dana is an accountant, Leslie is a chef, and Pat is a dentist.

YOU TRY IT 6

Mike, Clarissa, Roger, and Betty were recently elected as the new officers of the Wycliff Neighborhood Association. From the following statements, determine which position each holds.

1. Mike and the treasurer are next-door neighbors.
2. Clarissa and the secretary have lived in the neighborhood for 5 years, Roger has lived there for 8 years, and the president has lived there for 10 years.
3. Betty has lived in the neighborhood for fewer years than Mike.
4. The vice president has lived in the neighborhood for 5 years.

Solution See page S1. Mike: president, Clarissa: vice president, Roger: treasurer, Betty: secretary

1.1 EXERCISES Suggested Assignment: 7–53, odds

Topics for Discussion

1. List the four steps involved in Polya's problem-solving process.
   Understand the problem and state the goal, devise a strategy to solve the problem, solve the problem, and review the solution and check your work.
2. Discuss some of the strategies used by good problem solvers.
   Use the list provided on page 3 as a guideline.

3. When solving a problem, why is it important to write neatly?
   Answers may vary. For example, if the writing is not neat, a number might easily be misread.
4. What is inductive reasoning? Provide an example in which inductive reasoning is used.
   Inductive reasoning involves making generalizations from specific examples. Examples will vary.

5. What is deductive reasoning? Provide an example in which deductive reasoning is used.

Deductive reasoning involves drawing a conclusion that is based on given facts. Examples will vary.

■ Problem Solving

6. Find the units digit of 7^{97}.   7

7. A square floor is tiled with congruent square tiles. The tiles on the two diagonals of the floor are blue. The rest of the tiles are green. If 101 blue tiles are used, find the total number of tiles on the floor.   2601 tiles

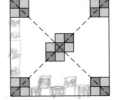

8. What is the least natural number greater than 1 that divides evenly into the sum $3^{11} + 5^{13}$?   2

9. How many of the first one hundred natural numbers are divisible by all of the numbers 2, 3, 4, and 5?   1

10. An ewok was visiting an island on which there lived knights, who only make true statements, and knaves, who only make false statements. The ewok needed to find a knight to be a trusty guide. While walking along the shore, the ewok came upon three natives, named Arthur, Bernard, and Charles. The ewok first asked Arthur, "Are Bernard and Charles both knights?" Arthur replied, "Yes." The ewok then asked, "Is Bernard a knight?" To his surprise, Arthur answered, "No." Who is a knight and who is a knave? [Modified from a puzzle by Raymond Smallyan.]

Arthur is a knave, Bernard is a knight, and Charles is a knave.

11. What are the next two letters of the sequence A, B, E, F, I, J, . . . ?   M, N

12. What is the 96th digit in the decimal equivalent of $\frac{1}{7}$?   7

13. One hundred college seniors were interviewed about their reading habits. The responses revealed that 63 read the *New York Times*, 41 read the *Wall Street Journal*, and 10 read both newspapers. How many students read neither newspaper?   6 students

14. What three-digit natural number is equal to 11 times the sum of its digits?   198

15. How many natural numbers greater than ten and less than one hundred are increased by nine when their digits are reversed?   8

16. A perfect number is one for which the sum of the proper divisors is equal to the number. (The proper divisors are the ones that are less than the number and divide evenly into the number.) For instance, 496 is a perfect number. The proper divisors of 496 are 1, 2, 4, 8, 16, 31, 62, 124, and 248. The sum of the divisors is

$$1 + 2 + 4 + 8 + 16 + 31 + 62 + 124 + 248 = 496.$$

Find a perfect number between 20 and 30.   28

17. A car has an odometer reading of 15951 miles, which is a palindrome. (A palindrome is a whole number that remains unchanged when its digits are written in reverse order.) After 2 hours of continuous driving at a constant speed, the odometer reading is the next palindrome. How fast, in miles per hour, was the car being driven during these 2 hours?   55 mph

18. If all of the digits must be different, how many three-digit odd numbers greater than 700 can be written using only the digits 1, 2, 3, 5, 6, 7?   12

19. A square is divided into a 100-by-100 grid of smaller squares. If 100 squares are shaded in the top row, 99 in the second row, 98 in the third row, and so on, what is the ratio of the squares shaded to the squares not shaded? Write the answer as a fraction in simplest form. $\frac{101}{99}$

20. Suppose you have a balance scale and 8 coins. One of the coins is counterfeit and weighs slightly more than the other 7 coins. Explain how you can find the counterfeit coin in two weighings.

The complete solution is given in the Solutions Manual. However, to start, place three coins on each pan of the balance scale.

21. How many children are there in a family wherein each girl has as many brothers as sisters, but each boy has twice as many sisters as brothers?   7 children

22. The natural numbers greater than 1 are arranged in five columns as shown below. In which column, 1, 2, 3, 4, or 5, will the number 1000 fall?

		2	3	4	5
	9	8	7	6	
		10	11	12	13
	17	16	15	14	
		18	19	20	21

Column 2

23. Which terms must be removed from $\frac{1}{2} + \frac{1}{4} + \frac{1}{6} + \frac{1}{8} + \frac{1}{10} + \frac{1}{12}$ if the sum of the remaining terms is to equal 1? $\frac{1}{8}$ and $\frac{1}{10}$

24. The number of a certain type of bacteria doubles every minute. If one of these bacteria is placed in a jar at 1:00 P.M. and starts doubling, the jar will be full in one hour. At what time was the jar half full?   1:59 P.M.

25. A new product, Super-Yeast, causes bread to double in volume each minute. If it takes one loaf of bread 30 minutes to fill an oven, how many minutes would it take two loaves to fill half the oven?   28 minutes

26. September 9, 1981 (9/9/81) was a square-root year date because both the month and the day are square roots of the last two digits of the year. How many square-root dates will there be during the 21st century?
9 square-root dates

■ Inductive and Deductive Reasoning

For Exercises 27 to 34, use inductive reasoning to predict the next term of the sequence.

27. 5, 11, 17, 23, 29, 35, . . . 41   **28.** 3, 5, 9, 15, 23, 33, . . . 45   **29.** 1, 8, 27, 64, 125, . . . 216

30. $\frac{3}{5}, \frac{5}{7}, \frac{7}{9}, \frac{9}{11}, \frac{11}{13}, \cdots$ $\frac{13}{15}$ **31.** 2, 3, 7, 16, 32, 57, . . . 93   **32.** $2, 7, -3, 2, -8, -3, -13, -8, \ldots$ −18

33. a, b, f, g, k, l, p, q, . . . u   **34.** Z, X, V, T, R, P, . . . N

35. Use a calculator to evaluate each of the following.

$$12{,}345{,}679 \cdot 9$$
$$12{,}345{,}679 \cdot 18$$
$$12{,}345{,}679 \cdot 27$$
$$12{,}345{,}679 \cdot 36$$
$$12{,}345{,}679 \cdot 45$$

Then use inductive reasoning to explain the pattern, and use your reasoning to evaluate

$$12{,}345{,}679 \cdot 54 \text{ and } 12{,}345{,}679 \cdot 63$$

without a calculator.
111,111,111; 222,222,222; 333,333,333; 444,444,444; 555,555,555;
Answers will vary; for example, multiplying a multiple of 9 times 12,345,679 results in a nine-digit number with repeating digits.
$12{,}345{,}679 \cdot 54 = 666{,}666{,}666$; $12{,}345{,}679 \cdot 63 = 777{,}777{,}777$

36. Use a calculator to evaluate 15^2, 25^2, 35^2, 45^2, 55^2, 65^2, and 75^2. Then use inductive reasoning to explain the pattern, and use your reasoning to evaluate 85^2 and 95^2 without a calculator.
225, 625, 1225, 2025, 3025, 4225, 5625.
Answers may vary; for example, each number ends in 25; the first digits are equal to the tens digit of the given number to be squared multiplied by 2, then 3, then 4, then 5, then 6, then 7, then 8. $85^2 = 7225$; $95^2 = 9025$

37. Draw the next figure in the sequence:

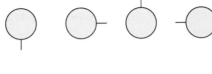

38. Draw the next figure in the sequence:

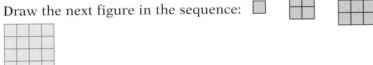

39. Given that $\triangledown\triangledown = \oplus\oplus\oplus$ and that $\oplus\oplus\oplus = \Lambda\Lambda\Lambda\Lambda$, how many Λ's equal $\triangledown\triangledown\triangledown\triangledown\triangledown\triangledown$? 12

40. Given that ⇑ = ◇◇ and ΩΩΩ = ◇◇, how many Ω's equal ⇑⇑⇑? 9

41. If ♠♠ = ♦♦♦♦♦, and ♦♦♦ = ♣♣, and ♣♣♣♣ = ♥, then how many ♥'s equal ♠♠♠♠♠♠? 3

42. If ⇓⇓⇓ = ΩΩΩΩΩΩ, and Ω = ◇◇◇, and ◇◇ = ⊕⊕⊕⊕⊕⊕, then how many ⊕'s equal ⇓⇓⇓? 30

43. Take the number 7654 and reverse the digits to form the number 4567. Now subtract the smaller number from the larger one ($7654 - 4567 = 3087$). Try this for other four-digit numbers whose digits are consecutive integers written in descending order (largest to smallest). Make a conjecture from your observations. The difference is always 3087.

44. There are four weights labeled A, B, C, and D. A weighs more than B, and B weighs more than D. B and D together weigh more than B and C together. Which weight is the lightest? C

45. The year 1998 was unusual in at least one respect: Friday the 13th occurred in two consecutive months. What were the months? (You should be able to do this problem without looking at a calendar for 1998.) February and March

For Exercises 46 to 50, determine whether the argument is an example of inductive or deductive reasoning.

46. All Mark Twain novels are worth reading. *The Adventures of Tom Sawyer* is a Mark Twain novel. Therefore, *The Adventures of Tom Sawyer* is worth reading. deductive reasoning

47. Every English setter likes to hunt. Duke is an English setter, so Duke likes to hunt. deductive reasoning

48. $2 \cdot 3 + 1 = 7$
$2 \cdot 3 \cdot 5 + 1 = 31$
$2 \cdot 3 \cdot 5 \cdot 7 + 1 = 211$
$2 \cdot 3 \cdot 5 \cdot 7 \cdot 11 + 1 = 2311$

Therefore, the product of the first n prime numbers increased by 1 is always a prime number. inductive reasoning

49. I have enjoyed each of Tom Clancy's novels. Therefore, I know that I will like his next novel. inductive reasoning

50. The Atlanta Braves have won eight games in a row. Therefore, the Atlanta Braves will win their next game. inductive reasoning

51. Each of four siblings (Anita, Tony, Maria, and Jose) is given $5000 to invest in the stock market. Each chooses a different stock. One chooses a utility stock, another an automotive stock, another a technology stock, and the other an oil stock. From the following statements, determine which sibling bought which stock.

a. Anita and the owner of the utility stock purchased their shares through an online brokerage, whereas Tony and the owner of the automotive stock did not.

b. The gain in value of Maria's stock is twice the gain in value of the automotive stock.

c. The technology stock is traded on NASDAQ, whereas the stock that Tony bought is traded on the New York Stock Exchange.

Maria: the utility stock; Jose: the automotive stock; Anita: the technology stock; Tony: the oil stock

52. The Changs, Steinbergs, Ontkeans, and Gonzaleses were winners in the All-State Cooking Contest. There was a winner in each of four categories: soup, entrée, salad, and dessert. From the following statements, determine in which category each family was the winner.

a. The soups were judged before the Ontkeans' winning entry.

b. This year's contest was the first for the Steinbergs and for the winner in the dessert category. The Changs and the winner in the soup category entered last year's contest.

c. The winning entrée took 2 hours to cook, whereas the Steinberg's entrée required no cooking at all.

Changs: entrée; Steinbergs: salad; Ontkeans: dessert; Gonzaleses: soup

53. The cities of Atlanta, Chicago, Philadelphia, and Seattle held conventions this summer for collectors of coins, stamps, comic books, and baseball cards. From the following statements, determine which collectors met in which city.

a. The comic book collectors convention was in August, as was the convention held in Chicago.

b. The baseball card collectors did not meet in Philadelphia, and the coin collectors did not meet in Seattle or Chicago.

c. The convention in Atlanta was held during the week of July 4, whereas the coin collectors convention was held the week after that.

d. The convention in Chicago had more collectors attending it than did the stamp collectors convention.

Atlanta: stamps; Chicago: baseball cards; Philadelphia: coins; Seattle: comic books

54. Each of the Little League teams in a small rural community is sponsored by a different local business. The names of the teams are the Dodgers, the Pirates, the Tigers, and the Giants. The businesses that sponsor the teams are the bank, the supermarket, the service station, and the drug store. From the following statements, determine which business sponsors each team.

a. The Tigers and the team sponsored by the service station have winning records this season.

b. The Pirates and the team sponsored by the bank are coached by parents of the players, whereas the Giants and the team sponsored by the drug store are coached by the director of the Community Center.

c. Jake is the pitcher for the team sponsored by the supermarket and coached by his father.

d. The game between the Tigers and the team sponsored by the drug store was rained out yesterday.

Dodgers: drug store; Pirates: supermarket; Tigers: bank; Giants: service station

Applying Concepts

55. It is a fact that the fourth power of any number ends with a 1, a 5, or a 6. On the basis of this, can you conclude that 134,512,357,186 is the fourth power of some number?   No

56. Predict the next term of the sequence 1, 5, 12, 22, 35, . . .   51

57. $1K31K4$ represents a six-digit number that is a multiple of 12 but not a multiple of 9. Find the value of K. *Note:* All K's represent the same digit.   6

58. Let x be the least of three natural numbers whose product is 720. Find the greatest possible value of x.   8

59. Find the least value of d that satisfies $a^2 + b^2 + c^2 = d^2$, where a, b, c, and d are natural numbers, not necessarily different.   3

60. During the spring campus clean-up day, four students (Daisy, Heather, Lily, and Rose) each did different chores (painting, pruning, raking, or washing). Each worked a different number of hours (5, 6, 7, or 8 hours). From the following statements, determine each student's chore and the length of time each worked. You might find it helpful to use the chart provided below the statements.

  a. Lily and the student who did the pruning worked the longest hours.

  b. Daisy and the student who did the washing started working at the same time, but Daisy worked 3 hours longer.

  c. Rose, Lily, and the student who did the washing all worked at the campus clean-up day last year.

  d. The student who did the raking worked 2 hours less than the student who did the pruning and 1 hour more than Heather.

	Painting	Pruning	Raking	Washing		5 hours	6 hours	7 hours	8 hours
Daisy									
Heather									
Lily									
Rose									
5 hours					
6 hours					
7 hours					
8 hours					

P

Daisy: pruning, 8 hours; Lily: painting, 7 hours; Rose: raking, 6 hours; Heather: washing, 5 hours

EXPLORATION

1. *The Game of Sprouts*   The mathematician John H. Conway has created several games that are easy to play but complex enough to be challenging. For instance, in 1967, Conway, along with Michael Paterson, created the two-person, paper-and-pencil game of Sprouts. After more than 30 years, the game has not been completely analyzed.

Here are the rules for Sprouts.

- Begin by drawing a few dots on a piece of paper. (Keep the number small to ensure that you can complete the game.)
- The players alternate turns. A turn consists of drawing an arc between two dots or drawing a curve that starts at a dot and ends at the same dot. The active player then draws a new dot at the midpoint of the new arc.
- No dot can have more than three arcs coming from it.
- No arc can cross itself or any previously drawn arc.
- The winner is the player to draw the last possible arc.

Here is an example of a game of Sprouts that begins with two dots.

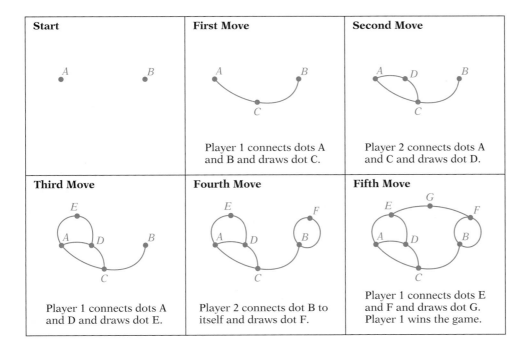

Start	First Move	Second Move
	Player 1 connects dots A and B and draws dot C.	Player 2 connects dots A and C and draws dot D.

Third Move	Fourth Move	Fifth Move
Player 1 connects dots A and D and draws dot E.	Player 2 connects dot B to itself and draws dot F.	Player 1 connects dots E and F and draws dot G. Player 1 wins the game.

A dot with no arc emanating from it is said to have three lives. A dot with one arc emanating from it has two lives. A dot with two arcs emanating from it has one life. A dot is dead and cannot be used if it has three arcs emanating from it.

Note in the game above that dot G has only two arcs emanating from it, so it has one life. But every other dot has three arcs emanating from it and is therefore dead. Thus there is no dot to connect to G, and the game is over. (Dot G cannot be connected to itself, because it would then have more than three arcs coming from it.)

a. Play a few games of one-spot Sprouts. How many moves are needed to determine a winner?
two moves

b. In a two-dot game, how many initial moves are possible?
three initial moves

c. Try to play out all possible two-dot games. How many moves are needed to determine a winner?
five moves

d. In a two-dot game, which player is guaranteed a win? Did you use inductive or deductive reasoning to answer this question?
Player 1; inductive reasoning

e. In a three-dot game, how many initial moves are possible?
six initial moves

f. Play several three-dot games. How many moves are needed to determine a winner?
eight moves

g. In a three-dot game, which player is guaranteed a win? Did you use inductive or deductive reasoning to answer this question?
Player 2; inductive reasoning

SECTION **1.2**   # Sets

■ Sets of Numbers
■ Union and Intersection of Sets
■ Interval Notation

■ Sets of Numbers

The tendency to group similar items seems to be a typical human trait. For instance, a botanist groups plants with similar characteristics in groups called species. Nutritionists classify foods according to food groups; for example, pasta, crackers, and rice are among the foods in the bread group.

Mathematicians place objects with similar properties in groups called sets. A **set** is a collection of objects. The objects in a set are called the **elements** of the set.

The **roster method** of writing sets is to enclose a list of the elements in braces. The set of sections within an orchestra is written {brass, percussion, string, woodwind}.

The numbers that we use to count objects, such as the number of students enrolled in a university or the number of stars in a constellation, are the natural numbers.

Natural numbers = {1, 2, 3, 4, 5, 6, 7, 8, 9, 10, . . .}

The set of natural numbers is an example of an **infinite set;** the pattern of numbers continues without end. It is impossible to list all the elements of an

▼ *Point of Interest*

Georg Cantor (1845–1918) was a German mathematician who developed many new concepts that dealt with the theory of sets. At the age of 15 he had decided that he wanted to become a mathematician, but his father coerced him into the field of engineering because it was a more lucrative profession. After
(continued)

▼ *Point of Interest (cont.)*

*a few years, Cantor's father real-
ized that his son was not suited
to engineering, and he permitted Georg to seek a career in
mathematics.*

   *Much of Cantor's work was
controversial. One of the simplest of the controversial concepts concerned points on a line
segment. For instance, consider
the line segment AB and the line
segment CD shown below.*

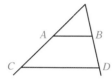

*Which line segment do you think
contains the most points? Cantor was able to show that they
both contain the same number
of points. In fact, he was able to
show that any line segment—no
matter how short—contains the
same number of points as a line,
or as a plane, or as all of three-
dimensional space.*

INSTRUCTOR NOTE
Exercise 10 can be used for
a similar in-class example.

Suggested Activity

Have students give examples
of finite sets, infinite sets, and
the empty set. Or ask students
to classify sets that you give
them. Here are a few
examples:
a. The set of odd negative integers greater than −8
b. The set of even negative
integers less than −8
c. The set of months in the
year with 41 days

infinite set. The set of even natural numbers less than 9 is written {2, 4, 6, 8}. This is an example of a **finite set;** all the elements of the set can be listed.

Each natural number greater than 1 is either a prime number or a composite number. A **prime number** is a natural number greater than 1 that is evenly divisible only by itself and 1. The first six prime numbers are 2, 3, 5, 7, 11, 13. A natural number greater than 1 that is not a prime number is a **composite number.** The numbers 4, 6, 8, 9, and 10 are the first five composite numbers.

❓ QUESTION   What is the 7th prime number?
             What is the 6th composite number?

The natural numbers do not have a symbol to denote the concept of none—for instance, the number of college students at Providence College that are under the age of 10. The whole numbers include zero and the natural numbers.

$$\textbf{Whole numbers} = \{0, 1, 2, 3, 4, 5, 6, 7, 8, 9, 10, \ldots\}$$

The whole numbers do not provide all the numbers that are useful in applications. For example, a meteorologist also needs numbers less than zero.

$$\textbf{Integers} = \{\ldots, -5, -4, -3, -2, -1, 0, 1, 2, 3, 4, 5, \ldots\}$$

The integers . . ., −5, −4, −3, −2, −1 are **negative integers.** The integers 1, 2, 3, 4, 5, . . . are **positive integers.** Note that the natural numbers and the positive integers are the same set of numbers. The integer zero is neither a positive nor a negative integer.

EXAMPLE 1

Use the roster method to write the set of whole numbers less than 7.

Solution   {0, 1, 2, 3, 4, 5, 6, 7}

YOU TRY IT 1

Use the roster method to write the set of positive odd integers less than 10.

Solution   See page S2.   {1, 3, 5, 7, 9}

Still other numbers are necessary to solve the variety of application problems that exist. For instance, a plumber may need to purchase drain pipe that has a diameter of $\frac{5}{8}$ inch. The set of numbers that include fractions are called rational numbers.

$$\textbf{Rational numbers} = \left\{\frac{p}{q}, \text{ where } p \text{ and } q \text{ are integers and } q \neq 0\right\}$$

Examples of rational numbers include $\frac{2}{3}$, $-\frac{9}{2}$, and $\frac{5}{1}$. Note that $\frac{5}{1} = 5$; all integers are rational numbers. The number $\frac{4}{\pi}$ is not a rational number because π is not an integer.

❓ ANSWER   The 7th prime number is 17. The 6th composite number is 12.

A fraction can be written in decimal notation by dividing the numerator by the denominator. For example, $\frac{7}{20} = 7 \div 20 = 0.35$ and $\frac{5}{9} = 5 \div 9 = 0.\overline{5}$. The number 0.35 is an example of a **terminating decimal.** $0.\overline{5}$ is an example of a **repeating decimal;** the bar over 5 indicates that this digit repeats. Every rational number can be written as either a terminating or a repeating decimal.

Some numbers cannot be written as terminating or repeating decimals; examples include 0.02002000200002 . . ., $\sqrt{7} = 2.645751 \ldots$, and $\pi = 3.1415926 \ldots$. These numbers have decimal representations that neither terminate nor repeat. They are called **irrational numbers.**

The rational numbers and the irrational numbers taken together are the **real numbers.**

The relationship among sets of numbers is shown in the figure below.

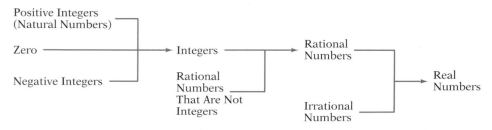

The **real number line** is used as a graphical representation of the real numbers. Although usually only integers are shown on the real number line, it represents the real numbers. The **graph of a real number** is made by placing a heavy dot on a number line directly above the number. The graphs of some real numbers are shown below.

It is common to designate a set by a capital letter. For instance, if A is the set of the first four letters of the alphabet, then $A = \{a, b, c, d\}$.

To refer to the elements of a set, the symbol $\in$ is used. This symbol is read "is an element of." The symbol $\notin$ means "is not an element of."

Given $B = \{1, 3, 5\}$, then $1 \in B$, $3 \in B$, and $6 \notin B$.

The **empty set,** or **null set,** is the set that contains no elements. The symbol $\varnothing$ or {} is used to represent the empty set. The set of people who have run a 2-minute mile is the empty set.

A second method of representing a set is **set-builder notation.** Set-builder notation can be used to describe almost any set, but it is especially useful for writing infinite sets. In set-builder notation, the set of integers greater than -4 is written

$\{x \mid x > -4, x \in \text{integers}\}$

INSTRUCTOR NOTE
Ask students which of the following numbers are in the set
$\{x \mid x > -4, x \in \text{integers}\}$:

$-9, -4, 0, 4, \frac{3}{8}, -2.61$

[Answer: 0, 4]
Ask students which of the following numbers are in the set $\{x \mid x < 5, x \in \text{real numbers}\}$:

$-7, -5, 0, 3, 5, \frac{4}{7}, -8.3, \pi, 5.62$

[Answer: $-7, -5, 0, 3, \frac{4}{7}, -8.3, \pi$]

This is read "the set of all x such that x is greater than -4 and x is an element of the integers." Recall that the symbol $>$ means "is greater than."

The set of real numbers less than 5 is written

$$\{x \mid x < 5, x \in \text{real numbers}\}$$

which is read "the set of all x such that x is less than 5 and x is an element of the real numbers." Recall that the symbol $<$ means "is less than."

INSTRUCTOR NOTE
Exercise 18 can be used for a similar in-class example.

EXAMPLE 2

Use set-builder notation to write the set of integers greater than -6.

Solution $\{x \mid x > -6, x \in \text{integers}\}$

YOU TRY IT 2

Use set-builder notation to write the set of real numbers greater than 19.

Solution   See page S2. $\{x \mid x > 19, x \in \text{real numbers}\}$

Suggested Activity

Let $A = \{0, 2, 4, 6\}$, $B = \{1, 2, 3, 4, 5\}$, and $C = \{1, 3, 5, 7\}$. Use the roster method to list the elements in the set that satisfy $\{x \mid x \in A \text{ and } x \notin B\}$. [Answer: $\{0, 6\}$]

The inequality symbols $>$ and $<$ are sometimes combined with the equality symbol.

$a \geq b$ is read "a is greater than or equal to b" and means $a > b$ or $a = b$.

$a \leq b$ is read "a is less than or equal to b" and means $a < b$ or $a = b$.

Sets described using set-builder notation and the inequality symbols $>$, $<$, $\geq$, and $\leq$ can be graphed on the real number line.

The graph of $\{x \mid x > -2, x \in \text{real numbers}\}$ is shown at the right. The set is the real numbers greater than -2. The parenthesis on the graph indicates that -2 is not included in the set.

The graph of $\{x \mid x \geq -2, x \in \text{real numbers}\}$ is shown at the right. The set is the real numbers greater than or equal to -2. The bracket at -2 indicates that -2 is included in the set.

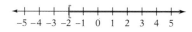

TAKE NOTE

A parenthesis is used to indicate that the number is not included in the set. A bracket is used to indicate that the number is included in the set.

For the remainder of this section, all variables will represent real numbers unless otherwise stated. Using this convention, the set above would be written $\{x \mid x \geq -2\}$.

EXAMPLE 3

Graph $\{x \mid x \leq 3\}$.

Solution   The set is the real numbers less than or equal to 3. Draw a right bracket at 3, and darken the number line to the left of 3.

INSTRUCTOR NOTE
Exercise 30 can be used for a similar in-class example.

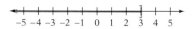

YOU TRY IT 3

Graph $\{x \mid x > -3\}$.

Solution See page S2.

■ Union and Intersection of Sets

Just as operations such as addition and multiplication are performed on real numbers, operations are performed on sets. Two operations performed on sets are union and intersection.

TAKE NOTE

When we list the elements of a set, the order is not important. Thus the set {2, 3, 4, 0, 1} is the same as the set {0, 1, 2, 3, 4}. However, numerical elements are generally listed in increasing order so that it is easier to read and compare sets.

The **union of two sets,** which is written $A \cup B$, is the set of all elements that belong to either A or B. In set-builder notation, this is written

$$A \cup B = \{x \mid x \in A \text{ or } x \in B\}$$

Given $A = \{2, 3, 4\}$ and $B = \{0, 1, 2, 3\}$, $A \cup B = \{0, 1, 2, 3, 4\}$. Note that an element that belongs to both sets is listed only once in their union.

INSTRUCTOR NOTE
Exercise 34 can be used for a similar in-class example.

EXAMPLE 4

Find $C \cup D$ given $C = \{1, 5, 9, 13, 17\}$ and $D = \{3, 5, 7, 9, 11\}$.

Solution $C \cup D = \{1, 3, 5, 7, 9, 11, 13, 17\}$

YOU TRY IT 4

Find $E \cup F$ given $E = \{-2, -1, 0, 1, 2\}$ and $F = \{-5, -1, 0, 1, 5\}$.

Solution See page S2. $E \cup F = \{-5, -2, -1, 0, 1, 2, 5\}$

Suggested Activity

See Section 1.2 of the *Student Activity Manual* for an activity on sets and graphing inequalities on a graphing calculator.

The set $\{x \mid x \leq -1\} \cup \{x \mid x > 3\}$ is the set of real numbers that are either less than or equal to -1 or greater than 3.

The set is written $\{x \mid x \leq -1 \text{ or } x > 3\}$.

The set $\{x \mid x > 2\} \cup \{x \mid x > 4\}$ is the set of real numbers that are either greater than 2 or greater than 4.

The set is written $\{x \mid x > 2\}$.

The union is the numbers greater than 2.

INSTRUCTOR NOTE
Exercise 44 can be used for
a similar in-class example.

▼ *Point of Interest*

*Some mathematics that con-
cerns sets has led to paradoxes.
For example, in 1902 Bertrand
Russell developed the following
paradox: "Is the set A of all sets
that are not elements of themselves an element of itself?"
Both the assumption that A is an
element of A and the assumption that A is not an element of
A lead to a contradiction.*

*Russell's paradox has been
popularized in the following
form: "The town barber shaves
all males who do not shave
themselves, and he shaves only
those males. The town barber is
a male who shaves. Who shaves
the barber?" The assumption
that the barber shaves himself
leads to a contradiction, and the
assumption that the barber does
not shave himself also leads to a
contradiction.*

INSTRUCTOR NOTE
Exercise 38 can be used for
a similar in-class example.

EXAMPLE 5

Graph $\{x \mid x \leq 0\} \cup \{x \mid x \geq 4\}$.

Solution   The set is the numbers less than or equal to 0 or greater than or
equal to 4.

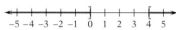

YOU TRY IT 5

Graph $\{x \mid x \geq 1\} \cup \{x \mid x \leq -3\}$.

Solution   See page S2.

The **intersection of two sets,** which is written $A \cap B$, is the set of all elements
that are common to both A and B. In set-builder notation, this is written

$$A \cap B = \{x \mid x \in A \text{ and } x \in B\}$$

Given $A = \{2, 3, 4\}$ and $B = \{0, 1, 2, 3\}$, $A \cap B = \{2, 3\}$.

EXAMPLE 6

a. Find $C \cap D$ given $C = \{3, 6, 9, 12\}$ and $D = \{0, 6, 12, 18\}$.
b. Find $E \cap F$ given $E = \{x \mid x \in \text{natural numbers}\}$ and $F = \{x \mid x \in \text{negative integers}\}$.

Solution   **a.** $C \cap D = \{6, 12\}$
b. There are no natural numbers that are also negative
integers.
$E \cap F = \varnothing$

YOU TRY IT 6

a. Find $A \cap B$ given $A = \{-2, -1, 0, 1, 2\}$ and $B = \{-10, -5, 0, 5, 10\}$.
b. Find $C \cap D$ given $C = \{x \mid x \in \text{odd integers}\}$ and $D = \{x \mid x \in \text{even integers}\}$.

Solution   See page S2.   **a.** $A \cap B = \{0\}$ **b.** $C \cap D = \varnothing$

The set $\{x \mid x > -2\} \cap \{x \mid x < 5\}$ is the set of real numbers that are greater than
-2 and less than 5.

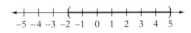   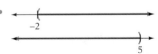

The intersection is the numbers
between -2 and 5.

The set can be written $\{x \mid x > -2 \text{ and } x < 5\}$. However, it is more commonly
written $\{x \mid -2 < x < 5\}$, which is read "the set of all x such that x is greater
than -2 and less than 5."

The set $\{x \,|\, x < 4\} \cap \{x \,|\, x < 5\}$ is the real numbers that are less than 4 and less than 5.

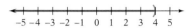

The set is written $\{x \,|\, x < 4\}$.

The intersection is the numbers less than 4.

INSTRUCTOR NOTE
Exercise 46 can be used for a similar in-class example.

EXAMPLE 7

Graph $\{x \,|\, x < 0\} \cap \{x \,|\, x > -3\}$.

Solution   The set is $\{x \,|\, -3 < x < 0\}$.

YOU TRY IT 7

Graph $\{x \,|\, x \leq 2\} \cap \{x \,|\, x \geq -1\}$.

Solution   See page S2.

■ Interval Notation

Some sets can also be expressed using **interval notation.** For example, the interval notation $(-3, 2]$ indicates the interval of all real numbers greater than -3 and less than or equal to 2. As on the graph of a set, the left parenthesis indicates that -3 is not included in the set. The right bracket indicates that 2 is included in the set.

INSTRUCTOR NOTE
Ask students which of the following numbers are in the set $(-3, 2)$:

$-4, -3, -2, 0, 3, 2, \frac{1}{5}, -2.67,$
1.9

[Answer: $-2, 0, \frac{1}{5}, -2.67,$
1.9]

Ask students which of the numbers listed above are in the set $[-3, 2]$.

[Answer: $-3, -2, 0, 2, \frac{1}{5},$
$-2.67, 1.9$]

An interval is said to be **closed** if it includes both endpoints; it is **open** if it does not include either endpoint. An interval is **half-open** if one endpoint is included and the other is not. In each of the following examples, -3 and 2 are the endpoints of the interval. In each case, the set notation, the interval notation, and the graph of the set are shown.

$\{x \,|\, -3 < x < 2\}$	$(-3, 2)$ Open interval	
$\{x \,|\, -3 \leq x \leq 2\}$	$[-3, 2]$ Closed interval	
$\{x \,|\, -3 \leq x < 2\}$	$[-3, 2)$ Half-open interval	
$\{x \,|\, -3 < x \leq 2\}$	$(-3, 2]$ Half-open interval	

To indicate an interval that extends forever in one or both directions using interval notation, we use the **infinity symbol** ∞ or the **negative infinity symbol** $-\infty$. The infinity symbol is not a number; it is simply used as a notation to indicate that the interval is unlimited. In interval notation, a parenthesis is always used to the right of an infinity symbol or to the left of a negative infinity symbol, as shown in the following examples.

Set-Builder Notation	Interval Notation	Graph
$\{x \mid x > 1\}$	$(1, \infty)$	
$\{x \mid x \geq 1\}$	$[1, \infty)$	
$\{x \mid x < 1\}$	$(-\infty, 1)$	
$\{x \mid x \leq 1\}$	$(-\infty, 1]$	
$\{x \mid -\infty < x < \infty\}$	$(-\infty, \infty)$	

INSTRUCTOR NOTE
Exercises 54 and 60 can be used for similar in-class examples.

EXAMPLE 8

a. Write $\{x \mid 0 < x \leq 5\}$ using interval notation.
b. Write $(-\infty, 9]$ using set-builder notation.

Solution **a.** The set is the real numbers greater than 0 and less than or equal to 5.
$(0, 5]$

b. The set is the real numbers less than or equal to 9.
$\{x \mid x \leq 9\}$

YOU TRY IT 8

a. Write $\{x \mid -8 \leq x < -1\}$ using interval notation.
b. Write $(-12, \infty)$ using set-builder notation.

Solution See page S2.   **a.** $[-8, -1)$ **b.** $\{x \mid x > -12\}$

INSTRUCTOR NOTE
Exercise 72 can be used for a similar in-class example.

EXAMPLE 9

Graph $(-\infty, 3) \cap [-1, \infty)$.

Solution $(-\infty, 3) \cap [-1, \infty)$ is the set of real numbers greater than or equal to -1 and less than 3.

YOU TRY IT 9

Graph $(-\infty, -2) \cup (-1, \infty)$.

Solution See page S2.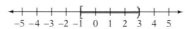

1.2 EXERCISES <u>Suggested Assignment: 7–81, odds</u>

Topics for Discussion

1. Explain the similarities and differences between rational and irrational numbers. Explanations will vary.

2. Explain the difference between the union of two sets and the intersection of two sets. Explanations will vary.

3. Explain the difference between $\{x \mid x < 5\}$ and $\{x \mid x \leq 5\}$.

   $\{x \mid x < 5\}$ does not include the element 5, whereas $\{x \mid x \leq 5\}$ does include the

   element 5.

4. Explain the similarities and differences between open intervals and closed intervals. Explanations will vary.

5. **a.** Is the intersection of two infinite sets always an infinite set? Explain your reasoning.
   b. Is the union of two infinite sets always an infinite set? Explain your reasoning.
   a. No. Explanations will vary. **b.** Yes. Explanations will vary.

▪ Sets of Numbers

Determine which of the numbers are **a.** natural numbers, **b.** whole numbers, **c.** integers, **d.** positive integers, **e.** negative integers, **f.** prime numbers. List all that apply.

6. $-14, 9, 0, 53, 7.8, -626$

   a. 9, 53 b. 0, 9, 53 c. -14, 9, 0, 53, -626 d. 9, 53
   e. $-14, -626$ f. 53

7. $31, -45, -2, 9.7, 8600, \dfrac{1}{2}$

   a. 31, 8600 b. 31, 8600 c. 31, -45, -2, 8600
   d. 31, 8600 e. -45, -2, f. 31

Determine which of the numbers are **a.** integers, **b.** rational numbers, **c.** irrational numbers, **d.** real numbers. List all that apply.

8. $-\dfrac{15}{2}, 0, -3, \pi, 2.\overline{33}, 4.232232223 \ldots, \dfrac{\sqrt{5}}{4}, \sqrt{7}$

   a. 0, -3 b. $-\frac{15}{2}$, 0, -3, $2.\overline{33}$

   c. π, $4.232232223 \ldots$, $\frac{\sqrt{5}}{4}$, $\sqrt{7}$ d. all

9. $-17, 0.3412, \dfrac{3}{\pi}, -1.010010001 \ldots, \dfrac{27}{91}, 6.1\overline{2}$

   a. -17 b. -17, 0.3412, $\frac{27}{91}$, $6.1\overline{2}$

   c. $\frac{3}{\pi}$, $-1.010010001 \ldots$ d. all

Use the roster method to list the elements of each set.

10. The integers between -3 and 5
    $\{-2, -1, 0, 1, 2, 3, 4\}$

11. The integers between -4 and 0
    $\{-3, -2, -1\}$

12. The even natural numbers less than or equal to 10 $\{2, 4, 6, 8, 10\}$

13. The odd natural numbers less than 15
    $\{1, 3, 5, 7, 9, 11, 13\}$

14. The letters in the word Mississippi $\{i, m, p, s\}$

15. The letters in the word banana $\{a, b, n\}$

16. The odd numbers evenly divisible by 2 $\varnothing$

17. The natural numbers less than 0 $\varnothing$

Use set-builder notation to write the set.

18. The integers greater than 7
$\{x \mid x > 7, x \in \text{integers}\}$

19. The integers less than -5
$\{x \mid x < -5, x \in \text{integers}\}$

20. The real numbers less than or equal to 0
$\{x \mid x \le 0\}$

21. The real numbers greater than or equal to -4
$\{x \mid x \ge -4\}$

22. The real numbers between -1 and 4
$\{x \mid -1 < x < 4\}$

23. The real numbers between -2 and 5
$\{x \mid -2 < x < 5\}$

For Exercises 24 to 29, answer True or False.

24. $7 \in \{2, 3, 5, 7, 9\}$
True

25. $4 \notin \{-8, -4, 0, 4, 8\}$
False

26. $\emptyset \in \{0, 1, 2, 4\}$
False

27. $\{a\} \in \{a, b, c, d, e\}$
False

28. $5 \in \{x \mid x \in \text{prime numbers}\}$
True

29. $0 \in \emptyset$
False

Graph.

30. $\{x \mid x < 2\}$

31. $\{x \mid x < -1\}$

32. $\{x \mid x \ge 1\}$

33. $\{x \mid x \le -2\}$

■ Union and Intersection of Sets

For Exercises 34 to 37, find $A \cup B$.

34. $A = \{1, 4, 9\}, B = \{2, 4, 6\}$
$\{1, 2, 4, 6, 9\}$

35. $A = \{2, 3, 5, 8\}, B = \{9, 10\}$
$\{2, 3, 5, 8, 9, 10\}$

36. $A = \{x \mid x \in \text{whole numbers}\}$
$B = \{x \mid x \in \text{positive integers}\}$
$\{x \mid x \in \text{whole numbers}\}$

37. $A = \{x \mid x \in \text{rational numbers}\}$
$B = \{x \mid x \in \text{real numbers}\}$
$\{x \mid x \in \text{real numbers}\}$

For Exercises 38 to 41, find $A \cap B$.

38. $A = \{6, 12, 18\}, B = \{3, 6, 9\}$ $\{6\}$

39. $A = \{2, 4, 6, 8, 10\}, B = \{4, 6\}$ $\{4, 6\}$

40. $A = \{x \mid x \in \text{rational numbers}\}$
$B = \{x \mid x \in \text{real numbers}\}$
$\{x \mid x \in \text{rational numbers}\}$

41. $A = \{x \mid x \in \text{rational numbers}\}$
$B = \{x \mid x \in \text{irrational numbers}\}$
$\emptyset$

42. Let $B = \{2, 4, 6, 8, 10\}$ and $C = \{2, 3, 5, 7\}$. Find $B \cup C$ and $B \cap C$. $B \cup C = \{2, 3, 4, 5, 6, 7, 8, 10\}; B \cap C = \{2\}$

43. Let $M = \{1, 4, 6, 8, 9, 10\}$ and $C = \{2, 3, 5, 7\}$. Find $M \cup C$ and $M \cap C$.
$M \cup C = \{1, 2, 3, 4, 5, 6, 7, 8, 9, 10\}; M \cap C = \emptyset$

Graph.

44. $\{x \mid x > 1\} \cup \{x \mid x < -1\}$

45. $\{x \mid x \le 2\} \cup \{x \mid x > 4\}$

46. $\{x \mid x \le 2\} \cap \{x \mid x \ge 0\}$

47. $\{x \mid x > -1\} \cap \{x \mid x \le 4\}$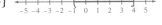

48. $\{x \mid 0 \le x \le 3\}$

49. $\{x \mid -1 < x < 5\}$

50. $\{x \mid 1 < x < 3\}$

51. $\{x \mid -1 \le x \le 1\}$

52. $\{x \mid x > 1\} \cap \{x \mid x \ge -2\}$

53. $\{x \mid x < -2\} \cup \{x \mid x < -4\}$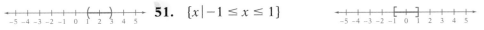

■ Interval Notation

For Exercises 54 to 59, write the interval in set-builder notation.

54. $(0, 8)$ $\{x | 0 < x < 8\}$

55. $[-5, 7]$ $\{x | -5 \le x \le 7\}$

56. $[-3, 6)$ $\{x | -3 \le x < 6\}$

57. $(-9, 5]$ $\{x | -9 < x \le 5\}$

58. $(-\infty, 4]$ $\{x | x \le 4\}$

59. $[-2, \infty)$ $\{x | x \ge -2\}$

For Exercises 60 to 68, write the set of real numbers in interval notation.

60. $\{x | -2 < x < 4\}$ $(-2, 4)$

61. $\{x | 0 \le x \le 3\}$ $[0, 3]$

62. $\{x | -4 \le x < -1\}$ $[-4, -1)$

63. $\{x | -2 \le x < 7\}$ $[-2, 7)$

64. $\{x | -10 < x \le -6\}$ $(-10, -6]$

65. $\{x | x \le -5\}$ $(-\infty, -5]$

66. $\{x | x < -2\}$ $(-\infty, -2)$

67. $\{x | x > 23\}$ $(23, \infty)$

68. $\{x | x \ge -8\}$ $[-8, \infty)$

Graph.

69. $[0, 3]$

70. $(-1, 4]$

71. $(1, \infty)$

72. $(-\infty, 2] \cup [4, \infty)$

73. $(-3, 4] \cup [-1, 5)$

74. $[-1, 2] \cap [0, 4]$

75. $[-5, 4) \cap (-2, \infty)$

76. $(2, \infty) \cup (-2, 4]$

77. $(-\infty, 2] \cup (4, \infty)$

Applying Concepts

78. Let $C = \{5, 12, 15, 17\}$. If $A = \{5, 12\}$ and $A \cap B = \{5\}$ and $A \cup B = C$, find the sum of the numbers that are elements of set B.   37

79. 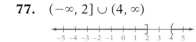 Some search engines on the World Wide Web make use of the operators "AND" and "OR." For instance, using the search engine Excite, a recent search for

"chocolate" produced 105,512 sites that mention the word *chocolate*
"dessert" produced 40,209 sites that mention the word *dessert*
"chocolate AND dessert" produced 9162 sites that mention the word *chocolate* and the word *dessert*

a. Explain this search in the context of the intersection of two sets.
b. Explain how a search engine might respond to a search for "chocolate OR dessert." How is this related to the union of two sets?
Explanations will vary.

80. Why are 2 and 5 the only prime numbers whose difference is 3?
Answers may vary. For example, 2 is the only even number that is a prime number. Since all other prime numbers are odd numbers, they differ by a multiple of 2.

81. What is a well-defined set? Provide examples of sets that are not well defined.

A set is well defined if it is possible to determine whether any given item is an element of the set. Examples will vary. For example, the set of great songs is not a well-defined set.

EXPLORATION

1. *Examining a Set of Positive Integers*   Let S be the set of positive integers that have the following property:

when divided by 6 leaves a remainder of 5
when divided by 5 leaves a remainder of 4
when divided by 4 leaves a remainder of 3
when divided by 3 leaves a remainder of 2
when divided by 2 leaves a remainder of 1

a. Find three elements of set S.

The LCM of 2, 3, 4, 5, and 6 is 60; one less than any multiple of 60 will generate an element of S. The form of these elements is $60k - 1$, where k is a natural number. Some possible values are 59, 119, 179, and 239.

b. Find the minimum value of S.

The minimum value occurs when $k = 1$. Therefore, the minimum value is 59.

c. Suppose the property is extended further to also include

when divided by 7 leaves a remainder of 6
when divided by n leaves a remainder of $n - 1$

Express in terms of n the least positive integer that satisfies this set of properties. $[\text{LCM} \{2, 3, 4, \ldots, n\}] - 1$

SECTION **1.3**

Evaluating Variable Expressions Using Integers

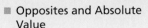
- Opposites and Absolute Value
- Evaluate Variable Expressions
- Addition and Subtraction of Integers
- Multiplication and Division of Integers
- Exponential Expressions
- The Order of Operations Agreement

■ Opposites and Absolute Value

Two numbers that are the same distance from zero on the number line but are on opposite sides of zero are **opposite numbers,** or **opposites.** The opposite of a number is also called its **additive inverse.**

The opposite of 5 is -5.
The opposite of -5 is 5.

The negative sign can be used to indicate "the opposite of."

$$-(2) = -2. \text{ The opposite of 2 is negative 2.}$$
$$-(-2) = 2. \text{ The opposite of } -2 \text{ is 2.}$$

▼ *Point of Interest*

The topic of this section is integers. Recall that the integers are the numbers
. . ., −3, −2, −1, 0, 1, 2, 3, . . .
    The word integer *comes directly from the Latin word* integer, *which means "whole, complete, perfect, entire."*
In fact, integer *and* entire *have the same word origin.*

INSTRUCTOR NOTE
In response to instructors' requests, the sections of this textbook are organized by topic, not by class presentation.

? QUESTION **a.** What is the opposite of −43?
b. What is the additive inverse of 51?
c. What is the opposite of 0?

The **absolute value** of a number is its distance from zero on the number line. The symbol for absolute value is two vertical bars, $||$.

The distance from 0 to 4 is 4. Therefore, the absolute value of 4 is 4.

$|4| = 4$

The distance from 0 to −4 is 4. Therefore, the absolute value of −4 is 4.

$|-4| = 4$

Note that because the absolute value of a number is its distance from zero on the number line, the absolute value of a number is nonnegative.

Absolute Value

The absolute value of a positive number is the number itself.

The absolute value of zero is zero.

The absolute value of a negative number is the opposite of the negative number.

TAKE NOTE
An alternative definition of absolute value is

$$|x| = \begin{cases} x, & x \geq 0 \\ -x, & x < 0 \end{cases}$$

To **evaluate an expression** means to determine what number the expression is equal to. Example 1 illustrates evaluating absolute value expressions.

EXAMPLE 1

Evaluate. **a.** $|-14|$ **b.** $|38|$ **c.** $-|-26|$

Solution **a.** $|-14| = 14$
b. $|38| = 38$
c. $-|-26| = -26$ • The negative sign *in front of* the absolute value sign means the opposite of $|-26|$.

YOU TRY IT 1

Evaluate. **a.** $|47|$ **b.** $|-50|$ **c.** $-|-89|$

Solution See page S2. **a.** 47 **b.** 50 **c.** −89

INSTRUCTOR NOTE
Exercise 28 can be used for a similar in-class example.

? ANSWERS **a.** The opposite of −43 is 43. **b.** The additive inverse of 51 is −51. **c.** The opposite of 0 is 0.

An expression containing absolute value can be evaluated by using the 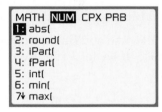 key. A check of the evaluation of the expressions in Example 1a and 1c above is shown below.

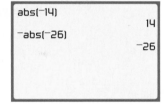

 See Appendix A:
Basic Operations

Note that on a graphing calculator there is a difference between the minus key ▬ and the negative key ⟨(-)⟩. The negative key was used above to enter negative numbers.

■ Evaluate Variable Expressions

Suppose that gasoline costs $1.50 per gallon. Then the amount you spend for gas for your car depends on how many gallons you purchase. Here are some examples.

A purchase of 5 gallons costs 1.50(5) = $7.50.

A purchase of 7 gallons costs 1.50(7) = $10.50.

A purchase of 11.3 gallons costs 1.50(11.3) = $16.95.

If you decide to fill the tank, you may not know how many gallons of gas will be required. In that case, you might use a letter, such as *g*, to represent the number of gallons that will be purchased.

A purchase of *g* gallons costs 1.50*g*.

A **variable** is a letter that represents a quantity that can change, or vary. The expression 1.50*g* (which means 1.50 times *g*) is a **variable expression.** The number 1.50 is the **coefficient** of the variable.

The expression 1.50*g* represents the cost to purchase *g* gallons of gas. As *g* (the number of gallons purchased) changes, the cost of the purchase changes.

INSTRUCTOR NOTE
Have students use the variable expression 1.50g to find the cost to purchase 9.5 gallons of gas. [Answer: $14.25]

➡ Use the variable expression 1.50*g* to find the cost to purchase 12.5 gallons of gas.

1.50*g*

1.50(12.5)   • Replace the variable *g* by 12.5.

= 18.75   • Multiply 1.50 times 12.5.

The cost for 12.5 gallons of gas is $18.75.   ⬅

Replacing a variable in a variable expression, as we did in Example 2, by a number and then simplifying the resulting numerical expression is called **evaluating the variable expression.** The number substituted for the variable (12.5) is called the **value of the variable.** The result (18.75) is called the **value of the variable expression.**

We could also use a variable expression to represent the total cost of a gasoline purchase. If we let T represent the total cost, in dollars, for a gasoline purchase, then

$$T = 1.50g$$

This *equation* shows the relationship between g, the number of gallons of gas purchased, and T, the total cost of the purchase.

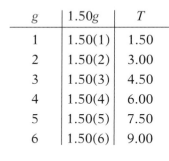

g	$1.50g$	T
1	1.50(1)	1.50
2	1.50(2)	3.00
3	1.50(3)	4.50
4	1.50(4)	6.00
5	1.50(5)	7.50
6	1.50(6)	9.00

We can also use the equation $T = 1.50g$ to prepare an **input/output table,** which shows how T changes as g changes. The input is g, the number of gallons of gas purchased. The output is T, the total cost of the purchase. For the input-output table at the left, we have chosen 1, 2, 3, 4, 5, and 6 as the values of g, but other values of g could have been used.

? QUESTION   In the table at the left, what is the meaning of the number 7.50?

See Appendix A:

A graphing calculator can be used to create input/output tables for equations such as $T = 1.50g$. This is accomplished by using the ⬤Y= editor screen. The output variable is designated as one of the calculator's Y variables. For this example, we will designate T as Y_1. The input variable is usually designated by X. Thus the equation would appear as $Y_1 = 1.50X$.

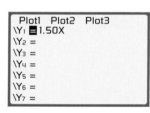

See Appendix A:
Table

To create the table, we use the TABLE feature of the calculator. Some typical screens for the gasoline purchase follow.

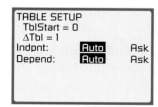

In the Table Setup screen, TblStart is the beginning value of X, and ΔTbl is the difference between successive values of X. The difference between any two successive X values is called the **change in X** or the **increment in X.** The symbol Δ is frequently used to represent the phrase *the change in*.

You can scroll down the screen by repeatedly pressing the down arrow key ⬤ to view greater values of X. Use the up arrow key to scroll up the screen to view lesser values of X.

X	Y_1	
5	7.5	
6	9	
7	10.5	
8	12	
9	13.5	
10	15	
11	16.5	
X = 11		

? ANSWER   7.50 is the output when the input is 5. The 7.50 means that it costs $7.50 for 5 gallons of gas.

INSTRUCTOR NOTE
Exercise 36 can be used for
a similar in-class example.

EXAMPLE 2

When a rock is dropped off a cliff, the distance the rock falls is given by the equation $d = 16t^2$, where d is the distance in feet that the rock has fallen and t is the time in seconds that the rock has been falling.

a. Create an input/output table for this equation. Use increments of 0.5 second, beginning with $t = 0$.

b. What is the meaning of the number 64 in the table?

Solution   a. The input variable is t, the number of seconds the rock has been falling. The output variable is d, the distance the rock falls.

ALGEBRAIC SOLUTION

t	d
0	0
0.5	4
1	16
1.5	36
2	64
2.5	100
3	144
3.5	196
4	256

GRAPHICAL CHECK

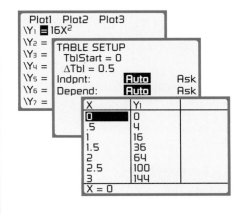

b. The number 64 is the output when the input is 2.

The number 64 means that the rock falls 64 feet in 2 seconds.

YOU TRY IT 2

Suppose that the average speed of an American Airlines flight from Los Angeles to Boston is 525 mph. Then the distance, in miles, that the plane is from Boston is given by the equation $d = 2650 - 525t$, where t is the number of hours since the plane left Los Angeles.

a. Create an input/output table for this equation. Use increments of 0.5 hour, beginning with $t = 0$.

b. What is the meaning of the number 1862.5 in the table?

Solution   See page S2.   See the solution on page S2.

■ Addition and Subtraction of Integers

▼ *Point of Interest*

Arrows are also used to represent numbers in engineering. Engineers call these arrows vectors and use them in many different situations.

A number can be represented by an arrow. A positive number is represented by an arrow pointing to the right, and a negative number is represented by an arrow pointing to the left. The magnitude (absolute value) of the number is represented by the length of the arrow. Arrows representing -6 and 4 are shown below.

? QUESTION   What number is represented by the arrow shown at the right?*

Addition is the process of finding the total of two numbers. The numbers being added are called **addends.** The total is called the **sum.** We can find rules for adding integers by using a number line and the arrow representation of numbers.

To add two integers, find the point on the number line that corresponds to the first addend. At that point, draw an arrow representing the second addend. The sum is the number directly below the tip of the arrow.

$3 + 5 = 8$

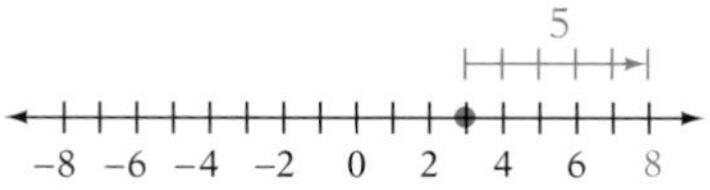

$-3 + (-5) = -8$

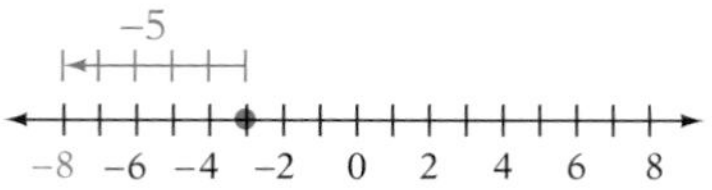

$3 + (-5) = -2$

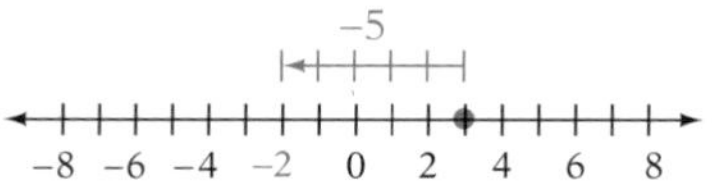

$-3 + 5 = 2$

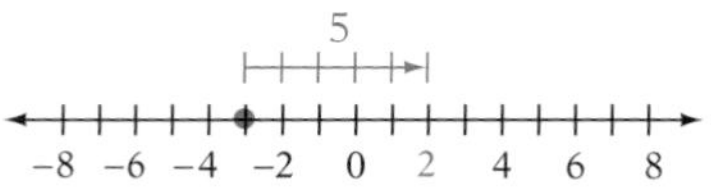

The arrow models for adding integers suggest the following rule.

Addition of Integers

Integers with the same sign

To add two numbers with the same sign, add the absolute values of the numbers. Then attach the sign of the addends.

Integers with different signs

To add two numbers with different signs, find the absolute value of each number. Then subtract the lesser of these absolute values from the greater. Attach the sign of the number with the greater absolute value.

? QUESTION   Which number has the greater absolute value, 37 or −62?†

INSTRUCTOR NOTE
Exercises 38 and 56 can be used for similar in-class examples.

EXAMPLE 3

  a. Find the sum of −14 and −48.
  b. Evaluate $a + b$ for $a = 23$ and $b = -51$.

? ANSWERS   *Because the arrow points to the left and is of length 5, the number is −5.
†$|37| = 37$ and $|-62| = 62$. Since $62 > 37$, −62 has a greater absolute value than 37.

Suggested Activity

There are a number of models of the addition of integers. Using arrows on the number line is just one of them. Another suggestion is to use a checking account. If there is a balance of $25 in a checking account, and a check is written for $30, the account will be overdrawn by $5 (−5).

Another model uses two colors of plastic chips, say blue for positive and red for negative, and the idea that a blue/red pair is equal to zero. To add −8 + 3, place 8 red and 3 blue chips in a circle. Make as many blue/red pairs as possible, and remove them from the region. There are 5 red chips remaining, or −5.

Solution   **a.** The word *sum* indicates addition.

$$-14 + (-48)$$

• The signs are the same. Find the absolute value of each number.

$$|-14| = 14, \; |-48| = 48$$

$$14 + 48 = 62$$

• Add the absolute values of the numbers.

$$-14 + (-48) = -62$$

• Both numbers are negative. The answer is negative.

b. $a + b$

$$23 + (-51)$$

• Replace a by 23 and b by -51.

$$|23| = 23, \; |-51| = 51$$

• The signs are different. Find the absolute value of each number.

$$51 - 23 = 28$$

• Subtract the lesser of the absolute values from the greater.

$$23 + (-51) = -28$$

• Because $|-51| > |23|$, the answer has the same sign as -51.

Check:

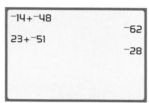

YOU TRY IT 3

a. What is the sum of -52 and 36?
b. Evaluate $c + d$ for $c = -18$ and $d = 9$.

Solution   See page S2.   a. -16 b. -9

Solution   See page S2.

The $-$ sign is used in two different ways: to mean *subtract*, as in $9 - 4$ (9 **minus** 4), and to mean *the opposite of*, as in -4 (the **opposite** of 4, or **negative** 4).

Look at the next four examples and be sure you understand the difference between *minus* (meaning subtract) and *negative* (meaning the opposite of).

$9 - 4$	positive 9 minus positive 4
$-9 - 4$	negative 9 minus positive 4
$9 - (-4)$	positive 9 minus negative 4
$-9 - (-4)$	negative 9 minus negative 4

❓ QUESTION   Write the expression in words.   **a.** $-6 - 4$ **b.** $8 - (-7)$

▼ *Point of Interest*

In mathematics manuscripts dating from the 1500s, an m was used to indicate minus. Some historians believe that, over time, the m went from m to ᴍ to ⌒ to —.

INSTRUCTOR NOTE
Provide students with several examples of expressions that require them to distinguish between a minus sign and a negative sign. Here are some examples:
$7 - 6$
$3 - (-9)$
$-4 - 1$
$-5 - (-2)$
$(-8) - 10$

❓ ANSWERS   **a.** negative 6 minus positive 4   **b.** positive 8 minus negative 7

Suggested Activity

A subtraction model based on blue and red chips similar to that of addition can be provided. Restrict the terms of the subtraction to, say, between -10 and 10, and start with 10 blue/red pairs in a circle. Because each blue/red pair is equal to zero, the circle contains 10 zeros. To model $-3 - (-7)$, place 3 more red chips in the circle and remove (subtract) any 7 red chips. Now pair as many blue and red chips as possible. There will be 4 blue chips left without a red chip. In other words, $-3 - (-7) = 4$.

Subtraction is the process of finding the difference between two numbers. Look at the following three problems. Opposites are used to rewrite subtraction problems as related addition problems. Notice below that the subtraction of whole numbers is the same as the addition of the opposite number.

Subtraction		Addition of the Opposite	
$8 - 4$	$=$	$8 + (-4)$	$= 4$
$7 - 5$	$=$	$7 + (-5)$	$= 2$
$9 - 2$	$=$	$9 + (-2)$	$= 7$

Subtraction of integers can be written as the addition of the opposite number. To subtract two integers, rewrite the subtraction expression as the first number plus the opposite of the second number. Some examples follow.

first number	$-$	second number	$=$	first number	$+$	opposite of the second number		
15	$-$	20	$=$	15	$+$	(-20)	$=$	-5
15	$-$	(-20)	$=$	15	$+$	20	$=$	35
-15	$-$	20	$=$	-15	$+$	(-20)	$=$	-35
-15	$-$	(-20)	$=$	-15	$+$	20	$=$	5

Subtraction of Integers

To subtract two numbers, add the opposite of the second number to the first number.

INSTRUCTOR NOTE
Exercises 46, 52, and 64 can be used for similar in-class examples.

CALCULATOR NOTE

Recall that on a graphing calculator, the ⬛ key is the "minus" key; it is used for subtraction. The $(-)$ key is the "negative" key; it is used to enter a negative sign. You will notice in the display that the negative sign is slightly higher and shorter than the minus sign.

See Appendix A:
Basic Operations

EXAMPLE 4

a. What is the difference between 9 and -17?
b. Evaluate: $2 - (-24) - 18 - (-3)$
c. Evaluate $a - b$ for $a = -43$ and $b = 25$.

Solution   **a.** The word *difference* indicates subtraction.

$$9 - (-17) = 9 + 17 \qquad \bullet \text{ Add the opposite of } -17 \text{ to } 9.$$
$$= 26$$

b. $2 - (-24) - 18 - (-3)$
$$= 2 + 24 + (-18) + 3 \qquad \bullet \text{ Rewrite each subtraction as}$$
$$= 26 + (-18) + 3 \qquad\qquad \text{addition of the opposite. Then}$$
$$= 8 + 3 \qquad\qquad\qquad\quad \text{add the numbers.}$$
$$= 11$$

c. $a - b$
$$-43 - 25 \qquad \bullet \text{ Replace } a \text{ by } -43 \text{ and } b \text{ by } 25.$$
$$= -43 + (-25) = -68 \qquad \bullet \text{ Add the opposite of } 25 \text{ to } -43.$$

Check:

```
9--17
                    26
2--24-18--3
                    11
-43-25
                    -68
```

YOU TRY IT 4

a. Find the difference between −8 and −26.
b. Evaluate: −15 − 12 − 9 − (−36)
c. Evaluate $a - b$ for $a = 46$ and $b = 72$.

Solution   See page S2.   a. 18   b. 0   c. −26

■ Multiplication and Division of Integers

One method of developing rules for multiplication of integers is to look for a pattern. Consider multiplying a decreasing sequence of integers by 5.

Decreasing sequence of integers

4	3	2	1	0	−1	−2	−3	−4
5(4)	5(3)	5(2)	5(1)	5(0)	5(−1)	5(−2)	5(−3)	5(−4)
20	15	10	5	0	?	?	?	?

Using inductive reasoning, it appears that each term of the sequence of products (20, 15, 10, 5, 0) is 5 less than the previous term. To continue this pattern, the question marks should be replaced by −5, −10, −15, and −20.

Decreasing sequence of integers

4	3	2	1	0	−1	−2	−3	−4
5(4)	5(3)	5(2)	5(1)	5(0)	5(−1)	5(−2)	5(−3)	5(−4)
20	15	10	5	0	−5	−10	−15	−20

This suggests that the product of a positive number and a negative number is a negative number.

Now consider multiplying a decreasing sequence of integers by −5.

Decreasing sequence of integers

4	3	2	1	0	−1	−2	−3	−4
−5(4)	−5(3)	−5(2)	−5(1)	−5(0)	−5(−1)	−5(−2)	−5(−3)	−5(−4)
−20	−15	−10	−5	0	?	?	?	?

Using inductive reasoning, it appears that each term of the sequence of products $(-20, -15, -10, -5, 0)$ is 5 more than the previous term. To continue this pattern, the question marks should be replaced by 5, 10, 15, and 20.

Decreasing sequence of integers

4	3	2	1	0	-1	-2	-3	-4
$-5(4)$	$-5(3)$	$-5(2)$	$-5(1)$	$-5(0)$	$-5(-1)$	$-5(-2)$	$-5(-3)$	$-5(-4)$
-20	-15	-10	-5	0	5	10	15	20

This suggests that the product of two negative numbers is a positive number.

Multiplication of Integers

Integers with the same sign

To multiply two numbers with the same sign, multiply the absolute values of the numbers. The product is positive.

Integers with different signs

To multiply two numbers with different signs, multiply the absolute values of the numbers. The product is negative.

INSTRUCTOR NOTE
Exercises 78 and 86 can be used for a similar in-class examples.

TAKE NOTE

When two or more variables are written together in an expression, the operation is multiplication. Thus mn in Example 5b means m times n.

EXAMPLE 5

a. Find the product of -4 and -27.
b. Evaluate mn for $m = 6$ and $n = -13$.

Solution   **a.** The word *product* indicates multiplication.

$$-4(-27) = 4 \cdot 27 = 108$$

• Multiply the absolute values. The signs are the same, so the product is positive.

b. mn

$$6(-13)$$

• Replace m by 6 and n by -13.

$$= -(6 \cdot 13) = -78$$

• Multiply the absolute values. The signs are different, so the product is negative.

Check:
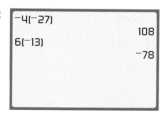

YOU TRY IT 5

a. What is -5 times 33?
b. Evaluate pr for $p = -18$ and $r = -21$.

Solution   See page S2.   **a.** -165 **b.** 378

For every division problem there is a related multiplication problem. For instance,

Division   Related Multiplication

$$\frac{12}{4} = 3 \quad \text{because} \quad 3 \cdot 4 = 12$$

Extending this to negative integers, we have

$$\frac{-12}{4} = -3 \quad \text{because} \quad -3 \cdot 4 = -12$$

$$\frac{-12}{-4} = 3 \quad \text{because} \quad 3 \cdot (-4) = -12$$

$$\frac{12}{-4} = -3 \quad \text{because} \quad -3 \cdot (-4) = 12$$

This suggests the following rules for dividing integers.

Division of Integers

Integers with the same sign

To divide two numbers with the same sign, divide the absolute values of the numbers. The quotient is positive.

Integers with different signs

To divide two numbers with different signs, divide the absolute values of the numbers. The quotient is negative.

INSTRUCTOR NOTE
To relate zero in division to a real situation, explain that 6 ÷ 3 means that if $6 is divided equally among 3 people, each person receives $2. If $0 is divided equally among 3 people, each person receives $0 (0 ÷ 3 = 0). Now, if $6 is divided equally among 0 people (6 ÷ 0), how many dollars does each person receive?

The relationship between division and multiplication can be used to illustrate principles of division involving zero and one.

$0 \div 3 = 0$ because $0 \cdot 3 = 0$.   **Zero divided by any number except zero is zero.**

$3 \div 3 = 1$ because $1 \cdot 3 = 3$.   **A number other than zero divided by itself is 1.**

$3 \div 1 = 3$ because $3 \cdot 1 = 3$.   **A number divided by 1 is the number.**

$3 \div 0 = ?$ $? \cdot 0 = 3$ What number can be multiplied by 0 to get 3? There is no number whose product with 0 is 3, because the product of a number and zero is 0. **Division by zero is undefined.**

INSTRUCTOR NOTE
Exercises 82 and 96 can be used for similar in-class examples.

EXAMPLE 6

a. What is the quotient of 36 and -12?

b. What is -18 divided by 1?

c. Evaluate $\dfrac{x}{y}$ for $x = -72$ and $y = -9$.

Solution **a.** The word *quotient* indicates division.

$$36 \div (-12) = -3$$

• Divide the absolute values. The signs are different, so the quotient is negative.

b. $-18 \div 1 = -18$

• A number divided by 1 is the number.

c. $\dfrac{x}{y}$

$$\dfrac{-72}{-9} = 8$$

• Replace x by -72 and y by -8. The fraction bar can be read "divided by." $\frac{-72}{-9}$ means -72 divided by -9. Divide the absolute values of the numbers. The signs are the same, so the quotient is positive.

Check:

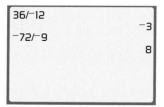

```
36/-12
              -3
-72/-9
              8
```

YOU TRY IT 6

a. Find the quotient of -121 and -11.
b. What is -24 divided by 0?
c. Evaluate $\dfrac{m}{n}$ for $m = 96$ and $n = -8$.

Solution See page S3. **a.** 11 **b.** undefined **c.** -12

■ Exponential Expressions

Genealogy is the study of the history of a family. A genealogist uses a family tree as a pictorial history of your ancestors. A sample of a family tree follows.

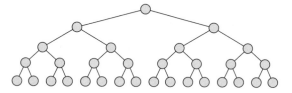

You
Parents
Grandparents
Great-grandparents
Great-great-grandparents

In each generation, you have twice as many ancestors as in the previous generation.

$$2 \text{ Parents}$$
$$2 \cdot 2 = 4 \text{ Grandparents}$$
$$2 \cdot 2 \cdot 2 = 8 \text{ Great-grandparents}$$
$$2 \cdot 2 \cdot 2 \cdot 2 = 16 \text{ Great-great-grandparents}$$

Instead of writing the number of ancestors as a product, in what is called **expanded form,** we could have written each product in **exponential form.**

Expanded form		Exponential form	Read as
2	=	2^1	"2 to the first power" or just "two."
$2 \cdot 2$	=	2^2	"2 squared" or "2 to the second power."
$2 \cdot 2 \cdot 2$	=	2^3	"2 cubed" or "2 to the third power."
$2 \cdot 2 \cdot 2 \cdot 2$	=	2^4	"2 to the fourth power."

In an exponential expression, the **exponent** indicates the number of times the **base** is used in a product.

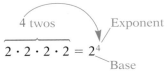

4 twos   Exponent

$2 \cdot 2 \cdot 2 \cdot 2 = 2^4$

Base

▼ *Point of Interest*

René Descartes (1596–1650) was the first mathematician to use exponential notation as it is used today.

Look at the following examples of exponential expressions involving negative signs.

$(-3)^2 = (-3)(-3) = 9$ The (-3) is squared. Multiply -3 by -3.
$-(3^2) = -(3 \cdot 3) = -9$ Read $-(3^2)$ as "the opposite of three squared." 3^2 is 9. The opposite of 9 is -9.

$-3^2 = -(3^2) = -9$ The expression -3^2 is the same as $-(3^2)$.

 QUESTION   How is each of the following expressions simplified?

a. -4^2 **b.** $(-4)^2$ **c.** $-(4^2)$

The caret key  on a graphing calculator is used to enter an exponent. The expressions -4^2, $(-4)^2$, and $-(4^2)$ are evaluated on the graphing calculator screen at the left.

There is a second method of squaring a number on a graphing calculator: press x^2. For example, to evaluate -4^2, press

(−)   4 x^2 ENTER

The x^2 key displays the exponent 2 to the right of the 4.

INSTRUCTOR NOTE
Exercises 114 and 120 can be used for similar in-class examples.

EXAMPLE 7

a. Evaluate -6^4.   **b.** Evaluate $b^3 c^2$ for $b = 2$ and $c = -3$.

Solution   **a.** $-6^4 = -(6 \cdot 6 \cdot 6 \cdot 6)$ • Write the exponential expression in expanded form. Then multiply.
$= -1296$

b. $b^3 c^2$
$(2)^3(-3)^2$ • Replace b by 2 and c by -3.
$= (2 \cdot 2 \cdot 2) \cdot [(-3)(-3)]$ • Write each exponential expression in expanded form.
$= 8 \cdot 9 = 72$ • Multiply.

 ANSWERS   **a.** $-4^2 = -(4 \cdot 4) = -16$ **b.** $(-4)^2 = (-4)(-4) = 16$ **c.** $-(4^2) = -(4 \cdot 4) = -16$

Check:

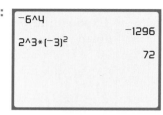

$$-6^4$$
$$-1296$$
$$2^3 * (-3)^2$$
$$72$$

YOU TRY IT 7

a. Evaluate $(-8)^4$.   **b.** Evaluate y^3z^2 for $y = 3$ and $z = 5$.

Solution   See page S3.   a. 4096   b. 675

Suggested Activity

See Section 1.3 of the *Student Activity Manual* for a lesson involving operations with integers.

▪ The Order of Operations Agreement

You buy a poster for $7 and two CDs, each priced at $18. The total cost of the items purchased is

$$7 + 2(18)$$

Note that the expression we use to find the total cost of the two purchases includes two arithmetic operations, addition and multiplication. Which operation should be performed first?

The correct answer is that we multiply first to find the cost of the two CDs. We then add the cost of the two CDs (36) to the cost of the poster (7) to determine that the total cost of the three items is $43.

Whenever an expression contains more than one operation, the operations must be performed in a specified order, as listed below in the Order of Operations Agreement.

The Order of Operations Agreement

Step 1   Perform operations inside grouping symbols. Grouping symbols include parentheses (), brackets [], braces {}, the absolute value symbol | |, and fraction bars.

Step 2   Simplify exponential expressions.

Step 3   Do multiplication and division as they occur from left to right.

Step 4   Do addition and subtraction as they occur from left to right.

❓ QUESTION   What operations are in the expression $67 - 3 \cdot 9$? Which operation must be performed first?

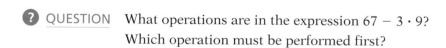

❓ ANSWER   The expression contains the operations of subtraction and multiplication. The multiplication must be performed first.

INSTRUCTOR NOTE
Exercise 138 can be used for
a similar in-class example.

EXAMPLE 8

a. Evaluate $\dfrac{a + b}{c} + a^2c$ for $a = 4$, $b = -8$, and $c = 2$.

b. Evaluate $a(b - c)^2 + a$ for $a = -2$, $b = 3$, and $c = 4$.

Solution

a. $\dfrac{a + b}{c} + a^2c$

$\dfrac{4 + (-8)}{2} + 4^2 \cdot 2$ • Replace each variable by its given value. $a = 4$, $b = -8$, and $c = 2$

$= \dfrac{-4}{2} + 4^2 \cdot 2$ • Simplify the expression in the numerator of the fraction.

$= \dfrac{-4}{2} + 16 \cdot 2$ • Simplify the exponential expression.

$= -2 + 16 \cdot 2$
$= -2 + 32$ • Do the multiplication and division from left to right. Recall that the fraction bar can be read "divided by."

$= 30$ • Add.

b. $a(b - c)^2 + a$

$-2(3 - 4)^2 + (-2)$ • Replace each variable by its given value. $a = -2$, $b = 3$, and $c = 4$

$= -2(-1)^2 + (-2)$ • Perform operations inside the parentheses.

$= -2(1) + (-2)$ • Simplify the exponential expression.

$= -2 + (-2)$ • Do the multiplication and division from left to right.

$= -4$ • Do the addition and subtraction from left to right.

Check:

Note: The numerator $(4 + {}^-8)$ in part a must be in parentheses so that the calculator understands that the entire expression, not just the number $^-8$, is in the numerator. This is related to Step 1 of the Order of Operations Agreement; the fraction bar is a grouping symbol.

```
(4+‾8)/2+4²*2
                    30
‾2(3‾4)²+‾2
                    ‾4
```

YOU TRY IT 8

a. Evaluate $a^2b \div c^3 - bc$ for $a = 4$, $b = 3$, and $c = 2$.

b. Evaluate $a(b - a)^2 - |c \div a|$ for $a = -4$, $b = -6$, and $c = 8$.

Solution   See page S3.   a. 0   b. −18

1.3 EXERCISES   Suggested Assignment: 13–29, every other odd; 31–55, odds; 57–157, every other odd

Topics for Discussion

For Exercises 1 to 6, determine whether the statement is always true, sometimes true, or never true. If the statement is sometimes true or never true, explain your answer.

1. A number and its opposite are different numbers.
   Sometimes true. (The opposite of 0 is 0.)
2. The absolute value of a number is positive.
   Sometimes true. (The absolute value of 0 is 0.)
3. The natural numbers plus the opposites of the natural numbers equals the integers.
   Never true. (The natural numbers do not include 0. The integers include 0.)
4. If two integers are added, the sum is greater than either of the two integers.
   Sometimes true.
5. If two integers are subtracted, the difference is less than either of the two integers.
   Sometimes true.
6. If two integers are multiplied, the product is greater than either of the two integers.
   Sometimes true.
7. What does it mean to evaluate a variable expression?
   It means to replace the variables in a variable expression with numbers and then simplify the resulting numerical expression.
8. Explain why division by zero is not allowed.
   Explanations will vary.
9. Explain why the absolute value of −24 is greater than the absolute value of 7.
   The distance from −24 to 0 on the number line is greater than the distance from 7 to 0.
10. **a.** Explain how to add two integers with the same sign.
    b. Explain how to add two integers with different signs.
    a. Add the absolute values of the numbers. Then attach the sign of the addends.
    b. Find the absolute value of each number. Then subtract the smaller of these absolute values from the larger one. Attach the sign of the number with the larger absolute value.

■ Opposites and Absolute Value

Find the opposite of the number.

11.	25	12.	42	13.	−34	14.	−45
	−25		−42		34		45
15.	0	16.	−7	17.	12	18.	−3
	0		7		−12		3

Evaluate.

19. $-(-16)$
16

20. $-(-30)$
30

21. $-(49)$
-49

22. $-(32)$
-32

23. $|16|$
16

24. $|25|$
25

25. $|-32|$
32

26. $|-21|$
21

27. $-|86|$
-86

28. $-|40|$
-40

29. $-|-54|$
-54

30. $-|-27|$
-27

▪ Evaluate Variable Expressions

For Exercises 31 to 34, create an input/output table for the equation using the given instructions. Use the input/output table to answer the questions.

31. $y = 3x + 2$, $\triangle$Tbl $= 1$, TblStart $= 0$

   a. What is the value of y when $x = 3$?
   b. What is the value of x when $y = 5$?

   a. 11   b. 1

Input, x	0	1	2	3	4	5
Output, $y = 3x + 2$	2	5	8	11	14	17

32. $y = 2x - 3$, $\triangle$Tbl $= 1$, TblStart $= 2$

   a. What is the value of y when $x = 6$?
   b. What is the value of x when $y = 5$?

   a. 9   b. 4

Input, x	2	3	4	5	6	7
Output, $y = 2x - 3$	1	3	5	7	9	11

33. $y = x^2 + 3x + 1$, $\triangle$Tbl $= 1$, TblStart $= 0$

   a. What is the value of y when $x = 3$?
   b. What is the value of x when $y = 29$?

   a. 19   b. 4

Input, x	0	1	2	3	4	5
Output, $y = x^2 + 3x + 1$	1	5	11	19	29	41

34. $y = x^2 + 5x - 2$, $\triangle$Tbl $= 1$, TblStart $= 1$

   a. What is the value of y when $x = 4$?
   b. What is the value of x when $y = 48$?

   a. 34   b. 5

Input, x	1	2	3	4	5	6
Output, $y = x^2 + 5x - 2$	4	12	22	34	48	64

35. *Aeronautics*   The altitude, or height above sea level, of a hot-air balloon is given by the equation $H = 100t + 1250$, where H is the altitude of the balloon t seconds after it has been released. Create an input/output table for this equation for increments of 1 second, beginning with $t = 0$.

   a. What is the altitude of the balloon 3 seconds after it is released?
   b. How many seconds after it is released is the balloon at an altitude of 1750 feet?

   a. 1550 ft   b. 5 s

36. *Sports*   The equation $h = -16t^2 + 64t + 5$ gives the height of a baseball thrown straight up, where h is the height of the baseball in feet and t is the time in seconds since the baseball was released. Create an input/output table for this equation for increments of 1 second, beginning with $t = 0$.

   a. What is the height of the baseball 2 seconds after it is released?
   b. The baseball is 53 feet above the ground at two different times, once on the way up and once on the way down. What are the two times?

   a. 69 ft   b. 1 s, 3 s

■ Addition and Subtraction of Integers

37. Find the sum of -8 and 11.
3

38. What is the total of -12 and -5?
-17

39. Find -9 added to -11.
-20

40. What is the sum of 17 and -21?
-4

41. Add -16, -8, and 14.
-10

42. Find the total of 32, -61, 17, and -44.
-56

43. Find the difference between 7 and 14.
-7

44. Subtract -24 from 24.
48

45. What is 32 minus -27?
59

46. What is the difference between -15 and 24?
-39

Evaluate.

47. $42 - (-30) - 65 - (-11)$
18

48. $12 - (-6) + 8$
26

49. $-8 - (-14) + 7$
13

50. $-4 + 6 - 8 - 2$
-8

51. $9 - 12 + 0 - 5$
-8

52. $11 - (-2) - 6 + 10$
17

53. $5 + 4 - (-3) - 7$
5

54. $-1 - 8 + 6 - (-2)$
-1

55. $-13 + 9 - (-10) - 4$
2

Evaluate the expression for the given values of the variables.

56. $x + y$ for $x = -5$ and $y = -7$
-12

57. $-a + b$ for $a = -8$ and $b = -3$
5

58. $a + b$ for $a = -8$ and $b = -3$
-11

59. $-x + y$ for $x = -5$ and $y = -7$
-2

60. $a + b + c$ for $a = -4$, $b = 6$, and $c = -9$
-7

61. $a + b + c$ for $a = -10$, $b = -6$, and $c = 5$
-11

62. $x + y + (-z)$ for $x = -3$, $y = 6$, and $z = -17$
20

63. $-x + (-y) + z$ for $x = -2$, $y = 8$, and $z = -11$
-17

64. $-x - y$ for $x = -3$ and $y = 9$
-6

65. $x - (-y)$ for $x = -3$ and $y = 9$
6

66. $-x - (-y)$ for $x = -3$ and $y = 9$
12

67. $a - (-b)$ for $a = -6$ and $b = 10$
4

68. $a - b - c$ for $a = 4$, $b = -2$, and $c = 9$
-3

69. $a - b - c$ for $a = -1$, $b = 7$, and $c = -15$
7

70. $x - y - (-z)$ for $x = -9$, $y = 3$, and $z = 30$
18

71. $-x - (-y) - z$ for $x = 8$, $y = 1$, and $z = -14$
7

Sports   The following table shows the top nine golfers in the 2001 Masters Golf Tournament. The golfers' scores in relation to par are given for the four rounds of play. Use this table for Exercises 72 and 73.

Name	Round 1	Round 2	Round 3	Round 4	Final Score
Calcavecchia, M.	0	–6	–4	0	–10
Duval, D.	–1	–6	–2	–5	–14
Els, E.	–1	–4	–4	0	–9
Furyk, J.	–3	–1	–2	–3	–9
Izawa, T.	–1	–6	2	–5	–10
Langer, B.	1	–3	–4	–3	–9
Michelson, P.	–5	–3	–3	–2	–13
Triplett, K.	–4	–2	–2	–1	–9
Woods, T.	–2	–6	–4	–4	–16

72. **a.** Complete the table by filling in the final score for each of the nine players. (Add the player's four scores.)
  b. Rank the players from lowest score to highest. For players with identical scores, list them alphabetically.
  Woods, Duval, Michelson, Calcavecchia, Izawa, Els, Furyk, Langer, Triplett

73. For the golf course on which the tournament was played, par is 72. If a player's score in relation to par for one round was −4, then he completed the course in 72 + (−4) = 68 strokes. For each player listed above, find the number of strokes taken throughout the four rounds of golf.
  Calcavecchia: 278; Duval: 274; Els: 279; Furyk: 279; Izawa: 278; Langer: 279; Michelson: 275; Triplett: 279; Woods: 272

74. *Economics*   The Bureau of Economic Analysis's web site, **www.bea.gov**, provides data on per capita personal income in the United States. **Per capita income** is total personal income divided by total population. For the year 2000, the per capita personal income in the United States was $29,676. Listed below is the dollar difference for some states from the national average. The dollar difference is a positive number if the per capita income for that state is above the national average. The dollar difference is negative if the per capita income for that state is below the national average. Calculate a state's per capita income by adding the dollar difference to the national average. Iowa's per capita personal income is calculated in the table. Find the per capita income for each of the other plains states listed.

Per Capita Personal Income for the United States in 2000: $29,676

State	Dollar Difference from National Average	Per Capita Personal Income
Iowa	–$2953	$29,676 + (–$2953) = $26,723
Kansas	–$1860	$27,816
Minnesota	$2425	$32,101
Missouri	–$2231	$27,445
Nebraska	–$1847	$27,829
North Dakota	–$3561	$26,115
South Dakota	–$4608	$25,068

■ Multiplication and Division of Integers

75. Find the product of 21, −4, and −3.
252

76. What is 12 times −7?
−84

77. What is 9 multiplied by −18?
−162

78. What is the product of 10 and −11?
−110

79. Multiply −3, −5, and 6.
90

80. Find the quotient of −168 and −7.
24

81. What is the quotient of 98 and −14?
−7

82. What is −84 divided by −6?
14

83. Find the quotient of −36 and 0.
undefined

84. Divide −28 by 1.
−28

Evaluate the expression for the given values of the variables.

85. xy for $x = -3$ and $y = -8$
24

86. $-xy$ for $x = -5$ and $y = -9$
−45

87. $x(-y)$ for $x = -2$ and $y = -10$
−20

88. $-xyz$ for $x = -6$, $y = 2$, and $z = -5$
−60

89. $-8a$ for $a = -24$
192

90. $-7n$ for $n = -51$
357

91. $5xy$ for $x = -9$ and $y = -2$
90

92. $8ab$ for $a = 7$ and $b = -1$
−56

93. $-4cd$ for $c = 25$ and $d = -8$
800

94. $-5st$ for $s = -40$ and $t = -8$
−1600

95. $a \div b$ for $a = -36$ and $b = -4$
9

96. $-a \div b$ for $a = -48$ and $b = -8$
−6

97. $a \div (-b)$ for $a = -72$ and $b = -9$
−8

98. $(-a) \div (-b)$ for $a = -100$ and $b = -25$
4

99. $\dfrac{x}{y}$ for $x = -42$ and $y = -7$
6

100. $\dfrac{-x}{y}$ for $x = -24$ and $y = -6$
−4

101. $\dfrac{x}{-y}$ for $x = -32$ and $y = -4$
−8

102. $\dfrac{-x}{-y}$ for $x = -81$ and $y = -9$
9

Mathematics   A **geometric sequence** is a list of numbers in which each number after the first is found by multiplying the preceding number in the list by the same number. For example, in the sequence 1, 3, 9, 27, 81, . . ., each number after the first is found by multiplying the preceding number in the list by 3. To find the multiplier in a geometric sequence, divide the second number in the sequence by the first number; for the example above, $3 \div 1 = 3$. Geometric sequences are given in Exercises 103 to 106. Find the next three numbers in each geometric sequence.

103. −4, 12, −36, . . .   108, −324, 972

104. 1, −4, 16, . . .   −64, 256, −1024

105. −3, −15, −75, . . .   −375, −1875, −9375

106. −1, −6, −36, . . .   −216, −1296, −7776

Business   The following table shows the net incomes for the first quarter of 2001 and the first quarter of 2000 for five Internet companies (*Source: www.wsj.com*). Profits are shown as positive numbers; losses are shown as negative numbers. One quarter of a year is 3 months. Use this table for Exercises 107 to 110.

Company	First-Quarter 2001 Net Income	First-Quarter 2000 Net Income
Adam.com	44,000	−4,187,000
Buy.com	−45,172,000	−32,846,000
Hotjobs.com	−13,600,000	−12,300,000
iVillage, Inc.	−12,175,000	−25,168,000
Tickets.com	−27,992,000	−21,610,000

107. If earnings were to continue throughout the year at the same level, what would the 2001 annual net income be for Buy.com?   −$180,688,000

108. If earnings were to continue throughout the year at the same level, what would the 2001 annual net income be for Tickets.com?   −$111,968,000

109. Find the difference between Adam.com's first-quarter net income for 2001 and its first-quarter net income for 2000. $4,231,000

110. For the first quarter of 2000, what was the average monthly net income for Hotjobs.com?   −$4,100,000

▪ Exponential Expressions

Evaluate.

111. -7^2
−49

112. -4^3
−64

113. $(-2)^3$
−8

114. $(-3)^4$
81

115. $(-3) \cdot 2^2$
−12

116. $(-5) \cdot 3^4$
−405

117. $(-4) \cdot (-2)^3$
32

118. $(-6) \cdot (-2)^2$
−24

Evaluate the expression for the given values of the variables.

119. $a^3 b^3 c$ for $a = 2$, $b = 3$, and $c = -4$
−864

120. $x^3 y^2 z$ for $x = -3$, $y = 5$, and $z = 10$
−6750

121. $-xy^2 z^2$ for $x = -7$, $y = 4$, and $z = 3$
1008

122. $-ab^3 c^2$ for $a = -2$, $b = -1$, and $c = -3$
−18

▪ The Order of Operations Agreement

Evaluate.

123. $16 - 2 \cdot 4^2$
−16

124. $27 - 18 \div (-3^2)$
29

125. $-2^2 + 4[16 \div (3 - 5)]$
−36

126. $24 \div \dfrac{3^2}{5 - 8} - (-5)$
−3

127. $|-10 \div 2| - 4^2 - (-3)^2$
−20

128. $18 \div |2^3 - 9| + (-3)$
15

129. $16 - 3(3 - 8)^2 \div (-5)$

31

130. $4(-8) \div [2(7 - 3)^2]$

−1

131. $16 - 4 \cdot \dfrac{3^3 - 7}{2^3 + 2} - (-2)^2$

4

Evaluate the variable expression for $a = 2$, $b = 3$, and $c = -4$.

132. $a - 2c$

10

133. $-3a + 4b$

6

134. $3b - 3c$

21

135. $-3c + 4$

16

136. $16 \div (2c)$

−2

137. $6b \div (-a)$

−9

138. $2a - (c + a)^2$

0

139. $(b - a)^2 + 4c$

−15

140. $(b - 2a)^2 + bc$

−11

Evaluate the variable expression for $a = -2$, $b = 4$, $c = -1$, and $d = 3$.

141. $\dfrac{b + c}{d}$

1

142. $\dfrac{d - b}{c}$

1

143. $\dfrac{2d + b}{-a}$

5

144. $\dfrac{b - d}{c - a}$

1

145. $2(b + c) - 2a$

10

146. $3(b - a) - bc$

22

147. $\dfrac{-4bc}{2a - b}$

−2

148. $\dfrac{abc}{b - d}$

8

149. $(b - a)^2 - (d - c)^2$

20

150. $(b + c)^2 + (a + d)^2$

10

151. $4ac + (2a)^2$

24

152. $3cd - (4c)^2$

−25

153. *Temperature*   The daily low temperatures, in degrees Celsius, in Fargo, North Dakota, for a 7-day period were $-4°$, $2°$, $7°$, $-5°$, $-4°$, $-2°$, and $-1°$. What was the average daily low temperature for this 7-day period?
−1°C

154. *Temperature*   The daily low temperatures, in degrees Celsius, in Billings, Montana, for a 5-day period were $-8°$, $-6°$, $-5°$, $2°$, and $-3°$. What was the average daily low temperature for this 5-day period?
−4°C

155. *Aptitude Tests*   The score on an aptitude test is the sum of 8 times the number of correct answers and -2 times the number of incorrect answers. Questions that are not answered are not counted in the score. What score does a person receive who answered 28 questions correctly, answered 5 questions incorrectly, and left 7 questions unanswered?
214 points

156. *Education*   To discourage guessing, a professor scores a multiple-choice exam by awarding 6 points for a correct answer, -2 points for an incorrect answer, and -1 point for a question that is not answered. What score does a student receive who answered 37 questions correctly, answered 5 questions incorrectly, and left 8 questions unanswered?
204 points

157. *Sports*   The scores, in relation to par, of the top 11 golfers in a recent PGA Seniors' Championship golf tournament are shown at the right. What was the average score of these 11 golfers?   −7

Final Scores from the PGA Seniors' Championship	
D. Tewell	−15
H. Irwin	−8
T. Kite	−8
D. Quigley	−8
L. Nelson	−8
H. Green	−7
V. Fernandez	−7
J. Mahaffey	−5
J. Ahern	−5
J. Bland	−3
K. Zarley	−3

Temperature   The following table shows the record low temperatures, in degrees Fahrenheit, in four cities in the United States for each of the first 6 months of the year (*Source:* **www.weather.com**). Use this table for Exercises 158 to 160.

Record Low Temperatures (in degrees Fahrenheit)

	January	*February*	*March*	*April*	*May*	*June*
Boston, MA	−12	−18	1	13	34	41
Chicago, IL	−24	−12	0	13	32	41
Jackson, WY	−46	−42	−25	−10	7	20
Minneapolis, MN	−34	−33	−32	2	18	33

158. Find Jackson's average record low temperature for the first 6 months of the year.

−16°F

159. For the city of Boston, what is the difference between the average record low temperature for the first 4 months of the year and the average record low temperature for the first 2 months of the year?

11°F

160. What is the difference between Chicago's average record low temperature for the first 3 months of the year and Minneapolis's average record low temperature for the first 3 months of the year?   21°F

Applying Concepts

161. a. On the number line, what number is 5 units to the right of −3?
 b. On the number line, what number is 6 units to the left of 4?

 a. 2   b. −2

162. a. Name two numbers that are 5 units from 1 on the number line.
 b. Name two numbers that are 6 units from 2 on the number line.

 a. −4 and 6   b. −4 and 8

163. *A* is a point on the number line halfway between −11 and 5. *B* is a point halfway between *A* and the graph of 1 on the number line. *B* is the graph of what number?

 −1

164. Write each expression in words.

 a. −7   **b.** −(−10)   **c.** −|9| **d.** −|−24|

 a. negative 7   b. the opposite of negative 10   c. the opposite of the absolute value of 9   d. the opposite of the absolute value of negative 24

165. Consider the numbers 5, −6, −3, 16, and −10. Find the greatest difference that can be obtained by subtracting one number in the list from a different number in the list. What is the least difference?   26; 3

166. Fill in the blank squares at the right with integers so that the sum of the integers along any row, column, or diagonal is zero. The resulting square is called a magic square.

 Answers will vary. For example, row 1: −3, 2, 1; row 2: 4, 0 −4; row 3: −1, −2, 3

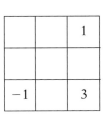

167. The sum of two negative integers is −9. Find the integers.

 Answers will vary. Possible answers include −1 and −8, −2 and −7, −3 and −6.

168. a. Find the greatest possible product of two negative integers whose sum is −18.

  b. Find the least possible sum of two negative integers whose product is 16.

  a. 81   b. −17

EXPLORATION

1. *Balance of Payments*   An **export** is a good or service produced in one's own country and sold for consumption in another country. An **import** is a good or service that is consumed in one's own country but was bought from another country. A nation's **balance of payments** or **balance of trade** is the difference between the value of its exports and the value of its imports during a particular period of time.

A **favorable balance of trade** exists when the value of the exports is greater than the value of the imports. In this case, the balance of trade is a positive number. An **unfavorable balance of trade** exists when the value of the imports is greater than the value of the exports. In this case, the balance of trade is a negative number. An unfavorable balance of trade is referred to as a **trade deficit.** A trade deficit is considered unfavorable because more money is going out of the country to pay for goods imported than is coming into the country to pay for goods exported.

The U.S. government provides data on international trade. On the Internet, go to **www.fedstats.gov**. Find tables that provide data on the value of U.S. imports and exports. All figures in the tables provided on this web site are in millions of dollars. The first two columns are annual figures. Subsequent columns are for quarters (3-month periods).

We located the following data for the first quarter of the year 2000.

        Exports: Total, all countries   183,659

        Imports: Total, all countries   289,699

We then calculated the balance of trade as follows:

        Balance of trade = value of exports − value of imports

$$= 183{,}659 - 289{,}699$$

$$= 183{,}659 + (-289{,}699)$$

$$= -106{,}040$$

The balance of trade for the first quarter of 2000 was −106,040 million dollars. This figure is provided in the table under "Balance: Total, all countries."

  a. Show the calculation of the balance of trade for each of the four quarters of last year. Use the calculation shown above as a model.

  b. Show the calculation of the annual balance of trade for last year.

  c. Show that the sum of the four quarterly figures is equal to the annual figure.

  Answers to parts a, b, and c will vary depending on the year.

SECTION **1.4**

- Addition and Subtraction of Rational Numbers
- Multiplication and Division of Rational Numbers
- The Order of Operations Agreement

Evaluating Variable Expressions Using Rational Numbers

■ Addition and Subtraction of Rational Numbers

Fractions with the same denominator are added by adding the numerators and placing the sum over the common denominator.

Suggested Activity

When two rational numbers are added, it is possible for the sum to be less than either addend, greater than either addend, or a number between the two addends. Give examples of each of these occurrences.

Possible examples:

$$-\frac{1}{2} + \left(-\frac{1}{4}\right) = -\frac{3}{4}; \frac{1}{2} + \frac{1}{4} = \frac{3}{4};$$
$$\frac{3}{4} + \left(-\frac{1}{4}\right) = \frac{1}{2}$$

> **Addition of Fractions**
> To add two fractions with the same denominator, add the numerators and place the sum over the common denominator.

For example, $\frac{2}{6} + \frac{1}{6} = \frac{2+1}{6} = \frac{3}{6} = \frac{1}{2}$.

Note that after adding the fractions, we write the sum in simplest form.

To add fractions with different denominators, first rewrite the fractions as equivalent fractions with a common denominator. Then add the fractions.

The least common denominator is the **least common multiple (LCM)** of the denominators. This is the least number that is a multiple of each of the denominators.

? QUESTION   What is the LCM of 6 and 8?

The sign rules for adding and subtracting fractions are the same as those for adding and subtracting integers.

INSTRUCTOR NOTE
Exercise 12 can be used for a similar in-class example.

 **CALCULATOR NOTE**

A graphing calculator can be used to find the LCM of two numbers. See Appendix A: Math.

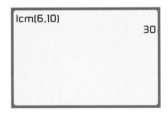

EXAMPLE 1

Add: $-\frac{5}{6} + \frac{3}{10}$

Solution   The LCM of 6 and 10 is 30. Rewrite the fractions as equivalent fractions with the denominator 30. Then add the fractions.

$$-\frac{5}{6} + \frac{3}{10} = -\frac{5}{6} \cdot \frac{5}{5} + \frac{3}{10} \cdot \frac{3}{3} = -\frac{25}{30} + \frac{9}{30} = \frac{-25+9}{30} = -\frac{16}{30} = -\frac{8}{15}$$

Note: You can find the LCM by multiplying the denominators and then dividing by the greatest common factor of the two denominators. In the

? ANSWER   The LCM of 6 and 8 is 24, because 24 is the least number that both 6 and 8 divide evenly into.

case of 6 and 10, $6 \cdot 10 = 60$. Now divide by 2, the common factor of 6 and 10. $60 \div 2 = 30$.

Check:

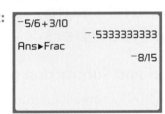

```
-5/6+3/10
                -.5333333333
Ans▶Frac
                        -8/15
```

- Note that the sum is first displayed as a decimal. Use the FRAC command in the MATH menu to convert the decimal to a fraction. (A **menu** gives a list of additional functions the calculator performs.)

See Appendix A: Math

YOU TRY IT 1

Add: $-\dfrac{3}{8} + \left(-\dfrac{1}{3}\right)$

Solution   See page S3. $-\dfrac{17}{24}$

Subtraction of Fractions

To subtract two fractions with the same denominator, subtract the numerators and place the difference over the common denominator.

INSTRUCTOR NOTE
Exercise 14 can be used for a similar in-class example.

EXAMPLE 2

Subtract: $-\dfrac{4}{9} - \dfrac{7}{12}$

Solution   The LCM of 9 and 12 is 36. Rewrite the fractions as equivalent fractions with the denominator 36. Then subtract the fractions.

$$-\frac{4}{9} - \frac{7}{12} = -\frac{16}{36} - \frac{21}{36} = \frac{-16 - 21}{36} = \frac{-37}{36} = -\frac{37}{36}$$

Check:

```
-4/9-7/12
                -1.027777778
Ans▶Frac
                        -37/36
```

YOU TRY IT 2

Subtract: $-\dfrac{3}{4} - \dfrac{3}{16}$

Solution   See page S3. $-\dfrac{15}{16}$

INSTRUCTOR NOTE
Exercise 24 can be used for
a similar in-class example.

EXAMPLE 3

Evaluate $a + b - c$ for $a = -\dfrac{3}{4}$, $b = \dfrac{1}{6}$, and $c = \dfrac{5}{8}$.

Solution $a + b - c$

$$-\frac{3}{4} + \frac{1}{6} - \frac{5}{8}$$ • Replace a by $-\dfrac{3}{4}$, b by $\dfrac{1}{6}$, and c by $\dfrac{5}{8}$.

$$= -\frac{18}{24} + \frac{4}{24} - \frac{15}{24}$$ • The LCM of 4, 6, and 8 is 24.

$$= \frac{-18 + 4 - 15}{24} = -\frac{29}{24}$$

Check:

```
−3/4+1/6−5/8
                −1.208333333
Ans▶Frac
                     −29/24
```

YOU TRY IT 3

Evaluate $x - y + z$ for $x = -\dfrac{7}{8}$, $y = \dfrac{5}{6}$, and $z = \dfrac{3}{4}$.

Solution   See page S3. $-\dfrac{23}{24}$

The sign rules for adding and subtracting decimals are the same as those for adding and subtracting integers.

INSTRUCTOR NOTE
Exercise 28 can be used for
a similar in-class example.

EXAMPLE 4

Evaluate $c - d$ for $c = 42.987$ and $d = 98.61$.

Solution $c - d$

$$42.987 - 98.61$$ • Replace c by 42.987 and d by 98.61.

$$= -55.623$$ • The signs are different. Find the difference between the absolute values of the numbers. Attach the sign of the number with the greater absolute value.

YOU TRY IT 4

Evaluate $a - b$ for $a = -16.127$ and $b = 67.91$.

Solution   See page S3. -84.037

Suggested Activity

When two rational numbers are multiplied, it is possible for the product to be less than either factor, greater than either factor, or a number between the two factors. Give examples of each of these occurrences.

Possible examples:

$$3\left(-\frac{1}{2}\right) = -\frac{3}{2};$$

$$-3\left(-\frac{1}{2}\right) = \frac{3}{2};$$

$$(-3)\frac{1}{2} = -\frac{3}{2}$$

INSTRUCTOR NOTE
Exercises 42 and 44 can be used for similar in-class examples.

■ Multiplication and Division of Rational Numbers

Fractions are multiplied and divided as described below.

> **Multiplication and Division of Fractions**
>
> To multiply two fractions, multiply the numerators and place the product over the product of the denominators.
>
> To divide fractions, rewrite the division as multiplication by the reciprocal of the second fraction. Then multiply the fractions.

The **reciprocal** of a fraction is the fraction with the numerator and denominator interchanged. The reciprocal of $\frac{4}{5}$ is $\frac{5}{4}$.

? QUESTION What is the reciprocal of $-\frac{7}{10}$?

EXAMPLE 5

a. Evaluate pq for $p = \frac{3}{8}$ and $q = \frac{4}{15}$.

b. Evaluate $v \div w$ for $v = \frac{3}{10}$ and $w = -\frac{18}{25}$.

Solution

a. pq

$$\frac{3}{8} \cdot \frac{4}{15} = \frac{3 \cdot 4}{8 \cdot 15} \qquad \text{• Multiply the numerators.}$$
$$\text{Multiply the denominators.}$$

$$= \frac{\overset{1}{\cancel{3}} \cdot \overset{1}{\cancel{4}}}{\underset{2}{\cancel{8}} \cdot \underset{5}{\cancel{15}}} \qquad \text{• Divide by the common factors.}$$

$$= \frac{1}{10} \qquad \text{• Write the answer in simplest form.}$$

Check:

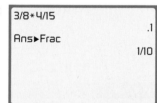

```
3/8*4/15
               .1
Ans▶Frac
            1/10
```

─────
? ANSWER The reciprocal of $-\frac{7}{10}$ is $-\frac{10}{7}$.

Suggested Activity
See Section 1.4 of the *Student Activity Manual* for an activity involving using a graphing calculator to evaluate expressions.

b. $v \div w$

$$\frac{3}{10} \div \left(-\frac{18}{25}\right) = -\left(\frac{3}{10} \div \frac{18}{25}\right)$$

- Because the signs are different, the quotient is negative.

$$= -\left(\frac{3}{10} \cdot \frac{25}{18}\right)$$

- Rewrite division as multiplication by the reciprocal.

$$= -\frac{3 \cdot 25}{10 \cdot 18}$$

- Multiply the fractions.

$$= -\frac{\overset{1}{3} \cdot \overset{5}{25}}{\underset{2}{10} \cdot \underset{6}{18}} = -\frac{5}{12}$$

- Divide by the common factors and write the answer in simplest form.

Check:

```
(3/10)/(-18/25)
              -.4166666667
Ans▶Frac
                    -5/12
```

YOU TRY IT 5

a. Evaluate st for $s = -\dfrac{3}{8}$ and $t = -\dfrac{5}{12}$.

b. Evaluate $a \div d$ for $a = -\dfrac{5}{8}$ and $d = -\dfrac{5}{40}$.

Solution   See page S3.   a. $\dfrac{5}{32}$ b. 5

As stated above, the sign rules are the same for decimals as for integers. These rules are used in Example 6 below to multiply and divide numbers written in decimal notation.

INSTRUCTOR NOTE
Exercises 50 and 54 can be used for similar in-class examples.

EXAMPLE 6

a. Evaluate bc for $b = -0.23$ and $c = 0.04$.

b. Evaluate $\dfrac{x}{y}$ for $x = 0.0527$ and $y = -0.27$. Round to the nearest hundredth.

Solution   **a.** bc

$(-0.23)(0.04) = -0.0092$

- The signs of the factors are different, so the product is negative.

b. $\dfrac{x}{y}$

$\dfrac{0.0527}{-0.27} \approx -0.20$

- The signs of the numbers are different, so the quotient is negative.

TAKE NOTE
The symbol $\approx$ is used to indicate that the quotient is an approximate value that has been rounded off.

YOU TRY IT 6

a. Evaluate $-cd$ for $c = 4.027$ and $d = 0.49$. Round to the nearest hundredth.

b. Evaluate $\dfrac{g}{h}$ for $g = -2.835$ and $h = -1.35$.

Solution   See page S3.   **a.** 1.97   **b.** 2.1

The table that follows shows the net incomes, in millions of dollars, of two U.S. companies for the first quarter of 2001 and the first quarter of 2000 (*Source:* **www.wsj.com**). Profits are shown as positive numbers, losses as negative numbers. One quarter of a year is 3 months. Use this table for Example 7 and You Try It 7.

Company	First-Quarter 2001 Net Income	First-Quarter 2000 Net Income
Cisco Systems, Inc.	−2,693	641
Friendly Ice Cream	−3.203	−18.510

INSTRUCTOR NOTE
Exercise 58 can be used for a similar in-class example.

EXAMPLE 7

If earnings were to continue throughout the year at the same level as in the first quarter, what would be the 2001 annual net income for Cisco Systems?

State the goal.   You need to find the annual net income for Cisco Systems for 2001, assuming the income were to continue throughout the year at the same level as in the first quarter.

Devise a strategy.   To find the annual net income, multiply the net income in the first quarter of 2001 (-2693) by the number of quarters in 1 year (4).

Solve the problem. $-2693(4) = -10,772$

Cisco System's annual net income for 2001 would be $-\$10,772$ million.

Check your work.   You should check to see that the answer is reasonable. For example, using estimation, $-2693 \approx -2700$, and -2700 times 4 is $-10,800$, which is close to our answer of $-10,772$.

YOU TRY IT 7

For the first quarter of 2001, what was the average monthly net income for Friendly Ice Cream? Round to the nearest thousand dollars.

Solution   See page S3.   −$1.068 million

■ The Order of Operations Agreement

Recall the Order of Operations Agreement from Section 1.3. This agreement is used for all numerical expressions.

INSTRUCTOR NOTE
When explaining that generally a variable that is a capital letter cannot be changed to lower case and a lower-case variable cannot be capitalized, you might cite the formula $A = \pi(R^2 - r^2)$, in which the capital R and the lower-case r represent different variables.

INSTRUCTOR NOTE
Exercise 78 can be used for a similar in-class example.

Suggested Activity

Define $a \, @ \, b$ as $a \cdot b + b$. Use this definition to find the value of $x \, @ \, (y \, @ \, z)$ when $x = 1.7$, $y = 2.3$, and $z = -1.8$. [Answer: -16.038]

The Order of Operations Agreement

Step 1 Perform operations inside grouping symbols. Grouping symbols include parentheses (), brackets [], braces {}, the absolute value symbol ||, and fraction bars.

Step 2 Simplify exponential expressions.

Step 3 Do multiplication and division as they occur from left to right.

Step 4 Do addition and subtraction as they occur from left to right.

EXAMPLE 8

Evaluate $(a + b)^2 \div c - a$ for $a = 1.9$, $b = -2.35$, and $c = 0.25$.

Solution $(a + b)^2 \div c - a$

$$[1.9 + (-2.35)]^2 \div 0.25 - 1.9$$ • Replace a by 1.9, b by -2.35, and c by 0.25.

$$= (-0.45)^2 \div 0.25 - 1.9$$ • Simplify the expression in the parentheses.

$$= 0.2025 \div 0.25 - 1.9$$ • Simplify the exponential expression.

$$= 0.81 - 1.9$$ • Do the division.

$$= 0.81 + (-1.9)$$ • Do the subtraction.

$$= -1.09$$

YOU TRY IT 8

Evaluate $x(x - y)^2 \div z$ for $x = 4.5$, $y = 6.2$, and $z = -0.5$.

Solution   See page S3. -26.01

A **formula** is a special type of equation that states a rule about measurements. For instance, the formula for the area of a rectangle is $A = LW$. This formula shows the relationship between the area, A, of a rectangle and its length, L, and width, W. A formula is used in Example 9 to find the surface area of a regular pyramid.

INSTRUCTOR NOTE
Exercise 94 can be used for a similar in-class example.

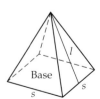

EXAMPLE 9

The formula for the surface area of a regular pyramid is $S = s^2 + 2sl$, where S is the surface area, s is the length of a side of the square base, and l is the slant height. Find the surface area of the regular pyramid shown at the right.

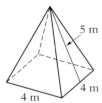

Solution $S = s^2 + 2sl$

$S = 4^2 + 2(4)(5)$ • Replace the variables s and l by their given values, $s = 4$ and $l = 5$.

$S = 16 + 2(4)(5)$ • Use the Order of Operations Agreement to simplify the numerical expression.

$S = 16 + 8(5)$

$S = 16 + 40$

$S = 56$

The surface area is 56 m².

See Appendix A:
Evaluating Variable
Expressions

Check: As an alternative method of checking the evaluation of a variable expression, store the value of each variable in the calculator. A typical screen for the evaluation of the expression $s^2 + 2sl$ in Example 9 follows.

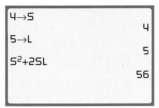

Note: Generally a variable that is a capital letter cannot be changed to lower case, and a lower-case variable cannot be capitalized. However, some graphing calculators cannot represent variables as lower-case letters.

YOU TRY IT 9

A rectangle has a length of 8.5 meters and a width of 3.5 meters. Find the perimeter of the rectangle. Use the formula $P = 2L + 2W$, where P is the perimeter, L is the length, and W is the width of a rectangle.

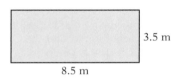

Solution   See page S3.   24 m

INSTRUCTOR NOTE
Exercise 108 can be used for a similar in-class example.

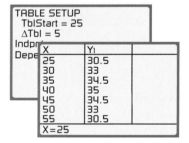

EXAMPLE 10

The fuel economy of a car is given by the equation $M = -0.02v^2 + 1.6v + 3$, where M is the fuel economy in miles per gallon (mpg) and v is the speed of the car in miles per hour (mph). Use a graphing calculator to create an input/output table for this equation with TblStart = 25 and $\triangle$Tbl = 5. Use the table to answer the following questions.

a. What is the fuel economy when the speed of the car is 50 mph?
b. What is the speed of the car when the fuel economy is 35 mpg?

Solution   The input variable is X, the speed of the car. The output variable is Y_1, the fuel economy of the car. The input/output table should look like the one at the left.

a. Because the question asks for the fuel economy (Y_1) when the speed is 50 mph (X), look in the table for an input value of 50. The corresponding output value is 33.

When the speed of the car is 50 mph, the fuel economy is 33 mpg.

b. Because the question asks for the speed (X) when the fuel economy is 35 mpg (Y_1), look in the table for an output value of 35. The corresponding input value is 40.

When the fuel economy is 35 mpg, the speed of the car is 40 mph.

YOU TRY IT 10

The amount of garbage generated by each person living in the United States is given by the equation $A = 0.05x - 95$, where A is the amount of garbage generated, in pounds per person per day, and x is the year. Use a graphing calculator to create an input/output table for this equation with TblStart = 1970 and $\triangle$Tbl = 5. Use the table to answer the following questions.

a. What was the amount of garbage generated per person per day in 1990?
b. In what year will the amount of garbage generated per person per day be 5.75 pounds?

Solution   See page S3.   **a.** 4.5 lb   **b.** 2015

1.4 EXERCISES

Suggested Assignment: 11–111, every other odd

Topics for Discussion

For Exercises 1 to 6, determine whether the statement is always true, sometimes true, or never true. If the statement is sometimes true or never true, explain your answer.

1. To add two fractions, find the sum of the numerators and place it over the sum of the denominators.   Never true

2. The sum of a number and its additive inverse is zero.   Always true

3. The rule for multiplying two fractions is to multiply the numerators and place the product over the least common multiple of the denominators.
    Never true

4. To multiply two fractions, you must first rewrite the fractions as equivalent fractions with a common denominator.   Never true

5. To divide two fractions, multiply the first fraction by the reciprocal of the second fraction.   Always true

6. The Order of Operations Agreement is used for natural numbers, integers, rational numbers, and real numbers.   Always true

7. Given both a fraction and a decimal, how can you determine which is the greater number?   Rewrite the fraction as a decimal. Then compare the two decimals.

8. **a.** Are there any integers that are not rational numbers?   **a.** No
    b. Are there any rational numbers that are not integers?   **b.** Yes

9. **a.** Is there a least positive rational number?   **a.** No
    b. Is there a greatest negative rational number?   **b.** No

10. Give some examples of how rational numbers are used in everyday experiences.   Answers will vary.

■ Addition and Subtraction of Rational Numbers

11. Find the sum of $-\dfrac{1}{2}$ and $\dfrac{3}{8}$. $-\dfrac{1}{8}$

12. Find the total of $\dfrac{2}{3}$, $-\dfrac{1}{2}$, and $\dfrac{5}{6}$.   1

13. Find the difference between $\dfrac{5}{6}$ and $\dfrac{11}{12}$. $-\dfrac{1}{12}$

14. What is 6.9027 minus 17.692? -10.7893

15. Evaluate $x - y$ for $x = \dfrac{5}{8}$ and $y = \dfrac{5}{6}$. $-\dfrac{5}{24}$

16. Evaluate $x - y$ for $x = \dfrac{1}{9}$ and $y = \dfrac{5}{27}$. $-\dfrac{2}{27}$

17. Evaluate $a - b$ for $a = -\dfrac{5}{12}$ and $b = \dfrac{3}{8}$. $-\dfrac{19}{24}$

18. Evaluate $p - r$ for $p = -\dfrac{3}{4}$ and $r = \dfrac{5}{16}$. $-\dfrac{17}{16}$

19. Evaluate $c - d$ for $c = -\dfrac{5}{6}$ and $d = \dfrac{5}{9}$. $-\dfrac{25}{18}$

20. Evaluate $c + d$ for $c = -\dfrac{7}{12}$ and $d = \dfrac{5}{8}$. $\dfrac{1}{24}$

21. Evaluate $v - w$ for $v = \dfrac{5}{8}$ and $w = -\dfrac{3}{4}$. $\dfrac{11}{8}$

22. Evaluate $s - t$ for $s = -\dfrac{5}{8}$ and $t = -\dfrac{11}{12}$. $\dfrac{7}{24}$

23. Evaluate $a + b - c$ for $a = -\dfrac{5}{16}$, $b = \dfrac{3}{4}$, and $c = \dfrac{7}{8}$. $-\dfrac{7}{16}$

24. Evaluate $a - b - c$ for $a = -\dfrac{1}{2}$, $b = \dfrac{3}{8}$, and $c = -\dfrac{1}{4}$. $-\dfrac{5}{8}$

25. Evaluate $x - y - z$ for $x = \dfrac{3}{4}$, $y = -\dfrac{7}{12}$, and $z = \dfrac{7}{8}$. $\dfrac{11}{24}$

26. Evaluate $x + y + z$ for $x = \dfrac{5}{16}$, $y = \dfrac{1}{8}$, and $z = -\dfrac{1}{2}$. $-\dfrac{1}{16}$

27. Evaluate $a - b$ for $a = -5.13$ and $b = 8.179$. -13.309

28. Evaluate $-a - b$ for $a = 32.1$ and $b = 6.7$. -38.8

29. Evaluate $x - y + z$ for $x = 2.09$, $y = 6.72$, and $z = -5.4$. -10.03

30. Evaluate $x - y - z$ for $x = -18.39$, $y = 4.9$, and $z = 23.7$. -46.99

31. Evaluate $a - b - c$ for $a = 19$, $b = -3.72$, and $c = 82.75$. -60.03

32. Evaluate $-a - b + c$ for $a = 3.09$, $b = 4.6$, and $c = 27.3$. 19.61

33. Which is greater, $\dfrac{5}{8} - \left(-\dfrac{5}{6}\right)$ or $-\dfrac{5}{6} - \dfrac{5}{9}$?

$\dfrac{5}{8} - \left(-\dfrac{5}{6}\right)$, because $\dfrac{5}{8} - \left(-\dfrac{5}{6}\right) = \dfrac{5}{8} + \dfrac{5}{6}$, so the difference is positive, whereas the difference $-\dfrac{5}{6} - \dfrac{5}{9}$ is negative.

34. Which is greater, $-\dfrac{1}{8} - \dfrac{3}{4}$ or $\dfrac{11}{12} - \left(-\dfrac{1}{4}\right)$?

$\dfrac{11}{12} - \left(-\dfrac{1}{4}\right)$, because $\dfrac{11}{12} - \left(-\dfrac{1}{4}\right) = \dfrac{11}{12} + \dfrac{1}{4}$, so the difference is positive, whereas the difference $-\dfrac{1}{8} - \dfrac{3}{4}$ is negative.

■ Multiplication and Division of Rational Numbers

35. What is $-\dfrac{1}{2}$ times $\dfrac{8}{9}$? $-\dfrac{4}{9}$

36. Find the product of $\dfrac{5}{12}$, $-\dfrac{8}{15}$, and $-\dfrac{1}{3}$. $\dfrac{2}{27}$

37. Find the quotient of $-\dfrac{3}{8}$ and $\dfrac{1}{4}$. $-\dfrac{3}{2}$

38. Divide -24.3 by 0.09. -270

39. Evaluate cd for $c = -\dfrac{2}{9}$ and $d = -\dfrac{3}{14}$. $\dfrac{1}{21}$

40. Evaluate ab for $a = -\dfrac{3}{8}$ and $b = -\dfrac{4}{15}$. $\dfrac{1}{10}$

41. Evaluate abc for $a = \dfrac{5}{8}$, $b = -\dfrac{7}{12}$, and $c = \dfrac{16}{25}$. $-\dfrac{7}{30}$

42. Evaluate wxy for $w = \dfrac{1}{2}$, $x = -\dfrac{3}{4}$, and $z = -\dfrac{5}{8}$. $\dfrac{15}{64}$

43. Evaluate $s \div t$ for $s = \dfrac{5}{6}$ and $t = -\dfrac{3}{4}$. $-\dfrac{10}{9}$

44. Evaluate $a \div b$ for $a = -\dfrac{5}{12}$ and $b = \dfrac{15}{32}$. $-\dfrac{8}{9}$

45. Evaluate $c \div d$ for $c = \dfrac{1}{8}$ and $d = -\dfrac{5}{12}$. $-\dfrac{3}{10}$

46. Evaluate $-y \div z$ for $y = \dfrac{4}{9}$ and $z = -\dfrac{2}{3}$. $\dfrac{2}{3}$

47. Evaluate xy for $x = 5.68$ and $y = 0.2$. 1.136

48. Evaluate ab for $a = 6.27$ and $b = 8$. 50.16

49. Evaluate xy for $x = -3.71$ and $y = 2.9$. -10.759

50. Evaluate cd for $c = -2.537$ and $d = -9.1$. 23.0867

51. Evaluate bcd for $b = 2.3$, $c = -0.6$, and $d = 0.8$. -1.104

52. Evaluate wxy for $w = 4.5$, $x = -0.22$, and $y = -0.8$. 0.792

53. Evaluate $\dfrac{c}{d}$ for $c = 10.15$ and $d = -2.9$. -3.5

54. Evaluate $\dfrac{s}{t}$ for $s = -1.24$ and $t = -0.31$. 4

55. Which is greater, $\left(-\dfrac{8}{9}\right)\left(-\dfrac{3}{4}\right)$ or $-\dfrac{5}{16} \div \dfrac{3}{8}$?

$\left(-\dfrac{8}{9}\right)\left(-\dfrac{3}{4}\right)$, because the product is positive, while the quotient $-\dfrac{5}{16} \div \dfrac{3}{8}$ is negative.

56. Which is greater, $-\dfrac{5}{6} \div (-5)$ or $-\dfrac{3}{4}\left(\dfrac{2}{9}\right)$?

$-\dfrac{5}{6} \div (-5)$, because the quotient is positive, while the product $-\dfrac{3}{4}\left(-\dfrac{2}{9}\right)$ is negative.

Business The following table shows the net incomes for the first quarter of 2001 and the first quarter of 2000 for three companies in the entertainment industry (*Source:* **www.wsj.com**). Figures are in millions of dollars. Profits are shown as positive numbers, losses as negative numbers. One quarter of a year is 3 months. Use this table for Exercises 57 to 62.

Company	First-Quarter 2001 Net Income	First-Quarter 2000 Net Income
Fox Entertainment	−9.0	19.0
Midway Games, Inc.	−25.852	−11.481
Six Flags, Inc.	−130.752	−113.892

57. If earnings were to continue throughout the year at the same level, what would the 2001 annual net income be for Midway Games?
−$103.408 million

58. If earnings were to continue throughout the year at the same level, what would the 2000 annual net income be for Six Flags? −$455.568 million

59. For the first quarter of 2001, what was the average monthly net income for Six Flags?
−$43.584 million

60. For the first quarter of 2000, what was the average monthly net income for Midway Games?
−$3.827 million

61. Find the difference between Fox Entertainment's first-quarter net income for 2000 and Midway Games's first-quarter net income for 2000.
$30.481 million

62. Find the difference between Fox Entertainment's first-quarter net income for 2001 and Six Flags's first-quarter net income for 2001.
$121.752 million

Commerce The table at the right shows the U.S. balance of trade, in billions of dollars, for the years 1970 to 2000 (*Source:* U.S. Dept. of Commerce). See page 52 for a discussion of balance of trade. Use the table for Exercises 63 to 70.

63. For which year was the trade balance lowest? For which was it highest?
2000; 1975

64. In which years did the trade balance increase from the previous year?
'73, '75, '79, '80, '81, '88, '89, '90, '91, '95

65. Calculate the difference between the trade balance in 1990 and that in 2000. $288.6 billion

66. What was the difference between the trade balance in 1970 and that in 2000? $372 billion

67. During which two consecutive years was the difference in the trade balance greatest? 1999–2000

68. How many times greater was the trade balance in 1990 than in 1980? Round to the nearest whole number. 4 times greater

69. Calculate the average trade balance per quarter for the year 2000.
−$92.425 billion

70. By examining the data, would you expect the trade balance to have increased or decreased from 1995 by the year 2000? Support your answer.
Answers will vary.

Year	Trade Balance
1970	2.3
1971	−1.3
1972	−5.4
1973	1.9
1974	−4.3
1975	12.4
1976	−6.1
1977	−27.2
1978	−29.8
1979	−24.6
1980	−19.4
1981	−16.2
1982	−24.2
1983	−57.8
1984	−109.2
1985	−122.1
1986	−140.6
1987	−153.3
1988	−115.9
1989	−92.2
1990	−81.1
1991	−30.7
1992	−35.7
1993	−68.9
1994	−97.0
1995	−95.9
1996	−102.1
1997	−104.7
1998	−166.9
1999	−265.0
2000	−369.7

■ **The Order of Operations Agreement**

Evaluate.

71. $0.3(1.7 − 4.8) + (−1.2)^2$

0.51

72. $(1.05 − 1.65)^2 ÷ 0.4 − 2$

−1.1

73. $-\dfrac{7}{12} + \dfrac{5}{6}\left(\dfrac{1}{6} - \dfrac{2}{3}\right)$

−1

74. $\left(-\dfrac{2}{3}\right)^2 + \left(-\dfrac{1}{6}\right) ÷ \dfrac{3}{8}$

0

75. $\left(\dfrac{1}{3} - \dfrac{5}{6}\right) + \dfrac{7}{8} ÷ \left(-\dfrac{1}{2}\right)^3$

$-\dfrac{15}{2}$

76. $\left(-\dfrac{1}{4}\right)^2 ÷ \left(\dfrac{1}{2} - \dfrac{3}{4}\right) + \dfrac{3}{8}$

$\dfrac{1}{8}$

Evaluate the variable expression when $a = 2.7$, $b = -1.6$, and $c = -0.8$.

77. $c^2 - ab$

4.96

78. $(a + b)^2 - c$

2.01

79. $\dfrac{b^3}{c} - 4a$

−5.68

Evaluate the variable expression when $a = \dfrac{2}{3}$ and $b = -\dfrac{3}{2}$.

80. $\dfrac{1}{3}a^3b^4$

$\frac{1}{2}$

81. $\dfrac{(2ab)^3}{2a^2b^2}$

−4

82. $|5ab - 8a^2b^2|$

13

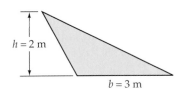 *Federal Budget*   The table at the right shows the surplus or deficit, in billions of dollars, for the federal budget every fifth year from 1945 to 1995 and every year from 1995 to 2000 (*Source:* U.S. Office of Management and Budget). The negative sign (−) indicates a deficit. Use this table for Exercises 83 to 90.

Year	Federal Budget Surplus or Deficit
1945	−47.533
1950	−3.119
1955	−2.993
1960	0.301
1965	−1.411
1970	−2.842
1975	−53.242
1980	−73.835
1985	−212.334
1990	−221.194
1995	−163.899
1996	−107.450
1997	−21.940
1998	69.246
1999	79.263
2000	117.305

83. Find the difference between the deficits in the years 1980 and 1985.
$138.499 billion

84. Calculate the difference between the surplus in 1960 and the deficit in 1955. $3.294 billion

85. How many times greater was the deficit in 1985 than in 1975? Round to the nearest whole number.   4 times greater

86. What was the average deficit, in billions of dollars, per month for the year 1985?   −$17.6945 billion

87. What was the average deficit, in millions of dollars, per quarter for the year 1970?   −$710.5 million

88. What was the average annual change in the deficit from 1980 to 1985? Round to the nearest million. $27,700 million

89. Find the average surplus or deficit for the years 1996 through 2000.
$27.2848 billion

90. Find the average surplus or deficit for the years 1995 through 2000. Round to the nearest billion.   −$5 billion

91. *Geometry*   The formula for the perimeter of a rectangle is $P = 2L + 2W$, where P is the perimeter, L is the length, and W is the width. Find the perimeter of the rectangle shown at the right.   220 cm

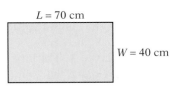

$L = 70$ cm

$W = 40$ cm

92. *Geometry*   The formula for the area of a rectangle is $A = LW$, where A is the area, L is the length, and W is the width. Find the area of the rectangle in Exercise 91.   2800 cm²

93. *Geometry*   The formula for the area of a triangle is $A = \frac{1}{2}bh$, where A is the area, b is the base, and h is the height. Find the area of the triangle shown at the right.   3 m²

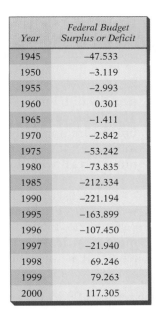

$h = 2$ m

$b = 3$ m

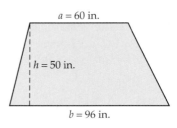

a = 60 in.

h = 50 in.

b = 96 in.

94. *Geometry*   The formula for the area of a trapezoid is $A = \frac{1}{2}h(a + b)$, where A is the area, h is the height, and a and b are the lengths of the bases. Find the area of the trapezoid shown at the right.   3900 in²

95. *Depreciation*   To determine the depreciated value of an X-ray machine, an accountant uses the formula $V = C - 5500t$, where V is the depreciated value of the machine in t years and C is the original cost. Find the depreciated value after 4 years of an X-ray machine that cost $70,000. $48,000

96. *Sports*   The world record time for a 1-mile race can be approximated by the formula $t = 17.08 - 0.0067y$, where y is the year of the race and t is the time, in minutes, of the race. Find the time predicted by this formula for the year 1954. Round to the nearest tenth.   4.0 min

97. *Automotive Technology*   Black ice is an ice covering on roads that is especially difficult to see and therefore extremely dangerous for motorists. The distance D, in feet, that a car traveling 30 mph will slide after its brakes are applied is related to the outside air temperature by the formula $D = 4C + 180$, where C is the Celsius temperature. Find the distance a car will slide on black ice when the outside temperature is $-11°C$.   136 ft

98. *Geometry*   Find the surface area of a rectangular solid that has a length of 5 meters, a width of 8 meters, and a height of 4 meters. Use the formula $S = 2LW + 2LH + 2WH$, where S is the surface area, L is the length, W is the width, and H is the height of the rectangular solid.
184 ft²

For Exercises 99 to 102, use a graphing calculator to create an input/output table for the equation using the given instructions. Use the input/output table to answer the questions.

99. $y = x^2$, TblStart = 1, ΔTbl = 0.5
   a.   What is the value of y when $x = 2$?
   b.   What is the value of x when $y = 12.25$?
   a. 4   b. 3.5

Input, x	1	1.5	2	2.5	3	3.5
Output, $y = x^2$	1	2.25	4	6.25	9	12.25

100. $y = x^2 - 1$, TblStart = 2, ΔTbl = 0.5
   a.   What is the value of y when $x = 4.5$?
   b.   What is the value of x when $y = 5.25$?
   a. 19.25   b. 2.5

Input, x	2	2.5	3	3.5	4	4.5
Output, $y = x^2 - 1$	3	5.25	8	11.25	15	19.25

101. $y = 2x^2 + 1$, TblStart = 1.5, ΔTbl = 0.25
   a.   What is the value of y when $x = 1.75$?
   b.   What is the value of x when $y = 16.125$?
   a. 7.125   b. 2.75

Input, x	1.5	1.75	2	2.25	2.5	2.75
Output, $y = 2x^2 + 1$	5.5	7.125	9	11.125	13.5	16.125

102. $y = 12 - x^2$, TblStart = 1, ΔTbl = 0.25
   a.   What is the value of y when $x = 1.5$?
   b.   What is the value of x when $y = 8.9375$?
   a. 9.75   b. 1.75

Input, x	1	1.25	1.5	1.75	2	2.25
Output, $y = 12 - x^2$	11	10.4375	9.75	8.9375	8	6.9375

103. *Architecture*   An architect charges a fee of $500 plus $2.65 per square foot to design a house. The equation that represents the architect's fee is $F = 2.65s + 500$, where F is the fee, in dollars, and s is the number of square feet in the house. Use a graphing calculator to create an input/output table for this equation for increments of 100 square feet, beginning with $s = 1200$. What is the meaning of the number 4740 in the table?

The architect charges a fee of $4740 to design a 1600-square-foot house.

104. *Business*   A rental car company charges a drop-off fee of $50 to return a car to a location different from that at which it was rented. It also charges a fee of $.18 per mile the car is driven. The equation that represents the total cost to rent a car from this company is $C = 0.18m + 50$, where C is the total cost, in dollars, and m is the number of miles the car is driven. Use a graphing calculator to create an input/output table for this equation for increments of 10 miles, beginning with $m = 100$. What is the meaning of the number 77 in the table?

If the rented car is driven 150 mi, the cost of the rental is $77.

105. *Temperature*   In June, the temperature at various elevations of the Grand Canyon can be approximated by the equation $T = -0.005x + 113.25$, where T is the temperature in Fahrenheit and x is the elevation, or distance above sea level. Use a graphing calculator to create an input/output table for this equation. Use TblStart = 2450 and ΔTbl = 500.

   a. According to this equation, what is the temperature at an elevation of 4450 feet (about halfway down the south rim of the canyon)?

   b. At Inner Gorge, or the bottom of the canyon, the temperature is 101°F. What is the elevation of Inner Gorge?

a. 91°F   b. 2450 ft

106. *Temperature*   The equation $F = \frac{9}{5}C + 32$ can be used to convert Celsius temperatures (C) to Fahrenheit temperatures (F). Use a graphing calculator to create an input/output table for this equation. Use TblStart = 0 and ΔTbl = 5.

   a. Find the Fahrenheit temperature when the Celsius temperature is 20°C.

   b. Find the Celsius temperature when the Fahrenheit temperature is 50°F.

a. 68°F   b. 10°C

107. *Geology*   Old Faithful is a geyser in Yellowstone National Park that was so named because of its regular eruptions for the past 100 years. An equation that can predict the approximate time until the next eruption is $T = 12.4L + 32$, where T is the time (in minutes) to the next eruption and L is the length of time (in minutes) of the last eruption. Use a graphing calculator to create an input/output table for this equation for increments of 0.5 minute, beginning with $L = 1$.

   a. According to this equation, how long will it be until the next eruption if the last eruption lasted 3.5 minutes?

   b. If the time between two eruptions is 63 minutes, what was the length of time of the last eruption?

a. 75.4 min   b. 2.5 min

108. *Aeronautics*   The altitude, or height above sea level, of a hot-air balloon is given by the equation $H = 100t + 1250$, where H is the altitude of the balloon in feet t seconds after it has been released. Use a graphing calculator to create an input/output table for this equation for increments of 0.5 second, beginning with $t = 0$.

   a. What is the altitude of the balloon 2.5 seconds after it is released?

   b. How many seconds after it is released is the balloon at an altitude of 1800 feet?

   a. 1500 ft   b. 5.5 s

109. *Geometry*   The formula for the perimeter of a square is $P = 4s$, where P is the perimeter and s is the length of one of the equal sides.

   a. Create an input/output table for this equation for increments of 2 inches, beginning with $s = 2$.

   b. What does the output value of 48 represent?

   a.

s	2	4	6	8	10	12	14
P	8	16	24	32	40	48	56

   b. When the length of a side of a square is 12 in., the perimeter is 48 in.

110. *Geometry*   The formula for the area of a square is $A = s^2$, where A is the area and s is the length of one of the equal sides.

   a. Create an input/output table for this equation for increments of 2 feet, beginning with $s = 2$.

   b. What does the output value 64 represent?

   a.

s	2	4	6	8	10	12	14
A	4	16	36	64	100	144	196

   b. When the length of a side of a square is 8 ft, the area of the square is 64 ft².

111. *Geometry*   The formula for the volume of a cube is $V = s^3$, where V is the volume and s is the length of a side of a cube.

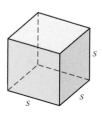

   a. Create an input/output table for this equation for increments of 1 meter, beginning with $s = 1$.

   b. Write a sentence that describes the meaning of the numbers 3 and 27.

   a.

s	1	2	3	4	5	6	7
A	1	8	27	64	125	216	343

   b. When the length of a side of a cube is 3 m, the volume of the cube is 27 m³.

112. *Geometry*   The formula for the surface area of a cube is $S = 6s^2$, where S is the surface area and s is the length of a side of a cube.

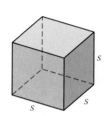

   a. Create an input/output table for this equation for increments of 1 centimeter, beginning with $s = 1$.

   b. Write a sentence that describes the meaning of the numbers 4 and 96.

   a.

s	1	2	3	4	5	6	7
A	6	24	54	96	150	216	294

   b. When the length of a side of a cube is 4 cm, the surface area of the cube is 96 cm².

Applying Concepts

113. If the same positive number is added to both the numerator and the denominator of $\frac{4}{7}$, is the new fraction less than, equal to, or greater than $\frac{4}{7}$?
greater than

114. A student simplified the expression $6 + 2(4 - 9)$ as follows:

$$6 + 2(4 - 9) = 6 + 2(-5)$$
$$= 8(-5)$$
$$= -40$$

Is this a correct simplification? Explain your answer.
No. In Step 2, addition was performed before multiplication.

115. A magic square is one in which the sum of the numbers in every row, column, and diagonal is the same number. Complete the magic square at the right.
Row 1: $-\frac{1}{6}$, 0; row 2: $-\frac{1}{2}$; row 3: $\frac{1}{3}, \frac{1}{2}$

$\frac{2}{3}$		
	$\frac{1}{6}$	$\frac{5}{6}$
		$-\frac{1}{3}$

116. Given any two distinct rational numbers, is it always possible to find a rational number between the two numbers? If so, explain how to find one.
Yes. Add the two numbers and then divide the sum by 2.

117. Suppose the numerator of a fraction is a fixed number—for instance, 5. How does the value of the fraction change as the denominator increases?
As the denominator increases, the value of the fraction decreases.

118. Explain why you need a common denominator when adding two fractions and why you don't need a common denominator when multiplying two fractions?
Answers will vary.

EXPLORATION

1. *Patterns in Mathematics*   For each of the following, determine the first natural number x, greater than 2, for which the second expression is greater than the first. On the basis of your answers, make a conjecture that appears to be true about the expressions x^n and n^x, where $n = 3, 4, 5, 6, 7, \ldots$ and x is a natural number greater than 2.
   a. $x^3, 3^x$ **b.** $x^4, 4^x$ **c.** $x^5, 5^x$ **d.** $x^6, 6^x$
   $n^x > x^n$ if $x \geq n + 1$

2. *Making Conjectures*   Consider the following expressions:
   $(a + b)^2, a^2 + b^2, (a + b)^3, a^3 + b^3$, and $(a + b)^4, a^4 + b^4$.
   a. By trying different values of a and b, determine whether $(a + b)^2 = a^2 + b^2$ is always true.
   b. By trying different values of a and b, determine whether $(a + b)^3 = a^3 + b^3$ is always true.
   c. By trying different values of a and b, determine whether $(a + b)^4 = a^4 + b^4$ is always true.
   d. Using inductive reasoning, make a conjecture about whether $(a + b)^n = a^n + b^n$ is always true when n is a natural number.
   a. No   b. No   c. No   d. No

SECTION **1.5** # Simplifying Variable Expressions

- Properties of Real Numbers
- Simplify Variable Expressions
- Translate Phrases into Variable Expressions

■ Properties of Real Numbers

In the last two sections, we *evaluated* variable expressions. That is, we replaced the variables by numbers and then simplified the resulting numerical expressions. Now we will look at *simplifying* variable expressions. This is accomplished by using the Properties of Real Numbers, presented below.

Note that numbers can be added in either order and the result is the same.

$$9 + (-12) = -3 \quad \text{and} \quad -12 + 9 = -3$$

This is the **Commutative Property of Addition,** which states that if a and b are any two numbers, then $a + b = b + a$.

When three numbers are added together, the numbers can be grouped in any order and the sum will be the same.

$$-6 + (3 + 7) = -6 + 10 = 4 \quad \text{and} \quad (-6 + 3) + 7 = -3 + 7 = 4$$

This is the **Associative Property of Addition,** which states that if a, b and c are any three numbers, then $a + (b + c) = (a + b) + c$.

Two other properties of addition are also important. The first says that the sum of a number and its opposite is zero.

$$-8 + 8 = 0 \quad \text{and} \quad 8 + (-8) = 0$$

This is the **Inverse Property of Addition,** which states that $a + (-a) = 0$ and $-a + a = 0$. Recall that a and $-a$ are opposites, or additive inverses, of each other.

The second of these two other properties expresses the fact that the sum of a number and zero is the number.

$$-3 + 0 = -3 \quad \text{and} \quad 0 + (-3) = -3$$

This is the **Addition Property of Zero,** which states that if a is any number, then $a + 0 = a$ and $0 + a = a$.

Note that numbers can be multiplied in either order and the result is the same.

$$9(-8) = -72 \quad \text{and} \quad (-8)9 = -72$$

This is the **Commutative Property of Multiplication,** which states that if a and b are any two numbers, then $ab = ba$.

When three numbers are multiplied together, the numbers can be grouped in any order and the product will be the same.

$$3(5 \cdot 2) = 3(10) = 30 \quad \text{and} \quad (3 \cdot 5)2 = 15 \cdot 2 = 30$$

This is the **Associative Property of Multiplication,** which states that if a, b, and c are any three numbers, then $a(bc) = (ab)c$.

Suggested Activity

A, B, C, and D are four distinct real numbers such that
$$A + B = A$$
$$B \cdot A = B$$
$$C + D = B$$
$$C(B + A) = A$$
$$C - D = A$$
Find the values of A, B, C, and D.
[Answer: $A = 2$, $B = 0$, $C = 1$, and $D = -1$.]

Two other properties of multiplication are also important. The first says that the product of a number and its reciprocal is 1.

$$\frac{1}{8} \cdot 8 = 1 \quad \text{and} \quad 8\left(\frac{1}{8}\right) = 1$$

This is the **Inverse Property of Multiplication,** which states that for $a \neq 0$, $a \cdot \frac{1}{a} = 1$ and $\frac{1}{a} \cdot a = 1$. The terms a and $\frac{1}{a}$ are **reciprocals.** They are also called **multiplicative inverses** of each other.

The second of these other two properties expresses the fact that the product of a number and 1 is the number.

$$9 \cdot 1 = 9 \quad \text{and} \quad 1 \cdot 9 = 9$$

This is the **Multiplication Property of One,** which states that if a is any number, then $a \cdot 1 = a$ and $1 \cdot a = a$.

> **TAKE NOTE**
>
> Here is a summary of the discussion at the right. If a coefficient is 1 or −1, the 1 is usually not written. For instance, we write $1x$ as x and write $-1xy$ as $-xy$.

Recall that the coefficient of a variable is the number that multiplies the variable. Note from the Multiplication Property of One that when the coefficient is 1, the 1 is not written. Thus we write x instead of $1x$ or $1 \cdot x$. A coefficient of -1 is treated in much the same way. For instance, we normally write $-1x$ and $-1 \cdot x$ as $-x$.

By the Order of Operations Agreement, the expression $6(4 + 7)$ is simplified by first adding the numbers inside the parentheses and then multiplying. However, we can multiply each number inside the parentheses by the number outside the parentheses and then add the products, and the result is the same.

$$6(4 + 7) = 6(11)$$
$$= 66$$
$$6(4 + 7) = 6(4) + 6(7)$$
$$= 24 + 42$$
$$= 66$$

This is an example of the **Distributive Property,** which states that if a, b, and c are any numbers, then $a(b + c) = ab + ac$.

INSTRUCTOR NOTE
Exercise 16 can be used for a similar in-class example.

EXAMPLE 1

a. Complete the statement by using the Commutative Property of Addition:

$$4 + x = ?$$

b. Complete the statement by using the Distributive Property:

$$4(6 + 9) = ?(6) + ?(9)$$

Solution   **a.** $4 + x = x + 4$ • The Commutative Property of Addition states that the order of addends can be interchanged.

b. $4(6 + 9)$
$= 4(6) + 4(9)$ • The Distributive Property states that each number inside the parentheses is multiplied by the number outside the parentheses.

a. Complete the statement by using the Associative Property of Multiplication:

$$4(3x) = ?$$

b. Complete the statement by using the Inverse Property of Addition.

$$12 + ? = 0$$

Solution See page S4. **a.** $(4 \cdot 3)x$ **b.** $12 + (-12) = 0$

▪ Simplify Variable Expressions

> **TAKE NOTE**
>
> Recall that we can rewrite subtraction as addition of the opposite. This step is rarely written when simplifying variable expressions but is always done mentally.
> $3x^2 + 3x - 6 - 7x + 9$
> equals
> $3x^2 + 3x + (-6) + (-7x) + 9$.

A variable expression is shown at the right. The expression can be rewritten by writing subtraction as addition of the opposite.

$$3x^2 - 4xy + 5z - 2$$
$$3x^2 + (-4xy) + 5z + (-2)$$

Note that the expression has 4 addends. The **terms** of a variable expression are the addends of the expression. The expression has 4 terms.

$$\overbrace{3x^2 \quad -4xy \quad +5z}^{\text{4 terms}} \quad -2$$
$$\text{Variable terms} \quad \text{Constant term}$$

The terms $3x^2$, $-4xy$, and $5z$ are **variable terms.** The term -2 is a **constant term,** or simply a **constant.**

Like terms of a variable expression are the terms that have the same variable parts. The terms $3x$ and $-7x$ are like terms. Constant terms are also like terms. Thus -6 and 9 are like terms. The terms $3x^2$ and $3x$ are not like terms, because $x^2 = x \cdot x$ and thus the variable parts are not the same.

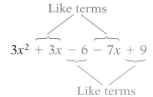

$$3x^2 + 3x - 6 - 7x + 9$$

Terms such as $4xy$ and $-7yx$ are like terms because, by the Commutative Property of Multiplication, $xy = yx$. The same is true for the like terms $-4abc$ and $12bca$.

> **TAKE NOTE**
>
> Combining like terms is an operation we do quite naturally, probably everyday. For instance,
> 4 apples + 7 apples
>   = 11 apples
> $5 + $2 = $7
> 6 pounds + 3 pounds
>   = 9 pounds
> Equally natural is the idea of *not* combining items that are not similar.
> 4 apples + 7 oranges
> 2 dogs + 5 cats

❓ QUESTION Which of the following pairs of terms are like terms?

    a. $3a$ and $3b$ **b.** $7z^2$ and $7z^3$ **c.** $6ab$ and $3a$ **d.** $-4c^2$ and $6c^2$

By using the Commutative Property of Multiplication, we can rewrite the Distributive Property as $ba + ca = (b + c)a$. This is sometimes called the *factoring form* of the Distributive Property. This form of the Distributive Property is used to **combine like terms** of a variable expression by adding their coefficients. For instance,

$$7x + 9x = (7 + 9)x \qquad \bullet \text{ Use the Distributive Property:}$$
$$= 16x \qquad\qquad\qquad ba + ca = (b + c)a$$

Combining like terms of a variable expression is referred to as **simplifying the variable expression.**

❓ ANSWER **a.** Not like terms; the variable parts are not the same. **b.** Not like terms; $z^2 = z \cdot z$ and $z^3 = z \cdot z \cdot z$. **c.** Not like terms; the variable parts are not the same. **d.** These are like terms; the variable parts are the same.

? QUESTION   What is the result when the expression $9y + 5y$ is simplified?

Because subtraction is defined as addition of the opposite, the Distributive Property also applies to subtraction. Thus, we can write $a(b - c) = ab - ac$ and $ac - bc = (a - b)c$. Here are the steps to simplify $8z - 12z$.

$$8z - 12z = (8 - 12)z \quad \bullet \text{ Use the Distributive Property: } ac - bc = (a - b)c$$
$$= -4z \quad \bullet \text{ Note: } 8 - 12 = 8 + (-12) = -4$$

Some variable expressions cannot be simplified. For instance, the variable expression $4a + 7b$ cannot be rewritten in a simpler form. The terms $4a$ and $7b$ do not have the same variable part. Therefore, the Distributive Property cannot be used. We say that $4a + 7b$ is in *simplest form*. As another example, $5x^2 + 8x$ is in simplest form because $5x^2$ and $8x$ are not like terms.

INSTRUCTOR NOTE
Exercise 66 can be used for a similar in-class example.

EXAMPLE 2

Simplify.   **a.** $2(-y)$ **b.** $-\frac{1}{3}(-3y)$

Solution   **a.** $2(-y) = 2(-1y)$ $\bullet$ Recall: $-y = -1 \cdot y$

$$= [2(-1)]y \quad \bullet \text{ Use the Associative Property of Multiplication to regroup factors.}$$
$$= -2y \quad \bullet \text{ Multiply.}$$

b. $-\frac{1}{3}(-3y) = \left[-\frac{1}{3}(-3)\right]y \quad \bullet \text{ Use the Associative Property of Multiplication to regroup factors.}$

$$= 1y \quad \bullet \text{ Use the Inverse Property of Multiplication.}$$
$$= y \quad \bullet \text{ Use the Multiplication Property of One.}$$

YOU TRY IT 2

Simplify.   **a.** $-5(-3a)$ **b.** $\left(-\frac{1}{2}c\right)2$

Solution   See page S4.   a. $15a$ b. $-c$

INSTRUCTOR NOTE
Exercise 54 can be used for a similar in-class example.

EXAMPLE 3

Simplify.   **a.** $5y + 3x - 5y$ **b.** $4x^2 + 5x - 6x^2 - 7x$

Solution   **a.** $5y + 3x - 5y$

$$= 3x + 5y - 5y \quad \bullet \text{ Use the Commutative Property of Addition to rearrange the terms.}$$
$$= 3x + (5y - 5y) \quad \bullet \text{ Use the Associative Property of Addition to group like terms.}$$
$$= 3x + 0 \quad \bullet \text{ Use the Inverse Property of Addition.}$$
$$= 3x \quad \bullet \text{ Use the Addition Property of Zero.}$$

? ANSWER $9y + 5y = (9 + 5)y = 14y$

TAKE NOTE

After simplifying an expression, it is customary to rewrite addition of the opposite as subtraction. That is why, in the solution to Example 3b, we write $-2x^2 + (-2x)$ as $-2x^2 - 2x$.

b. $4x^2 + 5x - 6x^2 - 7x$

$= 4x^2 - 6x^2 + 5x - 7x$ • Use the Commutative Property of Addition to rearrange the terms.

$= (4x^2 - 6x^2) + (5x - 7x)$ • Use the Associative Property of Addition to group like terms.

$= -2x^2 + (-2x)$ • Use the Distributive Property to combine like terms.

$= -2x^2 - 2x$ • Rewrite addition of the opposite as subtraction.

YOU TRY IT 3

Simplify. **a.** $3a - 2b + 5a$ **b.** $2z^2 - 5z - 3z^2 + 6z$

Solution See page S4. a. $8a - 2b$ b. $-z^2 + z$

? QUESTION Suppose you correctly simplify an expression and write the answer as $x + 7$, and another person writes the answer as $7 + x$. Are both answers correct?

Suggested Activity

See Section 1.5 of the *Student Activity Manual*, for an activity on using algebra tiles to represent variable expressions.

The Distributive Property also is used to remove parentheses from a variable expression. Here is an example.

$4(2x + 5z) = 4(2x) + 4(5z)$ • Use the Distributive Property: $a(b + c) = ab + ac$

$= (4 \cdot 2)x + (4 \cdot 5)z$ • Use the Associative Property of Multiplication to regroup factors.

$= 8x + 20z$ • Multiply $4 \cdot 2$ and $4 \cdot 5$.

When a negative number precedes the parentheses, be especially careful that all of the operations are performed correctly. Here are two examples.

$-5(3x - 7) = -5(3x) - (-5)(7)$ • Use the Distributive Property.

$= -15x - (-35)$ • Multiply.

$= -15x + 35$ • Rewrite subtraction of a negative number as addition of the opposite.

$-3(-7a + 4) = -3(-7a) + (-3)(4)$ • Use the Distributive Property.

$= 21a + (-12)$ • Multiply.

$= 21a - 12$ • Rewrite addition of a negative number as subtraction.

The Distributive Property can be extended to expressions containing more than two terms. For instance,

$$4(2x + 3y + 5z) = 4(2x) + 4(3y) + 4(5z)$$
$$= 8x + 12y + 20z$$

? ANSWER Yes. By the Commutative Property of Addition, $x + 7 = 7 + x$.

INSTRUCTOR NOTE
Exercise 80 can be used for a similar in-class example.

EXAMPLE 4

Simplify. **a.** $-3(2x + 4)$ **b.** $-(3z - 4)$
c. $(4a - 2c)5$ **d.** $6(3x - 4y + z)$

Solution **a.** $-3(2x + 4) = -3(2x) + (-3)(4)$ • Use the Distributive
$$= -6x - 12$$ Property.

b. $-(3z - 4) = -1(3z - 4)$ • Just as $-x = -1x$,
$-(3z - 4) = -1(3z - 4)$.

$$= -1(3z) - (-1)(4)$$ • Use the Distributive
$$= -3z + 4$$ Property.

c. $(4a - 2c)5 = (4a)(5) - (2c)(5)$ • Use the Distributive
$$= 20a - 10c$$ Property:
$(b + c)a = ba + ca$

d. $6(3x - 4y + z)$
$$= 6(3x) - 6(4y) + 6(z)$$ • Use the Distributive
$$= 18x - 24y + 6z$$ Property.

YOU TRY IT 4

Simplify. **a.** $-3(5y - 2)$ **b.** $-(6c + 5)$
c. $(3p - 7)(-3)$ **d.** $-2(4x + 2y - 6z)$

Solution See page S4. a. $-15y + 6$ b. $-6c - 5$
c. $-9p + 21$ d. $-8x - 4y + 12z$

TAKE NOTE

When we simplified $5 + 12x - 6$ (shown at the right), we wrote $12x - 1$. That is, the variable term was written first. Throughout the text, we use this convention of writing variable terms first and then the constant term. If there is more than one variable term, we arrange the variable terms alphabetically. There is no mathematical reason to do this. It is just a convention that developed over time.

To simplify the expression $5 + 3(4x - 2)$, use the Distributive Property to remove the parentheses. Remember that $3(4x - 2)$ means $3 \cdot (4x - 2)$. Thus, by the Order of Operations Agreement, perform the multiplication $3(4x - 2)$ before doing the addition.

$$5 + 3(4x - 2) = 5 + 3(4x) - 3(2)$$ • Use the Distributive Property
$$= 5 + 12x - 6$$
$$= 12x - 1$$ • Add the like terms 5 and -6.

In Example 5a and 5b below, the Distributive Property is used twice to simplify each expression.

INSTRUCTOR NOTE
Exercise 86 can be used for a similar in-class example.

EXAMPLE 5

Simplify. **a.** $3(2x - 4) - 5(3x + 2)$ **b.** $3a - 2[7a - 2(2a + 1)]$

Solution **a.** $3(2x - 4) - 5(3x + 2)$ • Use the Distributive Property
$$= 6x - 12 - 15x - 10$$ to remove parentheses.
$$= -9x - 22$$ • Combine like terms.

b. $3a - 2[7a - 2(2a + 1)]$

$= 3a - 2[7a - 4a - 2]$ • Use the Distributive Property to remove parentheses.

$= 3a - 2[3a - 2]$ • Combine like terms inside the brackets.

$= 3a - 6a + 4$ • Use the Distributive Property to remove the brackets.

$= -3a + 4$ • Combine like terms.

YOU TRY IT 5

Simplify.   **a.** $7(-3x - 4y) - 3(3x + y)$ **b.** $2y - 3[5 - 3(3 + 2y)]$

Solution   See page S4.   a. $-30x - 31y$ b. $20y + 12$

■ Translate Phrases into Variable Expressions

Creating a variable expression is an important goal in the applications of mathematics. Many application problems are given in verbal or written form and must be translated into a mathematical expression. A partial list of the verbal phrases used to indicate the different mathematical operations follows.

Addition	added to	7 added to z	$z + 7$
	more than	8 more than w	$w + 8$
	the sum of	the sum of z and 9	$z + 9$
	the total of	the total of r and s	$r + s$
	increased by	x increased by 7	$x + 7$
Subtraction	minus	t minus 3	$t - 3$
	less than	12 less than b	$b - 12$
	the difference between	the difference between x and 1	$x - 1$
	decreased by	17 decreased by a	$17 - a$
Multiplication	times	negative 2 times c	$-2c$
	the product of	the product of x and y	xy
	of	three fourths of m	$\frac{3}{4}m$
	twice	twice d	$2d$
	multiplied by	6 multiplied by y	$6y$
Division	divided by	v divided by 15	$\frac{v}{15}$
	the quotient of	the quotient of y and 3	$\frac{y}{3}$
Power	the square of *or* the second power of	the square of x	x^2
	the cube of *or* the third power of	the cube of r	r^3
	the fifth power of	the fifth power of a	a^5

TAKE NOTE

Note the translation of 12 *less than b* at the right as $b - 12$. It would be incorrect to write $12 - b$.

Ⓟ

Suggested Activity

Give students mathematical expressions and have them translate these into words. Here are some possibilities.

$5x$

$n + 9$

$\dfrac{7}{3b}$

$3y + 4$

$3(y + 4)$

Translating a phrase that contains the word *sum, difference, product,* or *quotient* can sometimes cause a problem. In the examples at the right, note that the operation symbol replaces the word *and.*

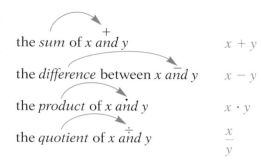

the *sum* of x and y $x + y$

the *difference* between x and y $x - y$

the *product* of x and y $x \cdot y$

the *quotient* of x and y $\dfrac{x}{y}$

TAKE NOTE

The phrase *difference between* indicates the mathematical operation of subtraction. The minus sign replaces the word *and.*

➡ Translate "the difference between three times a number and seven" into a variable expression.

Assign a variable, say x, to the unknown quantity.

Underline words that indicate mathematical operations.

Use the operations and the assigned variable (x) to write the variable expression.

the <u>difference between</u> three <u>times</u> a number and seven

$$\underbrace{\hspace{2cm}}_{3x} \quad - \quad 7$$

The variable expression is $3x - 7$. ⬅

INSTRUCTOR NOTE
Exercise 90 can be used for a similar in-class example.

EXAMPLE 6

Translate into a variable expression.

a. eight less than five times a number
b. the product of four times a number and the sum of the number and five

Solution   **a.** Let the unknown number be x.

eight <u>less than</u> five <u>times</u> a number

$5x - 8$

- Underline the words that indicate mathematical operations.
- Use the operations and the assigned variable to write the variable expression.

b. Let the unknown number be x.

the <u>product</u> of four <u>times</u> a number and the <u>sum</u> of the number and five

four times a number: $4x$
the sum of the number and 5: $x + 5$

$4x(x + 5)$

- Underline the words that indicate mathematical operations.
- Use the operations and the assigned variable to write variable expressions for internal phrases.
- Write the variable expression.

YOU TRY IT 6

Translate into a variable expression.

a. seven more than the product of a number and twelve
b. the total of eighteen and the quotient of a number and nine

Solution   See page S4.   a. $12x + 7$ b. $18 + \dfrac{x}{9}$

After translating a verbal expression into a variable expression, simplify the variable expression by using the Properties of Real Numbers.

INSTRUCTOR NOTE
Exercise 110 can be used for a similar in-class example.

EXAMPLE 7

Translate and simplify: "the total of four times an unknown number and twice the difference between the number and eight."

Solution   Let the unknown number be x.

the <u>total</u> of four <u>times</u> an unknown number and <u>twice</u> the <u>difference</u> <u>between</u> the number and eight
- Underline the words that indicate mathematical operations.

four times an unknown number: $4x$
twice the difference between the number and eight: $2(x - 8)$
- Use the operations and the assigned variable to write variable expressions for internal phrases.

$4x + 2(x - 8)$
- Write the variable expression.

$= 4x + 2x - 16$
$= 6x - 16$
- Simplify the variable expression.

YOU TRY IT 7

Translate and simplify: "a number minus the difference between the number and seventeen."

Solution   See page S4. $x - (x - 17)$; 17

▼ *Point of Interest*

The way in which expressions are symbolized has changed over time. Here are some expressions as they may have appeared in the early 16th century.
R p. 9 for x + 9. The symbol R was used for a variable to the first power. The symbol p. was used for plus.
R m. 3 for x − 3. The symbol R represented a variable. The symbol m. was used for minus.
The square of a variable was designated by Q, and the cube was designated by C. The expression x³ + x² was written C p. Q.

Many of the applications of mathematics require that you identify an unknown quantity, assign a variable to that quantity, and then express another unknown quantity in terms of that variable.

➡ A wire for a guitar is 12 feet long and is cut into two pieces. Use one variable to express the lengths of the two pieces.

Suppose that 4 feet are cut from the wire. Then the remaining piece would be

$$12 - 4 = 8 \text{ feet}$$

If 5 feet are cut from the wire, then the remaining piece would be

$$12 - 5 = 7 \text{ feet}$$

Extend this idea by letting x feet represent the length of the piece cut from the wire. Then the remaining piece would be

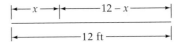

$$(12 - x) \text{ feet}$$

The lengths of the two pieces are x feet and $(12 - x)$ feet.

Note in this example that the sum of the two lengths of wire is 12, the length of the original wire.

$$x + (12 - x) = x + 12 - x = 12$$

INSTRUCTOR NOTE
Exercise 132 can be used for a similar in-class example.

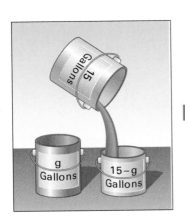

EXAMPLE 8

Fifteen gallons of paint were poured into two containers of different sizes. Express the amount of paint poured into the smaller container in terms of the amount poured into the larger container.

Solution   the number of gallons poured into the larger container: g

the number of gallons of paint poured into the smaller container: $15 - g$

YOU TRY IT 8

The sum of two numbers is 10. Express both numbers in terms of the same variable.

Solution   See page S4.   One number: x; the other number: $10 - x$

If you are having difficulty writing a variable expression for a problem, first try using numbers for the quantity that is changing. For instance, suppose the cost to rent a pair of skis is $10 plus $15 per day. Then the cost to rent the skis for 3 days is $10 + 15(3)$ or $55. The cost to rent the skis for 8 days is $10 + 15(8)$ or $130. Now replace the quantity that is changing with a variable. The variable expression is $10 + 15d$, where d is the number of days the skis are rented.

INSTRUCTOR NOTE
Exercise 136 can be used for a similar in-class example.

EXAMPLE 9

The cost to rent a car is $39.95 plus $.15 per mile driven. Express the cost of renting the car in terms of the number of miles driven.

Solution   Let m represent the number of miles driven.

3.95 + 0.15 for each mile driven:

$39.95 + 0.15m$

YOU TRY IT 9

A chef is paid $640 per week plus $32 for each hour of overtime worked. Express the chef's weekly pay in terms of the number of hours of overtime worked.

Solution   See page S4. $640 + 32h$

1.5 EXERCISES   Suggested Assignment: 15–43, odds; 45–135, every other odd

Topics for Discussion

For Exercises 1 to 12, determine whether the statement is always true, sometimes true, or never true. If the statement is sometimes true or never true, explain your answer.

1. The Multiplication Property of One states that multiplying a number by 1 does not change the number.
    Always true
2. The sum of a number and its additive inverse is zero.
    Always true
3. The product of a number and its multiplicative inverse is 1.
    Always true
4. The terms x and x^2 are like terms because both have a coefficient of 1.
    Never true
5. Like terms are terms with the same variables.
    Sometimes true
6. To add like terms, add the coefficients; the variable part remains unchanged.
    Always true
7. The expression $3x^2$ is a variable expression.
    Always true
8. In the expression $8y^2 - 4y$, the terms are $8y^3$ and $4y$.
    Never true
9. For the expression x^3, the value of x is 1.
    Sometimes true
10. For the expression $6a + 7b$, 7 is a constant term.
    Never true
11. If the sum of two numbers is 15 and one of the two numbers is x, then the other number can be expressed as $x - 15$.
    Never true
12. The expressions $7y - 8$ and $(7y) - 8$ are equivalent.
    Always true
13. Explain the difference between the Commutative and Associative Properties of Addition.   The Commutative Property says that two numbers can be added in either order. The Associative Property states that when three numbers are added together, the numbers can be grouped in any order.
14. Explain the difference between the Commutative and Associative Properties of Multiplication.   The Commutative Property says that two numbers can be multiplied in either order. The Associative Property states that when three numbers are multiplied, the numbers can be grouped in any order.

■ Properties of Real Numbers

Use the given property to complete the statement.

15. The Commutative Property of Multiplication
    $2 \cdot 5 = 5 \cdot ?$ 2

16. The Addition Property of Zero
    $? + x = x$ 0

17. The Commutative Property of Addition
    $9 + 17 = ? + 9$ 17

18. The Distributive Property
    $2(4 + 3) = 8 + ?$ 6

19. The Associative Property of Multiplication

$4(5x) = (? \cdot 5)x$ 4

20. The Multiplication Property of One

$? \cdot 1 = -4$ −4

21. The Associative Property of Addition

$(4 + 5) + 6 = ? + (5 + 6)$ 4

22. The Inverse Property of Addition

$8 + ? = 0$ −8

23. The Multiplication Property of Zero

$y \cdot ? = 0$ 0

24. The Inverse Property of Multiplication

$\left(-\frac{1}{5}\right)(-5) = ?$ 1

Identify the Property of Real Numbers that justifies the statement.

25. $1 \cdot a = a$
The Multiplication Property of One

26. $3(4x) = (3 \cdot 4)x$
The Associative Property of Multiplication

27. $0 + c = c$
The Addition Property of Zero

28. $z + (-z) = 0$
The Inverse Property of Addition

29. $\left(-\frac{2}{3}\right)\left(-\frac{3}{2}\right) = 1$
The Inverse Property of Multiplication

30. $3(4 + 7) = 12 + 21$
The Distributive Property

31. $2 + (4 + w) = (2 + 4) + w$
The Associative Property of Addition

32. $(-3 + 9)8 = -24 + 72$
The Distributive Property

33. $(3x)(4) = 4(3x)$
The Commutative Property of Multiplication

34. $(x + y) + z = z + (x + y)$
The Commutative Property of Addition

▪ Simplify Variable Expressions

Name the terms of the variable expression. Then underline the constant term.

35. $2x^2 + 5x - 8$ $2x^2$, $5x$, <u>−8</u>

36. $-3a^2 - 4a + 7$ $-3a^2$, $-4a$, <u>7</u>

37. $6 - n^4$ $-n^4$, <u>6</u>

Name the variable terms of the expression. Then underline the variable part of each term.

38. $9b^2 - 4ab + a^2$ $9\underline{b^2}$, $-4\underline{ab}$, $\underline{a^2}$

39. $7x^2y + 6xy^2 + 10$ $7\underline{x^2y}$, $6\underline{xy^2}$

40. $5 - 8n - 3n^2$ $-8\underline{n}$, $-3\underline{n^2}$

Name the coefficients of the variable terms.

41. $x^2 - 9x + 2$ 1, −9

42. $12a^2 - 8ab - b^2$ 12, −8, −1

43. $n^3 - 4n^2 - n + 9$ 1, −4, −1

Simplify each of the following. If the expression is already in simplest form, write "simplest form" as the answer.

44. $6x + 8x$ 14x

45. $12y + 9y$ 21y

46. $8b - 5b$ 3b

47. $4y - 10y$ −6y

48. $2a + 7$ simplest form

49. $x + y$ simplest form

50. $-12a + 17a$ 5a

51. $-12xy + 17xy$ 5xy

52. $3x + 5x + 3x$ 11x

53. $-5x^2 - 12x^2 + 3x^2$ $-14x^2$

54. $7x - 3y + 10x$ 17x − 3y

55. $3x - 8y - 10x + 4x$ −3x − 8y

56. $5a + 6a - 2a$ 9a

57. $-5x + 7x - 4x$ −2x

58. $2a - 5a + 3a$ 0

59. $12y^2 + 10y^2$ $22y^2$

60. $3x^2 - 15x^2$ $-12x^2$

61. $9z^2 - 9z^2$ 0

62. $\frac{3}{4}x - \frac{1}{4}x$ $\frac{1}{2}x$

63. $\frac{2}{5}y - \frac{3}{4}y$ $-\frac{7}{20}y$

64. $3x - 7 + 4x$ $7x - 7$

65. $4(3x)$ $12x$

66. $-2(-3y)$ $6y$

67. $(3a)(-2)$ $-6a$

68. $-5(3x^2)$ $-15x^2$

69. $\frac{1}{8}(8x)$ x

70. $\frac{12x}{5}\left(\frac{5}{12}\right)$ x

71. $\frac{1}{7}(14x)$ $2x$

72. $-\frac{5}{8}(24a^2)$ $-15a^2$

73. $(33y)\left(\frac{1}{11}\right)$ $3y$

74. $-(z + 2)$ $-z - 2$

75. $-2(a + 7)$ $-2a - 14$

76. $(5 - 3b)7$ $35 - 21b$

77. $3(5x^2 + 2x)$ $15x^2 + 6x$

78. $(-3x - 6)5$ $-15x - 30$

79. $-3(2y^2 - 7)$ $-6y^2 + 21$

80. $4(x^2 - 3x + 5)$
$4x^2 - 12x + 20$

81. $4(-3a^2 - 5a + 7)$
$-12a^2 - 20a + 28$

82. $5(2x^2 - 4xy - y^2)$
$10x^2 - 20xy - 5y^2$

83. $6a - (5a + 7)$ $a - 7$

84. $8 - (12 + 4y)$ $-4y - 4$

85. $6(2y - 7) - 3(3 - 2y)$ $18y - 51$

86. $2[x + 2(x + 7)]$ $6x + 28$

87. $-2[3x - (5x - 2)]$ $4x - 4$

88. $4a - 2[2b - (b - 2a)] + 3b$ b

■ Translate Phrases into Variable Expressions

Translate each phrase into a variable expression.

89. four divided by the difference between p and six
$\frac{4}{p - 6}$

90. the product of seven and the total of r and eight
$7(r + 8)$

91. three-eighths of the sum of t and fifteen
$\frac{3}{8}(t + 15)$

92. the total of nine times the cube of m and the square of m
$9m^3 + m^2$

93. thirteen less a number
$13 - x$

94. forty more than a number
$x + 40$

95. three-sevenths of a number
$\frac{3}{7}x$

96. the quotient of twice a number and five
$\frac{2x}{5}$

97. eight subtracted from the product of five and a number
$5x - 8$

98. the sum of four-ninths of a number and twenty
$\frac{4}{9}x + 20$

99. fourteen added to the product of seven and a number
$7x + 14$

100. the product of a number and ten more than the number
$x(x + 10)$

101. six less than the total of a number and the cube of the number
$(x + x^3) - 6$

102. the quotient of twelve and the sum of a number and two
$\frac{12}{x + 2}$

103. eleven plus one-half of a number
$11 + \frac{1}{2}x$

104. a number multiplied by the difference between the number and nine
$x(x - 9)$

105. eighty decreased by the product of thirteen and a number
$80 - 13x$

106. the difference between sixty and the quotient of a number and fifty
$60 - \frac{x}{50}$

107. four less than seven times the square of a number
$7x^2 - 4$

108. the sum of the square of a number and three times the number
$x^2 + 3x$

Translate into a variable expression. Then simplify the expression.

109. a number increased by the total of the number and ten
$x + (x + 10); 2x + 10$

110. a number added to the product of five and the number
$5x + x; 6x$

111. a number decreased by the difference between nine and the number
$x - (9 - x); 2x - 9$

112. eight more than the sum of a number and eleven
$(x + 11) + 8; x + 19$

113. the difference between one-fifth of a number and three-eighths of a number
$\frac{1}{5}x - \frac{3}{8}x; -\frac{7}{40}x$

114. the sum of one-eighth of a number and one-twelfth of the number
$\frac{1}{8}x + \frac{1}{12}x; \frac{5}{24}x$

115. four more than the total of a number and nine
$(x + 9) + 4; x + 13$

116. a number minus the sum of the number and fourteen
$x - (x + 14); -14$

117. twice the sum of three times a number and forty
$2(3x + 40); 6x + 80$

118. seven times the product of five and a number
$7(5x); 35x$

119. sixteen multiplied by one-fourth of a number
$16\left(\frac{1}{4}x\right); 4x$

120. the total of seventeen times a number and twice the number
$17x + 2x; 19x$

121. the difference between nine times a number and twice the number
$9x - 2x; 7x$

122. a number plus the product of the number and twelve
$x + 12x; 13x$

123. nineteen more than the difference between a number and five
$(x - 5) + 19; x + 14$

124. seven minus the sum of the number and two
$7 - (x + 2); -x + 5$

125. *Aviation*   The cruising speed of a jet plane is twice the cruising speed of a propeller-driven plane. Express the cruising speed of the jet plane in terms of the cruising speed of the propeller-driven plane.
Let p be the cruising speed of a propeller-driven plane; $2p$

126. *Sports*   In football, the number of points awarded for a touchdown is three times the number of points awarded for a safety. Express the number of points awarded for a touchdown in terms of the number of points awarded for a safety.
Let p be the number of points awarded for a safety; $3p$

127. *Food Mixtures*   A mixture contains four times as many peanuts as cashews. Express the amount of peanuts in the mixture in terms of the amount of cashews.
Let c be the amount of cashews; $4c$

128. *Coin Problem*   In a coin bank, there are ten more dimes than quarters. Express the number of dimes in the coin bank in terms of the number of quarters.
Let q be the number of quarters; $q + 10$

129. *Stamp Problem*   A 5¢ stamp in a stamp collection is 25 years older than an 8¢ stamp in the collection. Express the age of the 5¢ stamp in terms of the age of the 8¢ stamp.   Let a be the age of the 8¢ stamp; $a + 25$

130. *Geometry*   The length of a rectangle is five meters more than twice the width. Express the length of the rectangle in terms of the width.
Let *w* be the width; 2*w* + 5

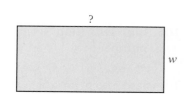

131. *Geometry*   In a triangle, the measure of the smallest angle is three degrees less than one-half the measure of the largest angle. Express the measure of the smallest angle in terms of the measure of the largest angle.
Let *L* be the measure of the largest angle; $\frac{1}{2}L - 3$

132. *Mathematics*   The sum of two numbers is twenty-three. Use one variable to represent the two numbers.
x and 23 − *x*

133. *Coin Problem*   A coin purse contains thirty-five coins in nickels and dimes. Use one variable to express the number of nickels and the number of dimes in the coin purse.
x and 35 − *x*

134. *Natural Resources*   Twenty gallons of oil was poured into two containers of different sizes. Use one variable to express the amount of oil poured into each container.
x and 20 − *x*

135. *Wages*   An employee is paid $640 per week plus $24 for each hour of overtime worked. Express the employee's weekly pay in terms of the number of hours of overtime worked.
640 + 24*h*

136. *Repair Bills*   An auto repair bill is $92 for parts and $45 for each hour of labor. Express the amount of the repair bill in terms of the number of hours of labor.   92 + 45*h*

Applying Concepts

Simplify.

137. $C - 0.7C$

0.3*C*

138. $\frac{1}{3}(3x + y) - \frac{2}{3}(6x - y)$

−3*x* + *y*

139. $-\frac{1}{4}[2x + 2(y - 6y)]$

$-\frac{1}{2}x + \frac{5}{2}y$

For each of the following, write a phrase that would translate into the given expression.

140. $2x + 3$ For example, the sum of twice a number and three

141. $5y - 4$ For example, four less than five times a number

142. $2(x + 3)$ For example, twice the sum of a number and three

143. $5(y - 4)$ For example, the product of five and four less than a number

144. *Travel*   Two cars start at the same point and travel in opposite directions and at different rates. Two hours later the cars are two hundred miles apart. Express the distance traveled by the slower car in terms of the distance traveled by the faster car.   Let *x* be the distance traveled by the faster car; 200 − *x*

145. *Coin Problem*   A coin bank contains nickels and dimes. Using n for the number of nickels in the bank and d for the number of dimes in the bank, write an expression for the value, in pennies, of the coins in the bank.
$5n + 10d$

146. *Chemistry*   Each molecule of octyl acetate, which gives air fresheners an orange scent, contains 10 carbon atoms, 20 hydrogen atoms, and 2 oxygen atoms. If x represents the number of atoms of oxygen in one gram of octyl acetate, express the number of carbon atoms in one gram of octyl acetate in terms of x. $5x$

147. *Chemistry*   Each molecule of glucose (sugar) contains 6 carbon atoms, 12 hydrogen atoms, and 6 oxygen atoms. If x represents the number of atoms of oxygen in a pound of sugar, express the number of hydrogen atoms in the pound of sugar in terms of x. $2x$

148. *Metalwork*   A wire whose length is given as x inches is bent into a square. Express the length of a side of the square in terms of x.
$\frac{1}{4}x$

EXPLORATION

1. *Investigation into Even and Odd Integers*   Complete each statement with the word *even* or *odd*.

   a. If k is an odd integer, then $k + 1$ is an _____ integer.   even
   b. If k is an odd integer, then $k - 2$ is an _____ integer.   odd
   c. If n is an integer, then $2n$ is an _____ integer.   even
   d. If m and n are even integers, then $m - n$ is an _____ integer.   even
   e. If m and n are even integers, then mn is an _____ integer.   even
   f. If m and n are odd integers, then $m + n$ is an _____ integer.   even
   g. If m and n are odd integers, then $m - n$ is an _____ integer.   even
   h. If m and n are odd integers, then mn is an _____ integer.   odd
   i. If m is an even integer and n is an odd integer, then $m - n$ is an _____ integer.   odd
   j. If m is an even integer and n is an odd integer, then $m + n$ is an _____ integer.   odd

2. *Investigation into Properties*   Determine whether the statement is true or false. If the statement is false, give an example that illustrates that it is false.

   a. Division is a commutative operation.
      False; for example, $8 \div 4 \neq 4 \div 8$
   b. Division is an associative operation.
      False; for example, $(8 \div 4) \div 2 \neq 8 \div (4 \div 2)$
   c. Subtraction is an associative operation.
      False; for example, $7 - (5 - 2) \neq (7 - 5) - 2$
   d. Subtraction is a commutative operation.
      False; for example, $7 - 4 \neq 4 - 7$
   e. Addition is a commutative operation.
      True; for example, $6 + 9 = 9 + 6$

CHAPTER **1** *SUMMARY*

Key Terms

absolute value [p. 30]
addends [p. 34]
addition [p. 34]
additive inverse [p. 29]
base [p. 41]
change in X [p. 32]
closed interval [p. 24]
coefficient [p. 31]
combine like terms [p. 72]
composite number [p. 19]
constant terms [p. 72]
deductive reasoning [p. 8]
difference [p. 36]
elements of a set [p. 18]
empty set [p. 20]
evaluate an expression [p. 30]
evaluate the variable expression [p. 32]
expanded form [p. 40]
exponent [p. 41]
exponential form [p. 40]
finite set [p. 19]
formula [p. 59]
graph of a real number [p. 20]
half-open interval [p. 24]
increment in X [p. 32]
inductive reasoning [p. 6]
infinite set [p. 18]
infinity symbol [p. 24]
input/output table [p. 32]
integers [p. 19]
intersection of two sets [p. 23]
interval notation [p. 24]
irrational numbers [p. 20]
least common multiple [p. 53]
like terms [p. 72]

multiplicative inverse [p. 71]
natural numbers [p. 18]
negative infinity symbol [p. 24]
negative integers [p. 19]
null set [p. 20]
open interval [p. 24]
opposites [p. 29]
positive integers [p. 19]
power [p. 76]
prime number [p. 19]
product [p. 38]
quotient [p. 40]
rational numbers [p. 19]
real number line [p. 20]
real numbers [p. 20]
reciprocal [pp. 56 and 71]
repeating decimal [p. 20]
roster method [p. 18]
sequence [p. 6]
set [p. 18]
set-builder notation [p. 20]
simplifying the variable expression [p. 72]
subtraction [p. 36]
sum [p. 34]
terminating decimal [p. 20]
term of a sequence [p. 6]
terms of a variable expression [p. 72]
union of two sets [p. 22]
value of the variable [p. 32]
value of the variable expression [p. 32]
variable [p. 31]
variable expression [p. 31]
variable terms [p. 72]
whole numbers [p. 19]

Essential Concepts

The four-step process in problem solving:
1. Understand the problem and state the goal.
2. Devise a strategy to solve the problem.
3. Solve the problem.
4. Review the solution and check your work. [pp. 2–3]

To add two numbers with the same sign, add the absolute values of the numbers. Then attach the sign of the addends. [p. 34]

To add two numbers with different signs, find the absolute value of each number. Then subtract the lesser of these absolute values from the greater one. Attach the sign of the number with the greater absolute value. **[p. 34]**

To subtract two numbers, add the opposite of the second number to the first number. **[p. 36]**

To multiply two numbers with the same sign, multiply the absolute values of the numbers. The product is positive. **[p. 38]**

To multiply two numbers with different signs, multiply the absolute values of the numbers. The product is negative. **[p. 38]**

To divide two numbers with the same sign, divide the absolute values of the numbers. The quotient is positive. **[p. 39]**

To divide two numbers with different signs, divide the absolute values of the numbers. The quotient is negative. **[p. 39]**

The Order of Operations Agreement

Step 1   Perform operations inside grouping symbols.
Step 2   Evaluate exponential expressions.
Step 3   Do multiplication and division as they occur from left to right.
Step 4   Do addition and subtraction as they occur from left to right. **[p. 42]**

Properties of Real Numbers
If a, b, and c are real numbers, then the following properties hold true.

Commutative Property of Addition	$a + b = b + a$
Associative Property of Addition	$a + (b + c) = (a + b) + c$
Inverse Property of Addition	$a + (-a) = -a + a = 0$
Addition Property of Zero	$a + 0 = 0 + a = a$
Commutative Property of Multiplication	$ab = ba$
Associative Property of Multiplication	$a(bc) = (ab)c$
Inverse Property of Multiplication	$a \cdot \dfrac{1}{a} = \dfrac{1}{a} \cdot a = 1, a \neq 0$
Multiplication Property of One	$a \cdot 1 = 1 \cdot a = a$
Distributive Property	$a(b + c) = ab + ac$ **[pp. 70–71]**

CHAPTER **1** *REVIEW EXERCISES*

1. I have one brother and two sisters. My mother's parents have 10 grand-children, while my father's parents have 11 grandchildren. If no divorces or remarriages occurred, how many first cousins do I have?

   13 first cousins

2. Use inductive reasoning to predict the next term in the sequence 1, 2, 4, 7, 11, 16, ...

   22

3. Given that ♠♠♠♠ = ♦♦, and ♦♦♦♦ = ♣♣, and ♣♣♣ = ♥♥♥♥♥♥, how many ♠'s equal ♥♥?

   4

4. Use the roster method to write the set of integers between -9 and -2.
{$-8, -7, -6, -5, -4, -3$}

5. Use set-builder notation to write the set of real numbers less than or equal to -10.
{$x | x \le -10$}

6. Find $A \cup B$ given $A = \{1, 3, 5, 7\}$ and $B = \{2, 4, 6, 8\}$.
{$1, 2, 3, 4, 5, 6, 7, 8$}

7. Find $C \cap D$ given $C = \{0, 1, 2, 3\}$ and $D = \{2, 3, 4, 5\}$.
{$2, 3$}

8. Write $[-2, 3]$ in set-builder notation.
{$x | -2 \le x \le 3$}

9. Write $\{x | x < -44\}$ in interval notation.
$(-\infty, -44)$

10. Graph: $(-2, 4]$

11. Graph: $\{x | x \le 3\} \cup \{x | x < -2\}$

12. Graph: $\{x | x < 3\} \cap \{x | x > -2\}$

13. Evaluate yz for $y = -\dfrac{1}{3}$ and $z = \dfrac{3}{7}$. $-\dfrac{1}{7}$

14. Evaluate $a - b$ for $a = -3.981$ and $b = -4.32$. 0.339

15. Evaluate $-x \div y$ for $x = -35.38$ and $y = 6.1$. 5.8

16. Find the sum of -247.8 and -193.4. -441.2

17. Evaluate $c - d$ for $c = \dfrac{7}{8}$ and $d = -\dfrac{5}{6}$. $\dfrac{41}{24}$

18. Evaluate $x \div y$ for $x = \dfrac{5}{9}$ and $y = -\dfrac{2}{3}$. $-\dfrac{5}{6}$

19. Evaluate $a - b(c + d)$ for $a = 6$, $b = 2$, $c = 1$, and $d = -7$. 18

20. Evaluate $ab^2 - c$ when $a = 4$, $b = -\dfrac{1}{2}$, and $c = \dfrac{5}{7}$. $\dfrac{2}{7}$

21. Find the temperature after an increase of $5°C$ from $-8°C$. $-3°C$

22. The boiling point of mercury is $356.58°C$. The melting point of mercury is $-38.87°C$. Find the difference between the boiling point and the melting point of mercury.
395.45°C

23. The formula for the perimeter of a rectangle is $P = 2L + 2W$, where P is the perimeter, L is the length, and W is the width. Find the perimeter of the rectangle with a length of 12.5 centimeters and a width of 6.25 centimeters.
37.5 cm

24. The pressure P, in pounds per square inch, at a certain depth in the ocean can be approximated by the equation $P = 15 + 0.5D$, where D is the depth in feet.

   a. Create an input/output table for this equation for increments of 2 feet, beginning with $D = 2$.

   b. Write a sentence that describes the meaning of the numbers 6 and 18.

a.

D	2	4	6	8	10	12	14
P	16	17	18	19	20	21	22

b. At a depth of 6 ft, the pressure is 18 lb/in².

25. Identify the property that justifies the statement: $-4(3) = 3(-4)$

The Commutative Property of Multiplication

26. Simplify: $(-6d)(-4)$

24d

27. Simplify: $7a^2 + 10a - 4a^2$

$3a^2 + 10a$

28. Simplify: $4(6a - 3) - (5a + 1)$

$19a - 13$

29. Translate and simplify: "eight times the quotient of twice a number and 16."

$8\left(\dfrac{2x}{16}\right);\ x$

30. The distance from Neptune to the Sun is thirty times the distance from Earth to the Sun. Express the distance from Neptune to the Sun in terms of the distance from Earth to the Sun.

Let d be the distance from Earth to the Sun; 30d

CHAPTER **1** *TEST*

1. Find the largest prime number between 210 and 220.   211

2. Determine whether the following argument is an example of inductive or deductive reasoning. "The product of an odd integer and an even integer is an odd integer. Therefore, the product of 17 and 42 is an odd integer."   deductive reasoning

3. Use the roster method to write the set of integers between -7 and 1.

$\{-6, -5, -4, -3, -2, -1, 0\}$

4. Use set-builder notation to write the set of real numbers greater than or equal to -2.

$\{x \mid x \geq -2\}$

5. Find $A \cap B$ given $A = \{-2, -1, 0, 1, 2, 3\}$ and $B = \{-1, 0, 1\}$.

$\{-1, 0, 1\}$

6. Write $[-4, 6]$ in set-builder notation.

$\{x \mid -4 \leq x \leq 6\}$

7. Write $\{x \mid x \geq -20\}$ in interval notation.

$[-20, \infty)$

8. Graph: $(-\infty, -1]$

9. Graph: $\{x \mid x \leq -3\} \ \cup \ \{x \mid x > 0\}$

10. Evaluate st for $s = -\dfrac{3}{8}$ and $t = -\dfrac{4}{15}$. $\dfrac{1}{10}$

11. Evaluate $c + d$ for $c = \dfrac{3}{4}$ and $d = -\dfrac{2}{3}$. $\dfrac{1}{12}$

12. Evaluate $x \div y$ for $x = \dfrac{1}{8}$ and $y = -\dfrac{5}{12}$. $-\dfrac{3}{10}$

13. Evaluate $m + n(p - q)^2$ for $m = -3, n = 4, p = 2,$ and $q = -1$.
    33

14. The daily high temperatures, in degrees Celsius, for a 4-day period were $-9°, -7°, 2°,$ and $-6°$. Find the average high temperature for the 4-day period. $-5°C$

15. A business analyst has determined that the cost per unit for a stereo amplifier is $127 and that the fixed costs per month are $20,000. Find the total cost during a month in which 147 amplifiers were produced. Use the formula $T = UN + F$, where T is the total cost, U is the cost per unit, N is the number of units produced, and F is the fixed cost. $38,669

16. The distance d, in feet, of a platform diver above the water t seconds after the dive begins is given by the equation $d = 50 - 16t^2$. Create an input/output table for this equation for increments of 0.25 second beginning with $t = 0$.
    a. How many feet above the water is the diver 0.25 second after the dive begins?
    b. After how many seconds is the diver 25 feet above the water?
    a. 49 ft   b. 1.25 s

17. Simplify: $-3y^2 + 9y - 5y^2$
    $-8y^2 + 9y$

18. Simplify: $3(4w - 1) - 7(w + 2)$
    $5w - 17$

19. Translate and simplify: "two more than a number added to the difference between the number and three."
    $(x - 3) + (x + 2); 2x - 1$

20. One car was driven 15 mph faster than a second car. Express the speed of the first car in terms of the speed of the second car.
    Let s be the speed of the second car; $s + 15$

2 Introduction to Functions and Relations

```
TABLE SETUP
 TblStart = 40
 ΔTbl = 5
Indpnt:    Auto    Ask
Depend:    Auto    Ask
```

Press **2nd** TBLSET to access the **TABLE SETUP** options.

*Officers investigating the scene of a traffic accident have a number of factors to analyze. For example, the skid marks, along with the type of street surface and the weather at the time of the accident, can give clues as to what happened. A quadratic function with variables representing the speed of the car and the length of the skid marks appears in **Exercise 30 on page 109**. Note that the length of the skid marks can potentially reveal whether or not the driver of the car was speeding.*

Need help? For online student resources, visit this web site:
Math.college.hmco.com

91

PREP TEST

1. Simplify: $-\dfrac{6}{3} + 4$ 2

2. Evaluate $-2x + 5$ for $x = -3$.   11

3. Evaluate $\dfrac{2r}{r-1}$ for $r = 5$.   2.5

4. Evaluate $2p^3 - 3p + 4$ for $p = -1$.   5

5. For what value of x is the expression $\dfrac{2x-5}{x}$ undefined?   0

6. Given $y = 6x - 5$, find the value of y when $x = -\dfrac{1}{3}$.   −7

INSTRUCTOR NOTE
The Prep Test is a means to test your students' mastery of *prerequisite* material that is assumed in the coming chapter. All answers, along with the section to review (if necessary), are provided in the Answers to Selected Exercises at the back of the book.

GO FIGURE

In a contest to guess how many jelly beans were in a jar, Hector guessed 223, Shannon guessed 215, Suki guessed 220, Deon guessed 221, Saul guessed 217, and Denise guessed 219. Two were off by 4, two were off by 2, one was off by 1, and one was correct. What was the correct number?

219

SECTION 2.1 Rectangular Coordinates and Graphs

- Introduction to Rectangular Coordinate Systems
- Input-Output Tables
- Graphs of Equations

Introduction to Rectangular Coordinate Systems

A cartographer (a person who makes maps) divides a city into little squares as shown on the map of Washington, DC below. Telling a visitor that the White House is located in square A3 enables the visitor to locate the White House within a small area of the map.

① Department of State
② FBI Building
③ Lincoln Memorial
④ National Air and Space Museum
⑤ National Gallery of Art
⑥ Vietnam Veterans Memorial
⑦ Washington Monument
⑧ White House

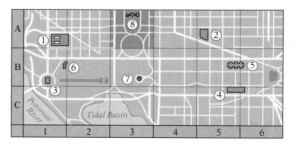

In mathematics we have a similar problem, that of locating a point in a plane. One way to solve the problem is to use a *rectangular coordinate system.*

A **rectangular coordinate system** is formed by two number lines, one horizontal and one vertical, that intersect at the zero point of each line. The point of intersection is called the **origin.** The two number lines are called the **coordinate axes** or simply the **axes.** Frequently, the horizontal axis is labeled the *x*-axis, and the vertical axis is labeled the *y*-axis. In this case, the axes form what is called the *xy*-**plane.**

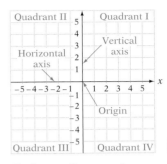

The two axes divide the plane into four regions called **quadrants** that are numbered counterclockwise, using Roman numerals I to IV, starting at the upper right.

Each point in the plane can be identified by a pair of numbers called an **ordered pair.** The first number of the ordered pair measures a horizontal change from the *y*-axis and is called the **abscissa, or *x*-coordinate.** The second number of the pair measures a vertical change from the *x*-axis and is called the **ordinate, or *y*-coordinate.** The ordered pair (*x, y*) associated with a point is also called the **coordinates** of the point.

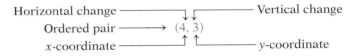

To **graph,** or **plot,** a point means to place a dot at the coordinates of the point. For example, to graph the ordered pair (4, 3), start at the origin. Move 4 units to the right and then 3 units up. Draw a dot. To graph (−3, −4), start at the origin. Move 3 units left and then 4 units down. Draw a dot.

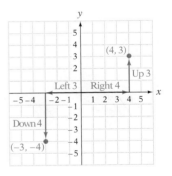

The **graph of an ordered pair** is the dot drawn at the coordinates of the point in the plane. The graphs of the ordered pairs (4, 3) and (−3, −4) are shown at the right.

The graphs of the points whose coordinates are (2, 3) and (3, 2) are shown at the right. Note that they are different points. **The order in which the numbers in an ordered pair appear is important.**

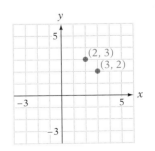

If the axes are labeled other than as *x* and *y*, then we refer to the ordered pair by the given labels. For instance, if the horizontal axis is labeled *t* and the vertical axis is labeled *d*, then the ordered pairs

▼ *Point of Interest*

The concept of a coordinate system developed over time, culminating in 1637 with the publication of Discourse on the Method for Rightly Directing One's Reason and Searching for Truth in the Sciences *by René Descartes (1596–1650) and* Introduction to Plane and Solid Loci *by Pierre de Fermat (1601–1665). Of the two mathematicians, Descartes is usually given more credit. In fact, he became so famous in Le Haye, the town in which he was born, that the town was renamed Le Haye-Descartes.*

Suggested Activity

Ask students to give the coordinates of any one point that satisfies each of the following given conditions.
1. A point in quadrant I.
2. A point on the positive *y*-axis.
3. A point in quadrant II.
4. A point on the negative *x*-axis.
5. A point in quadrant III.
6. A point on the negative *y*-axis.
7. A point in quadrant IV.
8. A point on the positive *x*-axis.

Suggested Activity

See Section 2.1 of the *Student Activity Manual* for an investigation on symmetry of points in a rectangular coordinate system.

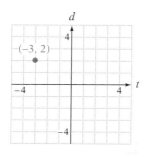

INSTRUCTOR NOTE
Exercise 8 can be used for
a similar in-class example.

are written as (t, d). In any case, we sometimes just refer to the first number in an ordered pair as the **first coordinate** of the ordered pair and to the second number as the **second coordinate** of the ordered pair.

EXAMPLE 1

Give the coordinates of the four points in the figure at the right.

Solution   A: $(-4, 2)$, B: $(4, 3)$, C: $(0, -1)$, D: $(2, 0)$, E: $(-1, -3)$.

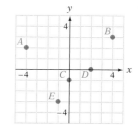

YOU TRY IT 1

Plot the points $A(-2, 4)$, $B(4, 0)$, $C(0, 3)$, and $D(-3, -4)$.

Solution   See page S4.

■ Input/Output Tables

One purpose of a coordinate system is to draw a picture of the solutions of an **equation in two variables.** Examples of equations in two variables are shown at the right.

$$y = 3x - 2$$
$$x^2 + y^2 = 25$$
$$s = t^2 - 4t + 1$$

A **solution of an equation in two variables** is an ordered pair that makes the equation a true statement. For instance, as shown below, $(2, 4)$ is a solution of $y = 3x - 2$ but $(3, -1)$ is not a solution of the equation.

TAKE NOTE

An ordered-pair solution of the equation $y = 3x - 2$ is of the form (x, y). The first number in an ordered pair is the x value; the second number is the y value.

$y = 3x - 2$	
4	$3(2) - 2$
4	$6 - 2$
$4 = 4$	

• $x = 2, y = 4$

• Checks.

$y = 3x - 2$	
-1	$3(3) - 2$
-1	$9 - 2$
$-1 \neq 7$	

• $x = 3, y = -1$

• Does not check.

There are many solutions of the equation $y = 3x - 2$. By choosing any value of x, we can calculate the corresponding value of y. The resulting ordered pair is a solution of the equation. For instance, we can choose $x = \frac{2}{3}$. Then, as shown at the right, $y = 0$. Thus $\left(\frac{2}{3}, 0\right)$ is also a solution of $y = 3x - 2$.

$$y = 3x - 2$$
$$y = 3\left(\frac{2}{3}\right) - 2$$
$$= 2 - 2$$
$$= 0$$

? QUESTION What is the solution of $y = 2x + 5$ corresponding to $x = 4$?

An **input/output table** shows some of the ordered-pair solutions of an equation in two variables. The values of x are the *inputs*; the values of y are the *outputs*.

Here is an input/output table for the equation $y = 3x - 2$. The graph of the ordered pairs is shown on the coordinate grid next to the table.

x	$3x - 2 = y$	(x, y)
-2	$3(-2) - 2 = -8$	$(-2, -8)$
-1	$3(-1) - 2 = -5$	$(-1, -5)$
0	$3(0) - 2 = -2$	$(0, -2)$
1	$3(1) - 2 = 1$	$(1, 1)$
2	$3(2) - 2 = 4$	$(2, 4)$
3	$3(3) - 2 = 7$	$(3, 7)$

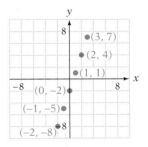

You may choose any values of x you wish. Generally, we do not choose fractional values for x because they are more difficult to graph.

The numbers do not have to be in sequence. For the input/output table above, we showed all the calculations for the ordered pairs. Normally an input/output table would show only the results, as illustrated below. The table can be displayed vertically or horizontally.

x	y
-2	-8
-1	-5
0	-2
1	1
2	4
3	7

x	-2	-1	0	1	2	3
y	-8	-5	-2	1	4	7

Note that in finding ordered-pair solutions of an equation, we determine a value of y after choosing a value of x. The value of y (the output) *depends* on the value of x (the input). Therefore, we say that y is the **dependent variable** and x is the **independent variable.**

? ANSWER Replace x in $y = 2x + 5$ by 4 and then simplify: $y = 2(4) + 5 = 8 + 5 = 13$. The solution is (4, 13).

INSTRUCTOR NOTE
Exercise 16 can be used for
a similar in-class example.

EXAMPLE 2

Create an input/output table for $y = x^2 - 1$ for $x = -2, -1, 0, 1, 2$, and 3. Graph the resulting ordered pairs.

Solution   Evaluate the expression $x^2 - 1$ for $x = -2, -1, 0, 1, 2$, and 3.

x	$x^2 - 1 = y$	(x, y)
-2	$(-2)^2 - 1 = 3$	$(-2, 3)$
-1	$(-1)^2 - 1 = 0$	$(-1, 0)$
0	$(0)^2 - 1 = -1$	$(0, -1)$
1	$(1)^2 - 1 = 0$	$(1, 0)$
2	$(2)^2 - 1 = 3$	$(2, 3)$
3	$(3)^2 - 1 = 8$	$(3, 8)$

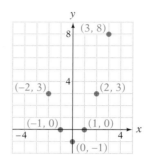

YOU TRY IT 2

Create an input/output table for $y = x^2 + 2x$ for $x = -4, -3, -2, -1, 0, 1$, and 2. Graph the resulting ordered pairs.

Solution   See page S4.

x	-4	-3	-2	-1	0	1	2
y	8	3	0	-1	0	3	8

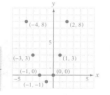

 See Appendix A:
$\boxed{\text{Y=}}$ and Table

A graphing calculator can be used to create input/output tables. The $\boxed{\text{Y=}}$ key is used to enter the variable expression, and the table capabilities are used to display the table. Some typical screens for the input/output table from Example 2 follow. By using the up arrow and down arrow keys, you can view additional inputs and outputs.

CALCULATOR NOTE

For the Table Setup screen shown at the right, ΔTbl = 1. This means the change between successive values of x is 1. In mathematics, the symbol Δ is frequently used to mean "change in."

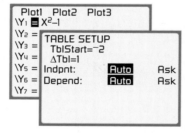

It is important to remember that **an input/output table gives ordered pairs that are solutions of an equation in two variables.** Because we entered $Y_1 = X^2 - 1$, the table at the right above shows the ordered-pair solutions of $y = x^2 - 1$. For instance, from the table, $(4, 15)$ is a solution of $y = x^2 - 1$.

❓ QUESTION   From the table above, what is the solution of $y = x^2 - 1$ corresponding to $x = 3$?

❓ ANSWER   The output value when $x = 3$ is 8. The solution is $(3, 8)$.

EXAMPLE 3

Without using a calculator, create an input/output table for $y = \frac{2}{3}x + 1$ for $x = -6, -3, 0, 3,$ and 6. Then check your table by using a graphing calculator. Graph the resulting ordered pairs.

INSTRUCTOR NOTE
Exercise 20 can be used for a similar in-class example.

Solution

ALGEBRAIC SOLUTION

The input/output table is created by evaluating $\frac{2}{3}x + 1$ for the given values of x.

x	$\frac{2}{3}x + 1 = y$	(x, y)
-6	$\frac{2}{3}(-6) + 1 = -3$	$(-6, -3)$
-3	$\frac{2}{3}(-3) + 1 = -1$	$(-3, -1)$
0	$\frac{2}{3}(0) + 1 = 1$	$(0, 1)$
3	$\frac{2}{3}(3) + 1 = 3$	$(3, 3)$
6	$\frac{2}{3}(6) + 1 = 5$	$(6, 5)$

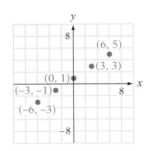

GRAPHICAL CHECK

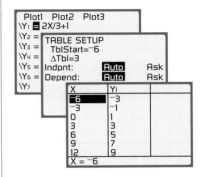

The calculator screens verify that our calculations are correct. Note that ΔTbl $= 3$ because the difference between successive input values is 3.

In this example, we chose input values that are multiples of 3 so that after we multiplied by $\frac{2}{3}$, the result would be an integer. This made graphing the points easier. However, we could have chosen any numbers for x.

YOU TRY IT 3

Without using a calculator, create an input/output table for $y = -\frac{x}{2} - 2$ for $x = -6, -4, -2, 0, 2,$ and 4. Check your results using a graphing calculator. Graph the resulting ordered pairs.

Solution   See page S5.

x	-6	-4	-2	0	2	4
y	1	0	-1	-2	-3	-4

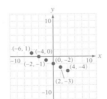

Input/output tables are also used in application problems.

INSTRUCTOR NOTE
Exercise 26 can be used for
a similar in-class example.

EXAMPLE 4

The height h, in feet, t seconds after a certain rocket used in fireworks celebrations is launched is given by $h = -16t^2 + 350t$.

a. Complete the input/output table below.

Input, time t (in seconds)	0	2	4	6	8	10
Output, height h (in feet)						

b. Write a sentence that explains the meaning of the ordered pair (6, 1524).

Solution

a. Evaluate the expression $-16t^2 + 350t$ for the given values of t.

Input, time t (in seconds)	0	2	4	6	8	10
Output, height h (in feet)	0	636	1144	1524	1776	1900

b. The ordered pair (6, 1524) means that 6 seconds after the rocket is launched it is 1524 feet above the ground.

⊞ CALCULATOR NOTE
This table can be created using a graphing calculator. Use the Y= editor to input Y₁ as
−16X² + 350X.
TblStart = 0, ΔTbl = 2
Note, in this case, that although the independent variable is time, *t*, the variable **X** is used with a graphing calculator.

YOU TRY IT 4

The temperature T, in degrees Fahrenheit, h hours after 4:00 P.M. one summer day was given by $T = \dfrac{960}{h + 12}$.

a. Complete the input/output table below. Round to the nearest tenth.

Input, time h (in hours)	0	0.5	1	1.5	2	2.5	3
Output, temperature T (in degrees Fahrenheit)	80	76.8	73.8	71.1	68.6	66.2	64

b. Write a sentence that explains the meaning of the ordered pair (2, 68.6).

Solution   See page S5.   **a.** See above.   **b.** At 6:00 P.M., the temperature was 68.6°F.

■ Graphs of Equations

The input/output table below was produced for the equation $y = 2x - 1$ using an increment of ΔTbl = 0.5. Scrolling through the table and graphing the ordered pairs produces the graph of the ordered pairs in the input/output table.

X	Y₁	
-2	-5	
-1.5	-4	
-1	-3	
-.5	-2	
0	-1	
.5	0	
1	1	
X = -2		

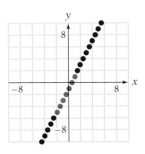

The points shown in red are from this table. The coordinates of the remaining points were found by scrolling through the table and then graphing the points displayed.

One observation we can make from the graph is that the points appear to lie on a straight line. If we change the increment to ΔTbl = 0.25, there will be even more points to graph.

X	Y₁	
-1	-3	
-.75	-2.5	
-.5	-2	
-.25	-1.5	
0	-1	
.25	-.5	
.5	0	
X = -1		

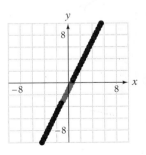

The points shown in red are from the table above. Plotting additional points seems to confirm our earlier observation that the graph is a straight line.

Note that as we use smaller and smaller increments, the graph of the ordered pairs of the input/output table begins to look more and more like a straight line. If we graph *all* of the ordered-pair solutions of $y = 2x - 1$, the resulting graph is a straight line. This line is called the **graph of the equation** and is shown at the right. Because the graph is a straight line, $y = 2x - 1$ is called a **linear equation in two variables.**

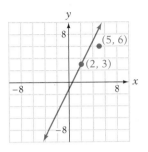

Suggested Activity

Draw a graph similar to the one below. (It does not have to be the graph of any particular equation.) Then ask students which of the points *A, B, C, D* and *E* are solutions of the equation.

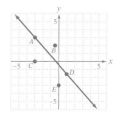

See Appendix A:
Graph

As shown below, $(2, 3)$ is a solution of $y = 2x - 1$, and note that the point $(2, 3)$ is on the graph of the equation. The ordered pair $(5, 6)$ is not a solution of the equation and is not a point on the graph.

$y = 2x - 1$		
3	$2(2) - 1$	• $x = 2, y = 3$
3	$4 - 1$	
$3 = 3$		• Checks.

$y = 2x - 1$		
6	$2(5) - 1$	• $x = 5, y = 6$
6	$10 - 1$	
$6 \neq 9$		• Does not check.

It is important to remember that **any ordered pair on a graph is a solution of the equation of the graph, and any ordered-pair solution of the equation is a point on the graph.**

A graphing calculator can be used to draw the graph of an equation by entering an expression in the Y= editor window. The portion of the rectangular coordinate grid that is shown on the calculator's screen is called the **viewing window** or just the **window.** All graphing calculators have some built-in viewing windows. One of these windows is called the **standard viewing window.** For the standard viewing window, the coordinate grid is shown for x-coordinates that are between -10 and 10 and y-coordinates that are between -10 and 10. The graph of $y = 2x - 1$ is shown below in the standard viewing window.

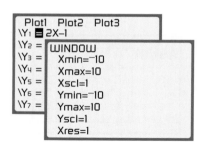

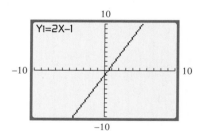

INSTRUCTOR NOTE
Exercise 32 can be used for a similar in-class example.

EXAMPLE 5

Let $y = -\dfrac{3}{4}x + 2$.

a. Complete the input/output table below.

x	-8	-4	0	4	8
y					

b. Graph the equation.
c. Use a graphing calculator to verify the input/output table and the graph.

Solution

a. Evaluate $-\dfrac{3}{4}x + 2$ for the given values of x.

x	-8	-4	0	4	8
y	8	5	2	-1	-4

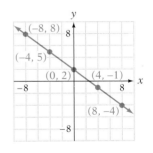

b. To graph the equation, graph the ordered pairs of the input/output table and draw a straight line through the points. The graph is shown at the right.

c. Some typical graphing calculator screens are shown below. Note that ΔTbl = 4 because the values of x (the input) in the table differ by 4. We are using the standard viewing window.

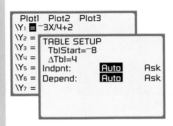

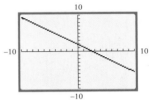

YOU TRY IT 5

Let $y = \dfrac{2}{3}x - 3$.

a. Complete the input/output table below.

x	-6	-3	0	3	6	9
y	-7	-5	-3	-1	1	3

b. Graph the equation.

b.

c. Use a calculator to verify the input/output table and the graph.

Solution   See page S5.   **a.**   **b.** See above   **c.**

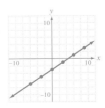

See Appendix A: Window and Trace

The graph of an equation is a drawing of all the ordered pair solutions of the equation. Besides using the TABLE feature, the TRACE feature of a graphing calculator can be used to find some of those solutions. To use this feature, enter the equation to be graphed using the Y= editor and graph the equation.

The graph of $y = -\frac{x}{2} + 1$ is shown at the right in the decimal viewing window. When the **TRACE** key is pressed, a small cursor is placed on the graph. **The coordinates of the point under the cursor are shown at the bottom of the screen. These coordinates are a solution of the graphed equation.**

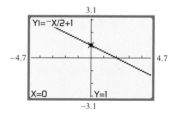

A solution is (0, 1).

Using the left and right arrow keys moves the cursor along the graph. As the arrow key is pressed, the coordinates of the point on which the cursor rests are shown at the bottom of the screen. Two examples follow.

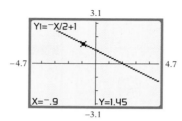

A solution is (−0.9, 1.45).

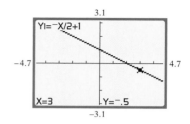

A solution is (3, −0.5).

The first shows that $(-0.9, 1.45)$ is a solution of the equation. The second shows $(3, -0.5)$ is a solution of the equation.

INSTRUCTOR NOTE
Exercise 48 can be used for a similar in-class example.

EXAMPLE 6

Let $y = 2x + 4$.

a. Graph the equation in the integer viewing window.
b. Trace along the graph to find the value of y when x is -6.
c. Trace along the graph to find the value of x when y is 16.

Solution

a. Enter $2x + 4$ for Y₁, select the integer viewing window, and press **ENTER**. Typical screens are shown below.

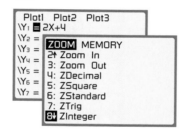

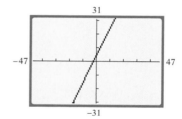

 See Appendix A: Window and Trace

b. To find the value of y when x is -6, press TRACE and use the left and right arrow keys until the value of x at the bottom of the screen is -6. The corresponding value of y is -8.

A solution of the equation is $(-6, -8)$.

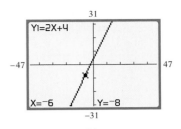

c. To find the value of x when y is 16, press TRACE and use the left and right arrow keys until the value of y at the bottom of the screen is 16. The corresponding value of x is 6.

A solution of the equation is $(6, 16)$.

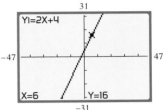

INSTRUCTOR NOTE
Examples 6 and 7 give students further experience with solutions and graphs of equations. They also teach students about the features of a graphing calculator that will be useful in later chapters.

YOU TRY IT 6

Let $y = -\dfrac{2}{3}x + 4$.

a. Graph the equation in the integer viewing window.
b. Trace along the graph to find the value of y when x is 9.
c. Trace along the graph to find the value of x when y is 8.

Solution   See page S5.   **a.**

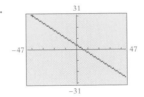

b. -2 **c.** -6

The TRACE feature can also be used to jump to any point on the graph in the viewing window by inputting the x-coordinate of that point. After you press the TRACE key, you can just enter the x value. The X = automatically appears on the screen. This procedure will work for any viewing window that contains the value of x that you input. If you input a value of x that is not in the viewing window, you will get an error message.

INSTRUCTOR NOTE
Exercise 56 can be used for a similar in-class example.

EXAMPLE 7

Let $y = -\dfrac{3}{2}x - 3$.

a. Use a graphing calculator to find the ordered-pair solution of the equation when $x = -\dfrac{13}{4}$.
b. Check the results algebraically.

See Appendix A:
Trace

Solution

a. Graph $y = -\frac{3}{2}x - 3$ in any viewing window that contains an x value of $-\frac{13}{4}$. We will use the standard viewing window. Use the TRACE feature to find the ordered pair whose x-coordinate is $-\frac{13}{4}$. The screens below show the result of graphing the equation and using the TRACE feature to enter the value $-\frac{13}{4}$.

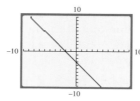 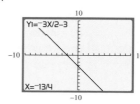 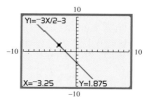

The ordered-pair solution is $(-3.25, 1.875)$.

Note that the calculator automatically converted the fraction $-\frac{13}{4}$ to a decimal. To write the ordered pair in fraction form, convert the decimals to fractions.

The ordered-pair solution in fraction form is $\left(-\frac{13}{4}, \frac{15}{8}\right)$.

b. To check the results algebraically, evaluate $-\frac{3}{2}x - 3$ when $x = -\frac{13}{4}$.

$$y = -\frac{3}{2}x - 3$$

$$y = -\frac{3}{2}\left(-\frac{13}{4}\right) - 3 \qquad \bullet \text{ Replace } x \text{ by } -\frac{13}{4}.$$

$$= \frac{39}{8} - 3 \qquad \bullet \text{ Multiply the fractions.}$$

$$= \frac{39}{8} - \frac{24}{8} \qquad \bullet \text{ Subtract the fractions. The common denominator is 8.}$$

$$= \frac{15}{8}$$

See Appendix A:
Operations

The result checks. The ordered-pair solution is $\left(-\frac{13}{4}, \frac{15}{8}\right)$.

YOU TRY IT 7

Let $y = \frac{1}{2}x + 2$.

a. Use a graphing calculator to find the ordered-pair solution of the equation when $x = -5$.

b. Check the results algebraically.

Solution   See page S5. $(-5, -0.5)$

INSTRUCTOR NOTE
The problem of finding x given y is discussed later in the text.

There is a similar problem of trying to find x for a given value of y. When tracing along the curve, the cursor may not stop at the particular value of y that you need. The solution to this problem is more complicated both with and without a calculator. We will discuss this in more detail later in the text.

2.1 EXERCISES   Suggested Assignment: 7–63, odds

Topics for Discussion

1. Describe a rectangular coordinate system. Include in your description the concepts of axes, ordered pairs, and quadrants.
   Answers will vary.
2. What is the graph of an ordered pair?
   The graph of an ordered pair is a dot drawn at the coordinates of the ordered pair.
3. What is a solution of an equation in two variables?
   A solution of an equation in two variables is an ordered pair that makes the equation a true statement.
4. What is an input/output table?
   An input/output table shows some of the ordered-pair solutions of an equation in two variables.
5. For the equation $s = 3t - 4$, what is the input variable? What is the output variable?
   The input variable is t. The output variable is s.
6. What is the graph of an equation?
   The graph of an equation is a drawing of all ordered-pair solutions of the equation.

▪ Introduction to Rectangular Coordinate Systems

7. Graph the ordered pairs $A(2, -3)$, $B(0, 3)$, $C(-2, 0)$, and $D(-3, -4)$.

8. Graph the ordered pairs $A(-3, 4)$, $B(3, 0)$, $C(0, -3)$, and $D(3, -2)$.

9. Name the x-coordinates for points A and B. Name the y-coordinates for points C and D.

$-2, 0; 0, -3$

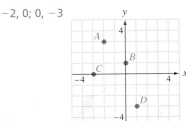

10. Name the x-coordinates for points A and B. Name the y-coordinates for points C and D.

$-4, -3; -4, 4$

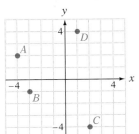

▪ Input/Output Tables

11. Is $(3, -4)$ a solution of $y = 2x - 10$?   Yes

12. Is $(-2, 3)$ a solution of $y = -2x - 2$?   No

13. Is $(-2, 5)$ a solution of $y = \dfrac{3}{2}x + 7$?   No

14. Is $(4, -2)$ a solution of $y = -\dfrac{3}{4}x + 1$?   Yes

15. Without using a calculator, create an input/output table for $y = -2x + 1$ for $x = -3, -2, -1, 0, 1, 2,$ and 3. Check your results using a graphing calculator. Graph the resulting ordered pairs.

x	-3	-2	-1	0	1	2	3
y	7	5	3	1	-1	-3	-5

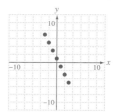

16. Without using a calculator, create an input/output table for $y = 2x - 3$ for $x = -3, -2, -1, 0, 1, 2,$ and 3. Check your results using a graphing calculator. Graph the resulting ordered pairs.

x	-3	-2	-1	0	1	2	3
y	-9	-7	-5	-3	-1	1	3

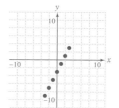

17. Without using a calculator, create an input/output table for $y = \frac{3}{4}x + 1$ for $x = -8, -4, 0, 4,$ and 8. Check your results using a graphing calculator. Graph the resulting ordered pairs.

x	-8	-4	0	4	8
y	-5	-2	1	4	7

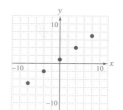

18. Without using a calculator, create an input/output table for $y = -\frac{2}{3}x - 3$ for $x = -9, -6, -3, 0, 3,$ and 6. Check your results using a graphing calculator. Graph the resulting ordered pairs.

x	-9	-6	-3	0	3	6
y	3	1	-1	-3	-5	-7

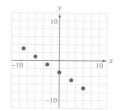

19. Without using a calculator, create an input/output table for $y = x^2 + 1$ for $x = -3, -2, -1, 0, 1, 2,$ and 3. Check your results using a graphing calculator. Graph the resulting ordered pairs.

x	-3	-2	-1	0	1	2	3
y	10	5	2	1	2	5	10

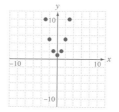

20. Without using a calculator, create an input/output table for $y = -x^2 + 2$ for $x = -3, -2, -1, 0, 1, 2,$ and 3. Check your results using a graphing calculator. Graph the resulting ordered pairs.

x	-3	-2	-1	0	1	2	3
y	-7	-2	1	2	1	-2	-7

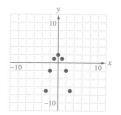

21. Without using a calculator, create an input/output table for $y = x^2 + 4x - 3$ for $x = -5, -4, -3, -2, -1, 0$, and 1. Check your results using a graphing calculator. Graph the resulting ordered pairs.

x	-5	-4	-3	-2	-1	0	1
y	2	-3	-6	-7	-6	-3	2

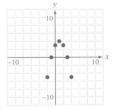

22. Without using a calculator, create an input/output table for $y = -x^2 + 2x + 3$ for $x = -2, -1, 0, 1, 2, 3$, and 4. Check your results using a graphing calculator. Graph the resulting ordered pairs.

x	-2	-1	0	1	2	3	4
y	-5	0	3	4	3	0	-5

23. *Sports*   If a jogger is running at a rate of 11 feet per second, then the distance d traveled by the jogger in t seconds is given by $d = 11t$.

a.   Complete the input/output table below.

Input, time t (in seconds)	0	5	10	15	20	25	30
Output, distance d (in feet)	0	55	110	165	220	275	330

b.   Write a sentence that explains the meaning of the ordered pair (20, 220).

In 20 s, the jogger runs 220 ft.

24. *Geometry*   Sand, dumped from a conveyor belt, is forming a cone-shaped mound. The relationship between the height h, in feet, of the cone and the diameter of the base b, in feet, is given by $h = \frac{4}{3}b$.

a.   Complete the input/output table below.

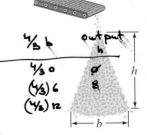

Input, diameter of base b (in feet)	0	6	12	18	24	30
Output, height h (in feet)	0	8	16	24	32	40

b.   Write a sentence that explains the meaning of the ordered pair (18, 24).

When the diameter of the base is 18 ft, the height of the mound is 24 ft.

25. *Physics*   Assuming no air resistance, the distance d, in feet, that an object will fall in t seconds is given by $d = 16t^2$.

a.   Complete the input/output table below.

Input, time t (in seconds)	0	0.5	1	1.5	2	2.5	3
Output, distance d (in feet)	0	4	16	36	64	100	144

b.   Write a sentence that explains the meaning of the ordered pair (1.5, 36).   In 1.5 s, the object will fall 36 ft.

26. *Food Mixtures*   Suppose a flavored drink contains 10% fruit juice. Then the quantity Q, in ounces, of fruit juice in a serving size of s ounces is given by $Q = 0.10s$.

    a.   Complete the input/output table below.

Input, serving size s (in ounces)	0	2	4	6	8	10	12	14
Output, quantity of fruit juice Q (in ounces)	0	0.2	0.4	0.6	0.8	1	1.2	1.4

    b.   Write a sentence that explains the meaning of the ordered pair (12, 1.2).

    In a 12-ounce serving of the flavored drink, there are 1.2 oz of fruit juice.

27. *Metallurgy*   Gold jewelry that is made with 18-carat gold contains 75% gold. The quantity Q, in grams, of gold in a piece of jewelry weighing w grams is given by $Q = 0.75w$.

    a.   Complete the input/output table below.

Input, weight of jewelry w (in grams)	0	5	10	15	20	25	30
Output, quantity of gold Q (in grams)	0	3.75	7.5	11.25	15	18.75	22.5

    b.   Write a sentence that explains the meaning of the ordered pair (15, 11.25).

    In a 15-gram piece of 18-carat gold jewelry, there are 11.25 g of gold.

28. *Fuel Consumption*   If a car averages 25 miles per gallon, then the number of miles m that a car can travel on g gallons of gasoline is given by $m = 25g$.

    a.   Complete the input/output table below.

Input, quantity of gas g (in gallons)	0	3	6	9	12	15	18	21
Output, distance traveled m (in miles)	0	75	150	225	300	375	450	525

    b.   Write a sentence that explains the meaning of the ordered pair (9, 225).

    The car can travel 225 mi on 9 gal of gasoline.

29. *Sports*   The height h of a baseball thrown upward at an initial velocity of 70 feet per second is given by $h = -16t^2 + 70t + 5$, where t is the time in seconds since the baseball was released.

    a.   Complete the input/output table below.

Input, time t (in seconds)	0	0.5	1	1.5	2	2.5	3
Output, height h (in feet)	5	36	59	74	81	80	71

    b.   Write a sentence that explains the meaning of the ordered pair (2.5, 80).

    The ball is 80 ft above the ground 2.5 s after it is released.

30. *Automotive Technology*   When the driver of a car is presented with a dangerous situation that requires braking, the distance the car will travel before stopping depends on the driver's reaction time and the speed of the car at the time the brakes are applied. The distance d, in feet, is given by $d = 0.05s^2 + 1.1s$, where s is the speed of the car in miles per hour.

   a.   Complete the input/output table below.

Input, speed s (in miles per hour)	40	45	50	55	60	65	70
Output, distance d (in feet)	124	150.75	180	211.75	246	282.75	322

   b.   Write a sentence that explains the meaning of the ordered pair (60, 246).

At a speed of 60 mph, the car will travel 246 ft before it stops.

Graphs of Equations

For Exercises 31 to 38, complete the input/output table and graph the equation without using a calculator. Then use a calculator to verify the table and graph.

31. $y = 2x - 4$

x	-2	-1	0	1	2
y	-8	-6	-4	-2	0

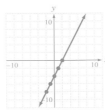

32. $y = -2x + 2$

x	-2	-1	0	1	2
y	6	4	2	0	-2

33. $y = \dfrac{x}{2} + 1$

x	-4	-2	0	2	4
y	-1	0	1	2	3

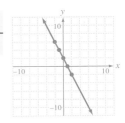

34. $y = \dfrac{2x}{3} - 2$

x	-3	0	3	6	9
y	-4	-2	0	2	4

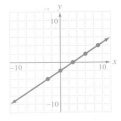

35. $y = \dfrac{-5x}{4}$

x	-8	-4	0	4	8
y	10	5	0	-5	-10

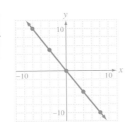

36. $y = -\dfrac{3}{2}x + 4$

x	-2	0	2	4	6
y	7	4	1	-2	-5

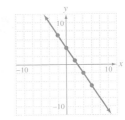

37. $y = \dfrac{3}{4}x - 4$

x	-8	-4	0	4	8
y	-10	-7	-4	-1	2

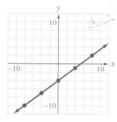

38. $y = -\dfrac{2}{3}x + 4$

x	-6	-3	0	3	6
y	8	6	4	2	0

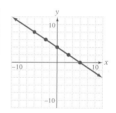

In Exercises 39 to 44, create your own input/output table for the equation and then graph the equation.

39. $y = 2x + 2$

40. $y = -x - 1$

41. $y = \dfrac{3}{2}x - 3$

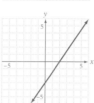

42. $y = -\dfrac{5}{2}x + 5$

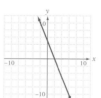

43. $y = -\dfrac{3}{4}x + 1$

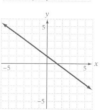

44. $y = \dfrac{2}{3}x - 4$

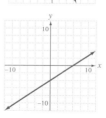

In Exercises 45 to 54, graph the equation in the integer viewing window. Then trace along the graph to find the requested values.

45. Let $y = 3x - 2$.

  a. Find y when $x = 4$.
  b. Find x when $y = 13$.
  a. 10   b. 5

46. Let $y = 3 - 2x$.

  a. Find y when $x = -2$.
  b. Find x when $y = 7$.
  a. 7   b. -2

47. Let $y = -\dfrac{x}{2} + 5$.

  a. Find y when $x = 10$.
  b. Find x when $y = -6$.
  a. 0   b. 22

48. Let $y = \dfrac{2}{3}x + 6$.

  a. Find y when $x = -12$.
  b. Find x when $y = -10$.
  a. -2   b. -24

49. Let $y = 5 - \dfrac{7x}{4}$.

  a. Find y when $x = -8$.
  b. Find x when $y = -9$.
  a. 19   b. 8

50. Let $y = \dfrac{3x}{2} - 5$.

  a. Find y when $x = -8$.
  b. Find x when $y = 10$.
  a. -17   b. 10

51. Let $y = -\dfrac{3}{4}x + 7$.

    a. Find y when $x = 24$.
    b. Find x when $y = -14$.
    a. -11 b. 28

52. Let $y = \dfrac{5}{7}x - 6$.

    a. Find y when $x = -14$.
    b. Find x when $y = 14$.
    a. -16 b. 28

53. Let $y = \dfrac{5}{3}x + 10$.

    a. Find y when $x = 0$.
    b. Find x when $y = 0$.
    a. 10 b. -6

54. Let $y = -\dfrac{2}{5}x - 8$.

    a. Find y when $x = 0$.
    b. Find x when $y = 0$.
    a. -8 b. -20

In Exercises 55 to 64, graph the equation in any viewing window that contains the given value of x. Use the TRACE feature of the calculator to find the ordered-pair solution of the equation for the given value of x. Check the results algebraically.

55. $y = 2x + 5$; $x = 7.5$ $(7.5, 20)$

56. $y = -3x + 2$; $x = -6.4$ $(-6.4, 21.2)$

57. $y = 2 - 3x$; $x = -6.3$ $(-6.3, 20.9)$

58. $y = 3 - 5x$; $x = 8.4$ $(8.4, -39)$

59. $y = \dfrac{3}{4}x + 3$; $x = -\dfrac{16}{3}$ $\left(-\dfrac{16}{3}, -1\right)$

60. $y = -\dfrac{3}{5}x - 5$; $x = -\dfrac{10}{3}$ $\left(-\dfrac{10}{3}, -3\right)$

61. $y = \dfrac{5}{3}x + 5$; $x = -\dfrac{57}{10}$ $\left(-\dfrac{57}{10}, -\dfrac{9}{2}\right)$

62. $y = -\dfrac{5}{4}x + 3$; $x = -\dfrac{16}{5}$ $\left(-\dfrac{16}{5}, 7\right)$

63. $y = -2.3x + 4.8$; $x = 12.1$ $(12.1, -23.03)$

64. $y = 1.95x - 4.5$; $x = -12.3$ $(-12.3, -28.485)$

Applying Concepts

65. What is the y-coordinate of a point at which a graph crosses the x-axis?
    0

66. What is the x-coordinate of a point at which a graph crosses the y-axis?
    0

67. Name any two points on a horizontal line that is 2 units above the x-axis.
    Answers will vary. For example, $(-3, 2)$ and $(5, 2)$.

68. Name any two points on a vertical line that is 3 units to the left of the y-axis.
    Answers will vary. For example, $(-3, -4)$ and $(-3, 1)$.

69. If $(-3, 4)$ and $(5, -1)$ are coordinates of two opposite vertices of a rectangle, what are the coordinates of the other two vertices?
    $(5, 4)$ and $(-3, -1)$

70. If $(0, -3)$ and $(-5, 4)$ are coordinates of two opposite vertices of a rectangle, what are the coordinates of the other two vertices?
    $(0, 4)$ and $(-5, -3)$.

es 71 to 76, determine the distance from the given point to **a.** the
). the *y*-axis.

71. (5, 3)
a. 5 b. 3

74. (−4, −8)
a. 8 b. 4

72. (−1, 6)
a. 6 b. 1

75. (−2, 0)
a. 0 b. 2

73. (7, −4)
a. 4 b. 7

76. (0, 6)
a. 6 b. 0

For Exercises 77 to 82, draw a graph in the *xy*-plane that satisfies the stated
conditions.

77. The *x*-coordinate is always −3;
the *y*-coordinate is any real
number.

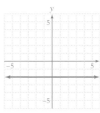

78. The *x*-coordinate is any real
number; the *y*-coordinate
always is 5.

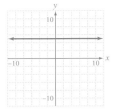

79. The *x*-coordinate is any real
number; the *y*-coordinate is
always −2.

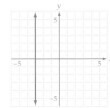

80. The *x*-coordinate is always 4;
the *y*-coordinate is any real
number.

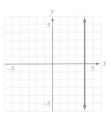

81. The *x*-coordinate always
equals the *y*-coordinate.

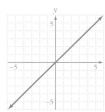

82. The *y*-coordinate is always
the opposite of the
x-coordinate.

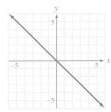

83.   There is a coordinate system on Earth that consists of *longitude* and *latitude*.
Write a report on how location is determined on the surface of Earth.

EXPLORATION

Properties of Graphs of Straight Lines

1. Using a graphing calculator, enter $Y_1 = 2x + 3$, $Y_2 = −x + 3$, $Y_3 = \frac{3}{4}x + 3$,
$Y_4 = −\frac{5}{3}x + 3$, and $Y_5 = x + 3$. Graph all of these in the standard viewing window.

a.   What are the similarities among the graphs?

b. What are the differences among the graphs?

c. Without graphing $y = −2.7x + 3$, determine through which point on
the *y*-axis the graph will pass.

    d. From the answers you have given and without graphing each equation, determine through which point on the y-axis each graph will pass.

      i. $\quad y = 2x + 4$ **ii.** $\quad y = -2x + 1$

      iii. $\quad y = \dfrac{1}{2}x - 3$ **iv.** $\quad y = -3.1x - 5$

    a. They all pass through the point (0, 3).   **b.** Each has a different slope (or slant or steepness).   **c.** (0, 3)   **d.** (0, 4), (0, 1), (0, −3), (0, −5)

2. Using a graphing calculator, enter $Y_1 = 2(x - 2) + 1$, $Y_2 = -\dfrac{1}{2}(x - 2) + 1$, $Y_3 = -3(x - 2) + 1$, $Y_4 = -(x - 2) + 1$, and $Y_5 = \dfrac{2}{3}(x - 2) + 1$. Graph all of these in the decimal viewing window.

    a.   What are the similarities among the graphs?

    b.   What are the differences among the graphs?

    c. Without graphing $y = 3.4(x - 2) + 1$, determine through which point in the rectangular coordinate system the graph will pass.

    d. From the answers you have given and without graphing each equation, determine through which point in the plane each graph will pass.

      i. $\quad y = 3(x - 1) + 2$ **ii.** $\quad y = -2(x - 3) + 4$

      iii. $\quad y = \dfrac{5}{2}(x - 5) + 1$ **iv.** $\quad y = -1.2(x - 4) + 3$

    a. They all pass through the point (2, 1).   **b.** Each has a different slope (or slant or steepness).   **c.** (2, 1)   **d.** (1, 2), (3, 4), (5, 1), (4, 3)

3.   Using a graphing calculator, enter $Y_1 = 2x + 3$, $Y_2 = 2x - 1$, $Y_3 = 2x + 2.5$, $Y_4 = 2x - 3.5$ and $Y_5 = 2x + 6$. Graph all of these in the standard viewing window.

    a. What are the similarities among the graphs?

    b. What are the differences among the graphs?

    c. How are the equations the same?

    d. How are they different?

    a. They all have the same slope (or slant or steepness).   **b.** Each has a different y-intercept. (Each goes through a different point on the y-axis.)   **c.** The coefficient of x in each equation is 2.   **d.** The constant in each equation is different.

4.   Using a graphing calculator, enter $Y_1 = -2x + 2$, $Y_2 = -2x - 2$, $Y_3 = -2x + 6$, $Y_4 = -2x - 4$, and $Y_5 = -2x + 3$. Graph all of these in the standard viewing window.

    a. What are the similarities among the graphs?

    b. What are the differences among the graphs?

    c. How are the equations the same?

    d. How are they different?

    a. They all have the same slope (or slant or steepness).   **b.** Each has a different y-intercept. (Each goes through a different point on the y-axis.)   **c.** The coefficient of x in each equation is −2.   **d.** The constant in each equation is different.

Relations and Functions

Introduction to Relations and Functions

Exploring relationships between known quantities frequently results in sets of ordered pairs. For instance, the amount of air resistance R, in pounds, experienced by a certain car is given by $R = 0.024s^2$, where s is the speed of the car in miles per hour. The table below shows how the resistance on the car depends on the speed of the car for various values of s.

Speed, s (in miles per hour)	10	15	25	35	55	70
Resistance, R (in pounds)	2.4	5.4	15	29.4	72.6	117.6

▼ *Point of Interest*

Automotive engineers use equations such as $R = 0.024s^2$ to study the flow of air over a car. By understanding the relationship between speed and resistance, engineers can design more fuel-efficient cars.

The numbers in the table can also be written as ordered pairs where the first coordinate of the ordered pair is the speed of the car and the second coordinate is the air resistance. The ordered pairs are (10, 2.4), (15, 5.4), (25, 15), (35, 29.4), (55, 72.6) and (70, 117.6). The ordered pairs from the table above are only some of the possible ordered pairs. Other possibilities include (31, 23.064), (52, 64.896), (65, 101.4) and many more.

A table is another way of describing the relationship between two quantities. The table at the right shows a grading scale for an exam. Some possible ordered pairs for the relationship between exam scores and letter grades are (91, A), (87, B), (78, C), (98, A), (68, D), and (85, B).

Test Score	Grade
90–100	A
80–89	B
70–79	C
60–69	D
0–59	F

A third way of describing a relationship between two quantities is a graph. The bar graph below, based on data from the U.S. Census Bureau, shows the increase of the median price of a house in the United States. The ordered pairs can be approximated by reading the graph as

(1996, 140,000), (1997, 146,000), (1998, 153,000), (1999, 161,000), (2000, 169,000), and (2001, 175,000).

For each of these situations, ordered pairs were used to show the relationship between two quantities. In mathematics, a set of ordered pairs is called a *relation*.

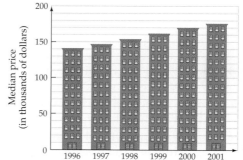

Definition of a Relation

A **relation** is any set of ordered pairs. The **domain** of the relation is the set of first coordinates of the ordered pairs. The **range** of the relation is the set of second coordinates of the ordered pairs.

For the relation between an exam score and a letter grade, the domain is all possible exam scores, and the range is all possible letter grades.

Domain = {0, 1, 2, 3, 4, . . ., 97, 98, 99, 100}   and   Range = {A, B, C, D, F}

Here is one more example of a relation. Consider the situation of a paleontologist trying to determine how fast a dinosaur could run. By studying the relation between the stride lengths of living creatures and the speed at which those creatures run, the scientist can infer the speed of a dinosaur. The table below shows the results of measuring the stride length and speed of various horses.

Stride length (in meters)	2.5	3.2	2.8	3.4	3.2	3.5	4.5	5.2
Speed (in meters per second)	2.6	2.7	2.5	2.8	2.6	2.8	3.2	3.0

This relation is the set of ordered pairs {(2.5, 2.6), (3.2, 2.7), (2.8, 2.5), (3.4, 2.8), (3.2, 2.6), (3.5, 2.8), (4.5, 3.2), (5.2, 3.0)}.

? QUESTION   What is the meaning of the ordered pair (3.5, 2.8) for the relation above?

Although relations are important in mathematics, the concept of *function* is especially useful in applications.

> **Definition of a Function**
> A **function** is a relation in which no two ordered pairs have the same first coordinate and different second coordinates.

The relation between stride length and speed given earlier is *not* a function because the ordered pairs (3.2, 2.7) and (3.2, 2.6) have the same first coordinate and different second coordinates.

Now consider the relation between exam scores and letter grades. If there were two ordered pairs with the same first coordinate and different second coordinates, it would mean that two students with the same exam score (first coordinate) would receive different letter grades (second coordinate). For example, one student with an exam score of 72 could receive a D, and another student with an exam score of 72 could receive an A. But this does not happen with a grading scale. Thus there are no two ordered pairs with the same first coordinate and different second coordinates. The relationship between exam scores and letter grades is a function.

? ANSWER   A horse with a stride length of 3.5 meters ran at a speed of 2.8 meters per second.

▼ *Point of Interest*

Studies of foot imprints of dinosaurs suggest they traveled between 2 mph and 7 mph. However, some dinosaurs could run about 25 mph. To put this in perspective, the fastest human can run about 23.5 mph. The fastest land animal (the cheetah) can run about 65 mph.

TAKE NOTE

The idea of *function* is one of the most important concepts in math. It is a concept you will encounter throughout this text.

TAKE NOTE

Note, however, that two students can receive different exam scores (first coordinates) and the same letter grade (second coordinates). For example, the ordered pairs (84, B) and (86, B) belong to the function.

INSTRUCTOR NOTE
Exercise 8 can be used for a
similar in-class example.

TAKE NOTE

Recall that when listing elements of a set, duplicate elements are listed only once.

EXAMPLE 1

Give the domain and range of the following relation. Is the relation a function?

$$\{(2, 4), (3, 6), (4, 7), (5, 4), (3, 2), (6, 8)\}$$

Solution   Domain $= \{2, 3, 4, 5, 6\}$ Range $= \{2, 4, 6, 7, 8\}$

The relation is not a function because there are two ordered pairs, $(3, 6)$ and $(3, 2)$, with the same first coordinate and different second coordinates.

YOU TRY IT 1

Give the domain and range of the following relation. Is the relation a function?

$$\{(1, 1), (2, 1), (3, 1), (4, 1), (5, 1), (6, 1), (7, 1)\}$$

Solution   See page S5.   Domain $= \{1, 2, 3, 4, 5, 6, 7\}$, Range $= \{1\}$; Yes

CALCULATOR NOTE

A graphing calculator can be used to find ordered pairs of $s = 16t^2$ for arbitrary values of t by using the *ASK* feature.

See Appendix A:
Table
Some typical screens
are shown below.

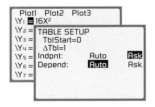

Although a function can be described in terms of ordered pairs, in a table, or by a graph, a major focus of this text will be functions defined by equations in two variables. For instance, when gravity is the only force acting on a falling body, a function that describes the distance s, in feet, that an object will fall in t seconds can be given by the equation $s = 16t^2$.

Given a value of t (time), the value of s (the distance the object falls) can be found. For instance, given $t = 3$, then

$$s = 16t^2$$
$$s = 16(3)^2 \qquad \bullet \text{ Replace } t \text{ by 3.}$$
$$= 16(9) = 144$$

In 3 seconds, an object falls 144 feet.

Because the distance the object falls depends on how long it has been falling, s is the dependent variable and t is the independent variable. We can find some of the ordered pairs of this function by evaluating $16t^2$ for various values of t. Here are the calculations for $t = 2$, $\frac{1}{2}$, and 4.5.

$s = 16t^2$	$s = 16t^2$	$s = 16t^2$
$s = 16(2)^2$	$s = 16\left(\dfrac{1}{2}\right)^2$	$s = 16(4.5)^2$
$= 64$	$= 4$	$= 324$
$(2, 64)$	$\left(\dfrac{1}{2}, 4\right)$	$(4.5, 324)$

The ordered pairs can be written as (t, s), where $s = 16t^2$. By substituting $16t^2$ for s, we can also write the ordered pairs as $(t, 16t^2)$.

Suggested Activity

Ask students to create a relation with five ordered pairs for each of the following situations.
1. The second coordinate is twice the first coordinate.
2. The second coordinate equals the first coordinate.
3. The second coordinate is 5, and the first coordinate is 1, 2, 3, 4, or 5.
4. The first coordinate is 5, and the second coordinate is 1, 2, 3, 4, or 5.
Now ask for the domain and range of each relation, and ask which of the relations are functions.

For the equation $s = 16t^2$, we say that "distance is a function of time."

Not all equations in two variables define a function. For instance, $y^2 = x^2 + 9$ is not an equation that defines a function. As shown below, the ordered pairs $(4, 5)$ and $(4, -5)$ belong to the equation.

$y^2 = x^2 + 9$	
5^2	$4^2 + 9$
25	$16 + 9$
$25 = 25$	

- Let $(x, y) = (4, 5)$.
  Replace x by 4 and y by 5.
- $(4, 5)$ checks.

$y^2 = x^2 + 9$	
$(-5)^2$	$4^2 + 9$
25	$16 + 9$
$25 = 25$	

- Let $(x, y) = (4, -5)$.
  Replace x by 4 and y by -5.
- $(4, -5)$ checks.

Thus there are two ordered pairs with the *same* first coordinate, 4, but *different* second coordinates, 5 and -5; the equation does not define a function.

The phrase "y is a function of x," or a similar phrase with different variables, is used to describe those equations in two variables that define functions.

■ Functional Notation

Functional notation is frequently used for those equations that define functions. Just as x is commonly used as a variable, the letter f is commonly used to name a function.

TAKE NOTE

The notation $f(x)$ does not mean f *times x*. The letter f stands for the name of the function, and $f(x)$ is the value of the function at x.

$$s(t) = 16t^2$$

To describe the relationship between a number and its square using functional notation, we can write $f(x) = x^2$. The symbol $f(x)$ is read "the *value* of f at x" or "f of x." The symbol $f(x)$ is the **value of the function** and represents the value of the dependent variable for a given value of the independent variable. We will often write $y = f(x)$ to emphasize the relationship between the independent variable, x, and the dependent variable, y. **Remember: y and $f(x)$ are different symbols for the same number.**

Also, the *name* of the function is f; the *value* of the function at x is $f(x)$. For instance, the equation $R = 0.024s^2$ discussed at the beginning of this section could be written as $R(s) = 0.024s^2$. The name of the function is R.

The letters used to represent a function are somewhat arbitrary. All of the following equations represent the same function.

$$f(x) = x^2 \qquad g(t) = t^2 \qquad P(v) = v^2$$

All three equations represent the square function.

The process of finding $f(x)$ for a given value of x is called **evaluating the function.** For instance, to evaluate $f(x) = x^2$ when x is 4, replace x by 4 and simplify.

$$f(x) = x^2$$
$$f(4) = 4^2 \qquad \bullet \text{ Replace } x \text{ by 4.}$$
$$= 16 \qquad \bullet \text{ Simplify.}$$

The *value* of the function is 16 when $x = 4$.

An ordered pair of the function is $(4, 16)$.

TAKE NOTE

To evaluate a function, you can use open parentheses for the variable. For instance,

$p(s) = s^2 - 4s - 1$
$p(\ ) = (\ )^2 - 4(\ ) - 1$

To evaluate the function, fill each of the parentheses with the same number and then simplify.

In many cases, you can think of a function as a machine that performs an operation on a number. For instance, you can think of the square function as taking an input (a number from the domain) and creating an output (a number in the range) that is the square of the input.

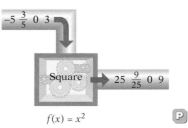

$f(x) = x^2$

? QUESTION   What is the value of the function $f(x) = x^2$ when $x = -5$?

➡ Evaluate $p(s) = s^2 - 4s - 1$ when $s = -2$.

$p(s) = s^2 - 4s - 1$

$p(-2) = (-2)^2 - 4(-2) - 1$ • Replace s by -2.

$= 11$ • Simplify.

The value of the function is 11 when $s = -2$.

See Appendix A:
Evaluating Functions

A graphing calculator can be used to check the evaluation of the function above. Some typical screens are shown at the right.

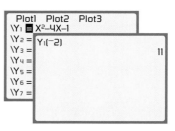

EXAMPLE 2

Let $g(t) = \dfrac{t^2}{t + 1}$.

a. Find $g(-2)$.
b. Find the value of g when $t = 3$.

Solution

INSTRUCTOR NOTE
Exercise 16 can be used for a similar in-class example.

ALGEBRAIC SOLUTION

$g(t) = \dfrac{t^2}{t + 1}$

a. $g(-2) = \dfrac{(-2)^2}{(-2) + 1}$ • Replace t by -2.

$= \dfrac{4}{-1} = -4$ • Simplify.

$g(-2) = -4$

GRAPHICAL CHECK

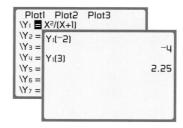

? ANSWER $f(-5) = (-5)^2 = 25$. The value of the function is 25 when $x = -5$.

b. To find the value of g when $t = 3$ means to evaluate the function when t is 3.

$$g(t) = \frac{t^2}{t + 1}$$

$$g(3) = \frac{(3)^2}{3 + 1}$$ • Replace t by 3.

$$= \frac{9}{4}$$ • Simplify.

The value of g when $t = 3$ is $\frac{9}{4}$.

Note the use of parentheses when inputting the function. If the parentheses were missing, the calculator, using the Order of Operations Agreement, would have interpreted the expression as $\frac{x^2}{x} + 1$, which is not correct. Also note that the decimal value was given for $Y_1(3)$. You can use the calculator to convert this to a fraction.

YOU TRY IT 2

Let $f(z) = 2z^3 - 4z$.

a. Find $f(-1)$.
b. Find the value of f when $z = -3$.

Solution See page S6. **a.** 2 **b.** −42

■ Graphs of Functions

The **graph of a function** is the graph of the ordered pairs that belong to the function. The graph of the speed–resistance function $R(s) = 0.024s^2$ given earlier is shown at the right. **The horizontal axis represents the domain of the function, or the independent variable (the speed of the car); the vertical axis represents the range of the function, or the dependent variable** (air resistance).

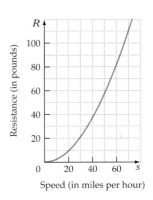

Resistance (in pounds)
Speed (in miles per hour)

The graph of a function can be drawn by finding ordered pairs of the function, plotting the ordered pairs, and then connecting the points with a curve.

➡ Graph $f(x) = x^2 - 4x - 2$ by completing the table below, plotting the ordered pairs, and then connecting the points with a curve.

TAKE NOTE

We are creating the graph of an equation in two variables as we did earlier. The only difference is the use of functional notation.

x	$f(x) = x^2 - 4x - 2$	(x, y)
−1	$f(-1) = (-1)^2 - 4(-1) - 2 = 3$	$(-1, 3)$
0		
1		
2		
3		
4		
5		

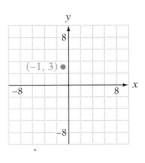

Complete the table by evaluating the function for the remaining values of x. Remember that $f(x)$ and y are two different symbols for the same number, the value of the dependent variable.

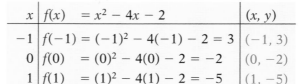

x	$f(x) = x^2 - 4x - 2$	(x, y)
-1	$f(-1) = (-1)^2 - 4(-1) - 2 = 3$	$(-1, 3)$
0	$f(0) = (0)^2 - 4(0) - 2 = -2$	$(0, -2)$
1	$f(1) = (1)^2 - 4(1) - 2 = -5$	$(1, -5)$
2	$f(2) = (2)^2 - 4(2) - 2 = -6$	$(2, -6)$
3	$f(3) = (3)^2 - 4(3) - 2 = -5$	$(3, -5)$
4	$f(4) = (4)^2 - 4(4) - 2 = -2$	$(4, -2)$
5	$f(5) = (5)^2 - 4(5) - 2 = 3$	$(5, 3)$

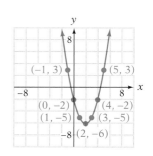

Plot the ordered pairs and draw a smooth curve through the points.

INSTRUCTOR NOTE
Exercise 30 can be used for a similar in-class example.

EXAMPLE 3

Let $f(x) = -\dfrac{3}{4}x + 2$.

a. Complete the following input/output table.

x	-8	-4	0	4	8
$f(x)$					

TAKE NOTE

In the previous section, we graphed $y = -\dfrac{3}{4}x + 2$. Notice that graphing $f(x) = -\dfrac{3}{4}x + 2$ results in the same graph. $f(x)$ and y are different symbols for the same quantity.

b. Graph the ordered pairs and then draw a line through the points.

Solution

a. Evaluate the function for the given values of x. The calculations for evaluating $f(-8)$ are shown at the right.

x	-8	-4	0	4	8
$f(x)$	8	5	2	-1	-4

$$f(x) = -\frac{3}{4}x + 2$$

$$f(-8) = -\frac{3}{4}(-8) + 2$$

$$= 6 + 2 = 8$$

b. Graph the ordered pairs $(-8, 8)$, $(-4, 5)$, $(0, 2)$, $(4, -1)$, and $(8, -4)$. Then draw a line through the points. The graph is shown at the right.

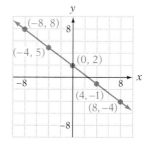

YOU TRY IT 3

Let $h(x) = 2x - 3$.

a. Complete the following input/output table.

x	-2	-1	0	1	2
$h(x)$	-7	-5	-3	-1	1

b. Graph the ordered pairs and then draw a line through the points.

Solution   See page S6.

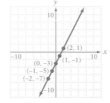

The equation $f(x) = 4$ is an example of a *constant function*. No matter what value of x is chosen, the value of the function is 4. For instance,

$$f(-3) = 4 \qquad f(0) = 4 \qquad f(2) = 4$$

The graph of $f(x) = 4$ is shown at the right. It is a horizontal line passing through $(0, 4)$.

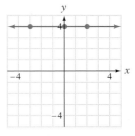

The **constant function** is written as $f(x) = c$, where c is a real number. No matter what value of x is chosen, the value of the constant function is c. The graph of the constant function $f(x) = c$ is a horizontal line passing through $(0, c)$.

INSTRUCTOR NOTE
Exercise 34 can be used for a similar in-class example.

EXAMPLE 4

Graph $h(x) = -1$.

Solution

The graph is a horizontal line through $(0, -1)$, as shown at the right.

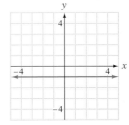

YOU TRY IT 4

Graph $g(x) = 2$.

Solution   See page S6.

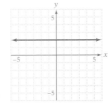

The graphs of many functions can be created by using a graphing calculator. The calculator uses the same procedure we would use by hand. That is, the calculator chooses values of *x*, evaluates the function for that value, plots the corresponding point, and then connects the points with a curve. The advantage of using a calculator is that we can produce the graph quickly and then investigate its properties.

The following typical graphing calculator screens could be used to graph the function in Example 3 in the standard viewing window.

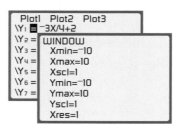

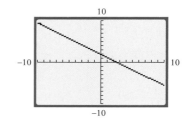

See Appendix A:
Trace

TAKE NOTE

In the previous section, we used the TRACE feature to find the value of *y* given a value of *x*. Here we are finding the value of *f(x)* given a value of *x*. The same procedure is used here because *f(x)* and *y* are different symbols for the same quantity.

In Example 2, we showed one method of evaluating a function by using a graphing calculator. It is also possible to evaluate a function using TRACE on a calculator.

➡ Graph $f(x) = x^3 - 2x + 5$ in the standard viewing. Use the TRACE feature to find $f(-3)$, the value of the function when $x = -3$. Verify the result algebraically.

Graph the function in the standard viewing window. Now use the TRACE feature to find the value of the function when $x = -3$.

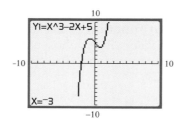

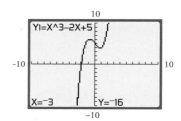

From the graph on the right, when $x = -3$, $y = -16$. Therefore, $f(-3) = -16$.

Algebraic check: $f(x) = x^3 - 2x + 5$
$$f(-3) = (-3)^3 - 2(-3) + 5$$
$$= -27 + 6 + 5$$
$$= -16$$

⬅

Suggested Activity

See Section 2.2 of the *Student Activity Manual* for an activity that involves matching the description of a function and its graph.

▪ Domain and Range

Recall that the domain of a function is the set of first coordinates of the ordered pairs of the function, and the range is the set of second coordinates. Another way of saying this is that **the domain is the set of all possible inputs, and the range is the set of corresponding outputs.** When the domain is a finite set of numbers, we can find the range by evaluating the function at each element of the domain.

INSTRUCTOR NOTE
Exercise 50 can be used for
a similar in-class example.

TAKE NOTE

Recall that when listing elements of a set, repeated elements are listed only once.

Suggested Activity

A function f is defined by
$\{(-4, -6), (-2, -2), (0, 2), (2, 6), (4, 10)\}$.
1. What is the domain of the function?
2. What is the range of the function?
3. Find $f(-2)$.
4. Find $f(4)$.
5. What is the input value when the output value is 6?
6. What is the input value when the output value is 2?
[Answers: **1.** $\{-4, -2, 0, 2, 4\}$
2. $\{-6, -2, 2, 6, 10\}$ **3.** -2
4. 10 **5.** 2 **6.** 0]

INSTRUCTOR NOTE
Exercise 60 can be used for
a similar in-class example.

EXAMPLE 5

Given $f(x) = x^2 - 4x$ with domain $\{-1, 0, 1, 2, 3, 4\}$, find the range of f.

Solution Evaluate the function for each element of the domain.

$$f(x) = x^2 - 4x$$
$$f(-1) = (-1)^2 - 4(-1) = 1 + 4 = 5$$
$$f(0) = 0^2 - 4(0) = 0 + 0 = 0$$
$$f(1) = 1^2 - 4(1) = 1 - 4 = -3$$
$$f(2) = 2^2 - 4(2) = 4 - 8 = -4$$
$$f(3) = 3^2 - 4(3) = 9 - 12 = -3$$
$$f(4) = 4^2 - 4(4) = 16 - 16 = 0$$

The range is $\{-4, -3, 0, 5\}$.

YOU TRY IT 5

Given $f(x) = -x^2 + 2x + 2$ with domain $\{-2, -1, 0, 1, 2, 3\}$, find the range of f.

Solution See page S6. $\{-6, -1, 2, 3\}$

In Example 5, the domain was given as a finite set. In many instances, we will assume that the domain of a function is all the real numbers for which the value of the function is a real number. For instance, the number 3 is not in the domain of $h(x) = \frac{2x}{x-3}$ because when $x = 3$, $h(3) = \frac{6}{3-3} = \frac{6}{0}$, which is not a real number.

EXAMPLE 6

For $f(x) = \frac{x}{x-2}$, which of the following numbers, if any, are *not* in the domain of f?

a. -4 **b.** 2   **c.** 0   **d.** 4

Solution When $x = -4$, $f(-4) = \dfrac{-4}{-4 - 2} = \dfrac{-4}{-6} = \dfrac{2}{3}$, a real number.

When $x = 2$, $f(2) = \dfrac{2}{2 - 2} = \dfrac{2}{0}$, which is not a real number.

When $x = 0$, $f(0) = \dfrac{0}{0 - 2} = \dfrac{0}{-2} = 0$, a real number.

When $x = 4$, $f(4) = \dfrac{4}{4 - 2} = \dfrac{4}{2} = 2$, a real number.

The number 2 is not in the domain of f.

> **YOU TRY IT 6**
>
> For $P(t) = \frac{t}{t^2 + 1}$, which of the following numbers, if any, are *not* in the domain of f?
>
> **a.** 4   **b.** -4 **c.** 0   **d.** 3
>
> **Solution**   See page S6.   All are included in the domain.

When $x = 2$ for the function $f(x) = \frac{x}{x - 2}$ in Example 6, the value of the function is undefined. In this case, given an input of 2, there is no output because division by zero is undefined. Therefore, 2 is *not* in the domain of the function. The domain of this function is all real numbers except 2.

The graph of $f(x) = \frac{x}{x - 2}$ is shown at the right.

Note that when we try to evaluate this function at 2 using the TRACE feature, there is no output shown for y. This indicates that 2 is not in the domain of the function. If you trace along the curve, there is an output value for other input values.

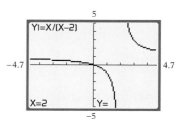

A portion of an input/output table for $f(x) = \frac{x}{x - 2}$ is shown at the right. Note that when $x = 2$, the output shows **ERROR**. This again demonstrates that 2 is not in the domain of f.

INSTRUCTOR NOTE
Exercise 78 can be used for a similar in-class example.

TAKE NOTE

Because of the way a graphing calculator graphs a function, use the viewing windows we suggest. Other windows can be used, but we have chosen these windows so that it is possible to trace to the necessary numbers.

EXAMPLE 7

Graph $f(x) = \frac{x}{x^2 - 4}$ using the decimal viewing window. Trace along the curve to find two numbers that are not in the domain of f. Algebraically verify that the value of the function is undefined at these two numbers.

Solution   Graph the function and trace along the graph to find where no value of y is shown.

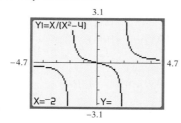

−2 is not in the domain of f.

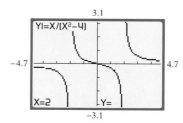

2 is not in the domain of f.

CALCULATOR NOTE

An input/output table for

$f(x) = \dfrac{x}{x^2 - 4}$ is shown

below.

Note the word **ERROR** in the Y_1 column for the input values of -2 and 2. This is another indication that these two numbers are not in the domain of f.

X	Y_1
-3	-.6
-2	ERROR
-1	.33333
0	0
1	-.3333
2	ERROR
3	.6

X = -3

We can algebraically verify that these numbers are not in the domain of f by attempting to evaluate the function for the two numbers.

$$f(x) = \frac{x}{x^2 - 4} \qquad\qquad f(x) = \frac{x}{x^2 - 4}$$

$$f(-2) = \frac{-2}{(-2)^2 - 4} \qquad\qquad f(2) = \frac{2}{2^2 - 4}$$

$$= \frac{-2}{0} \text{ Undefined} \qquad\qquad = \frac{2}{0} \text{ Undefined}$$

Because the value of the function is undefined when x equals -2 and when x equals 2, these numbers are not in the domain of f.

YOU TRY IT 7

Graph $f(x) = \dfrac{3}{x^2 - x - 6}$ using the decimal viewing window. Trace along the curve to find two numbers that are not in the domain of f. Algebraically verify that the value of the function is undefined at these two numbers.

Solution   See page S6. $-2, 3$

▪ Applications

Any letter or combination of letters can be used to name a function. In the next example, the letters *SA* are used to represent a *Surface Area* function. In this case, *SA* is the name of the function and does not mean *S* times *A*.

INSTRUCTOR NOTE
Exercise 84 can be used for a similar in-class example.

EXAMPLE 8

The surface area of a cube (the sum of the areas of each of the 6 faces of a cube) is given by $SA(s) = 6s^2$, where $SA(s)$ is the surface area of the cube and s is the length of one side of the cube. Find the surface area of a cube that has a side of 10 centimeters.

Solution $SA(s) = 6s^2$

$SA(10) = 6(10)^2$ • Replace s by 10.

$= 6(100)$ • Simplify.

$= 600$

The surface area is 600 square centimeters

YOU TRY IT 8

If m points are placed in the plane, no three of which are on the same line, then the number of different line segments that can be drawn between the points, $N(m)$, is given by $N(m) = \dfrac{m(m-1)}{2}$. Find the number of different line segments that can be drawn between 12 different points in the plane.

5 points
10 line segments

Solution   See page S7.   66 line segments

INSTRUCTOR NOTE
Exercise 88 can be used for
a similar in-class example.

EXAMPLE 9

A spherical snowball is melting in such a way that the volume of the snow ball can be approximated by $V(t) = 20\pi(2 - \frac{t}{90})^3$, where $V(t)$ is the volume, in cubic inches, of the snowball t minutes after it begins to melt.

a. Find the volume of the snowball when $t = 30$. Round to the nearest hundredth.

b. Explain why the domain of this function is the interval $[0, 180]$.

Solution

a. $V(t) = 20\pi\left(2 - \dfrac{t}{90}\right)^3$

$= 20\pi\left(2 - \dfrac{30}{90}\right)^3$ • Replace t by 30.

$= 20\pi\left(2 - \dfrac{1}{3}\right)^3 = 20\pi\left(\dfrac{5}{3}\right)^3$ • Simplify.

≈ 290.89

The volume is 290.89 cubic inches.

b. The snowball begins melting when $t = 0$, so the least value of t is 0. If t is greater than 180, then $\left(2 - \frac{t}{90}\right)^3$ is less than zero. This would mean the volume of the snowball was negative, which is not possible.

YOU TRY IT 9

An environmental study determined that the amount of carbon monoxide in the air surrounding a city depends on the population of the city and can be approximated by $C(p) = 0.15\sqrt{p^2 + 4p + 10}$, where p is the population in thousands and $C(p)$ is the concentration of carbon monoxide in parts per million (ppm).

a. Find the concentration of carbon monoxide in a city whose population is 100,000. Round to the nearest tenth.

b. What is the concentration of carbon monoxide when $p = 0$? Round to the nearest tenth. What is the significance of this number in the context of the problem?

Solution See page S7. **a.** 15.3 ppm **b.** 0.5 ppm. If there were no people in an area, the carbon monoxide concentration in the air would be approximately 0.5 ppm.

 See Appendix A:
Radical Expressions

2.2 **EXERCISES** Suggested Assignment: 9–91, odds

Topics for Discussion

1. Describe the concepts of relation and function. How are they the same? How are they different?
   A relation is a set of ordered pairs. A function is a relation in which no two ordered pairs have the same first coordinate and different second coordinates.

2. Are all functions relations? Are all relations functions?
   All functions are relations, but not all relations are functions.

3. What is the domain of a function? What is the range of a function?
   The domain is the set of first coordinates of the function; the range is the set of second coordinates of the function.

4. What is the value of a function?
   The value of a function is the output value for a given input value.

5. What does it mean to evaluate a function?
   To evaluate a function means to replace the independent variable by a given number and simplify the resulting expression.

6. Is it possible for a function to have the same output value for two different input values? Explain.
   Yes.

7. Is it possible for a function to have two different output values for the same input value? Explain.
   No.

■ Introduction to Relations and Functions

For Exercises 8 to 15, give the domain and range of the relation. Is the relation a function?

8. $\{(-2, 0), (2, 1), (4, 2), (6, 3), (8, 4)\}$
   D: {−2, 2, 4, 6, 8}, R: {0, 1, 2, 3, 4}, Yes

9. $\{(-3, -5), (-2, -7), (-1, -9), (0, -11), (1, -13)\}$
   D: {−3, −2, −1, 0, 1}, R: {−13, −11, −9, −7, −5}, Yes

10. $\{(1, 2), (2, 3), (3, 4), (3, 2), (2, 1)\}$
    D: {1, 2, 3}, R: {1, 2, 3, 4}, No

11. $\{(-4, 6), (-2, 8), (0, 10), (-4, 8), (2, 12)\}$
    D: {−4, −2, 0, 2}, R: {6, 8, 10, 12}, No

12. $\{(-1, 3), (0, 7), (1, 9), (4, 7), (7, 3)\}$
    D: {−1, 0, 1, 4, 7}, R: {3, 7, 9}, Yes

13. $\{(2, -6), (3, -3), (4, -6), (5, 6), (6, -6)\}$
    D: {2, 3, 4, 5, 6}, R: {−6, −3, 6}, Yes

14. $\{(-2, 5), (-1, 5), (0, 5), (1, 5), (2, 5)\}$
    D: {−2, −1, 0, 1, 2}, R: {5}, Yes

15. $\{(-4, 0), (-2, 0), (0, 0), (3, 0), (5, 0)\}$
    D: {−4, −2, 0, 3, 5}, R: {0}, Yes

■ Functional Notation

For Exercises 16 to 25, evaluate the function at the given value.

16. $f(x) = 2x + 7;\ x = -2$
    3

17. $y(x) = 1 - 3x;\ x = -4$
    13

18. $f(t) = t^2 - t - 3;\ t = 3$
    3

19. $P(n) = n^2 - 4n - 7;\ n = -3$
    14

20. $v(s) = s^3 + 3s^2 - 4s - 2;\ s = -2$
    10

21. $f(x) = 3x^3 - 4x^2 + 7;\ x = 2$
    15

22. $T(p) = \dfrac{p^2}{p - 2}; p = 0$

0

24. $r(x) = 2^x - x^2; x = 3$

−1

23. $s(t) = \dfrac{4t}{t^2 + 2}; t = 2$

$\dfrac{4}{3}$

25. $ABS(x) = |2x - 7|; x = -3$

13

■ Graphs of Functions

For Exercises 26 to 35, complete the input/output table, and graph the function without using a graphing calculator. Check your answer by using a graphing calculator.

26. $f(x) = 2x - 4$

x	−2	−1	0	1	2
y	−8	−6	−4	−2	0

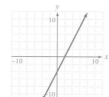

27. $f(x) = 2 - 2x$

x	−2	−1	0	1	2
y	6	4	2	0	−2

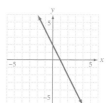

28. $f(x) = -\dfrac{1}{2}x + 3$

x	−4	−2	0	2	4
y	5	4	3	2	1

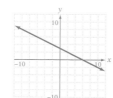

29. $f(x) = -\dfrac{2x}{3} + 4$

x	−6	−3	0	3	6
y	8	6	4	2	0

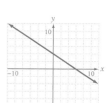

30. $f(x) = -x^2 + 2$

x	−3	−2	−1	0	1	2	3
y	−7	−2	1	2	1	−2	−7

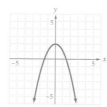

31. $f(x) = x^2 - 2$

x	−3	−2	−1	0	1	2	3
y	7	2	−1	−2	−1	2	7

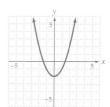

32. $f(x) = x^2 + 6x + 2$

x	−6	−5	−4	−3	−2	−1	0
y	2	−3	−6	−7	−6	−3	2

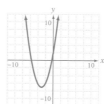

33. $f(x) = -x^2 + 2x - 1$

x	−2	−1	0	1	2	3	4
y	−9	−4	−1	0	−1	−4	−9

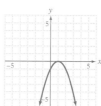

34. $f(x) = 3$

x	−4	−3	−2	−1	0	1	2
y	3	3	3	3	3	3	3

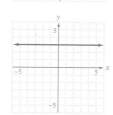

35. $f(x) = -2$

x	−2	−1	0	1	2	3	4
y	−2	−2	−2	−2	−2	−2	−2

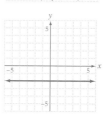

36. 🖎 Describe the graph of the function $G(x) = -4$.
The graph is a horizontal line through $(0, -4)$.

37. 🖎 Describe the graph of the function $H(x) = 1$.
The graph is a horizontal line through $(0, 1)$.

For Exercises 38 to 43, use a graphing calculator to graph each function in the integer viewing window. Then use the TRACE feature to find the x-coordinate for the given y-coordinate. Check your work by evaluating the function at the x-coordinate.

38. $f(x) = -3x + 6; y = 12$
-2

39. $f(x) = 2x - 7; y = 13$
10

40. $g(x) = -\dfrac{5}{4}x + 10; y = 15$
-4

41. $g(x) = \dfrac{2}{3}x + 8; y = -10$
-27

42. $F(x) = 5 - \dfrac{7}{5}x; y = 26$
-15

43. $y(x) = 9 - \dfrac{15}{8}x; y = -6$
8

For Exercises 44 to 49, graph the function in the standard viewing window. Use the TRACE feature to evaluate the function. Verify the results algebraically.

44. $f(x) = x^2 - 3x - 1; f(-2)$
9

45. $f(x) = x^2 + 2x - 5; f(3)$
10

46. $g(x) = x^3 - 4x - 1; g(2)$
-1

47. $g(x) = x^3 + 4x^2 - 2;$
$g(-3)$
7

48. $F(x) = x^3 - 3x^2 + x - 1;$
$F\left(\dfrac{3}{2}\right)$ $-\dfrac{23}{8}$

49. $F(x) = x^3 + 4x^2 - x - 3;$
$F\left(-\dfrac{5}{2}\right)$ $\dfrac{71}{8}$

■ Domain and Range

50. Given $f(x) = x^2 + 3x$ with domain $\{-4, -3, -2, -1, 0, 1, 2, 3, 4\}$, find the range of f. $\{-2, 0, 4, 10, 18, 28\}$

51. Given $g(x) = 10 - x^2$ with domain $\{-4, -3, -2, -1, 0, 1, 2, 3, 4\}$, find the range of g. $\{-6, 1, 6, 9, 10\}$

52. Given $s(t) = t^3 - t^2 + 3t - 5$ with domain $\{-4, -2, 0, 2, 4\}$, find the range of s. $\{-97, -23, -5, 5, 55\}$

53. Given $P(n) = \dfrac{n(n + 1)}{2}$ with domain $\{1, 2, 3, 4, 5, 6, 7\}$, find the range of P. $\{1, 3, 6, 10, 15, 21, 28\}$

54. Given $R(x) = \dfrac{1}{x + 2}$ with domain $\{-1, 0, 1, 2, 3, 4, 5\}$, find the range of R.
$\left\{1, \dfrac{1}{2}, \dfrac{1}{3}, \dfrac{1}{4}, \dfrac{1}{5}, \dfrac{1}{6}, \dfrac{1}{7}\right\}$

55. Given $v(s) = \frac{s}{s^2 + 1}$ with domain $\{-3, -2, -1, 0, 1, 2, 3\}$, find the range of v. Express the values of the range as fractions in simplest form.

$$\left\{ -\frac{3}{10}, -\frac{2}{5}, -\frac{1}{2}, 0, \frac{1}{2}, \frac{2}{5}, \frac{3}{10} \right\}$$

For Exercises 56 to 70, determine which, if any, of the given numbers are *not* in the domain of the function.

56. $h(x) = x^2$; $x = -3, 0, 4$

57. $f(t) = 2t - 4$; $t = -1, 0, 2$

58. $R(s) = \frac{5}{s - 3}$; $s = -3, 0, 3$

3

59. $f(x) = \frac{x}{x + 5}$; $x = -5, 0, 2$

-5

60. $g(v) = \frac{v + 1}{v^2 - 4}$; $v = -2, -1, 2$

$-2, 2$

61. $F(x) = \frac{x - 1}{x^2 - 2x - 3}$; $x = -1, 1, 3$

$-1, 3$

62. $h(x) = x^3$; $x = -2, 0, 3$

63. $f(t) = 3t + 1$; $t = -6, 0, 5$

64. $f(x) = \frac{x}{x - 5}$; $x = -3, 0, 5$

5

65. $g(v) = \frac{v + 1}{v + 4}$; $v = -4, -1, 0$

-4

66. $F(x) = \frac{x + 1}{x}$; $x = -1, 0, 1$

0

67. $F(x) = \frac{2x - 4}{x + 2}$; $x = -2, 0, 2$

-2

68. $R(s) = \frac{s^2 + s}{5}$; $s = -1, 0, 1$

69. $z(t) = \frac{t}{t^2 + 1}$; $t = -1, 0, 2$

70. $ABS(x) = |2x - 4|$; $x = -2, 0, 2$

For Exercises 71 to 76, graph the function in the decimal viewing window. Trace along the curve to find a number that is *not* in the domain of the function.

71. $f(x) = \frac{2}{x - 1}$

1

72. $f(x) = \frac{1}{x + 2}$

-2

73. $g(x) = \frac{x}{x + 3}$

-3

74. $g(x) = \frac{x}{x - 2}$

2

75. $F(x) = \frac{x}{2x - 3}$

1.5

76. $F(x) = \frac{x}{2x + 5}$

-2.5

For Exercises 77 to 82, graph the function in the decimal viewing window. Trace along the curve to find two numbers that are *not* in the domain of the function.

77. $f(x) = \frac{1}{x^2 - 1}$

$-1, 1$

78. $f(x) = \frac{1}{x^2 - x - 2}$

$-1, 2$

79. $g(x) = \frac{x}{x^2 - 2x - 3}$

$-1, 3$

80. $g(x) = \frac{x}{x^2 - 9}$

$-3, 3$

81. $F(x) = \frac{1}{2x^2 - x - 1}$

$-0.5, 1$

82. $F(x) = \frac{2}{2x^2 - x - 3}$

$-1, 1.5$

■ Applications

83. *Geometry*   The perimeter P of a square is a function of the length of one of its sides s and is given by $P(s) = 4s$.

   a. Find the perimeter of a square whose side is 4 meters.
   b. Find the perimeter of a square whose side is 5 feet.

   a. 16 m   b. 20 ft

84. *Geometry*   The area of a circle is a function of its radius and is given by $A(r) = \pi r^2$.

   a. Find the area of a circle whose radius is 3 inches. Round to the nearest tenth.
   b. Find the area of a circle whose radius is 12 centimeters. Round to the nearest tenth.

   a. 28.3 in²   b. 452.4 cm²

85. *Sports*   The height h, in feet, of a baseball that is released 4 feet above the ground with an initial upward velocity of 80 feet per second is a function of the time t, in seconds, and is given by $h(t) = -16t^2 + 80t + 4$.

   a. Find the height of the baseball above the ground 2 seconds after it is released.
   b. Find the height of the baseball above the ground 4 seconds after it is released.

   a. 100 ft   b. 68 ft

86. *Forestry*   The distance d, in miles, that a forest fire ranger can see from an observation tower is a function of the height h, in feet, of the tower above the ground and is given by $d(h) = 1.5\sqrt{h}$.

   a. Find the distance a ranger can see whose eye level is 20 feet above the ground. Round to the nearest tenth.
   b. Find the distance a ranger can see whose eye level is 35 feet above the ground. Round to the nearest tenth.

   a. 6.7 mi   b. 8.9 mi

87. *Physics*   The speed s, in feet per second, of sound in air depends on the temperature t of the air, in degrees Celsius, and is given by

$$s(t) = \frac{1087\sqrt{t + 273}}{16.52}.$$

   a. What is the speed of sound in air when the temperature is 25°C? Round to the nearest whole number.
   b. Does the speed of sound in air increase or decrease as temperature increases?

   a. 1136 ft/s   b. increases

88. *Business*   The number of personal digital assistants (PDAs) that a company can sell per month depends on the price of the PDA and is given by $D(p) = \frac{480,000}{100 + 5p}$, where p is price of the PDA, in dollars, and $D(p)$ is the number of PDAs that can be sold per month at that price.

   a. How many PDAs can be sold per month when the price is $300?
   b. Will the number of PDAs that can be sold per month increase or decrease as the price increases?

   a. 300 PDAs   b. decrease

89. *Sports*   In a softball league in which each team plays every other team three times, the number N of games that must be scheduled depends on the number of teams in the league and is given by $N(n) = \frac{3}{2}n^2 - \frac{3}{2}n$.

    a. How many games must be scheduled for a league that has 5 teams?
    b. How many games must be scheduled for a league that has 6 teams?

    a. 30 games   b. 45 games

90. *Solutions*   The percent concentration, P, of salt in a salt–water solution depends on the number of grams x of salt that is added to the solution and is given by $P(x) = \frac{100x + 100}{x + 10}$, where x is the number of grams of salt that is added.

    a. What is the percent concentration of salt after an additional 5 grams of salt is added to the solution?
    b. ✎ What is the percent concentration of salt when $x = 0$ and what is the significance of this result in the context of this problem?

    a. 40%   b. 10%; Evaluating the function at $x = 0$ gives the original concentration of salt.

91. *Physics*   The time T, in seconds, it takes a pendulum to make one swing depends on the length of the pendulum and is given by $T(L) = 2\pi\sqrt{\frac{L}{32}}$, where L is the length of the pendulum in feet.

    a. Find the time it takes the pendulum to make one swing if the length of the pendulum is 3 feet. Round to the nearest hundredth.
    b. Find the time it takes the pendulum to make one swing if the length of the pendulum is 9 inches. Round to the nearest hundredth.

    a. 1.92 s   b. 0.96 s

Applying Concepts

92. Create a set of five ordered pairs that is a function. Create a set of five ordered pairs that is a relation but not a function.

    Answers will vary.

93. a. Define a function that has domain = {1, 4, 6} and range = {2, 7, 9}.
    b. Define a function that has domain = {1, 2, 3} and range = {4}.

    Answers will vary. Possibilities: a. {(1, 2), (4, 7), (6, 9)}   b. {(1, 4), (2, 4), (3, 4)}

94. a. Suppose $f(2) = 7$ and $f(4) = 7$. Is it possible for f to be a function?
    b. Suppose $g(7) = 2$ and $g(7) = 4$. Is it possible for g to be a function?

    a. Yes   b. No

95. Create a function whose domain does not contain 5.

    Answers will vary. One possibility is $f(x) = \dfrac{1}{x - 5}$.

96. Modular functions have many different applications. One such application is in creating codes for secure communications so that, for instance, credit card information can be transmitted over the Internet. We define $a \equiv b \bmod n$ if a has remainder b when divided by n. (It is traditional to use $\equiv$ rather than $=$ for the mod function.) For instance,

$5 \equiv 2 \bmod 3$ because the remainder when 5 is divided by 3 is 2. On the other hand, $7 \not\equiv 2 \bmod 3$ because the remainder when 7 is divided by 3 is not 2.

a. Is $8 \equiv 3 \bmod 5$?
b. Is $7 \equiv 1 \bmod 3$?
c. Find three positive integers x for which $x \equiv 2 \bmod 5$.
d. Find the least positive integer x for which $2x \equiv 1 \bmod 11$.
e. Find the least positive integer x for which $x + 4 \equiv 0 \bmod 11$.

a. Yes b. Yes c. Answers will vary. Three possibilities are 2, 7, and 12. d. 6 e. 7

97. All books published in the United States have an ISBN (International Standard Book Number). Write a report that explains how mod 11 is used for ISBNs. (See Exercise 96.)

EXPLORATION

1. *Functions of More Than One Variable*   The value of some functions may depend on several variables. For instance, the perimeter of a rectangle depends on the length L and the width W. We can write this as

$$P(L, W) = 2L + 2W.$$

To evaluate this function, we need to be given the value of L and W. Here is an example: To find the perimeter of a rectangle whose length is 5 feet and whose width is 3 feet, evaluate $P(L, W)$ when $L = 5$ and $W = 3$.

$$P(L, W) = 2L + 2W$$
$$P(5, 3) = 2(5) + 2(3) \qquad \bullet \text{ Replace } L \text{ by 5 and } W \text{ by 3.}$$
$$= 10 + 6 = 16$$

The perimeter is 16 feet.

a. Evaluate $f(a, b) = 2a + 3b$ when $a = 3$ and $b = 4$.   18
b. Evaluate $R(s, t) = 2st - t^2$ when $s = -1$ and $t = 2$. -8
c. Although we normally do not think of addition as a function, it is a function of two variables. If we define the function *Add* as $Add(a, b) = a + b$, find $Add(3, 7)$.   10
d. Write the area of a triangle as a function of two variables.

$A(b, h) = \dfrac{1}{2}bh$

e. Give an example of a function whose value depends on more than one variable.

Examples will vary. One example is the calculation of simple interest as a function of principal, interest rate, and time: $I(p, r, t) = prt$.

SECTION **2.3**

Properties of Functions

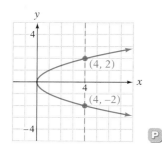

■ Vertical-Line Test

Consider the graph shown at the left. Note that two ordered pairs that belong to the graph are $(4, 2)$ and $(4, -2)$ and that these points lie on a vertical line. These two ordered pairs have the same first coordinates but different second coordinates, and therefore, the graph is not the graph of a function. With this observation in mind, we can give a quick method to determine whether a graph is the graph of a function.

> **Vertical-Line Test for the Graph of a Function**
>
> A graph defines a function if any vertical line intersects the graph at no more than one point.

This graphical interpretation of a function is often described by saying that each value in the domain of the function is paired with *exactly one* value in the range of the function.

TAKE NOTE

For the second graph on the right, note that there are values of *x* for which there is only one value of *y*. For instance, when $x = -5$, $y = 4$. In a function, however, *every* value of *x* in the domain of the function must have exactly one value of *y*. If there is even one value of *x* that has two or more values of *y*, the condition for a function is not met.

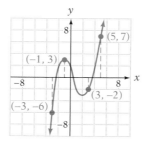

For each x there is exactly one value of y. *This is the graph of a function.*

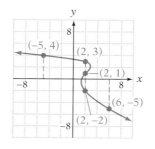

Some values of x can be paired with more than one value of y. For instance, 2 can be paired with −2, 1, and 3. *This is not the graph of a function.*

INSTRUCTOR NOTE
Exercise 10 can be used for a similar in-class example.

EXAMPLE 1

Use the vertical-line test to determine whether the graph is the graph of a function.

a.

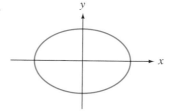

b.

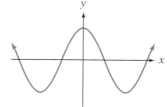

Solution

a. As shown at the right, there are vertical lines that intersect the graph at more than one point. Therefore, the graph is not the graph of a function.

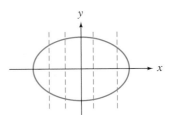

b. For the graph at the right, every vertical line intersects the graph at most once. Therefore, the graph is the graph of a function.

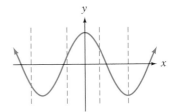

YOU TRY IT 1

Use the vertical-line test to determine whether the graph is the graph of a function.

a.

b.

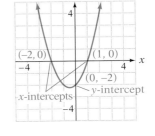

Solution   See page S7.   a. Yes   b. No

■ Intercepts

INSTRUCTOR NOTE
We present *x*-intercepts and intersections of graphs here as preparation for solving an equation by using a graphing calculator, which is discussed in the next chapter.
For this section, we focus on the value of the function and its graphical interpretation, rather than on the solution of an equation.

When choosing a window in which to graph a function, we usually choose the window so that the important characteristics of the graph are displayed. One important characteristic of a graph is its *intercepts*. A **y-intercept** is a point at which a graph crosses the *y*-axis. An **x-intercept** is a point at which a graph crosses the *x*-axis.

Consider the graph of $f(x) = x^2 + x - 2$ shown at the right. Note that **the *x*-coordinate of the y-intercept is zero.** To find the *y*-intercept, evaluate the function at $x = 0$.

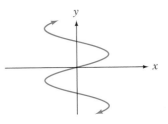

$$f(x) = x^2 + x - 2$$
$$f(0) = 0^2 + 0 - 2$$ • To find the *y*-intercept,
$$= -2$$ evaluate the function at 0.

The *y*-intercept is $(0, -2)$.

Look again at the graph of $f(x) = x^2 + x - 2$ and note that the x-coordinates of the x-intercepts are -2 and 1. Note also that **the y-coordinate of an x-intercept is zero.** We can verify this by evaluating the function at -2 and 1.

$$f(x) = x^2 + x - 2$$
$$f(-2) = (-2)^2 + (-2) - 2$$ • Evaluate f at -2.
$$= 4 - 2 - 2$$
$$= 0$$ • The value of the function is 0.

$$f(x) = x^2 + x - 2$$
$$f(1) = (1)^2 + (1) - 2$$ • Evaluate f at 1.
$$= 1 + 1 - 2$$
$$= 0$$ • The value of the function is 0.

Suggested Activity

Ask students to determine which of the numbers -1, 0, 2, and 5 are x-coordinates of the x-intercepts of the graph of $f(x) = x^2 - 4x - 5$.

INSTRUCTOR NOTE
Zeros of functions are discussed in greater depth later in the text.

The numbers -2 and 1 are called **zeros of the function** because the value of the function at these numbers is 0. From a graphical perspective, the x-coordinates of the x-intercepts of the graph of a function are zeros of the function. This idea is used by many graphing calculators to determine the x-intercepts of a graph.

To illustrate the procedure for using a graphing calculator to determine the x-intercepts of the graph of a function, we will use $f(x) = x^2 + x - 2$, the function discussed above. To begin, use the Y= editor to enter the expression for the function, and then graph the function in any viewing window that shows the x-intercepts. The standard viewing window is a good place to start. Once the graph is displayed, the ZERO feature of the calculator is used to find the x-intercepts. Some typical screens are shown below.

 See Appendix A:
Zero

CALCULATOR NOTE

The ZERO option on the graphing calculator calculates the value of x when y is zero. In other words, it finds the x-coordinates of the x-intercepts.

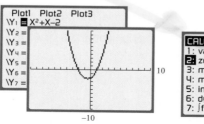
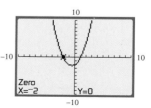

An x-intercept of the graph of $f(x) = x^2 + x - 2$ is $(-2, 0)$.

To find the second x-intercept, we repeat the process. The graph is shown at the right. A second x-intercept of the graph of $f(x) = x^2 + x - 2$ is $(1, 0)$.

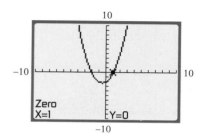

INSTRUCTOR NOTE
Exercise 16 can be used for
a similar in-class example.

Suggested Activity

See Section 2.3 of the *Student Activity Manual* for an activity that involves using technology to determine intercepts.

EXAMPLE 2

Find **a.** the y-intercept and **b.** the x-intercepts for the graph of $f(x) = x^2 + 2x - 2$. Round to the nearest hundredth.

Solution

a. To find the y-intercept, evaluate the function at zero.

$$f(x) = x^2 + 2x - 2$$
$$f(0) = 0^2 + 2(0) - 2 = -2$$

The y-intercept is $(0, -2)$.

b. To find the x-intercepts, we need to determine where the graph crosses the x-axis. The x-coordinate of an x-intercept is a zero of the function. Graph $f(x) = x^2 + 2x - 2$, and use the ZERO option of a graphing calculator to find the x-coordinates of the x-intercepts of the function.

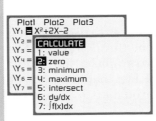 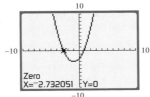 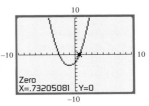

To the nearest hundredth, the x-coordinates of the x-intercepts are -2.73 and 0.73.

To the nearest hundredth, the x-intercepts are $(-2.73, 0)$ and $(0.73, 0)$.

YOU TRY IT 2

Find **a.** the y-intercept and **b.** the x-intercepts for the graph of $g(x) = 2x^2 - 5x + 2$.

Solution   See page S7.   **a.** $(0, 2)$ **b.** $(0.5, 0)$, $(2, 0)$

■ Intersections of the Graphs of Two Functions

In the last section, we discussed how we could determine an output for a given input. We can also solve the opposite problem of finding an input for a given output. Although there are algebraic methods for doing this, we will use a graphical approach here.

The graph of $f(x) = 2x + 1$ is shown at the right. To determine the value of x (the input) in the domain of f that will result in a given output of 7 in the range of f, draw a horizontal line from 7 on the y-axis to the graph of f. Now draw a vertical line to the x-axis. The x-coordinate (3, in this case) of the point where the vertical line touches the x-axis is the desired value. This can be verified by evaluating the function at this number.

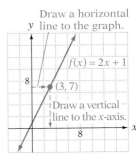

TAKE NOTE

The problem of finding the x-intercept is more difficult than finding the y-intercept because we must find the value of x for which $f(x)$ is zero rather than just evaluating the function at zero. In this chapter, we are using a graphical technique to solve this problem. In the next chapter, we will begin a discussion of algebraic techniques to solve the problem.

$$f(x) = 2x + 1$$
$$f(3) = 2(3) + 1 \quad \bullet \text{ Replace } x \text{ by 3.}$$
$$= 6 + 1$$
$$= 7 \quad \bullet \text{ The output is 7, the given number.}$$

The coordinates of the point at which the horizontal line through $(0, 7)$ intersects the graph is $(3, 7)$. This observation is the basis of a graphing calculator solution to this problem.

Example 3 illustrates how a graphing calculator can be used to find a number in the domain of a function that corresponds to a given output of that function.

INSTRUCTOR NOTE
Exercise 26 can be used for a similar in-class example.

 See Appendix A:
Intersect

EXAMPLE 3

Find the number in the domain of $f(x) = -\frac{2}{3}x + 2$ for which the output is 4.

Solution   The strategy is to use a calculator to graph both $f(x) = -\frac{2}{3}x + 2$ and a horizontal line through $(0, 4)$. The x-coordinate of the point of intersection of the two graphs is the desired number. To draw the horizontal line, we will use the fact that the graph of the constant function $g(x) = 4$ is a horizontal line through $(0, 4)$. Some typical screens are shown below.

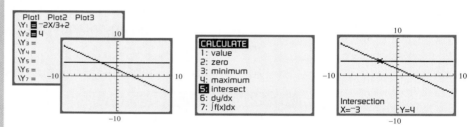

The point of intersection is $(-3, 4)$.

The number -3 in the domain of f produces an output of 4 in the range of f.

We can verify this algebraically as follows:

$$f(x) = -\frac{2}{3}x + 2$$

$$f(-3) = -\frac{2}{3}(-3) + 2 \quad \bullet \text{ Replace } x \text{ by } -3.$$

$$= 2 + 2$$

$$= 4 \quad \bullet \text{ The output is 4, the given number.}$$

YOU TRY IT 3

Find the number in the domain of $G(x) = 1 - 2x$ for which the output is -6.

Solution   See page S7.   3.5

In Example 3 and You Try It 3, the graphs of the functions were straight lines. We can solve similar problems for graphs that are not straight lines.

➡ Find the number in the domain of $f(x) = \sqrt{2x + 1}$ for which the output is 3. We proceed in a manner similar to Example 3. Draw the graph of $f(x) = \sqrt{2x + 1}$ and the graph of $g(x) = 3$, and then use a graphing calculator to find the point of intersection.

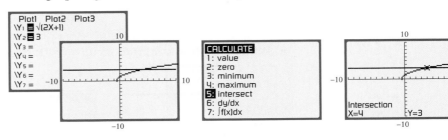

The point of intersection is (4, 3).

The number 4 in the domain of f produces an output of 3 in the range of f.   ⬅

TAKE NOTE

If $f(x) = x^2$, then because any nonzero real number squared is a positive number, it is impossible to have an input that will produce a negative output. This means not only that -3 is not in the range of f, but also that there are no negative numbers in the range of f. Graphically, any horizontal line passing through a point on the negative y-axis does not intersect the graph of f.

Suppose we want to determine an input that will result in an output of -3 for $f(x) = x^2$. The graph of $f(x) = x^2$ is shown at the right. Note that no matter in which direction a horizontal line through -3 on the y-axis is drawn, it will not intersect the graph of f. This means that there is no input for f that will produce an output of -3. Another way of saying this is that -3 is not in the range of f.

A problem that is related to finding the input for a given output is that of finding the input values for which the values of two functions are equal. The graphs of $f(x) = -2x + 1$ and $g(x) = x - 5$ are shown at the right. The two graphs intersect at the point $(2, -3)$. The values of the functions are equal at the x-coordinate of the point of intersection.

$$f(x) = -2x + 1 \qquad\qquad g(x) = x - 5$$
$$f(2) = -2(2) + 1 \quad \bullet \text{ Replace } x \text{ by 2.} \quad g(2) = 2 - 5 \quad \bullet \text{ Replace } x \text{ by 2.}$$
$$ = -3 \qquad\qquad\qquad\qquad = -3$$

From these evaluations of f and g, $f(2) = -3$ and $g(2) = -3$, which means $f(2) = g(2)$. That is, the value of the functions are equal when $x = 2$.

❓ QUESTION   For the graph at the right, is there a value of x for which the values of the functions f and g are equal?

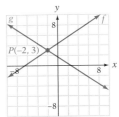

❓ ANSWER   Yes. When $x = -2$, $f(-2) = g(-2)$.

A graphing calculator can be used to find the point of intersection of the graphs of two functions.

INSTRUCTOR NOTE
Exercise 44 can be used for
a similar in-class example.

EXAMPLE 4

Find the element in the domain of $f(x) = 2x + 4$ and $g(x) = -3x - 1$ for which the values of the functions are equal. Verify the results algebraically.

Solution

Graph $f(x) = 2x + 4$ and $g(x) = -3x - 1$ in the same viewing window, and then determine the point of intersection.

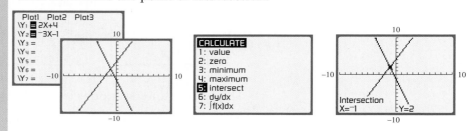

Suggested Activity

After completing Example 4,
or one similar to it, give students the functions
$f(x) = 2x - 3$ and
$g(x) = -x + 6$. Ask them how
they can determine, without
graphing, whether the values
of the functions are equal
when $x = -1$, 2, 3, or any
other number you might
give them.

The point of intersection is $(-1, 2)$.

The values of the functions are equal when $x = -1$.

To verify the results algebraically, evaluate the functions at the x-coordinate of the point of intersection.

$$f(x) = 2x + 4 \qquad\qquad g(x) = -3x - 1$$
$$f(-1) = 2(-1) + 4 \quad \bullet \text{ Replace } x \quad g(-1) = -3(-1) - 1 \quad \bullet \text{ Replace}$$
$$= 2 \qquad\qquad \text{by } -1. \qquad\qquad = 2 \qquad\qquad x \text{ by } -1.$$

The values of the functions are equal when $x = -1$.

YOU TRY IT 4

Find the element in the domain of $f(x) = 2x - 3$ and $g(x) = \frac{x}{2} + 3$ for which the values of the functions are equal. Verify the results algebraically.

Solution   See page S7.   4

There are many applications whose solutions use the techniques of Example 3 and Example 4.

INSTRUCTOR NOTE
Exercise 62 can be used for
a similar in-class example.

EXAMPLE 5

The distance a forest ranger can see from a lookout tower is given by $d(h) = 1.5\sqrt{h}$, where h is the height, in feet, of the ranger's binoculars above the ground and $d(h)$ is the distance, in miles, the ranger can see at a height of h feet. At what height above the ground must the ranger's eyes be to see 20 miles? Use a domain of $[0, 300]$ and a range of $[0, 30]$. Round to the nearest whole number.

CALCULATOR NOTE

It is important to choose a viewing window that includes the point of intersection. The viewing window we chose is shown at the right. However, other viewing windows could have been used.

See Appendix A:
Radical Expressions

Solution   To solve this problem, we need to determine the value of h for which $d = 20$. This is similar to Example 3. Graph $d(h) = 1.5\sqrt{h}$ and $g(h) = 20$, and determine the point of intersection.

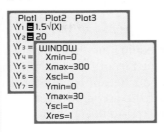
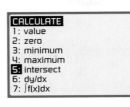
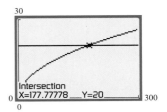

The ranger's eyes must be 178 feet above the ground to see 20 miles.

We can check this result by evaluating $d(h) = 1.5\sqrt{h}$ when $h = 178$.

$$d(h) = 1.5\sqrt{h}$$
$$d(178) = 1.5\sqrt{178}$$
$$\approx 20.012$$

The ranger can see approximately 20 miles at a height of 178 feet above the ground.

YOU TRY IT 5

The distance a marathon runner is from the finish line is given by $s(t) = 26 - 8t$, where t is the time in hours since the beginning of the race and $s(t)$ is the distance, in miles, from the finish line at time t. After how many hours will the runner be 10 miles from the finish line? Use a domain of $[0, 5]$ and a range of $[0, 30]$.

Solution   See page S8.   2 h

INSTRUCTOR NOTE
Exercise 66 can be used for a similar in-class example.

EXAMPLE 6

The value of an investment in t years is given by $V(t) = 400 + 50t$, and the value of a second investment in t years is given by $W(t) = 500 + 30t$. In how many years will the two investments have the same value? Use a domain of $[0, 7]$ and a range of $[0, 800]$.

Solution   To solve this problem, we need to determine the value of t for which $W(t) = V(t)$. This is similar to Example 4. Graph $V(t) = 400 + 50t$ and $W(t) = 500 + 30t$, and determine the point of intersection.

Suggested Activity

Have students algebraically verify that the investments in Example 6 are equal in 5 years.

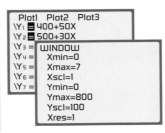
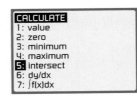
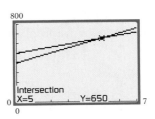

The point of intersection is (5, 650).

The investments will be equal in 5 years. At that time, each investment will be worth $650.

YOU TRY IT 6

A small plane is leaving an airport just as another plane is beginning a descent into the same airport. The height above the ground of the plane leaving the airport can be approximated by $h(t) = 115t$, where $h(t)$ is the height, in feet, of the airplane after t minutes. The height of the second plane is given by $f(t) = 2000 - 135t$, where $f(0)$ is the moment the second plane starts its descent. After how many minutes will the two planes be at the same height? Use a domain of [0, 10] and a range of [0, 2000].

Solution See page S8. 8 min

2.3 EXERCISES Suggested Assignment: 7–69, odds

Topics For Discussion

1. What is the vertical-line test?

   If every vertical line intersects a graph at most once, then the graph is the graph of a function.

2. Draw a graph that is the graph of a function. Draw a graph that is not the graph of a function.

   Answers will vary.

3. **a.** What is the value of the x-coordinate of any point on the y-axis?
   b. What is the value of the y-coordinate of any point on the x-axis?

   a. The value of the x-coordinate is 0.   b. The value of the y-coordinate is 0.

4. Can the graph of a function have more than one y-intercept? Explain.

   No. If a graph has more than one y-intercept, there are at least two ordered pairs with the same x-coordinate but different y-coordinates.

5. **a.** How are the independent and dependent variables of a function related to the domain and range of the function?
   b. How are the input and output values of a function related to the domain and range of the function?

   a. The domain of a function is the set of all values of the independent variable. The range of a function is the set of all values of the dependent variable.
   b. The domain of a function is the set of all input values. The range of a function is the set of all output values.

6. Draw a graph of a constant function.

   Answers will vary. The graph must be that of a horizontal line.

◼ Vertical-Line Test

In Exercises 7 to 12, use the vertical-line test to determine whether the graph is the graph of a function.

7.

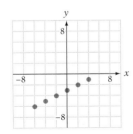

Yes

8.

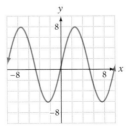

Yes

9.

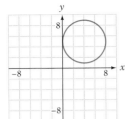

No

10.

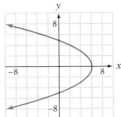

No

11.

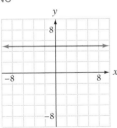

Yes

12.

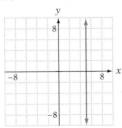

No

◼ Intercepts

For Exercises 13 to 24, find the x- and y-intercepts for the graph of the function.

13. $f(x) = 3x + 6$
   $(-2, 0); (0, 6)$

14. $y(x) = 2x - 5$
   $(2.5, 0); (0, -5)$

15. $h(x) = x^2 - x - 6$
   $(-2, 0), (3, 0); (0, -6)$

16. $g(x) = x^2 + 3x - 4$
   $(-4, 0), (1, 0); (0, -4)$

17. $G(x) = 2x^2 + 5x - 3$
   $(-3, 0), (0.5, 0); (0, -3)$

18. $H(x) = 2x^2 - 7x + 6$
   $(1.5, 0), (2, 0); (0, 6)$

19. $s(x) = -x^2 + 4x + 5$
   $(-1, 0), (5, 0); (0, 5)$

20. $F(x) = -x^2 + 4x$
   $(0, 0), (4, 0); (0, 0)$

21. $g(x) = x^3 - 4x^2 - 7x + 10$
   $(-2, 0), (1, 0), (5, 0); (0, 10)$

22. $f(x) = x^3 + 4x^2 - 7x - 10$
   $(-5, 0), (-1, 0), (2, 0); (0, -10)$

23. $P(x) = x^3 - x^2 - 12x$
   $(-3, 0), (0, 0), (4, 0); (0, 0)$

24. $h(x) = -x^3 - 3x^2 + 4x$
   $(-4, 0), (0, 0), (1, 0); (0, 0)$

■ Intersections of the Graphs of Two Functions

For Exercises 25 to 36, find the number in the domain of the function for which the output is the given number. If necessary, round to the nearest hundredth.

25. $f(x) = 3x - 4; 5$

3

26. $g(x) = 2x + 3; -1$

−2

27. $h(x) = -\dfrac{3}{2}x - 4; -7$

2

28. $s(t) = -\dfrac{3}{4}t + 2; -1$

4

29. $f(x) = \dfrac{5}{2}x - 2; 2$

1.6

30. $v(x) = \dfrac{5}{4}x - 2; 1$

2.4

31. $h(t) = 2.1t - 3; 3.3$

3

32. $g(x) = 3 - 2.4x; 5.4$

−1

33. $f(x) = \dfrac{6}{7}x - 4; -4$

0

34. $z(x) = 1 - \dfrac{5}{4}x; 1$

0

35. $f(x) = \dfrac{7}{3}x + 2; 6$

1.71

36. $g(x) = -\dfrac{6}{5}x + 3; -2$

4.17

For Exercises 37 to 42, find the number in the domain of the function for which the output is the given number. Use a domain of $[-10, 10]$ and a range of $[-10, 10]$.

37. $f(x) = \dfrac{8}{x + 1}; 2$

3

38. $g(x) = \dfrac{6}{x - 2}; 3$

4

39. $h(x) = \sqrt{x + 1}; 3$

8

40. $s(t) = \sqrt{8 - t}; 2$

4

41. $f(x) = x^3 + 0.5x + 2; -7$

−2

42. $v(s) = s^3 + s - 1; -3$

−1

For Exercises 43 to 54, find the input value for which the two functions have the same value. Verify the results algebraically.

43. $f(x) = 3x - 3, g(x) = -2x + 7$

2

44. $f(x) = -x + 8, g(x) = 2x - 1$

3

45. $f(x) = 2x + 8, g(x) = -3x - 2$

−2

46. $f(x) = \dfrac{2}{3}x - 6, g(x) = -2x + 2$

3

47. $f(x) = \dfrac{3}{2}x + 2, g(x) = -3x - 7$

−2

48. $f(x) = \dfrac{2x}{5} + 5, g(x) = \dfrac{-3x}{2} + 5$

0

49. $f(x) = 2x - 2, g(x) = -2x + 8$

2.5

50. $f(x) = \dfrac{3}{4}x, g(x) = -2x - 5.5$

−2

51. $f(x) = -2x - 3.25, g(x) = 3x + 8.5$

−2.35

52. $f(x) = -\dfrac{1}{4}x + 4, g(x) = \dfrac{7}{4}x + 10$

−3

53. $f(x) = 1.6x - 6, g(x) = -2.4x + 2.4$

2.1

54. $f(x) = 2.5x + 6, g(x) = -1.5x - 3.6$

−2.4

For Exercises 55 to 58, complete the ordered pairs that belong to the function.

55. Let $h(x) = 4x - 1$.

    a. $(-2, ?)$

    b. $(?, 3)$

    a. -9 b. 1

56. Let $g(x) = 5 - x$.

    a. $(-4, ?)$

    b. $(?, 6)$

    a. 9 b. -1

57. Let $F(x) = -\dfrac{x}{3} + 2$.

    a. $(-6, ?)$

    b. $(?, -1)$

    a. 4 b. 9

58. Let $H(x) = \dfrac{4}{3}x + 7$.

    a. $(-3, ?)$

    b. $(?, -5)$

    a. 3 b. -9

For Exercises 59 to 69, solve by using a graphing calculator.

59. *Metallurgy*   The amount of gold in an 18-carat gold necklace is given by $Q(w) = 0.75w$, where $Q(w)$ is the number of grams of pure gold in a necklace that weighs w grams. If a necklace contains 33 grams of pure gold, what is the weight of the necklace? Use a domain of $[0, 50]$ and a range of $[0, 40]$.

44 g

60. *Investments*   The value of an investment is given by $V(t) = 500 + 3.5t$, where $V(t)$ is the value of the investment, in dollars, after t months. In how many months will the investment be worth \$591? Use a domain of $[0, 30]$ and a range of $[0, 650]$.

26 months

This necklace is from Mycenae, a Greek city that flourished around 1500 B.C.

61. *Physics*   The height of a rock that was dropped from a cliff overlooking a ravine is given by $s(t) = 175 - 16t^2$, where $s(t)$ is the height, in feet, of the rock above the ocean t seconds after it is dropped.

    a.   Find the value of t for which $s(t) = 0$. Round to the nearest tenth.

    b.   Write a sentence that explains the meaning of the answer to part a in the context of this problem.

    a. 3.3 s   b. The rock hits the bottom of the ravine in 3.3 s.

62. *Automotive Technology*   When the driver of a car is presented with a dangerous situation that requires braking, the distance the car will travel before stopping depends on the driver's reaction time and the distance the car travels after the brakes are applied. The distance can be approximated by $d(s) = 0.05s^2 + 1.1s$, where $d(s)$ is the distance, in feet, the car travels before stopping when the car was traveling s miles per hour when the brakes were applied. Find the speed of a car that travels 253 feet after the brakes are applied. Round to the nearest whole number. Use a domain of $[0, 100]$ and a range of $[0, 500]$.

61 mph

63. *Population Growth*   The population of a city is increasing, and city planners have determined that the population can be modeled by $P(t) = 50 - \dfrac{30}{t + 1}$, where $P(t)$ is the population of the city, in thousands, t years from now. In how many years will the population be 45,000? Use a domain of $[0, 10]$ and a range of $[0, 50]$.

5 years

64. *Physics* The time T, in seconds, it takes a pendulum to make one swing depends on the length of the pendulum and is given by $T(L) = 2\pi\sqrt{\frac{L}{32}}$, where L is the length of the pendulum in feet. How long is a pendulum that takes 1 second to make one swing? Use a domain of $[0, 3]$ and a range of $[0, 3]$. Round to the nearest tenth.
0.8 ft

65. *Economics* For a particular computer game, the number of copies of the game that a company can sell per month depends on the price of the game and is given by $D(p) = \frac{250,000}{80 + 4p}$, where p is the price of the game in dollars and $D(p)$ is the number of games that can be sold per month at that price. At what price can 1500 games be sold per month? Round to the nearest whole number.
$22

66. *Investments* The value of an investment in t years is given by the function $V(t) = 600 + 48t$, and the value of a second investment in t years is given by $W(t) = 744 + 30t$. In how many years will the two investments have the same value? Use a domain of $[0, 10]$ and a range of $[0, 1200]$.
8 years

67. *Sports* Ramona starts walking along a hiking trail through a nature preserve, and 1 hour later Emily starts along the same trail trying to catch Ramona. The distance Ramona has traveled from the beginning of the trail is given by $g(t) = 3t + 3$, and the distance that Emily has traveled from the beginning of the trail is given by $f(t) = 4.5t$. In each case, t is the number of hours Emily has been walking. The distance each of them has walked is measured in miles. In how many hours will Emily catch up to Ramona? Use a domain of $[0, 4]$ and a range of $[0, 20]$.
2 h

68. *Economics* Suppose that the number of professional digital cameras that will be purchased by consumers each month is given by $n(p) = 8800 - 100p$, where $n(p)$ is the number of cameras that consumers will purchase at a selling price of p dollars. A company that manufactures digital cameras, however, is willing to manufacture the cameras only if the selling price is high enough to earn a profit. Suppose that the number of cameras a company is willing to produce each month is given by $c(p) = 10p$, where $c(p)$ is the number of cameras the manufacturer will produce at a selling price of p dollars. At what price will the number of cameras produced by the manufacturer equal the number of cameras purchased by consumers? Economists call this the *equilibrium price* of the commodity. Use a domain of $[0, 100]$ and a range of $[0, 10\ 000]$.
$80

69. *Population Growth* The population of a certain city t years after the year 2000 can be approximated by $P(t) = 53,000 + 1000t$, and the population of a second city can be approximated by $R(t) = 42,000 + 2000t$. In what year will the populations of the two cities be equal? Use a domain of $[0, 15]$ and a range of $[0, 75\ 000]$.
2011

Applying Concepts

In Exercises 25 to 36, you were asked to find the number in the domain of the function for which the output was the given number. Frequently in mathematics, we express directions such as these by using a combination of words and symbols. For instance, in Exercise 25, we could have written "Find the value of a in the domain of $f(x) = 3x - 4$ for which $f(a) = 5$." Recall that $f(a)$ is the output of a function for a given input a. Thus $f(a) = 5$ says the output is 5 for an input of a. We will use this terminology in Exercises 70 to 75.

70. Find the value of a in the domain of $f(x) = 2x + 5$ for which $f(a) = -3$.
   -4

71. Find the value of a in the domain of $g(x) = -2x - 1$ for which $g(a) = -5$.
   2

72. Find the value of a in the domain of $h(x) = \dfrac{3}{2}x - 2$ for which $h(a) = 4$.
   4

73. Find the value of a in the domain of $F(x) = -\dfrac{5}{4}x - 1$ for which $F(a) = 9$.
   -8

74. For $f(x) = x^3 - 4x - 1$, how many different values, a, in the domain of f satisfy the condition that $f(a) = 1$? (You do not have to find the values; just determine number of possible values for a.)
   3

75. For $f(x) = 0.01(x^4 - 49x^2 + 36x + 252)$, how many different values, a, in the domain of f satisfy the condition that $f(a) = 1$? (You do not have to find the values; just determine number of possible values for a.) 4

In Exercises 76 to 79, find two numbers in the domain of the functions for which the values of the functions are equal. *Hint:* Apply the method of finding the point of intersection of two graphs twice, once to each point of intersection.

76. $f(x) = x^2 + 4x - 1, g(x) = 3x + 5$
   $-3, 2$

77. $f(x) = x^2 - x - 1, g(x) = -3x + 2$
   $-3, 1$

78. $f(x) = 2x - 7, g(x) = x^2 - 4x - 2$
   $1, 5$

79. $f(x) = x + 4, g(x) = x^2 + 3x - 4$
   $-4, 2$

EXPLORATION

1. *Calculator Viewing Windows*   A graphing calculator screen consists of **pixels,**[2] which are small rectangles of light that can be turned on or off. When a calculator draws the graph of an equation, it is turning on the pixels that represent the ordered-pair solutions of the equation. The jagged appearance of the graph is a consequence of the solutions being approximations; the pixel nearest the ordered pair is turned on.

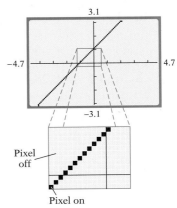

---

[2] Digital cameras are rated in pixels; some cameras have over 3 million pixels. A typical graphing calculator has just over 5800 pixels. Because of this, the graphs on these screens are not so sharp as an image in a digital camera.

Recall that Δ can be read "the change in." For a Texas Instruments TI-83 or TI-83 Plus, there are 94 horizontal pixels. Accordingly, the change in x, Δx, each time the left arrow key is pressed in TRACE mode is calculated by the formula

$$\Delta x = \frac{\text{Xmax} - \text{Xmin}}{94} \qquad \bullet \text{ Formula 1}$$

For the decimal viewing window, Xmin $= -4.7$ and Xmax $= 4.7$. Therefore,

$$\Delta x = \frac{\text{Xmax} - \text{Xmin}}{94}$$
$$= \frac{4.7 - (-4.7)}{94} = \frac{9.4}{94}$$
$$= 0.1$$

Thus each time the left or right arrow key is pressed, the value of x changes by 0.1.

The formula for Δx can be rewritten so that you can create a viewing window with your choice for Δx and for Xmin. This formula is

$$\textbf{Xmax} = \textbf{Xmin} + \textbf{94}(\boldsymbol{\Delta x}) \qquad \bullet \text{ Formula 2}$$

For instance, if we want to begin the graph with Xmin $= -20$ and have $\Delta x = 0.5$, then

$$\text{Xmax} = \text{Xmin} + 94(\Delta x)$$
$$= -20 + 94(0.5) = -20 + 47$$
$$= 27$$

Choosing Xmin $= -20$ and Xmax $= 27$ will result in a viewing window in which x will change by 0.5 each time the left or right arrow key is pressed when the calculator is in TRACE mode.

a. Using Formula 1, determine the change in x each time the left or right arrow key is pressed when the calculator uses the standard viewing window. Draw the graph of $y = 2x + 1$ in the standard viewing window and verify your answer.

$\dfrac{10}{47}$

b. Using Formula 1, verify that $\Delta x = 1$ for the integer viewing window.

$\dfrac{47 - (-47)}{94} = \dfrac{94}{94} = 1$

c. Using Formula 2, create a viewing window for which Xmin $= -10$ and $\Delta x = 0.25$.

Xmin $= -10$, Xmax $= 13.5$

CHAPTER **2** *SUMMARY*

Key Terms

abscissa [p. 93]
axes [p. 93]
constant function [p. 121]
coordinates [p. 93]
coordinate axes [p. 93]
dependent variable [p. 95]
domain [p. 122]
equation in two variables [p. 94]
evaluating a function [p. 117]
first coordinate [p. 94]
function [p. 115]
functional notation [p. 117]
graph of an equation [p. 99]
graph of a function [p. 119]
graph of an ordered pair [p. 93]
graph a point [p. 93]
independent variable [p. 95]
input/output table [p. 95]
intercepts [p. 135]
linear equation in two variables [p. 99]
ordered pair [p. 93]
ordinate [p. 93]

origin [p. 93]
pixel [p. 147]
plot a point [p. 93]
quadrants [p. 93]
range [p. 122]
rectangular coordinate system [p. 93]
relation [p. 114]
second coordinate [p. 94]
solution of an equation in two variables [p. 94]
standard viewing window [p. 100]
value of a function [p. 117]
viewing window [p. 100]
x-axis [p. 93]
x-coordinate [p. 93]
x-intercept [p. 135]
xy-plane [p. 93]
y-axis [p. 93]
y-coordinate [p. 93]
y-intercept [p. 135]
zero of a function [p. 136]

Essential Concepts

Vertical-Line Test for the Graph of a Function [p. 134]
A graph defines a function if any vertical line intersects the graph at no more than one point.

Graphing Calculator Techniques
Creating an input/output table [Appendix A: Table]
Adjusting the viewing window [Appendix A: Window]
Using the TRACE feature [Appendix A: Trace]
Evaluating a function [Appendix A: Evaluating Functions]
Finding the x-intercepts of a graph [Appendix A: Zero]
Finding the point of intersection of the graphs of two functions [Appendix A: Intersect]
Graphing a function [Appendix A: Graph]

CHAPTER **2** REVIEW EXERCISES

1. Without using a calculator, create an input/output table for $y = \frac{1}{2}x - 3$ for $x = -6, -4, -2, 0, 2$ and 4. Check your results using a graphing calculator. Graph the resulting ordered pairs.

x	−6	−4	−2	0	2	4
y	−6	−5	−4	−3	−2	−1

2. Without using a calculator, create an input/output table for the equation $y = x^2 + x - 3$ for $x = -3, -2, -1, 0, 1, 2$ and 3. Check your results using a graphing calculator. Graph the resulting ordered pairs.

x	−3	−2	−1	0	1	2	3
y	3	−1	−3	−3	−1	3	9

3. Complete the input/output table for $y = -\frac{1}{3}x + 2$ and graph the equation without using a calculator. Then use a calculator to verify the table and graph.

x	−9	−6	−3	0	3
y	5	4	3	2	1

For Exercises 4 and 5, graph the equation in the integer viewing window. Then trace along the graph to find the requested values.

4. Let $y = -\frac{4}{5}x - 3$.

    a. Find y when $x = 5$.
    b. Find x when $y = 1$.
    a. −7 b. −5

5. Let $y = \frac{5}{3}x - 1$.

    a. Find y when $x = 3$.
    b. Find x when $y = -6$.
    a. 4 b. −3

6. Find the domain and range of the relation $\{(-1, -1), (0, 1), (1, 3), (2, 5), (3, 3), (4, 1)\}$. Is the relation a function?
D: {−1, 0, 1, 2, 3, 4}, R: {−1, 1, 3, 5}; Yes

7. Find the range of $h(t) = -2t^2 + 5$ given the domain $\{-4, -3, -2, -1, 0, 1, 2, 3, 4\}$.
{−27, −13, −3, 3, 5}

For Exercises 8 and 9, determine which numbers, if any, must be excluded from the domain of the function.

8. $F(x) = \dfrac{2x - 4}{x + 2}, x = -2, 0, 2$

    −2

9. $G(x) = \dfrac{x}{x^2 - 9}, x = -3, 0, 3$

    −3, 3

For Exercises 10 to 12, complete the input/output table, and graph the function without using a calculator. Check your answer by using a graphing calculator.

10. $f(x) = -2x + 3$

x	-3	-2	-1	0	1
y	9	7	5	3	1

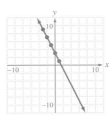

11. $f(x) = -x^2 + 4x - 1$

x	-1	0	1	2	3	4	5
y	-6	-1	2	3	2	-1	-6

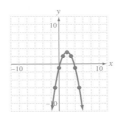

12. $f(x) = -4$

x	-2	-1	0	1	2
y	-4	-4	-4	-4	-4

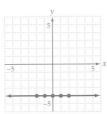

For Exercises 13 and 14, graph the function in the integer viewing window. Then use the TRACE feature to find the x-coordinate for the given y-coordinate. Check your work by evaluating the function at the x-coordinate.

13. $f(x) = -\dfrac{5}{4}x + 5; y = -15$

16

14. $g(x) = 3 - \dfrac{9}{5}x; y = 21$

-10

For Exercises 15 and 16, graph the function in the standard viewing window. Then use the TRACE feature to evaluate the function. Verify the results algebraically.

15. $f(x) = -x^3 - 4x^2 + 2; f(-3)$

-7

16. $F(x) = x^3 - x^2 + 2x - 2; F(-1.5)$

-10.625

For Exercises 17 and 18, graph the function in the decimal viewing window. Trace along the curve to find a number that is not in the domain of the function.

17. $G(x) = \dfrac{-2}{x + 1}$

-1

18. $F(x) = \dfrac{x}{2x - 4}$

2

19. Use the vertical-line test to determine whether the graph at the right is the graph of a function.

Yes, it is a function.

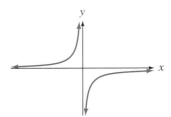

For Exercises 20 and 21, find the x- and y-intercepts of the graph of the function.

20. $F(x) = \frac{1}{2}x - 3$

(6, 0); (0, −3)

21. $H(x) = x^2 + 2x - 8$

(−4, 0); (2, 0); (0, −8)

22. Find the number in the domain of $f(x) = \frac{1}{2}x - 3$ for which the output is $-\frac{5}{4}$.   3.5

23. Find the number in the domain of $f(x) = -2x + 4$ and $g(x) = \frac{3}{2}x + 11$ for which the values of the functions are equal. Verify the results algebraically.   −2

24. The height of a plane leaving an airport can be given by $h(t) = 75t + 300$, where $h(t)$ is the height, in feet, of the plane above sea level t minutes after it takes off.

    a. In how many minutes will the plane be 750 feet above sea level? Use a domain of [0, 10] and a range of [0, 1000].

    b. What is the height of the airport above sea level?

    a. 6 min   b. 300 ft

25. The value, in dollars, of two investments can be given by $V(x) = 400 + 24x$ and $W(x) = 350 + 34x$, where x is the number of years after 2000. In what year will the values of the two investments be equal? Use a domain of [0, 20] and a range of [0, 1000].

2005

CHAPTER **2** *TEST*

1. Without using a calculator, create an input/output table for the equation $y = -x^2 + 4x + 5$ for $x = -2, -1, 0, 1, 2, 3, 4, 5,$ and 6. Graph the resulting ordered pairs.

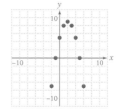

x	−2	−1	0	1	2	3	4	5	6
y	−7	0	5	8	9	8	5	0	−7

2. Complete the input/output table for $f(x) = \frac{2}{3}x - 2$ and then graph the equation.

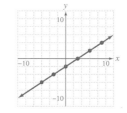

x	−6	−3	0	3	6	9
y	−6	−4	−2	0	2	4

3. Graph $y = -\frac{5}{2}x + 4$ in the integer viewing window. Then find the value of x for which $y = -6$.

    4

4. Find the domain and range of the relation $\{(-4, -2), (-2, -1), (0, 0), (2, -1), (4, -2)\}$. Is the relation a function?

    D: $\{-4, -2, 0, 2, 4\}$, R: $\{-2, -1, 0\}$; Yes

5. Given $P(x) = x^2 - 3x - 1$ with domain $\{-2, -1, 0, 1, 2, 3\}$, find the range of P.

    $\{-3, -1, 3, 9\}$

6. For $f(x) = \frac{2x + 4}{x - 2}$, which of the following numbers, if any, are *not* in the domain of f?

    a. -2 **b.** 0 **c.** 2

    2

For Exercises 7 and 8, complete the input/output table and graph the function.

7. $f(x) = -\frac{3}{2}x + 4$

x	-2	0	2	4	6
y	7	4	1	-2	-5

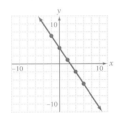

8. $f(x) = -x^2 - 2x + 3$

x	-4	-3	-2	-1	0	1	2
y	-5	0	3	4	3	0	-5

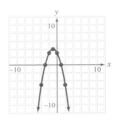

For Exercises 9 and 10, graph the function in the integer viewing window. Then use the TRACE feature to find the x-coordinate for the given y-coordinate. Check your work by evaluating the function at the x-coordinate.

9. $f(x) = -\frac{4}{3}x + 3$ when $y = 11$.

    -6

10. $g(x) = \frac{x}{2} - 5$ when $y = -11$.

    -12

11. Graph $f(x) = \frac{-2x}{x^2 - 9}$ in the decimal viewing window. Trace along the curve to find two numbers that are *not* in the domain of the function.

    $-3, 3$

12. Draw a graph that is not the graph of a function.

    Answers will vary.

For Exercises 13 and 14, find the x- and y-intercepts of the graph of the function.

13. $G(x) = 3 + \frac{3}{4}x$

    $(-4, 0)$; $(0, 3)$

14. $p(x) = x^2 - 2x - 3$

    $(-1, 0), (3, 0); (0, -3)$

15. Find the number in the domain of $f(x) = 2 + \frac{5}{4}x$ for which the output is -3. -4

16. Find the number in the domain of $f(x) = -\frac{5}{2}x + 4$ and $g(x) = \frac{3}{4}x - 9$ for which the values of the functions are equal. Verify the results algebraically.   4

17. The percent concentration, P, of acid in a solution depends on the number of grams x of acid that is added to the solution and is given by $P(x) = \frac{100x + 100}{x + 5}$, where x is the number of grams of acid that is added.
    a.   What is the percent concentration of acid after an additional 3 grams of acid is added to the solution?
    b.   What is the original percent concentration?
    a. 50%   b. 20%

18. The height of a roller coaster car above the ground as it is pulled to the highest point on the track is given by $H(t) = 2t + 40$, where $H(t)$ is the height of the car, in feet, after t seconds. In how many seconds will the car be 110 feet above the ground? Use a domain of $[0, 60]$ and a range of $[0, 200]$.   35 s

19. The height of a rock that was dropped from a cliff overlooking the ocean is given by $s(t) = 150 - 16t^2$, where $s(t)$ is the height, in feet, of the rock above the ocean t seconds after it is dropped. Find the value of t for which $s(t) = 0$. Round to the nearest tenth.   3.1 s

20. Because of changing economic conditions in two cities, Bradford and Candlewood, the populations of the cities are changing. The population of Bradford is given by $B(t) = 750t + 25,000$, and the population of Candlewood is given by $C(t) = 31,250 - 500t$, where t is the number of years after 2002. In which year will the two cities have the same population? Use a domain of $[0, 8]$ and a range of $[0, 32\ 000]$.   2007

◀ CUMULATIVE REVIEW EXERCISES

1. Write $[-2, 3]$ in set-builder notation.
   $\{x| -2 \leq x \leq 3\}$

2. True or false: If $x \in A \cap B$, then $x \in A$.
   True

3. Evaluate: $2 \cdot 3^2 - 4^3 \div (2 - 6)$
   34

4. Evaluate $-a^3b - 4ab^2$ when $a = -2$ and $b = 4$.
   160

5. Evaluate $3a + 2b - c^2$ when $a = \frac{5}{6}$, $b = -\frac{2}{3}$, and $c = -\frac{3}{2}$.
   $-\frac{13}{12}$

6. Simplify: $3 - 5(2a - 7)$
   $-10a + 38$

7. $3(a + b) = (a + b)3$ is an example of what Property of Real Numbers?
Commutative Property of Multiplication

8. Complete the following input/output table for $y = \frac{3}{2}x - 3$ and then graph the ordered pairs.

x	-4	-2	0	2	4	6
y	-9	-6	-3	0	3	6

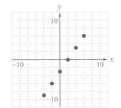

9. Federal government emission standards limit the grams of nitrous oxides (NO_x) that can be emitted from the exhaust of a car. The number of grams, g, of NO_x emitted into the atmosphere from the exhaust of a certain car is given by $g = 0.4m$, where m is the number of miles the car is driven.

a. Complete the input/output table below.

Input, distance driven m (in miles)	0	100	150	200	250	300	350
Output, NO_x g (in grams)	0	40	60	80	100	120	140

b. Write a sentence that explains the meaning of the ordered pair (150, 60).
When this car is driven 150 mi, it emits 60 g of NO_x.

10. Graph $y = 2 - \dfrac{x}{2}$.

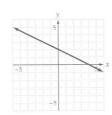

11. Graph $f(x) = -\dfrac{2}{3}x$.

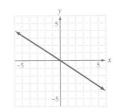

12. Complete the input/output table for $f(x) = x^2 - 6x + 1$ and then graph the function.

x	-1	0	1	2	3	4	5
y	8	1	-4	-7	-8	-7	-4

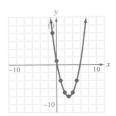

13. Complete the input/output table for $f(x) = -x^2 - 2x + 1$ and then graph the function.

x	-4	-3	-2	-1	0	1	2
y	-7	-2	1	2	1	-2	-7

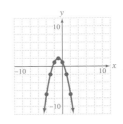

14. Graph $y = \frac{4}{3}x - 2$ in the integer viewing window. Then trace along the curve to find x when $y = -14$.   −9

15. Find the number in the domain of $f(x) = -\frac{3}{2}x + 3$ for which the output is -3.   4

16. For $g(x) = \frac{x + 3}{x - 2}$, which of the following numbers, if any, are *not* in the domain of g?

    a. −3   **b.** 2   **c.** −4

    2

17. Find the x- and y-intercepts for the graph of $f(x) = x^2 + 2x - 3$.
    (−3, 0), (1, 0); (0, −3)

18. Find the input value for which the values of $f(x) = 3x - 4$ and $g(x) = -x + 4$ are equal.   2

19. A biologist introduces a toxin into a culture of bacteria that causes the number of bacteria in the culture to decrease according to the population model $P(t) = \frac{20}{t + 2}$, where $P(t)$ is the number of bacteria, in thousands, t minutes after the toxin is introduced. In how many minutes will the population be 2000? Use a domain of $[0, 10]$ and a range of $[0, 20]$.   8 min

20. According to a certain demographic study, the population of the city of Goldcreek can be approximated by $G(n) = 0.75n + 8$, and the population of the city of Walnut Grove can be approximated by $W(n) = -0.5n + 13$, where the populations are in thousands and n is the number of years after the year 2000. In what year will the populations of the two cities be equal?   2004

3

First-Degree Equations and Inequalities

```
TEST  LOGIC
1: =
2: ≠
3: >
4: ≥
5: <
6: ≤
```

Press **2nd** TEST to access the inequality symbols in the TEST menu.

Making decisions is a part of life, and having a good understanding of the alternatives makes deciding between options much easier. For example, choosing which of two cell phone plans will be more economical may be easier if you know how many minutes you will use the phone each month. With that information, you can determine the "better deal" by writing and solving an inequality, such as the one required to solve **Exercise 54 on page 210.**

 Need help? For online student resources, visit this web site: Math.college.hmco.com

PREP TEST

For Exercises 1 to 4, add, subtract, multiply, or divide.

1. $8 - 12$ -4

2. $-9 + 3$ -6

3. $\dfrac{-18}{-6}$ 3

4. $-\dfrac{3}{4}\left(-\dfrac{4}{3}\right)$ 1

5. Simplify: $3x - 5 + 7x$ $10x - 5$

6. Simplify: $8x - 9 - 8x$ -9

7. Simplify: $6x - 3(6 - x)$ $9x - 18$

8. Evaluate: $|-8|$ 8

9. Twenty ounces of a snack mixture contains nuts and pretzels. Let n represents the number of ounces of nuts in the mixture. Express the number of ounces of pretzels in the mixture in terms of n. $20 - n$

GO FIGURE

A pair of perpendicular lines are drawn through the interior of a rectangle, dividing it into four smaller rectangles. The areas of the smaller rectangles are x, 2, 3, and 6. Find the possible values of x.   1, 4, 9

SECTION **3.1**

- Introduction to Solving Equations
- Solve Equations Using the Addition and Multiplication Properties
- Equations Containing Parentheses

Solving First-Degree Equations

▪ Introduction to Solving Equations

When the National Weather Service identifies a hurricane building up over the ocean, it tries to predict when the hurricane will reach land. Meteorologists do this by determining the direction and speed of the hurricane.

Suppose the National Weather Service locates at 9:00 A.M. a hurricane that is 150 miles from land and moving at a constant speed of 25 mph toward land. Then the distance d, in miles, that the hurricane is from land can be given by $d = 150 - 25t$, where t is the time in hours since 9:00 A.M. If the Weather Service wants to give a warning to residents along the coast when the hurricane is 100 miles from land, then they must find the input value for t that results in an output of 100 ($d = 100$).

$$d = 150 - 25t$$
$$100 = 150 - 25t \qquad \bullet \text{ We want an input value of } t \text{ for which } d = 100.$$

INSTRUCTOR NOTE
To help students understand the equation $d = 150 - 25t$, have them respond to the following:
1. What does $t = 0$ mean?
2. When $t = 1$, how far is the hurricane from land and what is the current time?
3. What does $d = 0$ mean?

CALCULATOR NOTE
The independent variable is t, which is input as X, and the dependent variable is d, which is Y_1. We are using a domain of $[0, 6]$ and a range of $[0, 150]$.

 See Appendix A: Intersect

This value of t can be found using a graphing calculator and its INTERSECT feature. Here are some typical screens.

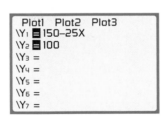

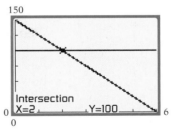

An input value of 2 results in an output value of 100. We can check this algebraically.

$$d = 150 - 25t$$
$$d = 150 - 25(2) \qquad \bullet \text{ Replace } t \text{ with 2.}$$
$$d = 150 - 50$$
$$d = 100 \qquad \bullet \text{ The input value 2 results in an output value of 100.}$$

The Weather Service should give a warning 2 hours after 9:00 A.M., or at 11:00 A.M.

Now suppose the Weather Service wants to issue another warning when the hurricane is 50 miles from land. The procedure above is repeated with $d = 50$. In this case, we want to find the input value of t that results in d equal to 50.

$$d = 150 - 25t$$
$$50 = 150 - 25t$$

Although saying we need to find the value of t that results in d equal to 50 is fine, this wording is usually rephrased as "Solve the equation $50 = 150 - 25t$ for t." Up to this point, we have accomplished this graphically. We will now begin an algebraic study of solving equations.

An **equation** expresses the equality of two mathematical expressions. The expressions can be either numerical or variable expressions. For the equation $50 = 150 - 25t$, the numerical expression 50 equals the variable expression $150 - 25t$. The expression to the left of the equals sign is the **left side of the equation,** and the expression to the right of the equals sign is the **right side of the equation.**

Suggested Activity

Students have a tendency to think of equations and expressions as the same thing. With students working in groups, ask them to determine which of the following are equations and which are expressions.
$3x + 1 = 5x^2 + 3$
$2y(3y + 4)$
$\dfrac{z}{z^2 + 1}$
$3b = 5$
Now have each group write down two equations and two expressions.

The equation $x + 3 = 8$ at the right is a **conditional equation.** The equation is true if the variable is replaced by 5. The equation is false if the variable is replaced by 4.

$$x + 3 = 8$$
$$5 + 3 = 8 \qquad \text{A true equation}$$
$$4 + 3 = 8 \qquad \text{A false equation}$$

The replacement values of the variable that will make an equation true are called the **roots,** or **solutions,** of the equation. The solution of the equation $x + 3 = 8$ is 5.

An **identity** is an equation for which any replacement for the variable will result in a true equation. For instance, the equation $2x = x + x$ is an identity.

Some equations have no solutions. For instance, the equation $n = n + 1$ has no solution. There is no number n that is equal to one more than itself.

Given a value for a variable, it is always possible to determine whether that value is a solution of an equation. **Replace the given value of the variable in the equation and then simplify. If the left and right sides of the equation are equal, the value of the variable is a solution of the equation.** As shown below, 4 is a solution of $50 = 150 - 25t$.

$$50 = 150 - 25t$$

50	$150 - 25(4)$	• Replace t by 4.
50	$150 - 100$	• Simplify.
$50 = 50$		• The left and right sides are equal.

Because the left and right sides of the equation are equal when $t = 4$, 4 is a solution of the equation. Four hours after 9:00 A.M. is 1:00 P.M., the time at which the hurricane will be 50 miles from land.

The equation $50 = 150 - 25t$ is a **first-degree equation in one variable.** It is called a first-degree equation because the exponent on the variable is 1. Several other examples of first-degree equations are given at the right.

$$2x - 7 = 15$$
$$3 - 4y = 8y + 1$$
$$6z = 2$$
$$3a - 2(4a + 1) = 7a$$

? QUESTION   Which of the following are first-degree equations?

    a. $5 = 3n + 7$ **b.** $6z^2 + 1 = 7$ **c.** $x = 4$ **d.** $x^3 = 8$

Solving an equation means finding a solution of the equation. The simplest equation to solve is an equation of the form *variable = constant* because the constant is the solution. If $x = 3$, then 3 is the solution of the equation because $3 = 3$ is a true equation. When solving an equation, the goal is to rewrite the given equation in the form

$$variable = constant$$

The Addition Property of Equations can be used to rewrite an equation in this form.

The Addition Property of Equations

If a, b, and c are algebraic expressions, then the equation $a = b$ has the same solutions as the equation $a + c = b + c$.

The Addition Property of Equations states that the same quantity can be added to each side of an equation without changing the solution of the equa-

---

? ANSWER   The equations in parts a and c are first-degree equations; the variable has an exponent of 1. The equations in parts b and d are not first-degree equations. In part b, the exponent on the variable is 2; in part d, the exponent on the variable is 3.

tion. **This property is used to remove a** *term* **from one side of the equation by adding the opposite of that term to each side of the equation.**

Note the effect of adding, to each side of the equation $x + 6 = 9$, the *opposite of the constant term* 6. After each side of the equation is simplified, the equation is in the form *variable = constant*. The solution is the constant.

$$x + 6 = 9$$
$$x + 6 + (-6) = 9 + (-6)$$
$$x + 0 = 3$$
$$x = 3$$
$$variable = constant$$

Because subtraction is defined in terms of addition, the Addition Property of Equations makes it possible to subtract the same number from each side of an equation without changing the solution of the equation.

➡ Solve: $y + \dfrac{3}{4} = \dfrac{2}{3}$

The goal is to write the equation in the form *variable = constant*.

Add the opposite of the constant term $\frac{3}{4}$ to each side of the equation. This is equivalent to subtracting $\frac{3}{4}$ from each side of the equation. After simplifying, the equation is in the form *variable = constant*.

$$y + \frac{3}{4} = \frac{2}{3}$$
$$y + \frac{3}{4} - \frac{3}{4} = \frac{2}{3} - \frac{3}{4}$$
$$y + 0 = \frac{8}{12} - \frac{9}{12}$$
$$y = -\frac{1}{12}$$

The solution is $-\dfrac{1}{12}$. ⬅

The goal of solving an equation can also be to write the equation as *constant = variable*. This is shown in Example 1.

TAKE NOTE

You should always check your solutions to an equation.

$$y + \frac{3}{4} = \frac{2}{3}$$
$$-\frac{1}{12} + \frac{3}{4} \;\Big|\; \frac{2}{3}$$
$$-\frac{1}{12} + \frac{9}{12} \;\Big|\; \frac{2}{3}$$
$$\frac{8}{12} \;\Big|\; \frac{2}{3}$$
$$\frac{2}{3} = \frac{2}{3}$$

A true equation

INSTRUCTOR NOTE
Exercise 8 can be used for a similar in-class example.

TAKE NOTE

You should always check your work. Here is the check for Example 1.

$$\frac{7 = x + 9}{7 \;\big|\; -2 + 9}$$
$$7 = 7 \;\checkmark$$

EXAMPLE 1

Solve: $7 = x + 9$

Solution

$$7 = x + 9$$
$$7 - 9 = x + 9 - 9 \qquad \bullet \text{ Subtract 9 from each side of the equation.}$$
$$-2 = x \qquad\qquad \bullet \text{ The equation is in the form } constant = variable.$$

The solution is -2.

YOU TRY IT 1

Solve: $9 + n = 4$

Solution   See page S8. -5

The Multiplication Property of Equations is also used to rewrite an equation in the form *variable = constant.*

> **The Multiplication Property of Equations**
>
> If a, b, and c are algebraic expressions, and $c \neq 0$, then the equation $a = b$ has the same solutions as the equation $ac = bc$.

The Multiplication Property of Equations states that we can multiply each side of an equation by the same nonzero number without changing the solutions of the equation. **This property is used to rewrite an equation so that the *coefficient* on the variable is 1. This is accomplished by multiplying each side of the equation by the reciprocal of the coefficient.**

➡ Solve: $\dfrac{2}{3}t = -\dfrac{1}{6}$

Note the effect of multiplying each side of the equation by $\dfrac{3}{2}$, the reciprocal of $\dfrac{2}{3}$.

$$\frac{2}{3}t = -\frac{1}{6}$$

$$\frac{3}{2}\left(\frac{2}{3}t\right) = \frac{3}{2}\left(-\frac{1}{6}\right)$$

$$1 \cdot t = -\frac{3}{12}$$

After simplifying, the equation is in the form *variable = constant.*

$$t = -\frac{1}{4}$$

Remember to check your solution.   The solution is $-\dfrac{1}{4}$. ⬅

Because division is defined in terms of multiplication, the Multiplication Property of Equations enables us to divide each side of an equation by the same nonzero number without changing the solution of the equation.

➡ Solve: $6x = -12$

Multiply each side of the equation by the reciprocal of 6. This is equivalent to dividing each side of the equation by 6.

$$6x = -12$$

$$\frac{6x}{6} = \frac{-12}{6}$$

After simplifying, the equation is in the form *variable = constant.*

$$x = -2$$

Remember to check your solution.   The solution is -2. ⬅

INSTRUCTOR NOTE
Exercise 12 can be used for
a similar in-class example.

EXAMPLE 2

Solve: $18 = -12z$

TAKE NOTE

Here is a check for Example 2.

$$\frac{18 = -12z}{}$$

$$18 \mid -12\left(-\frac{3}{2}\right)$$

$$18 = 18 \; \checkmark$$

Solution $18 = -12z$

$$\frac{18}{-12} = \frac{-12z}{-12}$$ • Divide each side of the equation by -12.

$$-\frac{3}{2} = z$$ • Write the answer in simplest form.

The solution is $-\frac{3}{2}$.

YOU TRY IT 2

Solve: $-4x = -20$

Solution   See page S8.   5

Solve Equations Using the Addition and Multiplication Properties

In the next example, we will apply both the Addition and the Multiplication Properties of Equations.

INSTRUCTOR NOTE
Exercise 22 can be used for a similar in-class example.

TAKE NOTE

The goal is to isolate the variable so that the equation can be written in the form *variable = constant*.

 CALCULATOR NOTE

The decimal approximation given at the right can be converted to a fraction.

 See Appendix A: Math

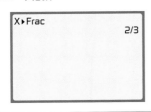

EXAMPLE 3

Solve: $8 = 6r + 4$

Solution $8 = 6r + 4$

$$8 - 4 = 6r + 4 - 4$$ • Subtract 4 from each side of the equation.

$$4 = 6r$$ • Simplify.

$$\frac{4}{6} = \frac{6r}{6}$$ • Divide each side of the equation by 6.

$$\frac{2}{3} = r$$ • Write the answer in simplest form.

The solution is $\frac{2}{3}$.

Graphical check:

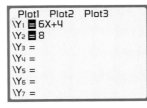

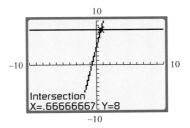

YOU TRY IT 3

Solve: $5 - 4z = 15$

Solution   See page S8. $-\frac{5}{2}$

Now consider an equation such as $2x + 5 = 4x - 1$. The solution of this equation is the value of x that results in the left side and the right side of the equation being equal. That is, we are trying to find the input value x so that the outputs of the functions $f(x) = 2x + 5$ and $g(x) = 4x - 1$ are equal.

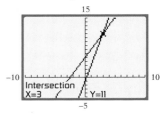

From the table and graph above, note that when $x = 3$, $Y_1 = Y_2$. The solution of $2x + 5 = 4x - 1$ is 3.

The equation $2x + 5 = 4x - 1$ can be solved by using the Addition and Multiplication Properties of Equations.

$$2x + 5 = 4x - 1$$

$$2x - 4x + 5 = 4x - 4x - 1 \qquad \bullet \text{ Subtract } 4x \text{ from each side of the equation.}$$

$$-2x + 5 = -1 \qquad \bullet \text{ Simplify.}$$

$$-2x + 5 - 5 = -1 - 5 \qquad \bullet \text{ Subtract 5 from each side of the equation.}$$

$$-2x = -6 \qquad \bullet \text{ Simplify.}$$

$$\frac{-2x}{-2} = \frac{-6}{-2} \qquad \bullet \text{ Divide each side of the equation by } -2.$$

$$x = 3 \qquad \bullet \text{ Simplify.}$$

The solution is 3. This algebraic solution confirms that the solution that we obtained from the table and graph is correct.

INSTRUCTOR NOTE
Exercise 26 can be used for a similar in-class example.

EXAMPLE 4

Solve: $2x + 5 = 5x - 1 - 7x$

Solution $2x + 5 = 5x - 1 - 7x$

$$2x + 5 = -2x - 1 \qquad \bullet \text{ Combine like terms.}$$

$$2x + 2x + 5 = -2x + 2x - 1 \qquad \bullet \text{ Add } 2x \text{ to each side of the equation.}$$

$$4x + 5 = -1 \qquad \bullet \text{ Simplify.}$$

$$4x + 5 - 5 = -1 - 5 \qquad \bullet \text{ Subtract 5 from each side of the equation.}$$

$$4x = -6 \qquad \bullet \text{ Simplify.}$$

$$\frac{4x}{4} = \frac{-6}{4} \qquad \bullet \text{ Divide each side of the equation by 4.}$$

$$x = -\frac{3}{2} \qquad \bullet \text{ Simplify.}$$

The solution is $-\dfrac{3}{2}$.

INSTRUCTOR NOTE
You might ask your students to find the solution using the TABLE feature of a graphing calculator. Suggest they use an increment of 0.5.

CALCULATOR NOTE

We entered $-2x - 1$ for Y_2
after simplifying $5x - 1 - 7x$.

Graphical check:

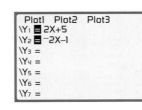

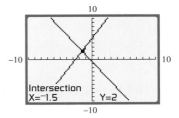

The solution is given as the decimal equivalent of $-\dfrac{3}{2}$.

YOU TRY IT 4

Solve: $6y - 3 + y = 2y + 7$

Solution   See page S9.   2

■ Equations Containing Parentheses

For an equation containing parentheses, the Distributive Property is used to remove the parentheses.

➡ Solve: $3 - x = 3 - 5(2x - 6)$

$3 - x = 3 - 5(2x - 6)$	
$3 - x = 3 - 10x + 30$	• Use the Distributive Property.
$3 - x = 33 - 10x$	• Add like terms on the right side of the equation.
$3 - x + 10x = 33 - 10x + 10x$	• Add $10x$ to each side of the equation.
$3 + 9x = 33$	• Simplify.
$3 - 3 + 9x = 33 - 3$	• Subtract 3 from each side of the equation.
$9x = 30$	• Simplify.
$\dfrac{9x}{9} = \dfrac{30}{9}$	• Divide each side of the equation by 9.
$x = \dfrac{10}{3}$	• Write the answer in simplest form.
The solution is $\dfrac{10}{3}$.	• Remember to check the solution. ⬅

This example illustrates the steps used to solve a first-degree equation.

Suggested Activity

See Section 3.1 of the *Student Activity Manual* for an activity involving using the TABLE feature on a graphing calculator to solve first-degree equations.

Steps in Solving First-Degree Equations
1. Use the Distributive Property to remove parentheses.
2. Combine like terms on each side of the equation.
3. Rewrite the equation with only one variable term.
4. Rewrite the equation with only one constant term.
5. Rewrite the equation so that the coefficient of the variable is 1.

INSTRUCTOR NOTE
Exercise 34 can be used for a similar in-class example.

Suggested Activity

The French club rented a bus to take all of its members to a French film festival. The cost per person was to be $18.00. However, those making the trip had to pay $24.00 because 10 members of the club canceled out at the last minute. How many French club members went to the film festival?
[Answer: 30 members]

TAKE NOTE

In Example 4, we simplified the expression on the right side of the equation before entering it. For Example 5, we did not simplify first. You may use either method.

EXAMPLE 5

Solve: $3 - 2(3x - 1) = 1 - 2x$

Solution

$$3 - 2(3x - 1) = 1 - 2x$$

$3 - 6x + 2 = 1 - 2x$	• Use the Distributive Property.
$5 - 6x = 1 - 2x$	• Simplify.
$5 - 6x + 2x = 1 - 2x + 2x$	• Add $2x$ to each side of the equation.
$5 - 4x = 1$	• Simplify.
$5 - 5 - 4x = 1 - 5$	• Subtract 5 from each side of the equation.
$-4x = -4$	• Simplify.
$\dfrac{-4x}{-4} = \dfrac{-4}{-4}$	• Divide each side of the equation by -4.
$x = 1$	

The solution is 1.

Graphical check:

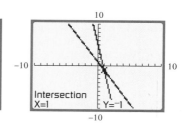

The solution checks.

YOU TRY IT 5

Solve: $2(3x + 1) = 4x + 8$

Solution   See page S9.   3

3.1 EXERCISES

Topics For Discussion

1. How does an equation differ from an expression?
   An equation has an equals sign; an expression does not have an equals sign.
2. What is the solution of an equation?
   The solution of an equation is a number that, when substituted for the variable, results in a true equation.
3. Explain the difference between solving an equation and simplifying an expression.
   The goal of solving an equation is to find its solutions. The goal of simplifying an expression is to combine like terms and write the expression in simplest form.
4. The solution of the equation $2x + 3 = 3$ is 0 and the equation $x = x + 1$ has no solution. Is there a difference between zero as a solution and no solution?
   Yes. No solution means that there are no values of the variable that make the equation a true statement. If the solution of the equation is zero, then there is a value of the variable (0) that results in a true equation.
5. What is the Addition Property of Equations? When is it used?
   The same number can be added to each side of an equation without changing the solution of the equation. This property is used to remove a term from one side of an equation.
6. What is the Multiplication Property of Equations? When is it used?
   Each side of an equation can be multiplied by the same nonzero number without changing the solution of the equation. This property is used to remove a coefficient other than 1 from a variable term.

■ **Introduction to Solving Equations**

Solve.

7. $x - 2 = 7$ 9

8. $a + 3 = -7$ −10

9. $3x = 12$ 4

10. $18 = 2a$ 9

11. $\frac{2}{3}y = 5$ $\frac{15}{2}$

12. $-\frac{5}{8}x = \frac{4}{5}$ $-\frac{32}{25}$

13. $-\frac{3}{5} = \frac{3b}{10}$ −2

14. $\frac{2}{3}y = 5$ $\frac{15}{2}$

15. $0.25x = 1.2$ 4.8

16. $-0.03z = 0.6$ −20

17. $12 = 3x + 5x$ $\frac{3}{2}$

18. $4t - 7t = 0$ 0

■ Solve Equations Using the Addition and Multiplication Properties

Solve.

19. $3x + 8 = 17$ 3

20. $2 + 5a = 12$ 2

21. $5 = 3x - 10$ 5

22. $4 = 3 - 5x$ $-\dfrac{1}{5}$

23. $\dfrac{2}{3}x + 5 = 3$ -3

24. $-\dfrac{1}{2}x + 4 = 1$ 6

25. $2x + 2 = 3x + 5$ -3

26. $2 - 3t = 3t - 4$ 1

27. $3b - 2b = 4 - 2b$ $\dfrac{4}{3}$

28. $\dfrac{1}{3} - 2b = 3$ $-\dfrac{4}{3}$

29. $d + \dfrac{1}{5}d = 2$ $\dfrac{5}{3}$

30. $\dfrac{5}{8}z - 3 = 12$ 24

■ Equations Containing Parentheses

Solve.

31. $4(x - 5) = 8$
7

32. $3(x - 2) = 21$
9

33. $5 - 2(2x + 3) = 11$
-3

34. $2 + 3(x - 5) = 20$
11

35. $5(2 - b) = -3(b - 3)$
$\dfrac{1}{2}$

36. $3 = 2 - 5(3y - 2)$
$\dfrac{3}{5}$

37. $4 - 3x = 7x - 2(3 - x)$
$\dfrac{5}{6}$

38. $-3x - 2(4 + 5x) = 14 - 3(2x - 3)$
$-\dfrac{31}{7}$

39. $3y = 2[5 - 3(2 - y)]$
$\dfrac{2}{3}$

40. $2[3 - 2(z + 4)] = 3(4 - z)$
-22

41. $3[x - (2 - x) - 2x] = 3(4 - x)$
6

42. $2 + 3[1 - 2(x + 3)] = -7(x + 1)$
6

Applying Concepts

43. If $3x - 5 = 9x + 4$, evaluate $6x - 3$.
-12

44. If $8 - 2(4x - 1) = 3x - 12$, evaluate $x^4 - x^2$.
12

45. Solve: $2[3(x + 4) - 2(x + 1)] = 5x + 3(1 - x)$
No solution

$3x - 5 = 9x + 4$
$-5 - 4 = 9x - 3x$
$-9 = 6x$
$-\dfrac{9}{6} = x$
$x = 1.5$

$6x - 3$
$6\left(-\dfrac{9}{6}\right) - 3$
$-9 - 3 = -12$

Solve.

46. $8 \div \dfrac{1}{x} = 3$

$\dfrac{3}{8}$

47. $\dfrac{\dfrac{1}{1}}{\dfrac{1}{x}} = 9$

9

48. $\dfrac{\dfrac{6}{7}}{a} = -18$

-21

49. $\dfrac{\dfrac{10}{3}}{x} - 5 = 4x$

$-\dfrac{15}{2}$

50. *Sports*   The speed v, in feet per second, of a foul tip that goes directly upward is given by $v = 100 - 32t$, where t is the number of seconds after the ball is hit. How many seconds after the ball is released will its speed be 8 feet per second?

2.875 s

51. *Monthly Income*   A sales executive for a software company can choose between two different equations that will determine the executive's monthly income, $I = 0.05x + 2500$ or $I = 0.02x + 4000$, where I is the monthly income for selling x dollars worth of software.

  a. For each option, determine the dollar amount the sales executive must sell in order to earn \$4500 in one month.

  b. For each option, determine the dollar amount the sales executive must sell in order to earn \$5500 in one month.

  c. Determine the dollar amount the sales executive must sell so that the monthly incomes from the two plans are equal.

  a. \$40,000; \$25,000   **b.** \$60,000; \$75,000   **c.** \$50,000

EXPLORATION

1. *Business Application*   Two people decide to open a business reconditioning toner cartridges for copy machines. They rent a building for \$7000 per year and estimate that building maintenance, taxes, and insurance will cost \$6500 per year. Each person wants to make \$12 per hour in the first year and will work 10 hours per day for 260 days of the year. Assume that it costs \$28 to restore a cartridge and that they can sell each restored cartridge for \$45.

  a. How many cartridges must they restore and sell annually to break even, not including the hourly wage they wish to earn?

  Approximately 794 cartridges

  b. How many cartridges must they restore and sell annually just to earn the hourly wage they desire?

  Approximately 3671 cartridges

  c. Suppose the entrepreneurs are successful in their business and are restoring and selling 25 cartridges each day of the 260 days they are open. What would be their hourly wage for the year?

  Approximately \$18.65 per hour

SECTION **3.2**

■ Value Mixture Problems
■ Uniform Motion Problems

Applications of First-Degree Equations

■ Value Mixture Problems

Suggested Activity

If a grocer blended peanuts costing $3.50 per pound with almonds costing $6.00 per pound and then sold the mixture for $3.50 per pound, would the grocer make money or lose money?

[Answer: lose money]

Suppose a coffee merchant blends Ethiopian Mocha Java coffee beans costing $7 per pound with Hawaiian Kona coffee beans costing $12 per pound. Will the coffee merchant always make a profit if the price of the blend is $13 per pound? Will the coffee merchant make a profit if the price of the blend is $10 per pound? Will the coffee merchant make a profit if the price of the blend is $6 per pound?

[Answers: yes; sometimes; no]

A **value mixture problem** involves combining two ingredients that have different prices into a single blend. For instance, a coffee manufacturer may blend two types of coffee into a single blend.

The solution of a value mixture problem is based on the equation $V = AC$, where V is the value of an ingredient, A is the amount of the ingredient, and C is the cost per unit of the ingredient.

For example, to find the value of 10 pounds of coffee costing $6.60 per pound, use the equation $V = AC$.

$$V = AC$$
$$V = (10)(6.60) \qquad \bullet \; A = 10, C = 6.60$$
$$V = 66$$

The value of the 10 pounds of coffee is $66.

? QUESTION   What is the value of 15 ounces of a silver alloy that costs $8 an ounce?

Now consider the situation of trying to determine how many pounds of peanuts that cost $2.25 per pound should be mixed with 40 pounds of cashews that cost $6.00 per pound to produce a mixture that has a value of $3.50 per pound. We will let x represent the number of pounds of peanuts that are being added to the 40 pounds of cashews. Then, using the equation $V = AC$, we have

Value of 40 Pounds of Cashews	Value of x Pounds of Peanuts
$V_1 = 40(6) = 240$	$V_2 = x(2.25) = 2.25x$

From the diagram at the left, note that

Total *amount* of the mixture = amount of peanuts + amount of cashews
$$= x + 40$$

Total *value* of the mixture = value of the peanuts + value of the cashews
$$= 2.25x + 40(6.00)$$
$$= 2.25x + 240$$

Peanuts + Cashews = Mixture

x	40	$x + 40$
2.25	6.00	3.50

$$2.25x + 40(6.00) = (x + 40)3.50$$

? ANSWER $V = AC = 15(8) = 120$. The value is $120.

To find the value of x, again use the equation $V = AC$, this time for the mixture. The unit cost of the mixture is $3.50 per pound.

$$V = AC$$
$$2.25x + 240 = (x + 40)3.50$$
$$2.25x + 240 = 3.50x + 140$$
$$-1.25x = -100$$
$$x = 80$$

- V is the total value of the mixture, $2.25x + 240$. A is the total amount of the mixture in pounds, $x + 40$. C is the unit cost of the mixture, $3.50.

To produce a mixture that has a value of $3.50 per pound, 80 pounds of peanuts should be added to the 40 pounds of cashews.

The unit cost, or cost per pound, of the cashew–peanut mixture changes as peanuts are added. Solving the equation $V = AC$ for C yields the unit cost of the blend as

$$C = \frac{V}{A} = \frac{2.25x + 240}{x + 40}$$

A graph of this equation is shown at the left, where X is the number of pounds of peanuts that have been added and Y_1 is the unit cost (C) of the mixture. By using the TRACE feature, we have positioned the cursor at X = 80. Note that when X = 80 (pounds), Y_1 is 3.5 ($3.50). This confirms the algebraic solution we found above.

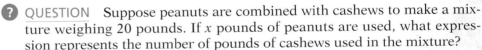

Note that the graph at the left is decreasing. This makes sense in the context of the problem: The peanuts have less value per pound than the cashews, so as peanuts are added, the cost per pound of the mixture decreases.

? QUESTION   Suppose peanuts are combined with cashews to make a mixture weighing 20 pounds. If x pounds of peanuts are used, what expression represents the number of pounds of cashews used in the mixture?

Suggested Activity
See Section 3.2 of the *Student Activity Manual* for an investigation involving value mixture problems.

INSTRUCTOR NOTE
Exercise 10 can be used for a similar in-class example.

EXAMPLE 1

A butcher wishes to combine hamburger that costs $3.00 per pound with hamburger that costs $1.80 per pound. How many pounds of each should be used to make 75 pounds of a mixture costing $2.20 per pound?

State the goal.   The goal is to determine how many pounds of hamburger that costs $3.00 per pound and how many pounds of hamburger that costs $1.80 per pound should be combined to form 75 pounds of hamburger costing $2.20 per pound.

Devise a strategy.   Let x = the number of pounds of $3.00 hamburger that are needed.

? ANSWER   Number of pounds of cashews = number of pounds in the mixture − number of pounds of peanuts = $20 - x$

Then $75 - x$ = the number of pounds of \$1.80 hamburger that are needed.

Find the value of each of the hamburger meats.

Value of the \$3.00 hamburger: $V = AC = x(3.00) = 3x$

Value of the \$1.80 hamburger: $V = AC = (75 - x)1.80 = 135 - 1.80x$

Value of the mixture (the \$2.20 hamburger):

$\qquad$ = value of the \$3.00 hamburger + value of the \$1.80 hamburger

$\qquad$ $= 3x + 135 - 1.80x$

$\qquad$ $= 1.2x + 135$

To find the value of x, use the equation $V = AC$ for the mixture (the \$2.20 hamburger). The amount A of the mixture is 75 pounds. The unit cost C of the mixture is \$2.20 per pound. Solve the equation for x.

Solve the problem.

$$V = AC$$

$$1.2x + 135 = 75(2.20)$$ • V is the value of the mixture, A is the amount of the mixture, and C is the unit cost of the mixture.

$$1.2x + 135 = 165$$

$$1.2x = 30$$

$$x = 25$$ • This is the number of pounds of the \$3.00 hamburger needed.

$$75 - x = 75 - 25 = 50$$ • Find the number of pounds of the \$1.80 hamburger needed by substituting the value of x into the expression for the amount of \$1.80 hamburger needed.

The butcher needs to combine 25 pounds of the hamburger that costs \$3.00 per pound with 50 pounds of the hamburger that costs \$1.80 per pound.

Check your work. One way to check the solution is to make sure that the value of the \$3.00 hamburger plus the value of the \$1.80 hamburger is equal to the value of the mixture.

Value of the \$3.00 hamburger: $V = AC = 25(3.00) = 75$

Value of the \$1.80 hamburger: $V = AC = 50(1.80) = 90$

Value of the two ingredients $= 75 + 90 = 165$

The value of the mixture is $V = AC = 75(2.20) = 165$ (\$165). The solution checks.

A second way to check the solution is to graph the variable expression for the value of the mixture: $V = 1.2x + 135$. Note in the graph at the left that the point (25, 165) is on the line.

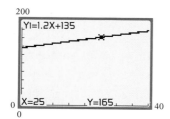

? QUESTION   The graph of the equation $V = 1.2x + 135$ is increasing. Why does this make sense in the context of the problem?

YOU TRY IT 1

How many ounces of gold that costs $320 per ounce must a jeweler mix with 100 ounces of an alloy that costs $100 per ounce to produce a new alloy that costs $160 per ounce?

Solution   See page S9.   37.5 oz

■ Uniform Motion Problems

A train that travels constantly in a straight line at 50 miles per hour is in *uniform motion*. **Uniform motion** means that the speed and direction of the object do not change.

The solution of a uniform motion problem is based on the equation $d = rt$. In this equation, d is the distance traveled by the object, r is the speed of the object, and t is the time the object spent traveling.

For instance, suppose a homing pigeon, on its return home, flies at a speed of 50 miles per hour. The distance flown by the pigeon in 2 hours is calculated using the uniform motion equation.

$$d = rt$$
$$d = 50(2) \qquad • \ r = 50, t = 2$$
$$d = 100$$

The pigeon flew 100 miles in 2 hours.

Substituting 50 for r in the equation $d = rt$ produces the equation $d = 50t$. This equation gives the distance d traveled by the pigeon in a certain amount of time t. The table and the graph below show the relationship between the time flying and the distance flown. The variable X represents t, the time in hours, and the variable Y_1 represents d, the distance in miles. The ordered pair $(4, 200)$ means that in 4 hours, the pigeon traveled 200 miles.

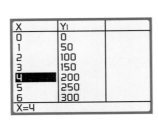

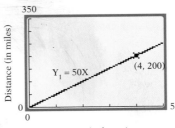

The graph above is called a **time–distance graph** and shows the relationship between the time traveled and the distance traveled. Time is on the horizontal axis, and distance is on the vertical axis.

▼ **Point of Interest**

Homing pigeons were first domesticated by the Egyptians around 5000 years ago.

? ANSWER   The $3 hamburger has more value per pound than the $1.80 hamburger, so as more $3 hamburger is added, the cost per pound of the mixture increases.

Suppose two brothers are going to race. Tom, who can run 7 meters per second, wants to give his younger brother Harry, who runs 5 meters per second, a 4-second head start in a 100-meter race. Let t represent the time Tom runs.

The distance Tom runs: $d = rt$

$$d = 7t \qquad \bullet \text{ Tom's rate is 7 meters per second.}$$

Harry had a 4-second head start, which means that he has been running 4 seconds longer than Tom. Therefore, the time that Harry has been running is $(t + 4)$ seconds.

The distance Harry runs: $d = rt$

$$d = 5(t + 4) \qquad \bullet \text{ Harry's rate is 5 meters per second.}$$
$$\text{Harry runs 4 seconds longer than Tom.}$$

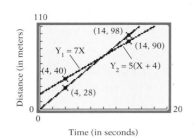

The time–distance graphs of the equations of the two brothers are shown at the left.

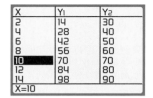

The graph shows that after Tom has been running for 4 seconds, he has traveled 28 meters and Harry has traveled 40 meters. The difference between the distances is $40 - 28 = 12$. This means that Harry is 12 meters ahead of Tom. You can also see this by looking at the table of values at the right.

The graph and the table show that Harry is ahead of Tom until Tom has been running for 10 seconds. At that time, Tom catches up to Harry and they have traveled the same distance, 70 meters. After 10 seconds, Tom is ahead of Harry. At the end of 14 seconds, Tom has traveled 98 meters and Harry has traveled 90 meters, so Tom is 8 meters ahead of Harry.

Another method of finding how long it takes Tom to catch up to his brother is to solve an equation. Again, let t be the time Tom is running. Use the fact that when Tom overtakes Harry, both have traveled the same distance.

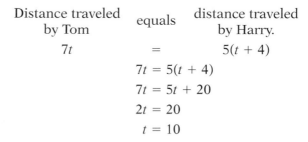

$$7t = 5(t + 4)$$
$$7t = 5t + 20$$
$$2t = 20$$
$$t = 10$$

Tom catches up to Harry in 10 seconds. This confirms the graphical approach discussed above.

? QUESTION   Suppose Car A is traveling 5 miles per hour faster than Car B. If the rate of Car B is represented by r, how can the rate of Car A be represented in terms of r?

? ANSWER $r + 5$

Suggested Activity

Tell students that you are going to show them the graphs of two runners on the same coordinate axis. Then graph $Y_1 = 6X$ and $Y_2 = 4X$ without showing students the equations for the graphs. Ask students to identity which runner has the faster speed. You can approach the answer to the question by moving the cursor to $X = 3$ and showing students that Y_1 is greater than Y_2. Thus the distance traveled by Y_1 is greater than the distance traveled by Y_2. Consequently, the rate of Y_1 is greater than the rate of Y_2. This activity will help prepare students for the concept of slope in the next chapter.

Suggested Activity

In preparation for Example 2, have students complete the following table.

Time to Monterey	Time back to Brown Field	Total Time
1	5	6
2	4	6
3	?	6
4	?	6
5	?	6
t	?	6

INSTRUCTOR NOTE
Exercise 28 can be used for a similar in-class example.

EXAMPLE 2

As part of flight training, a student pilot was required to fly from Brown Field to Monterey and then return. The average speed on the way to Monterey was 100 miles per hour, and the average speed returning was 150 miles per hour. Find the distance between the two airports if the total flying time was 6 hours.

State the goal.   The goal is to find the distance between the airports.

Devise a strategy.   If we can determine the time it took the pilot to fly to Monterey, then we can use the equation $d = rt$ to find the distance to that airport.

For instance, suppose it took 2 hours to fly to Monterey. Then the distance to the airport would be

Rate of the plane ———⌐   ⌐——— Time to the airport
$$d = 100(2) = 200$$
———— Distance to the airport

This suggests that we let t represent the time it takes the pilot to fly to Monterey.

The total time of the trip was 6 hours. Therefore, the time to return to Brown Field is the total time for the trip (6 hours) minus the time it took to fly to Monterey (t).

Time to return to Brown Field = 6 hours − the time flying to Monterey
$$= 6 - t$$

Use the equation $d = rt$ to write an equation for the distance traveled from Brown Field to Monterey and an equation for the distance traveled from Monterey to Brown Field.

Distance from Brown Field to Monterey: $d = rt$
$$d = 100t$$
Distance from Monterey to Brown Field: $d = rt$
$$d = 150(6 - t)$$

To assist in writing an equation, draw a diagram showing the distances traveled to and from Monterey.

$d = 100t$

Brown Field   Monterey

$d = 150(6 - t)$

Note that the distance the plane travels to Monterey is the same as the distance the plane travels returning from Monterey. **Translating this sentence into an equation, we have** $100t = 150(6 - t)$.

Solve the problem.

$100t = 150(6 - t)$ • The distance to Monterey equals the distance back
$100t = 900 - 150t$ to Brown.
$250t = 900$
   $t = 3.6$ • The time flying from Brown to Monterey was 3.6 hours.
$d = 100t$ • To find the distance between the airports, substitute 3.6
$d = 100(3.6)$ for t in the equation for the distance from Brown to
$d = 360$ Monterey.

The distance between the airports is 360 miles.

Check your work. It is important to note that the solution of the equation was 3.6 hours but that finding the answer to the question (What is the distance between the airports?) required substituting 3.6 into the equation $d = rt$.

As a check of your work, substitute 3.6 as the distance from Monterey to Brown.

$$d = 150(6 - t) = 150(6 - 3.6) = 150(2.4) = 360$$

This shows that the distance back is equal to the distance out, as it should be.

YOU TRY IT 2

Two cyclists, the second traveling 5 miles per hour faster than the first, start at the same time from the same point and travel in opposite directions. In 4 hours, they are 140 miles apart. Find the rate of each cyclist.

Solution   See page S9.   First cyclist: 15 mph; second cyclist: 20 mph

3.2 EXERCISES   Suggested Assignment: 11–37, odds

Topics for Discussion

1. Use the equation $V = AC$ to explain how to represent the value of x quarts of juice that costs $.85 per quart.
   $V = x(0.85) = 0.85x$

2. If a confectioner combines chocolate costing $5.50 per pound with caramel costing $4.00 per pound and then sells the candy at $6.00 per pound, would the confectioner make a profit or lose money?
   Make a profit

3. Suppose a merchant mixes apple juice costing $1.25 per quart with cranberry juice costing $2.00 per quart. Will the merchant always make a profit if the price of the mixture is $3.00 per quart? Will the merchant always make a profit if the price of the mixture is $1.50 per quart? Will the merchant ever make a profit if the price of the mixture is $1.00 per quart?

Yes. Sometimes; it depends on the ratio of the juices used. No.

4. Suppose you combined peanuts and raisins into a mixture that weighed a total of 5 pounds. Use one variable to express the number of pounds of peanuts and the number of pounds of raisins in the mixture.

Either x pounds of peanuts and $(5 - x)$ pounds of raisins, or x pounds of raisins and $(5 - x)$ pounds of peanuts.

5. Suppose a jogger starts on a 4-mile course. Two hours later a second jogger starts on the same course. If both joggers arrive at the finish line at the same time, which jogger is running faster?

The second jogger

6. A Boeing 757 airplane leaves San Diego, California, flying to Miami, Florida. One hour later, a Boeing 767 leaves San Diego taking the same route to Miami. Let t represents the time the Boeing 757 has been in the air. Use an expression involving t to represent the time the Boeing 767 has been in the air.

$(t - 1)$ h

7. If two objects started from the same point and are moving in opposite directions, how can the total distance between the two objects be expressed?

The total distance between the two objects =
the distance traveled by the first object + the distance traveled by the second object.

8. Two friends are standing 50 feet apart and begin walking toward each other on a straight sidewalk. When they meet, what is the total distance covered by the two friends?

50 ft

9. Suppose two planes are heading toward each other. One plane is traveling at 450 miles per hour, and the other plane is traveling at 350 miles per hour. What is the rate at which the distance between the planes is changing?

800 mph

■ Value Mixture Problems

10. *Food Mixtures*   A restaurant chef mixes 20 pounds of snow peas costing $1.99 per pound with 14 pounds of petite onions costing $1.19 per pound to make a vegetable medley for the evening meal. Find the cost per pound of the mixture.

$1.66

11. *Food Mixtures*   Find the cost per ounce of a salad dressing made from 64 ounces of olive oil that costs $.13 per ounce and 20 ounces of vinegar that costs $.09 per ounce. Round to the nearest cent.

$.12

12. *Food Mixtures* Forty pounds of cashews costing $5.60 per pound were mixed with 100 pounds of peanuts costing $1.89 per pound. Find the cost of the resulting mixture.
$2.95/lb

13. *Food Mixtures* A coffee merchant combined coffee costing $6 per pound with coffee costing $3.50 per pound. How many pounds of each were used to make 25 pounds of a blend costing $5.25 per pound?
17.5 lb of $6 coffee; 7.5 lb of $3.50 coffee

14. *Entertainment* Adult tickets for a play cost $5.00 and children's tickets cost $2.00. For one performance, 460 tickets were sold. Receipts for the performance were $1880. Find the number of adult tickets sold.
320 adult tickets

15. *Entertainment* Tickets for a school play sold for $2.50 for each adult and $1.00 for each child. The total receipts for 113 tickets sold were $221. Find the number of adult tickets sold.
72 adult tickets

16. *Food Mixtures* A breakfast cook mixes 5 liters of pure maple syrup that costs $9.50 per liter with imitation maple syrup that costs $4.00 per liter. How much imitation maple syrup is needed to make a mixture that costs $5.00 per liter?
22.5 L

17. *Food Mixtures* To make a flour mixture, a miller combined soybeans that cost $8.50 per bushel with wheat that cost $4.50 per bushel. How many bushels of each were used to make a mixture of 1000 bushels costing $5.50 per bushel?
250 bushels of soybeans; 750 bushels of wheat

18. *Metallurgy* A goldsmith combined pure gold that cost $400 per ounce with an alloy of gold that cost $150 per ounce. How many ounces of each were used to make 50 ounces of gold alloy costing $250 per ounce?
20 oz of pure gold; 30 oz of the alloy

19. *Metallurgy* A silversmith combined pure silver that cost $5.20 per ounce with 50 ounces of a silver alloy that cost $2.80 per ounce. How many ounces of the pure silver were used to make an alloy of silver costing $4.40 per ounce?
100 oz

20. *Food Mixtures* A tea mixture was made from 30 pounds of tea that cost $6.00 per pound and 70 pounds of tea that cost $3.20 per pound. Find the cost per pound of the tea mixture.
$4.04/lb

21. *Cosmetology* Find the cost per ounce of a face cream mixture made from 100 ounces of face cream that cost $3.46 per ounce and 60 ounces of face cream that cost $12.50 per ounce.
$6.85/oz

22. *Food Mixtures* A fruit stand owner combined cranberry juice that costs $4.20 per gallon with 50 gallons of apple juice that costs $2.10 per gallon. How much cranberry juice was used to make cranapple juice that costs $3.00 per gallon?
37.5 gal

23. *Food Mixtures*   Walnuts that cost $4.05 per kilogram were mixed with cashews that cost $7.25 per kilogram. How many kilograms of each were used to make a 50-kilogram mixture costing $6.25 per kilogram? Round to the nearest tenth.

15.6 kg of walnuts; 34.4 kg of cashews

■ Uniform Motion Problems

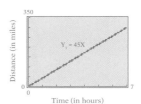

24. *Travel*   Write the equation for the distance a car traveling at a constant rate of 45 miles per hour travels in *t* hours. Use a graphing calculator to graph this equation with X as *t* and Y₁ as *d*. Use a domain of [0, 7] and a range of [0, 350].

25. *Transportation*   Write the equation for the distance a cyclist traveling at a constant rate of 12 miles per hour travels in *t* hours. Use a graphing calculator to graph this equation with X as *t* and Y₁ as *d*. Use a domain of [0, 5] and a range of [0, 60].

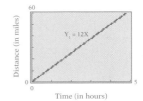

26. *Sports*   Jacob starts on a 16-mile hike on a path through a nature preserve. One hour later, Davadene starts on the same path, walking in the same direction as Jacob. The graphs of the two hikers are shown at the right, where *t* is the time that Davadene has been walking.
  a. After Davadene has been walking for 2 hours, has Jacob or Davadene traveled farther?
  b. Does Davadene ever pass Jacob on this hike? How can you tell?
  a. Davadene   b. Yes. Her graph intersects Jacob's graph.

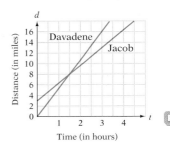

27. *Sports*   Imogene begins a 50-mile bicycle course. Two hours later, Alice starts on the same course, riding in the same direction as Imogene. The graphs of the two cyclists are shown at the right, where *t* is the time that Alice has been riding.
  a. After Alice has been riding for 4 hours, has Imogene or Alice traveled farther?
  b. Does Alice ever pass Imogene on this ride? How can you tell?
  a. Imogene   b. No. Her graph does not intersect Imogene's graph.

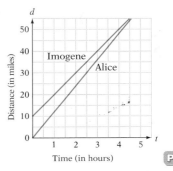

28. *Travel*   Two planes are 1380 miles apart and traveling toward each other. One plane is traveling 80 miles per hour faster than the other plane. The planes meet in 1.5 hours. Find the speed of each plane.

1st plane: 420 mph; 2nd plane: 500 mph

29. *Travel*   Two cars are 295 miles apart and traveling toward each other. One car travels 10 miles per hour faster than the other car. The cars meet in 2.5 hours. Find the speed of each car.

1st car: 54 mph; 2nd car: 64 mph

30. *Transportation*   A ferry leaves a harbor and travels to a resort island at an average speed of 18 miles per hour. On the return trip, the ferry travels at an average speed of only 12 miles per hour because of fog. The total time for the trip is 6 hours. How far is the island from the harbor?   43.2 mi

31. *Transportation* A commuter plane that provides transportation from an international airport to the surrounding cities averaged 210 miles per hour flying to a city and 140 miles per hour returning to the international airport. The total flying time was 4 hours. Find the distance between the two airports.   336 mi

32. *Travel* Two planes start from the same point and fly in opposite directions. The first plane is flying 50 miles per hour slower than the second plane. In 2.5 hours, the planes are 1400 miles apart. Find the rate of each plane.
1st plane: 255 mph; 2nd plane: 305 mph

33. *Sports* Two hikers start from the same point and hike in opposite directions around a lake whose shoreline is 13 miles long. One hiker walks 0.5 mile per hour faster than the other hiker. How fast did each hiker walk if they meet in 2 hours?
One hiker: 3 mph; other hiker: 3.5 mph

34. *Transportation* A student rode a bicycle to the repair shop and then walked home. The student averaged 14 miles per hour riding to the shop and 3.5 miles per hour walking home. The round trip took 1 hour. How far is it between the student's home and the bicycle shop?
2.8 mi

35. *Travel* A passenger train leaves a depot 1.5 hours after a freight train leaves the same depot. The passenger train is traveling 18 miles per hour faster than the freight train. Find the rate of each train if the passenger train overtakes the freight train 2.5 hours after leaving the depot.
Freight train: 30 mph; passenger train: 48 mph

36. *Travel* A plane leaves an airport at 3 P.M. At 4 P.M. another plane leaves the same airport traveling in the same direction at a speed 150 miles per hour faster than that of the first plane. Four hours after the first plane takes off, the second plane is 250 miles ahead of the first plane. How far did the second plane travel?
1050 mi

37. *Sports* A jogger and a cyclist set out at 9 A.M. from the same point headed in the same direction. The average speed of the cyclist is four times the average speed of the jogger. In 2 hours, the cyclist is 33 miles ahead of the jogger. How far did the cyclist ride?
44 mi

Applying Concepts

38. *Metallurgy* Find the cost per ounce of a mixture of 30 ounces of an alloy that costs $4.50 per ounce, 40 ounces of an alloy that costs $3.50 per ounce, and 30 ounces of an alloy that costs $3.00 per ounce.
$3.65

39. *Food Mixtures* A grocer combined walnuts that cost $2.60 per pound and cashews that cost $3.50 per pound with 20 pounds of peanuts that cost $2.00 per pound. Find the amount of walnuts and the amount of cashews used to make the 50-pound mixture costing $2.72 per pound.
Walnuts: 10 lb; cashews: 20 lb

40. *Food Mixtures*   A grocer creates a blend of two chocolate candies, chocolate mints and chocolate covered almonds. The grocer uses 30 pounds of chocolate mints costing $4.50 per pound and 20 pounds of chocolate covered almonds costing $6.00 per pound. At what price per pound should the grocer mark the blend to realize a $100 profit on the sale of the entire blend? $7.10 per pound

41. 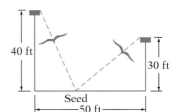 *Space Travel*   In 1999, astronomers confirmed the existence of planets orbiting a star other than the Sun. One of the planets, called Companion b, is approximately three-fourths the size of Jupiter. Companion c is approximately twice the size of Jupiter, and Companion d is four times the size of Jupiter. Each planet orbits the star Upsilon Andromedae, which is approximately 44 light years, or approximately 260 trillion (260,000,000,000,000) miles, from Earth. How many years (to the nearest hundred) after leaving Earth would it take a spacecraft traveling 18 million miles per hour to reach this star? (Note that 18 million miles per hour is about 1000 times faster than current spacecraft can travel.)
1600 years

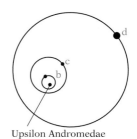

Upsilon Andromedae

The orbit of Companion c is approximately the same as the orbit of Earth around the Sun.

42. *Ancient Word Problem*   The following problem appears in a math text written around A.D. 1200. Two birds start flying from the tops of two towers 50 feet apart at the same time and at the same rate. One tower is 30 feet high, and the other tower is 40 feet high. The birds reach a grass seed on the ground at exactly the same time. How far is the grass seed from the 40-foot tower? *Note:* The solution requires use of the Pythagorean Theorem.
18 ft

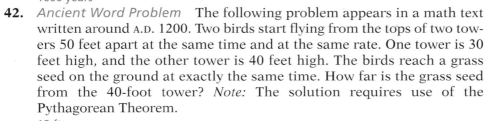

43. *Sports*   At 10 A.M., two campers left their campsite by canoe and paddled downstream at an average speed of 12 miles per hour. They then turned around and paddled back upstream at an average rate of 4 miles per hour. The total trip took 1 hour. At what time did the campers turn around downstream?
10:15 A.M.

EXPLORATION

1. *Challenging Uniform Motion Problems*
   a. If a parade 2 miles long is proceeding at 3 miles per hour, how long will it take a runner, jogging at 6 miles per hour, to travel from the end of the parade to the start of the parade?
   $\frac{2}{3}$ h

   b. Two cars are headed directly toward each other at rates of 40 miles per hour and 60 miles per hour. How many miles apart are they 2 minutes before impact?
   $3\frac{1}{3}$ mi

   c. A car travels a 1-mile track at an average speed of 30 miles per hour. At what average speed must the car travel the next mile so that the average speed for the 2 miles is 60 miles per hour?
   It is impossible to average 60 miles per hour.

d. Two horses, one mile apart, are running in a straight line toward each other. Each horse is running at a speed of 15 miles per hour. A bird, flying at 20 miles per hour, flies in a straight line back and forth from the nose of one horse to the nose of the other horse. How many miles will the bird fly before the horses reach one another?

$\frac{2}{3}$ mi

2. *Equations with No Solution*   Some equations have no solution. For instance, $x = x + 1$ has no solution. If we subtract x from each side of the equation, the result is $0 = 1$, which is not a true statement. One possible interpretation of this equation is "A number equals one more than itself." Because there is no number that is one more than itself, the equation has no solution. Now consider the equation $ax + b = cx + d$. Determine what conditions on a, b, c, and d will result in an equation with no solution.

$a = c, b \neq d$

3. *Modular Equations*   An equation of the form $ax + b \equiv c$ mod n, where a, b, c, and n are integers with $n > 1$, is one form of a *modular equation*. For instance $2x \equiv 3$ mod 5 is a modular equation ($a = 2$, $b = 0$, $c = 3$, and $n = 5$). These equations have many important applications. One use is to create codes so that sensitive or personal information can be transmitted electronically without someone other than the intended recipient being able to decode the message.

A solution of a modular equation is a number for which the expressions $ax + b$ and c have the same remainder when divided by n. For instance, 4 is a solution of $2x \equiv 3$ mod 5 because $\frac{2(4)}{5} = 1$ remainder 3 and $\frac{3}{5} = 0$ remainder 3. There are other solutions of $2x \equiv 3$ mod 5. For instance, 14 is a solution because $\frac{2(14)}{5} = 5$ remainder 3.

For Exercises a through d, find, by trial and error, only those solutions of the equation that are less than n.

a. $3x \equiv 2$ mod 4

2

b. $4x \equiv 1$ mod 3

1

c. $3x + 6 \equiv 1$ mod 11

2

d. $4x - 1 \equiv 3$ mod 5 (*Hint:* Try a decimal ending in five tenths.)

3.5

e. The modular equation $x^2 \equiv a$ mod p, where p is a prime number, is especially important in applications. Find the two solutions of $x^2 \equiv 9$ mod 11.

3 and 8

Applications to Geometry

■ Angles

A **ray** is part of a line that starts at a point, called the **endpoint** of the ray, but has no end. A ray is named by giving its endpoint and some other point on the ray. The ray below is named $\overrightarrow{PQ}$ or $\overrightarrow{PR}$.

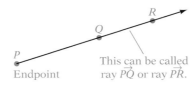

An **angle** is formed by two rays with a common endpoint. The endpoint is the **vertex** of the angle, and the rays are the **sides of the angle.** If A is a point on one ray of an angle, C is a point on the other ray, and B is the vertex, then the angle is called $\angle B$, $\angle ABC$, or $\angle CBA$, where $\angle$ is the symbol for angle. Note that the angle is named by the vertex, or when the angle is named by giving three points, the vertex is the second point listed.

TAKE NOTE

Although Greek letters are frequently used to designate an angle, other letters, such as x or y, can be used.

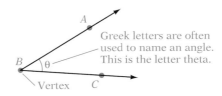

An angle can also be named by a letter inside the angle. The angle above can be named $\angle \theta$.

One unit of measure for an angle is **degrees** (°). One degree is equal in magnitude to $\frac{1}{360}$ of a complete revolution.

The measure of $\angle ABC$ is written $m\angle ABC$; the measure of $\angle B$ is written $m\angle B$.

A **right angle** has a measure of 90°. A right-angle symbol, ⌐, is frequently placed inside an angle to indicate a right angle.

Right angle

An angle whose measure is between 0° and 90° is called an **acute angle.** An angle whose measure is between 90° and 180° is called an **obtuse angle.**

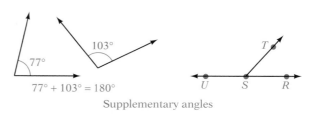

A **straight angle** has measure 180°.

Suggested Activity

See Section 3.3 of the *Student Activity Manual* for an investigation involving complementary and supplementary angles.

Two angles are **complements** of one another if the sum of the measures of the angles is 90°. The angles are called **complementary angles.** A 63° angle and a 27° angle are complementary angles. Angles *ABC* and *CBD* below are complementary angles.

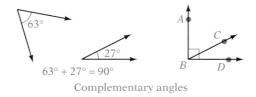

$$63° + 27° = 90°$$

Complementary angles

Two angles are **supplements** of one another if the sum of the measures of the angles is 180°. These angles are called **supplementary angles.** A 77° angle and a 103° angle are supplementary angles. Angles *RST* and *TSU* below are supplementary angles.

$$77° + 103° = 180°$$

Supplementary angles

INSTRUCTOR NOTE
Exercise 14 can be used for a similar in-class example.

EXAMPLE 1

One angle is 3° more than twice its supplement. Find the measure of each angle.

State the goal.   The goal is to find two supplementary angles such that one angle is 3° more than twice the other.

Devise a strategy.   Let x represent the measure of one angle.

The angles are supplements of one another. If the measure of one angle is x, then the measure of the supplementary angle is $180° - x$. Thus we have

Measure of one angle: x

Measure of the supplement: $180 - x$

Note that $x + (180 - x) = 180$, which shows that the angles x and $180 - x$ are supplementary angles.

Solve the problem. one angle = 3° more than twice its supplement

$$x = 2(180 - x) + 3$$
$$x = 360 - 2x + 3$$
$$x = 363 - 2x$$
$$3x = 363$$
$$x = 121$$

One angle is 121°.

To find the measure of the supplement, evaluate $180 - x$ when $x = 121$.

$$180 - x$$
$$180 - 121 = 59$$

The measure of the supplementary angle is 59°.

The measures of the two angles are 121° and 59°.

Check your work. Note that $121° + 59° = 180°$, so the two angles are supplements.

Also observe that $121 = 2(59) + 3$, so 121° is 3° more than twice 59°. This confirms that our solution is correct.

YOU TRY IT 1

One angle is 3° less than its complement. Find the two angles.

Solution See page S10. 43.5° and 46.5°

▼ *Point of Interest*

Many cities in the New World, unlike those in Europe, were designed using rectangular street grids. Washington, DC, was planned that way except that diagonal avenues were added to provide for quick troop movement in the event the city required defense. As an added precaution, monuments and statues were constructed at major intersections so that attackers would not have a straight shot down a boulevard.

■ Intersecting Lines

Four angles are formed by the intersection of two lines. If each of the four angles is a right angle, then the two lines are **perpendicular.** Line p is perpendicular to line q. This is written $p \perp q$, where $\perp$ is read "is perpendicular to."

If the two lines are not perpendicular, then two of the angles are acute angles and two of the angles are obtuse angles. The two acute angles are always opposite each other, and the two obtuse angles are always opposite each other. $\angle w$ and $\angle y$ are acute angles; $\angle x$ and $\angle z$ are obtuse angles.

Two angles that have the same vertex and share a common side are called **adjacent angles.** For the figure shown at the right, $\angle ABC$ and $\angle CBD$ are adjacent angles.

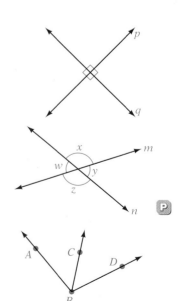

Suggested Activity

Suppose an obtuse angle of two intersecting lines is 135°. What are the measures of the three other angles? [Answer: 45°, 135°, 45°] Sketch a pair of nonperpendicular intersecting lines. Label one of the acute angles x. Label one of the obtuse angles $2x + 78°$. Find x. [Answer: 34°]

Adjacent angles of intersecting lines are supplementary angles. This is summarized by the following equations:

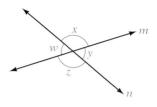

$$m\angle x + m\angle y = 180°$$
$$m\angle y + m\angle z = 180°$$
$$m\angle z + m\angle w = 180°$$
$$m\angle w + m\angle x = 180°$$

The nonadjacent angles formed when two lines intersect are called **vertical angles.** For the intersecting lines m and n above, $\angle x$ and $\angle z$ are vertical angles; $\angle w$ and $\angle y$ are also vertical angles. Vertical angles have the same measure. Thus,

$$m\angle x = m\angle z$$
$$m\angle w = m\angle y$$

INSTRUCTOR NOTE
Exercise 28 can be used for a similar in-class example.

EXAMPLE 2

In the figure at the right, what is the value of x?

State the goal. The goal is to find the value of x.

Devise a strategy. The labeled angles are vertical angles of intersecting lines. Therefore, the angles are equal.

Solve the problem.
$$3x - 50 = x + 50$$
$$2x - 50 = 50$$
$$2x = 100$$
$$x = 50$$

The value of x is 50.

Check your work. Replace x by 50 in the equation $3x - 50 = x + 50$ to ensure that the solution checks.

YOU TRY IT 2

The measures of two adjacent angles for a pair of intersecting lines are $2x + 20°$ and $3x + 50°$. Find the measure of the larger angle.

Solution See page S10. 116°

INSTRUCTOR NOTE
The definitions which follow are not dependent upon the lines being parallel.

Parallel lines never meet; the distance between them is always the same. The symbol for parallel lines is ∥.

A line that intersects two other lines at two distinct points is called a **transversal.** If the lines cut by a transversal are parallel lines and the transversal is perpendicular to the parallel lines, then all eight angles formed are right angles. For the diagram at the right, $p \parallel q$.

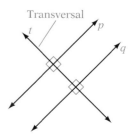

If the lines cut by a transversal are parallel lines and the transversal is not perpendicular to the parallel lines, then all four acute angles have the same measure and all four obtuse angles have the same measure. For the figure at the right:

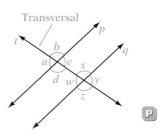

$$m\angle a = m\angle c = m\angle w = m\angle y$$
$$m\angle b = m\angle d = m\angle x = m\angle z$$

Alternate interior angles are two nonadjacent angles that are on opposite sides of the transversal and between the parallel lines. In the figure above, $\angle c$ and $\angle w$ are alternate interior angles; $\angle d$ and $\angle x$ are alternate interior angles. Alternate interior angles have the same measure.

Alternate interior angles have the same measure.

$$m\angle c = m\angle w$$
$$m\angle d = m\angle x$$

Alternate exterior angles are two nonadjacent angles that are on opposite sides of the transversal and outside the parallel lines. In the figure above, $\angle a$ and $\angle y$ are alternate exterior angles; $\angle b$ and $\angle z$ are alternate exterior angles. Alternate exterior angles have the same measure.

Alternate exterior angles have the same measure.

$$m\angle a = m\angle y$$
$$m\angle b = m\angle z$$

For two parallel lines cut by a transversal that is not perpendicular to the parallel lines, **corresponding angles** are two angles that are on the same side of the transversal and are both acute angles or are both obtuse angles. For the figure above, the following pairs of angles are corresponding angles: $\angle a$ and $\angle w$, $\angle d$ and $\angle z$, $\angle b$ and $\angle x$, $\angle c$ and $\angle y$. Corresponding angles have the same measure. (Note that if the transversal is perpendicular to the parallel lines, all corresponding angles are right angles.)

Corresponding angles have the same measure.

$$m\angle a = m\angle w$$
$$m\angle d = m\angle z$$
$$m\angle b = m\angle x$$
$$m\angle c = m\angle y$$

Suggested Activity

Sketch a pair of parallel lines l_1 and l_2 that are crossed by a transversal t. Label one of the exterior acute angles $x + 43°$. Label the exterior angle adjacent to the first angle $6x + 60°$. Find x.
[Answer: 11°]

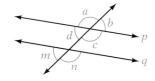

INSTRUCTOR NOTE
Exercise 34 can be used for a similar in-class example.

? QUESTION   In the figure at the left, $p \parallel q$. Which of the angles a, b, c, and d have the same measure as $\angle m$? Which angles have the same measure as $\angle n$?

EXAMPLE 3

Given that $p \parallel q$ in the figure at the right and that $m\angle 1 = x + 20°$ and $m\angle 2 = 3x$, find the value of x.

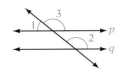

State the goal.   The goal is to find the value of x.

? ANSWER   The angles that have the same measure as angle $\angle m$ are $\angle b$ and $\angle d$. The angles that have the same measure as angle $\angle n$ are $\angle a$ and $\angle c$.

Devise a strategy. Because corresponding angles are equal, we can label ∠3 in the original diagram as $3x$.

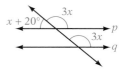

The sum of the measures of adjacent angles of intersecting lines is 180°. Therefore,

$$m\angle 3 + m\angle 1 = 180°$$

Solve the problem.
$$3x + (x + 20) = 180$$
$$4x + 20 = 180$$
$$4x = 160$$
$$x = 40$$

The value of x is 40.

Check your work. By replacing x by 40 in the equation $3x + (x + 20) = 180$, you can verify that the solution is correct.

YOU TRY IT 3

Given that $p \parallel q$, find the value of x in the diagram at the right.

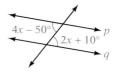

Solution   See page S10.   30°

■ Angles of a Triangle

If the lines cut by a transversal are not parallel lines, the three lines will intersect at three points. In the figure at the right, the transversal t intersects lines p and q. The three lines intersect at points A, B, and C. These three points define the three line segments $\overline{AB}$, $\overline{BC}$, and $\overline{AC}$. The plane figure formed by these line segments is a **triangle.**

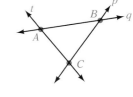

Each of the three points of intersection is the vertex of an angle of the triangle. The angles within the region enclosed by the triangle are called **interior angles.** In the figure at the right, ∠a, ∠b, and ∠c are interior angles. The sum of the measures of interior angles is 180°.

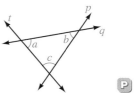

$$m\angle a + m\angle b + m\angle c = 180°$$

The Sum of the Measures of the Interior Angles of a Triangle

The sum of the measures of the interior angles of a triangle is 180°.

As an example of this, suppose the measures of two angles of a triangle are 25° and 47°. Let x be the measure of the third angle. Then

Suggested Activity

The measures of the angles of a triangle are consecutive integers. Find the measure of each angle.
[Answer: 59°, 60°, 61°]
Determine the measures of the angles of an equilateral triangle.
[Answer: 60°, 60°, 60°]

$$x + 25 + 47 = 180 \qquad \bullet \text{ The sum of the measures of the angles is } 180°.$$
$$x + 72 = 180 \qquad \bullet \text{ Add like terms.}$$
$$x = 108 \qquad \bullet \text{ Solve for } x.$$

The measure of the third angle is 108°.

An angle adjacent to an interior angle of a triangle is an **exterior angle** of the triangle. In the figure at the right, $\angle x$ and $\angle y$ are exterior angles for $\angle a$. The sum of the measures of an interior angle and the adjacent exterior angle of a triangle is 180°.

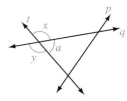

$$m\angle a + m\angle x = 180°$$
$$m\angle a + m\angle y = 180°$$

INSTRUCTOR NOTE
Exercise 38 can be used for a similar in-class example.

EXAMPLE 4

Given that $m\angle a = 110°$ and $m\angle c = 40°$ in the figure at the right, find the measure of $\angle b$.

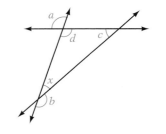

State the goal.   The goal is find the measure of $\angle b$.

Devise a strategy.   Note that $\angle b$ is an exterior angle for $\angle x$. Therefore, they are supplementary angles, and $m\angle b = 180° - m\angle x$. This means that we can find the measure of $\angle b$ by first finding the measure of $\angle x$.

$\angle d$ and $\angle a$ are vertical angles and therefore have the same measure.

That is, $m\angle d = m\angle a = 110°$.

We know the measure of $\angle d$ and the measure of $\angle c$. The measure of $\angle x$ can be found by using the fact that the sum of the interior angles of a triangle is 180°.

Solve the problem.
$$m\angle d + m\angle c + m\angle x = 180°$$
$$110° + 40° + m\angle x = 180°$$
$$150° + m\angle x = 180°$$
$$m\angle x = 30°$$
$$m\angle b = 180° - m\angle x = 180° - 30° = 150°$$

The measure of $\angle b$ is 150°.

Check your work.   Check over your work to be sure it is accurate.

YOU TRY IT 4

Given that $m\angle a = 112°$ in the figure at the right, find the $m\angle b$.

Solution   See page S10.   158°

■ Angles and Circles

The angles formed by rays that intersect circles have some special properties. In the diagram at the right, ∠*BOC* is a **central angle** because the vertex is at the center of the circle, *O*. The sides of the angle are radii of the circle. ∠*BAC* is an **inscribed angle** because its vertex is on the circumference of the circle and its sides are **chords,** or line segments whose endpoints lie on the circle.

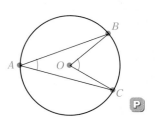

An **arc** is an unbroken part of a circle. In the diagram below, points *A* and *B* divide the circle into *minor arc* $\overset{\frown}{AB}$ and *major arc* $\overset{\frown}{ACB}$. Three points are always used to name a major arc. The **measure of an arc** is the measure of the central angle that intersects it. The measure of arc *AB* below, denoted as $m\overset{\frown}{AB}$, is 125°. Because a circle contains 360°, $m\overset{\frown}{ACB} = 360° - 125° = 235°$.

> **TAKE NOTE**
>
> If a chord is a diameter of a circle, then the central angle is a straight angle. Thus $m\overset{\frown}{ACB} = m\overset{\frown}{ADB} = 180°$.
>
>

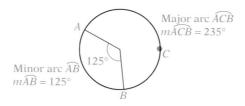

Major arc $\overset{\frown}{ACB}$
$m\overset{\frown}{ACB} = 235°$

Minor arc $\overset{\frown}{AB}$
$m\overset{\frown}{AB} = 125°$

? QUESTION   For the figure at the right, what is the measure of $\overset{\frown}{ABC}$?

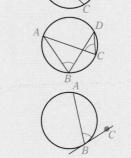

A **tangent** to a circle is a line that is in the same plane as the circle and is perpendicular to a radius of the circle at the point of intersection. See the diagram at the left.

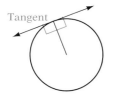

Tangent

The theorems below give some relationships between arcs and inscribed angles.

> **TAKE NOTE**
>
> In the second figure at the right, ∠*ABD* and ∠*ACD* are inscribed angles that intersect arc *AD*.

> **Inscribed-Angle Theorems**
>
> If ∠*ABC* is an inscribed angle of a circle, then $m\angle ABC = \frac{1}{2}(m\overset{\frown}{AC})$.
>
> Inscribed angles that intersect the same arc are equal. $m\angle ABD = m\angle ACD$
>
> The measure of an angle formed by a tangent and a chord is equal to one-half the measure of the intercepted arc. $m\angle ABC = \frac{1}{2}(m\overset{\frown}{AB})$

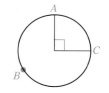

? ANSWER $m\overset{\frown}{ABC} = 360° - 90° = 270°$

INSTRUCTOR NOTE
Exercise 46 can be used for
a similar in-class example.

EXAMPLE 5

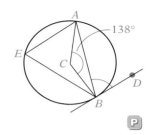

Find $m\angle ABD$ given that $\overline{BD}$ is tangent to the
circle at B and that $m\angle ACB = 138°$.

State the goal.   The goal is to find $m\angle ABD$.

Devise a strategy. $\angle ABD$ is formed by the
tangent $\overline{BD}$ and a chord of the circle. By the
theorem above, $m\angle ABD = \frac{1}{2}(m\widehat{AB})$. Thus we
can find $m\angle ABD$ by finding $m\widehat{AB}$.

Solve the problem.
$m\widehat{AB}$ = the measure of the central angle $ACB = 138°$
$m\angle ABD = \frac{1}{2}(m\widehat{AB}) = \frac{1}{2}(m\angle ACB) = \frac{1}{2}(138°) = 69°$

The measure of $\angle ABD = 69°$.

Check your work.   We can check the reasonableness of the answer by
noting that chord $\overline{AB}$ is not a diameter of the circle. Therefore, $m\angle ABD$
must be less than 90°.

YOU TRY IT 5

Use the diagram for Example 5, and find $m\angle AEB$.

Solution   See page S11.   69°

INSTRUCTOR NOTE
Exercise 50 can be used for
a similar in-class example.

EXAMPLE 6

If $\overline{AC}$ is a diameter of the circle at the right, show
that $\angle ABC$ is right angle.

State the goal.   The goal is to show that $\angle ABC$ is a right angle; that is,
that $m\angle ABC = 90°$.

Devise a strategy.   Because $\overline{AC}$ is a diameter, $m\widehat{ADC} = 180°$. From the
Inscribed-Angle Theorems, the measure of the inscribed angle ABC is
$\frac{1}{2}(m\widehat{ADC})$. Use this information to write and solve an equation.

Solve the problem. $m\angle ABC = \frac{1}{2}(m\widehat{ADC})$

$$= \frac{1}{2}(180°) = 90°$$

Because the measure of $\angle ABC$ is 90°, $\angle ABC$ is a right angle.

Check your work.   Be sure to check your work.

YOU TRY IT 6

Find the value of x in the figure at the right.

Solution See page S11. 50°

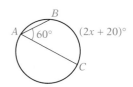

3.3 EXERCISES Suggested Assignment: 9–51, odds

Topics for Discussion

1. Describe each of the following: a right angle, an acute angle, an obtuse angle, a straight angle.

   A right angle is an angle whose measure is 90°. An acute angle is an angle whose measure is between 0° and 90°. An obtuse angle is an angle whose measure is between 90° and 180°. A straight angle is an angle whose measure is 180°.

2. What are complementary angles? What are supplementary angles?

   The sum of the measures of complementary angles is 90°. The sum of the measures of supplementary angles is 180°.

3. What are vertical angles?

   Vertical angles are nonadjacent angles formed by intersecting lines.

4. Draw a diagram with a transversal intersecting two parallel lines. Identify the corresponding angles, alternate interior angles, and alternate exterior angles.

   Answers will vary.

5. If a transversal cuts two lines p and q, and the alternate interior angles are not equal, what can be said about the lines p and q?

   They are not parallel.

6. What is a chord of a circle? Are all diameters chords of a circle? Are all chords diameters of a circle?

   A chord of a circle is a line segment whose endpoints are on the circle. Yes. No.

7. What is a central angle of a circle? How is the measure of a central angle related to the measure of the arc intercepted by the angle?

   A central angle is formed by two radii of the circle. The measure of the central angle is equal to the measure of the intercepted arc.

8. What is an inscribed angle of a circle? How is the measure of an inscribed angle related to the measure of the arc intercepted by the angle?

   An inscribed angle is an angle whose vertex is on the circumference of a circle and whose sides are chords of the circle. The measure of the inscribed angle is one-half the measure of the intercepted arc.

■ Angles

Solve.

9. Find the complement of a 43° angle.
   47°

10. Find the complement of a 53° angle.
    37°

11. Find the supplement of a 98° angle.
    82°

12. Find the supplement of a 33° angle.
    147°

13. Find two complementary angles such that one angle is 6 degrees more than twice the other.

28° and 62°

14. Find two complementary angles such that one angle is 15 degrees less than one-half the other.

20° and 70°

15. Find two supplementary angles such that one angle is 12 degrees less than three times the other.

48° and 132°

16. Find two supplementary angles such that one angle is 10 degrees less than two-thirds the other.

66° and 114°

Given that ∠ABC is a right angle, find the value of x.

17.

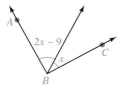

33°

18.

22.5°

19.

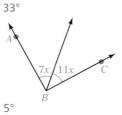

5°

20.

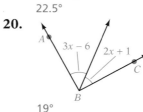

19°

Find the value of x.

21.

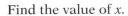

25°

22.

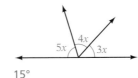

15°

23.

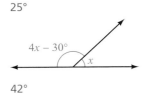

42°

24.

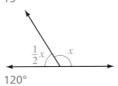

120°

Find the measure of ∠b.

25.

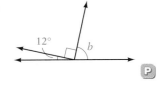

78°

26.

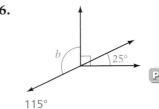

115°

■ Intersecting Lines

Find the value of *x*.

27.

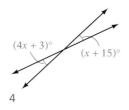

(4x + 3)°
(x + 15)°

4

28.

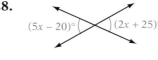

(5x − 20)° (2x + 25)°

15

Given that *p* ‖ *q*, find *m∠a* and *m∠b*.

29.

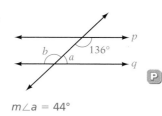

m∠a = 44°
m∠b = 136°

30.

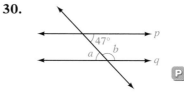

m∠a = 47°
m∠b = 133°

31.

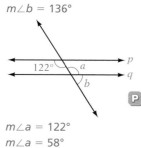

m∠a = 122°
m∠a = 58°

32.

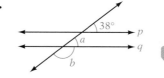

m∠b = 38°
m∠b = 142°

Given that *p* ‖ *q*, find the value of *x*.

33.

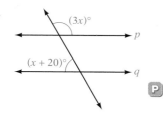

40

34.

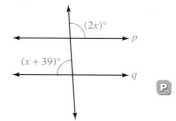

47

35.

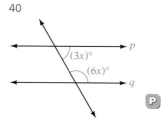

20

36.

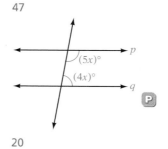

20

▪ Angles of a Triangle

37. Given that $m\angle a = 45°$ and $m\angle b = 100°$, find $m\angle x$ and $m\angle y$.

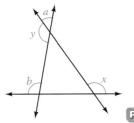

$m\angle x = 125°$
$m\angle y = 135°$

38. Given that $m\angle a = 80°$ and $m\angle b = 25°$, find $m\angle x$ and $m\angle y$.

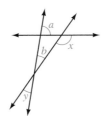

$m\angle x = 125°$
$m\angle y = 25°$

39. Given $m\angle a = 25°$, find $m\angle x$ and $m\angle y$.

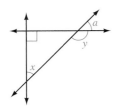

$m\angle x = 65°$
$m\angle y = 155°$

40. Given $m\angle a = 50°$, find $m\angle x$ and $m\angle y$.

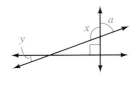

$m\angle x = 130°$
$m\angle y = 40°$

41. The measure of one of the acute angles of a right triangle is two degrees more than three times the measure of the other acute angle. Find the measure of each angle.
22° and 68°

42. The measure of the largest angle of a triangle is five times the measure of the smallest angle in the triangle. The measure of the third angle is three times the measure of the smallest angle. Find the measure of the largest angle.   100°

▪ Angles and Circles

Find the value of x.

43.

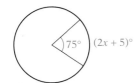

75° $(2x + 5)°$

35

44.

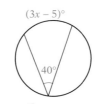

$(3x - 5)°$

40°

$28.\overline{3}$

45.

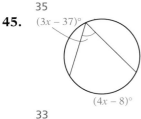

$(3x - 37)°$

$(4x - 8)°$

33

46.

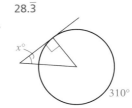

$x°$
y

310°

40

47.

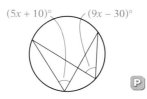

$(5x + 10)°$ $(9x − 30)°$

10

48.

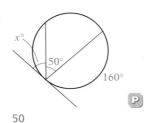

$x°$ $50°$ $160°$

50

49.

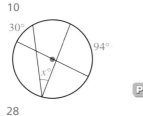

$30°$ $94°$ $x°$

28

50.

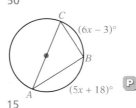

C $(6x − 3)°$ B A $(5x + 18)°$

15

51.

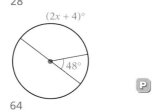

$(2x + 4)°$ $48°$

64

52.

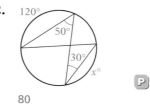

$120°$ $50°$ $30°$ $x°$

80

Applying Concepts

53. Cut out a paper triangle and then tear off two of the angles, as shown in the figure at the right. Position the pieces you tore off so that $\angle a$ is adjacent to $\angle b$ and $\angle c$ is adjacent to $\angle b$. Describe what you observe. What does this demonstrate?

The three angles form a straight angle. The sum of the measures of the interior angles of a triangle is 180°.

54. The measure of the supplement of the complement of $\angle a$ is 120°. What is the measure of $\angle a$?

30°

55. The measure of the complement of the supplement of $\angle a$ is 50°. What is the measure of $\angle a$?

140°

56. Determine whether the statement is always true, sometimes true, or never true.

a. Two lines that are parallel to a third line are parallel to each other.

b. A triangle contains two acute angles.

c. Vertical angles are complementary angles.

d. Adjacent angles are supplementary angles.

a. Always true **b.** Always true **c.** Sometimes true **d.** Sometimes true

EXPLORATION

1. 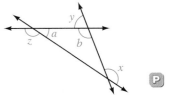 *Properties of Triangles*   For the figure at the right, explain why $m\angle a + m\angle b = m\angle x$. Write a rule that describes the relationship between an exterior angle of a triangle and the opposite interior angles.

   Let the third interior angle of the triangle be $\angle c$. Then $m\angle a + m\angle b + m\angle c = 180°$.
   $\angle c$ and $\angle x$ are adjacent angles of intersecting lines: $m\angle c + m\angle x = 180°$.
   Thus $m\angle a + m\angle b + m\angle c = m\angle c + m\angle x$.
   Subtract $m\angle c$ from each side: $m\angle a + m\angle b = m\angle x$.
   Rule: The measure of an exterior angle of a triangle is equal to the sum of the measures of the opposite interior angles of the triangle.

2. *Properties of Triangles*   For the figure at the right, find $m\angle x + m\angle y + m\angle z$.   360°

SECTION **3.4**

- The Addition and Multiplication Properties of Inequalities
- Compound Inequalities

Inequalities in One Variable

■ The Addition and Multiplication Properties of Inequalities

Recall that an equation contains an equals sign. An **inequality** contains the symbol $>$, $<$, $\geq$, or $\leq$. An inequality expresses the relative order of two mathematical expressions.

Here are some examples of inequalities in one variable:

$$\left.\begin{array}{l} 4x \geq 12 \\ 2x + 7 \leq 9 \\ x^2 + 1 > 3x \end{array}\right\} \quad \text{Inequalities in one variable}$$

A **solution of an inequality in one variable** is a number that, when substituted for the variable, results in a true inequality. For the inequality $x < 6$ shown below, 5, 0, and -4 are solutions of the inequality because replacing the variable by these numbers results in a true inequality.

$x < 6$		$x < 6$		$x < 6$	
$5 < 6$	True	$0 < 6$	True	$-4 < 6$	True

The number 7 is not a solution of the inequality $x < 6$ because $7 < 6$ is a false inequality.

Besides the numbers 5, 0, and -4, there are an infinite number of other solutions of the inequality $x < 6$. Any number less than 6 is a solution; for instance, -5.2, $\frac{5}{2}$, π, and 1 are also solutions of the inequality. The set of all the solutions of an inequality is called the **solution set of the inequality.** The solution set of the inequality $x < 6$ is written in set-builder notation as $\{x \mid x < 6\}$ and in interval notation as $(-\infty, 6)$.

Now consider the inequality $x - 1 < 4$. We can use a graphing calculator to visualize the solution set of this inequality by asking, "When is the graph of $y = x - 1$ less than 4?" To answer this question, graph $Y_1 = X - 1$ and $Y_2 = 4$. Use the INTERSECT feature to determine that the two lines intersect at $(5, 4)$. Observe that the graph of $Y_1 = X - 1$ is less than 4 (the graph is below 4) when $x < 5$.

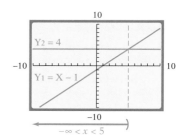

The graph of the solution set of $x - 1 < 4$ is usually displayed on a number line as shown at the right.

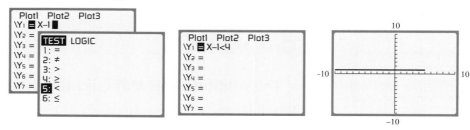

The solution set of the inequality $x - 1 < 4$ is $\{x \mid x < 5\}$ or $(-\infty, 5)$.

This solution set can be checked by using the TEST feature of a graphing calculator. Some typical screens follow.

In solving an inequality, the goal is to rewrite the given inequality in the form

$$variable < constant \quad \text{or} \quad variable > constant$$

The Addition Property of Inequalities is used to rewrite an inequality in this form.

> **The Addition Property of Inequalities**
>
> If $a > b$ and c is a real number, then the inequalities $a > b$ and $a + c > b + c$ have the same solution set.
>
> If $a < b$ and c is a real number, then the inequalities $a < b$ and $a + c < b + c$ have the same solution set.

The Addition Property of Inequalities states that the same number can be added to each side of an inequality without changing the solution set of the inequality. This property is also true for the symbols $\leq$ and $\geq$.

The Addition Property of Inequalities is used to remove a term from one side of an inequality by adding the additive inverse of that term to each side of the inequality. Because subtraction is defined in terms of addition, the same number can be subtracted from each side of an inequality without changing the solution set of the inequality.

INSTRUCTOR NOTE
It may be helpful for students to complete the activity for Section 1.2, Part B, in the *Student Activity Manual* before using a graphing calculator to check solutions to inequalities. The activity involves using a graphing calculator to create Boolean graphs of inequalities.

 See Appendix A: Test

 CALCULATOR NOTE
Note that the graphing calculator draws a horizontal line above the *x*-axis. The numbers below the line drawn are the elements of the solution set.

EXAMPLE 1

Solve and graph the solution set of $x + 5 > 3$. Write the solution set in set-builder notation and interval notation.

INSTRUCTOR NOTE
Exercise 8 can be used for a similar in-class example.

ALGEBRAIC SOLUTION

$x + 5 > 3$

$x + 5 - 5 > 3 - 5$ • Subtract 5 from each side of the inequality.

$x > -2$

The solution set is $\{x \mid x > -2\}$ or $(-2, \infty)$.

GRAPHICAL CHECK

Put the calculator in dot mode.

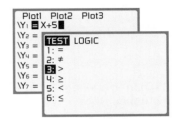

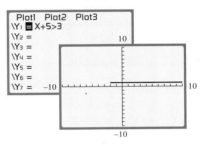

YOU TRY IT 1

Solve and graph the solution set of $x - 4 \le 1$. Write the solution set in set-builder notation and interval notation.

Solution   See page S11. $\{x \mid x \le 5\}$, $(-\infty, 5]$

As shown in the example that follows, the Addition Property of Inequalities applies to variable terms as well as to constants.

➡ Solve $3x - 4 \le 2x - 1$. Write the solution set in interval notation.

$3x - 4 \le 2x - 1$

$3x - 2x - 4 \le 2x - 2x - 1$ • Subtract $2x$ from each side of the inequality.

$x - 4 \le -1$ • Simplify.

$x - 4 + 4 \le -1 + 4$ • Add 4 to each side of the inequality.

$x \le 3$ • Simplify.

The solution set is $(-\infty, 3]$.

When we multiply or divide an inequality by a number, the inequality symbol may be reversed, depending on whether the number is positive or negative. Look at the following two examples.

$3 < 5$ • Multiply by positive 2. The inequality symbol remains the same.

$2(3) < 2(5)$

$6 < 10$ $6 < 10$ is a true statement.

$3 < 5$ • Multiply by negative 2. The inequality symbol is reversed in order to make the inequality a true statement.

$-2(3) > -2(5)$

$-6 > -10$

This is summarized in the Multiplication Property of Inequalities.

TAKE NOTE

$c > 0$ means c is a positive number. Note that an inequality symbol is not changed when multiplying by a positive number.

The Multiplication Property of Inequalities

Rule 1

If $a > b$ and $c > 0$, then the inequalities $a > b$ and $ac > bc$ have the same solution set.

If $a < b$ and $c > 0$, then the inequalities $a < b$ and $ac < bc$ have the same solution set.

Rule 2

If $a > b$ and $c < 0$, then the inequalities $a > b$ and $ac < bc$ have the same solution set.

If $a < b$ and $c < 0$, then the inequalities $a < b$ and $ac > bc$ have the same solution set.

TAKE NOTE

$c < 0$ means c is a negative number. Note that an inequality symbol is reversed when multiplying by a negative number.

Rule 1 states that when each side of an inequality is multiplied by a positive number, the inequality symbol remains the same. Rule 2 states that when each side of an inequality is multiplied by a negative number, the inequality symbol must be reversed.

Here are a few more examples of this property.

Rule 1		**Rule 2**	
$-4 < -2$	$5 > -3$	$1 < 7$	$-2 > -6$
$2(-4) < 2(-2)$	$3(5) > 3(-3)$	$-2(1) > -2(7)$	$-3(-2) < -3(-6)$
$-8 < -4$	$15 > -9$	$-2 > -14$	$6 < 18$

Use the Multiplication Property of Inequalities to remove a coefficient other than 1 from one side of an inequality so that the inequality can be written in the form *variable* $<$ *constant* or *variable* $>$ *constant*. The Multiplication Property of Inequalities is also true for the symbols $\leq$ and $\geq$.

Because division is defined in terms of multiplication, **when each side of an inequality is divided by a positive number, the inequality symbol remains the same. When each side of an inequality is divided by a negative number, the inequality symbol must be reversed.**

TAKE NOTE

Any time an inequality is multiplied or divided by a negative number, the inequality symbol must be reversed. Compare these two examples.

$2x < -4$ • Divide each side by *positive* 2. The inequality symbol *is not* reversed.
$\dfrac{2x}{2} < \dfrac{-4}{2}$
$x < -2$

$-2x < -4$ • Divide each side by *negative* 2. The inequality symbol *is* reversed.
$\dfrac{-2x}{-2} > \dfrac{-4}{-2}$
$x > 2$

➡ Solve $-3x < 9$. Write the solution set in interval notation.

$$-3x < 9$$

$$\dfrac{-3x}{-3} > \dfrac{9}{-3}$$ • Divide each side of the inequality by the coefficient -3 and reverse the inequality symbol.

$$x > -3$$ • Simplify.

$$(-3, \infty)$$ • Write the answer in interval notation.

The solution set is $(-3, \infty)$. ⬅

EXAMPLE 2

Solve and graph the solution set of $7x < -21$. Write the solution set in set-builder notation and interval notation.

INSTRUCTOR NOTE
Exercise 10 can be used for a similar in-class example.

ALGEBRAIC SOLUTION

$$7x < -21$$

$$\frac{7x}{7} < \frac{-21}{7}$$

- Divide each side of the inequality by 7. Because 7 is a positive number, the inequality symbol is not reversed.

$$x < -3$$

- Simplify.

The solution set is $\{x \mid x < -3\}$ or $(-\infty, -3)$.

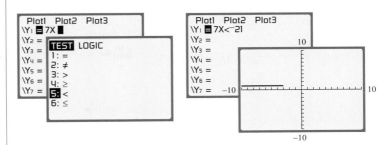

GRAPHICAL CHECK

YOU TRY IT 2

Solve and graph the solution set of $-3x \geq 6$. Write the solution set in set-builder notation and interval notation.

Solution   See page S11. $\{x \mid x \leq -2\}$, $(-\infty, -2]$

EXAMPLE 3

Solve $x + 3 > 4x + 6$. Write the solution set in set-builder notation.

INSTRUCTOR NOTE
Exercise 20 can be used for a similar in-class example.

ALGEBRAIC SOLUTION

$$x + 3 > 4x + 6$$

$$x - 4x + 3 > 4x - 4x + 6$$

- Subtract $4x$ from each side of the inequality.

$$-3x + 3 > 6$$

- Simplify.

$$-3x + 3 - 3 > 6 - 3$$

- Subtract 3 from each side of the inequality.

$$-3x > 3$$

- Simplify.

$$\frac{-3x}{-3} < \frac{3}{-3}$$

- Divide each side of the inequality by -3 and reverse the inequality symbol.

$$x < -1$$

- Simplify.

The solution set is $\{x \mid x < -1\}$.

- Write the solution set.

GRAPHICAL CHECK

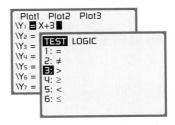

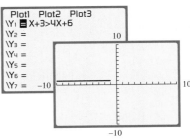

YOU TRY IT 3

Solve $3x - 1 \leq 5x - 7$. Write the solution set in set-builder notation.

Solution   See page S12. $\{x \mid x \geq 3\}$

When an inequality contains parentheses, often the first step in solving the inequality is to use the Distributive Property to remove the parentheses.

EXAMPLE 4

Solve $5(x - 2) \geq 9x - 3(2x - 4)$. Write the solution set in interval notation.

INSTRUCTOR NOTE
Exercise 32 can be used for a similar in-class example.

ALGEBRAIC SOLUTION

$5(x - 2) \geq 9x - 3(2x - 4)$

$5x - 10 \geq 9x - 6x + 12$
- Use the Distributive Property to remove parentheses.

$5x - 10 \geq 3x + 12$
- Combine like terms on the right side of the equation.

$2x - 10 \geq 12$
- Subtract $3x$ from each side of the equation.

$2x \geq 22$
- Add 10 to each side of the equation.

$x \geq 11$
- Divide each side of the equation by 2.

The solution set is $[11, \infty)$.
- Write the solution set.

GRAPHICAL CHECK

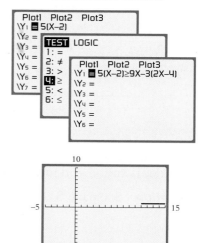

The graph is not visible in the standard viewing window. We had to choose a different window in order to display the graph of x ≥ 11.

YOU TRY IT 4

Solve $3 - 2(3x + 1) < 7 - 2x$. Write the solution set in interval notation.

Solution See page S12. $\left(-\frac{3}{2}, \infty\right)$

Suggested Activity

See Section 3.4 of the *Student Activity Manual* for an activity involving inequalities, their solution sets, and graphs of their solution sets.

Solving application problems requires recognition of the verbal phrases that translate into mathematical symbols. For instance, consider the sentence "To vote in an election for the president of the United States, a person must be *at least* 18 years old." This means the person must be 18 years old or older. If we let A represent a person's age, then this condition is represented mathematically as $A \geq 18$.

Now consider a plaque in a elevator that reads "The *maximum* allowable weight is 1050 pounds." This means that the total weight of all the occupants in the elevator must be 1050 pounds or less. If W represents the total weight of all the occupants, then this condition is expressed mathematically as $W \leq 1050$.

Suggested Activity

Have students translate the following sentences into inequalities.

1. x is at least 20.
2. The minimum value of y is 12.
3. w exceeds -3.
4. x is at most 7.
5. y is more than 15.
6. m is 17 or less.
7. The maximum value of n is 8.

[Answers:
1. $x \geq 20$
2. $y \geq 12$
3. $w > -3$
4. $x \leq 7$
5. $y > 15$
6. $m \leq 17$
7. $n \leq 8$]

INSTRUCTOR NOTE
Exercise 42 can be used for a similar in-class example.

There are additional words and phrases that translate into an inequality symbol. Here is a list of some the phrases used to indicate each of the four inequality symbols.

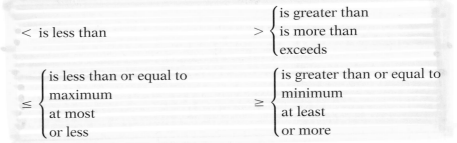

$<$ is less than	$>$	is greater than
		is more than
		exceeds

$\leq$	is less than or equal to	$\geq$	is greater than or equal to
	maximum		minimum
	at most		at least
	or less		or more

EXAMPLE 5

In 2002, the computer service America Online offered its customers the option of paying \$23.90 per month for unlimited use. Another option was a rate of \$4.95 per month with 3 free hours plus \$2.50 per hour thereafter (*Source:* AOL web site, March 2002). How many hours per month can you use this second option if it is to cost you less than the first option? Round to the nearest whole number.

State the goal.   The goal is to determine how many hours you can use the second option (\$4.95 per month plus \$2.50 per hour after the first 3 hours) if it is to cost you less than the first option (\$23.90 per month).

Devise a strategy.   Let x represent the number of hours per month you use the service. Then $x - 3$ represents the number of hours you would be paying \$2.50 per hour for service under the second option.

$$\text{Cost of first plan: } 23.90$$
$$\text{Cost of second plan: } 4.95 + 2.50(x - 3)$$

Write and solve an inequality that expresses that the second plan is less expensive (less than) the first plan.

Solve the problem.
$$4.95 + 2.50(x - 3) < 23.90$$
$$4.95 + 2.50x - 7.50 < 23.90$$
$$2.50x - 2.55 < 23.90$$
$$2.50x < 26.45$$
$$x < 10.58$$

The greatest whole number less than 10.58 is 10.

In order for the second option to cost you less than the first option, you can use the service for up to 10 hours per month.

Check your work.   One way to check the solution is to review the accuracy of our work. We can also make a partial check by ensuring that the answer makes sense. For instance, suppose you used the service for 11 hours per month. Then

Monthly cost = $4.95 + 2.50(11 - 3) = 4.95 + 2.50(8) = 4.95 + 20 = 24.95$

Because $24.95 > 23.90$, 11 is not in the solution set.

YOU TRY IT 5

The base of a triangle is 12 inches, and the height is $(x + 2)$ inches. Express as an integer the maximum height of the triangle when the area is less than 50 in².

Solution See page S12. 8 in.

■ Compound Inequalities

INSTRUCTOR NOTE
The compound inequalities in this section prepare the student to solve the absolute value inequalities that are given in the next section.

A **compound inequality** is formed by joining two inequalities with a connective word such as *and* or *or*. The inequalities shown below are compound inequalities.

$$2x - 1 \geq 4 \text{ and } x + 3 < 4$$
$$1 - 3x < 2 \text{ or } 5x - 7 > 3$$

The solution set of a compound inequality containing the word *and* is the intersection of the solution sets of the two inequalities.

EXAMPLE 6

Solve: $2x + 1 < 9$ and $2 - 3x < -4$

Write the solution set in set-builder notation.

INSTRUCTOR NOTE
Exercise 56 can be used for a similar in-class example.

ALGEBRAIC SOLUTION

Solve each inequality.

$$2x + 1 < 9 \qquad \text{and} \qquad 2 - 3x < -4$$
$$2x + 1 - 1 < 9 - 1 \qquad\qquad 2 - 2 - 3x < -4 - 2$$
$$2x < 8 \qquad\qquad -3x < -6$$
$$\frac{2x}{2} < \frac{8}{2} \qquad\qquad \frac{-3x}{-3} > \frac{-6}{-3}$$
$$x < 4 \qquad\qquad x > 2$$
$$\{x \mid x < 4\} \qquad\qquad \{x \mid x > 2\}$$

The solution of the compound inequality is the intersection of the solution sets of the individual inequalities.

$$\{x \mid x < 4\} \cap \{x \mid x > 2\} = \{x \mid 2 < x < 4\}$$

The solution set is $\{x \mid 2 < x < 4\}$.

GRAPHICAL CHECK

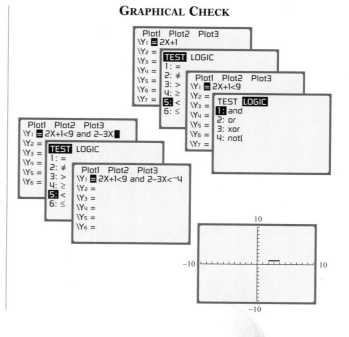

> **YOU TRY IT 6**
>
> Solve: $5x - 1 \geq -11$ and $4 - 6x > -14$.
>
> Write the solution set in set-builder notation.
>
> **Solution**   See page S12. $\{x \mid -2 \leq x < 3\}$

Some compound inequalities imply the use of the word *and* without actually stating it. This is illustrated in the following problem.

➡ Solve: $-3 \leq 2x + 1 < 7$
Write the solution set in set-builder notation.

This inequality is read "$2x + 1$ is greater than or equal to -3 and less than 7" and is equivalent to the compound inequality $-3 \leq 2x + 1$ and $2x + 1 < 7$.

Solve each inequality. Then find the intersection of the solution sets.

TAKE NOTE

The inequality $-2 \leq x$ is read, from right to left, "x is greater than or equal to -2," which is the inequality $x \geq -2$.

$$-3 \leq 2x + 1 \qquad \text{and} \qquad 2x + 1 < 7$$
$$-3 - 1 \leq 2x + 1 - 1 \qquad\qquad 2x + 1 - 1 < 7 - 1$$
$$-4 \leq 2x \qquad\qquad 2x < 6$$
$$\frac{-4}{2} \leq \frac{2x}{2} \qquad\qquad \frac{2x}{2} < \frac{6}{2}$$
$$-2 \leq x \qquad\qquad x < 3$$
$$\{x \mid x \geq -2\} \qquad\qquad \{x \mid x < 3\}$$

$$\{x \mid x \geq -2\} \cap \{x \mid x < 3\} = \{x \mid -2 \leq x < 3\}$$ • Find the intersection of the solution sets.

The solution set is $\{x \mid -2 \leq x < 3\}$.

The graph of the solution set is shown at the right.

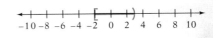

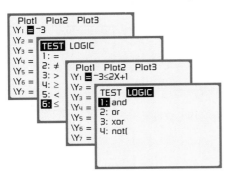
See Appendix A:
Test

Graphical Check:

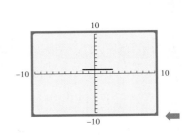

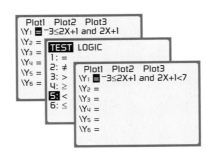

Here is an alternative method for solving the previous inequality.

➡ Solve: $-3 \le 2x + 1 < 7$

$$-3 \le 2x + 1 < 7$$

$$-3 - 1 \le 2x + 1 - 1 < 7 - 1$$ • Subtract 1 from each part of the inequality.

$$-4 \le 2x < 6$$

$$\frac{-4}{2} \le \frac{2x}{2} < \frac{6}{2}$$ • Divide each part of the inequality by 2.

$$-2 \le x < 3$$

The solution set is $\{x \mid -2 \le x < 3\}$.

The solution set of a compound inequality with the connective word *or* is the union of the solution sets of the two inequalities.

EXAMPLE 7

Solve and graph the solution set $2x + 3 > 7$ or $4x - 1 \le 3$. Write the solution set in interval notation.

ALGEBRAIC SOLUTION

Solve each inequality. Then find the union of the solution sets.

$$
\begin{array}{ccc}
2x + 3 > 7 & \text{or} & 4x - 1 \le 3 \\
2x + 3 - 3 > 7 - 3 & & 4x - 1 + 1 \le 3 + 1 \\
2x > 4 & & 4x \le 4 \\
\dfrac{2x}{2} > \dfrac{4}{2} & & \dfrac{4x}{4} \le \dfrac{4}{4} \\
x > 2 & & x \le 1 \\
(2, \infty) & & (-\infty, 1]
\end{array}
$$

The solution set is $(2, \infty) \cup (-\infty, 1]$. • The solution set is the union of the two intervals.

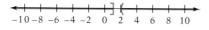

• Graph the solution set.

GRAPHICAL CHECK

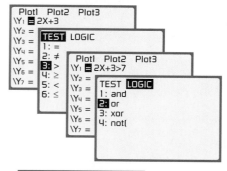

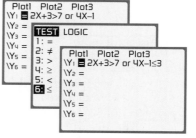

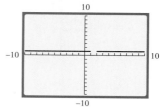

INSTRUCTOR NOTE
Exercise 62 can be used for a similar in-class example.

YOU TRY IT 7

Solve and graph the solution set $3 - 4x > 7$ or $4x + 5 > 9$

Write the solution set in interval notation.

Solution See page S12. $(-\infty, -1) \cup (1, \infty)$

Just as with equations, some inequalities may not have real number solutions. In this case the solution set is the empty set.

➡ Solve:

$$2x - 1 > 5 \quad \text{and} \quad 3x - 2 < 1$$
$$2x - 1 > 5 \quad \text{and} \quad 3x - 2 < 1$$
$$2x > 6 \qquad\qquad 3x < 3$$
$$x > 3 \qquad\qquad x < 1$$
$$\{x \mid x > 3\} \qquad\qquad \{x \mid x < 1\}$$
$$\{x \mid x > 3\} \cap \{x \mid x < 1\} = \varnothing \qquad\qquad ⬅$$

Another way of thinking about the solution of this inequality is to ask, "What number is greater than 3 and less than 1?" There is no such number, so the solution set is the empty set.

? QUESTION   Suppose the word *and* in the previous compound inequality is replaced with *or*. Does that change the solution set?

3.4 EXERCISES   Suggested Assignment: 7–69, odds

Topics for Discussion

1. How are the symbols), (,], and [used to distinguish the graphs of solution sets of inequalities?

) and (indicate that the endpoint of an interval is not included in the solution set;] and [indicate that the endpoint of an interval is included in the solution set.

2. State the Addition Property of Inequalities and give examples of its use.

   The Addition Property of Inequalities states that the same number can be added to each side of an inequality without changing the solution set of the inequality. Examples will vary.

3. State the Multiplication Property of Inequalities and give examples of its use.

   The Multiplication Property of Inequalities states that when each side of an inequality is multiplied by a positive number, the inequality symbol remains the same; when each side of an inequality is multiplied by a negative number, the inequality symbol must be reversed.

4. Which set operation is used when a compound inequality is combined with *or*? Which set operation is used when a compound inequality is combined with *and*?

   Union is used with *or*; intersection is used with *and*.

5. Explain why writing $-3 > x > 4$ does not make sense.

   A number cannot be less than -3 *and* greater than 4.

? ANSWER   Yes. The solution set is now the union of two sets, $\{x \mid x > 3\} \cup \{x \mid x < 1\}$. Another way of thinking about this is to note that we are looking for a number that is greater than 3 *or* less than 1.

■ The Addition and Multiplication Properties of Inequalities

Solve and graph the solution set. Write the solution set in set-builder notation.

6. $x + 1 < 3$
$\{x \mid x < 2\}$

7. $x - 5 > -2$
$\{x \mid x > 3\}$

8. $5 + n \geq 4$
$\{n \mid n \geq -1\}$

9. $-2 + n \geq 0$
$\{n \mid n \geq 2\}$

10. $8x \leq -24$
$\{x \mid x \leq -3\}$

11. $-4x < 8$
$\{x \mid x > -2\}$

12. $3n > 0$
$\{n \mid n > 0\}$

13. $-2n \leq -8$
$\{n \mid n \geq 4\}$

Solve. Write the answer in set-builder notation.

14. $x - 3 < 2$
$\{x \mid x < 5\}$

15. $4x \leq 8$
$\{x \mid x \leq 2\}$

16. $-2x > 8$
$\{x \mid x < -4\}$

17. $3x - 1 > 2x + 2$
$\{x \mid x > 3\}$

18. $2x - 1 > 7$
$\{x \mid x > 4\}$

19. $5x - 2 \leq 8$
$\{x \mid x \leq 2\}$

20. $6x + 3 > 4x - 1$
$\{x \mid x > -2\}$

21. $8x + 1 \geq 2x + 13$
$\{x \mid x \geq 2\}$

22. $4 - 3x < 10$
$\{x \mid x > -2\}$

23. $7 - 2x \geq 1$
$\{x \mid x \leq 3\}$

24. $-3 - 4x > -11$
$\{x \mid x < 2\}$

25. $4x - 2 < x - 11$
$\{x \mid x < -3\}$

Solve. Write the answer using interval notation.

26. $x + 7 \geq 4x - 8$
$(-\infty, 5]$

27. $3x + 2 \leq 7x + 4$
$\left[-\dfrac{1}{2}, \infty\right)$

28. $\dfrac{3}{5}x - 2 < \dfrac{3}{10} - x$
$\left(-\infty, \dfrac{23}{16}\right)$

29. $\dfrac{2}{3}x - \dfrac{3}{2} < \dfrac{7}{6} - \dfrac{1}{3}x$
$\left(-\infty, \dfrac{8}{3}\right)$

30. $\dfrac{1}{2}x - \dfrac{3}{4} < \dfrac{7}{4}x - 2$
$(1, \infty)$

31. $0.5x + 4 > 1.3x - 2.5$
$(-\infty, 8.125)$

32. $4(2x - 1) > 3x - 2(3x - 5)$
$\left(\dfrac{14}{11}, \infty\right)$

33. $2 - 5(x + 1) \geq 3(x - 1) - 8$
$(-\infty, 1]$

34. $3(4x + 3) \leq 7 - 4(x - 2)$
$\left(-\infty, \dfrac{3}{8}\right)$

35. $3 + 2(x + 5) \geq x + 5(x + 1) + 1$
$\left(-\infty, \dfrac{7}{4}\right]$

36. $3 - 4(x + 2) \leq 6 + 4(2x + 1)$
$\left[-\dfrac{5}{4}, \infty\right)$

37. $12 - 2(3x - 2) \geq 5x - 2(5 - x)$
$(-\infty, 2]$

38. *Mathematics*   Three-fifths of a number is greater than two-thirds. Find the least integer that satisfies the inequality.
2

39. *Income Tax*   One way a self-employed person can avoid a tax penalty is to pay at least 90% of her or his total annual income tax liability by April 15. What amount of income tax must be paid by April 15 by a person with an annual income tax liability of $3500?
$3150 or more

40. *Recycling*   A service organization will receive a bonus of $200 for collecting more than 1850 pounds of aluminum cans during its four collection drives. On the first three drives, the organization collected 505 pounds, 493 pounds, and 412 pounds. How many pounds of cans must the organization collect on the fourth drive to receive the bonus?
More than 440 lb

41. *Monthly Income*   A sales representative for a stereo store has the option of receiving either a monthly salary of $2000 or a 35% commission on the selling price of each item sold by the representative. What dollar amounts in sales will make the commission more attractive than the monthly salary?
More than $5714

42. *Monthly Income*   The sales agent for a jewelry company is offered a flat monthly salary of $3200 or a salary of $1000 plus an 11% commission on the selling price of each item sold by the agent. If the agent chooses the $3200 salary, what dollar amount does the agent expect to sell per month? $3200 < 1000 + \frac{11}{100}(x)$

32

$20,000 or less

43. *Telecommunications*   A computer bulletin board service charges either a flat fee of $10 per month or $4 per month plus $.10 for each minute the service is used. For how many minutes per month must a person use this service for the cost to exceed $10?
More than 60 min

44. *Food Mixtures*   For a product to be labeled orange juice, a state agency requires that at least 80% of the drink be real orange juice. How many ounces of artificial flavors can be added to 32 ounces of real orange juice if the drink is to be labeled orange juice?
8 or less ounces

45. *Transportation*   A shuttle service taking skiers to a ski area charges $8 per person each way. Four skiers are debating whether to take the shuttle bus or rent a car for $45 plus $.25 per mile. Assuming that the skiers will share the cost of the car and that they want the least expensive method of transportation, how far away is the ski area if they choose to take the shuttle service? $(8)(4)(2) < 45 + (0.25x)2$
$\frac{2(1?)}{}$
More than 38 mi

46. *Transportation*   Company A rents cars for $25 per day and $.08 per mile driven. Company B rents cars for $15 per day and $.14 per mile driven. You want to rent a car for one day. Find the maximum number of miles you can drive a Company B car if it is to cost you less than a Company A car. $\$25 + x(0.08) > \$15 + x(0.14)$
166 mi

47. *Telecommunications*   TopPage advertises local paging service for $6.95 per month for up to 400 pages, and $.10 per page thereafter. A competitor advertises service for $3.95 per month for up to 400 pages and $.15 per page thereafter. For what number of pages per month is the TopPage plan less expensive?
The TopPage plan is less expensive for more than 460 pages per month.

48. *Telecommunications*   During a weekday, to call a city 40 miles away from a certain pay phone costs $.70 for the first 3 minutes and $.15 for each additional minute. If you use a calling card, there is a $.35 fee and then the rates are $.196 for the first minute and $.126 for each additional minute. How long must a call last if it is to be cheaper to pay with coins rather than with a calling card?
A call must last 7 min or less.

49. *Temperature*   The temperature range for a week in a mountain town was between 0°C and 30°C. Find the temperature range in Fahrenheit degrees. The equation used to convert Fahrenheit to Celsius is $C = \frac{5(F - 32)}{9}$.
32°F to 86°F

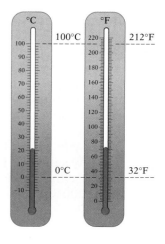

50. *Banking*   Heritage National Bank offers two different checking accounts. The first charges $3 per month and $.50 per check after the first 10 checks. The second account charges $8 per month with unlimited check writing. How many checks can be written per month if the first account is to be less expensive than the second account?
19 checks or less

51. *Banking*   Glendale Federal Bank offers a checking account to small businesses. The charge is $8 per month plus $.12 per check after the first 100 checks. A competitor is offering an account for $5 per month plus $.15 per check after the first 100 checks. If a business chooses the first account, how many checks does the business write each month if it is assumed that an account at the Glendale Federal Bank will cost less than the competitor's account?
more than 200 checks

52. *Education*   In a history class, an average score of 90 or above out of a possible 100 receives an A grade. You have grades of 95, 89, and 81 on three exams. Find the range of scores on the fourth exam that will give you an A for the course.
95 to 100

53. *Education*   An average of 70 to 79 in a mathematics class receives a C grade. A student has grades of 56, 91, 83, and 62 on four tests. If the maximum score on a fifth test is 100, find the range of scores on that test that will give the student a C for the course.
58 to 100

54. *Telecommunications*   The America's Choice℠ 550 plan offered by Verizon Wireless has a monthly rate of $55, which includes 550 minutes per month for calls. The rate for additional minutes over 550 is $.35 per minute. The America's Choice℠ plan has a monthly rate of $75 and includes 900 minutes per month for calls. The rate for additional minutes over 900 is $.35 per minute (*Source:* Verizon Wireless web site, April 2002). Assuming that a person is going to use at least 600 minutes per month but less than 900 minutes per month, how many additional minutes over 600 would the person have to use for the America's Choice℠ 550 plan to be less expensive? Round to the nearest whole number.
7 min or less

■ Compound Inequalities

Solve. Write the answer in set-builder notation.

55. $2x < 6$ or $x - 4 > 1$

$\{x | x < 3 \text{ or } x > 5\}$

56. $\dfrac{1}{2}x > -2$ and $5x < 10$

$\{x | -4 < x < 2\}$

57. $3x < -9$ and $x - 2 < 2$

$\{x | x < -3\}$

58. $7x < 14$ and $1 - x < 4$

$\{x | -3 < x < 2\}$

59. $6x - 2 < -14$ or $5x + 1 > 11$

$\{x | x < -2 \text{ or } x > 2\}$

60. $5 < 4x - 3 < 21$

$\{x | 2 < x < 6\}$

61. $3x - 5 > 10$ or $3x - 5 < -10$

$\left\{x \middle| x > 5 \text{ or } x < -\dfrac{5}{3}\right\}$

62. $6x - 2 < 5$ or $7x - 5 < 16$

$\{x | x < 3\}$

63. $5x + 12 \geq 2$ or $7x - 1 \leq 13$

$\{x | x \in \text{real numbers}\}$

64. $3 \leq 7x - 14 \leq 31$

$\left\{x \middle| \dfrac{17}{7} \leq x \leq \dfrac{45}{7}\right\}$

65. $6x + 5 < -1$ or $1 - 2x < 7$

$\{x | x \in \text{real numbers}\}$

66. $9 - x \geq 7$ and $9 - 2x < 3$

$\varnothing$

Applying Concepts

Use the roster method to list the positive integers that are solutions of the inequalities.

67. $7 - 2b \leq 15 - 5b$

$\{1, 2\}$

68. $13 - 8a \geq 2 - 6a$

$\{1, 2, 3, 4, 5\}$

69. $2(2c - 3) < 5(6 - c)$

$\{1, 2, 3\}$

70. $-6(2 - d) \geq 4(4d - 9)$

$\{1, 2\}$

Use the roster method to list the integers that are elements of the intersection of the solution sets of the two inequalities.

71. $5x - 12 \leq x + 8$
$3x - 4 \geq 2 + x$

$\{3, 4, 5\}$

72. $6x - 5 > 9x - 2$
$5x - 6 < 8x + 9$

$\{-4, -3, -2\}$

73. $4(x - 2) \leq 3x + 5$
$7(x - 3) \geq 5x - 1$

$\{10, 11, 12, 13\}$

74. $3(x + 2) < 2(x + 4)$
$4(x + 5) > 3(x + 6)$

$\{-1, 0, 1\}$

75. Determine whether the following statements are always true, sometimes true, or never true.
   a. If $a > b$, then $-a < -b$.
   b. If $a < b$ and $a \neq 0, b \neq 0$, then $\dfrac{1}{a} < \dfrac{1}{b}$.
   c. When dividing both sides of an inequality by an integer, we must reverse the inequality symbol.
   d. If $a < 1$, then $a^2 < a$.
   e. If $a < b < 0$ and $c < d < 0$, then $ac > bd$.

   a. Always true   b. Sometimes true   c. Sometimes true   d. Sometimes true
   e. Always true

76. Determine whether the following statements are always true, sometimes true, or never true. If a statement is sometimes true, find conditions that will make it always true.
  a. If $ax < bx$, then $a < b$.
  b. If $a < b$, then $a^2 < b^2$.
  c. If $a < b$, then $ax^2 < bx^2$.
  d. If $a < b$ and $a \neq 0$, $b \neq 0$, then $\frac{1}{a} < \frac{1}{b}$.

  e. If $a > b > 0$, then $\frac{1}{a} < \frac{1}{b}$.

  a. Sometimes true. $x > 0$ will make the statement always true.   **b.** Sometimes true. $0 < a < b$ will make the statement always true.   **c.** Always true   **d.** Sometimes true. $a < 0$, $b > 0$ will make the statement always true.   **e.** Always true.

EXPLORATION

1. *Measurements as Approximations*   Recall the rules for rounding a decimal, which are given at the right. Given these rules, some possible values of the number 2.7 that was rounded to the nearest tenth are 2.73, 2.68, 2.65, and 2.749. If V represents the exact value of 2.7 before it was rounded, then the inequality $2.65 \leq V < 2.75$ represents all possible values of 2.7 before it was rounded. This is read "V is greater than or equal to 2.65 and less than 2.75."

Now suppose a rectangle is measured as 3.4 meters by 4.8 meters, each measurement rounded to the nearest tenth of a meter. By using the least and greatest possible values of each measurement, we can find the possible values of the area, A.

$$3.35(4.75) \leq A < 3.45(4.85)$$
$$15.9125 \leq A < 16.7325$$

The area is greater than or equal to 15.9125 square meters and less than 16.7325 square meters.

  a. Suppose the length of a line is measured as 4.2 inches, rounded to the nearest tenth of an inch. Write an inequality that represents the possible lengths of the line.
    $4.15 \leq L < 4.25$, where L represents possible lengths
  b. The length of the side of a square was given as 6.4 centimeters, rounded to the nearest tenth of a centimeter. Write an inequality that represents the possible areas of the square.
    $40.3225 \leq A < 41.6025$, where A represents possible areas
  c. The base of a triangle was measured as 5.43 meters and the height as 2.47 meters, each measurement rounded to the nearest hundredth of a meter. Write an inequality that represents the possible areas of the triangle.
    $6.6863125 \leq A < 6.7258125$, where A represents possible areas
  d. A rectangle is measured as 3.0 meters by 4.0 meters, each measurement rounded to the nearest tenth of a meter. Write an inequality that represents the possible areas of the rectangle.
    $11.6525 \leq A < 12.3525$, where A represents possible areas

> **TAKE NOTE**
>
> If the digit to the right of the given place value is less than 5, that digit and all the digits to the right are dropped. For example, 2.73 rounded to the nearest tenth is 2.7.
>   If the digit to the right of the given place value is greater than or equal to 5, increase the number in the given place value by 1 and drop the remaining digits. For example, 2.65 rounded to the nearest tenth is 2.7.

2. *Inequalities and Logic*   Logical operators and inequalities are such an important part of mathematics that these functions are built into graphing calculators. Typical graphing calculator screens follow.

 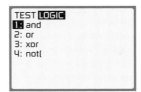

A graphing calculator will display a 1 when a statement is true and a 0 when the statement is false. For instance, the inequality $x > 2$ is true when $x = 3$ but false when $x = -1$. In these cases, the calculator assigns a *value* to the expression. The value of the expression $x > 2$ is 1 when $x = 3$; the value of $x > 2$ is 0 when $x = -1$.

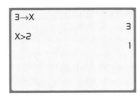 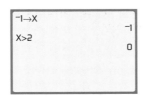

As shown at the right, a graphing calculator can be used to check the answer to a compound inequality. Enter 25 < 50 and 25 > 10. The calculator prints a 1 to the screen to mean that the statement is true. If we enter 30 > 50 or 20 < 10, a 0 is printed to the screen.

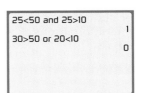

a. What is the value of $2x - 3 < 7$ when $x = 6$?

0

b. What is the value of $2x - 3 < 7$ when $x = 2$?

1

c. For what values of x is the value of $2x - 3 < 7$ equal to 1?

$x < 5$

d. What is the value of $4 - 3x \geq 13$ when $x = -4$?

1

e. What is the value of $4 - 3x \geq 13$ when $x = 2$?

0

f. For what values of x is the value of $4 - 3x \geq 13$ equal to 1?

$x \leq -3$

g. What is the value of the compound inequality $2x + 1 > 5$ and $3x - 2 < 13$ when $x = 3$?

1

h. What is the value of the compound inequality $2x + 1 > 5$ and $3x - 2 < 13$ when $x = 7$?

0

i. What is the value of the compound inequality $2x + 1 > 5$ and $3x - 2 < 13$ when $x = 1$?

0

j. For what values of x is the value of the compound inequality $2x + 1 > 5$ and $3x - 2 < 13$ equal to 1?

$2 < x < 5$

k. For what values of x is the value of the compound inequality $2x - 5 < 1$ or $4x + 1 > 21$ equal to 1?

$x < 3$ or $x > 5$

S E C T I O N **3.5**   **Absolute Value Equations and Inequalities**

- Absolute Value Equations
- Absolute Value Inequalities

Absolute Value Equations

The **absolute value** of a number is its distance from zero on the number line. Distance is always a positive number or zero. Therefore, the absolute value of a number is always a positive number or zero.

> **TAKE NOTE**
>
> Recall that the symbol for absolute value is $| \ |$.

The distance from 0 to 6 or from 0 to -6 is 6 units.

$$|6| = 6 \text{ and } |-6| = 6$$

An equation containing an absolute value symbol is called an **absolute value equation.** Examples of absolute value equations are shown at the right.

$$\left.\begin{array}{l} |x| = 3 \\ |x + 2| = 8 \\ |3x - 4| = 5x - 9 \end{array}\right\} \text{ Absolute value equations}$$

The solution of an absolute value equation is based on the following rule.

> **Solutions of Absolute Value Equations**
>
> If $a > 0$ and $|x| = a$, then $x = a$ or $x = -a$. If $a = 0$ and $|x| = a$, then $x = 0$.

> **TAKE NOTE**
>
> We require $a > 0$ because the absolute value of a number other than zero is positive. If $a < 0$, then $|x| = a$ has no solution.

For instance, if $|x| = 6$, then $x = 6$ or $x = -6$.

$$\begin{array}{cc} |x| = 6 & |x| = 6 \\ |6| = 6 \text{ True} & |-6| = 6 \text{ True} \end{array}$$

? QUESTION **a.** What are the solutions of the equation $|x| = 12$?

b. What are the solutions of the equation $|x| = -3$?

? ANSWERS **a.** 12 and -12 **b.** There are no solutions of the equation because the absolute value of a number is positive or zero.

TAKE NOTE

Be sure to check your solutions.

$\lvert x+3 \rvert = 7$		$\lvert x+3 \rvert = 7$	
$\lvert -10+3 \rvert$	7	$\lvert 4+3 \rvert$	7
$\lvert -7 \rvert$	7	$\lvert 7 \rvert$	7
7	= 7	7	= 7

The solutions check.

⇒ Solve: $\lvert x+3 \rvert = 7$

$$\lvert x+3 \rvert = 7$$

• If $x+3$ is positive, then $x+3 = 7$.
  If $x+3$ is negative, then $x+3 = -7$.

$$x+3 = 7 \qquad x+3 = -7$$
$$x = 4 \qquad x = -10$$

• Remove the absolute value sign and rewrite as two equations. Then solve each equation.

The solutions are -10 and 4.

INSTRUCTOR NOTE
Exercise 28 can be used for a similar in-class example.

EXAMPLE 1

Solve.   **a.** $\lvert 3x-5 \rvert = 10$ **b.** $6 - \lvert 1-4x \rvert = 1$

Solution   a.
$$\lvert 3x-5 \rvert = 10$$
$$3x-5 = 10 \qquad 3x-5 = -10$$
$$3x = 15 \qquad 3x = -5$$
$$x = 5 \qquad x = -\frac{5}{3}$$

• Remove the absolute value sign and rewrite as two equations.

The solutions are $-\dfrac{5}{3}$ and 5.

b.
$$6 - \lvert 1-4x \rvert = 1$$

• First solve for the absolute value expression.

$$-\lvert 1-4x \rvert = -5$$

• Subtract 6 from each side of the equation.

$$\lvert 1-4x \rvert = 5$$

• Multiply each side of the equation by -1.

$$1-4x = 5 \qquad 1-4x = -5$$

• Remove the absolute value sign and rewrite as two equations.

$$-4x = 4 \qquad -4x = -6$$
$$x = -1 \qquad x = \frac{3}{2}$$

The solutions are -1 and $\dfrac{3}{2}$.

Suggested Activity

Use absolute value to express the fact that there are two numbers 5 units from -2.
[Answer: $\lvert x+2 \rvert = 5$]
Use absolute value to express the fact that there are two numbers 8 units from 4.
[Answer: $\lvert x-4 \rvert = 8$]

YOU TRY IT 1

Solve.   **a.** $\lvert 5-6x \rvert = 1$ **b.** $\lvert 3x-7 \rvert + 4 = 2$

Solution   See page S12.   **a.** $\dfrac{2}{3}$ and 1   **b.** No solution

■ Absolute Value Inequalities

Recall that absolute value represents the distance between two points. For example, the solutions of the absolute value equation $|x - 3| = 5$ are the numbers whose distance from 3 is 5 units. Therefore, the solutions are -2 and 8.

The solutions of the absolute value inequality $|x - 3| < 5$ are the numbers whose distance from 3 is less than 5 units. Therefore, the solutions are the numbers greater than -2 and less than 8. The solution set is $\{x | -2 < x < 8\}$.

Another way to visualize the solution set of $|x - 3| < 5$ is to graph $Y_1 = \text{abs}(X - 3)$ and then draw the line $Y_2 = 5$. Using the INTERSECT feature of the calculator, we find that the graph of $Y_1 = \text{abs}(X - 3)$ intersects the graph of $Y_2 = 5$ at $(-2, 5)$ and $(8, 5)$. Observe that the graph of $Y_1 = \text{abs}(X - 3)$ is less than 5 (the graph is below 5) when $-2 < X < 8$.

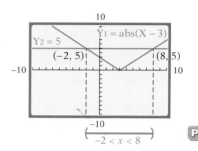

> ### Absolute Value Inequalities of the Form $|ax + b| < c$
>
> To solve an absolute value inequality of the form $|ax + b| < c$, $c > 0$, solve the equivalent compound inequality $-c < ax + b < c$.

For instance, if $|x| < 6$, then $-6 < x < 6$. The absolute value of all the numbers between -6 and 6 is less than 6.

? <u>QUESTION</u>   **a.** What are the solutions of $|x| < 12$?

    b. What are the solutions of $|x| < -3$?

In the previous statement on absolute value inequalities of the form $|ax + b| < c$, it is given that $c > 0$. If c is less than zero, then an absolute value inequality has no solution, as in part b of the Question.

INSTRUCTOR NOTE
Students can also check the solution to Example 2 on the next page by creating a Boolean graph on a graphing calculator. However, they may have difficulty reading $-\dfrac{7}{3}$ from the graph.

Also, the graphical check shown at the right of the algebraic solution may give them a greater understanding of absolute value inequalities because it is more visual.

***

? <u>ANSWER</u>   **a.** $-12 < x < 12$ **b.** There are no solutions because there are no numbers whose absolute value are less than -3; the absolute value of a number is positive or zero.

EXAMPLE 2

Solve $|3x + 2| < 5$. Write the solution set in set-builder notation.

INSTRUCTOR NOTE
Exercise 56 can be used for
a similar in-class example.

ALGEBRAIC SOLUTION

$$|3x + 2| < 5$$

$$-5 < 3x + 2 < 5$$ • Write an equivalent
inequality.

$$-5 - 2 < 3x + 2 - 2 < 5 - 2$$ • Subtract 2 from each part
of the inequality.

$$-7 < 3x < 3$$ • Simplify.

$$\frac{-7}{3} < \frac{3x}{3} < \frac{3}{3}$$ • Divide each part of the
inequality by 3.

$$-\frac{7}{3} < x < 1$$ • Simplify.

The solution set is $\left\{ x \mid -\dfrac{7}{3} < x < 1 \right\}$.

GRAPHICAL CHECK

Graph $Y_1 = \text{abs}(3X + 2)$ and $Y_2 = 5$.
Use the INTERSECT feature of the calculator to find the intersections of the two equations: $\left(-\frac{7}{3}, 5\right)$ and $(1, 5)$. Observe that the graph of $Y_1 = \text{abs}(3X + 2)$ is less than 5 (the graph is below 5) when $-\frac{7}{3} < X < 1$.

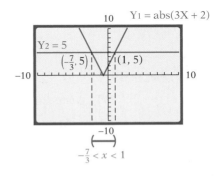

YOU TRY IT 2

Solve $|2x - 5| \le 7$. Write the solution set in set-builder notation.

Solution   See page S13. $\{x \mid -1 \le x \le 6\}$

The solutions of the absolute value inequality $|x + 2| > 5$ are the numbers whose distance from -2 is greater than 5 units. Therefore, the solutions are the numbers less than -7 *or* greater than 3. Because the word *or* is used, the solution set is the union of $\{x \mid x < -7\}$ and $\{x \mid x > 3\}$. The solution set is

$$\{x \mid x < -7\} \cup \{x \mid x > 3\} = \{x \mid x < -7 \text{ or } x > 3\}$$

Another way to visualize the solution set of $|x + 2| > 5$ is to graph $Y_1 = \text{abs}(X + 2)$ and $Y_2 = 5$. Using the INTERSECT feature of the calculator, we find that the graph of $Y_1 = \text{abs}(X + 2)$ intersects the graph of $Y_2 = 5$ at $(-7, 5)$ and $(3, 5)$. Observe that the graph of $Y_1 = \text{abs}(X + 2)$ is greater than 5 (the graph is above 5) when $X < -7$ or when $X > 3$.

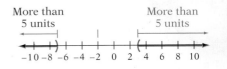

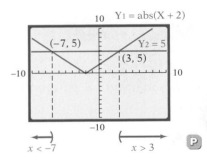

TAKE NOTE

Closely note the difference between solving an absolute value inequality of the form $|ax + b| < c$, discussed previously, and one of the form $|ax + b| > c$, discussed here.

Absolute Value Inequalities of the Form $|ax + b| > c$

To solve an absolute value inequality of the form $|ax + b| > c$, solve the equivalent compound inequality $ax + b > c$ or $ax + b < -c$.

Suggested Activity

See Section 3.5 of the *Student Activity Manual* for an activity involving inequalities and absolute value.

For instance, if $|x| > 6$, then $x < -6$ or $x > 6$.

? QUESTION   **a.** What are the solutions of $|x| > 12$?

b. What are the solutions of $|x| > -3$?

INSTRUCTOR NOTE
Exercise 58 can be used for a similar in-class example.

EXAMPLE 3

Solve $|1 - 2x| > 7$. Write the solution set in set-builder notation.

ALGEBRAIC SOLUTION

$$|1 - 2x| > 7$$

$1 - 2x > 7$	or	$1 - 2x < -7$	• Write an equivalent inequality.
$1 - 1 - 2x > 7 - 1$		$1 - 1 - 2x < -7 - 1$	• Subtract 1 from each side.
$-2x > 6$		$-2x < -8$	
$\dfrac{-2x}{-2} < \dfrac{6}{-2}$		$\dfrac{-2x}{-2} > \dfrac{-8}{-2}$	• Divide each side by -2. Because we are dividing by -2, the inequality symbol must be reversed.
$x < -3$		$x > 4$	
$\{x \mid x < -3\}$	or	$\{x \mid x > 4\}$	

The solution set is the union of the solution sets of the two inequalities.

$$\{x \mid x < -3\} \cup \{x \mid x > 4\} = \{x \mid x < -3 \text{ or } x > 4\}$$

The solution set is $\{x \mid x < -3 \text{ or } x > 4\}$.

GRAPHICAL CHECK

Graph $Y_1 = \text{abs}(1 - 2X)$ and $Y_2 = 7$. Use the INTERSECT feature of the calculator to find the intersections of the two equations: $(-3, 7)$ and $(4, 7)$. Observe that the graph of $Y_1 = \text{abs}(1 - 2X)$ is greater than 7 (the graph is above 7) when $X < -3$ or $X > 4$.

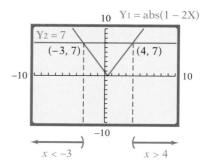

YOU TRY IT 3

Solve $|5x + 4| \geq 16$. Write the solution set in set-builder notation.

Solution   See page S13. $\left\{ x \mid x \leq -4 \text{ or } x \geq \dfrac{12}{5} \right\}$

? ANSWERS   a. $x < -12$ or $x > 12$ **b.** The numbers greater than -3 include 0 and all the positive numbers. The absolute value of any number is positive or zero. Therefore, the solutions are all the real numbers.

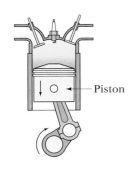

Piston

Suggested Activity

Have students write absolute value inequalities from data you give them. Here are a couple of examples.

a. An adult's normal body temperature is within 1°F of 98.6°F.

   [Answer: $|t - 98.6| \leq 1$]

b. The net weight of a cereal box labeled "20 oz" must contain within 0.45 oz of 20 oz of cereal.

   [Answer: $|c - 20| \leq 0.45$]

INSTRUCTOR NOTE
Exercise 72 can be used for a similar in-class example

The **tolerance** of a component, or part, is the acceptable amount by which the component may vary from a given measurement. For example, the diameter of a piston may vary from the given measurement of 9 centimeters by 0.001 centimeter. This is written as 9 centimeters ± 0.001 centimeter, which is read "9 centimeters plus or minus 0.001 centimeter." The maximum diameter, or **upper limit,** of the piston is 9 centimeters + 0.001 centimeter = 9.001 centimeters. The minimum diameter, or **lower limit,** of the piston is 9 centimeters − 0.001 centimeter = 8.999 centimeters.

The lower and upper limits of the diameter can also be found by solving the absolute value inequality $|d - 9| \leq 0.001$, where d is the diameter of the piston.

$$|d - 9| \leq 0.001$$
$$-0.001 \leq d - 9 \leq 0.001$$
$$-0.001 + 9 \leq d - 9 + 9 \leq 0.001 + 9$$
$$8.999 \leq d \leq 9.001$$

The lower and upper limits are 8.999 centimeters and 9.001 centimeters.

EXAMPLE 4

A doctor has prescribed 2 cc (cubic centimeters) of medication for a patient. The tolerance is 0.03 cc. Find the lower and upper limits of the amount of medication to be given.

State the goal.   The goal is to find the lower and upper limits of the amount of medication to be given.

Devise a strategy.   Let p represent the prescribed amount of medication, T the tolerance, and m the given amount of medication. Solve the absolute value inequality $|m - p| \leq T$ for m.

Solve the problem. $|m - p| \leq T$
$$|m - 2| \leq 0.03$$
$$-0.03 \leq m - 2 \leq 0.03$$
$$1.97 \leq m \leq 2.03$$

The lower and upper limits of the amount of medication to be given to the patient are 1.97 cc and 2.03 cc.

Check your work.   Be sure to check your work by doing a thorough check of your calculations. As an estimate, the answers appear reasonable in that the amounts of medication are close to 2 cc.

YOU TRY IT 4

A machinist must make a bushing that has a tolerance of 0.003 inch. The diameter of the bushing is 2.55 inch. Find the lower and upper limits of the diameter of the bushing.

Solution   See page S13.   2.547 in. and 2.553 in

INSTRUCTOR NOTE
Exercise 84 can be used for a similar in-class example.

EXAMPLE 5

A statistician is willing to state that a coin is fair if, when it is tossed 1000 times, the number of tails, t, satisfies the inequality $\left|\dfrac{t - 500}{15.81}\right| < 2.33$. Determine what values of t will cause the statistician to state that the coin is fair.

State the goal. The goal is to find the number of tails that must be obtained when tossing a coin 1000 times to suggest that the coin is fair.

Devise a strategy. Solve the inequality $\left|\dfrac{t - 500}{15.81}\right| < 2.33$ for t.

Solve the problem. $\dfrac{t - 500}{15.81} < 2.33$

$$-2.33 < \dfrac{t - 500}{15.81} < 2.33$$ • Write a compound inequality.

$$-36.8373 < t - 500 < 36.8373$$ • Multiply each part of the inequality by 15.81.

$$463.1627 < t < 536.8373$$ • Add 500 to each part of the inequality.

There must be between 464 tails and 536 tails, inclusive, for the statistician to state the coin is fair.

Check your work. Be sure to check your calculations.

> **TAKE NOTE**
>
> Note how the rounding was done for the answer. Since $t > 463.1527$ and t is a whole number, the lower bound is 464. Similarly, $t < 536.8373$, so the upper bound is 536.

YOU TRY IT 5

According to the registrar of a college, there is a 95% chance that a student applying for admission will have an SAT score, x, that satisfies the inequality $\left|\dfrac{x - 950}{98}\right| < 1.96$. Determine the values of x that the registrar expects from a student applicant.

Solution See page S13. $\{x \mid 757.92 < x < 1142.08\}$

3.5 EXERCISES

Suggested Assignment: 7–85, odds

Topics for Discussion

1. Determine whether the statement is always true, sometimes true, or never true.
   a. If $a > 0$ and $|x| = a$, then $x = a$ or $x = -a$.
   b. The absolute inequality $|x| < a$, $a > 0$, is equivalent to the compound inequality $x > a$ or $x < -a$.
   c. If $|x + 3| = 12$, then $|x + 3| = 12$ and $|x + 3| = -12$.
   d. An absolute value equation has two solutions.
   e. If $|x + b| < c$, then $x + b < -c$ or $x + b > c$.
   a. Always true   **b.** Never true   **c.** Never true   **d.** Sometimes true   **e.** Never true

2. If the absolute value of a number must be positive or zero, why can the solution of an absolute value equation be a negative number?

Answers will vary. For example, the absolute value of a number must be positive or zero, but the solution, which is a number (not the absolute value of a number) can be negative.

3. Which of the following inequalities have no solution? Which have all real numbers as the solution set?

a. $|3x + 5| \geq -9$
b. $|4x - 8| \leq -7$
c. $|2x + 1| > -6$
d. $|-x - 10| < -3$
e. $|7 - 5x| \geq 0$
f. $|9 - x| \leq -4$

Parts b, d, and f have no solution. For parts a, c, and e, the solution set is all real numbers.

4. The solution set of which inequality, $|ax + b| > c$ or $|ax + b| < c$, where in both cases $c > 0$, is the union of two solution sets and which is the intersection of two solution sets?

$|ax + b| > c$ is the union of the two solution sets; $|ax + b| < c$ is the intersection of the two solution sets.

5. How does the solution set of $|ax + b| < c, c > 0$, differ from the solution set of $|ax + b| \leq c$.

The solution set of $|ax + b| \leq c$ contains the endpoints of the interval. The solution set of $|ax + b| < c$ does not contain the endpoints.

▪ Absolute Value Equations

Solve.

6. $|x| = 7$
$-7, 7$

7. $|a| = 2$
$-2, 2$

8. $|-t| = 3$
$-3, 3$

9. $|-a| = 9$
$-9, 9$

10. $|-t| = -3$
No solution

11. $|-y| = -2$
No solution

12. $|x + 2| = 3$
$-5, 1$

13. $|x + 5| = 2$
$-7, -3$

14. $|y - 5| = 3$
$2, 8$

15. $|y - 8| = 4$
$4, 12$

16. $|a - 2| = 0$
2

17. $|a + 7| = 0$
-7

18. $|x - 2| = -4$

No solution

19. $|x + 8| = -2$

No solution

20. $|2x - 5| = 4$
$\dfrac{1}{2}, \dfrac{9}{2}$

21. $|4 - 3x| = 4$
$0, \dfrac{8}{3}$

22. $|2 - 5x| = 2$
$0, \dfrac{4}{5}$

23. $|2x - 3| = 0$
$\dfrac{3}{2}$

24. $|5x + 5| = 0$
-1

25. $|3x - 2| = -4$
No solution

26. $|2x + 5| = -2$
No solution

27. $|x - 2| - 2 = 3$

$-3, 7$

28. $|x - 9| - 3 = 2$

$4, 14$

29. $|3a + 2| - 4 = 4$

$-\dfrac{10}{3}, 2$

30. $|8 - y| - 3 = 1$

    4, 12

31. $|2x - 3| + 3 = 3$

    $\dfrac{3}{2}$

32. $|4x - 7| - 5 = -5$

    $\dfrac{7}{4}$

33. $|2x - 3| + 4 = -4$

    No solution

34. $|3x - 2| + 1 = -1$

    No solution

35. $|6x - 5| - 2 = 4$

    $\dfrac{11}{6}, -\dfrac{1}{6}$

36. $|4b + 3| - 2 = 7$

    $-3, \dfrac{3}{2}$

37. $|3t + 2| + 3 = 4$

    $-\dfrac{1}{3}, -1$

38. $|5x - 2| + 5 = 7$

    $\dfrac{4}{5}, 0$

39. $3 - |x - 4| = 5$

    No solution

40. $2 - |x - 5| = 4$

    No solution

41. $|2x - 8| + 12 = 2$

    No solution

42. $|3x - 4| + 8 = 3$

    No solution

43. $2 + |3x - 4| = 5$

    $\dfrac{7}{3}, \dfrac{1}{3}$

44. $5 + |2x + 1| = 8$

    $-2, 1$

45. $5 - |2x + 1| = 5$

    $-\dfrac{1}{2}$

46. $3 - |5x + 3| = 3$

    $-\dfrac{3}{5}$

47. $8 - |1 - 3x| = -1$

    $-\dfrac{8}{3}, \dfrac{10}{3}$

■ Absolute Value Inequalities

Solve.

48. $|x| > 3$

    $\{x \mid x < -3 \text{ or } x > 3\}$

49. $|x| < 5$

    $\{x \mid -5 < x < 5\}$

50. $|x + 1| > 2$

    $\{x \mid x < -3 \text{ or } x > 1\}$

51. $|x - 2| > 1$

    $\{x \mid x < 1 \text{ or } x > 3\}$

52. $|x - 5| \le 1$

    $\{x \mid 4 \le x \le 6\}$

53. $|x - 4| \le 3$

    $\{x \mid 1 \le x \le 7\}$

54. $|2 - x| \ge 3$

    $\{x \mid x \le -1 \text{ or } x \ge 5\}$

55. $|3 - x| \ge 2$

    $\{x \mid x \le 1 \text{ or } x \ge 5\}$

56. $|2x + 1| < 5$

    $\{x \mid -3 < x < 2\}$

57. $|3x - 2| < 4$

    $\left\{x \mid -\dfrac{2}{3} < x < 2\right\}$

58. $|5x + 2| > 12$

    $\left\{x \mid x < -\dfrac{14}{5} \text{ or } x > 2\right\}$

59. $|7x - 1| > 13$

    $\left\{x \mid x < -\dfrac{12}{7} \text{ or } x > 2\right\}$

60. $|4x - 3| \le -2$

    $\varnothing$

61. $|5x + 1| \le -4$

    $\varnothing$

62. $|2x + 7| > -5$

    $\{x \mid x \in \text{real numbers}\}$

63. $|3x - 1| > -4$

    $\{x \mid x \in \text{real numbers}\}$

64. $|4 - 3x| > 5$

    $\left\{x \mid x < -\dfrac{1}{3} \text{ or } x > 3\right\}$

65. $|7 - 2x| > 9$

    $\{x \mid x < -1 \text{ or } x > 8\}$

66. $|5 - 4x| \le 13$

    $\left\{x \mid -2 \le x \le \dfrac{9}{2}\right\}$

67. $|3 - 7x| < 17$

    $\left\{x \mid -2 < x < \dfrac{20}{7}\right\}$

68. $|6 - 3x| \le 0$

    $\{x \mid x = 2\}$

69. $|10 - 5x| \ge 0$

    $\{x \mid x \in \text{real numbers}\}$

70. $|2 - 9x| > 20$

    $\left\{x \mid x < -2 \text{ or } x > \dfrac{22}{9}\right\}$

71. $|5x - 1| < 16$

    $\left\{x \mid -3 < x < \dfrac{17}{5}\right\}$

72. *Mechanics* The diameter of a bushing is 1.75 inches. The bushing has a tolerance of 0.008 inch. Find the lower and upper limits of the diameter of the bushing.

    1.742 in., 1.758 in.

    ← 1.75 in. →

73. *Mechanics*   A machinist must make a bushing that has a tolerance of 0.004 inch. The diameter of the bushing is 3.48 inches. Find the lower and upper limits of the diameter of the bushing.
3.476 in., 3.484 in.

74. *Medicine*   A doctor has prescribed 2.5 cc of medication for a patient. The tolerance is 0.2 cc. Find the lower and upper limits of the amount of medication to be given.
2.3 cc, 2.7 cc

75. *Electricity*   A power strip is utilized on a computer to prevent the loss of programming by electrical surges. The power strip is designed to allow 110 volts plus or minus 16.5 volts. Find the lower and upper limits of voltage to the computer.
93.5 volts, 126.5 volts

76. *Electricity*   An electric motor is designed to run on 220 volts plus or minus 25 volts. Find the lower and upper limits of voltage on which the motor will run.
195 volts, 245 volts

77. *Mechanics*   A piston rod for an automobile is $10\frac{3}{8}$ inches long with a tolerance of $\frac{1}{32}$ inch. Find the lower and upper limits of the length of the piston rod.
$10\frac{11}{32}$ in., $10\frac{13}{32}$ in.

78. *Mechanics*   The diameter of a piston for an automobile is $3\frac{5}{16}$ inches with a tolerance of $\frac{1}{64}$ inch. Find the lower and upper limits of the diameter of the piston.
$3\frac{19}{64}$ in., $3\frac{21}{64}$ in.

79. *Electronics*   Find the lower and upper limits of a 29,000-ohm resistor with a 2% tolerance.
28,420 ohms, 29,580 ohms

80. *Electronics*   Find the lower and upper limits of a 15,000-ohm resistor with a 10% tolerance.
13,500 ohms, 16,500 ohms

81. *Electronics*   Find the lower and upper limits of a 25,000-ohm resistor with a 5% tolerance.
23,750 ohms, 26,250 ohms

82. *Electronics*   Find the lower and upper limits of a 56-ohm resistor with a 5% tolerance.
53.2 ohms, 58.8 ohms

83. *Fair Play*   In order to be reasonably sure that a coin is fair, the number of heads, h, in 500 tosses of the coin should satisfy the inequality $\frac{|h - 250|}{11.18} < 1.96$. Determine the values of h that would allow one to be reasonably sure that the coin is fair.
$228 < h < 272$

84. *Fair Play* The spinner for a game is shown at the right. If the spinner is fair (that is, the pointer is equally likely to land in each one of the sectors), then in 1000 spins, the number of times the spinner lands in sector 3 should satisfy the inequality $\frac{|x - 250|}{13.69} < 2.33$. What values of x will indicate that the spinner is fair?

218 < x < 282

85. *Agriculture* Pumpkins harvested from a certain farm had a mean diameter of 12.4 inches with a standard deviation of 2.6 inches. On the basis of these data, there is a 5% probability that a randomly selected pumpkin would have a diameter that satisfied the inequality $\frac{|x - 12.4|}{2.6} > 1.96$. Find the diameters of pumpkins that satisfy this condition. Round to the nearest tenth of an inch.

x < 7.3 in. or x > 17.5 in.

Applying Concepts

Solve.

86. $|2x + 1| = |x - 4|$
−5, 1

87. $|1 - 3x| = |x + 2|$
−0.25, 1.5

88. $|x| + |x - 1| = 3$
−1, 2

89. $|2x - 4| + |x| = 5$
$-\frac{1}{3}$, 3

90. $|3 - |x|| = 1$
−4, −2, 2, 4

91. $|x + 4| = 4x$
$\frac{4}{3}$

Solve. Write the answer in set-builder notation.

92. $|2x - 1| > |x + 2|$
$\left\{ x \mid x < -\frac{1}{3} \text{ or } x > 3 \right\}$

93. $|3x - 2| < |x - 3|$
$\{x \mid -0.5 < x < 1.25\}$

94. $x + |x| = 0$
$\{x \mid x \leq 0\}$

95. Solve each inequality for x. In each inequality, $a > 0$, $b > 0$.
   a. $|x + a| \leq b$
   b. $|x - a| > b$
   c. $|x + a| > a$
   d. $|x - a| \leq a$
   a. $-b - a \leq x \leq b - a$ b. $x < a - b$ or $x > a + b$ c. $x < -2a$ or $x > 0$
   d. $0 \leq x \leq 2a$

96. *Probability* In a survey of the heights of 1000 women college students, the mean height was 64 inches, and the standard deviation of the heights was 2.5 inches. Statisticians, using this information, have determined that there is approximately a 1% probability that a woman chosen from this group will have a height that satisfies the inequality $\frac{|x - 64|}{2.5} > 2.58$. What are the women's heights for which the probability is approximately 1%?

x < 57.55 in. or x > 70.45 in.

97. Express the fact that both −7 and 3 are 5 units from −2 using absolute value.

$|x + 2| = 5$

98. Use absolute value to represent the inequality $-3 \leq x \leq 5$.

$|x - 1| \leq 4$

99. Explain how the solution set of $|x - 4| \leq c$ changes for $c > 0$, $c = 0$, and $c < 0$.

For $c > 0$, the solution set is $-c \leq x - 4 \leq c$. For $c = 0$, the solution is 4. For $c < 0$, the solution set is the empty set.

EXPLORATION

1. *Graphing in Restricted Domains*   This Exploration builds on Question 2 of the Exploration in Section 3.4, in which we discussed the *value* of an expression as being 1 or 0, depending on whether the expression is true or false for a particular value of a variable.

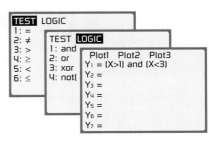

a. Consider the inequality $|x - 2| < 1$, which is equivalent to the compound inequality $x > 1$ and $x < 3$. When a graphing calculator evaluates an expression containing *and*, the calculator *multiplies* the values of the two expressions. Using the Y= editor, enter this expression into your calculator. Now create a table using a starting value of 0, and an increment of 0.25. Explain why the value of the expression is 0 for some values of x and is 1 for other values of x.

Because the calculator multiplies the values of the expressions joined by *and*, the values of 0 in the table represent the values of x for which one or the other or both of the expressions joined by *and* are false. The value of 1 corresponds to values of x for which both expressions joined by *and* are true.

b. Consider the inequality $|x - 2| > 1$, which is equivalent to the compound inequality $x < 1$ or $x > 3$. When a graphing calculator evaluates an expression containing *or*, the calculator *adds* the values of the two expressions. Using the Y= editor, enter this expression into your calculator. Now create a table using a starting value of 0, an ending value of 4, and an increment of 0.25. Explain why the value of the expression is 0 for some values of x and 1 for other values of x.

Because the calculator adds the values of the expressions joined by *or*, the values of 0 in the table represent the values of x for which both of the expressions joined by *or* are false. The value of 1 corresponds to values of x for which one or the other of the expressions joined by *or* is true.

c. Using the standard viewing window and the graph mode as dot rather than connected, enter and then graph $Y_1 = ((X > -2)$ and $(X < 5))(2X - 1)$. Explain why the graph displays as it does.

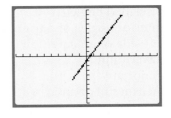

The graph is the graph of $y = 2x - 1$ only for the values of x between -2 and 5. This is because when x is less than -2 or greater than 5, the value of the compound inequality joined by *and* is 0, and $y = 0$ is plotted for these values of x. When x is between -2 and 5, the value of the compound inequality joined by *and* is 1, and the appropriate value of $y = 2x - 1$ is plotted for these values of x.

d. Using the standard viewing window and dot mode, graph $y = x^2$ for $0 < x < 3$.

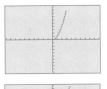

e. Using the standard viewing window and dot mode, graph $y = 2x + 2$ when $x < -2$ or when $x > 3$.

f.   Using the standard viewing window and dot mode, graph $y = |x + 1|$ for $-5 < x < 3$.

g.   Using Xmin $= -5$, Xmax $= 5$, Ymin $= -1$, Ymax $= 25$ and dot mode, graph $y = x^2$ when $x < -2$ or when $x > 3$.

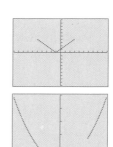

CHAPTER **3** *SUMMARY*

Key Terms

absolute value equation **[p. 214]**
absolute value inequality **[p. 216]**
acute angle **[p. 184]**
adjacent angles **[p. 185]**
angle **[p. 183]**
arc **[p. 190]**
central angle **[p. 190]**
chord **[p. 190]**
complementary angles **[p. 184]**
complements **[p. 184]**
compound inequality **[p. 204]**
conditional equation **[p. 159]**
corresponding angles **[p. 187]**
degrees **[p. 183]**
endpoint **[p. 183]**
equation **[p. 159]**
exterior angle of a triangle **[p. 189]**
first-degree equation in one variable
   [p. 160]
identity **[p. 159]**
inequality **[p. 197]**
inscribed angle **[p. 190]**
interior angles of a triangle **[p. 188]**
left side of an equation **[p. 159]**
lower limit **[p. 219]**
major arc **[p. 190]**
measure of an arc **[p. 190]**

minor arc **[p. 190]**
obtuse angle **[p. 184]**
parallel lines **[p. 186]**
perpendicular lines **[p. 185]**
ray **[p. 183]**
right angle **[p. 183]**
right side of an equation **[p. 159]**
root of an equation **[p. 159]**
sides of an angle **[p. 183]**
solution of an equation **[p. 159]**
solution of an inequality in one
   variable **[p. 197]**
solution set of an inequality **[p. 197]**
solving an equation **[p. 160]**
straight angle **[p. 184]**
supplementary angles **[p. 184]**
supplements **[p. 184]**
tangent **[p. 190]**
time-distance graph **[p. 173]**
tolerance **[p. 219]**
transversal **[p. 186]**
triangle **[p. 188]**
uniform motion **[p. 173]**
upper limit **[p. 219]**
value mixture problem **[p. 170]**
vertex **[p. 183]**
vertical angles **[p. 186]**

Essential Concepts

The Addition Property of Equations
If a, b, and c are algebraic expressions, then the equation $a = b$ has the same solutions as the equation $a + c = b + c$. **[p. 160]**

The Multiplication Property of Equations
If a, b, and c are algebraic expressions, and $c \neq 0$, then the equation $a = b$ has the same solutions as the equation $ac = bc$. **[p. 162]**

Steps in Solving First-Degree Equations
1. Use the Distributive Property to remove parentheses.
2. Combine like terms on each side of the equation.
3. Rewrite the equation with only one variable term.
4. Rewrite the equation with only one constant term.
5. Rewrite the equation so that the coefficient of the variable is 1. **[p. 166]**

A **value mixture problem** involves combining two ingredients that have different prices into a single blend. The solution of a value mixture problem is based on the equation $V = AC$, where V is the value of an ingredient, A is the amount of the ingredient, and C is the cost per unit of the ingredient. **[p. 170]**

Uniform motion means that the speed and direction of the object do not change. The solution of a uniform motion problem is based on the equation $d = rt$. In this equation, d is the distance traveled, r is the speed of the object, and t is the time spent traveling. **[p. 173]**

A **time–distance graph** shows the relationship between the time of travel and the distance traveled. Time is on the horizontal axis, and distance is on the vertical axis. **[p. 173]**

When a transversal intersects parallel lines, and the transversal is not perpendicular to the parallel lines, the following pairs of angles with equal measure are formed.
- **Alternate interior angles** are two nonadjacent angles that are on opposite sides of the transversal and between the parallel lines. **[p. 187]**
- **Alternate exterior angles** are two nonadjacent angles that are on opposite sides of the transversal and outside the parallel lines. **[p. 187]**
- **Corresponding angles** are two angles that are on the same side of the transversal and are both acute angles or are both obtuse angles. **[p. 187]**

The Sum of the Measures of the Interior Angles of a Triangle
The sum of the measures of the interior angles of a triangle is 180°. **[p. 188]**

Inscribed-Angle Theorems

If $\angle ABC$ is an inscribed angle of a circle, then $m\angle ABC = \frac{1}{2}(m\widehat{AC})$.

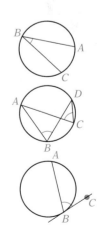

Inscribed angles that intersect the same arc are equal. $m\angle ABD = m\angle ACD$

The measure of an angle formed by a tangent and a chord is equal to one-half the measure of the intercepted arc. $m\angle ABC = \frac{1}{2}(m\widehat{AB})$. **[p. 190]**

The Addition Property of Inequalities
If $a > b$ and c is a real number, then the inequalities $a > b$ and $a + c > b + c$ have the same solution set.
If $a < b$ and c is a real number, then the inequalities $a < b$ and $a + c < b + c$ have the same solution set. **[p. 198]**

The Multiplication Property of Inequalities
Rule 1
If $a > b$ and $c > 0$, then the inequalities $a > b$ and $ac > bc$ have the same solution set.
If $a < b$ and $c > 0$, then the inequalities $a < b$ and $ac < bc$ have the same solution set.
Rule 2
If $a > b$ and $c < 0$, then the inequalities $a > b$ and $ac < bc$ have the same solution set.
If $a < b$ and $c < 0$, then the inequalities $a < b$ and $ac > bc$ have the same solution set. **[p. 199]**

Solutions of Absolute Value Equations
If $a > 0$ and $|x| = a$, then $x = a$ or $x = -a$. If $a = 0$ and $|x| = a$, then $x = 0$. **[p. 214]**

Absolute Value Inequalities of the Form $|ax + b| < c$
To solve an absolute value inequality of the form $|ax + b| < c$, $c > 0$, solve the equivalent compound inequality $-c < ax + b < c$. **[p. 216]**

Absolute Value Inequalities of the Form $|ax + b| > c$
To solve an absolute value inequality of the form $|ax + b| > c$, solve the equivalent compound inequality $ax + b < -c$ or $ax + b > c$. **[p. 218]**

CHAPTER **3** *REVIEW EXERCISES*

1. Solve: $m - \dfrac{3}{5} = -\dfrac{1}{4}$ $\dfrac{7}{20}$

2. Solve: $\dfrac{3}{2}y = 4$ $\dfrac{8}{3}$

3. Solve: $9 - 4b = 5b + 8$ $\dfrac{1}{9}$

4. Solve: $2[x + 3(4 - x) - 5x] = 6(x + 4)$ 0

5. Solve $5x - 3 < x + 9$. Write the answer in set-builder notation.
$\{x \mid x < 3\}$

6. Solve $-1 \le 3x + 5 \le 8$. Write the answer in interval notation.
$[-2, 1]$

7. Solve $5x - 2 > 8$ or $3x + 2 < -4$. Write the answer in interval notation.
$(-\infty, -2) \cup (2, \infty)$

8. Solve: $|6x + 4| - 3 = 7$
$1, -\dfrac{7}{3}$

9. Solve $|5 - 4x| > 3$. Write the answer in set-builder notation.
$\left\{ x \mid x < \dfrac{1}{2} \text{ or } x > 2 \right\}$

10. Solve $\left| \dfrac{3x - 1}{5} \right| < 4$. Write the answer in set-builder notation.
$\left\{ x \mid -\dfrac{19}{3} < x < 7 \right\}$

11. Find the value of x.

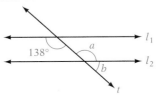

28

12. Given that $m\angle a = 103°$ and $m\angle b = 143°$, find $m\angle x$ and $m\angle y$.

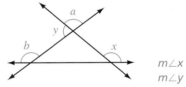

$m\angle x = 140°$
$m\angle y = 77°$

13. Given that $l_1 \parallel l_2$, find the measures of angles a and b.

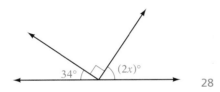

$m\angle a = 138°, m\angle b = 42°$

14. Find the value of x.

58

15. Find the supplement of a 32° angle.

148°

16. A statistics professor gives a pretest on the first day of a semester. On the basis of past experience, the professor knows that 95% of the students will have scores that satisfy the inequality $\left|\dfrac{x - 70}{10}\right| < 1.96$. Find the range of scores that 95% of the students taking the pretest will have.

$50.4 < x < 89.6$

17. At 1:00 P.M., two planes were 800 miles apart and flying toward each other at different altitudes. The rate of one plane was 320 miles per hour, and the rate of the second plane was 280 miles per hour. At what time will they pass each other?

2:20 P.M.

18. The diameter of a bushing is 2.75 inches. The bushing has a tolerance of 0.003 inch. Find the lower and upper limits of the diameter of the bushing.

2.747 in., 2.753 in

19. An average score of 80 to 90 in a psychology class receives a B grade. A student has grades of 92, 66, 72, and 88 on four exams. Find the range of scores on the fifth exam that will give the student a B in the course. Assume 100 is the highest possible score.

$82 \leq N \leq 100$, where N is the score on the fifth exam

20. How much apple juice that costs $3.20 per gallon must a grocer mix with 40 gallons of cranberry juice that costs $5.50 per gallon to make cranapple juice that costs $4.20 per gallon?

52 gal

CHAPTER **3** TEST

1. Solve: $\dfrac{2}{3} = b + \dfrac{3}{4}$

   $-\dfrac{1}{12}$

2. Solve: $\dfrac{2}{3}d = \dfrac{4}{9}$

   $\dfrac{2}{3}$

3. Solve: $4x + 1 = 7x - 7$

   $\dfrac{8}{3}$

4. Solve: $5x - 4(x + 2) = 7 - 2(3 - 2x)$

   -3

5. Solve $2 - 2(7 - 2x) \leq 4(5 - 3x)$. Write the answer in set-builder notation.

   $\{x \mid x \leq 2\}$

6. Solve $2 - 5(x + 1) \geq 3(x - 1) - 8$. Write the answer in set-builder notation.

   $\{x \mid x \leq 1\}$

7. Solve $-5 < 4x - 1 < 7$. Write the answer in interval notation.

   $(-1, 2)$

8. Solve $3x - 2 > -4$ or $7x - 5 < 3x + 3$. Write the answer in interval notation.

   $(-\infty, \infty)$

9. Solve $9x - 2 > 7$ and $3x - 5 < 10$. Write the answer in set-builder notation.

   $\{x \mid 1 < x < 5\}$

10. Solve: $|2x - 3| = 8$

    $-\dfrac{5}{2}, \dfrac{11}{2}$

11. Solve $|2x - 5| \leq 3$. Write the answer in set-builder notation.

    $\{x \mid 1 \leq x \leq 4\}$

12. Solve $|4x - 7| \geq 5$. Write the answer in set-builder notation.

    $\left\{ x \mid x \leq \dfrac{1}{2} \text{ or } x \geq 3 \right\}$

13. Solve: $|5x - 4| < -2$

    $\varnothing$

14. Find the value of x.

    $(3x + 5)^\circ$ $(x + 17)^\circ$

    6

15. Given that $l_1 \parallel l_2$, find the measures of angles a and b.

    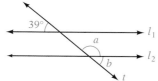

    $m\angle a = 141°, m\angle b = 39°$

16. Find the value of x.

    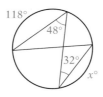

    82

17. The measure of one of the acute angles of a right triangle is twice the measure of the other acute angle. Find the measure of each angle.

    30°, 60°, 90°

18. Two planes are 1680 miles apart and traveling toward each other. One plane is traveling 80 mph faster than the other plane. The planes meet in 1.75 hours. Find the speed of each plane.

    440 mph, 520 mph

19. A tea mixture was made from 40 pounds of tea costing $6.40 per pound and 65 pounds of tea costing $3.80 per pound. Find the cost per pound of the tea mixture.

    $4.79 per pound

20. A doctor prescribed 5 milligrams of medication for a patient. The tolerance is 0.2 milligram. Find the lower and upper limits of the amount of medication to be given.

    4.8 mg; 5.2 mg

CUMULATIVE REVIEW EXERCISES

1. 1K31K4 represents a six-digit number that is a multiple of 12 but not a multiple of 9. Find the value of K.

   6

2. Give the order of the following statements so that the first two statements listed make it possible to use deductive reasoning to arrive at the third statement listed.
   (A) An eagle has feathers.
   (B) All birds have feathers.
   (C) An eagle is a bird.

   B, C, A

3. Use the roster method to list the odd natural numbers less than 9.

   {1, 3, 5, 7}

4. Let $E = \{0, 3, 6, 9, 12\}$ and $F = \{-6, -4, -2, 0, 2, 4, 6\}$. Find $E \cup F$ and $E \cap F$.

   $E \cup F = \{-6, -4, -2, 0, 2, 3, 4, 6, 9, 12\}$; $E \cap F = \{0, 6\}$

5. Evaluate the variable expression $2(b + c) - 2a^2$ for $a = -1$, $b = 3$, and $c = -2$.

   0

6. Find the temperature after a rise of 12°C from −8°C.

   4°C

7. Evaluate $y - z$ for $y = -3.597$ and $z = -4.826$.

   1.229

8. The formula for the area of a trapezoid is $A = \frac{1}{2} h(a + b)$, where A is the area, h is the height, and a and b are the lengths of the bases. Find the area of the trapezoid shown at the right.

   40 m²

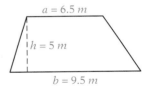

9. Simplify: $5(3y - 8) - 4(1 - 2y)$

   $23y - 44$

10. Translate "a number minus the sum of the number and twenty" into a variable expression. Then simplify the expression.

    $x - (x + 20)$; −20

11. If a car averages 28 miles per gallon, then the distance d, in miles, that a car can travel on g gallons of gasoline is given by $d = 28g$.
    a. Complete the following input/output table.

Input, quantity of gasoline g (in gallons)	0	2	4	6	8	10	12
Output, distance traveled d (in miles)	0	56	112	168	224	280	336

    b. Write a sentence that explains the meaning of the ordered pair (10, 280).

    The car can travel 280 mi on 10 gal of gasoline.

12. Evaluate $f(t) = t^2 + t - 4$ at $t = -2$.

    −2

13. Given $s(t) = 2t^3 - t^2 - 3t + 4$ with domain $\{-4, -2, 0, 2, 4\}$, find the range of s.

    {−128, −10, 4, 10, 104}

14. Find the x- and y-intercepts for the graph of $F(x) = x^2 + x - 6$.

    (−3, 0), (2, 0); (0, −6)

15.   Find the input value for which the two functions $f(x) = 4x - 2$ and $g(x) = -3x + 5$ have the same value.

1

16.   Solve: $3(2x - 3) + 1 = 2(1 - 2x)$

1

17.   Solve $3x - 2 \geq 6x + 7$. Write the solution set in interval notation.

$(-\infty, -3]$

18.   Solve: $3 - |2x - 3| = -8$

$-4, 7$

19.   Solve: $|3x - 5| \leq 4$

$\left\{ x \mid \dfrac{1}{3} \leq x \leq 3 \right\}$

20.   Two planes are 1400 miles apart and traveling toward each other. One plane is traveling 120 mph faster than the other plane. The planes meet in 2.5 hours. Find the speed of the faster plane.

340 mph

Linear Functions

Press **2nd** STAT PLOT **ENTER** to access the **STAT PLOTS** options.

*In order to monitor species that are or are becoming endangered, scientists need to determine the present population of that species. Scientists have tracked the increase in the population of the grey wolf in regions of Idaho since its reintroduction in the mid-1990s. This has led to the modeling of the size of the population with a linear function, as shown in **Exercise 9 on page 285.** From the equation, we can determine how fast the population is growing and make estimates about its future growth.*

Need help? For online student resources, visit this web site: Math.college.hmco.com

PREP TEST

1. Simplify: $-4(x - 3)$ $-4x + 12$

2. Simplify: $y - (-5)$ $y + 5$

3. Simplify: $\frac{1}{4}(3x - 16)$ $\frac{3}{4}x - 4$

4. Simplify: $\dfrac{3 - (-5)}{2 - 6}$ -2

5. Evaluate $8r + 240$ for $r = 0$.   240

6. Evaluate $\dfrac{a - b}{c - d}$ when $a = 3, b = -2, c = -3$, and $d = 2$. -1

7. Given $3x - 4y = 12$, find the value of x when $y = 0$.   4

8. Solve: $3x + 6 = 0$ -2

9. Solve: $y + 8 = 0$ -8

10. Which of the following are solutions of the inequality $-4 < x + 3$?
    a. 0   **b.** -3 **c.** 5   **d.** -7 **e.** -10 a, b, c

GO FIGURE

Two fractions are inserted between $\frac{1}{4}$ and $\frac{1}{2}$ so that the difference between any two successive fractions is the same. Find the sum of the four fractions. $\frac{3}{2}$

SECTION 4.1

■ Slope of a Line
■ Equations of the Form $Ax + By = C$

Slopes and Graphs of Linear Functions

■ Slope of a Line

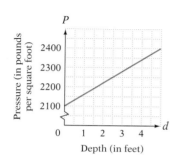

Pressure (in pounds per square foot)

2400
2300
2200
2100

0   1   2   3   4 d

Depth (in feet)

The graph at the left shows the pressure on a diver as the diver descends into the ocean. The graph of this equation can be represented by $P(d) = 64d + 2100$, where $P(d)$ is the pressure in pounds per square foot on a diver d feet below the surface of the ocean. By evaluating the function for various values of d, we can determine the pressure on the diver at that depth.

For instance, when $d = 2$, we have

$$P(d) = 64d + 2100$$
$$P(2) = 64(2) + 2100$$
$$= 128 + 2100$$
$$= 2228$$

The pressure on a diver 2 feet below the ocean surface is 2228 pounds per square foot.

The function $P(d) = 64d + 2100$ is an example of a *linear function.*

> **Linear Function**
> A **linear function** is one that can be written in the form $f(x) = mx + b$, where m is the coefficient of x and b is a constant.

Here are some examples of linear functions.

$$f(x) = 2x + 5 \qquad \bullet \ m = 2, b = 5$$
$$g(t) = \frac{2}{3}t - 1 \qquad \bullet \ m = \frac{2}{3}, b = -1$$
$$v(s) = -2s \qquad \bullet \ m = -2, b = 0$$
$$h(x) = 3 \qquad \bullet \ m = 0, b = 3$$
$$f(x) = 2 - 4x \qquad \bullet \ m = -4, b = 2$$

Note that for a linear function, the exponent on the variable is 1. Also note that, as shown above, different variables can be used to designate a linear function.

? QUESTION   Which of the following are linear functions?

a. $f(x) = 2x^2 + 5$ **b.** $g(x) = 1 - 3x$ **c.** $H(x) = \dfrac{1}{x}$

Evaluating at 0 the linear function that modeled the pressure on a diver, we have

$$P(d) = 64d + 2100$$
$$P(0) = 64(0) + 2100 = 2100$$

In this case, the P-intercept (the intercept on the vertical axis) is $(0, 2100)$. In the context of this application, this means that the pressure on a diver 0 feet below the ocean surface is 2100 pounds per square inch. Another way of saying "zero feet below the ocean surface" is to say "at sea level." Thus the pressure on a diver, or anyone else for that matter, is 2100 pounds per square foot at sea level.

Consider again the linear function $P(d) = 64d + 2100$ that models the pressure on a diver as the diver descends below the ocean surface. From the graph at the

---

? ANSWER   **a.** Because the exponent on the variable is 2, not 1, the function is not a linear function.   **b.** $1 - 3x = -3x + 1$. $g(x) = -3x + 1$ is a function of the form $f(x) = mx + b$. It is a linear function.   **c.** The variable is in the denominator. It is not a linear function.

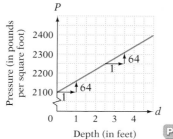

P

Pressure (in pounds per square foot)

2400
2300
2200
2100

0 1 2 3 4 → d

Depth (in feet) Ⓟ

left, note that as the depth of the diver increases by 1 foot, the pressure on the diver increases by 64 pounds per square foot. This can be verified algebraically.

$P(0) = 64(0) + 2100 = 2100$ • Pressure at sea level

$P(1) = 64(1) + 2100 = 2164$ • Pressure after descending 1 foot

$2164 - 2100 = 64$ • Change in pressure

If we choose two other depths that differ by 1 foot, such as 2.5 and 3.5 (as in the graph at the left), the change in pressure is the same.

$P(2.5) = 64(2.5) + 2100 = 2260$ • Pressure at 2.5 feet below surface

$P(3.5) = 64(3.5) + 2100 = 2324$ • Pressure at 3.5 feet below surface

$2324 - 2260 = 64$ • Change in pressure

The **slope** of a line is the change in the vertical direction caused by a 1-unit change in the horizontal direction. **In a linear function of the form $f(x) = mx + b$, m is the symbol used for slope.**

For the function $P(d) = 64d + 2100$, the slope is 64. In the context of this problem, the slope means that the pressure on a diver increases by 64 pounds per square foot for each additional foot the diver descends.

$$f(x) = mx + b$$
$$\updownarrow$$
$$P(d) = 64d + 2100$$

Suggested Activity

See Section 4.1 of the *Student Activity Manual* for an activity on investigating slope.

INSTRUCTOR NOTE
Exercise 42 can be used for a similar in-class example. We suggest you give the students the slope and ask for its meaning.

EXAMPLE 1

After a parachute is deployed, a function that models the height of the parachutist above the ground is $f(t) = -10t + 2800$, where $f(t)$ is the height, in feet, of the parachutist t seconds after the chute is deployed. What does the slope of this function mean in the context of this problem?

Solution   The function is of the form $f(x) = mx + b$, where m is the slope.

For the function $f(t) = -10t + 2800$, the slope m is -10.

The slope means that the height of the parachutist decreases 10 feet per second.

Note: The *negative* slope in $f(t) = -10t + 2800$ indicates a *decrease* in height; the *positive* slope in the function $P(d) = 64d + 2100$ indicates an *increase* in pressure.

YOU TRY IT 1

A function that models a certain small plane as it descends is given by $g(t) = -20t + 8000$, where $g(t)$ is the height, in feet, of the plane t seconds after it begins its descent. What does the slope of this function mean in the context of this problem?

Solution   See page S13.   The slope of -20 means the plane is descending 20 feet per second.

In general, for the linear function $f(x) = mx + b$, we define the slope as follows:

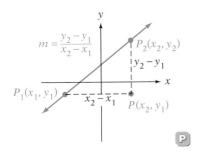

Slope of a Line

Let $P_1(x_1, y_1)$ and $P_2(x_2, y_2)$ be two points on a line. Then the **slope** m of the line through the two points is the ratio of the change in the y-coordinates to the change in the x-coordinates.

$$m = \frac{\text{change in } y}{\text{change in } x} = \frac{y_2 - y_1}{x_2 - x_1}, x_1 \neq x_2$$

? QUESTION   Why is the restriction $x_1 \neq x_2$ required in the definition of slope?

➡ Find the slope of the line between the points $P_1(-4, -3)$ and $P_2(-1, 1)$.

Let $(x_1, y_1) = (-4, -3)$ and $(x_2, y_2) = (-1, 1)$.

$$m = \frac{y_2 - y_1}{x_2 - x_1} = \frac{1 - (-3)}{-1 - (-4)} = \frac{4}{3}$$

The slope is $\frac{4}{3}$.

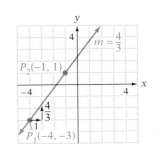

TAKE NOTE

$P_1(x_1, y_1)$ $P_2(x_2, y_2)$
  ↓ ↓   ↓ ↓
$P_1(-4, -3)$ $P_2(-1, 1)$
In the formula for slope, let
$x_1 = -4$,
$y_1 = -3$,
$x_2 = -1$, and
$y_2 = 1$

A *positive* slope indicates that the line slopes *upward* to the right.

For this line, the value of y *increases* by $\frac{4}{3}$ when x increases by 1.

➡ Find the slope of the line between the points $P_1(-2, 3)$ and $P_2(1, -3)$.

Let $(x_1, y_1) = (-2, 3)$ and $(x_2, y_2) = (1, -3)$.

$$m = \frac{y_2 - y_1}{x_2 - x_1} = \frac{-3 - 3}{1 - (-2)} = \frac{-6}{3} = -2$$

The slope is -2.

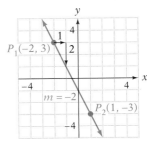

TAKE NOTE

It does not matter which point is named P_1 and which P_2; the slope will be the same. For the example at the right, let
$P_1 = (1, -3)$ and
$P_2 = (-2, 3)$. Then

$$m = \frac{y_2 - y_1}{x_2 - x_1}$$

$$= \frac{3 - (-3)}{-2 - 1}$$

$$= \frac{6}{-3} = -2$$

This is the same slope calculated using $P_1 = (-2, 3)$ and $P_2 = (1, -3)$.

A *negative* slope indicates that the line slopes *downward* to the right.

For this line, the value of y *decreases* by 2 when x increases by 1.

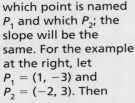

? ANSWER   If $x_1 = x_2$, then the difference $x_2 - x_1 = 0$. This would make the denominator 0, and division by 0 is undefined.

TAKE NOTE

Recall that the graph of the constant function $f(x) = c$, where c is a real number, is a horizontal line passing through $(0, c)$. The equation of the graph at the right is $f(x) = -3$.

➡ Find the slope of the line between the points $P_1(-1, -3)$ and $P_2(4, -3)$.

Let $(x_1, y_1) = (-1, -3)$ and $(x_2, y_2) = (4, -3)$.

$$m = \frac{y_2 - y_1}{x_2 - x_1} = \frac{-3 - (-3)}{4 - (-1)} = \frac{0}{5} = 0$$

The slope is 0.

A *zero* slope indicates that the line is *horizontal*.

For this particular line, the value of y stays the same when x increases by 1.

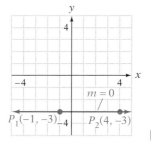

➡ Find the slope of the line between the points $P_1(4, 3)$ and $P_2(4, -1)$.

Let $(x_1, y_1) = (4, 3)$ and $(x_2, y_2) = (4, -1)$.

$$m = \frac{y_2 - y_1}{x_2 - x_1} = \frac{-1 - 3}{4 - 4} = \frac{-4}{0} \quad \text{undefined}$$

If the denominator of the slope formula is zero, the line has *no slope*. We say that the slope of the line is *undefined*.

The slope is undefined.

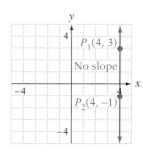

INSTRUCTOR NOTE
You might show students that the slope formula can also be written $m = \frac{y_1 - y_2}{x_1 - x_2}$. The slope will be the same. You may also want to describe slope as $\frac{\text{change in } y}{\text{change in } x}$.

INSTRUCTOR NOTE
Exercise 16 can be used for a similar in-class example.

EXAMPLE 2

Find the slope of the line between the two points.

a. $P_1(-6, 1)$ and $P_2(-4, 2)$ **b.** $P_1(-3, 3)$ and $P_2(4, 3)$

Solution   **a.** $(x_1, y_1) = (-6, 1)$, $(x_2, y_2) = (-4, 2)$

$$m = \frac{y_2 - y_1}{x_2 - x_1} = \frac{2 - 1}{-4 - (-6)} = \frac{1}{2}$$

The slope is $\frac{1}{2}$.

b. $(x_1, y_1) = (-3, 3)$, $(x_2, y_2) = (4, 3)$

$$m = \frac{y_2 - y_1}{x_2 - x_1} = \frac{3 - 3}{4 - (-3)} = \frac{0}{7} = 0$$

The slope is 0.

Suggested Activity

When you have finished discussing positive and negative slope, draw some random lines and ask students whether the slope is positive or negative. Then draw a line and ask students how they could approximate the slope of the line. Do this for both lines that have positive slope and lines that have negative slope.

YOU TRY IT 2

Find the slope of the line between the two points.

a. $P_1(-6, 5)$ and $P_2(4, -5)$ **b.** $P_1(-5, 0)$ and $P_2(-5, 7)$

Solution   See page S13.   **a.** -1 **b.** undefined

TAKE NOTE

Whether we write $f(t) = 6t$ or $d = 6t$, the equation represents a linear function. $f(t)$ and d are different symbols for the same quantity.

INSTRUCTOR NOTE
You might work with the students to create a list of some quantities that we use every day that are treated mathematically as slope—for instance, miles per gallon, miles per hour, and cost per unit. Explain that, in a sense, just as we translate the phrase *the sum of* as addition, a phrase that contains *per* can be translated as slope.

Suppose a jogger is running at a constant speed of 6 miles per hour. Then the linear function $d = 6t$ relates the time t spent running to the distance d traveled. Some of the entries in an input/output table are shown below.

Time t (in hours)	0	0.5	1	1.5	2	2.5
Distance d (in miles)	0	3	6	9	12	15

Because the equation $d = 6t$ represents a linear function, the slope of the graph is 6. This can be confirmed by choosing any two points on the graph at the right and finding the slope of the line between the two points. The points $(0.5, 3)$ and $(2, 12)$ are used here.

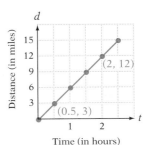

$$m = \frac{\text{change in } d}{\text{change in } t} = \frac{12 \text{ miles} - 3 \text{ miles}}{2 \text{ hours} - 0.5 \text{ hours}}$$

$$= \frac{9 \text{ miles}}{1.5 \text{ hours}} = 6 \text{ miles per hour}$$

This example demonstrates that the slope of the graph of an object in uniform motion is the speed of the object. In a more general way, we can say that anytime we discuss the speed of an object, we are discussing the slope of the graph that describes the relationship between the time the object travels and the distance it travels.

The value of the slope of a line gives the change in y for a *1-unit* change in x. For instance, a slope of -3 means that y changes by -3 as x changes by 1. We can write the slope as $\frac{-3}{1}$. A slope of $\frac{4}{3}$ means that y changes by $\frac{4}{3}$ as x changes by 1. Because it is difficult to graph a change of $\frac{4}{3}$, for fractional slopes it is easier to think of slope as integer changes in x and y.

For a slope of $\frac{4}{3}$, we have

$$m = \frac{\text{change in } y}{\text{change in } x} = \frac{4}{3}$$

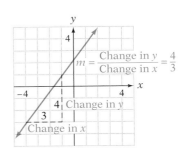

INSTRUCTOR NOTE
Exercise 32 can be used for
a similar in-class example.

INSTRUCTOR NOTE
We have chosen always to
place the negative sign of a
negative slope in the nu-
merator of the fraction. See
Example 3. Some students
may ask whether the nega-
tive sign can be placed in
the denominator. Show
these students that when
doing so, we move in the
positive y direction and the
negative x direction.

EXAMPLE 3

Graph the line that passes through $P(-2, 4)$ and has slope $-\dfrac{3}{4}$.

Solution   Rewrite the slope $-\dfrac{3}{4}$ as $\dfrac{-3}{4}$.

Draw a dot at $(-2, 4)$.

Starting at $(-2, 4)$, move 3 units down (the change
in y) and then 4 units to the right (the change in x).
Draw a dot at $(2, 1)$.

Draw a line through the two points.

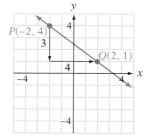

YOU TRY IT 3

Graph the line that passes through $P(2, 4)$ and has slope -1.

Solution   See page S14.

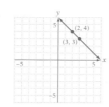

Suggested Activity

1. The line L in the xy-coordi-
   nate system has half the
   slope and twice the y-

   intercept of the line

   $y = \dfrac{2}{5}x + 6$.

   Find the equation of line L.
   Write the equation in
   slope–intercept form.

   [Answer: $y = \dfrac{1}{5}x + 12$]

2. Line A passes through
   points $(-1, 7)$ and $(3, 9)$.
   Line B passes through
   points $(1, -3)$ and $(6, -2)$.
   Which line is steeper, A or
   B? [Answer: Line A]

Recall that we can find the y-intercept of a linear equation by letting $x = 0$.

To find the y-intercept of $y = 3x + 4$, let $x = 0$.

$$y = 3x + 4$$
$$y = 3(0) + 4$$
$$y = 0 + 4$$
$$y = 4$$

The y-intercept is $(0, 4)$.

The constant term of $y = 3x + 4$ is the y-coordinate of the y-intercept.

In general, **for any equation of the form $y = mx + b$, the y-intercept is
$(0, b)$.**

Because the slope and the y-intercept can be determined directly from the
equation $f(x) = mx + b$, this equation is called the slope–intercept form of a
straight line.

TAKE NOTE

For a function of the
form $f(x) = mx + b$, m
is the slope and b is
the y-intercept.

Slope–Intercept Form of a Straight Line

The equation $y = mx + b$ is called the **slope–intercept form of a
straight line.** The slope of the line is m, the coefficient of x. The
y-intercept is $(0, b)$.

When an equation is in slope–intercept form, it is possible to draw a graph of
the function quickly.

INSTRUCTOR NOTE
Exercise 36 can be used for a similar in-class example.

INSTRUCTOR NOTE
Show students that $-\frac{2}{3}$ can also be written as $\frac{2}{-3}$. From the y-intercept, move up 2 and left 3. Place a dot at that location. Draw a line through the two points. The result is the same line shown in the graph at the right.

EXAMPLE 4

Graph $f(x) = -\frac{2}{3}x + 4$ by using the slope and y-intercept.

Solution   From the equation, the slope is $-\frac{2}{3}$ and the y-intercept is $(0, 4)$.

Rewrite the slope $-\frac{2}{3}$ as $\frac{-2}{3}$.

Place a dot at the y-intercept.

Starting at the y-intercept, move down 2 units (the change in y) and to the right 3 units (the change in x).

Place a dot at that location.

Draw a line through the two points.

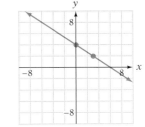

YOU TRY IT 4

Graph $y = \frac{3}{4}x - 1$ by using the slope and the y-intercept.

Solution   See page S14.

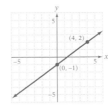

■ Equations of the Form Ax + By = C

TAKE NOTE

Note that $\dfrac{8 + 12}{4}$ can be simplified by first adding the terms in the numerator or by first dividing each term in the numerator by the denominator.

$$\frac{8 + 12}{4} = \frac{20}{4} = 5$$

$$\frac{8 + 12}{4} = \frac{8}{4} + \frac{12}{4}$$

$$= 2 + 3 = 5$$

It is this second method that is used to write the equation at the right in slope–intercept form.

Sometimes the equation of a line is written in the form $Ax + By = C$. This is called the **standard form of the equation of a line.** For instance, the equation $3x + 4y = 12$ is in standard form. For the equation $3x + 4y = 12$, $A = 3$, $B = 4$, and $C = 12$.

Here are two more examples of linear equations in standard form:

$$-2x + 5y = -10 \qquad (A = -2, B = 5, C = -10)$$
$$6x - y = 6 \qquad (A = 6, B = -1, C = 6)$$

Note that a linear equation written in standard form cannot be entered on a graphing calculator. The equation must be solved for y before it can be entered into the Y= editor of the calculator.

If an equation is in standard form, write the equation in slope–intercept form by solving the equation for y.

$$3x + 4y = 12 \qquad \bullet \text{ Standard form.}$$
$$3x - 3x + 4y = -3x + 12 \qquad \bullet \text{ Subtract } 3x \text{ from each side.}$$
$$4y = -3x + 12 \qquad \bullet \text{ Simplify.}$$
$$\frac{4y}{4} = \frac{-3x + 12}{4} \qquad \bullet \text{ Divide each side by 4.}$$
$$y = -\frac{3}{4}x + 3 \qquad \begin{array}{l} \bullet \text{ Divide each term of } -3x + 12 \text{ by 4.} \\ \text{ The equation is now in slope–intercept form.} \end{array}$$

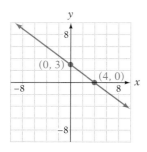

Once the equation is written in slope–intercept form, we can use the technique illustrated in Example 4 to graph the equation. The graph is shown at the left.

Another way to create the graph of a line when the equation is in standard form is to find the *x*- and *y*-intercepts (the points where the graph crosses the *x*- and *y*-axes). This is shown below for the equation $3x + 4y = 12$.

To find the *x*-intercept, let $y = 0$ and then solve for *x*.	To find the *y*-intercept, let $x = 0$ and then solve for *y*.
$3x + 4y = 12$	$3x + 4y = 12$
$3x + 4(0) = 12$	$3(0) + 4y = 12$
$3x = 12$	$4y = 12$
$x = 4$	$y = 3$
The *x*-intercept is (4, 0).	The *y*-intercept is (0, 3).

Plot the two intercepts. Then draw a line through the two points, as shown in the figure at the left above.

INSTRUCTOR NOTE
Exercise 72 can be used for a similar in-class example.

EXAMPLE 5

Write the equation $4x - 3y = 6$ in slope–intercept form. Then identify the slope and *y*-intercept.

Solution $4x - 3y = 6$ • The goal is to solve the equation for *y*.

$4x - 4x - 3y = -4x + 6$ • Subtract $4x$ from each side of the equation.

$-3y = -4x + 6$ • Simplify the left side of the equation.

$\dfrac{-3y}{-3} = \dfrac{-4x + 6}{-3}$ • Divide each side of the equation by -3.

$y = \dfrac{4}{3}x - 2$ • Simplify. On the right side, divide each term in the numerator by the denominator.

The slope is $\dfrac{4}{3}$. The *y*-intercept is (0, −2).

YOU TRY IT 5

Write the equation $3x + 2y = -6$ in slope–intercept form. Then identify the slope and *y*-intercept.

Solution See page S14. $y = -\dfrac{3}{2}x - 3$. The slope is $-\dfrac{3}{2}$. The *y*-intercept is (0, −3).

INSTRUCTOR NOTE
Exercise 82 can be used for
a similar in-class example.

EXAMPLE 6

Graph $2x - 5y = 10$ by finding the x- and y-intercepts.

Solution   To find the x-intercept,   To find the y-intercept,
let $y = 0$ and solve for x.   let $x = 0$ and solve for y.

$$2x - 5y = 10 \qquad\qquad 2x - 5y = 10$$
$$2x - 5(0) = 10 \qquad\qquad 2(0) - 5y = 10$$
$$2x = 10 \qquad\qquad -5y = 10$$
$$x = 5 \qquad\qquad y = -2$$

The x-intercept is $(5, 0)$.   The y-intercept is $(0, -2)$.

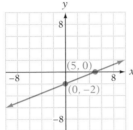

YOU TRY IT 6

Graph $3x + y = 6$ by finding the x- and y-intercepts.

Solution   See page S14.

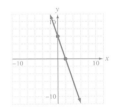

A linear equation in which one of the variables is missing has a graph that is either a horizontal or a vertical line. The equation $y = -2$ can be written in standard form as

$$0x + y = -2 \qquad \bullet \; A = 0, B = 1, \text{ and } C = -2$$

Because $0x = 0$ for all values of x, the value of y is -2 for all values of x.

Some of the possible ordered-pair solutions of $y = -2$ are given in the following table. The graph is shown at the left.

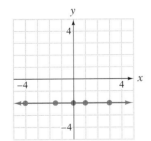

x	-4	-1.5	0	1	3
y	-2	-2	-2	-2	-2

The equation $y = -2$ represents a function. In functional notation we write $f(x) = -2$. Some of the ordered pairs of this function are $(-4, -2)$, $(-1.5, -2)$, $(0, -2)$, $(1, -2)$, and $(3, -2)$. This function is an example of a *constant function*. No matter what value of x is selected, $f(x) = -2$.

INSTRUCTOR NOTE
The constant function was
introduced in Chapter 2.
Students will probably need
to review this topic.

Definition of a Constant Function

A function given by $f(x) = b$, where b is a constant, is a **constant function**. The graph of a constant function is a horizontal line passing through $(0, b)$.

For each value in the domain of a constant function, the value of the function (the range) is the same (that is, it is constant). For instance, if $f(x) = 4$, then $f(-2) = 4, f(3) = 4, f(\pi) = 4, f(\sqrt{2}) = 4$, and so on. The value of $f(x)$ is 4 for all values of x.

? QUESTION   What is the value of $P(t) = 5$ when $t = 2$?

For the equation $y = -2$, the coefficient of x is zero. For the equation $x = 3$, the coefficient of y is zero. The equation $x = 3$ can be written in standard form as

$$x + 0y = 3 \qquad \bullet\ A = 1,\ B = 0,\ \text{and}\ C = 3$$

Because $0y = 0$ for all values of y, the value of x is 3 for all values of y.

Some of the possible ordered-pair solutions of $x = 3$ are given in the following table. The graph is shown at the left.

x	3	3	3	3	3
y	-3	-2	0	1	4

Because $(3, -3), (3, -2), (3, 0), (3, 1)$, and $(3, 4)$ are ordered pairs of this graph, the graph is not the graph of a function: there are ordered pairs with the same first coordinate and different second coordinates.

The graph of $x = a$ is not the graph of a function. It is a vertical line passing through the point $(a, 0)$.

EXAMPLE 7

Graph: $y + 1 = 0$

Solution   Solve the equation for y by subtracting 1 from each side of the equation.

$$y + 1 = 0$$
$$y = -1$$

This is a constant function. The graph of a constant function is a horizontal line passing through $(0, b)$.

The graph of $y = -1$ is a horizontal line through $(0, -1)$.

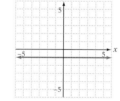

YOU TRY IT 7

Graph: $y - 5 = 0$

Solution   See page S14.

? ANSWER $P(t) = 5$ is a constant function. Therefore, $P(2) = 5$.

Suggested Activity

After completing the discussion of the graphs of $x = a$ and $y = b$, ask students to give the equations of the x-axis and y-axis.

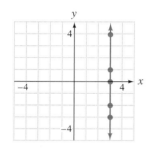

INSTRUCTOR NOTE
Exercise 94 can be used for a similar in-class example.

TAKE NOTE

Recall that a horizontal line has 0 slope.

INSTRUCTOR NOTE
Exercise 92 can be used for a similar in-class example.

EXAMPLE 8

Graph: $x = -7$.

Solution   This is an equation of the form $x = a$.

The graph of $x = a$ is a vertical line passing through the point $(a, 0)$.

The graph of $x = -7$ is a vertical line passing through the point $(-7, 0)$.

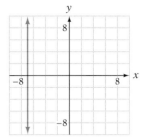

TAKE NOTE

Recall that the slope of a vertical line is undefined.

YOU TRY IT 8

Graph: $x = 1$

Solution   See page S14.

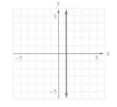

4.1 EXERCISES

Suggested Assignment: 9–95, odds

Topics for Discussion

1. Give an example of a linear equation in two variables that **a.** is in slope–intercept form and **b.** is in standard form.

   Answers will vary. For example,   **a.** $y = 3x - 4$,   **b.** $2x - 5y = 10$.

2. What is the formula for slope? Explain what each variable in the formula represents.

   $m = \dfrac{y_2 - y_1}{x_2 - x_1}$, $x_1 \neq x_2$, where m is the slope of a line, and (x_1, y_1) and (x_2, y_2) are two points on the line.

3. Explain the difference between zero slope and no slope.

   The graph of a line with zero slope is horizontal. The graph of a line with no slope is vertical.

4. Is the graph of $Ax + By = C$ always a straight line?

   Yes

5. Is the graph of a line always the graph of a function? If not, give an example of the graph of a line that is not the graph of a function.

   No. For instance, the graph of $x = 3$ is a line but is not the graph of a function.

6. What is a constant function? Give an example of a constant function.

   A constant function is a function of the form $f(x) = b$, where b is a constant. Examples will vary. For instance, $f(x) = 4$.

7. Describe the graph of $x = a$ and the graph of $y = b$.

   The graph of $x = a$ is a vertical line passing through $(a, 0)$.
   The graph of $y = b$ is a horizontal line passing through $(0, b)$.

8. For each of the following slopes of a straight line, discuss how the value of y changes when x changes by 1: $m = 2$, $m = \frac{2}{3}$, $m = -\frac{3}{4}$, $m = -3$.

   y increases by 2; y increases by $\frac{2}{3}$; y decreases by $\frac{3}{4}$; y decreases by 3

■ Slope of a Line

Is the function a linear function? Explain.

9. $f(x) = -\dfrac{3x}{4} + 1$

Yes. $-\dfrac{3x}{4} = -\dfrac{3}{4}x$. It is a function of the form $f(x) = mx + b$.

10. $f(c) = \dfrac{2}{5c} - 6$

No. The variable is in the denominator.

11. $f(a) = a^2 + 7$

No. The exponent on the variable is 2, not 1.

12. $F(x) = -10$

Yes. It is a function of the form $f(x) = mx + b$ in which $m = 0$.

Find the slope of the line containing the points P_1 and P_2.

13. $P_1(1, 3)$, $P_2(3, 1)$

-1

14. $P_1(2, 3)$, $P_2(5, 1)$

$-\dfrac{2}{3}$

15. $P_1(-1, 4)$, $P_2(2, 5)$

$\dfrac{1}{3}$

16. $P_1(3, -2)$, $P_2(1, 4)$

-3

17. $P_1(-1, 3)$, $P_2(-4, 5)$

$-\dfrac{2}{3}$

18. $P_1(-1, -2)$, $P_2(-3, 2)$

-2

19. $P_1(0, 3)$, $P_2(4, 0)$

$-\dfrac{3}{4}$

20. $P_1(-2, 0)$, $P_2(0, 3)$

$\dfrac{3}{2}$

21. $P_1(2, 4)$, $P_2(2, -2)$

Undefined

22. $P_1(4, 1)$, $P_2(4, -3)$

Undefined

23. $P_1(2, 5)$, $P_2(-3, -2)$

$\dfrac{7}{5}$

24. $P_1(4, 1)$, $P_2(-1, -2)$

$\dfrac{3}{5}$

25. $P_1(2, 3)$, $P_2(-1, 3)$

0

26. $P_1(3, 4)$, $P_2(0, 4)$

0

27. $P_1(0, 4)$, $P_2(-2, 5)$

$-\dfrac{1}{2}$

28. $P_1(-2, 3)$, $P_2(-2, 5)$

Undefined

29. $P_1(-3, -1)$, $P_2(-3, 4)$

Undefined

30. $P_1(-2, -5)$, $P_2(-4, -1)$

-2

31. Graph the line that passes through the point $(-1, -3)$ and has slope $\dfrac{4}{3}$.

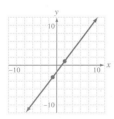

32. Graph the line that passes through the point $(-2, -3)$ and has slope $\dfrac{5}{4}$.

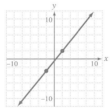

33. Graph the line that passes through the point $(-3, 0)$ and has slope -3.

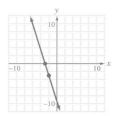

34. Graph the line that passes through the point $(2, 0)$ and has slope -1.

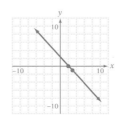

Graph by using the slope and the *y*-intercept.

35. $y = \dfrac{1}{2}x + 2$

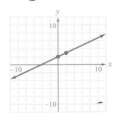

36. $y = \dfrac{2}{3}x - 3$

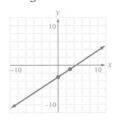

37. $y = -\dfrac{3}{2}x$

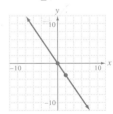

38. $y = \dfrac{3}{4}x$

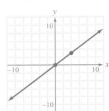

39. $y = \dfrac{1}{3}x - 1$

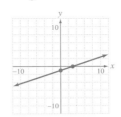

40. $y = -\dfrac{3}{2}x + 6$

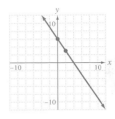

41.   *Telecommunications*   The graph below shows the total cost of a cellular phone call. Find the slope of the line. Write a sentence that states the meaning of the slope.

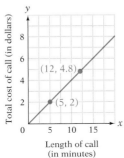

$m = 0.40$
The cellular call costs
$.40 per minute.

42.   *Aviation*   The graph below shows how the altitude of an airplane above the runway changes after takeoff. Find the slope of the line. Write a sentence that states the meaning of the slope.

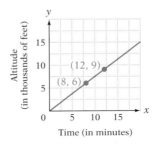

$m = 750$
The altitude of the plane is increasing at 750 ft/min.

43.   *Computers*   The graph below shows the relationship between the time, in seconds, it takes to download a file and the size of the file, in megabytes. Find the slope of the line. Write a sentence that states the meaning of the slope.

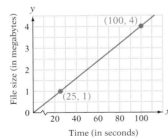

$m = 0.04$
Each second,
0.04 megabyte
is downloaded.

44.   *Computers*   The graph below shows the relationship between the size of a document and the time required to print the document using a laser printer. Find the slope of the line. Write a sentence that states the meaning of the slope.

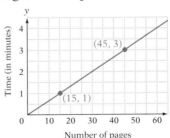

$m = 0.0\overline{6}$
Printing requires
$0.0\overline{6}$ minute per page.

45. *Temperature* The graph below shows the relationship between the temperature inside an oven and the time since the oven was turned off. Write a sentence that states the meaning of the slope.

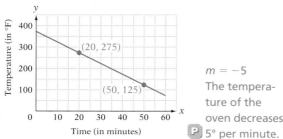

$m = -5$
The temperature of the oven decreases 5° per minute.

46. *Home Maintenance* The graph below shows the number of gallons of water remaining in a pool t minutes after a valve is opened to drain the pool. Find the slope of the line. Write a sentence that states the meaning of the slope.

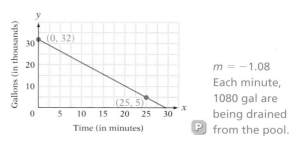

$m = -1.08$
Each minute, 1080 gal are being drained from the pool.

47. *Uniform Motion* The graph below shows the relationship between the distance traveled by a motorist and the time of travel. Find the slope of the line between the two points shown on the graph. Write a sentence that states the meaning of the slope.

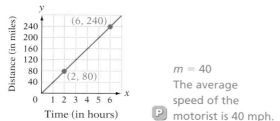

$m = 40$
The average speed of the motorist is 40 mph.

48. *Depreciation* The graph below shows the relationship between the value of a building and the depreciation allowed for income tax purposes. Find the slope of the line between the two points shown on the graph. Write a sentence that states the meaning of the slope.

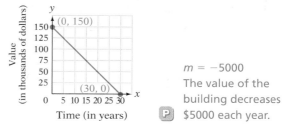

$m = -5000$
The value of the building decreases $5000 each year.

49. *Income Taxes* The graph below shows the relationship between the amount of tax and the amount of taxable income between $22,101 and $54,500. Find the slope of the line between the two points shown on the graph. Write a sentence that states the meaning of the slope.

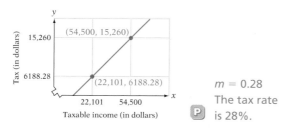

$m = 0.28$
The tax rate is 28%.

50. *Mortgages* The graph below shows the relationship between the payment on a mortgage and the amount of the mortgage. Find the slope of the line between the two points shown on the graph. Write a sentence that states the meaning of the slope.

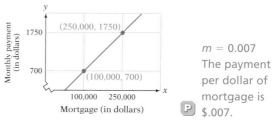

$m = 0.007$
The payment per dollar of mortgage is $.007.

51. 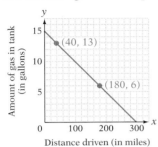 *Fuel Consumption* The graph below shows how the amount of gas in the tank of a car decreases as the car is driven. Find the slope of the line. Write a sentence that states the meaning of the slope.

$m = -0.05$
For each mile the car is driven, 0.05 gal of fuel is used.

52. 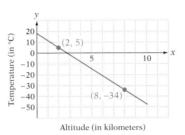 *Earth Science* The troposphere extends from the surface of Earth to an elevation of approximately 11 kilometers. The graph below shows the decrease in temperature of the troposphere as altitude increases. Find the slope of the line. Write a sentence that states the meaning of the slope.

$m = -6.5$
The temperature of the troposphere decreases 6.5°C per kilometer.

53. *Biology* There is a relationship between the number of times a cricket chirps per minute and the air temperature. A linear model of this relationship is given by $f(x) = 7x - 30$, where x is the temperature in degrees Celsius and $f(x)$ is the number of chirps per minute.
  a. Find and discuss the meaning of the x-intercept.
  b. What does the slope of this function mean in the context of this problem?

a. The x-intercept is $\left(\frac{30}{7}, 0\right)$. This means that when the temperature is $\frac{30}{7}$°C, the number of chirps per minute is 0. In other words, the cricket no longer chirps.   **b.** The slope of 7 means that the number of chirps per minute increases by 7 chirps for every 1°C increase in the temperature.

54. *Aviation* An approximate linear model that gives the remaining distance a plane must travel from Los Angeles to Paris is given by $s(t) = 6000 - 500t$, where $s(t)$ is the remaining distance t hours after the flight begins.
  a. Find and discuss the meaning of the intercepts on the vertical and horizontal axes.
  b. What does the slope of this function mean in the context of this problem?

a. The intercept on the vertical axis is (0, 6000). This means that when the trip begins, the plane's route between Los Angeles and Paris is 6000 miles. The intercept on the horizontal axis is (12, 0). This means that 12 hours after the trip begins, the plane reaches Paris.   **b.** The slope of -500 means that the remaining distance decreases by 500 mph.

55. *Refrigeration* The temperature of an object taken from a freezer gradually increases and can be modeled by $T(x) = 20x - 100$, where $T(x)$ is the Fahrenheit temperature of the object x hours after being removed from the freezer.
  a. Find and discuss the meaning of the intercepts on the vertical and horizontal axes.
  b. What does the slope of this function mean in the context of this problem?

a. The intercept on the vertical axis is (0, −100). This means that when the object was taken from the freezer, its temperature was −100°F. The intercept on the horizontal axis is (5, 0). This means that 5 hours after the object was removed from the freezer, its temperature was 0°F.   **b.** The slope of 20 means that the temperature of the object increases 20°F per hour.

56. *Investments* A retired botanist begins withdrawing money from a retirement account according to the linear model $A(t) = 100{,}000 - 2500t$, where $A(t)$ is the amount remaining in the account t months after withdrawals begin.
   a. Find and discuss the meaning of the intercepts on the vertical and horizontal axes.
   b. What does the slope of this function mean in the context of this problem?

   a. The intercept on the vertical axis is (0, 100,000). This means that initially there was $100,000 in the retirement account. The intercept on the horizontal axis is (40, 0). This means that 40 months after the withdrawals begin, there is no money left in the account.   b. The slope of −2500 means that the amount remaining in the account decreases $2500 per month.

57. *Temperature* The function $T(x) = -6.5x + 20$ approximates the temperature $T(x)$, in degrees Celsius, x kilometers above sea level. What is the slope of this function? Write a sentence that explains the meaning of the slope in the context of this problem.

   $m = -6.5$. The slope of −6.5 means that the temperature is decreasing 6.5°C for each 1-kilometer increase in height above sea level.

58. *Sound* The distance sound travels through air when the temperature is 75°F can be approximated by $d(t) = 1125t$, where d is the distance, in feet, that sound travels in t seconds. What is the slope of this function? What is the meaning of the slope in the context of this problem?

   $m = 1125$. The slope of 1125 means that sound travels 1125 ft/s when the temperature is 75°F.

59. *Biology* The distance that a homing pigeon can fly can be approximated by $d(t) = 50t$, where $d(t)$ is the distance, in miles, flown by the pigeon in t hours. What is the slope of this function? What is the meaning of the slope in the context of this problem?

   $m = 50$. The slope of 50 means that the pigeon flies 50 mph.

60. *Construction* The American National Standards Institute (ANSI) states that the slope for a wheelchair ramp must not exceed $\frac{1}{12}$.
   a. Does a ramp that is 6 inches high and 5 feet long meet the requirements of ANSI?
   b. Does a ramp that is 12 inches high and 170 inches long meet the requirements of ANSI?

   a. No   b. Yes

61. ⬤ *Construction* A certain ramp for a wheelchair must be 14 inches high. What is the minimum length of this ramp so that it will meet the ANSI requirements stated in Exercise 60.

168 in.

62. If $(2, 3)$ are the coordinates of a point on a line that has slope 2, what is the y-coordinate of the point on the line whose x-coordinate is 4?

7

63. If $(-1, 2)$ are the coordinates of a point on a line that has slope -3, what is the y-coordinate of the point on the line whose x-coordinate is 1?

-4

64. If $(1, 4)$ are the coordinates of a point on a line that has slope $\frac{2}{3}$, what is the y-coordinate of the point on the line whose x-coordinate is -2?

2

65. If $(-2, -1)$ are the coordinates of a point on a line that has slope $\frac{3}{2}$, what is the y-coordinate of the point on the line whose x-coordinate is -6?

-7

66. *Sports* Lois and Tanya start from the same place on a jogging course. Lois is jogging at 9 kilometers per hour, and Tanya is jogging at 6 kilometers per hour. The graphs below show the total distance traveled by each jogger and the total distance between Lois and Tanya. Which lines represent which distances?

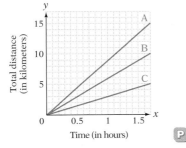

A - Lois
B - Tanya
Ⓟ   C - Distance between

67. *Gardening* A gardener is filling two cans from a faucet that releases water at a constant rate. Can 1 has a diameter of 20 millimeters, and Can 2 has a diameter of 30 millimeters. The depth of the water in each can is shown in the graph below. On the graph, which line represents the depth of the water for which can?

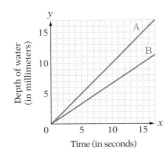

Can 1 - A
Ⓟ   Can 2 - B

■ Equations of the Form *Ax* + *By* = *C*

Write the equation in slope–intercept form.

68. $3x + y = 10$
  $y = -3x + 10$

69. $2x + y = 5$
  $y = -2x + 5$

70. $4x - y = 3$
  $y = 4x - 3$

71. $5x - y = 7$
  $y = 5x - 7$

72. $2x + 7y = 14$

  $y = -\dfrac{2}{7}x + 2$

73. $3x + 2y = 6$

  $y = -\dfrac{3}{2}x + 3$

74. $2x + 3y = 9$

  $y = -\dfrac{2}{3}x + 3$

75. $2x - 5y = 10$

  $y = \dfrac{2}{5}x - 2$

76. $5x - 2y = 4$

  $y = \dfrac{5}{2}x - 2$

77. $x + 3y = 6$

  $y = -\dfrac{1}{3}x + 2$

78. $x - 4y = 12$

  $y = \dfrac{1}{4}x - 3$

79. $6x - 5y = 10$

  $y = \dfrac{6}{5}x - 2$

Find the *x*- and *y*-intercepts and graph.

80. $x - 2y = -4$

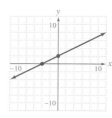

(−4, 0), (0, 2)

81. $3x + y = 3$

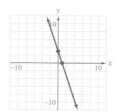

(1, 0), (0, 3)

82. $2x - 3y = 6$

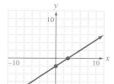

(3, 0), (0, −2)

83. $4x - y = 8$

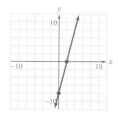

(2, 0), (0, −8)

84. $2x - y = 4$

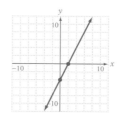

(2, 0), (0, −4)

85. $2x + y = 6$

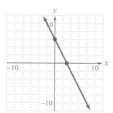

(3, 0), (0, 6)

86. $3x + 5y = 15$

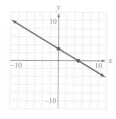

(5, 0), (0, 3)

87. $4x - 3y = 12$

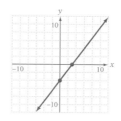

(3, 0), (0, −4)

88. $5x + 4y = 20$

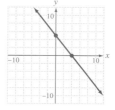

(4, 0), (0, 5)

89. $2x - 3y = 18$

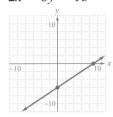

(9, 0), (0, −6)

90. $3x - 5y = 15$

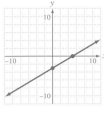

(5, 0), (0, −3)

91. $4x - 3y = 24$

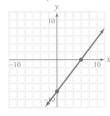

(6, 0), (0, −8)

Graph.

92. $x = 2$

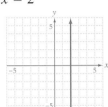

93. $y = -3$

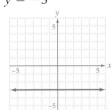

94. $y - 4 = 0$

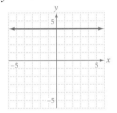

95. Given that f is a linear function for which $f(1) = -6$ and $f(-1) = -6$, determine $f(4)$.

−6

96. Given that f is a linear function for which $f(-3) = 2$ and $f(8) = 2$, determine $f(-5)$.   2

Applying Concepts

97. Explain how you can use the slope of a line to determine whether three given points lie on the same line. Then use your procedure to determine whether all of the following points lie on the same line.

a. (2, 5), (−1, −1), (3, 7)
b. (−1, 5), (0, 3), (−3, 4)

From the three points, *A, B,* and *C,* create three pairs of points, *A* and *B, A* and *C,* and *B* and *C.* Determine whether the lines containing each pair of points have the same slope.   **a.** The points lie on the same line.   **b.** The points do not lie on the same line.

98. a. What effect does increasing the coefficient of x have on the graph of $y = mx + b$?
b. What effect does decreasing the coefficient of x have on the graph of $y = mx + b$?
c. What effect does increasing the constant term have on the graph of $y = mx + b$?
d. What effect does decreasing the constant term have on the graph of $y = mx + b$?

a. Increases the slope   **b.** Decreases the slope   **c.** Increases the *y*-intercept   **d.** Decreases the *y*-intercept

99. Do the graphs of all straight lines have a y-intercept? If not, give an example of one that does not.

No; for example, $x = 2$

100. If two lines have the same slope and the same *y*-intercept, must the graphs of the lines be the same? If not, give an example.
Yes

101. A line with slope 3 passes through the point whose coordinates are (8, 12). If the ordered pair $(C, -3)$ belongs to the line, find C.
3

102. A secant is a line that passes through two points on a curve. To find the slope of the secant, the formula for slope is written in functional notation as $m = \dfrac{f(x_2) - f(x_1)}{x_2 - x_1}$, $x_1 \neq x_2$. For the graph at the right, $f(x) = x^2 - 2x - 1$, $x_1 = -1$, and $x_2 = 2$. Therefore,

$$m = \frac{f(2) - f(-1)}{2 - (-1)}$$

$$= \frac{[2^2 - 2(2) - 1] - [(-1)^2 - 2(-1) - 1]}{2 - (-1)}$$

$$= \frac{-1 - 2}{2 + 1} = \frac{-3}{3} = -1$$

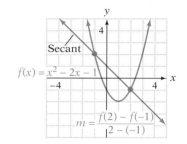

For each of the following, find the slope of the secant line for the given *x*-coordinates.

a. $f(x) = 2x - 1$, $x_1 = -2$, $x_2 = 3$ **b.** $f(x) = x^2$, $x_1 = -1$, $x_2 = 4$

c. $f(x) = x^2 - x$, $x_1 = 0$, $x_2 = 4$ **d.** $f(x) = \sqrt{x - 2}$, $x_1 = 3$, $x_2 = 6$

a. 2 b. 3 c. 3 d. $\dfrac{1}{3}$

103. A warning sign for drivers on a mountain road might read, "Caution: 6% down-grade next 2 miles." Explain this statement in the context of slope.
Answers will vary.

EXPLORATION

1. *Designing a Staircase* When you climb a staircase, the flat part of a stair that you step on is called the **tread** of the stair. The **riser** is the vertical part of the stair. The slope of a staircase is the quotient of the length of the riser and the length of the tread. Because the design of a staircase may affect safety, most cities have building codes that give rules for the design of a staircase.

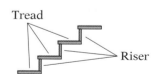

a. The traditional design of a staircase called for a 9-inch tread and a 8.25-inch riser. What is the slope of this staircase?
$0.91\overline{6}$

b. A newer design for a staircase uses an 11-inch tread and a 7-inch riser. What is the slope of this staircase?
$\dfrac{7}{11}$ or $0.\overline{63}$

c. An architect is designing a house with a staircase that is 8 feet high and 12 feet long. Is the architect using the traditional design given in part a or the newer design given in part b? Explain your answer.
newer design; $m = 0.\overline{6}$, which is closer to $0.\overline{63}$ (part b) than to $0.91\overline{6}$ (part a)

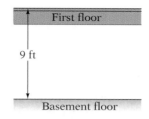

d. Staircases that have a slope between 0.5 and 0.7 are usually considered safer than those with a slope greater than 0.7. Design a safe staircase that goes from the first floor of a house to the basement, which is 9 feet below the first floor.

Answers will vary.

e. Measure the tread and riser for three staircases you encounter. Do these staircases match the traditional design in part a or the newer design in part b?

Answers will vary.

f. If there is an escalator in a building in your area, measure the tread and rise on a fully extended step of the escalator. Does it match the traditional design in part a or the newer design in part b?

Answers will vary.

2. *Intercept Form of a Straight Line*   We have discussed two equations whose graph is a straight line: $y = mx + b$ and $Ax + By = C$. There are other equations that represent straight lines. One such equation is $\frac{x}{a} + \frac{y}{b} = 1$, $a \neq 0, b \neq 0$. This is called the **intercept form of a straight line.**

a. Find the x- and y-intercepts of $\frac{x}{3} + \frac{y}{4} = 1$. Draw the graph of the equation.

(3, 0), (0, 4)

b. Find the x- and y-intercepts of $\frac{x}{2} - \frac{y}{5} = 1$. Draw the graph of the equation.

(2, 0), (0, −5)

c. Show that the x-intercept of $\frac{x}{a} + \frac{y}{b} = 1$ is $(a, 0)$ and that the y-intercept is $(0, b)$.

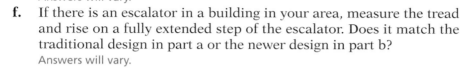

x-intercept: $\frac{x}{a} + \frac{y}{b} = 1, \frac{x}{a} + \frac{0}{b} = 1, \frac{x}{a} = 1, x = a$. The x-intercept is $(a, 0)$.

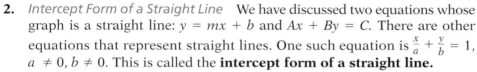

y-intercept: $\frac{x}{a} + \frac{y}{b} = 1, \frac{0}{a} + \frac{y}{b} = 1, \frac{y}{b} = 1, y = b$. The y-intercept is $(0, b)$.

d. Explain why this form of a linear equation is called the intercept form.

This is called the intercept form because the x-intercept is $(a, 0)$ and the y-intercept is $(0, b)$.

e. Write the equation $3x + 5y = 15$ in intercept form.

$\frac{x}{5} + \frac{y}{3} = 1$

f. Write the equation $y = 2x - 4$ in intercept form.

$\frac{x}{2} - \frac{y}{4} = 1$

g. Write the equation $3x - 4y = 8$ in intercept form.

$\frac{x}{8/3} - \frac{y}{2} = 1$

Finding Equations of Straight Lines

■ Find Equations of Lines Using $y = mx + b$

Suppose that a car uses 0.05 gallon of gas per mile driven and that the fuel tank, which holds 18 gallons of gas, is full. Using this information, we can determine a linear model for the amount of fuel remaining in the gas tank.

Recall that a linear function is one that can be written in the form $f(x) = mx + b$, where m is the slope of the line and b is the y-intercept. The slope is the rate at which the car is using fuel. Because the car is consuming the fuel, the amount of fuel in the tank is decreasing. Therefore, the slope is negative and we have $m = -0.05$.

The amount of fuel in the tank depends on the number of miles, x, the car is driven. Before the car starts (that is, when $x = 0$), there are 18 gallons of gas in the tank. The y-intercept is (0, 18).

Using this information, we can create the linear function by replacing m and b in the equation $f(x) = mx + b$ by their values.

$$f(x) = mx + b$$
$$f(x) = -0.05x + 18 \qquad \bullet \text{ Replace } m \text{ by } -0.05; \text{ replace } b \text{ by } 18.$$

The linear function that models the amount of fuel remaining in the tank is given by $f(x) = -0.05x + 18$, where $f(x)$ is the amount of fuel remaining after driving x miles. The graph of the function is shown at the right.

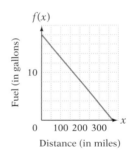

The x-intercept of this graph is the point at which $f(x) = 0$. For this application, $f(x) = 0$ means that there are 0 gallons of fuel remaining in the tank. Thus, replacing $f(x)$ by 0 in $f(x) = -0.5x + 18$ and solving for x will give the number of miles that can be driven before running out of gas.

$$f(x) = -0.05x + 18$$
$$0 = -0.05x + 18 \qquad \bullet \text{ Replace } f(x) \text{ by } 0.$$
$$-18 = -0.05x \qquad \bullet \text{ Subtract 18 from each side.}$$
$$360 = x \qquad \bullet \text{ Divide each side by } -0.05.$$

The car can travel 360 miles before running out of gas.

The domain of this function is the number of miles driven. Because the fuel tank is empty when the car has traveled 360 miles, the domain of this function is [0, 360]. The range of the function is the number of gallons of fuel in the tank. Therefore, the range is [0, 18].

TAKE NOTE

When we are creating a linear model, the slope will be the quantity that is expressed by using the word *per*. The car discussed at the right uses 0.05 gallon per mile. The slope is negative because the amount of fuel in the tank is decreasing.

Suggested Activity

Ask students to evaluate the function at the right for various values of x and to explain the meaning of the results.

TAKE NOTE

Recall that [0, 360] is interval notation and represents the set $\{x | 0 \le x \le 360\}$. The interval [0, 18] represents the set $\{x | 0 \le x \le 18\}$.

? QUESTION Why does it not make sense for the domain of $f(x) = -0.05x + 18$, discussed above, to exceed 360?

INSTRUCTOR NOTE
Exercise 40 can be used for a similar in-class example.

EXAMPLE 1

Suppose a 20-gallon gas tank contains 2 gallons when a motorist decides to fill up the tank. The gas pump fills the tank at a rate of 0.08 gallon per second. Find a linear function that models the amount of fuel in the tank x seconds after fueling begins.

Solution   Because there are 2 gallons of gas in the tank when fueling begins (at $x = 0$), the y-intercept is $(0, 2)$.

The slope is the rate at which fuel is being added to the tank. Because the amount of fuel in the tank is increasing, the slope is positive and we have $m = 0.08$.

To find the linear function, replace m and b in $f(x) = mx + b$ by their values.

$$f(x) = mx + b$$
$$f(x) = 0.08x + 2 \qquad \bullet \text{ Replace } m \text{ by } 0.08; \text{ replace } b \text{ by } 2.$$

The linear function is $f(x) = 0.08x + 2$, where $f(x)$ is the number of gallons of fuel in the tank x seconds after fueling begins.

YOU TRY IT 1

The boiling point of water at sea level is 100°C. The boiling point decreases 3.5°C for each 1-kilometer increase in altitude. Find a linear function that gives the boiling point of water as a function of altitude.

Solution   See page S14. $f(x) = -3.5x + 100$, where $f(x)$ is the boiling point of water x kilometers above sea level.

■ Find Equations of Lines Using the Point–Slope Formula

For each of the previous examples, the known point on the graph of the linear function was the y-intercept. This information enabled us to determine b for the linear function $f(x) = mx + b$. In some instances, a point other than the y-intercept is given. In this case, the *point–slope formula* is used to find the equation of the line.

INSTRUCTOR NOTE
Using parentheses may help some students with substituting into the point–slope formula.
$$y - y_1 = m(x - x_1)$$
$$y - (\ ) = (\ )(x - (\ ))$$

Point–Slope Formula of a Straight Line
Let $P_1(x_1, y_1)$ be a point on a line, and let m be the slope of the line. Then the equation of the line can be found using the point–slope formula

$$y - y_1 = m(x - x_1)$$

? ANSWER If $x > 360$, then $f(x) < 0$. This would mean that the tank contained negative gallons of gas. For instance, $f(400) = -2$.

⇒ Find the equation of the line that contains the point $(4, -1)$ and has slope $-\frac{3}{4}$.

$$y - y_1 = m(x - x_1)$$

• The slope and a point other than the y-intercept are given. Use the point–slope formula.

TAKE NOTE

$(x_1, y_1) = (4, -1)$. Substitute 4 for x_1 and -1 for y_1.

$$y - (-1) = -\frac{3}{4}(x - 4)$$

• $(x_1, y_1) = (4, -1)$ and $m = -\frac{3}{4}$.

$$y + 1 = -\frac{3}{4}x + 3$$

• Simplify the left side. Use the Distributive Property on the right side.

$$y = -\frac{3}{4}x + 2$$

• Subtract 1 from each side of the equation. The equation is now in the form $y = mx + b$.   ⬅

INSTRUCTOR NOTE
Exercise 18 can be used for a similar in-class example.

EXAMPLE 2

Find the equation of the line that passes through $P(1, -3)$ and that has slope -2.

Solution
$$y - y_1 = m(x - x_1)$$ • Use the point–slope formula.

$$y - (-3) = -2(x - 1)$$ • $m = -2$, $(x_1, y_1) = (1, -3)$

$$y + 3 = -2x + 2$$

$$y = -2x - 1$$

TAKE NOTE

Recall that $f(x)$ and y are different symbols for the same quantity, the value of the function at x.

In this example, we wrote the equation of the line as $y = -2x - 1$. We could have written the equation in functional notation as $f(x) = -2x - 1$.

YOU TRY IT 2

Find the equation of the line that passes through $P(-2, 2)$ and has slope $-\frac{1}{2}$.

Solution See page S14. $y = -\frac{1}{2}x + 1$

INSTRUCTOR NOTE
Exercise 22 can be used for a similar in-class example.

EXAMPLE 3

Find the equation of the line that passes through $P(5, -1)$ and has 0 slope.

Solution
$$y - y_1 = m(x - x_1)$$ • Use the point–slope formula.

$$y - (-1) = 0(x - 5)$$ • $m = 0$, $(x_1, y_1) = (5, -1)$

$$y + 1 = 0$$

$$y = -1$$

Recall that a line that has 0 slope is a horizontal line with equation $y = b$.

YOU TRY IT 3

Find the equation of the line that passes through $P(4, 3)$ and whose slope is undefined.

Solution   See page S14. $x = 4$

INSTRUCTOR NOTE
Exercise 42 can be used for a similar in-class example.

EXAMPLE 4

Judging on the basis of data from the Kelley Blue Book, the value of a certain car decreases approximately $250 per month. If the value of the car 2 years after it was purchased was $14,000, find a linear function that models the value of the car after x months of ownership. Use this function to find the value of the car after 3 years of ownership.

State the goal.   Find a linear model that gives the value of the car after x months of ownership. Then use the model to find the value of the car after 3 years.

Devise a strategy.   Because the function will predict the value of the car, let y represent the value of the car after x months.

Then $y = 14,000$ when $x = 24$ (2 years is 24 months).

The value of the car is decreasing $250 per month. Therefore, the slope is -250.

Use the point–slope formula to find the linear model.

To find the value of the car after 3 years (36 months), evaluate the function when $x = 36$.

Solve the problem.

$$y - y_1 = m(x - x_1)$$
$$y - 14,000 = -250(x - 24)$$
$$y - 14,000 = -250x + 6000$$
$$y = -250x + 20,000$$

A linear function that models the value of the car is $V(x) = -250x + 20,000$.

$$f(x) = -250x + 20,000$$
$$f(36) = -250(36) + 20,000 \qquad \bullet \text{ Evaluate the function at } x = 36.$$
$$= -9000 + 20,000$$
$$= 11,000$$

The value of the car is $11,000 after 36 months of ownership.

Check your work.   An answer of $11,000 seems reasonable. This value is less than $14,000, the value of the car after 2 years.

The graph of the function is shown at the right. Pressing ⟨TRACE⟩ 24 shows that the or-

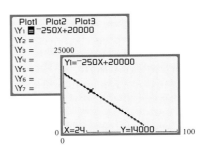

dered pair (24, 14,000) is on the graph. Pressing **TRACE** 36 shows that the ordered pair (36, 11,000) is also on the graph.

YOU TRY IT 4

In 1950, there were 13 million adults 65 years old or older in the United States. Data from the U.S. Census Bureau show that the population of these adults has been increasing at a constant rate of approximately 0.5 million per year. This rate of increase is expected to continue through the year 2010. Find a linear function that approximates the population of adults 65 years old or older in terms of the year. Use your equation to approximate the population of these adults in 2005.

Solution   See page S15. $f(x) = 0.5x - 962$; 40.5 million adults

There are instances in which it may be necessary **to find the equation of a line between two points.** The process involves two steps:

- Find the slope of the line between the two points.
- Use the point–slope formula to find the equation of the line.

Here is an example.

INSTRUCTOR NOTE
Exercise 30 can be used for a similar in-class example.

EXAMPLE 5

Find the equation of the line that contains the points (6, −4) and (3, 2).

Solution   Find the slope of the line between the two points.

$$m = \frac{y_2 - y_1}{x_2 - x_1} = \frac{2 - (-4)}{3 - 6} = \frac{6}{-3} = -2$$

Use the point–slope formula to find the equation of the line. Use either the point (6, −4) or the point (3, 2). The point (6, −4) is used here.

$$y - y_1 = m(x - x_1)$$
$$y - (-4) = -2(x - 6) \qquad \bullet \ m = -2, x_1 = 6, y_1 = -4$$
$$y + 4 = -2x + 12$$
$$y = -2x + 8$$

> **TAKE NOTE**
>
> The point (3, 2) could have been used instead of the point (6, −4). The result is the same.
>
> $$y - y_1 = m(x - x_1)$$
> $$y - 2 = -2(x - 3)$$
> $$y - 2 = -2x + 6$$
> $$y = -2x + 8$$

YOU TRY IT 5

Find the equation of the line containing the points (−2, 3) and (4, 1).

Solution   See page S15. $y = -\frac{1}{3}x + \frac{7}{3}$

INSTRUCTOR NOTE
Exercise 46 can be used for a similar in-class example.

EXAMPLE 6

Gabriel Daniel Fahrenheit invented the mercury thermometer in 1717. In terms of readings on this thermometer, water freezes at 32°F and boils at 212°F. In 1742 Anders Celsius invented the Celsius temperature scale. On this scale, water freezes at 0°C and boils at 100°C.

Determine a linear function that can be used to predict the Celsius temperature when the Fahrenheit temperature is known. Use the function to find the temperature in degrees Celsius when it is 70°F. Round to the nearest whole number.

State the goal.   Find a linear model that gives the temperature in degrees Celsius when the temperature in degrees Fahrenheit is given. Then use the model to find the temperature in degrees Celsius when the temperature is 70°F.

Devise a strategy.   Because the function will predict the temperature in degrees Celsius, let y represent the temperature in degrees Celsius. Then x represents the temperature in degrees Fahrenheit.

From the given data, two ordered pairs of the function are (32, 0) and (212, 100).

Use the two ordered pairs to find the slope of the line.

Use the point–slope formula to find the linear model.

To find the temperature in degrees Celsius when it is 70°F, evaluate the linear function at $x = 70$.

Solve the problem.   Let $(x_1, y_1) = (32, 0)$ and $(x_2, y_2) = (212, 100)$.

$$m = \frac{y_2 - y_1}{x_2 - x_1} = \frac{100 - 0}{212 - 32} = \frac{100}{180} = \frac{5}{9} \qquad \bullet \text{ The slope is } \frac{5}{9}.$$

$$y - y_1 = m(x - x_1) \qquad\qquad \bullet \text{ Use the point–slope formula.}$$

$$y - 0 = \frac{5}{9}(x - 32) \qquad\qquad \bullet \; m = \frac{5}{9}, x_1 = 32, y_1 = 0.$$

$$y = \frac{5}{9}x - \frac{160}{9}$$

The linear function is $f(x) = \dfrac{5}{9}x - \dfrac{160}{9}$.

$$f(x) = \frac{5}{9}x - \frac{160}{9}$$

$$f(70) = \frac{5}{9}(70) - \frac{160}{9} \qquad \bullet \text{ Evaluate the function at } x = 70.$$

$$\approx 21$$

When the temperature is 70°F, it is approximately 21°C.

Check your work.   The graph of the function is shown at the right. Use the TRACE feature to show that the ordered pairs (32, 0) and (212, 100) are on the graph. Use the same feature to show that when $x = 70$, $y \approx 21$.

See Appendix A: Trace

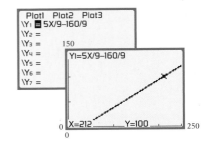

YOU TRY IT 6

There are approximately 126 calories in a 2-ounce serving of lean hamburger and approximately 189 calories in a 3-ounce serving. Write a linear equation for the number of calories in lean hamburger in terms of the size of the serving. Use your equation to estimate the number of calories in a 5-ounce serving of lean hamburger.

Solution   See page S15. $f(x) = 63x$; 315 calories

■ Parallel and Perpendicular Lines

The graph of $g(x) = 2x - 3$ and the graph of $f(x) = 2x + 4$ are shown at the right. Note from the equations that the lines have the same slope and different y-intercepts. Lines that have the same slope and different y-intercepts are **parallel lines**; the graphs of the lines never meet.

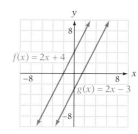

> **Parallel Lines**
>
> Two nonvertical lines with slopes m_1 and m_2 are **parallel lines** if and only if $m_1 = m_2$. Vertical lines are also parallel lines.

➡ Is the line that contains the points $P_1(-2, 1)$ and $P_2(-5, -1)$ parallel to the line that contains the points $Q_1(1, 0)$ and $Q_2(4, 2)$?

To determine whether the lines are parallel, find the slope of each line.

Slope of the line containing P_1 and P_2:

$$m_1 = \frac{y_2 - y_1}{x_2 - x_1} = \frac{-1 - 1}{-5 - (-2)} = \frac{-2}{-3} = \frac{2}{3}$$

Slope of the line containing Q_1 and Q_2: $m_2 = \dfrac{y_2 - y_1}{x_2 - x_1} = \dfrac{2 - 0}{4 - 1} = \dfrac{2}{3}$

$m_1 = m_2$: the slopes are equal. Therefore, the lines are parallel.   ⬅

? QUESTION   Is the graph of $y = -\frac{1}{2}x + 2$ parallel to the graph of $y = -x + 2$?

? ANSWER   No. The slope of one line is $-\frac{1}{2}$ and the slope of the other line is -1. The slopes are not equal, so the graphs are not parallel.

INSTRUCTOR NOTE
Exercise 62 can be used for
a similar in-class example.

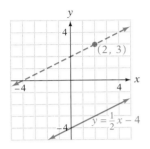

EXAMPLE 7

Find the equation of the line that is parallel to the graph of $y = \frac{1}{2}x - 4$ and contains the point $(2, 3)$.

Solution The slope of the given line is $\frac{1}{2}$.

Because parallel lines have the same slope, the slope of the unknown line is also $\frac{1}{2}$. We know the slope of the line and a point on the line. We can use the point–slope formula to find the equation of the line.

$$y - y_1 = m(x - x_1)$$

$$y - 3 = \frac{1}{2}(x - 2) \qquad \bullet \; m = \frac{1}{2}, x_1 = 2, y_1 = 3$$

$$y - 3 = \frac{1}{2}x - 1$$

$$y = \frac{1}{2}x + 2 \qquad \bullet \; \text{Check: The slope of this line is } \frac{1}{2}, \text{ and } (2, 3) \text{ is a}$$
$$\text{solution of this equation.}$$

YOU TRY IT 7

Find the equation of the line that is parallel to the graph of $y = -3x + 1$ and contains the point $(-5, -4)$.

Solution See page S15. $y = -3x - 19$

INSTRUCTOR NOTE
Exercise 64 can be used for
a similar in-class example

EXAMPLE 8

Find the equation of the line that is parallel to the graph of $2x - 3y = 12$ and passes through the point $(6, -1)$.

Solution Because the lines are parallel, the slope of the unknown line is the same as the slope of the given line. To determine the slope of the given line, write $2x - 3y = 12$ in slope–intercept form by solving the equation for y.

$$2x - 3y = 12$$

$$-3y = -2x + 12 \qquad \bullet \; \text{Subtract } 2x \text{ from each side of the equation.}$$

$$\frac{-3y}{-3} = \frac{-2x + 12}{-3} \qquad \bullet \; \text{Divide each side of the equation by } -3.$$

$$y = \frac{2}{3}x - 4 \qquad \bullet \; \text{The equation is in slope–intercept form.}$$

The slope of the given line is $\frac{2}{3}$.

Because parallel lines have the same slope, the slope of the unknown line is also $\frac{2}{3}$.

The slope of the line and a point on the line are known. Use the point–slope formula to find the equation of the line.

$$y - y_1 = m(x - x_1)$$

$$y - (-1) = \frac{2}{3}(x - 6) \qquad \bullet \; m = \frac{2}{3}, x_1 = 6, y_1 = -1$$

$$y + 1 = \frac{2}{3}x - 4$$

$$y = \frac{2}{3}x - 5 \qquad \bullet \; \text{Check: The slope of this line is } \frac{2}{3}, \text{ and } (6, -1) \text{ is a solution of this equation.}$$

YOU TRY IT 8

Find the equation of the line that is parallel to the graph of $3x + 5y = 15$ and passes through the point $P(-2, 3)$.

Solution   See page S16. $y = -\frac{3}{5}x + \frac{9}{5}$

For Example 8, we left the answer in slope–intercept form. However, we could have rewritten the equation in standard form.

$$y = \frac{2}{3}x - 5$$

$$3y = 3\left(\frac{2}{3}x - 5\right) \qquad \bullet \text{ Multiply each side by 3.}$$

$$3y = 2x - 15$$

$$-2x + 3y = -15 \qquad \bullet \text{ Subtract } 2x \text{ from each side. The equation is in standard form. } A = -2, B = 3, \text{ and } C = -15.$$

Two lines that intersect at right angles are **perpendicular** lines. A theorem allows us to determine whether the graphs of two lines are perpendicular.

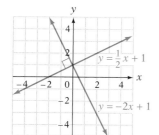

> ### Slopes of Perpendicular Lines
>
> If m_1 and m_2 are the slopes of two lines, neither of which is vertical, then the lines are perpendicular if and only if $m_1 \cdot m_2 = -1$.
>
> A vertical line is perpendicular to a horizontal line.

INSTRUCTOR NOTE
We suggest that after presenting the concept of perpendicular lines, you draw
two nonvertical perpendicular lines on a coordinate
grid. Construct a triangle
along one of the lines, using
the intersection of the lines
as a vertex. Mark on the triangle the rise and run of
the line. Ask the students
what happens when this
line is rotated 90° onto the
other line. You want the
students to see that the rotation results in the rise and
the run of the first line becoming the run and the rise,
respectively, of the second
line and that the run
changes direction.

Solving $m_1 \cdot m_2 = -1$ for m_1 gives $m_1 = -\dfrac{1}{m_2}$. This last equation states that the slopes of perpendicular lines are **negative reciprocals** of each other. The negative reciprocal of $-\dfrac{3}{4}$ is $\dfrac{4}{3}$. The negative reciprocal of 5 is $-\dfrac{1}{5}$.

? QUESTION **a.** What is the negative reciprocal of $\dfrac{7}{2}$?*
b. What is the negative reciprocal of -6?

➡ Is the line that contains the points $P_1(4, 2)$ and $P_2(-2, 5)$ perpendicular to the line that contains the points $Q_1(-4, 3)$ and $Q_2(-3, 5)$?

To determine whether the lines are perpendicular, find the slope of each line.

Slope of the line containing P_1 and P_2:

$$m_1 = \frac{y_2 - y_1}{x_2 - x_1} = \frac{5 - 2}{-2 - 4} = \frac{3}{-6} = -\frac{1}{2}$$

Slope of the line containing Q_1 and Q_2:

$$m_2 = \frac{y_2 - y_1}{x_2 - x_1} = \frac{5 - 3}{-3 - (-4)} = \frac{2}{1} = 2$$

m_1 is the negative reciprocal of m_2.

The lines are perpendicular. ⬅

? QUESTION Is the graph of $f(x) = -\dfrac{2}{3}x + 3$ perpendicular to the graph of $g(x) = \dfrac{2}{3}x - 3$? †

Graphs on a graphing calculator may give the impression that two lines are not perpendicular when in fact they are. This is due to the size of the pixels (picture elements). For instance, the graph of $y = -\dfrac{1}{2}x + 4$ and the graph of $y = 2x - 3$ are shown at the right in the standard viewing window. Although the graphs are perpendicular, they do not appear to be from the graph. This apparent distortion can be fixed, however, by using the square viewing window.

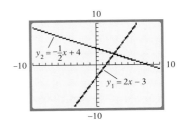

[📇] See Appendix A:
Window

? ANSWERS *a. $-\dfrac{2}{7}$ **b.** $\dfrac{1}{6}$ †No. The slope of f is $-\dfrac{2}{3}$ and the slope of g is $\dfrac{2}{3}$. The slopes are not negative reciprocals.

The graphs of $y = -\frac{1}{2}x + 4$ and $y = 2x - 3$ are shown at the right in the *square* viewing window. Note that the graphs now appear to be perpendicular. The main point of this discussion is that the appearance of a graph will change as the size of the viewing window changes.

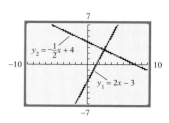

INSTRUCTOR NOTE
Exercise 70 can be used for a similar in-class example.

EXAMPLE 9

Find the equation of the line that is perpendicular to $y = \frac{2}{5}x + 1$ and passes through the point (5, 3).

Solution   Because the lines are perpendicular, the value of the slope of the unknown line is the negative reciprocal of the slope of the given line.

The slope of the given line is $\frac{2}{5}$. Therefore, the slope of the unknown perpendicular line is $-\frac{5}{2}$.

The slope of the line and a point on the line are known. Use the point–slope formula to find the equation of the line.

$$y - y_1 = m(x - x_1)$$

$$y - 3 = -\frac{5}{2}(x - 5) \quad \bullet \; m = -\frac{5}{2}, (x_1, y_1) = (5, 3).$$

$$y - 3 = -\frac{5}{2}x + \frac{25}{2}$$

$$y = -\frac{5}{2}x + \frac{31}{2} \quad \bullet \; \text{Check: The slope of the line is } -\frac{5}{2}, \text{ and} \\ (5, 3) \text{ is a solution of this equation.}$$

YOU TRY IT 9

Find the equation of the line that is perpendicular to the graph of $y = -\frac{4}{3}x - 1$ and passes through the point (−4, 3).

Solution   See page S16. $y = \frac{3}{4}x + 6$

INSTRUCTOR NOTE
Exercise 72 can be used for a similar in-class example.

EXAMPLE 10

Find the equation of the line that is perpendicular to $2x - y = -3$ and passes through the point (4, −5).

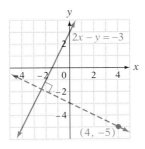

Solution   Because the lines are perpendicular, the slope of the unknown line is the negative reciprocal of the slope of the given line. To determine the slope of the given line, write $2x - y = -3$ in slope–intercept form by solving the equation for y.

$$2x - y = -3$$
$$-y = -2x - 3$$
$$y = 2x + 3$$

The slope of the given line is 2.

The slope of a perpendicular line is $-\dfrac{1}{2}$.

The slope of the line and a point on the line are known. Use the point–slope formula to find the equation of the line.

$$y - y_1 = m(x - x_1)$$

$$y - (-5) = -\frac{1}{2}(x - 4) \qquad \bullet \; m = -\frac{1}{2}, x_1 = 4, y_1 = -5$$

$$y + 5 = -\frac{1}{2}x + 2$$

$$y = -\frac{1}{2}x - 3 \qquad \bullet \; \text{Check: The slope of this line is } -\frac{1}{2}, \text{ and } (4, -5)$$
$$\text{is a solution of this equation.}$$

YOU TRY IT 10

Find the equation of the line that is perpendicular to the graph of $5x - 3y = 15$ and passes through the point $P(-5, -2)$.

Solution   See page S16. $y = -\dfrac{3}{5}x - 5$

Suggested Activity

See Section 4.2 of the *Student Activity Manual* for a lesson involving parallel and perpendicular lines, as well as equations in standard form.

There are many applications of the concept of perpendicular. We will consider one here.

Suppose a ball is being twirled on the end of a string. If the string breaks, the initial path of the ball is on a line that is perpendicular to the radius of the circle.

INSTRUCTOR NOTE
Exercise 74 can be used for a similar in-class example.

EXAMPLE 11

Suppose that a ball is being twirled on the end of a string and that the center of rotation is the origin of a coordinate system. See the figure at the right. If the string breaks when the ball is at the point whose coordinates are $P(6, 3)$, find the initial path of the ball.

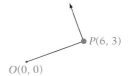

State the goal. The goal is to find the equation of the line that is perpendicular to the line containing the points (0, 0) and (6, 3) and goes through point (6, 3).

Devise a strategy. The initial path of the ball is perpendicular to the line through *OP*. Therefore, the slope of the initial path of the ball is the negative reciprocal of the slope of the line between *O* and *P*.

We need to find the slope of the line through *OP*.

The slope of the line we are looking for is the negative reciprocal of that slope.

We will then have the slope of the line and a point on the line. We can use the point–slope formula to find the equation of the line.

Solve the problem.

Slope of the line between *OP*: $m = \dfrac{y_2 - y_1}{x_2 - x_1} = \dfrac{3 - 0}{6 - 0} = \dfrac{1}{2}$

The slope of the line that is the initial path of the ball is the negative reciprocal of $\frac{1}{2}$. Therefore, the slope of a perpendicular line is -2.

$$y - y_1 = m(x - x_1) \qquad \bullet \text{ Use the point–slope formula.}$$
$$y - 3 = -2(x - 6) \qquad \bullet \ m = -2, x_1 = 6, y_1 = 3$$
$$y - 3 = -2x + 12$$
$$y = -2x + 15$$

The initial path of the ball is along the line whose equation is $y = -2x + 15$.

Check your work. The graphs of $f(x) = -2x + 15$ and $f(x) = \frac{1}{2}x$ are shown at the right in the square viewing window of a graphing calculator. The lines appear to be perpendicular. Use the TRACE feature to show that the ordered pair (6, 3) is on the graph.

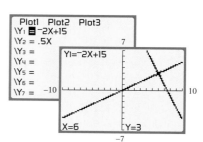

YOU TRY IT 11

Suppose that a ball is being twirled on the end of a string and that the center of rotation is the origin of a coordinate system. See the figure at the right. If the string breaks when the ball is at the point whose coordinates are (2, 8), find the initial path of the ball.

●*P*(2, 8)

O(0, 0)

Solution See page S16. $y = -\dfrac{1}{4}x + \dfrac{17}{2}$

4.2 EXERCISES Suggested Assignment: 7–71, odds

Topics For Discussion

1. Explain why the equation $y = mx + b$ is called the slope–intercept form of a straight line.

   The slope, m, and the y-intercept, $(0, b)$, can be read directly from the equation.

2. What is the point–slope formula and how is it used?

   The point–slope formula is $y - y_1 = m(x - x_1)$. The formula is used to find the equation of a line given a point on the line and its slope.

3. Suppose a biologist determined that there was a linear model that related the height of a tree, y, to its age, x. In the equation $y = mx + b$, would you expect m to be positive or negative? Why?

   m would be positive because as the age of the tree increases, the height of the tree increases.

4. Suppose a baked potato is taken from a hot oven and placed on a plate to cool. If y represents the temperature of the potato and x represents the time the potato has been out of the oven, would you expect m to be positive or negative in the equation $y = mx + b$? Why?

   m would be negative because as the time out of the oven increases, the temperature of the potato decreases.

5. **a.** Explain how to determine whether two lines are parallel.
   b. Explain how to determine whether two lines are perpendicular.

   a. Parallel lines have equal slopes.

   b. The product of the slopes of perpendicular lines is -1; that is, their slopes are negative reciprocals.

6. Are the lines $x = 0$ and $y = 0$ perpendicular? What is another name for these lines?

   Yes. The line $x = 0$ is the y-axis, and $y = 0$ is the x-axis.

▪ Writing Equations of Lines in the Form $y = mx + b$

7. Find the equation of the line that has slope -2 and y-intercept $(0, -1)$.

   $y = -2x - 1$

8. Find the equation of the line that has slope $\frac{5}{3}$ and y-intercept $(0, -6)$.

   $y = \frac{5}{3}x - 6$

9. Find the equation of the line that has slope $-\frac{1}{4}$ and passes through the point $(0, 2)$.

   $y = -\frac{1}{4}x + 2$

10. Find the equation of the line that has slope 5 and passes through the point $(0, -3)$.

    $y = 5x - 3$

11. Find the equation of the line that has slope $\frac{1}{6}$ and passes through the point $(0, 0)$.

    $y = \frac{1}{6}x$

12. Find the equation of the line that has slope -1 and passes through the point $(0, -8)$.

$y = -x - 8$

13. Find the equation of the line that has slope 2 and passes through the point $(0, 5)$.

$y = 2x + 5$

14. Find the equation of the line that has slope $\frac{1}{2}$ and passes through the point $(2, 3)$.

$y = \frac{1}{2}x + 2$

15. Find the equation of the line that has slope $-\frac{5}{4}$ and passes through the point $(4, 0)$.

$y = -\frac{5}{4}x + 5$

16. Find the equation of the line that has slope -3 and passes through the point $(-1, 7)$.

$y = -3x + 4$

17. Find the equation of the line that has slope 3 and passes through the point $(2, -3)$.

$y = 3x - 9$

18. Find the equation of the line that has slope $-\frac{2}{3}$ and passes through the point $(3, 5)$.

$y = -\frac{2}{3}x + 7$

19. Find the equation of the line that contains the point $(-2, -3)$ and has zero slope.

$y = -3$

20. Find the equation of the line that contains the point $(5, 8)$ and has zero slope.

$y = 8$

21. Find the equation of the line that contains the point $(3, -4)$ and whose slope is undefined.

$x = 3$

22. Find the equation of the line that contains the point $(-5, -1)$ and whose slope is undefined.

$x = -5$

23. Find the equation of the line that contains the points $(0, 2)$ and $(3, 5)$.

$y = x + 2$

24. Find the equation of the line that contains the points $(0, -3)$ and $(-4, 5)$.

$y = -2x - 3$

25. Find the equation of the line containing the points $(0, 0)$ and $(4, 3)$.

$y = \frac{3}{4}x$

26. Find the equation of the line containing the points $(2, -5)$ and $(0, 0)$.

$y = -\frac{5}{2}x$

27. Find the equation of the line that passes through the points $(-2, -3)$ and $(-1, -2)$.

$y = x - 1$

28. Find the equation of the line that passes through the points $(3, -1)$ and $(-2, 4)$.

$y = -x + 2$

29. Find the equation of the line that contains the points $(2, 0)$ and $(4, -3)$.

$y = -\dfrac{3}{2}x + 3$

30. Find the equation of the line that contains the points $(4, 1)$ and $(-4, -3)$.

$y = \dfrac{1}{2}x - 1$

31. Find the equation of the line that passes through the points $(-1, 3)$ and $(2, 4)$.

$y = \dfrac{1}{3}x + \dfrac{10}{3}$

32. Find the equation of the line that passes through the points $(2, 3)$ and $(5, 5)$.

$y = \dfrac{2}{3}x + \dfrac{5}{3}$

33. Find the equation of the line that contains the points $(3, -4)$ and $(-2, -4)$.

$y = -4$

34. Find the equation of the line that contains the points $(-3, 3)$ and $(-2, 3)$.

$y = 3$

35. Find the equation of the line that passes through the points $(-2, 5)$ and $(-2, -5)$.

$x = -2$

36. Find the equation of the line that passes through the points $(3, 2)$ and $(3, -4)$.

$x = 3$

37. *Aviation* The pilot of a Boeing 757 jet takes off from Boston's Logan Airport, which is at sea level, and climbs to a cruising altitude of 32,000 feet at a constant rate of 1200 feet per minute. Write a linear equation for the height of the plane in terms of the time after take-off. Use your equation to find the height of the plane 11 minutes after take-off.

$y = 1200x$; 13,200 ft

38. *Telecommunications* A cellular phone company offers several different options for using a cellular telephone. One option, for people who plan on using the phone only in emergencies, costs the user $4.95 per month plus $.59 per minute for each minute the phone is used. Write a linear equation for the monthly cost of the phone in terms of the number of minutes the phone is used. Use your equation to find the cost of using the cellular phone for 13 minutes in 1 month.

$y = 0.59x + 4.95$; $12.62

39. *Aviation* An Airbus 320 plane takes off from Denver International Airport in Denver, Colorado, which is 5200 feet above sea level, and climbs to 30,000 feet at a constant rate of 1000 feet per minute. Write a linear equation for the height of the plane in terms of the time after take-off. Use your equation to find the height of the plane 8 minutes after take-off.

$y = 1000x + 5200$; 13,200 ft

40. *Construction* A general building contractor estimates that the cost to build a new home is $30,000 plus $85 for each square foot of floor space in the house. Determine a linear function that will give the cost of building a house that contains a given number of square feet. Use this model to determine the cost to build a house that contains 1800 square feet.
$y = 85x + 30,000$; $183,000

41. *Boiling Points* At sea level, the boiling point of water is 100°C. At an altitude of 2 kilometers, the boiling point of water is 93°C. Write a linear equation for the boiling point of water in terms of the altitude above sea level. Use your equation to predict the boiling point of water on top of Mount Everest, which is approximately 8.85 kilometers above sea level. Round to the nearest degree.
$y = -3.5x + 100$; 69°C

42. *Uniform Motion* A plane travels 830 miles in 2 hours. Determine a linear model that will predict the number of miles the plane can travel in a given amount of time. Use this model to predict the distance the plane will travel in $4\frac{1}{2}$ hours.
$y = 415x$; 1867.5 mi

43. *Computers* According to the U.S. Department of Commerce, there were 24 million homes with computers in 1991. The average rate of growth in computers in homes was expected to increase by 2.4 million homes per year through 2005. Write a linear equation for the number of computers in homes in terms of the year. Let $x = 90$ represent 1990. Use your equation to find the number of computers expected to be in homes in 2004.
$y = 2.4x - 194.4$; 55.2 million

44. *Fuel Consumption* The gas tank of a certain car contains 16 gallons when the driver of the car begins a trip. Each mile driven by the driver decreases the amount of gas in the tank by 0.032 gallon per mile. Write a linear equation for the number of gallons of gas in the tank in terms of the number of miles driven. Use your equation to find the number of gallons in the tank after the car has been driven 150 miles.
$y = -0.032x + 16$; 11.2 gal

45. *Sound* Whales, dolphins, and porpoises communicate using high-pitched sounds that travel through the water. The speed at which the sound travels depends on many factors, one of which is the depth of the water. At approximately 1000 meters below sea level, the speed of sound is 1480 meters per second. Below 1000 meters, the speed of sound increases at a constant rate of 0.017 meters per second for each additional meter below 1000 meters. Write a linear equation for the speed of sound in terms of the number of meters below sea level. Use your equation to approximate the speed of sound 2500 meters below sea level. Round to the nearest meter per second.
$y = 0.017x + 1463$; 1506 m/s

46. *Compensation* An account executive receives a base salary plus a commission. On $20,000 in monthly sales, the account executive receives $1800. On $50,000 in monthly sales, the account executive receives $3000. Determine a linear function that will yield the compensation of the sales executive for a given amount of monthly sales. Use this model to determine the account executive's compensation for $85,000 in monthly sales.
$y = 0.04x + 1000$; $4400

47. *Business*   A manufacturer of pickup trucks has determined that 50,000 trucks per month can be sold at a price of $9000. At a price of $8750, the number of trucks sold per month would increase to 55,000. Determine a linear function that will predict the number of trucks that would be sold at a given price. Use this model to predict the number of trucks that would be sold at a price of $8500.

$y = -20x + 230{,}000$; 60,000 trucks

48. *Business*   A manufacturer of graphing calculators has determined that 10,000 calculators per week will be sold at a price of $95. At a price of $90, it is estimated that 12,000 calculators would be sold. Determine a linear function that will predict the number of calculators that would be sold at a given price. Use this model to predict the number of calculators per week that would be sold at a price of $75.

$y = -400x + 48{,}000$; 18,000 calculators

49. *Business*   The operator of a hotel estimates that 500 rooms per night will be rented if the room rate per night is $75. If the room rate per night is $85, then 494 rooms will be rented. Determine a linear function that will predict the number of rooms that will be rented for a given price per room. Use this model to predict the number of rooms that will be rented if the room rate is $100 per night.

$y = -\dfrac{3}{5}x + 545$; 485 rooms

50. *Boiling Points*   When sugar is added to water, the solution has a higher boiling point than pure water. The table at the right gives the boiling point (in degrees Celsius) of a certain quantity of water as various amounts of sugar are added to the water.

   a. Find a linear model for the boiling point of the solution in terms of the number of grams of sugar added.

   b. Write a sentence explaining the meaning of the slope of the line in the context of this problem.

   c. Using this model, determine the boiling point when 50 grams of sugar has been added to this quantity of pure water.

Sugar (in grams)	Boiling Point (in °C)
20	100.104
30	100.156
40	100.208
60	100.312
80	100.416

a. $y = 0.0052x + 100$ **b.** For every 1 g of sugar added, the boiling point increases 0.0052°C.   **c.** 100.26°C

51. *Freezing Points*   When sugar is added to water, the solution has a lower freezing point than pure water. The table at the right gives the freezing point (in degrees Celsius) of a certain quantity of water as various amounts of sugar are added to the water.

   a. Find a linear model for the freezing point of the solution in terms of the number of grams of sugar added.

   b. Write a sentence explaining the meaning of the slope of the line in the context of this problem.

   c. Using this model, determine the freezing point when 50 grams of sugar has been added to this quantity of pure water.

Sugar (in grams)	Freezing Point (in °C)
20	−0.372
30	−0.558
40	−0.744
60	−1.116
80	−1.488

a. $y = -0.0186x$ **b.** For every 1 g of sugar added, the freezing point decreases 0.0186°C.   **c.** −0.93°C.

52.  *Aviation* In 1927, Charles Lindbergh made history by making the first transatlantic flight from New York to Paris. It took Lindbergh approximately 33.5 hours to make the trip. In 1997, the Concorde could make the trip in approximately 3.3 hours. Write a linear equation for the time, in hours, it takes to cross the Atlantic in terms of the year the trip is made. Use your equation to predict how long a flight between the two cities would have taken in 1967. Round your answer to the nearest tenth. On the basis of your answer, do you think a linear model accurately predicts how the flying time between New York and Paris has changed? Explain your answer.

$y = -0.431x + 864$; 16.2 h. Answers will vary.

■ Parallel and Perpendicular Lines

53. Is the line $x = -2$ perpendicular to the line $y = 3$?

Yes

54. Is the line $y = \frac{1}{4}$ perpendicular to the line $y = -4$?

No

55. Is the line $x = -3$ parallel to the line $y = -3$?

No

56. Is the line $x = 4$ parallel to the line $x = -4$?

Yes

57. Is the line $y = -\frac{3}{2}x + 4$ parallel to the line $y = -\frac{3}{2}x - 1$?

Yes

58. Is the line $y = 5x - 6$ parallel to the line $y = -5x - 6$?

No

59. Is the line that contains the points $(3, 2)$ and $(1, 6)$ parallel to the line that contains the points $(-1, 3)$ and $(-1, -1)$?

No

60. Is the line that contains the points $(4, -3)$ and $(2, 5)$ parallel to the line that contains the points $(-2, -3)$ and $(-4, 1)$?

No

61. Find the equation of the line that is parallel to $y = -3x - 1$ and passes through $P(1, 4)$.

$y = -3x + 7$

62. Find the equation of the line that is parallel to $y = \frac{2}{3}x + 2$ and passes through $P(-3, 1)$.

$y = \frac{2}{3}x + 3$

63. Find the equation of the line that contains the point $(-2, -4)$ and is parallel to the line $2x - 3y = 2$.

$y = \frac{2}{3}x - \frac{8}{3}$

64. Find the equation of the line that contains the point $(3, 2)$ and is parallel to the line $3x + y = -3$.

$y = -3x + 11$

65. Is the line $y = -\frac{5}{2}x + 4$ perpendicular to the line $y = -\frac{2}{5}x - 1$?

No

66. Is the line $y = 4x - 8$ perpendicular to the line $y = -\frac{1}{4}x + 2$?

Yes

67. Is the line that contains the points $(-3, 2)$ and $(4, -1)$ perpendicular to the line that contains the points $(1, 3)$ and $(-2, -4)$?

Yes

68. Is the line that contains the points $(4, -1)$ and $(-4, 5)$ perpendicular to the line that contains the points $(6, -2)$ and $(-3, 6)$?

No

69. Find the equation of the line that contains the point $(4, 1)$ and is perpendicular to the line $y = -3x + 4$.

$y = \frac{1}{3}x - \frac{1}{3}$

70. Find the equation of the line that contains the point $(2, -5)$ and is perpendicular to the line $y = \frac{5}{2}x - 4$.

$y = -\frac{2}{5}x - \frac{21}{5}$

71. Find the equation of the line that contains the point $(-1, -3)$ and is perpendicular to the line $3x - 5y = 2$.

$y = -\frac{5}{3}x - \frac{14}{3}$

72. Find the equation of the line that contains the point $(-1, 3)$ and is perpendicular to the line $2x + 4y = -1$.

$y = 2x + 5$

73. *Physical Forces*   Suppose that a ball is being twirled on the end of a string and that the center of rotation is the origin of a coordinate system. If the string breaks when the ball is at the point whose coordinates are $(1, 9)$, find the initial path of the ball.

$y = -\frac{1}{9}x + \frac{82}{9}$

74. *Geometry*   A theorem from geometry states that a line passing through the center of a circle and through a point P on the circle is perpendicular to the tangent line at P. (See the figure at the right.) If the coordinates of P are $(5, 4)$ and the coordinates of C are $(3, 2)$, what is the equation of the tangent line?

$y = -x + 9$

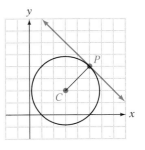

Applying Concepts

Is there a linear equation that contains all the given ordered pairs? If there is, find the equation.

75. $(5, 1), (4, 2), (0, 6)$

Yes; $y = -x + 6$

76. $(-2, -4), (0, -3), (4, -1)$

Yes; $y = \frac{1}{2}x - 3$

77. $(-1, -5), (2, 4), (0, 2)$

No

78. $(3, -1), (12, -4), (-6, 2)$

Yes, $y = -\frac{1}{3}x$

The given ordered pairs are solutions to the same linear equation. Find n.

79. $(0, 1), (4, 9), (3, n)$

7

80. $(2, 2), (-1, 5), (3, n)$

1

81. $(2, -2), (-2, -4), (4, n)$
-1

82. $(1, -2), (-2, 4), (4, n)$
-8

83. Suppose A_1, A_2, B_1, and B_2 are all not equal to zero. If the graphs of $A_1x + B_1y = C_1$ and $A_2x + B_2y = C_2$ are parallel, express $\frac{A_1}{B_1}$ in terms of A_2 and B_2.
$\frac{A_1}{B_1} = \frac{A_2}{B_2}$

84. Suppose A_1, A_2, B_1, and B_2 are all not equal to zero. If the graphs of $A_1x + B_1y = C_1$ and $A_2x + B_2y = C_2$ are perpendicular, express $\frac{A_1}{B_1}$ in term of A_2 and B_2.
$\frac{A_1}{B_1} = -\frac{B_2}{A_2}$

85. A line contains the points $(4, -1)$ and $(2, 1)$. Find the coordinates of three other points that are on this line.
Possible answers are (0, 3), (1, 2), and (3, 0).

86. Given that f is a linear function for which $f(1) = 3$ and $f(-1) = 5$, determine $f(4)$.
0

87. The graphs of $y = -\frac{1}{2}x + 2$ and $y = \frac{2}{3}x - 5$ intersect at the point whose coordinates are $(6, -1)$. Find the equation of a line whose graph intersects the graphs of the given lines to form a right triangle. (*Hint:* There is more than one answer to this question.)
Any equation of the form $y = 2x + b$, where $b \neq -13$, or of the form $y = -\frac{3}{2}x + c$, where $c \neq 8$.

88. A linear function includes the ordered pairs $(2, 4)$ and $(4, 10)$. Find the value of the function at $x = -1$.
-5

89. *Uniform Motion*   Assume the maximum speed your car will go varies linearly with the steepness of the hill it is climbing or descending. If the hill is 5° up, your car can go 77 kilometers per hour. If the hill is 2° down $(-2°)$, your car can go 154 kilometers per hour. When your top speed is 99 kilometers per hour, how steep is the hill? State your answer in degrees, and note whether it is up or down.
$x = 3$; 3° up

EXPLORATION

1. *Linear Parametric Equations*   Consider the situation of a person who can row at a rate of 3 miles per hour in calm water and is trying to cross a river for which there is a current, running perpendicular to the direction of rowing, of 4 miles per hour. See the figure at the right. Because of the current, the boat is being pushed downstream at the same time it is moving across the river. Because the boat is traveling 3 miles per hour in the x direction, its position after t hours is given by $x = 3t$. The current is pushing the boat in the negative y direction at 4 miles per hour. Therefore, its position after t hours is given by $y = -4t$, where we use -4 to indicate that the boat is moving down. The equations $x = 3t$ and $y = -4t$ are called **parametric equations,** and t is called the **parameter.**

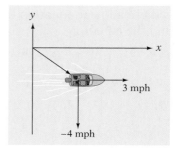

a. Assume the boat starts at the point (0, 0). What is the location of the boat after 15 minutes (0.25 hour)?
(0.75, −1)

b. If the river is 1 mile wide, how far down river will the boat be when it reaches the other shore? (*Suggestion:* Find the time it takes the boat to cross the river by solving $x = 3t$ for t when $x = 1$. Now replace t by this value in $y = -4t$ and simplify.)
1.3 mi

c. For the parametric equations $x = 3t$ and $y = -4t$, write y in terms of x by solving $x = 3t$ for t and then substituting this expression into $y = -4t$.
$y = -\dfrac{4}{3}x$

d. In the diagram at the right, a plane flying at 5000 feet above sea level begins a gradual ascent. Determine parametric equations for the path of the plane.
$x = 9000t$ and $y = 100t + 5000$

e. What is the altitude of the plane 5 minutes after it begins its ascent?
5500 ft

f. What is the altitude of the plane after it has traveled 12,000 feet in the positive x direction? Round to the nearest whole number.
5133 ft

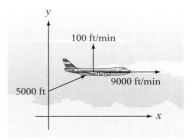

100 ft/min

9000 ft/min

5000 ft

SECTION **4.3**   # Linear Regression

■ Fit a Line to Data
■ Linear Regression

■ Fit a Line to Data

There are many instances when a linear function can be used to approximate collected data. For instance, the table below shows the recommended maximum exercise heart rates for individuals of various ages who exercise regularly.

Age, x	20	25	30	32	43	55	28	42	50	55	62
Heart rate, y	160	150	148	145	140	130	155	140	132	125	125

The graph at the right, called a **scatter diagram,** shows a graph of the ordered pairs given in the table. These ordered pairs suggest that the maximum exercise heart rate for an individual decreases as the person's age increases.

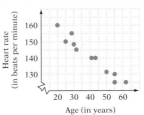

Heart rate (in beats per minute)

Age (in years)

Although these points do not lie on a straight line, it is possible to find a line that *approximately fits* the data. One way to do this is to select two data points. You want the line through the two points you select to be one that is close to all the data points. We chose $(25, 150)$ to be P_1 and $(55, 130)$ to be P_2. The slope of the line between these two points is

$$m = \frac{y_2 - y_1}{x_2 - x_1} = \frac{130 - 150}{55 - 25} = \frac{-20}{30} = -\frac{2}{3}$$

Suggested Activity

Ask students to bring data to class and then prepare a scatter diagram of the data. They can find data on the Internet or in the graphs printed in a newspaper such as *USA Today.* Be sure they label both the horizontal axis and the vertical axis. Also check that the distance between units is uniform.

You might ask students to determine whether there is a trend in the data graphed. For instance, in Example 1 on page 280, students can see that the sales of DVDs are increasing each year.

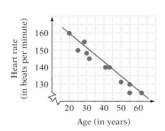

The graph of $y = -\dfrac{2}{3}x + \dfrac{500}{3}$

Now use the point–slope formula to find the equation of the line through the two points.

$$y - y_1 = m(x - x_1)$$

$$y - 150 = -\frac{2}{3}(x - 25)$$ • $m = -\dfrac{2}{3},\, x_1 = 25,\, y_1 = 150$

$$y - 150 = -\frac{2}{3}x + \frac{50}{3}$$ • Use the Distributive Property on the right side.

$$y = -\frac{2}{3}x + \frac{500}{3}$$ • Add 150 to each side of the equation.

The graph of $y = -\dfrac{2}{3}x + \dfrac{500}{3}$ is shown at the left. It *approximates* the data.

You may get a better approximation to the data if, rather than selecting two of the data points, you visually determine a line that closely fits the data and then use a straight-edge to draw that line. The goal is to draw on the graph the **line of best fit**—that is, the line that most closely approximates all the points. Here is an example.

INSTRUCTOR NOTE
Exercise 10 can be used for a similar in-class example.

EXAMPLE 1

The table below shows the actual and projected sales of DVDs, in billions of dollars, from 1998 through 2003 (*Source:* Adams Media Research). Sketch a line of best fit through a scatter diagram of the data. Then find a linear equation for the line you drew. Use decimals rather than fractions for the slope and *y*-intercept in the equation.

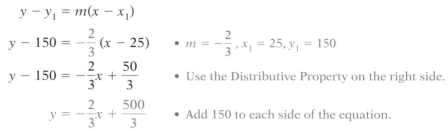

Year, x	1998	1999	2000	2001	2002	2003
DVD sales, y (in billions of dollars)	0.4	1.5	3.1	4.9	6.6	8.2

INSTRUCTOR NOTE
One way to motivate the idea of "best fit" is to draw some lines that are obviously not good approximations to the data. The diagram below shows three possibilities. Once students see these, they will be able to understand that some line will approximate the data better than other lines.

Solution   The data points are plotted in the scatter diagram at the right. Although the points do not lie on a straight line, they are close to linear.

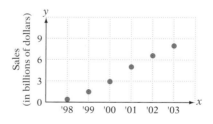

The goal is to draw on the graph the line of best fit. Use a straightedge to draw the line.

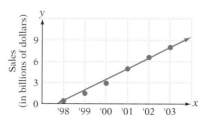

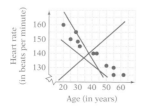

TAKE NOTE

We chose the points
(2001, 5) and (2003, 8)
because they are inte-
gers, making calcula-
tions easier. However,
the line appears to lie
close to the point
(1999, 1.5), and that
point could have been
used to find the equa-
tion of the line.

INSTRUCTOR NOTE
Students should understand
that the equation they
calculate may not be the
same as other students'
equations. However, the
slopes of the lines and the
y-intercepts should be "in
the same ball park."

You might ask the stu-
dents to explain the meaning of the slope of their
linear equation in the context of the problem. (It is
the annual decrease in the
average price of a DVD.)

Approximate the location of any two points on the line. We chose $(2001, 5)$ to be P_1 and $(2003, 8)$ to be P_2.

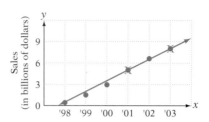

Use the two points selected to find the slope of the line.

$$m = \frac{y_2 - y_1}{x_2 - x_1} = \frac{8 - 5}{2003 - 2001} = \frac{3}{2} = 1.5$$

Now use the point–slope formula to find the equation of the line through the two points.

$$y - y_1 = m(x - x_1)$$
$$y - 5 = 1.5(x - 2001)$$
$$y - 5 = 1.5x - 3001.5$$
$$y = 1.5x - 2996.5$$

The equation for our line is $y = 1.5x - 2996.5$.

YOU TRY IT 1

The table below shows the actual and projected average price of a DVD, in dollars, from 1998 through 2003 (*Source:* Adams Media Research). Sketch a line of best fit through a scatter diagram of the data. (A scatter diagram is provided below the table.) Then find a linear equation for the line. Use decimals rather than fractions for the slope and y-intercept in the equation.

Year, x	1998	1999	2000	2001	2002	2003
Average price of a DVD, y (in dollars)	23.90	23.15	22.30	21.19	19.91	18.52

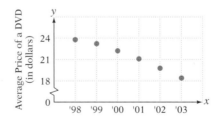

Solution See page S16. Answers will vary. For example, $y = -1.08x + 2182.07$.

■ Linear Regression

Consider again the heart rate data presented at the beginning of this section. The equation of the line we found by choosing two data points produces an approximate linear model to the data. If we had chosen different points, the result would have been a slightly different equation.

TAKE NOTE

The line of best fit, the regression line, is one that minimizes the *residuals.* A **residual** is the vertical distance between the actual *y* value and the *y* value the regression line predicts.

Among all the equations that could have been chosen, statisticians have determined that the line of best fit is the **regression line.** The equation of the regression line for the data presented there is $y = -0.827x + 174$. The graph of the regression line is shown at the right.

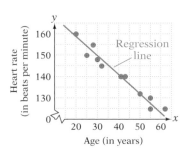

Using the equation of the regression line, an exercise physiologist can determine the recommended maximum exercise heart rate for an individual.

For instance, suppose an individual is 28 years old. Then the physiologist would replace *x* by 28 and determine the value of *y*.

$$y = -0.827x + 174$$
$$y = -0.827(28) + 174 \quad \bullet \text{ Replace } x \text{ by } 28.$$
$$y = 150.844$$

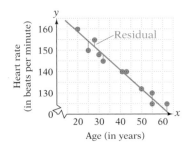

The recommended maximum exercise heart rate for a 28-year-old person is approximately 151 beats per minute.

In Example 1 above, we calculated a linear equation that modeled the data on sales of DVDs. We did this by first sketching a line through the data points and approximating the location of two points on this line. A graphing calculator, however, mathematically determines the line of best fit, or the regression line. The equation of the regression line for the DVD sales data is $y = 1.603x - 3202.399$.

The regression equation predicts that in 2005, sales of DVDs will be $11.616 billion.

$$y = 1.603x - 3202.399$$
$$y = 1.603(2005) - 3202.399$$
$$= 11.616$$

The example that follows illustrates the calculation of a regression line using a graphing calculator.

The table below shows the data recorded by a chemistry student who was trying to determine a relationship between the temperature (in degrees Celsius) and volume (in liters) at a constant pressure for 1 gram of oxygen. Chemists call this relationship Charles's Law.

See Appendix A: Regression

Temperature, T (in °C)	−100	−75	−50	−25	0	25	50
Volume, V (in liters)	0.43	0.5	0.57	0.62	0.7	0.75	0.81

A graphing calculator can be used to calculate the regression line. Some typical screens follow.

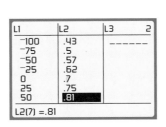

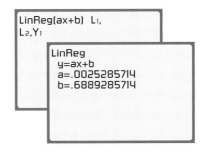

Note that the graphing calculator uses the variable a instead of m; the equation is of the form $y = ax + b$.

For this set of data, the regression equation is $V = 0.0025286T + 0.68892857$.

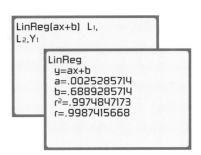

To determine the volume of 1 gram of oxygen when the temperature is $-30°C$, replace T in the regression equation by -30 and evaluate the expression. This can be done with your calculator. The volume will be displayed as approximately 0.61 liter.

You may have noticed the results of some other calculations on the screen when the regression equation was calculated. The variable r is the **correlation coefficient** and r^2 is the **coefficient of determination.** Statisticians use these numbers to determine how well the regression equation approximates the data. If $r = 1$, the data exactly fit a line of positive slope. If $r = -1$, the data exactly fit a line of negative slope. In general, the closer r^2 is to 1, the closer the data fit a linear model. For these data points, the r^2 value is 0.997, which is very close to 1; the regression line is quite a good fit to the data.

Y₁(⁻30) .6130714286

🖩 **CALCULATOR NOTE**

If your calculator does not display the values of r and r^2, see Appendix A: Correlation Coefficient

INSTRUCTOR NOTE
Exercise 14 can be used for a similar in-class example.

EXAMPLE 2

Sodium thiosulfate is used by photographers to develop some types of film. The amount of this chemical that will dissolve in water depends on the temperature of the water. The table that follows gives the number of grams of sodium thiosulfate that will dissolve in 100 milliliters of water for various temperatures.

Temperature, x (in °C)	20	35	50	60	75	90	100
Grams, y	50	80	120	145	175	205	230

a. Find the linear regression line for the data. Round decimals to the nearest thousandth.

INSTRUCTOR NOTE
You may want to discuss the meaning of the correlation coefficient, explaining that a high correlation indicates that the events occur together and not that one causes the other. For example, a study showed a high correlation between high academic achievement and families eating dinner together. Ask students whether eating dinner together *causes* high academic achievement. Have them discuss what the relationship might be.

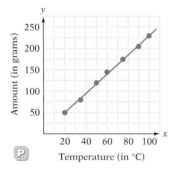

b. How many grams of sodium thiosulfate does the model predict will dissolve in 100 milliliters of water when the temperature is 70°C? Round to the nearest tenth.
c. What does the slope of the regression line mean in the context of the problem?
d. What does the *y*-intercept mean in the context of the problem?

Solution   **a.** Use a calculator to determine the regression line for the data.

The regression equation is $y = 2.252x + 5.248$.
b. Evaluate the regression equation when $x = 70$.

$$y = 2.252x + 5.248$$
$$= 2.252(70) + 5.248$$
$$= 162.888$$

Approximately 162.9 grams of sodium thiosulfate will dissolve when the temperature is 70°C.
c. The slope of 2.252 means that for every 1°C increase in temperature, an additional 2.252 grams of sodium thiosulfate will dissolve in the water.
d. The *y*-intercept of 5.248 means that at 0°C, 5.248 grams of sodium thiosulfate will dissolve in the water.

A scatter diagram of the data is shown at the left, along with the graph of the regression line. Note that the regression line is a very good fit for the data and that the r^2 value of 0.997959, shown in the graphing calculator screen at the left above, is very close to 1.

YOU TRY IT 2

The heights and weights of women swimmers on a college swim team are given in the table that follows.

Height, *x* (in inches)	68	64	65	67	62	67	65
Weight, *y* (in pounds)	132	108	108	125	102	130	105

a. Find the linear regression line for these data.
b. Use your regression equation to estimate the weight of a woman on a college swim team who is 63 inches tall. Round to the nearest whole number.
c. What does the slope of the regression line mean in the context of the problem?
d. What does the *y*-intercept mean in the context of the problem?

Solution   See page S17.   **a.** $y = 5.6\overline{3}x - 252.8\overline{6}$ **b.** 102 lb **c.** The slope indicates the increase in weight for every 1-inch increase in height.   **d.** A woman on a college swim team who is 0 in. tall is predicted to weigh −253 lb.

Suggested Activity

See Section 4.3 of the *Student Activity Manual* for an activity involving linear regression.

One important use of a scatter diagram is to determine the relationship between two variables before performing regression on the data. In this section, we are investigating two-variable data in which the relationship is linear. Therefore, the data values should lie close to a straight line.

The table from Example 2 is repeated below, and a scatter diagram for the data is shown at the left. Note that the points lie close to a straight line.

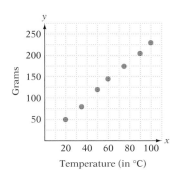

Temperature, x (in °C)	20	35	50	60	75	90	100
Grams, y	50	80	120	145	175	205	230

A graphing calculator can display data in a scatter diagram. Some typical screens follows.

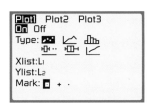
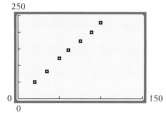

See Appendix A:
Stat Plot

Note that in the scatter diagram, all the points lie close to a straight line. Therefore, linear regression is appropriate for the data.

The following data are from a biology experiment in which the growth of bacteria was recorded. A scatter diagram of the data is shown at the left. Note that the points do not lie close to a straight line. Linear regression would not be appropriate for the data.

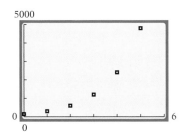

Time, x (in hours)	0	1	2	3	4	5
Population, y	150	300	600	1200	2400	4800

In the last section, we found the equation of a line between two points by first finding the slope of the line between the two points and then using the point–slope formula. Linear regression can also be used to find the equation of a line given two points on the line. Example 3 illustrates the procedure.

INSTRUCTOR NOTE
Exercise 22 can be used for a similar in-class example.

EXAMPLE 3

Use linear regression to find the equation of the line containing the points $(6, -4)$ and $(3, 2)$.

Solution   The points $(6, -4)$ and $(3, 2)$ are represented in the input/output table.

x	6	3
y	-4	2

```
LinReg
  y=ax+b
  a=-2
  b=8
  r²=1
  r=-1
```

Enter the x values in one list of a calculator and the y values in another. Then use the calculator to determine the regression line for the data.

The equation is $y = -2x + 8$.

YOU TRY IT 3

Use linear regression to find the equation of the line containing the points $(-2, 3)$ and $(4, 1)$.

Solution   See page S17. $y = -0.\overline{3}x + 2.\overline{3}$

? QUESTION   For the regression equation in Example 3, why is the value of r equal to -1?

4.3 EXERCISES   Suggested Assignment: 9–25 odds

Topics for Discussion

1. What does it mean to "fit a line to data"?
   Answers will vary.

2. What is a regression line?
   A line that approximates the data of a scatter diagram.

3. What might be the purpose of determining the regression equation for a set of data?
   Answers will vary. For example, we can use the equation to project possible future outcomes.

4. Determine whether the statement is always true, sometimes true, or never true.
   a. For linear regression to be performed on two-variable data, the points corresponding to the data values must lie on a straight line.
   b. If the correlation coefficient is equal to 1, the data exactly fit the regression line. If the correlation coefficient is equal to -1, the data do not fit the regression line.
   c. A scatter diagram is a graph of ordered pairs.
   a. Never true   b. Never true   c. Always true

5. Suppose an infant's weight is measured once a month for 2 years. If a regression equation were calculated for the weight of the infant in terms of the age of the infant, in months, would r be positive, zero, or negative? Why?
   r would be positive because as the number of months increases, the weight increases.

? ANSWER   The given points lie exactly on the line $y = -2x + 8$, and the slope of the line is negative.

6. Suppose a person purchases a used car. If data were collected giving the value of the car in terms of its age, would r be positive, zero, or negative? Why?

   r would be negative because as the age of the car increases, the value decreases.

7. Suppose that in a college history class, data are collected giving the height of a student and the student's score on an exam. If a regression equation were calculated on the data, would r be positive, zero, or negative? Why?

   r would be close to 0 because there is no correlation between height and history exam scores.

Fit a Line to Data

8. *Demography*   The table and scatter diagram show the projected number of U.S. high school graduates, in millions (*Source:* National Center for Education Statistics).

Year, x	'01	'03	'05	'07	'09
Number of graduates, y (in millions)	2.90	2.98	2.99	3.13	3.25

   a. Find the equation of a line that approximately fits the data by selecting two data points and finding the equation of the line through those two points.

   b. What does the slope of your line mean in the context of the problem?

   c. What does the y-intercept mean in the context of the problem?

   a. Answers will vary. The equation of the line through the points (1, 2.90) and (9, 3.25) is $y = 0.04375x + 2.85625$.   b. The slope of 0.04375 means that each year the number of U.S. high school graduates increases by 0.04375 million.
   c. The y-intercept of 2.85625 means that in 2000, there were 2.85625 million high school graduates.

9. *Zoology*   The table and scatter diagram show the increase in the grey wolf population in regions of Idaho after that species's re-introduction in the mid-1990s (*Source:* U.S. Fish and Wildlife Service).

   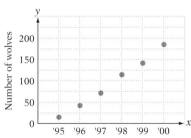

Year, x	'95	'96	'97	'98	'99	'00
Number of wolves, y	14	42	71	114	141	185

   a. Find the equation of a line that approximately fits the data by selecting two data points and finding the equation of the line through those two points.

   b. What does the slope of your line mean in the context of the problem?

   a. Answers will vary. The equation of the line through the points (95, 14) and (99, 141) is $y = 31.75x - 3002.25$.   b. The slope of 31.75 means that the wolf population increased by approximately 32 wolves per year from 1995 to 2000.

10. 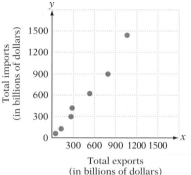 *Commerce*   The table shows the total U.S. exports and imports, in billions of dollars, for selected years (*Source:* U.S. Department of Commerce, Bureau of Economic Analysis). The scatter diagram represents the total exports along the horizontal axis and the total imports along the vertical axis.

Year	Total Exports	Total Imports
1970	56.6	54.4
1975	132.6	120.2
1980	271.8	291.2
1985	288.8	410.9
1990	537.2	618.4
1995	795.1	891.0
2000	1068.4	1438.1

a. Sketch a line of best fit through the scatter diagram of the data. Then find a linear equation for the line. Use decimals for the slope and *y*-intercept in the equation.

b. What does the slope of your line mean in the context of the problem?

a. Answers will vary. The regression equation is $y = 1.3045x - 40.8050$.   **b.** The slope of 1.3045 means that for each \$1 billion in exports, there is \$1.3045 billion in imports.

11. 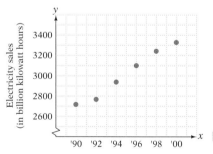 *Energy Consumption*   The table and scatter diagram show electricity sales, in billions of kilowatt hours, in the United States for selected years (*Source:* U.S. Department of Energy).

Year	Electricity Sales (in billion kilowatt hours)
1990	2713
1992	2763
1994	2935
1996	3098
1998	3240
2000	3325

a. Sketch a line of best fit through the scatter diagram of the data. Then find a linear equation for the line. Use decimals for the coefficient and constant in the equation.

b. What does the slope of your line mean in the context of the problem?

a. Answers will vary. The regression equation is $y = 66.4857x - 129{,}626.\overline{6}$ **b.** The slope of 66.4857 means that electricity sales increased 66.4857 billion kilowatt hours per year from 1990 to 2000.

Linear Regression

12. *Health*   A research hospital did a study on the relationship between stress and diastolic blood pressure. The results from 8 patients in the study are given in the table that follows.

Stress, x	55	62	58	78	92	88	75	80
Blood pressure, y	70	85	72	85	96	90	82	85

a. Find the regression line for the data.

b. Use the regression line to determine the blood pressure of a person whose stress test score was 85. Round to the nearest whole number.

c. Explain why the slope of the regression line indicates that blood pressure increases when stress increases.

a. $y = 0.563438438x + 41.71227477$ b. 90   c. The slope is positive; as x increases, y increases.

13. *Energy Consumption*   An automotive engineer studied the relationship between the speed of a car and the number of miles per gallon consumed at that speed. The results of the study are shown in the table that follows.

Speed, x (in miles per hour)	40	25	30	50	60	80	55	35	45
Consumption, y (in miles per gallon)	26	27	28	24	22	21	23	27	25

a. Find the regression line for the data.

b. Use the regression line to determine the expected number of miles per gallon for a car traveling 65 miles per hour. Round to the nearest whole number.

c. Explain why the slope of the regression line indicates that the number of miles per gallon decreases as speed increases.

a. $y = -0.1376741486 + 31.23316563$ b. 22 miles per gallon   c. The slope is negative; as x increases, y decreases.

14. *Earth Science*   A meteorologist studied the high temperature at various latitudes for January of a certain year. The results of the study are shown in the table that follows.

Latitude, x (in °N)	22	30	36	42	56	51	48
Temperature, y (in °F)	80	65	47	54	21	44	52

a. Find the regression line for this data.

b. Use the regression line to determine the expected temperature at a latitude of 45°N. Round to the nearest whole number.

c. What does the slope of the regression line mean in the context of the problem?

d. What does the y-intercept mean in the context of the problem?

a. $y = -1.353808752x + 106.9764992$ b. 46°F   c. The slope of approximately -1.35 indicates that the temperature decreases 1.35°F for every increase of 1°N in latitude.   d. The y-intercept of approximately 106.98 indicates that at 0°N, the temperature is 106.98°F.

15. *Zoology* A zoologist studied the running speed of animals in terms of the animal's body length. The results of the study are shown in the table that follows.

Body length, x (in centimeters)	1	9	15	16	24	25	60
Running speed, y (in meters per second)	1	2.5	7.5	5	7.4	7.6	20

a. Find the regression line for the data.

b. Use the regression line to determine the expected running speed of a deer mouse whose body length is 10 centimeters. Round to the nearest tenth.

c. ✎ What does the slope of the regression line mean in the context of the problem?

d. ✎ What does the *y*-intercept mean in the context of the problem?

a. $y = 0.3213184476x + 0.4003189793$ b. 3.6 m/s c. With an increase of 1 cm in body length, an animal's running speed increases approximately 0.32 m/s. d. The *y*-intercept of approximately 0.4 represents the running speed of an animal of length 0 cm.

16. *Paleontology* The data below shows the length, in centimeters, of the humerus and the total wingspan, in centimeters, of several pterosaurs, which are extinct flying reptiles of the order Pterosauria (*Source:* Southwest Educational Development Laboratory).

Pterosaur Data (in centimeters)

Humerus	Wingspan	Humerus	Wingspan
24	600	20	500
32	750	27	570
22	430	15	300
17	370	15	310
13	270	9	240
4.4	68	4.4	55
3.2	53	2.9	50
1.5	24		

a. Find the linear regression equation for the data.

b. On the basis of the linear regression model, what is the projected wingspan of the pterosaur *Quetzalcoatlus northropi*, which is thought to have been the largest of the prehistoric birds, if its humerus is 54 centimeters? Round to the nearest whole number.

a. $y = 23.55706665x - 24.4271215$ b. 1248 cm

17. *Demography* According to the U.S. Census Bureau, there were 69.9 million children under age 18 in the United States in 1998. In 1999 there were 70.2 million children under age 18 in the United States. It is projected that the number will increase to 77.2 million in 2020.

a. Let the year be the independent variable where $x = 98$ represents the year 1998, $x = 99$ represents the year 1999, and $x = 120$ represents the year 2020. Let the number of children, in millions, be the dependent variable. Find the regression line for the data.

b. ✎ What is the correlation coefficient? What does this mean about the fit of the data to the regression line?

c. Use the regression line to determine the expected number of children under age 18 in the United States in 2010. Round to the nearest hundred thousand.

d. ✎ What does the slope of the regression line mean in the context of the problem?

e. ✎ What does the *y*-intercept mean in the context of the problem?

a. $y = 0.3325054x + 37.2985961$ **b.** $r \approx 0.99999$; The fit of the data to the regression line is very good.   **c.** 73.9 million children   **d.** The slope of approximately 0.333 means that the number of children in the United States is increasing at a rate of about 0.333 million per year.   **e.** The *y*-intercept of approximately 37.299 means that in 1900 there were 37.299 million children in the United States.

18. *Air Density*   The density of air changes as temperature changes. These changes in density affect the resistance a cyclist experiences while riding. The table at the right gives the density of air, in kilograms per cubic meter, for various temperatures in degrees Celsius.

a. Find the linear regression equation of density in terms of temperature.

b. Use the regression equation to predict the density when the temperature is 27°C.

c. ✎ Write a sentence explaining the meaning of the slope of the repression line.

a. $y = -0.00411x + 1.28872$ **b.** 1.17777   **c.** The density of the air decreases by 0.00411 kg/m³ for each 1°C increase in temperature.

Temperature (in degrees Celsius)	Density of air (in kilograms per cubic meter)
0	1.292
15	1.225
20	1.204
25	1.184
30	1.165
35	1.146
40	1.127

19. *Electoral College* Listed at the right are the number of electoral college votes allotted to each of the 50 states and the District of Columbia (*Source:* Office of the Federal Register). Also listed is the population of each state (*Source:* U.S. Census Bureau).

a. Using population as the independent variable and number of electoral college votes as the dependent variable, find the regression line for the data.

b. ✎ What does the slope of the regression line mean in the context of the problem?

c. ✎ What does the *y*-intercept mean in the context of the problem?

d. ✎ Why is the *r* value not exactly equal to 1?

States	Population (in millions)	Electoral College Votes	States	Population (in millions)	Electoral College Votes
Alabama	4.4	9	Montana	0.9	3
Alaska	0.6	3	Nebraska	1.7	5
Arizona	5.1	10	Nevada	2.0	5
Arkansas	2.7	6	New Hampshire	1.2	4
California	33.9	55	New Jersey	8.4	15
Colorado	4.3	9	New Mexico	1.8	5
Connecticut	3.4	7	New York	19.0	31
Delaware	0.8	3	North Carolina	8.0	15
D.C.	0.6	3	North Dakota	0.6	3
Florida	16.0	27	Ohio	11.4	20
Georgia	8.2	15	Oklahoma	3.5	7
Hawaii	1.2	4	Oregon	3.4	7
Idaho	1.3	4	Pennsylvania	12.3	21
Illinois	12.4	21	Rhode Island	1.0	4
Indiana	6.1	11	South Carolina	4.0	8
Iowa	3.0	7	South Dakota	0.8	3
Kansas	2.7	6	Tennessee	5.7	11
Kentucky	4.0	8	Texas	20.9	34
Louisiana	4.5	9	Utah	2.2	5
Maine	1.3	4	Vermont	0.6	3
Maryland	5.3	10	Virginia	7.1	13
Massachusetts	6.3	12	Washington	5.9	11
Michigan	9.9	17	West Virginia	1.8	5
Minnesota	4.9	10	Wisconsin	5.4	10
Mississippi	2.8	6	Wyoming	0.5	3
Missouri	5.6	11			

a. $y = 1.551862378x + 1.986390721$ **b.** The slope of approximately 1.55 means that a state receives 1.55 electoral college votes per 1 million residents.
c. The *y*-intercept of approximately 1.986 means that a state with a population of 0 people would have 1.986 electoral college votes. (Note: Each state receives 2 electoral college votes for its 2 senators.)   **d.** The *r* value is not exactly 1 because the data are not completely linear. States cannot have a fractional part of a vote. The number of votes a state receives is rounded to a whole number.

20. Each screen shows a scatter diagram and the corresponding linear regression line. Match each screen with the corresponding correlation coefficient.

a.

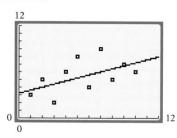

$y = 0.3\overline{69}x + 3.4\overline{6}$

b.

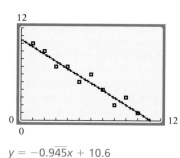

$y = -0.9\overline{45}x + 10.6$

i. $r = -0.515$

ii. $r = -0.970$

iii. $r = 0.515$

iv. $r = 0.970$

c.

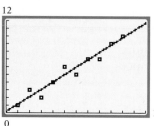

$y = 0.9\overline{45}x + 2$

d.

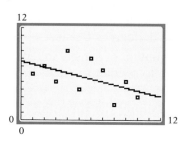

$y = -0.3\overline{69}x + 7.53$

a. iii   **b.** ii   **c.** iv   **d.** i

21. Each screen shows a scatter diagram and the corresponding linear regression line. Match each screen with the corresponding correlation coefficient.

a.

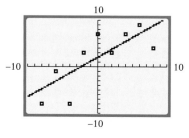

$y = 0.9x + 1.\overline{7}$

b.

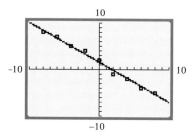

$y = -0.8\overline{6}x + 1.\overline{5}$

i. $r = -0.787$

ii. $r = 0.787$

iii. $r = -0.995$

iv. $r = 0.995$

c.

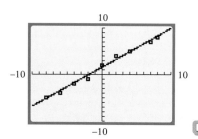

$y = 0.8\overline{6}x + 1.5$

d.

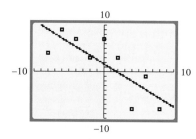

$y = -0.9x + 1.\overline{7}$

a. ii   **b.** iii   **c.** iv   **d.** i

22. Use linear regression to find the equation of the line containing the points $(1, 3)$ and $(5, -3)$.
$y = 1.5x + 4.5$

23. Use linear regression to find the equation of the line containing the points $(2, 4)$ and $(4, -1)$.
$y = -2.5x + 9$

24. *Demography*   Refer to the data on the number of high school graduates in Exercise 8.
   a. Find the regression line for these data. Use $x = 1$ for 2001, $x = 3$ for 2003, and so on.
   b. If the figure for 2009 were changed from 3.25 to 4.25, what would be the effect on the value of r^2? Why?
   a. $y = 0.0425x + 2.8375$ b. The r^2 value would decrease because the data values would not as closely fit a straight line.

25. *Zoology*   Refer to the data on the grey wolf population in Exercise 9.
   a. Find the regression line for these data. Use $x = 95$ for 1995, $x = 96$ for 1996, and so on.
   b. If the figure for 1998 were changed from 114 to 50, what would be the effect on the value of r^2? Why?
   a. $y = 34.142857x - 3234.4286$ b. The r^2 value would decrease because the data values would not as closely fit a straight line.

Applying Concepts

26. Use linear regression to find the equation of the line containing the points $(5, 1)$ and $(5, -2)$. Explain the result.
There is an error message; the slope of the line between the two points is undefined.

27. What is wrong with the statement "The r value is 1.35"?

The value of r is between -1 and 1; it cannot be greater than 1.

28. *Economics*   The following quote was printed on the web site **www.eia.doe.gov**. Provide an explanation of its meaning. "A strong correlation between economic growth and electricity use accounts for the variation in coal demand projections across the economic growth cases, with domestic coal consumption in 2020 projected to range from 1,245 to 1,426 million tons in the low and high economic growth cases, respectively."
Answers will vary. Students should note that there is a relationship between economic growth and electricity use: the greater the growth, the more electricity used. The projections for how much coal will be used in 2020 vary depending on whether low economic growth is expected (in which case it is anticipated that 1,245 million tons will be used) or high economic growth is expected (in which case it is anticipated that 1,426 million tons will be used).

Average SAT Scores

Verbal	Math
543	516
543	516
540	517
537	512
532	513
530	509
523	506
521	505
512	498
509	497
507	496
507	494
505	493
502	492
502	492
504	493
503	494
504	497
509	500
509	500
507	501
505	501
504	502
500	501
499	500
500	501
500	503
499	504
504	506
505	508
505	511
505	512
505	511
505	514

29. *Standardized Tests* The table at the right shows the average verbal SAT score and the average math SAT score for each year from 1967 through 2000 (*Source:* **www.collegeboard.org**).

   a. Let the average verbal SAT score be the independent variable and the average math SAT score be the dependent variable. Find the regression line for these data and the corresponding correlation coefficient.

   b. Let the average math SAT score be the independent variable and the average verbal SAT score be the dependent variable. Find the regression line for these data and the corresponding correlation coefficient. (*Note:* You do not need to reenter the data. You are able to instruct the calculator to treat the data in the second list as the independent variable and the data in the first list as the dependent variable. For assistance, see Appendix B: Regression.)

   c. Do you consider the average verbal SAT score to be a good predictor of the average math SAT score? Do you consider the average math SAT score to be a good predictor of the average verbal SAT score?

   a. $y = 0.36720042x + 315.62414$; $r \approx 0.6491$ **b.** $y = 1.1474246x - 66.269762$; $r = 0.6491$ **c.** No. The r values indicate that there is not a strong relationship between the two variables.

EXPLORATION

   1. *Elasticity Experiment*

   For this experiment, you will need:

   a styrofoam cup with two holes cut, on opposite sides of the cup, just below the top rim
   a piece of string several inches long
   a wide elastic band
   a large paper clip
   about 100 pennies
   heavy-duty tape
   a measuring tape or ruler with units in centimeters

   Thread the string through the two holes in the cup. Tie off the string on both sides of the cup. Attach the elastic band to the string. Attach the other end of the elastic band to the paper clip. Open the paper clip so that the cup can hang off the side of a desk or table. Tape the paper clip to the table.

   Use the tape measure to record the distance, in centimeters, from the top of the rubber band to the bottom of the cup. This is the initial distance. Record this trial in the table that follows. Put a number of pennies in the cup (anywhere from 10 to 25). Measure the distance to the bottom of the cup. Record this trial in the table. Continue adding pennies and recording the results in the table.

Number of pennies					
Distance to bottom of cup					

a. Make a scatter plot of the data. Describe any patterns you see.

Students should observe that the relationship between the variables is linear and that it is an increasing function.

b. Use your graphing calculator to generate a regression line.

Equations will vary.

c. What is the correlation coefficient of your line? What does this tell you?

Answers will vary. However, there should be a fairly high correlation between the two variables.

d. What does the slope of your line mean in the context of the experiment?

The slope represents the increase in distance measured, in centimeters, for every penny added to the cup.

e. What does the *y*-intercept of your line mean in the context of the experiment?

The *y*-intercept represents the distance measured when there are no pennies in the cup.

f. What is a reasonable domain of your equation? What is a reasonable range of your equation?

Answers will vary.

g. Use your model to predict what the distance from the top of the rubber band to the bottom of the cup would be if you placed 125 pennies in the cup.

Answers will vary.

h. If the elastic band were less elastic (tighter), what would you expect to happen to the slope of the line?

The line would be flatter; the slope would decrease.

i. If the elastic band were more elastic (stretchier), what would you expect to happen to the slope of the line?

The line would be steeper; the slope would be greater.

j. If the objects you were putting in the cup were lighter, how would that affect the slope of the line?

The line would be flatter; the slope would decrease.

k. If the objects you were putting in the cup were heavier, how would that affect the slope of the line?

The line would be steeper; the slope would be greater.

l. If instead of measuring from the top of the elastic to the bottom of the cup, you measured from the top of the paper clip to the bottom of the cup, how would that affect the equation of the line?

The *y*-intercept would be higher.

m. If instead of measuring from the top of the elastic to the bottom of the cup, you measured from the floor to the bottom of the cup, how would that affect the equation of the line?

The slope would be decreasing instead of increasing.

Linear Inequalities in Two Variables

■ Solution Sets of Inequalities in Two Variables

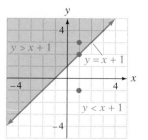

The graph of the linear equation $y = x + 1$ separates the plane into three sets: the set of points on the line, the set of points above the line, and the set of points below the line.

The point whose coordinates are $(1, 2)$ is a solution of $y = x + 1$ and is a point on the line.

The point whose coordinates are $(1, 3)$ is a solution of $y > x + 1$ and is a point above the line.

The point whose coordinates are $(1, -1)$ is a solution of $y < x + 1$ and is a point below the line.

The set of points on the line are the solutions of the equation $y = x + 1$. The set of points above the line are the solutions of the inequality $y > x + 1$. These points form a **half-plane.** The set of points below the line are solutions of the inequality $y < x + 1$. These points also form a half-plane.

? QUESTION Which ordered pairs are solutions of the inequality $y < 2x - 3$?

    a. $(-1, 4)$ **b.** $(0, 5)$ **c.** $(3, 1)$ **d.** $(-2, -7)$ **e.** $(2, -3)$

An inequality of the form $y > mx + b$ or $Ax + By > C$ is a **linear inequality in two variables.** (The inequality symbol could be replaced by $<$, $\leq$, or $\geq$.) The **solution set of a linear inequality in two variables** is a half-plane.

The following illustrates the procedure for graphing the solution set of a linear inequality in two variables.

➡ Graph the solution set of $2x - 5y > 10$.

Solve the inequality for y.

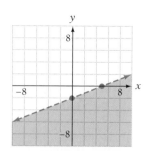

$$2x - 5y > 10$$
$$-5y > -2x + 10$$
$$\frac{-5y}{-5} < \frac{-2x + 10}{-5}$$
$$y < \frac{2}{5}x - 2$$

Suggested Activity

How would the graph at the right change if the inequality symbol in the original inequality were $\geq$ instead of $>$? What if it were $<$ instead of $>$? What if it were $\leq$ instead of $>$?

? ANSWER   **a.** No. $4 < -5$ is not true.   **b.** No. $5 < -3$ is not true.   **c.** Yes. $1 < 3$ is true.
d. No. $-7 < -7$ is not true.   **e.** Yes. $-3 < 1$ is true.

Change the inequality $y < \frac{2}{5}x - 2$ to the equality $y = \frac{2}{5}x - 2$, and graph the line.

After solving the equation for y, look at the inequality symbol. **If the inequality symbol is $\leq$ or $\geq$, the line belongs to the solution set and is shown by a solid line. If the inequality symbol is $<$ or $>$, the line is not part of the solution set and is shown by a dashed line.**

If the inequality contains $>$ or $\geq$, shade the upper half-plane. If the inequality contains $<$ or $\leq$, shade the lower half-plane. Every point in the shaded half-plane is in the solution set of the inequality.

As a check, use the ordered pair $(0, 0)$ to determine whether the correct region of the plane has been shaded. **If $(0, 0)$ is a solution of the inequality, then $(0, 0)$ should be in the shaded region. If $(0, 0)$ is not a solution of the inequality, then $(0, 0)$ should not be in the shaded region.**

To check whether $(0, 0)$ is a solution of the inequality, substitute 0 for x and 0 for y in the original inequality. For the example above,

$$\frac{2x - 5y > 10}{2(0) - 5(0) \ \big|\ 10}$$

$$0 > 10 \qquad \text{False. } (0, 0) \text{ is not a solution of the inequality.}$$
$$\qquad\qquad (0, 0) \text{ should not be in the shaded region.}$$

If the line passes through the point $(0, 0)$, another point, such as $(0, 1)$, must be used as a check. $\longleftarrow$

From the graph of $y < \frac{2}{5}x - 2$, note that for a given value of x, more than one value of y can be paired with the value of x. For instance, $(5, -1)$ and $(5, -2)$ are ordered pairs that are solutions of the inequality.

$y < \dfrac{2}{5}x - 2$			$y < \dfrac{2}{5}x - 2$	
-1	$\dfrac{2}{5}(5) - 2$		-2	$\dfrac{2}{5}(5) - 2$
-1	$2 - 2$		-2	$2 - 2$
$-1 < 0$	True		$-2 < 0$	True

Because there are ordered pairs with the same first component and different second components, the inequality does not represent a function. The inequality is a relation but not a function.

See Appendix A:
Graphing Linear
Inequalities

A graphing calculator can be used to graph the solution set of an inequality in two variables. Some typical screens for graphing the solution set of $y \leq 2x - 3$ on a graphing calculator follow.

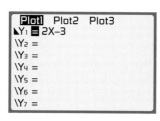

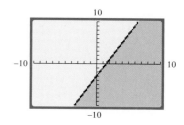

An inequality with the symbol ≥ or ≤ should be graphed as a solid line, and an inequality with the symbol > or < should be graphed as a dashed line. However, a graphing calculator does not distinguish between a solid line and a dashed line.

INSTRUCTOR NOTE
Exercise 12 can be used for a similar in-class example.

Suggested Activity

See Section 4.4 of the *Student Activity Manual* for a lesson involving linear inequalities in two variables.

EXAMPLE 1

Graph the solution set of $x + 2y \geq 6$. Is $(6, -4)$ in the solution set?

Solution Solve the inequality for y.

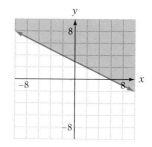

$$x + 2y \geq 6$$
$$2y \geq -x + 6$$
$$\frac{2y}{2} \geq \frac{-x + 6}{2}$$
$$y \geq -\frac{1}{2}x + 3$$

Graph $y = -\frac{1}{2}x + 3$ as a solid line.

Shade the upper half-plane.

From the graph, we can see that the point $(6, -4)$ is not in the solution set of the inequality.

GRAPHICAL CHECK:	**ALGEBRAIC CHECK:**

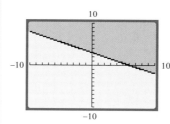

$$y \geq -\frac{1}{2}x + 3$$

-4	$-\frac{1}{2}(6) + 3$
-4	$-3 + 3$
$-4 \geq 0$	False.

The point $(6, -4)$ is not in the solution set.

YOU TRY IT 1

Graph the solution set of $2x - 3y < 12$. Is $(3, -1)$ in the solution set?

Solution See page S17.

Yes

Just as we can draw graphs of equations such as $x = 3$ and $y = -4$, we can draw graphs of similar inequalities.

INSTRUCTOR NOTE
Exercise 24 can be used for a similar in-class example.

EXAMPLE 2

Graph the solution set.

a. $x > -4$ **b.** $y \leq 2$

Solution **a.** Graph $x = -4$ as a dashed line.

$0 > -4$ is a true inequality.
The point $(0, 0)$ satisfies the inequality.
The point $(0, 0)$ should be in the shaded region.
Shade the half-plane to the right of the line.

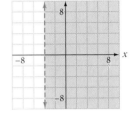

b. Graph $y = 2$ as a solid line.

$0 \leq 2$ is a true inequality.
The point $(0, 0)$ satisfies the inequality.
The point $(0, 0)$ should be in the shaded region.
Shade the half-plane below the line.

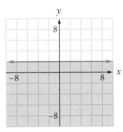

YOU TRY IT 2

Graph the solution set.

a. $x \geq 1$ **b.** $y < -5$

Solution See page S17. **a.** 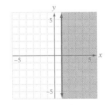 **b.**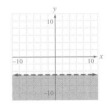

4.4 EXERCISES

Suggested Assignment: 5–25, odds

Topics for Discussion

1. Is it possible to write a linear inequality in two variables that has no solution?

   No.

2. What is a half-plane?

   A solution of a linear inequality in two variables.

3. Does a linear inequality in two variables define a function? Why or why not?

   No. There are ordered pairs with the same first component and different second components.

4. How does the graph of a linear inequality in two variables differ when $<$ is used from when $\leq$ is used?

When $<$ is used, the line is not part of the solution set and is drawn as a dashed line. When $\leq$ is used, the line is part of the solution set and is drawn as a solid line.

■ Solution Sets of Inequalities in Two Variables

Graph the solution set. State whether the given ordered pair is in the solution set.

5. $y \leq \dfrac{3}{2}x - 3$; $(4, -3)$

Yes

6. $y \geq \dfrac{4}{3}x - 4$; $(-6, 0)$

Yes

7. $y < -\dfrac{1}{3}x + 2$; $(-3, 6)$

No

8. $y < \dfrac{3}{5}x - 3$; $(-5, -2)$

No

9. $y < \dfrac{4}{5}x - 2$; $(5, -1)$

Yes

10. $y < -\dfrac{4}{3}x + 3$; $(3, -3)$

Yes

11. $x + 3y < 6$; $(1, 2)$

No

12. $2x - 5y \leq 10$; $(5, 0)$

Yes

13. $2x + 3y \geq 6$; $(6, -1)$

Yes

14. $3x + 2y < 4$; $(-4, -1)$

Yes

15. $-x + 2y > -8$; $(5, -3)$

No

16. $-3x + 2y > 2$; $(-2, -3)$

No

17. $y < 4$; $(0, 2)$

Yes

18. $y > 3$; $(-3, 4)$

Yes

19. $6x + 5y < 15$; $(-6, 0)$

Yes

20. $3x - 5y < 10$; $(0, -2)$
No

21. $-5x + 3y \geq -12$; $(2, -2)$
No

22. $3x + 4y \geq 12$; $(0, 3)$
Yes

23. $x \geq -2$; $(1, 1)$
Yes

24. $x < 3$; $(5, -1)$
No

25. $y \leq -2$; $(-4, 3)$
No

Applying Concepts

Graph the solution set.

26. $y - 5 < 4(x - 2)$

27. $y + 3 < 6(x + 1)$

28. $3x - 2(y + 1) \leq y - (5 - x)$

29. $2x - 3(y + 1) \geq y - (4 - x)$

Write the inequality given its graph.

30.
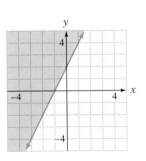

Ⓟ

$y \geq 2x + 2$

31.
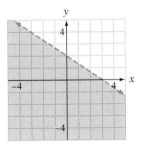

Ⓟ

$y < -\dfrac{2}{3}x + 2$

32. Are there any points whose coordinates satisfy both $y \le x + 3$ and $y \ge -\frac{1}{2}x + 1$? If so, give the coordinates of three such points. If not, explain why not.

Yes. Answers will vary.

33. Are there any points whose coordinates satisfy both $y \le x - 1$ and $y \ge x + 2$? If not, explain why not.

No. The solution sets do not intersect, which means that there are no ordered pairs that satisfy both inequalities.

EXPLORATION

INSTRUCTOR NOTE
You can ask students either to graph the inequalities on graph paper or to use a graphing calculator.

1. *Constraints* Linear inequalities are used as constraints (conditions that must be satisfied) for some application problems. For these problems, the focus of attention is only the first quadrant, so a solution outside the first quadrant is not considered. For each of the following, shade the region of the first quadrant that satisfies the constraints.

a. Suppose a manufacturer makes two types of computer monitors, 15-inch and 17-inch. Because of the production requirements for these monitors, the maximum number of monitors that can be produced in one day is 100. Shade the region of the first quadrant whose ordered pairs satisfy the constraint. (*Hint:* Let x = the number of 15-inch monitors and y = the number of 17-inch monitors. Write an inequality. Solve the inequality for y. Graph the inequality.)

b. A manufacturer makes two types of bicycle gears, standard and deluxe. It takes 4 hours of labor to produce a standard gear and 6 hours of labor to produce a deluxe gear. If there are a maximum of 480 hours of labor available, shade the region of the first quadrant whose ordered pairs satisfy the constraint. (*Hint:* Let x = the number of standard gears produced and y = the number of deluxe gears produced. Write an inequality for the number of hours of labor. Solve the inequality for y. Graph the inequality.)

c. Suppose a single tablet of the diet supplement SuperC contains 150 milligrams of calcium and that one tablet of the diet supplement CalcPlus contains 200 milligrams of calcium. A health care professional recommends that a patient take at least 2000 milligrams of calcium per day. Shade the region of the first quadrant whose ordered pairs satisfy these constraints. (*Hint:* Let x = the number of SuperC tablets and y = the number of CalcPlus tablets.)

d. A farmer is planning to raise wheat and barley. Each acre of wheat yields a profit of $50, and each acre of barley yields a profit of $70. The farmer wants to make a profit of at least $3500. Shade the region of the first quadrant whose ordered pairs satisfy the constraint. (*Hint:* Let x = the number of acres of wheat and y = the number of acres of barley.)

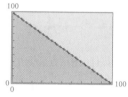

$y \le 100 - x$

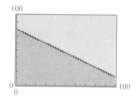

$y \le -\frac{2}{3}x + 80$

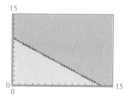

$y \ge -\frac{3}{4}x + 10$

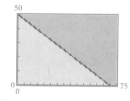

$y \ge -\frac{5}{7}x + 50$

CHAPTER **4** *SUMMARY*

Key Terms

coefficient of determination [p. 281]
constant function [p. 243]
correlation coefficient [p. 281]
half-plane [p. 294]
linear function [p. 235]
linear inequality in two variables
  [p. 294]
linear regression [p. 280]
line of best fit [p. 278]

negative reciprocals [p. 265]
parallel lines [p. 262]
perpendicular lines [p. 264]
regression line [p. 280]
scatter diagram [p. 277]
slope [p. 236]
solution set of a linear inequality in
  two variables [p. 294]

Essential Concepts

Slope of a Line

Let $P_1(x_1, y_1)$ and $P_2(x_2, y_2)$ be two points on a line. Then the slope m of the line through the two points is the ratio of the change in the y-coordinates to the change in the x-coordinates. [p. 237]

$$m = \frac{\text{change in } y}{\text{change in } x} = \frac{y_2 - y_1}{x_2 - x_1}, x_1 \neq x_2$$

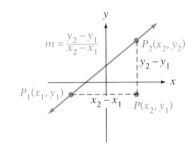

Slope–Intercept Form of a Straight Line

The equation $y = mx + b$ is called the slope–intercept form of a straight line. The slope of the line is m, the coefficient of x. The y-intercept is $(0, b)$. [p. 240]

Standard Form of the Equation of a Line

The equation of a line written in the form $Ax + By = C$ is called the standard form of the equation of a line. [p. 241]

Constant Function

A function given by $f(x) = b$, where b is a constant, is a constant function. The graph of a constant function is a horizontal line passing through $(0, b)$. [p. 243]

Point–Slope Formula of a Straight Line

Let $P_1(x_1, y_1)$ be a point on a line, and let m be the slope of the line. Then the equation of the line can be found by using the point–slope formula $y - y_1 = m(x - x_1)$. [p. 257]

To Find the Equation of a Line Between Two Points

Find the slope of the line between the two points. Then use the point–slope formula to find the equation of the line. [p. 260]

Parallel Lines

Two nonvertical lines with slopes m_1 and m_2 are parallel lines if and only if $m_1 = m_2$. Vertical lines are also parallel lines. [p. 262]

Perpendicular Lines

If m_1 and m_2 are the slopes of two lines, neither of which is vertical, then the lines are perpendicular if and only if $m_1 \cdot m_2 = -1$. A vertical line is perpendicular to a horizontal line. [p. 264]

CHAPTER **4** REVIEW EXERCISES

1. Find the slope of the line that contains the points $(-1, 3)$ and $(-2, 4)$.
   -1

2. Find the slope of the line that passes through the points $(-6, 5)$ and $(-6, 4)$.
   Undefined

3. Graph the line that has slope $\frac{1}{2}$ and passes through the point $(-2, 4)$.

4. Graph $y = -\frac{2}{3}x + 4$ by using the slope and y-intercept.

5. Graph $x + 2y = -4$.

6. Graph $y = 3$.

7. Find the equation of the line that has slope $-\frac{4}{3}$ and passes through the point $(0, -5)$.
   $y = -\frac{4}{3}x - 5$

8. Find the equation of the line that has slope 4 and passes through the point $(1, 2)$.
   $y = 4x - 2$

9. Find the equation of the line that passes through the points $(2, 6)$ and $(-4, 9)$.
   $y = -\frac{1}{2}x + 7$

10. Find the equation of the line that passes through the points $(3, -4)$ and $(-2, -4)$.
    $y = -4$

11. Is the line that contains the points $(4, 3)$ and $(6, 2)$ parallel to the line that contains the points $(3, 2)$ and $(1, 4)$?
    No

12. Find the equation of the line that is parallel to the line $3x + y = 4$ and contains the point $(3, -2)$.
    $y = -3x + 7$

13. Is the line that contains the points $(3, 5)$ and $(-3, 3)$ perpendicular to the line that contains the points $(2, -5)$ and $(-4, 4)$?
    No

14. Find the equation of the line that is perpendicular to $y = -\frac{2}{3}x + 6$ and contains the point $(2, 5)$.
    $y = \frac{3}{2}x + 2$

15. Graph the solution set of $y \le -\frac{3}{2}x + 6$.

16. A contractor estimates that it costs $60,000 plus $90 per square foot to build a house. Write a linear equation for the cost to build a house. Use your equation to find the cost to build a 2500-square-foot house.
    $y = 90x + 60,000$; $285,000

17. The graph at the right shows the relationship between the age of a person and the recommended maximum exercise heart rate for a person who exercises regularly. Find the slope of the line between the two points shown on the graph. Write a sentence that states the meaning of the slope in the context of this problem.

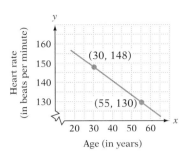

The slope is −0.72. The maximum recommended exercise heart rate decreases −0.72 beat per minute for every year older.

18. Water is being added to a pond that already contains 1000 gallons of water. The table at the right shows the number of gallons of water in the pond after selected times, in minutes.
  a. Find a linear model for the number of gallons of water in the tank after t minutes.
  b. Write a sentence that explains the meaning of the slope of the line in the context of this problem.
  c. Assuming that water continues to flow into the pond at the same rate, how many gallons of water will be in the tank after 6 hours?

Time (in minutes)	Water (in gallons)
30	1750
60	2500
80	3000
120	4000

a. $y = 25x + 1000$ b. Water is being added to the pond at a rate of 25 gal/min.
c. 10,000 gal

19. A linear model for a monthly cellular phone bill is given by $F(x) = 0.25x + 19.95$, where $F(x)$ is the monthly cellular phone bill and x is the number of minutes the phone was used during the month.
  a. Write a sentence that explains the meaning of the slope of this function in the context of this problem.
  b. Write a sentence that explains the meaning of the y-intercept in the context of this problem.

a. It costs $.25 per minute to use the phone.   b. The y-intercept is 19.95. When the phone is used for 0 minutes during the month, the phone bill is $19.95.

20. The "apparent temperature" takes into consideration not only the temperature but the relative humidity as well. The following table gives the apparent temperature when the actual temperature is 85°F for various humidities.
  a. Find a linear regression equation that gives apparent temperature in terms of the relative humidity.
  b. Use the regression equation to find the apparent temperature when the relative humidity is 75. Round to the nearest tenth.

Relative humidity, x (in percents)	30	40	50	60	70	80	90
Apparent temperature, y (in °F)	84	86	88	90	93	97	102

a. $y = 0.28928571x + 74.07142857$ b. 95.8°F

CHAPTER **4** *TEST*

1. Find the slope of the line that contains the points $(-2, 6)$ and $(-1, 4)$.

−2

2. Find the slope of the line that passes through the points $(-4, 3)$ and $(-8, 3)$.

0

3. Graph the line that has slope $-\frac{3}{2}$ and passes through the point $(-2, 3)$.

4. Graph $y = \frac{2}{3}x - 4$ by using the slope and y-intercept.

5. Graph $2x + y = 2$.

6. Graph $y = -2$.

7. Write the equation $3x - 4y = 8$ in slope–intercept form.

$y = \frac{3}{4}x - 2$

8. Graph the solution set of $6x - y > 6$.

9. Find the equation of the line that has slope $-\frac{5}{3}$ and passes through the point $(0, -4)$.

$y = -\frac{5}{3}x - 4$

10. Find the equation of the line that has slope $\frac{1}{2}$ and passes through the point $(3, -2)$.

$y = \frac{1}{2}x - \frac{7}{2}$

11. Find the equation of the line containing the points $(0, 1)$ and $(-1, 0)$.

$y = x + 1$

12. Find the equation of the line that passes through the points $(4, -2)$ and $(-4, 8)$.

$y = -\frac{5}{4}x + 3$

13. Is the line that contains the points $(3, 3)$ and $(-3, 7)$ parallel to the line that contains the points $(6, -5)$ and $(-6, 3)$?

Yes

14. Find the equation of the line that is parallel to the line $4x - y = 2$ and contains the point $(-3, 1)$.

$y = 4x + 13$

15. Is the line that contains the points $(4, 1)$ and $(-2, 4)$ perpendicular to the line that contains the points $(-1, -6)$ and $(3, 2)$?

Yes

16. Find the equation of the line that is perpendicular to $y = 3x + 2$ and contains the point $(3, 4)$.

$y = -\frac{1}{3}x + 5$

17. The table at the right shows the monthly profit for Sportlete.com for January, March, and May. Assume that profits continue in the same manner.
  a. Find a linear model for the data.
  b. Write a sentence that explains the meaning of the slope of this line in the context of this problem.
  c. Use the model to predict the profit in December.

Month	Profit
1 (January)	$20,000
3 (March)	$23,000
5 (May)	$26,000

  a. $y = 1500x + 18{,}500$ **b.** The slope of 1500 means that the profit is increasing by $1500 per month. **c.** The profit in December will be $36,500.

18. The increases in fuel prices for the first 6 months of a recent year are shown in the graph at the right. Find the slope of the line between the two points shown on the graph. Write a sentence that states the meaning of the slope in the context of this problem.

  $m = 7$. The prices are increasing 7 cents per month.

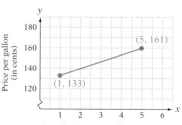

19. A molten piece of metal is allowed to cool in a controlled environment. The temperature, in degrees Fahrenheit, of the metal after it is removed from a smelter for various times t, in minutes, is shown at the right.

a. Find a linear model for the temperature of the metal after t minutes.

b. Write a sentence that explains the meaning of the slope of this line in the context of this problem.

c. Assuming that temperature continues to decrease at the same rate, what will be the temperature in 2 hours?

Time (in minutes)	Temperature (in degrees Fahrenheit)
15	2500
20	2400
30	2200

a. $y = -20x + 2800$ **b.** The slope of -20 means the metal is cooling 20°F per minute.   **c.** 400°F

20. The data in the following table show the curb weight, in pounds, and horsepower of the 10 best cars of 2001 as ranked by *Car and Driver* magazine (*Source: Car and Driver* web site, **www.10bestcars. com**).

a. Find a linear regression equation that gives horsepower in terms of weight.

b. Use the regression equation to find the horsepower when the curb weight is 2700 pounds. Round to the nearest whole number.

Curb Weight and Horsepower of 10 Selected Cars

Car Model	Weight	Horsepower	Car Model	Weight	Horsepower
Audi A6	3750	300	Ford Focus	2550	130
Audi TT	2900	225	Honda Accord	3000	200
BMW 3 Series	3250	225	Honda S2000	2800	240
BMW 5 Series	3450	282	Mazda MX-5	2400	155
Chrysler PT Cruiser	3150	150	Porsche Boxster	2800	250

a. $y = 0.098x - 80.079$ **b.** The equation predicts that a car that weighs 2700 lb has an engine that delivers approximately 186 hp.

◀ CUMULATIVE REVIEW EXERCISES

1. In how many different ways can a panel of four on–off switches be set if no two adjacent switches can be off?

8

2. Given the operation $a @ b = a + ab$, evaluate $(x @ y) @ z$ for $x = 2$, $y = 3$, $z = 4$.

40

3. Let $E = \{0, 5, 10, 15\}$ and $F = \{-10, -5, 0, 5, 10\}$. Find $E \cup F$ and $E \cap F$.

$E \cup F = \{-10, -5, 0, 5, 10, 15\}$; $E \cap F = \{0, 5, 10\}$

4. Evaluate the variable expression $2(a - d)^2 + (bc)^2$ for $a = -2$, $b = -1$, $c = 3$, and $d = -4$.

17

5. The formula for the volume of a regular pyramid is $V = \frac{1}{3}s^2h$, where V is the volume, s is the length of a side of the square base, and h is the height. Find the volume of the regular pyramid shown at the right.

15 ft³

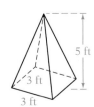

6. Simplify: $-3[4x - (2x - 1)]$

   $-6x - 3$

7. Translate "the product of twenty and one-fifth of a number" into a variable expression. Then simplify the expression.

   $20\left(\dfrac{1}{5}x\right)$; $4x$

8. Without using a calculator, create an input/output table for $y = x^2 - 2$ for $x = -2, -1, 0, 1,$ and 2. Check your results using a graphing calculator. Graph the resulting ordered pairs.

x	-2	-1	0	1	2
y	2	-1	-2	-1	2

9. Give the domain and range of the relation. Is the relation a function?
   $\{(-1, 0), (0, 1), (1, 2), (1, 0), (0, -1)\}$

   D: $\{-1, 0, 1\}$, R: $\{-1, 0, 1, 2\}$, No

10. Evaluate $p(s) = -2s^2 + 5s - 4$ when $s = -1$.

    -11

11. Find the x- and y-intercepts for the graph of the function $h(x) = x^2 - 2x - 3$.

    $(-1, 0), (3, 0); (0, -3)$

12. Use a graphing calculator to find the input value for which the two functions $f(x) = 3x - 4$ and $g(x) = 6x + 5$ have the same value. Verify the results algebraically.

    -3

13. Solve: $3t - 8t = 0$

    0

14. Solve: $8 - z = 6z - 3(4 - z)$

    2

15. A merchant combines coffee that costs $6 per pound with coffee that costs $4 per pound. How many pounds of each should be used to make 60 pounds of a blend that costs $4.50 per pound?

    15 lb of the $6 coffee; 45 lb of the $4 coffee

16. Given that $l_1 \| l_2$, find the measures of angles a and b.

    $m\angle a = 46°$, $m\angle b = 134°$

    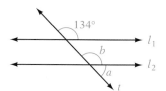

17. Solve $2 - 4(x + 1) \ge 11 + 3(2x - 6)$. Write the solution set in interval notation.

    $\left(-\infty, \dfrac{1}{2}\right]$

18. Solve: $2 + |3x + 1| = 7$

    $-2, \dfrac{4}{3}$

19. Graph $4x - 6y = 12$.

20. Find the equation of the line that contains the point $(2, 5)$ and is perpendicular to the line $y = -\dfrac{2}{3}x + 6$.

    $y = \dfrac{3}{2}x + 2$

5

Systems of Linear Equations and Inequalities

```
CALCULATE
1: value
2: zero
3: minimum
4: maximum
5: intersect
6: dy/dx
7: ∫f(x)dx
```

Press **2nd** CALC to access the intersect option.

*Pharmacists not only dispense pre-packaged pills and medications, they are also trained healthcare professionals who can offer advice about minor ailments and medicine usage. They can suggest over-the-counter treatments for symptoms of common ailments such as colds or allergies. Pharmacists are also called upon to mix different ingredients in proportions that satisfy certain criteria, thereby creating specific treatments and remedies for their customers. This is illustrated in **Exercise 37 on page 318,** in which a system of equations is used to determine the amount of each ingredient in the mixture.*

Need help? For online student resources, visit this web site:
Math.college.hmco.com

PREP TEST

1. Simplify: $10\left(\frac{3}{5}x + \frac{1}{2}y\right)$ 6x + 5y

2. Evaluate $3x + 2y - z$ for $x = -1$, $y = 4$, and $z = -2$.   7

3. Given $3x - 2z = 4$, find the value of x when $z = -2$.   0

4. Solve: $3x + 4(-2x - 5) = -5$ −3

5. Solve: $0.45x + 0.06(-x + 4000) = 630$ 1000

6. Graph: $y = \frac{1}{2}x - 4$

7. Graph: $3x - 2y = 6$

8. Graph: $y > -\frac{3}{5}x + 1$

GO FIGURE

I have two more sisters than brothers. Each of my sisters has two more sisters than brothers. How many more sisters than brothers does my youngest brother have?

four more sisters than brothers

SECTION 5.1

- Solve Systems of Equations by Graphing
- Solve Systems of Equations by the Substitution Method

Solving Systems of Linear Equations by Graphing and by the Substitution Method

■ Solve Systems of Equations by Graphing

Suppose Maria and Michael drove from California to Connecticut and that the total driving time was 48 hours. We now pose the question, "How long did Maria drive?" From the given information, it is impossible to tell. Maria may have driven 30 hours and Michael 18 hours; she may have driven 1 hour and Michael 47 hours; or many other possibilities. If we let x be the number of hours Michael drove and y the number of hours Maria drove, then the equation $x + y = 48$ expresses the fact that the total driving time was 48 hours. The graph of this equation is shown at the right.

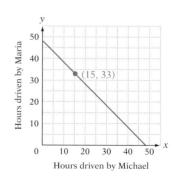

Any ordered pair on the graph represents possible driving times for Maria and Michael. For instance, the ordered pair $(15, 33)$ means that Michael drove 15 hours and Maria drove 33 hours.

Now suppose that we obtain the additional information that Maria drove twice as many hours as Michael. This can be expressed as the equation $y = 2x$. The graph of this equation is shown at the right, along with $x + y = 48$. The point of intersection $(16, 32)$ satisfies both conditions of the problem: The total number of hours driven were $48 [16 + 32 = 48]$ and Maria drove twice as many hours as Michael $[32 = 2(16)]$.

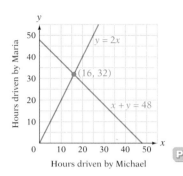

$$x + y = 48$$
$$y = 2x$$

A **system of equations** is two or more equations considered together. The system of equations for Maria and Michael is shown at the right. Because each equation of the system is a linear equation, this is a **system of linear equations in two variables.**

A **solution of a system of equations in two variables** is an ordered pair that is a solution of each equation of the system. For instance, as the following shows, $(16, 32)$ is a solution of the system of equations for the driving times of Michael and Maria.

$x + y = 48$	$y = 2x$	• Replace x by 16 and y by 32. Because
$16 + 32 \mid 48$	$32 \mid 2(16)$	the ordered pair is a solution of each
$48 = 48$	$32 = 32$	equation, it is a solution of the system of equations.

The solution of the system of equations is the ordered pair whose values of x and y simultaneously satisfy the conditions imposed by the equations.

A solution of a system of linear equations can be found by graphing the lines of the system on the same coordinate axes. **The coordinates of the point of intersection of the lines is the solution of the system of equations.**

Suggested Activity

See Section 5.1 of the *Student Activity Manual* for a lesson on using a graphing calculator to solve systems of linear equations.

EXAMPLE 1

Solve by graphing: $\begin{array}{l} 5x - 2y = 9 \\ 3x + 2y = -1 \end{array}$

Solve each equation for y.

$$5x - 2y = 9 \qquad\qquad 3x + 2y = -1$$
$$-2y = -5x + 9 \qquad\qquad 2y = -3x - 1$$
$$y = \frac{5}{2}x - \frac{9}{2} \qquad\qquad y = -\frac{3}{2}x - \frac{1}{2}$$

INSTRUCTOR NOTE
Exercise 8 can be used for a similar in-class example.

 See Appendix A:
Intersect

Enter the two equations into Y_1 and Y_2 and then graph them. If necessary, adjust the viewing window so that the point of intersection shows on the screen.

Use the INTERSECT feature of the calculator to find the point of intersection.

The solution is $(1, -2)$.

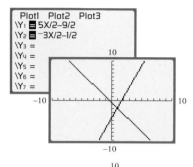

Algebraic Check:

Replace x by 1 and y by -2 in each equation of the system of equations.

$5x - 2y = 9$	
$5(1) - 2(-2)$	9
$5 + 4$	
$9 = 9$ √	

$3x + 2y = -1$	
$3(1) + 2(-2)$	-1
$3 - 4$	
$-1 = -1$ √	

The solution checks.

YOU TRY IT 1

Solve by graphing: $y = -\frac{2}{3}x + 1$
$2x + y = -3$

Solution   See page S17. $(-3, 3)$

When the graphs of a system of equations intersect at only one point, the system of equations is called an **independent system of equations.** The system of equations in Example 1 is an independent system of equations.

❓ QUESTION   Is the system of equations for Michael and Maria an independent system of the equations?

Recall that two lines with the same slope are parallel lines. If the y-intercepts are not equal, parallel lines do not intersect. Therefore, a system of equations that contains equations whose slopes are equal but whose y-intercepts are different has no solution. This is called an **inconsistent system of equations.**

➡ Show that the system of equations is inconsistent: $2x - 3y = 6$
$4x - 6y = -18$

❓ ANSWER   The graphs intersect at one point. Yes, the system of equations is independent.

CALCULATOR NOTE

Sometimes the graphs on a calculator may appear to be parallel when in fact they are not. The only way to be absolutely sure that the graphs are parallel is by solving each equation of the system for y and comparing the slopes.

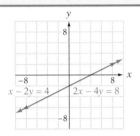

TAKE NOTE

Keep in mind the differences among independent, dependent, and inconsistent systems of equations. You should be able to express your understanding of these terms by using graphs.

Suggested Activity

To introduce the substitution method, you might ask students to solve the following systems of equations.

1. $3x - 2y = 4$
   $\qquad x = 2$ (2, 1)
2. $\qquad y = -2$
   $2x + 3y = 4$ (5, -2)

Ask students to explain why we can substitute 2 for x in the first example and -2 for y in the second example. Then ask them to use substitution to solve the following system.

3. $\qquad y = 2x - 1$
   $x + 2y = 3$ (1, 1)

Ask them to explain why we can substitute $2x - 1$ for y in this situation.

ALGEBRAIC SOLUTION

Solve each equation for y.

$$2x - 3y = 6 \qquad\qquad 4x - 6y = -18$$
$$-3y = -2x + 6 \qquad -6y = -4x - 18$$
$$y = \frac{2}{3}x - 2 \qquad\qquad y = \frac{2}{3}x + 3$$

The lines have the same slope and different y-intercepts. Therefore, the lines are parallel and do not intersect. The system of equations is inconsistent and has no solution.

GRAPHICAL CHECK

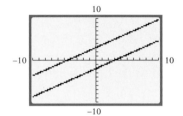

The lines are parallel. The system of equations has no solution.

Now consider the system of equations $\begin{aligned} x - 2y &= 4 \\ 2x - 4y &= 8 \end{aligned}$. Solving each equation for y, we have

$$x - 2y = 4 \qquad\qquad 2x - 4y = 8$$
$$-2y = -x + 4 \qquad -4y = -2x + 8$$
$$y = \frac{1}{2}x - 2 \qquad\qquad y = \frac{1}{2}x - 2$$

In this case, both the slopes and the y-intercepts are equal. Therefore, the equations represent the same line. This is a **dependent system of equations.**

When the equations of this system of equations are graphed, one line will graph on top of the other line. The solutions of this system of equations are the ordered pairs that satisfy each equation. Because both equations represent the same line, the solutions are the ordered pairs (x, y), where $y = \frac{1}{2}x - 2$. This is sometimes written $(x, \frac{1}{2}x - 2)$, where y has been replaced by the expression it is equal to.

■ Solve Systems of Equations by the Substitution Method

A graphical solution of a system of equations is based on approximating the coordinates of a point of intersection. An algebraic method called the **substitution method** can be used to find a solution of a system of equations. To use the substitution method, we must write one of the equations of the system in terms of x or in terms of y.

EXAMPLE 2

Solve by the substitution method: $\begin{aligned} (1) \quad 3x + y &= 5 \\ (2) \quad 4x + 5y &= 3 \end{aligned}$

Solution Solve Equation (1) for y. The result is labeled Equation (3).

$$3x + y = 5$$
$$(3) \qquad y = -3x + 5$$

INSTRUCTOR NOTE
Exercise 24 can be used for a similar in-class example.

 See Appendix A:
Intersect

INSTRUCTOR NOTE
When students evaluate a variable expression, they replace a variable by a constant. Here the student is replacing a variable with a variable expression. Mentioning this connection will help some students grasp the substitution method of solving a system of equations. Also, have students imagine parentheses around the variable. This will help ensure that they apply the Distributive Property correctly.

Substitute $-3x + 5$ for y in Equation (2) and solve for x.

$$4x + 5y = 3 \qquad \bullet \text{ This is Equation (2).}$$
$$4x + 5(-3x + 5) = 3 \qquad \bullet \text{ From Equation (3), replace } y \text{ by } -3x + 5.$$
$$4x - 15x + 25 = 3 \qquad \bullet \text{ Solve for } x.$$
$$-11x + 25 = 3$$
$$-11x = -22$$
$$x = 2$$

Replace x in Equation (3) by 2 and solve for y.

$$y = -3x + 5 \qquad \bullet \text{ This is Equation (3).}$$
$$= -3(2) + 5 \qquad \bullet \text{ Replace } x \text{ by 2.}$$
$$= -1$$

The solution is $(2, -1)$.

Graphical Check:

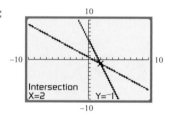

YOU TRY IT 2

Solve by the substitution method:
$$\begin{array}{ll} (1) & y = 2x + 3 \\ (2) & 2x + 3y = 17 \end{array}$$

Solution See page S18. $(1, 5)$

Here is an illustration of what happens when the substitution method is applied to an inconsistent system of equations.

➡ Solve:
$$\begin{array}{ll} (1) & 2x + 3y = 6 \\ (2) & x = -\dfrac{3}{2}y + 4 \end{array}$$

<table>
<tr><td colspan="1" align="center">**ALGEBRAIC SOLUTION**</td><td align="center">**GRAPHICAL CHECK**</td></tr>
</table>

ALGEBRAIC SOLUTION

Replace x in Equation (1) by $-\dfrac{3}{2}y + 4$ from Equation (2) and solve for y.

$$2x + 3y = 6 \qquad \bullet \text{ This is Equation (1).}$$
$$2\left(-\dfrac{3}{2}y + 4\right) + 3y = 6 \qquad \bullet \text{ Replace } x \text{ by } -\dfrac{3}{2}y + 4.$$
$$-3y + 8 + 3y = 6$$
$$8 = 6 \qquad \bullet \text{ This is not a true equation.}$$

The system of equations is inconsistent and has no solution.

GRAPHICAL CHECK

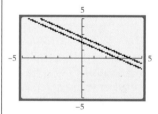

The lines are parallel. The system of equations has no solution.

TAKE NOTE

When a system of equations in two variables is inconsistent, the substitution method will always result in an equation that is not true. For the system of equations at the right, the false equation $8 = 6$ was reached.

Here is an example of a dependent system of equations.

Suggested Activity

Have students solve Equation (1) in Example 3 for y and verify that the slopes and y-intercepts are equal. Therefore, the graph of one line lies on top of the other.

EXAMPLE 3

Solve by the substitution method: $\quad$ (1) $\quad 3x + 4y = 12$

$\qquad\qquad\qquad\qquad\qquad\qquad\qquad\quad$ (2) $\qquad\quad y = -\dfrac{3}{4}x + 3$

Solution

Replace y in Equation (1) by $-\dfrac{3}{4}x + 3$ from Equation (2) and solve for x.

$$3x + 4y = 12$$
$$3x + 4\left(-\frac{3}{4}x + 3\right) = 12$$
$$3x - 3x + 12 = 12$$
$$12 = 12 \qquad \bullet \text{ This is a true equation.}$$

INSTRUCTOR NOTE

Exercise 30 can be used for a similar in-class example.

This means that if x is any real number and $y = -\dfrac{3}{4}x + 3$, then the ordered pair (x, y) is a solution of the system of equations. The solutions are the ordered pairs $\left(x, -\dfrac{3}{4}x + 3\right)$, where we have replaced y by $-\dfrac{3}{4}x + 3$.

Before leaving this problem, it is important to understand that **there are infinitely many solutions of this system of equations.** Because the graph of one equation lies on top of the graph of the other equation, the two lines intersect at an infinite number of points. The graph is shown at the left. To find some ordered-pair solutions, replace x in $\left(x, -\dfrac{3}{4}x + 3\right)$ by any real number and then simplify. This is shown at the right for selected values of x. Some of the ordered-pair solutions are $(-8, 9)$, $(4, 0)$, $(0, 3)$, and $\left(\dfrac{4}{3}, 2\right)$.

$$\left(x, -\frac{3}{4}x + 3\right)$$

$$(-8, 9) \qquad \bullet \; x = -8$$
$$(4, 0) \qquad \bullet \; x = 4$$
$$(0, 3) \qquad \bullet \; x = 0$$
$$\left(\frac{4}{3}, 2\right) \qquad \bullet \; x = \frac{4}{3}$$

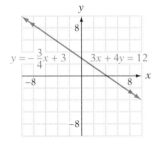

$y = -\dfrac{3}{4}x + 3 \qquad 3x + 4y = 12$

Suggested Activity

After students have completed You Try It 3, have them show that, for instance, if $x = 1$, then $y = -3(1) + 2 = -1$. $(1, -1)$ is a solution of the system of equations. If $x = -2$, then $y = -3(-2) + 2 = 8$. $(-2, 8)$ is a solution of the system of equations. Have them find two other ordered-pair solutions of the system of equations.

YOU TRY IT 3

Solve by the substitution method: $\quad$ (1) $\quad 3x + y = 2$
$\qquad\qquad\qquad\qquad\qquad\qquad\qquad$ (2) $\quad 9x + 3y = 6$

Solution   See page S18. $(x, -3x + 2)$

Application problems that contain two unknown quantities can be solved by using a system of equations.

EXAMPLE 4

A community theater sold 550 tickets for a benefit concert. Two types of tickets were sold. Orchestra tickets were $50 each, and loge tickets were $30 each. If income from the sales of tickets was $22,500, how many of each type were sold?

State the goal.   The goal is to determine the number of orchestra and loge tickets that were sold.

Devise a strategy.   Let x represent the number of orchestra tickets sold, and let y represent the number of loge tickets sold. A total of 550 tickets were sold. Therefore,

$$x + y = 550$$

The price of each orchestra ticket was $50. Therefore, $50x$ represents the income from the sale of orchestra seats. Similarly, the income from the sale of loge tickets was $30y$. Because the total income was $22,500, we have

$$50x + 30y = 22,500$$

Solve the system of equations formed from the two equations.

Solve the problem. $(1) \qquad x + y = 550$
$(2) \quad 50x + 30y = 22,500$

Solve Equation (1) for y.

$$x + y = 550$$
$$(3) \qquad\qquad y = -x + 550$$

Substitute $-x + 550$ for y in Equation (2) and solve for x.

$50x + 30y = 22,500$ • This is Equation (2).
$50x + 30(-x + 550) = 22,500$ • Replace y by $-x + 550$.
$50x - 30x + 16,500 = 22,500$
$20x + 16,500 = 22,500$
$20x = 6000$
$x = 300$

Substitute the value of x into Equation (3) and solve for y.

$$y = -x + 550$$
$$y = -300 + 550 = 250$$

There were 300 orchestra tickets and 250 loge tickets sold.

Check your work.   The sum of the orchestra tickets sold and the loge tickets sold is 550, which is the number sold by the theater. Also, the income from orchestra tickets is $300(50) = 15,000$, and the income from loge tickets is $250(30) = 7500$. The total income from the sale of the tickets is $15,000 + 7500 = 22,500$, which is the income received by the theater. The solution checks.

INSTRUCTOR NOTE
Exercise 32 can be used for a similar in-class example.

CALCULATOR NOTE
For a graphical check, choose a viewing window that contains the point of intersection.

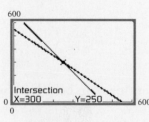

YOU TRY IT 4

In an isosceles triangle, the sum of the measures of the t
is equal to the measure of the third angle. Find the measu
(Recall that in an isosceles triangle, two angles have the
and that the sum of the measures of the interior angles of a triangle
is 180°.)

Solution   See page S18.   45°, 45°, 90°

INSTRUCTOR NOTE
Students may not realize
that investors may not always choose to put all their
money into the account
with the greatest interest
rate because that account
usually has the most risk.
Placing money in different
accounts allows the investor
to diversify.

Some problems that involve investing money can be solved using a system of
equations. The annual simple interest that an investment earns is given by the
equation $I = Pr$, where P is the principal, or the amount invested, r is the simple interest rate, and I is the simple interest earned.

For instance, if you invest $750 at an annual simple interest rate of 6%, then
the interest earned after 1 year is calculated as follows:

$$I = Pr$$
$$I = 750(0.06) \quad \bullet \text{ Replace } P \text{ by 750 and } r \text{ by 0.06.}$$
$$I = 45$$

The amount of interest earned is $45.

INSTRUCTOR NOTE
Exercise 38 can be used for
a similar in-class example.

EXAMPLE 5

Suppose an investor deposits a total of $5000 into two simple interest ac-
counts. On one account, the money market fund, the annual simple interest rate is 3.5%. On the second account, a bond fund, the annual simple
interest rate is 7.5%. If the investor wishes to earn $245 in interest from
the two investments, how much money should be placed in each account?

Solution

State the goal.   The goal is to find the amounts of money that should
be invested at 3.5% and at 7.5% so that the total interest earned is $245.

Devise a strategy.   Let x represent the amount invested at 3.5%, and let
y represent the amount invested at 7.5%. The total amount invested is
$5000. Therefore,

$$x + y = 5000$$

Using the equation $I = Pr$, we can determine the interest earned from each
account.

Interest earned at 3.5%: $0.035x$
Interest earned at 7.5%: $0.075y$

The total interest earned is $245. Therefore,

$$0.035x + 0.075y = 245$$

Solve the system of equations formed by the two equations.

Solve the problem.

$$(1) \qquad\qquad\qquad x + y = 5000$$
$$(2) \quad 0.035x + 0.075y = 245$$

Solve Equation (1) for y.

$$x + y = 5000$$
$$(3) \qquad\qquad\qquad y = -x + 5000$$

Substitute into Equation (2) and solve for x.

$$0.035x + 0.075y = 245 \qquad \bullet \text{ This is Equation (2).}$$
$$0.035x + 0.075(-x + 5000) = 245 \qquad \bullet \text{ Replace } y \text{ by } -x + 5000.$$
$$0.035 - 0.075x + 375 = 245$$
$$-0.04x + 375 = 245$$
$$-0.04x = -130$$
$$x = 3250$$

Substitute the value of x into Equation (3) and solve for y.

$$y = -x + 5000$$
$$y = -3250 + 5000 = 1750$$

The amount that should be invested at 3.5% is $3250. The amount that should be invested at 7.5% is $1750.

Check your work.   Note that $3250 + 1750 = 5000$. This confirms that the total amount invested is $5000. Also note that $0.035(3250) = 113.75$, that $0.075(1750) = 131.25$, and that $113.75 + 131.25$ is 245, the amount of interest to be earned. The solution checks.

YOU TRY IT 5

An investment club invests $13,600 in two simple interest accounts. On one account, the annual simple interest rate is 4.2%. On the other, the annual simple interest rate is 6%. How much should be invested in each account so that both accounts earn the same annual interest?

Solution   See page S18. $8000 at 4.2%, $5600 at 6%

5.1 EXERCISES   Suggested Assignment: 5–39, odds

Topics for Discussion

1. How is the solution of a system of linear equations in two variables represented?   Explanations will vary. For example, the solution is represented by an ordered pair (x, y).

2. For a system of two linear equations in two variables, explain, in graphical terms, each of the following: dependent system of equations, independent system of equations, and inconsistent system of equations.
    Explanations will vary. For example: the graphs of the equations in a dependent system of equations represent the same line, the graphs of the equations in an independent system of equations intersect at one point, and the graphs of the equations in an inconsistent system of equations do not intersect.

3. Explain how to determine, when solving a system of equations by the substitution method, whether the system of equations is dependent or inconsistent.   Explanations will vary. For example, for a dependent system, the resulting equation is true; for an inconsistent system, the resulting equation is false.

4. Can a system of two linear equations in two variables have exactly two solutions? Explain your answer.   No. Explanations will vary.

■ Solve Systems of Equations by Graphing

Solve by graphing.

5. $y = 2x - 1$
 $y = -x + 5$ (2, 3)

6. $y = x + 3$
 $y = -x + 5$ (1, 4)

7. $x + y = 1$
 $3x - y = -5$ (−1, 2)

8. $x - y = -2$
 $x + 2y = 10$ (2, 4)

9. $-3x + 2y = 11$
 $2x + 5y = 18$ (−1, 4)

10. $4x - 3y = 3$
 $2x + 5y = -31$ (−3, −5)

11. $2x - 5y = 10$
 $y = \frac{2}{5}x - 2$ $(x, \frac{2}{5}x - 2)$

12. $3x - 2y = 6$
 $4y - 6x = -12$ $(x, \frac{3}{2}x - 3)$

13. $x - 2y = 8$
 $y = \frac{1}{2}x - 2$ No solution

14. $2x + 3y = 6$
 $4x + 6y = 5$ No solution

15. $4x + 3y = -1$
 $2x - 2y = -11$ (−2.5, 3)

16. $3x + 2y = 15$
 $x - 4y = -16$ (2, 4.5)

■ Solve Systems of Equations by the Substitution Method

Solve by the substitution method.

17. $3x - 2y = 4$
 $x = 2$ (2, 1)

18. $2x + 3y = 4$
 $y = -2$ (5, −2)

19. $4x - 3y = 5$
 $y = 2x - 3$ (2, 1)

20. $x = 2y + 4$
 $4x + 3y = -17$ (−2, −3)

21. $5x + 4y = -1$
 $y = 2 - 2x$ (3, −4)

22. $7x - 3y = 3$
 $x = 2y + 2$ (0, −1)

23. $2x + 2y = 7$
 $y = 4x + 1$ $\left(\frac{1}{2}, 3\right)$

24. $3x + y = 5$
 $2x + 3y = 8$ (1, 2)

25. $x + 3y = 5$
 $2x + 3y = 4$ (−1, 2)

26. $3x + 4y = 14$
 $2x + y = 1$ (−2, 5)

27. $3x + 5y = 0$
 $x - 4y = 0$ (0, 0)

28. $5x - 3y = -2$
 $-x + 2y = -8$ (−4, −6)

29. $y = 3x + 2$
 $y = 2x + 3$ (1, 5)

30. $6x - 2y = 4$
 $y = 3x - 2$ $(x, 3x - 2)$

31. $y = -2x + 1$
 $6x + 3y = 3$ $(x, -2x + 1)$

32. *Carpentry*   A carpenter purchased 50 feet of redwood and 90 feet of pine for a total cost of $31.20. A second purchase, at the same prices, included 200 feet of redwood and 100 feet of pine for a total cost of $78. Find the cost per foot of redwood and of pine.   Pine: $.18/ft; redwood: $.30/ft

33. *Energy Consumption*   During one month, a homeowner used 400 units of electricity and 120 units of gas for a total cost of $147.20. The next month, 350 units of electricity and 200 units of gas were used for a total cost of $144. Find the cost per unit of gas. $.16

34. *Coin Problem*   The total value of the quarters and dimes in a coin bank is $6.90. If the quarters were dimes and the dimes were quarters, the total value of the coins would be $7.80. Find the number of quarters in the bank.
18 quarters

35. *Geometry*   In a right triangle, the measure of one acute angle is twice the measure of the second acute angle. Find the measure of the two acute angles. (Recall that a right triangle has a right angle whose measure is 90° and that the sum of the measures of the angles of a triangle is 180°.)
30°, 60°

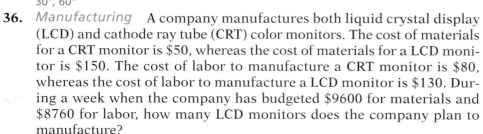

36. *Manufacturing*   A company manufactures both liquid crystal display (LCD) and cathode ray tube (CRT) color monitors. The cost of materials for a CRT monitor is $50, whereas the cost of materials for a LCD monitor is $150. The cost of labor to manufacture a CRT monitor is $80, whereas the cost of labor to manufacture a LCD monitor is $130. During a week when the company has budgeted $9600 for materials and $8760 for labor, how many LCD monitors does the company plan to manufacture?
60 LCD monitors

37. *Pharmacology*   A pharmacist has two vitamin-supplement powders. The first powder is 25% vitamin B1 and 15% vitamin B2. The second is 15% vitamin B1 and 20% vitamin B2. How many milligrams of each of the two powders should the pharmacist use to make a mixture that contains 117.5 milligrams of vitamin B1 and 120 milligrams of vitamin B2?
First powder: 200 mg; second powder: 450 mg

38. *Investments*   An investment of $12,000 is deposited into two simple interest accounts. On one account the annual simple interest rate is 5.5%. On the other, the annual simple interest rate is 6.5%. How much should be invested in each account so that both accounts earn the same interest? $6500 at 5.5%; $5500 at 6.5%

39. *Investments*   A total of $10,000 is deposited in two accounts. On one account the simple interest rate is 9.5%. On the other the annual simple interest rate is 7.5%. How much is invested in each account if the total annual interest earned is $870? $6000 at 9.5%; $4000 at 7.5%

Applying Concepts

For what values of k will the system of equations be inconsistent?

40. $2x - 2y = 5$
$kx - 2y = 3$ 2

41. $6x - 3y = 4$
$3x - ky = 1$ $\frac{3}{2}$

42. $6y + 6 = x$
$kx - 3y = 6$ $\frac{1}{2}$

43. $2y + 2 = x$
$kx - 8y = 2$ 4

Solve. (*Hint:* These equations are not linear equations. First rewrite the equations as linear equations by substituting x for $\frac{1}{a}$ and y for $\frac{1}{b}$. For example, rewrite $\frac{2}{a} + \frac{3}{b} = 4$ as $2x + 3y = 4$.)

44. $\frac{2}{a} + \frac{3}{b} = 4$
$\frac{4}{a} + \frac{1}{b} = 3$ (2, 1)

45. $\frac{2}{a} + \frac{1}{b} = 1$
$\frac{8}{a} - \frac{2}{b} = 0$ $\left(6, \frac{3}{2}\right)$

46. $\frac{1}{a} + \frac{3}{b} = 2$
$\frac{4}{a} - \frac{1}{b} = 3$ $\left(\frac{13}{11}, \frac{13}{5}\right)$

47. $\frac{3}{a} + \frac{4}{b} = -1$
$\frac{1}{a} + \frac{6}{b} = 2$ $(-1, 2)$

48. Write three different systems of equations:
  a. A system that has $(-3, 5)$ as its only solution
  b. A system for which there is no solution
  c. A dependent system of equations

Answers will vary. One possible answer is
a. $y = x + 8$
$y = -x + 2$
b. $y = 2x - 3$
$y = 2x + 4$
c. $y = 3x - 5$
$3x - y = 5$

49. If x and y are real numbers and $|x + y - 17| + |x - y - 5| = 0$, find the numerical value of y.   6

50. Find the equation of the line that passes through the solution of the system of equations $\begin{array}{l} 2x - 3y = 13 \\ x + 4y = -10 \end{array}$ and has slope 2. $y = 2x - 7$

EXPLORATION

1. *Ill-Conditioned Systems of Equations*   Solving systems of equations algebraically as we did in this chapter is not practical for systems of equations that contain a large number of variables. In those cases, a computer solution is the only hope. Computer solutions are not without some problems, however.

Consider the following system of equations.

$$0.24567x + 0.49133y = 0.73700$$
$$0.84312x + 1.68623y = 2.52935$$

It is easy to verify that the solution of this system of equations is (1, 1). However, change the constant 0.73700 to 0.73701 (add 0.00001) and the constant 2.52935 to 2.52936 (add 0.00001), and the solution is now (3, 0). Thus a very small change in the constant terms produced a dramatic change in the solution. A system of equations of this sort is said to be an *ill-conditioned* system.

These types of systems are important because computers generally cannot store numbers beyond a certain number of significant digits. Your calculator, for example, probably allows you to enter no more than 10 significant digits. If an exact number cannot be entered, then an approximation to that number is necessary. When a computer is solving an equation or system of equations, the hope is that approximations of the coefficients it uses will give reasonable approximations to the solutions. For ill-conditioned systems of equations, this is not always true.

In the system of equations, small changes in the constant terms caused a large change in the solution. It is possible that small changes in the coefficients of the variables will also cause large changes in the solution.

In the two systems of equations that follow, examine the effects on the solutions of approximating the fractional coefficients. Try approximating each fraction to the nearest hundredth, to the nearest thousandth, to the nearest ten-thousandth, and then to the limits of your calculator. The exact solution of the first system of equations is $(27, -192, 210)$. The exact solution of the second system of equations is $(-64, 900, -2520, 1820)$.

System 1

$$x + \frac{1}{2}y + \frac{1}{3}z = 1$$

$$\frac{1}{2}x + \frac{1}{3}y + \frac{1}{4}z = 2$$

$$\frac{1}{3}x + \frac{1}{4}y + \frac{1}{5}z = 3$$

System 2

$$x + \frac{1}{2}y + \frac{1}{3}z + \frac{1}{4}w = 1$$

$$\frac{1}{2}x + \frac{1}{3}y + \frac{1}{4}z + \frac{1}{5}w = 2$$

$$\frac{1}{3}x + \frac{1}{4}y + \frac{1}{5}z + \frac{1}{6}w = 3$$

$$\frac{1}{4}x + \frac{1}{5}y + \frac{1}{6}z + \frac{1}{7}w = 4$$

Note how the solutions change as the approximations change and thus how important it is to know whether a system of equations is ill-conditioned. For systems that are not ill-conditioned, approximations of the coefficients yield reasonable approximations of the solution. For ill-conditioned systems of equations, that is not always true.

System 1:

Nearest hundredth: $(266.6666667, -1433.333333, 1366.666667)$

Nearest thousandth: $(30.50502049, -209.9900227, 226.6966693)$

Nearest ten-thousandth: $(27.32317303, -193.65719, 211.5374196)$

To the limits of a calculator: $(27, -192, 210)$

System 2:

Nearest hundredth: $(55.98194131, -295.2595937, 190.9706546, 118.510158)$

Nearest thousandth: $(81.29464132, -708.9158728, 1335.912871, -682.7827636)$

Nearest ten-thousandth: $(-90.6440169, 1188.443724, -3201.682171, 2258.17129)$

To the limits of a calculator: $(-64, 900, -2520, 1820)$

Solving Systems of Linear Equations by the Addition Method

■ Solve Systems of Two Linear Equations in Two Variables by the Addition Method

The addition method is an alternative method for solving a system of equations. This method is based on the Addition Property of Equations and is the basis for solving systems of equations with more than two variables.

Note, for the system of equations at the right, the effect of adding Equation (2) to Equation (1). Because $-3y$ and $3y$ are additive inverses, adding the equations results in an equation with only one variable.

$$(1) \quad 5x - 3y = 14$$
$$(2) \quad \underline{2x + 3y = -7}$$
$$7x + 0y = 7$$
$$7x = 7$$
$$x = 1$$

The second component is found by substituting the value of x into Equation (1) or Equation (2) and then solving for y. Equation (1) is used here.

$$(1) \quad 5x - 3y = 14$$
$$5(1) - 3y = 14$$
$$5 - 3y = 14$$
$$-3y = 9$$
$$y = -3$$

The solution is $(1, -3)$.

Sometimes adding the two equations does not eliminate one of the variables. In this case, **use the Multiplication Property of Equations to rewrite one or both of the equations so that when the equations are added, one of the variables is eliminated.** To do this, first choose which variable to eliminate. The coefficients of that variable must be additive inverses. Multiply each equation by a constant that will produce coefficients that are additive inverses. This is illustrated in Example 1.

INSTRUCTOR NOTE
The addition method is not clear to many students. Try showing them that this method is really a variation of the substitution method. Using the system of equations at the right, add $2x + 3y$ to each side of $5x - 3y = 14$. The result is

$5x - 3y + (2x + 3y) =$
  $14 + (2x + 3y)$

Simplify the left side; substitute -7 for $2x + 3y$ on the right side. The result is $7x = 7$.

INSTRUCTOR NOTE
Exercise 18 can be used for a similar in-class example.

EXAMPLE 1

Solve by the addition method: $(1) \quad 3x + 4y = 2$
$(2) \quad 2x + 5y = -1$

Solution   We can choose to eliminate either x or y. We will eliminate x.

$$2 \diagdown (3x + 4y) = 2(2)$$
$$-3 \diagup (2x + 5y) = -3(-1)$$

• Multiply Equation (1) by 2 and multiply Equation (2) by -3.
• The negative sign is chosen so that the resulting coefficients are additive inverses.

$$6x + 8y = 4$$ • 2 times Equation (1).
$$\underline{-6x - 15y = 3}$$ • -3 times Equation (2).
$$-7y = 7$$ • Add the equations.
$$y = -1$$ • Solve for y.

▼ *Point of Interest*

There are records of Babylonian mathematicians solving systems of equations 3600 years ago. Here is a system of equations from that time (in our modern notation):

$$\frac{2}{3}x = \frac{1}{2}y + 500$$
$$x + y = 1800$$

We say modern notation for many reasons. Foremost is the fact that using variables did not become widespread until the 17th century. There are many other reasons, however. The equals sign had not been invented, 2 and 3 did not look like they do today, and zero had not even been considered as a possible number.

Suggested Activity

See Section 5.2 of the *Student Activity Manual* for a lesson on using a graphing calculator to investigate the addition method of solving systems of two linear equations in two variables.

INSTRUCTOR NOTE
Exercise 16 can be used for a similar in-class example.

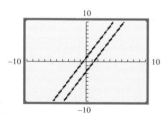

Substitute the value of y into one of the equations and solve for x. Equation (1) is used here.

$$3x + 4y = 2 \qquad \text{• This is Equation (1).}$$
$$3x + 4(-1) = 2 \qquad \text{• Replace } y \text{ by } -1.$$
$$3x - 4 = 2 \qquad \text{• Solve for } x.$$
$$3x = 6$$
$$x = 2$$

The solution is $(2, -1)$.

A graphical check (shown at the right) was created by solving each equation of the system for y, producing its graph, and then finding the point of intersection. For the equations of this system, we have $y = -\frac{3}{4}x + \frac{1}{2}$ and $y = -\frac{2}{5}x - \frac{1}{5}$.

[graphing calculator screen showing: 10, -10, 10, Intersection X=2, Y=-1, -10]

YOU TRY IT 1

Solve by the addition method: $\quad$ (1) $\quad 2x - 5y = 4$
$\qquad\qquad\qquad\qquad\qquad\qquad\qquad$ (2) $\quad 3x - 7y = 15$

Solution See page S19. (47, 18)

Example 2 shows the addition method applied to an inconsistent system of equations.

EXAMPLE 2

Solve by the addition method: $\quad$ (1) $\quad 4x - 2y = 5$
$\qquad\qquad\qquad\qquad\qquad\qquad\qquad$ (2) $\quad 6x - 3y = -3$

Solution We will choose to eliminate x. Multiply Equation (1) by 3 and Equation (2) by -2.

$$3(4x - 2y) = 3(5) \qquad \text{• 3 times Equation (1).}$$
$$-2(6x - 3y) = -2(-3) \qquad \text{• } -2 \text{ times Equation (2).}$$

$$12x - 6y = 15$$
$$\underline{-12x + 6y = 6}$$
$$0 = 21 \qquad \text{• Add the equations. This is not a true equation.}$$

The system of equations is inconsistent. The system does not have a solution.

A graphical check is shown at the left. **From the graphs, it appears that the lines are parallel and therefore do not intersect. The only way to be sure, however, is to solve each equation in the system of equations** for *y* and verify that the slopes are the same but the *y*-intercepts are different.

YOU TRY IT 2

Solve by the addition method: $\begin{array}{ll} (1) & x + 2y = 6 \\ (2) & 3x + 6y = 6 \end{array}$

Solution   See page S19.   No solution

In Example 3, the addition method is used to solve a dependent system of equations.

INSTRUCTOR NOTE
Exercise 14 can be used for a similar in-class example.

EXAMPLE 3

Solve by the addition method: $\begin{array}{ll} (1) & 6x + 2y = 12 \\ (2) & 3x + y = 6 \end{array}$

Solution   We will choose to eliminate *x*. Multiply Equation (2) by -2.

$$6x + 2y = 12 \qquad \bullet \text{ This is Equation (1).}$$
$$-2(3x + y) = -2(6) \qquad \bullet -2 \text{ times Equation (2).}$$

$$6x + 2y = 12$$
$$-6x - 2y = -12$$

$$0 = 0 \qquad \bullet \text{ Add the equations. This is a true equation.}$$

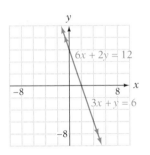

The system of equations is dependent. To find the ordered-pair solutions, solve one of the equations for *y*. Equation (2) is used here.

$$3x + y = 6$$
$$y = -3x + 6$$

The solutions are the ordered pairs $(x, -3x + 6)$.

A graphical check is shown at the left. Note that the graph of one line is on top of that of the other line.

YOU TRY IT 3

Solve by the addition method: $\begin{array}{ll} (1) & 2x + 5y = 10 \\ (2) & 8x + 20y = 40 \end{array}$

Solution   See page S19. $\left(x, -\frac{2}{5}x + 2\right)$

■ Solve Systems of Three Linear Equations in Three Variables by the Addition Method

An equation of the form $Ax + By + Cz = D$, where A, B, and C are coefficients and D is a constant, is a **linear equation in three variables.** Examples of these equations are shown at the right.

$$3x - 2y + z = 4$$
$$2x + y - 4z = 1$$

Suggested Activity

A good model of a three-dimensional coordinate system is the corner of the floor in a room. Have students identify the *xy*-plane as the floor, the *xz*-plane as one wall, and the *yz*-plane the other wall. Now give students various ordered triples (with positive coordinates) and have them indicate the location of the point in the room. Then ask where a point would be located that has some or all of its coordinates as negative numbers.

Finally, ask students to identify the region for which (1) $x = 0$, (2) $y = 0$, (3) $z = 0$, (4) *x* and *y* are both 0, (5) *x* and *z* are both 0, (6) *y* and *z* are both 0, and (7) *x*, *y*, and *z* are 0.

Graphing an equation in three variables requires a third coordinate axis perpendicular to the *xy*-plane. The third axis is commonly called the *z*-axis. The result is a three-dimensional coordinate system called the **xyz-coordinate system.** To help visualize a three-dimensional coordinate system, think of a corner of a room: The floor is the *xy*-plane, one wall is the *yz*-plane, and the other wall is the *xz*-plane. A three-dimensional coordinate system is shown at the right.

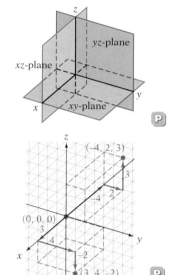

The graph of a point in an *xyz*-coordinate system is an **ordered triple** (x, y, z). Graphing an ordered triple requires three moves, the first along the *x*-axis, the second parallel to the *y*-axis, and the third parallel to the *z*-axis. The graphs of the points $(-4, 2, 3)$ and $(3, 4 - 2)$ are shown at the right.

The graph of a linear equation in three variables is a plane. That is, if all the solutions of a linear equation in three variables were plotted in an *xyz*-coordinate system, the graph would look like a large piece of paper with infinite extent. The graph of $x + y + z = 3$ is shown at the right.

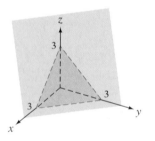

There are different ways in which three planes can be oriented in an *xyz*-coordinate system. The systems of equations represented by the planes below are inconsistent. There is no one point that lies on all three planes.

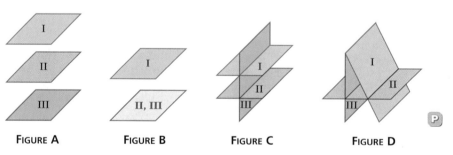

FIGURE A   FIGURE B   FIGURE C   FIGURE D

Graphs of Inconsistent Systems of Equations

For a system of three equations in three variables to have a solution, the graphs of the planes must intersect at a single point, they must intersect along a common line, or all equations must have a graph that is the same plane. Let's look at each of these situations.

The three planes shown in Figure E intersect at a point *P*. A system of equations represented by planes that intersect at a point is **independent.** The planes shown in Figures F and G intersect along a common line. The system of equations represented by the planes in Figure H has a graph that is the same plane. The systems of equations represented by Figures F, G, and H are **dependent.**

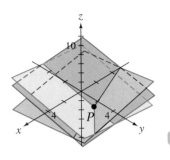

FIGURE E *An Independent System of Equations*

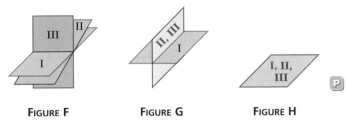

FIGURE F   **FIGURE G**   **FIGURE H**

Dependent Systems of Equations

CALCULATOR NOTE

Most graphing calculators cannot solve a system of three equations in three unknowns graphically and use an algebraic process instead. The method we show below forms a basis for the algebraic methods of a calculator.

Just as a solution of an equation in two variables is an ordered pair (x, y), a **solution of an equation in three variables** is an ordered triple (x, y, z). For example, $(2, -1, 3)$ is a solution of the equation $2x - 3y + 5z = 22$. The ordered triple $(1, 3, 2)$ is not a solution.

A **system of linear equations in three variables** is shown at the right. A **solution of a system of equations in three variables** is an ordered triple that is a solution of each equation of the system.

$$2x - y + z = 7$$
$$x + 2y + z = 12$$
$$x - 2y - z = -8$$

A system of linear equations in three variables can be solved by using the addition method. First, eliminate one variable from any two of the given equations. Then eliminate the same variable from any other two equations. The result will be a system of two equations in two variables. Solve this system by the addition method.

INSTRUCTOR NOTE

Show students that the first goal of solving a system of three equations in three variables is to replace it with an equivalent system of two equations in two variables.

➡ Solve:
(1) $2x - 3y + 2z = -7$
(2) $x + 4y - z = 10$
(3) $3x + 2y + z = 4$

Eliminate z from Equations (1) and (2) by multiplying Equation (2) by 2 and then adding to Equation (1).

$$2x - 3y + 2z = -7 \qquad \bullet \text{ This is Equation (1).}$$
$$\underline{2x + 8y - 2z = 20} \qquad \bullet \text{ 2 times Equation (2).}$$
(4) $\qquad\quad 4x + 5y = 13 \qquad \bullet \text{ Add the equations.}$

Eliminate z from Equations (2) and (3) by adding the two equations.

$$x + 4y - z = 10 \qquad \bullet \text{ This is Equation (2).}$$
$$\underline{3x + 2y + z = 4} \qquad \bullet \text{ This is Equation (3).}$$
(5) $\qquad 4x + 6y = 14 \qquad \bullet \text{ Add the equations.}$

Using Equations (4) and (5), solve the system of two equations in two variables.

(4) $\qquad 4x + 5y = 13$

(5) $\qquad 4x + 6y = 14$

Eliminate x by multiplying Equation (5) by -1 and then add to Equation (4).

$$4x + 5y = 13 \qquad \bullet \text{ This is Equation (4).}$$
$$\underline{-4x - 6y = -14} \qquad \bullet \text{ } -1 \text{ times Equation (5).}$$
$$-y = -1 \qquad \bullet \text{ Add the equations.}$$
$$y = 1$$

Substitute this value of y into Equation (4) or Equation (5) and solve for x. Equation (4) is used here.

$$4x + 5y = 13 \qquad \bullet \text{ This is Equation (4).}$$
$$4x + 5(1) = 13 \qquad \bullet \text{ } y = 1.$$
$$4x + 5 = 13 \qquad \bullet \text{ Solve for } x.$$
$$4x = 8$$
$$x = 2$$

Substitute the value of x and the value of y into one of the equations in the original system of equations and solve for z. Equation (1) is used here.

$$2x - 3y + 2z = -7 \qquad \bullet \text{ This is Equation (1).}$$
$$2(2) - 3(1) + 2z = -7 \qquad \bullet \text{ } x = 2, y = 1.$$
$$1 + 2z = -7 \qquad \bullet \text{ Solve for } z.$$
$$2z = -8$$
$$z = -4$$

The solution is $(2, 1, -4)$.

Just as a system of two equations in two variables may not have a solution, it is possible for a system of three equations in three variables not to have a solution. Here is an example:

$$\text{Solve: } \begin{array}{ll} (1) & 2x - 3y - z = -1 \\ (2) & x + 4y + 3z = 2 \\ (3) & 4x - 6y - 2z = 5 \end{array}$$

Eliminate x from Equations (1) and (3).

$$-4x + 6y + 2z = 2 \qquad \bullet \text{ } -2 \text{ times Equation (1).}$$
$$\underline{4x - 6y - 2z = 5} \qquad \bullet \text{ This is Equation (3).}$$
$$0 = 7 \qquad \bullet \text{ Add the equations.}$$

The equation $0 = 7$ is not a true equation. The system of equations is inconsistent and therefore has no solution. A graph of the system of equations is shown at the left. Note that two of the planes are parallel and therefore never intersect.

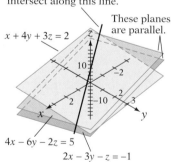

The top two planes intersect along this line.

These planes are parallel.

$x + 4y + 3z = 2$

$4x - 6y - 2z = 5$

$2x - 3y - z = -1$

INSTRUCTOR NOTE
Exercise 30 can be used for a similar in-class example.

EXAMPLE 4

Solve by the addition method: $\begin{array}{ll} (1) & 2x - y + z = 8 \\ (2) & x + 2y + z = -3 \\ (3) & x - 2y - z = 7 \end{array}$

Solution   You can choose any variable to eliminate first. We will choose x. We first eliminate x from Equation (1) and Equation (2) by multiplying Equation (2) by -2 and then adding to Equation (1).

$$2x - y + z = 8 \qquad \bullet \text{ This is Equation (1).}$$
$$-2(x + 2y + z) = -2(-3) \qquad \bullet -2 \text{ times Equation (2).}$$

$$\begin{array}{l} 2x - y + z = 8 \\ \underline{-2x - 4y - 2z = 6} \\ -5y - z = 14 \qquad (4) \quad \bullet \text{ Add the equations. This is Equation (4).} \end{array}$$

Eliminate x from Equation (2) and Equation (3) by multiplying Equation (3) by -1 and then adding it to Equation (2).

$$x + 2y + z = -3 \qquad \bullet \text{ This is Equation (2).}$$
$$-1(x - 2y - z) = -1(7) \qquad \bullet -1 \text{ times Equation (3).}$$

$$\begin{array}{l} x + 2y + z = -3 \\ \underline{-x + 2y + z = -7} \\ 4y + 2z = -10 \qquad (5) \quad \bullet \text{ Add the equations. This is Equation (5).} \end{array}$$

Now form a system of two equations in two variables using Equation (4) and Equation (5). We will solve this system of equations by multiplying Equation (4) by 2 and then adding to Equation (5).

$$-5y - z = 14 \qquad \bullet \text{ This is Equation (4).}$$
$$4y + 2z = -10 \qquad \bullet \text{ This is Equation (5).}$$

$$2(-5y - z) = 2(14) \qquad \bullet \text{ 2 times Equation (4).}$$
$$4y + 2z = -10$$

$$\begin{array}{l} -10y - 2z = 28 \\ \underline{4y + 2z = -10} \\ -6y = 18 \qquad \bullet \text{ Add the equations. Then solve for } y. \\ y = -3 \end{array}$$

Substitute -3 for y in Equation (4) or (5) and solve for z. We will use Equation (4).

$$-5y - z = 14 \qquad \bullet \text{ This is Equation (4).}$$
$$-5(-3) - z = 14 \qquad \bullet \text{ Replace } y \text{ by } -3.$$
$$15 - z = 14$$
$$-z = -1$$
$$z = 1$$

Now replace y by -3 and z by 1 in one of the original equations of the system. Equation (1) is used here.

Suggested Activity

Find a three-digit number such that: the sum of the digits is 7, the number is increased by 99 if the digits are reversed, and the hundreds digit is 3 less than the sum of the other two digits.
[Answer: 223]

$$2x - y + z = 8 \qquad \bullet \text{ This is Equation (1).}$$
$$2x - (-3) + 1 = 8 \qquad \bullet \text{ Replace } y \text{ by } -3 \text{ and replace } z \text{ by } 1.$$
$$2x + 4 = 8$$
$$2x = 4$$
$$x = 2$$

The solution of the system of equations is $(2, -3, 1)$.

YOU TRY IT 4

Solve by the addition method:
$$
\begin{array}{rl}
(1) & 3x - y - 2z = 11 \\
(2) & x - 2y + 3z = 12 \\
(3) & x + y - 2z = 5
\end{array}
$$

Solution   See page S19.   (9, 6, 5)

■ Rate-of-Wind and Rate-of-Current Problems

If a motorboat is on a river that is flowing at a rate of 4 mph, then the boat will float down the river at a speed of 4 mph even though the motor is not on. Now suppose the motor is turned on and the power adjusted so that the boat would travel 10 mph without the aid of the current. Then if the boat is moving with the current, its effective speed is the speed of the boat using power plus of the speed of the current: 10 mph + 4 mph = 14 mph. However, if the boat is moving in the direction opposite to the current, the current slows the boat down, and the effective speed of the boat is the speed of the boat using power minus the speed of the current: 10 mph − 4 mph = 6 mph.

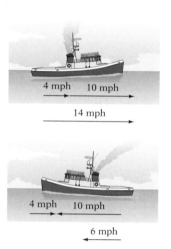

? QUESTION   The speed of a plane is 500 mph. There is a headwind of 50 mph. What is the speed of the plane relative to an observer on the ground?

INSTRUCTOR NOTE
Exercise 44 can be used for
a similar in-class example.

EXAMPLE 5

A motorboat traveling with the current can travel 24 miles in 2 hours. Against the current, it takes 3 hours to travel the same distance. Find the rate of the boat in calm water and the rate of the current.

State the goal.   The goal is to find the rate of the boat in calm water and the rate of the current.

? ANSWER   500 mph − 50 mph = 450 mph

Devise a strategy.   Let x represent the rate of the boat in calm water, and let y represent the rate of the current. Traveling with the current, the speed of the boat in calm water is increased by the rate of the current. Traveling against the current, the speed of the boat in calm water is decreased by the rate of the current. This can be expressed as follows:

Rate of boat with the current: $x + y$

Rate of boat against the current: $x - y$

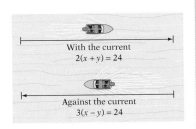

With the current
$2(x + y) = 24$

Against the current
$3(x - y) = 24$

Now use the equation $rt = d$ to express the distance traveled by the boat with the current and the distance traveled against the current in terms of the rate of the boat and the time traveled

Distance traveled with the current: $2(x + y) = 24$

Distance traveled against the current: $3(x - y) = 24$

These two equations form a system of equations.

Solve the problem. $2(x + y) = 24 \Rightarrow x + y = 12$ • Divide each side by 2.

$3(x - y) = 24 \Rightarrow \underline{x - y = 8}$ • Divide each side by 3.

$2x = 20$ • Add the equations.

$x = 10$ • Solve for x.

Substitute the value of x into one of the equations and solve for y. We will use $x + y = 12$.

$x + y = 12$
$10 + y = 12$
$y = 2$

The rate of the boat in calm water is 10 mph; the rate of the current is 2 mph.

Check your work.   With the current the boat can travel 10 mph + 2 mph = 12 mph, or 24 miles in 2 hours. Against the current the boat can travel 10 mph − 2 mph = 8 mph, or 24 miles in 3 hours. The answer is reasonable.

YOU TRY IT 5

Flying with the wind, a plane flew 1000 miles in 5 hours. Flying against the wind, the plane could fly only 500 miles in 5 hours. Find the rate of the plane in calm air and the rate of the wind.

Solution   See page S20.   Plane: 150 mph; wind: 50 mph

■ Percent Mixture Problems

The quantity of a substance in a solution can be given as a percent of the total solution. For instance, in an 8% salt–water solution, 8% of the total solution is salt. The remaining 92% of the solution is water.

The equation $Q = Ar$ relates the quantity, Q, of a substance in a solution to the amount, A, of solution and the percent concentration, r, of the solution. For example, suppose 40 ounces of a salt–water solution is 8% salt. Then the quantity of salt in the solution is

$$Q = Ar$$
$$= 40(0.08) \quad \bullet \text{ The total amount, } A, \text{ of solution is 40 ounces.}$$
$$= 3.2 \qquad\qquad \text{The percent concentration, } r, \text{ is } 8\% = 0.08.$$

There are 3.2 ounces of salt in the solution. Because the total amount of solution is 40 ounces, there are $40 - 3.2 = 36.8$ ounces of water in the solution.

Now suppose two different solutions are mixed together. For instance, suppose a 200-gram solution that is 15% sugar is mixed with 300 grams of a solution that is 25% sugar. Then the total amount of solution is the sum of the amounts in each solution.

$$200 + 300 = 500 \quad \bullet \text{ Grams of solution after mixing}$$

The quantity of sugar in the new solution is the sum of the quantities in the two original solutions. To find the quantity in the new solution, we use the equation $Q = Ar$.

$$200(0.15) + 300(0.25) = 30 + 75 = 105 \quad \bullet \text{ Grams of sugar after mixing}$$

Using the number of grams of solution after mixing the two solutions, the number of grams of sugar in the mixture, and the equation $Q = Ar$, we can find the percent concentration of sugar in the new solution.

$$Q = Ar$$
$$105 = 500r$$
$$\frac{105}{500} = \frac{500r}{500}$$
$$0.21 = r$$

The new solution is 21% sugar.

Problems involving mixtures expressed as percent are solved using the ideas discussed above.

TAKE NOTE

The quantity of sugar in the 15% solution:

$Q = Ar$
$Q = 200(0.15)$

The quantity of sugar in the 25% solution:

$Q = Ar$
$Q = 300(0.25)$

EXAMPLE 6

A chemist mixes an 11% acid solution with a 4% acid solution. How many milliliters of each solution should the chemist use to make a 700-milliliter solution that is 6% acid?

State the goal.   The goal is to find how many milliliters of each of the solutions must be mixed together to produce a 700-milliliter, 6% acid solution.

Devise a strategy.   Let x represent the number of milliliters of the 11% acid solution, and let y represent the number of milliliters of the 4% solution. After mixing, the new solution is 700 milliliters. Therefore,

$$x + y = 700 \quad \bullet \text{ This is Equation (1).}$$

The amount of acid in each solution can be found by using $Q = Ar$.

INSTRUCTOR NOTE
Exercise 52 can be used for a similar in-class example.

Quantity of acid in a solution: $Ar = rA$

Quantity of acid in the 11% solution: $0.11x$

Quantity of acid in the 4% solution: $0.04y$

Quantity of acid in the 6% solution (the mixture): $0.06(700)$

Write an equation using the fact that the sum of the quantity of acid in the 11% solution and the quantity of acid in the 4% solution equals the quantity of acid in the 6% solution.

$$0.11x + 0.04y = 0.06(700)$$
$$0.11x + 0.04y = 42 \qquad \bullet \text{ This is Equation (2).}$$

Equations (1) and (2) form a system of equations.

Solve the problem.

$$
\begin{aligned}
(1) && x + y &= 700 \\
(2) && 0.11x + 0.04y &= 42
\end{aligned}
$$

Eliminate x by multiplying Equation (1) by -0.11 and then adding it to Equation (2).

$$
\begin{aligned}
-0.11x - 0.11y &= -77 \qquad &\bullet \; -0.11 \text{ times Equation (1).} \\
\underline{0.11x + 0.04y} &= \underline{42} \qquad &\bullet \text{ This is Equation (2).} \\
-0.07y &= -35 \qquad &\bullet \text{ Add the equations.} \\
y &= 500 \qquad &\bullet \text{ Solve for } y.
\end{aligned}
$$

Substitute the value of y into Equation (1) and solve for x.

$$
\begin{aligned}
x + y &= 700 \\
x + 500 &= 700 \\
x &= 200
\end{aligned}
$$

Because x represents the number of milliliters of the 11% solution, the chemist must use 200 milliliters of the 11% solution. Because y represents the number of milliliters of the 4% solution, the chemist must use 500 milliliters of the 4% solution.

Check your work.   One way to check the solution is to calculate the percent concentration of the solution after mixing to be sure it is 6%.

Quantity of Acid in the 11% Solution	**Quantity of Acid in the 4% Solution**
$Q_1 = 0.11(200) = 22$	$Q_2 = 0.04(500) = 20$

The quantity of acid in the mixture is $22 + 20 = 42$ milliliters. The amount of mixture is 700 milliliters. To find the percent concentration of acid, solve $Q = Ar$ for r given that $Q = 42$ and $A = 700$.

$$
\begin{aligned}
Q &= Ar \\
42 &= 700r \\
\frac{42}{700} &= r \\
0.06 &= r
\end{aligned}
$$

The percent concentration is 6%. The solution checks.

> **YOU TRY IT 6**
>
> A hospital staff mixed a 55% disinfectant solution with a 15% disinfectant solution. How many liters of each were used to make 50 liters of a 25% disinfectant solution?
>
> **Solution**   See page S20.   12.5 L of the 55% solution; 37.5 L of the 15% solution

5.2 EXERCISES   Suggested Assignment: 11–63, odds

Topics for Discussion

1. When you solve a system of two linear equations in two variables by the addition method, how can you tell whether the system of equations is inconsistent?   The result of one of the steps will be an equation that is never true.

2. When you solve a system of two linear equations in two variables by the addition method, how can you tell whether the system of equations is dependent?   The result of one of the steps will be an equation that is always true.

3. What is a three-dimensional coordinate system?   A three-dimensional co-ordinate system is one formed by three mutually perpendicular axes.

4. Describe the graph of a linear equation in three variables. The graph is a plane.

5. Describe how the planes of an independent system of three linear equations in three variables intersect.   The planes intersect at one point.

6. Give an example of how the graphs of three planes would intersect for an inconsistent system of equations.   Answers may vary.

7. If a 10% apple juice solution is mixed with a 20% apple juice solution, is the resulting mixture less than 10% apple juice, between 10% and 20% apple juice, or greater than 20% apple juice?
   Between 10% and 20% apple juice

8. If a 50% gold alloy is mixed with pure gold, is the resulting alloy more than 50% gold or less than 50% gold?   More than 50% gold

9. Suppose you have a powerboat with the throttle set to move the boat at 8 miles per hour in calm water, and the rate of the current in the river the boat is on is 4 miles per hour. What is the speed of the boat when it is traveling with the current?   12 mph

10. If a cargo ship can travel 25 miles per hour in calm water, what speed can the cargo ship travel when moving against a 5-mile-per-hour river current?   20 mph

■ Solve Systems of Equations by the Addition Method

Solve by the addition method.

11. $x - y = 5$
$x + y = 7$ (6, 1)

12. $3x + y = 4$
$x + y = 2$ (1, 1)

13. $3x + y = 7$
$x + 2y = 4$ (2, 1)

14. $3x - y = 4$
$6x - 2y = 8$ $(x, 3x - 4)$

15. $2x + 5y = 9$
$4x - 7y = -16$ $\left(-\frac{1}{2}, 2\right)$

16. $4x - 6y = 5$
$2x - 3y = 7$ No solution

17. $3x - 5y = 7$
$x - 2y = 3$ $(-1, -2)$

18. $3x + 2y = 16$
$2x - 3y = -11$ (2, 5)

19. $4x + 4y = 5$
$2x - 8y = -5$ $\left(\frac{1}{2}, \frac{3}{4}\right)$

20. $5x + 4y = 0$
$3x + 7y = 0$ (0, 0)

21. $3x - 6y = 6$
$9x - 3y = 8$ $\left(\frac{2}{3}, -\frac{2}{3}\right)$

22. $5x + 2y = 2x + 1$
$2x - 3y = 3x + 2$ (1, −1)

23. $\frac{2}{3}x - \frac{1}{2}y = 3$
$\frac{1}{3}x - \frac{1}{4}y = \frac{3}{2}$ $\left(x, \frac{4}{3}x - 6\right)$

24. $\frac{2}{5}x - \frac{1}{3}y = 1$
$\frac{3}{5}x + \frac{2}{3}y = 5$ (5, 3)

25. $\frac{3}{4}x + \frac{2}{5}y = -\frac{3}{20}$
$\frac{3}{2}x - \frac{1}{4}y = \frac{3}{4}$ $\left(\frac{1}{3}, -1\right)$

26. $4x - 5y = 3y + 4$
$2x + 3y = 2x + 1$ $\left(\frac{5}{3}, \frac{1}{3}\right)$

27. $2x + 5y = 5x + 1$
$3x - 2y = 3y + 3$ No solution

28. $x - 3y = 2x - 5y$
$4x - y = x + y$ (0, 0)

29. $x + 2y - z = 1$
$2x - y + z = 6$ (2, 1, 3)
$x + 3y - z = 2$

30. $x + 3y + z = 6$
$3x + y - z = -2$ $(-1, 2, 1)$
$2x + 2y - z = 1$

31. $2x - y + 2z = 7$
$x + y + z = 2$ (1, −1, 2)
$3x - y + z = 6$

32. $x - 2y + z = 6$
$x + 3y + z = 16$ (6, 2, 4)
$3x - y - z = 12$

33. $3x - 2y + 3z = -4$
$2x + y - 3z = 2$ (0, 2, 0)
$3x + 4y + 5z = 8$

34. $3x + y = 5$
$3y - z = 2$ (1, 2, 4)
$x + z = 5$

35. $2x + 4y - 2z = 3$
$x + 3y + 4z = 1$ No solution
$x + 2y - z = 4$

36. $x - 3y + 2z = 1$
$x - 2y + 3z = 5$ No solution
$2x - 6y + 4z = 3$

37. $3x + 2y - 3z = 8$
$2x + 3y + 2z = 10$ (6, −2, 2)
$4x + 3y + 5z = 28$

38. $2x + 2y + 3z = 13$
$-3x + 4y - z = 5$ (2, 3, 1)
$5x - 3y + z = 2$

39. $2x - 3y + 7z = 0$
$4x - 5y + 2z = -11$ $(-2, 1, 1)$
$x - 2y + 3z = -1$

40. $3x - y + 2z = 2$
$4x + 2y - 7z = 0$ (1, 5, 2)
$2x + 3y - 5z = 7$

41. $2x + y - z = 5$
$x + 3y + z = 14$ (1, 4, 1)
$3x - y + 2z = 1$

42. $3x - 3y + 4z = 6$
$4x - 5y + 2z = 10$ (0, −2, 0)
$x - 2y + 3z = 4$

43. $5x + 3y - z = 5$
$3x - 2y + 4z = 13$ (1, 1, 3)
$4x + 3y + 5z = 22$

44. *Rate-of-Wind Problem* Flying with the wind, a small plane flew 320 miles in 2 hours. Against the wind, the plane could fly only 280 miles in the same amount of time. Find the rate of the plane in calm air and the rate of the wind. Plane: 150 mph; wind: 10 mph

45. *Rate-of-Wind Problem* A turbo-prop plane flying with the wind flew 600 miles between two cities in 2 hours. The return trip against the wind took 3 hours. Find the rate of the plane in calm air and the rate of the wind. Plane: 250 mph; wind: 50 mph

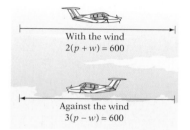

With the wind
$2(p + w) = 600$

Against the wind
$3(p - w) = 600$

46. *Rate-of-Current Problem* A cabin cruiser traveling with the current went 48 miles in 3 hours. Against the current, it took 4 hours to travel the same distance. Find the rate of the cabin cruiser in calm water and the rate of the current. Cabin cruiser: 14 mph; current: 2 mph

47. *Rate-of-Current Problem* A motorboat traveling with the current went 88 kilometers in 4 hours. Against the current, the boat could go only 64 kilometers in the same amount of time. Find the rate of the boat in calm water and the rate of the current. Boat: 19 kilometers per hour; current: 3 kilometers per hour

48. *Rate-of-Wind Problem* A plane flying with a tailwind flew 360 miles in 3 hours. Against the wind, the plane required 4 hours to fly the same distance. Find the rate of the plane in calm air and the rate of the wind. Plane: 105 mph; wind: 15 mph

49. *Rate-of-Current Problem* A motorboat traveling with the current went 54 miles in 3 hours. Against the current, it took 3.6 hours to travel the same distance. Find the rate of the boat in calm water and the rate of the current. Boat: 16.5 mph; current: 1.5 mph

50. *Percent Mixture Problem* A goldsmith mixed 10 grams of a 50% gold alloy with 40 grams of a 15% gold alloy. What is the percent concentration of the resulting alloy? 22%

51. *Percent Mixture Problem* A silversmith mixed 25 grams of a 70% silver alloy with 50 grams of a 15% silver alloy. What is the percent concentration of the resulting alloy? $33\frac{1}{3}$%

52. *Percent Mixture Problem* A butcher has some hamburger that is 20% fat and some hamburger that is 12% fat. How many pounds of each should be mixed to make 80 pounds of hamburger that is 17% fat? 50 lb of hamburger that is 20% fat; 30 lb of hamburger that is 12% fat

53. *Percent Mixture Problem* A chemist mixed a 3% hydrogen peroxide solution with a 12% hydrogen peroxide solution. The resulting 50-milliliter solution was 8.4% hydrogen peroxide. How many milliliters of each solution was used? 20 ml of 3% hydrogen peroxide; 30 ml of 12% hydrogen peroxide

54. *Percent Mixture Problem* A metallurgist mixed a 24% copper alloy with a 36% copper alloy to produce 300 pounds of an alloy that is 31% copper. Find the number of pounds of each alloy that was used. 125 lb of 24% alloy; 175 lb of 36% alloy

55. *Percent Mixture Problem*   A goldsmith mixed a 25% gold alloy with pure gold to produce a 120-gram alloy that was 75% gold. Find the amount of each substance used by the goldsmith. 40 g of 25% alloy; 80 g of pure gold

56. *Percent Mixture Problem*   A pharmacist wants to make a 200-milliliter salt solution that is 2% salt by mixing a 5% salt solution with pure water. How many milliliters of the two solutions are required? 80 ml of 5% solution; 120 ml of pure water

57. *Manufacturing*   On Monday, a computer manufacturing company sent out three shipments. The first order, which contained a bill for $114,000, was for 4 Model II, 6 Model VI and 10 Model IX computers. The second shipment, which contained a bill for $72,000, was for 8 Model II, 3 Model VI and 5 Model IX computers. The third shipment, which contained a bill for $81,000, was for 2 Model II, 9 Model VI, and 5 Model IX computers. What does the manufacturer charge for a Model VI computer? $4000

58. *Not-for-Profit Organizations*   A relief organization supplies blankets, cots, and lanterns to victims of fires, floods, and other natural disasters. One week the organization purchased 15 blankets, 5 cots, and 10 lanterns for a total cost of $1250. The next week, at the same prices, the organization purchased 20 blankets, 10 cots, and 15 lanterns for a total cost of $2000. The next week, at the same prices, the organization purchased 10 blankets, 15 cots, and 5 lanterns for a total cost of $1625. Find the cost of one blanket, the cost of one cot, and the cost of one lantern.
Blanket: $25; cot: $75; lantern: $50

59. *Investments*   An investor has a total of $18,000 deposited in three different accounts, which earn annual interest of 9%, 7%, and 5%. The amount deposited in the 9% account is twice the amount in the 5% account. If the three accounts earn total annual interest of $1340, how much money is deposited in each account?
$8000 at 9%, $6000 at 7%, $4000 at 5%

60. *Investments*   An investor has a total of $15,000 deposited in three different accounts, which earn annual interest of 9%, 6%, and 4%. The amount deposited in the 6% account is $2000 more than the amount in the 4% account. If the three accounts earn total annual interest of $980, how much money is deposited in each account? $5250 at 9%, $5875 at 6%, $3875 at 4%

61. *Investments*   A financial planner invested $33,000 of a client's money, part at 9%, part at 12%, and the remainder at 8%. The total annual income from these three investments was $3290. The amount invested at 12% was $5000 less than the combined amounts invested at 9% and 8%. Find the amount invested at each rate. $14,000 at 12%; $10,000 at 8%; $9000 at 9%

62. *Chemistry*   The following table shows the active chemical content of three different soil additives.

Additive	Ammonium Nitrate	Phosphorus	Iron
1	30%	10%	10%
2	40%	15%	10%
3	50%	5%	5%

A soil chemist wants to prepare two chemical samples. The first sample requires 380 grams of ammonium nitrate, 95 grams of phosphorus, and 85 grams of iron. The second sample requires 380 grams of ammonium nitrate, 110 grams of phosphorus, and 90 grams of iron. How many grams of each additive are required for sample 1, and how many grams of each additive are required for sample 2?

For sample 1, 500 g of additive 1, 200 g of additive 2, and 300 g of additive 3
For sample 2, 400 g of additive 1, 400 g of additive 2, and 200 g of additive 3

63. *Nutrition* The following table shows the carbohydrate, fat, and protein content of three food types.

Food Type	Carbohydrate	Fat	Protein
I	60%	10%	20%
II	10%	4%	60%
III	70%	0%	10%

A nutritionist must prepare two diets from these three food groups. The first diet must contain 220 grams of carbohydrate, 18 grams of fat, and 160 grams of protein. The second diet must contain 210 grams of carbohydrate, 28 grams of fat, and 170 grams of protein. How many grams of each food type are required for the first diet, and how many grams of each food type are required for the second diet?

For the first diet, 100 g of food type I, 200 g of food type II, and 200 g of food type III
For the second diet, 200 g of food type I, 200 g of food type II, and 100 g of food type III

Applying Concepts

64. The point of intersection of the graphs of the equations $Ax + 2y = 2$ and $2x + By = 10$ is $(2, -2)$. Find A and B. $A = 3, B = -3$

65. The point of intersection of the graphs of the equations $Ax - 4y = 9$ and $4x + By = -1$ is $(-1, -3)$. Find A and B. $A = 3, B = -1$

66. Given that the graphs of the equations $2x - y = 6$, $3x - 4y = 4$, and $Ax - 2y = 0$ all intersect at the same point, find A.   1

67. Given that the graphs of the equations $3x - 2y = -2$, $2x - y = 0$, and $Ax + y = 8$ all intersect at the same point, find A.   2

68. Find an equation such that the system of equations formed by your equation and $2x - 5y = 9$ will have $(2, -1)$ as a solution. Answers will vary.
$3x + 2y = 4$ is a possibility.

69. Let L be the line in which planes $2x + y - z = 13$ and $x - 2y + z = -4$ intersect. If the point $(x, 3, z)$ lies on L, find the value of $(x - z)$.   6

For Exercises 70 and 71, use the system of equations
$$\begin{array}{l} x - 3y - 2z = A^2 \\ 2x - 5y + Az = 9 \\ 2x - 8y + z = 18 \end{array}$$

70. Find all values of A for which the system has no solution.

The system of equations has no solution when $2A + 13 = 0$ or $A = -\frac{13}{2}$.

71. Find all values of A for which the system has a unique solution.

The system of equations has a unique solution when $2A + 13 \neq 0$ or $A \neq -\frac{13}{2}$.

72. *Rate-of-Wind Problem* A plane is flying the 3500 miles from New York City to London. The speed of the plane in calm air is 375 mph, and there is a 50-mph tailwind. The *point of no return* is the point at which the flight time required to return to New York City is the same as the flight time to travel on to London. For this flight, how far from New York is the point of no return? Round to the nearest whole number.

1517 mi

EXPLORATION

1. *Multiplying an Equation in a System of Equations by a Constant* When the addition method is used to solve a system of linear equations in two variables, sometimes it is necessary to multiply one or both equations by a constant and then add the equations. Multiplying and then adding equations does not change the solution of the system of equations. In this Exploration, you will investigate this fact.

a. Graph the equations of the system $\begin{array}{ll} (1) & 2x - 5y = -4 \\ (2) & 4x + 3y = 18 \end{array}$ and find the solution of the system of equations. (3, 2)

b. Multiply Equation (1) by 3 and add it to Equation (2). Call this Equation (3). Graph Equation (3) on the same coordinate grid as in part a. Does the system of equations made up of Equation (1) and Equation (3) have the same solution as the original system of equations? Explain. Yes. Explanations may vary.

c. Explain how the answer to part b illustrates that replacing an equation in a system of equations by the sum of that equation and a multiple of another equation does not change the solution of the system of equations. Explanations may vary.

d. Return to the original system of equations. Multiply Equation (1) by -2 and add it to Equation (2). Call this Equation (4). Graph Equation (4) on the same coordinate grid as in part a. Does the system of equations made up of Equation (1) and Equation (4) have the same solution as the original system of equations? Explain. Yes. Explanations may vary.

e. Explain why multiplying Equation (1) by -2 as in part d is better than multiplying it by 3 as in part b. The result is that the coefficients of x are opposite. Then when we add Equations (4) and (2), the variable x is eliminated. The sum is a linear equation in one variable that can easily be solved for y.

SECTION **5.3** # Solving Systems of Linear Equations Using Matrices

■ Elementary Row Operations

■ Solve Systems of Equations Using the Gaussian Elimination Method

■ Elementary Row Operations

A **matrix** is a rectangular array of numbers. Each number of a matrix is called an **element** of the matrix. The matrix at the right, with three rows and four columns, is called a 3×4 (read "3 by 4") matrix.

$$A = \begin{bmatrix} 2 & -3 & -6 & 0 \\ 7 & 4 & -2 & 5 \\ 1 & 6 & 0 & 3 \end{bmatrix}$$

A matrix of m rows and n columns is said to be of **order $m \times n$.** The order of matrix A on the previous page is 3×4. The notation a_{ij} refers to the element of the matrix in the ith row and jth column. For matrix A, $a_{23} = -2$ and $a_{31} = 1$.

? QUESTION For matrix A, what is a_{12}?

The elements a_{11}, a_{22}, a_{33}, . . ., a_{nn} form the **main diagonal** of a matrix. The elements 2, 4, and 0 form the main diagonal of matrix A given on the previous page.

▼ *Point of Interest*

Working with systems of equations is only one of the many ways in which we use matrices. Besides this application, matrices are used in such diverse fields as economics, biology, chemistry, and physics.

One application of matrices is solving a system of equations. For each system of equations, there is an associated matrix called an **augmented matrix.** This matrix consists of the coefficients of the variables and the constant terms.

System of Equations

$$\begin{aligned} 2x - y + 3z &= 5 \\ x + 4y \quad\ &= -2 \\ 4x + 3y - z &= 3 \end{aligned}$$

Augmented Matrix

$$\left[\begin{array}{ccc|c} 2 & -1 & 3 & 5 \\ 1 & 4 & 0 & -2 \\ 4 & 3 & -1 & 3 \end{array} \right]$$

• Typically, a vertical line is drawn between the coefficients of the variables and the constant terms.

Note that when a term is missing from one of the equations (there is no z term in the second equation), the coefficient of that term is 0, and 0 is entered in the matrix.

A system of equations can be written from an augmented matrix.

Augmented Matrix

$$\left[\begin{array}{ccc|c} 2 & 0 & -1 & 4 \\ 3 & -1 & 1 & 5 \\ 1 & 2 & -4 & -3 \end{array} \right]$$

System of Equations

$$\begin{aligned} 2x \quad\ - z &= 4 \\ 3x - y + z &= 5 \\ x + 2y - 4z &= -3 \end{aligned}$$

INSTRUCTOR NOTE
Writing the augmented matrix for the system of equations in Exercise 30 can serve as a similar in-class example.

EXAMPLE 1

Write the augmented matrix for $\begin{aligned} 2x - 3y &= 4 \\ x + 5y &= 0 \end{aligned}$.

Solution The augmented matrix is $\left[\begin{array}{cc|c} 2 & -3 & 4 \\ 1 & 5 & 0 \end{array} \right]$.

YOU TRY IT 1

Write the system of equations that corresponds to the augmented matrix
$$\left[\begin{array}{ccc|c} 2 & -3 & 1 & 4 \\ 1 & 0 & -2 & 3 \\ 0 & 1 & 2 & -3 \end{array} \right].$$

$$\begin{aligned} 2x - 3y + z &= 4 \\ x - 2z &= 3 \\ y + 2z &= -3 \end{aligned}$$

Solution See page S21.

? ANSWER $a_{12} = -3$

See Appendix A:
Matrix

A matrix can be entered into a graphing calculator using **EDIT** under the matrix key. There are 10 matrices with names A through J. By pressing the down arrow key, you can see the additional names.

A typical graphing calculator screen displaying the 2×3 matrix $\begin{bmatrix} 2 & -3 & 4 \\ 1 & 5 & 3 \end{bmatrix}$ is shown at the right.

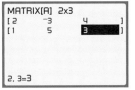

A system of equations can be solved by writing the system as an augmented matrix and then performing operations on the matrix similar to those performed on the equations of the system. These operations are called **elementary row operations.**

Elementary Row Operations

1. Interchange two rows.
2. Multiply all the elements in a row by the same nonzero number.
3. Replace a row by the sum of that row and a multiple of any other row.

TAKE NOTE

For each example of a row operation at the right, we are using the same system of equations. The solution of this system of equations is (2, 1). You should verify that (2, 1) is also a solution of the new system of equations. This means that these row operations have not changed the solution.

TAKE NOTE

For the third elementary row operation, the row being multiplied is *not* changed. For the example at the right, row 1 is not changed. The row being added to is replaced by the result of the operation. It is not absolutely necessary to follow this convention, but for consistency, we will always use this method.

Each of these elementary row operations has as its basis the operations that can be performed on a system of equations. These operations do not change the solution of the system of equations. Here are some examples of each row operation.

1. Interchange two rows.

Original System
$$2x - 3y = 1$$
$$4x + 5y = 13$$
$$\begin{bmatrix} 2 & -3 & | & 1 \\ 4 & 5 & | & 13 \end{bmatrix}$$

This notation means to interchange rows 1 and 2.
$$\overset{R_1 \longleftrightarrow R_2}{}$$

New System
$$\begin{bmatrix} 4 & 5 & | & 13 \\ 2 & -3 & | & 1 \end{bmatrix}$$
$$4x + 5y = 13$$
$$2x - 3y = 1$$

The solution is (2, 1).   The solution is (2, 1).

2. Multiply all the elements in a row by the same nonzero number.

Original System
$$2x - 3y = 1$$
$$4x + 5y = 13$$
$$\begin{bmatrix} 2 & -3 & | & 1 \\ 4 & 5 & | & 13 \end{bmatrix}$$

This notation means to multiply row 2 by 3.
$$\overset{3R_2}{} \longrightarrow$$

New System
$$\begin{bmatrix} 2 & -3 & | & 1 \\ 12 & 15 & | & 39 \end{bmatrix}$$
$$2x - 3y = 1$$
$$12x + 15y = 39$$

The solution is (2, 1).   The solution is (2, 1).

3. Replace a row by the sum of that row and a multiple of any other row.

Original System
$$2x - 3y = 1$$
$$4x + 5y = 13$$
$$\begin{bmatrix} 2 & -3 & | & 1 \\ 4 & 5 & | & 13 \end{bmatrix}$$

This notation means to replace row 2 by the sum of that row and -2 times row 1.
$$\overset{-2R_1 + R_2}{} \longrightarrow$$

New System
$$\begin{bmatrix} 2 & -3 & | & 1 \\ 0 & 11 & | & 11 \end{bmatrix}$$
$$2x - 3y = 1$$
$$11y = 11$$

The solution is (2, 1).   The solution is (2, 1).

Note that we replace the row that follows the plus sign. See the Take Note at the left.

INSTRUCTOR NOTE
Performing the following
operations on the matrix in
Exercise 12 can serve as a
similar in-class example:
$-2R_3$ and $-3R_1 + R_2$.

EXAMPLE 2

Let $A = \begin{bmatrix} 1 & 3 & -4 & 6 \\ 3 & 2 & 0 & -1 \\ -2 & -5 & 3 & 4 \end{bmatrix}$. Perform the following elementary row operations on A.

a. $R_1 \longleftrightarrow R_3$ **b.** $-2R_3$ **c.** $2R_3 + R_1$

Solution

a. $R_1 \longleftrightarrow R_3$ means to interchange row 1 and row 3.

$$\begin{bmatrix} 1 & 3 & -4 & 6 \\ 3 & 2 & 0 & -1 \\ -2 & -5 & 3 & 4 \end{bmatrix} \quad R_1 \longleftrightarrow R_3 \quad \begin{bmatrix} -2 & -5 & 3 & 4 \\ 3 & 2 & 0 & -1 \\ 1 & 3 & -4 & 6 \end{bmatrix}$$

b. $-2R_3$ means to multiply row 3 by -2.

$$\begin{bmatrix} 1 & 3 & -4 & 6 \\ 3 & 2 & 0 & -1 \\ -2 & -5 & 3 & 4 \end{bmatrix} \quad -2R_3 \longrightarrow \quad \begin{bmatrix} 1 & 3 & -4 & 6 \\ 3 & 2 & 0 & -1 \\ 4 & 10 & -6 & -8 \end{bmatrix}$$

c. $2R_3 + R_1$ means to multiply row 3 by 2 and then add the result to row 1. Only Row 1 will be changed.

$$\begin{bmatrix} 1 & 3 & -4 & 6 \\ 3 & 2 & 0 & -1 \\ -2 & -5 & 3 & 4 \end{bmatrix} \quad 2R_3 + R_1 \longrightarrow \quad \begin{bmatrix} -3 & -7 & 2 & 14 \\ 3 & 2 & 0 & -1 \\ -2 & -5 & 3 & 4 \end{bmatrix}$$

YOU TRY IT 2

Let $B = \begin{bmatrix} 1 & 8 & -2 & 3 \\ 2 & -3 & 4 & 1 \\ 3 & 5 & -7 & 3 \end{bmatrix}$. Perform the following elementary row operations on B.

a. $R_2 \longleftrightarrow R_3$ **b.** $3R_2$ **c.** $-3R_1 + R_3$

Solution   See page S21.

a. $\begin{bmatrix} 1 & 8 & -2 & 3 \\ 3 & 5 & -7 & 3 \\ 2 & -3 & 4 & 1 \end{bmatrix}$ **b.** $\begin{bmatrix} 1 & 8 & -2 & 3 \\ 6 & -9 & 12 & 3 \\ 3 & 5 & -7 & 3 \end{bmatrix}$ **c.** $\begin{bmatrix} 1 & 8 & -2 & 3 \\ 2 & -3 & 4 & 1 \\ 0 & -19 & -1 & -6 \end{bmatrix}$

See Appendix A:
Matrix

The elementary row operations can be performed using a graphing calculator. A typical screen from a graphing calculator is shown below.

Interchange rows ——

Multiply a row
by a constant ———

Multiply a row
by a constant
and then add to
another row

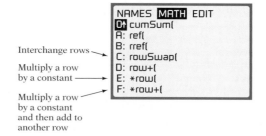

The operation row+(shown by D: is to add two rows. This is really the same as F: where the constant is 1.

Here are the calculator versions of the elementary row operations as they apply to Example 2.

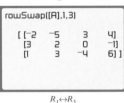

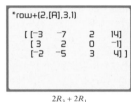

$R_1 \leftrightarrow R_3$ $2R_3$ $2R_3 + 2R_1$

Elementary row operations are used to solve a system of equations. **The goal is to use the elementary row operations to rewrite the augmented matrix with 1's down the main diagonal and 0's to the left of the 1's in all rows except the first.** This is called a **row echelon form** of the matrix. Examples of echelon form are shown below.

$$\begin{bmatrix} 1 & 3 & -2 \\ 0 & 1 & 3 \end{bmatrix} \quad \begin{bmatrix} 1 & -2 & 3 & 1 \\ 0 & 1 & 2.5 & -4 \\ 0 & 0 & 1 & 2 \end{bmatrix} \quad \begin{bmatrix} 1 & 4 & \frac{1}{2} & -3 \\ 0 & 1 & 3 & 0 \\ 0 & 0 & 1 & -\frac{2}{3} \end{bmatrix} \quad \begin{bmatrix} 1 & -2 & 3 & 4 \\ 0 & 1 & 5 & -6 \\ 0 & 0 & 0 & 5 \end{bmatrix}$$

We will follow a very definite procedure to rewrite an augmented matrix in row echelon form. For a 2 × 3 augmented matrix, use elementary row operations to

1. Change a_{11} to a 1.
2. Change a_{21} to a 0.
3. Change a_{22} to a 1.

$$\begin{bmatrix} a_{11} & a_{12} & a_{13} \\ a_{21} & a_{22} & a_{23} \end{bmatrix}$$

➡ Write the matrix $\begin{bmatrix} 3 & -6 & 12 \\ 2 & 1 & -3 \end{bmatrix}$ in row echelon form.

ALGEBRAIC SOLUTION

1. Change a_{11} to 1. One way to do this is to multiply row 1 by the reciprocal of a_{11}.

$$\begin{bmatrix} 3 & -6 & 12 \\ 2 & 1 & -3 \end{bmatrix} \xrightarrow{\frac{1}{3}R_1} \begin{bmatrix} 1 & -2 & 4 \\ 2 & 1 & -3 \end{bmatrix}$$

2. Change a_{21} to 0 by multiplying row 1 by the opposite of a_{21} and then adding to row 2.

$$\begin{bmatrix} 1 & -2 & 4 \\ 2 & 1 & -3 \end{bmatrix} \xrightarrow{-2R_1 + R_2} \begin{bmatrix} 1 & -2 & 4 \\ 0 & 5 & -11 \end{bmatrix}$$

3. Change a_{22} to 1 by multiplying by the reciprocal of a_{22}.

$$\begin{bmatrix} 1 & -2 & 4 \\ 0 & 5 & -11 \end{bmatrix} \xrightarrow{\frac{1}{5}R_2} \begin{bmatrix} 1 & -2 & 4 \\ 0 & 1 & -2.2 \end{bmatrix}$$

GRAPHICAL CHECK

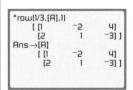

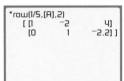

TAKE NOTE

Sometimes it is not possible, as shown in the fourth matrix at the right, to have all 1's on the main diagonal. In this case, we try as best we can to write the matrix with 1's on the main diagonal and rows with 0's following these. For instance, the following matrix is not in row echelon form.

$$\begin{bmatrix} 1 & -2 & 3 & 4 \\ 0 & 0 & 0 & 5 \\ 0 & 1 & 5 & -6 \end{bmatrix}$$

CALCULATOR NOTE

To rewrite a matrix in row echelon form, we make a series of changes to the matrix. After each step, we must replace [A] by the new matrix. The operation Ans->[A] replaces the matrix in [A] with the new matrix. If you need to keep the original matrix, you can make a copy of it and store it in another matrix, say [B].

INSTRUCTOR NOTE
A Texas Instruments calculator begins the sequence of steps to row echelon form by moving the row whose first term has the greatest absolute value to row 1, and then the row is multiplied by the reciprocal of the first element. For instance, in Example 3, the steps would be

$R_1 \longleftrightarrow R_3$, $-\frac{1}{3}R_1$. Once 0's are below a_{11}, and assuming a 3 × 4 matrix, the absolute values of a_{22} and a_{32} are examined to determine which is greater. The row with the first non-zero term with the greater absolute value is moved to row 2. This process is repeated until the matrix is in row echelon form. There is a little more to this, such as accounting for zeros, but the essence is as we have described. The main point is that a calculator-produced row echelon form may not be the same as one produced by a student.

A row echelon form of the matrix is $\begin{bmatrix} 1 & -2 & 4 \\ 0 & 1 & -2.2 \end{bmatrix}$.

The row echelon form of a matrix is not unique and depends on the elementary row operations that are used. For instance, suppose we again start with $\begin{bmatrix} 3 & -6 & 12 \\ 2 & 1 & -3 \end{bmatrix}$ and follow the elementary row operations below.

$$\begin{bmatrix} 3 & -6 & 12 \\ 2 & 1 & -3 \end{bmatrix} \xrightarrow{-1R_2 + R_1} \begin{bmatrix} 1 & -7 & 15 \\ 2 & 1 & -3 \end{bmatrix} \xrightarrow{-2R_1 + R_2}$$

$$\begin{bmatrix} 1 & -7 & 15 \\ 0 & 15 & -33 \end{bmatrix} \xrightarrow{\frac{1}{15}R_2} \begin{bmatrix} 1 & -7 & 15 \\ 0 & 1 & -2.2 \end{bmatrix}$$

In this case, we get $\begin{bmatrix} 1 & -7 & 15 \\ 0 & 1 & -2.2 \end{bmatrix}$ as the row echelon form rather than $\begin{bmatrix} 1 & -2 & 4 \\ 0 & 1 & -2.2 \end{bmatrix}$, which we got in the first case. Row echelon form is not unique.

The order in which the elements in a 3 × 4 matrix are changed is as follows:

1. Change a_{11} to a 1.
2. Change a_{21} and a_{31} to 0's.
3. Change a_{22} to a 1.
4. Change a_{32} to a 0.
5. Change a_{33} to a 1.

$$\begin{bmatrix} a_{11} & a_{12} & a_{13} & a_{14} \\ a_{21} & a_{22} & a_{23} & a_{24} \\ a_{31} & a_{32} & a_{33} & a_{34} \end{bmatrix}$$

EXAMPLE 3

INSTRUCTOR NOTE
Exercise 12 can be used for a similar in-class example.

Write $\begin{bmatrix} 2 & 1 & 3 & -1 \\ 1 & 3 & 5 & -1 \\ -3 & -1 & 1 & 2 \end{bmatrix}$ in row echelon form.

Solution

ALGEBRAIC SOLUTION

1. Change a_{11} to 1 by interchanging row 1 and row 2. *Note:*

   We could have chosen to multiply row 1 by $\frac{1}{2}$. The sequence of steps to get to row echelon form is not unique.

$$\begin{bmatrix} 2 & 1 & 3 & -1 \\ 1 & 3 & 5 & -1 \\ -3 & -1 & 1 & 2 \end{bmatrix} \xrightarrow{R_1 \longleftrightarrow R_2} \begin{bmatrix} 1 & 3 & 5 & -1 \\ 2 & 1 & 3 & -1 \\ -3 & -1 & 1 & 2 \end{bmatrix}$$

GRAPHICAL CHECK

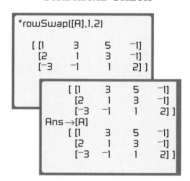

2. Change a_{21} to 0 by multiplying row 1 by the opposite of a_{21} and then adding to row 2.

$$\begin{bmatrix} 1 & 3 & 5 & -1 \\ 2 & 1 & 3 & -1 \\ -3 & -1 & 1 & 2 \end{bmatrix} \xrightarrow{-2R_1 + R_2} \begin{bmatrix} 1 & 3 & 5 & -1 \\ 0 & -5 & -7 & 1 \\ -3 & -1 & 1 & 2 \end{bmatrix}$$

Change a_{31} to 0 by multiplying row 1 by the opposite of a_{31} and then adding to row 3.

$$\begin{bmatrix} 1 & 3 & 5 & -1 \\ 0 & -5 & -7 & 1 \\ -3 & -1 & 1 & 2 \end{bmatrix} \xrightarrow{3R_1 + R_3} \begin{bmatrix} 1 & 3 & 5 & -1 \\ 0 & -5 & -7 & 1 \\ 0 & 8 & 16 & -1 \end{bmatrix}$$

3. Change a_{22} to 1 by multiplying row 2 by the reciprocal of a_{22}.

$$\begin{bmatrix} 1 & 3 & 5 & -1 \\ 0 & -5 & -7 & 1 \\ 0 & 8 & 16 & -1 \end{bmatrix} \xrightarrow{-\frac{1}{5}R_2} \begin{bmatrix} 1 & 3 & 5 & -1 \\ 0 & 1 & \dfrac{7}{5} & -\dfrac{1}{5} \\ 0 & 8 & 16 & -1 \end{bmatrix}$$

4. Change a_{32} to 0 by multiplying row 2 by the opposite of a_{32} and then adding to row 3.

$$\begin{bmatrix} 1 & 3 & 5 & -1 \\ 0 & 1 & \dfrac{7}{5} & -\dfrac{1}{5} \\ 0 & 8 & 16 & -1 \end{bmatrix} \xrightarrow{-8R_2 + R_3} \begin{bmatrix} 1 & 3 & 5 & -1 \\ 0 & 1 & \dfrac{7}{5} & -\dfrac{1}{5} \\ 0 & 0 & \dfrac{24}{5} & \dfrac{3}{5} \end{bmatrix}$$

5. Change a_{33} to 1 by multiplying row 3 by the reciprocal of a_{33}.

$$\begin{bmatrix} 1 & 3 & 5 & -1 \\ 0 & 1 & \dfrac{7}{5} & -\dfrac{1}{5} \\ 0 & 0 & \dfrac{24}{5} & \dfrac{3}{5} \end{bmatrix} \xrightarrow{\frac{5}{24}R_3} \begin{bmatrix} 1 & 3 & 5 & -1 \\ 0 & 1 & \dfrac{7}{5} & -\dfrac{1}{5} \\ 0 & 0 & 1 & \dfrac{1}{8} \end{bmatrix}$$

A row echelon form of the matrix is $\begin{bmatrix} 1 & 3 & 5 & -1 \\ 0 & 1 & \dfrac{7}{5} & -\dfrac{1}{5} \\ 0 & 0 & 1 & \dfrac{1}{8} \end{bmatrix}$.

```
*row+(-2,[A],1,2)
 [ [1     3     5    -1]
   [0    -5    -7     1]
   [-3   -1     1     2] ]
```

```
*row+(3,[A],1,3)

 [ [1     3     5    -1]
   [0    -5    -7     1]
   [0     8    16    -1] ]
```

```
*row(-1/5,[A],2)

 [ [1     3     5    -1]
   [0     1    1.4   -.2]
   [0     8    16    -1] ]
```

```
*row+(-8,[A],2,3)
 [ [1     3     5    -1]
   [0     1    1.4   -.2]
   [0     0    4.8    .6] ]
```

```
*row(5/24,[A],3)
 [ [1     3     5    -1]
   [0     1    1.4   -.2]
   [0     0     1    .125] ]
```

YOU TRY IT 3

Write $\begin{bmatrix} 1 & -3 & 2 & 1 \\ -4 & 14 & 0 & -2 \\ 2 & -5 & -3 & 16 \end{bmatrix}$ in row echelon form.

Solution See page S21. $\begin{bmatrix} 1 & -3 & 2 & 1 \\ 0 & 1 & 4 & 1 \\ 0 & 0 & 1 & -\dfrac{13}{11} \end{bmatrix}$

CALCULATOR NOTE

For the graphical check above, we have shown the result of interchanging rows. Remember that you must store the result in [A] after each step.

CALCULATOR NOTE

Your calculator probably shows the result of an elementary row operation using decimals rather than fractions. You can, however, convert the decimal numbers to fractions by using the ▶Frac command that can be found by pressing MATH .

See Appendix A: **Matrix** for assistance. A typical graphing calculator screen follows.

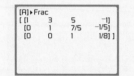

▼ *Point of Interest*

Johann Carl Friedrich Gauss (1777–1855) is considered one of the greatest mathematicians of all time. He contributed not only to mathematics but to astronomy and physics as well. The unit of magnetism called a gauss was named in his honor. The image of Gauss above appears on a German Deutsche mark note.

The **ref(** function on a graphing calculator performs all of the elementary row operations on a matrix and directly produces a row echelon form of a matrix. The abbreviation **ref** stands for **r**ow **e**chelon **f**orm. A typical screen is shown at the right for the matrix in Example 3, where we have also used the ▶Frac command to write the matrix with fractions rather than decimals. Again observe that this form is different from the form we created in Example 3. The echelon form that you produce by means of a calculator may not be the same one that you produce without a calculator.

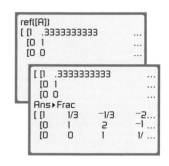

■ Solve Systems of Equations Using the Gaussian Elimination Method

If an augmented matrix is in row echelon form, the corresponding system of equations can be solved by substitution. For instance, consider the following matrix in row echelon form and the corresponding system of equations.

$$\begin{bmatrix} 1 & -3 & 4 & \vert & 7 \\ 0 & 1 & 3 & \vert & -6 \\ 0 & 0 & 1 & \vert & -1 \end{bmatrix} \qquad \begin{aligned} x - 3y + 4z &= 7 \\ y + 3z &= -6 \\ z &= -1 \end{aligned}$$

From the last equation of the system above, we have $z = -1$. Substitute this value into the second equation and solve for y. Thus $y = -3$.

$$\begin{aligned} y + 3z &= -6 \\ y + 3(-1) &= -6 \\ y - 3 &= -6 \\ y &= -3 \end{aligned}$$

Substitute $y = -3$ and $z = -1$ in the first equation of the system and solve for x. Thus $x = 2$.

$$\begin{aligned} x - 3y + 4z &= 7 \\ x - 3(-3) + 4(-1) &= 7 \\ x + 9 - 4 &= 7 \\ x + 5 &= 7 \\ x &= 2 \end{aligned}$$

The solution of the system of equations is $(2, -3, -1)$.

The process of solving a system of equations by using elementary row operations is called the **Gaussian elimination method.**

Suggested Activity

Find the ordered pair of numbers (x, y) that satisfies the system
$$123x + 321y = 345$$
$$321x + 123y = 543$$

$$\left[\text{Answer: } \left(\frac{3}{2}, \frac{1}{2} \right) \right]$$

EXAMPLE 4

Solve using the Gaussian elimination method: $\begin{aligned} 2x - 5y &= 19 \\ 3x + 4y &= -6 \end{aligned}$

Solution Write the augmented matrix and then use elementary row operations to rewrite the matrix in row echelon form.

$$\begin{bmatrix} 2 & -5 & \vert & 19 \\ 3 & 4 & \vert & -6 \end{bmatrix} \xrightarrow[\frac{1}{2}R_1]{\text{Change } a_{11} \text{ to 1.}} \begin{bmatrix} 1 & -\dfrac{5}{2} & \vert & \dfrac{19}{2} \\ 3 & 4 & \vert & -6 \end{bmatrix}$$

• Note that we multiplied R_1 by the reciprocal of a_{11}.

INSTRUCTOR NOTE
Exercise 20 can be used for
a similar in-class example.

$$\begin{bmatrix} 1 & -\dfrac{5}{2} & \bigg| & \dfrac{19}{2} \\ 3 & 4 & \bigg| & -6 \end{bmatrix} \xrightarrow[-3R_1 + R_2]{\text{Change } a_{21} \text{ to } 0.} \begin{bmatrix} 1 & -\dfrac{5}{2} & \bigg| & \dfrac{19}{2} \\ 0 & \dfrac{23}{2} & \bigg| & -\dfrac{69}{2} \end{bmatrix}$$

- Note that we multi-plied R_1 by the opposite of a_{21}.

$$\begin{bmatrix} 1 & -\dfrac{5}{2} & \bigg| & \dfrac{19}{2} \\ 0 & \dfrac{23}{2} & \bigg| & -\dfrac{69}{2} \end{bmatrix} \xrightarrow[\frac{2}{23} R_2]{\text{Change } a_{22} \text{ to } 1.} \begin{bmatrix} 1 & -\dfrac{5}{2} & \bigg| & \dfrac{19}{2} \\ 0 & 1 & \bigg| & -3 \end{bmatrix}$$

- This is row echelon form.

(1) $x - \dfrac{5}{2}y = \dfrac{19}{2}$

(2) $y = -3$

- Write the system of equations corre-sponding to the matrix that is in row echelon form.

$$x - \dfrac{5}{2}(-3) = \dfrac{19}{2}$$

$$x + \dfrac{15}{2} = \dfrac{19}{2}$$

$$x = 2$$

- Substitute -3 for y in Equation (1) and solve for x.

The solution is $(2, -3)$.

If a graphing calculator is used to find a row echelon form for $\begin{bmatrix} 2 & -5 & | & 19 \\ 3 & 4 & | & -6 \end{bmatrix}$, the result is as shown below. This gives a different corresponding system of equations. However, the final answer is the same.

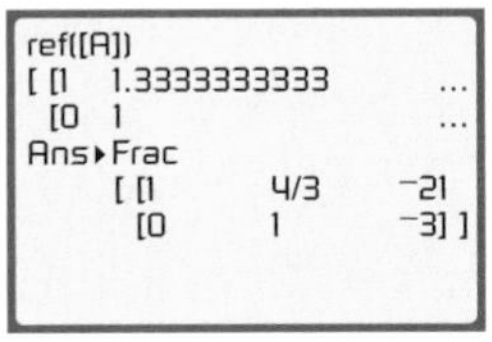

```
ref([A])
[ [1   1.3333333333    ...
  [0   1               ...
Ans▶Frac
    [ [1       4/3     -21
      [0       1       -3] ]
```

(1) $x + \dfrac{4}{3}y = -2$

(2) $y = -3$

Replace y in Equation (1) and solve for x.

$\xrightarrow{\hspace{2cm}}$

$x + \dfrac{4}{3}y = -2$

$x + \dfrac{4}{3}(-3) = -2$

$x - 4 = -2$

$x = 2$

The solution is $(2, -3)$.

YOU TRY IT 4

Solve by using the Gaussian elimination method: $\begin{array}{l} 4x - 5y = 17 \\ 3x + 2y = 7 \end{array}$

Solution   See page S21. $(3, -1)$

The Gaussian elimination method can be used with dependent and inconsistent systems of equations. Here is an example of a dependent system of equations.

➡ Solve by the Gaussian elimination method: $\begin{array}{l} x - 3y = 6 \\ -2x + 6y = -12 \end{array}$

$$\begin{bmatrix} 1 & -3 & | & 6 \\ -2 & 6 & | & -12 \end{bmatrix} \xrightarrow[2R_1 + R_2]{\begin{array}{c} a_{11} \text{ is 1. Change} \\ a_{21} \text{ to 0.} \end{array}} \begin{bmatrix} 1 & -3 & | & 6 \\ 0 & 0 & | & 0 \end{bmatrix}$$

- This is row echelon form.

- Write the system of equations corresponding to the matrix that is in row echelon form.

$$x - 3y = 6$$
$$0 = 0$$

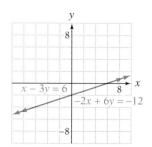

Because the equation $0 = 0$ is true, the solutions of the system of equations are the solutions of $x - 3y = 6$. Solving for y, we have $y = \frac{1}{3}x - 2$. The ordered-pair solutions are $\left(x, \frac{1}{3}x - 2\right)$. The graph of the system of equations is shown at the left. Note that the graphs are identical. The system of equations is dependent. ⬅

Here is an example of an inconsistent system of equations.

➡ Solve by the Gaussian elimination method:
$$\begin{array}{l} 4x + 2y = 6 \\ 2x + y = -4 \end{array}$$

$$\begin{bmatrix} 4 & 2 & | & 6 \\ 2 & 1 & | & -4 \end{bmatrix} \xrightarrow[\frac{1}{4}R_1]{\text{Change } a_{11} \text{ to 1.}} \begin{bmatrix} 1 & \frac{1}{2} & | & \frac{3}{2} \\ 2 & 1 & | & -4 \end{bmatrix}$$

$$\begin{bmatrix} 1 & \frac{1}{2} & | & \frac{3}{2} \\ 2 & 1 & | & -4 \end{bmatrix} \xrightarrow[-2R_1 + R_2]{\text{Change } a_{21} \text{ to 0.}} \begin{bmatrix} 1 & \frac{1}{2} & | & \frac{3}{2} \\ 0 & 0 & | & -7 \end{bmatrix}$$

- This is row echelon form.

$$x + \frac{1}{2}y = \frac{3}{2}$$
$$0 = -7$$

- Write the system of equations corresponding to the matrix that is in row echelon form.

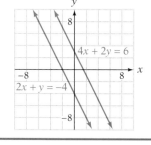

Because the equation $0 = -7$ is not true, the system of equations has no solution. The graphs of the lines are parallel and do not intersect. The system of equations is inconsistent. The graph is shown at the left. ⬅

The Gaussian elimination method can be extended to systems of equations with more than two variables.

Suggested Activity

Have students find a row echelon form for the system of equations in Example 5 by using a calculator and then solve the system of equations using the row echelon form produced by the calculator. Students should verify that although the row echelon forms of the two matrices are different, the solution of the system of equations is the same in each case.

EXAMPLE 5

Solve by using the Gaussian elimination method:

$$x + 2y - z = 9$$
$$2x - y + 2z = -1$$
$$-2x + 3y - 2z = 7$$

Solution

$$\begin{bmatrix} 1 & 2 & -1 & | & 9 \\ 2 & -1 & 2 & | & -1 \\ -2 & 3 & -2 & | & 7 \end{bmatrix} \xrightarrow[-2R_1 + R_2]{\begin{array}{c} a_{11} \text{ is 1. Change} \\ a_{21} \text{ to 0.} \end{array}} \begin{bmatrix} 1 & 2 & -1 & | & 9 \\ 0 & -5 & 4 & | & -19 \\ -2 & 3 & -2 & | & 7 \end{bmatrix}$$

INSTRUCTOR NOTE
Exercise 30 can be used for
a similar in-class example.

$$\begin{bmatrix} 1 & 2 & -1 & | & 9 \\ 0 & -5 & 4 & | & -19 \\ -2 & 3 & -2 & | & 7 \end{bmatrix} \xrightarrow[2R_1 + R_3]{\text{Change } a_{31} \text{ to } 0.} \begin{bmatrix} 1 & 2 & -1 & | & 9 \\ 0 & -5 & 4 & | & -19 \\ 0 & 7 & -4 & | & 25 \end{bmatrix}$$

$$\begin{bmatrix} 1 & 2 & -1 & | & 9 \\ 0 & -5 & 4 & | & -19 \\ 0 & 7 & -4 & | & 25 \end{bmatrix} \xrightarrow[-\frac{1}{5}R_2]{\text{Change } a_{22} \text{ to } 1.} \begin{bmatrix} 1 & 2 & -1 & | & 9 \\ 0 & 1 & -\frac{4}{5} & | & \frac{19}{5} \\ 0 & 7 & -4 & | & 25 \end{bmatrix}$$

$$\begin{bmatrix} 1 & 2 & -1 & | & 9 \\ 0 & 1 & -\frac{4}{5} & | & \frac{19}{5} \\ 0 & 7 & -4 & | & 25 \end{bmatrix} \xrightarrow[-7R_2 + R_3]{\text{Change } a_{32} \text{ to } 0.} \begin{bmatrix} 1 & 2 & -1 & | & 9 \\ 0 & 1 & -\frac{4}{5} & | & \frac{19}{5} \\ 0 & 0 & \frac{8}{5} & | & -\frac{8}{5} \end{bmatrix}$$

$$\begin{bmatrix} 1 & 2 & -1 & | & 9 \\ 0 & 1 & -\frac{4}{5} & | & \frac{19}{5} \\ 0 & 0 & \frac{8}{5} & | & -\frac{8}{5} \end{bmatrix} \xrightarrow[\frac{5}{8}R_3]{\text{Change } a_{33} \text{ to } 1.} \begin{bmatrix} 1 & 2 & -1 & | & 9 \\ 0 & 1 & -\frac{4}{5} & | & \frac{19}{5} \\ 0 & 0 & 1 & | & -1 \end{bmatrix}$$

• This is row echelon form.

(1) $x + 2y - z = 9$

(2) $y - \dfrac{4}{5}z = \dfrac{19}{5}$

(3) $z = -1$

• Write the system of equations corresponding to the matrix that is in row echelon form.

$$y - \dfrac{4}{5}(-1) = \dfrac{19}{5}$$

$$y + \dfrac{4}{5} = \dfrac{19}{5}$$

$$y = 3$$

• Substitute -1 for z in Equation (2) and solve for y.

$$x + 2y - z = 9$$

$$x + 2(3) - (-1) = 9$$

$$x + 7 = 9$$

$$x = 2$$

• Substitute -1 for z and 3 for y in Equation (1) and solve for x.

The solution is $(2, 3, -1)$.

If a graphing calculator is used to find a row echelon form for $\begin{bmatrix} 1 & 2 & -1 & | & 9 \\ 2 & -1 & 2 & | & -1 \\ -2 & 3 & -2 & | & 7 \end{bmatrix}$, the result is as shown on the next page. This gives a different corresponding system of equations. However, the final answer is the same.

Suggested Activity

Find the value of n given that n is a real number and that the system of equations has no solutions.

$$nx + y = 1$$
$$ny + z = 1$$
$$x + nz = 1$$

[Answer: -1]

```
ref([A])
[ [1    -.5   1    -.5...
  [0    1    -.8   3.8...
  [0    0    1    -1 ...
```

(1)　　$x - 0.5y + z = -0.5$
(2)　　　　$y - 0.8z = 3.8$
(3)　　　　　　　$z = -1$

$$y - 0.8(-1) = 3.8$$
$$y + 0.8 = 3.8$$
$$y = 3$$

- Replace z by -1 in Equation (2) and solve for y.

$$x - 0.5(3) + (-1) = -0.5$$
$$x - 2.5 = -0.5$$
$$x = 2$$

- Replace z by -1 and y by 3 in Equation (1) and solve for x.

The solution is $(2, 3, -1)$. This again illustrates that different row echelon forms of the same augmented matrix will yield the same solution of the system of equations.

YOU TRY IT 5

Solve by using the Gaussian elimination method:

$$2x + 3y + 3z = -2$$
$$x + 2y - 3z = 9$$
$$3x - 2y - 4z = 1$$

Solution　See page S21.　$(-1, 2, -2)$

Just as we can write the equation of a line in slope–intercept form as $y = mx + b$ or in standard form as $Ax + By = C$, we can write the equation of a plane in different ways. We can solve the equation in standard form of a plane, $Ax + By + Cz = D$, for z.

$$Ax + By + Cz = D$$
$$Cz = -Ax - By + D$$
$$z = -\frac{A}{C}x - \frac{B}{C}y + \frac{D}{C}$$

The last equation is usually written as $z = ax + by + c$, where $a = -\frac{A}{C}$, $b = -\frac{B}{C}$, and $c = \frac{D}{C}$. We will use the form $z = ax + by + c$ to find the equation of a plane.

INSTRUCTOR NOTE
Exercise 44 can be used for a similar in-class example.

EXAMPLE 6

Find the equation of the plane that passes through the points $P_1(1, 3, 6)$, $P_2(3, 2, 10)$, and $P_3(4, -1, 7)$.

State the goal.　The goal is to find the equation of the plane that contains the given points.

Devise a strategy.　To find the equation of the plane, we must determine the constants a, b, and c for the equation $z = ax + by + c$. Because the given ordered triples belong to a plane, they must satisfy that equation. Substitute the coordinates of each point into $z = ax + by + c$ and solve the resulting system of equations.

$$z = ax + by + c$$

$$6 = a(1) + b(3) + c \qquad \bullet \ P_1: x = 1, y = 3, z = 6$$

$$10 = a(3) + b(2) + c \qquad \bullet \ P_2: x = 3, y = 2, z = 10$$

$$7 = a(4) + b(-1) + c \qquad \bullet \ P_3: x = 4, y = -1, z = 7$$

Simplify and write the system of equations as an augmented matrix.

$$\begin{aligned} 6 &= a + 3b + c \\ 10 &= 3a + 2b + c \\ 7 &= 4a - b + c \end{aligned} \quad \xrightarrow{\text{Augmented matrix}} \quad \begin{bmatrix} 1 & 3 & 1 & 6 \\ 3 & 2 & 1 & 10 \\ 4 & -1 & 1 & 7 \end{bmatrix}$$

Solve the system of equations by using the Gaussian elimination method.

Solve the problem. A graphing calculator is used below to write the augmented matrix in row echelon form. We have used the ▶Frac command to write the augmented matrix with fractions.

```
ref([A])
[[1    -.25   .25     ...
 [0    1      .23076  ...
 [0    0      1       ...
      [[1    -.25   .25     ...
       [0    1      .23076  ...
       [0    0      1       ...
      Ans▶Frac
      [[1    -1/4   1/4    7...
       [0    1      3/13   1...
       [0    0      1      -...
```

$$(1) \quad a - \frac{1}{4}b + \frac{1}{4}c = \frac{7}{4}$$

$$(2) \quad b + \frac{3}{13}c = \frac{17}{13}$$

$$(3) \quad c = -3$$

Solve the resulting system of equations by substitution.

$$b + \frac{3}{13}(-3) = \frac{17}{13}$$ • Replace c by -3 in Equation (2) and solve for b.

$$b - \frac{9}{13} = \frac{17}{13}$$

$$b = 2$$

$$a - \frac{1}{4}(2) + \frac{1}{4}(-3) = \frac{7}{4}$$ • Replace c by -3 and b by 2 in Equation (1) and solve for a.

$$a - \frac{2}{4} - \frac{3}{4} = \frac{7}{4}$$

$$a = 3$$

We have $a = 3$, $b = 2$, and $c = -3$. The equation of the plane is $z = 3x + 2y - 3$.

Check your work. Verify that each given ordered triple is a solution of the equation by substituting into the equation of the plane. For instance, the check using P_1 is shown at the left. $P_1(1, 3, 6)$ checks. Now verify that the other points check.

$$\frac{z = 3x + 2y - 3}{6 \mid 3(1) + 2(3) - 3}$$
$$6 = 6$$

YOU TRY IT 6

Recall that a quadratic function can be written in the form $y = ax^2 + bx + c$. Find the equation of the quadratic function whose graph passes through $P_1(2, 3)$, $P_2(-1, 0)$, and $P_3(0, -3)$.

Solution See page S22. $y = 2x^2 - x - 3$

Suggested Activity

See Section 5.3 of the *Student Activity Manual* for an activity involving using matrices for problem solving.

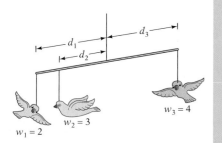

d_1, d_2, d_3

$w_1 = 2$ $w_2 = 3$ $w_3 = 4$

INSTRUCTOR NOTE
Exercise 48 can be used for a similar in-class example.

EXAMPLE 7

An artist is creating a mobile from which three objects will be suspended from a light rod that is 18 inches long, as shown below at the left. The weight, in ounces, of each object is shown in the diagram. For the mobile to balance, the objects must be positioned so that $w_1d_1 + w_2d_2 = w_3d_3$. The artist wants d_1 to be 1.5 times d_2. Find the distances d_1, d_2, and d_3 so that the mobile will balance.

State the goal.   The goal is to find the values of d_1, d_2, and d_3 so that the mobile will balance.

Devise a strategy.   There are three unknowns in this problem. Using the figure and information from the problem, write a system of three equations in three unknowns. The length of the rod is 18 inches. Therefore, $d_1 + d_3 = 18$. Because the artist wants d_1 to be 1.5 times d_2, we have $d_1 = 1.5d_2$. Using the equation $w_1d_1 + w_2d_2 = w_3d_3$, we have

$$w_1d_1 + w_2d_2 = w_3d_3$$
$$2d_1 + 3d_2 = 4d_3 \qquad \bullet \text{ From the diagram, } w_1 = 2, w_2 = 3, w_3 = 4.$$

Use the three equations to create a system of three equations in three unknowns.

$$\begin{aligned} d_1 \qquad\quad + d_3 &= 18 \\ d_1 - 1.5d_2 \qquad &= 0 \\ 2d_1 + 3d_2 - 4d_3 &= 0 \end{aligned} \qquad \underrightarrow{\text{Augmented matrix}} \qquad \begin{bmatrix} 1 & 0 & 1 & | & 18 \\ 1 & -1.5 & 0 & | & 0 \\ 2 & 3 & -4 & | & 0 \end{bmatrix}$$

Solve the system of equations by using the Gaussian elimination method.

Solve the problem.   A graphing calculator is used below to write the augmented matrix in row echelon form. We have used the ▶Frac command to write the augmented matrix with fractions.

```
ref([A])
[[1    1.5    -2      …
 [0    1     -.66666  …
 [0    0      1       …

 [[1    1.5    -2      …
  [0    1     -.66666  …
  [0    0      1       …
Ans▶Frac
 [[1    3/2    -2     0]
  [0    1     -2/3    0]
  [0    0      1      9]]
```

(1) $d_1 + \dfrac{3}{2}d_2 - 2d_3 = 0$

(2) $d_2 - \dfrac{2}{3}d_3 = 0$

(3) $d_3 = 9$

Solve the resulting system of equations by substitution.

$$d_2 - \frac{2}{3}(9) = 0 \qquad \bullet \begin{array}{l}\text{Replace } d_3 \text{ by} \\ \text{9 in Equation} \\ \text{(2) and solve} \\ \text{for } d_2.\end{array}$$
$$d_2 - 6 = 0$$
$$d_2 = 6$$

$$d_1 + \frac{3}{2}(6) - 2(9) = 0 \qquad \bullet \begin{array}{l}\text{Replace } d_3 \text{ by} \\ \text{9 and } d_2 \text{ by 6} \\ \text{in Equation} \\ \text{(1) and solve} \\ \text{for } d_1.\end{array}$$
$$d_1 + 9 - 18 = 0$$
$$d_1 = 9$$

The values are $d_1 = 9$ inches, $d_2 = 6$ inches, and $d_3 = 9$ inches.

Check your work.   You can check your solution by substituting the known values for w_1, w_2, and w_3 and the computed values for d_1, d_2, and d_3 into $w_1d_1 + w_2d_2 = w_3d_3$ and verifying that the solution checks.

> ### YOU TRY IT 7
>
> A science museum charges $10 for an admission ticket, but members receive a discount of $3, and students are admitted for half the regular admission price. Last Saturday, 750 tickets were sold for a total of $5400. If 20 more student tickets than full-price tickets were sold, how many of each type of ticket were sold?
>
> **Solution**   See page S22.   190 regular admission tickets, 350 member tickets, 210 student tickets

5.3 EXERCISES   Suggested Assignment: 5–49, odds

Topics for Discussion

1. What is a matrix?   A matrix is a rectangular array of numbers.
2. What is an augmented matrix?   An augmented matrix is one whose entries are the coefficients and constants from a system of equations.
3. What are the three elementary row operations on a matrix?
    1. Interchange two rows.   **2.** Multiply a row by a constant.   **3.** Replace a row by the sum of that row and a nonzero multiple of another row.
4. What is the next step toward writing the matrix $\begin{bmatrix} 1 & 3 & -5 \\ 4 & 3 & 2 \end{bmatrix}$ in row echelon form?   Multiply row 1 by -4 and then add to row 2.

▪ Elementary Row Operations

5. Which of the following matrices are not in row echelon form?

    a. $\begin{bmatrix} 1 & -2 & 0 \\ 0 & 0 & 3 \end{bmatrix}$ **b.** $\begin{bmatrix} 0 & 1 & 2 \\ 1 & 2 & 3 \end{bmatrix}$ **c.** $\begin{bmatrix} 1 & -1 & 3 & 0 \\ 0 & 1 & 4 & 0 \\ 0 & 0 & 1 & 0 \end{bmatrix}$ **d.** $\begin{bmatrix} 1 & -1 & -2 & 3 \\ 0 & 1 & 1 & 3 \\ 0 & 0 & 0 & 0 \end{bmatrix}$ b

Without using a calculator, write each matrix in row echelon form.

6. $\begin{bmatrix} 1 & -5 & 1 \\ 2 & -9 & 4 \end{bmatrix}$ $\begin{bmatrix} 1 & -5 & 1 \\ 0 & 1 & 2 \end{bmatrix}$

7. $\begin{bmatrix} 1 & 4 & -1 \\ -3 & -13 & 7 \end{bmatrix}$ $\begin{bmatrix} 1 & 4 & -1 \\ 0 & 1 & -4 \end{bmatrix}$

8. $\begin{bmatrix} 2 & -4 & 1 \\ 3 & -7 & -1 \end{bmatrix}$ $\begin{bmatrix} 1 & -2 & \frac{1}{2} \\ 0 & 1 & \frac{5}{2} \end{bmatrix}$

9. $\begin{bmatrix} 4 & 2 & -2 \\ 7 & 4 & -1 \end{bmatrix}$ $\begin{bmatrix} 1 & \frac{1}{2} & -\frac{1}{2} \\ 0 & 1 & 5 \end{bmatrix}$

10. $\begin{bmatrix} 5 & -2 & 3 \\ -7 & 3 & 1 \end{bmatrix}$ $\begin{bmatrix} 1 & -\frac{2}{5} & \frac{3}{5} \\ 0 & 1 & 26 \end{bmatrix}$

11. $\begin{bmatrix} 2 & 5 & -4 \\ 3 & 1 & 2 \end{bmatrix}$ $\begin{bmatrix} 1 & \frac{5}{2} & -2 \\ 0 & 1 & -\frac{16}{13} \end{bmatrix}$

12. $\begin{bmatrix} 1 & 4 & 1 & -2 \\ 3 & 11 & -1 & 2 \\ 2 & 3 & 1 & 4 \end{bmatrix}$ $\begin{bmatrix} 1 & 4 & 1 & -2 \\ 0 & 1 & 4 & -8 \\ 0 & 0 & 1 & -\frac{32}{19} \end{bmatrix}$

13. $\begin{bmatrix} 1 & 2 & 2 & -1 \\ -4 & -10 & -1 & 3 \\ 3 & 4 & 2 & -2 \end{bmatrix}$ $\begin{bmatrix} 1 & 2 & 2 & -1 \\ 0 & 1 & -\frac{7}{2} & \frac{1}{2} \\ 0 & 0 & 1 & -\frac{2}{11} \end{bmatrix}$

14. $\begin{bmatrix} 3 & 6 & -3 & 4 \\ -2 & -6 & -1 & 3 \\ 2 & 1 & 2 & 5 \end{bmatrix}$ $\begin{bmatrix} 1 & 2 & -1 & \frac{4}{3} \\ 0 & 1 & \frac{3}{2} & -\frac{17}{6} \\ 0 & 0 & 1 & -\frac{37}{51} \end{bmatrix}$ **15.** $\begin{bmatrix} -2 & 6 & -1 & 3 \\ 1 & -2 & 2 & 1 \\ 3 & -6 & 7 & 6 \end{bmatrix}$ $\begin{bmatrix} 1 & -3 & \frac{1}{2} & -\frac{3}{2} \\ 0 & 1 & \frac{3}{2} & \frac{5}{2} \\ 0 & 0 & 1 & 3 \end{bmatrix}$

16. $\begin{bmatrix} 2 & 6 & 10 & 3 \\ 3 & 8 & 15 & 0 \\ 1 & 2 & 3 & -1 \end{bmatrix}$ $\begin{bmatrix} 1 & 3 & 5 & \frac{3}{2} \\ 0 & 1 & 0 & \frac{9}{2} \\ 0 & 0 & 1 & -1 \end{bmatrix}$ **17.** $\begin{bmatrix} 4 & -6 & 9 & 4 \\ 2 & 2 & 1 & -5 \\ 3 & 3 & -5 & 1 \end{bmatrix}$ $\begin{bmatrix} 1 & -\frac{3}{2} & \frac{9}{4} & 1 \\ 0 & 1 & -\frac{47}{30} & -\frac{4}{15} \\ 0 & 0 & 1 & -\frac{17}{13} \end{bmatrix}$

■ Solve Systems of Equations Using the Gaussian Elimination Method

18. What is the solution of the system of equations that has $\begin{bmatrix} 1 & -1 & 3 & -2 \\ 0 & 1 & -1 & 1 \\ 0 & 0 & 1 & 3 \end{bmatrix}$

as the row echelon form of the augmented matrix for the system of equations? $(-7, 4, 3)$

19. What is the solution of the system of equations that has $\begin{bmatrix} 1 & -3 & 2 & 4 \\ 0 & 1 & -2 & 3 \\ 0 & 0 & 1 & -1 \end{bmatrix}$

as the row echelon form of the augmented matrix for the system of equations? $(9, 1, -1)$

Solve by using the Gaussian elimination method. Do not use a calculator.

20. $3x + y = 6$
$2x - y = -1$ $(1, 3)$

21. $2x + y = 3$
$x - 4y = 6$ $(2, -1)$

22. $x - 3y = 8$
$3x - y = 0$ $(-1, -3)$

23. $2x + 3y = 16$
$x - 4y = -14$ $(2, 4)$

24. $y = 4x - 10$
$2y = 5x - 11$ $(3, 2)$

25. $2y = 4 - 3x$
$y = 1 - 2x$ $(-2, 5)$

26. $2x - y = -4$
     $y = 2x - 8$
Inconsistent

27. $3x - 2y = -8$
     $y = \frac{3}{2}x - 2$
No solution

28. $4x - 3y = -14$
$3x + 4y = 2$
$(-2, 2)$

29. $5x + 2y = 3$
$3x + 4y = 13$ $(-1, 4)$

30. $5x + 4y + 3z = -9$
$x - 2y + 2z = -6$ $(0, 0\ -3)$
$x - y - z = 3$

31. $x - y - z = 0$
$3x - y + 5z = -10$ $(1, 3\ -2)$
$x + y - 4z = 12$

32. $5x - 5y + 2z = 8$
$2x + 3y - z = 0$
$x + 2y - z = 0$
$(1, -1, -1)$

33. $2x + y - 5z = 3$
$3x + 2y + z = 15$
$5x - y - z = 5$
$(2, 4, 1)$

34. $2x + 3y + z = 5$
$3x + 3y + 3z = 10$
$4x + 6y + 2z = 5$
Inconsistent

35. $x - 2y + 3z = 2$
$2x + y + 2z = 5$
$2x - 4y + 6z = -4$
No solution

36. $3x + 2y + 3z = 2$
$6x - 2y + z = 1$
$3x + 4y + 2z = 3$
$\left(\frac{1}{3}, \frac{1}{2}, 0\right)$

37. $2x + 3y - 3z = -1$
$2x + 3y + 3z = 3$
$4x - 4y + 3z = 4$
$\left(\frac{1}{2}, 0, \frac{2}{3}\right)$

38. $5x - 5y - 5z = 2$
 $5x + 5y - 5z = 6$ $\left(\frac{1}{5}, \frac{2}{5}, -\frac{3}{5}\right)$
 $10x + 10y + 5z = 3$

39. $3x - 2y + 2z = 5$
 $6x + 3y - 4z = -1$ $\left(\frac{2}{3}, -1, \frac{1}{2}\right)$
 $3x - y + 2z = 4$

40. $2x - y = 3$
 $3x + 2z = 7$ $(1, -1, 2)$
 $2y - 3z = -8$

41. $3y - 2z = -9$
 $2x + 3z = 13$ $(2, -1, 3)$
 $3x - y = 7$

42. $3x + y - 2z = 7$
 $2x - y = 2$ $(1, 0, -2)$
 $3x + 4z = -5$

43. $2y - 5z = 12$
 $3x + y - 4z = 9$ $(0, 1, -2)$
 $2x - 5z = 10$

44.   Find an equation of a plane that contains the points $(2, 1, 1)$, $(-1, 2, 12)$, and $(3, 2, 0)$. $z = -3x + 2y + 5$

45.   Find an equation of a plane that contains the points $(1, -1, 5)$, $(2, -2, 9)$, and $(-3, -1, -1)$. $z = \frac{3}{2}x - \frac{5}{2}y + 1$

46.   Find an equation of the form $y = ax^2 + bx + c$ whose graph passes through the points $(2, 3)$, $(-2, 7)$, and $(1, -2)$. $y = 2x^2 - x - 3$

47.   Find an equation of the form $y = ax^2 + bx + c$ whose graph passes through the points $(3, -4)$, $(2, -2)$, and $(1, -2)$. $y = -x^2 + 3x - 4$

48.   *Art*   A sculptor is creating a mobile from which three objects will be suspended from a light rod that is 15 inches long. The weight, in ounces, of each object is shown in the diagram at the right. For the mobile to balance, the objects must be positioned so that $w_1d_1 = w_2d_2 + w_3d_3$. The artist wants d_3 to be three times d_2. Find the distances d_1, d_2, and d_3 so that the mobile will balance. $d_1 = 6$ in., $d_2 = 3$ in., $d_3 = 9$ in.

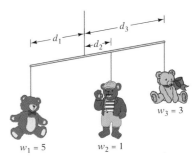

49.   *Art*   A mobile is made by suspending three objects from a light rod that is 20 inches long. The weight, in ounces, of each object is shown in the diagram at the right. For the mobile to balance, the objects must be positioned so that $w_1d_1 + w_2d_2 = w_3d_3$. The artist wants d_3 to be twice d_2. Find the distances d_1, d_2, and d_3 so that the mobile will balance. $d_1 = 8$ in., $d_2 = 6$ in., $d_3 = 12$ in.

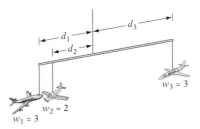

Applying Concepts

50.   *Biology*   Biologists use capture–recapture models to estimate how many animals live in a certain area. Say a sample of a certain number of fish is caught and tagged. When subsequent samples of fish are caught, a biologist can use a capture history matrix to record (with a 1) which, if any, of the fish in the original sample are caught again. The rows of the capture history matrix at the right represent particular fish (each has its own identification number), and the columns represent the number of the sample in which the fish was caught.

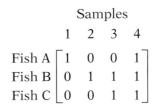

Samples

	1	2	3	4
Fish A	1	0	0	1
Fish B	0	1	1	1
Fish C	0	0	1	1

  a.   What is the meaning of the 1 in row A, column 4?   Fish A was caught again in sample 4.

  b.   Which fish was recaptured the most times?   Fish B

51. *Biology* Biologists can use a predator–prey matrix to study the relationships among animals in an ecosystem. Each row and each column represents an animal in that system. A 1 is used as an element in the matrix to indicate that the animal represented by that row preys on the animal in that column. A 0 is used to indicate that the animal in that row does not prey on the animal in that column. A simple predator–prey matrix is shown at the right. The abbreviations are H = hawk, R = rabbit, S = snake, C = coyote.

$$\begin{array}{c c c c c} & \text{H} & \text{R} & \text{S} & \text{C} \\ \text{H} & \begin{bmatrix} 0 & 1 & 1 & 0 \\ \text{R} & 0 & 0 & 0 & 0 \\ \text{S} & 1 & 1 & 0 & 0 \\ \text{C} & 0 & 1 & 1 & 0 \end{bmatrix} \end{array}$$

   a. What is the meaning of the 0 in row 2, column 1? Rabbits do not prey on hawks.

   b. What is the meaning of the 1 in row 3, column 2? Snakes prey on rabbits.

   c. What is the meaning of there being all zeros in column C? A coyote is not prey for hawks, rabbits, snakes, or coyotes.

   d. What is the meaning of all zeros in row R? A rabbit does not prey on hawks, rabbits, snakes, or coyotes.

52. The point of intersection of the graphs of $Ax + 3y = 6$ and $2x + By = -4$ is $(3, -2)$. Find A and B. $A = 4, B = 5$

53. The point of intersection of the graphs of $Ax + 3y + 2z = 8$, $2x + By - 3z = -12$, and $3x - 2y + Cz = 1$ is $(3, -2, 4)$. Find A, B, and C. $A = 2, B = 3, C = -3$

The following are not systems of linear equations. However, they can be solved by using a modification of the addition method. Solve each system of equations.

54. $\dfrac{1}{x} - \dfrac{2}{y} = 3$
    $\dfrac{2}{x} + \dfrac{3}{y} = -1$ $(1, -1)$

55. $\dfrac{1}{x} + \dfrac{2}{y} = 3$
    $\dfrac{1}{x} - \dfrac{3}{y} = -2$ $(1, 1)$

56. Suppose a system of equations contains three linear equations in two variables. Describe geometrically what must be true if the system of equations is to have a unique solution. The three lines must intersect at one point.

57. Describe the graph of each of the following equations in an xyz-coordinate system.
    a. $x = 3$ b. $y = 4$ c. $z = 2$ d. $y = x$ a. A plane parallel to the yz-plane passing through $x = 3$.   b. A plane parallel to the xz-plane passing through $y = 4$. c. A plane parallel to the xy-plane passing through $x = 2$.   d. A plane perpendicular to the xy-plane along the line $y = x$ in the xy-plane.

58. Solve the system and express the answer in the form (a, b, c, d).

$$a + b + c = 0$$
$$b + c + d = 1$$
$$a + c + d = 2$$
$$a + b + d = 3 \quad (1, 0, -1, 2)$$

EXPLORATION

1. *Reduced Row Echelon Form*   Another form in which an augmented matrix can be written is called *reduced row echelon form*. In this form, the matrix has 1's along the main diagonal and 0's above and below the main diagonal. The matrices at the right are in reduced row echelon form.

The advantage of having an augmented matrix in reduced row echelon form is that the corresponding system of equations is very easy to solve. The corresponding system of equations for each of the matrices at the right is shown below the matrix, along with the solution of the system of equations. The disadvantage, at least algebraically, is that it takes more steps to get the matrix in this form. However, a graphing calculator can be used to write a matrix in reduced row echelon form. Enter the augmented matrix into the calculator and then select the rref(function instead of ref(. Some sample screens are shown below.

$$\left[\begin{array}{cc|c} 1 & 0 & -2 \\ 0 & 1 & 3 \end{array}\right]$$

$$x = -2$$
$$y = 3$$

The solution is $(-2, 3)$.

$$\left[\begin{array}{ccc|c} 1 & 0 & 0 & 3 \\ 0 & 1 & 0 & -2 \\ 0 & 0 & 1 & 4 \end{array}\right]$$

$$x = 3$$
$$y = -2$$
$$z = 4$$

The solution is $(3, -2, 4)$.

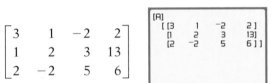

Enter the matrix.

Select **MATH** *under the* **MATRIX** *key.*

Scroll through the options to find rref(.

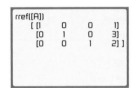

Find the reduced row echelon form for each matrix.

a. $\left[\begin{array}{cc} 3 & 4 & 25 \\ 2 & 1 & 10 \end{array}\right]$ $\left[\begin{array}{cc} 1 & 0 & 3 \\ 0 & 1 & 4 \end{array}\right]$

b. $\left[\begin{array}{cc} 3 & 2 & 16 \\ 2 & -3 & -11 \end{array}\right]$ $\left[\begin{array}{cc} 1 & 0 & 2 \\ 0 & 1 & 5 \end{array}\right]$

c. $\left[\begin{array}{ccc} 2 & 1 & -1 & 5 \\ 1 & 3 & 1 & 14 \\ 3 & -1 & 2 & 1 \end{array}\right]$ $\left[\begin{array}{ccc} 1 & 0 & 0 & 1 \\ 0 & 1 & 0 & 4 \\ 0 & 0 & 1 & 1 \end{array}\right]$

d. $\left[\begin{array}{ccc} 2 & -3 & 7 & 0 \\ 1 & 4 & -4 & -2 \\ 3 & 2 & 5 & 1 \end{array}\right]$ $\left[\begin{array}{ccc} 1 & 0 & 0 & -2 \\ 0 & 1 & 0 & 1 \\ 0 & 0 & 1 & 1 \end{array}\right]$

Solve the system of equations by finding the reduced row echelon form of the augmented matrix corresponding to the system of equations.

e. $2x - 5y = 13$
   $5x + 3y = 17$ $(4, -1)$

f. $4x + 4y = 5$
   $2x - 8y = -5$ $(0.5, 0.75)$

g. $5x + 3y - z = 5$
   $3x - 2y + 4z = 13$ $(1, 1, 3)$
   $4x + 3y + 5z = 22$

h. $3x - y - 2z = 11$
   $2x + y - 2z = 11$ $(2, 1, -3)$
   $x + 3y - z = 8$

Systems of Linear Inequalities and Linear Programming

■ Graph the Solution Set of a System of Linear Inequalities

Two or more inequalities considered together are called a **system of inequalities**. The **solution set of a system of inequalities** is the intersection of the solution sets of the individual inequalities. To graph the solution set of a system of inequalities, first graph the solution set of each inequality. The solution set of the system of inequalities is the region of the plane represented by the intersection of the shaded areas.

> **TAKE NOTE**
>
> You can use a test point to check that the correct region has been denoted as the solution set. We can see from the graph that the point (2, 4) is in the solution set, and as shown below, it is a solution of each inequality in the system. This indicates that the solution set as graphed is correct.
>
> $2x - y \le 3$
> $2(2) - (4) \le 3$
> $0 \le 3$ True
> $3x + 2y > 8$
> $3(2) + 2(4) > 8$
> $14 > 8$ True

➡ Graph the solution set: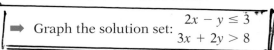
$$2x - y \le 3$$
$$3x + 2y > 8$$

Solve each inequality for y.

$2x - y \le 3$ $3x + 2y > 8$

$\quad -y \le -2x + 3$ $2y > -3x + 8$

$\quad\quad y \ge 2x - 3$ $y > -\dfrac{3}{2}x + 4$

Graph $y = 2x - 3$ as a solid line. Because the inequality is $\ge$, shade above the line.

Graph $y = -\dfrac{3}{2}x + 4$ as a dashed line. Because the inequality is $>$, shade above the line.

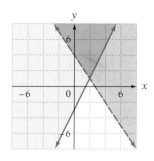

The solution set is the region of the plane represented by the intersection of the solution sets of the individual inequalities.

A graphing calculator can be used to draw the solution set of a system of inequalities. Begin as we have above by solving each equation for y.

$$y \ge 2x - 3$$
$$y > -\frac{3}{2}x + 4$$

 See Appendix A: Inequalities

Use the Y= editor window to enter the expressions, and then choose shading above or below the graph. The solution set is the intersection of the solution sets of the individual inequalities. Some typical graphing calculator screens are shown at the right.

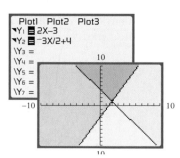

? QUESTION Is the point (3, 2) in the solution set of the system
$$3x + 4y \ge 12,$$
$$3x - 4y \ge 4$$
?

? ANSWER $3(3) + 4(2) = 17 \ge 12$; $3(3) - 4(2) = 1 \not\ge 4$. No, since (3, 2) is not a solution of each inequality in the system, it is not in the solution set of the system of inequalities.

▼ *Point of Interest*

Large systems of inequalities containing over 200 inequalities have been used to solve application problems in such diverse areas as providing health care, analyzing the economies of developing countries, and protecting nuclear silos.

Suggested Activity

The set of points that satisfies the system

$$2x - y < 0$$
$$x + y > 3$$

is contained entirely in which quadrants?
[Answer: I and II]

INSTRUCTOR NOTE
Exercise 8 can be used for a similar in-class example.

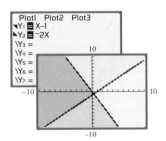

Suggested Activity

See Section 5.4 of the *Student Activity Manual* for an activity involving systems of linear inequalities.

➡ Graph the solution set: $-x + 2y \geq 4$
$$x - 2y \geq 6$$

Solve each inequality for y.

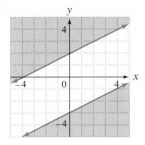

$$-x + 2y \geq 4 \qquad\qquad x - 2y \geq 6$$
$$2y \geq x + 4 \qquad\qquad -2y \geq -x + 6$$
$$y \geq \frac{1}{2}x + 2 \qquad\qquad y \leq \frac{1}{2}x - 3$$

Graph $y = \frac{1}{2}x + 2$ as a solid line. Because the inequality is $\geq$, shade above the line.

Graph $y = \frac{1}{2}x - 3$ as a solid line. Because the inequality is $\leq$, shade below the line.

Because the solution sets of the two inequalities do not intersect, the solution set of the system of inequalities is the empty set.   ⬅

EXAMPLE 1

Graph the solution set: $y \geq x - 1$
$$y < -2x$$

Solution   Shade the area above the solid line $y = x - 1$.

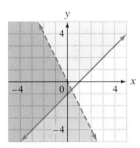

Shade the area below the dashed line $y = -2x$.

The solution of the system of inequalities is the intersection of the solution sets of the individual inequalities. A graphing calculator check is shown at the left.

YOU TRY IT 1

Graph the solution set: $y \geq 2x - 3$
$$y > -3x$$

Solution   See page S23.

■ Linear Programming

Consider a business analyst who is trying to maximize the profit from the production of a product, or an engineer who is trying to minimize the amount of energy an electrical circuit needs to operate. Generally, problems that seek to maximize or minimize a situation are called **optimization problems.** One strategy for solving certain of these problems was developed in the 1940s and is called **linear programming.**

A linear programming problem involves a **linear objective function,** which is the function that must be maximized or minimized. This objective function

▼ *Point of Interest*

George B. Dantzig and John von Neumann are generally given credit for the development of linear programming. During the early 1940s, Dantzig was trying to determine better methods of allocating resources to support soldiers fighting in World War II. Today, linear programming is used to schedule airline crews, select stocks for a mutual fund portfolio, explore methods to ensure the safety of nuclear missile sites, and solve many other types of problems.

In 1975, Leonid Kantorovich and Tjalling Koopmans shared the Nobel Prize in economics for demonstrating how linear programming could be used to allocate the resources of a developing country's economy.

Leonid Kantorovich

Tjalling Koopmans

is subject to some **constraints,** which are inequalities or equations that restrict the values of the variables. To illustrate these concepts, suppose a manufacturer produces two types of computer monitors: cathode ray tube (CRT) and liquid crystal display (LCD). Past sales experience shows that at least twice as many CRT monitors as LCD monitors are sold. Suppose further that the manufacturing plant is capable of producing 12 monitors per day. Let x represent the number of CRT monitors produced, and let y represent the number of LCD monitors produced. Then

$$x \geq 2y$$
$$x + y \leq 12$$

• These are the constraints.

These two inequalities place a constraint, or restriction, on the manufacturer. For example, the manufacturer cannot produce 5 LCD monitors, because that would require producing at least 10 CRT monitors, and $5 + 10 \nleq 12$.

Suppose a profit of $50 is earned on each CRT monitor sold and $75 is earned on each LCD monitor sold. Then the manufacturer's profit, P, is given by the equation

$$P = 50x + 75y \quad \text{• Objective function}$$

The equation $P = 50x + 75y$ defines the objective function. The goal of this linear programming problem is to determine how many of each monitor should be produced to maximize the manufacturer's profit and at the same time satisfy the constraints.

Because the manufacturer cannot produce fewer than zero units of either monitor, there are two other implied constraints, $x \geq 0$ and $y \geq 0$. Our linear programming problem now looks like

Objective function: $P = 50x + 75y$

Constraints: $\begin{cases} x - 2y \geq 0 \\ x + y \leq 12 \\ x \geq 0, y \geq 0 \end{cases}$

To solve this problem, graph the solution set of the constraints. The solution set of the constraints is called the **set of feasible solutions.** Ordered pairs in this set are used to evaluate the objective function to determine which ordered pair maximizes the profit.

For example, from the figure at the right, (5, 2), (8, 3), and (10, 1) are three ordered pairs in the set. For these ordered pairs, the profit would be

$$P = 50(5) + 75(2) = 400$$
$$P = 50(8) + 75(3) = 625$$
$$P = 50(10) + 75(1) = 575$$

It would be impossible to check every ordered pair in the set of feasible solutions to find which maximizes profit. Fortunately, we can find that ordered pair by solving the objective function $P = 50x + 75y$ for y.

$$y = -\frac{2}{3}x + \frac{P}{75}$$

In this form, the objective function is a linear equation whose graph has slope $-\frac{2}{3}$ and y-intercept $\frac{P}{75}$. If P is as great as possible (P a maximum), then the y-intercept will be as large as possible. Thus the maximum profit will occur on the line that has a slope of $-\frac{2}{3}$, has the greatest possible y-intercept, and intersects the set of feasible solutions.

From the figure at the left, the largest possible y-intercept occurs when the line passes through the point with coordinates $(8, 4)$. At this point, the profit is

$$P = 50(8) + 75(4) = 700$$

The manufacturer will maximize profit by producing 8 CRT monitors and 4 LCD monitors each day. The profit will be $700 per day.

P In general, the goal of any linear programming problem is to maximize or minimize the objective function, subject to the constraints. Minimization problems occur, for example, when a manufacturer wants to minimize the cost of operations.

Suppose that a minimization problem results in the following objective function and constraints.

Objective function: $C = 3x + 4y$

Constraints: $\begin{cases} x + y \geq 1 \\ 2x - y \leq 5 \\ x + 2y \leq 10 \\ x \geq 0, y \geq 0 \end{cases}$

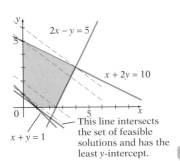

The figure at the left is the graph of the solution set of the constraints. The task is to find the ordered pair that satisfies all the constraints and gives the *least* value of C. This will occur when the line $y = -\frac{3}{4}x + \frac{1}{4}C$ passes through the point with coordinates $(1, 0)$. At this point, the cost is

$$C = 3(1) + 4(0) = 3$$

The smallest value of C that satisfies the constraints is 3.

P Finding a graphical solution as we have done is not practical for more complicated problems. Fortunately, there is a theorem from linear programming that can be used to find solutions to these types of problems.

Fundamental Linear Programming Theorem

If an objective function has an optimal solution, then that solution will be at a vertex of the set of feasible solutions.

Following is a list of the values of C at the vertices. The minimum value of the objective function occurs at the point whose coordinates are $(1, 0)$.

(x, y)	$C = 3x + 4y$
$(1, 0)$	$C = 3(1) + 4(0) = 3$ • Minimum

$$\left(\frac{5}{2}, 0\right) \quad C = 3\left(\frac{5}{2}\right) + 4(0) = 7.5$$

$(4, 3) \qquad C = 3(4) + 4(3) = 24 \qquad \bullet \text{ Maximum}$

$(0, 5) \qquad C = 3(0) + 4(5) = 20$

$(0, 1) \qquad C = 3(0) + 4(1) = 4$

The maximum value of the objective function can also be determined from the list. It occurs at (4, 3).

It is important to realize that the maximum or minimum value of an objective function depends on the objective function and on the set of feasible solutions. For example, using the same set of feasible solutions as in the foregoing figure but changing the objective function to $C = 2x + 5y$ changes the maximum value of C to 25 at the ordered pair (0, 5). You should verify this result by making a list similar to the one shown above.

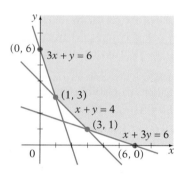

➡ Minimize the objective function $C = 4x + 7y$ with the constraints

$$\begin{cases} 3x + y \geq 6 \\ x + y \geq 4 \\ x + 3y \geq 6 \\ x \geq 0, y \geq 0 \end{cases}$$

Determine the set of feasible solutions by graphing the solution set of the inequalities. Note that in this instance, the set of feasible solutions is an unbounded set.

Find the vertices of the region by solving the following systems of equations. These systems are formed by the equations of the lines that intersect to form a vertex of the set of feasible solutions.

$$\begin{cases} 3x + y = 6 \\ x + y = 4 \end{cases} \qquad \begin{cases} x + 3y = 6 \\ x + y = 4 \end{cases}$$

The solutions of the two systems are (1, 3) and (3, 1), respectively. The points (6, 0) and (0, 6) are the vertices on the x- and y-axes.

Evaluate the objective function at each of the four vertices of the set of feasible solutions.

$(x, y) \qquad C = 4x + 7y$

$(0, 6) \qquad C = 4(0) + 7(6) = 42$

$(1, 3) \qquad C = 4(1) + 7(3) = 25$

$(3, 1) \qquad C = 4(3) + 7(1) = 19$

$(6, 0) \qquad C = 4(6) + 7(0) = 24$

The minimum value of the objective function is 19 at (3, 1). ⬅

Linear programming can be used to determine the best allocation of the resources available to a company. In fact, the word *programming* refers to a "program to allocate resources."

INSTRUCTOR NOTE
Exercise 40 can be used for
a similar in-class example.

EXAMPLE 2

A manufacturer of animal food makes two grain mixtures. Each kilogram of G_1 contains 300 grams of vitamins, 400 grams of protein, and 100 grams of carbohydrate. Each kilogram of G_2 contains 100 grams of vitamins, 300 grams of protein, and 200 grams of carbohydrate. Minimum nutritional guidelines require that a feed mixture made from these grains contain at least 900 grams of vitamins, 2200 grams of protein, and 800 grams of carbohydrate. G_1 costs \$2.00 per kilogram to produce, and G_2 costs \$1.25 per kilogram to produce. Find the number of kilograms of each grain mixture that should be combined to minimize cost.

State the goal.   The goal is to find the number of kilograms of each grain mixture that should be combined to minimize cost.

Devise a strategy.   Let x = the number of kilograms of G_1 and y = the number of kilograms of G_2.

The objective function is the function to be minimized. In this case, we are trying to minimize cost. Since G_1 costs \$2.00 per kilogram to produce and G_2 costs \$1.25 per kilogram to produce, the cost function is:

$$C = 2x + 1.25y$$

Next we must find the constraints. Because x kilograms of G_1 contains $300x$ grams of vitamins and y kilograms of G_2 contains $100y$ grams of vitamins, the total amount of vitamins contained in x kilograms of G_1 and y kilograms of G_2 is $300x + 100y$. At least 900 grams of vitamins must be included, so $300x + 100y \geq 900$. Following similar reasoning, we have the constraints

$$300x + 100y \geq 900$$
$$400x + 300y \geq 2200$$
$$100x + 200y \geq 800$$
$$x \geq 0, y \geq 0$$

Solve the problem.   Two of the vertices of the set of feasible solutions can be found by solving two systems of equations. These systems are formed by the equations of the lines that intersect to form a vertex of the set of feasible solutions. First use the Gaussian elimination method to solve the system of equations using the first two inequalities from the constraints.

System of Equations	**Augmented Matrix**	**Row Echelon Form**
$\begin{aligned} 300x + 100y &= 900 \\ 400x + 300y &= 2200 \end{aligned}$ $\longrightarrow$	$\begin{bmatrix} 300 & 100 & \mid & 900 \\ 400 & 300 & \mid & 2200 \end{bmatrix}$ $\longrightarrow$	$\begin{bmatrix} 1 & 0.75 & 5.5 \\ 0 & 1 & 6 \end{bmatrix}$

Solve the system of equations corresponding to the matrix that is in row echelon form.

(1) $x + 0.75y = 5.5$ • Write the system of equations corresponding
(2) $y = 6$ to the matrix that is in row echelon form.

$x + 0.75(6) = 5.5$ • Substitute 6 for y in Equation (1) and solve for x.

$x + 4.5 = 5.5$

$x = 1$

$(0, 9)$
$300x + 100y = 900$
$(1, 6)$
$100x + 200y = 800$
$(4, 2)$
Kilograms of G_2
0
$400x + 300y = 2200$
$(8, 0)$
Kilograms of G_1

The solution of this system of equations is $(1, 6)$. This is one vertex of the set of feasible solutions.

Form and solve a second system of equations using the second and third constraints.

System of Equations	Augmented Matrix	Row Echelon Form
$400x + 300y = 2200$	$\begin{bmatrix} 400 & 300 & \vert & 2200 \\ 100 & 200 & \vert & 800 \end{bmatrix}$	$\begin{bmatrix} 1 & 0.75 & 5.5 \\ 0 & 1 & 2 \end{bmatrix}$
$100x + 200y = 800$		

Solve the system of equations corresponding to the matrix that is in row echelon form.

(1) $x + 0.75y = 5.5$ • Write the system of equations corresponding
(2) $y = 2$ to the matrix that is in row echelon form.

$x + 0.75(2) = 5.5$ • Substitute 2 for y in Equation (1) and solve for x.

$x + 1.5 = 5.5$

$x = 4$

The solution is $(4, 2)$. This is another vertex of the set of feasible solutions.

The vertices on the x- and y-axes are the x- and y-intercepts $(8, 0)$ and $(0, 9)$. Substitute the coordinates of the vertices into the objective function.

(x, y)	$C = 2x + 1.25y$
$(0, 9)$	$C = 2(0) + 1.25(9) = 11.25$
$(1, 6)$	$C = 2(1) + 1.25(6) = 9.50$
$(4, 2)$	$C = 2(4) + 1.25(2) = 10.50$
$(8, 0)$	$C = 2(8) + 1.25(0) = 16.00$

The minimum value of the objective function is $9.50. It occurs when the company produces a feed mixture that contains 1 kilogram of G_1 and 6 kilograms of G_2.

Check your work. Be sure to check your work.

YOU TRY IT 2

A chemical firm produces two types of industrial solvents, S_1 and S_2. Each solvent is a mixture of three chemicals. Each kiloliter of S_1 requires 12 liters of chemical 1, 9 liters of chemical 2, and 30 liters of chemical 3. Each kiloliter of S_2 requires 24 liters of chemical 1, 5 liters of chemical 2, and 30 liters of chemical 3. The profit per kiloliter of S_1 is $100, and the profit per kiloliter of S_2 is $85. The inventory of the company shows 480 liters of chemical 1, 180 liters of chemical 2, and 720 liters of chemical 3. Assuming the company can sell all the solvent it makes, find the number of kiloliters of each solvent that the company should make to maximize profit.

Solution See page S23.   15 kiloliters of S_1, 9 kiloliters of S_2

5.4 EXERCISES Suggested Assignment: 5–41, odds

Topics for Discussion

1. Explain how to find the solution set of a system of linear inequalities.
   Graph each inequality. Then determine the intersection of the solution sets of the individual inequalities.

2. What is a constraint for a linear programming problem?
   A constraint is a condition or restriction on the variables of the problem.

3. What is the objective function of a linear programming problem?
   The objective function is the function that is to be maximized or minimized.

4. Explain how to solve a linear programming problem.
   Explanations may vary. Each explanation should include a reference to finding the vertices of the set of feasible solutions and then evaluating the objective function at these ordered pairs.

■ Graph the Solution Set of a System of Linear Inequalities

Which ordered pair is a solution of the system of inequalities?

5. $2x - y < 4$
   $x - 3y \geq 6$
   a. $(5, 1)$ **b.** $(-3, -5)$ b

6. $3x - 2y \geq 6$
   $x + y < 5$
   a. $(-2, 3)$ **b.** $(3, -2)$ b

Graph the solution set.

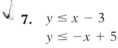

7. $y \leq x - 3$
   $y \leq -x + 5$

8. $y > 2x - 4$
   $y < -x + 5$

9. $y > 3x - 3$
   $y \geq -2x + 2$

10. $x + 2y \leq 6$
    $x - y \leq 3$

11. $2x + y \geq -2$
    $6x + 3y \leq 6$

12. $x + y \geq 5$
    $3x + 3y \leq 6$

no solution

13. $3x - 2y < 6$
    $y \leq 3$

14. $x \leq 2$
    $3x + 2y > 4$

15. $y > 2x - 6$
    $x + y < 0$

16. $x < 3$
$y < -2$

17. $x + 1 \geq 0$
$y - 3 \leq 0$

18. $5x - 2y \geq 10$
$3x + 2y \geq 6$

19. $2x + y \geq 4$
$3x - 2y < 6$

20. $3x - 4y < 12$
$x + 2y < 6$

21. $x - 2y \leq 6$
$2x + 3y \leq 6$

22. $x - 3y > 6$
$2x + y > 5$

23. $x - 2y \leq 4$
$3x + 2y \leq 8$
$x > -1$

24. $3x - 2y < 0$
$5x + 3y > 9$
$y < 4$

25. $2x + 3y \leq 15$
$3x - y \leq 6$
$y \geq 0$

26. $x + y \leq 6$
$x - y \leq 2$
$x \geq 0$

27. $x - y \leq 5$
$2x - y \geq 6$
$y \geq 0$

28. $x - 3y \leq 6$
$5x - 2y \geq 4$
$y \geq 0$

29. $2x - y \leq 4$
$3x + y < 1$
$y \leq 0$

30. $x - y \leq 4$
$2x + 3y > 6$
$x \geq 0$

■ Linear Programming

Solve the linear programming problem. Assume $x \geq 0$ and $y \geq 0$.

31. Minimize $C = 4x + 2y$ with the constraints
$$x + y \geq 7$$
$$4x + 3y \geq 24$$
$x \leq 10,\ y \leq 10$ The minimum is 16 at (0, 8).

32. Minimize $C = 5x + 4y$ with the constraints
$$3x + 4y \geq 32$$
$$x + 4y \geq 24$$
$x \leq 12,\ y \leq 15$ The minimum is 32 at (0, 8).

33. Maximize $C = 6x + 7y$ with the constraints
$$x + 2y \leq 16$$
$5x + 3y \leq 45$ The maximum is 71 at (6, 5).

34. Maximize $C = 6x + 5y$ with the constraints
$$2x + 3y \leq 27$$
$7x + 3y \geq 42$ The maximum is 81 at (13.5, 0).

35. Maximize $C = 2x + 7y$ with the constraints
$$x + y \leq 10$$
$$x + 2y \geq 16$$
$2x + y \leq 16$ The maximum is 70 at (0, 10).

36. Minimize $C = 4x + 3y$ with the constraints
$$2x + y \geq 8$$
$$2x + 3y \geq 16$$
$$x + 3y \geq 11$$
$x \leq 20,\ y \leq 20$ The minimum is 20 at (2, 4).

37. Minimize $C = 3x + 2y$ with the constraints
$$3x + y \geq 12$$
$$2x + 7y \geq 21$$
$x + y \geq 8$ The minimum is 18 at (2, 6).

38. Maximize $C = 2x + 6y$ with the constraints
$$x + y \leq 12$$
$$3x + 4y \leq 40$$
$x + 2y \leq 18$ The maximum is 54 at (0, 9).

39. *Agriculture*   A farmer is planning to raise wheat and barley. Each acre of wheat yields a profit of $50, and each acre of barley yields a profit of $70. To sow the crop, two machines, a tractor and tiller, are rented. The tractor is available for 200 hours, and the tiller is available for 100 hours. Sowing an acre of barley requires 3 hours of tractor time and 2 hours of tilling. Sowing an acre of wheat requires 4 hours of tractor time and 1 hour of tilling. How many acres of each crop should be planted to maximize the farmer's profit?   20 acres of wheat, 40 acres of barley

40. *Food Production*   An ice cream supplier has two machines that produce vanilla and chocolate ice cream. To meet one of its contractual obligations, the company must produce at least 60 gallons of vanilla ice cream and 100 gallons of chocolate ice cream per hour. One machine makes 4 gallons of vanilla and 5 gallons of chocolate ice cream per hour. The second machine makes 3 gallons of vanilla and 10 gallons of chocolate ice cream per hour. It costs $28 per hour to run machine 1 and $25 per hour to run machine 2. How many hours should each machine be operated to fulfill the contract at the least expense? Machine 1 for 12 h, machine 2 for 4 h

41. *Manufacturing*   A manufacturer makes two types of golf clubs, a starter model and a professional model. The starter model requires 4 hours in the assembly room and 1 hour in the finishing room. The professional model requires 6 hours in the assembly room and 1 hour in the finishing room. The total number of hours available per week in the assembly room is 108. There are 24 hours available per week in the finishing room. The profit for each starter model is $35, and the profit for each professional model is \$55. Assuming all the sets produced can be sold, find how many of each set should be manufactured per week to maximize profit.   0 starter sets, 18 pro sets

42.   *Manufacturing*   A company makes two types of telephone answering machines, the standard model and the deluxe model. Each machine passes through three processes: P_1, P_2, and P_3. One standard answering machine requires 1 hour in P_1, 1 hour in P_2, and 2 hours in P_3. One deluxe answering machine requires 3 hours in P_1, 1 hour in P_2, and 1 hour in P_3. Because of employee work schedules, P_1 is available for 24 hours per day, P_2 is available for 10 hours per day, and P_3 is available for 16 hours per day. If the profit is \$25 for each standard model and \$35 for each deluxe model, how many units of each type should the company produce per day to maximize profit?   3 standard models, 7 deluxe models

Applying Concepts

Write a system of inequalities to represent the shaded region.

43.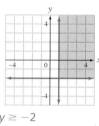

$y \geq -2$
$x \geq 1$

44.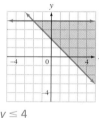

$y \leq 4$
$y \geq -x + 2$

45.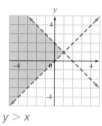

$y > x$
$y < -x + 2$

46.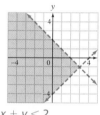

$x + y < 2$
$x - y < 4$

47.   *Nutrition*   A dietitian formulates a special diet from two food groups, A and B. Each ounce of food group A contains 3 units of vitamin A, 1 unit of vitamin C, and 1 unit of vitamin D. Each ounce of food group B contains 1 unit of vitamin A, 1 unit of vitamin C, and 3 units of vitamin D. Each ounce of food group A costs 40 cents, and each ounce of food group B costs 10 cents. The dietary constraints are such that at least 24 units of vitamin A, 16 units of vitamin C, and 30 units of vitamin D are required. Find the amount of each food group that should be used to minimize the cost. What is the minimum cost?
24 oz of food group B, 0 oz of food group A; \$2.40

48.   *Energy*   Among the many products it produces, an oil refinery makes two specialized petroleum distillates, Pymex A and Pymex B. Each distillate passes through three stages: S_1, S_2, and S_3. Each liter of Pymex A requires 1 hour in S_1, 3 hours in S_2, and 3 hours in S_3. Each liter of Pymex B requires 1 hour in S_1, 4 hours in S_2, and 2 hours in S_3. There are 10 hours available per week for S_1, 36 hours available per week for S_2, and 27 hours available per week for S_3. The profit per liter of Pymex A is \$12, and the profit per liter of Pymex B is \$9. How many liters of each distillate should be produced per week to maximize profit? What is the maximum profit?   7 L of Pymex A, 3 L of Pymex B; \$111

49. *Automotive Technology*   An engine reconditioning company works on 4- and 6-cylinder engines. Each 4-cylinder engine requires 1 hour for cleaning, 5 hours for overhauling, and 3 hours for testing. Each 6-cylinder engine requires 1 hour for cleaning, 10 hours for overhauling, and 2 hours for testing. The cleaning station is available for at most 9 hours per week, the overhauling equipment is available for at most 80 hours per week, and the testing equipment is available for at most 24 hours per week. For each reconditioned 4-cylinder engine, the company makes a profit of $150. A reconditioned 6-cylinder engine yields a profit of $250. The company can sell all the reconditioned engines it produces. How many of each type should be produced per week to maximize profit? What is the maximum profit? Two 4-cylinder engines, seven 6-cylinder engines; $2050

50. *Animal Science*   A producer of animal feed makes two food products: F_1 and F_2. The products contain three major ingredients: M_1, M_2, and M_3. Each ton of F_1 requires 200 pounds of M_1, 100 pounds of M_2, and 100 pounds of M_3. Each ton of F_2 requires 100 pounds of M_1, 200 pounds of M_2, and 400 pounds of M_3. There are at least 5000 pounds of M_1 available, at least 7000 pounds of M_2 available, and at least 10,000 pounds of M_3 available. Each ton of F_1 costs $450 to make, and each ton of F_2 costs $300 to make. How many tons of each food product should the feed producer make to minimize cost? What is the minimum cost? 10 tons of F_1, 30 tons of F_2; $13,500

EXPLORATION

Analyzing Graphs of Two Functions

1. *Sports*   Cara and Daren begin from the same point on a bicycle trail and return to that point some time later. The blue graph at the right shows Cara's distance, in miles, from the starting point t hours after starting the trip. The graph in red shows the same information for Daren.

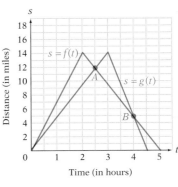

a.   In which intervals on the t-axis is $f(t) \le g(t)$?

b.   In which intervals on the t-axis is $f(t) \ge g(t)$?

c.   Based on your answer to part a, when is Daren closer to the starting point than Cara?

d.   Based on your answer to part b, when is Cara closer to the starting point than Daren?

e.   In the context of this problem, what is the significance of the points labeled A and B?

f.   How do points A and B differ in terms of Daren's and Cara's movement toward or away from the starting point?

g.   Who returns to the starting point first?

a. $2.5 \le t \le 4$ **b.** $0 \le t \le 2.5$ and $4 \le t \le 5$ **c.** $2.5 \le t \le 4$ **d.** $0 \le t \le 2.5$ and $4 \le t \le 5$ **e.** At A and B, Daren and Cara are exactly the same distance from the starting point.   **f.** At A, Daren is returning to the starting point while Cara is still moving away from it. At B, Cara and Daren are both returning to the starting point.

g. Cara.

2. *Business* When a company's income or revenue, *R*, exceeds its expenses or costs, *C*, the company has a profit. The graph at the right shows the revenue of a company and its costs to produce and sell *n* cell phones.

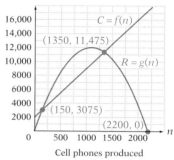

Cell phones produced

   a. In which intervals on the *n*-axis is $f(n) \leq g(n)$?
   b. In which intervals on the *n*-axis is $f(n) \geq g(n)$?
   c. Based on your answer to part a, how many cell phones should the company produce to be profitable?
   d. At which points is the company "breaking even"? That is, at which points does revenue equal cost?
   e. Is the *profit* greatest when $n = 250$, $n = 750$, or $n = 1000$?

   a. $150 \leq n \leq 1350$ **b.** $0 \leq n \leq 150$ and $1350 \leq n \leq 2200$ **c.** Between 150 and 1350   **d.** (150, 3075) and (1350, 11475)   **e.** $n = 750$

CHAPTER **5** *SUMMARY*

Key Terms

augmented matrix [**p. 338**]
constraint [**p. 358**]
dependent system of equations
   [**p. 311**]
element of a matrix [**p. 337**]
elementary row operations [**p. 339**]
Gaussian elimination method
   [**p. 344**]
inconsistent system of equations
   [**p. 310**]
independent system of equations
   [**p. 310**]
linear equation in three variables
   [**p. 323**]
linear objective function [**p. 357**]
linear programming [**p. 357**]
main diagonal [**p. 338**]
matrix [**p. 337**]
optimization problems [**p. 357**]
ordered triples [**p. 324**]
order $m \times n$ or dimension of a
   matrix [**p. 338**]

row echelon form [**p. 341**]
set of feasible solutions [**p. 358**]
solution of a system of equations in
   three variables [**p. 309**]
solution of a system of equations in
   two variables [**p. 325**]
solution of an equation in three
   variables [**p. 325**]
solution set of a system of inequalities [**p. 356**]
substitution method [**p. 311**]
system of equations [**p. 309**]
system of inequalities [**p. 356**]
system of linear equations in three
   variables [**p. 325**]
system of linear equations in two
   variables [**p. 309**]
three-dimensional coordinate
   system [**p. 324**]
xyz-coordinate system [**p. 324**]

Essential Concepts

Solve a System of Equations by Graphing
Graph each equation and graphically determine the point of intersection.
[**p. 309**]

Solve a System of Equations by the Substitution Method
Write one of the equations of the system in terms of *x* or *y*. Then substitute the variable expression for *x* or *y* into another equation of the system. [**p. 311**]

Solve a System of Equations by the Addition Method

Use the Multiplication Property of Equations to rewrite the equations in the system of equations so that the coefficients of one variable are additive inverses. Then add the two equations. **[p. 321]**

Elementary Row Operations on a Matrix

1. Interchange two rows.
2. Multiply all the elements in a row by the same nonzero constant.
3. Replace a row by the sum of that row and a multiple of any other row. **[p. 339]**

Solve a System of Equations by the Gaussian Elimination Method

Write the system of equations as an augmented matrix. Then use the elementary row operations to write the augmented matrix in row echelon row. Use the substitution method to solve the system of equations corresponding to row echelon form. **[p. 344]**

Fundamental Linear Programming Theorem

If an objective function has an optimal solution, then that solution will be at a vertex of the set of feasible solutions. **[p. 359]**

Solve a Linear Programming Problem

Graph each of the constraints and determine the points of intersection. Substitute the coordinates of the points of intersection into the objective function and determine the minimum or maximum value of that function. **[pp. 357–360]**

CHAPTER **5** *REVIEW EXERCISES*

1. Solve by substitution: $2x - 6y = 15$
$$x = 4y + 8$$
$(6, -\frac{1}{2})$

2. Solve by the addition method: $3x + 2y = 2$
$$x + y = 3$$
$(-4, 7)$

3. Solve by graphing: $x + y = 3$
$$3x - 2y = -6$$
$(0, 3)$

4. Solve by substitution: $2x - y = 4$
$$y = 2x - 4$$
$(x, 2x - 4)$

5. Solve by the addition method: $5x - 15y = 30$
$$x - 3y = 6$$
$(x, \frac{1}{3}x - 2)$

6. Solve by the addition method: $3x - 4y - 2z = 17$
$$4x - 3y + 5z = 5$$
$$5x - 5y + 3z = 14$$
$(3, -1, -2)$

7. Write the augmented matrix for the system of equations shown at the right.

$2x - 3y - z = 1$
$3x \quad\quad - 4z = -2$
$\quad\quad 4y - 5z = 0$

$$\begin{bmatrix} 2 & -3 & -1 & 1 \\ 3 & 0 & -4 & -2 \\ 0 & 4 & -5 & 0 \end{bmatrix}$$

8. Write the system of equations corresponding to the augmented matrix shown at the right.

$$\begin{bmatrix} 1 & 3 & 0 & -2 \\ 2 & -1 & 1 & 0 \\ 3 & 2 & -5 & 4 \end{bmatrix}$$

$x + 3y \quad\quad = -2$
$2x - y + \quad z = 0$
$3x + 2y - 5z = 4$

9. Write in row echelon form:

$$\begin{bmatrix} 3 & 6 & -3 & 9 \\ -3 & -5 & 1 & 4 \\ 2 & 3 & 5 & 2 \end{bmatrix} \qquad \begin{bmatrix} 1 & 2 & -1 & 3 \\ 0 & 1 & -2 & 13 \\ 0 & 0 & 1 & \frac{9}{5} \end{bmatrix}$$

10. Solve using the Gaussian elimination method:

$$2x + 5y = -1$$
$$3x - 4y = 10 \qquad (2, -1)$$

11. Solve using the Gaussian elimination method:

$$3x + 2y = 5$$
$$4x + 5y = 2 \qquad (3, -2)$$

12. Solve using the Gaussian elimination method:

$$x + 3y + z = 6$$
$$2x + y - z = 12 \qquad (2, 3, -5)$$
$$x + 2y - z = 13$$

13. Solve using the Gaussian elimination method:

$$x + y + z = 0$$
$$x + 2y + 3z = 5 \qquad (-1, -3, 4)$$
$$2x + y + 2z = 3$$

14. Minimize the objective function $P = 4x + y$ given the following constraints.

$$5x + 2y \geq 16$$
$$x + 2y \geq 8 \qquad \text{The minimum is 8 at } (0, 8).$$
$$x \leq 20, y \leq 20$$

15. Maximize the objective function $P = 2x + 2y$ given the following constraints.

$$x + 2y \leq 14$$
$$5x + 2y \leq 30$$
$$x \geq 0, y \geq 0$$

The maximum is 18 at (4, 5).

16. Graph the solution set: $x + 3y \leq 6$
$$2x - y \geq 4$$

17. Graph the solution set: $2x + 4y \geq 8$
$$x + y \leq 3$$

18. A cabin cruiser traveling with the current went 60 miles in 3 hours. Against the current, it took 5 hours to travel the same distance. Find the rate of the cabin cruiser in calm water and the rate of the current.

Cabin cruiser: 16 mph; current: 4 mph

19. At a movie theater, admission tickets are $5 for children and $8 for adults. The receipts for one Friday evening were $2500. The next day there were three times as many children as the preceding evening and half the number of adults as the night before; the receipts were $2500. Find the number of children who attended the Friday evening show.

100 children

20. A farmer has 160 acres available on which to plant oats and barley. It costs $15 per acre for oat seed and $13 per acre for barley seed. The labor cost is $15 per acre for oats and $20 per acre for barley. The farmer has $2200 available to purchase seed and has set aside $2600 for labor. The profit per acre for oats is $120, and the profit per acre for barley is $150. How many acres of oats should the farmer plant to maximize the profit on these crops?

$\frac{680}{7}$ acres

CHAPTER **5** *TEST*

 1. Solve by substitution: $3x + 2y = 4$
$$x = 2y - 1$$
$\left(\frac{3}{4}, \frac{7}{8}\right)$

2. Solve by the addition method: $4x - 6y = 5$
$$6x - 9y = 4$$
No solution

3. Solve by graphing: $3x + y = 7$ $(2, 1)$
$$2x - y = 3$$

4. Solve by substitution: $5x + 2y = -23$ $(-3, -4)$
$$2x + y = -10$$

 5. Solve by the addition method:
$4x - 12y = 12$ $\left(x, \frac{1}{3}x - 1\right)$
$$x - 3y = 3$$

6. Solve by the addition method:
$3x + 2y + 2z = 2$
$x - 2y - z = 1$ $(0, -2, 3)$
$2x - 3y - 3z = -3$

7. Write the augmented matrix for the system of equations shown at the right.
$3x - y + 2z = 4$
$x + 4y = -1$
$5y - z = 3$
$$\begin{bmatrix} 3 & -1 & 2 & 4 \\ 1 & 4 & 0 & -1 \\ 0 & 5 & -1 & 3 \end{bmatrix}$$

8. Write the system of equations corresponding to the augmented matrix shown at the right.
$$\begin{bmatrix} 2 & -1 & 3 & 4 \\ 1 & 5 & -2 & 6 \\ -3 & 0 & -4 & 1 \end{bmatrix}$$
$2x - y + 3z = 4$
$x + 5y - 2z = 6$
$-3x - 4z = 1$

9. Write in row echelon form:
$$\begin{bmatrix} 2 & 3 & -2 \\ 4 & 1 & 1 \end{bmatrix} \quad \begin{bmatrix} 1 & \frac{3}{2} & -1 \\ 0 & 1 & -1 \end{bmatrix}$$

10. Solve using the Gaussian elimination method:
$3x + 4y = -2$ $(-2, 1)$
$2x + 5y = 1$

11. Solve using the Gaussian elimination method:
$x - y = 3$
$2x + y = -4$ $\left(-\frac{1}{3}, -\frac{10}{3}\right)$

12. Solve using the Gaussian elimination method:
$x - y - z = 5$
$2x + z = 2$ $(2, -1, -2)$
$3y - 2z = 1$

13. Solve using the Gaussian elimination method:
$x - y + z = 10$
$2x - y - z = 5$ $(1, -6, 3)$
$x + 2y - 3z = -20$

14. Maximize the objective function $P = 4x + 5y$ given the following constraints.
$2x + 3y \leq 24$
$4x + 3y \leq 36$ The maximum is 44 at $(6, 4)$.

15. Minimize the objective function $P = 3x + 2y$ given the following constraints.
$x + 2y \geq 8$
$3x + y \geq 9$ The minimum is 12 at $(2, 3)$.
$x + 4y \geq 12$

 16. Graph the solution set: $3x - 2y \geq 4$
$$x + y < 3$$
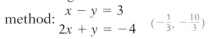

 17. Graph the solution set: $x + y > 2$
$$2x - y < -1$$

18. A plane flying with the wind went 350 miles in 2 hours. The return trip, flying against the wind, took 2.8 hours. Find the rate of the plane in calm air and the rate of the wind.   Plane: 150 mph; wind: 25 mph

19. A clothing manufacturer purchased 60 yards of cotton and 90 yards of wool for a total cost of $1800. Another purchase, at the same prices, included 80 yards of cotton and 20 yards of wool for a total cost of $1000. Find the cost per yard of the cotton and of the wool.
Cotton: $9; wool: $14

20. High Sports Company makes two types of baseball gloves, standard and professional. Each glove is produced at a station consisting of a machine and a person who finishes the gloves by hand. The standard glove requires 2 hours of machine time and 1 hour of finishing time. The professional glove requires 3 hours of machine time and 5 hours of finishing time. The profit on a standard glove is $50. The profit on a professional glove is $60. In an 8-hour workday, how many of each type of glove should be produced at each station to maximize the profit?
4 standard gloves, 0 professional gloves

◀ CUMULATIVE REVIEW EXERCISES

1. Let P represent the product of all the positive prime numbers less than 100. What is the units digit of the product P?   0

2. In a class election, one candidate received more than 94%, but less than 100%, of the votes cast. What is the least possible number of votes cast in the election?   17 votes

3. Evaluate $1 + 3$, $1 + 3 + 5$, $1 + 3 + 5 + 7$, and $1 + 3 + 5 + 7 + 9$. Then use inductive reasoning to explain the pattern and use your reasoning to determine $1 + 3 + 5 + 7 + 9 + 11$. $4 = 2^2$, $9 = 3^2$, $16 = 4^2$, $25 = 5^2$.
Answers may vary; for example, each number is the square of the number of odd consecutive integers added.
$1 + 3 + 5 + 7 + 9 + 11 = 36 = 6^2$.

4. Determine whether the following argument is an example of inductive or deductive reasoning. "All movies directed by Steven Spielberg are blockbusters. The movie *Saving Private Ryan* was directed by Steven Spielberg. Therefore, *Saving Private Ryan* was a blockbuster."
Deductive reasoning

5. Find the range of $g(x) = x^2 - 4x$ given the domain $\{-3, -2, -1, 0, 1, 2, 3\}$.
$\{-4, -3, 0, 5, 12, 21\}$

6. Use a graphing calculator to graph $H(x) = 2x^2 - 4x + 1$ using the viewing window $\mathsf{Xmin} = -4.7$, $\mathsf{Xmax} = 4.7$, $\mathsf{Xscl} = 1$, $\mathsf{Ymin} = -5$, $\mathsf{Ymax} = 5$, $\mathsf{Yscl} = 1$. Then use the TRACE feature to find the two x-coordinates for the y-coordinate 1.   0, 2

7. Find the x- and y-intercepts for the graph of the function $G(x) = x^3 + 5x^2 + 2x - 8$.   (−4, 0), (−2, 0), (1, 0), (0, −8)

8. An architect charges a fee of \$750 plus \$3.50 per square foot to design a house. The equation that represents the architect's fee is $F = 3.50s + 750$, where F is the fee, in dollars, and s is the number of square feet in the house. Create an input/output table for this equation for increments of 100 square feet, beginning with $s = 1500$ and ending with $s = 2100$.

s	1500	1600	1700	1800	1900	2000	2100
F	6000	6350	6700	7050	7400	7750	8100

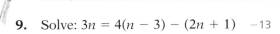

9. Solve: $3n = 4(n - 3) - (2n + 1)$ −13

10. Solve and write the answer in interval notation: $3(2t - 1) \geq 5t - 3(t + 9)$ [−6, ∞)

11. Solve and write the answer in set-builder notation: $7 < 4x - 5 < 19$ {x|3 < x < 6}

12. Solve: $|6 - 3x| - 5 = -2$ 1, 3

13. Find the equation of the line that passes through the points $(3, -2)$ and $(-1, 2)$. $y = -x + 1$

14. Solve by the substitution method:
$4x - y = 11$
$3x - 5y = 21$ (2, −3)

15. Solve by the addition method:
$3x - 5y = -1$
$4x + 3y = -11$ (−2, −1)

16. Solve by graphing: $4x - y = 8$
$2x + y = 4$

(2, 0)

17. Write in row echelon form: $x - y + z = 1$
$2x + 3y - z = 3$
$-x + 2y - 4z = 4$

$$\begin{bmatrix} 1 & -1 & 1 & | & 1 \\ 0 & 1 & -3 & | & 5 \\ 0 & 0 & 1 & | & -2 \end{bmatrix}$$

18. Solve using the Gaussian elimination method: $x - 2y + z = 5$
$3x - 2y - z = 3$
$4x + 5y - 4z = -9$ (1, −1, 2)

19. The percent of the world's population living in rural areas is shown in the table (*Source:* Food and Agricultural Organization of the United Nations).

Year, x	1950	1960	1970	1980	1990	2000	2010*
Percent, y	70	66	63	61	57	52	48

*Projection

a. Find the linear regression line for the data.

b. Use the regression line to determine the percent of the world's population expected to be living in rural areas in 2020. Round to the nearest percent.

c. Explain why the slope of the regression line indicates that the percent of the population living in rural areas is decreasing.

a. $y = -0.357142857x + 766.71429$ b. 45%   c. The slope is negative; as x increases, y decreases.

20. Traveling with the current, a cruise ship sailed between two islands, a distance of 90 miles, in 3 hours. The return trip against the current required 4 hours and 30 minutes. Find the rate of the cruise ship in calm water and the rate of the current.   Ship, 25 mph; current, 5 mph

6

Polynomials

```
CALCULATE
1: value
2: zero
3: minimum
4: maximum
5: intersect
6: dy/dx
7: ∫f(x)dx
```

Press **2nd** CALC to access the zero option.

Coordinating the position of each skydiver in a jump involves a lot of careful planning. Position within the 'pattern' is determined by when and where each chute is deployed, which is further dependent upon each diver's velocity during free fall. Factors such as the initial altitude of the jump, the size of the parachute, and the weight and body position of the diver affect velocity. The velocity of a falling object can be modeled by a quadratic equation, as shown in **Exercise 61 on page 455.**

Need help? For online student resources,
visit this web site:
Math.college.hmco.com

PREP TEST

1. Simplify: $-4(3y)$ $-12y$

2. Simplify: $(-2)^3$ -8

3. Simplify: $-4a - 8b + 7a$ $3a - 8b$

4. Simplify: $3x - 2[y - 4(x + 1) + 5]$
    $11x - 2y - 2$

5. Simplify: $-(x - y)$ $-x + y$

6. Write 40 as a product of prime numbers.
    $2 \cdot 2 \cdot 2 \cdot 5$

7. Find the GCF of 16, 20, and 24.   4

8. Evaluate $x^3 - 2x^2 + x + 5$ for $x = -2$. -13

9. Solve: $3x + 1 = 0$ $-\frac{1}{3}$

GO FIGURE

You are planning a large dinner party. If you seat 5 people at each table, you end up with only 2 people at the last table. If you seat 3 people at each table, you have 9 people left over with no place to sit. There are fewer than 10 tables. How many guests are coming to the dinner party?

27 guests

SECTION 6.1

- Addition and Subtraction of Monomials
- Multiplication of Monomials
- Division of Monomials
- Scientific Notation

Operations on Monomials and Scientific Notation

■ Addition and Subtraction of Monomials

The floorplan of a mobile home is shown below. All dimensions given are in feet. Not shown in the diagram is the fact that the height of each room is 8 feet.

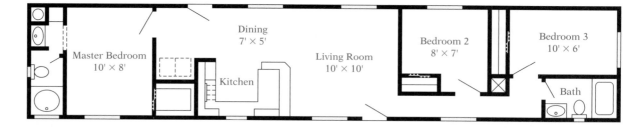

⟹ What is the combined length of the dining room and the living room?

The length of the dining room is 7 ft.

The length of the living room is 10 ft.

$$7 \text{ ft} + 10 \text{ ft} = 17 \text{ ft}$$

The combined length of the dining room and the living room is 17 ft. ⟸

➡ What is the combined area of the two smaller bedrooms?

The area of bedroom 3 is 10 ft × 6 ft = 60 ft².

The area of bedroom 2 is 8 ft × 7 ft = 56 ft².

$$60 \text{ ft}^2 + 56 \text{ ft}^2 = 116 \text{ ft}^2$$

The combined area of the two smaller bedrooms is 116 ft². ⬅

➡ What is the difference between the areas of the two smaller bedrooms?

$$60 \text{ ft}^2 - 56 \text{ ft}^2 = 4 \text{ ft}^2$$

The difference between the areas of the two smaller bedrooms is 4 ft². ⬅

As shown above, we can find the sum of two areas or the sum of two lengths. However, we cannot find the sum of an area and a length. For example, the area of the dining room is 7 ft × 5 ft = 35 ft². The length of the living room is 10 ft.

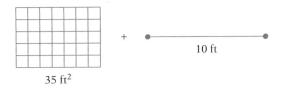

35 ft²   10 ft

INSTRUCTOR NOTE
Addition and subtraction of like terms has been developed previously. It is included here (1) so that students will see it as addition and subtraction of monomials, (2) for completeness in covering operations on monomials, and (3) to provide students with experience in differentiating addition of monomials from multiplication of monomials. (The exercise set provides mixed practice on the operations.)

The sum 35 ft² + 10 ft cannot be simplified.

Just as we cannot add square feet and feet, **we cannot add algebraic terms that do not have the same variable part. The same is true for subtraction.**

$$60x^2 + 56x^2 = 116x^2$$
$$60x^2 - 56x^2 = 4x^2$$

• Both $60x^2$ and $56x^2$ have the same variable part: x^2. Add or subtract the coefficients; the variable part stays the same.

$$35x^2 + 10x$$

• $35x^2$ and $10x$ do not have the same variable part. The terms cannot be combined.

When adding and subtracting like terms, we are actually adding and subtracting monomials. A **monomial** is a number, a variable, or a product of a number and variables. For instance.

7	b	$\dfrac{2}{3}a$	$12xy^2$
A number	A variable	A product of a number and a variable	A product of a number and variables

TAKE NOTE

x^3 can be written as the product $x \cdot x \cdot x$. a^2b can be written as the product $a \cdot a \cdot b$. $\sqrt{x}$ cannot be written as a product of variables.

The expression $3\sqrt{x}$ is not a monomial because $\sqrt{x}$ cannot be written as a product of variables.

The expression $\dfrac{2x}{y^2}$ is not a monomial because it is a quotient of variables.

TAKE NOTE

Addition of monomials involves using the distributive property:
$14x^3y^2z + 8x^3y^2z + 7x^3y^2z$
$= (14 + 8 + 7)x^3y^2z$
$= 29x^3y^2z$

INSTRUCTOR NOTE
Exercise 14 can be used for a similar in-class example.

INSTRUCTOR NOTE
Exercise 26 can be used for a similar in-class example.

▼ *Point of Interest*

A billion, which is 10^9, is too large a number for most of us to comprehend. If a computer were to start counting from 1 to 1 billion, writing to the screen one number every second of every day, it would take over 31 years for the computer to complete the task.

And if a billion is a large number, consider a googol. A googol is 1 with 100 zeros after it, or 10^{100}. Edward Kasner is the mathematician credited with thinking up this number, and his nine-year-old nephew is said to have thought up the name. The two then coined the word googolplex, which is 10^{googol}.

EXAMPLE 1

Simplify: $14x^3y^2z + 8x^3y^2z + 7x^3y^2z$

Solution $14x^3y^2z + 8x^3y^2z + 7x^3y^2z$

$= 29x^3y^2z$

• The terms have the same variable part. Add the coefficients. The variable part stays the same.

YOU TRY IT 1

Simplify: $16a^4b^3 + 10a^4b^3 + 5a^4b^3$

Solution   See page S24. $31a^4b^3$

EXAMPLE 2

Simplify: $29c^4d^5 - 6c^4d^5$

Solution $29c^4d^5 - 6c^4d^5$

$= 23c^4d^5$

• The terms have the same variable part. Subtract the coefficients. The variable part stays the same.

YOU TRY IT 2

Simplify: $37m^3n^2p - 14m^3n^2p$

Solution   See page S24. $23m^3n^2p$

■ Multiplication of Monomials

➡ What is the volume of air that must be heated in the master bedroom of the mobile home pictured at the beginning of this section? Note that the height of each room is 8 ft.

The volume of air in the master bedroom = 10 ft × 8 ft × 8 ft

= (10 ft × 8 ft) × 8 ft

= 80 ft² × 8 ft

= 640 ft³

There is 640 ft³ of air to heat in the master bedroom.   ⬅

This illustrates that we can multiply square feet by feet. The result is cubic feet, or volume.

Let's look at multiplication with monomials.

Recall that the exponential expression 3^4 means to multiply 3, the base, 4 times.

Therefore, $3^4 = 3 \cdot 3 \cdot 3 \cdot 3 = 81$. For the variable exponential expression x^6, x is the base and 6 is the exponent. **The exponent indicates the number of times the base occurs as a factor.** Therefore,

$$\overbrace{x^6 = x \cdot x \cdot x \cdot x \cdot x \cdot x}^{\text{Multiply } x \text{ 6 times}}$$

Suggested Activity

You might prefer to have students develop the rule of multiplying exponential expressions. Provide them with several expressions to simplify—for example,

$$x^4x^3$$
$$y^2y^6$$
$$a^5a^7$$

Ask them to simplify the expression by writing each expression in factored form and then writing the result with an exponent. After they have completed several examples, ask them to write a rule for x^mx^n. This same approach can be used for the Rule for Simplifying the Power of an Exponential Expression and the Rule for Dividing Exponential Expressions.

INSTRUCTOR NOTE
Exercise 6 can be used for a similar in-class example.

The product of exponential expressions with the *same* base can be simplified by writing each expression in factored form and writing the result with an exponent.

$$x^3 \cdot x^2 = \overbrace{(x \cdot x \cdot x)}^{\text{3 factors}} \cdot \overbrace{(x \cdot x)}^{\text{2 factors}}$$

$$\underbrace{}_{\text{5 factors}}$$

$$= x \cdot x \cdot x \cdot x \cdot x$$
$$= x^5$$

Note that adding the exponents results in the same product.

$$x^3 \cdot x^2 = x^{3+2} = x^5$$

This suggests the following rule for multiplying exponential expressions.

Rule for Multiplying Exponential Expressions

If m and n are positive integers, then $x^m \cdot x^n = x^{m+n}$.

EXAMPLE 3

Simplify: $a^4 \cdot a^5$

Solution $a^4 \cdot a^5 = a^{4+5} = a^9$ • The bases are the same. Add the exponents.

YOU TRY IT 3

Simplify: $t^3 \cdot t^8$

Solution   See page S24. t^{11}

INSTRUCTOR NOTE
Exercise 8 can be used for a similar in-class example.

EXAMPLE 4

Simplify: $c^3 \cdot c^4 \cdot c$

Solution $c^3 \cdot c^4 \cdot c = c^{3+4+1}$ • The bases are the same. Add the exponents. Note that $c = c^1$.
$$= c^8$$

YOU TRY IT 4

Simplify: $n^6 \cdot n \cdot n^2$

Solution   See page S24. n^9

? QUESTION   Why can the exponential expression x^5y^3 not be simplified?

? ANSWER   The bases are not the same. The Rule for Multiplying Exponential Expressions applies only to expressions with the *same* base.

INSTRUCTOR NOTE
Exercise 20 can be used for a similar in-class example.

EXAMPLE 5

Simplify: $(a^3b^2)(a^4)$

Solution $(a^3b^2)(a^4) = a^{3+4}b^2$ • Multiply variables with the same
                      $= a^7b^2$ base by adding the exponents.

YOU TRY IT 5

Simplify: $c^9(c^5d^8)$

Solution   See page S24. $c^{14}d^8$

INSTRUCTOR NOTE
Exercise 22 can be used for a similar in-class example.

EXAMPLE 6

Simplify: $(4x^3)(2x^6)$

Solution $(4x^3)(2x^6) = (4 \cdot 2)(x^3 \cdot x^6)$ • Use the Commutative and Associa-
                                         tive Properties of Multiplication to
                                         group the coefficients and variables
                                         with the same base.

$= 8x^{3+6}$ • Multiply the coefficients. Multiply
$= 8x^9$ variables with the same base by
                            adding the exponents.

> **TAKE NOTE**
>
> Note on page 378 that
> 10 ft $\times$ 8 ft $\times$ 8 ft
> $= (10 \cdot 8 \cdot 8)$ (ft $\cdot$ ft $\cdot$ ft)
> $= 640$ (ft^{1+1+1})
> $= 640$ ft^3

YOU TRY IT 6

Simplify: $(5y^4)(3y^2)$

Solution   See page S24. $15y^6$

INSTRUCTOR NOTE
Exercise 38 can be used for a similar in-class example.

EXAMPLE 7

Simplify: $(-2v^3z^5)(7v^2z^6)$

Solution $(-2v^3z^5)(7v^2z^6)$ • Multiply the coefficients of the
                                         monomials. Multiply variables
         $= [-2(7)](v^{3+2})(z^{5+6})$ with the same base by adding
         $= -14v^5z^{11}$ the exponents.

YOU TRY IT 7

Simplify: $(12p^4q^3)(-3p^5q^2)$

Solution   See page S24. $-36p^9q^5$

The expression $(x^4)^3$ is an example of a *power of a monomial*; the monomial x^4 is raised to a power of 3.

The power of a monomial can be simplified by writing the power in factored form and then using the Rule for Multiplying Exponential Expressions.

$$(x^4)^3 = x^4 \cdot x^4 \cdot x^4$$
$$= x^{4+4+4} = x^{12}$$

Note that multiplying the exponent inside the parentheses by the exponent outside the parentheses results in the same product.

$$(x^4)^3 = x^{4 \cdot 3} = x^{12}$$

This suggests the following rule for simplifying powers of monomials.

> **Rule for Simplifying the Power of an Exponential Expression**
> If m and n are positive integers, then $(x^m)^n = x^{m \cdot n}$.

? QUESTION   Which expression is the multiplication of two exponential expressions and which is the power of an exponential expression?*

a. $q^4 \cdot q^{10}$ **b.** $(q^4)^{10}$

INSTRUCTOR NOTE
Exercise 10 can be used for a similar in-class example.

EXAMPLE 8

Simplify: $(z^2)^5$

Solution $(z^2)^5 = z^{2 \cdot 5} = z^{10}$ • z^2 is raised to the power of 5. Simplify the power of an exponential expression by multiplying the exponents.

YOU TRY IT 8

Simplify: $(t^3)^6$

Solution   See page S24. t^{18}

TAKE NOTE

$(a^2b^3)^2$ is a *product* of exponential expressions raised to a power, whereas $(a^2 + b^3)^2$ is a *sum* of exponential expressions raised to a power. These two expressions are not simplified in the same manner.

The expression $(a^2b^3)^2$ is the *power of the product* of two exponential expressions, a^2 and b^3. The power of the product of exponential expressions can be simplified by writing the product in factored form and then using the Rule for Multiplying Exponential Expressions.

Write the exponential expression in factored form. Use the Rule for Multiplying Exponential Expressions.

$$(a^2b^3)^2 = (a^2b^3)(a^2b^3)$$
$$= a^{2+2}b^{3+3}$$
$$= a^4b^6$$

Note that multiplying each exponent inside the parentheses by the exponent outside the parentheses results in the same product.

$$(a^2b^3)^2 = a^{2 \cdot 2}b^{3 \cdot 2}$$
$$= a^4b^6$$

> **Rule for Simplifying Powers of Products**
> If m, n, and p are positive integers, then $(x^m y^n)^p = x^{m \cdot p} y^{n \cdot p}$.

? QUESTION   In the expression $(a^8 b^6)^5$, what is the product and what is the power?†

? ANSWERS   *a. This is the multiplication of two exponential expressions. q^4 is multiplied times q^{10}.   **b.** This is the power of an exponential expression. q^4 is raised to the 10th power.
† The product is $a^8 b^6$; a^8 is multiplied times b^6. The power is 5; $a^8 b^6$ is raised to the 5th power.

INSTRUCTOR NOTE
Exercise 12 can be used for
a similar in-class example.

EXAMPLE 9

Simplify: $(x^4y)^6$

Solution $(x^4y)^6 = x^{4\cdot6}y^{1\cdot6}$ • Multiply each exponent inside the parentheses by the exponent outside the parentheses. Remember that $y = y^1$.

$= x^{24}y^6$

YOU TRY IT 9

Simplify: $(bc^7)^8$

Solution   See page S24. b^8c^{56}

INSTRUCTOR NOTE
Use the following exercise
for a similar in-class example.
Simplify: $(6x^5)^4$

EXAMPLE 10

Simplify: $(5z^3)^2$

Solution $(5z^3)^2 = 5^{1\cdot2}z^{3\cdot2}$ • Multiply each exponent inside the parentheses by the exponent outside the parentheses. Note that $5 = 5^1$.

$= 5^2z^6$

$= 25z^6$ • Evaluate 5^2.

YOU TRY IT 10

Simplify: $(4y^6)^3$

Solution   See page S24. $64y^{18}$

INSTRUCTOR NOTE
Exercise 30 can be used for
a similar in-class example.

EXAMPLE 11

Simplify: $(3m^5p^2)^4$

Solution $(3m^5p^2)^4 = 3^{1\cdot4}m^{5\cdot4}p^{2\cdot4}$ • Multiply each exponent inside the parentheses by the exponent outside the parentheses.

$= 3^4m^{20}p^8$

$= 81m^{20}p^8$ • Evaluate 3^4.

YOU TRY IT 11

Simplify: $(2v^6w^9)^5$

Solution   See page S24. $32v^{30}w^{45}$

INSTRUCTOR NOTE
Exercise 24 can be used for
a similar in-class example.

EXAMPLE 12

Simplify: $(-a^5b^8)^6$

Solution $(-a^5b^8)^6 = (-1)^{1\cdot6}a^{5\cdot6}b^{8\cdot6}$ • Multiply each exponent inside the parentheses by the exponent outside the parentheses. Note that $-a^5b^8 = -1a^5b^8 = (-1)^1a^5b^8$.

$= (-1)^6a^{30}b^{48}$

$= a^{30}b^{48}$ • Evaluate $(-1)^6$. $(-1)^6 = 1$.

YOU TRY IT 12

Simplify: $(-2x^3y^7)^3$

Solution   See page S24. $-8x^9y^{21}$

In some products, it is necessary to use the Rule for Simplifying Powers of Products and the Rule for Multiplying Exponential Expressions.

➡ Simplify: $(3x^4)^2 (4x^3)$

$$(3x^4)^2(4x^3) = (3^{1 \cdot 2}x^{4 \cdot 2})(4x^3)$$

• Use the Rule for Simplifying Powers of Products to simplify $(3x^4)^2$.

$$= (3^2x^8)(4x^3)$$
$$= (9x^8)(4x^3)$$
$$= (9 \cdot 4)(x^8 \cdot x^3)$$

• Use the Rule for Multiplying Exponential Expressions.

$$= 36x^{8+3}$$
$$= 36x^{11}$$

⬅

INSTRUCTOR NOTE
Exercise 44 can be used for a similar in-class example.

EXAMPLE 13

Simplify: $(2a^2b)(2a^3b^2)^3$

Solution $(2a^2b)(2a^3b^2)^3$

$$= (2a^2b)(2^{1 \cdot 3}a^{3 \cdot 3}b^{2 \cdot 3})$$

• Use the Rule for Simplifying Powers of Products.

$$= (2a^2b)(2^3a^9b^6)$$
$$= (2a^2b)(8a^9b^6)$$
$$= (2 \cdot 8)(a^{2+9})(b^{1+6})$$

• Use the Rule for Multiplying Exponential Expressions.

$$= 16a^{11}b^7$$

YOU TRY IT 13

Simplify: $(-xy^4)(-2x^3y^2)^2$

Solution   See page S24. $-4x^7y^8$

■ Division of Monomials

The quotient of two exponential expressions with the *same* base can be simplified by writing each expression in factored form, dividing by the common factors, and then writing the result with an exponent.

$$\frac{x^6}{x^2} = \frac{\overset{1}{\cancel{x}} \cdot \overset{1}{\cancel{x}} \cdot x \cdot x \cdot x \cdot x}{\underset{1}{\cancel{x}} \cdot \underset{1}{\cancel{x}}} = x^4$$

Note that subtracting the exponents results in the same quotient.

$$\frac{x^6}{x^2} = x^{6-2} = x^4$$

This example suggests that to divide monomials with like bases, we subtract the exponents.

Rule for Dividing Exponential Expressions

If m and n are positive integers and $x \neq 0$, then $\dfrac{x^m}{x^n} = x^{m-n}$.

INSTRUCTOR NOTE
Exercise 18 can be used for a similar in-class example.

EXAMPLE 14

Simplify: $\dfrac{c^8}{c^5}$

Solution $\dfrac{c^8}{c^5} = c^{8-5}$ • The bases are the same. Subtract the exponents.

$= c^3$

YOU TRY IT 14

Simplify: $\dfrac{t^{10}}{t^4}$

Solution See page S24. t^6

❓ QUESTION   Why can the expression $\dfrac{x^8}{y^2}$ not be simplified?

INSTRUCTOR NOTE
Exercise 32 can be used for a similar in-class example.

EXAMPLE 15

Simplify: $\dfrac{x^5 y^7}{x^4 y^2}$

Solution $\dfrac{x^5 y^7}{x^4 y^2} = x^{5-4} \cdot y^{7-2}$ • Use the Rule for Dividing Exponential Expressions by subtracting the exponents of like bases. Note that $x^{5-4} = x^1$, but the exponent 1 is not written.

$= xy^5$

YOU TRY IT 15

Simplify: $\dfrac{a^7 b^6}{ab^3}$

Solution See page S24. $a^6 b^3$

Suggested Activity

1. Simplify: $\dfrac{2^{40}}{4^{20}}$

   [Answer: 1]

2. If $\dfrac{(a)(a)(a)}{a+a+a} = 3$, find the value of a^2.

   [Answer: 9]

The expression at the right has been simplified in two ways: by dividing by common factors, and by using the Rule for Dividing Exponential Expressions.

$$\dfrac{x^3}{x^3} = \dfrac{\overset{1}{\cancel{x}} \cdot \overset{1}{\cancel{x}} \cdot \overset{1}{\cancel{x}}}{\underset{1}{\cancel{x}} \cdot \underset{1}{\cancel{x}} \cdot \underset{1}{\cancel{x}}} = 1$$

Because $\dfrac{x^3}{x^3} = 1$ and $\dfrac{x^3}{x^3} = x^0$, 1 must equal x^0. Therefore, the following definition of zero as an exponent is used.

$$\dfrac{x^3}{x^3} = x^{3-3} = x^0$$

> **Zero as an Exponent**
>
> If $x \neq 0$, then $x^0 = 1$. The expression 0^0 is undefined.

❓ ANSWER   The bases are not the same. The Rule for Dividing Exponential Expressions applies only to expressions with the *same* base.

INSTRUCTOR NOTE
Exercise 34 can be used for
a similar in-class example.

EXAMPLE 16

Simplify: $(-15y^4)^0$, $y \neq 0$

Solution $(-15y^4)^0 = 1$ • Any nonzero expression to the zero power is 1.

YOU TRY IT 16

Simplify: $(-8x^2y^7)^0$

Solution   See page S24.   1

INSTRUCTOR NOTE
Exercise 16 can be used for
a similar in-class example.

EXAMPLE 17

Simplify: $-(6r^3t^2)^0$, $r \neq 0$, $t \neq 0$

Solution $-(6r^3t^2)^0 = -1$ • $(6r^3t^2)^0 = 1$. The negative sign in front of the parentheses can be read "the opposite of." The opposite of 1 is -1.

YOU TRY IT 17

Simplify: $-(9c^7d^4)^0$, $c \neq 0$, $d \neq 0$

Solution   See page S24. -1

The expression at the right has been simplified in two ways: by dividing by common factors, and by using the Rule for Dividing Exponential Expressions.

$$\frac{x^3}{x^5} = \frac{\overset{1}{\cancel{x}} \cdot \overset{1}{\cancel{x}} \cdot \overset{1}{\cancel{x}}}{\underset{1}{\cancel{x}} \cdot \underset{1}{\cancel{x}} \cdot \underset{1}{\cancel{x}} \cdot x \cdot x} = \frac{1}{x^2}$$

Because $\frac{x^3}{x^5} = \frac{1}{x^2}$ and $\frac{x^3}{x^5} = x^{-2}$, $\frac{1}{x^2}$ must equal x^{-2}. Therefore, the following definition of a negative exponent is used.

$$\frac{x^3}{x^5} = x^{3-5} = x^{-2}$$

Definition of Negative Exponents

If n is a positive integer and $x \neq 0$, then $x^{-n} = \frac{1}{x^n}$ and $\frac{1}{x^{-n}} = x^n$.

An exponential expression is in simplest form when there are no negative exponents in the expression.

➡ Simplify: y^{-7}

$y^{-7} = \frac{1}{y^7}$ • Use the Definition of Negative Exponents to rewrite the expression with a positive exponent.   ⬅

➡ Simplify: $\frac{1}{c^{-4}}$

$\frac{1}{c^{-4}} = c^4$ • Use the Definition of Negative Exponents to rewrite the expression with a positive exponent.   ⬅

▼ *Point of Interest*

In the 15th century, the expression $12^{\overline{2m}}$ was used to mean $12x^{-2}$. The use of $\overline{m}$ reflected an Italian influence. In Italy, m was used for minus and p was used for plus. It was understood that $\overline{2m}$ referred to an unnamed variable. Isaac Newton, in the 17th century, advocated the use of a negative exponent.

INSTRUCTOR NOTE
Exercise 60 can be used for
a similar in-class example.

? QUESTION: How are **a.** b^{-8} and **b.** $\frac{1}{w^{-5}}$ rewritten with positive exponents?

EXAMPLE 18

Simplify: $\dfrac{3n^{-5}}{4}$

Solution $\dfrac{3n^{-5}}{4} = \dfrac{3}{4}n^{-5} = \dfrac{3}{4} \cdot \dfrac{1}{n^5}$
- Use the Definition of Negative Exponents to rewrite the expression with a positive exponent.

$\qquad\qquad = \dfrac{3}{4n^5}$

YOU TRY IT 18

Simplify: $\dfrac{2}{c^{-4}}$

Solution See page S24. $2c^4$

TAKE NOTE

Note from the example at the right that 2^{-3} is a *positive* number. A negative exponent does not indicate a negative number.

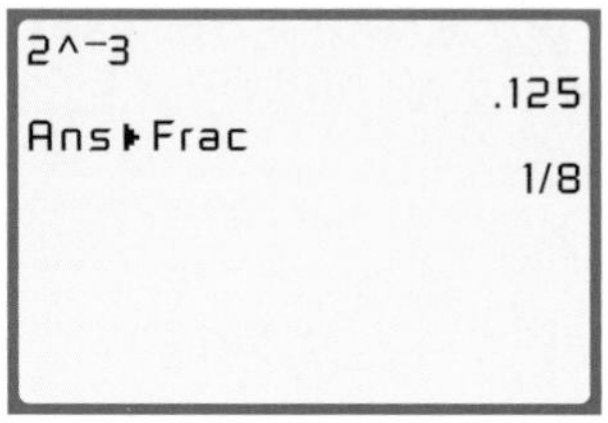

```
2^-3
            .125
Ans▶Frac
             1/8
```

A numerical expression with a negative exponent can be evaluated by first rewriting the expression with a positive exponent.

➡ Evaluate: 2^{-3}

$2^{-3} = \dfrac{1}{2^3}$
- Use the Definition of Negative Exponents to rewrite the expression with a positive exponent.

$\quad\; = \dfrac{1}{8}$
- Evaluate 2^3,

This answer can be checked using a calculator, as shown at the left. Note that $0.125 = \frac{1}{8}$. ⬅

Sometimes applying the Rule for Dividing Exponential Expressions results in a quotient that contains a negative exponent. If this happens, use the Definition of Negative Exponents to rewrite the expression with a positive exponent.

➡ Simplify: $\dfrac{6x^2}{8x^9}$

$\dfrac{6x^2}{8x^9} = \dfrac{3x^2}{4x^9} = \dfrac{3x^{2-9}}{4}$
- Divide the coefficients by their common factors. Then use the Rule for Dividing Exponential Expressions.

$\quad = \dfrac{3x^{-7}}{4} = \dfrac{3}{4} \cdot \dfrac{x^{-7}}{1} = \dfrac{3}{4} \cdot \dfrac{1}{x^7}$
- Rewrite the expression with only positive exponents.

$\quad = \dfrac{3}{4x^7}$ ⬅

? ANSWERS **a.** $b^{-8} = \frac{1}{b^8}$ **b.** $\frac{1}{w^{-5}} = w^5$

INSTRUCTOR NOTE
Exercise 76 can be used for a similar in-class example.

EXAMPLE 19

Simplify: $\dfrac{-35a^6b^{-2}}{25a^{-3}b^5}$

Solution $\dfrac{-35a^6b^{-2}}{25a^{-3}b^5} = -\dfrac{7a^6b^{-2}}{5a^{-3}b^5} = -\dfrac{7a^{6-(-3)}b^{-2-5}}{5} = -\dfrac{7a^9b^{-7}}{5} = -\dfrac{7a^9}{5b^7}$

YOU TRY IT 19

Simplify: $\dfrac{12x^{-8}y}{-16xy^{-3}}$

Solution See page S24. $-\dfrac{3y^4}{4x^9}$

The expression $\left(\dfrac{x^3}{y^4}\right)^2$ is the *power of the quotient* of two exponential expressions, x^3 and y^4. This expression can be simplified by squaring $\dfrac{x^3}{y^3}$—that is, by multiplying each exponent in the quotient by the exponent outside the parentheses.

$$\left(\dfrac{x^3}{y^4}\right) = \left(\dfrac{x^3}{y^4}\right)\left(\dfrac{x^3}{y^4}\right) = \dfrac{x^3 \cdot x^3}{y^4 \cdot y^4} = \dfrac{x^{3+3}}{y^{4+4}} = \dfrac{x^6}{x^8} \qquad \left(\dfrac{x^3}{y^4}\right)^2 = \dfrac{x^{3\cdot2}}{y^{4\cdot2}} = \dfrac{x^6}{y^8}$$

TAKE NOTE

The Rule for Simplifying Powers of Products states that we can multiply each exponent in the product by the exponent outside the parentheses.

$(x^my^n)^p = x^{m\cdot p}y^{n\cdot p}$

The Rule for Simplifying Powers of Quotients states that we can multiply each exponent in the quotient by the exponent outside the parentheses.

$\left(\dfrac{x^m}{y^n}\right)^p = \dfrac{x^{m\cdot p}}{y^{n\cdot p}}$

Rule for Simplifying Powers of Quotients

If m, n, and p are integers and $y \ne 0$, then $\left(\dfrac{x^m}{y^n}\right)^p = \dfrac{x^{m\cdot p}}{y^{n\cdot p}}$.

? QUESTION   In the expression $\left(\dfrac{c^4}{d^6}\right)^5$, what is the quotient and what is the power?

EXAMPLE 20

Simplify: $\left(\dfrac{a^4}{b^3}\right)^{-2}$

Solution $\left(\dfrac{a^4}{b^3}\right)^{-2} = \dfrac{a^{4(-2)}}{b^{3(-2)}} = \dfrac{a^{-8}}{b^{-6}}$ • Multiply each exponent inside the parentheses by the exponent outside the parentheses.

$= \dfrac{b^6}{a^8}$ • Rewrite the expression with positive exponents.

YOU TRY IT 20

Simplify: $\left(\dfrac{m^{-6}}{n^{-8}}\right)^3$

Solution See page S24. $\dfrac{n^{24}}{m^{18}}$

INSTRUCTOR NOTE
Exercise 82 can be used for a similar in-class example.

? ANSWER   The quotient is $\dfrac{c^4}{d^6}$; c^4 is divided by d^6. The power is 5; $\dfrac{c^4}{d^6}$ is raised to the 5th power.

The rules for simplifying exponential expressions and powers of exponential expressions apply to all integers. These rules are restated here.

Suggested Activity

See Section 6.1 of the *Student Activity Manual* for an investigation involving the rules of exponents.

Rules of Exponents

If m, n, and p are integers, then

$$x^m \cdot x^n = x^{m+n} \qquad (x^m)^n = x^{m \cdot n} \qquad (x^m y^n)^p = x^{m \cdot p} y^{n \cdot p}$$

$$\frac{x^m}{x^n} = x^{m-n}, x \neq 0 \qquad \left(\frac{x^m}{y^n}\right)^p = \frac{x^{m \cdot p}}{y^{n \cdot p}}, y \neq 0 \qquad x^{-n} = \frac{1}{x^n}, \frac{1}{x^{-n}} = x^n, x \neq 0$$

$$x^0 = 1, x \neq 0$$

Simplifying the expressions in Example 21 requires a combination of the rules of exponents.

INSTRUCTOR NOTE
Exercises 80 and 84 can be used for similar in-class examples.

EXAMPLE 21

Simplify.   **a.** $(-2x)(3x^{-2})^{-3}$ **b.** $\left(\dfrac{3a^2b^{-1}}{27a^{-3}b^{-4}}\right)^{-2}$

Solution

a. $(-2x)(3x^{-2})^{-3}$

$\qquad = (-2x)(3^{-3}x^6)$ • Use the Rule for Simplifying Powers of Products.

$\qquad = \dfrac{-2x \cdot x^6}{3^3}$ • Write the expression with positive exponents.

$\qquad = -\dfrac{2x^7}{27}$ • Use the Rule for Multiplying Exponential Expressions. Simplify 3^3.

b. $\left(\dfrac{3a^2b^{-1}}{27a^{-3}b^{-4}}\right)^{-2}$

$\qquad = \left(\dfrac{a^2b^{-1}}{9a^{-3}b^{-4}}\right)^{-2}$ • Simplify $\dfrac{3}{27}$.

$\qquad = \dfrac{a^{2(-2)}b^{(-1)(-2)}}{9^{1(-2)}a^{(-3)(-2)}b^{(-4)(-2)}}$ • Use the Rule for Simplifying Powers of Quotients.

$\qquad = \dfrac{a^{-4}b^2}{9^{-2}a^6b^8}$ • Simplify and rewrite the expression with positive exponents.

$\qquad = 9^2a^{-4-6}b^{2-8}$

$\qquad = 81a^{-10}b^{-6} = \dfrac{81}{a^{10}b^6}$

INSTRUCTOR NOTE
You might illustrate that the rules of exponents can be used in any order by having your students simplify the expression in Example 21b by first simplifying the expression inside the parentheses and then using the Rule for Simplifying Powers of Quotients.

$\left(\dfrac{3a^2b^{-1}}{27a^{-3}b^{-4}}\right)^{-2} = \left(\dfrac{a^5b^3}{9}\right)^{-2}$

$\qquad = \dfrac{a^{-10}b^{-6}}{9^{-2}}$

$\qquad = \dfrac{81}{a^{10}b^6}$

YOU TRY IT 21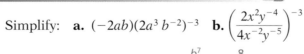

Simplify: **a.** $(-2ab)(2a^3 b^{-2})^{-3}$ **b.** $\left(\dfrac{2x^2y^{-4}}{4x^{-2}y^{-5}}\right)^{-3}$

Solution   See page S24.   a. $-\dfrac{b^7}{4a^8}$ b. $\dfrac{8}{x^{12}y^3}$

■ Scientific Notation

Very large and very small numbers are encountered in the fields of science and engineering. For example, the charge of an electron is

0.00000000000000000160 coulomb.

These numbers can be written more easily in scientific notation. **In scientific notation, a number is expressed as a product of two factors, one a number between 1 and 10 and the other a power of 10.**

TAKE NOTE

There are two steps in writing a number in scientific notation: (1) determine the number between 1 and 10, and (2) determine the exponent on 10.

To change a number written in decimal notation to one written in scientific notation, write it in the form $a \times 10^n$, where $1 \le a < 10$ and n is an integer.

For numbers greater than 10, move the decimal point to the right of the first digit. The exponent n is positive and equal to the number of places the decimal point has been moved.

$$240,000 = 2.4 \times 10^5$$

$$93,000,000 = 9.3 \times 10^7$$

For numbers less than 1, move the decimal point to the right of the first nonzero digit. The exponent n is negative. The absolute value of the exponent is equal to the number of places the decimal point has been moved.

$$0.00030 = 3.0 \times 10^{-4}$$

$$0.0000832 = 8.32 \times 10^{-5}$$

Look at the last example above: $0.0000832 = 8.32 \times 10^{-5}$. Using the Definition of Negative Exponents,

$$10^{-5} = \frac{1}{10^5} = \frac{1}{100,000} = 0.00001$$

Because $10^{-5} = 0.00001$, we can write

$$8.32 \times 10^{-5} = 8.32 \times 0.00001 = 0.0000832$$

which is the number we started with. We have not changed the value of the number; we have just written it in another form.

INSTRUCTOR NOTE
Exercises 100 and 102 can be used for similar in-class examples.

EXAMPLE 22

Write the number in scientific notation.

a. 824,300,000,000   **b.** 0.000000961

Solution

a. $824,300,000,000 = 8.243 \times 10^{11}$

• Move the decimal point 11 places to the left. The exponent on 10 is 11.

Graphical check:

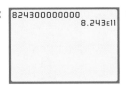

• Note how a graphing calculator writes a number in scientific notation. It displays the number between 1 and 10, followed by E, and then the exponent on 10.

b. $0.000000961 = 9.61 \times 10^{-7}$

• Move the decimal point 7 places to the right. The exponent on 10 is -7.

Graphical check:

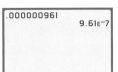

• 9.61E-7 represents 9.61×10^{-7}. The answer checks.

YOU TRY IT 22

Write the number in scientific notation.

a. 57,000,000,000   **b.** 0.000000017

Solution   See page S24.   **a.** 5.7×10^{10} **b.** 1.7×10^{-8}

Changing a number written in scientific notation to decimal notation also requires moving the decimal point.

When the exponent on 10 is positive, move the decimal point to the right the same number of places as the exponent.

$$3.45 \times 10^9 = 3,450,000,000$$
$$2.3 \times 10^8 = 230,000,000$$

When the exponent on 10 is negative, move the decimal point to the left the same number of places as the absolute value of the exponent.

$$8.1 \times 10^{-3} = 0.0081$$
$$6.34 \times 10^{-6} = 0.00000634$$

INSTRUCTOR NOTE
Exercises 112 and 114 can be used for similar in-class examples.

EXAMPLE 23

Write the number in decimal notation.

a. 7.329×10^6 **b.** 6.8×10^{-10}

Solution

a. $7.329 \times 10^6 = 7,329,000$

 See Appendix A:
Scientific Notation

Graphical check:
```
7.329*10^6
            7329000
```

• The exponent on 10 is positive. Move the decimal point 6 places to the right.

• Enter into a graphing calculator the number 7.329×10^6, in scientific notation. The calculator will return the number in decimal notation.

b. $6.8 \times 10^{-10} = 0.00000000068$

• The exponent on 10 is negative. Move the decimal point 10 places to the left.

Graphical check:
```
6.8E-10
                 6.8E-10
0.00000000068
                 6.8E-10
```

• If you enter 6.8E-10, the calculator returns 6.8E-10. Instead, enter the answer .00000000068 and check that the calculator displays the given expression, 6.8×10^{-10}.

YOU TRY IT 23

Write the number in decimal notation.

a. 5×10^{12} **b.** 4.0162×10^{-9}

Solution   See page S24.   **a.** 5,000,000,000,000   **b.** 0.0000000040162

? QUESTION   Is the expression written in scientific notation?

a. 2.84×10^{-4} **b.** 36.5×10^7 **c.** 0.91×10^{-12}

The rules for multiplying and dividing with numbers in scientific notation are the same as those for operating on algebraic expressions. The power of 10 corresponds to the variable part and the number between 1 and 10 corresponds to the coefficient of the variable.

	Algebraic Expressions	**Scientific Notation**
Multiplication	$(4x^{-3})(2x^5) = 8x^2$	$(4 \times 10^{-3})(2 \times 10^5) = 8 \times 10^2$
Division	$\dfrac{6x^5}{3x^{-2}} = 2x^{5-(-2)} = 2x^7$	$\dfrac{6 \times 10^5}{3 \times 10^{-2}} = 2 \times 10^{5-(-2)} = 2 \times 10^7$

INSTRUCTOR NOTE
Exercises 134 and 138 can be used for similar in-class examples.

EXAMPLE 24

Multiply or divide.   **a.** $(3.0 \times 10^5)(1.1 \times 10^{-8})$ **b.** $\dfrac{7.2 \times 10^{13}}{2.4 \times 10^{-3}}$

Solution

a. $(3.0 \times 10^5)(1.1 \times 10^{-8}) = 3.3 \times 10^{-3}$

• Multiply 3.0 times 1.1. Add the exponents on 10.

Graphical check: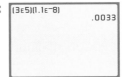

• The calculator returns the answer in decimal notation, .0033. The number .0033 in scientific notation is 3.3×10^{-3}. The answer checks.

b. $\dfrac{7.2 \times 10^{13}}{2.4 \times 10^{-3}} = 3 \times 10^{16}$

• Divide 7.2 by 2.4. Subtract the exponents on 10.

Graphical check: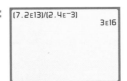

• 3E16 represents 3×10^{16}. The answer checks.

YOU TRY IT 24

Multiply or divide.

a. $(2.4 \times 10^{-9})(1.6 \times 10^3)$ **b.** $\dfrac{5.4 \times 10^{-2}}{1.8 \times 10^{-4}}$

Solution   See page S24.   **a.** 3.84×10^{-6} **b.** 3×10^2

? ANSWERS   **a.** 2.84×10^{-4} is written in scientific notation.   **b.** 36.5×10^7 is not written in scientific notation. 36.5 is not a number between 1 and 10.   **c.** 0.91×10^{-12} is not written in scientific notation. 0.91 is not a number between 1 and 10.

6.1 EXERCISES Suggested Assignment: 5–137, every other odd

Topics for Discussion

1. Explain why each of the following is or is not a monomial.

   a. $32a^3b$ This is a monomial because it is the product of a number, 32, and variables, a and b.

   b. $\dfrac{5n^3}{7}$

   This is a monomial because it is the product of a number, $\frac{5}{7}$, and a variable, n.

   c. $\dfrac{6c^4}{25d}$ This is not a monomial because there is a variable in the denominator.

2. Explain each of the following. Provide an example of each.

   a. The Rule for Multiplying Exponential Expressions
   To multiply exponential expressions with the same base, add the exponents.

   b. The Rule for Simplifying the Power of an Exponential Expression
   The power of a monomial is simplified by multiplying the exponents.

   c. The Rule for Simplifying Powers of Products
   To simplify a power of the product of exponential expressions, multiply each exponent inside the parentheses by the exponent outside the parentheses.

   d. The Rule for Dividing Exponential Expressions
   To divide two exponential expressions with the same base, subtract the exponents.

   e. The Rule for Simplifying Powers of Quotients
   To simplify a power of the quotient of exponential expressions, multiply each exponent inside the parentheses by the exponent outside the parentheses.

3. Explain the error in each of the following. Then correct the error.

   a. $6x^{-3} = \dfrac{1}{6x^3}$ The exponent on 6 is positive; it should not be moved to the denominator. $6x^{-3} = \frac{6}{x^3}$

   b. $xy^{-2} = \dfrac{1}{xy^2}$ The exponent on x is positive; it should not be moved to the denominator. $xy^{-2} = \frac{x}{y^2}$

   c. $\dfrac{1}{8a^{-4}} = 8a^4$ The exponent on 8 is positive; it should not be moved to the numerator. $\frac{1}{8a^{-4}} = \frac{a^4}{8}$

   d. $\dfrac{1}{b^{-5}c} = b^5c$ The exponent on c is positive; it should not be moved to the numerator. $\frac{1}{b^{-5}c} = \frac{b^5}{c}$

4. In your own words, explain how you know that a number is written in scientific notation. Answers will vary.

■ Operations on Monomials

Simplify.

5. $a^4 \cdot a^5$ a^9 **6.** $y^5 \cdot y^8$ y^{13} **7.** $z^3 \cdot z \cdot z^4$ z^8 **8.** $b \cdot b^2$

9. $(x^3)^5$ x^{15} **10.** $(b^2)^4$ b^8 **11.** $(x^2y^3)^6$ $x^{12}y^{18}$ **12.** $(m^4n^2)^3$ $m^{12}n^6$

13. $12s^4t^3 + 5s^4t^3$ $17s^4t^3$ **14.** $8b^6c^5 + 9b^6c^5$ $17b^6c^5$ **15.** 27^0 1 **16.** $-(17)^0$ -1

17. $\dfrac{a^8}{a^2}$ a^6 **18.** $\dfrac{c^{12}}{c^5}$ c^7 **19.** $(-m^3n)(m^6n^2)$ $-m^9n^3$ **20.** $(-r^4t^3)(r^2t^9)$ $-r^6t^{12}$

21. $(2x)(3x^2)(4x^4)$ $24x^7$ **22.** $(5a^2)(4a)(3a^5)$ $60a^8$ **23.** $(-2a^2)^3$ $-8a^6$ **24.** $(-3b^3)^2$ $9b^6$

25. $11p^4q^5 - 7p^4q^5$ $4p^4q^5$ **26.** $16c^2d^3 - 9c^2d^3$ $7c^2d^3$ **27.** $(6r^2)(-4r)$ $-24r^3$ **28.** $(7v^3)(-2v)$ $-14v^4$

29. $(2a^3bc^2)^3$ $8a^9b^3c^6$ **30.** $(4xy^3z^2)^2$ $16x^2y^6z^4$ **31.** $\dfrac{m^4n^7}{m^3n^5}$ mn^2 **32.** $\dfrac{a^5b^6}{a^3b^2}$ a^2b^4

33. $(3x)^0$ 1 **34.** $(2a)^0$ 1 **35.** $\dfrac{-16a^7}{24a^6}$ $-\dfrac{2a}{3}$ **36.** $\dfrac{18b^5}{-45b^4}$ $-\dfrac{2b}{5}$

37. $(9mn^4p)(-3mp^2)$ $-27m^2n^4p^3$ **38.** $(-3v^2wz)(-4vz^4)$ $12v^3wz^5$ **39.** $(-xy^5)(3x^2)(5y^3)$ $-15x^3y^8$ **40.** $(-6m^3n)(-mn^2)(m)$ $6m^5n^3$

41. $\dfrac{x^4}{x^9}$ $\dfrac{1}{x^5}$ **42.** $\dfrac{b}{b^5}$ $\dfrac{1}{b^4}$ **43.** $(-2n^2)(-3n^4)^3$ $54n^{14}$ **44.** $(-3m^3n)(-2m^2n^3)^3$ $24m^9n^{10}$

45. $\dfrac{14x^4y^6z^2}{16x^3y^9z}$ $\dfrac{7xz}{8y^3}$ **46.** $\dfrac{25x^4y^7z^2}{20x^5y^9z^{11}}$ $\dfrac{5}{4xy^2z^9}$ **47.** $(-2x^3y^2)^3(-xy^2)^4$ $-8x^{13}y^{14}$ **48.** $(-m^4n^2)^5(-2m^3n^3)^3$ $8m^{29}n^{19}$

Simplify. Remember that an exponential expression is in simplest form when it contains only positive exponents.

49. w^{-8} $\dfrac{1}{w^8}$ **50.** m^{-9} $\dfrac{1}{m^9}$ **51.** $\dfrac{1}{a^{-5}}$ a^5 **52.** $\dfrac{1}{c^{-6}}$ c^6

53. 4^{-3} $\dfrac{1}{64}$ **54.** 5^{-2} $\dfrac{1}{25}$ **55.** $\dfrac{1}{3^{-5}}$ 243 **56.** $\dfrac{1}{2^{-4}}$ 16

57. $4x^{-7}$ $\dfrac{4}{x^7}$ **58.** $-6y^{-1}$ $-\dfrac{6}{y}$ **59.** $\dfrac{2x^{-2}}{y^4}$ $\dfrac{2}{x^2y^4}$ **60.** $\dfrac{a^3}{4b^{-2}}$ $\dfrac{a^3b^2}{4}$

61. $x^{-4}x^4$ 1 **62.** $x^{-3}x^{-5}$ $\dfrac{1}{x^8}$ **63.** $\dfrac{x^{-3}}{x^2}$ $\dfrac{1}{x^5}$ **64.** $\dfrac{x^4}{x^{-5}}$ x^9

65. $(3x^{-2})^2$ $\dfrac{9}{x^4}$

66. $(5x^2)^{-3}$ $\dfrac{1}{125x^6}$

67. $\dfrac{1}{3x^{-2}}$ $\dfrac{x^2}{3}$

68. $\dfrac{2}{5c^{-6}}$ $\dfrac{2c^6}{5}$

69. $(x^2\,y^{-4})^3$ $\dfrac{x^6}{y^{12}}$

70. $(x^3\,y^5)^{-4}$ $\dfrac{1}{x^{12}y^{20}}$

71. $(3x^{-1}\,y^{-2})^2$ $\dfrac{9}{x^2y^4}$

72. $(5xy^{-3})^{-2}$ $\dfrac{y^6}{25x^2}$

73. $(2x^{-1})(x^{-3})$ $\dfrac{2}{x^4}$

74. $(-2x^{-5})(x^7)$ $-2x^2$

75. $\dfrac{3x^{-2}y^2}{6xy^2}$ $\dfrac{1}{2x^3}$

76. $\dfrac{2x^{-2}y}{8xy}$ $\dfrac{1}{4x^3}$

77. $\dfrac{2x^{-1}y^{-4}}{4xy^2}$ $\dfrac{1}{2x^2y^6}$

78. $\dfrac{3a^{-2}\,b}{ab}$ $\dfrac{3}{a^3}$

79. $(x^{-2}\,y)^2\,(xy)^{-2}$ $\dfrac{1}{x^6}$

80. $(x^{-1}\,y^2)^{-3}\,(x^2\,y^{-4})^{-3}$ $\dfrac{y^6}{x^3}$

81. $\left(\dfrac{x^2y^{-1}}{xy}\right)^{-4}$ $\dfrac{y^8}{x^4}$

82. $\left(\dfrac{x^{-2}y^{-4}}{x^{-2}y}\right)^{-2}$ y^{10}

83. $\left(\dfrac{4a^{-2}b}{8a^3b^{-4}}\right)^2$ $\dfrac{b^{10}}{4a^{10}}$

84. $\left(\dfrac{6ab^{-2}}{3a^{-2}b}\right)^{-2}$ $\dfrac{b^6}{4a^6}$

85. *Geometry*   Find the length of line segment *AC*.

11*xy*

86. *Geometry*   Find the length of line segment *DF*.

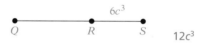

16y^2

87. *Geometry*   The length of line segment *LN* is $27a^2b$. Find the length of line segment *MN*.

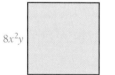

15a^2b

88. *Geometry*   The length of line segment *QS* is $18c^3$. Find the length of line segment *QR*.

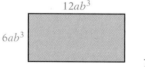

12c^3

89. *Geometry*   Find the area of the square. The dimension given is in meters.

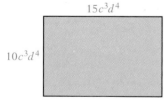

$64x^4y^2$ m²

90. *Geometry*   Find the area of the rectangle. The dimensions given are in feet.

$72a^2b^6$ ft²

91. *Geometry*   Find the perimeter of the rectangle. The dimensions given are in miles.

15c^3d^4

10c^3d^4

$50c^3d^4$ mi

92. *Geometry*   Find the perimeter of the square. The dimension given is in centimeters.

50x^4y

$200x^4y$ cm

93. *Geometry*   Find the area of the rectangle. The dimensions given are in kilometers.

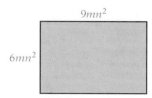

9mn^2

6mn^2

$54m^2n^4$ km²

94. *Geometry*   Find the area of the parallelogram. The dimensions given are in inches.

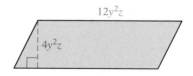

12y^2z

4y^2z

$48y^4z^2$ in²

95. *Geometry* The area of the rectangle is $24a^3b^5$ square yards. Find the length of the rectangle. $6a^2b^3$ yd

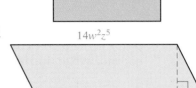

$4ab^2$

96. *Geometry* The area of the parallelogram is $56w^4z^6$ square meters. Find the height of the parallelogram. $4w^2z$ m

$14w^2z^5$

97. The product of a monomial and $4b$ is $12a^2b$. Find the monomial. $3a^2$

98. The product of a monomial and $8y^2$ is $32x^2y^3$. Find the monomial. $4x^2y$

■ Scientific Notation

Write the number in scientific notation.

99. 2,370,000
2.37×10^6

100. 75,000
7.5×10^4

101. 0.00045
4.5×10^{-4}

102. 0.000076
7.6×10^{-5}

103. 309,000
3.09×10^5

104. 819,000,000
8.19×10^8

105. 0.000000601
6.01×10^{-7}

106. 0.00000000096
9.6×10^{-10}

107. 57,000,000,000
5.7×10^{10}

108. 934,800,000,000
9.348×10^{11}

109. 0.000000017
1.7×10^{-8}

110. 0.0000009217
9.217×10^{-7}

Write the number in decimal notation.

111. 7.1×10^5
710,000

112. 2.3×10^7
23,000,000

113. 4.3×10^{-5}
0.000043

114. 9.21×10^{-7}
0.000000921

115. 6.71×10^8
671,000,000

116. 5.75×10^9
5,750,000,000

117. 7.13×10^{-6}
0.00000713

118. 3.54×10^{-8}
0.0000000354

119. 5×10^{12}
5,000,000,000,000

120. 1.0987×10^{11}
109,870,000,000

121. 8.01×10^{-3}
0.00801

122. 4.0162×10^{-9}
0.0000000040162

Solve.

123. *Physics* Light travels approximately 16,000,000,000 miles in one day. Write this number in scientific notation.
1.6×10^{10}

124. *Geology* The mass of the planet Earth is approximately 5,980,000,000,000,000,000,000,000 kilograms. Write this number in scientific notation. 5.98×10^{24}

125. *The Arts* The graph at the right shows the box office income, in millions, for three movies through the first two weekends after the release of each film (*Source:* AC-Nielsen EDI). Write in scientific notation the dollar amount of income for *Star Wars, Episode I: The Phantom Menace*.
$\$1.317 \times 10^8$

126. *Light* The length of an infrared light wave is approximately 0.0000037 meters. Write this number in scientific notation. 3.7×10^{-6}

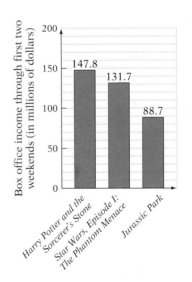

127. *Electricity* The charge on an electron is 0.00000000000000000016 coulomb. Write this number in scientific notation. 1.6×10^{-19}

128. *Computers* A unit used to measure the speed of a computer is the picosecond. One picosecond is 0.000000000001 second. Write this number in scientific notation. 1×10^{-12}

129. *Chemistry* Avogadro's number is used in chemistry. Its value is approximately 602,300,000,000,000,000,000,000. Write this number in scientific notation. 6.023×10^{23}

130. 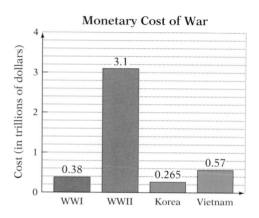 *The Military* The graph at the right shows the monetary cost of four wars (*Source:* Congressional Research Service, using numbers from the *Statistical Abstract of the United States*). Write the monetary cost of World War II in scientific notation. 3.1×10^{12}

Monetary Cost of War

Cost (in trillions of dollars)

131. *Astronomy* A parsec is a distance measurement that is used by astronomers. One parsec is 3,086,000,000,000,000,000 centimeters. Write this number in scientific notation. 3.086×10^{18}

132. *Astronomy* One light year is the distance traveled by light in one year. One light year is 5,880,000,000,000 miles. Write this number in scientific notation. 5.88×10^{12}

Simplify.

133. $(1.9 \times 10^{12})(3.5 \times 10^7)$ 6.65×10^{19}

134. $(4.2 \times 10^7)(1.8 \times 10^{-5})$ 7.56×10^2

135. $(2.3 \times 10^{-8})(1.4 \times 10^{-6})$ 3.22×10^{-14}

136. $(3 \times 10^{-20})(2.4 \times 10^9)$ 7.2×10^{-11}

137. $\dfrac{6.12 \times 10^{14}}{1.7 \times 10^9}$
3.6×10^5

138. $\dfrac{6 \times 10^{-8}}{2.5 \times 10^{-2}}$
2.4×10^{-6}

139. $\dfrac{5.58 \times 10^{-7}}{3.1 \times 10^{11}}$
1.8×10^{-18}

140. $\dfrac{9.03 \times 10^6}{4.3 \times 10^{-5}}$
2.1×10^{11}

Applying Concepts

141. Evaluate.
   a. $8^{-2} + 2^{-5}$ $\dfrac{3}{64}$ **b.** $9^{-2} + 3^{-3}$ $\dfrac{4}{81}$

142. Determine whether the statement is always true, sometimes true, or never true.
   a. The phrase *a power of a monomial* means a monomial is the base of an exponential expression.   Always true
   b. To multiply $x^m \cdot x^n$, multiply the exponents.   Never true
   c. The rules of exponents can be applied to expressions that contain an exponent of zero or contain negative exponents.   Always true
   d. The expression 3^{-2} represents the reciprocal of 3^2.   Always true

143. Evaluate 2^x and 2^{-x} when $x = -2, -1, 0, 1,$ and 2. $\frac{1}{4}, \frac{1}{2}, 1, 2, 4; 4, 2, 1, \frac{1}{2}, \frac{1}{4}$

144. Write in decimal notation.
   a. 2^{-4} 0.0625
   b. 25^{-2} 0.0016

145. If $m = n + 1$ and $a \neq 0$, then $\dfrac{a^m}{a^n} = $ _____. a

146. Solve: $(-8.5)^x = 1$ 0

EXPLORATION

1. *Scientific Notation and Order Relations*
   a. Place the correct symbol, $<$ or $>$, between the two numbers.
       (i) 5.23×10^{18} ? 5.23×10^{17}
       (ii) 3.12×10^{13} ? 3.12×10^{12}
       (iii) 3.45×10^{-14} ? 3.45×10^{-15}
       (iv) 4.2×10^8 ? 9.7×10^9
       (v) 2.7×10^{-11} ? 6.8×10^{-10}
       (i) > (ii) > (iii) > (iv) < (v) <

   b. Write a rule for ordering two numbers written in scientific notation.
       The number with the larger power of 10 is the larger number.

2. *Expressions with Negative Exponents*

   a. If x is a nonzero real number, is x^{-2} always positive, always negative, or positive or negative depending on whether x is positive or negative? Explain your answer.
       The expression x^{-2} is positive for all nonzero real numbers x. The reason is that $x^{-2} = \frac{1}{x^2}$, and x^2 is positive for all nonzero real values of x.

   b. If x is a nonzero real number, is x^{-3} always positive, always negative, or positive or negative depending on whether x is positive or negative? Explain your answer.
       The expression x^{-3} is positive or negative, depending on whether x is positive or negative. This is because $x^{-3} = \frac{1}{x^3}$, and x^3 is positive if x is positive and negative if x is negative.

3. *Negative Exponents on Fractional Expressions*
   a. Simplify each of the following expressions.

   (i) $\left(\dfrac{a^2}{b^3}\right)^{-2}$ (ii) $\left(\dfrac{x^4}{y}\right)^{-3}$ (iii) $\left(\dfrac{c^5}{d^2}\right)^{-4}$ (iv) $\left(\dfrac{2^3}{3^4}\right)^{-1}$

   (i) $\dfrac{b^6}{a^4}$ (ii) $\dfrac{y^3}{x^{12}}$ (iii) $\dfrac{d^8}{c^{20}}$ (iv) $\dfrac{81}{8}$

   b. Write a rule for rewriting with a positive exponent a fraction raised to a negative exponent.

   If $a \neq 0$, $b \neq 0$, and n is a positive integer, then $\left(\dfrac{a}{b}\right)^{-n} = \left(\dfrac{b}{a}\right)^n$.

Addition and Subtraction of Polynomials

■ Introduction to Polynomials

Some forecasters predicted that revenue generated by business on the Internet from 1997 to 2002 could be approximated by the function

$$R(t) = 15.8t^2 - 17.2t + 10.2,$$

where R is the annual revenue in billions of dollars and t is the time in years, with $t = 0$ corresponding to the year 1997. Use this function to approximate the annual revenue in the year 2000.

Because $t = 0$ corresponds to 1997, $t = 3$ corresponds to the year 2000. Evaluate the given function for $t = 3$.

$$\begin{aligned} R(t) &= 15.8t^2 - 17.2t + 10.2 \\ R(3) &= 15.8(3)^2 - 17.2(3) + 10.2 \\ &= 15.8(9) - 17.2(3) + 10.2 \\ &= 142.2 - 51.6 + 10.2 \\ &= 100.8 \end{aligned}$$

According to this function, in the year 2000, the revenue generated by business conducted on the Internet was approximately $100.8 billion.

In the function $R(t) = 15.8t^2 - 17.2t + 10.2$, the variable expression

$$15.8t^2 - 17.2t + 10.2$$

is a polynomial. A **polynomial** is a variable expression in which the terms are monomials. The polynomial $15.8t^2 - 17.2t + 10.2$ has three terms: $15.8t^2$, $-17.2t$, and 10.2. Note that each of these three terms is a monomial.

A polynomial of *one* term is a **monomial.** $-7x^2$ is a monomial.

A polynomial of *two* terms is a **binomial.** $4y + 3$ is a binomial.

A polynomial of *three* terms is a **trinomial.** $6b^2 + 5b - 8$ is a trinomial.

? QUESTION   Is the expression a polynomial? If it is a polynomial, is it a monomial, a binomial, or a trinomial?

a. $16a^2 - 9b^2$ **b.** $-\dfrac{2}{3}xy$ **c.** $x^2 + 2xy - 8$ **d.** $\dfrac{3}{x} - 5$

The terms of a polynomial in one variable are usually arranged so that the exponents of the variable decrease from left to right. This is called **descending order.** The polynomials at the right are written in descending order.

$2x^3 - 3x^2 + 6x - 1$

$5y^4 - 9y^3 + y^2 - 7y + 8$

$t - 4$

? ANSWERS   **a.** It is a polynomial. It is a binomial because it has two terms ($16a^2$ and $-9b^2$). **b.** It is a polynomial. It is a monomial because it has one term.   **c.** It is a polynomial. It is a trinomial because it has three terms (x^2, $2xy$, and -8).   **d.** $\dfrac{3}{x}$ is not a monomial, so $\dfrac{3}{x} - 5$ is not a polynomial.

INSTRUCTOR NOTE
An analogy may help students understand these terms. Dogs, lions, and monkeys are specific types of animals, just as monomials, binomials, and trinomials are specific types of polynomials.

? QUESTION   Is the polynomial written in descending order?

    a. $3a^2 - 2a^3 + 4a$ **b.** $6d^5 + 4d^3 - 7$

The **degree of a polynomial in one variable** is its greatest exponent.

The degree of $t - 4$ is 1. It is a first-degree or **linear polynomial.**
The degree of $6b^2 + 5b - 8$ is 2. It is a second-degree or **quadratic polynomial.**
The degree of $2x^3 - 3x^2 + 6x - 1$ is 3. It is a third-degree or **cubic polynomial.**
The degree of $5y^4 - 9y^3 + y^2 - 7y + 8$ is 4. It is a fourth-degree polynomial.

▼ *Point of Interest*

The dimples on a golf ball have a dramatic effect on its flight. A golf ball with a well-designed dimple pattern will travel 2 to 3 times farther than a ball with no dimples.

Suggested Activity

Have students investigate the graphs on this page by answering such questions as
1. Does the golf ball reach a maximum height?
2. Are there different distances from where the ball was struck at which it is the same height above the ground?
3. Are there different times when the ball is the same distance above the ground?
4. How can you find the time it takes the ball to be 90 feet from where it was struck?

Polynomial functions are used to model many different situations. For instance, ignoring forces other than that of gravity, the height, h, of a golf ball x feet from the point from where it was hit is given by the polynomial function $h(x) = -0.0133x^2 + 1.7321x$. By evaluating this function, we can determine the height of the ball at various distances from the point at which it was struck. The graph above shows that the ball is 48.159 feet high at a distance of 90 feet from the point where it was hit. Evaluating the polynomial when $x = 90$, we have

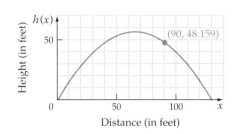

$$h(x) = -0.0133x^2 + 1.7321x$$
$$h(90) = -0.0133(90)^2 + 1.7321(90) \quad \bullet \text{ Replace } x \text{ by } 90.$$
$$= 48.159$$

The polynomial function above gave the height of the ball in terms of its *distance* from where it was hit. That is, the height depended on distance. A different polynomial function can be written that will give the height of the ball in terms of the amount of *time* the ball is in flight. Using this function, you can determine the height of the ball at various times during its flight. This is shown in Example 1.

INSTRUCTOR NOTE
Exercise 10 can be used for a similar in-class example.

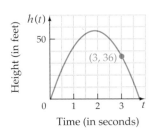

EXAMPLE 1

The height h, in feet, of a golf ball t seconds after it has been struck is given by $h(t) = -16t^2 + 60t$. Determine the height of the ball 3 seconds after it is hit.

Solution $h(t) = -16t^2 + 60t$

           $h(3) = -16(3)^2 + 60(3)$ $\bullet$ Replace t by 3.

                $= 36$

The golf ball is 36 feet high 3 seconds after being hit. See the graph at the left.

? ANSWER   **a.** No. The exponents on a (2 and 3) do not decrease from left to right. $-2a^3 + 3a^2 + 4a$ is the same polynomial written in descending order.   **b.** Yes. The exponents on d (5 and 3) decrease from left to right.

INSTRUCTOR NOTE
Exercise 20 can be used for a similar in-class example.

Suggested Activity

See Section 6.2 of the *Student Activity Manual* for an investigation in which visual models are used to add polynomials.

> ### YOU TRY IT 1
>
> If $2000 is deposited into an individual retirement account (IRA), then the value, V, of that investment 3 years later is given by the cubic polynomial function $V(r) = 2000r^3 + 6000r^2 + 6000r + 2000$, where r is the interest rate (as a decimal) earned on the investment. Determine the value after 3 years of $2000 deposited in an IRA that earns an interest rate of 7%.
>
> **Solution**   See page S24. $2450.09

■ Addition and Subtraction of Polynomials

Polynomials can be added by combining like terms.

> ### EXAMPLE 2
>
> Add: $(3x^2 - 7x + 4) + (5x^2 + 2x - 8)$
>
> **Solution** $(3x^2 - 7x + 4) + (5x^2 + 2x - 8)$
>
> $\qquad = (3x^2 + 5x^2) + (-7x + 2x) + (4 - 8)$ • Use the Properties of Addition to rearrange and group like terms.
>
> $\qquad = 8x^2 - 5x - 4$ • Combine like terms. Write the polynomial in descending order.

> ### YOU TRY IT 2
>
> Add: $(-4d^2 - 3d + 2) + (3d^2 - 4d)$
>
> **Solution**   See page S25. $-d^2 - 7d + 2$

For Example 2, you can use a graphing calculator to check your work. Here are some typical graphing calculator screens that you might use to produce the graphs.

CALCULATOR NOTE

Note the blackened equals signs for Y₃ and Y₄. They indicate that Y₃ and Y₄ will be graphed. The clear equals signs for Y₁ and Y₂ indicate that they have been "turned off"; they will not be graphed. For assistance, see Appendix A: **Y=** .

Enter the polynomials to be added.

Enter the sum.
Enter your answer.

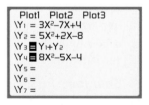

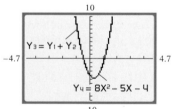

When the graphs above are produced, the graph of Y₃ lies on top of the graph of Y₄. This means that the two graphs are the same. It does not ensure that the addition is correct; however, if the two graphs do not match exactly, the addition is definitely incorrect.

Recall that **the definition of subtraction is addition of the opposite.**

$$a - b = a + (-b)$$

This definition holds true for polynomials. Polynomials can be subtracted by adding the opposite of the second polynomial to the first. The **opposite of a polynomial** is the polynomial with the sign of every term changed.

The opposite of the polynomial $x^2 - 2x + 3$ is $-x^2 + 2x - 3$.
The opposite of the polynomial $-4y^3 + 5y - 8$ is $4y^3 - 5y + 8$.

? QUESTION   What is the opposite of the polynomial $5d^4 - 6d^2 + 9$?

INSTRUCTOR NOTE
Exercise 28 can be used for a similar in-class example.

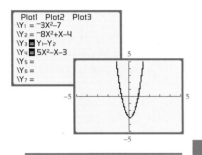

Suggested Activity

1. Which of the following have the same sum?
   a. $(4x^2 - 5x + 1) + (-2x^2 + x - 6)$
   b. $(x^2 - 8x - 2) + (x^2 + 4x - 3)$
   c. $(-3x^2 + 6x + 7) + (5x^2 - 10x - 12)$
   [Answer: They all have the sum $2x^2 - 4x - 5$.]
2. Which of the following have the same difference?
   a. $(6x^2 + 3x + 7) - (3x^2 + 2x + 5)$
   b. $(x^2 - 7x - 4) - (-2x^2 - 8x - 6)$
   c. $(-2x^2 - 6x) - (-5x^2 - 7x - 2)$
   [Answer: They all have the difference $3x^2 + x + 2$.]

INSTRUCTOR NOTE
Exercise 44 can be used for a similar in-class example.

EXAMPLE 3

Subtract: $(-3a^2 - 7) - (-8a^2 + a - 4)$

Solution $(-3a^2 - 7) - (-8a^2 + a - 4)$

$= (-3a^2 - 7) + (8a^2 - a + 4)$ • Rewrite subtraction as addition of the opposite.

$= (-3a^2 + 8a^2) - a + (-7 + 4)$ • Use the Properties of Addition to rearrange and group like terms.

$= 5a^2 - a - 3$ • Combine like terms. Write the polynomial in descending order.

A graphical check is shown at the left.

YOU TRY IT 3

Subtract: $(5x^2 - 3x + 4) - (-6x^3 - 2x + 8)$

Solution   See page S25. $6x^3 + 5x^2 - x - 4$

A company's **revenue** is the money the company earns by selling its products. A company's **cost** is the money it spends to manufacture and sell its products. A company's **profit** is the difference between its revenue and cost. This relationship is expressed by the formula $P = R - C$, where P is the profit, R is the revenue, and C is the cost. This formula is used in Example 4 and You Try It 4.

EXAMPLE 4

A company manufactures and sells woodstoves. The total monthly cost, in dollars, to produce n woodstoves is $30n + 2000$. The company's revenue, in dollars, obtained from selling all n woodstoves is $-0.4n^2 + 150n$. Express in terms of n the company's monthly profit.

State the goal.   Our goal is to write a variable expression for the company's profit from manufacturing and selling n woodstoves.

Devise a strategy.   Use the formula $P = R - C$. Substitute the given polynomials for R and C. Then subtract the polynomials.

? ANSWER   The opposite of $5d^4 - 6d^2 + 9$ is $-5d^4 + 6d^2 - 9$.

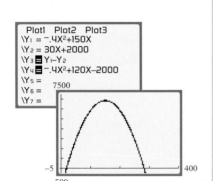

Solve the problem.

$$P = R - C$$

$$P = (-0.4n^2 + 150n) - (30n + 2000) \qquad \bullet \; R = -0.4n^2 + 150n, \; C = 30n + 2000$$

$$P = (-0.4n^2 + 150n) + (-30n - 2000) \qquad \bullet \; \text{Rewrite subtraction as addition of the opposite.}$$

$$P = -0.4n^2 + (150n - 30n) - 2000 \qquad \bullet \; \text{Combine like terms.}$$

$$P = -0.4n^2 + 120n - 2000$$

The company's monthly profit, in dollars, is $-0.4n^2 + 120n - 2000$.

Check your work.   A graphical check of the solution is shown at the left.

YOU TRY IT 4

A company's total monthly cost, in dollars, for manufacturing and selling n videotapes per month is $35n + 2000$. The company's monthly revenue, in dollars, from selling all n videotapes is $-0.2n^2 + 175n$. Express in terms of n the company's monthly profit.

Solution   See page S25. $(-0.2n^2 + 140n - 2000)$ dollars

6.2 EXERCISES   Suggested Assignment: 1–47, odds

Topics for Discussion

1. State whether the polynomial is a monomial, a binomial, or a trinomial. Explain your answer.
   a. $8x^4 - 6x^2$ This is a binomial. It contains two terms, $8x^4$ and $-6x^2$.
   b. $4a^2b^2 + 9ab + 10$
      This is a trinomial. It contains three terms, $4a^2b^2$, $9ab$, and 10.
   c. $7x^3y^4$ This is a monomial. It is one term, $7x^3y^4$. (*Note:* It is a product of a number and variables. There is no addition or subtraction operation in the expression.)
2. Explain each of the following terms. Give an example of each.
   a. Polynomial   A polynomial is a variable expression in which the terms are monomials. Examples will vary.
   b. Monomial   A polynomial of one term is a monomial. Examples will vary.
   c. Binomial   A polynomial of two terms is a binomial. Examples will vary.
   d. Trinomial   A polynomial of three terms is a trinomial. Examples will vary.

3. State whether the expression is a polynomial. Explain your answer.
   a. $\frac{1}{5}x^3 + \frac{1}{2}x$ Yes. Both $\frac{1}{5}x^3$ and $\frac{1}{2}x$ are monomials. (*Note:* The coefficients of variables can be fractions.)
   b. $\frac{1}{5x^2} + \frac{1}{2x}$ No. A polynomial does not have a variable in the denominator of a fraction.
   c. $x + \sqrt{5}$ Yes. Both x and $\sqrt{5}$ are monomials. (*Note:* The variable is not under a radical sign.)
4. Determine whether the statement is always true, sometimes true, or never true.
   a. The terms of a polynomial are monomials.   Always true
   b. Subtraction is addition of the opposite.   Always true

Introduction to Polynomials

5. *Geometry*   As sand that is very fine is poured into a pile, the volume, V, of the cone-shaped pile is given by $V(h) = \frac{8}{3}\pi h^3$, where h is the height of the cone. Find the volume of a sand pile that is 2 feet high. Round to the nearest hundredth.   67.02 ft³

6. *Geometry*   The area, A, of a rectangle with a perimeter of 100 meters is given by $A(w) = 50w - w^2$, where w is the width of the rectangle. What is the area of this rectangle when the width is 10 meters?   400 m²

10 m

7. *Oceanography*   The wavelength L, in meters, of a deep-water wave can be approximated by the function $L(v) = 0.6411v^2$, where v is the wave speed in meters per second. Find the length of a deep-water wave that has a speed of 30 meters per second.   576.99 m

8. *Polygonal Numbers*   In the diagram at the right, the total number of circles, T, when there are n rows is given by $T = 0.5n^2 + 0.5n$. Verify the formula for the four figures shown. What is the total number of circles when there are 10 rows?   For $n = 1$, $T = 1$. For $n = 2$, $T = 3$. For $n = 3$, $T = 6$. For $n = 4$, $T = 10$. For $n = 10$, $T = 55$.

9. *Sports*   The height h, in feet, of a cliff diver t seconds after beginning a dive can be modeled by the function $h(t) = -16t^2 + 5t + 50$. How high is the cliff from which the diver is jumping? (*Hint:* Determine the value of t before the diver starts the dive.)   50 ft

10. *Oceanography*   The height H, in feet, of the tide at a certain beach in Encinitas, California, can be approximated by

$$H(t) = -0.00013t^5 + 0.00839t^4 - 0.186t^3 + 1.635t^2 - 4.8t + 3.40$$

where t is the number of hours after midnight. Find the height of the tide at 9:00 A.M. Round to the nearest tenth.   4.4 ft

11. *Physics*   The amount of force F, in pounds, on one side of a triangular trough is given by $F(x) = -14.2x^3 + 63.9x^2$, where x is the height of the water in feet. What is the force on one side of this trough when the water is 3 feet deep?   191.7 lb

12. *Probability*   The probability, P, that all three security lights in a garage will not fail is given by the function $P(x) = 1 - 3x + 3x^2 - x^3$, where x is the probability that one light will fail. Find the probability that all three lights will not fail if the probability that one light will fail is 0.01. Write the answer to the nearest tenth of a percent.   97.0%

13. *Energy*   The amount of energy E, in foot-pounds, that is required to pump the water out of a certain conical tank 12 feet tall that contains x feet of water is given by $E(x) = 62.4\pi\left(-\frac{x^4}{64} + \frac{5x^3}{16}\right)$. How much energy is required to pump out the water when the depth of the water is 8 feet? Round to the nearest whole number.   18,819 foot-pounds

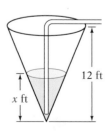

12 ft

x ft

14. *Pollution* One source of the pollutant carbon monoxide is our exhaling the air we breathe. Suppose the amount of carbon monoxide due to exhaling, in parts per million, is given by

$$P(x) = -0.02x^4 + 0.2x^3 + 1,$$

where x is the population in hundred thousands. Evaluate this polynomial when $x = 2$. Write a sentence that explains the meaning of the value
of the polynomial when $x = 2$.   2.28; A population of 200,000 people con-
tributes 2.28 parts per million of carbon monoxide to the air.

■ Addition and Subtraction of Polynomials

Add or subtract.

15. $(x^2 + 7x) + (-3x^2 - 4x)$ $-2x^2 + 3x$

16. $(3y^2 - 2y) + (5y^2 + 6y)$ $8y^2 + 4y$

17. $(x^2 - 6x) - (x^2 - 10x)$ $4x$

18. $(y^2 + 4y) - (y^2 + 10y)$ $-6y$

19. $(4b^2 - 5b) + (3b^2 + 6b - 4)$ $7b^2 + b - 4$

20. $(2c^2 - 4) + (6c^2 - 2c + 4)$ $8c^2 - 2c$

21. $(2y^2 - 4y) - (-y^2 + 2)$ $3y^2 - 4y - 2$

22. $(-3a^2 - 2a) - (4a^2 - 4)$ $-7a^2 - 2a + 4$

23. $(2a^2 - 7a + 10) + (a^2 + 4a + 7)$ $3a^2 - 3a + 17$

24. $(-6x^2 + 7x + 3) + (3x^2 + x + 3)$ $-3x^2 + 8x + 6$

25. $(x^2 - 2x + 1) - (x^2 + 5x + 8)$ $-7x - 7$

26. $(3x^2 + 2x - 2) - (5x^2 - 5x + 6)$ $-2x^2 + 7x - 8$

27. $(-2x^3 + x - 1) - (-x^2 + x - 3)$ $-2x^3 + x^2 + 2$

28. $(2x^2 + 5x - 3) - (3x^3 + 2x - 5)$ $-3x^3 + 2x^2 + 3x + 2$

29. $(x^3 - 7x + 4) + (2x^2 + x - 10)$
$x^3 + 2x^2 - 6x - 6$

30. $(3y^3 + y^2 + 1) + (-4y^3 - 6y - 3)$
$-y^3 + y^2 - 6y - 2$

31. $(5x^3 + 7x - 7) + (10x^2 - 8x + 3)$
$5x^3 + 10x^2 - x - 4$

32. $(3y^3 + 4y + 9) + (2y^2 + 4y - 21)$
$3y^3 + 2y^2 + 8y - 12$

33. $(2y^3 + 6y - 2) - (y^3 + y^2 + 4)$
$y^3 - y^2 + 6y - 6$

34. $(-2x^2 - x + 4) - (-x^3 + 3x - 2)$
$x^3 - 2x^2 - 4x + 6$

35. $(4y^3 - y - 1) - (2y^2 - 3y + 3)$
$4y^3 - 2y^2 + 2y - 4$

36. $(3x^2 - 2x - 3) - (2x^3 - 2x^2 + 4)$
$-2x^3 + 5x^2 - 2x - 7$

37. *Geometry* Find the length of line segment AC.

$\overset{3x^2 - 4x + 5}{} \quad \overset{8x^2 + 6x - 1}{}$
$\underset{A}{\bullet} \quad \underset{B}{\bullet} \quad \underset{C}{\bullet}$

$11x^2 + 2x + 4$

38. *Geometry* Find the length of line segment DF.

$\overset{5y^2 - y}{} \quad \overset{7y^2 + 4}{}$
$\underset{D}{\bullet} \quad \underset{E}{\bullet} \quad \underset{F}{\bullet}$

$12y^2 - y + 4$

39. *Geometry* The length of line segment LN is
$7a^2 + 4a - 3$. Find the length of line segment
MN.

$\overset{2a^2 + a + 6}{}$
$\underset{L}{\bullet} \quad \underset{M}{\bullet} \quad \underset{N}{\bullet}$

$5a^2 + 3a - 9$

40. *Geometry* The length of line segment QS is
$12c^3 + 4c^2 - 6$. Find the length of line segment
QR.

$\overset{5c^3 - 3c + 9}{}$
$\underset{Q}{\bullet} \quad \underset{R}{\bullet} \quad \underset{S}{\bullet}$

$7c^3 + 4c^2 + 3c - 15$

41. *Geometry* Find the perimeter of the rectangle. The dimensions given are in kilometers.

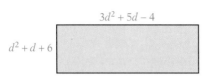

$3d^2 + 5d - 4$

$d^2 + d + 6$

$(8d^2 + 12d + 4)$ km

42. *Geometry* Find the perimeter of the rectangle. The dimensions given are in meters.

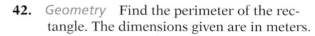

$4b^2 - 8b + 6$

$b^2 + 5b - 1$

$(10b^2 - 6b + 10)$ m

43. *Business* The total monthly cost, in dollars, for a company to produce and sell n guitars per month is $240n + 1200$. The company's monthly revenue, in dollars, from selling all n guitars is $-2n^2 + 400n$. Express in terms of n the company's monthly profit. Use the formula $P = R - C$.
$(-2n^2 + 160n - 1200)$ dollars

44. *Business* A company's total monthly cost, in dollars, for manufacturing and selling n cameras per month is $40n + 1800$. The company's monthly revenue, in dollars, from selling all n cameras is $-n^2 + 250n$. Express in terms of n the company's monthly profit. Use the formula $P = R - C$.
$(-n^2 + 210n - 1800)$ dollars

45. What polynomial must be added to $3x^2 - 4x - 2$ so that the sum is $-x^2 + 2x + 1$? $\quad -4x^2 + 6x + 3$

46. What polynomial must be added to $-2x^3 + 4x - 7$ so that the sum is $x^2 - x - 1$? $\quad 2x^3 + x^2 - 5x + 6$

47. What polynomial must be subtracted from $6x^2 - 4x - 2$ so that the difference is $2x^2 + 2x - 5$? $\quad 4x^2 - 6x + 3$

48. What polynomial must be subtracted from $2x^3 - x^2 + 4x - 2$ so that the difference is $x^3 + 2x - 8$? $\quad x^3 - x^2 + 2x + 6$

Applying Concepts

49. Determine whether the statement is always true, sometimes true, or never true.

a. Like terms have the same coefficient and the same variable part.
Sometimes true

b. The opposite of the polynomial
$ax^3 - bx^2 + cx - d$ is $-ax^3 + bx^2 - cx + d$. Always true

c. A binomial is a polynomial of degree 2. Sometimes true

50. Is it possible to add two polynomials, each of degree 3, and have the sum be a polynomial of degree 2? If so, give an example. If not, explain why not.
Yes. For example, $(3x^3 - 2x^2 + 3x - 4) + (-3x^3 - 4x^2 + 6x - 5) = -6x^2 + 9x - 9$

51. For what value of k is the given equation an identity?

a. $(2x^3 + 3x^2 + kx + 5) - (x^3 + 2x^2 + 3x + 7) = x^3 + x^2 + 5x - 2$
$k = 8$

b. $(6x^3 + kx^2 - 2x - 1) - (4x^3 - 3x^2 + 1) = 2x^3 - x^2 - 2x - 2$
$k = -4$

EXPLORATION

1. *Sums of Consecutive Integers* Given that x is an integer, $x + 1$ is the next consecutive integer. Three consecutive integers can be represented as x, $x + 1$, and $x + 2$.

    a. Represent the sum of two consecutive integers. Simplify.
    b. Show that the sum of any two consecutive integers is an odd number.
    c. Represent the sum of three consecutive integers. Simplify.
    d. Show that the sum of any three consecutive integers is divisible by 3.
    e. Represent the sum of four consecutive integers. Simplify.
    f. Show that the sum of any four consecutive integers is an even number.
    g. Represent the sum of five consecutive integers. Simplify.
    h. What is true about the sum of any five consecutive integers?
    i. Is the sum of any six consecutive numbers an even or an odd number?

    a. $x + (x + 1)$; $2x + 1$ **b.** Given x is an integer, $2x$ is an even number, so $2x + 1$ is an odd number. **c.** $x + (x + 1) + (x + 2)$; $3x + 3$ **d.** Given x is an integer, $3x$ is a multiple of 3, so $3x$ is divisible by 3; 3 is divisible by 3; the sum of two numbers that are divisible by 3 is divisible by 3; so $3x + 3$ is divisible by 3. It can also be stated that $3x + 3 = 3(x + 1)$, a multiple of 3; so $3x + 3$ is divisible by 3.   **e.** $x + (x + 1) + (x + 2) + (x + 3)$; $4x + 6$ **f.** Given x is an integer, $4x$ is an even number; 6 is an even number; the sum of two even numbers is an even number; so $4x + 6$ is an even number. It can also be stated that $4x + 6 = 2(2x + 3)$, which is an even number.   **g.** $x + (x + 1) + (x + 2) + (x + 3) + (x + 4)$; $5x + 10$ **h.** The sum of any five consecutive integers is divisible by 5.   **i.** The sum of any six consecutive numbers, $6x + 15$, is an odd number because $6x$ is even and 15 is odd. The sum of an even number and an odd number is odd.

SECTION 6.3

Multiplication and Division of Polynomials

- Multiplication of Polynomials
- Division of Polynomials
- Synthetic Division
- The Remainder Theorem
- The Factor Theorem

■ Multiplication of Polynomials

To multiply a polynomial by a monomial, use the Distributive Property and the Rule for Multiplying Exponential Expressions.

The monomial $-2x$ is multiplied by the trinomial $x^2 - 4x - 3$ as follows:

$-2x(x^2 - 4x - 3)$
$= -2x(x^2) - (-2x)(4x) - (-2x)(3)$ • Use the Distributive Property.
$= -2(x^{1+2}) - (-2 \cdot 4)(x^{1+1}) - (-2 \cdot 3)x$ • Use the Rule for Multiplying Exponential Expressions.
$= -2x^3 + 8x^2 + 6x$

TAKE NOTE

Distribute $-2x$ over each term inside the parentheses.

INSTRUCTOR NOTE
Exercise 10 can be used for a similar in-class example.

EXAMPLE 1

Multiply. **a.** $(5y + 4)(-2y)$ **b.** $x^3(2x^2 - 3x + 2)$

Solution **a.** $(5y + 4)(-2y) = 5y(-2y) + 4(-2y)$
$= -10y^2 - 8y$

b. $x^3(2x^2 - 3x + 2) = x^3(2x^2) - x^3(3x) + x^3(2)$
$= 2x^5 - 3x^4 + 2x^3$

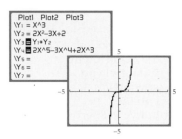

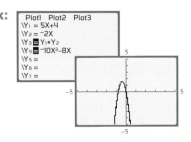

CALCULATOR NOTE

When the graphs at the right are produced, the graph of Y₃ lies on top of the graph of Y₄. This means that the two graphs are the same. It does not ensure that the product is correct; however, if the two graphs do not match exactly, the multiplication is definitely incorrect.

Suggested Activity

1. Replace the ? to make a true statement.
$(4x^2 + 3x + 6)(x + 2) =$
$4x^2(?) + 3x(?) + 6(?)$
[Answer: $x + 2$]
2. How many terms are in the product
$(a^2 - ab + b^2)(a + b)$?
[Answer: 2 (The product is $a^3 + b^3$.)]

INSTRUCTOR NOTE
Before doing an example similar to the one at the right, provide an illustration of multiplying two whole numbers—for example, 473×28. Compare the procedure for multiplying two polynomials to this example.

INSTRUCTOR NOTE
Exercise 14 can be used for a similar in-class example.

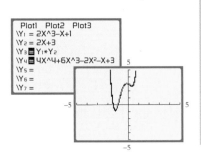

Check:

YOU TRY IT 1

Multiply.   **a.** $(-2d + 3)(-4d)$ **b.** $-a^3(3a^2 + 2a - 7)$

Solution   See page S25.   a. $8d^2 - 12d$ b. $-3a^5 - 2a^4 + 7a^3$

Multiplication of two polynomials requires the repeated application of the Distributive Property.

Shown below is the binomial $y - 2$ multiplied by the trinomial $y^2 + 3y + 1$.

$(y - 2)(y^2 + 3y + 1)$

$= (y - 2)(y^2) + (y - 2)(3y) + (y - 2)(1)$ • Use the Distributive Property to multiply $y - 2$ times each term of the trinomial.

$= y^3 - 2y^2 + 3y^2 - 6y + y - 2$ • Use the Distributive Property.

$= y^3 + y^2 - 5y - 2$ • Combine like terms.

Two polynomials can also be multiplied using a vertical format similar to that used for multiplication of whole numbers. Note that the factors in the multiplication below are the same as those used in the previous example.

$$
\begin{array}{r}
y^2 + 3y + 1 \\
y - 2 \\
\hline
-2y^2 - 6y - 2 \\
y^3 + 3y^2 + y \\
\hline
y^3 + y^2 - 5y - 2
\end{array}
$$

• Multiply each term in the trinomial by -2.
• Multiply each term in the trinomial by y. Like terms must be written in the same column.
• Add the terms in each column.

EXAMPLE 2

Multiply: $(2b^3 - b + 1)(2b + 3)$

Solution

$$
\begin{array}{r}
2b^3 - b + 1 \\
2b + 3 \\
\hline
6b^3 \quad - 3b + 3 \\
4b^4 \quad - 2b^2 + 2b \\
\hline
4b^4 + 6b^3 - 2b^2 - b + 3
\end{array}
$$

• This is $3(2b^3 - b + 1)$.

• This is $2b(2b^3 - b + 1)$. Like terms are in the same column.

• Add the terms in each column.

Check: A graphical check is shown at the left.

TAKE NOTE

The FOIL method is not really a different way of multiplying two polynomials. It is based on the Distributive Property.

$(2x + 3)(x + 5)$
$= (2x + 3)x + (2x + 3)5$
$= 2x^2 + 3x + 10x + 15$
$= 2x^2 + 13x + 15$

Note that the terms $2x^2 + 3x + 10x + 15$ are the same products we found using the FOIL method.

INSTRUCTOR NOTE
Exercise 30 can be used for a similar in-class example.

TAKE NOTE

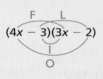

 CALCULATOR NOTE

The products in Examples 3a and 3b can be checked using the graphing calculator approach used in Examples 2a and 2b.

Suggested Activity

See Section 6.3 of the *Student Activity Manual* for an activity involving multiplying polynomials.

YOU TRY IT 2

Multiply: $(3c^3 - 2c^2 + c - 3)(2c + 5)$

Solution   See page S25. $6c^4 + 11c^3 - 8c^2 - c - 15$

The product of two binomials can be found by using a method called **FOIL**, which is based on the Distributive Property. The letters of FOIL stand for **F**irst, **O**uter, **I**nner, and **L**ast.

To multiply $(2x + 3)(x + 5)$:

Multiply the First terms.	$(2x + 3)(x + 5)$	$2x \cdot x = 2x^2$
Multiply the Outer terms.	$(2x + 3)(x + 5)$	$2x \cdot 5 = 10x$
Multiply the Inner terms.	$(2x + 3)(x + 5)$	$3 \cdot x = 3x$
Multiply the Last terms.	$(2x + 3)(x + 5)$	$3 \cdot 5 = 15$

$$ \overset{\text{F}}{} \quad \overset{\text{O}}{} \quad \overset{\text{I}}{} \quad \overset{\text{L}}{}$$

Add the products. $(2x + 3)(x + 5)$ $= 2x^2 + 10x + 3x + 15$
Combine like terms. $= 2x^2 + 13x + 15$

EXAMPLE 3

Multiply.   **a.** $(4x - 3)(3x - 2)$ **b.** $(3x - 2y)(x + 4y)$

Solution   **a.** This is the product of two binomials. Use the FOIL method.

$(4x - 3)(3x - 2)$
$= 4x(3x) + (4x)(-2) + (-3)(3x) + (-3)(-2)$
$= 12x^2 - 8x - 9x + 6$
$= 12x^2 - 17x + 6$

b. This is the product of two binomials. Use the FOIL method.

$(3x - 2y)(x + 4y)$
$= 3x(x) + (3x)(4y) + (-2y)(x) + (-2y)(4y)$
$= 3x^2 + 12xy - 2xy - 8y^2$
$= 3x^2 + 10xy - 8y^2$

YOU TRY IT 3

Multiply. **a.** $(4y - 5)(3y - 3)$ **b.** $(3a + 2b)(3a - 5b)$

Solution   See page S25.   **a.** $12y^2 - 27y + 15$ **b.** $9a^2 - 9ab - 10b^2$

The expression $(a + b)^2$ is the **square of a binomial.** We are squaring $(a + b)$, which means that we are multiplying it times itself.

$$(a + b)^2 = (a + b)(a + b)$$

$(a + b)(a + b)$ is the product of two binomials. Use the FOIL method to multiply.

$$\boldsymbol{(a + b)^2} = (a + b)(a + b) = a^2 + ab + ab + b^2 \boldsymbol{= a^2 + 2ab + b^2}$$

INSTRUCTOR NOTE
Exercise 48 can be used for
a similar in-class example.

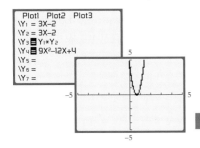

EXAMPLE 4

Multiply: $(3x - 2)^2$

Solution $(3x - 2)^2$ • This is the square of a binomial.

$\quad\quad\quad = (3x - 2)(3x - 2)$ • Multiply $(3x - 2)$ times itself.

$\quad\quad\quad = 9x^2 - 6x - 6x + 4$ • Use the FOIL method.

$\quad\quad\quad = 9x^2 - 12x + 4$ • Combine like terms.

Check: A graphical check is shown at the left.

YOU TRY IT 4

Multiply: $(3x + 2y)^2$

Solution   See page S25. $9x^2 + 12xy + 4y^2$

INSTRUCTOR NOTE
Exercise 94 can be used for
a similar in-class example.

EXAMPLE 5

A rectangular piece of cardboard measures 12 inches by 16 inches. An open box is formed by cutting four squares that measure x inches on a side from the corners of the cardboard and then folding up the sides, as shown in the figure at the left. Determine the volume of the box in terms of x. Use your equation to find the volume of the box when x is 2 inches.

State the goal.   The goal is to determine the volume of the box in terms of x and then to find the volume when $x = 2$.

Devise a strategy.   To determine the volume of the box in terms of x, use the formula for the volume of a box, $V = LWH$. Substitute variable expressions for L, W, and H. Then multiply.

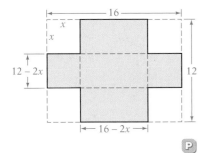

To determine the volume when $x = 2$, substitute 2 for x in the equation. Then simplify the numerical expression.

Solve the problem.

$V = LWH$

$V = (16 - 2x)(12 - 2x)x$ • $L = 16 - 2x$, $W = 12 - 2x$, $H = x$

$V = (4x^2 - 56x + 192)x$ • Multiply $(16 - 2x)(12 - 2x)$. Write the
                                               terms in descending order.

$V = 4x^3 - 56x^2 + 192x$ • Multiply the trinomial by x.

The volume of the box in terms of x is $(4x^3 - 56x^2 + 192x)$ cubic inches.

$V = 4x^3 - 56x^2 + 192x$

$V = 4(2)^3 - 56(2)^2 + 192(2)$ • Replace x by 2.

$V = 192$

When x is 2 inches, the volume of the box is 192 cubic inches.

Check your work.   A graphical check of the product $4x^3 - 56x^2 + 192x$ can be performed on a graphing calculator.

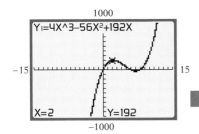

YOU TRY IT 5

The radius of a circle is $(x - 4)$ feet. Find the area of the circle in terms of the variable x. Leave the answer in terms of π.

Solution   See page S25. $(\pi x^2 - 8\pi x + 16\pi)$ ft^2

■ Division of Polynomials

As shown below, $\frac{8 + 4}{2}$ can be simplified by first adding the terms in the numerator and then dividing the result by the denominator. It can also be simplified by first dividing each term in the numerator by the denominator and then adding the results.

$$\frac{8 + 4}{2} = \frac{12}{2} = 6 \qquad \frac{8 + 4}{2} = \frac{8}{2} + \frac{4}{2} = 4 + 2 = 6$$

It is this second method that is used **to divide a polynomial by a monomial: Divide each term in the numerator by the denominator, and then write the sum of the quotients.**

> **TAKE NOTE**
>
> Recall that the fraction bar can be read "divided by."

To divide $\frac{6x^2 + 4x}{2x}$, divide each term of the polynomial $6x^2 + 4x$ by the monomial $2x$. Then simplify each quotient.

$$\frac{6x^2 + 4x}{2x} = \frac{6x^2}{2x} + \frac{4x}{2x}$$
$$= 3x + 2$$

INSTRUCTOR NOTE
Exercise 66 can be used for a similar in-class example.

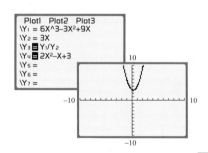

INSTRUCTOR NOTE
Ask students how the domains of $\frac{6x^3 - 3x^2 + 9x}{3x}$ and $2x^2 - x + 3$ differ.

EXAMPLE 6

Divide: $\dfrac{6x^3 - 3x^2 + 9x}{3x}$

Solution $\dfrac{6x^3 - 3x^2 + 9x}{3x}$

$$= \frac{6x^3}{3x} - \frac{3x^2}{3x} + \frac{9x}{3x} \qquad \bullet \text{ Divide each term in the numerator by the denominator.}$$

$$= 2x^2 - x + 3 \qquad \bullet \text{ Simplify each quotient.}$$

Check: A graphical check is shown at the left.

YOU TRY IT 6

Divide: $\dfrac{4x^3y + 8x^2y^2 - 4xy^3}{2xy}$

Solution   See page S25. $2x^2 + 4xy - 2y^2$

The method illustrated above is appropriate only when the divisor is a monomial. To divide two polynomials in which the divisor is not a monomial, we use a method similar to that used for division of whole numbers.

To divide $(x^2 - 5x + 8) \div (x - 3)$:

INSTRUCTOR NOTE
It may help students if you start with the division algorithm for whole numbers and show them that a similar procedure is used to divide polynomials.

Step 1

$$\begin{array}{r} x \\ x - 3 \overline{) x^2 - 5x + 8} \\ \underline{x^2 - 3x} \\ -2x + 8 \end{array}$$

Think: $x \overline{)x^2} = \dfrac{x^2}{x} = x$

Multiply: $x(x - 3) = x^2 - 3x$
Subtract: $(x^2 - 5x) - (x^2 - 3x) = -2x$
Bring down the $+8$.

Step 2

$$\begin{array}{r} x - 2 \\ x - 3\overline{)x^2 - 5x + 8} \\ \underline{x^2 - 3x} \\ -2x + 8 \\ \underline{-2x + 6} \\ 2 \end{array}$$

Think: $x\overline{)-2x} = \dfrac{-2x}{x} = -2$

Multiply: $-2(x - 3) = -2x + 6$
Subtract: $(-2x + 8) - (-2x + 6) = 2$
The remainder is 2.

The same equation we use to check division of whole numbers is used to check polynomial division.

(Quotient × Divisor) + Remainder = Dividend

Check: $(x - 2)(x - 3) + 2 = x^2 - 3x - 2x + 6 + 2 = x^2 - 5x + 8$

$(x^2 - 5x + 8) \div (x - 3) = x - 2 + \dfrac{2}{x - 3}$

INSTRUCTOR NOTE
Exercise 70 can be used for a similar in-class example.

> ### TAKE NOTE
> Note that the remainder is used to write a fraction with the remainder over the divisor. This is similar to arithmetic, in which the answer to $15 \div 4$ is written $3\frac{3}{4}$.

> ### TAKE NOTE
> Inserting the term $0x^2$ is similar to using zero as a placeholder in division.
>
> $45\overline{)702}$ • The 0 represents zero tens.

EXAMPLE 7

Divide: $(6x + 2x^3 + 26) \div (x + 2)$

Solution   Arrange the terms of the dividend in descending order. There is no x^2 term in $2x^3 + 6x + 26$. Insert $0x^2$ for the missing term so that like terms will be in columns.

$$\begin{array}{r} 2x^2 - 4x + 14 \\ x + 2\overline{)2x^3 + 0x^2 + 6x + 26} \\ \underline{2x^3 + 4x^2} \\ -4x^2 + 6x \\ \underline{-4x^2 - 8x} \\ 14x + 26 \\ \underline{14x + 28} \\ -2 \end{array}$$

Check: $(x + 2)(2x^2 - 4x + 14) - 2 = 2x^3 + 6x + 28 - 2 = 2x^3 + 6x + 26$

$(6x + 2x^3 + 26) \div (x + 2) = 2x^2 - 4x + 14 - \dfrac{2}{x + 2}$

YOU TRY IT 7

Divide: $(x^3 - 7 - 2x) \div (x - 2)$

Solution   See page S25. $x^2 + 2x + 2 - \dfrac{3}{x - 2}$

INSTRUCTOR NOTE
Exercise 108 can be used for a similar in-class example.

EXAMPLE 8

Given that $x + 1$ is a factor of $x^3 + x^2 + 4x + 4$, find another factor of $x^3 + x^2 + 4x + 4$.

Solution   Recall that a factor of a number divides that number evenly. **A factor of a polynomial divides that polynomial evenly. The quotient is another factor of the polynomial.**

Suggested Activity

Consider asking your students to perform this more difficult division problem.

$$(3x^3 - 5x^2 - 6x + 8)$$
$$\div (2x - 4)$$

[Answer: $\frac{3}{2}x^2 + \frac{1}{2}x - 2$]

$$\begin{array}{r} x^2 + 4 \\ x + 1 \overline{\smash{)}\,x^3 + x^2 + 4x + 4} \\ \underline{x^3 + x^2} \\ 0 + 4x + 4 \\ \underline{4x + 4} \\ 0 \end{array}$$

Check: $(x + 1)(x^2 + 4) = x^3 + 4x + x^2 + 4 = x^3 + x^2 + 4x + 4$

Another factor of $x^3 + x^2 + 4x + 4$ is $x^2 + 4$.

YOU TRY IT 8

Given that $x + 5$ is a factor of $x^4 + 5x^3 + 2x + 10$, find another factor of $x^4 + 5x^3 + 2x + 10$.

Solution See page S25. $x^3 + 2$

INSTRUCTOR NOTE
Exercise 98 can be used for a similar in-class example.

$x - 3$

EXAMPLE 9

The area of a rectangle is $(2x^2 - 3x - 9)$ square meters. The width of the rectangle is $(x - 3)$ meters. Find the length of the rectangle in terms of the variable x. Use the formula $L = \frac{A}{W}$, where L is the length, A is the area, and W is the width of a rectangle.

State the goal. The goal is to write a variable expression for the length of a rectangle that has an area of $(2x^2 - 3x - 9)$ square meters and a width of $(x - 3)$ meters.

Devise a strategy. Using the formula $L = \frac{A}{W}$, substitute the given polynomials for A and W. Then divide the polynomials.

Solve the problem.

$$L = \frac{A}{W} = \frac{2x^2 - 3x - 9}{x - 3}$$

$$\begin{array}{r} 2x + 3 \\ x - 3 \overline{\smash{)}\,2x^2 - 3x - 9} \\ \underline{2x^2 - 6x} \\ 3x - 9 \\ \underline{3x - 9} \\ 0 \end{array}$$

The length of the rectangle is $(2x + 3)$ meters.

Check your work. $(x - 3)(2x + 3) = 2x^2 + 3x - 6x - 9 = 2x^2 - 3x - 9$

YOU TRY IT 9

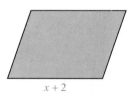

$x + 2$

The area of a parallelogram is $(3x^2 + 2x - 8)$ square feet. The length of the base is $(x + 2)$ feet. Find the height of the parallelogram in terms of the variable x. Use the formula $h = \frac{A}{b}$, where h is the height, A is the area, and b is the length of the base of a parallelogram.

Solution See page S25. $(3x - 4)$ ft

■ Synthetic Division

Synthetic division is a shorter method of dividing a polynomial by a binomial of the form $x - a$. This method of dividing uses only the coefficients of the variable terms.

Both long division and synthetic division are used below to divide the polynomial $3x^2 - 4x + 6$ by $x - 2$.

LONG DIVISION

Compare the coefficients in this problem worked by long division with the coefficients in the same problem worked by synthetic division.

$$
\begin{array}{r}
3x + 2 \\
x - 2 \overline{\smash{)}3x^2 - 4x + 6} \\
\underline{3x^2 - 6x} \\
2x + 6 \\
\underline{2x - 4} \\
10
\end{array}
$$

$$(3x^2 - 4x + 6) \div (x - 2) = 3x + 2 + \frac{10}{x - 2}$$

SYNTHETIC DIVISION

$x - a = x - 2; a = 2$

Value of a	Coefficients of the dividend		
2	3	-4	6

Bring down the 3.

	3		

Multiply $2 \cdot 3$ and add the product (6) to -4.

2	3	-4	6
		6	
	3	2	

Multiply $2 \cdot 2$ and add the product (4) to 6.

2	3	-4	6
		6	4
	3	2	10

Coefficients of the quotient   Remainder

The degree of the first term of the quotient is one degree less than the degree of the first term of the dividend.

$$(3x^2 - 4x + 6) \div (x - 2) = 3x + 2 + \frac{10}{x - 2}$$

Check: $(3x + 2)(x - 2) + 10$

$$= 3x^2 - 6x + 2x - 4 + 10$$
$$= 3x^2 - 4x + 6$$

? QUESTION   Suppose you are going to divide $2x^3 + 13x^2 + 15x - 5$ by $x + 5$ using synthetic division.

a. What are the coefficients of the dividend?
b. What is the value of a?
c. What is the degree of the first term of the quotient?

? ANSWERS   **a.** The coefficients of the dividend are 2, 13, 15, and -5.
b. $x - a = x + 5 = x - (-5); a = -5$ **c.** The degree of the first term of the dividend is 3; therefore, the degree of the first term of the quotient is 2.

INSTRUCTOR NOTE
To check their understanding, ask students why synthetic division uses addition rather than subtraction.

➡ Divide: $(2x^3 + 3x^2 - 4x + 8) \div (x + 3)$

$x - a = x + 3 = x - (-3); a = -3$
Write down the value of a and the coefficients of the dividend. Bring down the 2. Multiply $-3 \cdot 2$ and add the product (-6) to 3. Continue until all the coefficients have been used.

$$
\begin{array}{r|rrrr}
-3 & 2 & 3 & -4 & 8 \\
& & -6 & 9 & -15 \\
\hline
& 2 & -3 & 5 & -7
\end{array}
$$

Coefficients of   Remainder
the quotient

Write the quotient. The degree of the quotient is one less than the degree of the dividend.

$(2x^3 + 3x^2 - 4x + 8) \div (x + 3)$

$= 2x^2 - 3x + 5 - \dfrac{7}{x + 3}$ ⬅

INSTRUCTOR NOTE
Exercise 116 can be used for a similar in-class example.

EXAMPLE 10

Divide. **a.** $(5x^2 - 3x + 7) \div (x - 1)$ **b.** $(3x^4 - 8x^2 + 2x + 1) \div (x + 2)$

Solution   **a.**
$$
\begin{array}{r|rrr}
1 & 5 & -3 & 7 \\
& & 5 & 2 \\
\hline
& 5 & 2 & 9
\end{array}
$$
• $x - a = x - 1; a = 1$

$(5x^2 - 3x + 7) \div (x - 1) = 5x + 2 + \dfrac{9}{x - 1}$

b.
$$
\begin{array}{r|rrrrr}
-2 & 3 & 0 & -8 & 2 & 1 \\
& & -6 & 12 & -8 & 12 \\
\hline
& 3 & -6 & 4 & -6 & 13
\end{array}
$$
• Insert a zero for the missing cubic term. $x - a = x + 2$; $a = -2$

$(3x^4 - 8x^2 + 2x + 1) \div (x + 2) = 3x^3 - 6x^2 + 4x - 6 + \dfrac{13}{x + 2}$

TAKE NOTE

You can check the answer to a synthetic division problem in the same way that you check an answer to a long division problem.

YOU TRY IT 10

Divide. **a.** $(6x^2 + 8x - 5) \div (x + 2)$ **b.** $(2x^4 - 3x^3 - 8x^2 - 2) \div (x - 3)$

Solution   See page S26.   **a.** $6x - 4 + \dfrac{3}{x + 2}$ **b.** $2x^3 + 3x^2 + x + 3 + \dfrac{7}{x - 3}$

■ The Remainder Theorem

A polynomial can be evaluated by using synthetic division. Consider the polynomial $P(x) = 2x^4 - 3x^3 + 4x^2 - 5x + 1$. One way to evaluate the polynomial when $x = 2$ is to replace x by 2 and then simplify the numerical expression.

$$
\begin{aligned}
P(x) &= 2x^4 - 3x^3 + 4x^2 - 5x + 1 \\
P(2) &= 2(2)^4 - 3(2)^3 + 4(2)^2 - 5(2) + 1 \\
&= 2(16) - 3(8) + 4(4) - 5(2) + 1 \\
&= 32 - 24 + 16 - 10 + 1 \\
&= 15
\end{aligned}
$$

Now use synthetic division to divide $2x^4 - 3x^3 + 4x^2 - 5x + 1$ by $x - 2$.

$$
\begin{array}{r|rrrrr}
2 & 2 & -3 & 4 & -5 & 1 \\
  &   & 4 & 2 & 12 & 14 \\
\hline
  & 2 & 1 & 6 & 7 & 15 \\
\end{array}
$$

Coefficients of the quotient   Remainder

Note that the remainder is 15, which is the same value as $P(2)$. This is not a coincidence. The following theorem states that this situation is always true.

> ### Remainder Theorem
> If the polynomial $P(x)$ is divided by $x - a$, the remainder is $P(a)$.

➡ Evaluate $P(x) = x^4 - 3x^2 + 4x - 5$ when $x = -2$ by using the Remainder Theorem.

$$
\begin{array}{r|rrrrr}
-2 & 1 & 0 & -3 & 4 & -5 \\
   &   & -2 & 4 & -2 & -4 \\
\hline
   & 1 & -2 & 1 & 2 & -9 \\
\end{array}
$$

• A 0 is inserted for the missing cubic term.

↑ The remainder

$P(-2) = -9$

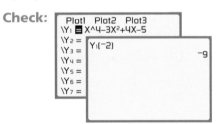

Check:

See Appendix A: Evaluating Expressions

INSTRUCTOR NOTE
Exercise 136 can be used for a similar in-class example.

EXAMPLE 11

Use the Remainder Theorem to evaluate $P(-2)$ when $P(x) = x^3 - 3x^2 + x + 3$.

Solution Use synthetic division with $a = -2$.

$$
\begin{array}{r|rrrr}
-2 & 1 & -3 & 1 & 3 \\
   &   & -2 & 10 & -22 \\
\hline
   & 1 & -5 & 11 & -19 \\
\end{array}
$$

By the Remainder Theorem, $P(-2) = -19$.

YOU TRY IT 11

Use the Remainder Theorem to evaluate $P(3)$ when $P(x) = 2x^3 - 4x - 5$.

Solution   See page S26. $P(3) = 37$

■ The Factor Theorem

The Factor Theorem is a result of the Remainder Theorem.

The Factor Theorem
A polynomial $P(x)$ has a factor $(x - c)$ if and only if $P(c) = 0$.

This theorem states that a remainder of zero means that the divisor is a factor of the dividend.

➡ Determine whether $x + 1$ is a factor of $P(x) = 2x^3 + 3x^2 - x - 2$.

Use synthetic division to divide $2x^3 + 3x^2 - x - 2$ by $x + 1$. $a = -1$.

$$
\begin{array}{r|rrrr}
-1 & 2 & 3 & -1 & -2 \\
   &   & -2 & -1 & 2 \\
\hline
   & 2 & 1 & -2 & 0 \\
\end{array}
$$

↑ Remainder

The remainder is zero. $x + 1$ is a factor of $P(x) = 2x^3 + 3x^2 - x - 2$. ⬅

INSTRUCTOR NOTE
Exercise 148 can be used for a similar in-class example.

EXAMPLE 12

Determine whether $x - 2$ is a factor of $P(x) = x^4 - x^3 - 3x^2 + x + 1$.

Solution Use synthetic division to divide $x^4 - x^3 - 3x^2 + x + 1$ by $x - 2$. $a = 2$.

$$
\begin{array}{r|rrrrr}
2 & 1 & -1 & -3 & 1 & 1 \\
  &   & 2 & 2 & -2 & -2 \\
\hline
  & 1 & 1 & -1 & -1 & -1 \\
\end{array}
$$

The remainder is not zero.
$x - 2$ is not a factor of $P(x) = x^4 - x^3 - 3x^2 + x + 1$.

See Appendix A:
Evaluating Expressions

Check:

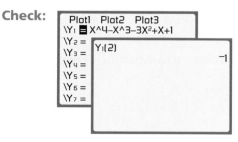

YOU TRY IT 12

Determine whether $x - 1$ is a factor of $P(x) = -x^3 + 4x^2 - 5x + 2$.

Solution See page S26. Yes

6.3 EXERCISES   Suggested Assignment: 9–153, every other odd

Topics for Discussion

1. When is the FOIL method used?
    The FOIL method is used to multiply two binomials.

2. Why is $(a + b)^2$ not equal to $a^2 + b^2$.
    Because $(a + b)^2 = (a + b)(a + b) = a^2 + 2ab + b^2 \neq a^2 + b^2$.

3. Given that $\frac{x^3 + 1}{x + 1} = x^2 - x + 1$. Name two factors of $x^3 + 1$.
    $x + 1$ and $x^2 - x + 1$

4. If a polynomial of degree 3 is multiplied by a polynomial of degree 2, what is the degree of the resulting polynomial?   Degree 5

5. Determine whether the statement is always true, sometimes true, or never true.
    a. The FOIL method is used to multiply two polynomials.   Sometimes true
    b. Using the FOIL method, the terms $3x$ and 5 are the "First" terms in $(3x + 5)(2x + 7)$.   Never true
    c. To square a binomial means to multiply it times itself.   Always true

6. What is synthetic division?
    Answers may vary. For example: It is a method of dividing a polynomial by a binomial of the form $x - a$; it uses only the coefficients of the variable terms.

7. When synthetic division is used to divide a polynomial by a binomial of the form $x - a$, how is the degree of the quotient related to the degree of the dividend?
    The degree of the first term of the quotient is one degree less than the degree of the first term of the dividend.

8. What does the Remainder Theorem state?
    If the polynomial $P(x)$ is divided by $x - a$, the remainder is $P(a)$.

■ **Multiplication and Division of Polynomials**

Multiply.

9. $-b(5b^2 + 7b - 35)$ $-5b^3 - 7b^2 + 35b$

10. $x^2(3x^4 - 3x^2 - 2)$ $3x^6 - 3x^4 - 2x^2$

11. $y^3(-4y^3 - 6y + 7)$ $-4y^6 - 6y^4 + 7y^3$

12. $2y^2(-3y^2 - 6y + 7)$ $-6y^4 - 12y^3 + 14y^2$

13. $(-2b^2 - 3b + 4)(b - 5)$ $-2b^3 + 7b^2 + 19b - 20$

14. $(-a^2 + 3a - 2)(2a - 1)$ $-2a^3 + 7a^2 - 7a + 2$

15. $(x^3 - 3x + 2)(x - 4)$ $x^4 - 4x^3 - 3x^2 + 14x - 8$

16. $(y^3 + 4y^2 - 8)(2y - 1)$ $2y^4 + 7y^3 - 4y^2 - 16y + 8$

17. $(y + 2)(y^3 + 2y^2 - 3y + 1)$
    $y^4 + 4y^3 + y^2 - 5y + 2$

18. $(2a - 3)(2a^3 - 3a^2 + 2a - 1)$
    $4a^4 - 12a^3 + 13a^2 - 8a + 3$

19. $(x + 1)(x + 3)$ $x^2 + 4x + 3$

20. $(y + 2)(y + 5)$ $y^2 + 7y + 10$

21. $(a - 3)(a + 4)$ $a^2 + a - 12$

22. $(b - 6)(b + 3)$ $b^2 - 3b - 18$

23. $(y - 7)(y - 3)$ $y^2 - 10y + 21$

24. $(a - 8)(a - 9)$ $a^2 - 17a + 72$

25. $(2x + 1)(x + 7)$ $\quad 2x^2 + 15x + 7$

26. $(y + 2)(5y + 1)$ $\quad 5y^2 + 11y + 2$

27. $(3x - 1)(x + 4)$ $\quad 3x^2 + 11x - 4$

28. $(7x - 2)(x + 4)$ $\quad 7x^2 + 26x - 8$

29. $(4x - 3)(x - 7)$ $\quad 4x^2 - 31x + 21$

30. $(2x - 3)(4x - 7)$ $\quad 8x^2 - 26x + 21$

31. $(3y - 8)(y + 2)$ $\quad 3y^2 - 2y - 16$

32. $(5y - 9)(y + 5)$ $\quad 5y^2 + 16y - 45$

33. $(7a - 16)(3a - 5)$ $\quad 21a^2 - 83a + 80$

34. $(5a - 12)(3a - 7)$ $\quad 15a^2 - 71a + 84$

35. $(3b + 13)(5b - 6)$ $\quad 15b^2 + 47b - 78$

36. $(x + y)(2x + y)$ $\quad 2x^2 + 3xy + y^2$

37. $(2a + b)(a + 3b)$ $\quad 2a^2 + 7ab + 3b^2$

38. $(3x - 4y)(x - 2y)$ $\quad 3x^2 - 10xy + 8y^2$

39. $(2a - b)(3a + 2b)$ $\quad 6a^2 + ab - 2b^2$

40. $(5a - 3b)(2a + 4b)$ $\quad 10a^2 + 14ab - 12b^2$

41. $(d - 6)(d + 6)$ $\quad d^2 - 36$

42. $(y - 5)(y + 5)$ $\quad y^2 - 25$

43. $(2x + 3)(2x - 3)$ $\quad 4x^2 - 9$

44. $(4x - 7)(4x + 7)$ $\quad 16x^2 - 49$

45. $(x + 1)^2$ $\quad x^2 + 2x + 1$

46. $(y - 3)^2$ $\quad y^2 - 6y + 9$

47. $(3a - 5)^2$ $\quad 9a^2 - 30a + 25$

48. $(6x - 5)^2$ $\quad 36x^2 - 60x + 25$

Divide.

49. $\dfrac{2x + 2}{2}$ $\quad x + 1$

50. $\dfrac{5y + 5}{5}$ $\quad y + 1$

51. $\dfrac{10a - 25}{5}$ $\quad 2a - 5$

52. $\dfrac{16b - 40}{8}$ $\quad 2b - 5$

53. $\dfrac{3a^2 + 2a}{a}$ $\quad 3a + 2$

54. $\dfrac{6y^2 + 4y}{y}$ $\quad 6y + 4$

55. $\dfrac{4b^3 - 3b}{b}$ $\quad 4b^2 - 3$

56. $\dfrac{12x^2 - 7x}{x}$ $\quad 12x - 7$

57. $\dfrac{3x^2 - 6x}{3x}$ $\quad x - 2$

58. $\dfrac{10y^2 - 6y}{2y}$ $\quad 5y - 3$

59. $\dfrac{5x^2 - 10x}{-5x}$ $\quad -x + 2$

60. $\dfrac{3y^2 - 27y}{-3y}$ $\quad -y + 9$

61. $\dfrac{x^3 + 3x^2 - 5x}{x}$ $\quad x^2 + 3x - 5$

62. $\dfrac{a^3 - 5a^2 + 7a}{a}$ $\quad a^2 - 5a + 7$

63. $\dfrac{x^6 - 3x^4 - x^2}{x^2}$ $\quad x^4 - 3x^2 - 1$

64. $\dfrac{a^8 - 5a^5 - 3a^3}{a^2}$ $\quad a^6 - 5a^3 - 3a$

65. $\dfrac{5x^2y^2 + 10xy}{5xy}$ $\quad xy + 2$

66. $\dfrac{8x^2y^2 - 24xy}{8xy}$ $\quad xy - 3$

67. $(b^2 - 14b + 49) \div (b - 7)$ $\quad b - 7$

68. $(x^2 - x - 6) \div (x - 3)$ $\quad x + 2$

69. $(2x^2 + 5x + 2) \div (x + 2)$ $\quad 2x + 1$

70. $(2y^2 - 13y + 21) \div (y - 3)$ $\quad 2y - 7$

71. $(x^2 + 1) \div (x - 1)$ $\quad x + 1 + \dfrac{2}{x - 1}$

72. $(x^2 + 4) \div (x + 2)$ $\quad x - 2 + \dfrac{8}{x + 2}$

73. $(6x^2 - 7x) \div (3x - 2)$ $\quad 2x - 1 - \dfrac{2}{3x - 2}$

74. $(6y^2 + 2y) \div (2y + 4)$ $\quad 3y - 5 + \dfrac{20}{2y + 4}$

75. $(a^2 + 5a + 10) \div (a + 2)$ $\quad a + 3 + \dfrac{4}{a + 2}$

76. $(b^2 - 8b - 9) \div (b - 3)$ $\quad b - 5 - \dfrac{24}{b - 3}$

77. $(2y^2 - 9y + 8) \div (2y + 3)$ $\quad y - 6 + \dfrac{26}{2y + 3}$

78. $(3x^2 + 5x - 4) \div (x - 4)$ $\quad 3x + 17 + \dfrac{64}{x - 4}$

79. $(8x + 3 + 4x^2) \div (2x - 1)$ $2x + 5 + \frac{8}{2x - 1}$

80. $(10 + 21y + 10y^2) \div (2y + 3)$ $5y + 3 + \frac{1}{2y + 3}$

81. $(x^3 + 3x^2 + 5x + 3) \div (x + 1)$ $x^2 + 2x + 3$

82. $(x^3 - 6x^2 + 7x - 2) \div (x - 1)$ $x^2 - 5x + 2$

83. $(x^4 - x^2 - 6) \div (x^2 + 2)$ $x^2 - 3$

84. $(x^4 + 3x^2 - 10) \div (x^2 - 2)$ $x^2 + 5$

85. *Geometry*   Find the area of the square. The dimension given is in meters.

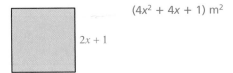

$(4x^2 + 4x + 1)$ m^2

$2x + 1$

86. *Geometry*   Find the area of the square. The dimension given is in yards.

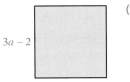

$(9a^2 - 12a + 4)$ yd^2

$3a - 2$

87. *Geometry*   Find the area of the rectangle. The dimensions given are in miles.

$5x$

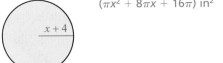

$(10x^2 - 35x)$ mi^2

$2x - 7$

88. *Geometry*   Find the area of the rectangle. The dimensions given are in feet.

$2x + 3$

$(2x^2 - 9x - 18)$ ft^2

$x - 6$

89. *Geometry*   The radius of a circle is $(x + 4)$ inches. Find the area of the circle in terms of the variable x. Leave the answer in terms of π.

$(\pi x^2 + 8\pi x + 16\pi)$ in^2

$x + 4$

90. *Geometry*   The radius of a circle is $(x - 3)$ centimeters. Find the area of the circle in terms of the variable x. Leave the answer in terms of π.

$(\pi x^2 - 6\pi x + 9\pi)$ cm^2

$x - 3$

91. *Geometry*   The length of a side of a cube is $(4x + 1)$ inches. Find the volume of the cube in terms of the variable x. $(64x^3 + 48x^2 + 12x + 1)$ in^3

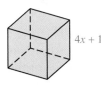

$4x + 1$

92. *Geometry*   A rectangular box has a length of $(5x + 3)$ centimeters, a width of $(2x - 1)$ centimeters, and a height of $4x$ centimeters. Find the volume of the box in terms of the variable x. $(40x^3 + 4x^2 - 12x)$ cm^3

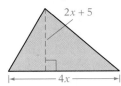

$2x + 5$

$4x$

93. *Geometry*   The base of a triangle is $4x$ meters and the height is $(2x + 5)$ meters. Find the area of the triangle in terms of the variable x.
$(4x^2 + 10x)$ m^2

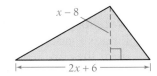

$x - 8$

$2x + 6$

94. *Geometry*   The base of a triangle is $(2x + 6)$ inches and the height is $(x - 8)$ inches. Find the area of the triangle in terms of the variable x.
$(x^2 - 5x - 24)$ in^2

95. *Geometry*   The width of a rectangle is $(3x + 1)$ inches. The length of the rectangle is twice the width. Find the area of the rectangle in terms of the variable x. $(18x^2 + 12x + 2)$ in^2

96. *Geometry*   The width of a rectangle is $(4x - 3)$ centimeters. The length of the rectangle is twice the width. Find the area of the rectangle in terms of the variable x. $(32x^2 - 48x + 18)$ cm^2

97. *Geometry*   The area of a rectangle is $(3x^2 - 22x - 16)$ square feet. The width of the rectangle is $(x - 8)$ feet. Find the length of the rectangle in terms of the variable x. Use the formula $L = \frac{A}{W}$, where L is the length, A is the area, and W is the width of a rectangle. $(3x + 2)$ ft

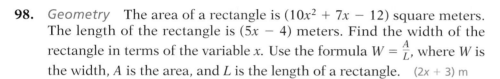
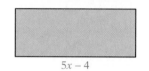

98. *Geometry*   The area of a rectangle is $(10x^2 + 7x - 12)$ square meters. The length of the rectangle is $(5x - 4)$ meters. Find the width of the rectangle in terms of the variable x. Use the formula $W = \frac{A}{L}$, where W is the width, A is the area, and L is the length of a rectangle. $(2x + 3)$ m

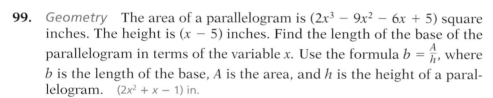

99. *Geometry*   The area of a parallelogram is $(2x^3 - 9x^2 - 6x + 5)$ square inches. The height is $(x - 5)$ inches. Find the length of the base of the parallelogram in terms of the variable x. Use the formula $b = \frac{A}{h}$, where b is the length of the base, A is the area, and h is the height of a parallelogram. $(2x^2 + x - 1)$ in.

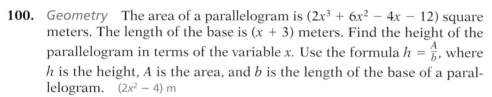
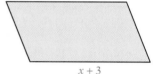

100. *Geometry*   The area of a parallelogram is $(2x^3 + 6x^2 - 4x - 12)$ square meters. The length of the base is $(x + 3)$ meters. Find the height of the parallelogram in terms of the variable x. Use the formula $h = \frac{A}{b}$, where h is the height, A is the area, and b is the length of the base of a parallelogram. $(2x^2 - 4)$ m

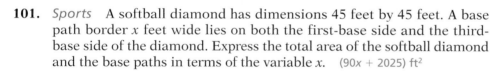
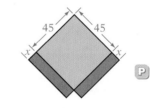

101. *Sports*   A softball diamond has dimensions 45 feet by 45 feet. A base path border x feet wide lies on both the first-base side and the third-base side of the diamond. Express the total area of the softball diamond and the base paths in terms of the variable x. $(90x + 2025)$ ft^2

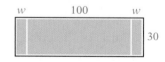

102. *Sports*   An athletic field has dimensions 30 yards by 100 yards. An end zone that is w yards wide borders each end of the field. Express the total area of the field and the end zones in terms of the variable w.
$(60w + 3000)$ yd^2

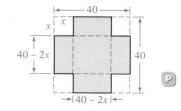

103. *Packaging*   An open box is made from a square piece of cardboard that measures 40 inches on each side. To construct the box, squares that measure x inches on a side are cut from each corner. Express the volume of the box in terms of x. What is the volume of the box when x is 3 inches? $(4x^3 - 160x^2 + 1600x)$ in^3; 3468 in^3

104. *Metallurgy*   A sheet of tin 50 centimeters wide and 200 centimeters long is made into a trough by bending up two sides, each of length x, until they are perpendicular to the bottom. Express the volume of the trough in terms of x. What is the volume of the trough when x is 10 centimeters? $(10{,}000x - 400x^2)$ cm^3; 60,000 cm^3

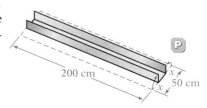

105. What polynomial has quotient $x^2 + 2x - 1$ when divided by $x + 3$?
$x^3 + 5x^2 + 5x - 3$

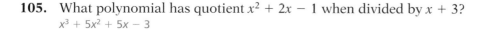

106. What polynomial has quotient $3x - 4$ when divided by $4x + 5$?
$12x^2 - x - 20$

107. Given that $x + 5$ is a factor of $x^3 + 12x^2 + 36x + 5$, find another factor
of $x^3 + 12x^2 + 36x + 5$. $x^2 + 7x + 1$

108. Given that $x - 2$ is a factor of $x^3 + 2x^2 - 9x + 2$, find another factor of
$x^3 + 2x^2 - 9x + 2$. $x^2 + 4x - 1$

109. Subtract $4x^2 - x - 5$ from the product of $x^2 + x + 3$ and $x - 4$.
$x^3 - 7x^2 - 7$

110. Add $x^2 + 2x - 3$ to the product of $2x - 5$ and $3x + 1$. $7x^2 - 11x - 8$

111. The quotient of a polynomial and $2x + 1$ is $2x - 4 + \dfrac{7}{2x + 1}$. Find the
polynomial. $4x^2 - 6x + 3$

112. The quotient of a polynomial and $x - 3$ is $x^2 - x + 8 + \dfrac{22}{x - 3}$. Find the
polynomial. $x^3 - 4x^2 + 11x - 2$

113. Let $f(x) = \dfrac{x^3 - 10x^2 + 33x - 36}{x^2 - 6x + 9}$, $x \neq 3$. Simplify the expression

$\dfrac{x^3 - 10x^2 + 33x - 36}{x^2 - 6x + 9}$ and then graph it using a graphing calculator. What

is the relationship between the quotient and the graph?

The quotient is $x - 4$, and the graph is $y = x - 4$, except when $x = 3$.

■ Synthetic Division

Divide by using synthetic division.

114. $(x^3 - 6x^2 + 11x - 6) \div (x - 3)$ $x^2 - 3x + 2$ **115.** $(x^3 - 4x^2 + x + 6) \div (x + 1)$ $x^2 - 5x + 6$

116. $(2x^3 - x^2 + 6x + 9) \div (x + 1)$ $2x^2 - 3x + 9$ **117.** $(3x^3 + 10x^2 + 6x - 4) \div (x + 2)$ $3x^2 + 4x - 2$

118. $(6x - 3x^2 + x^3 - 9) \div (x + 2)$ **119.** $(5 - 5x + 4x^2 + x^3) \div (x - 3)$

$x^2 - 5x + 16 - \dfrac{41}{x + 2}$ $x^2 + 7x + 16 + \dfrac{53}{x - 3}$

120. $(x^3 + x - 2) \div (x + 1)$ $x^2 - x + 2 - \dfrac{4}{x + 1}$ **121.** $(x^3 + 2x + 5) \div (x - 2)$ $x^2 + 2x + 6 + \dfrac{17}{x - 2}$

122. $(3x^2 - 4) \div (x - 1)$ $3x + 3 - \dfrac{1}{x - 1}$ **123.** $(4x^2 - 8) \div (x - 2)$ $4x + 8 + \dfrac{8}{x - 2}$

124. $\dfrac{16x^2 - 13x^3 + 2x^4 - 9x + 20}{x - 5}$ $2x^3 - 3x^2 + x - 4$ **125.** $\dfrac{3 - 13x - 5x^2 + 9x^3 - 2x^4}{x - 3}$ $-2x^3 + 3x^2 + 4x - 1$

126. $\dfrac{3x^4 + 3x^3 - x^2 + 3x + 2}{x + 1}$ $3x^3 - x + 4 - \dfrac{2}{x + 1}$ **127.** $\dfrac{4x^4 + 12x^3 - x^2 - x + 2}{x + 3}$ $4x^3 - x + 2 - \dfrac{4}{x + 3}$

128. $\dfrac{2x^4 - x^2 + 2}{x - 3}$ $2x^3 + 6x^2 + 17x + 51 + \dfrac{155}{x - 3}$ **129.** $\dfrac{x^4 - 3x^3 - 30}{x + 2}$ $x^3 - 5x^2 + 10x - 20 + \dfrac{10}{x + 2}$

130. *Geometry*   A rectangular box has a volume of $(x^3 + 11x^2 + 38x + 40)$
cubic inches. The height of the box is $(x + 2)$ inches. The length of
the box is $(x + 5)$ inches. Find the width of the box in terms of x.
$(x + 4)$ in.

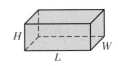

131. *Geometry* The volume of a right circular cylinder is $\pi(x^3 + 7x^2 + 15x + 9)$ cubic centimeters. The height of the cylinder is $(x + 1)$ centimeters. Find the area of the base of the cylinder in terms of x.
$\pi(x^2 + 6x + 9)$ cm^2

132. Three linear factors of $x^4 + x^3 - 7x^2 - x + 6$ are $x - 1$, $x - 2$, and $x + 3$. Find the other linear factor of $x^4 + x^3 - 7x^2 - x + 6$. $x + 1$

133. Three linear factors of $x^4 + 3x^3 - 8x^2 - 12x + 16$ are $x + 2$, $x - 1$, and $x + 4$. Find the other linear factor of $x^4 + 3x^3 - 8x^2 - 12x + 16$. $x - 2$

■ **The Remainder Theorem**

Use the Remainder Theorem to evaluate the polynomial.

134. $P(z) = 2z^3 - 4z^2 + 3z - 1; P(-2)$ -39

135. $R(t) = 3t^3 + t^2 - 4t + 2; R(-3)$ -58

136. $Q(x) = x^4 + 3x^3 - 2x^2 + 4x - 9; Q(2)$ 31

137. $Y(z) = z^4 - 2z^3 - 3z^2 - z + 7; Y(3)$ 4

138. $F(x) = 2x^4 - x^3 - 2x - 5; F(-3)$ 190

139. $Q(x) = x^4 - 2x^3 + 4x - 2; Q(-2)$ 22

140. $R(t) = 4t^4 - 3t^2 + 5; R(-3)$ 302

141. $P(z) = 2z^4 + z^2 - 3; P(-4)$ 525

142. $Q(x) = x^5 - 4x^3 - 2x^2 + 5x - 2; Q(2)$ 0

143. $T(x) = 2x^5 - 4x^4 - x^2 + 4; T(3)$ 157

144. Suppose you know that when $f(x)$ is divided by $x - 4$, the remainder is zero. What is $f(4)$? 0

145. Find the remainder when $P(x) = 37x^{50} - 3x^{35} + 2x^{17} - 21x^{10} + x^5 - 5$ is divided by $x + 1$. 11

146. What is the remainder when $x^{51} + 51$ is divided by $x + 1$? 50

■ **The Factor Theorem**

Use synthetic division and the Factor Theorem to determine whether the given polynomial is a factor of $P(x)$.

147. $P(x) = x^3 + 6x^2 - x - 30; x - 3$ No

148. $P(x) = -x^3 + 4x^2 - 5x + 2; x - 1$ Yes

149. $P(x) = 3x^3 + 7x^2 - 4x + 6; x + 3$ Yes

150. $P(x) = 2x^3 - 5x^2 + 3x - 8; x - 4$ No

151. $P(x) = 6x^4 + 15x^3 + 28x^2 + 3; x + 3$ No

152. $P(x) = 3x^4 - 25x^2 - 18; x - 3$ Yes

153. $P(x) = 2x^4 - 5x^3 - 3x^2 - x; x - 4$ No

154. $P(x) = x^5 - 4x^3 - 3x^2 + 2x + 16; x + 2$ Yes

Applying Concepts

155. Determine whether the statement is always true, sometimes true, or never true.

   a. To multiply two polynomials, multiply each term of one polynomial by the other polynomial. Always true

   b. The square of a binomial is a trinomial. Always true

156. Is it possible to multiply a polynomial of degree 2 by a polynomial of degree 2 and have the product be a polynomial of degree 3? If so, give an example. If not, explain why not.   No. Two polynomials of degree 2 will have terms ax^2 and bx^2, $a \neq 0$, $b \neq 0$. Multiplying these terms yields abx^4, where $ab \neq 0$. Therefore, the product will have an x^4 term and will be of degree 4.

157. *Packaging*   An open box is made from a square piece of cardboard that measures 20 inches on each side. To construct the box, squares that measure x inches on a side are cut from each corner. Express the volume of the box in terms of x. Can x be 10 inches? Explain your answer. $(4x^3 - 80x^2 + 400x)$ in³; No; Explanations will vary.

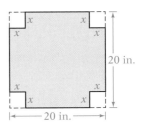

20 in.

20 in.

For what value of k will the remainder be zero?

158. $(x^3 - 3x^2 - x + k) \div (x - 3)$ 3

159. $(x^3 - 2x^2 + x + k) \div (x - 2)$ −2

160. $(x^2 + kx - 6) \div (x - 3)$ −1

161. $(x^3 + kx + k - 1) \div (x - 1)$ 0

162. When $x^2 + x + 2$ is divided by a polynomial, the quotient is $x + 4$, and the remainder is 14. Find the polynomial. $x - 3$

163. Find the value of t given that $x + 1$ is a factor of $3x^3 - 2x^2 + tx - 4$.   −9

164. When a polynomial $P(x)$ is divided by a polynomial $d(x)$, it produces a quotient $q(x)$ and remainder $r(x)$. This can be stated mathematically as

$$P(x) = d(x) \cdot q(x) + r(x)$$

Suppose $P(x) = x^2 + 5x + 8$, and $d(x) = x + 1$. Find possible polynomials for $q(x)$ and $r(x)$.   Answers will vary. For example, $q(x) = x + 4$ and $r(x) = 4$.

165. Find the ordered pair of numbers (a, b) for which $x - 3$ is a factor of both $x^2 - (a + b)x + 3b$ and $(a - 1)x^2 + bx + a$. $(3, -7)$

166. **a.** Is $x - 3$ a factor of $x^2 - x - 6$?   Yes
    b. For $f(x) = x^2 - x - 6$, find $f(3)$.   0
    c. If $f(x)$ is a polynomial that has $x - a$ as a factor, what is the value of $f(a)$?   0

167. A polynomial $P(x)$ has remainder 3 when divided by $x - 1$ and remainder 5 when divided by $x - 3$. Find the remainder when $P(x)$ is divided by $(x - 1)(x - 3)$. $x + 2$

EXPLORATION

1. *Patterns in Products of Polynomials*

    a. Multiply: $(x + 1)(x - 1)$
    b. Multiply: $(x + 1)(-x^2 + x - 1)$
    c. Multiply: $(x + 1)(x^3 - x^2 + x - 1)$
    d. Multiply: $(x + 1)(-x^4 + x^3 - x^2 + x - 1)$

    e. Use the pattern of the answers to parts a through d to multiply $(x + 1)(x^5 - x^4 + x^3 - x^2 + x - 1)$.

    f. Use the pattern of the answers to parts a through e to multiply $(x + 1)(-x^6 + x^5 - x^4 + x^3 - x^2 + x - 1)$.

a. $x^2 - 1$ b. $-x^3 - 1$ c. $x^4 - 1$ d. $-x^5 - 1$ e. $x^6 - 1$ f. $-x^7 - 1$

2. *Patterns in Quotients of Polynomials*

Part I

    a. Divide each polynomial given below by $x - y$.

$$x^3 - y^3 \qquad x^5 - y^5 \qquad x^7 - y^7 \qquad x^9 - y^9$$

    b. Explain the pattern, and use the pattern to write the quotient of $(x^{11} - y^{11}) \div (x - y)$.

Part II: Determine whether the second polynomial is a factor of the first.

c. $x^3 + 8; x + 2$	**d.** $x^3 - 8; x + 2$	**e.** $x^3 + 8; x - 2$	**f.** $x^3 - 8; x - 2$
g. $x^4 + 16; x + 2$	**h.** $x^4 - 16; x + 2$	**i.** $x^4 + 16; x - 2$	**j.** $x^4 - 16; x - 2$

Use your answers to parts c through j to determine whether the statement is true or false.

    k. For $n > 0$, $x - y$ is a factor of $(x^n - y^n)$.

    l. For $n > 0$ and n an even integer, $x + y$ is a factor of $(x^n - y^n)$.

    m. For $n > 0$ and n an odd integer, $x + y$ is a factor of $(x^n - y^n)$.

    n. For $n > 0$ and n an even integer, $x + y$ is a factor of $(x^n + y^n)$.

    o. For $n > 0$ and n an odd integer, $x + y$ is a factor of $(x^n + y^n)$.

a. $x^2 + xy + y^2$; $x^4 + x^3y + x^2y^2 + xy^3 + y^4$; $x^6 + x^5y + x^4y^2 + x^3y^3 + x^2y^4 + xy^5 + y^6$; $x^8 + x^7y + x^6y^2 + x^5y^3 + x^4y^4 + x^3y^5 + x^2y^6 + xy^7 + y^8$

b. The base of the first term is x, and the base of the last term is y. The exponent on each of these terms is one less than the exponent on that term of the dividend. Every other term has both variables. The exponents on x decrease by one in each successive term, while the exponents on y increase by one in each successive term. $x^{10} + x^9y + x^8y^2 + x^7y^3 + x^6y^4 + x^5y^5 + x^4y^6 + x^3y^7 + x^2y^8 + xy^9 + y^{10}$

c. Yes   d. No   e. No   f. Yes   g. No   h. Yes   i. No   j. Yes   k. True   l. True   m. False   n. False   o. True

3. *Diagramming the Square of a Binomial*

    a. Explain why the diagram at the right represents $(a + b)^2 = a^2 + 2ab + b^2$.

    b. Draw diagrams representing each of the following.

$$(x + 3)^2 = x^2 + 6x + 9$$
$$(y + 5)^2 = y^2 + 10y + 25$$
$$(x + y)^2 = x^2 + 2xy + y^2$$

See the Solutions Manual for the complete solution.

Factoring Polynomials

■ Common Monomial Factors

In the last section, we multiplied polynomials. For example, here the Distributive Property is used to multiply a monomial times a trinomial.

$$2x(x^2 + 3x - 5) = 2x^3 + 6x^2 - 10x$$

Here the FOIL method is used to multiply two binomials.

$$(x + 4)(x - 7) = x^2 - 7x + 4x - 28 = x^2 - 3x - 28$$

In this section we will write polynomials in factored form. A polynomial is in **factored form** when it is written as a product of other polynomials. It can be thought of as the reverse of multiplication. In the examples above, the factored form is on the left and the polynomial is on the right. In this section, we will be given polynomials to write in factored form.

Polynomial		**Factored Form**
$2x^3 + 6x^2 - 10x$	$=$	$2x(x^2 + 3x - 5)$
$x^2 - 3x - 28$	$=$	$(x + 4)(x - 7)$

Factoring enables us to write a more complicated polynomial as the product of simpler polynomials.

> **? QUESTION** Is the expression written in factored form? In other words, is it written as a product of polynomials?
>
> **a.** $a^3(4b + 9)$ **b.** $2y^2 - y + 1$ **c.** $(5c + 6)(c - 8)$

To factor out a common monomial from the terms of a polynomial, first find the greatest common factor (GCF) of the terms.

The GCF of two or more monomials is the product of the GCF of the coefficients and the common variable factors.

$$6x^3y = 2 \cdot 3 \cdot x \cdot x \cdot x \cdot y$$
$$8x^2y^2 = 2 \cdot 2 \cdot 2 \cdot x \cdot x \cdot y \cdot y$$
$$2 \cdot x \cdot x \cdot y = 2x^2y$$

Note that **the exponent of each variable in the GCF is the same as the *least* exponent of that variable in either of the monomials.**

The GCF of $6x^3y$ and $8x^2y^2$ is $2x^2y$.

> **TAKE NOTE**
>
> 2 is the GCF of 6 and 8 because 2 is the greatest integer that divides evenly into both 6 and 8.

> **? ANSWERS** **a.** Yes, this is in factored form. It is a monomial (a^3) times a binomial ($4b + 9$). **b.** No, this is not in factored form. **c.** Yes, this is in factored form. It is a binomial ($5c + 6$) times a binomial ($c - 8$).

INSTRUCTOR NOTE
Exercise 16 can be used for a similar in-class example.

TAKE NOTE

Division of a polynomial by a monomial is presented in Section 6.3.

TAKE NOTE

Note that to factor
$16x^4y^5 + 8x^4y^2 - 12x^3y$
as
$4x^3y(4xy^4 + 2xy - 3)$,
we are using the Distributive Property in reverse. Recall that the Distributive Property is

$a(b + c) = ab + ac$

The Distributive Property in reverse is

$ab + ac = a(b + c)$

EXAMPLE 1

Factor: $16x^4y^5 + 8x^4y^2 - 12x^3y$

Solution The GCF is $4x^3y$.

$$\frac{16x^4y^5 + 8x^4y^2 - 12x^3y}{4x^3y}$$

$$= 4xy^4 + 2xy - 3$$

$$16x^4y^5 + 8x^4y^2 - 12x^3y$$

$$= 4x^3y(4xy^4 + 2xy - 3)$$

$$4x^3y(4xy^4 + 2xy - 3)$$

$$= 16x^4y^5 + 8x^4y^2 - 12x^3y$$

- Find the GCF of the terms.

- Divide the polynomial by the GCF.

- Write the polynomial as a product of the GCF and the quotient found above.

- Check by using the Distributive Property.

YOU TRY IT 1

Factor: $6x^4y^2 - 9x^3y^2 + 12x^2y^4$

Solution See page S26. $3x^2y^2(2x^2 - 3x + 4y^2)$

■ Factor by Grouping

In the example below, the common monomial factor of a is factored out of the binomial.

$$2xa + 5ya = a(2x + 5y)$$

If we replace the a with $(a + b)$ in $2xa + 5ya$, the result is

$$2x(a + b) + 5y(a + b)$$

Now instead of a common monomial factor, there is a **common binomial factor.** The Distributive Property is used to factor a common binomial factor from an expression.

$$2x(a + b) + 5y(a + b) = (a + b)(2x + 5y)$$

INSTRUCTOR NOTE
For a similar in-class example, have students factor

$$y(x - 5) + 3(x - 5)$$

EXAMPLE 2

Factor: $y(x + 2) + 3(x + 2)$

Solution $y(x + 2) + 3(x + 2)$ • The common binomial factor is $x + 2$.

$$= (x + 2)(y + 3)$$

YOU TRY IT 2

Factor: $a(b - 7) + b(b - 7)$

Solution See page S26. $(b - 7)(a + b)$

Consider the following simplification of $-(a - b)$.

$$-(a - b) = -1(a - b) = -a + b = b - a$$

Thus,

$$\mathbf{-(a - b) = b - a}$$

This equivalence is sometimes used to factor a common binomial factor from an expression.

➡ Factor: $2x(x - y) + 5(y - x)$

$2x(x - y) + 5(y - x)$

$\quad = 2x(x - y) - 5(x - y)$ • $5(y - x) = 5(-1)(x - y) = -5(x - y)$

$\quad = (x - y)(2x - 5)$ • The common binomial factor is $x - y$. ⬅

? QUESTION   Which of the following expressions is equal to $(d - 3c)$?

    a. $(3c - d)$ **b.** $(-3c - d)$ **c.** $-(3c - d)$

Some polynomials can be factored by grouping the terms so that a common binomial factor is found. This is illustrated in the example that follows.

➡ Factor: $2x^3 - 3x^2 + 4x - 6$

$2x^3 - 3x^2 + 4x - 6$ • Group the first two terms and the last two terms. (Put them in parentheses.)

$\quad = (2x^3 - 3x^2) + (4x - 6)$

$\quad = x^2(2x - 3) + 2(2x - 3)$ • Factor out the GCF from each group.

$\quad = (2x - 3)(x^2 + 2)$ • Write the expression as a product of factors.

$(2x - 3)(x^2 + 2)$ • Check the factorization by multiplying the binomials. Use the FOIL method.

$\quad = 2x^3 + 4x - 3x^2 - 6$

$\quad = 2x^3 - 3x^2 + 4x - 6$ ⬅

Suggested Activity

Replace the ? to make a true statement.
1. $d - 6 = ?(6 - d)$
2. $3 - (a - b) = 3 + (?)$
3. $2x + (5a - b) = 2x - (?)$
[Answers: **1.** -1 **2.** $b - a$ **3.** $b - 5a$]

CALCULATOR NOTE

These factoring problems can also be checked using a graphing calculator. Here are screens used to check the factoring at the right.

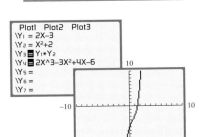

INSTRUCTOR NOTE
Exercise 22 can be used for a similar in-class example.

TAKE NOTE

By the Commutative Property of Addition, $3y^3 - 6y - 4y^2 + 8 = 3y^3 - 4y^2 - 6y + 8$. The factorization checks.

EXAMPLE 3

Factor: $3y^3 - 4y^2 - 6y + 8$

Solution $3y^3 - 4y^2 - 6y + 8$

$\quad = (3y^3 - 4y^2) - (6y - 8)$ • Group the first two terms and the last two terms. Note that $-6y + 8 = -(6y - 8)$.

$\quad = y^2(3y - 4) - 2(3y - 4)$ • Factor out the GCF from each group.

$\quad = (3y - 4)(y^2 - 2)$ • Write the expression as a product of two binomials.

$(3y - 4)(y^2 - 2)$ • Check by using the FOIL method.

$\quad = 3y^3 - 6y - 4y^2 + 8$

? ANSWER $(d - 3c)$ is equal to part c. Note that $-(3c - d) = -3c + d = d - 3c$.

INSTRUCTOR NOTE
Exercise 28 can be used for
a similar in-class example.

YOU TRY IT 3

Factor: $y^5 - 5y^3 + 4y^2 - 20$

Solution   See page S26. $(y^2 - 5)(y^3 + 4)$

EXAMPLE 4

Factor: $4ab - 6 + 3b - 2ab^2$

Solution $4ab - 6 + 3b - 2ab^2$

$= (4ab - 6) + (3b - 2ab^2)$ • Group the first two terms and
the last two terms.

$= 2(2ab - 3) + b(3 - 2ab)$ • Factor out the GCF from each
group.

$= 2(2ab - 3) - b(2ab - 3)$ • $b(3 - 2ab) = b(-1)(2ab - 3)$
$\qquad\qquad\qquad\qquad = -b(2ab - 3)$

$= (2ab - 3)(2 - b)$ • Write the expression as a product of two binomials.

$(2ab - 3)(2 - b)$ • Check by using the FOIL
method.

$= 4ab - 2ab^2 - 6 + 3b$

$= 4ab - 6 + 3b - 2ab^2$

YOU TRY IT 4

Factor: $3xy - 6y - 8 + 4x$

Solution   See page S26. $(x - 2)(3y + 4)$

Suggested Activity

Complete the table by finding
two integers whose product is
given in the column headed
ab and whose sum is given in
the column headed *a* + *b*.
Assume *a* ≤ *b*.

ab	a + b	a	b
100	20		
40	13		
−42	−11		
−72	−1		
75	−20		
44	−15		

[Answer: Column a: 10, 5,
−14, −9, −15, −11.
Column b: 10, 8, 3, 8, −5, −4.]

■ **Factor Trinomials of the Form $x^2 + bx + c$**

A **quadratic trinomial** is a trinomial of the form $ax^2 + bx + c$, where a and b are coefficients and c is a constant. The degree of a quadratic trinomial is 2.

Examples of quadratic trinomials follow.

$$x^2 + 9x + 14 \qquad a = 1, b = 9, c = 14$$
$$x^2 - 2x - 15 \qquad a = 1, b = -2, c = -15$$
$$3x^2 - x + 4 \qquad a = 3, b = -1, c = 4$$
$$4x^2 - 16 \qquad a = 4, b = 0, c = -16$$

❓ QUESTION   What are the values of a, b, and c in $4x^2 + 5x - 8$?

To **factor a quadratic trinomial** means to express the trinomial as the product of two binomials. For example,

Trinomial		**Factored Form**
$2x^2 - x - 1$	=	$(2x + 1)(x - 1)$
$y^2 - 3y + 2$	=	$(y - 1)(y - 2)$

We will begin by factoring trinomials of the form $x^2 + bx + c$, where $a = 1$.

❓ ANSWER   In $4x^2 + 5x - 8$, $a = 4$, $b = 5$, $c = -8$.

The method by which factors of a trinomial are found is based on FOIL. Consider the following binomial products, noting the relationship between the constant term of the binomials and the terms of the trinomial.

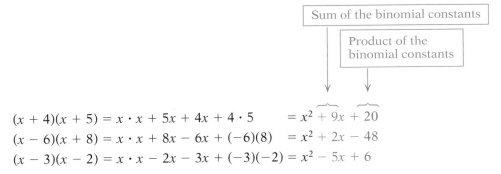

$$(x + 4)(x + 5) = x \cdot x + 5x + 4x + 4 \cdot 5 \quad = x^2 + 9x + 20$$
$$(x - 6)(x + 8) = x \cdot x + 8x - 6x + (-6)(8) \quad = x^2 + 2x - 48$$
$$(x - 3)(x - 2) = x \cdot x - 2x - 3x + (-3)(-2) = x^2 - 5x + 6$$

INSTRUCTOR NOTE
Students sometimes see factoring as unrelated to multiplication. Remind students that the relationships between the binomial factors and the terms of the trinomial are based on multiplying binomials.

> ### Important Relationships in Factoring Quadratic Trinomials
>
> 1. When the constant term of the trinomial is positive, the constant terms of the binomials have the same sign. They are both positive when the coefficient of the x term in the trinomial is positive. They are both negative when the coefficient of the x term in the trinomial is negative.
> 2. When the constant term of the trinomial is negative, the constant terms of the binomials have different signs.
> 3. In the trinomial, the coefficient of x is the sum of the constant terms of the binomials.
> 4. In the trinomial, the constant term is the product of the constant terms of the binomials.

➡ Factor: $x^2 - 7x + 10$

Because the constant term is positive ($+10$) and the coefficient of x is negative (-7), both of the binomial constants will be negative. Find two negative factors of 10 whose sum is -7. The results can be recorded in a table.

Negative Factors of 10	Sum
$-1, -10$	-11
$-2, \; -5$	**-7**

• These are the correct factors.

TAKE NOTE
Always check your proposed factorization to ensure accuracy.

$$x^2 - 7x + 10 = (x - 2)(x - 5)$$

• Write the trinomial as a product of the factors.

Check the proposed factorization by multiplying the two binomials.

Check: $(x - 2)(x - 5) = x^2 - 5x - 2x + 10$
$$= x^2 - 7x + 10$$

⬅

➡ Factor: $x^2 - 9x - 36$

The constant term is negative (-36). The binomial constants will have different signs. Find two factors of -36 whose sum is -9.

Factors of -36	Sum
$+1, -36$	-35
$-1, 36$	35
$+2, -18$	-16
$-2, 18$	16
$+3, -12$	**-9**

• These are the correct factors. Once the correct factors are found, it is not necessary to try the remaining factors.

$x^2 - 9x - 36 = (x + 3)(x - 12)$ • Write the trinomial as a product of the factors.

Check: $(x + 3)(x - 12) = x^2 - 12x + 3x - 36$
$$= x^2 - 9x - 36$$

INSTRUCTOR NOTE
The phrase *nonfactorable over the integers* may require additional examples. Explain that it does not mean the polynomial does not factor. It just does not factor if integers are used. An analogy to numbers may help. For instance, the only ways to write 7 as a product *involving integers* are $1 \cdot 7$ or $(-1)(-7)$. However, $\frac{21}{5} \cdot \frac{5}{3} = 7$. The ability to factor depends on the numbers that can be used.

➡ Factor: $x^2 + 7x + 8$

Because the constant term is positive ($+8$) and the coefficient of x is positive ($+7$), both of the binomial constants will be positive. Find two positive factors of 8 whose sum is 7.

Positive Factors of 8	Sum
1, 8	9
2, 4	6

There are no positive integer factors of 8 whose sum is 7. The trinomial $x^2 + 7x + 8$ is said to be **nonfactorable over the integers.**

Just as 17 is a prime number, $x^2 + 7x + 8$ is a **prime polynomial.** Binomials of the forms $x - a$ and $x + a$ are also prime polynomials.

INSTRUCTOR NOTE
Exercise 34 can be used for a similar in-class example.

EXAMPLE 5

Factor. **a.** $x^2 - 8x + 15$ **b.** $x^2 + 6x - 27$

Solution   **a.** Two negative factors of 15 whose sum is -8 are -3 and -5.

$$x^2 - 8x + 15 = (x - 3)(x - 5)$$

Check: $(x - 3)(x - 5) = x^2 - 5x - 3x + 15 = x^2 - 8x + 15$

b. Two factors of -27 whose sum is 6 are -3 and 9.

$$x^2 + 6x - 27 = (x - 3)(x + 9)$$

Check: $(x - 3)(x + 9) = x^2 + 9x - 3x - 27 = x^2 + 6x - 27$

CALCULATOR NOTE

These factoring problems can also be checked using a graphing calculator. Here are screens used to check Example 5a.

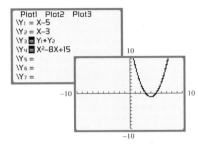

INSTRUCTOR NOTE

Factoring by using trial factors is presented here. This presentation is followed by a discussion of factoring by grouping. You may either skip one of these methods or do both.

YOU TRY IT 5

Factor: **a.** $x^2 + 9x + 20$ **b.** $x^2 + 7x - 18$

Solution   See page S26.   **a.** $(x + 4)(x + 5)$ **b.** $(x + 9)(x - 2)$

Factor Trinomials of the Form $ax^2 + bx + c$

There are various methods of factoring trinomials of the form $ax^2 + bx + c$, where $a \neq 1$. We will discuss both factoring by using trial factors and factoring by grouping. Factoring by using trial factors is illustrated first.

To use the trial factor method, use the factors of a and the factors of c to write all of the possible binomial factors of the trinomial. Then use FOIL to determine the correct factorization. To reduce the number of trial factors that must be considered, remember the following guidelines.

1. Use the signs of the constant term and the coefficient of x in the trinomial to determine the signs of the binomial factors. If the constant term is positive, the signs of the binomial factors will be the same as the sign of the coefficient of x in the trinomial. If the sign of the constant term is negative, the constant terms in the binomials will have different signs.
2. If the terms of the trinomial do not have a common factor, then the terms in either one of the binomial factors will not have a common factor.

➡ Factor by using trial factors: $2x^2 - 7x + 3$

Because the constant term is positive ($+3$) and the coefficient of x is negative (-7), the binomial constants will be negative. Use negative factors of 3. The results can be recorded in a table.

Positive Factors of 2 ($a = 2$)	Negative Factors of 3 ($c = 3$)
1, 2	−1, −3

Write trial factors. Use the **O**uter and **I**nner products of FOIL to determine the middle term, $-7x$, of the trinomial.

Trial Factors	Middle Term	
$(x - 1)(2x - 3)$	$-3x - 2x = -5x$	
$(x - 3)(2x - 1)$	$-x - 6x = -7x$	• $-7x$ is the middle term.

Write the factors of the trinomial.

$$2x^2 - 7x + 3 = (x - 3)(2x - 1)$$

Check: $(x - 3)(2x - 1) = 2x^2 - x - 6x + 3 = 2x^2 - 7x + 3$ ⬅

Suggested Activity

See Section 6.4 of the *Student Activity Manual* for an activity involving factoring trinomials of the form $ax^2 + bx + c$.

➡ Factor by using trial factors: $5x^2 + 22x - 15$

The constant term is negative (-15). The binomial constants will have different signs.

Positive Factors of 5 ($a = 5$)	Factors of -15 ($c = -15$)
1, 5	$-1, 15$
	$1, -15$
	$-3, 5$
	$3, -5$

Write trial factors. Use the **O**uter and **I**nner products of FOIL to determine the middle term, $22x$, of the trinomial. Note that **it is not necessary to test trial factors that have a common factor; if the trinomial does not have a common factor, then its factors cannot have a common factor.**

Trial Factors	Middle Term	
$(x - 1)(5x + 15)$	common factor of 5 in second binomial	
$(x + 15)(5x - 1)$	$-x + 75x = 74x$	
$(x + 1)(5x - 15)$	common factor of 5 in second binomial	
$(x - 15)(5x + 1)$	$x - 75x = -74x$	
$(x - 3)(5x + 5)$	common factor of 5 in second binomial	
$(x + 5)(5x - 3)$	$-3x + 25x = 22x$	• $22x$ is the middle term.
$(x + 3)(5x - 5)$	common factor of 5 in second binomial	
$(x - 5)(5x + 3)$	$3x - 25x = -22x$	

Write the factors of the trinomial.

$$5x^2 + 22x - 15 = (x + 5)(5x - 3)$$

Check: $(x + 5)(5x - 3) = 5x^2 - 3x + 25x - 15 = 5x^2 + 22x - 15$ ⬅

In the previous example, all the trial factors were listed. Once the correct factors have been found, however, the remaining trial factors need not be checked.

INSTRUCTOR NOTE
Exercise 52 can be used for a similar in-class example.

EXAMPLE 6

Factor by using trial factors: $3x^2 + x - 2$

Solution

Positive factors of 3	Factors of -2
1, 3	$1, -2$
	$-1, 2$

CALCULATOR NOTE

Remember that these factoring problems can be checked using a graphing calculator. Here are screens used to check Example 6.

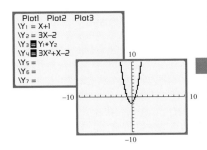

Trial Factors	Middle Term	
$(x + 1)(3x - 2)$	$-2x + 3x = x$	• x is the middle term.
$(x - 2)(3x + 1)$	$x - 6x = -5x$	
$(x - 1)(3x + 2)$	$2x - 3x = -x$	
$(x + 2)(3x - 1)$	$-x + 6x = 5x$	

$3x^2 + x - 2 = (x + 1)(3x - 2)$

Check: $(x + 1)(3x - 2) = 3x^2 - 2x + 3x - 2 = 3x^2 + x - 2$

YOU TRY IT 6

Factor by using trial factors: $2x^2 - x - 3$

Solution   See page S26. $(x + 1)(2x - 3)$

We have described factoring trinomials of the form $ax^2 + bx + c$ by using trial factors. These trinomials can also be factored by grouping.

To factor $ax^2 + bx + c$ by grouping, first find two factors of the product $a \cdot c$ whose sum is b. For instance, for the trinomial $3x^2 + 11x + 8$, $a = 3$, $b = 11$, and $c = 8$. The product $a \cdot c$ is $3(8) = 24$. Find two factors of 24 whose sum is 11. The factorization of this trinomial follows.

INSTRUCTOR NOTE

Show students that $3x^2 + 11x + 8$ can also be written as $3x^2 + 8x + 3x + 8$ and then factored by grouping. The result is the same.

$3x^2 + 8x + 3x + 8$
$= (3x^2 + 8x) + (3x + 8)$
$= x(3x + 8) + 1(3x + 8)$
$= (3x + 8)(x + 1)$

➡ Factor by grouping: $3x^2 + 11x + 8$

The product $a \cdot c = 3(8) = 24$.

Find two positive factors of 24 whose sum is 11 ($b = 11$).

Positive Factors of 24	Sum	
1, 24	25	
2, 12	14	
3, 8	**11**	• $11x$ is the middle term.
4, 6	10	

TAKE NOTE

$3x^2 + 3x + 8x + 8 =$
$3x^2 + 11x + 8$
We have not changed the original polynomial.

Use the factors 3 and 8 to write $11x$ as $3x + 8x$. Then factor by grouping.

$$3x^2 + 11x + 8 = 3x^2 + 3x + 8x + 8$$
$$= (3x^2 + 3x) + (8x + 8)$$
$$= 3x(x + 1) + 8(x + 1)$$
$$= (x + 1)(3x + 8)$$

Check: $(x + 1)(3x + 8) = 3x^2 + 8x + 3x + 8 = 3x^2 + 11x + 8$ ⬅

INSTRUCTOR NOTE
Exercise 54 can be used for a similar in-class example.

INSTRUCTOR NOTE
You might ask students to factor $3x^2 - 2x - 4$ by grouping. For this trinomial, $a \cdot c = 3(-4) = -12$. There are no integer factors of -12 that have a sum of -2. The trinomial is nonfactorable over the integers.

TAKE NOTE
First of all,
$2x^2 - x + 20x - 10 = 2x^2 + 19x - 10$.
We have not changed the original polynomial.
Second, if you write $19x$ as $20x - x$ rather than as $-x + 20$, the result is the same.
Factor
$2x^2 + 20x - x - 10$
by grouping to show the factors are the same.

EXAMPLE 7

Factor by grouping: $2x^2 + 19x - 10$

Solution   The product $a \cdot c = 2(-10) = -20$.

Find two factors of -20 whose sum is 19 ($b = 19$).

Factors of -20	Sum
-1, 20	**19**
1, -20	-19
-2, 10	8
2, -10	-8
-4, 5	1
4, -5	-1

• $19x$ is the middle term.

Use the factors -1 and 20 to write $19x$ as $-x + 20x$. Then factor by grouping.

$$2x^2 + 19x - 10 = 2x^2 - x + 20x - 10$$
$$= (2x^2 - x) + (20x - 10)$$
$$= x(2x - 1) + 10(2x - 1)$$
$$= (2x - 1)(x + 10)$$

Check: $(2x - 1)(x + 10) = 2x^2 + 20x - x - 10 = 2x^2 + 19x - 10$

YOU TRY IT 7

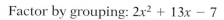

Factor by grouping: $2x^2 + 13x - 7$

Solution   See page S26. $(2x - 1)(x + 7)$

6.4 EXERCISES   Suggested Assignment: 7–75, odds

Topics for Discussion

1. Explain the meaning of "a factor" and the meaning of "to factor."
    A factor is a number or expression in a multiplication. To factor means to write a polynomial as a product of other polynomials.

2. Determine whether the statement is always true, sometimes true, or never true.

    a. To factor a polynomial means to write it as a product of factors.
        Always true
    b. A binomial is factorable.   Sometimes true
    c. The terms of a trinomial are monomials.   Always true
    d. A binomial is a polynomial of degree 2.   Sometimes true
    e. The value of the coefficient of x in the trinomial $x^2 - 3x + 5$ is 3.
        Never true

3. Explain why the statement is true.
   a. The terms of the binomial $3x - 9$ have a common factor.
      The common factor is 3.
   b. The expression $3x^2 + 15$ is not in factored form.
      $3x^2 + 15$ is a sum; it is not a product. It can be factored as $3(x^2 + 5)$.
   c. $2x - 1$ is a factor of $x(2x - 1)$.
      In $x(2x - 1)$, x is multiplied times $2x - 1$. Both x and $2x - 1$ are factors.
   d. The trinomial $x^2 + 3x + 5$ is a prime polynomial.
      It cannot be expressed as the product of two binomials.
   e. The factored form of $2x^2 - 7x - 15$ is $(2x + 3)(x - 5)$.
      $(2x + 3)(x - 5)$ is a product, both $2x + 3$ and $x - 5$ are nonfactorable, and
      when the two binomials are multiplied, the result is the trinomial.
4. What does it mean to factor a trinomial of the form $ax^2 + bx + c$?
   It means to express the trinomial as the product of other polynomials.
5. When factoring a trinomial of the form $ax^2 + bx + c$ by using trial factors, how do we determine the signs of the last terms of the two binomial factors?   When the constant term of the trinomial is positive, the constant terms of the binomials have the same sign. When the constant term of the trinomial is negative, the constant terms of the binomials have different signs.

■ Factoring Polynomials

Factor.

6. $8x + 12$ $4(2x + 3)$

7. $12y^2 - 5y$ $y(12y - 5)$

8. $10x^4 - 12x^2$ $2x^2(5x^2 - 6)$

9. $10x^2yz^2 + 15xy^3z$ $5xyz(2xz + 3y^2)$

10. $x^3 - 3x^2 - x$ $x(x^2 - 3x - 1)$

11. $5x^2 - 15x + 35$ $5(x^2 - 3x + 7)$

12. $3x^3 + 6x^2 + 9x$ $3x(x^2 + 2x + 3)$

13. $3y^4 - 9y^3 - 6y^2$ $3y^2(y^2 - 3y - 2)$

14. $x^3y - 3x^2y^2 + 7xy^3$ $xy(x^2 - 3xy + 7y^2)$

15. $x^4y^4 - 3x^3y^3 + 6x^2y^2$ $x^2y^2(x^2y^2 - 3xy + 6)$

16. $4x^5y^5 - 8x^4y^4 + x^3y^3$ $x^3y^3(4x^2y^2 - 8xy + 1)$

17. $16x^2y - 8x^3y^4 - 48x^2y^2$ $8x^2y(2 - xy^3 - 6y)$

18. $x^3 + 4x^2 + 3x + 12$ $(x + 4)(x^2 + 3)$

19. $x^3 - 4x^2 - 3x + 12$ $(x - 4)(x^2 - 3)$

20. $2y^3 + 4y^2 + 3y + 6$ $(y + 2)(2y^2 + 3)$

21. $3y^3 - 12y^2 + y - 4$ $(y - 4)(3y^2 + 1)$

22. $ab + 3b - 2a - 6$ $(a + 3)(b - 2)$

23. $yz + 6z - 3y - 18$ $(y + 6)(z - 3)$

24. $x^2a - 2x^2 - 3a + 6$ $(a - 2)(x^2 - 3)$

25. $x^2y + 4x^2 + 3y + 12$ $(y + 4)(x^2 + 3)$

26. $x^2 - 3x + 4ax - 12a$ $(x - 3)(x + 4a)$

27. $t^2 + 4t - st - 4s$ $(t + 4)(t - s)$

28. $10xy^2 - 15xy + 6y - 9$ $(2y - 3)(5xy + 3)$

29. $10a^2b - 15ab - 4a + 6$ $(2a - 3)(5ab - 2)$

30. $x^2 + 5x + 6$ $(x + 2)(x + 3)$

31. $x^2 + x - 2$ $(x - 1)(x + 2)$

32. $x^2 + x - 6$ $(x + 3)(x - 2)$

33. $a^2 + a - 12$ $(a + 4)(a - 3)$

34. $a^2 - 2a - 35$ $(a + 5)(a - 7)$

35. $a^2 - 3a + 2$ $(a - 1)(a - 2)$

36. $a^2 - 5a + 4$ $(a - 1)(a - 4)$ **37.** $b^2 + 7b - 8$ $(b + 8)(b - 1)$ **38.** $y^2 + 6y - 55$ $(y + 11)(y - 5)$

39. $z^2 - 4z - 45$ $(z + 5)(z - 9)$ **40.** $y^2 - 8y + 15$ $(y - 3)(y - 5)$ **41.** $z^2 - 14z + 45$ $(z - 5)(z - 9)$

42. $p^2 + 12p + 27$ **43.** $b^2 + 9b + 20$ **44.** $y^2 - 8y + 32$
$(p + 3)(p + 9)$ $(b + 4)(b + 5)$ Nonfactorable over the integers
45. $y^2 - 9y + 81$ **46.** $p^2 + 24p + 63$ **47.** $x^2 - 15x + 56$
Nonfactorable over the integers $(p + 3)(p + 21)$ $(x - 7)(x - 8)$
48. $5x^2 + 6x + 1$ $(x + 1)(5x + 1)$ **49.** $2y^2 + 7y + 3$ $(y + 3)(2y + 1)$ **50.** $2a^2 - 3a + 1$ $(a - 1)(2a - 1)$

51. $3a^2 - 4a + 1$ $(a - 1)(3a - 1)$ **52.** $4x^2 - 3x - 1$ $(x - 1)(4x + 1)$ **53.** $2x^2 - 5x - 3$ $(x - 3)(2x + 1)$

54. $6t^2 - 11t + 4$ $(2t - 1)(3t - 4)$ **55.** $10t^2 + 11t + 3$ $(2t + 1)(5t + 3)$ **56.** $8x^2 + 33x + 4$ $(x + 4)(8x + 1)$

57. $10z^2 + 3z - 4$ $(2z - 1)(5z + 4)$ **58.** $3x^2 + 14x - 5$ $(x + 5)(3x - 1)$ **59.** $3z^2 + 95z + 10$
  Nonfactorable over the integers
60. $8z^2 - 36z + 1$ **61.** $2t^2 - t - 10$ $(t + 2)(2t - 5)$ **62.** $2t^2 + 5t - 12$ $(t + 4)(2t - 3)$
Nonfactorable over the integers
63. $12y^2 + 19y + 5$ $(3y + 1)(4y + 5)$ **64.** $5y^2 - 22y + 8$ $(y - 4)(5y - 2)$ **65.** $11a^2 - 54a - 5$ $(a - 5)(11a + 1)$

66. $4z^2 + 11z + 6$ $(z + 2)(4z + 3)$ **67.** $6b^2 - 13b + 6$ $(2b - 3)(3b - 2)$ **68.** $6x^2 + 35x - 6$ $(x + 6)(6x - 1)$

Geometry   Write an expression in factored form for the shaded portion of the diagram.

69.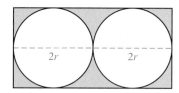

70.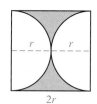

71.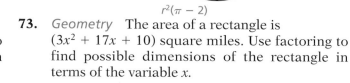

$2r^2(4 - \pi)$ $r^2(4 - \pi)$ $r^2(\pi - 2)$

72. *Geometry*   The area of a rectangle is $(2x^2 + 9x + 4)$ square inches. Use factoring to find possible dimensions of the rectangle in terms of the variable x.

$A = 2x^2 + 9x + 4$

$(2x + 1)$ in. by $(x + 4)$ in.

73. *Geometry*   The area of a rectangle is $(3x^2 + 17x + 10)$ square miles. Use factoring to find possible dimensions of the rectangle in terms of the variable x.

$A = 3x^2 + 17x + 10$

$(3x + 2)$ mi by $(x + 5)$ mi

74. *Geometry*   The area of a parallelogram is $(30x^2 + 23x + 3)$ square yards. Use factoring to find possible dimensions of the parallelogram in terms of the variable x.

$A = 30x^2 + 23x + 3$

$(6x + 1)$ yd by $(5x + 3)$ yd

75. *Geometry*   The area of a parallelogram is $(4x^2 + 23x + 15)$ square feet. Use factoring to find possible dimensions of the parallelogram in terms of the variable x.

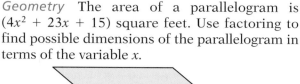

$A = 4x^2 + 23x + 15$

$(4x + 3)$ ft by $(x + 5)$ ft

Applying Concepts

76. Match equivalent expressions.

   a. $5y(x - 3) - 2(x - 3)$ **i.** $5y(x - 3) + 2(x - 3)$
   b. $5y(x - 3) + 2(3 - x)$ **ii.** $5y(x - 3) - 2(3 + x)$
   c. $5y(x - 3) - 2(3 - x)$ **iii.** $5y(x - 3) - 2(x - 3)$
   d. $5y(x - 3) + 2(x - 3)$ **iv.** $5y(x - 3) + 2(x + 3)$
   e. $5y(x - 3) + 2(3 + x)$ **v.** $5xy - 15y - 2x + 6$
   f. $5y(x - 3) - 2(x + 3)$ **vi.** $5xy - 15y + 2x - 6$

   a matches iii and v; b matches iii and v; c matches i and vi; d matches i and vi; e matches iv; f matches ii

77. In the expression $P = 2L + 2W$, what is the effect on P when the quantity $L + W$ doubles?   *P doubles*

78. Find all integers k such that the trinomial can be factored over the integers.

   a. $x^2 + kx + 35$ −36, 36, −12, 12
   b. $x^2 + kx + 18$ −19, 19, −11, 11, −9, 9
   c. $x^2 - kx + 21$ −22, 22, −10, 10
   d. $x^2 - kx + 14$ −15, 15, −9, 9

79. Find all integers k such that the trinomial can be factored over the integers.

   a. $2x^2 + kx + 3$ −7, 7, −5, 5
   b. $2x^2 + kx - 3$ −5, 5, −1, 1
   c. $3x^2 + kx + 2$ −7, 7, −5, 5
   d. $3x^2 + kx - 2$ −5, 5, −1, 1
   e. $2x^2 + kx + 5$ −11, 11, −7, 7
   f. $2x^2 + kx - 5$ −9, 9, −3, 3

80. Determine the positive integer values of k for which the following polynomials are factorable over the integers.

   a. $y^2 + 4y + k$ 3, 4
   b. $z^2 + 7z + k$ 6, 10, 12
   c. $a^2 - 6a + k$ 5, 8, 9
   d. $c^2 - 7c + k$ 6, 10, 12
   e. $x^2 - 3x + k$ 2
   f. $y^2 + 5y + k$ 4, 6

81. Exercise 80 included the requirement that $k > 0$. If k is allowed to be any integer, how many different values of k are possible for each polynomial? Explain your answer.

   An infinite number of different values of k are possible. Explanations will vary.

82. Given that $x + 2$ is a factor of $x^3 - 2x^2 - 5x + 6$, factor $x^3 - 2x^2 - 5x + 6$ completely. $(x + 2)(x - 3)(x - 1)$

83. Can a third-degree polynomial have factors $x - 1$, $x + 1$, $x - 3$, and $x + 4$? Why or why not?

   No. The leading term of the product of these four factors will be x^4, not x^3.

84. *Geometry* The area of a rectangle is $(3x^2 + x - 2)$ square feet. Find the dimensions of the rectangle in terms of the variable x. Given that $x > 0$, specify the dimension that is the length and the dimension that is the width. Can $x < 0$? Can $x = 0$?

$A = 3x^2 + x - 2$

The dimensions are $3x - 2$ and $x + 1$. If $x = 1.5$, then the rectangle is a square. If $x < 1.5$, then the length is $(x + 1)$ ft and the width is $(3x - 2)$ ft. If $x > 1.5$, then the width is $(x + 1)$ ft and the length is $(3x - 2)$ ft. If $x < 0$, then $3x^2 + x - 2 < 0$, which is not possible. Therefore, x cannot be less than 0. If $x = 0$, then $3x - 2$ is negative, which is not possible. Therefore, x cannot be equal to 0.

EXPLORATION

1. *The Grouping of Terms in Factoring by Grouping*

   a. Factor $2x^2 + 6x + 5x + 15$ and $2x^2 + 5x + 6x + 15$.
   b. Factor $3x^2 + 3xy - xy - y^2$ and $3x^2 - xy + 3xy - y^2$.
   c. Factor $2a^2 - 2ab - 3ab + 3b^2$ and $2a^2 - 3ab - 2ab + 3b^2$.
   d. Compare your answers in part a, in part b, and in part c. Do different groupings of the terms in a polynomial affect the binomial factoring?

   a. $(x + 3)(2x + 5)$; $(2x + 5)(x + 3)$ **b.** $(x + y)(3x - y)$; $(3x - y)(x + y)$ **c.** $(a - b)(2a - 3b)$; $(2a - 3b)(a - b)$ **d.** No

---

SECTION 6.5 **Special Factoring**

- Factor the Difference of Two Perfect Squares or a Perfect-Square Trinomial
- Factor the Sum or the Difference of Two Perfect Cubes
- Factor a Trinomial That Is Quadratic in Form
- Factor Completely

■ Factor the Difference of Two Perfect Squares or a Perfect-Square Trinomial

The product of a term and itself is called a **perfect square.** The exponents on variables of perfect squares are always even numbers.

Term		Perfect Square
5	$5 \cdot 5 =$	25
x	$x \cdot x =$	x^2
$3y^4$	$3y^4 \cdot 3y^4$	$9y^8$
x^n	$x^n \cdot x^n$	x^{2n}

The **square root** of a perfect square is one of the two equal factors of the perfect square. The symbol $\sqrt{}$ is the symbol for square root. Note that we are taking the square roots of the perfect squares listed above.

$$\sqrt{25} = 5$$
$$\sqrt{x^2} = x$$
$$\sqrt{9y^8} = 3y^4$$
$$\sqrt{x^{2n}} = x^n$$

Note that we can find the exponent of the square root of a variable term by dividing the exponent by 2.

? QUESTION   Which of the following are perfect squares? If the expression is a perfect square, find its square root.*

a. 36   **b.** x^{10} **c.** $100y^6$

The **difference of two perfect squares,** $a^2 - b^2$, is the product of the sum and difference of two terms, $(a + b)(a - b)$. The two terms, a and b, are the square roots of the perfect squares, a^2 and b^2.

TAKE NOTE

Recall that the product $(a + b)(a - b)$ can be found using the FOIL method.
$(a + b)(a - b)$
$= a^2 - ab + ab - b^2$
$= a^2 - b^2$

Factors of the Difference of Two Perfect Squares
$$a^2 - b^2 = (a + b)(a - b)$$

? QUESTION $(3x + 4)(3x - 4)$ is the product of the sum and difference of two terms. What are the two terms? Which factor is the sum of the two terms? Which factor is the difference of the two terms?†

The **sum of two perfect squares,** $a^2 + b^2$, is nonfactorable over the integers.

➡ Factor: $4x^2 - 81y^2$

$4x^2 - 81y^2 = (2x)^2 - (9y)^2$ • $4x^2 - 81y^2$ is the difference of two perfect squares.
$= (2x + 9y)(2x - 9y)$ • The factors are the sum $(2x + 9y)$ and difference $(2x - 9y)$ of the square roots of the perfect squares $4x^2$ and $81y^2$.

Check:
$(2x + 9y)(2x - 9y) = 4x^2 - 18xy + 18xy - 81y^2 = 4x^2 - 81y^2$ ⬅

INSTRUCTOR NOTE
Exercise 14 can be used for a similar in-class example.

EXAMPLE 1

Factor: $25x^2 - 1$

Solution $25x^2 - 1 = (5x)^2 - (1)^2$ • This is the difference of two perfect squares.
$= (5x + 1)(5x - 1)$

Check: $(5x + 1)(5x - 1) = 25x^2 - 5x + 5x - 1 = 25x^2 - 1$

Suggested Activity

See Section 6.5 of the *Student Activity Manual*, for an activity involving factoring the difference of two perfect squares and perfect-square trinomials.

YOU TRY IT 1

Factor: $x^2 - 36y^4$

Solution   See page S26. $(x + 6y^2)(x - 6y^2)$

? ANSWERS   ***a.** 36 is a perfect square because $6^2 = 36$. $\sqrt{36} = 6$.   **b.** x^{10} is a perfect square because the exponent on x is an even number. $\sqrt{x^{10}} = x^5$.   **c.** $100y^6$ is a perfect square because $10^2 = 100$ and the exponent on y is an even number, $\sqrt{100y^6} = 10y^3$.
†The two terms are $3x$ and 4. $(3x + 4)$ is the sum of the two terms. $(3x - 4)$ is the difference of the two terms.

TAKE NOTE

Recall that the square of a binomial can be found using the FOIL method.

$(a + b)^2$
  $= (a + b)(a + b)$
  $= a^2 + ab + ab + b^2$
  $= a^2 + 2ab + b^2$
$(a - b)^2$
  $= (a - b)(a - b)$
  $= a^2 - ab - ab + b^2$
  $= a^2 - 2ab + b^2$

$a^2 + 2ab + b^2$ and $a^2 - 2ab + b^2$ are called perfect-square trinomials because each is a trinomial that is the result of squaring a binomial.

A **perfect-square trinomial** is the square of a binomial.

Factors of a Perfect-Square Trinomial
$$a^2 + 2ab + b^2 = (a + b)^2$$
$$a^2 - 2ab + b^2 = (a - b)^2$$

? QUESTION   Which of the following is the square of a binomial?

a. $(3x - 2)$ **b.** $(4y + 7)$ **c.** $(x + 5)^2$ **d.** $(x + y + 1)^2$

In factoring a perfect-square trinomial, remember that the terms of the binomial are the square roots of the perfect squares of the trinomial. The sign in the binomial is the sign of the middle term of the trinomial.

➡ Factor: $4x^2 + 12x + 9$

$4x^2$ is a perfect square $[4x^2 = (2x)^2]$, and 9 is a perfect square $[9 = 3^2]$.

Try factoring $4x^2 + 12x + 9$ as the square of a binomial.

$$4x^2 + 12x + 9 = (2x + 3)^2$$

Check: $(2x + 3)^2 = (2x + 3)(2x + 3) = 4x^2 + 6x + 6x + 9 = 4x^2 + 12x + 9$

The check verifies that $4x^2 + 12x + 9 = (2x + 3)^2$.   ⬅

TAKE NOTE

The square roots of the perfect squares $4x^2$ and 9 are $2x$ and 3. $2x$ and 3 are the terms of the binomial factor. The sign of the middle term in the trinomial $4x^2 + 12x + 9$ is +. The sign in the binomial is +.

EXAMPLE 2

Factor: $9x^2 + 12x + 4$

Solution $9x^2$ is a perfect square $[9x^2 = (3x)^2]$, and 4 is a perfect square $[4 = 2^2]$. Try factoring $9x^2 + 12x + 4$ as the square of a binomial.

$$9x^2 + 12x + 4 = (3x + 2)^2$$

Check: $(3x + 2)^2 = (3x + 2)(3x + 2)$
$$= 9x^2 + 6x + 6x + 4 = 9x^2 + 12x + 4$$

The check verifies that $9x^2 + 12x + 4 = (3x + 2)^2$.

INSTRUCTOR NOTE
Exercise 24 can be used for a similar in-class example.

YOU TRY IT 2

Factor: $4x^2 - 20x + 25$

Solution   See page S26. $(2x - 5)^2$

Suggested Activity

To emphasize the importance of checking a proposed factorization, have students factor $x^2 + 13x + 36$. Although the first and last terms are perfect squares, it is not a perfect-square trinomial. It factors as $(x + 4)(x + 9)$. Another example of what appears to be a perfect-square trinomial, but does not factor, is $4x^2 - 17x + 25$.

? ANSWER   **a.** $(3x - 2)$ is a binomial, but it is not squared.   **b.** $(4y + 7)$ is a binomial, but it is not squared.   **c.** $(x + 5)^2$ is the square of a binomial.   **d.** $(x + y + 1)^2$ is a trinomial squared.

■ Factor the Sum or the Difference of Two Perfect Cubes

The product of the same three factors is called a **perfect cube.** The exponents on variables of perfect cubes are always divisible by 3.

Term		Perfect Cube
2	$2 \cdot 2 \cdot 2 =$	8
$3y$	$3y \cdot 3y \cdot 3y =$	$27y^3$
x^2	$x^2 \cdot x^2 \cdot x^2 =$	x^6

The **cube root** of a perfect cube is one of the three equal factors of the perfect cube. The symbol $\sqrt[3]{}$ is the symbol for cube root. Note that we are taking the cube roots of the perfect cubes listed above.

$$\sqrt[3]{8} = 2$$
$$\sqrt[3]{27y^3} = 3y$$
$$\sqrt[3]{x^6} = x^2$$

Note that we can find the exponent of the cube root of a variable term by dividing the exponent by 3.

? QUESTION   Which of the following are perfect cubes? If the expression is a perfect cube, find its cube root.*

a. 64   **b.** x^{12} **c.** $9y^6$

The following rules are used to factor the sum and difference of two perfect cubes.

> **Factors of the Sum or Difference of Two Perfect Cubes**
> $$a^3 + b^3 = (a + b)(a^2 - ab + b^2)$$
> $$a^3 - b^3 = (a - b)(a^2 + ab + b^2)$$

? QUESTION   Which of the following is the sum of two cubes? Which is the difference of two cubes?†

a. $x^6 - 27$ **b.** $y^3 + 12$ **c.** $c^3 + 1$

? ANSWERS   *a. 64 is a perfect cube because $4^3 = 64$, $\sqrt[3]{64} = 4$.   **b.** x^{12} is a perfect cube because the exponent on x is a divisible by 3. $\sqrt[3]{x^{12}} = x^4$.   **c.** $9y^6$ is not a perfect cube because 9 is not a perfect cube. †**a.** $x^6 - 27$ is the difference of two cubes.   **b.** $y^3 + 12$ is neither; 12 is not a perfect cube.   **c.** $c^3 + 1$ is the sum of two cubes.

CALCULATOR NOTE

The factorization can be checked by multiplying the two factors. It can also be checked using a graphing calculator.

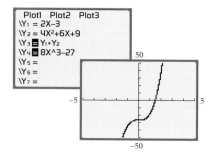

INSTRUCTOR NOTE
Exercise 30 can be used for a similar in-class example.

⇒ Factor: $8x^3 - 27$

Write the binomial as the difference of two perfect cubes. Then write the binomial factor and the trinomial factor.

$$8x^3 - 27 = (2x)^3 - (3)^3$$
$$= (2x - 3)(4x^2 + 6x + 9)$$

Square of the first term ——————
Opposite of the product
   of the two terms $2x$ and -3 ——————
Square of the last term ——————

Note that the terms of the binomial factor, $2x - 3$, are the cube roots of the perfect cubes $8x^3$ and -27. The sign of the binomial factor is the same sign as in the given binomial. The trinomial factor is obtained from the terms in the binomial factor.

EXAMPLE 3

Factor: $a^3 + 64y^3$

Solution $a^3 + 64y^3$

$= a^3 + (4y)^3$ • Write the binomial as the sum of two perfect cubes.

$= (a + 4y)(a^2 - 4ay + 16y^2)$ • Factor.

Check: $(a + 4y)(a^2 - 4ay + 16y^2)$
$= a^3 - 4a^2y + 16ay^2 + 4a^2y - 16ay^2 + 64y^3$
$= a^3 + 64y^3$

YOU TRY IT 3

Factor: $x^3y^3 - 1$

Solution   See page S26. $(xy - 1)(x^2y^2 + xy + 1)$

■ Factor a Trinomial That Is Quadratic in Form

Certain trinomials that are not quadratic can be expressed as quadratic trinomials by making suitable variable substitutions. A trinomial is **quadratic in form** if it can be written as $au^2 + bu + c$.

Each of the trinomials shown below is quadratic in form.

$$x^4 + 5x^2 + 6$$
$$(x^2)^2 + 5(x^2) + 6$$

Let $u = x^2$. $u^2 + 5u + 6$ • A quadratic trinomial

$$2x^2y^2 + 3xy - 9$$
$$2(xy)^2 + 3(xy) - 9$$

Let $u = xy$. $2u^2 + 3u - 9$ • A quadratic trinomial

As shown on page 442, $x^4 + 5x^2 + 6$ can be made quadratic in form by letting $u = x^2$. This trinomial is factored in the example that follows.

⇒ Factor: $x^4 + 5x^2 + 6$

$$x^4 + 5x^2 + 6 = u^2 + 5u + 6 \qquad \text{• Let } u = x^2.$$
$$= (u + 3)(u + 2) \qquad \text{• Factor the quadratic trinomial.}$$
$$= (x^2 + 3)(x^2 + 2) \qquad \text{• Replace } u \text{ with } x^2.$$

Check: $(x^2 + 3)(x^2 + 2) = x^4 + 2x^2 + 3x^2 + 6 = x^4 + 5x^2 + 6$ ⬅

INSTRUCTOR NOTE
Exercise 44 can be used for a similar in-class example.

EXAMPLE 4

Factor: $6x^2y^2 - xy - 12$

Solution $6x^2y^2 - xy - 12$

$$= 6u^2 - u - 12 \qquad \text{• Let } u = xy.$$
$$= (3u + 4)(2u - 3)$$
$$= (3xy + 4)(2xy - 3) \qquad \text{• Replace } u \text{ with } xy.$$

Check: $(3xy + 4)(2xy - 3) = 6x^2y^2 - 9xy + 8xy - 12 = 6x^2y^2 - xy - 12$

YOU TRY IT 4

Factor: $3x^4 + 4x^2 - 4$

Solution   See page S26. $(3x^2 - 2)(x^2 + 2)$

▪ Factor Completely

A polynomial is factored completely when it is written as a product of factors that are nonfactorable over the integers.

When factoring a polynomial completely, ask the following questions about the polynomial.

TAKE NOTE   Ⓟ

The first step in factoring a polynomial is to factor out the common factor if there is one. The last step is to check that each factor is nonfactorable over the integers.

Questions to Ask When Factoring a Polynomial

1. Is there a common factor? If so, factor out the GCF.
2. If the polynomial is a binomial, is it the difference of two perfect squares, the sum of two perfect cubes, or the difference of two perfect cubes? If so, factor.
3. If the polynomial is a trinomial, is it a perfect-square trinomial or the product of two binomials? If so, factor.
4. If the polynomial has four terms, can it be factored by grouping? If so, factor.
5. Is each factor nonfactorable over the integers? If not, factor.

INSTRUCTOR NOTE
Exercise 62 can be used for
a similar in-class example.

TAKE NOTE

Note that $2y$ and $10y$ were placed in the binomials. The check shows that this was necessary because of the y^2 in the term $100y^2$.

EXAMPLE 5

Factor: $5x^2 + 60xy + 100y^2$

Solution $5x^2 + 60xy + 100y^2$

$$= 5(x^2 + 12xy + 20y^2)$$ • There is a common factor, 5. Factor out the GCF.

$$= 5(x + 2y)(x + 10y)$$ • Factor $x^2 + 12xy + 20y^2$. The two factors of 20 whose sum is 12 are 2 and 10.

Check: $5(x + 2y)(x + 10y) = (5x + 10y)(x + 10y)$

$$= 5x^2 + 50xy + 10xy + 100y^2$$

$$= 5x^2 + 60xy + 100y^2$$

YOU TRY IT 5

Factor: $4a^3 - 4a^2b - 24ab^2$

Solution   See page S27. $4a(a - 3b)(a + 2b)$

INSTRUCTOR NOTE
Exercise 66 can be used for
a similar in-class example.

EXAMPLE 6

Factor: $x^2y + 2x^2 - y - 2$

Solution $x^2y + 2x^2 - y - 2$

$$= (x^2y + 2x^2) - (y + 2)$$ • There are four terms. Try factoring by grouping. Note that $-y - 2 = -(y + 2)$.

$$= x^2(y + 2) - 1(y + 2)$$

$$= (y + 2)(x^2 - 1)$$

$$= (y + 2)(x + 1)(x - 1)$$ • Factor the difference of two squares.

Check: $(y + 2)(x + 1)(x - 1) = (y + 2)(x^2 - 1)$

$$= x^2y - y + 2x^2 - 2$$

$$= x^2y + 2x^2 - y - 2$$

YOU TRY IT 6

Factor: $4x - 4y - x^3 + x^2y$

Solution   See page S27. $(x - y)(2 + x)(2 - x)$

INSTRUCTOR NOTE
Exercise 86 can be used for
a similar in-class example.

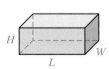

EXAMPLE 7

The volume of a box is $(2xy^2 + 12xy + 10x)$ cubic inches. Use factoring to find possible dimensions of the box in terms of the variables x and y.

State the goal.   The goal is to find the length, width, and height of a box (a rectangular solid) that has a volume of $(2xy^2 + 12xy + 10x)$ cubic inches. We are looking for three expressions that when multiplied equal $2xy^2 + 12xy + 10x$.

> **Devise a strategy.**   Factor the polynomial $2xy^2 + 12xy + 10x$.
>
> **Solve the problem.** $2xy^2 + 12xy + 10x = 2x(y^2 + 6y + 5)$
> $$= 2x(y + 1)(y + 5)$$
>
> The dimensions of the box are $2x$ inches by $(y + 1)$ inches by $(y + 5)$ inches.
>
> **Check your work.** $2x(y + 1)(y + 5) = (2xy + 2x)(y + 5)$
> $$= 2xy^2 + 10xy + 2xy + 10x$$
> $$= 2xy^2 + 12xy + 10x$$

YOU TRY IT 7

The volume of a box is $(3x^2y + 21xy + 36y)$ cubic inches. Use factoring to find possible dimensions of the box in terms of the variables x and y.

Solution   See page S27. $3y$ in. by $(x + 3)$ in. by $(x + 4)$ in.

6.5 EXERCISES

Suggested Assignment: 7–87, odds

Topics for Discussion

1. Determine whether the statement is always true, sometimes true, or never true.
   a. The expression $x^2 - 12$ is an example of the difference of two squares.   Never true
   b. The expression $(y + 8)(y - 8)$ is the product of the sum and difference of two terms. The two terms are y and 8.   Always true

2. Is $x^2 + 9$ factorable? Why or why not?   No. Explanations will vary. For example, the sum of two perfect squares is nonfactorable over the integers.

3. Provide three examples of the product of the sum and difference of two terms. For each example, state the two terms, the sum of the two terms, the difference of the two terms, and how the product is represented.
   Examples will vary.

4. Is the product of the sum and difference of the same two terms always equal to a binomial? Why or why not?   Yes. Explanations will vary.

5. Provide an example of each of the following.   Examples will vary.
   a. The difference of two perfect squares
   b. A perfect-square trinomial
   c. The square of a binomial
   d. The sum of two perfect squares
   e. The sum of two perfect cubes
   f. The difference of two perfect cubes
   g. A prime polynomial

■ **Special Factoring**

Factor.

6. $a^2 - 81$ $(a + 9)(a - 9)$ **7.** $a^2 - 49$ $(a + 7)(a - 7)$ **8.** $4x^2 - 1$ $(2x + 1)(2x - 1)$

9. $9x^2 - 1$ $(3x + 1)(3x - 1)$ **10.** $1 - 49x^2$ $(1 + 7x)(1 - 7x)$ **11.** $1 - 64x^2$ $(1 + 8x)(1 - 8x)$

12. $t^2 + 36$
Nonfactorable over the integers
13. $x^2 + 64$
Nonfactorable over the integers
14. $25 - a^2b^2$
$(5 + ab)(5 - ab)$

15. $64 - x^2y^2$ $(8 + xy)(8 - xy)$ **16.** $a^{2n} - 1$ $(a^n + 1)(a^n - 1)$ **17.** $b^{2n} - 16$ $(b^n + 4)(b^n - 4)$

18. $x^2 - 12x + 36$ $(x - 6)^2$ **19.** $y^2 - 6y + 9$ $(y - 3)^2$ **20.** $16x^2 - 40x + 25$ $(4x - 5)^2$

21. $49x^2 + 28x + 4$ $(7x + 2)^2$ **22.** $4a^2 + 4a - 1$ Nonfactorable over the integers

23. $9x^2 + 12x - 4$ Nonfactorable over the integers **24.** $x^2 + 6xy + 9y^2$ $(x + 3y)^2$

25. $4x^2y^2 + 12xy + 9$ $(2xy + 3)^2$ **26.** $25a^2 - 40ab + 16b^2$ $(5a - 4b)^2$

27. $4a^2 - 36ab + 81b^2$ $(2a - 9b)^2$ **28.** $x^{2n} + 6x^n + 9$ $(x^n + 3)^2$

29. $y^{2n} - 16y^n + 64$ $(y^n - 8)^2$ **30.** $y^3 + 125$ $(y + 5)(y^2 - 5y + 25)$

31. $x^3 - 27$ $(x - 3)(x^2 + 3x + 9)$ **32.** $8x^3 - 1$ $(2x - 1)(4x^2 + 2x + 1)$

33. $64a^3 + 27$ $(4a + 3)(16a^2 - 12a + 9)$ **34.** $27a^3 + b^3$ $(3a + b)(9a^2 - 3ab + b^2)$

35. $x^3 - 8y^3$ $(x - 2y)(x^2 + 2xy + 4y^2)$ **36.** $1 - 125b^3$ $(1 - 5b)(1 + 5b + 25b^2)$

37. $x^3y^3 + 64$ $(xy + 4)(x^2y^2 - 4xy + 16)$ **38.** $16x^3 - y^3$ Nonfactorable over the integers

39. $8x^3 - 9y^3$ Nonfactorable over the integers **40.** $x^{3n} + 8$ $(x^n + 2)(x^{2n} - 2x^n + 4)$

41. $a^{3n} + 64$ $(a^n + 4)(a^{2n} - 4a^n + 16)$ **42.** $x^2y^2 - 8xy + 15$ $(xy - 3)(xy - 5)$

43. $x^2y^2 - 8xy - 33$ $(xy + 3)(xy - 11)$ **44.** $x^4 - 9x^2 + 18$ $(x^2 - 3)(x^2 - 6)$

45. $y^4 - 6y^2 - 16$ $(y^2 + 2)(y^2 - 8)$ **46.** $x^4y^4 - 8x^2y^2 + 12$ $(x^2y^2 - 2)(x^2y^2 - 6)$

47. $a^4b^4 + 11a^2b^2 - 26$ $(a^2b^2 + 13)(a^2b^2 - 2)$ **48.** $x^{2n} + 3x^n + 2$ $(x^n + 1)(x^n + 2)$

49. $a^{2n} - a^n - 12$ $(a^n + 3)(a^n - 4)$ **50.** $3x^2y^2 - 14xy + 15$ $(3xy - 5)(xy - 3)$

51. $5x^2y^2 - 59xy + 44$ $(5xy - 4)(xy - 11)$ **52.** $2x^4 - 13x^2 - 15$ $(2x^2 - 15)(x^2 + 1)$

53. $3x^4 + 20x^2 + 32$ $(3x^2 + 8)(x^2 + 4)$ **54.** $4x^2y^2 + 12xy + 9$ $(2xy + 3)^2$

55. $x^{2n} + 6x^n + 9$ $(x^n + 3)^2$ **56.** $y^{2n} - 16y^n + 64$ $(y^n - 8)^2$

57. $3x^2 + 15x + 18$ $3(x + 2)(x + 3)$ **58.** $3a^2 + 3a - 18$ $3(a + 3)(a - 2)$

59. $ab^2 + 7ab - 8a$ $a(b + 8)(b - 1)$ **60.** $12x^2y - 36xy + 27y$ $3y(2x - 3)^2$

61. $8x^4 - 40x^3 + 50x^2$ $2x^2(2x - 5)^2$

62. $3y^3 - 15y^2 + 18y$ $3y(y - 2)(y - 3)$

63. $2y^4 - 26y^3 - 96y^2$ $2y^2(y + 3)(y - 16)$

64. $3y^4 + 54y^3 + 135y^2$ $3y^2(y + 3)(y + 15)$

65. $x^3 - 2x^2 - x + 2$ $(x + 1)(x - 1)(x - 2)$

66. $x^3 - 2x^2 - 4x + 8$ $(x - 2)^2(x + 2)$

67. $x^4 - 5x^2 - 4$ Nonfactorable over the integers

68. $a^4 - 25a^2 - 144$ Nonfactorable over the integers

69. $2x^3 - 11x^2 + 5x$ $x(x - 5)(2x - 1)$

70. $2x^3 + 3x^2 - 5x$ $x(x - 1)(2x + 5)$

71. $10t^2 - 5t - 50$ $5(t + 2)(2t - 5)$

72. $16t^2 + 40t - 96$ $8(t + 4)(2t - 3)$

73. $6p^3 + 5p^2 + p$ $p(2p + 1)(3p + 1)$

74. $2x^2 - 18$ $2(x + 3)(x - 3)$

75. $4x^2y^2 - 4x^2 - 9y^2 + 9$
$(2x + 3)(2x - 3)(y + 1)(y - 1)$

76. $4x^4 - x^2 - 4x^2y^2 + y^2$
$(x + y)(x - y)(2x + 1)(2x - 1)$

77. $50 - 2x^2$ $2(5 + x)(5 - x)$

78. $72 - 2x^2$ $2(6 + x)(6 - x)$

79. $b^4 - a^2b^2$ $b^2(b + a)(b - a)$

80. $2x^4y^2 - 2x^2y^2$ $2x^2y^2(x + 1)(x - 1)$

81. *Geometry* The area of a square is $(4x^2 + 12x + 9)$ square centimeters. Use factoring to find a possible length of a side of the square in terms of the variable x.

$A = 4x^2 + 12x + 9$

$(2x + 3)$ cm

82. *Geometry* The area of a square is $(9x^2 + 6x + 1)$ square meters. Use factoring to find a possible length of a side of the square in terms of the variable x.

$A = 9x^2 + 6x + 1$

$(3x + 1)$ m

83. *Geometry* Find the dimensions of a rectangle that has the same area as the shaded region in the diagram at the right. Write the dimensions in terms of the variable x. $(2x - 3)$ by $(2x + 3)$

84. *Geometry* Find the dimensions of a rectangle that has the same area as the shaded region in the diagram at the right. Write the dimensions in terms of the variable a. $(a + 4)$ by $(a - 4)$

85. *Geometry* Find the dimensions of a rectangle that has the same area as the shaded region in the diagram at the right. Write the dimensions in terms of the variable a. $(3a + 5)$ by $(3a - 5)$

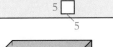

86. *Geometry* The volume of a box is $(3xy^2 + 21xy + 18x)$ cubic centimeters. Find the dimensions of the box in terms of the variables x and y.
$3x$ cm by $(y + 1)$ cm by $(y + 6)$ cm

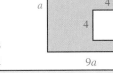

87. *Geometry* The volume of a box is $(4x^2y + 32xy + 60y)$ cubic inches. Find the dimensions of the box in terms of the variables x and y.
$4y$ in. by $(x + 5)$ in. by $(x + 3)$ in.

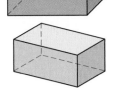

Applying Concepts

88. What is the least positive integer by which 252 should be multiplied to obtain a perfect cube?   294

89. Factor: $ax^3 - b + bx^3 - a$ $(x - 1)(x^2 + x + 1)(a + b)$

90. Factor.
   a. $x^4 + 64$ [*Hint*: Add and subtract $16x^2$ so that the expression becomes $(x^4 + 16x^2 + 64) - 16x^2$. Now factor by grouping.]
   $(x^2 - 4x + 8)(x^2 + 4x + 8)$
   b. $x^4 + x^2y^2 + y^4$ [*Hint*: Use the strategy in part a. Add and subtract x^2y^2.] $(x^2 + xy + y^2)(x^2 - xy + y^2)$

91. *Geometry*   The area of a square is $(16x^2 + 24x + 9)$ square feet. Find the dimensions of the square in terms of the variable x. Can $x = 0$? What are the possible values of x? $(4x + 3)$ ft by $(4x + 3)$ ft; yes; $x > -\frac{3}{4}$

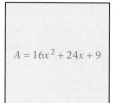

$A = 16x^2 + 24x + 9$

EXPLORATION

1. Select any odd integer greater than 1, square it, and then subtract 1. Is the result evenly divisible by 8? Prove that this procedure always produces a number divisible by 8. (*Suggestion:* Any odd integer greater than 1 can be expressed as $2n + 1$, where n is a natural number.)
$(2n + 1)^2 - 1 = (2n + 1)(2n + 1) - 1 = (4n^2 + 4n + 1) - 1 = 4n^2 + 4n = 4n(n + 1)$.
Because n or $n + 1$ is an even number, $4n(n + 1)$ is divisible by 8.

SECTION **6.6**   ## Solving Equations by Factoring

■ Solve Equations by Factoring

■ Solve Equations by Factoring

Earlier in this chapter, a quadratic trinomial was defined as a trinomial of the form $ax^2 + bx + c$, where a and b are coefficients and c is a constant. A **quadratic equation** is an equation of the form $ax^2 + bx + c = 0$, where a and b are coefficients, c is a constant, and $a \neq 0$. Here are some examples of quadratic equations.

$$3x^2 - x + 2 = 0, \qquad a = 3, \qquad b = -1, \qquad c = 2$$
$$-x^2 + 4 = 0, \qquad a = -1, \qquad b = 0, \qquad c = 4$$
$$6x^2 - 5x = 0, \qquad a = 6, \qquad b = -5, \qquad c = 0$$

A quadratic equation is in **standard form** when the polynomial is in descending order and equal to 0. Each of the three quadratic equations above is in standard form.

Because the degree of the polynomial $ax^2 + bx + c$ is 2, a quadratic equation is also called a **second-degree equation.**

? QUESTIONS **1.** What are the values of a, b, and c in the second-degree equation $5x^2 + x - 9 = 0$?

2. Which of the following are second-degree equations written in standard form?

a. $3y^2 + 5y - 2 = 0$ **b.** $8p - 4p^2 + 7 = 0$
c. $z^3 - 6z + 9 = 0$ **d.** $4r^2 + r - 1 = 6$
e. $v^2 - 16 = 0$

In this section, we will be solving quadratic equations by factoring.

Consider the equation $ab = 0$. If a is not zero, then b is zero. Conversely, if b is not zero, then a must be zero. This is stated in the Principle of Zero Products.

Principle of Zero Products

If the product of two factors is zero, then at least one of the factors must be zero.

If $ab = 0$, then $a = 0$ or $b = 0$.

The Principle of Zero Products is used to solve equations.

➡ Solve: $(x + 1)(x - 4) = 0$

By the Principle of Zero Products, if $(x + 1)(x - 4) = 0$, then $x + 1 = 0$ or $x - 4 = 0$.

$$x + 1 = 0 \qquad x - 4 = 0$$
$$x = -1 \qquad x = 4 \qquad \bullet \text{ Solve each equation for } x.$$

Check:

$$
\begin{array}{c|c}
(x + 1)(x - 4) = 0 & \\
\hline
(-1 + 1)(-1 - 4) & 0 \\
(0)(-5) & 0 \\
0 = 0\,\checkmark &
\end{array}
\qquad
\begin{array}{c|c}
(x + 1)(x - 4) = 0 & \\
\hline
(4 + 1)(4 - 4) & 0 \\
(5)(0) & 0 \\
0 = 0\,\checkmark &
\end{array}
$$

The solutions are -1 and 4.   ⬅

It is important to note that before we can use the Principle of Zero Products to solve a quadratic equation, the equation must be written in standard form. In Example 1 below, writing the equation in standard form is the first step of the solution.

? ANSWERS **1.** a is the coefficient of x^2; $a = 5$. b is the coefficient of x; $b = 1$. c is the constant; $c = -9$. **2.** The equations in parts a and e are second-degree equations in standard form. The equation in part b is not in standard form because $8p - 4p^2 + 7$ is not written in descending order. The equation in part c is not a second-degree equation because there is an exponent of 3 on the variable. The equation in part d is not in standard form because it is not equal to 0.

INSTRUCTOR NOTE
Exercise 34 can be used for
a similar in-class example.

EXAMPLE 1

Solve: $2x^2 - x = 1$

Solution

$$2x^2 - x = 1$$
$$2x^2 - x - 1 = 0$$

- First write the equation in standard form. Subtract 1 from each side of the equation.

$$(2x + 1)(x - 1) = 0$$

- Factor the left side of the equation.

$$2x + 1 = 0 \qquad x - 1 = 0$$

- Use the Principle of Zero Products.

$$2x = -1 \qquad x = 1$$

- Solve each equation for x.

$$x = -\frac{1}{2}$$

The solutions are $-\frac{1}{2}$ and 1.

 See Appendix A:
Intersection

Graphical check: Perform a graphical check by graphing $Y_1 = 2X^2 - X$ and $Y_2 = 1$ and then using the intersect feature of the calculator. The x-coordinates of the points of intersection are the solutions of the equation.

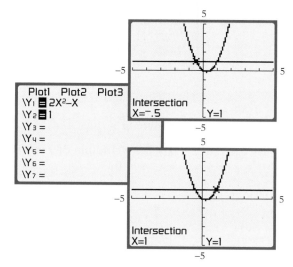

> ### TAKE NOTE
>
> An algebraic check in-volves substituting the proposed solutions for x in the original equa-tion. Here is the check for the solution $x = 1$.
>
> $$\begin{array}{c|c} 2x^2 - x = 1 \\ \hline 2(1)^2 - 1 & 1 \\ 2(1) - 1 & 1 \\ 2 - 1 & 1 \\ 1 = 1 \ \checkmark \end{array}$$

YOU TRY IT 1

Solve by factoring: $3x^2 + 5x = 2$

Solution See page S27. -2 and $\frac{1}{3}$

INSTRUCTOR NOTE
Exercise 52 can be used for
a similar in-class example.

EXAMPLE 2

Solve: $(x + 4)(x - 3) = 8$

Solution

$$(x + 4)(x - 3) = 8$$
$$x^2 + x - 12 = 8$$

- Write the equation in standard form. Multiply the binomials.

$$x^2 + x - 20 = 0$$

- Subtract 8 from each side of the equation.

$$(x + 5)(x - 4) = 0$$

- Factor the trinomial.

$$x + 5 = 0 \qquad x - 4 = 0$$ • Use the Principle of Zero Products.

$$x = -5 \qquad x = 4$$ • Solve each equation for x.

The solutions are -5 and 4.

Graphical check:

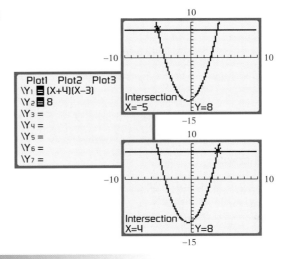

YOU TRY IT 2

Solve by factoring: $(x - 2)(x + 5) = 8$

Solution   See page S27. -6 and 3

The solution of Example 2 illustrates the steps involved in solving a second-degree equation by factoring.

Suggested Activity

See Section 6.6 of the *Student Activity Manual* for an investigation involving solving equations by factoring.

> **Solving a Second-Degree Equation by Factoring**
> 1. Write the equation in standard form.
> 2. Factor the polynomial.
> 3. Use the Principle of Zero Products to set each factor of the polynomial equal to zero.
> 4. Solve each equation for the variable.

The Principle of Zero Products can be extended to more than two factors. For example, if $abc = 0$, then $a = 0$, $b = 0$, or $c = 0$.

➡ Solve: $x^3 - x^2 - 4x + 4 = 0$

$$x^3 - x^2 - 4x + 4 = 0$$

$$(x^3 - x^2) - (4x - 4) = 0$$ • Factor by grouping.

$$x^2(x - 1) - 4(x - 1) = 0$$

$$(x - 1)(x^2 - 4) = 0$$

$$(x - 1)(x + 2)(x - 2) = 0$$

$$x - 1 = 0 \quad x + 2 = 0 \quad x - 2 = 0$$ • Use the Principle of Zero Products.

$$x = 1 \qquad x = -2 \qquad x = 2$$ • Solve each equation for x.

-2, 1, and 2 check as solutions.

The solutions are -2, 1, and 2.

EXAMPLE 3

Solve: $x^3 - x^2 - 25x + 25 = 0$

Solution

$$x^3 - x^2 - 25x + 25 = 0$$
$$(x^3 - x^2) - (25x - 25) = 0 \qquad \bullet \text{ Factor by grouping.}$$
$$x^2(x - 1) - 25(x - 1) = 0$$
$$(x - 1)(x^2 - 25) = 0$$
$$(x - 1)(x + 5)(x - 5) = 0$$
$$x - 1 = 0 \quad x + 5 = 0 \quad x - 5 = 0 \qquad \bullet \text{ Set each factor equal to 0.}$$
$$x = 1 \qquad x = -5 \qquad x = 5 \qquad \bullet \text{ Solve each equation for } x.$$

The solutions are -5, 1, and 5.

Graphical check:

To check these solutions graphically, we need to find all three points at which the graphs of $Y_1 = X^3 - X^2 - 25X + 25$ and $Y_2 = 0$ intersect.

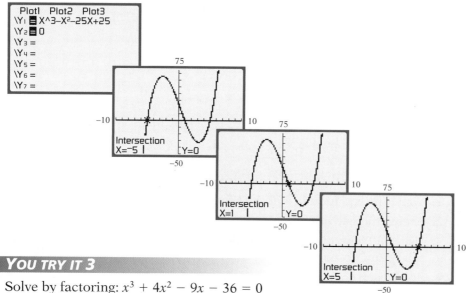

INSTRUCTOR NOTE
Exercise 54 can be used for a similar in-class example.

YOU TRY IT 3

Solve by factoring: $x^3 + 4x^2 - 9x - 36 = 0$

Solution See page S27. $-4, -3,$ and 3

INSTRUCTOR NOTE
Exercise 66 can be used for a similar in-class example.

Suggested Activity

1. What is the sum of the solutions of $x^2 + 5x - 24 = 0$? What is the product of the solutions?
   [Answer: -5; -24]
2. One solution of the equation $2x^2 - 5x + c = 0$ is 3. Find the other solution.

[Answer: $-\frac{1}{2}$]

EXAMPLE 4

An object is released from the top of a building 320 feet high. The initial velocity is 16 feet per second. How many seconds later will the object hit the ground? Use the formula $d = vt + 16t^2$, where d is the distance in feet, v is the initial velocity in feet per second, and t is the time in seconds.

State the goal. The goal is to find the time it takes for the object to hit the ground.

Devise a strategy. The distance d is 320 feet. The initial velocity v is 16 feet per second. Substitute these values into the given formula. Then solve for t, the time.

Solve the problem.

$$d = vt + 16t^2$$

$$320 = 16t + 16t^2$$

$16t^2 + 16t - 320 = 0$ • Write the equation in standard form.

$16(t^2 + t - 20) = 0$ • Factor out the common monomial factor.

$t^2 + t - 20 = 0$ • Divide each side of the equation by 16.

$(t + 5)(t - 4) = 0$ • Factor the trinomial.

$t + 5 = 0 \quad t - 4 = 0$ • Set each factor equal to zero.

$t = -5 \quad\quad t = 4$ • Solve each equation for x.

The time it takes for the object to reach the ground cannot be a negative number, so the solution -5 is not possible.

The object will hit the ground 4 seconds after it is released.

Check your work. $320 = 16t + 16t^2$

$$
\begin{array}{c|c}
320 & 16(4) + 16(4)^2 \\
320 & 16(4) + 16(16) \\
320 & 64 + 256 \\
320 = 320 & \sqrt{}\ \text{The solution checks.}
\end{array}
$$

You can also perform a graphical check.

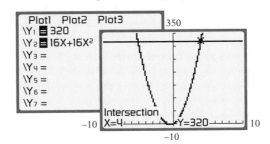

YOU TRY IT 4

How many consecutive natural numbers beginning with 1 will give a sum of 78? Use the formula $2S = n^2 + n$, where S is the sum of the first n natural numbers.

Solution   See page S27.   12

6.6 EXERCISES   Suggested Assignment: 7–69, odds

Topics for Discussion

1. What is a quadratic equation? How does it differ from a linear equation? Give an example of each type of equation.

    A quadratic equation is an equation of the form $ax^2 + bx + c = 0$, $a \neq 0$. A linear equation is an equation of the form $ax + b = 0$, $a \neq 0$. A quadratic equation has a term of degree 2. In a linear equation, the highest exponent on a variable is 1. Examples will vary. $3x^2 - 4x + 5 = 0$ is an example of a quadratic equation. $6x + 1 = 0$ is an example of a linear equation.

2. What is the difference between a quadratic trinomial and a quadratic equation?

A quadratic trinomial is a trinomial of the form $ax^2 + bx + c$; the values of a, b, and c are nonzero. A quadratic equation is an equation of the form $ax^2 + bx + c = 0$; only a must be nonzero.

3. What does the Principle of Zero Products state?

The Principle of Zero Products is based on the Multiplication Property of Zero, which states that the product of a number and zero is zero. The Principle of Zero Products begins with the conclusion of the Multiplication Property of Zero, stating that if the product of two or more factors is zero, then at least one of the factors must be equal to zero.

4. Why is it possible to solve some quadratic equations by using the Principle of Zero Products?

When the polynomial $ax^2 + bx + c$ is factorable and $ax^2 + bx + c = 0$, we can factor the polynomial, thereby rewriting the polynomial as a product. The equation now states that a product is equal to zero, so we can apply the Principle of Zero Products.

5. Consider the equation $(x + 5)(x - 6) = 0$. Can the Principle of Zero Products be used to solve this equation? Why? If so, how?

Yes, because it is the product of two factors, $x + 5$ and $x - 6$, and their product is zero. It can be solved by setting each factor equal to zero and solving the resulting equations for x.

■ Solve Equations by Factoring

Is the equation a quadratic equation?

6. $3x^2 - 5 = 0$ Yes

7. $3x - 5 = 0$ No

8. $x^2 = 5x$ Yes

Write the equation in standard form.

9. $x^2 + 6 = 5x$ $x^2 - 5x + 6 = 0$

10. $x + x^2 = 12$ $x^2 + x - 12 = 0$

11. $2x^2 = -3x + 5$ $2x^2 + 3x - 5 = 0$

Can the equation be solved by using the Principle of Zero Products without first rewriting the equation?

12. $8x(3x + 5) = 0$ Yes

13. $3x(x - 7) - 1 = 0$ No

14. $0 = (3x - 4)(2x + 5)$ Yes

15. $(x - 9)(y + 6) = 0$ Yes

16. $0 = (4x - 1)x + 2$ No

17. $0 = (5x - 3)(x + 7)$ Yes

Solve.

18. $(y + 4)(y + 6) = 0$ $-6, -4$

19. $(a - 5)(a - 2) = 0$ $2, 5$

20. $x(x - 7) = 0$ $0, 7$

21. $b(b + 8) = 0$ $-8, 0$

22. $3z(2z + 5) = 0$ $-\frac{5}{2}, 0$

23. $4y(3y - 2) = 0$ $0, \frac{2}{3}$

24. $(2x + 3)(x - 7) = 0$ $-\frac{3}{2}, 7$

25. $(4a - 1)(a + 9) = 0$ $-9, \frac{1}{4}$

26. $4z^2 - 1 = 0$ $-\frac{1}{2}, \frac{1}{2}$

27. $9t^2 - 16 = 0$ $-\frac{4}{3}, \frac{4}{3}$

28. $x^2 + x - 6 = 0$ $-3, 2$

29. $y^2 + 4y - 5 = 0$ $-5, 1$

30. $t^2 - 8t = 0$ $0, 8$

31. $x^2 - 9x = 0$ $0, 9$

32. $2y^2 - 10y = 0$ $0, 5$

33. $3a^2 - 12a = 0$ 0, 4

34. $b^2 - 4b = 32$ −4, 8

35. $z^2 - 3z = 28$ −4, 7

36. $2x^2 - 5x = 12$ $-\frac{3}{2}, 4$

37. $3t^2 + 13t = 10$ $-5, \frac{2}{3}$

38. $4y^2 - 19y = 5$ $-\frac{1}{4}, 5$

39. $5b^2 - 17b = -6$ $\frac{2}{5}, 3$

40. $6a^2 + a = 2$ $-\frac{2}{3}, \frac{1}{2}$

41. $8x^2 - 10x = 3$ $-\frac{1}{4}, \frac{3}{2}$

42. $z(z - 1) = 20$ −4, 5

43. $y(y - 2) = 35$ −5, 7

44. $t(t + 1) = 42$ −7, 6

45. $x(x - 12) = -27$ 3, 9

46. $x(2x - 5) = 12$ $-\frac{3}{2}, 4$

47. $y(3y - 2) = 8$ $-\frac{4}{3}, 2$

48. $2b^2 - 6b = b - 3$ $\frac{1}{2}, 3$

49. $3a^2 - 4a = 20 - 15a$ $-5, \frac{4}{3}$

50. $(y + 5)(y - 7) = -20$ −3, 5

51. $(x + 2)(x - 6) = 20$ −4, 8

52. $(b + 5)(b + 10) = 6$ −11, −4

53. $(a - 9)(a - 1) = -7$ 2, 8

54. $x^3 + 7x^2 - 9x - 63 = 0$
−7, −3, 3

55. $x^3 - x^2 - 16x + 16 = 0$
−4, 1, 4

56. $x^3 + 4x^2 - x - 4 = 0$
−4, −1, 1

57. $2x^3 + x^2 - 8x - 4 = 0$
$-2, -\frac{1}{2}, 2$

58. $2x^3 + x^2 - 50x - 25 = 0$
$-5, -\frac{1}{2}, 5$

59. $12x^3 - 8x^2 - 3x + 2 = 0$
$-\frac{1}{2}, \frac{1}{2}, \frac{2}{3}$

60. *Physics*   A stone is thrown into a well with an initial velocity of 8 feet per second. The well is 440 feet deep. How many seconds later will the stone hit the bottom of the well? Use the formula $d = vt + 16t^2$, where d is the distance in feet, v is the initial velocity in feet per second, and t is the time in seconds.   5 s

61. *Physics*   An object is released from a plane at an altitude of 1600 feet. The initial velocity is 0 feet per second. How many seconds later will the object hit the ground? Use the formula $d = vt + 16t^2$, where d is the distance in feet, v is the initial velocity in feet per second, and t is the time in seconds.   10 s

62. *Sports*   A team has 28 games scheduled. How many teams are in the league if each team plays every other team once? Use the formula $2N = t^2 - t$, where N is the number of football games that must be scheduled in a league with t teams when each team plays every other team once.
8 teams

63. *Sports*   A team has 45 games scheduled. How many teams are in the league if each team plays every other team once? Use the formula $2N = t^2 - t$, where N is the number of football games that must be scheduled in a league with t teams when each team plays every other team once.
10 teams

64. *Number Problems*   How many consecutive natural numbers beginning with 1 will give a sum of 55? Use the formula $2S = n^2 + n$, where S is the sum of the first n natural numbers.   10

65. *Geometry*   The number of possible diagonals D in a polygon with n sides is given by $2D = n(n - 3)$. Find the number of sides for a polygon with 20 diagonals.   8 sides

66. *Sports*   A baseball player hits a "Baltimore chop," which means the ball bounces off home plate after the batter hits it. The ball leaves home plate with an initial velocity of 32 feet per second. How many seconds after the ball hits home plate will the ball be 16 feet above the ground? Use the formula $h = vt - 16t^2$, where h is the height in feet that an object will attain (neglecting air resistance) in t seconds, and v is the initial velocity in feet per second.   1 s

67. *Sports*   A golf ball is thrown onto a cement surface and rebounds straight up. The initial velocity of the rebound is 96 feet per second. How many seconds later will the golf ball return to the ground? Use the formula $h = vt - 16t^2$, where h is the height in feet that an object will attain (neglecting air resistance) in t seconds, and v is the initial velocity in feet per second.   6 s

68. *Physics*   A model of the height above the ground of an arrow projected into the air with an initial velocity of 120 ft/s is $h = -16t^2 + 120t + 5$, where h is the height, in feet, of the arrow t seconds after it is released from the bow. Determine at what times the arrow is 181 ft above the ground.   After 2 s and after 5.5 s

69. *Physics*   The height of a projectile fired upward is given by the formula $s = vt - 16t^2$, where s is the height in feet, v is the initial velocity in feet per second, and t is the time in seconds. Find the time for a projectile to return to Earth if it has an initial velocity of 200 ft/s.   12.5 s

Applying Concepts

70. Solve for the greatest positive root of the equation $2x^3 + x^2 - 8x = 4$.   2

71. Find $3n^2 + 2n - 1$ if $n(n + 6) = 16$.   175 or 15

72. Find two consecutive integers whose cubes differ by sixty-one.
    4 and 5 or −5 and −4

73. The sum of the squares of three consecutive even integers is fifty-six. Find the three integers.
    2, 4, and 6 or −2, −4, and −6

74. The sum of the squares of three consecutive odd integers is eighty-three. Find the three integers.   3, 5, and 7 or −3, −5, and −7

EXPLORATION

1.   Show that the solutions of the equation $ax^2 + bx = 0$ are 0 and $-\frac{b}{a}$.

    Factor the left side: $x(ax + b) = 0$; then $x = 0$ or $ax + b = 0$; solve $ax + b = 0$ for x:
    $x = -\frac{b}{a}$.

2. The following seems to show that $1 = 2$. Explain the error.

$$a = b$$
$$a^2 = ab$$ • Multiply each side of the equation by a.
$$a^2 - b^2 = ab - b^2$$ • Subtract b^2 from each side of the equation.
$$(a - b)(a + b) = b(a - b)$$ • Factor.
$$a + b = b$$ • Divide each side by $a - b$.
$$b + b = b$$ • Because $a = b$, substitute b for a.
$$2b = b$$
$$2 = 1$$ • Divide both sides by b.

$a - b$ is equal to zero, and division by zero is undefined.

CHAPTER **SUMMARY**

Key Terms

binomial **[p. 398]**
common binomial factor **[p. 426]**
common monomial factor **[p. 426]**
cube root **[p. 441]**
cubic polynomial **[p. 399]**
degree of a polynomial in one
    variable **[p. 399]**
descending order **[p. 398]**
factor **[p. 425]**
factor completely **[p. 443]**
factor a quadratic trinomial **[p. 428]**
factored form **[p. 425]**
linear polynomial **[p. 399]**
monomial **[pp. 377, 398]**
nonfactorable over the integers
    [p. 430]
opposite of a polynomial **[p. 401]**

perfect cube **[p. 441]**
perfect square **[p. 438]**
polynomial **[p. 398]**
power of a monomial **[p. 380]**
prime polynomial **[p. 430]**
quadratic equation **[p. 448]**
quadratic in form **[p. 442]**
quadratic polynomial **[p. 399]**
quadratic trinomial **[p. 428]**
second-degree equation **[p. 499]**
square of a binomial **[p. 408]**
square root **[p. 438]**
standard form of a quadratic
    equation **[p. 448]**
synthetic division **[p. 413]**
trinomial **[p. 398]**

Essential Concepts

Rule for Multiplying Exponential Expressions [p. 379]
$x^m \cdot x^n = x^{m+n}$

Rule for Simplifying the Power of an Exponential Expression [p. 381]
$(x^m)^n = x^{m \cdot n}$

Rule for Simplifying Powers of Products [p. 381]
$(x^m y^n)^p = x^{m \cdot p} y^{n \cdot p}$

Rule for Dividing Exponential Expressions
For $x \neq 0$, $\dfrac{x^m}{x^n} = x^{m-n}$. **[p. 383]**

Zero as an Exponent

For $x \neq 0$, $x^0 = 1$. [**p. 384**]

Definition of Negative Exponents

For $x \neq 0$, $x^{-n} = \dfrac{1}{x^n}$ and $\dfrac{1}{x^{-n}} = x^n$. [**p. 385**]

Rule for Simplifying Powers of Quotients

For $y \neq 0$, $\left(\dfrac{x^m}{y^n}\right)^p = \dfrac{x^{m \cdot p}}{y^{n \cdot p}}$. [**p. 387**]

Addition of Polynomials

To add polynomials, add the coefficients of the like terms. [**p. 400**]

Subtraction of Polynomials

To subtract two polynomials, add the opposite of the second polynomial to the first. [**p. 401**]

Multiplication of Polynomials

To multiply two polynomials, multiply each term of one polynomial times each term of the other polynomial. [**p. 407**]

The FOIL Method

To multiply two binomials, add the products of the **F**irst terms, the **O**uter terms, the **I**nner terms, and the **L**ast terms. [**p. 408**]

Division of Polynomials

To divide a polynomial by a monomial, divide each term of the polynomial by the monomial. If the divisor is not a monomial, use the long-division method similar to that used for division of whole numbers. [**pp. 410–411**]

Equation Used to Check Division [**p. 411**]

(Quotient $\times$ Divisor) + Remainder = Dividend

The Remainder Theorem

If the polynomial $P(x)$ is divided by $x - a$, the remainder is $P(a)$. [**p. 415**]

The Factor Theorem

A polynomial $P(x)$ has a factor $(x - c)$ if and only if $P(c) = 0$. [**p. 416**]

Common Monomial Factor

To factor a monomial from a polynomial, use the Distributive Property in reverse: $ab + ac = a(b + c)$. [**p. 425**]

Scientific Notation

To express a number in scientific notation, write it in the form $a \times 10^n$, where a is a number between 1 and 10 and n is an integer. If the number is greater than 10, the exponent on 10 will be positive. If the number is less than 1, the exponent on 10 will be negative.

$$367,000,000 = 3.67 \times 10^8$$
$$0.0000059 = 5.9 \times 10^{-6}$$

To change a number written in scientific notation to decimal notation, move the decimal point to the right if the exponent on 10 is positive and to the left if the exponent on 10 is negative. Move the decimal point the same number of places as the absolute value of the exponent on 10. **[p. 389]**

$$2.418 \times 10^7 = 24,180,000$$
$$9.06 \times 10^{-5} = 0.0000906$$

Factor by Grouping
To factor a polynomial with four terms, group the first two terms and the last two terms (put them in parentheses). Factor out the GCF from each group. Write the expression as a product of factors by factoring out the common binomial factor. **[p. 427]**

Factor a Trinomial by Grouping
To factor a trinomial by grouping, find the product of the first and last terms. Find two factors of this product whose sum is the middle term of the trinomial. Replace the middle term of the trinomial with the sum of these two factors. The resulting polynomial has four terms. Factor it by grouping. **[p. 433]**

Factors of the Difference of Two Perfect Squares **[p. 439]**

$a^2 - b^2 = (a + b)(a - b)$

Factors of a Perfect-Square Trinomial **[p. 440]**

$a^2 + 2ab + b^2 = (a + b)^2$
$a^2 - 2ab + b^2 = (a - b)^2$

Factors of the Sum or Difference of Two Perfect Cubes **[p. 441]**

$a^3 + b^3 = (a + b)(a^2 - ab + b^2)$
$a^3 - b^3 = (a - b)(a^2 + ab + b^2)$

Questions to Ask When Factoring a Polynomial
1. Is there a common factor? If so, factor out the GCF.
2. If the polynomial is a binomial, is it the difference of two perfect squares, the sum of two perfect cubes, or the difference of two perfect cubes? If so, factor.
3. If the polynomial is a trinomial, is it a perfect-square trinomial or the product of two binomials? If so, factor.
4. If the polynomial has four terms, can it be factored by grouping? If so, factor.
5. Is each factor nonfactorable over the integers? If not, factor. **[p. 443]**

Principle of Zero Products
If the product of two factors is zero, then at least one of the factors must be zero.
If $ab = 0$, then $a = 0$ or $b = 0$. **[p. 449]**

Solving a Second-Degree Equation by Factoring
1. Write the equation in standard form.
2. Factor the polynomial.
3. Use the Principle of Zero Products to set each factor of the polynomial equal to zero.
4. Solve each equation for the variable. **[p. 451]**

CHAPTER 6 REVIEW EXERCISES

1. Subtract: $47a^2b^3c - 23a^2b^3c$
   $24a^2b^3c$

2. Add: $(3x^3 - 2x^2 - 4) + (8x^2 - 8x + 7)$
   $3x^3 + 6x^2 - 8x + 3$

3. Multiply: $(5xy^2)(-4x^2y^3)$ $-20x^3y^5$

4. Simplify: $\dfrac{12x^2}{-3x^{-4}}$ $-4x^6$

5. Factor: $4a^2 - 12a + 9$ $(2a - 3)^2$

6. Factor: $5x^2 - 45x - 15$ $5(x^2 - 9x - 3)$

7. Simplify: $(2ab^{-3})(3a^{-2}b^4)$ $\dfrac{6b}{a}$

8. Divide: $\dfrac{16x^5 - 8x^3 + 20x}{4x}$ $4x^4 - 2x^2 + 5$

9. Factor: $a^2 - 19a + 48$ $(a - 3)(a - 16)$

10. Factor: $x^3 + 2x^2 - 15x$ $x(x + 5)(x - 3)$

11. Multiply: $-3y^2(-2y^2 + 3y - 6)$
    $6y^4 - 9y^3 + 18y^2$

12. Simplify: $(2x - 5)^2$
    $4x^2 - 20x + 25$

13. Factor: $6x^2 + 19x + 8$ $(2x + 1)(3x + 8)$

14. Factor: $2b^2 - 32$ $2(b + 4)(b - 4)$

15. Simplify: $(-3a^2 b^{-3})^2$
    $\dfrac{9a^4}{b^6}$

16. Write 0.0000029 in scientific notation.
    2.9×10^{-6}

17. Factor: $ab + 6a - 3b - 18$ $(b + 6)(a - 3)$

18. Multiply: $(4y - 3)(4y + 3)$ $16y^2 - 9$

19. Factor: $16x^2 - 25$
    $(4x + 5)(4x - 5)$

20. Multiply: $(2a - 7)(5a^2 - 2a + 3)$
    $10a^3 - 39a^2 + 20a - 21$

21. Factor: $2x^2 + 4x - 5$
    Nonfactorable over the integers

22. Solve: $z^2 + 5z = 14$
    $-7, 2$

23. Simplify: $\dfrac{-2a^2b^3}{8a^4b^8}$ $-\dfrac{1}{4a^2b^5}$

24. Divide: $(8x^2 + 4x - 3) \div (2x - 3)$
    $4x + 8 + \dfrac{21}{2x - 3}$

25. Write 3.5×10^{-8} in decimal notation.
    0.000000035

26. Factor: $27p^3 - 8$
    $(3p - 2)(9p^2 + 6p + 4)$

27. Factor: $2y^4 - 14y^3 - 16y^2$ $2y^2(y + 1)(y - 8)$

28. Solve: $(y + 3)(2y + 3) = 5$ $-4, -\dfrac{1}{2}$

29. Subtract: $(3y^2 - 5y + 8) - (-2y^2 + 5y + 8)$ $5y^2 - 10y$

30. The length of the side of a square is $(2x + 3)$ meters. Find the area of the square in terms of the variable x. $(4x^2 + 12x + 9)$ m²

2x + 3

31. The mass of the moon is 8.103×10^{19} tons. Write this number in standard form. 81,030,000,000,000,000,000

32. The height h, in feet, of a golf ball t seconds after it has been struck is given by $h(t) = -16t^2 + 60t$. Determine the height of the ball 2 seconds after it is hit. 56 ft

33. The area of a rectangle is $(4x^2 + 13x + 3)$ square inches. Use factoring to find possible dimensions of the rectangle in terms of the variable x.
    $(4x + 1)$ in. by $(x + 3)$ in.

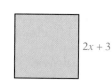

$A = 4x^2 + 13x + 3$

CHAPTER **6** *TEST*

1. Add: $(12y^2 + 17y - 4) + (9y^2 - 13y + 3)$
$21y^2 + 4y - 1$

2. Multiply: $(6a^2b^5)(-3a^6b)$ $-18a^8b^6$

3. Multiply: $4x^2(3x^3 + 2x - 7)$ $12x^5 + 8x^3 - 28x^2$

4. Simplify: $\dfrac{-6x^{-2}y^4}{3xy}$ $-\dfrac{2y^3}{x^3}$

5. Factor: $9a^2 - 30a + 25$ $(3a - 5)^2$

6. Factor: $6x^3 - 8x^2 + 10x$ $2x(3x^2 - 4x + 5)$

7. Simplify: $(5a^{-1}b^{-4})(-2a^2b^3)$ $-\dfrac{10a}{b}$

8. Divide: $\dfrac{12b^7 + 36b^5 - 3b^3}{3b^3}$ $4b^4 + 12b^2 - 1$

9. Factor: $a^2 - 9a - 36$ $(a + 3)(a - 12)$

10. Factor: $12x^3 + 12x^2 - 45x$ $3x(2x + 5)(2x - 3)$

11. Simplify: $(-2a^4b^{-5})^3$ $-\dfrac{8a^{12}}{b^{15}}$

12. Write 78,000,000,000 in scientific notation.
7.8×10^{10}

13. Factor: $2ax + 4bx - 3ay - 6by$ $(a + 2b)(2x - 3y)$

14. Multiply: $(6y - 5)(6y + 5)$ $36y^2 - 25$

15. Factor: $81x^2 - 1$ $(9x + 1)(9x - 1)$

16. Multiply: $(2a + 3)(3a^2 + 4a - 7)$
$6a^3 + 17a^2 - 2a - 21$

17. Solve: $6b^2 = b + 1$ $-\dfrac{1}{3}, \dfrac{1}{2}$

18. Factor: $64a^3 - b^3$ $(4a - b)(16a^2 + 4ab + b^2)$

19. Subtract: $(6x^3 - 7x^2 + 6x - 7) - (4x^3 - 3x^2 + 7)$
$2x^3 - 4x^2 + 6x - 14$

20. Divide: $(x^3 - 5x^2 + 5x + 5) \div (x - 3)$
$x^2 - 2x - 1 + \dfrac{2}{x - 3}$

21. Factor: $36y^8 - 36y^4 + 5$ $(6y^4 - 5)(6y^4 - 1)$

22. Solve: $p(p - 8) = -15$ $3, 5$

23. The length of a rectangle is $(5x + 3)$ centimeters. The width is $(2x - 7)$ centimeters. Find the area of the rectangle in terms of the variable x.
$(10x^2 - 29x - 21)$ cm^2

24. Light from the sun supplies Earth with 2.4×10^{14} horsepower. Write this number in standard form.   240,000,000,000,000

25. The number of possible diagonals D in a polygon with n sides is given by $2D = n(n - 3)$. Find the number of sides for a polygon with 54 diagonals.   12 sides

◀ *CUMULATIVE REVIEW EXERCISES*

1. If $5 \le a \le 10$ and $20 \le b \le 30$, find the maximum value of $\dfrac{a}{b}$. $\dfrac{1}{2}$

2. Evaluate $-2a^2 \div (2b) - c$ when $a = -4$, $b = 2$, and $c = -1$. -7

3. Identify the property that justifies the statement.
$$(3 + 8) + 7 = 3 + (8 + 7)$$
The Associative Property of Addition

4. Multiply: $-\dfrac{3}{4}(-24x^2)$ $18x^2$

5. Find the domain and range of the relation $\{(-5, -4), (-3, -2), (-1, 0), (1, 2), (3, 4)\}$. Is the relation a function?   D: $\{-5, -3, -1, 1, 3\}$;
R: $\{-4, -2, 0, 2, 4\}$; Yes

6. Find the range of the function given by the equation $f(x) = \dfrac{4}{5}x - 3$ if the domain is $\{-10, -5, 0, 5, 10\}$. $\{-11, -7, -3, 1, 5\}$

7. Solve: $4 + 3(x - 2) = 13$

5

8. Solve $-4x - 2 \geq 10$. Write the solution set in interval notation.

$(-\infty, -3]$

9. Graph $3x - 4y = 12$ by using the x- and y-intercepts.

10. Graph $f(x) = -3x - 3$.

11. Graph the solution set of $-3x + 2y < 6$.

12. Find the equation of the line that contains the points $(-5, 2)$ and $(5, 6)$.

$y = \frac{2}{5}x + 4$

13. Solve by the addition method: $2x - 3y = -4$
$5x + y = 7$

$(1, 2)$

14. Simplify: $(-2x^{-4}y^2)^3$

$-\frac{8y^6}{x^{12}}$

15. Subtract: $(3y^3 - 5y^2 - 6) - (2y^2 - 8y + 1)$

$3y^3 - 7y^2 + 8y - 7$

16. Divide: $(8x^2 + 4x - 3) \div (2x - 3)$

$4x + 8 + \frac{21}{2x - 3}$

17. Factor: $25a^2 - 36b^2$

$(5a + 6b)(5a - 6b)$

18. Solve: $3x^2 + 11x = 20$

$-5, \frac{4}{3}$

19. How many ounces of pure gold that costs $360 per ounce must be mixed with 80 ounces of an alloy that costs $120 per ounce to make a mixture that costs $200 per ounce?

40 oz

20. The graph shows the relationship between the distance traveled, in miles, and the time of travel, in hours. Find the slope of the line between the two points on the graph. Write a sentence that states the meaning of the slope.

$m = 50$. A slope of 50 means the average speed was 50 mph.

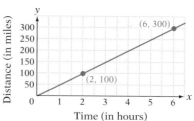

7 Rational Expressions and Equations

```
CALCULATE
1: value
2: zero
3: minimum
4: maximum
5: intersect
6: dy/dx
7: ∫f(x)dx
```

Press **2nd** CALC to access the minimum and maximum options.

Manufacturers who package their products in cans must consider a number of aspects of the packaging. They want the product to be attractive and comfortable for the consumer to hold. They would also like the can designed such that a minimum amount of aluminum is needed. A rational function that relates the surface area of a can (the amount of aluminum needed) to the radius of the bottom of the can is given in **Exercise 72 on page 487.**

Need help? For online student resources, visit this web site:
Math.college.hmco.com

PREP TEST

1. Find the LCM of 10 and 25.   50

For Exercises 2 to 5, add, subtract, multiply, or divide.

2. $-\dfrac{3}{8} \cdot \dfrac{4}{9}$ $-\dfrac{1}{6}$

3. $-\dfrac{4}{5} \div \dfrac{8}{15}$ $-\dfrac{3}{2}$

4. $-\dfrac{5}{6} + \dfrac{7}{8}$ $\dfrac{1}{24}$

5. $-\dfrac{3}{8} - \left(-\dfrac{7}{12}\right)$ $\dfrac{5}{24}$

6. Evaluate $\dfrac{2x - 3}{x^2 - x + 1}$ for $x = 2$. $\dfrac{1}{3}$

7. Solve: $4(2x + 1) = 3(x - 2)$ -2

8. Solve: $10\left(\dfrac{t}{2} + \dfrac{t}{5}\right) = 10(1)$ $\dfrac{10}{7}$

9. Two planes start from the same point and fly in opposite directions. The first plane is flying 20 mph slower than the second plane. In 2 h, the planes are 480 mi apart. Find the rate of each plane.   110 mph, 130 mph

GO FIGURE

Two children are on their way from school to home. Carla runs half the time and walks half the time. James runs half the distance and walks half the distance. If Carla and James walk at the same speed and run at the same speed, which child arrives home first?   Carla

---

SECTION 7.1

- Rational Functions
- Simplify Rational Expressions

Introduction to Rational Expressions

■ Rational Functions

Recall that a value mixture problem involves combining two ingredients that have different prices into a single blend. The solution of a value mixture problem is based on the equation $V = AC$, where V is the value of the blend, A is the amount of the ingredient, and C is the cost per unit of the ingredient.

The equation $V = AC$ can be solved for C by dividing each side of the equation by A. The result is

$$C = \frac{V}{A}$$

Suppose we have 10 pounds of cashews that cost $6 per pound and we want to add peanuts that cost $3 per pound to the cashews. The cost of the mixture

TAKE NOTE

The rational expression $\frac{3p + 60}{p + 10}$ represents

$$\frac{\text{value of} \atop \text{peanuts} + \text{value of} \atop \text{cashews}}{\text{amount of} \atop \text{peanuts} + \text{amount of} \atop \text{cashews}}$$

will depend on the number of pounds of peanuts added to the cashews and can be given by

$$C(p) = \frac{3p + 60}{p + 10}$$

where p is the number of pounds of peanuts added to the cashews.

To find the cost of the mixture when 20 pounds of peanuts are added to the cashews, evaluate the function for $p = 20$.

$$C(20) = \frac{3(20) + 60}{20 + 10} = \frac{120}{30} = 4$$

The cost of the mixture when 20 pounds of peanuts are added to the cashews is \$4 per pound.

The expression $\frac{3p + 60}{p + 10}$ is a rational expression. A **rational expression** is one in which the numerator and denominator are polynomials. Further examples of rational expressions follow.

$$\frac{9}{z} \qquad \frac{3x + 4}{2x^2 + 1} \qquad \frac{x^3 - x + 1}{x^2 - 3x - 5}$$

The expression $\frac{\sqrt{x} + 3}{x}$ is not a rational expression because $\sqrt{x} + 3$ is not a polynomial.

? **QUESTION**   Which of the following are not rational expressions?

a. $\dfrac{7x^{1/2}}{3x - 8}$ 　　　　 **b.** $\dfrac{9x}{2x - 5}$ 　　　　 **c.** $\dfrac{4x}{6x^{-2} + x^{-1} - 10}$

A function that is written in terms of a rational expression is a **rational function.** Each of the following represents a rational function.

$$f(x) = \frac{x^2 + 3}{2x - 1} \qquad g(t) = \frac{3}{t^2 - 4} \qquad R(z) = \frac{z^2 + 3z - 1}{z^2 + z - 12}$$

Because division by zero is not defined, **the domain of a rational function must exclude those numbers for which the value of the polynomal in the denominator is zero.** For the function $C(p) = \frac{3p + 60}{p + 10}$, the value of p cannot be -10.

The graph of $C(p) = \frac{3p + 60}{p + 10}$ is shown at the left as it would be graphed on a graphing calculator that is in the CONNECTED mode. The vertical line at X $= -10$ appears to be a part of the graph, but it is not. The calculator is "connecting" the plotted points to the left of X $= -10$ with the plotted points to the right of X $= -10$.

INSTRUCTOR NOTE
For students who have not learned the definition of *polynomial*, it would be helpful to review the definition at this time.

Suggested Activity

The average speed for a round trip is given by the expression $\frac{2v_1 v_2}{v_1 + v_2}$, where v_1 is the average speed on the way to your destination and v_2 is the average speed on your return trip. Find the average speed for a round trip if $v_1 = 50$ mph and $v_2 = 40$ mph.

[Answer: $44.\overline{4}$ mph or $44\frac{4}{9}$ mph]

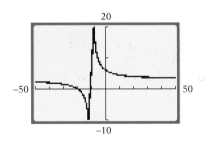

? ANSWER   **a.** This is not a rational expression because $7x^{1/2}$ is not a polynomial.   **c.** This is not a rational expression because $6x^{-2} + x^{-1} - 10$ is not a polynomial.

We can avoid this problem by putting the graphing calculator in the DOT mode. The graph will then appear as shown at the right. There is no point on the graph at X = −10.

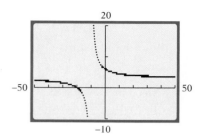

➡ Determine the domain of $g(x) = \dfrac{x^2 + 4}{3x - 6}$.

The domain of g must exclude values of x for which the denominator is zero.

$3x - 6 = 0$ • Set the denominator equal to zero.

$3x = 6$ • Solve for x.

$x = 2$ • This value must be *excluded* from the domain.

The domain of g is $\{x \mid x \neq 2\}$. ⬅

INSTRUCTOR NOTE
Exercise 24 can be used for a similar in-class example.

EXAMPLE 1

Find the domain of $f(x) = \dfrac{2x - 6}{x^2 - 3x - 4}$.

Solution $x^2 - 3x - 4 = 0$ • The domain must exclude values of x for which $x^2 - 3x - 4 = 0$. Solve this equation for x.

$(x + 1)(x - 4) = 0$

$x + 1 = 0 \qquad x - 4 = 0$

$x = -1 \qquad x = 4$ • When $x = -1$ and $x = 4$ the value of the denominator is zero. Therefore, these values must be excluded from the domain of f.

The domain is $\{x \mid x \neq -1, 4\}$.

The graph of this function is shown at the left. Note that the graph never intersects the lines $x = -1$ and $x = 4$ (shown as dashed lines). These are the two values of x excluded from the domain of f.

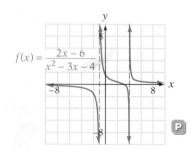

$f(x) = \dfrac{2x - 6}{x^2 - 3x - 4}$

YOU TRY IT 1

Find the domain of $g(x) = \dfrac{5 - x}{x^2 - 4}$.

$(a + b)(a - b) = a^2 - b^2$

$(x - 2)(x + 2) = \varnothing$

Solution See page S27. $\{x \mid x \neq -2, 2\}$

INSTRUCTOR NOTE
Exercise 22 can be used for a similar in-class example.

EXAMPLE 2

Find the domain of $f(x) = \dfrac{3x + 2}{x^2 + 1}$.

Solution The domain must exclude values of x for which $x^2 + 1 = 0$. It is not possible for $x^2 + 1 = 0$, because $x^2 \geq 0$, and a positive number added to a number equal to or greater than zero cannot equal zero. Therefore, there are no real numbers that must be excluded from the domain of f.

The domain is $\{x \mid x \in \text{real numbers}\}$.

The graph of this function is shown at the right. Because the domain of f is all real numbers, there are points on the graph for all real numbers.

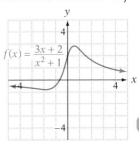

$f(x) = \dfrac{3x + 2}{x^2 + 1}$

YOU TRY IT 2

Find the domain of $p(x) = \dfrac{6x}{x^2 + 4}$.

Solution   See page S27. $\{x \mid x \in \text{real numbers}\}$

TAKE NOTE

In order to draw the graph of a discontinuous function, at some point you need to lift your pencil from the paper and put it back down in a different place.

A function is **continuous** if its graph has no breaks or undefined range values. The function in Example 2 is continuous. A function is **discontinuous** if there is a break in the graph. The function in Example 1 is discontinuous.

Look again at the graph of the function in Example 1. Note that as x gets closer and closer to -1 from the left, the y value gets smaller and smaller without bound. We say the value decreases without bound. As x gets closer and closer to -1 from the right, the y value gets larger and larger without bound. We say the value increases without bound. Also, as x gets closer and closer to 4 from the left, the y value decreases without bound, and as x gets closer and closer to 4 from the right, the y value increases without bound.

TAKE NOTE

As x gets closer to -1 from the left, y gets smaller and smaller. It decreases without bound.

x	-1.5	-1.2	-1.1	-1.01
y	-3.3	-8.1	-16.1	-160.1

As x gets closer to -1 from the right, y gets larger and larger. It increases without bound.

x	-0.5	-0.8	-0.9	-0.99
y	3.1	7.9	15.9	159.9

When the values of y increase or decrease without bound as the value of x approaches a number a, then the vertical line $x = a$ is called a **vertical asymptote** of the graph of the function. The vertical lines $x = -1$ and $x = 4$ are vertical asymptotes of the graph of $f(x) = \dfrac{2x - 6}{x^2 - 3x - 4}$.

Vertical Asymptotes of a Rational Function

The graph of $f(x) = \dfrac{p(x)}{q(x)}$, where $p(x)$ and $q(x)$ have no common factors, has a vertical asymptote at $x = a$ if a is a real number and a is a zero of the denominator $q(x)$.

EXAMPLE 3

Find the vertical asymptotes of the graph of the rational function.

a. $g(x) = \dfrac{x}{x^2 - x - 6}$ **b.** $h(x) = \dfrac{5x}{x^2 + 1}$

INSTRUCTOR NOTE
Exercises 42 and 44 can be used for similar in-class examples.

Solution

a. Find the zeros of the denominator.

$$x^2 - x - 6 = 0$$
$$(x + 2)(x - 3) = 0$$
$$x + 2 = 0 \qquad x - 3 = 0$$
$$x = -2 \qquad x = 3$$

Graphical check:

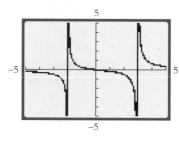

- The graph of g has vertical asymptotes at $x = -2$ and $x = 3$.

The numerator and denominator have no common factors, so both $x = -2$ and $x = 3$ are vertical asymptotes of the graph.

The lines $x = -2$ and $x = 3$ are vertical asymptotes of the graph of g.

b. $x^2 + 1 = 0$

Find the zeros of the denominators.

There are no real zeros of the denominator, so the graph has no vertical asymptotes.

The graph of h has no vertical asymptotes.

Graphical check:

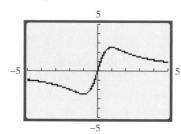

- The graph of h has no vertical asymptotes.

YOU TRY IT 3

Find the vertical asymptotes of the graph of the rational function.

a. $g(x) = \dfrac{3x^2 + 5}{x^2 - 25}$ **b.** $h(x) = \dfrac{4}{x^2 + 9}$

[handwritten:] $x^2 - 25 \Rightarrow (x+5)(x-5) = 0$
$x = 5 \quad x = -5$

Solution See page S28. **a.** $x = -5$, $x = 5$ **b.** none

Suggested Activity

See Section 7.1 of the *Student Activity Manual* for an activity on exploring vertical asymptotes and on simplifying rational expressions.

Suggested Activity

Have students evaluate $\dfrac{x^2 - 25}{x^2 + 13x + 40}$ and $\dfrac{x - 5}{x + 8}$ and determine that, except for $x = -8$ and $x = -5$, the expressions are equal.

■ Simplify Rational Expressions

The definition of a vertical asymptote of a rational function given above states that for the rational expression $\dfrac{p(x)}{q(x)}$, $p(x)$ and $q(x)$ have no common factors. A rational expression is in **simplest form** when the numerator and the denominator have no common factors other than 1.

The Multiplication Property of One is used to write a rational expression in simplest form. This is illustrated in the example that follows.

➡ Simplify: $\dfrac{x^2 - 25}{x^2 + 13x + 40}$

$$\frac{x^2 - 25}{x^2 + 13x + 40} = \frac{(x - 5)(x + 5)}{(x + 8)(x + 5)} = \frac{(x - 5)}{(x + 8)} \cdot \frac{(x + 5)}{(x + 5)} = \frac{x - 5}{x + 8} \cdot 1 = \frac{x - 5}{x + 8},$$
$$x \neq -8, -5$$

In this example, the requirement $x \neq -8, -5$ must be included because division by zero is not allowed.

The simplification above is usually shown with slashes to indicate that a common factor has been removed:

$$\frac{x^2 - 25}{x^2 + 13x + 40} = \frac{(x - 5)\overset{1}{\cancel{(x + 5)}}}{(x + 8)\underset{1}{\cancel{(x + 5)}}} = \frac{x - 5}{x + 8}, \qquad x \neq -8, -5 \quad \longleftarrow$$

We will show a simplification with slashes. We will also omit the restrictions that prevent division by zero. Nonetheless, those restrictions *always* are implied.

INSTRUCTOR NOTE
Exercises 64 and 72 can be used for similar in-class examples.

TAKE NOTE

Recall that
$(b - a) = -(a - b)$
Therefore,
$(4 - x) = -(x - 4)$
In general,

$$\frac{b - a}{a - b} = \frac{-\overset{1}{\cancel{(a - b)}}}{\underset{1}{\cancel{b - a}}}$$

$$= -1$$

EXAMPLE 4

Simplify.

a. $\dfrac{12 + 5x - 2x^2}{2x^2 - 3x - 20}$ **b.** $\dfrac{x^{2n} + x^n - 2}{x^{2n} - 1}$

Solution

a. $\dfrac{12 + 5x - 2x^2}{2x^2 - 3x - 20} = \dfrac{(4 - x)(3 + 2x)}{(x - 4)(2x + 5)}$

- Factor the numerator and the denominator.

$$= \frac{\overset{-1}{\cancel{(4 - x)}}(3 + 2x)}{\underset{1}{\cancel{(x - 4)}}(2x + 5)}$$

- Divide by the common factors. Remember: $4 - x = -(x - 4)$. Therefore, $\frac{4 - x}{x - 4} = \frac{-(x - 4)}{x - 4} = \frac{-1}{1} = -1$.

$$= -\frac{2x + 3}{2x + 5}$$

- Write the answer in simplest form.

b. $\dfrac{x^{2n} + x^n - 2}{x^{2n} - 1} = \dfrac{(x^n - 1)(x^n + 2)}{(x^n - 1)(x^n + 1)}$

- Factor the numerator and the denominator.

$$= \frac{\overset{1}{\cancel{(x^n - 1)}}(x^n + 2)}{\underset{1}{\cancel{(x^n - 1)}}(x^n + 1)}$$

- Divide by the common factors.

$$= \frac{x^n + 2}{x^n + 1}$$

- Write the answer in simplest form.

YOU TRY IT 4

Simplify.

a. $\dfrac{6x^4 - 24x^3}{12x^3 - 48x^2}$ **b.** $\dfrac{20x - 15x^2}{15x^3 - 5x^2 - 20x}$ **c.** $\dfrac{x^{2n} + x^n - 12}{x^{2n} - 3x^n}$

$$\frac{(x^n + 4)(x^n - 3)}{x^n(x^n - 3)}$$

Solution See page S28. **a.** $\dfrac{x}{2}$ **b.** $-\dfrac{1}{x + 1}$ **c.** $\dfrac{x^n + 4}{x^n}$

7.1 EXERCISES   Suggested Assignment: 7–71, odds

Topics for Discussion

1. What is a rational expression? Provide two examples of rational expressions.   A rational expression is one in which the numerator and denominator are polynomials. Examples will vary.

2. What is a rational function? Provide two examples of rational functions. A rational function is a function that is written in terms of a rational expression. Examples will vary.

3. The denominator of a rational function is $x^2 + 5$. Why are no real numbers excluded from the domain of this function?   Explanations will vary. For example: Because $x^2 \geq 0$, $x^2 + 5$ is positive, and there are no values of x for which the denominator is equal to zero.

4. Explain how to determine whether the graph of a rational function has a vertical asymptote.   Explanations will vary. For example: If the numerator and denominator have no common factors, and the denominator of the function has a real number zero, then the graph of the function has a vertical asymptote.

5. Explain how to simplify a rational expression.   Explanations will vary. For example: Factor the numerator and denominator, divide by the common factors, and then write the answer in simplest form.

6. Determine whether the statement is always true, sometimes true, or never true.

   a. A rational expression with a variable in the denominator will have restrictions on the value of that variable.

   b. When simplifying a rational expression, divide the numerator and denominator by the greatest common factor of the numerator and denominator.

   c. $\dfrac{a + 4}{a^2 + 6a + 8} = \dfrac{a + 4}{(a + 4)(a + 2)} = \dfrac{\cancel{a + 4}}{\cancel{(a + 4)}(a + 2)} = \dfrac{0}{a + 2} = 0$

   d. $\dfrac{a(a + 4)}{a} = a + 4$

   a. Sometimes true   b. Always true   c. Never true   d. Sometimes true

■ Rational Functions

7. Given $f(x) = \dfrac{x - 2}{x + 4}$, find $f(-2)$. -2

8. Given $f(x) = \dfrac{x - 3}{2x - 1}$, find $f(3)$. 0

9. Given $f(x) = \dfrac{1}{x^2 - 2x + 1}$, find $f(-2)$. $\dfrac{1}{9}$

10. Given $f(x) = \dfrac{-3}{x^2 - 4x + 2}$, find $f(-1)$. $-\dfrac{3}{7}$

11. Given $f(x) = \dfrac{x - 2}{2x^2 + 3x + 8}$, find $f(3)$. $\dfrac{1}{35}$

12. Given $f(x) = \dfrac{x^2}{3x^2 - 3x + 5}$, find $f(4)$. $\dfrac{16}{41}$

13. Given $f(x) = \dfrac{x^2 - 2x}{x^3 - x + 4}$, find $f(-1)$. $\dfrac{3}{4}$

14. Given $f(x) = \dfrac{8 - x^2}{x^3 - x^2 + 4}$, find $f(-3)$. $\dfrac{1}{32}$

Find the domain of the function.

15. $f(x) = \dfrac{x}{x + 4}$ $\{x \mid x \neq -4\}$

16. $g(x) = \dfrac{3x}{x - 5}$ $\{x \mid x \neq 5\}$

17. $R(x) = \dfrac{5x}{3x + 9}$ $\{x \mid x \neq -3\}$

18. $p(x) = \dfrac{-2x}{6 - 2x}$ $\{x \mid x \neq 3\}$

19. $f(x) = \dfrac{x^2 + 1}{x}$ $\{x \mid x \neq 0\}$

20. $g(x) = \dfrac{2x^3 - x - 1}{x^2}$ $\{x \mid x \neq 0\}$

21. $k(x) = \dfrac{x + 1}{x^2 + 1}$
$\{x \mid x \in \text{real numbers}\}$

22. $P(x) = \dfrac{2x + 3}{2x^2 + 3}$
$\{x \mid x \in \text{real numbers}\}$

23. $G(x) = \dfrac{3 - 4x}{x^2 + 4x - 5}$
$\{x \mid x \neq -5, 1\}$

24. $A(x) = \dfrac{5x + 2}{x^2 + 2x - 24}$
$\{x \mid x \neq -6, 4\}$

25. $f(x) = \dfrac{x^2 + 8x + 4}{2x^3 + 9x^2 - 5x}$
$\left\{x \mid x \neq -5, 0, \dfrac{1}{2}\right\}$

26. $H(x) = \dfrac{x^4 - 1}{2x^3 + 2x^2 - 24x}$
$\{x \mid x \neq -4, 0, 3\}$

27. **a.** Complete the input/output tables for the function $f(x) = \dfrac{x + 4}{x - 2}$.

x	1	1.5	1.75	1.95	1.999
$f(x)$					

x	3	2.5	2.25	2.05	2.001
$f(x)$					

b. Describe the graph of the function $f(x) = \dfrac{x + 4}{x - 2}$ as x approaches 2 from the left.

c. Describe the graph of the function $f(x) = \dfrac{x + 4}{x - 2}$ as x approaches 2 from the right.

a. First table: $-5, -11, -23, -119, -5999$; second table: 7, 13, 25, 121, 6001
b. It decreases.   **c.** It increases.

28. **a.** Complete the input/output tables for the function $f(x) = \dfrac{2x}{x + 3}$.

x	-4	-3.5	-3.2	-3.05	-3.001
$f(x)$					

x	-2	-2.5	-2.8	-2.9	-2.999
$f(x)$					

b. Describe the graph of the function $f(x) = \dfrac{2x}{x + 3}$ as x approaches -3 from the left.

c. Describe the graph of the function $f(x) = \dfrac{2x}{x + 3}$ as x approaches -3 from the right.

a. First table: 8, 14, 32, 122, 6002; second table: $-4, -10, -28, -58, -5998$
b. It increases.   **c.** It decreases.

29. Evaluate $h(x) = \frac{x + 2}{x - 3}$ when $x = 2.9, 2.99, 2.999$ and 2.9999. On the basis of your evaluations, complete the following sentence. As x becomes closer to 3 from the left, the value of $h(x)$ _____. decreases

30. Evaluate $h(x) = \frac{x + 2}{x - 3}$ when $x = 3.1, 3.01, 3.001$ and 3.0001. On the basis of your evaluations, complete the following sentence. As x becomes closer to 3 from the right, the value of $h(x)$ _____. increases

State whether the function is continuous or discontinuous.

31. $\dfrac{2x - 1}{x^2 + 3x}$

Discontinuous

32. $\dfrac{x^2 - 4}{x^2 + 16}$

Continuous

33. $\dfrac{8x}{x^4 + 1}$

Continuous

34. $\dfrac{x + 7}{x^2}$

Discontinuous

Use an algebraic method to find the vertical asymptote(s) of the graph of the function. Verify your answer by graphing the function.

35. $H(x) = \dfrac{4}{x - 3}$ $x = 3$

36. $G(x) = \dfrac{-2}{x + 2}$ $x = -2$

37. $V(x) = \dfrac{x^2}{(2x + 5)(3x - 6)}$ $x = -\dfrac{5}{2}, x = 2$

38. $F(x) = \dfrac{x^2 - 1}{(4x + 8)(3x - 1)}$ $x = -2, x = \dfrac{1}{3}$

39. $q(x) = \dfrac{3 - x}{x^2 - 2x - 8}$ $x = -2, x = 4$

40. $h(x) = \dfrac{2x + 1}{x^2 + 6x + 5}$ $x = -5, x = -1$

41. $f(x) = \dfrac{2x - 1}{x^2 + x - 6}$ $x = -3, x = 2$

42. $h(x) = \dfrac{3x}{x^2 - 4}$ $x = -2, x = 2$

43. $f(x) = \dfrac{4x - 7}{3x^2 + 12}$ No vertical asymptotes

44. $g(x) = \dfrac{x^2 + x + 1}{5x^2 + 1}$ No vertical asymptotes

45. $G(x) = \dfrac{x^2 + 1}{6x^2 - 13x + 6}$ $x = \dfrac{2}{3}, x = \dfrac{3}{2}$

46. $A(x) = \dfrac{5x - 7}{x(x - 2)(x - 3)}$ $x = 0, x = 2, x = 3$

47. *Water Purification*   The cost C, in dollars, to remove $p\%$ of the salt in a tank of sea water is given by the rational function $C(p) = \dfrac{2000p}{100 - p}$, $0 \leq p < 100$.

 a. Find the cost of removing 40% of the salt.
 b. Find the cost of removing 80% of the salt.
 c. As the percent of salt removed from the sea water increases, does the cost per percent removed increase or decrease?
 d. The point whose coordinates are (50, 2000) is on the graph of the function. Write a sentence that describes the meaning of this ordered pair.

e. 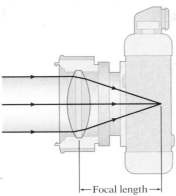 Use the function to explain why it is not possible to remove 100% of the salt from the sea water.

a. $1333.33   b. $8000   c. The cost increases.   d. The cost to remove 50% of the salt in a tank of sea water is $2000.   e. When $p = 100$, the denominator of the rational function is 0, and division by zero is undefined.

48. *Temperature*   The temperature F, in degrees Fahrenheit, of a dessert placed in a freezer for t hours is given by the rational function $F(t) = \dfrac{60}{t^2 + 2t + 1}, t \geq 0.$

   a.   According to the function, what is the temperature of the dessert before it is placed in the freezer?

   b.   Find the temperature of the dessert after it has been in the freezer for 1 hour.

   c.   Find the temperature of the dessert after 4 hours.

   d. The point whose coordinates are $(5, 1.\overline{6})$ is on the graph of the function. Write a sentence that describes the meaning of this ordered pair.

   e. Use the function to explain why it is not possible for the temperature of the dessert to be negative.

   a. 60°   b. 15°   c. 2.4°   d. Five hours after the dessert is placed in the freezer, its temperature is 1.$\overline{6}$°F.   e. It is given that $t \geq 0$. For $t \geq 0$, $t^2 + 2t + 1 > 0$. 60 divided by a positive number is a positive number. Therefore, the temperature cannot be negative.

49. *Traffic Flow*   The flow of traffic on a freeway F, measured in cars per hour past a reference point, is influenced by many factors. Under certain conditions, the flow can be modeled by the function $F(s) = \dfrac{s}{0.004 + 0.0008s}$, where s is the speed of the cars in miles per hour.

   a.   Find the flow of traffic when the speed is 50 miles per hour. Round to the nearest whole number.

   b.   Is the increase in flow when the speed changes from 20 miles per hour to 30 miles per hour the same as the increase in flow when the speed changes from 30 miles per hour to 40 miles per hour?

   c. The point whose coordinates are $(45, 1125)$ is on the graph of the function. Write a sentence that describes the meaning of this ordered pair.

   a. 1136 cars per hour   b. No   c. The flow of traffic is 1125 cars per hour when the speed of the cars is 45 mph.

50. *Photography*   The focal length of a camera lens is the distance from the lens to the point where parallel rays of light come to a focus.

←Focal length→

The relationship among the focal length (F), the distance between the object and the lens (x), and the distance between the lens and the film (y) is given by $\frac{1}{F} = \frac{1}{x} + \frac{1}{y}$. A camera used by a professional photographer has a dial that allows the focal length to be set at a constant value. Suppose a photographer chooses a focal length of 50 millimeters. Substituting this value into the equation, solving for y, and using the notation $y = f(x)$ yields the equation $f(x) = \frac{50x}{x - 50}$.

a. Graph this equation for $50 < x \le 6000$.

b. The point whose coordinates are (2000, 51), to the nearest integer, is on the graph of the function. Give an interpretation of this ordered pair.

c. Give a reason for choosing the domain so that $x > 50$.

d. Photographers refer to depth of field as a range of distances in which an object remains in focus. Use the graph to explain why the depth of field is larger for objects that are far from the lens than for objects that are close to the lens.

a.

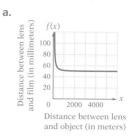

b. When the distance between the object and the lens is 2000 m, the distance between the lens and the film is 51 mm.   **c.** For $x = 50$, the expression $\frac{50x}{x - 50}$ is undefined. For $0 < x \le 50$, $f(x)$ is negative, and the distance cannot be negative. Therefore, the domain is $x > 50$.   **d.** For $x > 1000$, $f(x)$ changes very little for large changes in x.

■ Simplify Rational Expressions

Simplify.

51. $\dfrac{6x^2 - 2x}{2x}$ $3x - 1$

52. $\dfrac{3y - 12y^2}{3y}$ $1 - 4y$

53. $\dfrac{2x - 6}{3x - x^2}$ $-\dfrac{2}{x}$

54. $\dfrac{3a^2 - 6a}{12 - 6a}$ $-\dfrac{a}{2}$

55. $\dfrac{6x^3 - 15x^2}{12x^2 - 30x}$ $\dfrac{x}{2}$

56. $\dfrac{-36a^2 - 48a}{18a^3 + 24a^2}$ $-\dfrac{2}{a}$

57. $\dfrac{3x^{3n} - 9x^{2n}}{12x^{2n}}$ $\dfrac{x^n - 3}{4}$

58. $\dfrac{8a^n}{4a^{2n} - 8a^n}$ $\dfrac{2}{a^n - 2}$

59. $\dfrac{x^2 - 7x + 12}{x^2 - 9x + 20}$ $\dfrac{x - 3}{x - 5}$

60. $\dfrac{x^2 - x - 20}{x^2 - 2x - 15}$ $\dfrac{x + 4}{x + 3}$

61. $\dfrac{x^2 - xy - 2y^2}{x^2 - 3xy + 2y^2}$ $\dfrac{x + y}{x - y}$

62. $\dfrac{2x^2 + 7xy - 4y^2}{4x^2 - 4xy + y^2}$ $\dfrac{x + 4y}{2x - y}$

63. $\dfrac{3x^2 + 10x - 8}{8 - 14x + 3x^2}$ $-\dfrac{x + 4}{4 - x}$

64. $\dfrac{14 - 19x - 3x^2}{3x^2 - 23x + 14}$ $\dfrac{x + 7}{x - 7}$

65. $\dfrac{x^2 + x - 12}{x^2 - x - 12}$ Simplest form

66. $\dfrac{a^2 - 7a + 10}{a^2 + 9a + 14}$ Simplest form

67. $\dfrac{x^4 + 3x^2 + 2}{x^4 - 1}$ $\dfrac{x^2 + 2}{(x + 1)(x - 1)}$

68. $\dfrac{x^4 - 2x^2 - 3}{x^4 + 2x^2 + 1}$ $\dfrac{x^2 - 3}{x^2 + 1}$

69. $\dfrac{x^2y^2 + 4xy - 21}{x^2y^2 - 10xy + 21}$ $\dfrac{xy + 7}{xy - 7}$

70. $\dfrac{6x^2y^2 + 11xy + 4}{9x^2y^2 + 9xy - 4}$ $\dfrac{2xy + 1}{3xy - 1}$

71. $\dfrac{a^{2n} - a^n - 2}{a^{2n} + 3a^n + 2}$ $\dfrac{a^n - 2}{a^n + 2}$

72. $\dfrac{a^{2n} + a^n - 12}{a^{2n} - 2a^n - 3}$ $\dfrac{a^n + 4}{a^n + 1}$

Applying Concepts

73. A function f has $x = 6$ and $x = -2$ as vertical asymptotes. Write a possible expression for $f(x)$.   Answers will vary. For example, $\frac{x}{x^2 - 4x - 12}$.

74. If $a : b : c = 1 : 3 : 5$, find the value of $\dfrac{a + 3b + 5c}{a}$.   35

75. If $\dfrac{x^n(x^{n+1}y^{2n-1})^3}{(x^{2n-1}y^{3n-2})^2} = x^r y^s$, find the value of r.   5

76. For $x > 0$, let $f(x) = \dfrac{x}{1 + x}$. Find the least possible value of $B - A$ where $A < f(x) < B$ and A and B are integers.   1

77. Suppose that $F(x) = \dfrac{g(x)}{h(x)}$ and that, for some real number a, $g(a) = 0$ and $h(a) = 0$. Is $F(x)$ in simplest form? Explain your answer.
No. Explanations may vary. For example: Because $g(a) = h(a) = 0$, $x - a$ is a factor of both $g(x)$ and $h(x)$; therefore, $F(x)$ is not in simplest form.

78. Why can the numerator and denominator of a rational expression be divided by their common factors? What conditions must be placed on the value of the variables when a rational expression is simplified?   Answers will vary.

EXPLORATION

1. *Sets of Functions*

   a. For each of the functions below, state whether the function is a linear function, a quadratic function, a polynomial function, a rational function, or none of these types.

   1. $f(x) = \dfrac{2}{3}x^2$ **2.** $g(x) = -x$ **3.** $h(x) = \sqrt{3x} + 4$

   4. $k(x) = \dfrac{1}{x}$ **5.** $j(x) = \sqrt[3]{x} - 1$ **6.** $m(x) = \dfrac{1 - x^2}{x}$

   7. $p(x) = 2x^4 - x^2 + 7$ **8.** $q(x) = |2x - 3|$ **9.** $r(x) = 5$

   10. $t(x) = |x + 5|$ **11.** $s(x) = \dfrac{\sqrt{x} + 1}{\sqrt{x} - 1}$ **12.** $v(x) = x^5$

   Linear: 2, 3, 9; quadratic: 1; polynomial: 1, 2, 3, 7, 9, 12; rational: 1, 2, 3, 4, 6, 7, 9, 12; none: 5, 8, 10, 11.

   b. Are all polynomial functions also rational functions? Why or why not?   Yes. Rational functions are defined as being the ratio of two polynomials. If the polynomial in the denominator is the constant 1, then the rational function is also a polynomial function.

   c. Why are the graphs of rational functions often discontinuous, whereas those of polynomial functions are not?   Explanations may vary. For example, rational functions are often discontinuous because they are the ratio of two polynomial functions, whereas polynomial functions are not ratios. Or students may explain that in a rational function, a value of the variable may result in the denominator being zero, in which case the function is not defined for that value of the variable.

Operations on Rational Expressions

Multiply Rational Expressions

The product of two fractions is a fraction whose numerator is the product of the numerators of the two fractions and whose denominator is the product of the denominators of the two fractions.

$$\frac{a}{b} \cdot \frac{c}{d} = \frac{ac}{bd}$$

For example: $\dfrac{5}{a+2} \cdot \dfrac{b-3}{3} = \dfrac{5(b-3)}{(a+2)3} = \dfrac{5b-15}{3a+6}$

The product of two rational expressions can often be simplified by factoring the numerator and the denominator.

> ➡ Simplify: $\dfrac{x^2 - 2x}{2x^2 + x - 15} \cdot \dfrac{2x^2 - x - 10}{x^2 - 4}$

TAKE NOTE

Note that the factorizations are performed prior to the multiplication. If the numerators and denominators were multiplied first, factoring the resulting polynomials would be very difficult.

INSTRUCTOR NOTE
You might show students that the technique used to multiply rational expressions is the same as that used to multiply arithmetic fractions.

INSTRUCTOR NOTE
Exercises 8 and 12 can be used for similar in-class examples.

$$\frac{x^2 - 2x}{2x^2 + x - 15} \cdot \frac{2x^2 - x - 10}{x^2 - 4}$$

Factor the numerator and the denominator of each fraction.

$$= \frac{x(x-2)}{(x+3)(2x-5)} \cdot \frac{(x+2)(2x-5)}{(x+2)(x-2)}$$

Multiply.

$$= \frac{x(x-2)(x+2)(2x-5)}{(x+3)(2x-5)(x+2)(x-2)}$$

Divide by the common factors.

$$= \frac{x \cancel{(x-2)}^1 \cancel{(x+2)}^1 \cancel{(2x-5)}^1}{(x+3)\cancel{(2x-5)}_1 \cancel{(x+2)}_1 \cancel{(x-2)}_1}$$

Write the answer in simplest form.

$$= \frac{x}{x+3}$$

⬅

EXAMPLE 1

Multiply. **a.** $\dfrac{2x^2 - 6x}{3x - 6} \cdot \dfrac{6x - 12}{8x^3 - 12x^2}$ **b.** $\dfrac{6x^2 + x - 2}{6x^2 + 7x + 2} \cdot \dfrac{2x^2 + 9x + 4}{4 - 7x - 2x^2}$

Solution

a. $\dfrac{2x^2 - 6x}{3x - 6} \cdot \dfrac{6x - 12}{8x^3 - 12x^2} = \dfrac{2x(x - 3)}{3(x - 2)} \cdot \dfrac{6(x - 2)}{4x^2(2x - 3)}$

$$= \frac{12x(x-3)\cancel{(x-2)}^1}{12x^2\cancel{(x-2)}_1(2x-3)} = \frac{x-3}{x(2x-3)}$$

b. $\dfrac{6x^2 + x - 2}{6x^2 + 7x + 2} \cdot \dfrac{2x^2 + 9x + 4}{4 - 7x - 2x^2} = \dfrac{(2x - 1)(3x + 2)}{(3x + 2)(2x + 1)} \cdot \dfrac{(2x + 1)(x + 4)}{(1 - 2x)(4 + x)}$

$$= \dfrac{\overset{-1}{\cancel{(2x - 1)}}\,\overset{1}{\cancel{(3x + 2)}}\,\overset{1}{\cancel{(2x + 1)}}\,\overset{1}{\cancel{(x + 4)}}}{\underset{1}{\cancel{(3x + 2)}}\,\underset{1}{\cancel{(2x + 1)}}\,\underset{1}{\cancel{(1 - 2x)}}\,\underset{1}{\cancel{(x + 4)}}}$$

$$= -1$$

YOU TRY IT 1

Multiply.

a. $\dfrac{12 + 5x - 3x^2}{x^2 + 2x - 15} \cdot \dfrac{2x^2 + x - 45}{3x^2 + 4x}$ **b.** $\dfrac{2x^2 - 13x + 20}{x^2 - 16} \cdot \dfrac{2x^2 + 9x + 4}{6x^2 - 7x - 5}$

Solution   See page S28.   a. $-\dfrac{2x - 9}{x}$ b. $\dfrac{2x - 5}{3x - 5}$

■ Divide Rational Expressions

TAKE NOTE

The reciprocal of an expression is also called the **multiplicative inverse**.

The **reciprocal of a rational expression** is the rational expression with the numerator and denominator interchanged.

$$\text{Rational expression} \left\{ \begin{array}{cc} \dfrac{a}{b} & \dfrac{b}{a} \\[2ex] \dfrac{a^2 - 2y}{4} & \dfrac{4}{a^2 - 2y} \end{array} \right\} \text{Reciprocal}$$

? QUESTION   What is the reciprocal of $\dfrac{3x^2 - 5}{x + 7}$?

To divide two rational expressions, multiply by the reciprocal of the divisor.

$$\dfrac{a}{b} \div \dfrac{c}{d} = \dfrac{a}{b} \cdot \dfrac{d}{c} = \dfrac{ad}{bc}$$

For example: $\dfrac{2}{a} \div \dfrac{5}{b} = \dfrac{2}{a} \cdot \dfrac{b}{5} = \dfrac{2b}{5a}$

INSTRUCTOR NOTE
Exercise 20 can be used for a similar in-class example.

EXAMPLE 2

Divide.

a. $\dfrac{12x^2y^2 - 24xy^2}{5z^2} \div \dfrac{4x^3y - 8x^2y}{3z^4}$ **b.** $\dfrac{3y^2 - 10y + 8}{3y^2 + 8y - 16} \div \dfrac{2y^2 - 7y + 6}{2y^2 + 5y - 12}$

$$\dfrac{12xy^2(x - 2)}{5z^2}$$

? ANSWER $\dfrac{x + 7}{3x^2 - 5}$

Solution

a.
$$\frac{12x^2y^2 - 24xy^2}{5z^2} \div \frac{4x^3y - 8x^2y}{3z^4} = \frac{12x^2y^2 - 24xy^2}{5z^2} \cdot \frac{3z^4}{4x^3y - 8x^2y}$$

$$= \frac{12xy^2(x - 2)}{5z^2} \cdot \frac{3z^4}{4x^2y(x - 2)}$$

$$= \frac{36xy^2z^4(x - 2)}{20x^2yz^2(x - 2)} = \frac{9yz^2}{5x}$$

b.
$$\frac{3y^2 - 10y + 8}{3y^2 + 8y - 16} \div \frac{2y^2 - 7y + 6}{2y^2 + 5y - 12} = \frac{3y^2 - 10y + 8}{3y^2 + 8y - 16} \cdot \frac{2y^2 + 5y - 12}{2y^2 - 7y + 6}$$

$$= \frac{(y - 2)(3y - 4)}{(3y - 4)(y + 4)} \cdot \frac{(y + 4)(2y - 3)}{(y - 2)(2y - 3)}$$

$$= \frac{(y - 2)(3y - 4)(y + 4)(2y - 3)}{(3y - 4)(y + 4)(y - 2)(2y - 3)}$$

$$= 1$$

YOU TRY IT 2

Divide.

a. $\dfrac{6x^2 - 3xy}{10ab^4} \div \dfrac{16x^2y^2 - 8xy^3}{15a^2b^2}$ **b.** $\dfrac{6x^2 - 7x + 2}{3x^2 + x - 2} \div \dfrac{4x^2 - 8x + 3}{5x^2 + x - 4}$

Solution   See page S28.   **a.** $\dfrac{9a}{16b^2y^2}$ **b.** $\dfrac{5x - 4}{2x - 3}$

■ Add and Subtract Rational Expressions

When adding rational expressions in which the denominators are the same, add the numerators. The denominator of the sum is the common denominator. Write the answer in simplest form.

$$\frac{a}{c} + \frac{b}{c} = \frac{a + b}{c}$$

For example: $\dfrac{4x}{15} + \dfrac{8x}{15} = \dfrac{4x + 8x}{15} = \dfrac{12x}{15} = \dfrac{4x}{5}$

When subtracting rational expressions with the same denominators, subtract the numerators. The denominator of the difference is the common denominator. Write the answer in simplest form.

$$\frac{a}{c} - \frac{b}{c} = \frac{a - b}{c}$$

For example: $\dfrac{y}{y - 3} - \dfrac{3}{y - 3} = \dfrac{y - 3}{y - 3} = \dfrac{(y - 3)}{(y - 3)} = 1$

TAKE NOTE

Note in the example at the right that we must subtract the entire numerator of the second expression.
$(7x - 12) - (3x - 6)$
$= 7x - 12 - 3x + 6$

Here is another example of subtracting rational expressions with the same denominator.

$$\frac{7x - 12}{2x^2 + 5x - 12} - \frac{3x - 6}{2x^2 + 5x - 12}$$

$$= \frac{(7x - 12) - (3x - 6)}{2x^2 + 5x - 12} = \frac{7x - 12 - 3x + 6}{2x^2 + 5x - 12} = \frac{4x - 6}{2x^2 + 5x - 12}$$

$$= \frac{2(2x - 3)}{(2x - 3)(x + 4)} = \frac{\overset{1}{2(2x - 3)}}{\underset{1}{(2x - 3)}(x + 4)} = \frac{2}{x + 4}$$

Before two rational expressions with different denominators can be added or subtracted, both rational expressions must be expressed in terms of a common denominator. This common denominator is the least common multiple (LCM) of the denominators of the rational expressions.

The LCM of two or more polynomials is the simplest polynomial that contains the factors of each polynomial. To find the LCM, first factor each polynomial completely. The LCM is the product of each factor the greatest number of times it occurs in any one factorization.

To find the LCM of $x^2 + 5x$ and $x^4 + 4x^3 - 5x^2$, factor each polynomial.

$$x^2 + 5x = x(x + 5)$$
$$x^4 + 4x^3 - 5x^2 = x^2(x^2 + 4x - 5) = x^2(x - 1)(x + 5)$$

The LCM is the product of each factor the greatest number of times it occurs in any one factorization.

$$LCM = x^2(x - 1)(x + 5)$$

➡ Add the rational expressions: $\dfrac{x}{2x - 3} + \dfrac{x + 2}{2x^2 + x - 6}$

$2x^2 + x - 6 = (2x - 3)(x + 2)$

The LCM of the denominators is $(2x - 3)(x + 2)$.

$$\frac{x}{2x - 3} + \frac{x + 2}{2x^2 + x - 6}$$

Rewrite each fraction in terms of the LCM of the denominators.

$$= \frac{x}{2x - 3} \cdot \frac{x + 2}{x + 2} + \frac{x + 2}{(2x - 3)(x + 2)}$$

$$= \frac{x^2 + 2x}{(2x - 3)(x + 2)} + \frac{x + 2}{(2x - 3)(x + 2)}$$

Add the fractions.

$$= \frac{(x^2 + 2x) + (x + 2)}{(2x - 3)(x + 2)}$$

$$= \frac{x^2 + 3x + 2}{(2x - 3)(x + 2)}$$

Factor the numerator to determine whether there are common factors in the numerator and denominator.

$$= \frac{(x + 2)(x + 1)}{(2x - 3)(x + 2)}$$

$$= \frac{\overset{1}{(x + 2)}(x + 1)}{(2x - 3)\underset{1}{(x + 2)}} = \frac{x + 1}{2x - 3}$$

INSTRUCTOR NOTE

One reason why adding and subtracting rational expressions is difficult for students is the number of steps. Encourage students to develop an outline similar to the following that they can use for the exercises in this section.

• Find a common denominator.
• Express each fraction in terms of the common denominator.
• Add (or subtract) the rational expressions.
• Simplify.

INSTRUCTOR NOTE
Exercise 32 can be used for a similar in-class example.

EXAMPLE 3

Subtract: $\dfrac{x}{x-2} - \dfrac{4}{x^2 - 2x}$

Solution $\dfrac{x}{x-2} - \dfrac{4}{x^2 - 2x}$

$= \dfrac{x}{x-2} \cdot \dfrac{x}{x} - \dfrac{4}{x(x-2)}$ • Write the fractions in terms of the LCM. The LCM is $x(x-2)$.

$= \dfrac{x^2}{x(x-2)} - \dfrac{4}{x(x-2)}$

$= \dfrac{x^2 - 4}{x(x-2)}$ • Subtract the fractions.

$= \dfrac{(x+2)(x-2)}{x(x-2)}$ • Factor the numerator.

$= \dfrac{(x+2)\cancel{(x-2)}}{x\cancel{(x-2)}} = \dfrac{x+2}{x}$ • Divide by the common factors.

> **TAKE NOTE**
>
> We are multiplying the expression $\frac{x}{x-2}$ by $\frac{x}{x}$, which equals 1.
>
> Multiplying an expression by 1 does not change the value of the expression.

YOU TRY IT 3

Subtract: $\dfrac{5}{y-3} - \dfrac{2}{y+1}$

Solution See page S28. $\dfrac{3y+11}{(y-3)(y+1)}$

INSTRUCTOR NOTE
Exercise 34 can be used for a similar in-class example

EXAMPLE 4

Add: $y + \dfrac{3}{5y}$

Solution $y + \dfrac{3}{5y} = \dfrac{y}{1} + \dfrac{3}{5y}$ • Rewrite y as a fraction.

$= \dfrac{y}{1} \cdot \dfrac{5y}{5y} + \dfrac{3}{5y}$ • The LCM of 1 and $5y$ is $5y$.

$= \dfrac{5y^2}{5y} + \dfrac{3}{5y}$

$= \dfrac{5y^2 + 3}{5y}$ • Add the fractions. This fraction is in simplest form.

YOU TRY IT 4

Subtract: $x - \dfrac{5}{6x}$

Solution See page S28. $\dfrac{6x^2 - 5}{6x}$

▼ *Point of Interest*

Complex fractions occur in formulas used in many fields of study. For example, the total resistance, R_T, of three resistors in parallel is

$$R_T = \cfrac{1}{\dfrac{1}{R_1} + \dfrac{1}{R_2} + \dfrac{1}{R_3}}$$

■ Complex Fractions

A **complex fraction** is a fraction whose numerator or denominator contains one or more fractions. Examples of complex fractions follow.

$$\cfrac{5}{2 + \dfrac{1}{2}} \qquad \cfrac{5 + \dfrac{1}{y}}{5 - \dfrac{1}{y}} \qquad \cfrac{x + 4 + \dfrac{1}{x + 2}}{x - 2 + \dfrac{1}{x + 2}}$$

One method of simplifying a complex fraction is to multiply the numerator and denominator of the complex fraction by the LCM of the denominators. Here is an example.

➡ Simplify: $\cfrac{\dfrac{1}{x} + \dfrac{1}{y}}{\dfrac{1}{x} - \dfrac{1}{y}}$

TAKE NOTE

The Distributive Property is used to multiply xy times each term in the numerator and each term in the denominator.

Multiply the numerator and denominator of the complex fraction by the LCM of the denominators. The LCM of x and y is xy.

$$\cfrac{\dfrac{1}{x} + \dfrac{1}{y}}{\dfrac{1}{x} - \dfrac{1}{y}} = \cfrac{\dfrac{1}{x} + \dfrac{1}{y}}{\dfrac{1}{x} - \dfrac{1}{y}} \cdot \dfrac{xy}{xy} = \cfrac{\dfrac{1}{x} \cdot xy + \dfrac{1}{y} \cdot xy}{\dfrac{1}{x} \cdot xy - \dfrac{1}{y} \cdot xy} = \dfrac{y + x}{y - x}$$

Note that after the numerator and denominator of the complex fraction have been multiplied by the LCM of the denominators, no fraction remains in the numerator or denominator.

INSTRUCTOR NOTE
Exercise 58 can be used for a similar in-class example.

EXAMPLE 5

Simplify. **a.** $\cfrac{2 - \dfrac{11}{x} + \dfrac{15}{x^2}}{3 - \dfrac{5}{x} - \dfrac{12}{x^2}}$ **b.** $\cfrac{2x - 1 + \dfrac{7}{x + 4}}{3x - 8 + \dfrac{17}{x + 4}}$

Solution

TAKE NOTE

We are using the Distributive Property to multiply x^2 times each term in the numerator and each term in the denominator.

a. $\cfrac{2 - \dfrac{11}{x} + \dfrac{15}{x^2}}{3 - \dfrac{5}{x} - \dfrac{12}{x^2}} = \cfrac{2 - \dfrac{11}{x} + \dfrac{15}{x^2}}{3 - \dfrac{5}{x} - \dfrac{12}{x^2}} \cdot \dfrac{x^2}{x^2}$ • The LCM is x^2.

$$= \cfrac{2 \cdot x^2 - \dfrac{11}{x} \cdot x^2 + \dfrac{15}{x^2} \cdot x^2}{3 \cdot x^2 - \dfrac{5}{x} \cdot x^2 - \dfrac{12}{x^2} \cdot x^2}$$

$$= \dfrac{2x^2 - 11x + 15}{3x^2 - 5x - 12}$$

$$= \dfrac{(2x - 5)(x - 3)}{(3x + 4)(x - 3)} = \dfrac{2x - 5}{3x + 4}$$

b. $\dfrac{2x - 1 + \dfrac{7}{x + 4}}{3x - 8 + \dfrac{17}{x + 4}} = \dfrac{2x - 1 + \dfrac{7}{x + 4}}{3x - 8 + \dfrac{17}{x + 4}} \cdot \dfrac{x + 4}{x + 4}$ • The LCM is $x + 4$.

$$= \dfrac{2x(x + 4) - 1(x + 4) + \dfrac{7}{x + 4}(x + 4)}{3x(x + 4) - 8(x + 4) + \dfrac{17}{x + 4}(x + 4)}$$

$$= \dfrac{2x^2 + 8x - x - 4 + 7}{3x^2 + 12x - 8x - 32 + 17}$$

$$= \dfrac{2x^2 + 7x + 3}{3x^2 + 4x - 15}$$

$$= \dfrac{(2x + 1)(x + 3)}{(3x - 5)(x + 3)} = \dfrac{2x + 1}{3x - 5}$$

Suggested Activity

Have students find the reciprocal, in simplest form, of the complex fraction $\dfrac{\dfrac{x}{2x-5}}{\dfrac{x^2}{4x^2 - 25}}$ and show that the product of the complex fraction and its reciprocal is 1.

Note: The reciprocal is $\dfrac{x}{2x + 5}$.

YOU TRY IT 5

Simplify. **a.** $\dfrac{3 + \dfrac{16}{x} + \dfrac{16}{x^2}}{6 + \dfrac{5}{x} - \dfrac{4}{x^2}}$ **b.** $\dfrac{2x + 5 + \dfrac{14}{x - 3}}{4x + 16 + \dfrac{49}{x - 3}}$

Solution See page S28. **a.** $\dfrac{x + 4}{2x - 1}$ **b.** $\dfrac{x - 1}{2x + 1}$

An alternative method of simplifying a complex fraction is to rewrite the numerator and denominator of the complex fraction as a single fraction and then divide the numerator by the denominator. Here, the same complex fraction presented on page 481 is simplified using this method.

$$\dfrac{\dfrac{1}{x} + \dfrac{1}{y}}{\dfrac{1}{x} - \dfrac{1}{y}} = \dfrac{\dfrac{1}{x} \cdot \dfrac{y}{y} + \dfrac{1}{y} \cdot \dfrac{x}{x}}{\dfrac{1}{x} \cdot \dfrac{y}{y} - \dfrac{1}{y} \cdot \dfrac{x}{x}} = \dfrac{\dfrac{y}{xy} + \dfrac{x}{xy}}{\dfrac{y}{xy} - \dfrac{x}{xy}} = \dfrac{\dfrac{y + x}{xy}}{\dfrac{y - x}{xy}}$$

$$= \dfrac{y + x}{xy} \div \dfrac{y - x}{xy} = \dfrac{y + x}{xy} \cdot \dfrac{xy}{y - x} = \dfrac{(y + x)xy}{xy(y - x)} = \dfrac{y + x}{y - x}$$

INSTRUCTOR NOTE
Exercise 56 can be used for a similar in-class example.

EXAMPLE 6

Simplify using the alternative method illustrated in the above example:

$$\dfrac{2 + \dfrac{5}{x} - \dfrac{3}{x^2}}{3 + \dfrac{11}{x} + \dfrac{6}{x^2}}$$

Solution

$$\frac{2 + \dfrac{5}{x} - \dfrac{3}{x^2}}{3 + \dfrac{11}{x} + \dfrac{6}{x^2}} = \frac{\dfrac{2}{1} \cdot \dfrac{x^2}{x^2} + \dfrac{5}{x} \cdot \dfrac{x}{x} - \dfrac{3}{x^2}}{\dfrac{3}{1} \cdot \dfrac{x^2}{x^2} + \dfrac{11}{x} \cdot \dfrac{x}{x} + \dfrac{6}{x^2}} = \frac{\dfrac{2x^2 + 5x - 3}{x^2}}{\dfrac{3x^2 + 11x + 6}{x^2}}$$

$$= \frac{2x^2 + 5x - 3}{x^2} \div \frac{3x^2 + 11x + 6}{x^2}$$

$$= \frac{2x^2 + 5x - 3}{x^2} \cdot \frac{x^2}{3x^2 + 11x + 6}$$

$$= \frac{(2x - 1)(x + 3)}{x^2} \cdot \frac{x^2}{(3x + 2)(x + 3)}$$

$$= \frac{(2x - 1)(x + 3) \cdot x^2}{x^2(3x + 2)(x + 3)} = \frac{2x - 1}{3x + 2}$$

Suggested Activity

See Section 7.2 of the *Student Activity Manual* for an activity on evaluating complex fractions using technology. A focus of the activity is the formula for a monthly loan payment.

YOU TRY IT 6

Simplify using the alternative method illustrated in Example 6:

$$\frac{\dfrac{1}{x + 2} + \dfrac{4}{x - 3}}{\dfrac{2}{x - 3} - \dfrac{7}{x + 2}}$$

Solution   See page S29. $-\dfrac{x + 1}{x - 5}$

7.2 EXERCISES   Suggested Assignment 7–59, odds

Topics for Discussion

1. Explain how to multiply two rational expressions.   Explanations may vary. For example: Multiply the numerators and denominators, divide by the common factors, and write the answer in simplest form.

2. Explain how to divide two rational expressions.   Explanations may vary. For example: Change the division to multiplication, write the reciprocal of the divisor, and then multiply the two expressions.

3. Why must rational expressions have the same denominator before they can be added or subtracted?   Answers will vary. Students might make an analogy to monomials: Monomials that are like terms can be added or subtracted; otherwise they cannot be combined. In order for us to add or subtract fractions, they must be "like terms," which means they must have the same denominator.

4. What is a complex fraction?   A complex fraction is a fraction whose numerator or denominator contains one or more fractions.

5. What is the general goal of simplifying a complex fraction?
The goal is to rewrite the expression with no fractions in the numerator
or denominator.

■ Operations on Rational Expressions

Add, subtract, multiply, or divide.

6. $\dfrac{27a^2b^5}{16xy^2} \cdot \dfrac{20x^2y^3}{9a^2b}$ $\dfrac{15b^4xy}{4}$

7. $\dfrac{15x^2y^4}{24ab^3} \cdot \dfrac{28a^2b^4}{35xy^4}$ $\dfrac{abx}{2}$

8. $\dfrac{3x - 15}{4x^2 - 2x} \cdot \dfrac{20x^2 - 10x}{15x - 75}$ 1

9. $\dfrac{2x^2 + 4x}{8x^2 - 40x} \cdot \dfrac{6x^3 - 30x^2}{3x^2 + 6x}$ $\dfrac{x}{2}$

10. $\dfrac{x^2y^3}{x^2 - 4x - 5} \cdot \dfrac{2x^2 - 13x + 15}{x^4y^3}$ $\dfrac{2x - 3}{x^2(x + 1)}$

11. $\dfrac{2x^2 - 5x + 3}{x^6y^3} \cdot \dfrac{x^4y^4}{2x^2 - x - 3}$ $\dfrac{y(x - 1)}{x^2(x + 1)}$

12. $\dfrac{x^2 - 3x + 2}{x^2 - 8x + 15} \cdot \dfrac{x^2 + x - 12}{8 - 2x - x^2}$ $-\dfrac{x - 1}{x - 5}$

13. $\dfrac{x^2 + x - 6}{12 + x - x^2} \cdot \dfrac{x^2 + x - 20}{x^2 - 4x + 4}$ $-\dfrac{x + 5}{x - 2}$

14. $\dfrac{x^{n+1} + 2x^n}{4x^2 - 6x} \cdot \dfrac{8x^2 - 12x}{x^{n+1} - x^n}$ $\dfrac{2(x + 2)}{x - 1}$

15. $\dfrac{x^{2n} + 2x^n}{x^{n+1} + 2x} \cdot \dfrac{x^2 - 3x}{x^{n+1} - 3x^n}$ 1

16. $\dfrac{x^{2n} - x^n - 6}{x^{2n} + x^n - 2} \cdot \dfrac{x^{2n} - 5x^n - 6}{x^{2n} - 2x^n - 3}$ $\dfrac{x^n - 6}{x^n - 1}$

17. $\dfrac{x^{2n} + 3x^n + 2}{x^{2n} - x^n - 6} \cdot \dfrac{x^{2n} + x^n - 12}{x^{2n} - 1}$ $\dfrac{x^n + 4}{x^n - 1}$

18. $\dfrac{6x^2y^4}{35a^2b^5} \div \dfrac{12x^3y^3}{7a^4b^5}$ $\dfrac{a^2y}{10x}$

19. $\dfrac{12a^4b^7}{13x^2y^2} \div \dfrac{18a^5b^6}{26xy^3}$ $\dfrac{4by}{3ax}$

20. $\dfrac{3x^2 - 10x - 8}{6x^2 + 13x + 6} \div \dfrac{2x^2 - 9x + 10}{4x^2 - 4x - 15}$ $\dfrac{x - 4}{x - 2}$

21. $\dfrac{x^2 - 8x + 15}{x^2 + 2x - 35} \div \dfrac{x^2 + 2x - 15}{x^2 + 9x + 14}$ $\dfrac{x + 2}{x + 5}$

22. $\dfrac{2x^{2n} - x^n - 6}{x^{2n} - x^n - 2} \div \dfrac{2x^{2n} + x^n - 3}{x^{2n} - 1}$ 1

23. $\dfrac{x^{4n} - 1}{x^{2n} + x^n - 2} \div \dfrac{x^{2n} + 1}{x^{2n} + 3x^n + 2}$ $(x^n + 1)^2$

24. $-\dfrac{3}{4x^2} + \dfrac{8}{4x^2} - \dfrac{3}{4x^2}$ $\dfrac{1}{2x^2}$

25. $\dfrac{x}{x^2 - 3x + 2} - \dfrac{2}{x^2 - 3x + 2}$ $\dfrac{1}{x - 1}$

26. $\dfrac{3x}{3x^2 + x - 10} - \dfrac{5}{3x^2 + x - 10}$ $\dfrac{1}{x + 2}$

27. $\dfrac{3}{2x^2y} - \dfrac{8}{5x} - \dfrac{9}{10xy}$ $\dfrac{15 - 16xy - 9x}{10x^2y}$

28. $\dfrac{2}{5ab} - \dfrac{3}{10a^2b} + \dfrac{4}{15ab^2}$ $\dfrac{12ab - 9b + 8a}{30a^2b^2}$

29. $\dfrac{2x - 1}{12x} - \dfrac{3x + 4}{9x}$ $-\dfrac{6x + 19}{36x}$

30. $\dfrac{3x - 4}{6x} - \dfrac{2x - 5}{4x}$ $\dfrac{7}{12x}$

31. $\dfrac{2x}{x - 3} - \dfrac{3x}{x - 5}$ $-\dfrac{x^2 + x}{(x - 3)(x - 5)}$

32. $\dfrac{3a}{a - 2} - \dfrac{5a}{a + 1}$ $-\dfrac{2a^2 - 13a}{(a - 2)(a + 1)}$

33. $x + \dfrac{8}{5y}$ $\dfrac{5xy + 8}{5y}$

34. $\dfrac{6}{7x} + y$ $\dfrac{7xy + 6}{7x}$

35. $\dfrac{10}{x} - \dfrac{2}{x - 4}$ $\dfrac{8(x - 5)}{x(x - 4)}$

36. $\dfrac{6a}{a - 3} + \dfrac{3}{a}$ $\dfrac{3(2a + 3)(a - 1)}{a(a - 3)}$

37. $\dfrac{2x - 3}{x + 5} - \dfrac{x^2 - 4x - 19}{x^2 + 8x + 15}$ $\dfrac{x + 2}{x + 3}$

38. $\dfrac{-3x^2 + 8x + 2}{x^2 + 2x - 8} - \dfrac{2x - 5}{x + 4}$ $-\dfrac{5x^2 - 17x + 8}{(x + 4)(x - 2)}$

Solve.

39. Use $x = 3$ and $y = 5$ to show that $\dfrac{1}{x} + \dfrac{1}{y} \neq \dfrac{1}{x + y}$. $\dfrac{8}{15} \neq \dfrac{1}{8}$

40. Use $x = 3$ and $y = 5$ to show that $\dfrac{1}{x} - \dfrac{1}{y} \neq \dfrac{1}{x - y}$. $\dfrac{2}{15} \neq -\dfrac{1}{2}$

41. Find the rational expression in simplest form that represents the sum of the reciprocals of the consecutive integers x and $x + 1$. $\dfrac{2x + 1}{x(x + 1)}$

Rewrite the expression as the sum of two fractions in simplest form.

42. $\dfrac{3x + 6y}{xy}$ $\dfrac{3}{y} + \dfrac{6}{x}$

43. $\dfrac{5a + 8b}{ab}$ $\dfrac{5}{b} + \dfrac{8}{a}$

44. $\dfrac{4a^2 + 3ab}{a^2b^2}$ $\dfrac{4}{b^2} + \dfrac{3}{ab}$

45. Complete the equation: $\dfrac{2x - 6}{6x^2 - 15x} \div ? = \dfrac{3}{2}$ $\dfrac{4x - 12}{18x^2 - 45x}$

46. Complete the equation: $\dfrac{1}{2x - 3} - ? = \dfrac{5}{2x}$ $\dfrac{-8x + 15}{2x(2x - 3)}$

■ Complex Fractions

Simplify.

47. $\dfrac{1 + \dfrac{1}{x}}{1 - \dfrac{1}{x^2}}$ $\dfrac{x}{x - 1}$

48. $\dfrac{\dfrac{1}{y^2} - 1}{1 + \dfrac{1}{y}}$ $\dfrac{1 - y}{y}$

49. $\dfrac{\dfrac{1}{a^2} - \dfrac{1}{a}}{\dfrac{1}{a^2} + \dfrac{1}{a}}$ $-\dfrac{a - 1}{a + 1}$

50. $\dfrac{\dfrac{1}{b} + \dfrac{1}{2}}{\dfrac{4}{b^2} - 1}$ $\dfrac{b}{2(2 - b)}$

51. $\dfrac{2 - \dfrac{4}{x + 2}}{5 - \dfrac{10}{x + 2}}$ $\dfrac{2}{5}$

52. $\dfrac{4 + \dfrac{12}{2x - 3}}{5 + \dfrac{15}{2x - 3}}$ $\dfrac{4}{5}$

53. $\dfrac{\dfrac{x}{x + 1} - \dfrac{1}{x}}{\dfrac{x}{x + 1} + \dfrac{1}{x}}$ $\dfrac{x^2 - x - 1}{x^2 + x + 1}$

54. $\dfrac{\dfrac{2a}{a - 1} - \dfrac{3}{a}}{\dfrac{1}{a - 1} + \dfrac{2}{a}}$ $\dfrac{2a^2 - 3a + 3}{3a - 2}$

55. $\dfrac{1 - \dfrac{1}{x} - \dfrac{6}{x^2}}{1 - \dfrac{4}{x} + \dfrac{3}{x^2}}$ $\dfrac{x + 2}{x - 1}$

56. $\dfrac{1 - \dfrac{3}{x} - \dfrac{10}{x^2}}{1 + \dfrac{11}{x} + \dfrac{18}{x^2}}$ $\dfrac{x-5}{x+9}$

57. $\dfrac{x - 1 + \dfrac{2}{x-4}}{x + 3 + \dfrac{6}{x-4}}$ $\dfrac{x-2}{x+2}$

58. $\dfrac{x - 5 - \dfrac{18}{x+2}}{x + 7 + \dfrac{6}{x+2}}$ $\dfrac{x-7}{x+5}$

59. Find the sum of the reciprocals of three consecutive even integers.

$\dfrac{3n^2 + 12n + 8}{n(n+2)(n+4)}$

60. *Electricity*   The total resistance, R_T, of three resistors in parallel is given by $R_T = \dfrac{1}{\dfrac{1}{R_1} + \dfrac{1}{R_2} + \dfrac{1}{R_3}}$. Find the total parallel resistance when $R_1 = 2$ ohms, $R_2 = 4$ ohms, and $R_3 = 8$ ohms. Round to the nearest hundredth.

1.14 ohms

61. *Harmonic Mean*   A number h is the harmonic mean of the numbers a and b if the reciprocal of h is equal to the average of the reciprocals of a and b.

   a.   Write an expression for the harmonic mean of a and b.
   b.   Find the harmonic mean of 10 and 15.

a. $\dfrac{2ab}{a+b}$ **b.** 12

62. *Uniform Motion Problem*   A plane flew from St. Louis to Boston, a distance of d miles, at an average rate of 400 mph. Because of prevailing winds, on the return trip the plane flew at an average rate of 500 mph.

   a.   Write an expression for the total flying time.
   b.   Find the average rate for the entire round trip.

a. $\dfrac{9d}{2000}$ **b.** $444.\overline{4}$ mph

63. *Interest Rates*   The interest rate on a loan to purchase a car affects the monthly payment. The function that relates the monthly payment for a 5-year (60-month) loan to the monthly interest rate is given by $P(x) = \dfrac{Cx}{1 - \dfrac{1}{(x+1)^{60}}}$, where x is the monthly interest rate as a decimal, C is the loan amount for the car, and $P(x)$ is the monthly payment.

   a.   Simplify the complex fraction.
   b.   Graph this equation for $0 < x \le 0.025$ and $C = \$20,000$.
   c.   What is the interval of *annual* interest rates for the domain in part b?
   d.   The point whose coordinates are approximately $(0.006, 198.96)$ is on the graph of this equation. Write a sentence that gives an interpretation of this ordered pair.
   e.   Use a graphing calculator to determine the monthly payment for a car with a loan amount of $\$10,000$ and an annual interest rate of 8%. Round to the nearest dollar.

a. $P(x) = \dfrac{Cx(x+1)^{60}}{(x+1)^{60} - 1}$ **c.** 0% to 22.8%   **d.** The ordered pair $(0.006, 198.96)$ means that when the monthly interest rate on a car loan is 0.6%, the monthly payment on the loan is $\$198.96$.   **e.** $\$203$

b.

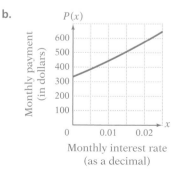

Monthly interest rate (as a decimal)

64. *Mass of a Moving Object*   According to the theory of relativity, the mass of a moving object is given by an equation that contains a complex fraction. The equation is $m = \dfrac{m_0}{\sqrt{1 - \frac{v^2}{c^2}}}$, where m is the mass of the moving object, m_0 is the mass of the object at rest, v is the speed of the object, and c is the speed of light.

   a. Evaluate the expression at speeds of $0.5c$, $0.75c$, $0.90c$, $0.95c$, and $0.99c$ when the mass of the object at rest is 10 grams.

   b. Explain how m changes as the speed of the object becomes close to the speed of light.

   c. Explain how this equation can be used to support the statement that an object cannot travel at the speed of light.

   a. 11.547, 15.119, 22.942, 32.026, 70.888   **b.** As the velocity v of the mass approaches the velocity of light c, the mass of the object increases.   **c.** According to the theory of relativity, the mass of an object at the speed of light would be infinite and would require an infinite force to accelerate it. Because this is impossible, the theory suggests that no object of any mass can attain the speed of light.

Applying Concepts

Simplify.

65. $\left(\dfrac{2m}{3}\right)^2 \div \left(\dfrac{m^2}{6} + \dfrac{m}{2}\right)$ $\dfrac{8m}{3(m + 3)}$

66. $\dfrac{b + 3}{b - 1} \div \dfrac{b + 3}{b - 2} \cdot \dfrac{b - 1}{b + 4}$ $\dfrac{b - 2}{b + 4}$

67. $\left(\dfrac{1}{3} - \dfrac{2}{a}\right) \div \left(\dfrac{3}{a} - 2 + \dfrac{a}{4}\right)$ $\dfrac{4}{3(a - 2)}$

68. $\left(\dfrac{x + 1}{2x - 1} - \dfrac{x - 1}{2x + 1}\right) \cdot \left(\dfrac{2x - 1}{x} - \dfrac{2x - 1}{x^2}\right)$ $\dfrac{6(x - 1)}{x(2x + 1)}$

69. $\dfrac{x^{-1} + y^{-1}}{x^{-1} - y^{-1}}$ $\dfrac{y + x}{y - x}$

70. $\dfrac{\dfrac{1}{x + h} - \dfrac{1}{x}}{h}$ $-\dfrac{1}{x(x + h)}$

71. For adding and subtracting fractions, any common denominator will do. Explain the advantages and disadvantages of using the LCM of the denominators.   Answers will vary.

72. *Manufacturing*   Manufacturers who package their product in cans would like to design the can so that the minimum amount of aluminum is needed. If a soft drink can contains 12 ounces (355 cubic centimeters), the function that relates the surface area of the can (the amount of aluminum needed) to the radius of the bottom of the can is given by the equation $f(r) = 2\pi r^2 + \dfrac{710}{r}$, where r is measured in centimeters.

   a. Express the right side of this equation with a common denominator.
   b. Graph the equation for $0 < r \le 7.15$.
   c. The point whose coordinates are (7, 409), to the nearest integer, is on the graph of f. Write a sentence that gives an interpretation of this ordered pair.

d. Use a graphing calculator to determine the radius of the can that has a minimum surface area. (For assistance, see Appendix B: Min and Max.) Round to the nearest tenth.

e. The height of the can is determined by $h = \frac{355}{\pi r^2}$. Use the answer to part d to determine the height of the can that has a minimum surface area. Round to the nearest tenth.

f. Determine the minimum surface area. Round to the nearest tenth.

a. $f(r) = \frac{2\pi r^3 + 710}{r}$ **c.** The ordered pair (7, 409) means that when the radius of the can is 7 cm, the surface area of the can is 409 cm². **d.** 3.8 cm **e.** 7.8 cm
f. 277.5 cm²

b.

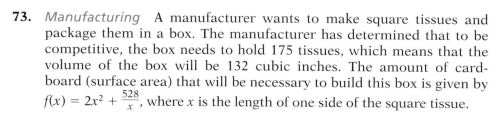

73. *Manufacturing* A manufacturer wants to make square tissues and package them in a box. The manufacturer has determined that to be competitive, the box needs to hold 175 tissues, which means that the volume of the box will be 132 cubic inches. The amount of cardboard (surface area) that will be necessary to build this box is given by $f(x) = 2x^2 + \frac{528}{x}$, where x is the length of one side of the square tissue.

a. Express the right side of this equation with a common denominator.
b. Graph the equation for $1.8 < r \le 10$.
c. The point whose coordinates are (4, 164) is on the graph. Write a sentence that explains the meaning of this point.
d. Use a graphing calculator to determine, to the nearest tenth, the height of the box that uses the minimum amount of cardboard. (For assistance, see Appendix B: Min and Max.)
e. Determine the minimum amount of cardboard. Round to the nearest tenth.

a. $f(x) = \frac{2x^3 + 528}{x}$ **c.** When the height of the box is 4 in., 164 in² of cardboard will be needed. **d.** 5.1 in. **e.** 155.5 in²

b.

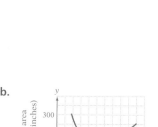

EXPLORATION

1. *Patterns in Mathematics* A student incorrectly tried to add the fractions $\frac{1}{5}$ and $\frac{2}{3}$ by adding the numerators and the denominators. The procedure was shown as

$$\frac{1}{5} + \frac{2}{3} = \frac{1+2}{5+3} = \frac{3}{8}$$

Write the fractions $\frac{1}{5}$, $\frac{2}{3}$, and $\frac{3}{8}$ in order from smallest to largest. Now take any two other fractions, add the numerators and the denominators, and then write the fractions in order from smallest to largest. Do you see a pattern? If so, explain it. If not, try a few more examples until you find a pattern and can explain it.

The value of the fraction that is the result of adding the numerators and denominators is always between the values of the two original fractions.

SECTION **7.3**

■ Solve Rational Equations
■ Work Problems
■ Uniform Motion Problems

Rational Equations

■ Solve Rational Equations

In this section, we will be solving two types of application problems: work problems and uniform motion problems. Each of these types of problems involves solving equations that contain fractions, so we will first look at solving these types of equations.

To solve an equation containing fractions, **clear denominators** by multiplying each side of the equation by the LCM of the denominators. Then solve for the variable.

➡ Solve: $\dfrac{3x}{x-5} = 5 - \dfrac{5}{x-5}$

<table>
<tr><td>

Multiply each side of the equation by the LCM of the denominators.

</td><td>

$$\dfrac{3x}{x-5} = 5 - \dfrac{5}{x-5}$$

$$(x-5)\left(\dfrac{3x}{x-5}\right) = (x-5)\left(5 - \dfrac{5}{x-5}\right)$$

</td></tr>
<tr><td>

Use the Distributive Property.

</td><td>

$$3x = (x-5)5 - (x-5)\left(\dfrac{5}{x-5}\right)$$

$$3x = 5x - 25 - 5$$

$$3x = 5x - 30$$

$$-2x = -30$$

$$x = 15$$

</td></tr>
<tr><td>

A graphical check is shown at the left.

</td><td>

15 checks as a solution.
The solution is 15.　⬅

</td></tr>
</table>

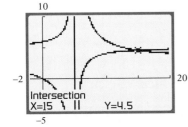

The x-coordinate of the point of intersection of the graphs of $Y_1 = \dfrac{3x}{x-5}$ and $Y_2 = 5 - \dfrac{5}{x-5}$ is 15.

Suggested Activity

Find the slope of all lines through (1, 3) where the slope of the line is $-\frac{1}{2}$ of the value of the x-intercept.

$\left[\text{Answer: } -\dfrac{3}{2} \text{ and } 1\right]$

Occasionally, a value of the variable that appears to be a solution will make one of the denominators zero. In this case, the equation has no solution for that value of the variable. For instance, look at the solution of $\dfrac{3x}{x-3} = 2 + \dfrac{9}{x-3}$ shown below.

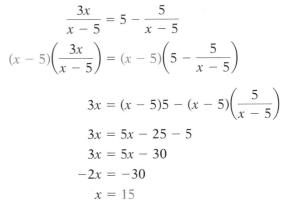

$$\dfrac{3x}{x-3} = 2 + \dfrac{9}{x-3}$$

Multiply each side of the equation by the LCM of the denominators.

$$(x-3)\left(\dfrac{3x}{x-3}\right) = (x-3)\left(2 + \dfrac{9}{x-3}\right)$$

$$3x = (x-3)2 + (x-3)\left(\dfrac{9}{x-3}\right)$$

$$3x = 2x - 6 + 9$$

$$x = 3$$

CALCULATOR NOTE

Use a graphing calculator to

graph $Y_1 = \dfrac{3x}{x-3}$ and

$Y_2 = 2 + \dfrac{9}{x-3}$. Then use the

intersect feature to find the

point of intersection. You will

get an error message be-

cause the two graphs do not

intersect.

Substituting 3 into the equation results in division by zero. Because division by zero is not defined, the equation has no solution.

$$\dfrac{3x}{x-3} = 2 + \dfrac{9}{x-3}$$

$$\begin{array}{c|c} \dfrac{3(3)}{3-3} & 2 + \dfrac{9}{3-3} \\[2mm] \dfrac{9}{0} & 2 + \dfrac{9}{0} \end{array}$$

Multiplying each side of an equation by a variable expression may produce an equation with different solutions from those of the original equation. Thus, **any time you multiply each side of an equation by a variable expression, you must check the resulting solution.**

EXAMPLE 1

Solve.   **a.** $\dfrac{1}{4} = \dfrac{5}{x+5}$ **b.** $\dfrac{1}{r} + \dfrac{1}{r+1} = \dfrac{3}{2}$

INSTRUCTOR NOTE
Exercises 14 and 16 can be used for similar in-class examples.

Solution

a.
$$\dfrac{1}{4} = \dfrac{5}{x+5}$$

$$4(x+5)\,\dfrac{1}{4} = 4(x+5)\,\dfrac{5}{x+5}$$

• Multiply each side of the equation by the LCM of the denominators.

$$x + 5 = 4(5)$$

$$x + 5 = 20$$

$$x = 15$$

ALGEBRAIC CHECK

$$\dfrac{1}{4} = \dfrac{5}{x+5}$$

$$\begin{array}{c|c} \dfrac{1}{4} & \dfrac{5}{15+5} \\[2mm] \dfrac{1}{4} & \dfrac{5}{20} \\[2mm] \dfrac{1}{4} & \dfrac{1}{4} \end{array}$$

• True

The solution checks.

The solution is 15.

GRAPHICAL CHECK

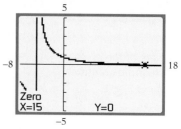

The zero of $f(x) = \dfrac{5}{x+5} - \dfrac{1}{4}$ is 15. Alternatively, find the x-coordinate of the intersection of the graphs of $Y_1 = \dfrac{1}{4}$ and $Y_2 = \dfrac{5}{x+5}$.

b.

$$\frac{1}{r} + \frac{1}{r + 1} = \frac{3}{2}$$

$$2r(r + 1)\left(\frac{1}{r} + \frac{1}{r + 1}\right) = 2r(r + 1) \cdot \frac{3}{2}$$

• Multiply each side by the LCM of the denominators.

$$2(r + 1) + 2r = r(r + 1) \cdot 3$$

$$2r + 2 + 2r = 3r^2 + 3r$$

$$4r + 2 = 3r^2 + 3r$$

$$0 = 3r^2 - r - 2$$

• Write the quadratic equation in standard form.

$$0 = (3r + 2)(r - 1)$$

• Solve for r by factoring.

$$3r + 2 = 0 \qquad\qquad r - 1 = 0$$

$$3r = -2 \qquad\qquad r = 1$$

$$r = -\frac{2}{3}$$

ALGEBRAIC CHECK

$$\frac{\dfrac{1}{r} + \dfrac{1}{r + 1} = \dfrac{3}{2}}{\dfrac{1}{-\dfrac{2}{3}} + \dfrac{1}{-\dfrac{2}{3} + 1} \;\Bigg|\; \dfrac{3}{2}}$$

$$-\frac{3}{2} + 3 \;\Bigg|\; \frac{3}{2}$$

$$\frac{3}{2} = \frac{3}{2} \qquad \text{• True}$$

$$\frac{\dfrac{1}{r} + \dfrac{1}{r + 1} = \dfrac{3}{2}}{\dfrac{1}{1} + \dfrac{1}{1 + 1} \;\Bigg|\; \dfrac{3}{2}}$$

$$1 + \frac{1}{2} \;\Bigg|\; \frac{3}{2}$$

$$\frac{3}{2} = \frac{3}{2} \qquad \text{• True}$$

GRAPHICAL CHECK

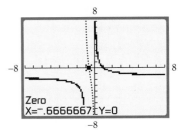

The zeros of $f(r) = \dfrac{1}{r} + \dfrac{1}{r + 1} - \dfrac{3}{2}$ are $-\dfrac{2}{3}$ and 1.

The solutions check.

The solutions are $-\dfrac{2}{3}$ and 1.

YOU TRY IT 1

Solve.   **a.** $\dfrac{5}{2x - 3} = \dfrac{-2}{x + 1}$ **b.** $3y + \dfrac{25}{3y - 2} = -8$

Solution   See page S29.   **a.** $\dfrac{1}{9}$ **b.** -1

▼ Point of Interest

The following problem was recorded in the Jiuzhang, a Chinese text that dates to the Han dynasty (about 200 B.C. to A.D. 200). "A reservoir has 5 channels bringing water to it. The first can fill the reservoir in $\frac{1}{3}$ day, the second in 1 day, the third in $2\frac{1}{2}$ days, the fourth in 3 days, and the fifth in 5 days. If all channels are open, how long does it take to fill the reservoir?" This is the earliest known work problem.

Suggested Activity

Have your students solve the problem presented in the Point of Interest. The answer is $\frac{15}{74}$ day.

■ Work Problems

If a mason can build a retaining wall in 12 hours, then in 1 hour the mason can build $\frac{1}{12}$ of the wall. The mason's rate work is $\frac{1}{12}$ of the wall each hour. The **rate of work** is that part of the task that is completed in 1 unit of time. If an apprentice can build the wall in x hours, then the rate of work for the apprentice is $\frac{1}{x}$ of the wall each hour.

In solving a work problem, the goal is to determine the time it takes to complete a task. The basic equation that is used to solve work problems is

Rate of work × Time worked = Part of task completed

For example, if a pipe can fill a tank in 5 hours, then in 2 hours the pipe will fill $\frac{1}{5} \times 2 = \frac{2}{5}$ of the tank. In t hours, the pipe will fill $\frac{1}{5} \times t = \frac{t}{5}$ of the tank.

➡ A mason can build a wall in 10 hours. An apprentice can build a wall in 15 hours. How long would it take them to build the wall if they worked together?

State the goal. We want to find the amount of time it will take to build the wall if the mason and the apprentice work together.

Devise a strategy.

- Let t represent the amount of time it takes for the mason and the apprentice to build the wall if they work together.
- Write an equation using the fact that the sum of the part of the task completed by the mason and the part of the task completed by the apprentice equals 1, the complete task. Solve this equation for t.

Solve the problem.

$$\begin{array}{l}\text{Part of task completed} \\ \text{by mason}\end{array} = \text{Rate of work} \cdot \text{Time worked} = \frac{1}{10} \cdot t = \frac{t}{10}$$

$$\begin{array}{l}\text{Part of task completed} \\ \text{by apprentice}\end{array} = \text{Rate of work} \cdot \text{Time worked} = \frac{1}{15} \cdot t = \frac{t}{15}$$

The sum of the part of the task completed by the mason and the part of the task completed by the apprentice is 1.

$$\frac{t}{10} + \frac{t}{15} = 1$$

$$30\left(\frac{t}{10} + \frac{t}{15}\right) = 30(1) \qquad \bullet \text{ Multiply each side of the equation by the LCM of the denominators, 30.}$$

$$3t + 2t = 30$$

$$5t = 30$$

$$t = 6$$

Working together, they would build the wall in 6 hours.

Check your work.   The mason completes $\dfrac{t}{10} = \dfrac{6}{10} = \dfrac{3}{5}$ of the job.

The apprentice completes $\dfrac{t}{15} = \dfrac{6}{15} = \dfrac{2}{5}$ of the job.

$\dfrac{3}{5} + \dfrac{2}{5} = \dfrac{5}{5} = 1$, the complete job.

The mason should complete a larger fraction of the job, and $\dfrac{3}{5} > \dfrac{2}{5}$.

A graphical check is shown at the left. The zero of $f(t) = \dfrac{t}{10} + \dfrac{t}{15} - 1$ is 6.

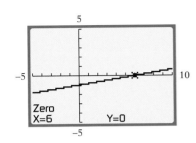

Zero
X=6   Y=0

INSTRUCTOR NOTE
Exercise 36 can be used for
a similar in-class example.

EXAMPLE 2

An electrician requires 12 hours to wire a house. The electrician's apprentice can wire a house in 16 hours. After working alone on one job for
4 hours, the electrician leaves, and the apprentice completes the task.
How long does it take the apprentice to finish wiring the house?

State the goal.   The goal is to determine how much time it will take the
apprentice to complete the job after the electrician has worked on it for
4 hours.

Devise a strategy.

- Let t be the time required for the apprentice to finish wiring the house.

- Write an equation using the fact that the sum of the part of the task
  completed by the electrician and the part of the task completed by the
  apprentice equals 1, the complete task. Solve this equation for t.

Solve the problem.

$$\text{Part of task completed by electrician} = \text{Rate of work} \cdot \text{Time worked} = \frac{1}{12} \cdot 4 = \frac{4}{12} = \frac{1}{3}$$

$$\text{Part of task completed by apprentice} = \text{Rate of work} \cdot \text{Time worked} = \frac{1}{16} \cdot t = \frac{t}{16}$$

The sum of the part of the task completed by the electrician and the part
of the task completed by the apprentice is 1.

$$\frac{1}{3} + \frac{t}{16} = 1$$

$$48\left(\frac{1}{3} + \frac{t}{16}\right) = 48(1)$$

$$16 + 3t = 48$$

$$3t = 32$$

$$t = \frac{32}{3}$$

Suggested Activity

See Section 7.3 of the *Student
Activity Manual* for an activity
on work problems.

It takes the apprentice $10\frac{2}{3}$ hours to finish wiring the house.

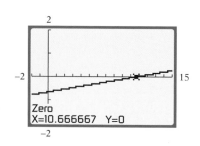

Zero
X=10.666667 Y=0

Check your work. The apprentice completes $\dfrac{t}{16} = \dfrac{\frac{32}{3}}{16} = \dfrac{2}{3}$ of the job.

$\dfrac{1}{3} + \dfrac{2}{3} = \dfrac{3}{3} = 1$, the complete job.

A graphical check is shown at the left. The zero of $(t) = \dfrac{1}{3} + \dfrac{t}{16} - 1$ is $10\dfrac{2}{3}$.

❓ QUESTION Could the answer to Example 2 be more than 16 hours? Why or why not?

YOU TRY IT 2

Two water pipes can fill a tank with water in 6 hours. The larger pipe working alone can fill the tank in 9 hours. How long would it take the smaller pipe, working alone, to fill the tank?

Solution   See page S29.   18 h

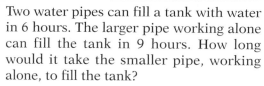

Fills tank in x hours   Fills tank in 9 hours

Fills $\frac{1}{x}$ of the tank each hour   Fills $\frac{1}{9}$ of the tank each hour

■ Uniform Motion Problems

A car that travels constantly in a straight line at 55 mph is in uniform motion. **Uniform motion** means that the speed of an object does not change. The basic equation used to solve uniform motion problems is

$$\textbf{Distance} = \textbf{Rate} \times \textbf{Time}$$

An alternative form of this equation can be written by solving the equation for time. This form of the equation is used to solve the following problem.

$$\frac{\textbf{Distance}}{\textbf{Rate}} = \textbf{Time}$$

TAKE NOTE

Suppose you drive 150 miles at a speed of 50 mph. The time it takes you to travel the 150 miles is

$$\text{Time} = \frac{\text{Distance}}{\text{Rate}} = \frac{150}{50}$$
$$= 3 \text{ hours}$$

➡ A motorist drove 150 miles on country roads before driving 50 miles on mountain roads. The rate of speed on the country roads was three times the rate on the mountain roads. The time spent traveling the 200 miles was 5 hours. Find the rate of the motorist on the country roads.

State the goal. We want to find the rate at which the motorist traveled on the country roads.

Devise a strategy.

- Let r represent the rate on the mountain roads. Then the rate on the country roads is $3r$.
- Write an equation using the fact that the time spent driving on country roads plus the time spent driving on mountain roads equals 5 hours. Solve this equation for r.
- Substitute the value of r into the expression $3r$ to find the rate on the country roads.

❓ ANSWER The apprentice can complete the entire job in 16 hours. Because the electrician has already completed part of the job, it must take the apprentice less than 16 hours to finish it.

Solve the problem.

Time spent traveling country roads: $\dfrac{\text{Distance}}{\text{Rate}} = \dfrac{150}{3r}$

Time spent traveling mountain roads: $\dfrac{\text{Distance}}{\text{Rate}} = \dfrac{50}{r}$

The time spent traveling the country roads plus the time spent traveling the mountain roads is 5 hours, the total time for the trip.

$$\frac{150}{3r} + \frac{50}{r} = 5$$

$$\frac{50}{r} + \frac{50}{r} = 5 \qquad \bullet \ \text{Simplify } \frac{150}{3r}.$$

$$r\left(\frac{50}{r} + \frac{50}{r}\right) = r(5) \qquad \bullet \ \text{Multiply each side by the LCM of the denominator.}$$

$$50 + 50 = 5r$$

$$100 = 5r$$

$$20 = r \qquad \bullet \ \text{This is the rate on the mountain roads.}$$

$$3r = 3(20) = 60 \qquad \bullet \ \text{The rate on the country roads was } 3r. \\ \text{Replace } r \text{ with 20 and evaluate.}$$

The rate of speed on the country roads was 60 mph.

Check your work.   The sum of the times is 5:

$$\frac{150}{3r} + \frac{50}{r} = \frac{150}{3(20)} + \frac{50}{20} = \frac{150}{60} + \frac{50}{20} = \frac{5}{2} + \frac{5}{2} = 5$$

A graphical check shows that the zero of $f(t) = \frac{150}{3r} + \frac{50}{r} - 5$ is 20. Note that this is the value of r, the speed on the mountain roads, not $3r$, the speed on the country roads.

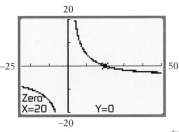

INSTRUCTOR NOTE
Exercise 56 can be used for
a similar in-class example.

EXAMPLE 3

A marketing executive traveled 810 miles on a corporate jet in the same amount of time that it took to travel an additional 162 miles by helicopter. The rate of the jet was 360 mph greater than the rate of the helicopter. Find the rate of the jet.

State the goal.   We want to find the rate of the jet.

Devise a strategy.

• Let r represent the rate of the helicopter. Then the rate of the jet is $r + 360$.

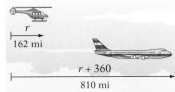

- Write an equation using the fact that the time spent on the jet equals the time spent on the helicopter. Solve this equation for r.
- Substitute the value of r into the expression $r + 360$ to find the rate of the jet.

Solve the problem.

Time spent on the jet: $\dfrac{\text{Distance}}{\text{Rate}} = \dfrac{810}{r + 360}$

Time spent on the helicopter: $\dfrac{\text{Distance}}{\text{Rate}} = \dfrac{162}{r}$

The time traveled by jet is equal to the time traveled by helicopter.

$$\frac{810}{r + 360} = \frac{162}{r}$$

$$r(r + 360)\left(\frac{810}{r + 360}\right) = r(r + 360)\left(\frac{162}{r}\right)$$ • Multiply each side by the LCM of the denominators.

$$810r = (r + 360)162$$
$$810r = 162r + 58{,}320$$
$$648r = 58{,}320$$
$$r = 90$$ • This is the rate of the helicopter.

$$r + 360 = 90 + 360 = 450$$ • The rate of the jet is $r + 360$. Replace r with 90 and evaluate.

The rate of the jet was 450 mph.

Check your work.

The times are equal: $\dfrac{810}{r + 360} = \dfrac{810}{90 + 360} = \dfrac{810}{450} = 1.8$

$$\frac{162}{r} = \frac{162}{90} = 1.8$$

A graphical check shows that the zero of $f(t) = \dfrac{810}{r + 360} - \dfrac{162}{r}$ is 90. Note that this is the value of r, the rate of the helicopter, not $r + 360$, the rate of the jet.

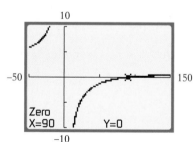

YOU TRY IT 3

A plane can fly at a rate of 150 mph in calm air. Traveling with the wind, the plane flew 700 miles in the same amount of time it took to fly 500 miles against the wind. Find the rate of the wind.

Solution See page S30. 25 mph

7.3 EXERCISES   Suggested Assignment: 7–75, odds

Topics for Discussion

1. What is a rational equation? Provide an example of a rational equation.
    A rational equation is an equation that contains fractions. Examples will vary.

2. Explain why it is necessary to check the solution of a rational equation.
    Explanations may vary. For example: A proposed solution might make one of the
    denominators equal to zero.

3. If a gardener can mow a lawn in 20 minutes, what fraction of the lawn
    can the gardener mow in x minutes? $\frac{x}{20}$ of the lawn

4. If one person can complete a task in 2 hours and another person can
    complete the same task in 3 hours, will it take more or less than 2 hours
    to complete the task when both people are working? Explain your an-
    swer.   It will take less than 2 hours. Explanations will vary.

5. Only two people worked on a job, and together they completed it. One
    person completed $\frac{t}{10}$ of the job and the other person completed $\frac{t}{15}$ of the
    job. Write an equation to express the fact that together they completed
    the whole job. $\frac{t}{10} + \frac{t}{15} = 1$

6. A plane flies 350 mph in calm air, and the rate of the wind is r mph.
    a.   Write an expression for the rate of the plane flying with the wind.
    b.   Write an expression for the rate of the plane flying against the wind.
    a. $350 + r$ **b.** $350 - r$

■ **Solve Rational Equations**

Solve.

7. $1 - \dfrac{3}{y} = 4$ -1

8. $7 + \dfrac{6}{y} = 5$ -3

9. $\dfrac{8}{2x - 1} = 2$ $\dfrac{5}{2}$

10. $3 = \dfrac{18}{3x - 4}$ $\dfrac{10}{3}$

11. $\dfrac{x - 2}{5} = \dfrac{1}{x + 2}$ $-3, 3$

12. $\dfrac{x + 4}{10} = \dfrac{6}{x - 3}$ $-9, 8$

13. $\dfrac{3}{x - 2} = \dfrac{4}{x}$ 8

14. $\dfrac{5}{x} = \dfrac{2}{x + 3}$ -5

15. $\dfrac{3}{x - 4} + 2 = \dfrac{5}{x - 4}$ 5

16. $\dfrac{5}{y + 3} - 2 = \dfrac{7}{y + 3}$
    -4

17. $5 + \dfrac{8}{a - 2} = \dfrac{4a}{a - 2}$
    No solution

18. $\dfrac{-4}{a - 4} = 3 - \dfrac{a}{a - 4}$
    No solution

19. $\dfrac{x}{2} + \dfrac{20}{x} = 7$ $4, 10$

20. $3x = \dfrac{4}{x} - \dfrac{13}{2}$ $-\dfrac{8}{3}, \dfrac{1}{2}$

21. $\dfrac{x}{x + 2} = \dfrac{6}{x + 5}$ $-3, 4$

22. $\dfrac{x}{x - 2} = \dfrac{3}{x - 4}$ $1, 6$

23. $\dfrac{y - 1}{y + 2} + y = 1$ $-3, 1$

24. $\dfrac{2p - 1}{p - 2} + p = 8$ $3, 5$

25. $\dfrac{16}{z-2} + \dfrac{16}{z+2} = 6$ $-\dfrac{2}{3}, 6$ **26.** $\dfrac{5}{2p-1} + \dfrac{4}{p+1} = 2$ $-\dfrac{1}{4}, 3$ **27.** $\dfrac{2v}{v+2} + \dfrac{6}{v+4} = 2$ 2

28. $\dfrac{x+3}{x+1} - \dfrac{x+7}{x+3} = 0$ 1 **29.** $\dfrac{x-2}{x+1} + x = 2$ $-2, 2$ **30.** $\dfrac{4x-1}{2x+1} + x = 1$ $-2, \dfrac{1}{2}$

31. The sum of a number and twice its reciprocal is $\dfrac{33}{4}$. Find the number. $\dfrac{1}{4}, 8$

32. The numerator of a fraction is 3 less than the denominator. The sum of the fraction and four times its reciprocal is $\dfrac{17}{2}$. Find the fraction. $\dfrac{3}{6}$

33. If the ratio of $(2x - y)$ to $(x + y)$ is $2 : 3$ what is the ratio of x to y? $5 : 4$

34. *Sports*   A baseball team won 50 games out of 70 games played. How many more games must the team win in succession to raise its record to 80%?   30 games

35. *Metallurgy*   If a cube of metal 2 inches on an edge weighs 3 pounds, how many pounds will a cube that is made of the same metal weigh if its edges are 4 inches?   24 lb

■ Work Problems

36. *Data Processing*   A large biotech firm uses two computers to process the daily results of its research studies. One computer can process data in 2 hours; the other computer takes 3 hours to do the same job. How long would it take to process the data if both computers were used?
1.2 h

37. *Business*   Two college students have started their own business building computers from kits. Working alone, one student can build a computer in 20 hours. When the second student helps, they can build a computer in 7.5 hours. How long would it take the second student, working alone, to build the computer?   12 h

38. *Energy Consumption*   One solar heating panel can raise the temperature of water 1 degree in 30 minutes. A second solar heating panel can raise the temperature of the water 1 degree in 45 minutes. How long would it take to raise the temperature of the water 1 degree with both solar panels operating?   18 min

39. *Home Maintenance*   After flood waters along the Mississippi began to recede, a young family was faced with pumping the water from the basement. One pump they were using could dispose of 9000 gallons in 3 hours. A second pump could dispose of the same number of gallons in 4.5 hours. How many hours would it take to dispose of 9000 gallons if both pumps were working?   1.8 h

40. *Energy Consumption*   A heat wave in Texas during a recent summer forced even small businesses to run their air conditioners 24 hours a day. In the office of one such business, there were two air conditioners, one

older than the other. The newer one was able to cool the room by 2 degrees in 8 minutes. With both running, the room could be cooled by the same number of degrees in 4.8 minutes. How long would it take the older air conditioner, working alone, to cool the room by 2 degrees?
12 min

41. *Business*   A new machine can package transistors four times as fast as an older machine. Working together, the machines can package the transistors in 8 hours. How long would it take the new machine, working alone, to package the transistors?   10 h

42. *Business*   The larger of two printers being used to print the payroll for a major corporation requires 40 minutes to print the payroll. After both printers have been operating for 10 minutes, the larger printer malfunctions. The smaller printer requires 50 more minutes to complete the payroll. How long would it take the smaller printer, working alone, to print the payroll?   80 min

43. *Construction*   A roofer requires 12 hours to shingle a roof. After the roofer and an apprentice work on a roof for 3 hours, the roofer moves on to another job. The apprentice requires 12 more hours to finish the job. How long would it take the apprentice, working alone, to do the job?
20 h

44. *Masonry*   An experienced bricklayer can work twice as fast as an apprentice brick layer. After they worked together on a job for 8 hours, the experienced bricklayer left. The apprentice required 12 more hours to finish the job. How long would it take the experienced bricklayer, working alone, to do the job?   18 h

45. *Construction*   A welder requires 25 hours to do a job. After the welder and an apprentice work on a job for 10 hours, the welder quits. The apprentice finishes the job in 17 hours. How long would it take the apprentice, working alone, to do the job?   45 h

46. *Business*   A New York City pizza parlor hired three part-time employees to make pizzas. After a short training period of 2 days, the first employee was able to make a large pepperoni pizza in 3.5 minutes. The second employee took 2.5 minutes, and the third took 3.0 minutes to create the same large pizza. To the nearest minute, how long would it take to make the pizza if all three employees worked on it at the same time?   1 min

47. *Business*   Three machines fill soda bottles. The machines can fill the daily quota of soda bottles in 12 hours, 15 hours, and 20 hours, respectively. How long would it take to fill the daily quota of soda bottles with all three machines working?   5 h

48. *Parenting*   With both hot and cold water running, a bathtub can be filled in 10 minutes. The drain will empty the tub in 15 minutes. A child turns both faucets on and leaves the drain open. How long will it be before the bathtub starts to overflow?   30 min

49. *Reservoirs* The inlet pipe can fill a water tank in 30 minutes. The outlet pipe can empty the tank in 20 minutes. How long would it take to empty a full tank with both pipes open? 60 min

50. *Oil Tanks* An oil tank has two inlet pipes and one outlet pipe. One inlet pipe can fill the tank in 12 hours, and the other inlet pipe can fill the tank in 20 hours. The outlet pipe can empty the tank in 10 hours. How long would it take to fill the tank with all three pipes open? 30 h

51. *Irrigation* Water from a tank is being used for irrigation at the same time as the tank is being filled. The two inlet pipes can fill the tank in 6 hours and 12 hours, respectively. The outlet pipe can empty the tank in 24 hours. How long would it take to fill the tank with all three pipes open? 4.8 h

52. *Gasoline Tanks* A small pipe can fill a gasoline tank in 6 minutes more time than it takes a larger pipe to fill the same tank. Working together, both pipes can fill the tank in 4 minutes. How long would it take each pipe, working alone, to fill the tank? Smaller pipe: 12 min; larger pipe: 6 min

53. *Energy Consumption* A small heating unit takes 8 hours longer to melt a piece of iron than does a larger unit. Working together, the heating units can melt the iron in 3 hours. How long would it take each heating unit, working alone, to melt the iron? Smaller unit: 12 h; larger unit: 4 h

54. *Science* A science experiment requires that a vacuum be created in a chamber. A small vacuum pump requires 15 seconds longer than does a second, larger pump to evacuate the chamber. Working together, the pumps can evacuate the chamber in 4 seconds. Find the time required for the larger vacuum pump, working alone, to evacuate the chamber. 5 s

55. *U.S. Postal Service* An old mechanical sorter takes 21 minutes longer to sort a batch of mail than does a second, newer model. With both sorters working, a batch of mail can be sorted in 10 minutes. How long would it take each sorter, working alone, to sort the batch of mail? Newer sorter: 14 min; older sorter: 35 min

■ **Uniform Motion Problems**

56. *Recreation* Two skaters take off for an afternoon of roller blading in Central Park. The first skater can cover 15 miles in the same time it takes the second skater, traveling 3 mph slower than the first skater, to cover 12 miles. Find the rate of each roller blader. 15 mph; 12 mph

57. *Aviation* A commercial jet travels 1620 miles in the same amount of time it takes a corporate jet to travel 1260 miles. The rate of the commercial jet is 120 mph greater than the rate of the corporate jet. Find the rate of each jet. Commercial: 540 mph; corporate: 420 mph

58. *Railroads* A passenger train travels 295 miles in the same amount of time it takes a freight train to travel 225 miles. The rate of the passenger train is 14 mph greater than the rate of the freight train. Find the rate of each train.   Passenger train: 59 mph; freight train: 45 mph

59. *Athletics*   The rate of a bicyclist is 7 mph greater than the rate of a long-distance runner. The bicyclist travels 30 miles in the same amount of time it takes the runner to travel 16 miles. Find the rate of the runner.
8 mph

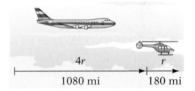

60. *Athletics*   A cyclist rode 40 miles before having a flat tire and then walking 5 miles to a service station. The cycling rate was four times the walking rate. The time spent cycling and walking was 5 hours. Find the rate at which the cyclist was riding.   12 mph

61. *Travel*   A sales executive traveled 32 miles by car and then an additional 576 miles by plane. The rate of the plane was nine times the rate of the car. The total time of the trip was 3 hours. Find the rate of the plane.
288 mph

62. *Travel*   A motorist drove 72 miles before running out of gas and then walking 4 miles to a gas station. The driving rate of the motorist was twelve times the walking rate. The time spent driving and walking was 2.5 hours. Find the rate at which the motorist walks.   4 mph

63. *Travel*   To assess the damage done by a fire, a forest ranger traveled 1080 miles by jet and then an additional 180 miles by helicopter. The rate of the jet was four times the rate of the helicopter. The entire trip took 5 hours. Find the rate of the jet.   360 mph

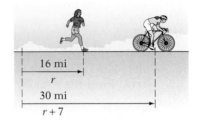

64. *Travel*   A business executive can travel the 480 feet between two terminals of an airport by walking on a moving sidewalk in the same time required to walk 360 feet without using the moving sidewalk. If the rate of the moving sidewalk is 2 feet per second, find the rate at which the executive can walk.   6 ft/s

65. *Athletics*   A cyclist and a jogger start from a town at the same time and head for a destination 18 miles away. The rate of the cyclist is twice the rate of the jogger. The cyclist arrives 1.5 hours before the jogger. Find the rate of the cyclist.   12 mph

66. *Travel*   A single-engine plane and a commercial jet leave an airport at 10 A.M. and head for an airport 960 miles away. The rate of the jet is four times the rate of the single-engine plane. The single-engine plane arrives 4 hours after the jet. Find the rate of each plane.
Jet: 720 mph; single-engine plane: 180 mph

67. *Athletics*   Marlys can row a boat 3 mph faster than she can swim. She is able to row 10 miles in the same time it takes her to swim 4 miles. Find rate at which Marlys swims.   2 mph

68. *Travel*   A cruise ship can sail 28 mph in calm water. Sailing with the Gulf Stream, the ship can sail 170 miles in the same amount of time it takes to sail 110 miles against the Gulf Stream. Find the rate of the Gulf Stream.   6 mph

69. *Travel*   A commercial jet can fly 500 mph in calm air. Traveling with the jet stream, the plane flew 2420 miles in the same amount of time it takes to fly 1580 miles against the jet stream. Find the rate of the jet stream.
105 mph

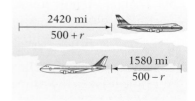

70. *Travel*   A tour boat used for river excursions can travel 7 mph in calm water. The amount of time it takes to travel 20 miles with the current is the same as the amount of time it takes to travel 8 miles against the current. Find the rate of the current.   3 mph

71. *Athletics*   A canoe can travel 8 mph in still water. Traveling with the current of a river, the canoe can travel 15 miles in the same amount of time it takes to travel 9 miles against the current. Find the rate of the current.
2 mph

72. *Travel*   A cruise ship made a trip of 100 miles in 8 hours. The ship traveled the first 40 miles at a constant rate before increasing its speed by 5 mph. Another 60 miles was traveled at the increased speed. Find the rate of the cruise ship for the first 40 miles.   10 mph

73. *Athletics*   A cyclist traveled 60 miles at a constant rate before reducing the speed by 2 mph. Another 40 miles was traveled at the reduced speed. The total time for the 100-mile trip was 9 hours. Find the rate during the first 60 miles.   12 mph

74. *Sports*   The rate of a river's current is 2 mph. A rowing crew can row 16 miles down this river and back in 6 hours. Find the rowing rate of the crew in calm water.   6 mph

75. *Sports*   A fishing boat traveled 30 miles down a river and then returned. The total time for the round trip was 4 hours, and the rate of the river's current was 4 mph. Find the rate of the boat in still water.   16 mph

Applying Concepts

76. *Water Tanks*   One pipe can fill a water tank in 3 hours, a second pipe can fill the tank in 4 hours, and a third pipe can fill the tank in 6 hours. How long would it take to fill the tank with all three pipes operating?
$1\frac{1}{3}$ h

77. *Business*   One printer can print a company's paychecks in 24 minutes, a second printer can print the checks in 16 minutes, and a third printer can do the job in 12 minutes. How long would it take to print the checks with all three printers operating? $5\frac{1}{3}$ min

78. *Business*   If 6 machines can fill 12 boxes of cereal in 7 minutes, how many boxes of cereal can be filled by 14 machines in 12 minutes?
48 boxes

79. *Travel*   By increasing your speed by 10 mph, you can drive the 200-mile trip to your hometown in 40 minutes less time than the trip usually takes you. How fast do you usually drive?   50 mph

80. *Travel*   Because of weather conditions, a bus driver reduced the usual speed along a 165-mile bus route by 5 mph. The bus arrived only 15 minutes later than its usual arrival time. How fast does the bus usually travel?   60 mph

81. *Recreation*   If a pump can fill a pool in *A* hours and a second pump can fill the pool in *B* hours, find a formula, in terms of *A* and *B*, for the time it takes both pumps, working together, to fill the pool. $t = \dfrac{AB}{A + B}$

82. *Recreation*   If a parade is 1 mile long and is proceeding at 3 mph, how long will it take a runner, jogging at 5 mph, to run from the beginning of the parade to the end and then back to the beginning? $\dfrac{5}{8}$ h

83. If $\dfrac{A}{x + 2} + \dfrac{B}{2x - 3} = \dfrac{5x - 11}{2x^2 + x - 6}$, find the values of *A* and *B*.
$A = 3, B = -1$

EXPLORATION

1. *Golden Rectangles*   The ancient Greeks defined a rectangle as a "golden rectangle" if its length *L* and its width *W* satisfied the equation

$$\frac{L}{W} = \frac{W}{L - W}$$

a. This equation can be solved for *W*. The result is $W = \dfrac{L(-1 + \sqrt{5})}{2}$.
Write this equation in function notation.

b. Evaluate the function from part a at 101. Round to the nearest tenth. Explain what the value of the function at 101 means in the context of a golden rectangle.

c. Find applications of the golden rectangle to art and architecture.

a. $f(L) = \dfrac{L(-1 + \sqrt{5})}{2}$ **b.** 62.4. If the length *L* of a golden rectangle measures 101 units, the width of the rectangle is approximately 62.4 units.   **c.** Answers will vary. For example, portions of Leonardo da Vinci's Mona Lisa, the Greek architecture of the Parthenon, and the United Nations Building in New York City.

2. *Applications to Clock Movements*

a. How many minutes does it take a clock's hour hand to move through one degree of revolution?

b. Bill has designed an eight-hour clock. The minute hand still takes 60 minutes to complete one revolution, but for the complete revolution of the minute hand, the hour hand goes one-eighth of a revolution. At how many minutes after one o'clock will the minute hand and the hour hand coincide? Round to the nearest hundredth.

a. 2 min   **b.** 8.57 min

SECTION **7.4**

- Solve Proportions
- Similar Triangles
- Dimensional Analysis

Proportions and Similar Triangles

Solve Proportions

The following table shows the automobile fatality rate per 100,000 people in the United States for the years 1990 through 2000. Also shown is the U.S. population for those years (*Sources:* National Highway Traffic Safety Administration; U.S. Bureau of the Census).

Year	*Fatality Rate per 100,000 Population*	*U.S. Population*
1990	17.88	249,000,000
1991	16.46	252,000,000
1992	15.39	255,000,000
1993	15.58	258,000,000
1994	15.64	260,000,000
1995	15.91	263,000,000
1996	15.86	265,000,000
1997	15.69	268,000,000
1998	15.36	270,000,000
1999	15.30	273,000,000
2000	15.23	275,000,000

What was the number of automobile fatalities in the United States in 2000? Was this more or less than the number in 1995? We can answer these questions by writing and solving proportions. But first we need to define *rate* and *ratio*.

A **rate** is the quotient of two quantities that have different units. The 1990 automobile fatality rate of 17.88 per 100,000 population can be written

$$\frac{17.88 \text{ deaths}}{100,000 \text{ people}}$$

A **ratio** is the quotient of two quantities that have the same unit. The ratio of the number of home-schooled students in Florida during the 2001–2002 school year to the number of home-schooled students in Florida during the 1991–1992 school year is

$$\frac{41,128 \text{ home-schooled students}}{10,039 \text{ home-schooled students}} = \frac{41,128}{10,039}$$

Note that units are written as part of a rate, but units are not written as part of a ratio.

A **proportion** is the equality of two rates or ratios. The following are examples of proportions.

$$\frac{200 \text{ miles}}{4 \text{ hours}} = \frac{50 \text{ miles}}{1 \text{ hour}}$$

$$\frac{7}{14} = \frac{1}{2}$$

Note that the units in the numerators (miles) are the same and the units in the denominators (hours) are the same.

▼ *Point of Interest*

Here is an example of a ratio from the field of geometry: If we solve the literal equation $C = \pi d$ *for d, the result is*
$d = \frac{C}{\pi}$. *This means that the diameter of a circle is equal to the ratio of the circumference of the circle to* π.

Suggested Activity

Here is a challenging problem involving rates.
For the first 5 miles of a 10-mile race, Ray's rate was 10 mph. For the last 5 miles, his rate slowed to 8 mph. How long did it take Ray to complete the race? What was his average rate for the race?

[Answer: 1 1/8 hours; 8 8/9 mph]

The definition of a proportion can be stated as follows:

> If $\dfrac{a}{b}$ and $\dfrac{c}{d}$ are equal ratios or rates, then $\dfrac{a}{b} = \dfrac{c}{d}$ is a proportion.

Each of the four members in a proportion is called a **term.** These terms are numbered as follows:

$$\text{First term} \longrightarrow \quad \dfrac{a}{b} = \dfrac{c}{d} \quad \longleftarrow \text{Third Term}$$
$$\text{Second Term} \longrightarrow \qquad \qquad \longleftarrow \text{Fourth Term}$$

The second and third terms of the proportion are called the **means,** and the first and fourth terms are called the **extremes.**

If we multiply both sides of the proportion by the LCM of the denominators, bd, we obtain the following result:

$$\frac{a}{b} = \frac{c}{d}$$

$$bd\left(\frac{a}{b}\right) = bd\left(\frac{c}{d}\right)$$

$$ad = bc$$

In a proportion, **the product of the means equals the product of the extremes.** This is sometimes phrased "the cross products are equal."

As shown below, in the proportion $\dfrac{2}{3} = \dfrac{8}{12}$, the cross products are equal.

$$\text{Product of the means} = \text{Product of the extremes}$$
$$3 \cdot 8 = 2 \cdot 12$$
$$24 = 24$$

? QUESTION Consider the proportion $\dfrac{3}{8} = \dfrac{9}{24}$.

    a. Which are the first and third terms?
    b. Which terms are the means?
    c. What is the product of the extremes?

When three of the four numbers in a proportion are known, we can solve the proportion for the unknown number.

EXAMPLE 1

Solve: $\dfrac{8}{3} = \dfrac{12}{x}$

? <u>ANSWERS</u> **a.** The first term is 3. The third term is 9. **b.** The 8 and the 9 are the means.
c. The product of the extremes is $3 \cdot 24 = 72$.

TAKE NOTE

b and c are the means
a and d are the extremes.

$$\frac{a}{b} = \frac{c}{d}$$

INSTRUCTOR NOTE
Exercise 10 can be used for a similar in-class example.

Solution $\dfrac{8}{3} = \dfrac{12}{x}$

$3 \cdot 12 = 8 \cdot x$ • The product of the means equals the product of the extremes.

$36 = 8x$

$\dfrac{36}{8} = \dfrac{8x}{8}$ • Divide both sides of the equation by the coefficient of x.

$4.5 = x$

Check: $\dfrac{8}{3}$ $\dfrac{12}{4.5}$

$3(12) = 8(4.5)$

$36 = 36$

The solution is 4.5.

YOU TRY IT 1

Solve: $\dfrac{n}{5} = \dfrac{12}{25}$

Solution See page S30.   2.4

Now let's return to the questions related to automobile fatalities. The table of rates and populations is repeated here (*Sources:* National Highway Safety Administration; U.S. Bureau of the Census).

Year	Fatality Rate per 100,000 Population	U.S. Population
1990	17.88	249,000,000
1991	16.46	252,000,000
1992	15.39	255,000,000
1993	15.58	258,000,000
1994	15.64	260,000,000
1995	15.91	263,000,000
1996	15.86	265,000,000
1997	15.69	268,000,000
1998	15.36	270,000,000
1999	15.30	273,000,000
2000	15.23	275,000,000

To approximate the number of automobile fatalities in the United States in 2000, we can write a proportion using any variable to represent the number of fatalities. The variable N is used in the proportion below.

$$\dfrac{15.23 \text{ deaths}}{100{,}000 \text{ people}} = \dfrac{N \text{ deaths}}{275{,}000{,}000 \text{ people}}$$

$$100{,}000 \cdot N = 15.23(275{,}000{,}000)$$

$$100{,}000N = 4{,}188{,}250{,}000$$

$$N \approx 41{,}883$$

There were approximately 41,883 automobile fatalities in the United States in 2000.

TAKE NOTE

The fatality rate (the number of fatalities per 100,000 people) is equal to the rate of the total number of fatalities to the total population.

TAKE NOTE

The same unit is in both numerators (deaths), and the same unit is in both denominators (people).

Suggested Activity

See Section 7.4 of the *Student Activity Manual* for an investigation that involves using proportions to solve problems.

To answer the question "Was this more or less than the number in 1995?" we need to solve another proportion.

$$\frac{15.91 \text{ deaths}}{100,000 \text{ people}} = \frac{N \text{ deaths}}{263,000,000 \text{ people}}$$

$$100,000 \cdot N = 15.91(263,000,000)$$

$$100,000N = 4,184,330,000$$

$$N \approx 41,843$$

$$41,883 > 41,843$$

The number of automobile fatalities in 2000 was greater than the number in 1995.

It is important to remember, in setting up a proportion, to keep the same units in the numerators and the same units in the denominators. In the example above, "deaths" is the unit in the numerators and "people" is the units in the denominators. The following proportion, with the unit "people" in the numerators and "deaths" in the denominators, could also be used to solve the problem.

$$\frac{100,000 \text{ people}}{15.91 \text{ deaths}} = \frac{263,000,000 \text{ people}}{N \text{ deaths}}$$

? QUESTION   Write two proportions that can be used to determine the number of automobile fatalities in 1996. Use the data in the table on page 506.

INSTRUCTOR NOTE
Exercise 30 can be used for a similar in-class example.

EXAMPLE 2

The following table shows the marriage rate per 1000 people aged 15 years and over in the United States during selected years. The U.S. population for those years is also given (*Sources:* National Center for Health Statistics; U.S. Bureau of the Census). Use these data to find the difference between the number of marriages in 1950 and 1990. Round to the nearest hundred thousand.

Year	Marriage Rate per 1000 Population	U.S. Population
1950	11.1	152,271,000
1960	8.5	180,671,000
1970	10.6	205,052,000
1980	10.6	227,726,000
1990	9.8	249,913,000

? ANSWER   Two possibilities are the proportion $\frac{15.86 \text{ deaths}}{100,000 \text{ people}} = \frac{N \text{ deaths}}{265,000,000 \text{ people}}$, which has the unit "deaths" in the numerators, and the proportion $\frac{100,000 \text{ people}}{15.86 \text{ deaths}} = \frac{265,000,000 \text{ people}}{N \text{ deaths}}$, which has the unit "deaths" in the denominators.

State the goal. The goal is to find the difference between the number of marriages in 1950 and the number of marriages in 1990.

Devise a strategy.

- Write and solve a proportion to find the number of marriages in 1950.
- Write and solve a proportion to find the number of marriages in 1990.
- Subtract to find the difference.

Solve the problem.

$$\frac{11.1 \text{ marriages}}{1000 \text{ people}} = \frac{M \text{ marriages}}{152{,}271{,}000 \text{ people}}$$ • Find the number of marriages in 1950.

$$1000 \cdot M = 11.1(152{,}271{,}000)$$
$$1000M = 1{,}690{,}208{,}100$$
$$M \approx 1{,}690{,}208$$

$$\frac{9.8 \text{ marriages}}{1000 \text{ people}} = \frac{M \text{ marriages}}{249{,}913{,}000 \text{ people}}$$ • Find the number of marriages in 1990.

$$1000 \cdot M = 9.8(249{,}913{,}000)$$
$$1000M = 2{,}449{,}147{,}400$$
$$M \approx 2{,}449{,}147$$

$$2{,}449{,}147 - 1{,}690{,}208 = 758{,}939 \approx 800{,}000$$ • Subtract to find the difference.

There were approximately 800,000 more marriages in the United States in 1990 than in 1950.

Check your work. Be sure to check the solution.

▼ *Point of Interest*

Note that the population and the number of marriages increased from 1950 to 1990 but the marriage rate decreased during that time.

YOU TRY IT 2

A major concern in regard to the Social Security system is the fact that fewer and fewer workers are paying into the system to support more and more beneficiaries.

a. In 1960 the ratio of workers to beneficiaries was 5 to 1. If the number of beneficiaries in 1960 was 14 million, approximately how many workers were there in 1960?

b. The ratio of the number of workers to beneficiaries is expected to be 2 to 1 in 2030. How many Social Security beneficiaries are there expected to be in 2030 if the number of workers in 2030 is expected to be 167 million?

Sources: Social Security Administration; Census Bureau.

Solution See page S30. **a.** 70 million workers **b.** 83.5 million beneficiaries

■ Similar Triangles

Proportions have applications to the field of geometry. We will present here their application to similar triangles.

Similar objects have the same shape but not necessarily the same size. A tennis ball is similar to a basketball. A model airplane is similar to an actual airplane.

Similar objects have corresponding parts. For example, the wing on a model airplane corresponds to the wing on the actual airplane. The relationship between the sizes of each of the corresponding parts can be written as a ratio, and all these ratios will be the same. If the wing on the model airplane is $\frac{1}{100}$ the size of the wing on the actual airplane, then the model fuselage is $\frac{1}{100}$ the size of the actual fuselage, the wheels of the landing gear on the model are $\frac{1}{100}$ of the size of the wheels on the actual airplane, and so on.

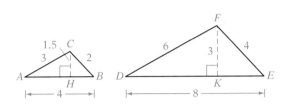

The two triangles *ABC* and *DEF* shown at the left are similar. Side *AB* corresponds to side *DE*, side *AC* corresponds to side *DF*, and side *BC* corresponds to side *EF*. Height *CH* corresponds to height *FK*. The ratios of corresponding parts are equal.

$$\frac{AB}{DE} = \frac{4}{8} = \frac{1}{2}, \qquad \frac{AC}{DF} = \frac{3}{6} = \frac{1}{2}, \qquad \frac{BC}{EF} = \frac{2}{4} = \frac{1}{2}, \qquad \frac{CH}{FK} = \frac{1.5}{3} = \frac{1}{2}$$

Because the ratios of corresponding parts are equal, three proportions can be formed using the sides of the triangles.

$$\frac{AB}{DE} = \frac{AC}{DF}, \qquad \frac{AB}{DE} = \frac{BC}{EF}, \qquad \text{and} \qquad \frac{AC}{DF} = \frac{BC}{EF}$$

Three proportions can also be formed by using the sides and height of the triangles.

$$\frac{AB}{DE} = \frac{CH}{FK}, \qquad \frac{AC}{DF} = \frac{CH}{FK}, \qquad \text{and} \qquad \frac{BC}{EF} = \frac{CH}{FK}$$

The corresponding angles in similar triangles are equal. Therefore,

$$m\angle A = m\angle D, \qquad m\angle B = m\angle E, \text{ and} \qquad m\angle C = m\angle F$$

➡ Triangles *ABC* and *DEF* shown below are similar. Find the area of triangle *ABC*.

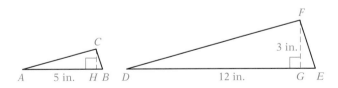

State the goal.   The goal is to find the area of triangle *ABC*.

Devise a strategy.

• Write and solve a proportion to find *CH*, the height of triangle *ABC*.

• Use the formula for the area of a triangle: $A = \frac{1}{2}bh$.

Solve the problem.

$$\frac{AB}{DE} = \frac{CH}{FG}$$

$$\frac{5}{12} = \frac{CH}{3}$$

$$12(CH) = 5(3)$$

$$12(CH) = 15$$

$$CH = 1.25$$

$$A = \frac{1}{2}bh$$

$$A = \frac{1}{2}(5)(1.25) = 3.125$$

The area of triangle *ABC* is 3.125 square inches.

Check your work.   Be sure to check the solution.   ⬅

It is also true that if the three angles of one triangle are equal, respectively, to the three angles of another triangle, then the two triangles are similar.

In the triangle at the right, line segment *DE* is drawn parallel to the base *AB*. $m\angle x = m\angle r$ and $m\angle y = m\angle n$ because corresponding angles are equal, and $m\angle C = m\angle C$. Therefore, the three angles of triangle *DEC* are equal, respectively, to the three angles of triangle *ABC*. Triangle *DEC* is similar to triangle *ABC*.

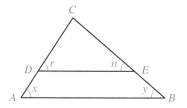

The sum of the three angles of a triangle is 180°. If two angles of one triangle are equal to two angles of another triangle, then the third angles must be equal. Thus we can say that **if two angles of one triangle are equal to two angles of another triangle, then the two triangles are similar.**

➡ Line segments *AB* and *CD* intersect at point *O* in the figure at the right. Angles *C* and *D* are right angles. Find the length of *DO*.

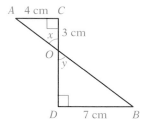

State the goal.   The goal is to find the length of *DO*.

Devise a strategy.

• First determine whether triangle *AOC* is similar to triangle *BOD*: $m\angle C = m\angle D$ because they are right angles. $m\angle x = m\angle y$ because they are vertical angles. Therefore, triangle *AOC* is similar to triangle *BOD* because two angles of one triangle are equal to two angles of the other triangle.

• Use a proportion to find the length of *DO*.

Solve the problem. $\dfrac{AC}{DB} = \dfrac{CO}{DO}$

$$\dfrac{4}{7} = \dfrac{3}{DO}$$

$$7(3) = 4(DO)$$

$$21 = 4(DO)$$

$$5.25 = DO$$

The length of *DO* is 5.25 centimeters.

Check your work.   The length of *AC* is greater than the length of *CO*, so the length of *BD* should be greater than the length of *DO*; $7 > 5.25$. The length of *BD* is greater than the length of *AC*, so the length of *DO* should be greater than the length of *CO*; $5.25 > 3$. A length of 5.25 for *DO* is reasonable.   ⬅

INSTRUCTOR NOTE
Exercise 46 can be used for a similar in-class example.

EXAMPLE 3

In the figure at the right, *AB* is parallel to *CD*, and angles *B* and *D* are right angles. *AB* = 12 meters, *CD* = 4 meters, and *AC* = 18 meters. Find the length of *CO*.

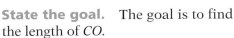

State the goal.   The goal is to find the length of *CO*.

Devise a strategy.   Triangle *AOB* is similar to triangle *COD*. Solve a proportion to find the length of *CO*. Let *x* represent the length of *CO*. Then $18 - x$ represents the length of *AO*.

Solve the problem. $\dfrac{CD}{AB} = \dfrac{CO}{AO}$

$$\dfrac{4}{12} = \dfrac{x}{18 - x}$$

$$12(x) = 4(18 - x)$$

$$12x = 72 - 4x$$

$$16x = 72$$

$$x = 4.5$$

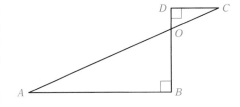

The length of *CO* is 4.5 meters.

Check your work.   Be sure to check the solution.

YOU TRY IT 3

In the figure at the right, *AB* is parallel to *CD*, and angles *A* and *D* are right angles. *AB* = 10 centimeters, *CD* = 4 centimeters, and *DO* = 3 centimeters. Find the area of triangle *AOB*.

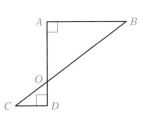

Solution   See page S30.   37.5 cm²

Surveyors use similar triangles to find the measures of distances that cannot be measured directly. This is illustrated in Example 4 and You Try It 4.

INSTRUCTOR NOTE
Exercise 50 can be used for
a similar in-class example.

EXAMPLE 4

The diagram at the right represents a river of width *CD*. Triangles *AOB* and *DOC* are similar. The distances *AB*, *BO*, and *CO* were measured and found to have the lengths given in the diagram. Find the width of the river.

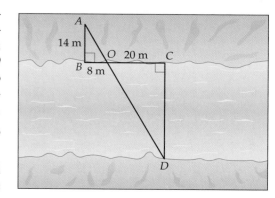

State the goal.   The goal is to find *CD*, the width of the river.

Devise a strategy.   Write and solve a proportion to find the length of side *CD*.

Solve the problem. $\dfrac{AB}{CD} = \dfrac{BO}{CO}$

$$\frac{14}{CD} = \frac{8}{20}$$

$$CD(8) = 14(20)$$

$$CD(8) = 280$$

$$CD = 35$$

The width of the river is 35 meters.

Check your work.   Be sure to check the solution.

YOU TRY IT 4

The diagram below shows how surveyors laid out similar triangles along the Winnepaugo River. Find the width, *w*, of the river.

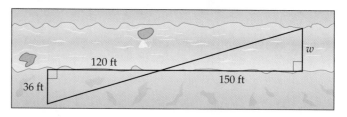

Solution   See page S31.   45 ft

■ Dimensional Analysis

In application problems, many numbers have *units of measurement* associated with the number, such as

9 **feet**   32 **pounds** $\dfrac{3}{4}$ **gallons**   23.8 **meters**

When a problem contains two different units, say feet and inches, it is usually necessary to convert one of the measurements so that both units are the same. This is accomplished by using **conversion factors.**

For instance, because 1 foot = 12 inches, we have the two conversion factors

$$\frac{1 \text{ foot}}{12 \text{ inches}} \quad \text{and} \quad \frac{12 \text{ inches}}{1 \text{ foot}}$$

We can show that each of these conversion factors is equal to 1 by replacing one of the measurements by its equivalent measurement. You can think of dividing the numerator and denominator by the common unit "inches" in one case and "foot" in the other case. It is important to remember that **all conversion factors are equal to 1.**

$$\frac{1 \text{ foot}}{12 \text{ inches}} = \frac{\overset{1}{\cancel{12 \text{ inches}}}}{\underset{1}{\cancel{12 \text{ inches}}}} = 1$$

$$\frac{12 \text{ inches}}{1 \text{ foot}} = \frac{\overset{1}{\cancel{1 \text{ foot}}}}{\underset{1}{\cancel{1 \text{ foot}}}} = 1$$

INSTRUCTOR NOTE
Exercise 56 can be used for a similar in-class example.

EXAMPLE 5

Convert 5 pints to quarts.

Solution   To convert from one unit to another, it is necessary to use the appropriate conversion factor. If you do not know the factor, look it up in a reference book.

TAKE NOTE

When converting from one unit to a second unit, use the conversion factor that has the second unit in the numerator.

For pints and quarts, the conversion factors are $\frac{2 \text{ pints}}{1 \text{ quart}}$ and $\frac{1 \text{ quart}}{2 \text{ pints}}$. Because we are converting from pints to quarts, we use the conversion factor that contains quart in the numerator. We use the abbreviations pt for pints and qt for quarts.

$$5 \text{ pt} = 5 \text{ pt} \cdot \boxed{1} = \frac{5 \cancel{\text{pt}}}{1} \cdot \boxed{\frac{1 \text{qt}}{2 \cancel{\text{pt}}}} = \frac{5}{2} \text{ qt} = 2.5 \text{ quarts}$$

YOU TRY IT 5

Convert 29 feet to yards.

Solution   See page S31. $9\frac{2}{3}$ yd

TAKE NOTE

We are using the conversion factors $\frac{1 \text{ hour}}{3600 \text{ seconds}} = 1$ and $\frac{5280 \text{ feet}}{1 \text{ mile}} = 1$.

Conversions between units may require using more than one conversion factor. Here is an example.

➡ Convert 60 miles per hour to feet per second.

$$\frac{60 \text{ miles}}{1 \text{ hour}} = \frac{60 \cancel{\text{ miles}}}{1 \cancel{\text{ hour}}} \cdot \frac{1 \cancel{\text{ hour}}}{3600 \text{ seconds}} \cdot \frac{5280 \text{ feet}}{1 \cancel{\text{ mile}}}$$

$$= \frac{60 \cdot 5280 \text{ feet}}{3600 \text{ seconds}} = 88 \text{ feet/second}$$

←

INSTRUCTOR NOTE
Exercise 68 can be used for
a similar in-class example.

EXAMPLE 6

A caterer has $7\frac{1}{2}$ gallons of iced tea for a party. How many $\frac{2}{3}$-cup servings of iced tea is it possible to make?

State the goal.   We must determine the number of $\frac{2}{3}$-cup servings in $7\frac{1}{2}$ gallons.

Devise a strategy.   To find the number of servings, we need to find how many $\frac{2}{3}$ cups are in $7\frac{1}{2}$ gallons. This requires that we divide $7\frac{1}{2}$ gallons by $\frac{2}{3}$ cups. Because the units, gallons and cups, are not the same, part of the solution must be to change gallons to cups. It may require some research to find that there are 16 cups in one gallon.

Solve the problem. $7\frac{1}{2}\text{ gal} = 7\frac{1}{2}\text{ gal} \cdot \dfrac{16\text{ c}}{1\text{ gal}} = \dfrac{15}{2} \cdot \dfrac{\overset{8}{16}\text{ c}}{\underset{1}{1}} = 120\text{ c}$

There are 120 cups in $7\frac{1}{2}$ gallons. Find the number of $\frac{2}{3}$-cup servings in 120 cups.

$$120 \div \frac{2}{3} = 120 \cdot \frac{3}{2} = 180$$

There are 180 $\frac{2}{3}$-cup servings in $7\frac{1}{2}$ gallons.

Check your work.   An important part of this solution was recognizing that we must work with the same units. You should also consider alternative solutions.

? QUESTION   In Example 6, would it have been possible to find the number of gallons in $\frac{2}{3}$ cup and then to divide $7\frac{1}{2}$ gallons by that number?

YOU TRY IT 6

The Meridian Group purchased $8\frac{1}{2}$ acres of land to be used for a housing development. How many housing lots of 10,000 square feet each are possible?

Solution   See page S31.   37 housing lots (1 acre = 43,560 ft²)

? ANSWER   Yes. $\dfrac{2}{3}\text{ c} = \dfrac{2}{3}\overset{1}{\text{c}} \cdot \dfrac{1\text{ gal}}{\underset{8}{16}\text{ c}} = \dfrac{1}{24}\text{ gal}; 7\dfrac{1}{2} \div \dfrac{1}{24} = \dfrac{15}{2} \div \dfrac{1}{24} = \dfrac{15}{2} \cdot \dfrac{24}{1} = 180$ servings

7.4 EXERCISES Suggested Assignment: 9–75, odds

Topics for Discussion

1. **a.** What is a rate? **b.** What is a ratio? **c.** What is a proportion?
   Answers may vary. For example: **a.** A rate is the quotient of two quantities that have different units. **b.** A ratio is the quotient of two quantities that have the same unit. **c.** A proportion is the equality of two rates or ratios.

2. Provide two examples of proportions, one involving rates and one involving ratios. Examples will vary.

3. Explain what the means and the extremes in a proportion are.
   Answers will vary. For example, the second and third terms are the means. The first and fourth terms are the extremes.

4. Explain why the product of the means in a proportion is equal to the product of the extremes. Answers will vary. For example, it is equivalent to multiplying both sides of the proportion by the LCM of the denominators.

5. Determine whether the statement is always true, sometimes true, or never true.
   a. If an acute angle of a right triangle is equal to an acute angle of another right triangle, then the triangles are similar. Always true
   b. Two isosceles triangles are similar triangles. Sometimes true
   c. Two equilateral triangles are similar triangles. Always true
   d. Two squares are similar. Always true
   e. Two rectangles are similar. Sometimes true

6. Explain how proportions are related to similar triangles. Answers will vary.

7. Explain why we can multiply an expression by the conversion factor $\frac{5280 \text{ feet}}{1 \text{ mile}}$. Answers will vary. For example, the conversion factor is equal to 1, and multiplying an expression by 1 does not change its value.

■ Solving Proportions

Solve.

8. $\frac{4}{5} = \frac{12}{x}$ 15

9. $\frac{6}{x} = \frac{2}{3}$ 9

10. $\frac{20}{9} = \frac{64}{x}$ 28.8

11. $\frac{n}{12} = \frac{5}{8}$ 7.5

12. $\frac{8}{n+3} = \frac{4}{n}$ 3

13. $\frac{3}{x-2} = \frac{4}{x}$ 8

14. $\frac{6}{x+4} = \frac{12}{5x-13}$ 7

15. $\frac{2}{3x-1} = \frac{3}{4x+1}$ 5

16. $\frac{2}{x+3} = \frac{6}{5x+5}$ 2

17. $\frac{5}{n+3} = \frac{3}{n-1}$ 7

18. $\frac{5}{2x-3} = \frac{10}{x+3}$ 3

19. $\frac{4}{5y-1} = \frac{2}{2y-1}$ −1

20. $\frac{x}{x-1} = \frac{8}{x+2}$ 2, 4

21. $\frac{x}{x+12} = \frac{1}{x+5}$ −6, 2

22. $\frac{2x}{x+4} = \frac{3}{x-1}$ $-\frac{3}{2}$, 4

23. $\frac{5}{3y-8} = \frac{y}{y+2}$ $-\frac{2}{3}$, 5

24. *Fuel* A cord is a quantity of cut wood, to be used for fuel, that is equal to 128 cubic feet in a stack measuring 4 feet by 4 feet by 8 feet. Cutting 8 cords of wood produces 1 cord of sawdust. At this rate, how much sawdust is produced by cutting 14 cords of wood? 1.75 cords

25. *Media* In a city of 25,000 homes, a survey was taken to determine the number with cable television. Of the 300 homes surveyed, 210 had cable television. Based upon the sample, estimate the number of homes in the city that have cable television. 17,500 homes

26. *Power* The lighting for some billboards is provided by using solar energy. If 3 small solar energy panels can generate 10 watts of power, how many panels are necessary to provide 600 watts of power? 180 panels

27. *Taxes* The sales tax on a car that sold for $12,000 is $780. At this rate, what is the sales tax on a car that sells for $18,500? $1202.50

28. *Architecture* To conserve energy and still allow for as much natural lighting as possible, an architect suggests that the ratio of the area of window to the area of the total wall surface be 5 to 12. Using this ratio, determine the recommended area of a window to be installed in a wall that measures 8 feet by 12 feet. 40 ft²

29. *Conservation* As part of a conservation effort for a lake, 40 fish are caught, tagged, and then released. Later 80 fish are caught. Four of these 80 fish are found to have tags. Estimate the number of fish in the lake.
800 fish

30. *Birth Rates* The following table shows the states with the highest birth rates. Also shown is the population for those states (*Sources:* Center for Disease Control and Prevention; U.S. Bureau of the Census).

State	Birth Rate per 1000 Residents	Population
Utah	21.7	2,233,000
Texas	17.3	20,852,000
Arizona	17.0	5,131,000
Georgia	16.3	8,186,000
Nevada	16.2	1,998,000
Alaska	16.1	630,000
Idaho	15.9	1,294,000

a. Find the number of births in Georgia.
b. How many more births were there in Nevada than in Alaska? Round to the nearest whole number.
c. Write an explanation for providing statistics on the birth rate per 1000 residents rather than the number of births.
a. 133,432 births b. 22,225 more births c. Answers will vary.

31. *Divorce Rates* The following table shows the divorce rates for the United States. Also shown is the population of the United States for each of the years given (*Sources:* National Center for Health Statistics; U.S. Bureau of the Census).

Year	Divorce Rate per 1000 Population	U.S. Population
1960	2.2	181,000,000
1970	3.5	205,000,000
1980	5.2	228,000,000
1990	4.7	250,000,000
2000	4.2	275,000,000

a. Find the number of divorces in the United States in 1960.
b. During which year were there more divorces, 1980 or 1990?
c. Find the difference between the number of divorces in 2000 and the number in 1970.
d. Find the percent increase in the number of divorces from 1970 to 1980. Round to the nearest percent.
e. Find the percent decrease in the divorce rate per 1000 residents from 1990 to 2000. Round to the nearest percent.

a. 398,200 divorces   b. 1980   c. 437,500 divorces   d. 65%   e. 11%

32. *Law Enforcement*   The following table shows the cities in the United States with the highest number of full-time police officers per 10,000 residents. Also shown is the population of each city (*Sources:* Local Police Departments; the Bureau of Justice Statistic; U. S. Bureau of the Census).

City	Full-Time Officers per 10,000 Residents	Population
Washington, DC	67	572,000
New York	52	8,008,000
Newark, NJ	52	274,000
Chicago	49	2,896,000

a. Find the number of full-time officers in Washington, DC.
b. Which city has more full-time officers, New York or Newark? Explain why you can determine the answer without performing any calculations.
c. Find the difference between the number of full-time officers in New York and in Newark.

a. 3832 officers   b. New York   c. 40,217 officers

33. *Traffic Accidents*   The following graph shows the motor vehicle crash rate per 100,000 drivers in different age groups in the United States (*Source:* National Safety Council).

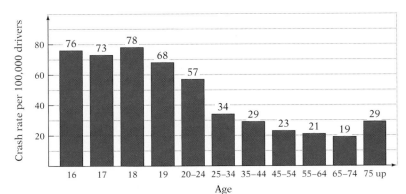

a. The U.S. population of 25- to 34-year-olds is 38,979,000. Find the number of motor vehicle crashes this age group was involved in.

b. The U.S. population of 20- to 24-year-olds is 17,633,000. The U.S. population of 35- to 44-year-olds is 44,353,000. Which age group was involved in more motor vehicle crashes? How many more?

c. If ages 16–19 were combined, rather than shown separately, would the combined rate for this group of drivers be greater than or less than the individual rate shown for 17-year olds?

d. How are statistics such as these used by automobile insurance companies to determine premiums for drivers of different ages?

a. 13,253 crashes   **b.** 35- to 44-year-olds; 2811 more   **c.** Greater than   **d.** Answers will vary.

▪ Similar Triangles

Triangles *ABC* and *DEF* in Exercises 34 to 41 are similar. Round answers to the nearest tenth.

34. Find side *AC*.

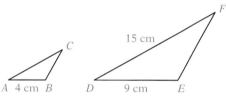

6.7 cm

35. Find side *DE*.

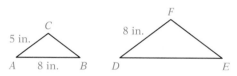

12.8 in.

36. Find the height of triangle *ABC*.

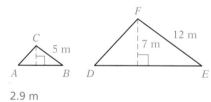

2.9 m

37. Find the height of triangle *DEF*.

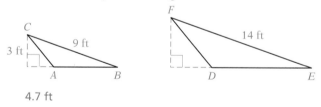

4.7 ft

38. Find the perimeter of triangle *DEF*.

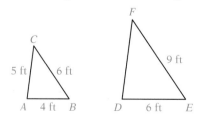

22.5 ft

39. Find the perimeter of triangle *ABC*.

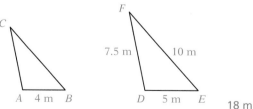

18 m

40. Find the area of triangle *ABC*.

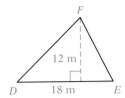

48 m²

41. Find the area of triangle *ABC*.

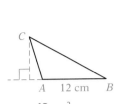

48 cm²

42. Given *BD*∥*AE*, *BD* measures 5 centimeters, *AE* measures 8 centimeters, and *AC* measures 10 centimeters, find the length of *BC*.

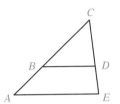

6.25 cm

43. Given *AC*∥*DE*, *BD* measures 8 meters, *AD* measures 12 meters, and *BE* measures 6 meters, find the length of *BC*.

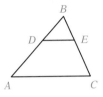

15 m

44. Given *DE*∥*AC*, *DE* measures 6 inches, *AC* measures 10 inches, and *AB* measures 15 inches, find the length of *DA*.

6 in.

45. Given *MP* and *NQ* intersect at *O*, *NO* measures 25 feet, *MO* measures 20 feet, and *PO* measures 8 feet, find the length of *QO*.

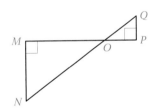

10 ft

46. Given *MP* and *NQ* intersect at *O*, *NO* measures 24 centimeters, *MN* measures 10 centimeters, *MP* measures 39 centimeters, and *QO* measures 12 centimeters, find the length of *OP*.

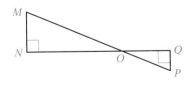

13 cm

47. Given *MQ* and *NP* intersect at *O*, *NO* measures 12 meters, *MN* measures 9 meters, *PQ* measures 3 meters, and *MQ* measures 20 meters, find the perimeter of triangle *OPQ*.

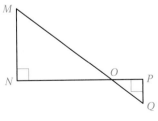

12 m

48. *Measurement* The sun's rays cast a shadow as shown in the diagram at the right. Find the height of the flagpole. Write the answer in terms of feet.   14.375 ft

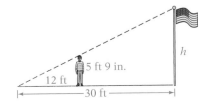

49. *Surveying*   The diagram at the right represents a river of width *CD*. The distances *AB*, *BO*, and *OC* were measured and found to have the lengths given in the diagram. Find the width of the river.   36 m

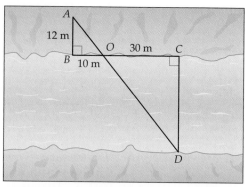

50. *Surveying*   The diagram at the right shows how surveyors laid out similar triangles along a ravine. Find the width, *w*, of the ravine.   82.5 ft

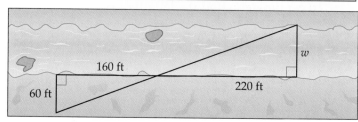

■ **Dimensional Analysis**

51. Convert 28 inches to feet. $2\frac{1}{3}$ ft

52. Convert 9 feet to inches.   108 in.

53. Convert 3 gallons to quarts.   12 qt

54. Convert 4 pints to quarts.   2 qt

55. Convert $4\frac{1}{2}$ feet to inches.   54 in.

56. Convert $\frac{1}{2}$ cup to fluid ounces.   4 fl oz

57. Convert 28 ounces to pounds. $1\frac{3}{4}$ lb

58. Convert 2500 pounds to tons. $1\frac{1}{4}$ tons

59. Convert 8 fluid ounces to pints. $\frac{1}{2}$ pt

60. Convert 12 pints to quarts.   6 qt

61. Convert 44 feet per second to miles per hour.   30 mph

62. Convert 66 feet per second to miles per hour.   45 mph

63. Convert 12 quarts per minute to gallons per second. $\frac{1}{20}$ gal/s

64. Convert 15 miles per hour to feet per second.   22 ft/s

65. *Time*   When a person reaches the age of 35, for how many seconds has that person lived?   1,103,760,000 s

66. *Recreation*   Five students are going backpacking in the desert. Each student requires 2 quarts of water per day. How many gallons of water should they take for a 3-day trip? $7\frac{1}{2}$ gal

67. *Recreation*   A hiker is carrying 5 quarts of water. Water weighs $8\frac{1}{3}$ pounds per gallon. Find the weight of the water carried by the hiker. $10\frac{5}{12}$ lb

68. *Food Industry*   A can of cranberry juice contains 25 fluid ounces. How many quarts of cranberry juice are in a case of 24 cans?   18.75 qt

69. *Sound*   The speed of sound is about 1100 feet per second. Find the speed of sound in miles per hour.   750 mph

70. *Real Estate*   A $\frac{1}{2}$-acre commercial lot is selling for $3 per square foot. Find the price of the commercial lot. $65,340

71. *Interior Decorating*   Each pleat in a drape requires 8 inches of fabric. How many pleats can be made from a drape material that is 4 yards long?   18 pleats

72. *Construction*   Studs for the support of a wall are required at each end and 16 inches apart. How many studs are required for a wall that is 13 feet long?   11 studs

73. *Business*   A candy store purchases 12 pounds of candy for $6.50 and then repackages the candy in 4-ounce packages. The selling price for each 4-ounce package is $.60. Find the profit earned on each 4-ounce package that is sold. Round to the nearest cent. $.46

74. *Business*   A mechanic purchases a 40-gallon container of oil for $128 and charges customers $1.50 per quart for an oil change. Ignoring the cost of labor, find the profit the mechanic makes on one quart of oil. $.70

75. *Fund Raising*   A charity group is trying to raise money by recycling aluminum cans. The charity has found a recycler that will pay $.50 per pound for the 10,000 cans the group has collected. If four aluminum cans weigh 3 ounces, how much money will the recycler pay to the charity? $234.38

76. *Interior Decorating*   Wall-to-wall carpeting is to be laid in the living room of the home shown in the following floor plan. At $24 per square yard, how much will the carpeting cost? (Note: The dining room dimensions of 12 × 15 mean 12 ft by 15 ft.) $800

FIRST FLOOR

Applying Concepts

77. *Divorce Rates* The divorce rate per 1000 people in the United States in 1950 was 2.6; 395,000 people were divorced that year. The divorce rate per 1000 people in the United States in 1960 was 2.2; 398,200 people were divorced that year (*Sources:* National Center for Health Statistics; U.S. Bureau of the Census). Discuss the fact that in 1950 the rate was higher than in 1960 but the number of people divorced was lower. Answers will vary. For example, the population of the United States was greater in 1960.

78. *Sports* A basketball player has made 5 out of every 6 foul shots attempted. If 48 foul shots were missed in the player's career, how many foul shots were made in the player's career? 240 foul shots

79. *Photography* The "sitting fee" for school pictures is $8. If 10 photos cost $20, including the sitting fee, what would 24 photos cost, including the sitting fee? $36.80

80. *Lotteries* Three people put their money together to buy lottery tickets. The first person put in $20, the second person put in $25, and the third person put in $30. One of their tickets was a winning ticket. If they won $15 million, what was the first person's share of the jackpot? $4 million

81. *Sports* After the skiing-related deaths of Michael Kennedy and Sonny Bono, the following statistics were published (*Source:* National Ski Areas Association).

Ski Season	1996–97	1995–96	1994–95	1993–94	1992–93	1991–92	1990–91
Fatalities	36	45	49	41	42	35	38
Rate per Million Skier Days	0.69	0.65	0.93	0.75	0.78	0.69	0.60

   a. How many skier days were there during the 1996–1997 ski season?

   b. On the basis of the data in the table and your answer to part a, what do you think the definition of a "skier day" is?

   c. Judging on the basis of the data, how dangerous do you consider the sport of skiing to be?

   a. 52,173,913 skier days   **b.** Answers will vary.   **c.** Answers will vary.

82. The height of a right triangle is drawn from the right angle to the hypotenuse. (See the accompanying diagram.) Explain why the two triangles formed are similar to the original triangle and similar to each other. Explanations may vary.

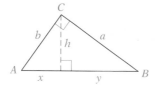

EXPLORATION

1. *Topology*   In this section, we discussed similar figures—that is, figures with the same shape. The branch of geometry called **topology** is interested in even more basic properties of figures than their size and shape. For example, look at the figures below. We could take a rubber band and stretch it into any one of these shapes.

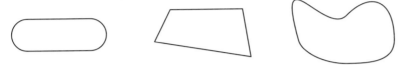

All three of these figures are different shapes, but each can be turned into one of the others by stretching the rubber band.

In topology, figures that can be stretched, molded, or bent into the same shape without puncturing or cutting belong to the same family. They are called **topologically equivalent.**

Rectangles, triangles, and circles are topologically equivalent.

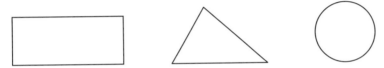

Line segments and wavy curves are topologically equivalent.

Note that the figures formed from a rubber band and a line segment are not topologically equivalent; to form a line segment from a rubber band, we would have to cut the rubber band.

In the following plane figures, the lines are joined where they cross. They are topologically equivalent. They are not topologically equivalent to any of the figures shown above.

A topologist (a person who studies topology) is interested in identifying and describing different families of equivalent figures. This applies to solids as well as to plane figures. For example, a topologist considers a brick, a potato, and a cue ball to be equivalent to each other. Think of using modeling clay to form each of these shapes.

For parts a, b, and c, which of the figures listed is not topologically equivalent to the others?

a. parallelogram   square   ray   trapezoid
b. wedding ring   doughnut   fork   sewing needle
c. A D O P T
d. Make a list of three objects that are topologically equivalent.

a. ray   **b.** fork   **c.** T   **d.** Answers will vary.

SECTION **7.5**

Variation

■ Variation Problems

■ Variation Problems

As illustrated in the last section, many types of problems can be solved by setting up a proportion. In this section we will begin by defining a special type of proportion called a direct proportion.

Two quantities are **directly proportional** if an increase in one quantity leads to a proportional increase in the other quantity. Let's look at an example:

The number of filters purchased is directly proportional to the cost; the more filters purchased, the greater the cost. If 8 filters cost $39.60, what is the cost of 6 filters?

We can write the equality of two ratios: the ratio of 6 filters to 8 filters, and the ratio of the cost of 6 filters to the cost of 8 filters.

As the number of filters increases,   the cost increases.

$$\frac{6 \text{ filters}}{8 \text{ filters}} = \frac{C}{39.60}$$

$$8C = 6(39.60)$$
$$8C = 237.60$$
$$C = 29.70$$

The cost of 6 filters is $29.70.

? QUESTION   One gallon of paint is required for every 600 square feet of wall to be painted. Are these two quantities directly proportional?

Many important relationships in business, science, and engineering involve quantities that are directly proportional. Often these relationships can be written in the form $y = kx$, where k is a constant. The constant k is called the **constant of variation** or the **constant of proportionality.** The equation $y = kx$ is a **direct variation equation** and is read "y varies directly as x" or "y is directly proportional to x."

? ANSWER   Yes. As the number of square feet of wall to be painted increases, the number of gallons of paint required increases.

In the previous problem, the cost of 1 filter is \$4.95 (\$29.70 ÷ 6 = \$4.95, or \$39.60 ÷ 8 = \$4.95). Therefore, the problem could be described by the equation $y = 4.95x$, where x is the number of filters purchased and y is the total cost of x filters. We can find the cost of 6 filters as follows:

$y = 4.95x$ • The number 4.95 is the constant of proportionality.

$y = 4.95(6)$ • Replace x, the number of filters, by 6.

$y = 29.70$

The cost of 6 filters is \$29.70, the same answer that was given above.

The distance traveled by a car driven at a constant rate of 55 mph is represented by the equation $y = 55x$, where x is the number of hours and y is the total distance traveled. Suppose you travel 4 hours at 55 mph.

$y = 55x$ • The number 55 is the constant of proportionality.

$y = 55(4)$ • Replace x, the number of hours, by 4.

$y = 220$

If you travel 4 hours at 55 mph, you travel a total of 220 miles.

INSTRUCTOR NOTE
Exercise 10 can be used for a similar in-class example.

EXAMPLE 1

Find the constant of variation if y varies directly as x, and $y = 35$ when $x = 5$.

State the goal.   The goal is to determine the value of k in the equation $y = kx$ when $y = 35$ and $x = 5$.

Devise a strategy.   In the equation $y = kx$, replace y with 35 and x with 5. Solve for k.

Solve the problem. $y = kx$ • Use the direct variation equation.

$35 = k \cdot 5$ • Replace y by 35 and x by 5.

$\dfrac{35}{5} = \dfrac{k \cdot 5}{5}$ • Divide both sides of the equation by 5.

$7 = k$

The constant of variation is 7.

Check your work.   Check by substituting 35 for y, 7 for k, and 5 for x in the direct variation equation.

$$\frac{y = kx}{35 \mid 7 \cdot 5}$$

$35 = 35$ • A true equation.

YOU TRY IT 1

Find the constant of variation if y varies directly as x, and $y = 120$ when $x = 8$.

Solution   See page S31.   15

Suggested Activity

The monthly interest charged on the unpaid balance on a credit card varies directly as the unpaid balance. If the interest on $475 is $6.65, what is the constant of variation? [Answer: 0.014] What does the constant of variation in this situation represent? [Answer: The monthly interest rate, 1.4%.] For what unpaid balance would the monthly interest charge be $4.06? [Answer: $290]

 See Appendix A: Trace

➡ The distance, d, that sound travels varies directly as the time, t, it travels. If sound travels 8920 feet in 8 seconds, find the distance that sound travels in 3 seconds.

$$d = kt$$
• This is a direct variation. k is the constant of proportionality.

$$8920 = k \cdot 8$$
• Replace d by 8920 and t by 8.

$$\frac{8920}{8} = \frac{k \cdot 8}{8}$$
• Solve for k. Divide both sides by 8.

$$1115 = k$$

$$d = 1115t$$
• Write the direct variation equation. $k = 1115$.

$$d = 1115 \cdot 3$$
• Replace t with 3 to find d when t is 3.

$$d = 3345$$

Sound travels 3345 feet in 3 seconds.

The distance 3345 can also be found by graphing the direct variation equation $d = 1115t$ and finding d when t is 3.

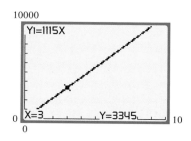

❓ QUESTION   How could the above problem be set up as a proportion? Is the solution of the proportion 3345 feet?

INSTRUCTOR NOTE
Exercise 20 can be used for a similar in-class example.

EXAMPLE 2

The amount, A, of medication prescribed for a person is directly related to the person's weight, W. For a 50-kilogram person, 2 milliliters of medication are prescribed. How many milliliters of medication are required for a person who weighs 75 kilograms?

State the goal.   We want to find the amount of medication to be given to a 75-kilogram person.

Devise a strategy.
• This is a direct variation. To find the value of k, write the basic direct variation equation, replace the variables by the given values, and solve for k.
• Write the direct variation equation, replacing k by its value. Substitute 75 for W and solve for A.

❓ ANSWER   Here are two possibilities: $\dfrac{8920 \text{ ft}}{d \text{ ft}} = \dfrac{8 \text{ s}}{3 \text{ s}}$ and $\dfrac{d \text{ ft}}{8920 \text{ ft}} = \dfrac{3 \text{ s}}{8 \text{ s}}$. The solution of either proportion is 3345 feet.

Solve the problem.

$$A = kW$$ • Use the direct variation equation.

$$2 = k \cdot 50$$ • Replace A by 2 and W by 50.

$$\frac{2}{50} = \frac{k \cdot 50}{50}$$ • Solve for k. Divide both sides by 50.

$$0.04 = k$$

$$A = 0.04W$$ • Write the direct variation equation. $k = 0.04$.

$$A = 0.04(75)$$ • Replace W with 75 to find A when W is 75.

$$A = 3$$

The required amount of medication is 3 milliliters.

Check your work.

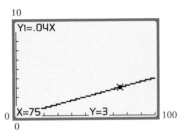

YOU TRY IT 2

A nurse's total wage, w, is directly proportional to the number of hours, h, worked. If the nurse earns \$264 for working 12 hours, what is the nurse's total wage for working 18 hours?

Solution   See page S31.   \$396

Some direct variation equations are written in the form $y = kx^n$. For example, the equation $y = kx^2$ is read "y varies directly as the square of x" or "y is directly proportional to the square of x." Example 3 and You Try It 3 illustrate this type of direct variation.

INSTRUCTOR NOTE
Exercise 24 can be used for a similar in-class example.

EXAMPLE 3

The load, L, that a horizontal beam can safely support is directly proportional to the square of the depth, d, of the beam. A beam with a depth of 8 inches can support 800 pounds. Find the load that a beam with a depth of 6 inches can support.

State the goal.   We want to determine the amount of weight a beam that has a depth of 6 inches can support.

Devise a strategy.

• This is a direct variation. To find the value of k, write the basic direct variation equation, replace the variables by the given values, and solve for k.
• Write the direct variation equation, replacing k by its value. Substitute 6 for d and solve for L.

Solve the problem.

$$L = kd^2$$ • Use the direct variation equation.

$$800 = k \cdot 8^2$$ • Replace L by 800 and d by 8.

$$800 = k \cdot 64$$ • Square 8.

$$\frac{800}{64} = \frac{k \cdot 64}{64}$$ • Solve for k. Divide each side by 64.

$$12.5 = k$$

$$L = 12.5d^2$$ • Write the direct variation equation. $k = 12.5$.

$$L = 12.5 \cdot 6^2$$ • Replace d with 6 to find L when d is 6.

$$L = 12.5 \cdot 36$$ • Square 6.

$$L = 450$$

The beam can support a load of 450 pounds.

Check your work.

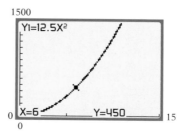

YOU TRY IT 3

The distance, s, a body falls from rest varies directly as the square of the time, t, of the fall. An object falls 64 feet in 2 seconds. How far will it fall in 5 seconds?

Solution See page S31. 400 ft

Two quantities are **inversely proportional** if an increase in one quantity leads to a proportional decrease in the other quantity, or if a decrease in one leads to a proportional increase in the other.

An **inverse variation** is one that can be written in the form $y = \frac{k}{x}$, where k is a constant. The equation $y = \frac{k}{x}$ is read "y varies inversely as x" or "y is inversely proportional to x."

In an automobile cylinder, the volume, V, of a gas is inversely proportional to the pressure, P, given that the temperature does not change. The inverse variation equation is written.

$$V = \frac{k}{P}$$

➡ If the volume of gas in the cylinder is 300 cubic centimeters when the pressure is 20 pounds per square inch, what is the volume when the pressure is increased to 80 pounds per square inch?

▼ *Point of Interest*

The time required to travel a given distance is inversely proportional to the speed of travel. The faster you travel, the shorter the time to reach the destination. The more slowly you travel, the longer it takes to reach the destination.

$$V = \frac{k}{P}$$

• Use the inverse variation equation shown above. k is the constant.

$$300 = \frac{k}{20}$$

• Replace V by 300 and P by 20.

$$20 \cdot 300 = 20 \cdot \frac{k}{20}$$

• Solve for k. Multiply both sides by 20.

$$6000 = k$$

$$V = \frac{6000}{P}$$

• Write the inverse variation equation. $k = 6000$.

$$V = \frac{6000}{80}$$

• Replace P with 80 to find V when P is 80.

$$V = 75$$

When the pressure is 80 pounds per square inch, the volume is 75 cubic centimeters. ⬅

Because this is an *inverse* variation, we cannot write a proportion as we did with direct variations. Look at the two ratios we would form:

$$\frac{300 \text{ cm}^2}{75 \text{ cm}^2} \qquad\qquad \frac{20 \text{ psi}}{80 \text{ psi}}$$

This ratio simplified to 4.   This ratio simplifies to $\frac{1}{4}$.

These ratios are not equal. Therefore, we cannot set them equal to each other to write a proportion. Note, however, that the simplified ratios are reciprocals of each other, or multiplicative inverses. This is always the case with inverse variations: One ratio is the multiplicative inverse of the other.

❓ QUESTION   The loudness of the music heard on a stereo speaker is 5 decibels at a distance of 20 feet from the speaker. Are these two quantities inversely proportional?

EXAMPLE 4

A company that produces personal computers has determined that the number of computers it can sell, s, is inversely proportional to the price, P, of the computer. (This means that the higher the price, the lower the number of computers sold; the lower the price, the higher the number of computers sold.) Two thousand computers can be sold when the price is $2500. How many computers can be sold when the price is $2000?

State the goal.   The goal is to find the number of computers that can be sold at a price of $2000 each.

❓ ANSWER   Yes. As the distance from the speaker increases, the decibels decrease; or as the decibels increase, the distance from the speaker decreases.

Suggested Activity
See Section 7.5 of the *Student Activity Manual* for an investigation that involves inverse variation.

INSTRUCTOR NOTE
Exercise 32 can be used for a similar in-class example.

Devise a strategy.
- This is an inverse variation. To find the value of k, write the basic inverse variation equation, replace the variables by the given values, and solve for k.
- Write the inverse variation equation, replacing k by its value. Substitute 2000 for P and solve for s.

Solve the problem.

$$s = \frac{k}{P}$$ • Use the inverse variation equation.

$$2000 = \frac{k}{2500}$$ • Replace s by 2000 and P by 2500.

$$2500 \cdot 2000 = 2500 \cdot \frac{k}{2500}$$ • Solve for k. Multiply both sides by 2500.

$$5{,}000{,}000 = k$$

$$s = \frac{5{,}000{,}000}{P}$$ • Write the inverse variation equation. $k = 5{,}000{,}000$.

$$s = \frac{5{,}000{,}000}{2000}$$ • Replace P with 2000 to find s when P is 2000.

$$s = 2500$$

At a price of $2000, 2500 computers can be sold.

Check your work.

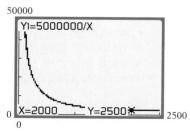

YOU TRY IT 4

At an assembly plant, the number of hours, h, it takes to complete the daily quota is inversely proportional to the number of assembly machines, m, operating. If five assembly machines can complete the daily quota in 9 hours, how many hours does it take for four assembly machines to complete the daily quota?

Solution   See page S32.   11.25 h

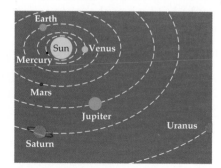

Some inverse variation equations are written in the form $y = \frac{k}{x^n}$. For example, the gravitational force, F, between two planets is inversely proportional to the square of the distance, d, between the planets. This inverse equation is written $F = \frac{k}{d^2}$ and is read "F varies inversely as the square of d" or "F is inversely proportional to the square of d." Note that this means that the farther apart the planets, the smaller the gravitational force between them.

INSTRUCTOR NOTE
Exercise 34 can be used for
a similar in-class example.

EXAMPLE 5

The resistance, R, to the flow of electric current in a wire of fixed length varies inversely as the square of the diameter, d, of the wire. A wire of diameter 0.01 centimeter has a resistance of 0.5 ohm. Find the resistance in a wire that is 0.02 centimeter in diameter.

State the goal.   The goal is to find the resistance in a wire that has a diameter of 0.02 centimeter.

Devise a strategy.

- This is an inverse variation. To find the value of k, write the basic inverse variation equation, replace the variables by the given values, and solve for k.
- Write the inverse variation equation, replacing k by its value. Substitute 0.02 for d and solve for R.

Solve the problem.

$$R = \frac{k}{d^2}$$
- Use the inverse variation equation.

$$0.5 = \frac{k}{0.01^2}$$
- Replace R by 0.5 and d by 0.01.

$$0.5 = \frac{k}{0.0001}$$
- Square 0.01.

$$0.0001 \cdot 0.5 = 0.0001 \cdot \frac{k}{0.0001}$$
- Solve for k. Multiply both sides by 0.0001.

$$0.00005 = k$$

$$R = \frac{0.00005}{d^2}$$
- Write the inverse variation equation. $k = 0.00005$.

$$R = \frac{0.00005}{0.02^2}$$
- Replace d with 0.02 to find R when d is 0.02.

$$R = \frac{0.00005}{0.0004}$$

$$R = 0.125$$

The resistance is 0.125 ohm.

Check your work.

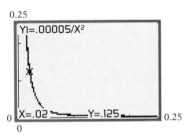

YOU TRY IT 5

The intensity, I, of a light source is inversely proportional to the square of the distance, d, from the source. If the intensity is 20 footcandles at a distance of 8 feet, what is the intensity when the distance is 5 feet?

Solution   See page S32.   51.2 footcandles

7.5 EXERCISES   Suggested Assignment: 7–37, odds

Topics for Discussion

1. **a.** When are two quantities directly proportional?
    b. When are two quantities inversely proportional?

    a. When an increase in one quantity leads to a proportional increase in the other quantity. **b.** When an increase in one quantity leads to a proportional decrease in the other quantity.

2. State whether the two quantities are directly proportional or inversely proportional. Explain your answer.

    a. One acre planted in wheat will produce 45 bushels of wheat.
    b. Traveling at an average speed of 30 mph, a trip took 2 hours.
    c. A truck travels 17 miles on 1 gallon of fuel.

    a. Directly proportional **b.** Inversely proportional **c.** Directly proportional (Explanations will vary.)

3. What is a constant of variation? Answers will vary.

4. Explain the relationship between direct variation and proportion.
    Answers will vary.

5. Determine whether the statement is true or false.

    a. In the direct variation equation $y = kx$, if x increases, then y increases.
    b. In the inverse variation equation $y = \frac{k}{x}$, if x increases, then y increases.
    c. In the direct variation equation $y = kx^2$, if x doubles, then y doubles.

    a. True   **b.** False   **c.** False

▪ Variation Problems

For Exercises 6 and 7, solve using a proportion and using a direct variation equation.

6. If y varies directly as x, and $x = 10$ when $y = 4$, find y when $x = 15$.   6

7. Given that L varies directly as P, and $L = 24$ when $P = 21$, find P when $L = 80$.   70

8. Find the constant of variation when y varies directly as x, and $y = 15$ when $x = 2$.   7.5

9. Find the constant of variation when n varies directly as the square of m, and $n = 64$ when $m = 2$.   16

10. Find the constant of proportionality when T varies inversely as S, and $T = 0.2$ when $S = 8$.   1.6

11. Find the constant of variation when W varies inversely as the square of V, and $W = 5$ when $V = 0.5$.   1.25

12. Given that P varies directly as R, and $P = 20$ when $R = 5$, find P when $R = 6$.   24

13. Given that M is directly proportional to P, and $M = 15$ when $P = 30$, find M when $P = 20$.   10

14. Given that W is directly proportional to the square of V, and $W = 50$ when $V = 5$, find W when $V = 12$.   288

15. If A varies directly as the square of r, and $A = \frac{22}{7}$ when $r = 1$, find A when $r = 7$.   154

16. If y varies inversely as x, and $y = 500$ when $x = 4$, find y when $x = 10$.   200

17. If L varies inversely as the square of d, and $L = 25$ when $d = 2$, find L when $d = 5$.   4

18. *Compensation*   A worker's wage, w, is directly proportional to the number of hours, h, worked. If \$82 is earned for working 8 hours, how much is earned for working 30 hours?   \$307.50

19. *Physics*   The distance, d, a spring will stretch varies directly as the force, F, applied to the spring. If a force of 12 pounds is required to stretch a spring 3 inches, what force is required to stretch the spring 5 inches?   20 lb

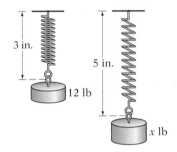

20. *Scuba Diving*   The pressure, P, on a diver in the water varies directly as the depth, d. If the pressure is 2.25 pounds per square inch when the depth is 5 feet, what is the pressure when the depth is 12 feet?   5.4 psi

21. *Clerical Work*   The number of words typed, w, is directly proportional to the time, t, spent typing. A typist can type 260 words in 4 minutes. Find the number of words typed in 15 minutes.   975 words

22. *Electricity*   The current, I, varies directly as the voltage, V, in an electric circuit. If the current is 4 amperes when the voltage is 100 volts, find the current when the voltage is 75 volts.   3 amperes

23. *Travel*   The distance traveled, d, varies directly as the time, t, of travel, assuming that the speed is constant. If it takes 45 minutes to travel 50 miles, how many hours would it take to travel 180 miles?   2.7 h

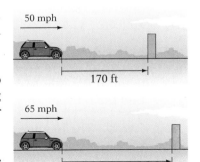

24. *Automotive Technology*   The distance, d, required for a car to stop varies directly as the square of the velocity, v, of the car. If a car traveling 50 mph requires 170 feet to stop, find the stopping distance for a car traveling 65 mph.   287.3 ft

25. *Physics*   The distance, d, an object falls is directly proportional to the square of the time, t, of the fall. If an object falls a distance of 8 feet in 0.5 second, how far will the object fall in 5 seconds?   800 ft

26. *Physics*   The distance, *s*, that a ball rolls down an inclined plane is directly proportional to the square of the time, *t*. If the ball rolls 5 feet in 1 second, how far will it roll in 4 seconds?   80 ft

27. *Consumerism*   The number of items, *N*, that can be purchased for a given amount of money is inversely proportional to the cost, *C*, of the item. If 390 items can be purchased when the cost per item is $.50, how many items can be purchased when the cost per item is $.20?   975 items

28. *Geometry*   The length, *L*, of a rectangle of fixed area varies inversely as the width, *W*. If the length of the rectangle is 8 feet when the width is 5 feet, find the length of the rectangle when the width is 4 feet.   10 ft

29. *Travel*   The time, *t*, of travel of an automobile trip varies inversely as the speed, *v*. Traveling at an average speed of 65 mph, a trip took 4 hours. The return trip took 5 hours. Find the average speed of the return trip. 52 mph

30. *Electricity*   The current, *I*, in an electric circuit is inversely proportional to the resistance, *R*. If the current is 0.25 ampere when the resistance is 8 ohms, find the resistance when the current is 1.2 amperes. $1.\overline{6}$ ohms

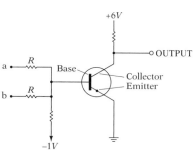

31. *Physics*   For a constant temperature, the pressure, *P*, of a gas varies inversely as the volume, *V*. If the pressure is 25 pounds per square inch when the volume is 400 cubic feet, find the pressure when the volume is 150 cubic feet. $66.\overline{6}$ psi

32. *Physics*   The volume, *V*, of a gas varies inversely as the pressure, *P*, on the gas. If the volume of the gas is 12 cubic feet when the pressure is 15 pounds per square foot, find the volume of the gas when the pressure is 4 pounds per square foot.   45 cubic feet

33. *Business*   A computer company that produces personal computers has determined that the number of computers it can sell, *S*, is inversely proportional to the price, *P*, of the computer. Eighteen hundred computers can be sold if the price is $1800. How many computers can be sold if the price is $1500?   2160 computers

34. *Light*   The intensity, *I*, of a light source is inversely proportional to the square of the distance, *d*, from the source. If the intensity is 80 footcandles at a distance of 4 feet, what is the intensity when the distance is 10 feet?   12.8 footcandles

35. *Magnetism*   The repulsive force, *f*, between the north poles of two magnets is inversely proportional to the square of the distance, *d*, between them. If the repulsive force is 18 pounds when the distance is 3 inches, find the repulsive force when the distance is 1.2 inches.   112.5 lb

36. *Sound*   The loudness, *L*, measured in decibels, of a stereo speaker is inversely proportional to the square of the distance, *d*, from the speaker. The loudness is 20 decibels at a distance of 10 feet. What is the loudness at a distance of 6 feet from the speaker? $55.\overline{5}$ decibels.

37. *Mechanics*   The speed of a gear varies inversely as the number of teeth. If a gear that has 40 teeth makes 15 revolutions per minute, how many revolutions per minute will a gear that has 32 teeth make?   18.75 rpm

Applying Concepts

38. *Optics*   The distance, *d*, a person can see to the horizon from a point above the surface of Earth varies directly as the square root of the height, *H*. If, for a height of 500 feet, the horizon is 19 miles away, how far is the horizon from a point that is 800 feet high? Round to the nearest hundredth.   24.03 mi

39. *Physics*   The period, *p*, of a pendulum, or the time it takes for the pendulum to make one complete swing, varies directly as the square root of the length, *L*, of the pendulum. If the period of a pendulum is 1.5 seconds when the length is 2 feet, find the period when the length is 4.5 feet. Round to the nearest hundredth.   2.25 s

40.   Explain why the formula for the area of a circle is a direct variation.
Answers will vary.

41. Determine whether the statement is always true, sometimes true, or never true.
  a. If *x* varies inversely as *y*, then when *x* is doubled, *y* is doubled.
  b. If *a* varies inversely as *b*, then *ab* is a constant.
  c. If the length of a rectangle is held constant, then the area of the rectangle varies directly as the width.
  d. If the area of a rectangle is held constant, then the length varies directly as the width.
  a. Never true   b. Always true   c. Always true   d. Never true

42. *Unit Pricing*   Explain how proportion may be used in pricing large quantities of a purchase as compared to small quantities of a purchase. Is the unit price of a large purchase always smaller than the unit price of a small purchase?   Answers will vary.

43. **a.** The variable *y* varies directly as the cube of *x*. If *x* is doubled, by what factor is *y* increased?
  b. The variable *y* varies inversely as the cube of *x*. If *x* is doubled, by what factor is *y* decreased?   a. 8   b. $\frac{1}{8}$

EXPLORATION

1. *Determining a Variation*   You order an extra large pizza, cut into 12 slices, to be delivered to your home.
  a. If there are 6 people to share the pizza, how many slices does each person get?
  b. If two people leave before the pizza arrives, how many slices does each person get?
  c. Is the variation direct or inverse?

d. What is the constant of variation?
e. Write the equation that represents the variation.

a. 2 slices **b.** 3 slices **c.** inverse variation **d.** 12 **e.** $s = \frac{k}{p}$, where s is the
number of slices per person and p is the number of persons.

2. *Joint Variation*   A variation may involve more than two variables. If a
quantity varies directly as the product of two or more variables, it is
known as a **joint variation.**

➡ The weight of a rectangular metal box is directly proportional to the
volume of the box, which is given by length times width times
height. The variation equation is written

$$\text{Weight} = kLWH$$

The weight of a box with $L = 24$ inches, $W = 12$ inches, and $H = 12$
inches is 72 pounds. Find the weight of another box with $L = 18$
inches, $W = 9$ inches, and $H = 18$ inches.

12 in.

24 in.   12 in.

$$\text{Weight} = kLWH$$
$$72 = k(24)(12)(12) \qquad \bullet \text{ Use the values of the first box.}$$
$$\frac{72}{(24)(12)(12)} = \frac{k(24)(12)(12)}{(24)(12)(12)} \qquad \bullet \text{ Solve for } k.$$
$$\frac{1}{48} = k$$
$$\text{Weight} = \frac{1}{48} LWH \qquad \bullet \text{ Write the variation equation.}$$
$$\qquad\qquad\qquad\qquad k = \frac{1}{48}.$$
$$\text{Weight} = \frac{1}{48}(18)(9)(18) \qquad \bullet \text{ Use the values of the second box.}$$
$$\text{Weight} = 60.75$$

The weight of the other box is 60.75 pounds.   ⬅

The force on a flat surface that is perpendicular to the wind is directly
proportional to the product of the area of the surface and the square of
the speed of the wind.
a. Write the joint variation.
b. What effect does doubling the area have on the force of the wind?
c. What effect does doubling the speed of the wind have on the force of
the wind?
d. When the wind is blowing at 30 mph, the force on a 10-square-foot
area is 45 pounds. Find the force on this area when the wind is
blowing at 60 miles per hour.

The power, P, in an electric circuit is directly proportional to the product
of the current, I, and the square of the resistance, R.
e. If the power is 100 watts when the current is 4 amperes and the resistance is 5 ohms, find the power when the current is 2 amperes
and the resistance is 10 ohms.

The pressure, p, in a liquid varies directly as the product of the depth, d,
and the density, D, of the liquid.

f. If the pressure is 37.5 pounds per square inch when the depth is 100 inches and the density is 1.2, find the pressure when the density remains the same and the depth is 60 inches.

a. $f = kaw^2$ **b.** It doubles the force of the wind. **c.** It quadruples the force of the wind. **d.** 180 lb **e.** 200 watts **f.** 22.5 psi

CHAPTER **7** *SUMMARY*

Key Terms

clear denominators [**p. 489**]
complex fraction [**p. 481**]
constant of proportionality [**p. 524**]
constant of variation [**p. 524**]
continuous [**p. 467**]
conversion factor [**p. 513**]
dimensional analysis [**p. 513**]
directly proportional [**p. 524**]
direct variation [**p. 524**]
direct variation equation [**p. 524**]
discontinuous [**p. 467**]
extremes [**p. 505**]
inversely proportional [**p. 528**]
inverse variation [**p. 528**]
joint variation [**p. 536**]
means [**p. 505**]
multiplicative inverse [**p. 477**]

proportion [**p. 504**]
rate [**p. 504**]
rate of work [**p. 492**]
ratio [**p. 504**]
rational equation [**p. 489**]
rational expression [**p. 465**]
rational function [**p. 465**]
reciprocal of a rational expression [**p. 477**]
similar objects [**p. 508**]
simplest form of a rational expression [**p. 468**]
term of a proportion [**p. 505**]
topology [**p. 523**]
uniform motion [**p. 494**]
unit of measurement [**p. 512**]
vertical asymptote [**p. 467**]

Essential Concepts

Vertical Asymptotes of a Rational Function
The graph of $f(x) = \frac{p(x)}{q(x)}$, where $p(x)$ and $q(x)$ have no common factors, has a vertical asymptote at $x = a$ if a is a real number and a is a zero of the denominator $q(x)$. [**p. 467**]

Multiplication of Rational Expressions
The product of two fractions is a fraction whose numerator is the product of the numerators of the two fractions and whose denominator is the product of the denominators of the two fractions. [**p. 476**]
$$\frac{a}{b} \cdot \frac{c}{d} = \frac{ac}{bd}$$

Division of Rational Expressions
To divide two rational expressions, multiply by the reciprocal of the divisor. [**p. 477**]
$$\frac{a}{b} \div \frac{c}{d} = \frac{a}{b} \cdot \frac{d}{c} = \frac{ad}{bc}$$

Addition of Rational Expressions [**p. 478**]
$$\frac{a}{c} + \frac{b}{c} = \frac{a+b}{c}$$

Subtraction of Rational Expressions [p. 478]

$$\frac{a}{c} - \frac{b}{c} = \frac{a-b}{c}$$

Equation for Work Problems [p. 492]

Rate of work × Time worked = Part of task completed

Equation for Uniform Motion Problems [p. 494]

$$\text{Distance} = \text{Rate} \times \text{Time} \quad \text{or} \quad \frac{\text{Distance}}{\text{Rate}} = \text{Time}$$

Proportions

In a proportion, the product of the means equals the product of the extremes. This is sometimes phrased "The cross products are equal." [p. 505]

Direct Variation

Two quantities are **directly proportional** if an increase in one quantity leads to a proportional increase in the other quantity. The equation $y = kx$ is a **direct variation equation** and is read "y varies directly as x" or "y is directly proportional to x." [p. 524]

Inverse Variation

Two quantities are **inversely proportional** if an increase in one quantity leads to a proportional decrease in the other quantity, or if a decrease in one leads to a proportional increase in the other. An **inverse variation** is one that can be written in the form $y = \frac{k}{x}$, where k is a constant. The equation $y = \frac{k}{x}$ is read "y varies inversely as x" or "y is inversely proportional to x." [p. 528]

CHAPTER **7** *REVIEW EXERCISES*

1. Given $f(x) = \dfrac{x^2 - 2}{3x^2 - 2x + 5}$, find $f(-2)$.

$\dfrac{2}{21}$

2. Find the domain of $f(x) = \dfrac{2x - 7}{3x^2 + 3x - 18}$.

$\{x \mid x \neq -3, 2\}$

3. Simplify: $\dfrac{x^2 - 16}{x^3 - 2x^2 - 8x}$

$\dfrac{x + 4}{x(x + 2)}$

4. Divide: $\dfrac{x^{2n} - 5x^n + 4}{x^{2n} - 2x^n - 8} \div \dfrac{x^{2n} - 4x^n + 3}{x^{2n} + 8x^n + 12}$

$\dfrac{x^n + 6}{x^n - 3}$

5. Subtract: $\dfrac{3x^2 + 2}{x^2 - 4} - \dfrac{9x - x^2}{x^2 - 4}$ $\dfrac{4x - 1}{x + 2}$

6. Add: $\dfrac{5}{3a^2b^3} + \dfrac{7}{8ab^4}$ $\dfrac{21a + 40b}{24a^2b^4}$

7. Simplify: $\dfrac{x + \dfrac{3}{x - 4}}{3 + \dfrac{x}{x - 4}}$ $\dfrac{x - 1}{4}$

8. Multiply: $\dfrac{a^6b^4 + a^4b^6}{a^5b^4 - a^4b^4} \cdot \dfrac{a^2 - b^2}{a^4 - b^4}$ $\dfrac{1}{a - 1}$

9. Solve: $\dfrac{2}{x - 4} + 3 = \dfrac{x}{2x - 3}$ 2, 3

10. Solve: $\dfrac{3x + 7}{x + 2} + x = 3$ −1

11. A car uses 4 tanks of fuel to travel 1800 miles. At this rate, how many tanks of fuel would be required for a trip of 3000 miles?   6.$\overline{6}$ tanks

12. State whether the function $F(x) = \dfrac{x^2 - x}{3x^2 + 4}$ is continuous or discontinuous.

Continuous

13. If the reciprocal of $x + 2$ is $-x$, what is the value of x?   −1

14. Use an algebraic method to find the vertical asymptote(s) of the graph of $g(x) = \dfrac{2x}{x - 3}$. Verify your answer by graphing the function using a graphing calculator. $x = 3$

15. Given that T varies directly as the square of S, and $T = 50$ when $S = 5$, find T when $S = 120$.

28,800

16. The denominator of a fraction is 4 more than the numerator. If both the numerator and the denominator of the fraction are increased by 3, the new fraction is $\frac{5}{6}$. Find the original fraction.

$\frac{17}{21}$

17. One member of a gardening team can landscape a new lawn in 36 hours. The other member of the team can do the job in 45 hours. How long would it take to landscape the lawn if both gardeners worked together?

20 h

18. A car travels 200 miles. A second car, traveling 10 mph faster than the first car, makes the same trip in 1 hour less time. Find the speed of each car.   First car: 40 mph; second car: 50 mph

19. On a certain map, 2.5 inches represents 10 miles. How many miles would be represented by 12 inches?   48 mi

20. The inlet pipe can fill a tub in 4 minutes. The drain pipe can empty the tub in 15 minutes. How long would it take to fill an empty tub with both pipes open?   5.$\overline{45}$ min

21. A canoeist can travel 10 mph in calm water. The amount of time it takes to travel 60 miles with the current is the same amount of time it takes to travel 40 miles against the current. Find the rate of the current.   2 mph

22. The distance, d, that sound travels varies directly as the time, t, it travels. If sound travels 5575 feet in 5 seconds, find the distance that sound travels in 9 seconds.   10,035 ft

23. The illumination, I, produced by a light varies inversely as the square of the distance, d, from the light. If the illumination produced 10 feet from a light is 12 footcandles, find the illumination 2 feet from the light.

300 footcandles

24. Triangles ABC and DEF are similar triangles. Determine the length of side BC. Round to the nearest hundredth.   4.27 ft

25. Find the speed in feet per second of a baseball pitched at 87 miles per hour.   127.6 ft/s

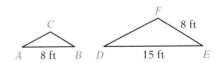

CHAPTER **7** *TEST*

1. Given $P(x) = \dfrac{3 - x^2}{x^3 - 2x^2 + 4}$, find $P(-1)$.

2

2. Find the domain of $f(x) = \dfrac{3x^2 - x + 1}{x^2 - 9}$.

$\{x \mid x \neq -3, 3\}$

3. Simplify: $\dfrac{v^3 - 4v}{2v^2 - 5v + 2}$ $\dfrac{v(v + 2)}{2v - 1}$

4. Divide: $\dfrac{2x^2 - x - 3}{2x^2 - 5x + 3} \div \dfrac{3x^2 - x - 4}{x^2 - 1}$ $\dfrac{x + 1}{3x - 4}$

5. Add: $\dfrac{5}{3x-4} + \dfrac{4}{2x+3}$ $\dfrac{22x-1}{(3x-4)(2x+3)}$

6. Subtract: $\dfrac{5x}{2x+4} - \dfrac{x}{2x+4}$ $\dfrac{2x}{x+2}$

7. Simplify: $\dfrac{\dfrac{5}{x-1} - \dfrac{3}{x+3}}{\dfrac{6}{x+3} + \dfrac{2}{x-1}}$ $\dfrac{x+9}{4x}$

8. Multiply: $\dfrac{3x^2+4x-15}{x^2-11x+28} \cdot \dfrac{x^2-5x-14}{3x^2+x-10}$ $\dfrac{x+3}{x-4}$

9. Solve: $\dfrac{x+8}{x+4} = 1 + \dfrac{5}{x+4}$ No solution

10. Solve: $\dfrac{6}{x-7} = \dfrac{8}{x-6}$ 10

11. An investment of $8000 earns $520 in dividends. At the same rate, how much money must be invested to earn $780 in dividends? $12,000

12. State whether the function $F(x) = \dfrac{2x+3}{x^2+4}$ is continuous or discontinuous.

Continuous

13. Use an algebraic method to find the vertical asymptote(s) of the graph of $g(x) = \dfrac{x+1}{x^2-x-6}$. Verify your answer by graphing the function using a graphing calculator. $x = -2, x = 3$

14. A brick mason can construct a patio in 3 hours. If the mason works with an apprentice, they can construct the patio in 2 hours. How long would it take the apprentice, working alone, to construct the patio? 6 h

15. The rate of a jet plane is 400 mph in calm air. Traveling with the wind, the jet can fly 2100 miles in the same amount of time as it takes to fly 1900 miles against the wind. Find the rate of the wind.
20 mph

16. A gardener uses 4 ounces of insecticide to make 2 gallons of garden spray. At this rate, how much insecticide is necessary to make 10 gallons of the garden spray? 20 oz

17. A car travels 315 miles in the same amount of time it takes a bus to travel 245 miles. The rate of the car is 10 mph greater than that of the bus. Find the rate of the car. 45 mph

18. Hooke's Law states that the distance, d, a spring will stretch is directly proportional to the weight, w, on the spring. A weight of 5 pounds will stretch the spring 2 inches. How far will a weight of 28 pounds stretch the spring? 11.2 in.

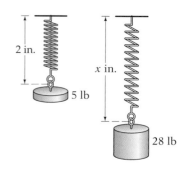

19. Triangles ABC and DEF are similar triangles. Determine the perimeter of triangle ABC. 24 in.

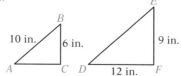

20. Find the speed in feet per second of a soccer ball kicked at 52 mph. Round to the nearest hundredth. 76.27 ft/s

CHAPTER **7** *CUMULATIVE REVIEW EXERCISES*

1. Find the y- and x-intercepts of the graph of $f(x) = x^2 + 2x - 8$.
   (0, −8); (−4, 0) and (2, 0)

2. Solve and write the solution in interval notation: $2 - 5(x + 1) \geq 3(x - 1) - 8$
   (−∞, 1]

3. Solve: $\dfrac{5}{8}x + 2 < -3$ or $2 - \dfrac{3}{5}x < -7$
   $\{x \mid x < -8 \text{ or } x > 15\}$

4. Solve: $|5 - 3x| \geq 4$
   $\{x \mid x \leq \dfrac{1}{3} \text{ or } x \geq 3\}$

5. Solve: $3x - 2y = 1$
            $5x - 3y = 3$ (3, 4)

6. Solve: $2x + 3y - z = 5$
            $x - 2y + z = 1$ (2, 0, −1)
            $3x + y + 2z = 4$

7. Simplify: $(p^{-10}q^5)^2 \dfrac{q^{10}}{p^{20}}$

8. Solve: $w^2 + 4w = -4$ −2

9. Add: $\dfrac{3x}{x - 2} + \dfrac{4}{x + 2}$ $\dfrac{(3x - 2)(x + 4)}{(x + 2)(x - 2)}$

10. Divide by using long division: $4x + 7 + \dfrac{10}{3x - 2}$
    $(12x^2 + 13x - 4) \div (3x - 2)$

11. Divide by using synthetic division:
    $(12 - 3x^2 + x^3) \div (x + 3)$
    $x^2 - 6x + 18 - \dfrac{42}{x + 3}$

12. Solve: $\dfrac{x}{x + 2} - \dfrac{4x}{x + 3} = 1$ $-\dfrac{3}{2}, -1$

13. Graph: $f(x) = \dfrac{3}{5}x - 2$

14. Divide: $\dfrac{x^2 - y^2}{14x^2y^4} \div \dfrac{x^2 + 2xy + y^2}{7xy^3}$ $\dfrac{x - y}{2xy(x + y)}$

15. Determine whether the following argument is an example of inductive or deductive reasoning. "Ron got an A on each of his first four math exams, so he will get an A on the next math exam."   Inductive reasoning

16. Translate and simplify: "a number minus the difference between ten and twice the number."
    $x - (10 - 2x)$; $3x - 10$

17. Is the graph shown at the right the graph of a function?   Yes

18. The measures of two adjacent angles of a pair of intersecting lines are $(3x + 10)°$ and $(2x + 25)°$. Find the measure of the larger angle.   97°

19. How many pounds of almonds that cost \$5.40 per pound must be mixed with 50 pounds of peanuts that cost \$2.60 per pound to make a mixture that costs \$4.00 per pound?   50 lb

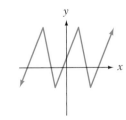

20. Find the equation of the line that passes through the point $(6, -2)$ and is perpendicular to the line that contains the points $(-3, 1)$ and $(5, -5)$.
$y = \frac{4}{3}x - 10$

21. The operator of a hotel estimates that 200 rooms per night will be rented if the room rate per night is $90. For each $5 increase in the price of a room, 4 fewer rooms will be rented. Determine a linear function that will predict the number of rooms that will be rented for a given price per room. Use this model to predict the number of rooms that will be rented if the room rate is $120 per night. $y = -\frac{4}{5}x + 272$; 176 rooms

22. Given $R(x) = \dfrac{4 - x^2}{x^3 - 2x^2 + 4}$, find $R(1)$.   1

23. For the function $f(x) = \dfrac{3}{x^2 - 3x - 4}$, find **a.** the domain and **b.** the vertical asymptote(s) of its graph.

a. $\{x \mid x \neq -1, 4\}$ b. $x = -1, x = 4$

24. A car travels 120 miles. A second car, traveling 10 mph faster than the first car, makes the same trip in 1 hour less time. Find the speed of each car.   First car: 30 mph; second car: 40 mph

25. One member of a telephone crew can wire new telephone lines in 5 hours. It takes 7.5 hours for the other member of the crew to do the job. How long would it take to wire new telephone lines if both members of the crew worked together?   3 h

Radical Expressions and Rational Exponents

```
EDIT CALC TESTS
5↑ QuadReg
6: CubicReg
7: QuartReg
8: LinReg(a+bx)
9: LnReg
0: ExpReg
A↓ PwrReg
```

Press **STAT** ▶ to access the PwrReg option in the CALC menu.

*Real data does not provide just the past and present conditions of a situation; it can also give a possible glimpse into the future. As shown in **Exploration 1 on page 564,** regression can be performed on data of sales of personal computers. The resulting radical equation is used to predict future sales. This information is important to computer manufacturers who need to determine how many computers to produce. Manufacturing more computers than will be purchased by consumers puts a company's financials in jeopardy, and manufacturing fewer computers than will be puchased by customers means lost revenue.*

Need help? For online student resources, visit this web site: Math.college.hmco.com

PREP TEST

1. Complete: $48 = ? \cdot 3$ 16

For Exercises 2 to 7, simplify.

2. 2^5 32

3. $6\left(\dfrac{3}{2}\right)$ 9

4. $\dfrac{1}{2} - \dfrac{2}{3} + \dfrac{1}{4}$ $\dfrac{1}{12}$

5. $(3 - 7x) - (4 - 2x)$ $-5x - 1$

6. $\dfrac{3x^5y^6}{12x^4y}$ $\dfrac{xy^5}{4}$

7. $(3x - 2)^2$ $9x^2 - 12x + 4$

For Exercises 8 and 9, multiply.

8. $(2 + 4x)(5 - 3x)$ $-12x^2 + 14x + 10$

9. $(6x - 1)(6x + 1)$ $36x^2 - 1$

10. Solve: $x^2 - 14x - 5 = 10$ $-1, 15$

GO FIGURE

If $x + y$, xy, and $\dfrac{x}{y}$ all equal the same number, find the values of x and y. $x = \dfrac{1}{2}, y = -1$

SECTION 8.1

Rational Exponents and Radical Expressions

- **Expressions with Rational Exponents**
- **Exponential Expressions and Radical Expressions**

■ Expressions with Rational Exponents

When the Rules of Exponents were presented earlier in this text, we operated on expressions with integer exponents. In this section, we begin by assuming that these rules apply also to exponents that are rational numbers. Recall that a rational number is one of the form $\dfrac{p}{q}$, where p and q are integers and $q \neq 0$.

Use your calculator to evaluate $(9^{1/2})^2$. The display should read 9.

Use your calculator to evaluate $(16^{1/2})^2$. The display should read 16.

Note that we are evaluating the power of an exponential expression. Therefore, by the Rules of Exponents, we can simplify the expressions by multiplying the exponents.

$$(9^{1/2})^2 = 9^{\frac{1}{2} \cdot 2} = 9^1 = 9 \qquad \text{This is the same result obtained above.}$$
$$(16^{1/2})^2 = 16^{\frac{1}{2} \cdot 2} = 16^1 = 16 \qquad \text{This is the same result obtained above.}$$

Use your calculator to evaluate $(8^{1/3})^3$. The display should read 8.

Use your calculator to evaluate $(27^{1/3})^3$. The display should read 27.

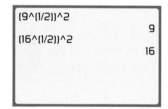

INSTRUCTOR NOTE
You may prefer to have your students discover the meaning of rational exponents by having them do the Exploration exercises at the end of this section.

We can also simplify these expressions by multiplying the exponents.

$(8^{1/3})^3 = 8^{\frac{1}{3}\cdot3} = 8^1 = 8$ This is the same result obtained above.

$(27^{1/3})^3 = 27^{\frac{1}{3}\cdot3} = 27^1 = 27$ This is the same result obtained above.

The pattern developed above can be stated as a rule. For $a > 0$ and $n > 0$,

$$(a^{1/n})^n = a$$

? QUESTION

a. As shown above, the square of $9^{1/2}$ is 9. What positive number, when squared, is equal to 9?*

b. As shown above, the square of $16^{1/2}$ is 16. What positive number, when squared, is equal to 16?

c. As shown above, the cube of $8^{1/3}$ is 8. What positive number, when cubed, is equal to 8?

d. As shown above, the cube of $27^{1/3}$ is 27. What positive number, when cubed, is equal to 27?

Look at your answers to the questions above. Note that

The square of $9^{1/2}$ is 9, and the square of **3** is 9. $9^{1/2} = 3$

The square of $16^{1/2}$ is 16, and the square of **4** is 16. $16^{1/2} = 4$

The cube of $8^{1/3}$ is 8, and the cube of **2** is 8. $8^{1/3} = 2$

The cube of $27^{1/3}$ is 27, and the cube of **3** is 27. $27^{1/3} = 3$

Because $(a^{1/n})^n = a$, $a^{1/n}$ is the number whose nth power is a.

$$25^{1/2} = 5 \text{ because } 5^2 = 25.$$
$$64^{1/3} = 4 \text{ because } 4^3 = 64.$$

In the expression $a^{1/n}$, if a is a negative number and n is a positive even integer, then $a^{1/n}$ is not a real number.

$(-9)^{1/2}$ is not a real number because there is no real number that when squared equals -9.

When n is a positive odd integer, a can be a positive or a negative number.

$(-8)^{1/3} = -2$ because $(-2)^3 = -8$.

? QUESTION What integer is each of the following equal to?†

a. $49^{1/2}$ **b.** $(-125)^{1/3}$ **c.** $16^{1/4}$ **d.** $(-81)^{1/2}$

———

? ANSWERS   *a. $3^2 = 9$ **b.** $4^2 = 16$ **c.** $2^3 = 8$ **d.** $3^3 = 27$ †a. $49^{1/2} = 7$ because $7^2 = 49$. **b.** $(-125)^{1/3} = -5$ because $(-5)^3 = -125$.   **c.** $16^{1/4} = 2$ because $2^4 = 16$.   **d.** $(-81)^{1/2}$ is not a real number because there is no number that when squared equals -81.

We will now define any exponential expression that contains a rational exponent.

INSTRUCTOR NOTE
Reinforce the concept that in the expression $a^{1/n}$, a must be a positive number when n is an even integer by asking your students why the definition states that $a^{1/n}$ is a real number.

> **Definition of $a^{m/n}$**
>
> If m and n are positive integers and $a^{1/n}$ is a real number, then $a^{m/n} = (a^{1/n})^m$.

As $(-9)^{1/2}$ demonstrates, expressions that contain rational exponents do not always represent real numbers when the base of the exponential expression is a negative number. For this reason, **all variables in this chapter represent positive numbers unless otherwise stated.**

INSTRUCTOR NOTE
Exercises 8, 10, and 12 can be used for similar in-class examples.

EXAMPLE 1

Simplify. **a.** $27^{4/3}$ **b.** $32^{-3/5}$ **c.** $(-16)^{-3/4}$

Solution

a. $27^{4/3} = (3^3)^{4/3}$ • Rewrite 27 as 3^3.

$= 3^4$ • Simplify the power of an exponential expression by multiplying the exponents.

$= 81$ • Evaluate the exponential expression.

b. $32^{-3/5} = (2^5)^{-3/5}$ • Rewrite 32 as 2^5.

$= 2^{-3}$ • Simplify the power of an exponential expression by multiplying the exponents.

$= \dfrac{1}{2^3}$ • Rewrite the expression with a positive exponent.

$= \dfrac{1}{8}$ • Evaluate the exponential expression.

c. $(-16)^{-3/4}$ • The base of the exponential expression, -16, is a negative number, and the denominator of the exponent is a positive even number.

$(-16)^{-3/4}$ is not a real number.

TAKE NOTE

An expression with a negative exponent must be rewritten with a positive exponent before it can be evaluated.

Recall that $a^{-n} = \dfrac{1}{a^n}$.

Suggested Activity

If $P = 2^{1996} + 2^{-1996}$ and $Q = 2^{1996} - 2^{-1996}$, find the value of $P^2 - Q^2$.
[Answer: 4]

YOU TRY IT 1

Simplify. **a.** $16^{3/4}$ **b.** $64^{-2/3}$ **c.** $(-100)^{3/4}$

Solution See page S32. a. 8 b. $\dfrac{1}{16}$ c. Not a real number

In Example 1, numerical expressions with rational exponents were simplified. In Example 2, variable expressions with rational exponents are simplified.

INSTRUCTOR NOTE
Exercises 14, 16, and 26 can
be used for similar in-class
examples.

TAKE NOTE

For Example 2a, recall
the Rule for Multiply-
ing Exponential
Expressions:

$$x^m \cdot x^n = x^{m+n}$$

For Example 2b, recall
the Rule for Simplify-
ing Powers of Prod-
ucts:

$$(x^m y^n)^p = x^{m \cdot p} y^{n \cdot p}$$

For Example 2c, recall
the Rule for Dividing
Exponential Expres-
sions and the Rule for
Simplifying Powers of
Quotients:

$$\frac{x^m}{x^n} = x^{m-n}$$

$$\left(\frac{x^m}{y^n}\right)^p = \frac{x^{m \cdot p}}{y^{n \cdot p}}$$

EXAMPLE 2

Simplify. **a.** $b^{1/2}(b^{2/3})(b^{-1/4})$ **b.** $(x^4 y^6)^{3/2}$ **c.** $\left(\dfrac{3a^3 b^{-4}}{24a^{-9}b^2}\right)^{2/3}$

Solution

a. $b^{1/2}(b^{2/3})(b^{-1/4})$

$= b^{1/2+2/3-1/4}$

$= b^{6/12+8/12-3/12}$

$= b^{11/12}$

- Multiply exponential expressions with the same base by adding the exponents.

b. $(x^4 y^6)^{3/2} = x^{4(3/2)} y^{6(3/2)}$

$= x^6 y^9$

- Simplify the power of an exponential expression by multiplying the exponents.

c. $\left(\dfrac{3a^3 b^{-4}}{24a^{-9}b^2}\right)^{2/3} = \left(\dfrac{a^{12}b^{-6}}{8}\right)^{2/3}$

- Simplify inside the parenthesis. Divide exponential expressions with the same base by subtracting the exponents.

$= \left(\dfrac{a^{12}}{2^3 b^6}\right)^{2/3}$

- Rewrite the expression with positive exponents. Rewrite 8 as 2^3.

$= \dfrac{a^8}{2^2 b^4}$

- Simplify the power of an exponential expression by multiplying the exponents.

$= \dfrac{a^8}{4b^4}$

- Evaluate 2^2.

YOU TRY IT 2

Simplify. **a.** $p^{3/4}(p^{-1/8})(p^{1/2})$ **b.** $(a^{5/3}b^{1/6})^6$ **c.** $\left(\dfrac{2a^{-2}b}{50a^6 b^{-3}}\right)^{1/2}$

Solution See page S33. a. $p^{9/8}$ b. $a^{10}b$ c. $\dfrac{b^2}{5a^4}$

INSTRUCTOR NOTE
Exercise 28 can be used for
a similar in-class example.

EXAMPLE 3

The Federal Reserve System provides data on the average life span of different denominations of paper currency, up to the $100 bill. The function that approximately models the data is $f(x) = 1.2x^{2/5}$, where x is the denomination of the bill and $f(x)$ is its average life span in years.

a. Use the model to approximate the life span of a $20 bill. Round to the nearest whole number.
b. Use the model to approximate the difference between the average life span of a $10 bill and that of a $100 bill. Round to the nearest whole number.
c. What is the domain of the function $f(x) = 1.2x^{2/5}$?

Solution

a. $f(x) = 1.2x^{2/5}$

$f(20) = 1.2(20)^{2/5} \approx 4$ • Evaluate the function at $x = 20$.

The life span of a $20 bill is approximately 4 years.

```
1.2(20^.4)
                3.977344821
1.2(10^.4)
                3.014263718
1.2(100^.4)
                7.571488134
```

▼ **Point of Interest**

*You can learn more about U.S. currency at the web site of the Bureau of Engraving and Printing: **www.moneyfactory.com**.*

b. $f(x) = 1.2x^{2/5}$

$f(10) = 1.2(10)^{2/5} \approx 3$ • Evaluate the function at $x = 10$.

$f(100) = 1.2(100)^{2/5} \approx 8$ • Evaluate the function at $x = 100$.

$8 - 3 = 5$ • Subtract to find the difference.

The life span of a $100 bill is approximately 5 years greater than the life span of a $10 bill.

c. The domain of the function is the values of the U.S. bills, up to 100, that are in circulation.

The domain is {1, 2, 5, 10, 20, 50, 100}.

YOU TRY IT 3

T. Rowe Price Associates has provided data on how much money parents must save each month in order to have enough money set aside for their child's college expenses, assuming that no money has been saved up to this point. The function that approximates the data for a child who will attend a public college and pay in-state tuition is $f(x) = 3605x^{-39/40}$, where x is the number of years before the child enters college and $f(x)$ is the monthly savings.

a. Use the model to approximate the monthly savings for a child who will be entering college in 12 years. Round to the nearest dollar.

b. Use the model to approximate the difference between the monthly savings for a child who is 15 years from entering college and a child who will be going to college in 5 years. Round to the nearest dollar.

c. Given that the domain of the function is $\{x \mid 1 \le x \le 20, x \in \text{integers}\}$, find the range of the function.

Solution See page S33. **a.** $320 **b.** $494 **c.** $\{y \mid 194 \le y \le 3605, y \in \text{integers}\}$

```
√(9)
                3
√(16)
                4
```

▤ **Suggested Activity**

See Section 8.1 of the *Student Activity Manual* for an activity involving the use of technology to investigate rational exponents and radical expressions.

■ Exponential Expressions and Radical Expressions

Use your calculator to evaluate $\sqrt{9}$. The display should read 3.

Use your calculator to evaluate $\sqrt{16}$. The display should read 4.

$$9^{1/2} = 3 \text{ and } \sqrt{9} = 3. \qquad 9^{1/2} = \sqrt{9}$$
$$16^{1/2} = 4 \text{ and } \sqrt{16} = 4. \qquad 16^{1/2} = \sqrt{16}$$

The expression $a^{1/n}$ is the **nth root of a**. The expression $\sqrt[n]{a}$ is another symbol for the nth root of a.

Alternative Notation for the nth root of a

If $a^{1/n}$ is a real number and n is a positive integer, then $a^{1/n} = \sqrt[n]{a}$.

In the expression $\sqrt[n]{a}$, the symbol $\sqrt{}$ is called a **radical sign,** n is the **index** of the radical, and a is the **radicand.** When $n = 2$, the radical expression represents a square root, and the index 2 is usually not written.

An exponential expression with a rational exponent can be written as a radical expression.

INSTRUCTOR NOTE
Show students that although they can simplify an expression by using $(a^m)^{1/n}$, it is usually easier to simplify $(a^{1/n})^m$. For example, simplifying $(27^{1/3})^2$ is easier than simplifying $(27^2)^{1/3}$.

> **Definition of the nth root of a^m**
>
> If $a^{1/n}$ is a real number, then
> $$a^{m/n} = a^{(1/n)m} = (\sqrt[n]{a})^m$$
> $a^{m/n}$ can also be written $a^{m/n} = a^{m(1/n)} = \sqrt[n]{a^m}$.

INSTRUCTOR NOTE
Exercises 32, 34, and 36 can be used for similar in-class examples.

EXAMPLE 4

Rewrite the exponential expression as a radical expression.

a. $y^{2/3}$　　　**b.** $(5x)^{3/4}$　　　**c.** $-6x^{4/5}$

Solution

a. $y^{2/3} = (y^2)^{1/3}$
$\quad\quad = \sqrt[3]{y^2}$

- The denominator of the rational exponent is the index of the radical. The numerator is the power of the radicand.

b. $(5x)^{3/4} = \sqrt[4]{(5x)^3}$
$\quad\quad\quad = \sqrt[4]{125x^3}$

- The denominator of the rational exponent is the index of the radical. The numerator is the power of the radicand.

c. $-6x^{4/5} = -6(x^4)^{1/5}$
$\quad\quad\quad = -6\sqrt[5]{x^4}$

- Only x is raised to the 4/5 power; -6 is not. The denominator is the index of the radical. The numerator is the power of the radicand.

YOU TRY IT 4

Rewrite the exponential expression as a radical expression.

a. $b^{3/7}$　　　**b.** $(3y)^{2/5}$　　　**c.** $-9d^{5/8}$

Solution　See page S33.　a. $\sqrt[7]{b^3}$　b. $\sqrt[5]{9y^2}$　c. $-9\sqrt[8]{d^5}$

INSTRUCTOR NOTE
Exercises 44 and 50 can be used for similar in-class examples.

EXAMPLE 5

Rewrite the radical expression as an exponential expression.

a. $\sqrt[3]{z^8}$　　　**b.** $\sqrt{19}$　　　**c.** $\sqrt[4]{x^4 + y^4}$

Solution

a. $\sqrt[3]{z^8} = (z^8)^{1/3}$
$\quad\quad = z^{8/3}$

- The index of the radical is the denominator of the rational exponent. The power of the radicand is the numerator of the rational exponent.

b. $\sqrt{19} = (19)^{1/2}$

$= 19^{1/2}$

• The index of the radical is the denominator of the rational exponent.

c. $\sqrt[4]{x^4 + y^4}$

$= (x^4 + y^4)^{1/4}$

• Note that $\sqrt[4]{x^4 + y^4} \neq x + y.$

YOU TRY IT 5

Rewrite the radical expression as a exponential expression.

a. $\sqrt[5]{p^9}$ **b.** $\sqrt[3]{26}$ **c.** $\sqrt[3]{c^3 + d^3}$

Solution See page S33. **a.** $p^{9/5}$ **b.** $26^{1/3}$ **c.** $(c^3 + d^3)^{1/3}$

? QUESTION We used the function $f(x) = 1.2x^{2/5}$ in Example 3. How can the expression $1.2x^{2/5}$ be written as a radical expression?

Now that the relationship between radical expressions and expressions with rational exponents has been presented, we can look at alternative methods of using a graphing calculator to evaluate these expressions. To illustrate, we will evaluate the expression $\sqrt[5]{32^4}$ first as the radical expression $\sqrt[5]{32^4}$ and then as the exponential expression $32^{4/5}$.

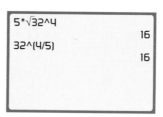

In either case, the display reads 16.

INSTRUCTOR NOTE
Exercise 52 can be used for a similar in-class example.

EXAMPLE 6

Use a graphing calculator to evaluate $\sqrt[6]{64^5}$ first as a radical expression and then as an exponential expression.

Solution $\sqrt[6]{64^5} = 32$

$64^{5/6} = 32$

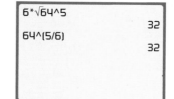

See Appendix A:
Radical Expressions

YOU TRY IT 6

Use a graphing calculator to evaluate $\sqrt[5]{243^4}$ first as a radical expression and then as an exponential expression.

Solution See page S33. 81; 81

? ANSWER $1.2x^{2/5}$ written as a radical expression is $1.2\sqrt[5]{x^2}$.

8.1 EXERCISES   Suggested Assignment: 7–55, odds

Topics for Discussion

1. Explain why $a^{1/2}$ is not a real number when a is a negative number.
    There is no real number that, when squared, equals a negative number.

2. Write two expressions that represent the nth root of a. For each expression, name the term that describes each part of the expression.
    $a^{1/n}$ (a is the base, $1/n$ is the exponent) and $\sqrt[n]{a}$ (n is the index, $\sqrt{}$ is the radical sign; a is the radicand)

3. Write an exponential expression of the form $a^{m/n}$. Explain how to rewrite it as a radical expression.   Answers will vary.

4. Write a radical expression of the form $\sqrt[n]{a^m}$. Explain how to rewrite it as an exponential expression.   Answers will vary.

5. Explain why $\sqrt[3]{x^3 + y^3} \neq x + y$.   Because $(x + y)^3 \neq x^3 + y^3$.

■ **Expressions with Rational Exponents**

Simplify.

6. $9^{3/2}$ 27

7. $25^{3/2}$ 125

8. $32^{2/5}$ 4

9. $64^{-2/3}$ $\dfrac{1}{16}$

10. $27^{-2/3}$ $\dfrac{1}{9}$

11. $16^{5/4}$ 32

12. $(-25)^{5/2}$ Not a real number

13. $(-36)^{3/4}$ Not a real number

14. $a^{1/3}(a^{3/4})(a^{-1/2})$ $a^{7/12}$

15. $(t^{-1/6})(t^{2/3})(t^{1/2})$ t

16. $(x^8y^2)^{5/2}$ $x^{20}y^5$

17. $(a^3b^9)^{2/3}$ a^2b^6

18. $(x^4y^2z^6)^{3/2}$ $x^6y^3z^9$

19. $(a^8b^4c^{12})^{3/4}$ $a^6b^3c^9$

20. $(x^{-3}y^6)^{-1/3}$ $\dfrac{x}{y^2}$

21. $(a^2b^{-6})^{-1/2}$ $\dfrac{b^3}{a}$

22. $\left(\dfrac{x^{1/2}}{y^2}\right)^4$ $\dfrac{x^2}{y^8}$

23. $\left(\dfrac{b^{-3/4}}{a^{-1/2}}\right)^8$ $\dfrac{a^4}{b^6}$

24. $\left(\dfrac{x^{1/2}y^{-5/4}}{y^{-3/4}}\right)^{-4}$ $\dfrac{y^2}{x^2}$

25. $\left(\dfrac{2a^3b^{-5}}{72ab^{-7}}\right)^{1/2}$ $\dfrac{ab}{6}$

26. $\left(\dfrac{40x^7y^{-2}}{5x^4y^{-8}}\right)^{2/3}$ $4x^2y^4$

27. Birth Rates   Statistics on birth rates in the United States are provided by the U.S. Census Bureau. The function that approximately models the data is $f(x) = 16.7x^{-1/16}$, where x is the year, with 1990 corresponding to $x = 1$, and $f(x)$ is the annual birth rate per 1000 people.

    a. Use the model to approximate the birth rate in 2000. Round to the nearest tenth.

    b. Use the model to approximate the difference between the birth rate in 1990 and the birth rate in 2005. Round to the nearest tenth.

c. Does the function indicate that the birth rate in the United States is increasing or decreasing?

a. 14.4 births per 1000 people per year   b. 2.7 births per 1000 people per year
c. Decreasing

28. *Demographics*   The U.S. Census Bureau provides projections of resident populations in the United States. The function that approximately models a projection for the population of children ages 5 through 13 in the years 1997 to 2050 is $f(x) = 43,000x^{-1/10}$, where x is the year, with 1997 corresponding to $x = 7$, and $f(x)$ is the population in thousands.

   a. Use the model to estimate the population of children ages 5 through 13 in 2010. Round to the nearest thousand.
   b. Use the model to estimate the difference in this population group between 2050 and 2000. Round to the nearest thousand.
   c. Who might be interested in statistics on the population of this age group?

   a. 31,869,000 children   b. 5,603,000 children   c. Answers will vary. For example, schools and manufacturers of children's clothing.

29. *Traffic Accidents*   The Insurance Institute for Highway Safety has released data on the number of car accidents in which motorists of different ages are involved. The function that approximately models the data is $f(x) = 6434x^{-4/3}$, where x is the age of the driver and $f(x)$ is the number of crashes per 1000 licensed drivers.

   a. Use the model to approximate the accident rate per 1000 drivers for 18-year-olds. Round to the nearest tenth.
   b. Use the model to approximate the difference between the accident rate for 16-year-olds and the accident rate for 60-year-olds. Round to the nearest tenth.
   c. Provide an explanation for the decrease in the accident rate as age increases.

   a. 136.4 crashes per 1000 licensed drivers   b. 132.2 crashes per 1000 licensed drivers   c. Answers will vary. For example, older drivers have more driving experience.

30. *Health*   The per capita consumption of cigarettes for the years 1963 through 1999 can be approximated by the function $f(x) = 955,954x^{-32/25}$, where x is the year, with 1963 corresponding to $x = 63$, and $f(x)$ is the per capita consumption of cigarettes (*Sources:* Tobacco Institute, Economic Research Service, Agriculture Department).

   a. Use the function to approximate the per capita consumption in 1985. Round to the nearest whole number.
   b. Use the function to approximate the difference in per capita consumption of cigarettes between 1963 and 1995. Round to the nearest whole number.
   c. Provide an explanation for the decrease in per capita consumption of cigarettes in the United States during the last few decades of the 1900s.

   a. 3242 cigarettes   b. 1945 cigarettes   c. Answers will vary. For example, increased awareness of its health risks and increased emphasis on healthful lifestyles.

31.  *Telecommunications*   The Yankee Group and Technology Futures, Inc., provided data on the average monthly price of cellular phone use in the United States at the end of the 20th century. The function that approximately models the data is $f(x) = 84.7x^{-1/4}$, where x is the year, with 1995 corresponding to $x = 0$, and $f(x)$ is the average monthly price.

    a. Use the model to approximate the monthly price in 1997. Round to the nearest dollar.

    b. Use the model to approximate the difference between the monthly price in 1998 and the monthly price in 1999. Round to the nearest dollar.

    c. What is the range of the function given that the domain is the years 1996 through 2000? Round the elements in the range to the nearest dollar.

    a. $71   b. $4   c. {85, 71, 64, 60, 57}

▪ Exponential Expressions and Radical Expressions

Rewrite the exponential expression as a radical expression.

32. $a^{3/2}$ $\sqrt{a^3}$ **33.** $b^{4/3}$ $\sqrt[3]{b^4}$ **34.** $(2t)^{5/2}$ $\sqrt{32t^5}$ **35.** $(3x)^{2/3}$ $\sqrt[3]{9x^2}$ **36.** $-2x^{2/3}$ $-2\sqrt[3]{x^2}$

37. $-3a^{2/5}$ $-3\sqrt[5]{a^2}$ **38.** $(4x-3)^{3/4}$ $\sqrt[4]{(4x-3)^3}$ **39.** $(3x-2)^{1/3}$ $\sqrt[3]{3x-2}$ **40.** $x^{-2/3}$ $\dfrac{1}{\sqrt[3]{x^2}}$ **41.** $b^{-3/4}$ $\dfrac{1}{\sqrt[4]{b^3}}$

Rewrite the radical expression as an exponential expression.

42. $\sqrt[3]{x}$ $x^{1/3}$ **43.** $\sqrt[4]{y}$ $y^{1/4}$ **44.** $\sqrt[3]{c^4}$ $c^{4/3}$ **45.** $\sqrt[4]{d^3}$ $d^{3/4}$ **46.** $\sqrt[3]{2x^2}$ $(2x^2)^{1/3}$

47. $\sqrt[5]{4y^7}$ $(4y^7)^{1/5}$ **48.** $3x\sqrt[3]{y^2}$ $3xy^{2/3}$ **49.** $2p\sqrt[5]{r}$ $2pr^{1/5}$ **50.** $\sqrt{n^2-2}$ $(n^2-2)^{1/2}$ **51.** $\sqrt[6]{a^2+5}$ $(a^2+5)^{1/6}$

Use a graphing calculator to evaluate the expression as both a radical expression and an exponential expression.

52. $\sqrt[5]{243^4}$ 81 **53.** $\sqrt[7]{128^3}$ 8 **54.** $\sqrt[4]{625^3}$ 125 **55.** $343^{2/3}$ 49 **56.** $256^{5/8}$ 32

Applying Concepts

57. Find the value of the expression $(27^{2/3} + 64^{2/3})^{3/2} - 10^2$.   25

58. Simplify the product: $(x^{1/4} + y^{1/4})(x^{3/4} - x^{1/2}y^{1/4} + x^{1/4}y^{1/2} - y^{3/4})$ $x - y$

59. If a and b are real numbers, and $3^{a/b} \cdot 3^{b/a} = 3^{5/2}$, find the value of $9^{|a/b - b/a|}$.
27

60. ✎ Provide an explanation for the concept that $25^{1/2} = 5$ and $\sqrt{25} = 5$.

Answers will vary. For example, $(25^{1/2})^2 = 25$ and $5^2 = 25$, so $25^{1/2} = 5$. $\sqrt{25} = 25^{1/2}$ and $25^{1/2} = 5$, so $\sqrt{25} = 5$.

61. If m and n have a common factor, does $x^{m/n} = x^{p/q}$, where p and q have no common factors? For example, does $x^{4/6} = x^{2/3}$ for all real numbers x? Explain your answer.   No. Explanations will vary.

EXPLORATION

1. *Developing Rules for Rational Exponents*

a. Use a calculator to evaluate each expression given below.

$\dfrac{1}{2^{-1}}$ 2 $\dfrac{1}{3^{-1}}$ 3 $\dfrac{1}{4^{-1}}$ 4 $\dfrac{1}{5^{-1}}$ 5 $\dfrac{1}{6^{-1}}$ 6

$\dfrac{1}{2^{-2}}$ 4 $\dfrac{1}{3^{-2}}$ 9 $\dfrac{1}{4^{-2}}$ 16 $\dfrac{1}{5^{-2}}$ 25 $\dfrac{1}{6^{-2}}$ 36

$\dfrac{1}{2^{-3}}$ 8 $\dfrac{1}{3^{-3}}$ 27 $\dfrac{1}{4^{-3}}$ 64 $\dfrac{1}{5^{-3}}$ 125 $\dfrac{1}{6^{-3}}$ 216

b. Use the pattern of the answers to the exercises in part a to determine a rule for $\dfrac{1}{x^{-n}}$, $x \neq 0$. $\frac{1}{x^{-n}} = x^n$

Use the rule to evaluate each of the following.

$\dfrac{1}{2^{-4}}$ 16 $\dfrac{1}{3^{-4}}$ 81 $\dfrac{1}{7^{-2}}$ 49 $\dfrac{1}{8^{-2}}$ 64 $\dfrac{1}{2^{-5}}$ 32

c. Use a calculator to evaluate each expression given below.

$4^{1/2}$ 2 $9^{1/2}$ 3 $16^{1/2}$ 4 $25^{1/2}$ 5 $36^{1/2}$ 6

d. Use the pattern of your answers to the exercises in part c to determine a rule for $x^{1/2}$. $x^{1/2} = \sqrt{x}$

Use the rule to evaluate each of the following.

$49^{1/2}$ 7 $64^{1/2}$ 8 $81^{1/2}$ 9 $100^{1/2}$ 10 $144^{1/2}$ 12

e. Use a graphing calculator to graph $y = x^{1/2}$. Check that points on the graph confirm your answers to part d.

f. Use a calculator to evaluate each expression given below.

$8^{1/3}$ 2 $(-8)^{1/3}$ −2 $27^{1/3}$ 3 $(-27)^{1/3}$ −3 $64^{1/3}$ 4

g. Use the pattern of your answers to the exercises in part f to determine a rule for $x^{1/3}$. $x^{1/3} = \sqrt[3]{x}$

Use the rule to evaluate $125^{1/3}$.   5

h. Use a graphing calculator to graph $y = x^{1/3}$. Check that points on the graph confirm your answers to part g.

i. On the basis of the pattern of your answers, how would you define $x^{1/n}$? $x^{1/n} = \sqrt[n]{x}$

Use your rule to evaluate each of the following.

$16^{1/4}$ 2 $81^{1/4}$ 3 $32^{1/5}$ 2 $64^{1/6}$ 2 $(-32)^{1/5}$ −2

j. Use a graphing calculator to graph the function $y = (x^3)^{1/3}$. Write a linear function that has the same graph as $y = (x^3)^{1/3}$. Why do these two functions have the same graph? Is the graph of the function $y = (x^{1/3})^3$ also the same graph? Why?

$y = x$. They are the same graph because $(x^3)^{1/3} = x$. Yes. They are the same graph because $(x^{1/3})^3 = x$.

k. Use a graphing calculator to graph the function $y = (x^5)^{1/5}$. Write a linear function that has the same graph as $y = (x^5)^{1/5}$. Why do these two functions have the same graph? Is the graph of the function $y = (x^{1/5})^5$ also the same graph? Why?

$y = x$. They are the same graph because $(x^5)^{1/5} = x$. Yes. They are the same graph because $(x^{1/5})^5 = x$.

l. Use a graphing calculator to graph the function $y = (x^2)^{1/2}$. Why is this not the same as the graph of $y = (x^3)^{1/3}$?

Because the square root of a negative number is not defined.

2. *Exploring Values of Expressions with Rational Exponents*

a. Evaluate the expression 3^n for $n = 0, 1, 2, 3$, and 4.

b. What values would you expect $3^{1.5}$ to be between?

c. What values would you expect $3^{5/2}$ to be between?

d. What values would you expect $3^{3.5}$ to be between?

e. Graph $y = 3^n$. Approximate $3^{1.5}$, $3^{5/2}$, and $3^{3.5}$. Compare the estimates with the answers to parts b–d. Do they confirm what you expected in parts b–d?

f. Use a calculator to evaluate $\sqrt{3^5}$ and $(\sqrt{3})^5$. Compare your results with the value of $3^{5/2}$ found in part e. What do you notice?

g. Using the patterns suggested in part f, write and evaluate two radical expressions for $3^{1.5}$. Compare the values with the graphical approximation of $3^{1.5}$ found in part e. Are the results the same?

a. 1, 3, 9, 27, 81 **b.** between 3 and 9 **c.** between 9 and 27 **d.** between 27 and 81 **e.** $3^{1.5} \approx 5.2$, $3^{5/2} \approx 15.6$, $3^{3.5} \approx 46.8$ **f.** $\sqrt{3^5} \approx 15.6$, $(\sqrt{3})^5 \approx 15.6$.

g. $\sqrt{3^3} \approx 5.2$, $(\sqrt{3})^3 \approx 5.2$

SECTION **8.2** # Simplifying Radical Expressions

- Simplify Radical Expressions That Are Roots of Perfect Powers
- Decimal Approximations of Radical Expressions
- Radical Expressions in Simplest Form

■ Simplify Radical Expressions That Are Roots of Perfect Powers

The formula $R = 1.4\sqrt{h}$ is used to determine the distance a person looking through a submarine periscope can see. In this formula, R is the distance in miles and h is the height in feet of the periscope above the surface of the water. When a periscope is 9 feet above the surface of the water, how far can a lookout see?

$$R = 1.4\sqrt{h}$$

$$R = 1.4\sqrt{9} \qquad \bullet \text{ Replace } h \text{ in the given formula by 9.}$$

$$R = 1.4(3) \qquad \bullet \text{ Take the square root of 9.}$$

$$R = 4.2 \qquad \bullet \text{ Multiply 1.4 by 3.}$$

When the periscope is 9 feet above the surface of the water, the lookout can see a distance of 4.2 miles.

In the solution above, we rewrote $\sqrt{9}$ as 3. Actually, every positive number has two square roots, one a positive number and one a negative number. For example, because $(3)^2 = 9$ and $(-3)^2 = 9$, there are two square roots of 9: 3 and -3. The symbol $\sqrt{}$ is used to indicate the positive or **principal square root.** To indicate the negative square root of a number, a negative sign is placed in front of the radical sign.

$$\sqrt{9} = 3 \qquad -\sqrt{9} = -3$$

The square root of 0 is 0.

$$\sqrt{0} = 0$$

The square root of a negative number is not a real number because the square of a real number must be positive.

$$\sqrt{-9} \text{ is not a real number.}$$

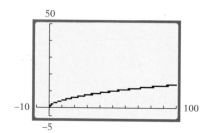

The graph of $R = 1.4\sqrt{h}$ is shown at the left. For no point on the graph is the h value less than 0 because the square root of a negative number is not a real number; the domain of $R = 1.4\sqrt{h}$ is $\{h \mid h \geq 0\}$. For no point on the graph is the R value less than 0; the range of the function is $\{R \mid R \geq 0\}$. This is reasonable in the context of the application: it is not possible for the lookout to see a negative distance.

The square root of a negative number is not a real number, but the square root of a squared negative number is a positive number. Let's look at an example.

$$\text{If } a = 3, \text{ then } \sqrt{a^2} = \sqrt{3^2} = \sqrt{9} = 3 = a.$$
$$\text{If } a = -3, \text{ then } \sqrt{a^2} = \sqrt{(-3)^2} = \sqrt{9} = 3 = -a.$$

TAKE NOTE

This is similar to absolute value:

$|a| = a$ if $a = 3$
$|a| = -a$ if $a = -3$

Thus we have

$$\sqrt{a^2} = a \text{ if } a = 3.$$
$$\sqrt{a^2} = -a \text{ if } a = -3.$$

Using these ideas, we can state $\sqrt{a^2} = |a|$.

Besides square roots, we can also determine cube roots, fourth roots, and so on:

$$\text{If } a = 2, \text{ then } \sqrt[3]{a^3} = \sqrt[3]{2^3} = \sqrt[3]{8} = 2.$$
$$\text{If } a = -2, \text{ then } \sqrt[3]{a^3} = \sqrt[3]{(-2)^3} = \sqrt[3]{-8} = -2.$$

TAKE NOTE

$\sqrt[3]{8} = 2$ because
$2^3 = 8$.
$\sqrt[3]{-8} = -2$ because
$(-2)^3 = -8$.
$\sqrt[4]{16} = 2$ because
$2^4 = 16$.
$\sqrt[5]{32} = 2$ because
$2^5 = 32$.
$\sqrt[5]{-32} = -2$ because
$(-2)^5 = -32$.

Note that the cube root of a positive number is positive, and the cube root of a negative number is negative. We can state $\sqrt[3]{a^3} = a$.

$$\text{If } a = 2, \text{ then } \sqrt[4]{a^4} = \sqrt[4]{2^4} = \sqrt[4]{16} = 2.$$
$$\text{If } a = -2, \text{ then } \sqrt[4]{a^4} = \sqrt[4]{(-2)^4} = \sqrt[4]{16} = 2.$$

From this example, $\sqrt[4]{a^4} = |a|$.

$$\text{If } a = 2, \text{ then } \sqrt[5]{a^5} = \sqrt[5]{2^5} = \sqrt[5]{32} = 2.$$
$$\text{If } a = -2, \text{ then } \sqrt[5]{a^5} = \sqrt[5]{(-2)^5} = \sqrt[5]{-32} = -2.$$

From this example, $\sqrt[5]{a^5} = a$.

TAKE NOTE

Note that when the index is an even natural number, the nth root requires absolute value symbols.

$\sqrt[6]{d^6} = |d|$ but $\sqrt[5]{y^5} = y$

Because we stated that variables within radicals represent positive numbers, we will omit the absolute value symbols when writing an answer.

The following properties hold true for finding the nth root of a real number.

If n is an even integer, then $\sqrt[n]{a^n} = |a|$ and $-\sqrt[n]{a^n} = -|a|$.

If n is an odd integer, then $\sqrt[n]{a^n} = a$.

For example,

$$\sqrt[6]{d^6} = |d| \qquad -\sqrt[14]{y^{14}} = -|y| \qquad \sqrt[5]{c^5} = c$$

Because we have stated that all variables in this chapter represent positive numbers unless otherwise stated, it is not necessary to use the absolute value signs.

In simplifying radical expressions, we use *perfect powers*. For example:

TAKE NOTE

Perfect squares and perfect cubes were discussed in the section Special Factoring.

The square of a term is a **perfect square.** The exponents on variables of perfect squares are even numbers.

	Perfect Square
$5^2 =$	25
$(x^4)^2 =$	x^8

The cube of a term is a **perfect cube.** The exponents on variables of perfect cubes are multiples of 3.

	Perfect Cube
$5^3 =$	125
$(y^7)^3 =$	y^{21}

The radicand of the radical expression $\sqrt[3]{x^6y^9}$ is a perfect cube because the exponents on the variables are multiples of 3. To simplify this expression, write the radical expression as an exponential expression.

$$\sqrt[3]{x^6y^9} = (x^6y^9)^{1/3}$$

Simplify the power of an exponential expression by multiplying the exponents.

$$= x^2y^3$$

Note that **a variable expression is a perfect power if the exponents on the factors are evenly divisible by the index of the radical.**

The chart below shows roots of perfect powers. Knowledge of these roots is very helpful in simplifying radical expressions.

Square Roots		**Cube Roots**	**Fourth Roots**	**Fifth Roots**
$\sqrt{1} = 1$	$\sqrt{36} = 6$	$\sqrt[3]{1} = 1$	$\sqrt[4]{1} = 1$	$\sqrt[5]{1} = 1$
$\sqrt{4} = 2$	$\sqrt{49} = 7$	$\sqrt[3]{8} = 2$	$\sqrt[4]{16} = 2$	$\sqrt[5]{32} = 2$
$\sqrt{9} = 3$	$\sqrt{64} = 8$	$\sqrt[3]{27} = 3$	$\sqrt[4]{81} = 3$	$\sqrt[5]{243} = 3$
$\sqrt{16} = 4$	$\sqrt{81} = 9$	$\sqrt[3]{64} = 4$	$\sqrt[4]{256} = 4$	
$\sqrt{25} = 5$	$\sqrt{100} = 10$	$\sqrt[3]{125} = 5$	$\sqrt[4]{625} = 5$	

P

TAKE NOTE

From the chart, $\sqrt[5]{243} = 3$, which means that $3^5 = 243$. From this we know that $(-3)^5 = -243$, which means that $\sqrt[5]{-243} = -3$.

➡ Simplify: $\sqrt[5]{-243x^5y^{15}}$

$$\sqrt[5]{-243x^5y^{15}} = (-243x^5y^{15})^{1/5}$$

$$= -3xy^3$$

• From the chart, 243 is a perfect fifth power, and each exponent is divisible by 5. Therefore, the radicand is a perfect fifth power. $\sqrt[5]{-243} = -3$. Divide each exponent by 5.

INSTRUCTOR NOTE
Exercises 6, 12, and 14 can be used for similar in-class examples.

EXAMPLE 1

Simplify.   **a.** $\sqrt{49x^2y^{12}}$ **b.** $-\sqrt[4]{16a^4b^8}$ **c.** $\sqrt[5]{c^{10}d^5}$

Solution

a. $\sqrt{49x^2y^{12}} = 7xy^6$

- Each exponent is divisible by 2. The radicand is a perfect square. $\sqrt{49} = 7$. Divide each exponent by 2.

b. $-\sqrt[4]{16a^4b^8} = -2ab^2$

- Each exponent is divisible by 4. The radicand is a perfect fourth power. $\sqrt[4]{16} = 2$. Divide each exponent by 4.

c. $\sqrt[5]{c^{10}d^5} = c^2d$

- The radicand is a perfect fifth power. Divide each exponent by 5.

YOU TRY IT 1

Simplify.   **a.** $\sqrt{121x^{10}y^4}$ **b.** $\sqrt[3]{-8x^{12}y^3}$ **c.** $-\sqrt[4]{81a^{12}b^8}$

Solution   See page S33.   **a.** $11x^5y^2$ **b.** $-2x^4y$ **c.** $-3a^3b^2$

Suggested Activity

Have students answer the following questions:
By what factor must you multiply a number in order to double its square root? [4] triple its square root? [9] double its cube root? [8] triple its cube root? [27]
Have students explain their answers.

■ Decimal Approximations of Radical Expressions

The formula $R = 1.4\sqrt{h}$ has been used to determine that a person looking through a submarine periscope that is 9 feet above the surface of the water can see a distance of 4.2 miles. How far can the lookout see when the periscope is 6 feet above the surface of the water?

$$R = 1.4\sqrt{h}$$
$$R = 1.4\sqrt{6}$$ • Replace h in the given formula by 6.
$$R \approx 3.4$$

In this situation, 6 is not a perfect square; the square root of 6 is not an integer. We used a calculator to approximate $\sqrt{6}$ and multiplied the result by 1.4. To the nearest tenth of a mile, the lookout can see a distance of 3.4 miles when the periscope is 6 feet above the surface of the water.

If a number is not a perfect power, its root can only be approximated. For example,

$$\sqrt{6} = 2.449489743\ldots \qquad \sqrt[3]{5} = 1.709975947\ldots$$

These numbers are **irrational numbers**. Their decimal representations never terminate or repeat.

▼ *Point of Interest*

The Latin expression for irrational number was numerus surdus, *which literally means "inaudible number." A prominent 16th-century mathematician wrote of irrational numbers, ". . . just as an infinite number is not a number, so an irrational number is not a true number, but lies in some sort of cloud of infinity." In 1872 Richard Dedekind wrote a paper that established the first logical treatment of irrational numbers.*

? QUESTION   Which of the following represent irrational numbers?

   a. $\sqrt{18}$ **b.** $\sqrt[3]{6}$ **c.** $\sqrt[4]{81}$

? ANSWERS   **a.** 18 is not a perfect square. $\sqrt{18}$ is an irrational number.   **b.** 6 is not a perfect cube. $\sqrt[3]{6}$ is an irrational number.   **c.** 81 is a perfect fourth power because $3^4 = 81$. $\sqrt[4]{81}$ is not an irrational number.

INSTRUCTOR NOTE
Exercise 36 can be used for
a similar in-class example.

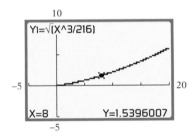

EXAMPLE 2

Weather satellites can measure the diameter of a storm. The duration of the storm can then be determined by using the formula $t = \sqrt{\dfrac{d^3}{216}}$, where t is the duration of the storm in hours and d is the diameter of the storm in miles. Find the duration of a storm that has an 8-mile diameter. Round to the nearest tenth.

Solution $t = \sqrt{\dfrac{d^3}{216}}$

$t = \sqrt{\dfrac{8^3}{216}}$ • Replace d with 8.

$t = \sqrt{\dfrac{512}{216}}$ • Evaluate 8^3.

$t \approx 1.5$ • Use a calculator to evaluate the radical expression.

A storm that has a diameter of 8 miles will last 1.5 hours.

The graph of $t = \sqrt{\dfrac{d^3}{216}}$ is shown at the left. Note that the domain of the function is $\{d \,|\, d \geq 0\}$, and the range is $\{t \,|\, t \geq 0\}$. This is reasonable in the context of the problem: Neither the time nor the duration of the storm can be negative.

Note that when $d = 8$, $t \approx 1.5$. This is the answer we calculated in Example 2.

YOU TRY IT 2

The time t, in hours, needed to cook a pot roast that weighs p pounds can be approximated by the formula $t = 0.9 \sqrt[5]{p^3}$. Use this formula to find the time required to cook a 12-pound pot roast. Round to the nearest tenth.

Solution   See page S33.   4.0 h

■ Radical Expressions in Simplest Form

Sometimes we are not interested in an approximation of the root of a number but, rather, in the exact value in simplest form.

A radical expression is in simplest form when the radicand contains no factor, other than 1, that is a perfect power. The Product Property of Radicals is used to simplify radical expressions whose radicands are not perfect powers.

INSTRUCTOR NOTE
Provide numerical examples
of the Product Property of
Radicals. For example, show
that $\sqrt{4 \cdot 9} = \sqrt{4} \cdot \sqrt{9}$ and
$\sqrt[3]{8 \cdot 64} = \sqrt[3]{8} \cdot \sqrt[3]{64}$.

Product Property of Radicals
If $\sqrt[n]{a}$ and $\sqrt[n]{b}$ are real numbers, then
$\sqrt[n]{ab} = \sqrt[n]{a} \cdot \sqrt[n]{b}$ and $\sqrt[n]{a} \cdot \sqrt[n]{b} = \sqrt[n]{ab}$.

TAKE NOTE

16 is the greatest perfect-square factor of 48.

$48 \div 16 = 3$

The factor 3 does not contain a perfect square.

Suggested Activity

See Section 8.2 of the *Student Activity Manual* for an investigation involving geometry and the simplifying of radical expressions.

INSTRUCTOR NOTE
Exercises 24 and 30 can be used for similar in-class examples.

To simplify $\sqrt{48}$, write the radicand as the product of a perfect square and a factor that does not contain a perfect square. Use the Product Property of Radicals to write the expression as a product. Then simplify $\sqrt{16}$.

$$\sqrt{48} = \sqrt{16 \cdot 3}$$
$$= \sqrt{16} \cdot \sqrt{3}$$
$$= 4\sqrt{3}$$

Note that 48 must be written as the product of a perfect square and *a factor that does not contain a perfect square*. Therefore, it would not be correct to rewrite $\sqrt{48}$ as $\sqrt{4 \cdot 12}$ and simplify the expression as shown at the right. Although 4 is a perfect-square factor of 48, 12 contains a perfect square ($12 = 4 \cdot 3$) so $\sqrt{12}$ can be simplified. Remember to **find the largest perfect power that is a factor of the radicand.**

$$\sqrt{48} = \sqrt{4 \cdot 12}$$
$$= \sqrt{4} \cdot \sqrt{12}$$
$$= 2\sqrt{12}$$

Not in simplest form

? <u>QUESTION</u>   Is the radical expression $\sqrt[3]{16}$ in simplest form?

EXAMPLE 3

Simplify.   **a.** $\sqrt[4]{x^9}$ **b.** $\sqrt{18x^2y^3}$ **c.** $\sqrt[3]{-27a^5b^{12}}$

Solution

a. $\sqrt[4]{x^9} = \sqrt[4]{x^8 \cdot x}$

   • Write the radicand as the product of a perfect fourth power and a factor that does not contain a perfect fourth power.

$$= \sqrt[4]{x^8} \cdot \sqrt[4]{x}$$

   • Use the Product Property of Radicals to write the expression as a product.

$$= x^2 \sqrt[4]{x}$$

   • Simplify $\sqrt[4]{x^8}$.

b. $\sqrt{18x^2y^3} = \sqrt{9x^2y^2 \cdot 2y}$

   • Write the radicand as the product of a perfect square and factors that do not contain a perfect square.

$$= \sqrt{9x^2y^2} \cdot \sqrt{2y}$$

   • Use the Product Property of Radicals to write the expression as a product.

$$= 3xy\sqrt{2y}$$

   • Simplify.

c. $\sqrt[3]{-27a^5b^{12}} = \sqrt[3]{-27a^3b^{12} \cdot a^2}$
$$= \sqrt[3]{-27a^3b^{12}} \cdot \sqrt[3]{a^2}$$
$$= -3ab^4\sqrt[3]{a^2}$$

YOU TRY IT 3

Simplify.   **a.** $\sqrt[5]{x^7}$ **b.** $\sqrt[4]{32x^{10}}$ **c.** $\sqrt[3]{-64c^8d^{18}}$

Solution   See page S33.   a. $x\sqrt[5]{x^2}$ b. $2x^2\sqrt[4]{2x^2}$ c. $-4c^2d^6\sqrt[3]{c^2}$

? <u>ANSWER</u>   No, $\sqrt[3]{16}$ is not in simplest form because the radicand, 16, contains a factor (8) that is a perfect cube.

8.2 EXERCISES   Suggested Assignment: 5–37, odds

Topics for Discussion

1. Which of the following represent irrational numbers and why?
   a. $\sqrt{24}$ **b.** $\sqrt[3]{9}$ **c.** $\sqrt[5]{32}$
   Parts a and b are irrational numbers. Explanations will vary.

2. Explain how to determine whether a radical expression is the root of a perfect power.   Explanations will vary.

3. Explain how to write $\sqrt{32a^5}$ in simplest form.   Explanations will vary.

4. Determine whether the statement is always true, sometimes true, or never true.
   a. For real numbers, $\sqrt{x^2} = x$.   Sometimes true
   b. If a is a real number, then $\sqrt{a}$ represents a real number.
        Sometimes true
   c. $\sqrt{(-2)^2} = -2$.   Never true
   d. $\sqrt[3]{(-3)^3} = -3$.   Always true
   e. If b is a real number, then $\sqrt[3]{b}$ is a real number.   Always true
   f. The nth root of a negative number is a negative number.
        Sometimes true

■ Simplify Radical Expressions

Simplify.

5. $\sqrt{16a^4b^{12}}$
   $4a^2b^6$

6. $\sqrt{25x^8y^2}$
   $5x^4y$

7. $\sqrt{-16x^4y^2}$
   Not a real number

8. $\sqrt{-9a^6b^8}$
   Not a real number

9. $\sqrt[3]{27x^9y^{12}}$
   $3x^3y^4$

10. $\sqrt[3]{8a^{21}b^6}$
   $2a^7b^2$

11. $\sqrt[3]{-64c^9d^{12}}$
   $-4c^3d^4$

12. $\sqrt[3]{-27x^3y^{15}}$
   $-3xy^5$

13. $-\sqrt[4]{x^8y^{12}}$
   $-x^2y^3$

14. $-\sqrt[4]{a^{16}b^4}$
   $-a^4b$

15. $\sqrt[4]{81p^{16}q^4}$
   $3p^4q$

16. $\sqrt[4]{16a^8b^{20}}$
   $2a^2b^5$

17. $\sqrt[5]{32w^5x^{10}}$
   $2wx^2$

18. $\sqrt[5]{-32y^{15}z^{20}}$
   $-2y^3z^4$

19. $\sqrt{98}$
   $7\sqrt{2}$

20. $\sqrt{128}$
   $8\sqrt{2}$

21. $\sqrt[3]{72}$
   $2\sqrt[3]{9}$

22. $\sqrt[3]{16}$
   $2\sqrt[3]{2}$

23. $\sqrt{8c^3d^8}$
   $2cd^4\sqrt{2c}$

24. $\sqrt{24x^9y^6}$
   $2x^4y^3\sqrt{6x}$

25. $\sqrt{45x^2y^3z^5}$
   $3xyz^2\sqrt{5yz}$

26. $\sqrt{60ab^7c^{12}}$
   $2b^3c^6\sqrt{15ab}$

27. $\sqrt[3]{-125c^2d^4}$
   $-5d\sqrt[3]{c^2d}$

28. $\sqrt[3]{-216x^5y^9}$
   $-6xy^3\sqrt[3]{x^2}$

29. $\sqrt[3]{16p^8q^{11}r^{15}}$
   $2p^2q^3r^5\sqrt[3]{2p^2q^2}$

30. $\sqrt[3]{54a^5b^8c^6}$
   $3ab^2c^2\sqrt[3]{2a^2b^2}$

31. $\sqrt[4]{32x^9y^5}$
   $2x^2y\sqrt[4]{2xy}$

32. $\sqrt[4]{64y^8z^{10}}$
   $2y^2z^2\sqrt[4]{4z^2}$

33. If $\dfrac{\sqrt[3]{x}}{3}$ is an even integer, what is a possible value of x?

Answers may vary. For example, 216, 1728, or 5832.

34. If $\sqrt[4]{4x - 12}$ represents an integer, what is a possible value of x?

Answers may vary. For example, $\dfrac{13}{4}$, 7, or $\dfrac{93}{4}$.

35. *Optics*   The percent, *P,* of light that will pass through a certain translucent material is given by the equation $P = \sqrt[5]{\dfrac{1}{10}}$. Find the percent of light that will pass through the material. Round to the nearest tenth of a percent.   63.1%

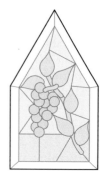

36. *Demographics*   The data below show the number of married couples, in millions, in the United States for selected years (*Source:* U.S. Bureau of the Census). The equation that approximately models the data is $y = 8.1\sqrt[5]{x^2}$, where y is the number of married couples, in millions, in year x, and $x = 0$ for 1900. Use the equation to predict, to the nearest tenth of a million, the number of married couples in the United States in **a.** 1975 and **b.** 2000.

Year	1985	1990	1995	2000
Married Couples (in millions)	51.1	53.3	54.9	54.5

a. 45.6 million married couples   **b.** 51.1 million married couples

37. *Crime*   The table below shows the number of property crimes, in millions, in the United States for selected years (*Source:* U.S. Federal Bureau of Investigation, Crime in the United States, annual). The equation that approximately models the data is $y = 9\sqrt[10]{x}$, where y is the number of property crimes, in millions, in year x, and $x = 5$ represents the year 1975. Use the equation to predict, to the nearest tenth of a million, the number of property crimes in the United States in **a.** 1971, **b.** 1988, and **c.** 2005. **d.** Are the numbers for these years reasonable when compared to the data in the table? Why or why not? **e.** There were 10.2 million property crimes in the United States in 2000. What is the difference between the number of property crimes the model predicts for 2000 and the actual number? Round to the nearest tenth of a million.

> **▼ Point of Interest**
>
> *Property crimes include burglary, larceny-theft, motor vehicle theft, and arson.*

Year	1975	1985	1995
Property Crimes (in millions)	10.3	11.1	12.1

a. 9.0 million property crimes   **b.** 12.0 million property crimes   **c.** 12.8 million property crimes   **d.** Answers will vary.   **e.** 2.4 million property crimes

38. *Crime*   The table on the next page shows the number of violent crimes, in millions, in the United States for selected years (*Source:* U.S. Federal Bureau of Investigation, Crime in the United States, annual). The equation that approximately models the data is

> **▼ Point of Interest**
>
> *Violent crimes include murder and nonnegligent manslaughter, forcible rape, robbery, and aggravated assault.*

$y = 0.54 \sqrt[5]{x^2}$, where y is the number of violent crimes, in millions, in year x, and $x = 5$ represents the year 1975. Use the equation to predict, to the nearest tenth of a million, the number of violent crimes in the United States in **a.** 1977, **b.** 1988, and **c.** 2005. **d.** Are the numbers for these years reasonable when compared to the data in the table? Why or why not? **e.** There were 1.4 million violent crimes in the United States in 2000. What is the difference between the number of violent crimes the model predicts for 2000 and the actual number? Round to the nearest tenth of a million.

Year	1975	1985	1995
Violent Crimes (in millions)	1.0	1.3	1.8

a. 1.2 million violent crimes   **b.** 1.7 million violent crimes   **c.** 2.2 million violent crimes   **d.** Answers will vary.   **e.** 0.7 million violent crimes.

Applying Concepts

39. Prove **a.** $\sqrt{a^2 + b^2} \neq a + b$ and **b.** $\sqrt[3]{a^3 + b^3} \neq a + b$ by finding a counter example.   **a.** Answers will vary.   **b.** Answers will vary.

40. Write in exponential form and then simplify.
   a. $\sqrt{16^{1/2}}$ **b.** $\sqrt[3]{4^{3/2}}$ **c.** $\sqrt[4]{32^{-4/5}}$ **d.** $\sqrt{243^{-4/5}}$

   a. $16^{1/4}$; 2   **b.** $4^{1/2}$; 2   **c.** $32^{-1/5}$; $\frac{1}{2}$ **d.** $243^{-2/5}$; $\frac{1}{9}$

41. If $A \triangle B$ means A^B and $A \triangledown B$ means $\sqrt[B]{A}$, what is the value of the expression $[(2 \triangle 6) \triangledown 3] \triangle 2$?   16

42. For how many real numbers x will the expression $\sqrt{-(x + 1)^2}$ be a real number?   1 (-1)

43. If $\sqrt[5]{x} = 4$, what is the value of $\sqrt{x}$?   32

44. *Roller Coasters*   The formula $v = (32r)^{1/2}$ is used to estimate the speed at which a roller coaster car must travel in order to stay on a vertical loop of track. In this formula, v is the speed in feet per second and r is the radius of the loop in feet.

   a.   How fast must a roller coaster car travel on a vertical loop of track that has a radius of 25 feet? Round to the nearest foot per second.
   b.   As the radius of the roller coaster loop increases, does the speed at which the roller coaster car must travel increase or decrease?
   c.   As the speed at which the roller coaster car must travel increases, does the radius of the roller coaster loop increase or decrease?

   a. 28 ft/s   **b.** Increase   **c.** Increase

EXPLORATION

1. *Mathematical Models of Data: Regression Analysis* A graphing calculator can be used to create mathematical models of equations containing radicals. To do this, use the calculator to perform regression analysis, using power regression. Convert the exponent, which is given in decimal form, to fraction form. Then rewrite the exponential expression as a radical expression. Here is an example.

The data given below are actual and projected worldwide sales of PCs (*Source:* John Gantz, research director, IDC).

Year	2000	2001	2002	2003	2004
Personal Computer Sales (in billions of dollars)	205	206	214	225	236

For a TI-83, enter the data for the years in L1, with 2000 corresponding to $x = 10$, 2001 corresponding to $x = 11$, and so on. Enter the data on computer sales, in billions of dollars, in L2.

L1	L2	L3	2
10	205	------	
11	206		
12	214		
13	225		
14	236		

L2(3) =214

Perform power regression on the data. See Appendix A: Regression. The screen shown at the right will appear. This means that the equation that approximately models the data is $y = 73.948385x^{0.4342389749}$ or approximately $y = 74x^{0.45}$. Because $0.45 = 9/20$, we can write the equation as $y = 74x^{9/20}$ or $y = 74\sqrt[20]{x^9}$.

PwrReg
y=a*x^b
a=73.948385
b=.4342389749
r²=.9204678616
r=.9594101634

a. How closely does the equation $y = 74\sqrt[20]{x^9}$ predict sales of PCs for 2000 and 2004? Are the results close to the data in the table? What is the prediction for 2010? Round to the nearest billion.

The prediction for 2000 is \$209 billion, for 2004 is \$243 billion, and for 2010 is \$285 billion.

The table below provides public approval, in percent, of home schooling (*Source:* 33rd Annual Phi Delta Kappa/Gallup Poll).

Year	1985	1988	1997	2001
Public Approval of Home Schooling (in percent)	16	28	36	41

b. Use a graphing calculator to write a mathematical model of the data. Use $x = 5$ for 1985, $x = 8$ for 1988, and so on. For the values of a and b in the equation $y = ax^b$, round to the nearest tenth. Write a paragraph describing how well your model predicts public approval of home schooling.

$y = 6.8\sqrt[5]{x^3}$. Paragraphs will vary.

c. Use data of your own choosing to write a mathematical model in the form of a radical expression. A lot of data are available at the web site **www.fedstats.gov**. Mathematical models will vary.

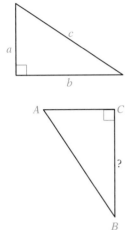

The Pythagorean Theorem

■ The Pythagorean Theorem

A **right triangle** contains one right angle. The side opposite the right angle is called the **hypotenuse.** The other two sides are called **legs.**

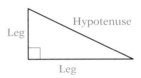

? QUESTION   For the triangle at the left which side is the hypotenuse?*

The angles in a right triangle are usually labeled with capital letters A, B, and C, with C reserved for the right angle. The side opposite angle A is side a, the side opposite angle B is side b, and the hypotenuse is side c.

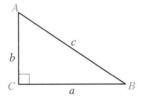

? QUESTION   Using the convention stated in the above paragraph, what letter should replace the question mark in the triangle at the left?†

The Pythagorean Theorem states an important relationship that exists among the sides of a right triangle. The theorem is named after Pythagoras, a Greek mathematician and philosopher who lived from about 572 to 501 B.C. Although Pythagoras is generally credited with the discovery, there is evidence that it was known years before his birth. There is reference to this theorem in Chinese writings from about 1100 B.C., and it is believed that the Egyptians and Babylonians knew of it as early as 2000 B.C.

The **Pythagorean Theorem** states that the square of the hypotenuse of a right triangle is equal to the sum of the squares of the two legs.

In the figure below is a right triangle with legs measuring 3 units and 4 units and a hypotenuse measuring 5 units. Each side of the triangle is also the side of a square. The number of square units in the area of the largest square is equal to the sum of the areas of the smaller squares.

▼ *Point of Interest*

Historians believe that surveyors in ancient Egypt used stretched ropes with 12 equally spaced knots to measure right angles when laying out land boundaries. The surveyors knew that a triangle with sides measuring 3, 4, and 5 units was a right triangle.

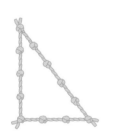

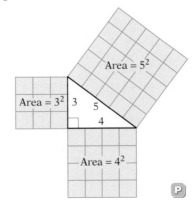

Square of the hypotenuse   =   sum of the squares of the two legs

$$5^2 = 3^2 + 4^2$$
$$25 = 9 + 16$$
$$25 = 25$$

─────────

? ANSWERS   *Side c is the hypotenuse because it is the side opposite the right angle.   †The letter a (lowercase a) should replace the question mark because it is the side opposite angle A.

Suggested Activity

See Section 8.3 of the *Student Activity Manual* for an activity involving the Pythagorean Theorem.

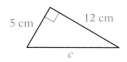

5 cm   12 cm

c

Pythagorean Theorem

If a and b are the lengths of the legs of a right triangle and c is the length of the hypotenuse, then $c^2 = a^2 + b^2$.

If the lengths of two sides of a right triangle are known, the Pythagorean Theorem can be used to find the length of the third side.

Consider a right triangle with legs that measure 5 cm and 12 cm. Use the Pythagorean Theorem, with $a = 5$ and $b = 12$, to find the length of the hypotenuse. (If you let $a = 12$ and $b = 5$, the result is the same.)

$$c^2 = a^2 + b^2$$
$$c^2 = 5^2 + 12^2$$
$$c^2 = 25 + 144$$
$$c^2 = 169$$

This equation states that the square of c is 169. Because $13^2 = 169$, $c = 13$, and the length of the hypotenuse is 13 cm. We can find c by taking the square root of 169. $\sqrt{169} = 13$. This suggests the following property.

The Principal Square Root Property

If $r^2 = s$, then $r = \sqrt{s}$, and r is called the square root of s.

The Principal Square Root Property and its application are illustrated as follows:

Because $4^2 = 16$, $4 = \sqrt{16}$. Therefore, if $c^2 = 16$, $c = \sqrt{16} = 4$.
Because $8^2 = 64$, $8 = \sqrt{64}$. Therefore, if $c^2 = 64$, $c = \sqrt{64} = 8$.

? QUESTION   If $c^2 = 81$, what is the value of c?

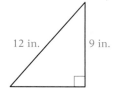

12 in.   9 in.

➡ The length of one leg of a right triangle is 9 inches. The hypotenuse is 12 inches. Find the length of the other leg. Round to the nearest tenth.

$$a^2 + b^2 = c^2$$ • Use the Pythagorean Theorem.
$$9^2 + b^2 = 12^2$$ • $a = 9$ and $c = 12$.
$$81 + b^2 = 144$$ • Square 9 and square 12.
$$81 - 81 + b^2 = 144 - 81$$ • We want to solve for b^2. Subtract
$$b^2 = 63$$ 81 from each side of the equation.
$$b = \sqrt{63}$$ • Use the Principal Square Root Property.
$$b \approx 7.9$$ • Use a calculator to approximate $\sqrt{63}$.

The length of the other leg is approximately 7.9 inches.   ⬅

TAKE NOTE

In this example, if you let $b = 9$ and solve for a^2, the result is the same.

? ANSWER $c = \sqrt{81} = 9$

INSTRUCTOR NOTE
Exercise 10 can be used for
a similar in-class example.

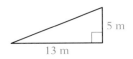

EXAMPLE 1

The two legs of a right triangle measure 5 meters and 13 meters. Find the hypotenuse of the right triangle. Round to the nearest hundredth.

Solution $c^2 = a^2 + b^2$ • Use the Pythagorean Theorem.

$c^2 = 5^2 + 13^2$ • $a = 5$ and $b = 13$.

$c^2 = 25 + 169$ • Square 5 and square 13.

$c^2 = 194$

$c = \sqrt{194}$ • Use the Principal Square Root Property.

$c \approx 13.93$

The length of the hypotenuse is 13.93 meters.

YOU TRY IT 1

The two legs of a right triangle measure 7 meters and 14 meters. Find the hypotenuse of the right triangle. Round to the nearest hundredth.

Solution   See page S33.   15.65 m

INSTRUCTOR NOTE
Exercise 12 can be used for
a similar in-class example.

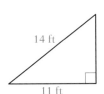

EXAMPLE 2

The hypotenuse of a right triangle measures 14 feet, and one leg measures 11 feet. Find the measure of the other leg. Round to the nearest tenth.

Solution $a^2 + b^2 = c^2$ • Use the Pythagorean Theorem.

$11^2 + b^2 = 14^2$ • $a = 11$ and $c = 14$.

$121 + b^2 = 196$ • Square 11 and square 14.

$b^2 = 75$ • Solve for b^2. Subtract 121 from each side of the equation.

$b = \sqrt{75}$ • Use the Principal Square Root Property.

$b \approx 8.7$

The measure of the other leg is approximately 8.7 feet.

YOU TRY IT 2

The hypotenuse of a right triangle measures 16 feet, and one leg measures 5 feet. Find the measure of the other leg. Round to the nearest hundredth.

Solution   See page S33.   15.20 ft

The hypotenuse is always the longest side of a right triangle. This fact can be used as a quick check when you are solving a right triangle for the length of one of the sides. For example, in Example 2, we found that the length of a leg was 8.7 feet. Because 8.7 < 14 (the hypotenuse), we know that the answer is "in the right ball-park."

? QUESTION   Two legs of a right triangle measure 3 meters and 9 meters. Is the length of the hypotenuse less than 3 meters, between 3 and 9 meters, or greater than 9 meters?

INSTRUCTOR NOTE
Exercise 14 can be used for a similar in-class example.

EXAMPLE 3

High-definition television (HDTV) gives consumers a wider viewing area, more like a film seen in a movie theater. A regular television with a 27-inch diagonal measurement has a screen 16.2 inches tall. An HDTV screen with the same 16.2-inch height has a diagonal measuring 33 inches. How many inches wider is the HDTV screen? Round to the nearest tenth.

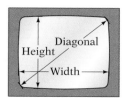

State the goal.   The goal is to find the difference between the width of an HDTV screen and a regular television screen given that both have a height of 16.2 inches, the diagonal of the HDTV screen is 33 inches, and the diagonal of the regular TV is 27 inches.

Devise a strategy.
- Use the Pythagorean Theorem to find the width of an HDTV screen that has a 16.2-inch height and a 33-inch diagonal.
- Use the Pythagorean Theorem to find the width of a regular TV screen that has a 16.2-inch height and a 27-inch diagonal.
- Subtract the width of the regular television screen from the width of the HDTV screen.

Solve the problem.

$$a^2 + b^2 = c^2$$
$$(16.2)^2 + b^2 = 33^2 \qquad \text{• For the HDTV, } a = 16.2 \text{ and } c = 33.$$
$$262.44 + b^2 = 1089$$
$$b^2 = 826.56 \qquad \text{• Solve for } b^2. \text{ Subtract 262.44 from each side.}$$
$$b = \sqrt{826.56} \qquad \text{• Use the Product Property of Square Roots.}$$
$$b \approx 28.7 \qquad \text{• The width of the HDTV screen is 28.7 inches.}$$

$$a^2 + b^2 = c^2$$
$$(16.2)^2 + b^2 = 27^2 \qquad \text{• For the regular TV, } a = 16.2 \text{ and } c = 27.$$
$$262.44 + b^2 = 729$$
$$b^2 = 466.56 \qquad \text{• Solve for } b^2. \text{ Subtract 262.44 from each side.}$$
$$b = \sqrt{466.56} \qquad \text{• Use the Product Property of Square Roots.}$$
$$b = 21.6 \qquad \text{• The width of the regular TV screen is 21.6 inches.}$$

$$28.7 - 21.6 = 7.1 \qquad \text{• Find the difference between the two widths.}$$

The width of the HDTV screen is 7.1 inches greater than the width of the regular TV screen.

Check your work.   √

? ANSWER   The length of the hypotenuse is longer than either leg, so the length is greater than 9 meters.

YOU TRY IT 3

Dave Marshall needs to clean the gutters of his home. The gutters are 28 feet above the ground. For safety, the distance a ladder reaches up a wall should be four times the distance from the bottom of the ladder to the base of the side of the house. Therefore, the ladder must be 7 feet from the base of the house. Will a 30-foot ladder be long enough to reach the gutters?

Solution   See page S33.   Yes.

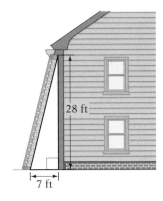

■ The Distance Formula

In the rectangular coordinate system at the right, points are graphed at (1, 3) and (5, 3). A line segment is drawn between the two points. We can determine the length of the line segment by counting the number of units between the points; it is 4 units. We can also determine the length by finding the absolute value of the difference between the *x*-coordinates of the two points.

$$|5 - 1| = |4| = 4$$

It does not matter in what order the two numbers are subtracted; the result is the same.

$$|1 - 5| = |-4| = 4$$

In the rectangular coordinate grid at the right, points are graphed at (3, 0) and (3, 4). A line segment is drawn between the two points. We can determine the length of the line segment by counting the number of units between the points; it is 4 units. We can also determine the length by finding the absolute value of the difference between the *y*-coordinates of the two points.

$$|4 - 0| = |4| = 4$$

It does not matter in what order the two numbers are subtracted; the result is the same.

$$|0 - 4| = |-4| = 4$$

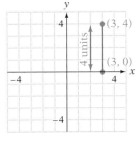

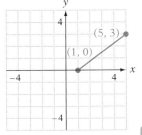

In the coordinate grid at the left, points are graphed at (1, 0) and (5, 3). A line segment is drawn between the two points. Because the line segment is neither horizontal nor vertical, we cannot determine its length by counting the number of units between the points. We can, however, draw a right triangle (as shown on the next page) in which the line segment drawn is the hypotenuse. The vertex of the right angle is at the point (5, 0). One leg of the right triangle is a horizontal line segment, and the other leg is a vertical line segment. We can determine the length of each leg and then use the Pythagorean Theorem to calculate the length of the hypotenuse.

Find the length of each leg.

$$a = |5 - 1| = |4| = 4$$
$$b = |3 - 0| = |3| = 3$$

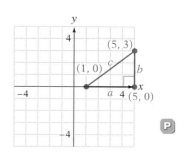

Use the Pythagorean Theorem.

$$c^2 = a^2 + b^2$$
$$c^2 = 4^2 + 3^2$$
$$c^2 = 16 + 9$$
$$c^2 = 25$$
$$c = \sqrt{25}$$
$$c = 5$$

The length of the line segment between the points (1, 0) and (5, 3) is 5 units.

We can apply this method of using the Pythagorean Theorem to find the distance between the two points (x_1, y_1) and (x_2, y_2).

The vertical distance between P_1 and Q is $|y_1 - y_2|$. The horizontal distance between P_2 and Q is $|x_1 - x_2|$.

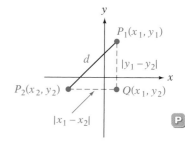

Apply the Pythagorean Theorem to the right triangle.

Because the square of a number cannot be negative, the absolute value signs are not necessary.

$$d^2 = |x_1 - x_2|^2 + |y_1 - y_2|^2$$

$$d^2 = (x_1 - x_2)^2 + (y_1 - y_2)^2$$

Use the Principal Square Root Property to solve the equation for d. This equation is known as the distance formula.

$$d = \sqrt{(x_1 - x_2)^2 + (y_1 - y_2)^2}$$

The Distance Formula

If (x_1, y_1) and (x_2, y_2) are two points in the plane, then the distance d between the two points is given by

$$d = \sqrt{(x_1 - x_2)^2 + (y_1 - y_2)^2}$$

INSTRUCTOR NOTE
Exercise 26 can be used for
a similar in-class example.

TAKE NOTE

If $(x_1, y_1) = (-3, 2)$,
then $x_1 = -3$ and
$y_1 = 2$. If
$(x_2, y_2) = (4, -1)$, then
$x_2 = 4$ and $y_2 = -1$.

Suggested Activity

Have the students solve Example 4 using $(x_1, y_1) = (4, -1)$
and $(x_2, y_2) = (-3, 2)$ to show
that the answer is the same.

EXAMPLE 4

Find the distance between the points $(-3, 2)$ and $(4, -1)$. Round to the
nearest tenth.

Solution

$$d = \sqrt{(x_1 - x_2)^2 + (y_1 - y_2)^2}$$ • Use the distance formula.

$$d = \sqrt{(-3 - 4)^2 + [2 - (-1)]^2}$$ • Let $(x_1, y_1) = (-3, 2)$ and $(x_2, y_2) = (4, -1)$.

$$d = \sqrt{(-7)^2 + (3)^2}$$

$$d = \sqrt{49 + 9}$$

$$d = \sqrt{58}$$

$$d \approx 7.6$$

The distance between the points is approximately 7.6 units.

YOU TRY IT 4

Find the distance between the points $(5, -2)$ and $(-4, 3)$. Round to the
nearest tenth.

Solution   See page S34.   10.3 units

8.3 EXERCISES   Suggested Assignment: 7–35, odds

Topics for Discussion

1. Label the right triangle shown at the right. Include the right angle symbol, the three angles, and the three sides.
   The right angle symbol must be at the 90° angle. The right angle must be labeled C
   and the <u>hypotenuse</u> labeled c. One acute angle should be labeled A with the side
   opposite it labeled a. The other acute angle should be labeled B with the side opposite it labeled b.

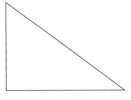

2. What does the Pythagorean Theorem state?
   The square of the hypotenuse of a right triangle is equal to the sum of the squares
   of the two legs.

3. Can the Pythagorean Theorem be used to find the length of side c of the
   triangle at the right? If so, find c. If not, explain why the theorem cannot
   be used.   No. The triangle is not a right triangle.

4. What does the Principal Square Root Property state?
   If $r^2 = s$, then $r = \sqrt{s}$, and r is called the square root of s.

5. What does the distance formula state?
   If (x_1, y_1) and (x_2, y_2) are two points in the plane, then the distance between the two
   points is given by $d = \sqrt{(x_1 - x_2)^2 + (y_1 - y_2)^2}$.

■ The Pythagorean Theorem

Determine whether the triangle is a right triangle.

6.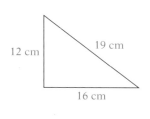

12 cm   19 cm

16 cm

No

7.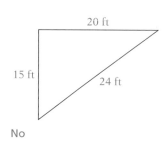

20 ft

15 ft   24 ft

No

8.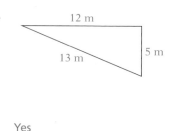

12 m

13 m   5 m

Yes

9. *Geometry* The two legs of a right triangle measure 5 centimeters and 9 centimeters. Find the length of the hypotenuse. Round to the nearest tenth.   10.3 cm

5 cm

9 cm

10. *Geometry* The two legs of a right triangle measure 8 inches and 4 inches. Find the length of the hypotenuse. Round to the nearest tenth.   8.9 in.

11. *Geometry* The hypotenuse of a right triangle measures 12 feet. One leg of the triangle measures 7 feet. Find the length of the other leg of the triangle. Round to the nearest hundredth.   9.75 ft

12 ft   7 ft

12. *Geometry* The hypotenuse of a right triangle measures 20 centimeters. One leg of the triangle measures 16 centimeters. Find the length of the other leg of the triangle.   12 cm

13. *Geometry* The diagonal of a rectangle is a line drawn from one vertex to the opposite vertex. Find the length of the diagonal in the rectangle shown at the right. Round to the nearest tenth.   13.9 mi

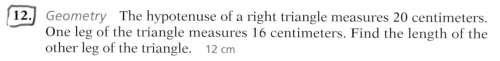

13 mi

5 mi

14. *Sports* The infield of a baseball diamond is a square. The distance between successive bases is 90 feet. The pitcher's mound is on the diagonal between home plate and second base at a distance of 60.5 feet from home plate. Is the pitcher's mound more or less than halfway between home plate and second base?   Less than halfway

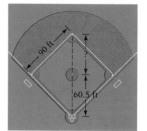

90 ft

60.5 ft

15. *Sports* The infield of a softball diamond is a square. The distance between successive bases is 60 feet. The pitcher's mound is on the diagonal between home plate and second base at a distance of 46 feet from home plate. Is the pitcher's mound more or less than halfway between home plate and second base?   More than halfway

16. *Parks* An L-shaped sidewalk from the parking lot to a memorial is shown in the figure at the right. The distance directly across the grass to the memorial is 650 feet. The distance to the corner is 600 feet. Find the distance from the corner to the memorial.   250 ft

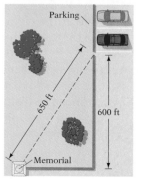

Parking

650 ft   600 ft

Memorial

17. *Travel* A commuter plane leaves an airport traveling due south at 400 mph. Another plane leaving at the same time travels due east at 300 mph. Find the distance between the two planes after 2 hours.   1000 mi

18. *Television* The measure of a television screen is given by the length of a diagonal across the screen. A 36-inch television has a width of 28.8 inches. Find the height of the screen. Round to the nearest tenth.   21.6 in.

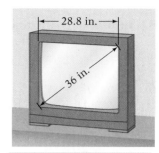

19. *Television* The measure of a television screen is given by the length of a diagonal across the screen. A 33-inch television has a width of 26.4 inches. Find the height of the screen. Round to the nearest tenth.   19.8 in.

20. *Engineering* A guy wire is to be attached to a telephone pole at a point 20 meters above the ground. The wire is anchored to the ground at a point 8 meters from the base of the telephone pole. How long a guy wire is required? Round to the nearest tenth.   21.5 m

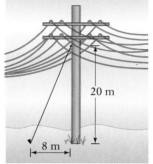

21. *Communications* Melissa leaves a dock in her sailboat and sails 4 miles due east. She then tacks and sails 2.5 miles due south. The walkie-talkie Melissa has on board has a range of 5 miles. Will she be able to call a friend on the dock from her location using the walkie-talkie?   Yes

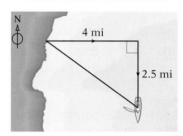

■ **The Distance Formula**

Find the distance between the two points. Give the exact value.

22. $(3, 5)$ and $(5, 1)$ $2\sqrt{5}$

23. $(-2, 3)$ and $(4, -1)$ $2\sqrt{13}$

24. $(0, 3)$ and $(-2, 4)$ $\sqrt{5}$

Find the distance between the two points. Round to the nearest tenth.

25. $(6, -1)$ and $(-3, -2)$ 9.1

26. $(-3, -5)$ and $(2, -4)$ 5.1

27. $(-7, -5)$ and $(-2, -1)$ 6.4

28. $(5, -2)$ and $(-2, 5)$ 9.9

29. $(3, -6)$ and $(6, 0)$ 6.7

30. $(-5, 5)$ and $(2, -5)$ 12.2

31. *Geometry* A triangle has vertices at $(3, -2)$, $(-3, -2)$, and $(-3, 2)$. Find **a.** the perimeter and **b.** the area of the triangle. Round to the nearest tenth.   **a.** 17.2 units   **b.** 12 square units

32. *Geometry* A triangle has vertices at $(1, 3)$, $(1, -3)$, and $(-6, -3)$. Find **a.** the perimeter and **b.** the area of the triangle. Round to the nearest tenth.   **a.** 22.2 units   **b.** 21 square units

33. *Geometry* A parallelogram has vertices at $(-4, 1)$, $(2, 1)$, $(4, -3)$, and $(-2, -3)$. Find **a.** the perimeter and **b.** the area of the parallelogram. Round to the nearest tenth.   **a.** 20.9 units   **b.** 24 square units

34. *Geometry* A parallelogram has vertices at $(4, 6)$, $(-1, 6)$, $(-3, -1)$, and $(2, -1)$. Find **a.** the perimeter and **b.** the area of the parallelogram. Round to the nearest tenth.   **a.** 24.6 units   **b.** 35 square units

35. *Geometry* A trapezoid has vertices at (4, 1), (−6, 1), (−2, −5), and (2, −5). Find **a.** the perimeter and **b.** the area of the trapezoid. Round to the nearest tenth.   **a.** 27.5 units   **b.** 42 square units

36. *Geometry* A trapezoid has vertices at (−2, 2), (6, 5), (6, −5), and (−2, −4). Find **a.** the perimeter and **b.** the area of the trapezoid. Round to the nearest tenth.   **a.** 32.6 units   **b.** 64 square units

Applying Concepts

37. *Geometry* Write an expression in factored form for the shaded region in the diagram at the right. $r^2(\pi - 2)$

38. *Geometry* The hypotenuse of a right triangle is $5\sqrt{2}$ centimeters, and one leg is $3\sqrt{2}$ centimeters. Find **a.** the perimeter and **b.** the area of the triangle.   **a.** $12\sqrt{2}$ cm   **b.** 12 cm²

39. Explain why, when using the distance formula, it makes no difference which point you designate as (x_1, y_1) and which you designate as (x_2, y_2).   Answers will vary.

40. *Geometry* A triangle has vertices at (4, 5), (−3, 9), and (1, 3). Is the triangle scalene, isosceles, or equilateral? Explain your reasoning.
Scalene. Explanations will vary.

41. *Geometry* A triangle has vertices at (4, 5), (−3, 9), and (1, 3). Is the triangle a right triangle? Explain your reasoning.   Yes. Explanations will vary.

42. For what positive value of y will the distance between the points (5, −2) and (−3, y) be 10 units?   4

43. Determine whether the statement is always true, sometimes true, or never true.
  a. $(x_1 - x_2) = |x_1 - x_2|$ Sometimes true
  b. $(x_1 - x_2) = (x_2 - x_1)$ Sometimes true
  c. $|x_1 - x_2| = |x_2 - x_1|$ Always true
  d. $|x_1 - x_2|^2 = (x_1 - x_2)^2$ Always true
  e. $(x_1 - x_2)^2 = (x_2 - x_1)^2$ Always true

EXPLORATION

1. *Pythagorean Triples* The Pythagorean Theorem states that if a and b are the legs of a right triangle and c is the hypotenuse, then $a^2 + b^2 = c^2$. For instance, the triangle with legs 3 and 4 and hypotenuse 5 is a right triangle because $3^2 + 4^2 = 5^2$. The numbers 3, 4, and 5 are called a Pythagorean triple because they are natural numbers that satisfy the equation of the Pythagorean Theorem.

  a. Determine whether the numbers are a Pythagorean triple.
    i. 5, 7, and 9   No   **ii.** 8, 15, and 17   Yes
    iii. 1, 60, and 61   No   **iv.** 64, 120, and 136   Yes

b. Determine whether multiples of Pythagorean triples are also Pythagorean triples. Do fractional multiples of Pythagorean triples satisfy the Pythagorean Theorem? Explain how you arrived at your conclusions.   Yes. Yes. Explanations will vary.

2. *Pythagorean Relationships*   Complete the statement using the symbol $<$, $=$, or $>$. Explain how you determined which symbol to use.
Explanations will vary.

    a. For an acute triangle with side c the longest side, $a^2 + b^2 \,\square\, c^2$. $>$
    b. For a right triangle with side c the longest side, $a^2 + b^2 \,\square\, c^2$. $=$
    c. For an obtuse triangle with side c the longest side, $a^2 + b^2 \,\square\, c^2$. $<$

SECTION **8.4**

- Addition and Subtraction of Radical Expressions
- Multiplication of Radical Expressions
- Division of Radical Expressions

Operations on Radical Expressions

■ Addition and Subtraction of Radical Expressions

Triangle ABC in the rectangular coordinate system at the right has vertices at $(0, 3)$, $(4, -1)$, and $(-3, 0)$. Find the perimeter of triangle ABC.

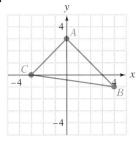

Use the distance formula to find the lengths of the three sides of the triangle.

$$AB = \sqrt{(0 - 4)^2 + [3 - (-1)]^2} = \sqrt{16 + 16} = \sqrt{32} = 4\sqrt{2}$$
$$BC = \sqrt{[4 - (-3)]^2 + (-1 - 0)^2} = \sqrt{49 + 1} = \sqrt{50} = 5\sqrt{2}$$
$$AC = \sqrt{[0 - (-3)]^2 + (3 - 0)^2} = \sqrt{9 + 9} = \sqrt{18} = 3\sqrt{2}$$

The perimeter of triangle ABC is the sum of the lengths of the three sides.

$$AB + BC + AC = 4\sqrt{2} + 5\sqrt{2} + 3\sqrt{2} = (4 + 5 + 3)\sqrt{2} = 12\sqrt{2}$$

The perimeter of triangle ABC is $12\sqrt{2}$ units.

Note from this example that the Distributive Property is used to simplify the sum of radical expressions with the same index and the same radicand. It is also used to simplify the difference of radical expressions with the same index and the same radicand.

INSTRUCTOR NOTE
Mention to students that adding and subtracting radicals is similar to combining like terms.

$$8\sqrt[3]{5x} + 7\sqrt[3]{5x} = (8 + 7)\sqrt[3]{5x} = 15\sqrt[3]{5x}$$
$$2\sqrt[4]{3y} - 9\sqrt[4]{3y} = (2 - 9)\sqrt[4]{3y} = -7\sqrt[4]{3y}$$

Radical expressions that are in simplest form and have unlike radicands or different indices cannot be simplified by the Distributive Property. The following expressions cannot be simplified by the Distributive Property.

$$3\sqrt[4]{2} - 6\sqrt[4]{3}$$ • The radicands are different.
$$2\sqrt[4]{3x} + 5\sqrt[3]{3x}$$ • The indices are different.

Suggested Activity

See the *Student Activity Manual*, Section 8.4, for an activity involving operations on radical expressions.

? **QUESTION** Which of the following expressions cannot be simplified?

a. $\sqrt[4]{8y} + \sqrt[5]{8y}$ **b.** $\sqrt[3]{a^2b} + \sqrt[3]{ab^2}$ **c.** $\sqrt{x+y} + \sqrt{x+y}$

To simplify $3\sqrt{32x^2} - 2x\sqrt{2} + \sqrt{128x^2}$, first simplify each term. Then combine like terms by using the Distributive Property.

$$3\sqrt{32x^2} - 2x\sqrt{2} + \sqrt{128x^2}$$
$$= 3\sqrt{16x^2 \cdot 2} - 2x\sqrt{2} + \sqrt{64x^2 \cdot 2}$$
$$= 3\sqrt{16x^2} \cdot \sqrt{2} - 2x\sqrt{2} + \sqrt{64x^2} \cdot \sqrt{2}$$
$$= 3 \cdot 4x\sqrt{2} - 2x\sqrt{2} + 8x\sqrt{2} = 12x\sqrt{2} - 2x\sqrt{2} + 8x\sqrt{2} = 18x\sqrt{2}$$

INSTRUCTOR NOTE
Exercise 10 can be used for a similar in-class example.

EXAMPLE 1

Subtract: $5b\sqrt[4]{32a^7b^5} - 2a\sqrt[4]{162a^3b^9}$

Solution

$$5b\sqrt[4]{32a^7b^5} - 2a\sqrt[4]{162a^3b^9} = 5b\sqrt[4]{16a^4b^4 \cdot 2a^3b} - 2a\sqrt[4]{81b^8 \cdot 2a^3b}$$
$$= 5b\sqrt[4]{16a^4b^4} \cdot \sqrt[4]{2a^3b} - 2a\sqrt[4]{81b^8} \cdot \sqrt[4]{2a^3b}$$
$$= 5b \cdot 2ab\sqrt[4]{2a^3b} - 2a \cdot 3b^2\sqrt[4]{2a^3b}$$
$$= 10ab^2\sqrt[4]{2a^3b} - 6ab^2\sqrt[4]{2a^3b}$$
$$= 4ab^2\sqrt[4]{2a^3b}$$

YOU TRY IT 1

Subtract: $3xy\sqrt[3]{81x^5y} - \sqrt[3]{192x^8y^4}$

Solution See page S34. $5x^2y\sqrt[3]{3x^2y}$

■ Multiplication of Radical Expressions

Rectangle *ABCD* in the rectangular coordinate system at the right has vertices at $(-1, 2)$, $(4, -3)$, $(2, -5)$, and $(-3, 0)$. To find the area of rectangle *ABCD*, use the distance formula to find the length and the width of the rectangle. We used *AB* for the length and *BC* for the width.

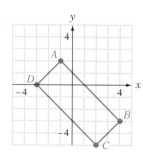

$$AB = \sqrt{(-1-4)^2 + [2-(-3)]^2} = \sqrt{25+25} = \sqrt{50}$$
$$BC = \sqrt{(4-2)^2 + [-3-(-5)]^2} = \sqrt{4+4} = \sqrt{8}$$

? ANSWER Parts a and b (In part a, the indices are different. In part b, the radicands are different.)

CALCULATOR NOTE

You can verify this answer by using a calculator to multiply the two radical expressions.

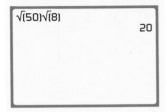

The area of rectangle $ABCD$ is the product of the length and the width.

$$A = LW = (\sqrt{50})(\sqrt{8}) = \sqrt{50 \cdot 8} = \sqrt{400} = 20$$

The area of rectangle $ABCD$ is 20 square units.

The product $(\sqrt{50})(\sqrt{8})$ was simplified by using the Product Property of Radicals.

? QUESTION The Product Property of Radicals was presented earlier. What is the Product Property of Radicals?

To multiply $\sqrt[3]{2a^5} \cdot \sqrt[3]{16a^2}$, use the Product Property of Radicals to multiply the radicands. Then simplify.

$$\sqrt[3]{2a^5} \cdot \sqrt[3]{16a^2} = \sqrt[3]{32a^7} = \sqrt[3]{8a^6 \cdot 4a} = \sqrt[3]{8a^6} \cdot \sqrt[3]{4a} = 2a^2 \sqrt[3]{4a}$$

INSTRUCTOR NOTE
Exercise 34 can be used for a similar in-class example.

EXAMPLE 2

Multiply: $\sqrt{3x}(\sqrt{27x^2} - \sqrt{3x})$

Solution

$$\sqrt{3x}(\sqrt{27x^2} - \sqrt{3x}) = \sqrt{81x^3} - \sqrt{9x^2}$$

• Use the Distributive Property and the Product Property of Radicals.

$$= \sqrt{81x^2 \cdot x} - \sqrt{9x^2}$$

• Simplify.

$$= \sqrt{81x^2} \cdot \sqrt{x} - \sqrt{9x^2}$$
$$= 9x\sqrt{x} - 3x$$

YOU TRY IT 2

Multiply: $\sqrt{5b}(\sqrt{3b} - \sqrt{10})$

Solution See page S34. $b\sqrt{15} - 5\sqrt{2b}$

INSTRUCTOR NOTE
Exercise 36 can be used for a similar in-class example.

Suggested Activity

Verify that $\sqrt{5} - \sqrt{3}$ is a square root of $8 - 2\sqrt{15}$.
[Answer: $(\sqrt{5} - \sqrt{3})^2 = 8 - 2\sqrt{15}$]

EXAMPLE 3

Multiply: $(2\sqrt[3]{x} - 3)(3\sqrt[3]{x} - 4)$

Solution

$$(2\sqrt[3]{x} - 3)(3\sqrt[3]{x} - 4) = 6\sqrt[3]{x^2} - 8\sqrt[3]{x} - 9\sqrt[3]{x} + 12$$

• Use the FOIL method.

$$= 6\sqrt[3]{x^2} - 17\sqrt[3]{x} + 12$$

• Combine like terms.

YOU TRY IT 3

Multiply: $(2\sqrt[3]{2x} - 3)(\sqrt[3]{2x} - 5)$

Solution See page S34. $2\sqrt[3]{4x^2} - 13\sqrt[3]{2x} + 15$

? ANSWER If $\sqrt[n]{a}$ and $\sqrt[n]{b}$ are positive real numbers, then $\sqrt[n]{ab} = \sqrt[n]{a} \cdot \sqrt[n]{b}$ and $\sqrt[n]{a} \cdot \sqrt[n]{b} = \sqrt[n]{ab}$.

TAKE NOTE

The concept of conjugate is used in a number of different ways. Make sure you understand this idea.
The conjugate of $\sqrt{3} - 4$ is $\sqrt{3} + 4$.
The conjugate of $\sqrt{5a} + \sqrt{b}$ is $\sqrt{5a} - \sqrt{b}$.

The expressions $a + b$ and $a - b$ are **conjugates** of each other. **The product of conjugates, $(a + b)(a - b)$, is $a^2 - b^2$.**

$$(a + b)(a - b) = a^2 - b^2$$

This equivalence is used to multiply conjugate radical expressions. For example,

$$(\sqrt{11} - 3)(\sqrt{11} + 3) = (\sqrt{11})^2 - 3^2 = 11 - 9 = 2$$

? QUESTION   What is the conjugate of $\sqrt{2y} + 7$?[3]

EXAMPLE 4

Multiply: $(2\sqrt{x} - \sqrt{2y})(2\sqrt{x} + \sqrt{2y})$

Solution $(2\sqrt{x} - \sqrt{2y})(2\sqrt{x} + \sqrt{2y}) = (2\sqrt{x})^2 - (\sqrt{2y})^2$
$= 4x - 2y$

YOU TRY IT 4

Multiply: $(\sqrt{a} - 3\sqrt{y})(\sqrt{a} + 3\sqrt{y})$

Solution   See page S34. $a - 9y$

INSTRUCTOR NOTE
Exercise 38 can be used for a similar in-class example.

■ Division of Radical Expressions

The Quotient Property of Radicals is used to divide radical expressions with the same index.

The Quotient Property of Radicals

If $\sqrt[n]{a}$ and $\sqrt[n]{b}$ are real numbers and $b \neq 0$, then $\sqrt[n]{\dfrac{a}{b}} = \dfrac{\sqrt[n]{a}}{\sqrt[n]{b}}$ and $\dfrac{\sqrt[n]{a}}{\sqrt[n]{b}} = \sqrt[n]{\dfrac{a}{b}}$.

This property is used in Example 5.

INSTRUCTOR NOTE
Exercise 44 can be used for a similar in-class example.

EXAMPLE 5

Simplify.   **a.** $\sqrt[3]{\dfrac{81x^5}{y^6}}$ **b.** $\dfrac{\sqrt{5a^4b^7c^2}}{\sqrt{ab^3c}}$

Solution

a. $\sqrt[3]{\dfrac{81x^5}{y^6}} = \dfrac{\sqrt[3]{81x^5}}{\sqrt[3]{y^6}}$

$= \dfrac{\sqrt[3]{27x^3 \cdot 3x^2}}{\sqrt[3]{y^6}} = \dfrac{\sqrt[3]{27x^3} \cdot \sqrt[3]{3x^2}}{\sqrt[3]{y^6}}$

$= \dfrac{3x\sqrt[3]{3x^2}}{y^2}$

- Use the Quotient Property of Radicals.

- Simplify each radical expression. 27 is the largest perfect-cube factor of 81. x^3 is the largest perfect-cube factor of x^5. y^6 is a perfect cube.

? ANSWER   The conjugate of $\sqrt{2y} + 7$ is $\sqrt{2y} - 7$.

b. $\dfrac{\sqrt{5a^4b^7c^2}}{\sqrt{ab^3c}} = \sqrt{\dfrac{5a^4b^7c^2}{ab^3c}}$ • Use the Quotient Property of Radicals.

$= \sqrt{5a^3b^4c} = \sqrt{a^2b^4 \cdot 5ac}$ • Simplify the radicand.

$= \sqrt{a^2b^4} \cdot \sqrt{5ac} = ab^2\sqrt{5ac}$

YOU TRY IT 5

Simplify. **a.** $\sqrt{\dfrac{48p^7}{q^4}}$ **b.** $\dfrac{\sqrt[3]{54y^8z^4}}{\sqrt[3]{2y^5z}}$

Solution See page S34. a. $\dfrac{4p^3\sqrt{3p}}{q^2}$ b. $3yz$

A radical expression is in simplest form when there is no fraction as part of the radicand and no radical remains in the denominator of the radical expression. The procedure used to remove a radical from the denominator is called **rationalizing the denominator.**

To simplify $\dfrac{5}{\sqrt{2}}$, multiply the expression by $\dfrac{\sqrt{2}}{\sqrt{2}}$, which equals 1. Then simplify.

$$\dfrac{5}{\sqrt{2}} = \dfrac{5}{\sqrt{2}} \cdot 1 = \dfrac{5}{\sqrt{2}} \cdot \dfrac{\sqrt{2}}{\sqrt{2}} = \dfrac{5\sqrt{2}}{(\sqrt{2})^2} = \dfrac{5\sqrt{2}}{2}$$

To simplify $\dfrac{3x}{\sqrt[3]{4}}$, multiply the expression by $\dfrac{\sqrt[3]{2}}{\sqrt[3]{2}}$, which equals 1. Then simplify. (*Note:* We chose the cube root of 2 because we must multiply 4 by a number that will result in a product that is a perfect cube. $4 \cdot 2 = 8$, a perfect cube.)

$$\dfrac{3x}{\sqrt[3]{4}} = \dfrac{3x}{\sqrt[3]{4}} \cdot 1 = \dfrac{3x}{\sqrt[3]{4}} \cdot \dfrac{\sqrt[3]{2}}{\sqrt[3]{2}} = \dfrac{3x\sqrt[3]{2}}{\sqrt[3]{8}} = \dfrac{3x\sqrt[3]{2}}{2}$$

EXAMPLE 6

Simplify. **a.** $\dfrac{5}{\sqrt{5x}}$ **b.** $\dfrac{3}{\sqrt[4]{2x}}$

Solution **a.** $\dfrac{5}{\sqrt{5x}} = \dfrac{5}{\sqrt{5x}} \cdot 1 = \dfrac{5}{\sqrt{5x}} \cdot \dfrac{\sqrt{5x}}{\sqrt{5x}} = \dfrac{5\sqrt{5x}}{(\sqrt{5x})^2} = \dfrac{5\sqrt{5x}}{5x} = \dfrac{\sqrt{5x}}{x}$

b. $\dfrac{3}{\sqrt[4]{2x}} = \dfrac{3}{\sqrt[4]{2x}} \cdot 1 = \dfrac{3}{\sqrt[4]{2x}} \cdot \dfrac{\sqrt[4]{8x^3}}{\sqrt[4]{8x^3}} = \dfrac{3\sqrt[4]{8x^3}}{\sqrt[4]{16x^4}} = \dfrac{3\sqrt[4]{8x^3}}{2x}$

YOU TRY IT 6

Simplify. **a.** $\dfrac{b}{\sqrt{3b}}$ **b.** $\dfrac{3}{\sqrt[3]{3y^2}}$

Solution See page S34. a. $\dfrac{\sqrt{3b}}{3}$ b. $\dfrac{\sqrt[3]{9y}}{y}$

INSTRUCTOR NOTE
Assure students that the value of the expression has not changed by having them evaluate both $\dfrac{5}{\sqrt{2}}$ and $\dfrac{5\sqrt{2}}{2}$ with their calculators.

TAKE NOTE

Multiplying $\dfrac{3x}{\sqrt[3]{4}}$ by $\dfrac{\sqrt[3]{4}}{\sqrt[3]{4}}$ will not rationalize the denominator:
$\dfrac{3x}{\sqrt[3]{4}} \cdot \dfrac{\sqrt[3]{4}}{\sqrt[3]{4}} = \dfrac{3x\sqrt[3]{4}}{\sqrt[3]{16}}$, and 16 is not a perfect cube.

INSTRUCTOR NOTE
Exercises 48 and 52 can be used for similar in-class examples.

To simplify a fraction that has a radical expression with two terms in the denominator, multiply the numerator and denominator by the conjugate of the denominator. Then simplify. For example,

$$\frac{\sqrt{x} - \sqrt{y}}{\sqrt{x} + \sqrt{y}} = \frac{\sqrt{x} - \sqrt{y}}{\sqrt{x} + \sqrt{y}} \cdot \frac{\sqrt{x} - \sqrt{y}}{\sqrt{x} - \sqrt{y}}$$

$$= \frac{(\sqrt{x})^2 - \sqrt{xy} - \sqrt{xy} + (\sqrt{y})^2}{(\sqrt{x})^2 - (\sqrt{y})^2} = \frac{x - 2\sqrt{xy} + y}{x - y}$$

> **TAKE NOTE**
>
> This is an example of using a conjugate to simplify a radical expression. The FOIL method is used to multiply the conjugates.

Suggested Activity

Have students explain why the product of two conjugates produces an expression without a radical.

INSTRUCTOR NOTE
Exercises 58 and 60 can be used for similar in-class examples.

EXAMPLE 7

Simplify.　**a.** $\dfrac{3}{5 - 2\sqrt{3}}$　**b.** $\dfrac{2 - \sqrt{2}}{3 + \sqrt{2}}$

Solution　**a.** $\dfrac{3}{5 - 2\sqrt{3}} = \dfrac{3}{5 - 2\sqrt{3}} \cdot \dfrac{5 + 2\sqrt{3}}{5 + 2\sqrt{3}} = \dfrac{15 + 6\sqrt{3}}{5^2 - (2\sqrt{3})^2}$

$$= \frac{15 + 6\sqrt{3}}{25 - 12} = \frac{15 + 6\sqrt{3}}{13}$$

b. $\dfrac{2 - \sqrt{2}}{3 + \sqrt{2}} = \dfrac{2 - \sqrt{2}}{3 + \sqrt{2}} \cdot \dfrac{3 - \sqrt{2}}{3 - \sqrt{2}} = \dfrac{6 - 2\sqrt{2} - 3\sqrt{2} + 2}{9 - 2}$

$$= \frac{8 - 5\sqrt{2}}{7}$$

YOU TRY IT 7

Simplify.　**a.** $\dfrac{6}{5 - \sqrt{7}}$　**b.** $\dfrac{3 + \sqrt{6}}{2 - \sqrt{6}}$

Solution　See page S34.　**a.** $\dfrac{5 + \sqrt{7}}{3}$　**b.** $-\dfrac{12 + 5\sqrt{6}}{2}$

8.4 EXERCISES　Suggested Assignment: 7–67, odds

Topics for Discussion

1. Why must radical expressions have the same index and the same radicand before they can be added or subtracted?　Answers will vary.

2. Why is it not necessary for two radical expressions that are to be multiplied to have the same radicand?　Answers will vary.

3. Must two radical expressions have the same index if they are to be multiplied or divided? If so, explain why. If not, give an example.

    No. Examples will vary. For example, $\dfrac{\sqrt{x}}{\sqrt[3]{x}} = \dfrac{x^{1/2}}{x^{1/3}} = x^{1/6} = \sqrt[6]{x}$.

4. Explain how to write the conjugate of $\sqrt{a} + \sqrt{b}$.

    Write the same expression except change the plus sign to a minus sign.

5. Explain what it means to rationalize the denominator of a radical expression and how to do so.　Explanations will vary.

■ Operations on Radical Expressions

Simplify.

6. $\sqrt{128x} - \sqrt{98x}$ $\sqrt{2x}$

7. $\sqrt{48x} + \sqrt{147x}$ $11\sqrt{3x}$

8. $2\sqrt{2x^3} + 4x\sqrt{8x}$ $10x\sqrt{2x}$

9. $5y\sqrt{8y} + 2\sqrt{50y^3}$ $20y\sqrt{2y}$

10. $x\sqrt{75xy} - \sqrt{27x^3y}$ $2x\sqrt{3xy}$

11. $3\sqrt{8x^2y^3} - 2x\sqrt{32y^3}$ $-2xy\sqrt{2y}$

12. $7b\sqrt{a^5b^3} - 2ab\sqrt{a^3b^3}$ $5a^2b^2\sqrt{ab}$

13. $2a\sqrt{27ab^5} + 3b\sqrt{3a^3b}$ $6ab^2\sqrt{3ab} + 3ab\sqrt{3ab}$

14. $\sqrt[3]{128} + \sqrt[3]{250}$ $9\sqrt[3]{2}$

15. $\sqrt[3]{16} - \sqrt[3]{54}$ $-\sqrt[3]{2}$

16. $2\sqrt[3]{3a^4} - 3a\sqrt[3]{81a}$ $-7a\sqrt[3]{3a}$

17. $2b\sqrt[4]{16b^2} + \sqrt[4]{128b^5}$ $8b\sqrt[4]{2b^2}$

18. $3\sqrt[3]{x^5y^7} - 8xy\sqrt[3]{x^2y^4}$ $-5xy^2\sqrt[3]{x^2y}$

19. $3\sqrt[4]{32a^5} - a\sqrt[4]{162a}$ $3a\sqrt[4]{2a}$

20. $2\sqrt{50} - 3\sqrt{125} + \sqrt{98}$ $17\sqrt{2} - 15\sqrt{5}$

21. $3\sqrt{108} - 2\sqrt{18} - 3\sqrt{48}$ $6\sqrt{3} - 6\sqrt{2}$

22. $5a\sqrt{3a^3b} + 2a^2\sqrt{27ab} - 4\sqrt{75a^5b}$ $-9a^2\sqrt{3ab}$

23. $\sqrt[3]{54xy^3} - 5\sqrt[3]{2xy^3} + \sqrt[3]{128xy^3}$ $2y\sqrt[3]{2x}$

24. $\sqrt{2x^3y} \cdot \sqrt{32xy}$ $8x^2y$

25. $\sqrt{5x^3y} \cdot \sqrt{10x^3y^4}$ $5x^3y^2\sqrt{2y}$

26. $\sqrt[3]{x^2y} \cdot \sqrt[3]{16x^4y^2}$ $2x^2y\sqrt[3]{2}$

27. $\sqrt[3]{4a^2b^3} \cdot \sqrt[3]{8ab^5}$ $2ab^2\sqrt[3]{4b^2}$

28. $\sqrt[4]{12ab^3} \cdot \sqrt[4]{4a^5b^2}$ $2ab\sqrt[4]{3a^2b}$

29. $\sqrt[4]{36a^2b^4} \cdot \sqrt[4]{12a^5b^3}$ $2ab\sqrt[4]{27a^3b^3}$

30. $2\sqrt{14xy} \cdot 4\sqrt{7x^2y} \cdot 3\sqrt{8xy^2}$ $672x^2y^2$

31. $\sqrt[3]{8ab} \cdot \sqrt[3]{4a^2b^3} \cdot \sqrt[3]{9ab^4}$ $2ab^2\sqrt[3]{36ab^2}$

32. $\sqrt{3}(\sqrt{27} - \sqrt{3})$ 6

33. $\sqrt{10}(\sqrt{10} - \sqrt{5})$ $10 - 5\sqrt{2}$

34. $\sqrt{2x}(\sqrt{8x} - \sqrt{32})$ $4x - 8\sqrt{x}$

35. $\sqrt{3a}(\sqrt{27a^2} - \sqrt{a})$ $9a\sqrt{a} - a\sqrt{3}$

36. $(\sqrt{2} - 3)(\sqrt{2} + 4)$ $-10 + \sqrt{2}$

37. $(\sqrt{5} - 5)(2\sqrt{5} + 2)$ $-8\sqrt{5}$

38. $(\sqrt{2x} - 3\sqrt{y})(\sqrt{2x} + 3\sqrt{y})$ $2x - 9y$

39. $(2\sqrt{3x} - \sqrt{y})(2\sqrt{3x} + \sqrt{y})$ $12x - y$

40. $(\sqrt[3]{a} + 2)(\sqrt[3]{a} + 3)$ $\sqrt[3]{a^2} + 5\sqrt[3]{a} + 6$

41. $(\sqrt[3]{x} - 4)(\sqrt[3]{x} + 5)$ $\sqrt[3]{x^2} + \sqrt[3]{x} - 20$

42. $\dfrac{\sqrt{32x^2}}{\sqrt{2x}}$ $4\sqrt{x}$

43. $\dfrac{\sqrt{60y^4}}{\sqrt{12y}}$ $y\sqrt{5y}$

44. $\dfrac{\sqrt{42a^3b^5}}{\sqrt{14a^2b}}$ $b^2\sqrt{3a}$

45. $\sqrt[3]{\dfrac{49m^5}{n^{12}}}$ $\dfrac{m\sqrt[3]{49m^2}}{n^4}$

46. $\sqrt{\dfrac{32y^8}{z^6}}$ $\dfrac{4y^4\sqrt{2}}{z^3}$

47. $\dfrac{5}{\sqrt{5x}}$ $\dfrac{\sqrt{5x}}{x}$

48. $\dfrac{9}{\sqrt{3a}}$ $\dfrac{3\sqrt{3a}}{a}$

49. $\sqrt{\dfrac{x}{5}}$ $\dfrac{\sqrt{5x}}{5}$

50. $\sqrt{\dfrac{y}{2}}$ $\dfrac{\sqrt{2y}}{2}$

51. $\dfrac{3}{\sqrt[3]{4x^2}}$ $\dfrac{3\sqrt[3]{2x}}{2x}$

52. $\dfrac{5}{\sqrt[3]{3y}}$ $\dfrac{5\sqrt[3]{9y^2}}{3y}$

53. $\dfrac{\sqrt{40x^3y^2}}{\sqrt{80x^2y^3}}$ $\dfrac{\sqrt{2xy}}{2y}$

54. $\dfrac{\sqrt{15a^2b^5}}{\sqrt{30a^5b^3}}$ $\dfrac{b\sqrt{2a}}{2a^2}$

55. $\dfrac{3}{\sqrt[4]{8x^3}}$ $\dfrac{3\sqrt[4]{2x}}{2x}$

56. $\dfrac{a}{\sqrt[5]{81a^4}}$ $\dfrac{\sqrt[5]{3a}}{3}$

57. $\dfrac{2}{\sqrt{5} + 2}$ $2\sqrt{5} - 4$

58. $\dfrac{5}{2 - \sqrt{7}}$ $-\dfrac{10 + 5\sqrt{7}}{3}$

59. $\dfrac{\sqrt{2} - \sqrt{3}}{\sqrt{2} + \sqrt{3}}$ $-5 + 2\sqrt{6}$

60. $\dfrac{\sqrt{2} + \sqrt{3}}{\sqrt{3} - \sqrt{2}}$ $5 + 2\sqrt{6}$

61. $\dfrac{3 - \sqrt{x}}{3 + \sqrt{x}}$ $\dfrac{9 - 6\sqrt{x} + x}{9 - x}$

62. $\dfrac{\sqrt{a} + 5}{\sqrt{a} - 5}$ $\dfrac{a + 10\sqrt{a} + 25}{a - 25}$

63. *Geometry*   Find **a.** the perimeter and **b.** the area of the rectangle with vertices at $(-4, 3)$, $(2, 5)$, $(4, -1)$, and $(-2, -3)$.
a. $8\sqrt{10}$ units   **b.** 40 square units

64. *Geometry*   Find **a.** the perimeter and **b.** the area of the rectangle with vertices at $(0, 4)$, $(3, 2)$, $(-1, -4)$, and $(-4, -2)$.
a. $6\sqrt{13}$ units   **b.** 26 square units

65. *Geometry*   Find **a.** the perimeter and **b.** the area of the triangle with vertices at $(-3, 6)$, $(5, 2)$, and $(1, -6)$.
a. $8\sqrt{5} + 4\sqrt{10}$ units   **b.** 40 square units

66. *Geometry*   A triangle has vertices at $(0, 1)$, $(3, -2)$, and $(-4, -3)$. Find **a.** the perimeter and **b.** the area of the triangle.
a. $12\sqrt{2}$ units   **b.** 12 square units

67. *Geometry*   Two vertices of a rectangle are at $(-5, 0)$ and $(0, -5)$. The length of the line segment joining these two vertices represents the length of the rectangle. The area of the rectangle is 40 square units. **a.** Find the width of the rectangle. **b.** Name the other two vertices of the rectangle given that one vertex lies in quadrant II and one lies in quadrant IV.   **a.** $4\sqrt{2}$ units   **b.** $(-1, 4)$ and $(4, -1)$

Applying Concepts

Simplify.

68. $(\sqrt{8} - \sqrt{2})^2$ 2

69. $(\sqrt{27} - \sqrt{3})^2$ 12

70. $(\sqrt{2} - 3)^3$ $-45 + 29\sqrt{2}$

71. $(\sqrt{5} + 2)^3$ $38 + 17\sqrt{5}$

72. $(\sqrt[3]{a} + \sqrt[3]{b})(\sqrt[3]{a^2} - \sqrt[3]{ab} + \sqrt[3]{b^2})$ $a + b$

73. $\dfrac{3}{\sqrt{y + 1} + 1}$ $\dfrac{3\sqrt{y + 1} - 3}{y}$

74. $\dfrac{3}{\sqrt{x + 4} + 2}$ $\dfrac{3\sqrt{x + 4} - 6}{x}$

75. $\dfrac{\sqrt{b + 9} - 3}{\sqrt{b + 9} + 3}$ $\dfrac{b + 18 - 6\sqrt{b + 9}}{b}$

EXPLORATION

1. *Comparing Radical Expressions with Polynomial Expressions*

a. Write a paragraph that compares adding two monomials to adding two radical expressions. For example, compare the addition of $7y + 9y$ to the addition of $7\sqrt{y} + 9\sqrt{y}$.

b. Write a paragraph that compares simplifying a variable expression such as $7x + 9y$ to simplifying a radical expression such as $7\sqrt{x} + 9\sqrt{y}$.

c. Write a paragraph that compares multiplying two monomials to multiplying two radical expressions. For example, compare the multiplication $(5x)(8x)$ to $(5\sqrt{x})(8\sqrt{x})$.
Answers will vary.

Radical Functions

■ Graphs of Radical
  Functions

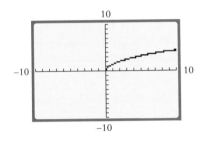

■ Graphs of Radical Functions

Earlier in this chapter, we used the formula $R = 1.4\sqrt{h}$, where R is the distance, in miles, that a person looking through a submarine periscope can see, and h is the height, in feet, of the periscope above the surface of the water. The graph of the equation $R = 1.4\sqrt{h}$ is shown at the left, graphed on a graphing calculator and using the standard viewing window.

As stated previously, for no point on the graph of $R = 1.4\sqrt{h}$ is the h value less than 0 because the square root of a negative number is not a real number; the domain of $R = 1.4\sqrt{h}$ is $\{h \mid h \geq 0\}$. For no point on the graph is the R value less than 0; the range of the function is $\{R \mid R \geq 0\}$. This is reasonable in the context of the application: It is not possible for the lookout to see a negative distance.

Note that the graph of this equation passes the vertical-line test for a function: Any vertical line intersects the graph no more than once. Therefore, $R = 1.4\sqrt{h}$ represents a function. We can emphasize this by writing the equation in functional notation as

$$f(h) = 1.4\sqrt{h}$$

$f(h) = 1.4\sqrt{h}$ is an example of a radical function. A **radical function** is a function that contains a variable under a radical sign or contains a variable raised to a fractional exponent. Further examples of radical functions are

$$g(x) = 4\sqrt[3]{x^2} - 6$$
$$h(x) = 3x - 2x^{1/2} + 5$$

? QUESTION

INSTRUCTOR NOTE
Writing exponential expressions as radical expressions and radical expressions as exponential expressions was a topic in the first section of this chapter.

  a. How is the function $g(x) = 4\sqrt[3]{x^2} - 6$ rewritten with a fractional exponent rather than with a radical expression?

  b. How is the function $h(x) = 3x - 2x^{1/2} + 5$ rewritten with a radical expression rather than with a fractional exponent?

The domain of a radical function is the set of real numbers for which the radical expression is a real number. For example, -8 is one number that would be excluded from the domain of $f(x) = \sqrt{x + 6}$ because

$$f(-8) = \sqrt{-8 + 6} = \sqrt{-2}$$

which is not a real number. We can determine the domain of $f(x) = \sqrt{x + 6}$ algebraically. The value of the expression $\sqrt{x + 6}$ is a real number when $x + 6$ is greater than or equal to zero:

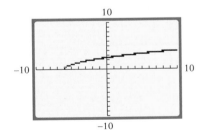

$$x + 6 \geq 0$$
$$x \geq -6 \qquad \text{• Subtract 6 from each side of the inequality.}$$

The domain of $f(x) = \sqrt{x + 6}$ is $\{x \mid x \geq -6\}$. This is confirmed by the graph of the function, shown at the left. Note that no value of x is less than -6.

—————

? ANSWERS   **a.** $g(x) = 4x^{2/3} - 6$ **b.** $h(x) = 3x - 2\sqrt{x} + 5$

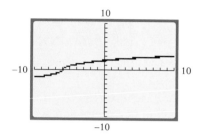

As shown above, -8 is not an element in the domain of $f(x) = \sqrt{x + 6}$ because $\sqrt{x + 6}$ is not a real number when the radicand $x + 6$ is negative.

Now consider $F(x) = \sqrt[3]{x + 6}$. Because the cube root of a negative number is a real number, the radicand $x + 6$ can be negative. For example, -14 is in the domain of x because

$$F(-14) = \sqrt[3]{-14 + 6} = \sqrt[3]{-8} = -2$$

which is a real number. The expression $\sqrt[3]{x + 6}$ is a real number for all values of x. Therefore, the domain of F is $\{x \mid x \in \text{real numbers}\}$. This is confirmed by the graph of the function, shown above at the left. There are no values of x for which $\sqrt[3]{x + 6}$ is not a real number.

TAKE NOTE

If the index of a radical expression is 2, 4, 6, 8, . . ., the radicand must be greater than or equal to zero. If the index is 3, 5, 7, 9, . . ., the radicand is a real number for any value of the variable.

These last two examples suggest the following about the domain of a radical function:

- **If a radical expression contains an even root, the radicand must be greater than or equal to zero to ensure that the value of the expression will be a real number.**
- **If a radical expression contains an odd root, the radicand may be a positive or a negative number.**

EXAMPLE 1

State the domain of each function in interval notation. Confirm your answer by graphing the function on a graphing calculator.

a. $g(x) = \sqrt[4]{8 - 2x}$ **b.** $h(x) = \sqrt[5]{4x + 3}$

INSTRUCTOR NOTE
Exercises 4 and 12 can be used for similar in class examples.

Solution

a. $g(x) = \sqrt[4]{8 - 2x}$

- g contains an even root. The radicand must be greater than or equal to zero.

$$8 - 2x \geq 0$$
$$-2x \geq -8$$
$$x \leq 4$$

The domain is $(-\infty, 4]$.

b. $h(x) = \sqrt[5]{4x + 3}$

- h contains an odd root. The radicand can be positive or negative. x can be any real number.

The domain is $(-\infty, \infty)$.

GRAPHICAL CHECK

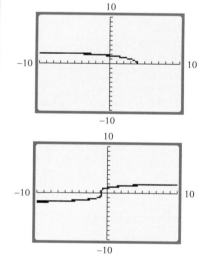

YOU TRY IT 1

State the domain of each function in interval notation. Confirm your answer by graphing the function on a graphing calculator.

a. $f(x) = 2\sqrt[3]{6x}$ **b.** $F(x) = (5x - 10)^{1/2}$

Solution See page S34.   **a.** $(-\infty, \infty)$ **b.** $[2, \infty)$

INSTRUCTOR NOTE
Exercise 22 can be used for
a similar in-class example.

Suggested Activity

Have your students do the
exercises in the Exploration at
the end of this section. It involves translations of graphs
of radical functions.

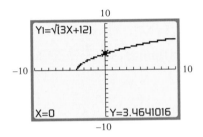

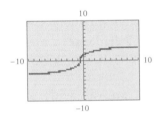

See Appendix A:
Trace

EXAMPLE 2

a. Graph $f(x) = \sqrt{3x + 12}$.
b. State the domain and range of the function in set-builder notation.
c. To the nearest tenth, find $f(0)$.

Solution

a.

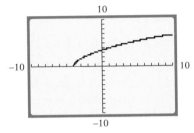

b. For no point on the graph is the x value less than -4.
The domain is $\{x \mid x \geq -4\}$.
For no point on the graph is the y value less than 0.
The range of the function is $\{y \mid y \geq 0\}$.

c. $f(0) \approx 3.5$ • Use the TRACE feature to find the y
value of the function when $x = 0$.

YOU TRY IT 2

a. Graph $g(x) = \sqrt[3]{5x + 2}$.
b. State the domain and range of the function in set-builder notation.
c. To the nearest tenth, find $f(0)$.

Solution   See page S35. **a.** See the graph at the left. **b.** D: $\{x \mid x \in \text{real numbers}\}$;
R: $\{y \mid y \in \text{real numbers}\}$ **c.** 1.3

INSTRUCTOR NOTE
Exercise 32 can be used for
a similar in-class example.

Suggested Activity

See the *Student Activity
Manual*, Section 8.5, for an
extended activity involving
pendulums.

EXAMPLE 3

The period of a pendulum is the time, T, it takes the pendulum to complete one swing from left to right and then back again. For a pendulum
near the surface of Earth, $T = 2\pi \sqrt{\dfrac{L}{32}}$, where T is measured in seconds
and L is the length of the pendulum in feet.

a. Find the period of a pendulum that has a length of 2 feet. Round to the
nearest tenth.
b. Find the length of a pendulum that has a period of 4 seconds. Round
to the nearest tenth.
c. What would you consider to be a reasonable domain for this function?
What would you consider to be a reasonable range?

Solution   The graph of the function $T = 2\pi \sqrt{\dfrac{L}{32}}$ is shown on the next
page.

See Appendix A:
Trace or Table

▼ *Point of Interest*

Here is an excerpt from Edgar Allan Poe's The Pit and the Pendulum.

"The vibration of the pendulum was at right angles to my length. I saw that the crescent was designed to cross the region of the heart. It would fray the serge of my robe; it would return and repeat its operations—again—and again. Notwithstanding its terrifically wide sweep (some thirty feet or more) and the hissing vigour of its descent, sufficient to sunder these very walls of iron, still the fraying of my robe would be all that, for several minutes, it would accomplish."

Suggested Activity

Have student calculate the period of the pendulum described in the Point of Interest above. Use a length of 30 feet.
[Answer: 6.08 seconds, to the nearest hundredth]

a. Use the TRACE feature or the TABLE feature to find T when $L = 2$. When $L = 2$, $T \approx 1.6$.
The period of a pendulum that has a length of 2 feet is 1.6 seconds.

b. Use the TRACE feature or the TABLE feature to find L when $T = 4$. When $T = 4$, $L \approx 13.0$.
The length of a pendulum that has a period of 4 seconds is 13.0 feet.

c. The domain of the function $T = 2\pi \sqrt{\dfrac{L}{32}}$ is all real numbers for which $\dfrac{L}{32}$ is greater than or equal to zero. Solving this for L, we get $L \geq 0$. But it would not be reasonable to have a pendulum of 0 feet, so $L > 0$. Many museums exhibit pendulums that are very large. However, we can probably assume an upper limit of about 50 feet.
A reasonable domain is (0, 50].
When $L = 0$, $T = 0$, so the range is greater than 0. When $L = 50$, $T \approx 7.9$.
Given a domain of (0, 50], the range is (0, 7.9].

Check your work.

a. To check part a, substitute 2 for L in the equation and solve for T.
b. To check part b, substitute 13 for L in the equation and solve for T.
c. To check the reasonableness of part c, check the internet or a resource book to find the lengths of actual pendulums.

Graph: $Y_1 = 2\pi\sqrt{(X/32)}$, X=2, Y=1.5707963

YOU TRY IT 3

The speed of a rider on a merry-go-round is given by the formula $v = \sqrt{12r}$, where v is the speed, in feet per second, of a rider on a merry-go-round and r is the distance in feet from the center of the merry-go-round to the rider.

a. Find the speed of a rider who is sitting 6 feet from the center of a merry-go-round. Round to the nearest tenth.
b. The speed of a rider on a merry-go-round is 10 feet per second. Find the distance between the rider and the center of the merry-go-round. Round to the nearest tenth.
c. What would you consider to be a reasonable domain for this function? What would you consider to be a reasonable range?

Solution See page S35. **a.** 8.5 ft/s **b.** 8.3 ft **c.** Answers will vary.

8.5 EXERCISES Suggested Assignment: 5–37, odds

Topics for Discussion

1. Which of the following are radical functions? Why?
   a. $f(x) = \sqrt{x} + 7$ **b.** $g(x) = \sqrt{2x - 5}$
   c. $h(x) = x + \sqrt{6}$ **d.** $F(x) = \sqrt{4x}$

   Parts a and b are radical functions because the radicand contains a variable expression. No variable appears within a radical in part c or d.

2. **a.** Explain why 8 is not in the domain of $f(x) = \sqrt{3 - x}$.
    b. Explain why 8 is in the domain of $f(x) = \sqrt[3]{3 - x}$.
    a. The expression $\sqrt{3 - x}$ is not a real number when $x = 8$.
    b. The expression $\sqrt[3]{3 - x}$ is a real number when $x = 8$.

3. Explain how to use an algebraic method to find the domain of
$f(x) = \sqrt{4x + 16}$.
Write the inequality $4x + 16 \geq 0$. Then solve for x.

■ **Graphs of Radical Functions**

State the domain of each function in set-builder notation. Confirm your answer by graphing the function on a graphing calculator.

4. $f(x) = 2x^{1/3}$ $\{x|x \in \text{real numbers}\}$

5. $g(x) = -3\sqrt[5]{2x}$ $\{x|x \in \text{real numbers}\}$

6. $h(x) = -2\sqrt{x + 1}$ $\{x|x \geq -1\}$

7. $r(x) = 3x^{1/4} - 2$ $\{x|x \geq 0\}$

8. $F(x) = 2x\sqrt{x} - 3$ $\{x| x \geq 0\}$

9. $G(x) = -3\sqrt[3]{5 + x}$ $\{x|x \in \text{real numbers}\}$

10. $C(x) = 6\sqrt[5]{x^2} + 7$ $\{x| x \in \text{real numbers}\}$

11. $H(x) = -3x^{3/4} + 1$ $\{x|x \geq 0\}$

For Exercises 12 to 19, state the domain of each function in interval notation. Confirm your answer by graphing the function on a graphing calculator.

12. $f(x) = -2(4x - 12)^{1/2}$ $[3, \infty)$

13. $g(x) = 2(2x - 10)^{2/3}$ $(-\infty, \infty)$

14. $h(x) = 4 - (3x - 3)^{2/3}$ $(-\infty, \infty)$

15. $F(x) = x - \sqrt{12 - 4x}$ $(-\infty, 3]$

16. $G(x) = -6 + \sqrt{6 - x}$ $(-\infty, 6]$

17. $f(x) = 3\sqrt[4]{(x - 2)^3}$ $[2, \infty)$

18. $H(x) = \frac{2}{3}\sqrt[4]{(4 - x)^3}$ $(-\infty, 4]$

19. $V(x) = x - (4 - 6x)^{1/2}$ $\left(-\infty, \frac{2}{3}\right]$

20. **a.** Graph $f(x) = -\sqrt[3]{x}$.
    b. State the domain and range in set-builder notation.
    c. To the nearest tenth, find $f(4)$.
    a.

21. **a.** Graph $f(x) = \sqrt[3]{x} + 1$.
    b. State the domain and range in set-builder notation.
    c. To the nearest tenth, find $f(-4)$.
    a.

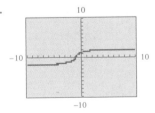

    b. Domain: $\{x|x \in \text{real numbers}\}$
       Range: $\{y|y \in \text{real numbers}\}$
    c. $f(4) \approx -1.6$

    b. Domain: $\{x|x \in \text{real numbers}\}$
       Range: $\{y|y \in \text{real numbers}\}$
    c. $f(-4) \approx -1.4$

22. a. Graph $f(x) = -\sqrt[4]{x}$.
   b. State the domain and range in set-builder notation.
   c. To the nearest tenth, find $f(5)$.

   a.

   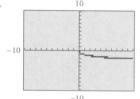

   b. Domain: $\{x \mid x \geq 0\}$;
   Range: $\{y \mid y \leq 0\}$
   c. $f(5) \approx -1.5$

23. a. Graph $f(x) = (x + 2)^{1/4}$.
   b. State the domain and range in set-builder notation.
   c. To the nearest tenth, find $f(6)$.

   a.

   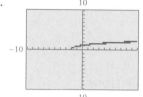

   b. Domain: $\{x \mid x \geq -2\}$;
   Range: $\{y \mid y \geq 0\}$
   c. $f(6) \approx 1.7$

24. a. Graph $f(x) = (x - 3)^{1/3}$.
   b. State the domain and range in interval notation.
   c. To the nearest tenth, find $f(-7)$.

   a.

   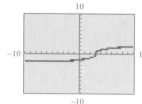

   b. Domain: $(-\infty, \infty)$;
   Range: $(-\infty, \infty)$
   c. $f(-7) \approx -2.2$

25. a. Graph $f(x) = \sqrt[3]{-x}$.
   b. State the domain and range in interval notation.
   c. To the nearest tenth, find $f(-9)$.

   a.

   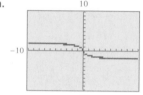

   b. Domain: $(-\infty, \infty)$;
   Range: $(-\infty, \infty)$
   c. $f(-9) \approx 2.1$

26. a. Graph $f(x) = 2x^{2/5} - 1$.
   b. State the domain and range in interval notation.
   c. To the nearest tenth, find $f(-7)$.

   a.

   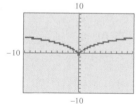

   b. Domain: $(-\infty, \infty)$;
   Range: $[-1, \infty)$
   c. $f(-7) \approx 3.4$

27. a. Graph $f(x) = 3\sqrt[5]{x^2} + 2$.
   b. State the domain and range in interval notation.
   c. To the nearest tenth, find $f(-5)$.

   a.

   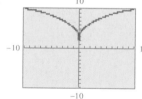

   b. Domain: $(-\infty, \infty)$;
   Range: $[2, \infty)$
   c. $f(-5) \approx 7.7$

Graph.

28. $f(x) = 3 - (5 - 2x)^{1/2}$

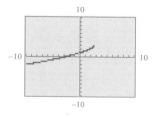

29. $g(x) = x\sqrt{3x - 9}$

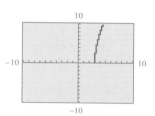

30. $h(x) = 1 + \sqrt{4 - 8x}$

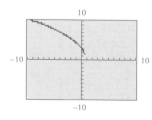

31. $F(x) = 3x\sqrt{4x + 8}$

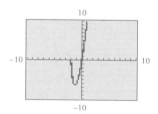

32. *Automotive Technology*   Under certain conditions, the length *L*, in feet, of skid marks left on dry concrete by a vehicle traveling *r* miles per hour is given by the equation $r = \sqrt{24L}$. Round answers to the nearest tenth.

   a. Find the speed of a vehicle that left skid marks 20 feet long.

   b. Find the length of the skid marks left by a vehicle traveling 25 mph.

   c. What would you consider to be a reasonable domain for this function? What would you consider to be a reasonable range?

   a. 21.9 mph   **b.** 26.0 ft   **c.** Answers will vary.

33. *Automotive Technology*   Under certain conditions, the length *L*, in feet, of skid marks left on wet concrete by a vehicle traveling *r* miles per hour is given by the equation $r = \sqrt{12L}$. Round answers to the nearest tenth.

   a. Find the speed of a vehicle that left skid marks 20 feet long.

   b. Find the length of the skid marks left by a vehicle traveling 25 mph.

   c. What would you consider to be a reasonable domain for this function? What would you consider to be a reasonable range?

   a. 15.5 mph   **b.** 52.1 ft   **c.** Answers will vary.

34. *Automotive Technology*   Compare the answers to Exercises 32 and 33. What conclusions can you draw regarding the relationship between speed and the length of a skid in dry vs. wet concrete?

Answers will vary.

35. Match each graph with its equation. In each graph shown, Xscl = 1 and Yscl = 1. (It might be helpful to do Exercise 1 in the Exploration at the end of this section before attempting this exercise.)

I = E, II = D, III = A, IV = F, V = G, VI = H, VII = C, VIII = B

I.

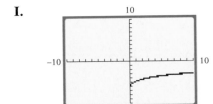

II.

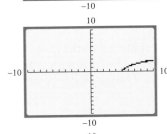

A. $f(x) = \sqrt{x} + 3$

B. $f(x) = \dfrac{1}{2}\sqrt{x}$

III.

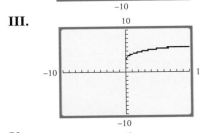

IV.

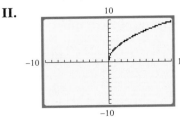

C. $f(x) = 4\sqrt{-x}$

D. $f(x) = 3\sqrt{x}$

V.

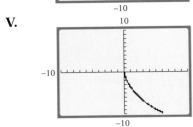

VI.

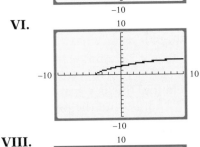

E. $f(x) = \sqrt{x} - 6$

F. $f(x) = \sqrt{x - 5}$

VII.

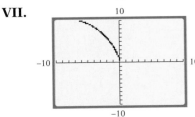

VIII.

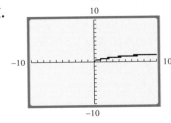

G. $f(x) = -4\sqrt{x}$

H. $f(x) = \sqrt{x + 4}$

36. *Geometry* The radius r of a sphere of volume V is given by the equation $r = \sqrt[3]{\dfrac{3V}{4\pi}}$. Find the radius of a sphere that has a volume of 8 cubic centimeters. Round to the nearest tenth.   1.2 cm

37. *Geometry* A container 16 centimeters high is in the shape of a right circular cone with the vertex at the bottom. A valve at the vertex can be opened to allow the container to be emptied. The time T, in seconds, it takes to empty the container is given by $T = 0.04[1024 - (16 - h)^{5/2}]$, where h is the height, in centimeters, of the water in the container.

  a. How long will it take to empty the container when $h = 12$ centimeters? Round to the nearest tenth.
  b. What is the domain of this function?   a. 39.7 s   b. $\{h \mid 0 \le h \le 16\}$

38. *Sound* The speed of sound in different air temperatures is calculated using $v = \dfrac{1087\sqrt{t + 273}}{16.52}$, where v is the speed in feet per second and t is the temperature in degrees Celsius.

  a. What must the temperature be in order for sound to travel at a speed of 1250 feet per second? Round to the nearest whole number.
  b. What might be a reasonable domain for this function? What might be a reasonable range?   a. 88°C   b. Answers will vary.

Applying Concepts

Use a graphing calculator to find the zeros of each function.

39. $f(x) = x - 3\sqrt{x} + 2$ 1, 4

40. $f(x) = 2x - 3\sqrt{x} + 1$ 0.25, 1

41. $f(x) = 3x - 5\sqrt{x} + 6$ There are no zeros.

42. $f(x) = 4x - 6\sqrt{x} + 5$ There are no zeros.

43. $f(x) = \sqrt[3]{x^2} + 2\sqrt[3]{x} - 8$ -64, 8

44. $f(x) = \sqrt[3]{x^2} - \sqrt[3]{x} - 2$ -1, 8

45. *Sports* Many new major league baseball parks have a symmetrical design as shown in the figure at the right. One question that the designer must decide is the shape of the outfield. One possible design uses the function

$$f(x) = k + (400 - k)\sqrt{1 - \dfrac{x^2}{a^2}}$$

to determine the shape of the outfield.

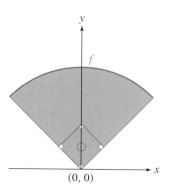

  a. Graph this equation for $k = 0$, $a = 287$, and $-240 \le x \le 240$.
  b. What is the maximum value of this function for the given interval?
  c. The equation of the right-field foul line is $y = x$. Where does the foul line intersect the graph of f? That is, find the point on the graph of f for which $y = x$.
  d. If the units on the axes are feet, what is the distance from home plate to the base of the right-field wall?

  a.   **b.** 400   **c.** Approximately (233, 233)   **d.** 330 ft

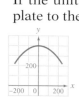

EXPLORATION

1.   *Translations of Graphs of Radical Functions*

   a. Graph each of the following functions. Describe how each differs from the position of the graph of $f(x) = \sqrt{x}$.

   $f(x) = \sqrt{x} + 4 \quad f(x) = \sqrt{x} - 5 \quad f(x) = \sqrt{x} + 1 \quad f(x) = \sqrt{x} - 2$

   Write a description of the graph of $f(x) = \sqrt{x} + c$, where c is a constant.

   b. Graph each of the following functions. Describe how each differs from the position of the graph of $f(x) = \sqrt{x}$.

   $f(x) = \sqrt{x+3} \quad f(x) = \sqrt{x-6} \quad f(x) = \sqrt{x+2} \quad f(x) = \sqrt{x-4}$

   Write a description of the graph of $f(x) = \sqrt{x+c}$, where c is a constant.

   c. Graph each of the following functions. Describe how each differs from the shape of the graph of $f(x) = \sqrt{x}$.

   $f(x) = 2\sqrt{x} \quad f(x) = \frac{1}{2}\sqrt{x} \quad f(x) = 4\sqrt{x} \quad f(x) = \frac{1}{3}\sqrt{x}$

   Write a description of the graph of $f(x) = c\sqrt{x}$, where c is a constant and $c > 0$.

   d. Graph each of the following functions. Describe how each differs in position and/or shape from the graph of $f(x) = \sqrt{x}$.

   $f(x) = -\sqrt{x} \quad f(x) = -4\sqrt{x} \quad f(x) = -3\sqrt{x} \quad f(x) = -\frac{1}{2}\sqrt{x}$

   Write a description of the graph of $f(x) = c\sqrt{x}$, where c is a constant and $c < 0$.

   e. Graph each of the following functions. Describe how each differs in position and/or shape from the graph of $f(x) = \sqrt{x}$.

   $f(x) = \sqrt{-x} \quad f(x) = 5\sqrt{-x} \quad f(x) = 2\sqrt{-x} \quad f(x) = \frac{1}{2}\sqrt{-x}$

   Write a description of the graph of $f(x) = c\sqrt{-x}$, where c is a constant and $c > 0$.

   f. Describe the graph of $f(x) = 4\sqrt{x+2}$.

   g. Describe the graph of $f(x) = \sqrt{x-5} - 6$.

   h. Describe the graph of $f(x) = -3\sqrt{x} + 1$.

   a. A vertical shift of the graph of $f(x) = \sqrt{x}$, c units upward for $c > 0$ or c units downward for $c < 0$ **b.** A horizontal shift of the graph of $f(x) = \sqrt{x}$, c units to the left for $c > 0$ or c units to the right for $c < 0$ **c.** The graph of $f(x) = \sqrt{x}$ scaled by the value of c **d.** A reflection of the graph of $f(x) = |c|\sqrt{x}$ over the x-axis   **e.** A reflection of the graph of $f(x) = c\sqrt{x}$ over the y-axis   **f.** The graph of $f(x) = \sqrt{x}$ shifted 2 units to the left and scaled by a factor of 4   **g.** The graph of $f(x) = \sqrt{x}$ shifted right 5 units and downward 6 units   **h.** The graph of $f(x) = \sqrt{x}$ scaled by a factor of 3, reflected over the x-axis, and shifted upward 1 unit

2. *Graphs of Radical Functions*   Graph all four of the radical functions given below on the same graphing calculator screen.

   $$y = \sqrt{\frac{1}{2}x} \qquad y = \sqrt{x} \qquad y = \sqrt{2x} \qquad y = \sqrt{3x}$$

   a. For $x = 4$, which function has the least y value?

   b. For $x = 3$, which function has the greatest y value?

c. As the coefficient of x increases, is the graph higher or lower in the rectangular coordinate grid?

d. For a given positive value of x, which function has the least y value?

e. For a given positive value of x, which function has the greatest y value?

f. Describe where you would expect the graph of $y = \sqrt{\frac{1}{4}x}$ to lie relative to the other graphs. Check your conjecture by graphing this function.

g. Describe where you would expect the graph of $y = \sqrt{4x}$ to lie relative to the other graphs. Check your conjecture by graphing this function.

h. Describe where the graph of $y = x^{1/2}$ would lie relative to the other graphs.

i. Describe where the graph of $y = (3x)^{1/2}$ would lie relative to the other graphs.

Now consider the four functions shown below.

$$y = \sqrt[4]{\frac{1}{2}x} \qquad y = \sqrt[4]{x} \qquad y = \sqrt[4]{2x} \qquad y = \sqrt[4]{3x}$$

j. For a given positive value of x, which function would you assume has the least y value? the greatest y value? Check your conjecture by graphing these functions.

k. Describe where you would expect the graph of $y = \sqrt[4]{6x}$ to lie relative to the other graphs. Check your conjecture by graphing this function.

l. Describe where the graph of $y = (2x)^{1/4}$ would lie relative to the other graphs.

a. $y = \sqrt{\frac{1}{2}x}$ b. $y = \sqrt{3x}$ c. Higher d. $y = \sqrt{\frac{1}{2}x}$ e. $y = \sqrt{3x}$ f. Below $y = \sqrt{\frac{1}{2}x}$ g. Above $y = \sqrt{3x}$ h. It is the same as the graph of $y = \sqrt{x}$. i. It is the same as the graph of $y = \sqrt{3x}$ j. $y = \sqrt[4]{\frac{1}{2}x}$; $y = \sqrt[4]{3x}$ k. Above $y = \sqrt[4]{3x}$ l. It is the same as the graph of $y = \sqrt[4]{2x}$.

SECTION **8.6** **Solving Radical Equations**

■ Solve Equations Containing Radical Expressions

■ Solve Equations Containing Radical Expressions

Earlier in this chapter, we used the formula $R = 1.4\sqrt{h}$ to determine the distance a person looking through a submarine periscope can see. Recall that in this formula, R is the distance, in miles, that a person can see and h is the height, in feet, of the periscope above the surface of the water. Given the height of the periscope, we found the distance the person could see.

Now suppose the lookout on a submarine wants to be able to see a ship 3.5 miles away. How far must the periscope be above the surface? (*Note:* In this problem, the distance we want to be able to see is given, and we are asked to find what height the periscope must be above the surface of the water.)

State the goal. We want to find the height of the periscope above the surface of the water when the lookout can see a distance of 3.5 miles.

Devise a strategy.   Replace the variable R in the formula $R = 1.4\sqrt{h}$ by 3.5. Then solve the equation for h.

Solve the problem. $R = 1.4\sqrt{h}$

$$3.5 = 1.4\sqrt{h}$$ • Replace R by 3.5.

$$\frac{3.5}{1.4} = \sqrt{h}$$ • Solve the equation for $\sqrt{h}$. Divide each side by 1.4.

$$\left(\frac{3.5}{1.4}\right)^2 = (\sqrt{h})^2$$ • We want to solve the equation for h. The square of $\sqrt{h}$ is h, so square each side of the equation.

$$6.25 = h$$

The periscope must be 6.25 feet above the surface of the water.

Check your work.   Earlier in this chapter, we found that when the periscope is 9 feet above the surface of the water, the lookout can see a distance of 4.2 miles. To see a shorter distance (3.5 miles), the periscope would need to be less than 9 feet above the surface, and $6.25 < 9$.

The graph of $R = 1.4\sqrt{h}$ is shown at the left, along with the graph of $R = 3.5$ (the distance the lookout wants to see). Note that the coordinates of the point of intersection of the two graphs confirm that a lookout can see 3.5 miles when the periscope is 6.25 feet above the surface of the water.

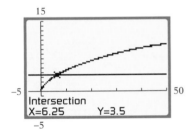

In solving the equation $\frac{3.5}{1.4} = \sqrt{h}$, we used the Property of Raising Both Sides of an Equation to a Power to square both sides of the equation.

Property of Raising Both Sides of an Equation to a Power

If two numbers are equal, then the same powers of the numbers are equal. If $a = b$, then $a^n = b^n$.

EXAMPLE 1

Solve.   **a.** $\sqrt[3]{2x - 1} = -3$ **b.** $\sqrt{3x - 2} - 4 = 3$

Solution

a. $\sqrt[3]{2x - 1} = -3$

$$(\sqrt[3]{2x - 1})^3 = (-3)^3$$ • Because $(\sqrt[3]{a})^3 = a$, cube each side of the equation.
• Solve the resulting equation.

$$2x - 1 = -27$$

$$2x = -26$$

$$x = -13$$

INSTRUCTOR NOTE
Exercises 14 and 16 can be used for similar in-class examples.

ALGEBRAIC CHECK

$$\sqrt[3]{2x - 1} = -3$$

$\sqrt[3]{2(-13) - 1}$	-3
$\sqrt[3]{-26 - 1}$	-3
$\sqrt[3]{-27}$	-3
$-3 = -3$	• True

The solution checks.

The solution is -13.

b. $\sqrt{3x - 2} - 4 = 3$
  • We want to rewrite the equation with the radical alone on one side of the equation.

$\sqrt{3x - 2} = 7$
  • Add 4 to each side of the equation.

$(\sqrt{3x - 2})^2 = 7^2$
  • Because $(\sqrt{a})^2 = a$, square each side of the equation.

$3x - 2 = 49$
  • Solve the resulting equation.

$3x = 51$

$x = 17$

ALGEBRAIC CHECK

$$\sqrt{3x - 2} - 4 = 3$$

$\sqrt{3(17) - 2} - 4$	3
$\sqrt{51 - 2} - 4$	3
$\sqrt{49} - 4$	3
$7 - 4$	3
$3 = 3$	• True

The solution checks.

The solution is 17.

GRAPHICAL CHECK

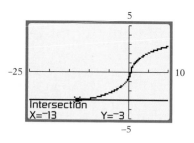

The solution of $\sqrt[3]{2x - 1} = -3$ is the x-coordinate of the intersection of the graphs of $Y_1 = \sqrt[3]{2x - 1}$ and $Y_2 = -3$.

GRAPHICAL CHECK

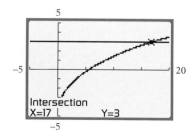

The solution of $\sqrt{3x - 2} - 4 = 3$ is the x-coordinate of the intersection of the graphs of $Y_1 = \sqrt{3x - 2} - 4$ and $Y_2 = 3$.

YOU TRY IT 1

Solve.   **a.** $\sqrt[4]{2x - 9} = 3$ **b.** $\sqrt{4x + 5} - 12 = -5$

Solution   See page S35.   a. 45   b. 11

Suggested Activity

See Section 8.6 of the *Student Activity Manual* for an activity in which technology is used to investigate radical equations.

INSTRUCTOR NOTE
Exercise 28 can be used for a similar in-class example.

The Property of Raising Both Sides of an Equation to a Power states that if $a = b$, then $a^n = b^n$. The converse of this property (If $a^n = b^n$, then $a = b$) is not true. For example, let $a = -4$ and $b = 4$. Then $(-4)^2 = (4)^2$, but $-4 \neq 4$. Because the converse of this property is not true, using this property may lead to extraneous solutions. Therefore, **whenever you raise both sides of an equation to an even power, it is necessary to check the solutions of the equation because the resulting equation may have a solution that is not a solution of the original equation.**

INSTRUCTOR NOTE
Ask students to explain why Step 3 in the solution of the equation in Example 2 indicates that there is no solution to the equation. [There is no solution of the equation $\sqrt{x - 1} = -3$ because the square root of a number is nonnegative.]
You might also ask what the solution of the equation in Example 2 would be if the equation were changed to $7 + 2\sqrt{x - 1} = 13$. [Because the algebraic check shows that when $x = 10$, $7 + 2\sqrt{x - 1} = 13$, the solution of $7 + 2\sqrt{x - 1} = 13$ is 10.]

EXAMPLE 2

Solve: $7 + 2\sqrt{x - 1} = 1$

Solution $7 + 2\sqrt{x - 1} = 1$ • We want to rewrite the equation with the radical expression alone on one side of the equation.

$2\sqrt{x - 1} = -6$ • Subtract 7 from each side of the equation.

$\sqrt{x - 1} = -3$ • Divide each side of the equation by 2.

$(\sqrt{x - 1})^2 = (-3)^2$ • Square each side of the equation.

$x - 1 = 9$

$x = 10$ • Solve for x.

ALGEBRAIC CHECK

$$7 + 2\sqrt{x - 1} = 1$$

$7 + 2\sqrt{10 - 1}$	1
$7 + 2\sqrt{9}$	1
$7 + 2(3)$	1
$7 + 6$	1
$13 \neq 1$	

The solution does not check.

There is no solution.

GRAPHICAL CHECK

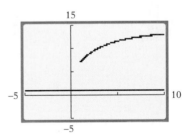

The graphs of $Y_1 = 7 + 2\sqrt{x - 1}$ and $Y_2 = 1$ do not intersect. There is no value of x for which $7 + 2\sqrt{x - 1} = 1$.

YOU TRY IT 2

Solve: $8 + 3\sqrt{x + 2} = 5$

Solution   See page S35.   No solution

Suggested Activity

Explain how solving the equation $3\sqrt{x-8} - 5 = 7$ is similar to solving the equation $3x - 5 = 7$.

In Examples 1 and 2, each equation contained only one radical. Example 3 below illustrates the procedure for solving a radical equation containing two radical expressions. Note that the process of squaring both sides of the equation is performed twice.

EXAMPLE 3

Solve: $\sqrt{x+7} - \sqrt{x} = 1$

INSTRUCTOR NOTE
Exercise 34 can be used for a similar in-class example.

Solution

$\sqrt{x+7} - \sqrt{x} = 1$ • We want to rewrite the equation with one of the radical expressions alone on one side of the equation.

$\sqrt{x+7} = \sqrt{x} + 1$ • Add $\sqrt{x}$ to each side of the equation.

$(\sqrt{x+7})^2 = (\sqrt{x} + 1)^2$ • Square each side of the equation.

$x + 7 = x + 2\sqrt{x} + 1$ • $(\sqrt{x} + 1)^2 = (\sqrt{x} + 1)(\sqrt{x} + 1) = x + \sqrt{x} + \sqrt{x} + 1$

$7 = 2\sqrt{x} + 1$ • Subtract x from each side of the equation.

$6 = 2\sqrt{x}$ • Subtract 1 from each side of the equation.

$3 = \sqrt{x}$ • Divide each side of the equation by 2.

$3^2 = (\sqrt{x})^2$ • Square each side of the equation.

$9 = x$

ALGEBRAIC CHECK

$\sqrt{x+7} - \sqrt{x} = 1$	
$\sqrt{9+7} - \sqrt{9}$	1
$\sqrt{16} - \sqrt{9}$	1
$4 - 3$	1
$1 = 1$	• True

The solution checks.

The solution is 9.

GRAPHICAL CHECK

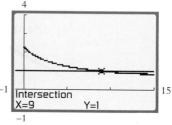

The solution of $\sqrt{x+7} - \sqrt{x} = 1$ is the x-coordinate of the intersection of the graphs of $Y_1 = \sqrt{x+7} - \sqrt{x}$ and $Y_2 = 1$.

? QUESTION   In Example 3, why is the first step in solving the equation not to square each side of the equation?

YOU TRY IT 3

Solve: $\sqrt{x+5} + \sqrt{x} = 5$

Solution   See page S35.   4

INSTRUCTOR NOTE
You might ask students this same question in regard to Example 1b and Example 2.

? ANSWER   Squaring the left side of the equation, $\sqrt{x+7} - \sqrt{x}$, would not eliminate either of the radical expressions.

INSTRUCTOR NOTE
Exercise 38 can be used for
a similar in-class example.

EXAMPLE 4

The perimeter of the rectangle shown
on the right is 32 meters. Find the
value of x.

$(\sqrt{x} - 7)$ m

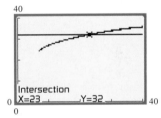

12 m

State the goal.   The goal is to determine the value of x in the expression
$\sqrt{x} - 7$.

Devise a strategy.   We are given the perimeter of the rectangle. Therefore, we need to use the formula for the perimeter of a rectangle to write
an equation. We can do this by substituting, in the formula, 32 for P (the
perimeter), 12 for L (the length), and $\sqrt{x} - 7$ for W (the width). We can
then solve the equation for x.

Solve the problem.

$P = 2L + 2W$
   • This is the formula for the perimeter of a rectangle.

$32 = 2(12) + 2(\sqrt{x} - 7)$
   • Substitute 32 for P, 12 for L, and $\sqrt{x} - 7$ for W.

$32 = 24 + 2(\sqrt{x} - 7)$
   • We want to get the radical expression alone on one side of the equation.

$8 = 2(\sqrt{x} - 7)$

$4 = \sqrt{x} - 7$

$4^2 = (\sqrt{x} - 7)^2$
   • Square each side of the equation.

$16 = x - 7$

$23 = x$

ALGEBRAIC CHECK

$$32 = 2(12) + 2(\sqrt{x} - 7)$$

32	$24 + 2(\sqrt{23} - 7)$
32	$24 + 2(\sqrt{16})$
32	$24 + 2(4)$
32	$24 + 8$

$32 = 32$

The value of x is 23.

GRAPHICAL CHECK

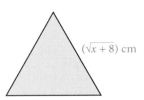

The solution of $32 = 24 + 2(\sqrt{x} - 7)$
is the x-coordinate of the intersection
of the graphs of $Y_1 = 24 + 2(\sqrt{x} - 7)$
and $Y_2 = 32$.

Check your work.   When $x = 23$, $\sqrt{x} - 7 = \sqrt{23} - 7 = \sqrt{16} = 4$.

We are given that the length of the rectangle is 12. If the width is 4, then
the perimeter is $12 + 4 + 12 + 4 = 32$. This is the perimeter we are given
in the problem statement. The solution checks.

YOU TRY IT 4

The perimeter of the equilateral triangle shown
at the right is 15 centimeters. Find the value of x.

$(\sqrt{x} + 8)$ cm

Solution   See page S35.   17

8.6 EXERCISES Suggested Assignment: 7–49, odds

Topics for Discussion

1. The graph of the equation $Y_1 = 1.4\sqrt{X}$ is shown above Example 1 in this section. In this equation, Y_1 is the distance, in miles, that a person can see and X is the height, in feet, of the periscope above the surface of the water. The point (20, 6.26) is on the graph. What is the meaning of this ordered pair in the context of the given formula?   When the periscope is 20 ft above the surface of the water, the lookout can see 6.26 mi.

2. What does the Property of Raising Both Sides of an Equation to a Power state?   If a and b are real numbers and $a = b$, then $a^n = b^n$.

3. When both sides of an equation are raised to an even power, why is it necessary to check the solutions?   The resulting equation may have a solution that is not a solution of the original equation.

4. Explain how to solve an equation containing two radical expressions. Answers will vary.

5. Suppose you solve the equation $\sqrt{2x + 1} + 3 = 6$ algebraically, and the result is $x = 4$. Describe two methods by which the solution can be checked.   Answers will vary. For example; Algebraically: Substitute 4 for x in the equation; evaluate the left side of the equation; if the result is 6, the solution checks. Graphically: On a graphing calculator, graph the equations $Y_1 = \sqrt{2x + 1} + 3$ and $Y_2 = 6$; find the point of intersection; if the point of intersection is (4, 6), the solution checks.

■ Solve Equations Containing Radical Expressions

Solve.

6. $\sqrt{3 - 2x} = 7$ −23

7. $\sqrt{9 - 4x} = 4$ $-\dfrac{7}{4}$

8. $\sqrt[3]{4x - 1} = 2$ $\dfrac{9}{4}$

9. $\sqrt[3]{1 - 2x} = -3$ 14

10. $\sqrt[4]{4x + 1} = 2$ $\dfrac{15}{4}$

11. $\sqrt[4]{2x - 9} = 3$ 45

12. $\sqrt{3x + 9} - 12 = 0$ 45

13. $\sqrt{4x - 3} - 5 = 0$ 7

14. $\sqrt{2x - 1} - 8 = -5$ 5

15. $\sqrt{7x + 2} - 10 = -7$ 1

16. $\sqrt[3]{2x - 3} + 5 = 2$ −12

17. $\sqrt[3]{x - 4} + 7 = 5$ −4

18. $\sqrt[3]{4x - 3} - 2 = 3$ 32

19. $\sqrt[3]{1 - 3x} + 5 = 3$ 3

20. $1 - \sqrt{4x + 3} = -5$ $\dfrac{33}{4}$

21. $7 - \sqrt{3x + 1} = -1$ 21

22. $\sqrt{x^2 + 3x - 2} - x = 1$ 3

23. $\sqrt{x^2 + 4x - 1} + 3 = 5$ −5, 1

24. $\sqrt[4]{2x + 8} - 2 = 0$ 4

25. $\sqrt[4]{x - 1} - 1 = 0$ 2

26. $4\sqrt{x + 1} - 5 = 11$ 15

27. $3\sqrt{x - 2} + 6 = 15$ 11

28. $\sqrt{2x - 3} + 5 = 1$ No solution

29. $\sqrt{9x + 1} + 6 = 2$ No solution

30. $\sqrt[4]{2x - 8} + 7 = 5$ No solution

31. $\sqrt[4]{3x + 4} + 5 = 3$ No solution

32. $\sqrt{3x + 4} = 7 - \sqrt{3x - 3}$ 4

33. $\sqrt{x + 1} = 2 - \sqrt{x}$ $\dfrac{9}{16}$

34. $\sqrt{2x + 4} + \sqrt{2x} = 3$ $\dfrac{25}{72}$

35. $\sqrt{4x + 1} - \sqrt{4x - 2} = 1$ $\dfrac{3}{4}$

36. *Visibility*   How high a hill must you climb in order to be able to see a distance of 45 kilometers? Use the formula $d = \sqrt{12h}$, where d is the distance in kilometers to the horizon from a point h meters above Earth's surface.   a 168.75-meter hill

37. *Oceanography*   A tsunami is a great sea wave produced by underwater earthquakes or volcanic eruption. Find the depth in feet of the water when the velocity of a tsunami reaches 20 feet per second. Use the formula $v = 3\sqrt{d}$, where v is the velocity in feet per second of a tsunami as it approaches land and d is the depth in feet of the water. Round to the nearest tenth.   44.4 ft

38. *Sports*   The equation $s = 16.97\sqrt[9]{n}$ can be used to predict the maximum speed s, in feet per second, of n rowers on a scull.

   a.   How many rowers are needed to travel at 20 feet per second? Round to the nearest whole number.

   b.   Does doubling the number of rowers double the maximum speed of the scull?

   a. 4 rowers   b. No

39. *Sports*   Find the distance required for a car to reach a velocity of 60 meters per second when the acceleration is 10 meters per second squared. Use the equation $v = \sqrt{2as}$, where v is the velocity, a is the acceleration, and s is the distance.   180 m

40. *Physics*   The time it takes for an object to fall a distance of d feet on the moon is given by the formula $t = \sqrt{\dfrac{d}{2.75}}$, where t is the time in seconds. If an astronaut drops an object on the moon, how far will it fall in 8 seconds?   176 ft

41. *Astronautics*   The weight of an object is related to its distance above the surface of Earth. A formula for this relationship is

$$d = 4000\sqrt{\frac{E}{S}} - 4000,$$

where E is the object's weight on the surface of Earth and S is the object's weight at a distance of d miles above Earth's surface. An astronaut weighs 24 pounds when she is 5000 miles above Earth's surface. How much does the astronaut weigh on Earth's surface?   121.5 lb

42. *Sound*   The speed of sound in different air temperatures is calculated using the formula $v = \dfrac{1087\sqrt{t + 273}}{16.52}$, where v is the speed in feet per second and t is the temperature in degrees Celsius. What must the temperature be in order for sound to travel at a speed of 1100 feet per second? Round to the nearest tenth.   6.5°C

43. *Geometry*   The perimeter of a rectangle that has a width of $(\sqrt{5x + 1})$ meters and a length of 14 meters is 36 meters. Find the value of x.   3

44. *Meteorology*   The sustained wind velocity v, in meters per second, in a hurricane is given by $v = 6.3 \sqrt{1013 - p}$, where p is the air pressure in millibars.

  a. If the velocity of the wind in a hurricane is 64 meters per second, what is the air pressure? Round to the nearest tenth.
  b. What happens to wind speed in a hurricane as air pressure decreases?

  a. 909.8 mb   b. Increases

45. *Astronomy*   The time T, in days, that it takes a planet to revolve around the sun can be approximated by the equation $T = 0.407 \sqrt{d^3}$, where d is the mean distance of the planet from the sun in millions of miles. It takes Venus approximately 226 days to complete one revolution of the sun. What is the mean distance of Venus from the sun? Round to the nearest million.   68 million miles

46. *Astronomy*   The time T, in days, that it takes a moon of Saturn to revolve around Saturn can be approximated by the equation $T = 0.373\sqrt{d^3}$, where d is the mean distance of the moon from Saturn in units of 100,000 kilometers. It takes the moon Tethys approximately 1.89 days to complete one revolution of Saturn. What is the mean distance of Tethys from Saturn? Round to the nearest thousand.   295,000 km

47. *Demographics*   In Section 8.2, data related to the number of married couples in the United States were given. We provided an equation that approximately models the data: $y = 8.1\sqrt[5]{x^2}$, where y is the number of married couples, in millions, in year x, and $x = 0$ represents 1900. Use the equation to predict the years in which there were **a.** 45 million married couples and **b.** 55 million married couples in the United States. **c.** Are these years reasonable when compared to the data in the table in Exercise 36 at the end of Section 8.2? Why or why not?

  a. 1973   b. 2020   c. Answers will vary.

48. *Crime*   In Section 8.2, we presented data on the total number of property crimes in the United States and an equation that approximately models the data: $y = 9 \sqrt[10]{x}$, where y is the number of property crimes, in millions, in year x, and $x = 5$ represents the year 1975. Use the equation to predict the years in which there were **a.** 10 million property crimes and **b.** 12 million property crimes in the United States. **c.** Are these years reasonable when compared to the data in the table in Exercise 37 at the end of Section 8.2? Why or why not?

  a. 1973   b. 1988   c. Answers will vary.

49. *Animal Science*   The number of calories an animal uses per day (called the metabolic rate of the animal) can be approximated by $M = 126.4 \sqrt[4]{W^3}$, where M is the metabolic rate and W is the weight, in pounds, of the animal. Find the weight of an elephant whose metabolic rate is 60,000 calories per day. Round to the nearest hundred.   3700 lb

Applying Concepts

Solve.

50. $x^{3/4} = 8$ 16

51. $x^{2/3} = 9$ 27

52. $x^{5/4} = 32$ 16

53. Find two positive numbers whose sum is 20 and whose arithmetic mean is 2 more than the geometric mean. (*Hint:* The geometric mean of two positive numbers p and q is $\sqrt{pq}$.)   4 and 16

54. When does $\sqrt[3]{a^3 + b^3} = a^3 + b^3$? (*Hint:* Cube both sides of the equation.)
The expressions are equal when $a = -b$ or $b = -a$.

55. Solve for x: $\sqrt{\dfrac{1}{9} + \dfrac{1}{3} + \dfrac{5}{9}} = \sqrt{\dfrac{1}{9} + \dfrac{1}{3} + \sqrt{\dfrac{x}{9}}}$ 1

56. In the figure at the right, the area of the small square is one-third of the total area of the large square. Calculate the ratio $y : x$. $\dfrac{1}{\sqrt{3} - 1}$

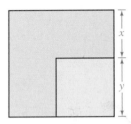

EXPLORATION

1. *Hydroplaning* Hydroplaning occurs when, rather than gripping the road's surface, a tire slides on the surface of water that is on pavement. The equation $v = 8.6\sqrt{p}$ gives the relationship between v, the minimum hydroplaning speed, in miles per hour, and p, the tire pressure in pounds per square inch.

    a. As the tire pressure increases, does the minimum hydroplaning speed increase or decrease? How did you determine this?
Increases. Explanations will vary.

    b. As the minimum hydroplaning speed increases, does the tire pressure increase or decrease? How did you determine this?
Increases. Explanations will vary.

    c. Is there more danger of hydroplaning when the tire pressure is low or when the tire pressure is high? How did you determine this?
Low. Explanations will vary.

    d. What implications does this formula have for drivers with respect to checking the tires on their vehicles?
Drivers should check the tire pressure. Low tire pressure increases the danger of hydroplaning.

SECTION **8.7** # Complex Numbers

- Simplify Complex Numbers
- Addition and Subtraction of Complex Numbers
- Multiplication of Complex Numbers
- Division of Complex Numbers

Simplify Complex Numbers

The radical expression $\sqrt{-4}$ is not a real number because there is no real number whose square is -4. However, the solution of an algebraic equation is sometimes the square root of a negative number.

For example, the equation $x^2 + 1 = 0$ does not have a real number solution because there is no real number whose square is a negative number.

$$x^2 + 1 = 0$$
$$x^2 = -1$$

INSTRUCTOR NOTE
Students will be solving quadratic equations with complex number solutions later in the text.

Around the 17th century, a new number, called an **imaginary number,** was defined so that a negative number would have a square root. The letter i was chosen to represent the number whose square is -1.

$$i^2 = -1$$

An imaginary number is defined in terms of i.

▼ *Point of Interest*

In the 17th century, René Descartes called square roots of negative numbers imaginary numbers, in contrast to the numbers everyone understood, which he called real numbers. In his book De Formulis Differentialibus Angularibus, he wrote, "In the following I shall denote the expression $\sqrt{-1}$ by the letter i so that $i \cdot i = -1$."

Note that Descartes wrote ii instead of i^2. He also wrote xx instead of x^2.

Principal Square Root of a Negative Number

If a is a positive real number, then the principal square root of negative a is the imaginary number $i\sqrt{a}$.

$$\sqrt{-a} = i\sqrt{a}$$

Here are some examples of imaginary numbers.

$$\sqrt{-25} = i\sqrt{25} = 5i \qquad \sqrt{-18} = i\sqrt{18} = 3i\sqrt{2}$$
$$\sqrt{-17} = i\sqrt{17} \qquad \sqrt{-1} = i\sqrt{1} = i$$

It is customary to write i in front of a radical sign to avoid confusing $\sqrt{a}\, i$ with $\sqrt{ai}$.

The real numbers and the imaginary numbers make up the complex numbers. A **complex number** is a number of the form $a + bi$, where a and b are real numbers and $i = \sqrt{-1}$. The number a is the **real part** of $a + bi$, and b is the **imaginary part.**

TAKE NOTE

The imaginary part of $5 + 4i$ is 4.
The imaginary part of $6 - 9i$ is -9.

Here are some examples of complex numbers.

Real part ⟍　　　 ⟋Imaginary part

$$5 + 4i$$
$$6 - 9i$$
$$-2 + 7i$$
$$-3 - 8i$$

? QUESTION   What is **a.** the real part and **b.** the imaginary part of $-1 + 10i$?

When a complex number is entered into a graphing calculator, the calculator will return the real part or the imaginary part of the complex number. Some typical screens are shown below.

See Appendix A:
Complex Numbers

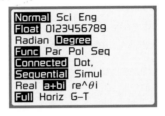

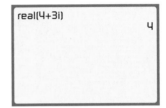

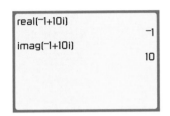

A graphing calculator can be used to verify the real and imaginary parts of $-1 + 10i$ in the Question above. The screen is shown at the left.

The following diagram illustrates that all real numbers are complex numbers and all imaginary numbers are complex numbers. No real number is an imaginary number; no imaginary number is a real number. For example:

2 is a real number and a complex number.

$3i$ is an imaginary number and a complex number.

$2 + 3i$ is a complex number.

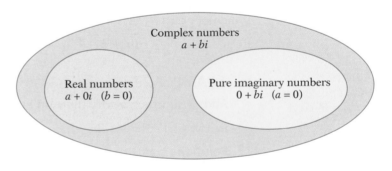

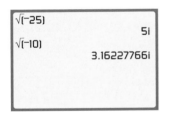

A graphing calculator can simplify complex numbers written in radical form. For example, with the graphing calculator in complex number mode, enter $\sqrt{-25}$ and press ENTER. The calculator will print **5i** to the screen.

If the absolute value of the radicand is not a perfect square, a graphing calculator will return a decimal approximation of the complex number. For example, enter $\sqrt{-10}$. The calculator will print **3.16227766i** to the screen.

See Appendix A:
Complex Numbers

? ANSWERS   **a.** -1 **b.** 10

Sometimes we want an exact value of a complex number rather than a decimal approximation. Example 1 illustrates simplifying complex numbers that are written as radical expressions.

INSTRUCTOR NOTE
Exercises 10 and 18 can be
used for similar in-class
examples.

EXAMPLE 1

Simplify.   **a.** $\sqrt{-50}$ **b.** $\sqrt{20} + \sqrt{-45}$

Solution

a. $\sqrt{-50} = i\sqrt{50} = i\sqrt{25 \cdot 2} = 5i\sqrt{2}$

b. $\sqrt{20} + \sqrt{-45} = \sqrt{20} + i\sqrt{45}$
   - Write the complex number in the form $a + bi$.

$$= \sqrt{4 \cdot 5} + i\sqrt{9 \cdot 5}$$
   - Use the Product Property of Radicals to simplify each radical.

$$= 2\sqrt{5} + 3i\sqrt{5}$$

YOU TRY IT 1

Simplify.   **a.** $\sqrt{-60}$ **b.** $\sqrt{40} - \sqrt{-80}$

Solution   See page S36.   **a.** $2i\sqrt{15}$ **b.** $2\sqrt{10} - 4i\sqrt{5}$

■ Addition and Subtraction of Complex Numbers

To add two complex numbers, add the real parts and add the imaginary parts.

$$(a + bi) + (c + di) = (a + c) + (b + d)i$$

To subtract two complex numbers, subtract the real parts and subtract the imaginary parts.

$$(a + bi) - (c + di) = (a - c) + (b - d)i$$

INSTRUCTOR NOTE
Exercises 22, 24, and 28 can
be used for similar in-class
examples.

EXAMPLE 2

Add or subtract. Verify the sum or difference using a graphing calculator.

a. $(6 - 3i) + (-4 + 2i)$ **b.** $(-8 + 5i) - (7 - i)$ **c.** $(9 + 3i) + (-9 - 3i)$

Solution

```
(6-3i)+(-4+2i)
                    2-i
(-8+5i)-(7-i)
                  -15+6i
(9+3i)+(-9-3i)
                    0
```

a. $(6 - 3i) + (-4 + 2i)$

$$= [6 + (-4)] + (-3 + 2)i$$
   - Add the real parts and add the imaginary parts.

$$= 2 - i$$

b. $(-8 + 5i) - (7 - i)$
   - Subtract the real parts and subtract the imaginary parts.

$$= (-8 - 7) + [5 - (-1)]i$$

$$= -15 + 6i$$

c. $(9 + 3i) + (-9 - 3i)$
   - Add the real parts and add the imaginary parts.

$$= 0 + 0i = 0$$

YOU TRY IT 2

Add or subtract. Verify the sum or difference using a graphing calculator.

a. $(-10 + 6i) + (9 - 4i)$ **b.** $(3 + i) - (8i)$ **c.** $(4 - 2i) + (-4 + 2i)$

Solution   See page S36.   **a.** $-1 + 2i$ **b.** $3 - 7i$ **c.** 0

■ Multiplication of Complex Numbers

When multiplying complex numbers, the term i^2 is frequently a part of the product. Recall that $i^2 = -1$. Note how this equivalence is used in Example 3 to multiply two imaginary numbers.

INSTRUCTOR NOTE
Exercises 34 and 40 can be used for similar in-class examples.

EXAMPLE 3

Multiply. Verify the product using a graphing calculator.

a. $5i \cdot 3i$ **b.** $-6i(4 + 3i)$

Solution

a. $5i \cdot 3i = 15i^2$ • Multiply the complex numbers.

$\qquad = 15(-1)$ • Replace i^2 by -1.

$\qquad = -15$ • Simplify.

b. $-6i(4 + 3i) = -24i - 18i^2$ • Use the Distributive Property.

$\qquad = -24i - 18(-1)$ • Replace i^2 by -1.

$\qquad = 18 - 24i$ • Simplify and write the complex number in the form $a + bi$.

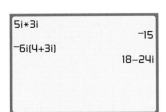

YOU TRY IT 3

Multiply. Verify the product using a graphing calculator.

a. $-7i \cdot 2i$ **b.** $5i(2 - 8i)$

Solution   See page S36.   a. 14   b. $40 + 10i$

▼ *Point of Interest*

The imaginary unit i *is important in the field of electricity. However, because the letter* i *is used to represent electrical current, engineers use the variable* j *for the imaginary unit.*

When multiplying square roots of negative numbers, first rewrite the radical expressions using i.

For example, to multiply $\sqrt{-6} \cdot \sqrt{-24}$, $\sqrt{-6} \cdot \sqrt{-24}$

write each radical as the product of a real number and i. $= i\sqrt{6} \cdot i\sqrt{24}$

Then multiply the imaginary numbers. $= i^2\sqrt{144}$

Replace i^2 with -1. $= -1\sqrt{144}$

Simplify $\sqrt{144}$. $= -12$

Note from this example that it would have been incorrect to multiply the radicands of the two radical expressions. To illustrate:

$$\sqrt{-6} \cdot \sqrt{-24} = \sqrt{(-6)(-24)} = \sqrt{144} = 12, \text{ } not \text{ } -12$$

❓ QUESTION   What is the product of $\sqrt{-2}$ and $\sqrt{-8}$?

❓ ANSWER $\sqrt{-2} \cdot \sqrt{-8} = i\sqrt{2} \cdot i\sqrt{8} = i^2\sqrt{16} = -1\sqrt{16} = -4$

The product of two complex numbers can be found by using the FOIL method. This is illustrated in Example 4.

INSTRUCTOR NOTE
Exercises 42 and 56 can be used for similar in-class examples.

EXAMPLE 4

Multiply. Verify the product using a graphing calculator.

a. $(2 + 4i)(3 - 5i)$ **b.** $(3 - i)\left(\dfrac{3}{10} + \dfrac{1}{10}i\right)$

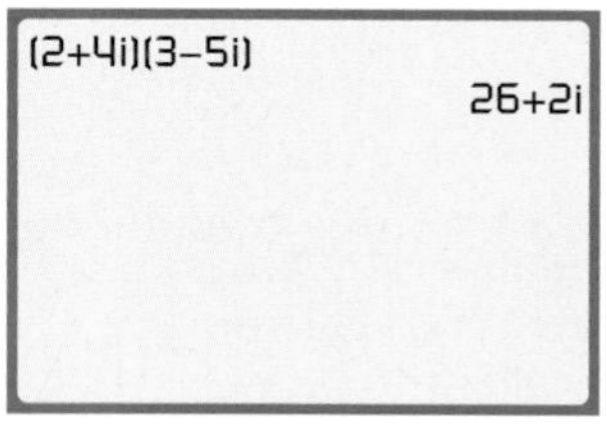

Solution   **a.** $(2 + 4i)(3 - 5i)$

$= 6 - 10i + 12i - 20i^2$ • Use the FOIL method.

$= 6 + 2i - 20i^2$ • Combine like terms.

$= 6 + 2i - 20(-1)$ • Replace i^2 by -1.

$= 6 + 2i + 20$ • Simplify and write the complex number in the form $a + bi$.

$= 26 + 2i$

b. $(3 - i)\left(\dfrac{3}{10} + \dfrac{1}{10}i\right)$

$= \dfrac{9}{10} + \dfrac{3}{10}i - \dfrac{3}{10}i - \dfrac{1}{10}i^2$ • Use the FOIL method.

$= \dfrac{9}{10} - \dfrac{1}{10}i^2$ • Combine like terms.

$= \dfrac{9}{10} - \dfrac{1}{10}(-1)$ • Replace i^2 by -1.

$= \dfrac{9}{10} + \dfrac{1}{10} = 1$ • Simplify.

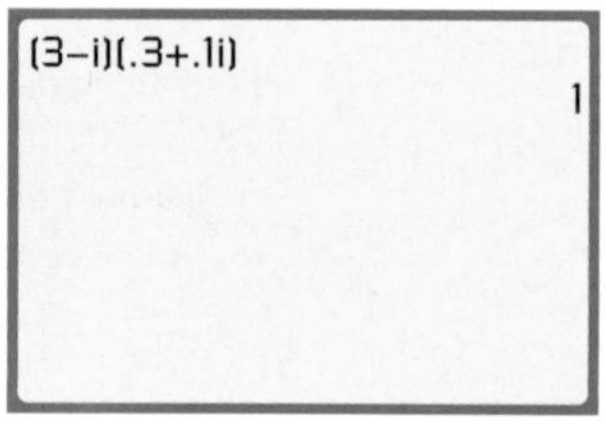

(2+4i)(3−5i)
        26+2i

(3−i)(.3+.1i)
        1

YOU TRY IT 4

Multiply. Verify the product using a graphing calculator.

a. $(3 - 4i)(2 + 5i)$ **b.** $\left(\dfrac{9}{10} + \dfrac{3}{10}i\right)\left(1 - \dfrac{1}{3}i\right)$

Solution   See page S36.   **a.** $26 + 7i$ **b.** 1

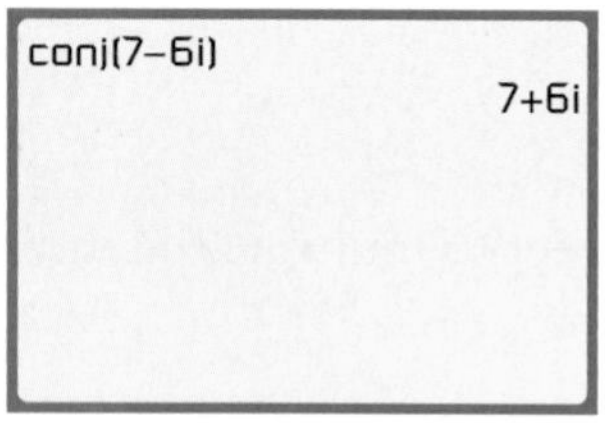

The conjugate of $a + bi$ is $a - bi$. For example, the conjugate of $8 - 10i$ is $8 + 10i$.

? QUESTION   What is the conjugate of $7 - 6i$?

The conjugate of $7 - 6i$ can be verified using a graphing calculator. A typical screen is shown at the left.

conj(7−6i)
        7+6i

Note the product when we multiply conjugates of the form $(a + bi)(a - bi)$.

$$(a + bi)(a - bi) = a^2 - b^2i^2 = a^2 - b^2(-1) = a^2 + b^2$$

See Appendix A:
Complex Numbers

? ANSWER $7 + 6i$

> **The Product of Conjugates of the Form $(a + bi)(a - bi)$**
> The product of conjugates of the form $(a + bi)(a - bi)$ is $a^2 + b^2$.

For example, $(2 + 3i)(2 - 3i) = 2^2 + 3^2 = 4 + 9 = 13$.

Note that **the product of a complex number and its conjugate is a real number.**

INSTRUCTOR NOTE
Exercise 50 can be used for
a similar in-class example.

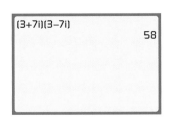

EXAMPLE 5

Multiply $(3 + 7i)(3 - 7i)$. Verify the product using a graphing calculator.

Solution $(3 + 7i)(3 - 7i) = 3^2 + 7^2$ • The product of conjugates of the
form $(a + bi)(a - bi)$ is $a^2 + b^2$.

$$= 9 + 49$$

$$= 58$$

YOU TRY IT 5

Multiply $(6 - 5i)(6 + 5i)$. Verify the product using a graphing calculator.

Solution   See page S36.   61

■ Division of Complex Numbers

A fraction containing one or more complex numbers is in simplest form when no imaginary number remains in the denominator. This is illustrated in Example 6.

INSTRUCTOR NOTE
Exercise 62 can be used for
a similar in-class example.

EXAMPLE 6

Simplify $\dfrac{2 - 3i}{2i}$. Verify the quotient using a graphing calculator.

Solution $\dfrac{2 - 3i}{2i} = \dfrac{2 - 3i}{2i} \cdot \dfrac{i}{i}$ • Multiply the expression by 1 in the
form $\frac{i}{i}$.

$$= \frac{2i - 3i^2}{2i^2}$$ • Multiply the numerators.
Multiply the denominators.

$$= \frac{2i - 3(-1)}{2(-1)}$$ • Replace i^2 by -1.

$$= \frac{3 + 2i}{-2}$$ • Simplify.

$$= -\frac{3}{2} - i$$ • Write the number in the form $a + bi$.

(2−3i)/(2i)
−1.5−i

YOU TRY IT 6

Simplify $\dfrac{4 + 5i}{3i}$. Verify the quotient using a graphing calculator.

Solution   See page S36. $\dfrac{5}{3} - \dfrac{4}{3}i$

To simplify a fraction that has a complex number in the denominator, multiply the numerator and denominator by the conjugate of the complex number in the denominator. This is illustrated in Example 7.

INSTRUCTOR NOTE
Exercise 64 can be used for a similar in-class example.

Suggested Activity

Have students find the reciprocal of $2 - 5i$, which is $\frac{2}{29} + \frac{5}{29}i$. Then ask them to show that the product of $2 - 5i$ and its reciprocal is 1.

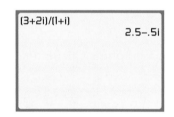

Suggested Activity

See Section 8.7 of the *Student Activity Manual* for an activity involving using technology to iterate functions.

EXAMPLE 7

Simplify $\dfrac{3 + 2i}{1 + i}$. Verify the quotient using a graphing calculator.

Solution

$$\frac{3 + 2i}{1 + i} = \frac{3 + 2i}{1 + i} \cdot \frac{1 - i}{1 - i}$$

- The conjugate of the denominator is $1 - i$. Multiply the expression by $\frac{1 - i}{1 - i}$.

$$= \frac{3 - 3i + 2i - 2i^2}{1^2 + 1^2}$$

- In $1 + i$, $a = 1$ and $b = 1$. $a^2 + b^2 = 1^2 + 1^2$.

$$= \frac{3 - i - 2(-1)}{1 + 1}$$

- Simplify. Replace i^2 by -1.

$$= \frac{5 - i}{2}$$

$$= \frac{5}{2} - \frac{1}{2}i$$

- Write the number in the form $a + bi$.

YOU TRY IT 7

Simplify $\dfrac{5 - 3i}{4 + 2i}$. Verify the quotient using a graphing calculator.

Solution See page S36. $\dfrac{7}{10} - \dfrac{11}{10}i$

8.7 EXERCISES   Suggested Assignment: 7–73, odds

Topics for Discussion

1. What does the letter i represent?
   The letter i represents the number whose square is -1.

2. What is a complex number?   A complex number is a number of the form $a + bi$, where a and b are real numbers and $i = \sqrt{-1}$.

3. Explain why the real numbers are complex numbers.
   Explanations may vary. For example: A real number is a complex number in which $b = 0$.

4. Explain why the imaginary numbers are complex numbers.
   Explanations may vary. For example: An imaginary number is a complex number in which $a = 0$.

5. Explain the error in the following calculation.

$$\sqrt{-8} \cdot \sqrt{-50} = \sqrt{(-8)(-50)} = \sqrt{400} = 20$$

The radical expressions must first be rewritten in terms of i before the radicands can be multiplied.

6. Determine whether the following statements are always true, sometimes true, or never true.

  a. The product of two imaginary numbers is a real number. Always true
  b. The sum of two complex numbers is a real number.   Sometimes true
  c. The product of two complex numbers is a real number. Sometimes true
  d. The product of a complex number and its conjugate is a real number.   Always true

▪ Simplify Complex Numbers

Simplify.

7. $\sqrt{-4}$ $2i$

8. $\sqrt{-64}$ $8i$

9. $\sqrt{-98}$ $7i\sqrt{2}$

10. $\sqrt{-72}$ $6i\sqrt{2}$

11. $\sqrt{-27}$ $3i\sqrt{3}$

12. $\sqrt{-75}$ $5i\sqrt{3}$

13. $\sqrt{-9a^2}$ $3ai$

14. $\sqrt{-16b^6}$ $4b^3i$

15. $\sqrt{-49x^{12}}$ $7x^6i$

16. $\sqrt{-32x^3y^2}$ $4xyi\sqrt{2x}$

17. $\sqrt{-18a^{10}b^9}$ $3a^5b^4i\sqrt{2b}$

18. $\sqrt{16} + \sqrt{-81}$ $4 + 9i$

19. $\sqrt{25} + \sqrt{-9}$ $5 + 3i$

20. $\sqrt{12} - \sqrt{-18}$ $2\sqrt{3} - 3i\sqrt{2}$

21. $\sqrt{60} + \sqrt{-48}$ $2\sqrt{15} + 4i\sqrt{3}$

▪ Operations on Complex Numbers

Add or subtract. Verify the sum or difference using a graphing calculator.

22. $(2 + 4i) + (6 - 5i)$ $8 - i$

23. $(6 - 9i) + (4 + 2i)$ $10 - 7i$

24. $(-2 - 4i) - (6 - 8i)$ $-8 + 4i$

25. $(3 - 5i) - (8 - 2i)$ $-5 - 3i$

26. $(5 - 3i) + 2i$ $5 - i$

27. $(6 - 8i) + 4i$ $6 - 4i$

28. $(7 + 2i) + (-7 - 2i)$ 0

29. $(8 - 3i) + (-8 + 3i)$ 0

30. $(9 + 4i) + 6$ $15 + 4i$

31. $(4 + 6i) + 7$ $11 + 6i$

32. $8 - (2 + 4i)$ $6 - 4i$

33. $5 - (-11 - 7i)$ $16 + 7i$

Multiply. Verify the product using a graphing calculator.

34. $(7i)(-9i)$ 63

35. $(-6i)(-4i)$ -24

36. $\sqrt{-2} \cdot \sqrt{-8}$ -4

37. $\sqrt{-5} \cdot \sqrt{-45}$ -15

38. $6(3 - 8i)$ $18 - 48i$

39. $-10(7 + 4i)$ $-70 - 40i$

40. $2i(6 + 2i)$ $-4 + 12i$

41. $-3i(4 - 5i)$ $-15 - 12i$

42. $(5 - 2i)(3 + i)$ $17 - i$

43. $(2 - 4i)(2 - i)$ $-10i$

44. $(6 + 5i)(3 + 2i)$ $8 + 27i$

45. $(4 - 7i)(2 + 3i)$ $29 - 2i$

46. $(1 - i)\left(\frac{1}{2} + \frac{1}{2}i\right)$ 1

47. $(2 - i)\left(\frac{2}{5} + \frac{1}{5}i\right)$ 1

48. $\left(\frac{4}{5} - \frac{2}{5}i\right)\left(1 + \frac{1}{2}i\right)$ 1

49. $\left(\dfrac{6}{5} + \dfrac{3}{5}i\right)\left(\dfrac{2}{3} - \dfrac{1}{3}i\right)$ 1

50. $(4 - 3i)(4 + 3i)$ 25

51. $(8 - 5i)(8 + 5i)$ 89

52. $(3 - i)(3 + i)$ 10

53. $(7 - i)(7 + i)$ 50

Multiply. Find the exact product.

54. $\sqrt{-3} \cdot \sqrt{-6}$
$-3\sqrt{2}$

55. $\sqrt{-5} \cdot \sqrt{-10}$
$-5\sqrt{2}$

56. $\sqrt{-8} \cdot \sqrt{-4}$
$-4\sqrt{2}$

57. $\sqrt{-12} \cdot \sqrt{-2}$
$-2\sqrt{6}$

Simplify. Verify the quotient using a graphing calculator.

58. $\dfrac{3}{i}$ $-3i$

59. $\dfrac{4}{5i}$ $-\dfrac{4}{5}i$

60. $\dfrac{-6}{i}$ $6i$

61. $\dfrac{2 - 3i}{-4i}$ $\dfrac{3}{4} + \dfrac{1}{2}i$

62. $\dfrac{16 + 5i}{-3i}$ $-\dfrac{5}{3} + \dfrac{16}{3}i$

63. $\dfrac{5 + 2i}{3i}$ $\dfrac{2}{3} - \dfrac{5}{3}i$

64. $\dfrac{1 - 3i}{3 + i}$ $-i$

65. $\dfrac{3 + 5i}{1 - i}$ $-1 + 4i$

Simplify.

66. $\dfrac{4}{5 + i}$ $\dfrac{10}{13} - \dfrac{2}{13}i$

67. $\dfrac{6}{5 + 2i}$ $\dfrac{30}{29} - \dfrac{12}{29}i$

68. $\dfrac{2}{2 - i}$ $\dfrac{4}{5} + \dfrac{2}{5}i$

69. $\dfrac{5}{4 - i}$ $\dfrac{20}{17} + \dfrac{5}{17}i$

70. $\dfrac{2 - 3i}{3 + i}$ $\dfrac{3}{10} - \dfrac{11}{10}i$

71. $\dfrac{2 + 12i}{5 + i}$ $\dfrac{11}{13} + \dfrac{29}{13}i$

72. $\dfrac{4 - 5i}{3 - i}$ $\dfrac{17}{10} - \dfrac{11}{10}i$

73. $\dfrac{5 - i}{6 - 2i}$ $\dfrac{4}{5} + \dfrac{1}{10}i$

Iteration   **Fractal geometry** is the study of nonlinear dimensions. Fractal images are generated by substituting an initial value into a complex function, calculating the output, and then using the output as the next value to substitute into the function. This second output is then substituted into the function, and the process is repeated. This continual recycling of outputs is called **iteration,** and each output is called an **iterate.** Complex numbers are usually symbolized by the variable z, so we will use z in the functions in Exercises 74 to 76.

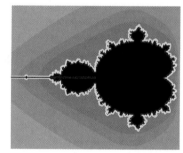

74. Let $f(z) = z + 4 + 3i$. Begin with the initial value $z = -2 + i$. Determine the first four iterates of the function. $2 + 4i$, $6 + 7i$, $10 + 10i$, $14 + 13i$

75. Let $f(z) = 3iz$. Begin with the initial value $z = 4 + 3i$. Determine the first three iterates of the function. $-9 + 12i$, $-36 - 27i$, $81 - 108i$

76. Let $f(z) = iz$. Begin with the initial value $z = 1 - i$. Determine the first four iterates of the function. Predict the next four iterates of $f(z)$ and explain your reasoning. $1 + i$, $-1 + i$, $-1 - i$, $1 - i$. The next four iterates will be $1 + i$, $-1 + i$, $-1 - i$, $1 - i$. Explanations will vary.

Applying Concepts

77. **a.** Is $3 + i$ a solution of $x^2 - 6x + 10 = 0$?
b. Is $3 - i$ a solution of $x^2 - 6x + 10 = 0$?
a. Yes   b. Yes

78. **a.** Is $-3 + 2i$ a solution of $x^2 + 6x + 13 = 0$?
b. Is $-3 - 2i$ a solution of $x^2 + 6x + 13 = 0$?
a. Yes   b. Yes

79. Simplify and express as a single term: $\sqrt{\dfrac{-3}{2}} + \sqrt{\dfrac{-2}{3}}$ $\dfrac{5\sqrt{6}}{6}i$

80. The sum of two complex numbers is $1 + 3i$. Their difference is $7 - 5i$. Find the product of the two numbers. $-8 + 19i$

81. For how many integers n is $(n + i)^4$ an integer?   Two integers (0 and 1)

82. Find a complex number z such that $3z = 4iz - 10$. Express z in the form $a + bi$. $-\dfrac{6}{5} - \dfrac{8}{5}i$

83. **a.** Find the reciprocal of $a + bi$.
    b. Write the additive inverse of $a + bi$. (*Hint:* See Example 4 in Section 8.7.)
    c. Find the multiplicative inverse of $a + bi$.

a. $\dfrac{a}{a^2 + b^2} - \dfrac{b}{a^2 + b^2}i$ b. $-a - bi$ c. $\dfrac{a}{a^2 + b^2} - \dfrac{b}{a^2 + b^2}i$

84. Show that $\sqrt{i} = \dfrac{\sqrt{2}}{2} + \dfrac{\sqrt{2}}{2}i$ by simplifying $\left(\dfrac{\sqrt{2}}{2} + \dfrac{\sqrt{2}}{2}i\right)^2$.

The complete solution is in the Solutions Manual.

The property that the product of conjugates of the form $(a + bi)(a - bi)$ is equal to $a^2 + b^2$ can be used to factor the sum of two perfect squares over the set of complex numbers. For example, $x^2 + y^2 = (x + yi)(x - yi)$. Factor the binomials over the set of complex numbers.

85. $x^2 + 25$
$(x + 5i)(x - 5i)$

86. $4b^2 + 9$
$(2b + 3i)(2b - 3i)$

87. $16x^2 + y^2$
$(4x + yi)(4x - yi)$

88. $9a^2 + 64$
$(3a + 8i)(3a - 8i)$

EXPLORATION

1. *Powers of i*   Note the pattern that emerges when successive powers of i are simplified.

$i^1 = i$ $i^5 = i \cdot i^4 = i(1) = i$
$i^2 = -1$ $i^6 = i^2 \cdot i^4 = -1$
$i^3 = i^2 \cdot i = -i$ $i^7 = i^3 \cdot i^4 = -i$
$i^4 = i^2 \cdot i^2 = (-1)(-1) = 1$ $i^8 = i^4 \cdot i^4 = 1$

   a. When the exponent on i is a multiple of 4, the power equals _____.

Use the pattern above to simplify the power of i.

   b. i^{57} **c.** i^{65} **d.** i^{122} **e.** i^{460}
   f. i^{-6} **g.** i^{-34} **h.** i^{-58} **i.** i^{-180}

   a. 1   b. i c. i d. -1 e. 1   f. -1 g. -1 h. -1 i. 1

2. *Graphs of Complex Numbers*   Real numbers are graphed as points on a number line. Complex numbers are graphed on a coordinate plane called an **Argand diagram** or the **complex plane.** The horizontal axis of the complex plane is called the **real axis,** and the vertical axis is called the **imaginary axis.**

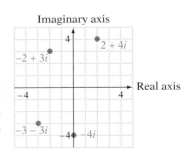

Graph a complex number $a + bi$ on the complex plane as you would graph the ordered pair (a, b) on the rectangular coordinate system. The complex numbers $2 + 4i$, $-2 + 3i$, $-3 - 3i$, and $-4i$ are graphed on the complex plane on the previous page.

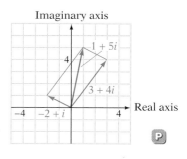

An Argand diagram can be used to find the sum of two complex numbers. For example, to find the sum of $-2 + i$ and $3 + 4i$, graph each complex number. Draw a line segment from each point to the origin. Then complete the parallelogram in which the line segments are two of the sides. The fourth vertex represents the sum $1 + 5i$. See the figure at the right.

a. Graphically add the complex numbers: $(4 - 5i) + (-1 + 6i)$
b. Graphically add the complex numbers: $(-3i) + (5 + 4i)$

An Argand diagram is also used to show a geometric explanation of multiplying by i. Consider the complex number $3 + 4i$. Multiply this number by i.

$$i(3 + 4i) = -4 + 3i$$

The two complex numbers are shown in the graph at the right.

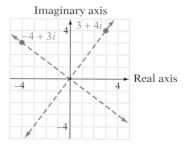

c. Find the slope of the line that passes through the origin and the point $(3, 4)$.
d. Find the slope of the line that passes through the origin and the point $(-4, 3)$.
e. Find the product of the slopes of the two lines. What does this product tell you?
f. Select another complex number, multiply it by i, graph each point, and follow the procedure described in parts c through e.
g.  What conclusion can be drawn about the graphical representation of a complex number and the graph of the product of that number and i?   **a.** $3 + i$ **b.** $5 + i$ **c.** $\frac{4}{3}$ **d.** $-\frac{3}{4}$ **e.** -1. The lines are perpendicular. **f.** Answers will vary.   **g.** Multiplying a number by i rotates that number 90°.

3. *Graphs of Powers of Complex Numbers*

a. Graph each of the six complex numbers on the complex plane. (*Hint:* You will need to simplify the powers of $1 + i$ before graphing them.)

$$1 + i \quad (1 + i)^2 \quad (1 + i)^3 \quad (1 + i)^4 \quad (1 + i)^5 \quad (1 + i)^6$$

b. Predict where the graphs of $(1 + i)^7$ and $(1 + i)^8$ will be. Verify your predictions.
a. Points at $(1, 1)$, $(0, 2)$, $(-2, 2)$, $(-4, 0)$, $(-4, -4)$, $(0, -8)$ **b.** $(8, -8)$, $(16, 0)$

CHAPTER **8** *SUMMARY*

Key Terms

complex number [p. 602]
conjugates [p. 578]
fractal geometry [p. 610]
hypotenuse [p. 565]
imaginary number [p. 602]

imaginary part of a complex number [p. 602]
index [p. 549]
iterate [p. 610]
iteration [p. 610]

Essential Concepts

Definition of $a^{m/n}$

If m and n are positive integers and $a^{1/n}$ is a real number, then $a^{m/n} = (a^{1/n})^m$.
[p. 546]

Alternative Notation for the nth root of a

If $a^{1/n}$ is a real number and n is a positive integer, then $a^{1/n} = \sqrt[n]{a}$. [p. 548]

Definition of the nth root of a^m

If $a^{1/n}$ is a real number, then $a^{m/n} = a^{(1/n)m} = (\sqrt[n]{a})^m$.
$a^{m/n}$ can also be written $a^{m/n} = a^{m(1/n)} = \sqrt[n]{a^m}$. [p. 549]

Exponents on Perfect Powers

A variable expression is a perfect power if the exponents on the factors are evenly divisible by the index of the radical. [p. 557]

Product Property of Radicals

If $\sqrt[n]{a}$ and $\sqrt[n]{b}$ are real numbers, then $\sqrt[n]{ab} = \sqrt[n]{a} \cdot \sqrt[n]{b}$ and $\sqrt[n]{a} \cdot \sqrt[n]{b} = \sqrt[n]{ab}$.
[p. 559]

Pythagorean Theorem

If a and b are the lengths of the legs of a right triangle and c is the length of the hypotenuse, then $c^2 = a^2 + b^2$. [p. 566]

The Principal Square Root Property

If $r^2 = s$, then $r = \sqrt{s}$, and r is called the square root of s. [p. 566]

The Distance Formula

If (x_1, y_1) and (x_2, y_2) are two points in the plane, then the distance d between the two points is given by $d = \sqrt{(x_1 - x_2)^2 + (y_1 - y_2)^2}$. [p. 570]

The Quotient Property of Radicals

If $\sqrt[n]{a}$ and $\sqrt[n]{b}$ are real numbers and $b \neq 0$, then $\sqrt[n]{\dfrac{a}{b}} = \dfrac{\sqrt[n]{a}}{\sqrt[n]{b}}$ and $\dfrac{\sqrt[n]{a}}{\sqrt[n]{b}} = \sqrt[n]{\dfrac{a}{b}}$.

[p. 578]

Domain of a Radical Function

The domain of a radical function is the set of real numbers for which the radical expression is a real number.

- If a radical expression contains an even root, the radicand must be greater than or equal to zero to ensure that the value of the expression will be a real number.
- If a radical expression contains an odd root, the radicand may be a positive or a negative number. [p. 584]

Property of Raising Both Sides of an Equation to a Power

If two numbers are equal, then the same powers of the numbers are equal.
If $a = b$, then $a^n = b^n$. **[p. 593]**

Principal Square Root of a Negative Number

If a is a positive real number, then the principal square root of negative a is
the imaginary number $i\sqrt{a}$. $\sqrt{-a} = i\sqrt{a}$ **[p. 602]**

Addition of Complex Numbers
$(a + bi) + (c + di) = (a + c) + (b + d)i$ **[p. 604]**

Subtraction of Complex Numbers
$(a + bi) - (c + di) = (a - c) + (b - d)i$ **[p. 604]**

The Product of Conjugates of the Form $(a + bi)(a - bi)$
The product of conjugates of the form $(a + bi)(a - bi)$ is $a^2 + b^2$. **[p. 607]**

Simplest Form of a Fraction Containing a Complex Number
A fraction containing one or more complex numbers is in simplest form when
no imaginary number remains in the denominator. **[p. 607]**

Simplifying a Fraction That Has a Complex Number in the Denominator
To simplify a fraction that has a complex number in the denominator, multiply the numerator and denominator by the conjugate of the complex number in the denominator. **[p. 608]**

CHAPTER 8 REVIEW EXERCISES

Simplify.

1. $81^{3/4}$ 27

2. $b^{2/3}(b^{5/6})(b^{-1/2})$ b

3. $(x^{-9}y^6)^{-2/3}$ $\dfrac{x^6}{y^4}$

4. $\left(\dfrac{a^{3/4}}{b^2}\right)^8$ $\dfrac{a^6}{b^{16}}$

5. $\sqrt[4]{81x^8y^{12}}$ $3x^2y^3$

6. $\sqrt[5]{-64a^8b^{12}}$ $-2ab^2\sqrt[5]{2a^3b^2}$

7. $\dfrac{8}{\sqrt{3y}}$ $\dfrac{8\sqrt{3y}}{3y}$

8. $\dfrac{x+2}{\sqrt{x}+\sqrt{2}}$ $\dfrac{x\sqrt{x} - x\sqrt{2} + 2\sqrt{x} - 2\sqrt{2}}{x-2}$

9. $\sqrt{-50}$ $5i\sqrt{2}$

10. $\dfrac{7}{2-i}$ $\dfrac{14}{5} + \dfrac{7}{5}i$

In Exercises 11–17, add, subtract, or multiply.

11. $\sqrt{50a^4b^3} - ab\sqrt{18a^2b}$ $2a^2b\sqrt{2b}$

12. $\sqrt{3x}(3 + \sqrt{3x})$ $3x + 3\sqrt{3x}$

13. $(\sqrt{3} + 8)(\sqrt{3} - 2)$ $-13 + 6\sqrt{3}$

14. $(20 - 3i) + (-15 + 4i)$ $5 + i$

15. $(-8 + 3i) - (4 - 7i)$ $-12 + 10i$

16. $-9i(10i)$ 90

17. $(6 - 5i)(4 + 3i)$ $39 - 2i$

18. Rewrite $3x^{3/8}$ as a radical expression. $3\sqrt[8]{x^3}$

Solve.

19. $\sqrt{3x - 5} - 5 = 3$ 23

20. $\sqrt[3]{2x - 2} + 4 = 2$ -3

21. $\sqrt{x + 12} - \sqrt{x} = 2$ 4

22.  The Transportation Department has provided statistics on passenger reports of lost, damaged, or delayed baggage on domestic flights. The function that approximately models the data is $f(x) = 6.447x^{-1/10}$, where x is the year, with $1991 = 1$, and $f(x)$ is the number of reports per 1000 passengers.

    a. Use the model to approximate the number of reports per 1000 passengers in 2005. Round to the nearest hundredth.

    b. Does the function indicate that the annual number of reports is increasing or decreasing?

    a. 4.92 reports per 1000 passengers   b. decreasing

23. Find **a.** the perimeter and **b.** the area of the triangle with vertices at $(-6, 2)$, $(-4, 4)$, and $(6, -6)$.

    a. $12\sqrt{2} + 4\sqrt{13}$ units   b. 20 square units

24. How far from the center of a merry-go-round is a child sitting when the child is traveling at a speed of 6 feet per second? Use the formula $v = \sqrt{12r}$, where v is the speed in feet per second and r is the distance in feet from the center of the merry-go-round to the rider.   3 ft

25. Find the distance between the two points $(3, -4)$ and $(-1, 5)$. Give the exact value. $\sqrt{97}$

26. State the domain of the function $F(x) = \sqrt[3]{3 - 6x} + 4$ in interval notation. Confirm your answer by graphing the function on a graphing calculator. $(-\infty, \infty)$

    a.
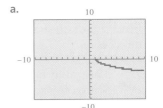

27. **a.** Graph $f(x) = -(x - 1)^{1/2}$.

    b. State the domain and range in set-builder notation.

    c. To the nearest tenth, find $f(4)$.

    b. Domain: $\{x \mid x \geq 1\}$; Range: $\{y \mid y \leq 0\}$ c. $f(4) \approx -1.7$

28. 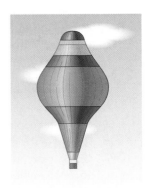 The distance you can see while flying in an airplane is a function of the altitude of the plane. This is given by the equation $d = \sqrt{1.5a}$, where d is the viewing distance to the horizon in miles and a is the altitude in feet. On day 13 of Steve Fossett's around-the-world balloon flight (July 1, 2002), the ballon achieved a speed of 197 miles per hour and an altitude of 30,400 feet (*Source:* Reuters).

    a. To the nearest whole number, what was the distance that Steve Fossett could see from that altitude?

    b. To the nearest whole number, at what altitude was his balloon flying when the viewing distance was 230 miles?

    a. 214 mi   b. 35,267 ft

CHAPTER **8** *TEST*

Simplify.

1. $16^{-5/4}$ $\dfrac{1}{32}$

2. $(c^8 d^{12})^{3/4}$ $c^6 d^9$

3. $\left(\dfrac{6x^{-2}y^4}{24x^{-8}y^{10}}\right)^{1/2}$ $\dfrac{x^3}{2y^3}$

4. $\sqrt[3]{-8a^6 b^{12}}$ $-2a^2 b^4$

5. $\sqrt[4]{x^6 y^8 z^{10}}$ $xy^2 z^2 \sqrt[4]{x^2 z^2}$

6. $\dfrac{\sqrt{125x^6}}{\sqrt{5x^3}}$ $5x\sqrt{x}$

7. $\dfrac{3}{\sqrt{y} + 1}$ $\dfrac{3\sqrt{y} - 3}{y - 1}$

8. $\sqrt{49} - \sqrt{-16}$ 7 − 4i   **9.** $\dfrac{6 + 4i}{2i}$ 2 − 3i   **10.** $\dfrac{5 + 9i}{1 - i}$ −2 + 7i

Add, subtract, or multiply.

11. $\sqrt{54} + \sqrt{24}$
  $5\sqrt{6}$

12. $\sqrt[3]{16x^4} \cdot \sqrt[3]{4x}$
  $4x\sqrt[3]{x^2}$

13. $(\sqrt{3} + 5)(\sqrt{3} - 4)$
  $-17 + \sqrt{3}$

14. $(5 + 2i) + (4 - 3i)$ 9 − i

15. $(9 - 2i) - (6 + 7i)$ 3 − 9i

16. $6i(-8i)$ 48

17. $i(3 - 7i)$ 7 + 3i

18. $\sqrt{-12} \cdot \sqrt{-6}$ $-6\sqrt{2}$

19. $(4 - 7i)(2 + i)$ 15 − 10i

20. Rewrite $7y\sqrt[5]{z^2}$ as an exponential expression.
  $7yz^{2/5}$

Solve.

21. $\sqrt{2x - 7} + 3 = 4$ 4

22. $\sqrt{x + 7} - \sqrt{x} = 2$ $\dfrac{9}{16}$

23. The perimeter of a square that has a side of length $(\sqrt{7x + 8})$ inches is 24 inches. Find the value of x.   4

24. The length of one leg of a right triangle is 7 inches. The hypotenuse is 16 inches. Find the length of the other leg. Round to the nearest tenth.
  14.4 in.

25. Find the distance between the points $(-2, 5)$ and $(1, -4)$. Round to the nearest tenth.   9.5 units

26. 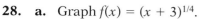 The table below shows the population, in millions, in the United States for selected years (*Source:* U.S. Bureau of the Census). The equation that approximately models the data is $y = 146.7\sqrt[5]{x}$, where y is the population, in millions, in year x, and $x = 10$ represents the year 1980. Use the equation to predict, to the nearest hundred thousand, the population of the United States in **a.** 1978, **b.** 1992, and **c.** 2010. **d.** Are the numbers for these years reasonable when compared to the data in the table?

Year	1980	1985	1990	1995	2000
U.S. Population (in millions)	228	238	250	263	281

a. 222.4 million people   **b.** 272.2 million people
c. 306.8 million people   **d.** Answers will vary.

27. Exercise 26 shows the population in the United States for selected years and the equation that approximately models the data: $y = 146.7\sqrt[5]{x}$, where y is the population, in millions, in year x, and $x = 10$ represents the year 1980. Use the equation to predict the years in which the population of the United States was **a.** 220 million people and **b.** 265 million people.   **a.** 1978   **b.** 1989

28. **a.** Graph $f(x) = (x + 3)^{1/4}$.
  b. State the domain and range in interval notation.
  c. Find $f(13)$.
  b. Domain: $[-3, \infty)$; Range: $[0, \infty)$ **c.** $f(13) = 2$

a.
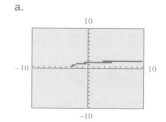

◀ CUMULATIVE REVIEW EXERCISES

1. Use inductive reasoning to predict the next term in the sequence a, c, d, g, h, i, m, n, o, p, . . .   u

2. If ♣♣ = ▽▽▽▽ and ▽▽ = ∞∞∞∞, then ♣♣♣ equal how many ∞?   9

3. Graph: $\{x\,|\,x \le -3\} \cup \{x\,|\,x > 0\}$

$$\overset{-5\ -4\ -3\ -2\ -1\ \ 0\ \ 1\ \ 2\ \ 3\ \ 4\ \ 5}{\longleftrightarrow}$$

4. Evaluate $m + n(p - q)^2$ for $m = -3, n = 2, p = -1$, and $q = 4$.   47

5. Subtract: $(3b^3 - 4b^2 + b - 8) - (b^3 - 5b + 7)$
$2b^3 - 4b^2 + 6b - 15$

6. Write 3.04×10^{11} in decimal notation.
304,000,000,000

7. Find the range of the function $F(x) = 3x^2 - 4$ if the domain is $\{-2, -1, 0, 1, 2\}$. $\{-4, -1, 8\}$

8. Find the slope of the line between the points $(-4, 1)$ and $(-1, -5)$. -2

9. What numbers must be excluded from the domain of $f(x) = \frac{x + 1}{x + 2}$? -2

10. Find the y- and x-intercepts for the graph of $f(x) = x^2 + 3x - 4$. $(0, -4); (-4, 0)$ and $(1, 0)$

11. Solve: $5 - \frac{2}{3}x = 4$ $\frac{3}{2}$

12. Solve: $2(4x - 2) = 4(1 - x)$ $\frac{2}{3}$

13. Solve: $5 < 2x - 3 < 7$
Write the answer in set-builder notation.
$\{x\,|\,4 < x < 5\}$

14. Solve: $|7 - 3x| > 1$
Write the answer in set-builder notation.
$\left\{x \,|\, x < 2 \text{ or } x > \frac{8}{3}\right\}$

15. Write the equation $3x - 2y = -6$ in slope–intercept form. Then identify the slope and y-intercept.
$y = \frac{3}{2}x + 3;\ m = \frac{3}{2};\ y$-intercept: $(0, 3)$

16. Simplify: $32^{-6/5}$
$\frac{1}{64}$

17. Multiply: $(2\sqrt{3} + 4)(3\sqrt{3} - 1)$
$14 + 10\sqrt{3}$

18. Add: $(9 - 4i) + (5 + 6i)$
$14 + 2i$

19. Evaluate $-st$ for $s = -\frac{2}{3}$ and $t = -\frac{9}{10}$. $-\frac{3}{5}$

20. Simplify: $4(3y - 7) - (2y - 9)$ $10y - 19$

21. Simplify: $\frac{8a^{-3}b^5}{2a^4b^{-7}}$ $\frac{4b^{12}}{a^7}$

22. Factor: $3x^2 - 11x - 4$ $(3x + 1)(x - 4)$

23. The radius of a circle is $(x + 2)$ meters. Find the area of the circle in terms of the variable x. Leave the answer in terms of π.
$(x^2\pi + 4\pi x + 4\pi)$ m²

24. Find the length of the line segment with endpoints $(-2, 4)$ and $(3, 5)$.
$\sqrt{26}$

25. A pilot flies 70 miles west from Bradford Airport to Murdock Airport and then 50 miles south from Murdock Airport to Plimpton Airfield. Find the distance of the return flight straight back to Bradford from Plimpton. Round to the nearest mile.   86 mi

26. How many ounces of pure gold that costs \$360 per ounce must be mixed with 80 ounces of an alloy that costs \$120 per ounce to make a mixture that costs \$200 per ounce?   40 oz

27. Graph $y = -2x + 3$.

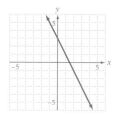

28. Find the equation of the line that contains the point (2, 5) and is perpendicular to the line $y = -\frac{2}{3}x + 6$. $y = \frac{3}{2}x + 2$

29. Solve by the substitution method: $x - 2y = 3$
$$-2x + y = -3 \quad (1, -1)$$

30. Solve by the addition method: $x - y + z = 0$
$$2x + y - 3z = -7$$
$$-x + 2y + 2z = 5 \quad \left(-\frac{9}{7}, \frac{2}{7}, \frac{11}{7}\right)$$

CHAPTER

9

Quadratic Functions and Quadratic Inequalities

```
EDIT CALC TESTS
4↑LinReg(ax+b)
5:QuadReg
6:CubicReg
7:QuartReg
8:LinReg(a+bx)
9:LnReg
0↓ExpReg
```

Press **STAT** to access the QuadReg option in the CALC menu.

*This retired couple wants to enjoy their retirement in the decades ahead. They want to maintain their health and have sufficient income to meet their expenses. Many couples in their senior years depend heavily upon Social Security to make ends meet. Several sources provide data on the expected average annual retirement income from Social Security. The data from one such source is provided in **Exercise 92 on page 683**. Regression is used to determine a quadratic model for this data, and the model used to predict annual Social Security benefits in the years ahead, important information for a retired couple planning their future.*

Need help? For online student resources, visit this web site:
Math.college.hmco.com

PREP TEST

1. Simplify: $\sqrt{18}$ $3\sqrt{2}$

2. Simplify: $\sqrt{-9}$ $3i$

3. Simplify: $\dfrac{3x-2}{x-1} - 1$ $\dfrac{2x-1}{x-1}$

4. Evaluate $b^2 - 4ac$ when $a = 2$, $b = -4$, and $c = 1$. 8

5. Is $4x^2 + 28x + 49$ a perfect-square trinomial? Yes

6. Factor: $4x^2 - 4x + 1$ $(2x-1)^2$

7. Factor: $9x^2 - 4$ $(3x+2)(3x-2)$

8. Graph: $\{x \mid x < -1\} \cap \{x \mid x < 4\}$

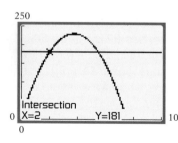

9. Solve: $x(x - 1) = x + 15$ $-3, 5$

10. Solve: $\dfrac{4}{x-3} = \dfrac{16}{x}$ 4

GO FIGURE

In a school election, one candidate for class president received more than 94%, but less than 100%, of the votes cast. What is the least possible number of votes cast? 17 votes

---

SECTION **9.1**

■ Solve Quadratic Equations by Factoring

■ Write a Quadratic Equation Given Its Solutions

Introduction to Quadratic Equations

■ Solve Quadratic Equations by Factoring

A model of the height above the ground of an arrow projected into the air with an initial velocity of 120 feet per second is $h = -16t^2 + 120t + 5$, where h is the height, in feet, of the arrow t seconds after it is released from the bow. Using a graphing calculator, we can determine at what times the arrow would be, say, 181 feet above the ground.

$$h = -16t^2 + 120t + 5$$
$$181 = -16t^2 + 120t + 5$$

Graph both $Y_1 = -16X^2 + 120X + 5$ and $Y_2 = 181$. Then find the point of intersection of the two graphs. As shown at the left, one point of intersection is (2, 181). A second point of intersection is (5.5, 181).

The arrow will be 181 feet above the ground twice. Once on its way up, 2 seconds after it leaves the bow, and once on its way down, 5.5 seconds after it leaves the bow.

The equation $181 = -16t^2 + 120t + 5$ is an example of a quadratic equation. A **quadratic equation** is an equation of the form $ax^2 + bx + c = 0$, where

a and b are coefficients, c is a constant, and $a \neq 0$. A quadratic equation is in **standard form** when the polynomial is in descending order and equal to zero.

The equation $181 = -16t^2 + 120t + 5$ is not in standard form but can be written in standard form by subtracting 181 from each side of the equation.

$$181 = -16t^2 + 120t + 5$$
$$181 - 181 = -16t^2 + 120t + 5 - 181 \quad \bullet \text{ Subtract 181 from each side.}$$
$$0 = -16t^2 + 120t - 176 \quad \bullet \text{ The equation is now in standard form.}$$

? <u>QUESTION</u>   What are the values of a, b, and c in the equation $0 = -16t^2 + 120t - 176$?

➡ Rewrite the quadratic equations in standard form.

a. $x^2 = 2x + 7$ **b.** $3x - 7 = 4x^2$

a. $\quad\quad\quad\quad x^2 = 2x + 7$ $\bullet$ Subtract $2x$ and subtract 7 from
$\quad\quad x^2 - 2x - 7 = 0$ each side of the equation.

b. $3x - 7 = 4x^2$ $\bullet$ Subtract $3x$ from each side and
$\quad\quad 0 = 4x^2 - 3x + 7$ add 7 to each side of the equation. ⬅

An equation of the form $y = ax^2 + bx + c$, $a \neq 0$, is a **quadratic equation in two variables.** Below are three examples of quadratic equations in two variables.

$$y = 4x^2 - x + 1$$
$$y = -x^2 + 5$$
$$y = 3x^2 + 2x$$

For these equations, y is a function of x, and we can write $f(x) = ax^2 + bx + c$. This represents a **quadratic function.**

The graph of a quadratic function is a **parabola.** The graphs below are examples of parabolas.

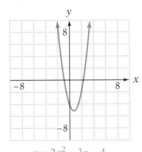
$y = 2x^2 - 3x - 4$

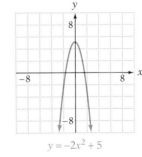
$y = -2x^2 + 5$

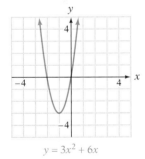
$y = 3x^2 + 6x$

The graph of the equation $h = -16t^2 + 120t + 5$ shown at the beginning of this section is a parabola.

We solved the equation $181 = -16t^2 + 120t + 5$ using a graphing calculator. We can also solve this equation by factoring. Before doing so, we will present the Principle of Zero Products.

? <u>ANSWER</u> $a = -16$, $b = 120$, $c = -176$

TAKE NOTE

A quadratic equation is in standard form whether it is written as $ax^2 + bx + c = 0$ or as $0 = ax^2 + bx + c$.

▼ *Point of Interest*

If you consider the roadway of the Golden Gate Bridge as the x-axis, then each main suspension cable of the bridge can be approximated by the quadratic function

$f(x) = \frac{1}{9000}x^2 + 5.$

Consider the equation $ab = 0$. If a is not zero, then b is zero. Conversely, if b is not zero, then a must be zero. This is stated in the Principle of Zero Products.

INSTRUCTOR NOTE
The material on solving a quadratic equation by factoring was presented earlier in the text. It is reviewed here for completeness.

> **Principle of Zero Products**
>
> If the product of two factors is zero, then at least one of the factors must be zero.
>
> If $ab = 0$, then $a = 0$ or $b = 0$.

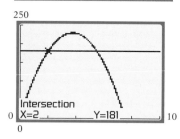

Suggested Activity

See Section 9.1 of the *Student Activity Manual* for an investigation in which technology is used to investigate quadratic equations, factors, and zeros, as well as to determine quadratic equations from known zeros.

The Principle of Zero Products states that, for instance, if $(x + 1)(x - 4) = 0$, then $(x + 1) = 0$ or $(x - 4) = 0$.

Now let's return to the problem presented at the beginning of this section: We needed to solve the equation $181 = -16t^2 + 120t + 5$ to determine when an arrow projected into the air would be 181 feet above the ground. This time we will solve the equation by factoring.

$$0 = -16t^2 + 120t - 176$$ • Write the equation in standard form.

$$0 = -8(2t^2 - 15t + 22)$$ • Factor out -8 on the right side.

$$0 = 2t^2 - 15t + 22$$ • Divide both sides of the equation by -8.

$$0 = (2t - 11)(t - 2)$$ • Factor the trinomial.

$$2t - 11 = 0 \qquad t - 2 = 0$$ • Use the Principle of Zero Products to set each factor equal to zero. Then solve each equation for t.
$$2t = 11 \qquad\qquad t = 2$$
$$t = 5.5$$

The arrow will be 181 feet above the ground after 2 seconds and after 5.5 seconds. This is the same answer obtained by solving the equation by graphing.

INSTRUCTOR NOTE
Exercise 24 can be used for a similar in-class example.

EXAMPLE 1

Solve by factoring: $(x - 3)(x - 5) = 35$

Solution $(x - 3)(x - 5) = 35$

$$x^2 - 8x + 15 = 35$$ • First write the equation in standard form.

$$x^2 - 8x - 20 = 0$$

$$(x + 2)(x - 10) = 0$$ • Factor the left side of the equation.

$$x + 2 = 0 \qquad x - 10 = 0$$ • Use the Principle of Zero Products.

$$x = -2 \qquad\qquad x = 10$$

The solutions are -2 and 10.

YOU TRY IT 1

Solve by factoring: $(x + 3)(x - 7) = 11$

Solution   See page S36. $-4, 8$

The Principle of Zero Products is used to find elements in the domain of a quadratic function that correspond to a given element in the range.

INSTRUCTOR NOTE
Exercise 28 can be used for a similar in-class example.

EXAMPLE 2

Given that 1 is in the range of the function defined by $f(x) = x^2 + x - 5$, find two values of c for which $f(c) = 1$.

Solution $f(c) = 1$

$$c^2 + c - 5 = 1 \qquad \bullet \; f(c) = c^2 + c - 5.$$
$$c^2 + c - 6 = 0 \qquad \bullet \; \text{Write the quadratic equation in standard form.}$$
$$(c + 3)(c - 2) = 0 \qquad \bullet \; \text{Factor the left side of the equation.}$$
$$c + 3 = 0 \qquad c - 2 = 0 \qquad \bullet \; \text{Use the Principle of Zero Products.}$$
$$c = -3 \qquad c = 2 \qquad \bullet \; \text{Solve each equation for } c.$$

The values of c are -3 and 2.

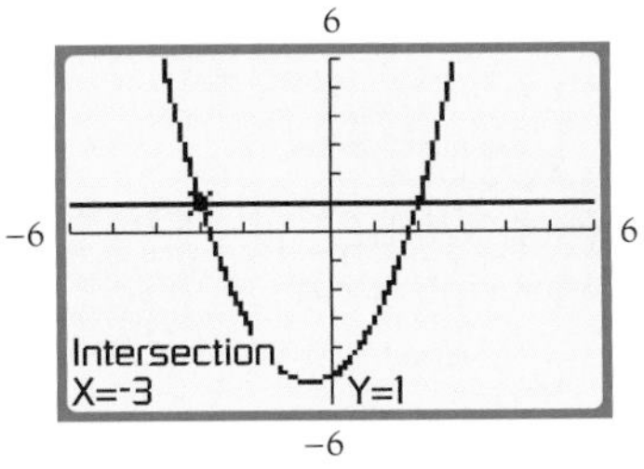

Note: There are two values in the domain that can be paired with the range element 1: -3 and 2. Two ordered pairs that belong to the function are $(-3, 1)$ and $(2, 1)$. The graph of $f(x) = x^2 + x - 5$ is shown at the left, along with the graph of $y = 1$. The two points of intersection are $(-3, 1)$ and $(2, 1)$.

YOU TRY IT 2

Given that 4 is in the range of the function defined by $s(t) = t^2 - t - 2$, find two values of c for which $s(c) = 4$.

Solution   See page S36. $-2, 3$

INSTRUCTOR NOTE
Exercise 36 can be used for a similar in-class example.

EXAMPLE 3

The base of a triangle is 2 inches less than four times the height. The area of the triangle is 45 square inches. Find the height and the length of the base of the triangle.

State the goal.   The goal is to determine the height and the length of the base of the triangle.

Devise a strategy.

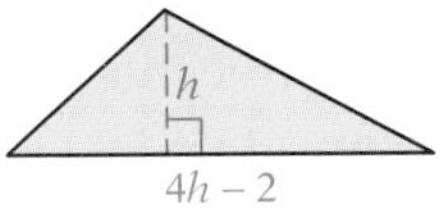

- Let h be the height of the triangle. Then the base is $4h - 2$.
- Use the formula for the area of a triangle, $A = \frac{1}{2} bh$.

  Substitute 45 for A, substitute $4h - 2$ for b, and solve for h.

- After determining the value for h, substitute that value in the expression $4h - 2$ to find the length of the base of the triangle.

Solve the problem. $A = \dfrac{1}{2}bh$ • Use the formula for the area of a triangle.

$45 = \dfrac{1}{2}(4h - 2)h$ • Substitute 45 for A and $4h - 2$ for b.

$45 = (2h - 1)h$ • Write the equation in standard form.

$45 = 2h^2 - h$

$0 = 2h^2 - h - 45$

$0 = (2h + 9)(h - 5)$ • Factor the trinomial.

$2h + 9 = 0 \qquad h - 5 = 0$ • Use the Principle of Zero Products.

$2h = -9 \qquad\quad h = 5$

$h = -\dfrac{9}{2}$ • The solution $-\dfrac{9}{2}$ is not possible because the height cannot be a negative number. The solution is $h = 5$.

$4h - 2$

$4(5) - 2 = 20 - 2 = 18$ • Substitute 5 for h in the expression for the length of the base of the triangle.

The height of the triangle is 5 inches, and the length of the base is 18 inches.

Check your work.

ALGEBRAIC CHECK

$A = \dfrac{1}{2}bh$

$A = \dfrac{1}{2}(18)(5) = 9(5) = 45$

GRAPHICAL CHECK

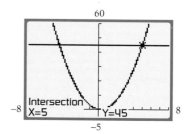

For the function $y = \dfrac{1}{2}(4x - 2)x$, when $y = 45$, the value of x is 5.

YOU TRY IT 3

The height of a projectile fired upward is given by the formula $s = v_0t - 16t^2$, where s is the height, v_0 is the initial velocity, and t is the time. Find the time for a projectile to return to Earth if it has an initial velocity of 200 feet per second.

Solution   See page S37.   12.5 s

■ Write a Quadratic Equation Given Its Solutions

As shown below, the solutions of the equation $(x - r_1)(x - r_2) = 0$ are r_1 and r_2.

$$(x - r_1)(x - r_2) = 0$$

$x - r_1 = 0 \qquad x - r_2 = 0$ • Use the Principle of Zero Products.

$\qquad x = r_1 \qquad\qquad x = r_2$ • Solve each equation for x.

Check:

$(x - r_1)(x - r_2) = 0$	
$(r_1 - r_1)(r_1 - r_2)$	0
$(0)(r_1 - r_2)$	0
	$0 = 0$

$(x - r_1)(x - r_2) = 0$	
$(r_2 - r_1)(r_2 - r_2)$	0
$(r_2 - r_1)(0)$	0
	$0 = 0$

Suggested Activity

A quadratic equation has two roots, one of which is twice the other. If the sum of the roots is 30, find the equation and express it in the form $ax^2 + bx + c = 0$, where a, b, and c are integers and $c > 0$.
[Answer: $x^2 - 30x + 200 = 0$]

Using the equation $(x - r_1)(x - r_2) = 0$ and the fact that r_1 and r_2 are solutions of this equation, it is possible to write a quadratic equation given its solutions.

➡ Write a quadratic equation that has solutions 3 and -5.

$(x - r_1)(x - r_2) = 0$

$(x - 3)[x - (-5)] = 0$ • Replace r_1 by 3 and r_2 by -5.

$(x - 3)(x + 5) = 0$ • Simplify $[x - (-5)]$.

$x^2 + 2x - 15 = 0$ • Multiply the binomials. ⬅

EXAMPLE 4

Write a quadratic equation that has integer coefficients and has solutions $-\frac{2}{3}$ and $\frac{1}{2}$.

INSTRUCTOR NOTE
Exercise 62 can be used for a similar in-class example.

Solution

ALGEBRAIC SOLUTION

$(x - r_1)(x - r_2) = 0$

$\left[x - \left(-\dfrac{2}{3}\right)\right]\left(x - \dfrac{1}{2}\right) = 0$ • Replace r_1 by $-\dfrac{2}{3}$ and r_2 by $\dfrac{1}{2}$.

$\left(x + \dfrac{2}{3}\right)\left(x - \dfrac{1}{2}\right) = 0$ • Simplify $\left[x - \left(-\dfrac{2}{3}\right)\right]$.

$x^2 + \dfrac{1}{6}x - \dfrac{1}{3} = 0$ • Multiply the binomials.

$6\left(x^2 + \dfrac{1}{6}x - \dfrac{1}{3}\right) = 6 \cdot 0$ • Multiply each side of the equation by the LCM of the denominators.

$6x^2 + x - 2 = 0$ • Check that this is correct by solving the equation by factoring.

The solutions are $-\dfrac{2}{3}$ and $\dfrac{1}{2}$.

GRAPHICAL CHECK

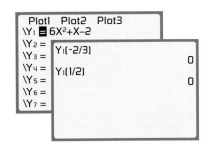

> ### YOU TRY IT 4
>
> Write a quadratic equation that has integer coefficients and has solutions $-\frac{1}{6}$ and $\frac{2}{3}$.
>
> **Solution**   See page S37. $18x^2 - 9x - 2 = 0$

9.1 EXERCISES

Suggested Assignment: 7–27, every other odd: 29-65, odds

Topics for Discussion

1. **a.**   What is a quadratic equation?
    b.   Provide two examples of quadratic equations.
    a. A quadratic equation is an equation of the form $ax^2 + bx + c = 0$, $a \neq 0$.
    b. Examples will vary.

2. Explain why the restriction that $a \neq 0$ is given in the definition of a quadratic equation.   If $a = 0$ in $ax^2 + bx + c = 0$, then there is no x^2 term, and it is the x^2 term that makes it a quadratic equation.

3. **a.**   Describe a quadratic equation in standard form.
    b.   Explain how to write the equation $6x - 3 + 4x^2 = 0$ in standard form.
    c.   Provide two examples of quadratic equations in standard form.
    a. The polynomial $ax^2 + bx + c$ is in descending order and equal to zero.
    b. Explanations may vary. For example, use the Commutative Property to rearrange the terms: $4x^2 + 6x - 3 = 0$.   **c.** Examples will vary.

4. Which of the following are graphs of quadratic functions? Explain your answers.

    a. 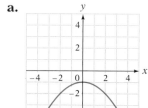   **b.** 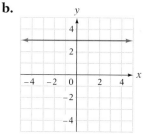   **c.**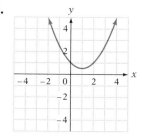

    Parts a and c are graphs of quadratic functions. Explanations will vary.

5. **a.**   What does the Principle of Zero Products state?
    b.   When is the Principle of Zero Products used?
    a. If the product of two factors is zero, then at least one of the factors must be zero.
    b. It is used to solve some quadratic equations.

■ Solve Quadratic Equations by Factoring

What are the values of a, b, and c in the quadratic equation?

6. $2x^2 - 3x + 5 = 0$ $a = 2, b = -3, c = 5$

7. $-x^2 + x - 8 = 0$ $a = -1, b = 1, c = -8$

8. $4x^2 - 6 = 0$ $a = 4, b = 0, c = -6$

9. $x^2 - 7x = 0$ $a = 1, b = -7, c = 0$

Is the equation a quadratic equation? If it is, write the equation in standard form.

10. $2d^2 - 8 + 7d = 0$

Yes; $2d^2 + 7d - 8 = 0$

11. $3p = 4 - 5p^2$

Yes; $5p^2 + 3p - 4 = 0$ or $-5p^2 - 3p + 4 = 0$

12. $10 - n = 9n$ No

13. $z(z + 1) = 5$ Yes; $z^2 + z - 5 = 0$

Solve by factoring.

14. $d^2 + 10 = 7d$ 2, 5

15. $v^2 - 16 = 15v$ −1, 16

16. $4t^2 = 9t - 2$ $\frac{1}{4}$, 2

17. $2z^2 = 9z - 9$ $\frac{3}{2}$, 3

18. $3s^2 + 11s = 4$ $-4, \frac{1}{3}$

19. $2w^2 + w = 6$ $-2, \frac{3}{2}$

20. $6r^2 = 23r + 18$ $-\frac{2}{3}, \frac{9}{2}$

21. $6x^2 = 7x - 2$ $\frac{1}{2}, \frac{2}{3}$

22. $9d^2 - 18d = 0$ 0, 2

23. $4t^2 + 20t = 0$ −5, 0

24. $4z(z + 3) = z - 6$ $-2, -\frac{3}{4}$

25. $3w(w - 2) = 11w + 6$ $-\frac{1}{3}$, 6

26. $u^2 - 2u + 4 = (2u - 3)(u + 2)$ −5, 2

27. $(3r - 4)(r + 4) = r^2 - 3r - 28$ $-4, -\frac{3}{2}$

Find the values c in the domain of f for which $f(c)$ is the indicated value.

28. $f(x) = x^2 - 3x + 3; f(c) = 1$ 1, 2

29. $f(x) = x^2 + 4x - 2; f(c) = 3$ −5, 1

30. $f(x) = 2x^2 - x - 5; f(c) = -4$ $-\frac{1}{2}$, 1

31. $f(x) = 6x^2 - 5x - 9; f(c) = -3$ $-\frac{2}{3}, \frac{3}{2}$

32. $f(x) = 4x^2 - 4x + 3; f(c) = 2$ $\frac{1}{2}$

33. $f(x) = x^2 - 6x + 12; f(c) = 3$ 3

34. *Geometry*   The length of a rectangle is 2 feet less than three times the width of the rectangle. The area of the rectangle is 65 square feet. Find the length and width of the rectangle.   Length: 13 ft; width: 5 ft

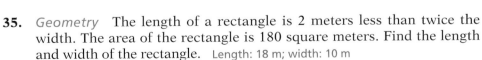

35. *Geometry*   The length of a rectangle is 2 meters less than twice the width. The area of the rectangle is 180 square meters. Find the length and width of the rectangle.   Length: 18 m; width: 10 m

36. *Geometry*   The length of each side of a square is extended 2 centimeters. The area of the resulting square is 64 square centimeters. Find the length of a side of the original square.   6 cm

37. *Geometry*   The length of each side of a square is extended 4 meters. The area of the resulting square is 64 square meters. Find the length of a side of the original square.   4 m

38. *Integer Problems*   The sum of the squares of two consecutive odd integers is thirty-four. Find the two integers. (*Hint:* Odd integers differ by 2. Two consecutive odd integers can be represented by x and $x + 2$.)
3 and 5 or −3 and −5

39. *Integer Problems*   The sum of the squares of three consecutive even integers is fifty-six. Find the three integers. (*Hint:* Odd integers differ by 2. Three consecutive odd integers can be represented by x, $x + 2$, and $x + 4$.)   2, 4, and 6 or −6, −4, and −2

40. *Integer Problems*   Find two consecutive integers whose cubes differ by 127. (*Hint:* Two consecutive integers can be represented by x and $x + 1$.)
6 and 7 or −7 and −6

41. *Integer Problems*   Find two consecutive even integers whose cubes differ by 488. (*Hint:* Even integers differ by 2. Two consecutive even integers can be represented by x and $x + 2$.)   8 and 10 or −10 and −8

42. *Geometry*   The length of a rectangle is 7 centimeters, and the width is 4 centimeters. If both the length and the width are increased by equal amounts, the area of the rectangle is increased by 42 square centimeters. Find the length and width of the larger rectangle.
Length: 10 cm; width: 7 cm

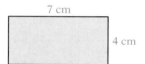

Use the formula $d = vt + 16t^2$, where d is the distance in feet, v is the initial velocity in feet per second, and t is the time in seconds.

43. *Physics*   An object is released from a plane at an altitude of 1600 feet. The initial velocity is 0 feet per second. How many seconds later will the object hit the ground?   10 s

44. *Physics*   An object is released from the top of a building 320 feet high. The initial velocity is 16 feet per second. How many seconds later will the object hit the ground?   4 s

Use the formula $S = \dfrac{n^2 + n}{2}$, where S is the sum of the first n natural numbers.

45. *Mathematics*   How many consecutive natural numbers beginning with 1 will give a sum of 78?   12

46. *Mathematics*   How many consecutive natural numbers beginning with 1 will give a sum of 120?   15

Use the formula $N = \frac{t^2 - t}{2}$, where N is the number of football games that must be scheduled in a league with t teams if each team is to play every other team once.

47. *Sports*   A league has 21 games scheduled. How many teams are in the league if each team plays every other team once?   7 teams

48. *Sports*   A league has 36 games scheduled. How many teams are in the league if each team plays every other team once?   9 teams

For Exercises 49 and 50, use the formula $h = vt - 16t^2$, where h is the height an object will attain (neglecting air resistance) in t seconds and v is the initial velocity.

49. *Sports*   A baseball player hits a "Baltimore chop," meaning the ball bounces off home plate after he hits it. The ball leaves home plate with an initial upward velocity of 64 feet per second. How many seconds after the ball hits home plate will the ball be 48 feet above the ground?   1 s and 3 s

50. *Sports*   A golf ball is thrown onto a cement surface and rebounds straight up. The initial velocity of the rebound is 80 feet per second. How many seconds later will the golf ball return to the ground?   5 s

51. *Geometry*   A rectangular piece of cardboard is 10 inches longer than it is wide. Squares 2 inches on a side are to be cut from each corner, and then the sides will be folded up to make an open box with a volume of 192 cubic inches. Find the length and width of the piece of cardboard.
Length: 20 in.; width: 10 in.

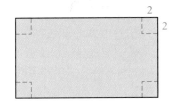

52. *Geometry*   A rectangular piece of cardboard has a length that is 8 inches more than its width. An open box is formed from the piece of cardboard by cutting squares whose sides are 2 inches in length from each corner and then folding up the sides. Find the dimensions of the box if its volume is 256 square inches.   2 in. by 8 in. by 16 in.

■ Write a Quadratic Equation Given Its Solutions

Write a quadratic equation that has integer coefficients and has as solutions the given pair of numbers.

53. 6 and -1 $x^2 - 5x - 6 = 0$

54. -2 and 5 $x^2 - 3x - 10 = 0$

55. 3 and -3 $x^2 - 9 = 0$

56. 5 and -5 $x^2 - 25 = 0$

57. 0 and 5 $x^2 - 5x = 0$

58. 0 and -2 $x^2 + 2x = 0$

59. 2 and $\frac{2}{3}$ $3x^2 - 8x + 4 = 0$

60. $-\frac{1}{2}$ and 5 $2x^2 - 9x - 5 = 0$

61. $\frac{6}{5}$ and $-\frac{1}{2}$ $10x^2 - 7x - 6 = 0$

62. $\frac{3}{4}$ and $-\frac{3}{2}$ $8x^2 + 6x - 9 = 0$

63. $-\frac{1}{4}$ and $-\frac{1}{2}$ $8x^2 + 6x + 1 = 0$

64. $-\frac{5}{6}$ and $-\frac{2}{3}$ $18x^2 + 27x + 10 = 0$

65. *Physics*   An arrow is shot upward with an initial velocity of 128 feet per second. The equation $16t(t - a) = 0$, where t is the time in seconds and a is a constant, describes the times at which the arrow is on the ground. Suppose you know that the arrow is on the ground at 0 seconds and 8 seconds. What is the value of a?   8

Applying Concepts

Solve for x.

66. $x^2 - 9ax + 14a^2 = 0$ 2a, 7a

67. $x^2 + 9xy - 36y^2 = 0$ −12y, 3y

68. $3x^2 - 4cx + c^2 = 0$ $\frac{c}{3}$, c

69. $2x^2 + 3bx + b^2 = 0$ $-\frac{b}{2}$, −b

Write a quadratic equation that has as solutions the given pair of numbers.

70. $\sqrt{5}$ and $-\sqrt{5}$ $x^2 - 5 = 0$

71. $2i$ and $-2i$ $x^2 + 4 = 0$

72. $2\sqrt{2}$ and $-2\sqrt{2}$ $x^2 - 8 = 0$

73. $2\sqrt{3}$ and $-2\sqrt{3}$ $x^2 - 12 = 0$

74. $i\sqrt{2}$ and $-i\sqrt{2}$ $x^2 + 2 = 0$

75. $2i\sqrt{3}$ and $-2i\sqrt{3}$ $x^2 + 12 = 0$

76.   Show that the solutions of the equation $ax^2 + bx = 0$ are 0 and $-\frac{b}{a}$.

Factor the left side of the equation: $x(ax + b) = 0$; then $x = 0$ or $ax + b = 0$; solve $ax + b = 0$ for x: $x = -\frac{b}{a}$.

77.   Explain the error made in solving the equation at the right. Solve the equation correctly.

$x^2 = x$

$\dfrac{x^2}{x} = \dfrac{x}{x}$

$x = 1$

The equation should first be written in standard form: $x^2 - x = 0$. Solve by factoring: $x(x - 1) = 0$. $x = 0$ or $x = 1$.

78.   *Geometry*   The volumes of two spheres differ by 372π cubic centimeters. The radius of the larger sphere is 3 centimeters more than the radius of the smaller sphere. Find the radius of the larger sphere. The formula for the volume of a sphere is $V = \frac{4}{3}\pi r^3$.   7 cm

79.   Find all real values of x for which $(x^2 - 5x + 5)^{x^2 - 9x + 20} = 1$.   1, 2, 3, 4, 5

80.   Find both values of x for which $x\%$ of $x\%$ equals $x\%$.   0 and 100

EXPLORATION

1.   Solutions of a Quadratic Equation and Zeros of a Quadratic Function

a.   Solve each of the following quadratic equations by factoring.

$x^2 + 6x + 5 = 0$ −5, −1

$x^2 + x - 12 = 0$ −4, 3

$x^2 - 5x - 14 = 0$ −2, 7

$x^2 - 9x + 8 = 0$ 1, 8

b. Use a graphing calculator to find the zeros of each of the following quadratic functions. A **zero** of a function is a value of x for which $f(x) = 0$. Therefore, it is the x-coordinate of an x-intercept of the graph of the function.

$f(x) = x^2 + 6x + 5$ $-5, -1$
$f(x) = x^2 + x - 12$ $-4, 3$
$f(x) = x^2 - 5x - 14$ $-2, 7$
$f(x) = x^2 - 9x + 8$ $1, 8$

c.   Note the similarity between the quadratic equations in part a and the quadratic functions in part b. Explain the connection between the solutions of the quadratic equations in part a and the zeros of the functions in part b.
The solutions in part a are the same as the zeros in part b.

d. Write two quadratic equations that can be solved by factoring. For each quadratic equation, write the corresponding quadratic function. What do you expect to be the zeros of the quadratic functions? Verify the zeros by using a graphing calculator.   Equations will vary.

e. Solve each of the following equations by factoring.

$x^2 + 6x + 9 = 0$ -3
$x^2 - 8x + 16 = 0$ 4
$4x^2 - 20x + 25 = 0$ $\dfrac{5}{2}$

When a quadratic equation has two solutions that are the same number, the solution is called a **double root** of the equation.

f. Use a graphing calculator to find the zeros of each of the following quadratic functions.

$f(x) = x^2 + 6x + 9$ -3
$f(x) = x^2 - 8x + 16$ 4
$f(x) = 4x^2 - 20x + 25$ $\dfrac{5}{2}$

g.   Note the similarity between the quadratic equations in part e and the quadratic functions in part f. Describe the graph of a quadratic function when the corresponding quadratic equation has a double root. How many x-intercepts does the graph have?
Descriptions may vary. For example, the graph just touches the x-axis. The graph has only one x-intercept.

h. Use the zeros of the graphs of the quadratic functions shown at the right to write the quadratic functions.
$f(x) = x^2 - 8x + 12$
$f(x) = x^2 - 4x - 5$

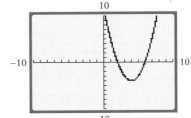
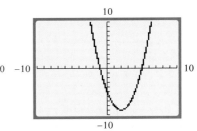

i.   Each of the quadratic equations at the left below is nonfactorable over the integers. Graph the corresponding quadratic functions at the right. Describe the zeros of the functions. Are any of them integers?

$x^2 + 2x - 5 = 0$ $f(x) = x^2 + 2x - 5$
$x^2 - 6x + 1 = 0$ $f(x) = x^2 - 6x + 1$
$x^2 - 3x - 7 = 0$ $f(x) = x^2 - 3x - 7$
None of the zeros of the functions are integers.

Solving Quadratic Equations by Taking Square Roots and by Completing the Square

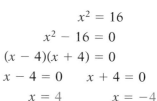

■ Solve Quadratic Equations by Taking Square Roots

INSTRUCTOR NOTE
Another way to demonstrate that $x^2 = a$ implies that $x = \pm\sqrt{a}$ is to rely on the fact that $\sqrt{x^2} = |x|$. Thus $x^2 = a$ implies that $|x| = \sqrt{a}$, or that $x = \pm\sqrt{a}$.

The solution of the quadratic equation $x^2 = 16$ is shown at the right.

$$x^2 = 16$$
$$x^2 - 16 = 0$$
$$(x - 4)(x + 4) = 0$$
$$x - 4 = 0 \qquad x + 4 = 0$$
$$x = 4 \qquad\qquad x = -4$$

Note that the solution is the positive or negative square root of 16, 4 or -4.

The solution can also be found by taking the square root of each side of the equation and writing the positive and negative square roots of the number.

$$x^2 = 16$$
$$\sqrt{x^2} = \sqrt{16}$$
$$x = \pm\sqrt{16}$$
$$x = \pm 4$$

The notation $x = \pm 4$ means $x = 4$ or $x = -4$.

The solutions are 4 and -4.

TAKE NOTE

When a quadratic equation has complex number solutions, the corresponding quadratic function has no x-intercepts.
     The equation $2x^2 + 18 = 0$ has complex number solutions. The graph of $y = 2x^2 + 18$, as shown below, has no x-intercepts.

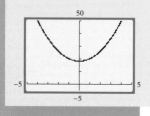

? QUESTION   What are the solutions of the equation $x^2 = 25$?

➡ Solve by taking square roots: $3x^2 = 54$

$$3x^2 = 54$$
$$x^2 = 18 \qquad\qquad$$ • Solve for x^2.
$$\sqrt{x^2} = \sqrt{18} \qquad$$ • Take the square root of each side of the equation.
$$x = \pm\sqrt{18}$$
$$x = \pm 3\sqrt{2} \qquad$$ • Simplify.

$3\sqrt{2}$ and $-3\sqrt{2}$ check as solutions.

The solutions are $3\sqrt{2}$ and $-3\sqrt{2}$.   • Write the solutions.   ⬅

Solving a quadratic equation by taking the square root of each side of the equation can lead to solutions that are complex numbers.

➡ Solve by taking square roots: $2x^2 + 18 = 0$

$$2x^2 + 18 = 0$$
$$2x^2 = -18 \qquad$$ • Solve for x^2.
$$x^2 = -9$$
$$\sqrt{x^2} = \sqrt{-9} \qquad$$ • Take the square root of each side of the equation. Simplify.
$$x = \pm\sqrt{-9}$$
$$x = \pm 3i$$

$3i$ and $-3i$ check as solutions.

The solutions are $3i$ and $-3i$.   • Write the solutions.   ⬅

? ANSWER   If $x^2 = 25$, then $\sqrt{x^2} = \sqrt{25}$, so $x = \pm\sqrt{25} = \pm 5$. The solutions are 5 and -5.

An equation containing the square of a binomial can be solved by taking square roots. This is illustrated in Example 1.

EXAMPLE 1

Solve by taking square roots: $3(x - 2)^2 + 12 = 0$

Solution

$$3(x - 2)^2 + 12 = 0$$

$$3(x - 2)^2 = -12$$

$$(x - 2)^2 = -4 \qquad \bullet \text{ Solve for } (x - 2)^2.$$

$$\sqrt{(x - 2)^2} = \sqrt{-4} \qquad \bullet \text{ Take the square root of each side of the equation. Then simplify.}$$

$$x - 2 = \pm\sqrt{-4}$$

$$x - 2 = \pm 2i$$

$$x - 2 = 2i \qquad x - 2 = -2i \qquad \bullet \text{ Solve for } x.$$

$$x = 2 + 2i \qquad x = 2 - 2i$$

The solutions are $2 + 2i$ and $2 - 2i$. $\qquad \bullet \text{ You should always check the solutions.}$

> **TAKE NOTE**
>
> $(x - 2)^2$ is the square of a binomial.

INSTRUCTOR NOTE
Exercise 22 can be used for a similar in-class example.

Handwritten notes:
$3(x-2)^2 - 12 = 0$
$3(x-2)^2 = 12$
$(x-2)^2 = 4$

YOU TRY IT 1

Solve by taking square roots: $2(x + 1)^2 + 24 = 0$

Solution   See page S37. $-1 + 2i\sqrt{3}, \ -1 - 2i\sqrt{3}$

INSTRUCTOR NOTE
Exercise 30 can be used for a similar in-class example.

EXAMPLE 2

The equation $d = 0.071v^2$ can be used to approximate the distance d required for a car traveling v miles per hour to stop after its brakes are applied. An officer investigating an auto accident noted that the vehicle involved required 144 feet to stop. At what speed was the vehicle traveling before its brakes were applied? Round to the nearest whole number.

State the goal.   The goal is to determine the speed of the car before its brakes were applied.

Devise a strategy.   Substitute 144 for d in the equation $d = 0.071v^2$. Then solve the equation for v.

Solve the problem. $d = 0.071v^2$

$$144 = 0.071v^2 \qquad \bullet \text{ Substitute 144 for } d.$$

$$2028.169 \approx v^2 \qquad \bullet \text{ Solve for } v^2. \text{ Divide each side by } 0.071.$$

$$\sqrt{2028.169} \approx \sqrt{v^2} \qquad \bullet \text{ Take the square root of each side of the equation.}$$

$$45 \approx v \qquad \bullet \text{ The speed cannot be negative. Take the positive square root.}$$

The speed of the vehicle before its brakes were applied was 45 mph.

Check your work.

<table>
<tr><td colspan="2" align="center">**ALGEBRAIC CHECK**</td></tr>
</table>

$$d = 0.071v^2$$
$$d = 0.071(45)^2$$
$$d = 143.775$$
$$d \approx 144$$

GRAPHICAL CHECK

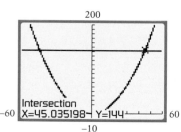

YOU TRY IT 2

An artist wants to paint a circle that has an area of 64 square inches. What should the diameter of the circle be? Round to the nearest tenth.

Solution   See page S37.   9.0 in.

■ Solve Quadratic Equations by Completing the Square

A perfect-square trinomial is the square of a binomial. Some examples of perfect–square trinomials are shown below.

Perfect-Square Trinomial		Square of a Binomial
$x^2 + 8x + 16$	$=$	$(x + 4)^2$
$x^2 - 10x + 25$	$=$	$(x - 5)^2$
$x^2 + 2ax + a^2$	$=$	$(x + a)^2$

For each perfect-square trinomial, the square of one-half the coefficient of x equals the constant term.

$$\left(\frac{1}{2} \text{ coefficient of } x\right)^2 = \text{Constant term}$$

$$x^2 + 8x + 16, \qquad \left(\frac{1}{2} \cdot 8\right)^2 = 16$$

$$x^2 - 10x + 25, \qquad \left[\frac{1}{2}(-10)\right]^2 = 25$$

$$x^2 + 2ax + a^2, \qquad \left(\frac{1}{2} \cdot 2a\right)^2 = a^2$$

This relationship can be used to write the constant term for a perfect-square trinomial. Adding to a binomial the constant term that makes it a perfect-square trinomial is called **completing the square.** In general, to complete the square on $x^2 + bx$, add $\left(\frac{1}{2}b\right)^2$ to $x^2 + bx$.

❓ QUESTION   To complete the square on $x^2 + 6x$, what number must be added to the expression?

❓ ANSWER $\left(\frac{1}{2} \cdot 6\right)^2 = 3^2 = 9$. To complete the square on $x^2 + 6x$, add 9 to the expression.

▼ **Point of Interest**

Early attempts to solve quadratic equations were primarily geometric. The Persian mathematician al-Khwarizmi (c. A.D. 800) essentially completed a square of $x^2 + 12x$ *as shown below.*

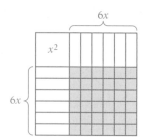

➡ Complete the square on $x^2 - 12x$. Write the resulting perfect-square trinomial as the square of a binomial.

$$\left[\frac{1}{2}(-12)\right]^2 = (-6)^2 = 36 \qquad \bullet \text{ Find the constant term.}$$

$$x^2 - 12x + 36 \qquad\qquad\qquad \bullet \text{ Complete the square on } x^2 - 12x$$
$$\text{by adding the constant term.}$$

$$x^2 - 12x + 36 = (x - 6)^2 \qquad \bullet \text{ Write the resulting perfect-square}$$
$$\text{trinomial as the square of a binomial.} \quad ⬅$$

➡ Complete the square on $z^2 + 3z$. Write the resulting perfect-square trinomial as the square of a binomial.

$$\left(\frac{1}{2} \cdot 3\right)^2 = \left(\frac{3}{2}\right)^2 = \frac{9}{4} \qquad \bullet \text{ Find the constant term.}$$

$$z^2 + 3z + \frac{9}{4} \qquad\qquad\qquad \bullet \text{ Complete the square on } z^2 + 3z \text{ by}$$
$$\text{adding the constant term.}$$

$$z^2 + 3z + \frac{9}{4} = \left(z + \frac{3}{2}\right)^2 \qquad \bullet \text{ Write the resulting perfect-square}$$
$$\text{trinomial as the square of a binomial.} \quad ⬅$$

Though not all quadratic equations can be solved by factoring, any quadratic equation can be solved by completing the square. Add to each side of the equation the term that completes the square. Rewrite the equation in the form $(x + a)^2 = b$. Then take the square root of each side of the equation.

➡ Solve by completing the square: $x^2 - 4x - 14 = 0$

$$x^2 - 4x - 14 = 0$$

$$x^2 - 4x = 14 \qquad \bullet \text{ Add 14 to each side of the equation.}$$

$$x^2 - 4x + 4 = 14 + 4 \qquad \bullet \text{ Add the constant term that}$$
$$\text{completes the square on } x^2 - 4x$$
$$\text{to each side of the equation.}$$

$$\left[\frac{1}{2}(-4)\right]^2 = 4$$

$$(x - 2)^2 = 18 \qquad \bullet \text{ Factor the perfect-square trinomial.}$$
$$\sqrt{(x - 2)^2} = \sqrt{18} \qquad \bullet \text{ Take the square root of each side}$$
$$\text{of the equation.}$$

$$x - 2 = \pm\sqrt{18} \qquad \bullet \text{ Simplify.}$$
$$x - 2 = \pm 3\sqrt{2}$$

$$x - 2 = 3\sqrt{2} \qquad x - 2 = -3\sqrt{2} \qquad \bullet \text{ Solve for } x.$$
$$x = 2 + 3\sqrt{2} \qquad x = 2 - 3\sqrt{2}$$

ALGEBRAIC CHECK

$$x^2 - 4x - 14 = 0 \qquad\qquad x^2 - 4x - 14 = 0$$

$(2 + 3\sqrt{2})^2 - 4(2 + 3\sqrt{2}) - 14$	0	$(2 - 3\sqrt{2})^2 - 4(2 - 3\sqrt{2}) - 14$	0
$4 + 12\sqrt{2} + 18 - 8 - 12\sqrt{2} - 14$	0	$4 - 12\sqrt{2} + 18 - 8 + 12\sqrt{2} - 14$	0
	$0 = 0$		$0 = 0$

The solutions are $2 + 3\sqrt{2}$ and $2 - 3\sqrt{2}$.

Suggested Activity

See Section 9.2 of the *Student Activity Manual* for activities involving completing the square geometrically and completing the square algebraically.

▼ *Point of Interest*

Mathematicians have studied quadratic equations for centuries. Many of the initial equations were a result of trying to solve a geometry problem. One of the most famous, which dates from around 500 B.C., is "squaring the circle." The question was "Is it possible to construct a square whose area is that of a given circle?" For these early mathematicians, to construct meant to draw with only a straightedge and a compass. It was approximately 2300 years later that mathematicians were able to prove that such a construction was impossible.

Suggested Activity

Marc and Elena tried to solve the same equation, one of the form $ax^2 + bx + c = 0$. Unfortunately, each made just one copying error. The solutions of Marc's equation were -2 and $-\frac{3}{2}$ but his value of a was incorrect. The solutions of Elena's equations were 2 and 3 but her value of b was incorrect. Find the solutions of the correct equation.
Answer: -6 and -1

The solutions of this equation can also be checked using a graphing calculator.

First find the decimal approximations of the solutions:

$$2 + 3\sqrt{2} \approx 6.2426$$
$$2 - 3\sqrt{2} \approx -2.2426$$

Then graph $y = x^2 - 4x - 14$, and find the x-coordinates of the x-intercepts:

The x-intercept $(-2.2426, 0)$ is shown in the graph at the left.
The x-intercepts are approximately $(-2.2426, 0)$ and $(6.2426, 0)$.
The solutions check.

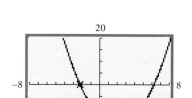

When a, the coefficient of the x^2 term, is not 1, divide each side of the equation by a before completing the square.

➡ Solve by completing the square: $2x^2 - x = 2$

$$2x^2 - x = 2$$

$$\frac{2x^2 - x}{2} = \frac{2}{2}$$
• Divide each side of the equation by the coefficient of x^2.

$$x^2 - \frac{1}{2}x = 1$$
• The coefficient of the x^2 term is now 1.

$$x^2 - \frac{1}{2}x + \frac{1}{16} = 1 + \frac{1}{16}$$
• Add the term that completes the square on $x^2 - \frac{1}{2}x$ to each side of the equation.

$$\left(x - \frac{1}{4}\right)^2 = \frac{17}{16}$$
• Factor the perfect-square trinomial.

$$\sqrt{\left(x - \frac{1}{4}\right)^2} = \sqrt{\frac{17}{16}}$$
• Take the square root of each side of the equation. Then simplify.

$$x - \frac{1}{4} = \pm\sqrt{\frac{17}{16}}$$

$$x - \frac{1}{4} = \pm\frac{\sqrt{17}}{4}$$

$$x - \frac{1}{4} = \frac{\sqrt{17}}{4} \qquad x - \frac{1}{4} = -\frac{\sqrt{17}}{4}$$
• Solve for x.

$$x = \frac{1}{4} + \frac{\sqrt{17}}{4} \qquad x = \frac{1}{4} - \frac{\sqrt{17}}{4}$$

$\dfrac{1 + \sqrt{17}}{4}$ and $\dfrac{1 - \sqrt{17}}{4}$ check as solutions.

The solutions are $\dfrac{1 + \sqrt{17}}{4}$ and $\dfrac{1 - \sqrt{17}}{4}$.   • Write the solutions.

INSTRUCTOR NOTE
To say that students will find completing the square a daunting task is an understatement. The positive side is that the procedure is the same each time.

• Isolate $ax^2 + bx$.
• Multiply by $\dfrac{1}{a}$.
• Complete the square.
• Factor.
• Take the square root.
• Solve for x.

TAKE NOTE

The solution
$\dfrac{1 + \sqrt{17}}{4} \approx 1.2808$.
The solution
$\dfrac{1 - \sqrt{17}}{4} \approx -0.7808$.
The graph of
$y = 2x^2 - x - 2$ is shown below. The x-coordinates of the x-intercepts are approximately -0.7808 and 1.2808.

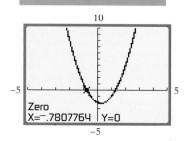

EXAMPLE 3

Solve by completing the square: $4x^2 - 8x + 1 = 0$

INSTRUCTOR NOTE
Exercise 52 can be used for
a similar in-class example.

Solution $4x^2 - 8x + 1 = 0$

$$4x^2 - 8x = -1$$

• Subtract 1 from each side of the equation.

$$\frac{4x^2 - 8x}{4} = \frac{-1}{4}$$

• The coefficient of the x^2 term must be 1. Divide each side by 4.

$$x^2 - 2x = -\frac{1}{4}$$

$$x^2 - 2x + 1 = -\frac{1}{4} + 1$$

• Complete the square. $\left[\frac{1}{2}(-2)\right]^2 = 1$.

$$(x - 1)^2 = \frac{3}{4}$$

• Factor the perfect-square trinomial.

$$\sqrt{(x - 1)^2} = \sqrt{\frac{3}{4}}$$

• Take the square root of each side of the equation. Then simplify.

$$x - 1 = \pm\sqrt{\frac{3}{4}}$$

$$x - 1 = \pm\frac{\sqrt{3}}{2}$$

$$x - 1 = \frac{\sqrt{3}}{2} \qquad x - 1 = -\frac{\sqrt{3}}{2}$$

• Solve for x.

$$x = 1 + \frac{\sqrt{3}}{2} \qquad x = 1 - \frac{\sqrt{3}}{2}$$

$$x = \frac{2 + \sqrt{3}}{2} \qquad x = \frac{2 - \sqrt{3}}{2}$$

ALGEBRAIC CHECK

$$4x^2 - 8x + 1 = 0$$

$$4\left(\frac{2 + \sqrt{3}}{2}\right)^2 - 8\left(\frac{2 + \sqrt{3}}{2}\right) + 1 \;\Big|\; 0$$

$$4 + 4\sqrt{3} + 3 - 8 - 4\sqrt{3} + 1 \;\Big|\; 0$$

$$0 = 0$$

$$4x^2 - 8x + 1 = 0$$

$$4\left(\frac{2 - \sqrt{3}}{2}\right)^2 - 8\left(\frac{2 - \sqrt{3}}{2}\right) + 1 \;\Big|\; 0$$

$$4 - 4\sqrt{3} + 3 - 8 + 4\sqrt{3} + 1 \;\Big|\; 0$$

$$0 = 0$$

The solutions are $\dfrac{2 + \sqrt{3}}{2}$ and $\dfrac{2 - \sqrt{3}}{2}$.

GRAPHICAL CHECK

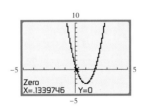
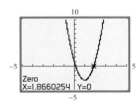

The x-coordinates of the x-intercepts of $y = 4x^2 - 8x + 1$ are approximately 1.8660 and 0.1340.

$$\frac{2 + \sqrt{3}}{2} \approx 1.8660 \text{ and } \frac{2 - \sqrt{3}}{2} \approx 0.1340.$$

YOU TRY IT 3

Solve by completing the square: $4x^2 - 4x - 1 = 0$

Solution   See page S37. $\dfrac{1 + \sqrt{2}}{2}, \dfrac{1 - \sqrt{2}}{2}$.

EXAMPLE 4

The height of an arrow shot upward can be given by the formula $s = v_0t - 16t^2$, where s is the height, v_0 is the initial velocity, and t is the time. Find the time it takes for the arrow to reach a height of 64 feet if it has an initial velocity of 128 feet per second. Round to the nearest hundredth.

INSTRUCTOR NOTE
Exercise 62 can be used for a similar in-class example.

State the goal.   The goal is to find out how long it takes an arrow to reach a height of 64 feet after it has been shot into the air.

Devise a strategy.   Substitute 128 for v_0 and 64 for s in the equation $s = v_0t - 16t^2$. Then solve the equation for t.

Solve the problem. $s = v_0t - 16t^2$

$$64 = 128t - 16t^2$$

• Substitute 64 for s and 128 for v_0.

$$16t^2 - 128t + 64 = 0$$

• Write the quadratic equation in standard form.

$$16(t^2 - 8t + 4) = 0$$

• Factor out the GCF from the trinomial.

$$t^2 - 8t + 4 = 0$$

• Divide each side by 16.

$$t^2 - 8t = -4$$

• $t^2 - 8t + 4$ is nonfactorable over the integers. Solve by completing the square. Subtract 4 from each side.

$$t^2 - 8t + 16 = -4 + 16$$

• Complete the square

$$\left[\frac{1}{2}(-8)\right]^2 = 16$$

$$(t - 4)^2 = 12$$

• Factor the perfect-square trinomial.

$$\sqrt{(t - 4)^2} = \sqrt{12}$$

• Take the square root of each side.

$$t - 4 = \pm 2\sqrt{3}$$

• Simplify.

$$t - 4 = 2\sqrt{3} \qquad t - 4 = -2\sqrt{3}$$

• Solve for t.

$$t = 4 + 2\sqrt{3} \qquad t = 4 - 2\sqrt{3}$$

$$t \approx 7.46 \qquad t \approx 0.54$$

The arrow will be at a height of 64 feet after 0.54 second and after 7.46 seconds.

Check your work.

ALGEBRAIC CHECK

$$
\begin{array}{c|c}
64 = 128t - 16t^2 \\
\hline
64 & 128(0.54) - 16(0.54)^2 \\
64 \approx 64.4544
\end{array}
$$

$$
\begin{array}{c|c}
64 = 128t - 16t^2 \\
\hline
64 & 128(7.46) - 16(7.46)^2 \\
64 \approx 64.4544
\end{array}
$$

GRAPHICAL CHECK

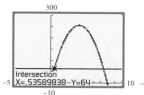
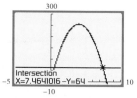

The x-coordinates of the points of intersection of $y = 128t - 16t^2$ and $y = 64$ are approximately 0.54 and 7.46.

YOU TRY IT 4

A rectangular corral is constructed in a pasture. The length of the rectangle is 12 feet longer than the width. The area of the rectangle is 620 square feet. What are the length and width of the rectangle? Round to the nearest hundredth.

Solution   See page S38.   Length: 31.61 ft; width: 19.61 ft

9.2 EXERCISES   Suggested Assignment: 5–65, odds

Topics for Discussion

1. If $x^2 = 16$, then $x = \pm4$. Explain why the $\pm$ sign is necessary.
    Answers will vary.
2. Determine whether the statement is always true, sometimes true, or never true.
    a. A quadratic equation can be solved by factoring.
    b. A quadratic equation can be solved by taking square roots.
    c. A quadratic equation can be solved by completing the square.
    d. A quadratic equation has two different real roots.
    e. To complete the square on $3x^2 + 6x$, add 9 to the expression.
    a. Sometimes true   b. Sometimes true   c. Always true   d. Sometimes true   e. Never true
3. Explain how to complete the square of $x^2 + bx$.   Explanations will vary.
    For example, add the square of one-half of b to the expression.
4. Describe the steps used to solve a quadratic equation by completing the square.   Descriptions may vary. For example: (1) Isolate $ax^2 + bx$. (2) Multiply both sides by the reciprocal of a. (3) Complete the square on $x^2 + \frac{b}{a}x$. (4) Factor the trinomial. (5) Take the square root of each side. (6) Solve for x.

■ Solve Quadratic Equations by Taking Square Roots

Solve by taking square roots.

5. $4x^2 - 81 = 0$ $\pm\frac{9}{2}$

6. $9x^2 - 16 = 0$ $\pm\frac{4}{3}$

7. $y^2 + 49 = 0$ $\pm7i$

8. $z^2 + 16 = 0$ $\pm4i$

9. $v^2 - 48 = 0$ $\pm4\sqrt{3}$

10. $s^2 - 32 = 0$ $\pm4\sqrt{2}$

11. $z^2 + 18 = 0$ $\pm3i\sqrt{2}$

12. $t^2 + 27 = 0$ $\pm3i\sqrt{3}$

13. $(x - 1)^2 = 36$ $-5, 7$

14. $(x + 2)^2 = 25$ $-7, 3$

15. $3(y + 3)^2 = 27$ $-6, 0$

16. $4(s - 2)^2 = 36$ $-1, 5$

17. $5(x + 2)^2 = -125$ $-2 \pm 5i$

18. $3(x - 9)^2 = -27$ $9 \pm 3i$

19. $\left(x - \dfrac{2}{5}\right)^2 = \dfrac{9}{25}$ $-\dfrac{1}{5}, 1$

20. $\left(y + \dfrac{1}{3}\right)^2 = \dfrac{4}{9}$ $-1, \dfrac{1}{3}$

21. $3\left(x - \dfrac{5}{3}\right)^2 = \dfrac{4}{3}$ $\dfrac{7}{3}, 1$

22. $2\left(x + \dfrac{3}{5}\right)^2 = \dfrac{8}{25}$ $-1, -\dfrac{1}{5}$

23. $(x + 5)^2 - 6 = 0$ $-5 \pm\sqrt{6}$

24. $(t - 1)^2 - 15 = 0$ $1 \pm\sqrt{15}$

25. $(z + 1)^2 = 12$ $-1 \pm 2\sqrt{3}$

26. $(r - 2)^2 + 28 = 0$ $2 \pm 2i\sqrt{7}$

27. $\left(z - \dfrac{1}{2}\right)^2 - 20 = 0$ $\dfrac{1 \pm 4\sqrt{5}}{2}$

28. $\left(r - \dfrac{3}{2}\right)^2 + 48 = 0$ $\dfrac{3}{2} \pm 4i\sqrt{3}$

e Technology   On a certain type of road surface, the equation
^{2} can be used to approximate the distance d a car traveling
hour will slide when its brakes are applied. After applying the
., ... owner of a car involved in an accident skidded 40 feet. Did
the traffic officer investigating the accident issue the driver a ticket for
speeding if the speed limit was 30 miles per hour?   No ($v \approx 27.0$ mph)

30. *Investments*   The value P of an initial investment of A dollars after
2 years is given by $P = A(1 + r)^2$, where r is the annual percentage rate
earned by the investment. If an initial investment of $5000 grew to a
value of $5724.50 in 2 years, what was the annual percentage rate?   7%

31. *Energy*   The kinetic energy, E, of a moving body is given by $E = \frac{1}{2}mv^2$,
where m is the mass of the moving body and v is the velocity in meters
per second. What is the velocity of a moving body whose mass is 25 kilograms and whose kinetic energy is 250 newton-meters? Round to the
nearest hundredth.   4.47 m/s

32. *Geometry*   The circles shown are concentric. The inner circle has a radius of 4 centimeters, and the area of the shaded region is 20π square
centimeters. What is the diameter of the larger circle?   12 cm

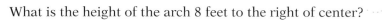

33. *Travel*   Two sisters head off on their bicycles starting from the same
point at the same time. One is riding east at 2.5 miles per hour, and the
other is riding north at 6 miles per hour. After how many hours will they
be 26 miles apart?   4 h

34. *Geometry*   A solid cylinder has a conical core removed so that the base
of the cylinder is the base of the cone, and the vertex of the cone is on
the base of the cylinder. If the height of the cylinder is 10 inches and the
volume of the part of the cylinder that is left after the cone is removed is
20π square inches, what is the radius of the cylinder? The formula for
the volume of a cylinder is $V = \pi r^2 h$. The formula for the volume of a
cone is $V = \frac{1}{3}\pi r^2 h$. $\sqrt{3}$ in.

35. *Mathematics*   Given that A and B are real numbers such that $A + B = A(B)$
and $A - B = 2$, find the value of B. $\sqrt{2}$ or $-\sqrt{2}$

36. *Construction*   The height of an arch is given by the equation
$h(x) = -\frac{3}{64}x^2 + 27$, where $-24 \le x \le 24$ and $|x|$ is the distance
in feet from the center of the arch.
   a.   What is the height of the arch 8 feet to the right of center?
   b.   To the nearest hundredth, how far from the center is the
      arch 8 feet tall?
   a. 24 ft   b. 20.13 ft

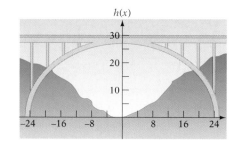

■ Solve Quadratic Equations by Completing the Square

Solve by completing the square.

37. $r^2 + 4r - 7 = 0$ $-2 \pm \sqrt{11}$

38. $s^2 + 6s - 1 = 0$ $-3 \pm \sqrt{10}$

39. $x^2 - 6x + 7 = 0$ $3 \pm \sqrt{2}$

40. $y^2 + 8y + 13 = 0$ $-4 \pm \sqrt{3}$

41. $z^2 - 2z + 2 = 0$ $1 \pm i$

42. $t^2 - 4t + 8 = 0$ $2 \pm 2i$

First try to solve by factoring. If you are unable to solve the equation by factoring, solve the equation by completing the square.

43. $t^2 - t - 1 = 0$ $\dfrac{1 \pm \sqrt{5}}{2}$

44. $u^2 - u - 7 = 0$ $\dfrac{1 \pm \sqrt{29}}{2}$

45. $p^2 + 6p = -13$ $-3 \pm 2i$

46. $x^2 + 4x = -20$ $-2 \pm 4i$

47. $y^2 - 2y = 17$ $1 \pm 3\sqrt{2}$

48. $x^2 + 10x = 7$ $-5 \pm 4\sqrt{2}$

49. $2y^2 + 3y + 1 = 0$ $-1, -\dfrac{1}{2}$

50. $2t^2 + 5t - 3 = 0$ $-3, \dfrac{1}{2}$

51. $4x^2 - 4x + 5 = 0$ $\dfrac{1}{2} \pm i$

52. $4t^2 - 4t + 17 = 0$ $\dfrac{1}{2} \pm 2i$

53. $2s^2 = 4s + 5$ $\dfrac{2 \pm \sqrt{14}}{2}$

54. $3u^2 = 6u + 1$ $\dfrac{3 \pm 2\sqrt{3}}{3}$

Solve by completing the square. Approximate the solutions to the nearest thousandth.

55. $z^2 + 2z = 4$ $-3.236, 1.236$

56. $t^2 - 4t = 7$ $-1.317, 5.317$

57. $2x^2 = 4x - 1$ $0.293, 1.707$

58. $3y^2 = 5y - 1$
$0.232, 1.434$

59. $4z^2 + 2z - 1 = 0$
$-0.809, 0.309$

60. $4w^2 - 8w = 3$
$-0.323, 2.323$

61. *Mathematics*   What number is equal to one less than its square?
$\dfrac{1 + \sqrt{5}}{2}$ or $\dfrac{1 - \sqrt{5}}{2}$

62. *Physics*   A rock is tossed upward from the top of a cliff that is 74 feet above the ocean. The height h, in feet, of the rock above the ocean t seconds after it has been released is given by the equation $h = -16t^2 + 64t + 74$. How many seconds after the rock is released is it 10 feet above the ocean? Round to the nearest hundredth.   4.83 s

63. *Day Care Management*   The area of a rectangular playground in a child care center is 250 square feet. The width of the playground is 20 feet less than twice the length. Fencing must be purchased by the whole foot and costs $6.95 per foot. Find the cost of fencing purchased to surround the playground. $444.80

64. *Forestry*   Will Harris, the owner of the Evergreen Tree Farm, always arranges his trees in a square array. This year, after forming a square array, Will finds there are still 46 trees unplanted. He buys 13 more trees and then adds all the trees to the original square, thus forming a square array that is one row larger than the previous square. Find the total number of trees Will planted this year.   900 trees

74 ft

65. *Sports*   The height, h, in feet, of a baseball above the ground t seconds after it is hit can be approximated by the equation $h = -16t^2 + 70t + 4$.

4 ft

    a. Use the given equation to determine when the ball will hit the ground. Round to the nearest thousandth.

    b. After a baseball is hit, there are two quantities that can be considered. One is the equation in part a. The second is the distance s, in feet, the ball is from home plate t seconds after it is hit. A model of this situation is given by $s = 44.5t$. Using these equations, determine whether the ball will clear a fence 325 feet from home plate. Round to the nearest tenth.

    a. 4.431 s   b. No. The ball will have gone only 197.2 ft when it hits the ground.

66. If $\dfrac{2}{x} - x = 2$, find the value of $\dfrac{4}{x^2} + x^2$.   8

Applying Concepts

Solve for x.

67. $2a^2x^2 = 32b^2$ $\pm\dfrac{4b}{a}$

68. $5y^2x^2 = 125z^2$ $\pm\dfrac{5z}{y}$

69. $(x + a)^2 - 4 = 0$ $-a \pm 2$

70. $2(x - y)^2 - 8 = 0$ $y \pm 2$

71. $(2x - 1)^2 = (2x + 3)^2$ $-\dfrac{1}{2}$

72. $(x - 4)^2 = (x + 2)^2$ 1

73. Show that the solutions of the equation $ax^2 + c = 0$, $a > 0$, $c > 0$, are $\dfrac{\sqrt{ca}}{a}i$ and $-\dfrac{\sqrt{ca}}{a}i$.   The complete solution is in the Solutions Manual.

74. If $3x^2 - 7x + 6 = a(x - 2)^2 + b(x - 2) + c$ is true for all values of x, what is the value of $a + b + c$?   12

75. *Chemistry*   A chemical reaction between carbon monoxide and water vapor is used to increase the proportion of hydrogen gas in certain gas mixtures. In the process, carbon dioxide is also formed. For a certain reaction, the concentration of carbon dioxide x, in moles per liter, is given by the equation $0.58 = \dfrac{x^2}{(0.02 - x)^2}$. Solve this equation for x. Round to the nearest ten-thousandth.   0.0086 mole per liter

76. *Geometry*   A perfectly spherical scoop of fudge ripple ice cream is placed in a cone as shown at the right. How far is the bottom of the scoop of ice cream from the bottom of the cone? Round to the nearest tenth. (*Hint:* A line segment from the center of the scoop of ice cream to the point at which the ice cream touches the cone is perpendicular to the edge of the cone.)   2.3 in.

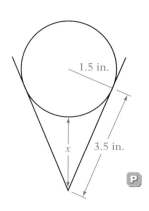

1.5 in.

x 3.5 in.

77. *Astronomy*   You have been hired by an observatory to track meteorites and determine whether they will strike Earth. The equation of Earth's path is $x^2 + y^2 = 40$. The first meteorite you observe is moving along a path whose equation is $18x - y^2 = -144$. Will the meteorite strike Earth?

No (When $18x + 144$ is substituted for y^2 in $x^2 + y^2 = 40$, the solutions of the resulting equation are complex numbers.)

EXPLORATION

1. *Construction Materials* Building materials such as steel, aluminum, concrete, and brick expand as a result of increases in temperature. This is why, for example, fillers are placed between the cement slabs of a sidewalk.

Suppose you have a roof truss that is 100 feet long and is securely fastened at both ends. We will assume that the buckle is linear. (Although the buckle is not linear, the linear assumption suffices as a reasonable approximation.) Let the height of the buckle be x feet, as shown in the figure below.

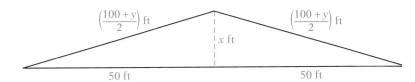

Let the percent increase due to swelling be y. Then, for one-half the 100-foot length,

Length after buckling = length before buckling + increase in length

$$= 50 + \left(\frac{y}{100}\right) 50$$

$$= 50 + \frac{y}{2}$$

$$= \frac{100 + y}{2}$$

a. Using the figure above and the Pythagorean Theorem, show how to derive the equation $y^2 + 200y - 4x^2 = 0$.

Begin with $x^2 + 50^2 = \left(\frac{100 + y}{2}\right)^2$. The complete derivation is in the Solutions Manual.

b. Solve the equation in part a for x. Then calculate the amount of buckling for each of the following materials. Round answers to the nearest tenth.

$$\text{Steel: } y = 0.06$$
$$\text{Aluminum: } y = 0.12$$
$$\text{Concrete: } y = 0.05$$
$$\text{Brick: } y = 0.03$$

$x = \pm\dfrac{\sqrt{y^2 + 200y}}{2}$; Steel: 1.7 ft; Aluminum: 2.5 ft; Concrete: 1.6 ft; Brick: 1.2 ft

c. Which of the four materials in part b expands the most? The least?

The most: aluminum; the least: brick

Solving Quadratic Equations by Using the Quadratic Formula and by Graphing

▼ *Point of Interest*

Although mathematicians have studied quadratic equations since about 500 B.C., it was not until the 18th century that the formula was written as it is today.

Of further note, the word quadratic *has the same Latin root as does the word* square.

INSTRUCTOR NOTE
You may want to make a case for needing another method of solving quadratic equations, in addition to those already presented. For example, if $a = 2.7$ in $ax^2 + bx + c = 0$, one might want to use a method other than completing the square.

Also, you may want the first quadratic equation you solve using the quadratic formula to be one that has integer solutions (one that would have factored), illustrating to your students that the quadratic formula works!

INSTRUCTOR NOTE
One of the difficulties students have with the quadratic formula is correct substitution. Have them first make sure that the equation is in standard form. Then circling a, b, and c may help.

■ Solve Quadratic Equations by Using the Quadratic Formula

In the previous section, the following problem was solved by completing the square:

The height of an arrow shot upward can be given by the formula $s = v_0 t - 16t^2$, where s is the height, v_0 is the initial velocity, and t is the time since the arrow was released. Find the time it takes for the arrow to reach a height of 64 feet if it has an initial velocity of 128 feet per second. Round to the nearest hundredth.

This problem can also be solved by using the quadratic formula as shown later in Example 4. The quadratic formula can be derived by applying the method of completing the square to the standard form of a quadratic equation. (See Exercise 70 in the exercise set accompanying this section.) This formula, which can be used to solve any quadratic equation, is given below.

> **The Quadratic Formula**
> The solutions of $ax^2 + bx + c = 0$, $a \neq 0$, are
> $$\frac{-b + \sqrt{b^2 - 4ac}}{2a} \quad \text{and} \quad \frac{-b - \sqrt{b^2 - 4ac}}{2a}$$
> The quadratic formula is frequently written in the form
> $$x = \frac{-b \pm \sqrt{b^2 - 4ac}}{2a}$$

➡ Solve by using the quadratic formula: $4x^2 = 8x - 13$

$$4x^2 = 8x - 13$$
$$4x^2 - 8x + 13 = 0$$ • Write the equation in standard form.
$$a = 4, b = -8, c = 13$$ • Find the values of a, b, and c.
$$x = \frac{-b \pm \sqrt{b^2 - 4ac}}{2a}$$ • Replace a, b, and c in the quadratic formula by their values.
$$= \frac{-(-8) \pm \sqrt{(-8)^2 - 4 \cdot 4 \cdot 13}}{2 \cdot 4}$$
$$= \frac{8 \pm \sqrt{64 - 208}}{8} = \frac{8 \pm \sqrt{-144}}{8}$$ • Simplify.
$$= \frac{8 \pm 12i}{8} = \frac{2 \pm 3i}{2}$$
$$= 1 \pm \frac{3}{2}i$$ • Write the answer in the form $a + bi$.

They must remember that the sign that precedes a number is the sign of the number. For the equation at the right, we have

$$\boxed{4}x^2 \boxed{-8}x \boxed{+13} = 0$$

with a, b, c labeled above.

🖩 *Suggested Activity*

See Section 9.3 of the *Student Activity Manual* for an activity on using technology to evaluate the quadratic formula. Also included is an activity on solving cubic equations by graphing.

Algebraic Check

$$4x^2 = 8x - 13$$

$$4\left(1 + \frac{3}{2}i\right)^2 \,\bigg|\, 8\left(1 + \frac{3}{2}i\right) - 13$$

$$4\left(1 + 3i - \frac{9}{4}\right) \,\bigg|\, 8 + 12i - 13$$

$$4\left(-\frac{5}{4} + 3i\right) \,\bigg|\, -5 + 12i$$

$$-5 + 12i = -5 + 12i$$

$$4x^2 = 8x - 13$$

$$4\left(1 - \frac{3}{2}i\right)^2 \,\bigg|\, 8\left(1 - \frac{3}{2}i\right) - 13$$

$$4\left(1 - 3i - \frac{9}{4}\right) \,\bigg|\, 8 - 12i - 13$$

$$4\left(-\frac{5}{4} - 3i\right) \,\bigg|\, -5 - 12i$$

$$-5 - 12i = -5 - 12i$$

The solutions check. The solutions are $1 + \frac{3}{2}i$ and $1 - \frac{3}{2}i$. ⬅

❓ QUESTION   Why can't the solutions be checked using a graphing calculator to graph $y = 4x^2 - 8x + 13$ and finding the x-intercepts?

EXAMPLE 1

Solve by using the quadratic formula: $2x^2 - x = 5$

Solution $2x^2 - x = 5$

$$2x^2 - x - 5 = 0$$

$$x = \frac{-b \pm \sqrt{b^2 - 4ac}}{2a}$$

$$= \frac{-(-1) \pm \sqrt{(-1)^2 - 4(2)(-5)}}{2 \cdot 2}$$

$$= \frac{1 \pm \sqrt{1 + 40}}{4}$$

$$= \frac{1 \pm \sqrt{41}}{4}$$

INSTRUCTOR NOTE
Exercise 14 can be used for a similar in-class example.

• Write the equation in standard form. $a = 2, b = -1, c = -5$

• Replace a, b, and c in the quadratic formula by their values. Then simplify.

ALGEBRAIC CHECK

$$2x^2 - x = 5$$

$$2\left(\frac{1 + \sqrt{41}}{4}\right)^2 - \left(\frac{1 + \sqrt{41}}{4}\right) \,\bigg|\, 5$$

$$2\left(\frac{1 + 2\sqrt{41} + 41}{16}\right) - \left(\frac{1 + \sqrt{41}}{4}\right) \,\bigg|\, 5$$

$$\frac{2\sqrt{41} + 42}{8} - \frac{2 + 2\sqrt{41}}{8} \,\bigg|\, 5$$

$$\frac{40}{8} \,\bigg|\, 5$$

$$5 = 5$$

$$2x^2 - x = 5$$

$$2\left(\frac{1 - \sqrt{41}}{4}\right)^2 - \left(\frac{1 - \sqrt{41}}{4}\right) \,\bigg|\, 5$$

$$2\left(\frac{1 - 2\sqrt{41} + 41}{16}\right) - \left(\frac{1 - \sqrt{41}}{4}\right) \,\bigg|\, 5$$

$$\frac{-2\sqrt{41} + 42}{8} - \frac{2 - 2\sqrt{41}}{8} \,\bigg|\, 5$$

$$\frac{40}{8} \,\bigg|\, 5$$

$$5 = 5$$

❓ ANSWER   When a quadratic equation has complex number solutions, the corresponding quadratic function has no x-intercepts.

GRAPHICAL CHECK

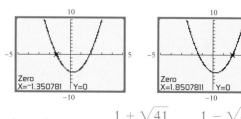

The solutions are $\dfrac{1 + \sqrt{41}}{4}$ and $\dfrac{1 - \sqrt{41}}{4}$.

The x-coordinates of the x-intercepts of $y = 2x^2 - x - 5$ are approximately 1.8508 and -1.3508.

$$\dfrac{1 + \sqrt{41}}{4} \approx 1.8508 \text{ and } \dfrac{1 - \sqrt{41}}{4} \approx -1.3508.$$

YOU TRY IT 1

Solve by using the quadratic formula: $4x^2 = 4x - 1$

Solution   See page S38. $\dfrac{1}{2}$

For the equation $4x^2 = 8x - 13$, solved before Example 1, the solutions are two different complex numbers. In Example 1, the solutions are two different real numbers. In You Try It 1, the equation has two solutions that are the same number; this is called a **double root.**

In the quadratic formula, the quantity $b^2 - 4ac$ is called the **discriminant.** When a, b, and c are real numbers, the discriminant determines whether a quadratic equation will have a double root, two real number solutions that are not equal, or two complex number solutions.

INSTRUCTOR NOTE
Rather than presenting students with this material on the discriminant, you may prefer to have them do the Exploration at the end of the exercise set.

The Effect of the Discriminant on the Solutions of a Quadratic Equation

1. If $b^2 - 4ac = 0$, the equation has one real number solution, a double root.
2. If $b^2 - 4ac > 0$, the equation has two real number solutions that are not equal.
3. If $b^2 - 4ac < 0$, the equation has two complex number solutions.

The equation $x^2 - 4x - 5 = 0$ has two real number solutions because the discriminant is greater than zero.

$a = 1, b = -4, c = -5$
$b^2 - 4ac = (-4)^2 - 4(1)(-5)$
$\qquad\qquad = 16 + 20$
$\qquad\qquad = 36$
$\quad 36 > 0$

INSTRUCTOR NOTE
Exercise 32 can be used for a similar in-class example.

EXAMPLE 2

Use the discriminant to determine whether $4x^2 - 2x + 5 = 0$ has one real number solution, two real number solutions, or two complex number solutions.

Solution $4x^2 - 2x + 5 = 0$ • $a = 4, b = -2, c = 5$

$b^2 - 4ac = (-2)^2 - 4(4)(5)$

$= 4 - 80$

$= -76$

$-76 < 0$ • The discriminant is less than 0.

The equation has two complex number solutions.

YOU TRY IT 2

Use the discriminant to determine whether $3x^2 - x - 1 = 0$ has one real number solution, two real number solutions, or two complex number solutions.

Solution   See page S38.   two real number solutions

Because there is a connection between the solutions of $ax^2 + bx + c = 0$ and the x-intercepts of the graph of $y = ax^2 + bx + c$, the discriminant can be used to determine the number of x-intercepts of a parabola.

Suggested Activity Ⓟ

Have students draw examples of graphs of parabolas in which $b^2 - 4ac = 0$, $b^2 - 4ac > 0$, and $b^2 - 4ac < 0$ for (1) $a > 0$ and (2) $a < 0$.

The Effect of the Discriminant on the Number of x-Intercepts of a Parabola

1. If $b^2 - 4ac = 0$, the parabola has one x-intercept.
2. If $b^2 - 4ac > 0$, the parabola has two x-intercepts.
3. If $b^2 - 4ac < 0$, the parabola has no x-intercepts.

The graph of the equation $y = 2x^2 - x + 2$ has no x-intercepts because the discriminant is less than zero.

$a = 2, b = -1, c = 2$

$b^2 - 4ac = (-1)^2 - 4(2)(2)$

$= 1 - 16$

$= -15$

$-15 < 0$

INSTRUCTOR NOTE
Exercise 46 can be used for a similar in-class example.

EXAMPLE 3

Use the discriminant to determine the number of x-intercepts of the parabola whose equation is $y = x^2 - 6x + 9$.

Solution $y = x^2 - 6x + 9$ • $a = 1, b = -6, c = 9$

$b^2 - 4ac = (-6)^2 - 4(1)(9)$

$= 36 - 36$

$= 0$ • The discriminant is equal to 0.

The parabola has one x-intercept.

YOU TRY IT 3

Use the discriminant to determine the number of x-intercepts of the parabola whose equation is $y = x^2 - x - 6$.

Solution   See page S38.   two x-intercepts

In Example 4, we use the quadratic formula to solve the problem presented at the beginning of the section.

EXAMPLE 4

The height of an arrow shot upward can be given by the formula $s = v_0 t - 16t^2$, where s is the height, v_0 is the initial velocity, and t is the time since the arrow was released. Find the time it takes for the arrow to reach a height of 64 feet if it has an initial velocity of 128 feet per second. Round to the nearest hundredth.

State the goal.   The goal is to find out how long it takes an arrow to reach a height of 64 feet after it has been shot into the air.

Devise a strategy.   Substitute 128 for v_0 and 64 for s in the equation $s = v_0 t - 16t^2$. Then solve the equation for t.

Solve the problem.

$$s = v_0 t - 16t^2$$

$$64 = 128t - 16t^2 \qquad \bullet \text{ Substitute 64 for } s \text{ and 128 for } v_0.$$

$$16t^2 - 128t + 64 = 0 \qquad \bullet \text{ Write the quadratic equation in standard form.}$$

$$16(t^2 - 8t + 4) = 0 \qquad \bullet \text{ Factor out the GCF from the trinomial.}$$

$$t^2 - 8t + 4 = 0 \qquad \bullet \text{ Divide each side by 16.}$$

$$t = \frac{-b \pm \sqrt{b^2 - 4ac}}{2a} \qquad \begin{array}{l}\bullet\ t^2 - 8t + 4 \text{ is nonfactorable over}\\ \text{the integers. Solve by using the}\\ \text{quadratic formula.}\end{array}$$

$$= \frac{-(-8) \pm \sqrt{(-8)^2 - 4 \cdot 1 \cdot 4}}{2 \cdot 1} \qquad \bullet\ a = 1, b = -8, c = 4$$

$$= \frac{8 \pm \sqrt{64 - 16}}{2} = \frac{8 \pm \sqrt{48}}{2}$$

$$= \frac{8 \pm 4\sqrt{3}}{2} = 4 \pm 2\sqrt{3}$$

$$4 + 2\sqrt{3} \approx 7.46 \qquad 4 - 2\sqrt{3} \approx 0.54$$

The arrow will be at a height of 64 feet after 0.54 second and after 7.46 seconds.

Check your work.

ALGEBRAIC CHECK

$$\frac{64 = 128t - 16t^2}{64 \mid 128(0.54) - 16(0.54)^2}$$

$$64 \approx 64.4544$$

$$\frac{64 = 128t - 16t^2}{64 \mid 128(7.46) - 16(7.46)^2}$$

$$64 \approx 64.4544$$

GRAPHICAL CHECK

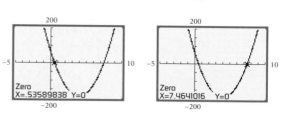

The x-coordinates of the x-intercepts of $y = 16t^2 - 128t + 64$ are approximately 0.54 and 7.46.

Note that these are the same solutions we obtained in Example 4 of the previous section when we solved the equation by completing the square.

INSTRUCTOR NOTE
Exercise 50 can be used for a similar in-class example.

TAKE NOTE

The equation can be written in standard form as
$0 = -16t^2 + 128t - 64.$
The solutions will be the same.

Suggested Activity

In this graphical check, we graphed the equation $y = 16t^2 - 128t + 64$ and found the x-coordinates of the x-intercepts. In the previous section, we graphed $Y_1 = 128t - 16t^2$ and $Y_2 = 64$ and found the x-coordinates of the points of intersection. Have students discuss why the two methods produce the same x-coordinates.

YOU TRY IT 4

A piece of fabric is to be cut in the shape of a triangle to make a kite. The height of the triangle is to be 10 inches less than the base of the triangle. The area of the triangle is to be 150 square inches. Find the base and the height of the triangle to be cut from the fabric. Round to the nearest thousandth.

Solution See page S38.   Base: 23.028 in.; height: 13.028 in.

■ Solve Quadratic Equations Graphically

We have presented different methods of solving quadratic equations: by factoring, by taking square roots, by completing the square, and by using the quadratic formula. Generally, if a quadratic equation is factorable, that is the easiest method to use. Taking square roots is used when $b = 0$; for example, $4x^2 - 6 = 0$. Completing the square is used in other situations in mathematics and is a method used to derive the quadratic formula. The quadratic formula is used for those quadratic equations that are nonfactorable over the integers or for which it is not easy to determine the factors.

Now let's look at a situation where the most reasonable approach may be to use a graphing calculator.

INSTRUCTOR NOTE
Exercise 58 can be used for a similar in-class example.

EXAMPLE 5

According to the *Keenan Report #1*, "Exchanges in the Internet Economy," the actual and projected consumer-to-consumer transactions, in billions, from online auctions, such as eBay, are as shown in the chart below. An equation that models these data is $y = 0.6144x^2 - 7.7658x + 23.9343$, where y is the consumer-to-consumer transactions, in billions of dollars, in year x, and $x = 10$ represents the year 2000. During which year does the model predict that transactions will reach $30 billion?

Year	1995	1996	1997	1998	1999	2000	2001
Transactions (in billions)	$0	$.03	$.13	$.75	$3.5	$7.6	$13.1

State the goal. The goal is to determine the year in which transactions will be $30 billion. This means that we want to find the value of x when y is $30 billion.

Devise a strategy. Use a graphing calculator to graph $Y_1 = 0.6144x^2 - 7.7658x + 23.9343$ and $Y_2 = 30$. Find the point of intersection.

Solve the problem.

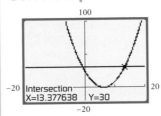

We want the rightmost point of intersection, because we are asking a question about the years after 2001.

The x-coordinate of the intersection of $Y_1 = 0.6144x^2 - 7.7658x + 23.9343$ and $Y_2 = 30$ is approximately 13.38. Because $x = 10$ corresponds to 2000, $x = 13$ corresponds to 2003.

The model predicts that transactions will reach $30 billion in 2003.

CALCULATOR NOTE
Note that our strategy could be to solve the quadratic equation $30 = 0.6144x^2 - 7.7658x + 23.9343$ by using the quadratic formula. However, given the number of digits in the coefficients and constant, using a graphing calculator may be less "messy."

> **Check your work.** One method is to substitute 13.377638 for x in $0.6144x^2 - 7.7658x + 23.9343$ and evaluate. The result should be close to 30. Another method is to use the quadratic formula to solve the equation $30 = 0.6144x^2 - 7.7658x + 23.9343$.

YOU TRY IT 5

During which year does the model given in Example 5 predict that consumer-to-consumer transactions will reach $50 billion?

Solution See page S38.   2005

9.3 EXERCISES   Suggested Assignment: 7–63, odds

Topics for Discussion

1. **a.** Write the quadratic formula.
    b. What is the quadratic formula used for?
    c. What does each variable in the quadratic formula represent?
    a. $x = \dfrac{-b \pm \sqrt{b^2 - 4ac}}{2a}$ **b.** It is used to solve quadratic equations.
    c. a is the coefficient of x^2, b is the coefficient of x, and c is the constant in the quadratic equation $ax^2 + bx + c = 0$.

2. **a.** If a quadratic equation is solved using the quadratic formula, and the result is $x = \dfrac{1 \pm \sqrt{23}}{3}$, what are the solutions of the equation?
    b. If a quadratic equation is solved using the quadratic formula, and the result is $x = \dfrac{2 \pm 6}{4}$, what are the solutions of the equation?
    a. $\dfrac{1 + \sqrt{23}}{3}$ and $\dfrac{1 - \sqrt{23}}{3}$ **b.** -1 and 2

3. What is a double root of a quadratic equation?
    When a quadratic equation has two solutions that are the same number, it is called a double root of the equation.

4. **a.** What is the discriminant?
    b. What formula is the discriminant taken from?
    c. What can the discriminant be used to determine?
    a. For an equation of the form $ax^2 + bx + c = 0$, the discriminant is the quantity $b^2 - 4ac$.   **b.** It is the quantity under the radical sign in the quadratic formula.
    c. It can be used to determine the nature of the solutions of a quadratic equation or the number of x-intercepts of a parabola.

5. Suppose you have just solved a quadratic equation by using the quadratic formula. Describe two methods by which you can check that your solutions are correct.
    Answers will vary. For example, students might provide descriptions of substituting the solutions back into the original equation, using a graphing calculator, or solving the equation by completing the square.

■ Solve Quadratic Equations by Using the Quadratic Formula

Solve by using the quadratic formula.

6. $x^2 - 4x - 2 = 0$ $2 \pm \sqrt{6}$

7. $y^2 - 8y - 1 = 0$ $4 \pm \sqrt{17}$

8. $z^2 - 3z - 40 = 0$ $-5, 8$

9. $y^2 + 5y - 36 = 0$ $-9, 4$ **10.** $v^2 = 4v + 8$ $2 \pm 2\sqrt{3}$ **11.** $w^2 = 8w + 72$ $4 \pm 2\sqrt{22}$

First try to solve by factoring. If you are unable to solve the equation by factoring, solve the equation by using the quadratic formula.

12. $t^2 - 2t - 11 = 0$ $1 \pm 2\sqrt{3}$ **13.** $u^2 - 2u - 7 = 0$ $1 \pm 2\sqrt{2}$ **14.** $2p^2 - 2p = 1$ $\dfrac{1 \pm \sqrt{3}}{2}$

15. $4x^2 - 4x = 1$ $\dfrac{1 \pm \sqrt{2}}{2}$ **16.** $z^2 + 2z + 2 = 0$ $-1 \pm i$ **17.** $y^2 - 4y + 5 = 0$ $2 \pm i$

18. $6w^2 = 19w - 10$ $\dfrac{2}{3}, \dfrac{5}{2}$ **19.** $4t^2 + 8t + 3 = 0$ $-\dfrac{3}{2}, -\dfrac{1}{2}$ **20.** $x^2 - 2x + 5 = 0$ $1 \pm 2i$

21. $t^2 + 6t + 13 = 0$ $-3 \pm 2i$ **22.** $2s^2 + 6s + 5 = 0$ $-\dfrac{3}{2} \pm \dfrac{1}{2}i$ **23.** $2u^2 + 2u + 13 = 0$ $-\dfrac{1}{2} \pm \dfrac{5}{2}i$

Solve by using the quadratic formula. Approximate the solutions to the nearest thousandth.

24. $z^2 + 6z = 6$ $-6.873, 0.873$ **25.** $t^2 = 8t - 3$ $0.394, 7.606$ **26.** $r^2 - 2r - 4 = 0$ $-1.236, 3.236$

27. $w^2 + 4w - 1 = 0$
$-4.236, 0.236$

28. $3t^2 = 7t + 1$
$-0.135, 2.468$

29. $2y^2 = y + 5$
$-1.351, 1.851$

Use the discriminant to determine whether the quadratic equation has one real number solution, two real number solutions, or two complex number solutions.

30. $2z^2 - z + 5 = 0$ Two complex   **31.** $5t^2 + 2 = 0$ Two complex   **32.** $9x^2 - 12x + 4 = 0$ One real

33. $4y^2 + 20y + 25 = 0$ One real   **34.** $2v^2 - 3v - 1 = 0$ Two real   **35.** $3w^2 + 3w - 2 = 0$ Two real

36. *Sports* The height h, in feet, of a baseball above the ground t seconds after it is hit by a Little Leaguer can be approximated by the equation $h = -0.01t^2 + 2t + 3.5$. Does the ball reach a height of 100 feet?
Yes. (The discriminant is greater than zero.)

37. *Sports* The height h, in feet, of an arrow shot upward can be given by the equation $h = 128t - 16t^2$, where t is the time in seconds. Does the arrow reach a height of 275 feet?   No. (The discriminant is less than zero.)

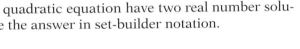

3.5 ft

For what values of p does the quadratic equation have two real number solutions that are not equal? Write the answer in set-builder notation.

38. $x^2 - 6x + p = 0$
$\{p \mid p < 9, p \in \text{real numbers}\}$

39. $x^2 + 10x + p = 0$
$\{p \mid p < 25, p \in \text{real numbers}\}$

For what values of p does the quadratic equation have two complex number solutions? Write the answer in set-builder notation.

40. $x^2 - 2x + p = 0$
$\{p \mid p > 1, p \in \text{real numbers}\}$

41. $x^2 + 4x + p = 0$
$\{p \mid p > 4, p \in \text{real numbers}\}$

Use the discriminant to determine the number of x-intercepts of the graph of the parabola.

42. $y = 2x^2 + 2x - 1$ Two

43. $y = -x^2 - x + 3$ Two

44. $y = x^2 - 8x + 16$ One

45. $y = x^2 - 10x + 25$ One

46. $y = -3x^2 - x - 2$ None

47. $y = 2x^2 + x + 4$ None

48. Find all values of x that satisfy the equation $x^2 + ix + 2 = 0$. $i, -2i$

49. *Geometry*   The base of a triangle is 4 meters less than the height of the triangle. The area is 25 square meters. Find the height and the base of the triangle. Round to the nearest thousandth.
Height: 9.348 m, base: 5.348 m

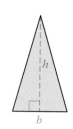

50. *Sports*   A ball is thrown straight up at a velocity of 44 feet per second (30 miles per hour). After t seconds the height h, in feet, of the ball is given by the equation $h = 44t - 16t^2$. If the ball is released from a point 5 feet above the ground, then the height h of the ball above the ground after t seconds is $h = 44t - 16t^2 + 5$. Use the equation $h = 44t - 16t^2 + 5$ to find the times at which the height of the ball will be 25 feet above the ground. Round to the nearest hundredth.   0.57 s and 2.18 s

51. *Sports*   An arrow is shot in the air with an initial upward velocity of 50 meters per second. The height h, in meters, is given by $h = 50t - 5t^2$, where t is the number of seconds since the arrow was released. Find the interval of time when the arrow will be more than 100 meters high. Round to the nearest hundredth.   Between 2.76 s and 7.24 s

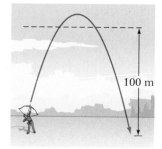

100 m

52. *Physics*   The bridge over the Royal Gorge of the Arkansas River in Colorado is 1053 feet above the water. The height h, in feet, of a rock thrown upward from this bridge with an initial velocity of 64 feet per second is given by the equation $h = -16t^2 + 64t + 1053$. For how many seconds will the rock be more than 1053 feet above the ground?   4 s

53. *Business*   A small manufacturer of watches determines that the daily revenue R, in dollars, from selling x watches is $800x - 50x^2$. The daily cost C, in dollars, to manufacture the x watches is $10x^2 + 80x + 1200$. The daily profit P is determined by using the equation $P = R - C$.

a. How many watches must be manufactured in order to make a profit of more than $800 per day?

b. How many watches must be manufactured in order to make a profit?

c. ✏ What is the smallest possible element in the domain of the profit function? What is the corresponding value of the range for this element in the domain? Write a sentence to explain the meaning of these numbers.

a. 5, 6 or 7 watches   b. 3 to 9 watches   c. 0; −1200. No less than 0 watches can be manufactured. When $x = 0$, there are 0 watches manufactured and the manufacturer experiences a loss of $1200.

■ **Solve Quadratic Equations Graphically**

54. Find all real numbers x that satisfy the equation

$$0.0764x^2 - 19.67x - 203.3 = 0.$$

Round to the nearest thousandth.   −9.951, 267.412

55. *Traffic Patterns*   The number of cars per minute entering a highway during morning rush hour can be approximated by

$$N(t) = -0.0025t^2 + 0.44t + 5,$$

where t is the number of minutes after 6:00 A.M. and $N(t)$ is the number of cars entering the highway each minute.   **a.** How many cars per minute are entering the highway at 7:30 A.M.? Round to the nearest whole number.   **b.** At what time will 3 cars enter the highway per minute? Round to the nearest minute.   **a.** 24 cars per minute   **b.** 9:00 A.M.

56. *Gardening*   If air resistance is neglected, the path of a stream of water from a garden hose can be approximated by $H(x) = -0.21x^2 + 1.8x + 3.5$, where $H(x)$ is the height, in feet, of the water x feet from the gardener holding the hose.   **a.** Find the height of the water 6 feet from the gardener.   **b.** When the height of the water is 3 feet, what is its distance from the gardener? Round to the nearest hundredth.   **a.** 6.74 ft   **b.** 8.84 ft

57. *Online Shopping*   According to Jupiter Communications, actual and projected total purchases, in millions of dollars, by U.S. consumers for online grocery shopping were as shown in the chart below.

Year	1997	1998	1999	2000	2001	2002
Purchases (in millions)	$63	$148	$350	$553	$1,332	$3,529

An equation that approximately models the data is

$$y = 229.8x^2 - 3763.5x + 15,340.7,$$

where y is the total purchases, in millions of dollars, in year x, and $x = 10$ represents the year 2000. Use the model to predict the year in which purchases will total $8 billion.   2004

58. *Internet Radio Stations*   The table below shows actual and projected revenue, in millions of dollars, from Internet radio stations (*Source:* Paul Kagan Associates).

Year	1998	1999	2000	2001	2002	2003	2004
Revenue (in millions)	$21	$76	$190	$384	$678	$1,095	$2,004

An equation that approximately models the data is

$$y = 71.25x^2 - 1264.82x + 5642.21,$$

where y is the revenue, in millions of dollars, in year x, and $x = 10$ represents the year 2000. In what year does the model predict that Internet radio stations' revenue will be $4 billion?   2006

59. *Agriculture*   The data below show the U.S. peppermint crop in 1990, 1997, and 2001 (*Source:* National Agricultural Statistics Service).

Year	1990	1997	2001
Harvest of Peppermint (in millions of pounds)	7	10	6.3

An equation that approximates the data is $y = -0.1231x^2 + 1.290x + 7$, where y is the annual harvest, in millions of pounds, in year x, and $x = 10$ represents the year 2000. According to the model, in what years was the annual harvest of peppermint 8 million pounds? Round to the nearest year.   1991 and 2000

60. *Agriculture*   The data below show the average price per gallon of maple syrup in 1995, 1998, and 2001 (*Source:* National Agricultural Statistics Service).

Year	1995	1998	2001
Average Price per Gallon	$26.20	$27.80	$26.90

An equation that approximately models the data is

$$y = -0.1389x^2 + 2.3389x + 17.9778,$$

where y is the average price per gallon of maple syrup in year x, and $x = 10$ represents the year 2000. According to the model, in what years was the average price per gallon of maple syrup $25?
1994 and 2003

61. *Banking*   The actual and projected amounts of money, in trillions of dollars, deposited in the world's private banks, are shown in the chart below (*Sources:* Gemini Consulting; Private Banker International).

Year	1986	1997	2000
Deposits (in trillions)	$4.3	$10	$13.6

An equation that approximately models the data is

$$y = 0.0487x^2 - 8.3942x + 366.0026,$$

where y is the deposits, in trillions of dollars, in year x, and $x = 100$ represents the year 2000.

a.   Use the model to predict the amount that will be deposited in the world's private banks in 2005. Round to the nearest trillion.
b.   In what year does the model predict that $50 trillion will be deposited in the world's private banks?
a. $22 trillion   **b.** 2017

62. *Business*   The actual and projected fees paid by small businesses represents online services are shown in the chart below (*Source:* Keenan Vision).

Year	2000	2002	2004	2006
Fees Paid (in billions)	$2.8	$12.8	$30.6	$47.3

An equation that models the data is $y = 0.41875x^2 + 5.0525x + 2.355$, where y represents the fees paid, in billions of dollars, in year x, and $x = 0$ represents the year 2000.

a. Use the model to predict the fees that will be paid by small businesses for online services in 2010. Round to the nearest billion.

b. In what year does the model predict that $60 billion will be paid in fees?   a. $95 billion   b. 2007

63. *Aeronautics* The Gateway Arch in St. Louis is a curve given by the equation $y = -0.028(x - 81)^2 + 183.75$, where x and y are measured in meters. An airplane with a wingspan of 40 meters, and with its wings parallel to the ground, tries to fly through the arch at an altitude of 170 meters. If the plane makes it, how much room does it have to spare? If the plane does not make it, how much more room does it need? Round to the nearest tenth.   4.3 m to spare

Applying Concepts

64. Find the difference between the larger root and the smaller root of $x^2 - px + \frac{(p^2 - 1)}{4} = 0$.   1

65. For what values of k does the equation $2x^2 - kx + x + 8 = 0$ have two equal and real roots?   −7 and 9

66. If the sum of the squares of the roots of a quadratic equation is equal to twice the product of those roots, what is the value of the discriminant of the equation?   0

67. a. Show that if r_1 and r_2 are solutions of $ax^2 + bx + c = 0$, then $r_1 + r_2 = -\frac{b}{a}$ and $r_1 r_2 = \frac{c}{a}$.

b. Show how these relationships can be used to check the solutions of a quadratic equation.   a. Answers may vary. For example, from the quadratic formula, $r_1 = \frac{-b + \sqrt{b^2 - 4ac}}{2a}$ and $r_2 = \frac{-b - \sqrt{b^2 - 4ac}}{2a}$. Use these expressions to show that $r_1 + r_2 = -\frac{b}{a}$ and $r_1 r_2 = \frac{c}{a}$.   b. Answers will vary. For example, use the solutions $r_1 = 1 + \sqrt{3}$ and $r_2 = 1 - \sqrt{3}$ of the equation $x^2 - 2x - 2 = 0$ to show that $r_1 + r_2 = 2 = -\frac{b}{a}$ and $r_1 r_2 = -2 = \frac{c}{a}$.

68. What is the least integral value of K such that $2x(Kx - 4) - x^2 + 6 = 0$ has no real roots?   2

69. Show that the equation $x^2 + bx - 1 = 0$ always has real number solutions regardless of the value of b.
$b^2 - 4ac = b^2 + 4 > 0$ for any real number b.

70. Derive the quadratic formula by applying the method of completing the square to the standard form of a quadratic equation, $ax^2 + bx + c = 0$. You may want to perform each of the steps listed below.

Subtract the constant term from each side of the equation.
Divide each side of the equation by a, the coefficient of x^2.
Complete the square by adding $\left(\frac{1}{2} \cdot \frac{b}{a}\right)^2$ to each side of the equation.
Simplify the right side of the equation so that it is written as a fraction with a denominator of $4a^2$.
Factor the perfect-square trinomial on the left side of the equation.
Take the square root of each side of the equation.
Solve the equation for x.

The complete derivation is in the Solutions Manual.

EXPLORATION

1. The Effect of the Discriminant on Solutions of Quadratic Equations and x-Intercepts of Parabolas

As you work through this project, record your results in a table such as the one shown here.

$y = ax^2 + bx + c$	Number of x-intercepts of the graph	Solutions of the equation $ax^2 + bx + c = 0$	Nature of the solutions of $ax^2 + bx + c = 0$	Value of $b^2 - 4ac$
$y = 4x^2 - 4x + 1$	1	0.5, 0.5	Double root	0
$y = 4x^2 - 4x + 3$	0	$\frac{1}{2} \pm \frac{\sqrt{2}}{2}i$	Two complex roots	-32
$y = 4x^2 - 4x - 2$	2	$-0.366, 1.366$	Two real roots	48
$y = -x^2 + 6x - 9$	1	3, 3	Double root	0
$y = -x^2 + 6x - 10$	0	$3 \pm i$	Two complex roots	-4
$y = -x^2 + 6x - 5$	2	1, 5	Two real roots	16

a. For each quadratic equation in two variables given in the first column, use a graphing calculator to determine the number of x-intercepts of the graph of the equation (column 2).

b. Calculate the solutions of the corresponding quadratic equation in one variable (column 3).

c. Record the nature of the solutions to the quadratic equation (column 4). Is it a double root? Two real number solutions? Two complex number solutions?

d. The quantity $b^2 - 4ac$ is called the discriminant. It is the quantity under the radical sign in the quadratic formula. Record in the last column the value of the determinant of each quadratic equation.

e. Make a conjecture as to the effect of the determinant on the solutions of a quadratic equation and the number of x-intercepts of a parabola. Test your conjecture on other quadratic equations and determine whether your conjecture is correct for these equations. If not, modify your conjecture and try a few more equations.

SECTION **9.4** **Equations That Are Quadratic in Form**

■ Solve Equations That Are Quadratic in Form

■ Solve Equations That Are Quadratic in Form

Certain equations that are not quadratic equations can be rewritten as quadratic equations by making suitable substitutions. An equation is **quadratic in form** if it can be written as $au^2 + bu + c = 0$.

To see that the equation at the right is quadratic in form, let $x^2 = u$. Replace x^2 by u. The equation is in quadratic form.

$$x^4 - 4x^2 - 5 = 0$$
$$(x^2)^2 - 4(x^2) - 5 = 0$$
$$u^2 - 4u - 5 = 0$$

▼ *Point of Interest*

The book Mathematical
Treatise by Ch'in Chiu-Shao
was published in 1245. The
book contains equations of
degree higher than 3. The mathematician Li Yeh published texts
in 1248 and 1259 that contain
equations from the first through
the sixth degree.

Suggested Activity

See Section 9.4 of the
Student Activity Manual for
an activity on recognizing
equations that are quadratic
in form. Also included is an
investigation of graphs of
equations that are quadratic
in form.

TAKE NOTE

Recall that a solution
that does not check in
the original equation
is an extraneous
solution.

To see that the equation at the right is quadratic in form, let $y^{1/2} = u$. Replace $y^{1/2}$ by u. The equation is in quadratic form.

$$y - y^{1/2} - 6 = 0$$
$$(y^{1/2})^2 - (y^{1/2}) - 6 = 0$$
$$u^2 - u - 6 = 0$$

The key to recognizing equations that are quadratic in form is that the exponent on one variable term is one-half the exponent on the other variable term.

? QUESTION   Is the equation $t^6 - 4t^3 + 4 = 0$ quadratic in form?

➡ Solve: $z + 7z^{1/2} - 18 = 0$

$$z + 7z^{1/2} - 18 = 0$$
$$(z^{1/2})^2 + 7(z^{1/2}) - 18 = 0 \qquad \bullet \text{ The equation is quadratic in form.}$$
$$u^2 + 7u - 18 = 0 \qquad \bullet \text{ To solve this equation, let } z^{1/2} = u.$$
$$(u - 2)(u + 9) = 0 \qquad \bullet \text{ Solve for } u \text{ by factoring.}$$

$u - 2 = 0$	$u + 9 = 0$
$u = 2$	$u = -9$
$z^{1/2} = 2$	$z^{1/2} = -9$
$(z^{1/2})^2 = 2^2$	$(z^{1/2})^2 = (-9)^2$
$z = 4$	$z = 81$

• Replace u by $z^{1/2}$.

• Solve for z by squaring each side of the equation.

Note: We must check the solutions. **When each side of an equation has been squared, the resulting equation may have a solution that is not a solution of the original equation.**

ALGEBRAIC CHECK

$z + 7z^{1/2} - 18 = 0$		$z + 7z^{1/2} - 18 = 0$	
$4 + 7(4)^{1/2} - 18$	0	$81 + 7(81)^{1/2} - 18$	0
$4 + 7 \cdot 2 - 18$	0	$81 + 7 \cdot 9 - 18$	0
$4 + 14 - 18$	0	$81 + 63 - 18$	0
	$0 = 0$		$126 \neq 0$

4 checks as a solution. 81 does not.
The solution is 4.

GRAPHICAL CHECK

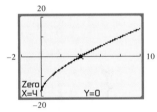

The graph of
$Y = X + 7X^{0.5} - 18$
has only one x-intercept,
at $X = 4$.   ⬅

EXAMPLE 1

Solve.   **a.** $x^4 + x^2 - 12 = 0$ **b.** $x^{2/3} - 2x^{1/3} - 3 = 0$

Solution

a.
$$x^4 + x^2 - 12 = 0$$
$$(x^2)^2 + (x^2) - 12 = 0 \qquad \bullet \text{ The equation is quadratic in form.}$$
$$u^2 + u - 12 = 0 \qquad \bullet \text{ Let } x^2 = u.$$
$$(u - 3)(u + 4) = 0 \qquad \bullet \text{ Solve for } u \text{ by factoring.}$$

INSTRUCTOR NOTE
Exercises 8 and 10 can be
used for similar in-class
examples.

? ANSWER   The exponent 3 is one-half the exponent 6. The equation is quadratic in form.
$(t^3)^2 - 4(t^3) + 4 = 0$

$$u - 3 = 0 \qquad u + 4 = 0$$
$$u = 3 \qquad u = -4$$
$$x^2 = 3 \qquad x^2 = -4 \qquad \text{• Replace } u \text{ by } x^2.$$
$$\sqrt{x^2} = \sqrt{3} \qquad \sqrt{x^2} = \sqrt{-4} \qquad \text{• Solve for } x \text{ by taking square roots.}$$
$$x = \pm\sqrt{3} \qquad x = \pm 2i$$

ALGEBRAIC CHECK		**GRAPHICAL CHECK**

$$\begin{array}{r|l} x^4 + x^2 - 12 = 0 & \\ \hline (\sqrt{3})^4 + (\sqrt{3})^2 - 12 & 0 \\ 9 + 3 - 12 & 0 \\ 0 = 0 \end{array} \qquad \begin{array}{r|l} x^4 + x^2 - 12 = 0 & \\ \hline (-\sqrt{3})^4 + (-\sqrt{3})^2 - 12 & 0 \\ 9 + 3 - 12 & 0 \\ 0 = 0 \end{array}$$

$$\begin{array}{r|l} x^4 + x^2 - 12 = 0 & \\ \hline (2i)^4 + (2i)^2 - 12 & 0 \\ 16 - 4 - 12 & 0 \\ 0 = 0 \end{array} \qquad \begin{array}{r|l} x^4 + x^2 - 12 = 0 & \\ \hline (-2i)^4 + (-2i)^2 - 12 & 0 \\ 16 - 4 - 12 & 0 \\ 0 = 0 \end{array}$$

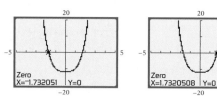

The x-coordinates of the x-intercepts of the graph of $Y = X^4 + X^2 - 12$ are -1.732051 and 1.732051.

$$\sqrt{3} \approx 1.732051; \ -\sqrt{3} \approx -1.732051$$

The complex number solutions cannot be checked graphically.

The solutions are $\sqrt{3}$, $-\sqrt{3}$, $2i$, and $-2i$.

b. $\qquad x^{2/3} - 2x^{1/3} - 3 = 0$

$$(x^{1/3})^2 - 2(x^{1/3}) - 3 = 0 \qquad \text{• The equation is quadratic in form.}$$
$$u^2 - 2u - 3 = 0 \qquad \text{• Let } x^{1/3} = u.$$
$$(u - 3)(u + 1) = 0 \qquad \text{• Solve for } u \text{ by factoring.}$$

$$u - 3 = 0 \qquad u + 1 = 0$$
$$u = 3 \qquad u = -1$$
$$x^{1/3} = 3 \qquad x^{1/3} = -1 \qquad \text{• Replace } u \text{ by } x^{1/3}.$$
$$(x^{1/3})^3 = 3^3 \qquad (x^{1/3})^3 = (-1)^3 \qquad \begin{array}{l} \text{• Solve for } x \text{ by cubing both sides of} \\ \text{the equation.} \end{array}$$
$$x = 27 \qquad x = -1$$

ALGEBRAIC CHECK		**GRAPHICAL CHECK**

$$\begin{array}{r|l} x^{2/3} - 2x^{1/3} - 3 = 0 & \\ \hline 27^{2/3} - 2 \cdot 27^{1/3} - 3 & 0 \\ 9 - 6 - 3 & 0 \\ 0 = 0 \end{array} \qquad \begin{array}{r|l} x^{2/3} - 2x^{1/3} - 3 = 0 & \\ \hline (-1)^{2/3} - 2 \cdot (-1)^{1/3} - 3 & 0 \\ 1 + 2 - 3 & 0 \\ 0 = 0 \end{array}$$

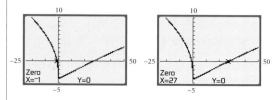

The x-coordinates of the x-intercepts of the graph of $Y = X^{2/3} - 2X^{1/3} - 3$ are -1 and 27.

The solutions check.
The solutions are 27 and -1.

YOU TRY IT 1

Solve.   **a.** $x - 5x^{1/2} + 6 = 0$ **b.** $4x^4 + 35x^2 - 9 = 0$

Solution   See page S39.   **a.** $4, 9$ **b.** $-3i, 3i, -\dfrac{1}{2}, \dfrac{1}{2}$

Suggested Activity

Solve the equation

$$(x^2 - 7)^{1/2} = (x - 1)^{1/2}$$

for all real values of x.
[Answer: 3]

Recall that the Property of Raising Both Sides of an Equation to a Power was used to solve radical equations. Sometimes after this property is applied to a radical equation, the result is a quadratic equation. This is illustrated in Example 2.

 QUESTION   What does the Property of Raising Both Sides of an Equation to a Power state?

Remember that **when each side of an equation is raised to an even power, the resulting equation may have a solution that is not a solution of the original equation. Therefore, the solutions must be checked.**

EXAMPLE 2

Solve: $\sqrt{x + 2} + 4 = x$

INSTRUCTOR NOTE
Exercise 30 can be used for a similar in-class example.

Solution

$$\sqrt{x + 2} + 4 = x$$
$$\sqrt{x + 2} = x - 4 \qquad \bullet \text{ Solve for the radical expression.}$$
$$(\sqrt{x + 2})^2 = (x - 4)^2 \qquad \bullet \text{ Square each side of the equation.}$$
$$x + 2 = x^2 - 8x + 16 \qquad \bullet \text{ Simplify.}$$
$$0 = x^2 - 9x + 14 \qquad \bullet \text{ Write the equation in standard form.}$$
$$0 = (x - 7)(x - 2) \qquad \bullet \text{ Solve for } x \text{ by factoring.}$$

$$x - 7 = 0 \qquad x - 2 = 0$$
$$x = 7 \qquad\quad x = 2$$

ALGEBRAIC CHECK

$$\begin{array}{c|c} \sqrt{x + 2} + 4 = x & \\ \hline \sqrt{7 + 2} + 4 & 7 \\ \sqrt{9} + 4 & 7 \\ 7 = 7 & \end{array} \qquad \begin{array}{c|c} \sqrt{x + 2} + 4 = x & \\ \hline \sqrt{2 + 2} + 4 & 2 \\ \sqrt{4} + 4 & 2 \\ 6 \neq 2 & \end{array}$$

GRAPHICAL CHECK

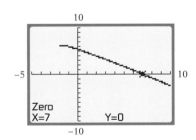

The graph of $Y = \sqrt{X + 2} + 4 - X$ has only one x-intercept, at $X = 7$.

7 checks as a solution, but 2 does not.
The solution is 7.

YOU TRY IT 2

Solve: $\sqrt{2x + 1} + x = 7$

Solution   See page S39.   4

Now let's look at an application of an equation that is quadratic in form.

 ANSWER   If $a = b$, then $a^n = b^n$.

➡ Suppose we want to find two points on the line $y = 4$ that are 5 units from the point $(3, 1)$.

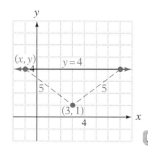

State the goal. We want to find two points in the rectangular coordinate system that are a distance of 5 units from the point $(3, 1)$.

Devise a strategy. Let a desired point be (x, y). Use the distance formula.

$$d = \sqrt{(x_1 - x_2)^2 + (y_1 - y_2)^2}$$

The distance d is 5. Let $(x_1, y_1) = (x, y)$ and $(x_2, y_2) = (3, 1)$. Because the point (x, y) is on the line $y = 4$, the value of y is 4. Solve the equation for x.

Solve the problem.

$d = \sqrt{(x_1 - x_2)^2 + (y_1 - y_2)^2}$ • Use the distance formula.

$5 = \sqrt{(x - 3)^2 + (y - 1)^2}$ • Substitute 5 for d. $(x_1, y_1) = (x, y)$ and $(x_2, y_2) = (3, 1)$.

$5 = \sqrt{(x - 3)^2 + (4 - 1)^2}$ • The point (x, y) is on the line $y = 4$. Substitute 4 for y.

$5 = \sqrt{(x - 3)^2 + 9}$

$25 = (x - 3)^2 + 9$ • Square both sides of the equation.

$16 = (x - 3)^2$ • Subtract 9 from each side of the equation.

$\sqrt{16} = \sqrt{(x - 3)^2}$ • Solve by taking square roots.

$\pm 4 = x - 3$

$x - 3 = 4 \qquad x - 3 = -4$ • Solve for x.

$x = 7 \qquad\quad x = -1$

The points $(7, 4)$ and $(-1, 4)$ are points on the line $y = 4$ that are 5 units from the point $(3, 1)$.

Check your work. Check the reasonableness of the answer. Review all the steps in the solution. ⬅

The depth s from the opening of a well to the water can be determined by measuring the total time between the instant you drop a stone and the time you hear it hit the water. The time, in seconds, it takes the stone to hit the water is given by $\frac{\sqrt{s}}{4}$, where s is measured in feet. The time, also in seconds, required for the sound of the impact to travel up to your ears is given by $\frac{s}{1100}$. Thus the total time T, in seconds, between the instant you drop a stone and the moment you hear its impact is

$$T = \frac{\sqrt{s}}{4} + \frac{s}{1100}$$

Rewrite the right side of the equation as a single fraction with a denominator of 1100.

$$T = \frac{275\sqrt{s}}{1100} + \frac{s}{1100}$$

$$T = \frac{275\sqrt{s} + s}{1100}$$

Multiply both sides of the equation by 1100. This equation is used in Example 3.

$$1100T = 275\sqrt{s} + s$$

INSTRUCTOR NOTE
Exercise 50 can be used for
a similar in-class example.

EXAMPLE 3

The total time between the instant you drop a stone into a well and the moment you hear its impact is 5 seconds. Find the depth to the water in the well.

Use the equation $1100T = 275\sqrt{s} + s$, where T is the time in seconds and s is the depth to the water in feet. Round to the nearest thousandth.

State the goal.   The goal is to find the distance to the water in the well.

Devise a strategy.   Substitute 5 for T in the equation $1100T = 275\sqrt{s} + s$. Then solve the equation for s.

Solve the problem.

$$1100T = 275\sqrt{s} + s$$

$$1100(5) = 275\sqrt{s} + s \qquad \text{• Substitute 5 for } T.$$

$$5500 = 275\sqrt{s} + s$$

$$0 = s + 275\sqrt{s} - 5500 \qquad \text{• The equation is quadratic in form.}$$

$$0 = u^2 + 275u - 5500 \qquad \text{• Let } \sqrt{s} = u.\ u^2 + 275u - 5500 \text{ is}$$
$$\text{nonfactorable over the integers.}$$
$$\text{Use the quadratic formula.}$$

$$u = \frac{-b \pm \sqrt{b^2 - 4ac}}{2a}$$

$$= \frac{-275 \pm \sqrt{(275)^2 - 4(1)(-5500)}}{2 \cdot 1} \qquad \text{• } a = 1, b = 275, c = -5500$$

$$= \frac{-275 \pm \sqrt{97,625}}{2}$$

$$u = \frac{-275 + \sqrt{97,625}}{2} \qquad u = \frac{-275 - \sqrt{97,625}}{2}$$

$$u \approx 18.724998 \qquad u \approx -293.724998$$

$$\sqrt{s} \approx 18.724998 \qquad \sqrt{s} \approx -293.724998 \qquad \text{• Replace } u \text{ by } \sqrt{s}.$$

$$(\sqrt{s})^2 \approx (18.724998)^2 \qquad (\sqrt{s}) = (-293.724998)^2 \qquad \text{• Square each side}$$
$$s = 350.62555 \qquad s \approx 86,274.37445 \qquad \text{of the equation.}$$

The answer 86,274.37445 is not reasonable because 86,274.37445 feet is over 16 miles, and wells are not that deep.

The distance to the water in the well is 350.626 feet.

Check your work.   Substitute 5 for T and 350.626 for s in the equation $1100T = 275\sqrt{s} + s$. After simplifying each side of the equation, the results should be approximately equal.

The solution can also be checked by using a graphing calculator.

Suggested Activity

If $5^{2x} = 12 - 5^x$, find the value of 5^{x+2}.
[Answer: 75]

YOU TRY IT 3

The sum of the cube of a number and the product of the number and twelve is equal to seven times the square of the number. Find the number.

Solution   See page S39.   0, 3, 4

9.4 EXERCISES   Suggested Assignment: 5–53, odds

Topics for Discussion

1. **a.** Is the equation $x + 3\sqrt{x} - 8 = 0$ quadratic in form? Explain why it is or is not.
    b. Is the equation $\sqrt[4]{x} + 2\sqrt[3]{3x} - 8 = 0$ quadratic in form? Explain why it is or is not.
    a. Yes. Explanations will vary.   b. No. Explanations will vary.

2. Show that the equation $x^8 - 2x^4 - 15 = 0$ is quadratic in form.
    Let $x^4 = u$. Then the equation can be written as $u^2 - 2u - 15 = 0$.

3. Determine whether the statement is always true, sometimes true, or never true.

    a. An equation that is quadratic in form can be solved by using the quadratic formula.
    b. Squaring both sides of a radical equation produces an extraneous root.
    c. When the Property of Raising Both Sides of an Equation to a Power is used to solve an equation, the solutions must be checked.
    a. Always true   b. Sometimes true   c. Always true

4. Write two equations that are not quadratic equations but can be written in quadratic form. Then write them in quadratic form.
    Answers will vary.

■ Solve Equations That Are Quadratic in Form

Solve.

5. $x^4 - 13x^2 + 36 = 0$ $\pm2, \pm3$

6. $y^4 - 5y^2 + 4 = 0$ $\pm1, \pm2$

7. $z^4 - 6z^2 + 8 = 0$ $\pm2, \pm\sqrt{2}$

8. $t^4 - 12t^2 + 27 = 0$ $\pm3, \pm\sqrt{3}$

9. $p - 3p^{1/2} + 2 = 0$ 1, 4

10. $v - 7v^{1/2} + 12 = 0$ 9, 16

11. $x - x^{1/2} - 12 = 0$ 16

12. $w - 2w^{1/2} - 15 = 0$ 25

13. $z^4 + 3z^2 = 4$ $\pm1, \pm2i$

14. $y^4 + 5y^2 = 36$ $\pm2, \pm3i$

15. $x^4 + 12x^2 - 64 = 0$ $\pm2, \pm4i$

16. $x^4 - 81 = 0$ $\pm3, \pm3i$

17. $p + 2p^{1/2} = 24$ 16

18. $v + 3v^{1/2} = 4$ 1

19. $y^{2/3} - 9y^{1/3} + 8 = 0$ 1, 512

20. $z^{2/3} - z^{1/3} - 6 = 0$
    $-8, 27$

21. $x^6 - 9x^3 + 8 = 0$
    $1, 2, -1 \pm i\sqrt{3}, -\dfrac{1}{2} \pm \dfrac{\sqrt{3}}{2}i$

22. $y^6 + 9y^3 + 8 = 0$
    $-2, -1, 1 \pm i\sqrt{3}, \dfrac{1}{2} \pm \dfrac{\sqrt{3}}{2}i$

23. $z^8 = 17z^4 - 16$
    $\pm1, \pm2, \pm i, \pm2i$

24. $v^4 - 15v^2 = 16$
    $\pm4, \pm i$

25. $p^{2/3} + 2p^{1/3} = 8$
    $-64, 8$

26. $w^{2/3} + 3w^{1/3} = 10$ $-125, 8$

27. $2x = 3x^{1/2} - 1$ $\dfrac{1}{4}, 1$

28. $3y = 5y^{1/2} + 2$ 4

29. $x^{2/5} + 6 = 5x^{1/5}$ 32, 243

30. $\sqrt{x + 1} + x = 5$ 3

31. $\sqrt{x - 4} + x = 6$ 5

32. $x = \sqrt{x} + 6$ 9

33. $\sqrt{2y - 1} = y - 2$ 5

34. $\sqrt{3w + 3} = w + 1$ $-1, 2$

35. $\sqrt{2s + 1} = s - 1$ 4

36. $\sqrt{4y + 1} - y = 1$ 0, 2

37. $\sqrt{3s + 4} + 2s = 12$ 4

38. $\sqrt{10x + 5} - 2x = 1$ $-\dfrac{1}{2}$, 2

39. $\sqrt{t + 8} = 2t + 1$ 1

40. $\sqrt{p + 11} = 1 - p$ −2

41. $x - 7 = \sqrt{x - 5}$ 9

42. $\sqrt{2x - 1} = 1 - \sqrt{x - 1}$ 1

43. $\sqrt{t + 3} + \sqrt{2t + 7} = 1$ −3

44. $\sqrt{5 - 2x} = \sqrt{2 - x} + 1$ ±2

45. $\sqrt{x^4 - 2} = x$ $\sqrt{2}$, i

46. $\sqrt{x^4 + 4} = 2x$ $\sqrt{2}$

47. *Mathematics*   The fourth power of a number is twenty-five less than ten times the square of the number. Find the number. $\sqrt{5}$ or $-\sqrt{5}$

48. *Mathematics*   The difference between sixteen times the square of a number and sixty-four is equal to the fourth power of the number. Find the number. $2\sqrt{2}$ and $-2\sqrt{2}$

49. *Geometry*   The width of a rectangle is twice the square root of the length. The diagonal of the rectangle is 12 inches. Find the length of the rectangle. Round to the nearest tenth.   10.2 in.

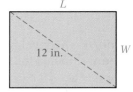

50. *Geometry*   The longer leg of a right triangle is four times the square root of the shorter leg. Find the lengths of the two legs if the hypotenuse is 6 feet. Round to the nearest tenth.   2 ft and 5.7 ft

51. *Mathematics*   The sum of the cube of a number and the product of the number and seven is equal to eight times the square of the number. Find the number.   0, 1, or 7

52. *Mathematics*   The sum of the cube of a number and twice the square of the number is equal to ninety-nine times the number. Find the number.
−11, 0, or 9

53. *Sound*   The total time between the instant you drop a stone into a well and the moment you hear its impact is 5.5 seconds. Use the equation $1100T = 275\sqrt{s} + s$, where T is the time in seconds and s is the depth to the water in feet, to find the distance to the water in the well. Round to the nearest tenth.   419.2 ft

54. *Sound*   The total time between the instant you drop a rock into an abandoned mine shaft and the moment you hear its impact is 4.5 seconds. Find the depth of the mine shaft. Use the equation $1100T = 275\sqrt{s} + s$, where T is the time in seconds and s is the depth of the mine shaft in feet. Round to the nearest tenth.   287.5 ft

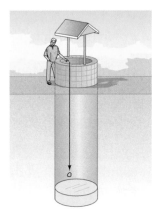

55. *Distance*   Find two points on the line $y = 6$ that are 7 units from the point (5, 2). Give both the exact values and approximations to the nearest hundredth. $(5 - \sqrt{33}, 6)$ and $(5 + \sqrt{33}, 6)$, or $(-0.74, 6)$ and $(10.74, 6)$

56. *Distance*   Find two points on the line $x = -4$ that are 6 units from the point (−1, 3). Give both the exact values and approximations to the nearest hundredth.
$(-4, 3 - 3\sqrt{3})$ and $(-4, 3 + 3\sqrt{3})$ or $(-4, -2.20)$ and $(-4, 8.20)$

57. *Mathematics* Given that $x^{2000} - x^{1998} = x^{1999} - x^{1997}$, find the value of x.
−1, 0, 1

Applying Concepts

58. Solve: $|x - 3|^2 - 9|x - 3| = -18$ −3, 0, 6, 9   **59.** Solve for x: $2^{2x} + 32 = 12(2^x)$ 2, 3

60. Solve: $(\sqrt{x} - 2)^2 - 5\sqrt{x} + 14 = 0$ (*Hint:* Let $u = \sqrt{x} - 2$.)   9, 36

61. 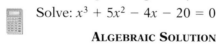 Why does the equation $x^2 + 5x - 6 = 0$ have two solutions whereas $x + 5\sqrt{x} - 6 = 0$ has only one solution?   Explanations will vary.

62. One real root of $x^8 + x^6 + x^4 + x^2 = 340$ is 2. Find the only other real root of this equation.   −2

63. Given that $x^{400} = 400^{400}$ and $x \neq 400$, find the real value of x.   −400

64. *Sports* According to the *Compton's Interactive Encyclopedia,* the minimum dimensions of a football used in the National Football Association games are 10.875 inches long and 20.75 inches in circumference at the center. A possible <u>model</u> for the cross section of a football is given by $y = \pm 3.3041\sqrt{1 - \dfrac{x^2}{29.7366}}$, where x is the distance from the center of the football and y is the radius of the football at x.

a. What is the domain of the equation?

b. Graph $y = 3.3041\sqrt{1 - \dfrac{x^2}{29.7366}}$ and $y = -3.3041\sqrt{1 - \dfrac{x^2}{29.7366}}$ on the same coordinate axes. Explain why the $\pm$ symbol occurs in the equation.

c. Determine the radius of the football when x is 3 inches. Round to the nearest ten-thousandth.
a. $\{x|-\sqrt{29.7366} \leq x \leq \sqrt{29.7366}\}$ **b.** The $\pm$ symbol occurs in the equation so that the graph pictures the entire shape of the football. **c.** 2.7592 in.

EXPLORATION

1. *Solutions of Higher-Degree Equations*
   Solve: $x^3 + 5x^2 - 4x - 20 = 0$

ALGEBRAIC SOLUTION	**GRAPHICAL SOLUTION**

ALGEBRAIC SOLUTION

$$x^3 + 5x^2 - 4x - 20 = 0$$
$$(x^3 + 5x^2) - (4x + 20) = 0$$
$$x^2(x + 5) - 4(x + 5) = 0$$
$$(x + 5)(x^2 - 4) = 0$$
$$(x + 5)(x + 2)(x - 2) = 0$$

$x + 5 = 0$ $x + 2 = 0$ $x - 2 = 0$
$x = -5$ $x = -2$ $x = 2$

The solutions are −5, −2, and 2.

GRAPHICAL SOLUTION

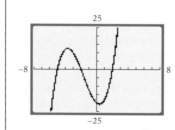

The x-coordinates of the x-intercepts of the graph of $Y = X^3 + 5X^2 - 4X - 20$ are −5, −2, and 2.

a. Solve the equation $x^3 - 6x^2 - 9x + 54 = 0$ by factoring and by using a graphing calculator.

b. Solve the equation $x^3 + 2x^2 - 15x = 0$ by factoring and by using a graphing calculator.

c. Solve the equation $x^3 + 3x^2 + x = 0$. What algebraic methods can you use to solve the equation? Can you check all the solutions using a graphing calculator?

d. Solve the equation $x^3 + 3x^2 + 4x = 0$. What algebraic methods can you use to solve the equation? Can you check all the solutions using a graphing calculator?

e. Solve the equation $x^4 + 5x^3 - 16x^2 - 80x = 0$ using a graphing calculator. From the solutions, determine the factorization of $x^4 + 5x^3 - 16x^2 - 80x$.

f. Solve the equation $x^4 + 4x^3 - 11x^2 - 30x = 0$ using a graphing calculator. From the solutions, determine the factorization of $x^4 + 4x^3 - 11x^2 - 30x$.

g. Find the zeros of the function $f(x) = 0.5x(x - 4)(x - 3)(x + 2)(x + 1)$. Check your answers using a graphing calculator.

h. Find the zeros of the function $f(x) = 0.25x(x + 5)(x + 3)(x - 4)(x - 2)$. Check your answers using a graphing calculator.

i. What are the zeros of the function
$f(x) = 0.3x(x + a)(x - b)(x - c)(x + d)$?

a. −3, 3, 6   b. −5, 0, 3   c. 0, −2.62, −0.38; factoring and completing the square or quadratic formula; yes   d. 0, $\frac{-3 \pm i\sqrt{7}}{2}$; factoring and completing the square or quadratic formula; no   e. −5, −4, 0, 4; $x(x + 5)(x + 4)(x - 4)$ f. −5, −2, 0, 3; $x(x + 5)(x + 2)(x - 3)$ g. −2, −1, 0, 3, 4   h. −5, −3, 0, 2, 4   i. 0, −a, b, c, −d

---

SECTION **9.5** # Quadratic Functions

■ Properties of Quadratic Functions

Recall that a linear function is one of the form $f(x) = mx + b$. Its graph has certain characteristics: it is a straight line with slope m and y-intercept $(0, b)$.

A **quadratic function** is a function of the form $f(x) = ax^2 + bx + c$, $a \neq 0$. The graph of this function, which is called a **parabola,** also has certain characteristics.

 QUESTION

a. State whether the function is a linear function or a quadratic function.
  i. $f(x) = -3x + 1$ **ii.** $f(x) = x^2 - 4x - 5$
b. What are the values of a, b, and c in the quadratic function
$f(x) = -3x^2 + 5x - 8$?

? ANSWERS   **a. i.** $f(x) = 3x + 1$ is an equation of the form $f(x) = mx + b$. It is a linear function.
ii. $f(x) = x^2 - 4x - 5$ is an equation of the form $f(x) = ax^2 + bx + c$. It is a quadratic function.
b. $a = -3, b = 5, c = -8$

INSTRUCTOR NOTE
The graph of a quadratic function is discussed in this section, along with some of its properties and applications. A general discussion of parabolas as one of the conic sections appears in Additional Topics in Algebra, which follows Chapter 10.

The graphs of two quadratic functions are shown below.

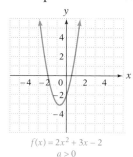

$f(x) = 2x^2 + 3x - 2$
$a > 0$

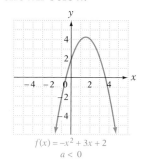

$f(x) = -x^2 + 3x + 2$
$a < 0$

For the figure on the left, the value of a is *positive* ($a = 2$). The graph opens up. For the figure on the right, the value of a is *negative* ($a = -1$). The graph opens down.

This can be stated in general terms: **A parabola opens up when $a > 0$ and opens down when $a < 0$.**

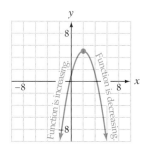

? QUESTION   State whether the graph of the equation is a parabola that opens up or opens down.

    a. $y = 3x^2 - 2x - 4$ **b.** $y = -\dfrac{1}{2}x^2 + 5$

Consider the graph of the quadratic function at the left. As a point on the graph moves from left to right, the values of y are decreasing for $x < 1$. For $x > 1$, the values of y are increasing. The function is said to be *decreasing* for $x < 1$, and the function is said to be *increasing* when $x > 1$. The point at which the graph changes from decreasing to increasing is called a **minimum** of the function.

As a point on the graph of the quadratic function at the right moves from left to right, the values of y are increasing for $x < 2$. For $x > 2$, the values of y are decreasing. The function is said to be *increasing* for $x < 2$, and the function is said to be *decreasing* when $x > 2$. The point at which the graph changes from increasing to decreasing is called a **maximum** of the function.

The point at which the graph of a parabola has a minimum or maximum is called the **vertex** of the parabola. Note that **the vertex of a parabola is the point with the least y-coordinate when $a > 0$ and is the point with the greatest y-coordinate when $a < 0$.**

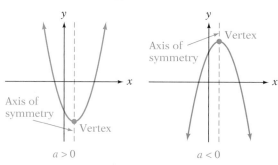

$a > 0$ $a < 0$

▼ *Point of Interest*

Recall that the suspension cables for some bridges, such as the Golden Gate Bridge, have the shape of a parabola. Parabolic shapes are also used for mirrors in telescopes and in the design of certain antennas.

? ANSWERS   **a.** $a > 0$ ($a = 3$). The graph is a parabola that opens up.   **b.** $a < 0$ ($a = -\frac{1}{2}$). The graph is a parabola that opens down.

TAKE NOTE

The axis of symmetry is a vertical line. The vertex of the parabola lies on the axis of symmetry.

 See Appendix A: Min and Max

Suggested Activity

1. Draw the graphs of two different quadratic functions that have the line $x = 3$ as their line of symmetry.
2. Draw the graphs of two different quadratic functions that have the line $x = -2$ as their line of symmetry.

The **axis of symmetry** of the graph of a quadratic function is a line that passes through the vertex of the parabola and is parallel to the *y*-axis. To understand the axis of symmetry, think of folding the graph along that line. The two portions of the graph will match up.

The vertex and axis of symmetry can be found by using the MINIMUM or MAXIMUM operation of a graphing calculator.

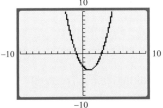

→ Find the vertex and axis of symmetry for the graph of $f(x) = x^2 - 2x - 3$.

For this quadratic function, $a > 0$ ($a = 1$), so the parabola opens up and the function has a minimum. Enter the expression for the function in Y₁ and display its graph in the standard viewing window. Use the CALCULATE feature to find the minimum.

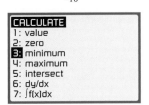

The vertex is $(1, -4)$.

Because the axis of symmetry is a vertical line and passes through the vertex, its equation can be determined once the vertex is determined. Its equation is $x = $ constant, where the constant is the *x*-coordinate of the vertex. The *x*-coordinate of the vertex $(1, -4)$ is 1.

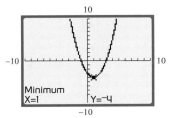

The axis of symmetry is $x = 1$.

? QUESTION

a. The axis of symmetry of a parabola is the line $x = -6$. What is the *x*-coordinate of the vertex of the parabola?

b. The vertex of a parabola is $(3, -2)$. What is the equation of the axis of symmetry of the parabola?

Because the value of $f(x) = x^2 - 2x - 3$ graphed above is a real number for all values of *x*, the domain of *f* is all real numbers. Since the vertex is the point at the minimum of the function, all values of $y \geq -4$. Thus the range is $\{y \mid y \geq -4\}$. The range can also be determined algebraically, as shown below, by completing the square.

Group the variable terms.

Complete the square on $x^2 - 2x$. Add 1 to and subtract 1 from $x^2 - 2x$.

$$f(x) = x^2 - 2x - 3$$
$$f(x) = (x^2 - 2x) - 3$$
$$f(x) = (x^2 - 2x + 1) - 1 - 3$$

? ANSWERS **a.** -6 **b.** $x = 3$

TAKE NOTE

In completing the square, 1 is both added and subtracted. Because $1 - 1 = 0$, the expression $x^2 - 2x - 3$ is not changed. Note that
$$(x - 1)^2 - 4$$
$$= (x^2 - 2x + 1) - 4$$
$$= x^2 - 2x - 3,$$
which is the original equation.

Factor and combine like terms. $f(x) = (x - 1)^2 - 4$

The square of a number is always positive. Therefore, the expression $(x - 1)^2$ is positive. Subtract 4 from each side of the inequality so that the left side is equal to $f(x)$.

$$(x - 1)^2 \geq 0$$

$$(x - 1)^2 - 4 \geq -4$$

Replace $(x - 1)^2 - 4$ with $f(x)$.

$$f(x) \geq -4$$
$$y \geq -4$$

From the last inequality, the range is $\{y \,|\, y \geq -4\}$.

By following the process illustrated in the last example and completing the square of $f(x) = ax^2 + bx + c$, we can find a formula for the coordinates of the vertex of a parabola. This formula will allow us to determine the vertex and axis of symmetry without having to graph the function.

> **The Vertex and Axis of Symmetry of a Parabola**
>
> Let $f(x) = ax^2 + bx + c$ be the equation of a parabola. The coordinates of the vertex are $\left(-\dfrac{b}{2a}, f\left(-\dfrac{b}{2a} \right) \right)$. The equation of the axis of symmetry is $x = -\dfrac{b}{2a}$.

EXAMPLE 1

Find the vertex and the axis of symmetry of the parabola whose equation is $y = -3x^2 + 6x + 1$.

INSTRUCTOR NOTE
Exercise 20 can be used for a similar in-class example.

Solution

$$x = -\frac{b}{2a} = -\frac{6}{2(-3)} = 1$$

- Find the x-coordinate of the vertex. $a = -3$ and $b = 6$.

$$y = -3x^2 + 6x + 1$$
$$y = -3(1)^2 + 6(1) + 1$$
$$y = 4$$

- Find the y-coordinate of the vertex by replacing x by 1 and solving for y.

The vertex is $(1, 4)$.

The axis of symmetry is the line $x = 1$.

- The axis of symmetry is the line $x = -\dfrac{b}{2a}$.

GRAPHICAL CHECK

See Appendix A: Min and Max

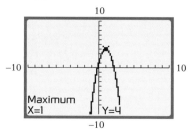

YOU TRY IT 1

Find the vertex and the axis of symmetry of the parabola whose equation is $y = x^2 - 2$.

Solution   See page S39.   Vertex: $(0, -2)$; axis of symmetry: $x = 0$

INSTRUCTOR NOTE
Exercise 36 can be used for
a similar in-class example.

EXAMPLE 2

State the domain and range of $f(x) = -2x^2 - 4x + 3$.

Solution The graph of f is a parabola that opens down $(a = -2)$.

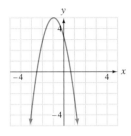

The x-coordinate of the vertex is $x = -\dfrac{b}{2a} = -\dfrac{-4}{2(-2)} = -1$.

The y-coordinate of the vertex is $f(-1) = -2(-1)^2 - 4(-1) + 3 = 5$.

The vertex is $(-1, 5)$.

Because $f(x) = -2x^2 - 4x + 3$ is a real number for all values of x, the domain of the function is $\{x \mid x \in \text{real numbers}\}$. The vertex of the parabola is the highest point on the graph. Because the y-coordinate at that point is 5, the range is $\{y \mid y \le 5\}$.

YOU TRY IT 2

State the domain and range of $g(x) = x^2 + 4x - 2$.

Solution See page S39. Domain: $\{x \mid x \in \text{real numbers}\}$; Range: $\{y \mid y \ge -6\}$

TAKE NOTE

If the coordinates of
the vertex are known,
the range of the qua-
dratic function can be
determined.

■ Intercepts of Quadratic Functions

Recall that a point at which a graph crosses the x- or y-axis is called an *intercept* of the graph. **The x-intercepts of the graph of an equation occur when $y = 0$; the y-intercepts occur when $x = 0$.**

The graph of $y = x^2 + x - 6$ is shown at the left. The points whose coordinates are $(-3, 0)$ and $(2, 0)$ are x-intercepts of the graph. The y-intercept is $(0, -6)$.

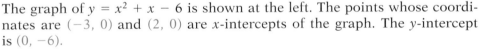

x-intercepts
$(-3, 0)$ $(2, 0)$
$(0, -6)$
y-intercept
$y = x^2 + x - 6$

➡ Find the x-intercepts for the parabola whose equation is $y = 4x^2 - 4x + 1$.

$$y = 4x^2 - 4x + 1$$
$$0 = 4x^2 - 4x + 1$$ • To find x-intercepts, let $y = 0$.
$$0 = (2x - 1)(2x - 1)$$ • Solve for x by factoring and using the Principle of Zero Products.

$$2x - 1 = 0 \qquad 2x - 1 = 0$$
$$2x = 1 \qquad 2x = 1$$
$$x = \frac{1}{2} \qquad x = \frac{1}{2}$$

The x-intercept is $\left(\dfrac{1}{2}, 0\right)$. ⬅

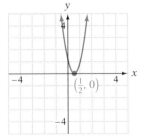

$\left(\frac{1}{2}, 0\right)$
$y = 4x^2 - 4x + 1$

In this example, the parabola has only one x-intercept. In this case, the parabola is said to be *tangent* to the x-axis at $x = \frac{1}{2}$.

⇒ Find the x-intercepts of $y = 2x^2 - x - 6$.

$y = 2x^2 - x - 6$

$0 = (2x + 3)(x - 2)$ • To find the x-intercepts, let $y = 0$.

$2x + 3 = 0$ $x - 2 = 0$ • Solve for x by factoring and using the Principle of Zero Products.

$x = -\dfrac{3}{2}$ $x = 2$

> ### TAKE NOTE
>
> Note that $-\frac{3}{2}$ and 2 are *zeros* of the function. The *points* $(-\frac{3}{2}, 0)$ and (2, 0) are *x-intercepts* of the graph of the function.

The x-intercepts are $\left(-\dfrac{3}{2}, 0\right)$ and $(2, 0)$.

If the equation in this example, $y = 2x^2 - x - 6$, were written in functional notation as $f(x) = 2x^2 - x - 6$, then to find the x-intercepts you would let $f(x) = 0$ and solve for x. A value of x for which $f(x) = 0$ is a **zero of the function.** Thus $-\dfrac{3}{2}$ and 2 are zeros of $f(x) = 2x^2 - x - 6$.

❓ QUESTION   The zeros of the function $f(x) = x^2 - 2x - 3$ are -1 and 3. What are the x-intercepts of the graph of the equation $y = x^2 - 2x - 3$?

EXAMPLE 3

Find the x-intercepts of the parabola given by the equation.

a. $y = x^2 + 2x - 2$ **b.** $y = 4x^2 + 4x + 1$

INSTRUCTOR NOTE
Exercises 48 and 54 can be used for similar in-class examples.

Solution

a. $y = x^2 + 2x - 2$

$0 = x^2 + 2x - 2$ • Let $y = 0$.

$x = \dfrac{-b \pm \sqrt{b^2 - 4ac}}{2a}$ • The trinomial $x^2 + 2x - 2$ is non-factorable over the integers. Use the quadratic formula to solve for x. $a = 1$, $b = 2$, $c = -2$.

$= \dfrac{-2 \pm \sqrt{2^2 - 4(1)(-2)}}{2(1)}$

$= \dfrac{-2 \pm \sqrt{4 + 8}}{2}$

$= \dfrac{-2 \pm \sqrt{12}}{2} = \dfrac{-2 \pm 2\sqrt{3}}{2}$

$= -1 \pm \sqrt{3}$

The x-intercepts are $(-1 + \sqrt{3}, 0)$ and $(-1 - \sqrt{3}, 0)$.

Graphical Check

See Appendix A: Zero

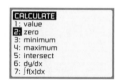

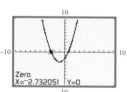
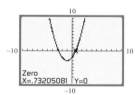

The x-coordinates of the x-intercepts of the graph of $Y = X^2 + 2X - 2$ are -2.732051 and 0.73205081.

$-1 - \sqrt{3} \approx -2.73205$;

$-1 + \sqrt{3} \approx 0.73205$

❓ ANSWER   The x-intercepts are $(-1, 0)$ and $(3, 0)$.

b. $y = 4x^2 + 4x + 1$

$0 = 4x^2 + 4x + 1$ • Let $y = 0$.

$0 = (2x + 1)(2x + 1)$ • Solve for x by factoring.

$2x + 1 = 0 \qquad 2x + 1 = 0$

$\qquad 2x = -1 \qquad\qquad 2x = -1$

$\qquad x = -\dfrac{1}{2} \qquad\qquad x = -\dfrac{1}{2}$ • The equation has a double root.

The x-intercept is $\left(-\dfrac{1}{2}, 0\right)$.

Graphical Check

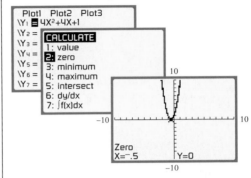

The x-coordinate of the x-intercept of the graph of $Y = 4X^2 + 4X + 1$ is -0.5.

YOU TRY IT 3

Find the x-intercepts of the parabola given by the equation.

a. $y = 2x^2 - 5x + 2$ **b.** $y = x^2 + 4x + 4$

Solution   See page S40.   **a.** $\left(\dfrac{1}{2}, 0\right)$, $(2, 0)$ **b.** $(-2, 0)$

INSTRUCTOR NOTE
This is difficult for students. Giving additional simple examples may help.
Find the x-intercept of
$y = 2x - 6$.
Find the zero of
$f(x) = 2x - 6$.
Solve $2x - 6 = 0$.

Suggested Activity

See Section 9.5 of the *Student Activity Manual* for an activity in which technology is used to investigate quadratics.

The preceding examples suggest that there is a relationship among the x-intercepts of the graph of a function, the zeros of a function, and the solutions of an equation. In fact, these three concepts are different ways of discussing the same number. The choice depends on the focus of the discussion.

• If we are discussing graphing, then intercept is our focus.
• If we are discussing functions, then the zero of the function is our focus.
• If we are discussing equations, then the solution of the equation is our focus.

Recall from Section 9.3 that the graph of a quadratic function may not have x-intercepts. The graph of $y = -x^2 + 2x - 2$ is shown at the right. The graph does not pass through the x-axis and thus there are no x-intercepts. This means that there are no real number zeros of the function $f(x) = -x^2 + 2x - 2$ and that there are no real number solutions of the equation $-x^2 + 2x - 2 = 0$. However, using the quadratic formula, we find that the solutions of $-x^2 + 2x - 2 = 0$ are the complex numbers $1 - i$ and $1 + i$. Thus the zeros of $f(x) = -x^2 + 2x - 2$ are the complex numbers $1 - i$ and $1 + i$.

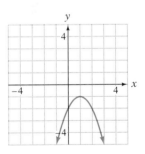

■ The Minimum or Maximum of a Quadratic Function

The graph of $f(x) = x^2 - 2x + 3$ is shown at the right. Because a is positive, the parabola opens up. **The vertex of the parabola is the lowest point on the parabola. It is the point that has the minimum y-coordinate. Therefore, the value of the function at this point is a minimum.**

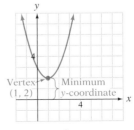

The graph of $f(x) = -x^2 + 2x + 1$ is shown at the right. Because a is negative, the parabola opens down. **The vertex of the parabola is the highest point on the parabola. It is the point that has the maximum y-coordinate. Therefore, the value of the function at this point is a maximum.**

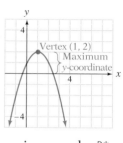

? QUESTION   Does the function have a minimum or a maximum value?*

a. $f(x) = -x^2 + 6x - 1$ **b.** $f(x) = 2x^2 - 4$ **c.** $f(x) = -5x^2 + x$

TAKE NOTE

The maximum value of a function is the y-coordinate of the vertex. See Example 4.

As described earlier, the maximum or minimum value of a quadratic function can be found by using a graphing calculator. An algebraic method of finding the maximum or minimum value involves first finding the x-coordinate of the vertex and then evaluating the function at that value. This is illustrated in Example 4.

EXAMPLE 4

Find the maximum value of $f(x) = -2x^2 + 4x + 3$.

Solution

$x = -\dfrac{b}{2a} = -\dfrac{4}{2(-2)} = 1$

- Find the x-coordinate of the vertex. $a = -2$ and $b = 4$

$f(x) = -2x^2 + 4x + 3$

$f(1) = -2(1)^2 + 4(1) + 3$

- Find the y-coordinate of the vertex by replacing x by 1 and solving for y.

$f(1) = 5$

The maximum value of the function is 5.

INSTRUCTOR NOTE
Exercise 68 can be used for a similar in-class example.

GRAPHICAL CHECK

See Appendix A: Min and Max

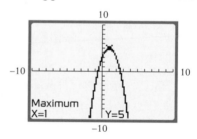

YOU TRY IT 4

Find the minimum value of $f(x) = 2x^2 - 3x + 1$.

Solution   See page S40. $-\dfrac{1}{8}$

? QUESTION   The vertex of a parabola that opens up is $(-4, 7)$. What is the minimum value of the function?†

The concepts surrounding increasing, decreasing, maximum, and minimum play an important role in the application of mathematics.

———

? ANSWERS   *a. A maximum value ($a = -1$)   **b.** A minimum value ($a = 2$)   **c.** A maximum value ($a = -5$)   †The minimum value is 7 (the y-coordinate of the vertex).

Suggested Activity

Have students determine the point at which the graph at the right intercepts the *t*-axis. Once they have found that point, ask them to explain its relevance in the context of this problem. This activity will connect the abstract concept of intercept with a real situation.

Suppose a ball is tossed in the air with an initial velocity of 64 feet per second by a softball player. The function that describes the distance $s(t)$, in feet, of the ball above the ground t seconds after it is released at a point 5 feet above the ground is given by $s(t) = -16t^2 + 64t + 5$. (This assumes no air resistance.)

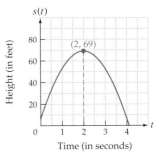

The graph shows that the function is increasing from $t = 0$ to $t = 2$. Thinking about this in terms of the flight of the ball, this means that the distance between the softball player and the ball is increasing. After 2 seconds have elapsed, the *maximum* distance between the softball player and the ball is achieved.

From 2 seconds to just after 4 seconds, the function is decreasing. In terms of the flight of the ball, this means that the distance between the softball player and the ball is decreasing. This corresponds to our experience: Once a ball reaches its maximum height, the ball starts down and gets closer and closer to us.

INSTRUCTOR NOTE
Exercise 76 can be used for a similar in-class example.

EXAMPLE 5

A mining company has determined that the cost c, in dollars per ton, of mining a certain mineral is given by $c(x) = 0.2x^2 - 2x + 12$, where x is the number of tons of the mineral that is mined. Find how many tons of the mineral should be mined to minimize the cost. What is the minimum cost?

State the goal.   The goal is to find the number of tons of the mineral that should be mined to minimize the cost and then to determine the minimum cost.

Devise a strategy.

- Find the x-coordinate of the vertex.
- Evaluate the function at the x-coordinate of the vertex.

Solve the problem.

$$x = -\frac{b}{2a} = -\frac{-2}{2(0.2)} = 5$$ • $a = 0.2, b = -2$

To minimize the cost, 5 tons should be mined.

$$c(x) = 0.2x^2 - 2x + 12$$
$$c(5) = 0.2(5)^2 - 2(5) + 12 = 5 - 10 + 12 = 7$$ • Evaluate the function at 5.

The minimum cost per ton is $7.

Check your work.

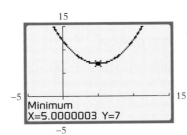

- Use a graphing calculator to find the minimum value of the function. This will give the vertex, (5, 7). Note that a graphing calculator may give an approximation to the exact value. Here the value 5 is returned as 5.0000003.

See Appendix A:
Min and Max

YOU TRY IT 5

The height s, in feet, of a ball thrown straight up is given by $s(t) = -16t^2 + 64t$, where t is the time in seconds. Find the time it takes the ball to reach its maximum height. What is the maximum height?

Solution   See page S40.   2s; 64 ft

INSTRUCTOR NOTE
Exercise 90 can be used for
a similar in-class example.

EXAMPLE 6

Find two numbers whose difference is 10 and whose product is a minimum.

State the goal.   We want to find two numbers that when subtracted equal 10 and whose product is the lowest possible number.

Devise a strategy.   Let x represent one number. Because the difference between the two numbers is 10, $x + 10$ represents the other number. Then their product is represented by $x(x + 10) = x^2 + 10x$.

- To find one of the two numbers, find the x-coordinate of the vertex of $f(x) = x^2 + 10x$.
- To find the other number, replace x in $x + 10$ by the x-coordinate of the vertex and evaluate.

Solve the problem.

$$x = -\frac{b}{2a} = -\frac{10}{2(1)} = -5$$

$$x + 10 = -5 + 10 = 5$$

The numbers are -5 and 5.

Check your work.   The difference of -5 and 5 is 10: $5 - (-5) = 10$.

The product of -5 and 5 is -25.

Try several numbers whose difference is 10 (for example, 6 and -4, 7 and -3, 8 and -2) to see whether the products are less than the product of -5 and 5.

YOU TRY IT 6

A mason is forming a rectangular floor for a storage shed. The perimeter of the rectangle is 44 feet. What dimensions will give the floor a maximum area?

Solution   See page S40.   11 ft by 11 ft

■ Quadratic Regression

Given data that approximate a quadratic function, you can use a graphing calculator to perform a regression and determine a quadratic model. An example follows.

INSTRUCTOR NOTE
Exercise 92 can be used for
a similar in-class example.

EXAMPLE 7

The fuel efficiency of an automobile engine varies with the speed at which the car is driven. Because the fuel efficiency depends on the speed, fuel efficiency is a function of the speed of the car. The results of an automotive study comparing the speed x, in miles per hour, and the average fuel efficiency y, in miles per gallon, are shown in the table below.

x	15	20	25	30	35	40	45	50	55	60	65	70
y	22.1	25.3	27.3	28.2	28.6	28.8	29.7	30.0	30.2	28.6	27.2	25.3

a. Use regression to determine a quadratic model for these data.

b. Use the equation to predict the fuel efficiency when the car is driven at 48 miles per hour.

c. Use the equation to predict the speeds at which the fuel efficiency of a car is 27 miles per gallon.

d. ✎ Find the vertex of the graph of the equation. Round coordinates to the nearest tenth. Explain the meaning of the vertex in the context of the data.

Solution

a. Use a graphing calculator to perform the regression.

 See Appendix A: Regression

The quadratic regression equation is given in the form $y = ax^2 + bx + c$.

To the nearest millionth, the equation is $y = -0.007635x^2 + 0.702502x + 13.818057$.

 See Appendix A: Trace

b. Graph the regression equation. Use the TRACE feature to find the value of y when $x = 48$.

The coordinates printed at the bottom of the screen are approximately (48, 29.95).

The equation predicts that the fuel efficiency when the car is driven at 48 miles per hour is approximately 29.95 miles per gallon.

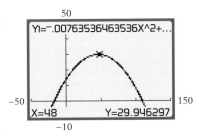

See Appendix A:
Intersect

c. The regression equation is already stored in Y_1. Enter 27 for Y_2. Find the points of intersection of Y_1 and Y_2 to determine the x values when $y = 27$.

The coordinates printed at the bottom of the screen are approximately (26.26, 27). At the second point of intersection, the coordinates are approximately (65.75, 27).

The equation predicts that the speeds at which the fuel efficiency of a car is 27 miles per gallon are approximately 26.26 miles per hour and 65.75 miles per hour.

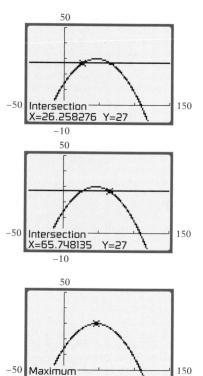

 See Appendix A:
Min and Max

d. Because the parabola opens down, the vertex is the highest point on the parabola; the value of the function at this point is a maximum. To find the vertex, use a graphing calculator to find the maximum.

The coordinates printed at the bottom of the screen are approximately (46.0, 30.0).

The vertex is (46.0, 30.0).

The vertex represents the speed at which the fuel efficiency is greatest. At a speed of 46 miles per hour, the fuel efficiency of a car is 30 miles per gallon.

YOU TRY IT 7

The distance traveled by a baseball varies with the angle at which the ball is hit. The table below shows the distance y, in meters, traveled by a baseball hit at various angles x, in degrees. For these data, the initial speed of the ball off the bat is 40 meters per second, and the ball is hit with backspin.

x	10	15	30	36	42	45	48	54	60
y	61.2	83.0	130.4	139.4	143.2	142.7	140.7	132.8	119.7

INSTRUCTOR NOTE
Exercises 94 and 95 provide data on baseballs hit with no spin and with topspin, respectively. You might have students compare the data for the three situations. For example, ask them what the differences in the vertices represent.

a. Use regression to determine a quadratic model for these data. Round values to the nearest hundredth.
b. Use the equation to predict the distance the ball travels when it is hit at an angle of 40°.
c. Use the equation to predict the angle at which the ball must be hit to travel a distance of 120 meters.

d. Find the vertex of the graph of the equation. Round coordinates to the nearest hundredth. Explain the meaning of the vertex in the context of the data.

Solution See page S40. **a.** $y = -0.8x^2 + 6.58x + 2.46$ **b.** 142.33 m **c.** 25.53° or 59.91° **d.** Vertex: (42.68, 142.88). The ball will travel a maximum distance of 142.88 m when it is hit at an angle of 42.68°.

9.5 EXERCISES

Suggested Assignment: 9–27, odds; 34; 37–95, odds

Topics for Discussion

1. What is a quadratic function?
   A quadratic function is a function of the form $f(x) = ax^2 + bx + c, a \neq 0$.

2. Describe the graph of a parabola.
   Descriptions will vary.

3. What is the vertex of a parabola? When $a > 0$, it is the point with the least y-coordinate. When $a < 0$, it is the point with the greatest y-coordinate.

4. What is the axis of symmetry of the graph of a parabola? It is a line that passes through the vertex of a parabola of the form $y = ax^2 + bx + c$ and is parallel to the y-axis.

5. Describe an algebraic method of finding the vertex of a parabola. Descriptions may vary. For example: Use the equation $x = -\frac{b}{2a}$ to find the x-coordinate of the vertex. Then find the y-coordinate of the vertex by replacing x in the equation of the parabola by the value found for x.

6. **a.** What is an x-intercept of a graph of a parabola?
   b. What is a y-intercept of a graph of a parabola?
   a. It is a point at which the graph crosses the x-axis. **b.** It is a point at which the graph crosses the y-axis.

7. What is the minimum value or the maximum value of a quadratic function? Answers may vary. For example, it is the value of the function at the vertex of the graph of the function.

8. Describe how to find the minimum or maximum value of a quadratic function. Descriptions may vary. For example, find the y-coordinate of the vertex of the graph of the function.

▪ Properties of Quadratic Functions

9. The axis of symmetry of a parabola is the line $x = -5$. What is the x-coordinate of the vertex of the parabola? -5

10. The axis of symmetry of a parabola is the line $x = 8$. What is the x-coordinate of the vertex of the parabola? 8

11. The vertex of a parabola is $(7, -9)$. What is the axis of symmetry of the parabola? $x = 7$

12. The vertex of a parabola is $(-4, 10)$. What is the axis of symmetry of the parabola? $x = -4$

Find the vertex and axis of symmetry of the parabola given by the equation.
Verify your answers using a graphing calculator.

13. $y = x^2 - 2$
Vertex: $(0, -2)$
Axis of symmetry: $x = 0$

14. $y = x^2 + 2$
Vertex: $(0, 2)$
Axis of symmetry: $x = 0$

15. $y = -x^2 + 3$
Vertex: $(0, 3)$
Axis of symmetry: $x = 0$

16. $y = -x^2 - 1$
Vertex: $(0, -1)$
Axis of symmetry: $x = 0$

17. $y = \dfrac{1}{2} x^2$
Vertex: $(0, 0)$
Axis of symmetry: $x = 0$

18. $y = -\dfrac{1}{2} x^2 + 2$
Vertex: $(0, 2)$
Axis of symmetry: $x = 0$

19. $y = 2x^2 - 1$
Vertex: $(0, -1)$
Axis of symmetry: $x = 0$

20. $y = x^2 - 2x$
Vertex: $(1, -1)$
Axis of symmetry: $x = 1$

21. $y = x^2 + 2x$
Vertex: $(-1, -1)$
Axis of symmetry: $x = -1$

22. $y = -2x^2 + 4x$
Vertex: $(1, 2)$
Axis of symmetry: $x = 1$

23. $y = -\dfrac{1}{2} x^2 - x$
Vertex: $\left(-1, \dfrac{1}{2} \right)$
Axis of symmetry: $x = -1$

24. $y = x^2 - x - 2$
Vertex: $\left(\dfrac{1}{2}, -\dfrac{9}{4} \right)$
Axis of symmetry: $x = \dfrac{1}{2}$

25. $y = x^2 - 3x + 2$
Vertex: $\left(\dfrac{3}{2}, -\dfrac{1}{4} \right)$
Axis of symmetry: $x = \dfrac{3}{2}$

26. $y = 2x^2 - x - 5$
Vertex: $\left(\dfrac{1}{4}, -\dfrac{41}{8} \right)$
Axis of symmetry: $x = \dfrac{1}{4}$

27. $y = 2x^2 - x - 3$
Vertex: $\left(\dfrac{1}{4}, -\dfrac{25}{8} \right)$
Axis of symmetry: $x = \dfrac{1}{4}$

To answer Exercises 28 to 31, use the graphs you produced to verify your answers in Exercises 13 to 27.

28. What effect does increasing the coefficient of x^2 have on the graph of $y = ax^2 + bx + c$, $a > 0$?   The graph becomes thinner.

29. What effect does decreasing the coefficient of x^2 have on the graph of $y = ax^2 + bx + c$, $a > 0$?   The graph becomes wider.

30. What effect does increasing the constant term have on the graph of $y = ax^2 + bx + c$, $a \neq 0$?
The graph is higher on the rectangular coordinate system.

31. What effect does decreasing the constant term have on the graph of $y = ax^2 + bx + c$, $a \neq 0$?
The graph is lower on the rectangular coordinate system.

32. What is the value of k if the vertex of the parabola $y = x^2 + 10x + k$ is a point on the x-axis?   25

33. What is the value of k if the vertex of the parabola $y = x^2 - 8x + k$ is a point on the x-axis?   16

34. Match each of the graphs below with one of the equations listed. Explain the reasons for your answers.

$f(x) = 2x^2 - 4x - 3$ b   Explanations will vary.

$g(x) = -2x^2 + 4x + 3$ f

$h(x) = 0.5x^2 - 2x - 4$ d

$j(x) = 2x - 4$ a

$k(x) = -0.5x^2 + 2x + 4$ e

$L(x) = 2x^2 - 4x + 1$ c

a.
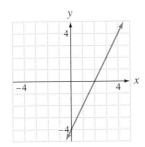

b.
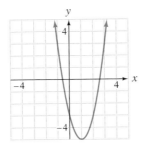

c.
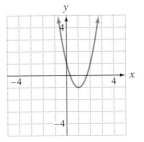

d.
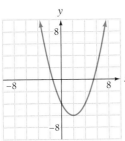

e.
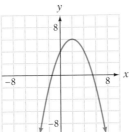

f.
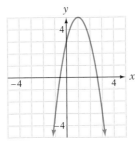

P

35. Sketch a graph of a function of the form $f(x) = ax^2 + bx + c$ in which:

  a. $a > 0, c > 0$, and the function has no real zeros

  b. $a < 0, c < 0$, and the function has no real zeros

  c. $a > 0, c > 0$, and the function has two real zeros

  d. $a < 0, c < 0$, and the function has two real zeros

  e. $a > 0, c < 0$, and the function has two real zeros

  f. $a < 0, c > 0$, and the function has two real zeros

  g. $a > 0, c > 0$, and the function has one real zero

  h. $a < 0, c < 0$, and the function has one real zero

  i. $a > 0, c < 0$, and the function has no real zeros

  j. $a < 0, c > 0$, and the function has one real zero

Sketches for parts a through h will vary. Examples are provided in the Solutions Manual. Parts i and j are not possible.

State the domain and the range of the function.

For Exercises 36 to 44, the domain is $\{x \mid x \in \text{real numbers}\}$.

36. $f(x) = 2x^2 - 4x - 5$
Range: $\{y \mid y \geq -7\}$

37. $f(x) = 2x^2 + 8x + 3$
Range: $\{y \mid y \geq -5\}$

38. $f(x) = -2x^2 - 3x + 2$
Range: $\{y \mid y \leq \frac{25}{8}\}$

39. $f(x) = -x^2 + 6x - 9$
Range: $\{y \mid y \leq 0\}$

40. $f(x) = x^2 - 4x + 4$
Range: $\{y \mid y \geq 0\}$

41. $f(x) = x^2 + 4x - 3$
Range: $\{y \mid y \geq -7\}$

42. $f(x) = -x^2 - 4x - 5$
Range: $\{y \mid y \leq -1\}$

43. $f(x) = -x^2 + 4x + 1$
Range: $\{y \mid y \leq 5\}$

44. $f(x) = x^2 - 2x - 2$
Range: $\{y \mid y \geq -3\}$

▪ Intercepts of Quadratic Functions

Find the x-intercepts of the parabola given by the equation. Verify the intercepts using a graphing calculator.

45. $y = 2x^2 - 4x$

(0, 0), (2, 0)

46. $y = 3x^2 + 6x$

(0, 0), (−2, 0)

47. $y = 2x^2 - 5x - 3$

(3, 0), $\left(-\dfrac{1}{2}, 0\right)$

48. $y = 4x^2 + 11x + 6$

$\left(-\dfrac{3}{4}, 0\right)$, (−2, 0)

49. $y = x^2 - 2$

$(\sqrt{2}, 0)$, $(-\sqrt{2}, 0)$

50. $y = 9x^2 - 2$

$\left(\dfrac{\sqrt{2}}{3}, 0\right)$, $\left(-\dfrac{\sqrt{2}}{3}, 0\right)$

51. $y = x^2 + 2x - 1$

$(-1 + \sqrt{2}, 0)$, $(-1 - \sqrt{2}, 0)$

52. $y = x^2 + 4x - 3$

$(-2 + \sqrt{7}, 0)$, $(-2 - \sqrt{7}, 0)$

53. $y = x^2 + 6x + 10$

No x-intercepts

54. $y = -x^2 - 4x - 5$

No x-intercepts

55. $y = -x^2 - 2x + 1$

$(-1 + \sqrt{2}, 0)$, $(-1 - \sqrt{2}, 0)$

56. $y = -x^2 + 4x + 1$

$(2 + \sqrt{5}, 0)$, $(2 - \sqrt{5}, 0)$

Find the real zeros of the function. Round to the nearest tenth.

57. $y = x^2 + 3x - 1$

−3.3, 0.3

58. $y = x^2 - 2x - 4$

−1.2, 3.2

59. $y = -2x^2 + 3x + 7$

−1.3, 2.8

60. $y = -2x^2 - x + 2$

−1.3, 0.8

61. $y = x^2 + 6x + 12$

No real zeros

62. $y = -x^2 + 3x - 9$

No real zeros

▪ The Minimum or Maximum of a Quadratic Function

Find the minimum or maximum value of each quadratic function. Verify your answers using a graphing calculator.

63. $f(x) = x^2 - 2x + 3$

Minimum: 2

64. $f(x) = 2x^2 + 4x$

Minimum: −2

65. $f(x) = -2x^2 + 4x - 3$

Maximum: −1

66. $f(x) = -2x^2 + 4x - 5$

Maximum: −3

67. $f(x) = 3x^2 + 3x - 2$

Minimum: $-\dfrac{11}{4}$

68. $f(x) = 3x^2 + 5x + 2$

Minimum: $-\dfrac{1}{12}$

69. $f(x) = -x^2 - x + 2$

Maximum: $\dfrac{9}{4}$

70. $f(x) = -3x^2 + 4x - 2$

Maximum: $-\dfrac{2}{3}$

71. $f(x) = -2x^2 - 3x$

Maximum: $\dfrac{9}{8}$

72. Which of the following parabolas has the highest minimum value?

a. $y = x^2 - 2x - 3$ **b.** $y = x^2 - 10x + 20$ **c.** $y = 4x^2 - 3$

c

73. Which of the following parabolas has the lowest maximum value?

a. $y = -2x^2 + 4x - 1$ **b.** $y = -x^2 - 3x + 5$ **c.** $y = -\dfrac{1}{2}x^2 - 6$

c

74. *Physics*   The height s, in feet, of a rock thrown upward at an initial speed of 64 feet per second from a cliff 50 feet above an ocean beach is given by the function $s(t) = -16t^2 + 64t + 50$, where t is the time in seconds. Find the maximum height above the beach that the rock will attain.   114 ft

50 ft

75. *Physics*   The height s, in feet, of a ball thrown upward at an initial speed of 80 feet per second from a platform 50 feet high is given by the function $s(t) = -16t^2 + 80t + 50$, where t is the time in seconds. Find the maximum height above the ground that the ball will attain.   150 ft

76. *Business*   A manufacturer of microwave ovens believes that the revenue R, in dollars, that the company receives is related to the price P, in dollars, of an oven by the function $R(P) = 125P - 0.25P^2$. What price will give the maximum revenue? $250

77. *Business*   A manufacturer of camera lenses estimated that the average monthly cost, C, of a lens is given by the function $C(x) = 0.1x^2 - 20x + 2000$, where x is the number of lenses produced each month. Find the number of lenses the company should produce in order to minimize the average cost.   100 lenses

78. *Chemistry*   A pool is treated with a chemical to reduce the amount of algae. The amount of algae in the pool t days after the treatment can be approximated by the function $A(t) = 40t^2 - 400t + 500$. How many days after treatment will the pool have the least amount of algae?   5 days

79. *Structural Engineering*   The suspension cable that supports a small footbridge hangs in the shape of a parabola. The height h, in feet, of the cable above the bridge is given by the function $h(x) = 0.25x^2 - 0.8x + 25$, where x is the distance in feet from one end of the bridge. What is the minimum height of the cable above the bridge?   24.36 ft

80. *Income*   The net annual income I, in dollars, of a family physician can be modeled by the equation $I(x) = -290(x - 48)^2 + 148,000$, where x is the age of the physician and $27 \le x \le 70$. Find **a.** the age at which the physician's income will be a maximum and **b.** the maximum income.
   a. 48 years   **b.** $148,000

81. *Physics*   Karen is throwing an orange to her brother Saul, who is standing on the balcony of their home. The height h, in feet, of the orange above the ground t seconds after it is thrown is given by $h(t) = -16t^2 + 32t + 4$. If Saul's outstretched arms are 18 feet above the ground, will the orange ever be high enough so that he can catch it?   Yes

82. *Sports*   Some football fields are built in a parabolic mound shape so that water will drain off the field. A model for the shape of the field is given by $h(x) = -0.00023475x^2 + 0.0375x$, where h is the height of the field, in feet, at a distance of x feet from the sideline. What is the maximum height? Round to the nearest tenth.   1.5 ft

83. *Art*   The Buckingham Fountain in Chicago shoots water from a nozzle at the base of the fountain. The height h, in feet, of the water above the ground t seconds after it leaves the nozzle is given by the equation $h(t) = -16t^2 + 90t + 15$. What is the maximum height of the water spout to the nearest tenth of a foot?   141.6 ft

84. *Automotive Engineering* On wet concrete, the stopping distance s, in feet, of a car traveling v miles per hour is given by $s(v) = 0.055v^2 + 1.1v$. At what speed could a car be traveling and still stop at a stop sign 44 feet away? 20 mph

85. *Automotive Engineering* The fuel efficiency of an average car is given by the equation $E(v) = -0.018v^2 + 1.476v + 3.4$, where E is the fuel efficiency in miles per gallon and v is the speed of the car in miles per hour. What speed will yield the maximum fuel efficiency? What is the maximum fuel efficiency? 41 mph; 33.658 mpg

86. *Business* A manufacturer has determined that the profit received from producing and selling x cans of paint is given by $P(x) = -\frac{1}{10}x^2 + 50x - 800$.

   a. Graph this function for $0 \le x \le 550$.

   b. Find the intervals on which the function is increasing or decreasing. Write a sentence that explains the meaning of these intervals in the context of this problem.

   c. How many cans of paint should the manufacturer produce and sell to maximize profit? What is the maximum profit?

   b. The function is increasing on $0 < x < 250$. This means that the manufacturer's profit is increasing as more cans of paint are produced and sold. The function is decreasing on $250 < x < 550$. This means that profit is decreasing as manufacturing exceeds 250 cans of paint. **c.** From the graph, the manufacturer should produce and sell 250 cans of paint to maximize profit. The maximum profit is $5450.

87. *Business* The number of calories burned by a small bird is a function of the speed of the bird and can be modeled by $C(x) = 0.7x^2 - 29.8x + 387.4$, where $C(x)$ is the number of calories burned by the bird when it is flying x miles per hour.

   a. Graph this function for $0 \le x \le 50$.

   b. At what speed does the bird burn the least number of calories? How many calories are burned at this speed? Round to the nearest tenth.

   c. To the nearest whole number, find the intervals on which the function is increasing or decreasing. Write a sentence that explains the meaning of these intervals in the context of this problem.

   b. 21.3 mph; 70 calories per hour **c.** The function is decreasing on $0 < x < 21$. This means that the number of calories burned is decreasing as the speed increases. The function is increasing on $21 < x < 50$. This means that the number of calories burned is increasing as the speed increases.

88. *Mathematics* Find two numbers whose sum is 20 and whose product is a maximum. 10 and 10

89. *Mathematics* Find two numbers whose difference is 14 and whose product is a minimum. 7 and −7

90. *Farming* A rancher has 200 feet of fencing to build a rectangular corral alongside an existing fence. Determine the dimensions of the corral that will maximize the enclosed area. Length: 100 ft; width: 50 ft

a.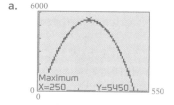

a.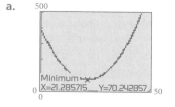

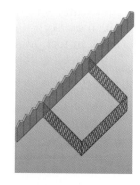

■ Quadratic Regression

91. *Automotive Engineering*   The distance it takes a driver to stop a car is a function of the speed at which the driver is traveling. The table shows speed x, in miles per hour, and the corresponding stopping distance y, in feet, that is required when traveling on dry, level asphalt.

x	20	25	30	35	40	45	50	55	60
y	63	85	109	136	164	196	229	265	304

a. Use regression to determine a quadratic model for these data.
b. Use the equation to predict the distance required to stop a car traveling at 43 miles per hour. Round to the nearest tenth.
c. Skid marks on a straight roadway measured 420 feet. Assuming that this represents the total stopping distance, predict the speed at which the car had been traveling. Round to the nearest tenth.

a. To the nearest ten-thousandth: $y = 0.0475x^2 + 2.2107x - 0.0701$ **b.** 182.9 ft
c. 73.6 mph

92. Social Security   The table below shows expected average annual retirement income from Social Security for a married couple (*Source:* **www.taxfoundation.org**).

Year	2012	2015	2019	2023	2027	2031
Social Security Benefits	\$39,484	\$45,631	\$55,339	\$64,961	\$77,256	\$93,896

a. Use regression to determine a quadratic model for these data. Use $x = 0$ for the year 2000.
b. Use the equation to predict the annual Social Security benefit in 2005. Round to the nearest dollar.
c. Use the equation to predict the year during which the annual Social Security benefit will be \$100,000. Round to the nearest year.

a. To the nearest hundredth: $y = 62.32x^2 + 126.24x + 29,458.14$ **b.** \$31,647
c. 2033

93. Labor Unions   The table below shows the percentage of workers who are members of unions (*Source:* Bureau of Labor Statistics).

Year	1985	1990	1995	2000
Percentage of Workers in Unions	18	16	14.9	13.5

a. Use regression to determine a quadratic model for these data. Use $x = 85$ for the year 1985.
b. Use the equation to predict the percent of workers who will be members of unions in 2010. Round to the nearest tenth of a percent.
c. Use the equation to predict the year before 2000 in which 14% of workers were members of unions.
d. After what year does the equation predict that the percentage will begin to increase?

a. $y = 0.006x^2 - 1.402x + 93.76$ **b.** 12.1%   **c.** 1998   **d.** 2016

94. *Sports*   The distance traveled by a baseball varies with the angle at which the ball is hit. The table below shows the distance y, in meters, traveled by a baseball hit at various angles x, in degrees. In these data, the initial speed of the ball off the bat is 40 meters per second and the ball is hit with no spin.

x	10	15	30	36	42	45	48	54	60
y	58.3	79.7	126.7	136.6	140.6	140.9	139.3	132.5	120.5

  a.   Use regression to determine a quadratic model for these data.

  b.   What is the correlation coefficient for your model? What does that imply about the fit of the data to the graph of the equation?

  c.   Use the equation to predict the distance the ball travels when it is hit at an angle of 20°. Round to the nearest tenth.

  d.   Use the equation to predict the angle at which the ball must be hit to travel a distance of 130 meters. Round to the nearest tenth.

  e.   Find the vertex of the graph of the equation. Round coordinates to the nearest tenth. Explain the meaning of the vertex in the context of the data.

a. To the nearest hundred-thousandth: $y = -0.07374x^2 + 6.42409x + 0.73657$ **b.** $r \approx 0.999$. This is very close to 1, which means that the data points lie very close to the graph of the regression equation.   **c.** 99.7 m   **d.** 31.5° and 55.6°
e. The vertex is (43.6, 140.6). The vertex represents the angle at which the ball must be hit in order to travel the maximum distance. When hit at an angle of 43.6°, the ball travels 140.6 m.

95. *Sports*   The distance traveled by a baseball varies with the angle at which the ball is hit. The table below shows the distance y, in meters, traveled by a baseball hit at various angles x, in degrees. In these data, the initial speed of the ball off the bat is 40 meters per second and the ball is hit with topspin.

x	10	15	30	36	42	45	48	54	60
y	56.1	76.3	122.8	133.2	138.3	139.0	137.8	132.1	120.9

  a.   Use regression to determine a quadratic model for these data.

  b.   What is the correlation coefficient for your model? What does that imply about the fit of the data to the graph of the equation?

  c.   Use the equation to predict the distance the ball travels when it is hit at an angle of 65°. Round to the nearest tenth.

  d.   Use the equation to predict the angle at which the ball must be hit to travel a distance of 110 meters. Round to the nearest tenth.

  e.   Find the vertex of the graph of the equation. Round coordinates to the nearest tenth. Explain the meaning of the vertex in the context of the data.

a. To the nearest hundred-thousandth: $y = -0.06998x^2 + 6.22845x - 0.26229$ **b.** $r \approx 0.999$. This is very close to 1, which means that the data points lie very close to the graph of the regression equation.   **c.** 108.9 m   **d.** 24.4° and 64.6°   **e.** The vertex is (44.5, 138.3). The vertex represents that angle at which the ball must be hit in order to travel the maximum distance. When hit at an angle of 44.5°, the ball travels 138.3 m.

96. *Sports* The table below shows the effect of wind on a runner's performance in the 200-meter dash. Wind speed is given as x, in meters per second. Positive wind speeds correspond to tailwinds; negative wind speeds correspond to headwinds. The values of y are changes in the finishing times, in seconds. Positive changes in finishing times correspond to longer running times; negative changes in finishing times correspond to shorter running times.

x	-6	-4	-2	0	2	4	6
y	2.3	1.4	0.7	0	-0.6	-1.1	-1.4

a. Use regression to determine a quadratic model for these data.
b. Use the equation to predict the change in finishing time when there is a tailwind of 3 meters per second. Round to the nearest tenth.
c. Use the equation to predict the change in finishing time when there is a headwind of 3 meters per second. Round to the nearest tenth.
d. Find the vertex of the graph of the equation. Round coordinates to the nearest tenth. Explain the meaning of the vertex in the context of the data.

a. To the nearest ten-thousandth: $y = 0.0125x^2 - 0.3107x - 0.0143$ **b.** -0.8 s
c. 1.0   **d.** The vertex is $(12.4, -1.9)$. The vertex represents the wind speed that results in the greatest decrease in finishing time. With a 12.4-meters-per-second tailwind, the change in the finishing time is -1.9.

Applying Concepts

97. The axis of symmetry of a parabola is the line $x = 0$. The point $(-2, -3)$ lies on the parabola. Use the symmetry of a parabola to find a second point on the graph. $(2, -3)$

98. The axis of symmetry of a parabola is the line $x = 1$. The point $(3, 0)$ lies on the parabola. Use the symmetry of a parabola to find a second point on the graph. $(-1, 0)$

99. The axis of symmetry of a parabola is the line $x = 2$. The point $(4, -4)$ lies on the parabola. Use the symmetry of a parabola to find a second point on the graph. $(0, -4)$

100. The axis of symmetry of a parabola is the line $x = -1$. The point $(1, -1)$ lies on the parabola. Use the symmetry of a parabola to find a second point on the graph. $(-3, -1)$

Find the value of k such that the graph of the equation contains the given point.

101. $y = x^2 - 3x + k$; $(2, 5)$ 7

102. $y = x^2 + 2x + k$; $(-3, 1)$ -2

103. $y = 2x^2 + kx - 3$; $(4, -3)$ -8

104. $y = 3x^2 + kx - 6$; $(-2, 4)$ 1

105. Suppose a quadratic function was entered into a graphing calculator, and it was displayed on the viewing window as shown at the right.

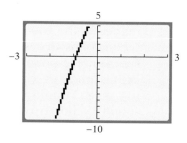

    a. Based on the portion of the graph you can see, is the vertex above or below the x-axis?

    b. Does the function have real number zeros?

    c. Does the graph have a minimum value or a maximum value?

    a. Above **b.** Yes **c.** Maximum

106. The point (x_1, y_1) lies in quadrant II and is on the graph of the equation $y = -2x^2 - 3x + 3$. Given $y_1 = 1$, find x_1. -2

107. The point (x_1, y_1) lies in quadrant III and is a solution of the equation $y = -x^2 + 2x + 3$. Given $y_1 = -5$, find x_1. -2

108. The graph of a quadratic function passes through the points $(-1, 12)$, $(0, 5)$, and $(2, -3)$. Find the value of $a + b + c$. 0

109. One root of the quadratic equation $2x^2 - 5x + k = 0$ is 4. What is the other root? $-\frac{3}{2}$

110. *Metallurgy* Squares are cut from the corners of a rectangular sheet of metal that measures 8 inches by 14 inches. The metal is then folded up to make an open box; it has no lid. Find the maximum volume of the box. Round to the nearest whole number. 83 in³

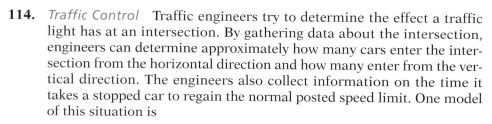

14 in.

8 in.

111. For what values of k will the roots of the equation $7x^2 + 4x + k = 0$ be reciprocals? 7

112. The roots of the function $f(x) = mx^2 + nx + 1$ are -2 and 3. What are the roots of the function $g(x) = nx^2 + mx - 1$? $-2, 3$

113. The figure at the right is a parabolic arch. M and C are the midpoints of line segment AB and arch AB, respectively. Line segment $AB = 40$ units, line segment $MC = 16$ units, and line segment $MX = 5$ units. Find the length of line segment XY. 15 units

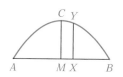

114. *Traffic Control* Traffic engineers try to determine the effect a traffic light has at an intersection. By gathering data about the intersection, engineers can determine approximately how many cars enter the intersection from the horizontal direction and how many enter from the vertical direction. The engineers also collect information on the time it takes a stopped car to regain the normal posted speed limit. One model of this situation is

$$T = \left(\frac{H + V}{2}\right)R^2 + (0.08H - 1.08V)R + 0.58V$$

where H is the number of cars arriving at the intersection from the horizontal direction, V is the number of cars arriving at the intersection

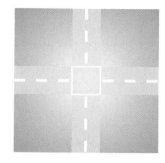

from the vertical direction, and R is the percent of time the light is red in the horizontal direction. T is the total delay time for all cars and is measured as the number of times the traffic light changes from red to green and back to red.

a.   Graph this equation for $H = 100$, $V = 150$, and $0 \le R \le 1$.

b.   Write a sentence that explains why the graph is drawn only for $0 \le R \le 1$.

c.   What percent of the time should the traffic light remain red in the horizontal direction to minimize T? Round to the nearest whole percent.

a.

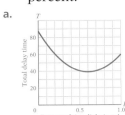

b. The percent cannot be less than 0% or more than 100%.

c. 62%

EXPLORATION

1.   *Alternative Form of the Equation of a Parabola*   An equation of the form $y = ax^2 + bx + c$ can be written in the form $y = a(x - h)^2 + k$, where h and k are constants.

a.   Find the vertex of each equation given below. Then use the process of completing the square to rewrite the equation in the form $y = a(x - h)^2 + k$.

$$y = x^2 - 4x + 7$$
$$y = x^2 - 2x - 2$$
$$y = x^2 - 6x + 5$$
$$y = x^2 + x + 2$$

b.   Based on your answers to part a, for a parabola of the form $y = a(x - h)^2 + k$, what does the ordered pair (h, k) represent?

c.   Use the equation $y = a(x - h)^2 + k$ to find the equation of the parabola that has its vertex at $(1, 2)$ and passes through the point $(2, 5)$.

d.   Use the equation $y = a(x - h)^2 + k$ to find the equation of the parabola that has its vertex at $(0, -3)$ and passes through the point $(3, -2)$.

a. Vertex: $(2, 3)$, $y = (x - 2)^2 + 3$; Vertex: $(1, -3)$; $y = (x - 1)^2 - 3$; Vertex: $(3, -4)$;

$y = (x - 3)^2 - 4$; Vertex: $\left(-\frac{1}{2}, \frac{7}{4}\right)$; $y = \left(x + \frac{1}{2}\right)^2 + \frac{7}{4}$ **b.** The ordered pair (h, k)

represents the vertex of the parabola.   **c.** $y = 3(x - 1)^2 + 2$ **d.** $y = \frac{1}{9}x^2 - 3$

2.   *Local Maximum and Local Minimum*   The graph at the right shows a function that increases, then decreases, and then increases again. The point at which a graph changes from increasing to decreasing is called a **local maximum** for the function. The point at which a function changes from decreasing to increasing is called a **local minimum** for the function. The word **extrema** is used to refer to either a maximum or a minimum.

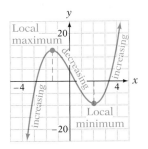

For most functions, finding the local maximum or the local minimum requires techniques that are studied in calculus. We can, however, estimate local maximums and minimums by using a graphing calculator.

a. Use a graphing calculator to find the local maximum and the local minimum for $f(x) = 2x^3 - 3x^2 - 12x + 1$. *Note:* It may take some experimenting to find a viewing window that will show the local maximum and local minimum. For this exercise, use a domain of $[-5, 5]$ and a range of $[-25, 25]$.

b. Use a graphing calculator to find the local maximum and the local minimum for $f(x) = -2x^3 - 3x^2 + 12x + 1$.

After finding the local maximum and minimum for a function, we can use that information to determine on which intervals a function is increasing or decreasing. For instance, for the function in part a, using the fact that $(-1, 8)$ is a local maximum and that $(2, -19)$ is a local minimum, we have the following:

The function is increasing on the interval $(-\infty, -1)$.
The function is decreasing on the interval $(-1, 2)$.
The function is increasing on the interval $(2, \infty)$.

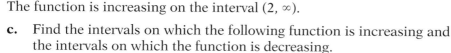

$f(x) = 2x^3 - 3x^2 - 12x + 1$

c. Find the intervals on which the following function is increasing and the intervals on which the function is decreasing.

$$f(x) = 3x^4 - 8x^3 - 66x^2 + 144x + 25$$

d. Find the intervals on which the following function is increasing and the intervals on which the function is decreasing.

$$f(x) = -3x^4 - 16x^3 + 24x^2 + 192x + 10$$

a. Local maximum: $(-1, 8)$; local minimum: $(2, -19)$ **b.** Local maximum: $(1, 8)$; local minimum: $(-2, -19)$ **c.** Decreasing on $(-\infty, -3)$, increasing on $(-3, 1)$, decreasing on $(1, 4)$, and increasing on $(4, \infty)$ **d.** Increasing on $(-\infty, -4)$, decreasing on $(-4, -2)$, increasing on $(-2, 2)$, and decreasing on $(2, \infty)$

SECTION **9.6** **Quadratic Inequalities**

■ Solve Quadratic Inequalities

■ Solve Quadratic Inequalities

A **quadratic inequality in one variable** is one that can be written in the form $ax^2 + bx + c < 0$ or $ax^2 + bx + c > 0$, where $a \neq 0$. The symbols $\leq$ and $\geq$ can also be used.

? QUESTION Is 4 an element in the solution set of the quadratic inequality $x^2 - 6x + 5 > 0$?

Quadratic inequalities can be solved by algebraic means. However, it is often easier to use a graphical method to solve these inequalities. The graphical method is used in the example that follows.

———

? ANSWER $4^2 - 6(4) + 5 = 16 - 24 + 5 = -3$, and $-3 \not> 0$. No, 4 is not an element of the solution set.

Suggested Activity

Ask students questions
such as
For what values of x is $x - 5$
positive? [Answer: $x > 5$]
For what values of x is $x - 5$
negative? [Answer: $x < 5$]
For what values of x is $x + 6$
positive? [Answer: $x > -6$]
For what values of x is $x + 6$
negative? [Answer: $x < -6$]

TAKE NOTE

The product of the
factors $(x - 3)(x + 2)$
will be negative if one
factor is positive and
the other factor is
negative.

Suggested Activity

See Section 9.6 of the
Student Activity Manual for
an activity in which technology is used to visualize the
solution set of a quadratic
inequality. Also included is an
activity involving higher-order
inequalities.

TAKE NOTE

For each factor,
choose a number in
each region. For ex-
ample, when $x = -4$,
$x - 2$ is negative;
when $x = 1$, $x - 2$ is
negative; when $x = 3$,
$x - 2$ is positive; and
when $x = 5$, $x - 2$ is
positive.

➡ Solve and graph the solution set of $x^2 - x - 6 < 0$.

Factor the trinomial.

$$x^2 - x - 6 < 0$$
$$(x - 3)(x + 2) < 0$$

On a number line, draw vertical lines
at the numbers that make each factor
equal to zero.

$$x - 3 = 0 \qquad x + 2 = 0$$
$$x = 3 \qquad\quad x = -2$$

For each factor, **place plus signs
above the number line for those regions where the factor is positive
and negative signs where the factor
is negative.**

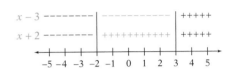

$$x - 3 \text{ is positive for } x > 3.$$
$$x + 2 \text{ is positive for } x > -2.$$

Because $x^2 - x - 6 < 0$, **the solution set will be the regions where one
factor is positive and the other factor is negative.**

Write the solution set. $\qquad\qquad \{x \mid -2 < x < 3\}$

The graph of the solution set of the
inequality $x^2 - x - 6 < 0$ is shown at
the right.

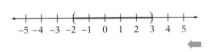

❓ QUESTION   To satisfy the inequality, must both factors be positive, both
factors negative, or one factor positive and one factor negative?†

a. $(x - 4)(x + 5) > 0$ **b.** $(x - 3)(x - 6) < 0$

This method of solving quadratic inequalities can be used on any polynomial
that can be factored into linear factors.

➡ Solve and graph the solution set of $x^3 - 4x^2 - 4x + 16 > 0$.

Factor the polynomial by grouping.

$$x^3 - 4x^2 - 4x + 16 > 0$$
$$x^2(x - 4) - 4(x - 4) > 0$$
$$(x^2 - 4)(x - 4) > 0$$
$$(x - 2)(x + 2)(x - 4) > 0$$

On a number line, identify for each
factor the regions where the factor is
positive and where the factor is
negative.

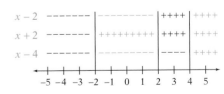

$$x - 2 = 0 \qquad x + 2 = 0 \qquad x - 4 = 0$$
$$x = 2 \qquad\quad x = -2 \qquad\quad x = 4$$

❓ ANSWER   **a.** Either both factors are positive or both factors are negative.   **b.** One factor is
positive and one factor is negative.

There are two regions where **the product of the three factors is positive.**

Write the solution set. $\{x \mid -2 < x < 2 \text{ or } x > 4\}$

The graph of the solution set of the inequality $x^3 - 4x^2 - 4x + 16 > 0$ is shown at the right.

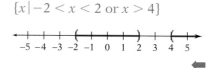

INSTRUCTOR NOTE
Exercise 10 can be used for a similar in-class example.

EXAMPLE 1

Solve and graph the solution set of $2x^2 - x - 3 \geq 0$.

Solution $2x^2 - x - 3 \geq 0$

$(2x - 3)(x + 1) \geq 0$

$2x - 3 = 0 \qquad x + 1 = 0$

$x = \dfrac{3}{2} \qquad\qquad x = -1$

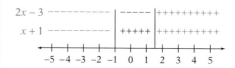

See Appendix A:
Test

$\left\{ x \mid x \leq -1 \text{ or } x \geq \dfrac{3}{2} \right\}$

GRAPHICAL CHECK

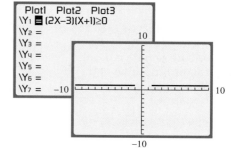

YOU TRY IT 1

Solve and graph the solution set of $2x^2 - x - 10 \leq 0$.

Solution   See page S41. $\left\{ x \mid -2 \leq x \leq \dfrac{5}{2} \right\}$

9.6 EXERCISES   Suggested Assignment: 5–29, odds

Topics for Discussion

1. If $(x - 3)(x - 5) > 0$, what must be true of the values of $x - 3$ and $x - 5$?
   Both $x - 3$ and $x - 5$ are positive or both $x - 3$ and $x - 5$ are negative.

2. How does the solution set of $(x - 5)(x + 2) > 0$ differ from the solution set of $(x - 5)(x + 2) \geq 0$?
   5 and -2 are elements of the solution set of $(x - 5)(x + 2) \geq 0$, but not of the solution set of $(x - 5)(x + 2) > 0$.

3. Why is the first step in solving the quadratic inequality $x^2 - 3x - 10 > 0$ to factor the trinomial?
   The trinomial must be written as a product so that we can determine what the signs of the factors must be in order for their product to be greater than zero.

4. Determine whether the statement is always true, sometimes true, or never true.
   a. The solution set of $x^2 - 4x - 5 < 0$ includes the elements 0 and 4.
   b. The solution set of $(x - 3)(x - 2)(x + 2) > 0$ is $\{-2, 2, 3\}$.
   c. The endpoints of a solution set of a quadratic inequality are not included in the solution set.
   a. Always true   b. Never true   c. Sometimes true

■ Solve Quadratic Inequalities

Solve and graph the solution set.

5. $(x - 4)(x + 2) > 0$
   $\{x \,|\, x < -2 \text{ or } x > 4\}$

6. $(x + 1)(x - 3) > 0$
   $\{x \,|\, x < -1 \text{ or } x > 3\}$

7. $x^2 - 3x + 2 \geq 0$
   $\{x \,|\, x \leq 1 \text{ or } x \geq 2\}$

8. $x^2 + 5x + 6 > 0$
   $\{x \,|\, x < -3 \text{ or } x > -2\}$

9. $x^2 - x - 12 < 0$
   $\{x \,|\, -3 < x < 4\}$

10. $x^2 + x - 20 < 0$
    $\{x \,|\, -5 < x < 4\}$

11. $(x - 1)(x + 2)(x - 3) < 0$
    $\{x \,|\, x < -2 \text{ or } 1 < x < 3\}$

12. $(x + 4)(x - 2)(x - 1) \geq 0$
    $\{x \,|\, -4 \leq x \leq 1 \text{ or } x \geq 2\}$

Solve.

13. $x^2 - 16 > 0$
    $\{x \,|\, x < -4 \text{ or } x > 4\}$

14. $x^2 - 4 \geq 0$
    $\{x \,|\, x \leq -2 \text{ or } x \geq 2\}$

15. $x^2 - 4x + 4 > 0$
    $\{x \,|\, x \neq 2\}$

16. $x^2 + 6x + 9 > 0$
    $\{x \,|\, x \neq -3\}$

17. $x^2 - 9x \leq 36$
    $\{x \,|\, -3 \leq x \leq 12\}$

18. $x^2 + 4x > 21$
    $\{x \,|\, x < -7 \text{ or } x > 3\}$

19. $2x^2 - 5x + 2 \geq 0$
    $\left\{x \,\middle|\, x \leq \dfrac{1}{2} \text{ or } x \geq 2\right\}$

20. $4x^2 - 9x + 2 < 0$
    $\left\{x \,\middle|\, \dfrac{1}{4} < x < 2\right\}$

21. $4x^2 - 8x + 3 < 0$
    $\left\{x \,\middle|\, \dfrac{1}{2} < x < \dfrac{3}{2}\right\}$

22. $2x^2 + 11x + 12 \geq 0$
    $\left\{x \,\middle|\, x \leq -4 \text{ or } x \geq -\dfrac{3}{2}\right\}$

23. $(x - 6)(x + 3)(x - 2) \leq 0$
    $\{x \,|\, x \leq -3 \text{ or } 2 \leq x \leq 6\}$

24. $(x + 5)(x - 2)(x - 3) > 0$
    $\{x \,|\, -5 < x < 2 \text{ or } x > 3\}$

25. $(2x - 1)(x - 4)(2x + 3) > 0$
    $\left\{x \,\middle|\, -\dfrac{3}{2} < x < \dfrac{1}{2} \text{ or } x > 4\right\}$

26. $(x - 2)(3x - 1)(x + 2) \leq 0$
    $\left\{x \,\middle|\, x \leq -2 \text{ or } \dfrac{1}{3} \leq x \leq 2\right\}$

27. $x^3 + 3x^2 - x - 3 \leq 0$
    $\{x \,|\, x \leq -3 \text{ or } -1 \leq x \leq 1\}$

28. $x^3 + x^2 - 9x - 9 < 0$
    $\{x \,|\, x < -3 \text{ or } -1 < x < 3\}$

29. $x^3 - x^2 - 4x + 4 \geq 0$
    $\{x \,|\, -2 \leq x \leq 1 \text{ or } x \geq 2\}$

30. $2x^3 + 3x^2 - 8x - 12 \geq 0$
    $\left\{x \,\middle|\, -2 \leq x \leq -\dfrac{3}{2} \text{ or } x \geq 2\right\}$

Applying Concepts

Graph the solution set.

31. $(x + 2)(x - 3)(x + 1)(x + 4) > 0$

32. $(x - 1)(x + 3)(x - 2)(x - 4) \geq 0$

33. $(x^2 + 2x - 8)(x^2 - 2x - 3) < 0$

34. $(x^2 + 2x - 3)(x^2 + 3x + 2) \geq 0$

35. $(x^2 + 1)(x^2 - 3x + 2) > 0$

36. $(x^2 - 9)(x^2 + 5x + 6) \leq 0$

37. $x < x^2$

38. $x^3 > x$

39. How many integers satisfy the inequality $x^2 + 48 < 16x$?   7

40. Find all values of x that satisfy both $x^2 - 8 \leq 2x$ and $x^2 - 2x \geq 8$. $-2, 4$

41. Solve: $(x - 4)^2 > -2$ All real numbers

42. *Sports*   You shoot an arrow into the air with an initial velocity of 70 meters per second. The distance up, in meters, is given by $d = rt - 5t^2$, where t is the number of seconds since the arrow was shot and r is the initial velocity. Find the interval of time when the arrow will be more than 200 meters high. $4 < t < 10$; between 4 s and 10 s

EXPLORATION

1. *Rational Inequalities*   The graphical method used in this section can be used to solve rational inequalities.

Solve: $\dfrac{2x - 5}{x - 4} \leq 1$

$\dfrac{2x - 5}{x - 4} \leq 1$

$\dfrac{2x - 5}{x - 4} - 1 \leq 0$ • Rewrite the inequality so that 0 appears on the right side of the inequality.

$\dfrac{2x - 5}{x - 4} - \dfrac{x - 4}{x - 4} \leq 0$ • Rewrite the left side so that there is one fraction.

$\dfrac{x - 1}{x - 4} \leq 0$

• On a number line, identify, for each factor of the numerator and each factor of the denominator, the regions where the factor is positive and where the factor is negative.

• The region where the quotient of the two factors is negative is between 1 and 4.

$\{x \mid 1 \leq x < 4\}$ • Write the solution set.

Note that **1 is part of the solution set but that 4 is not part of the solution set, because the denominator of the rational expression is zero when $x = 4$.**

Solve and graph the solution set.

a. $\dfrac{x-4}{x+2} > 0$ $\{x \mid x < -2 \text{ or } x > 4\}$

b. $\dfrac{x+2}{x-3} > 0$ $\{x \mid x < -2 \text{ or } x > 3\}$

c. $\dfrac{x-3}{x+1} \le 0$ $\{x \mid -1 < x \le 3\}$

d. $\dfrac{x-1}{x} > 0$ $\{x \mid x < 0 \text{ or } x > 1\}$

e. $\dfrac{(x-1)(x+2)}{x-3} \le 0$ $\{x \mid x \le -2 \text{ or } 1 \le x < 3\}$

f. $\dfrac{(x+3)(x-1)}{x-2} \ge 0$ $\{x \mid -3 \le x \le 1 \text{ or } x > 2\}$

Solve.

g. $\dfrac{3x}{x-2} > 1$ $\{x \mid x < -1 \text{ or } x > 2\}$

h. $\dfrac{2x}{x+1} < 1$ $\{x \mid -1 < x < 1\}$

i. $\dfrac{2}{x+1} \ge 2$ $\{x \mid -1 < x \le 0\}$

j. $\dfrac{3}{x-1} < 2$ $\left\{x \mid x < 1 \text{ or } x > \dfrac{5}{2}\right\}$

k. $\dfrac{x}{(x-1)(x+2)} \ge 0$ $\{x \mid -2 < x \le 0 \text{ or } x > 1\}$

l. $\dfrac{x-2}{(x+1)(x-1)} \le 0$ $\{x \mid x < -1 \text{ or } 1 < x \le 2\}$

CHAPTER **9** SUMMARY

Key Terms

axis of symmetry [**p. 667**]
discriminant [**p. 646**]
double root [**p. 646**]
intercept [**p. 669**]
maximum value of a function
[**p. 666**]
minimum value of a function
[**p. 666**]
parabola [**pp. 621 and 665**]
quadratic equation [**p. 620**]
quadratic equation in two
variables [**p. 621**]

quadratic function
[**pp. 621 and 665**]
quadratic inequality in one
variable [**p. 688**]
quadratic in form [**p. 656**]
quadratic regression [**p. 674**]
standard form of a quadratic
equation [**p. 621**]
vertex [**p. 666**]
x-intercept [**p. 669**]
y-intercept [**p. 669**]
zero of a function [**p. 670**]

Essential Concepts

The Principle of Zero Products
If the product of two factors is zero, then at least one of the factors must be zero. If $ab = 0$, then $a = 0$ or $b = 0$. [**p. 622**]

To solve a quadratic equation by factoring:
- Write the equation in standard form.
- Factor the polynomial.
- Use the Principle of Zero Products to set each factor equal to zero.
- Solve each equation for the variable. [**p. 623**]

Writing a Quadratic Equation Given Its Solutions
In the equation $(x - r_1)(x - r_2) = 0$, substitute one solution for r_1 and the other solution for r_2. Then rewrite the equation in standard form. **[p. 625]**

Completing the Square

To complete the square on $x^2 + bx$, add $\left(\frac{1}{2}b\right)^2$ to $x^2 + bx$. **[p. 634]**

To solve a quadratic equation by completing the square:
- Solve the equation for $ax^2 + bx$.
- Multiply each side of the equation by the reciprocal of a.
- Add to each side of the equation the term that completes the square.
- Factor the perfect-square trinomial.
- Take the square root of each side of the equation.
- Solve for x. **[p. 636]**

The Quadratic Formula

The solutions of $ax^2 + bx + c = 0$, $a \neq 0$, are $x = \dfrac{-b \pm \sqrt{b^2 - 4ac}}{2a}$. **[p. 644]**

The Effect of the Discriminant on the Solutions of a Quadratic Equation
1. If $b^2 - 4ac = 0$, the equation has one real number solution, a double root.
2. If $b^2 - 4ac > 0$, the equation has two real number solutions that are not equal.
3. If $b^2 - 4ac < 0$, the equation has two complex number solutions. **[p. 646]**

The Effect of the Discriminant on the Number of x-Intercepts of a Parabola
1. If $b^2 - 4ac = 0$, the parabola has one x-intercept.
2. If $b^2 - 4ac > 0$, the parabola has two x-intercepts.
3. If $b^2 - 4ac < 0$, the equation has no x-intercepts. **[p. 647]**

The Vertex and Axis of Symmetry of a Parabola
Let $f(x) = ax^2 + bx + c$ be the equation of a parabola. The coordinates of the vertex are $\left(-\frac{b}{2a}, f\left(-\frac{b}{2a}\right)\right)$. The equation of the axis of symmetry is $x = -\frac{b}{2a}$. **[p. 668]**

Properties of Quadratic Functions
If $a > 0$, the parabola opens up and the function has a minimum value at the vertex.
If $a < 0$, the parabola opens down, and the function has a maximum value at the vertex. **[p. 666]**

Finding the Minimum or Maximum of a Quadratic Function
Find the x-coordinate of the vertex. Then evaluate the function at that value. **[p. 672]**

CHAPTER **9** *REVIEW EXERCISES*

1. Solve: $x + 18 = x(x - 6)$ $-2, 9$

2. Solve: $r^2 - 75 = 0$ $\pm 5\sqrt{3}$

3. Solve: $5(z + 2)^2 = 125$
   $-7, 3$

4. Solve by completing the square: $r^2 = 3r - 1$
   $\dfrac{3 \pm \sqrt{5}}{2}$

5. Solve by completing the square: $x^2 + 13 = 2x$
$1 \pm 2i\sqrt{3}$

6. Solve by using the quadratic formula:
$4x^2 - 4x = 7$
$\dfrac{1 \pm 2\sqrt{2}}{2}$

7. Solve by using the quadratic formula:
$t^2 = 6t - 10$
$3 \pm i$

8. Solve: $x^4 - 4x^2 - 5 = 0$
$\pm\sqrt{5}, \pm i$

9. Solve: $2x^{2/3} + 3x^{1/3} - 2 = 0$ $-8, \dfrac{1}{8}$

10. Solve: $\sqrt{3x - 2} + 4 = 3x$ 2

11. Find the maximum value of the function
$f(x) = -x^2 + 8x - 7$. 9

12. Find the zeros of $g(x) = x^2 + 3x - 8$.
$\dfrac{-3 \pm \sqrt{41}}{2}$

13. Solve: $2x^2 + x < 15$ $\left\{x \mid -3 < x < \dfrac{5}{2}\right\}$

14. Find the x-intercepts of the parabola whose equation is $y = 2x^2 + 5x - 12$.
$(-4, 0), \left(\dfrac{3}{2}, 0\right)$

15. Find the vertex and axis of symmetry of the parabola whose equation is
$y = x^2 - 2x + 3$.   Vertex: $(1, 2)$; Axis of symmetry: $x = 1$

16. State the domain and range of $f(x) = x^2 + 2x - 4$.
Domain: $\{x \mid x \in \text{real numbers}\}$; Range: $\{y \mid y \geq -5\}$

17. Given that 6 is in the range of the function defined by $f(t) = t^2 + 5t - 8$,
find two values of c for which $f(c) = 6$.   2 and -7

18. Write a quadratic equation that has integer coefficients and has solutions -3 and $\dfrac{1}{3}$. $3x^2 + 8x - 3 = 0$

19. Use the discriminant to determine whether $2x^2 + 5x + 1 = 0$ has one real number solution, two real number solutions, or two complex number solutions.   Two real

20. Use the discriminant to determine the number of x-intercepts of the graph of $y = 2x^2 + x + 1$.   None

21. A car with good tire tread can stop in less distance than a car with poor tread. The formula for the stopping distance d, in feet, of a car with good tread on dry cement is approximated by $d = 0.04v^2 + 0.5v$, where v is the speed of the car. If the driver must be able to stop within 60 feet, what is the maximum safe speed, to the nearest mile per hour, of the car?   33 mph

22. A ball is thrown straight up at a velocity of 40 feet per second. After t seconds the height h, in feet, of the ball is given by the equation $h = 5 + 40t - 16t^2$.
   a. Find the times at which the ball will be 20 feet above the ground. Round to the nearest hundredth.
   b. What is the maximum height reached by the ball?
   c. When will it reach the maximum height?
   a. 0.46 s and 2.04 s   **b.** 30 ft   **c.** after 1.25 s

23. The total time between the instant you drop a stone into a certain well and the moment you hear its impact is 4.8 seconds. Use the equation $1100T = 275\sqrt{s} + s$, where T is the time in seconds and s is the depth to the water in feet, to find the distance to the water in the well. Round to the nearest tenth.   324.7 ft

24. The point (x_1, y_1) lies in quadrant I and is a solution of the equation $y = 3x^2 - 2x - 1$. Given $y_1 = 5$, find x.
$\dfrac{1 + \sqrt{19}}{3}$

25. ⊘ The table below shows the actual and projected number of U.S. households using electronic bill payment (*Source:* Yankee Group).

Year	1998	1999	2000	2001	2002
E-payments (in millions)	1.7	4.1	5.2	7.3	9.9

  a. Use regression to determine a quadratic model for these data. Use $x = 100$ for the year 2000.
  b. Use the equation to predict how many U.S. households will use electronic bill payment in 2010.
  c. Use the equation to predict the year during which the number of U.S. households using electronic bill payments will be 25 million. Round to the nearest year.
  a. $y = 0.1x^2 - 18.04x + 809.44$ b. 35.04 million U.S. households   c. 2007

CHAPTER **9** *TEST*

1. Solve: $2x^2 + 9x = 5$ $-5, \dfrac{1}{2}$

2. Solve: $t^2 - 48 = 0$ $\pm 4\sqrt{3}$

3. Solve: $\left(p + \dfrac{1}{2}\right)^2 + 4 = 0$
$-\dfrac{1}{2} \pm 2i$

4. Solve by completing the square: $x^2 - 6x - 2 = 0$
$3 \pm \sqrt{11}$

5. Solve by completing the square:
$x^2 + 6x + 10 = 0$
$-3 \pm i$

6. Solve by using the quadratic formula:
$2x^2 - 2x = 1$
$\dfrac{1 \pm \sqrt{3}}{2}$

7. Solve by using the quadratic formula:
$3t^2 = 4t - 2$
$\dfrac{2}{3} \pm \dfrac{\sqrt{2}}{3}i$

8. Solve: $x^4 - 6x^2 + 8 = 0$

$\pm 2, \pm\sqrt{2}$

9. Solve: $2x + 7x^{1/2} - 4 = 0$ $\dfrac{1}{4}$

10. Solve: $\sqrt{3x + 1} + 2 = 4x$ 1

11. Find the maximum value of the function
$f(x) = -2x^2 + 4x + 1$.
3

12. Find the zeros of $h(x) = 3x^2 + 2x + 2$.
$-\dfrac{1}{3} \pm \dfrac{\sqrt{5}}{3}i$

16. There are three different digits such that any two of them, written in any order, are the digits of a two-digit prime number. Find all three of these digits.   1, 3, 7

17. Find the distance between the points $(-2, 3)$ and $(2, 5)$. $2\sqrt{5}$

18. A pilot flew from Westerly Airfield to Rockingham and then back again. The average speed on the way to Rockingham was 140 miles per hour, and the average speed returning was 100 miles per hour. Find the distance between the two airports if the total flying time was 6 hours.
350 mi

19. A chemist mixes a 12% acid solution with a 6% acid solution. How many milliliters of each should the chemist use to make a 900-milliliter solution that is 10% acid?   600 ml of the 12% solution; 300 ml of the 6% solution

20. How many ounces of a silver alloy that costs $225 per ounce must a jeweler mix with 150 ounces of an alloy that costs $150 per ounce to produce a new alloy that costs $175 per ounce?   75 oz

21. A building contractor estimates that the cost to build a new home is $32,000 plus $75 for each square foot of floor space in the house. Determine a linear function that will give the cost to build a house that contains a given number of square feet. Use this model to determine the cost of building a house that contains 2200 square feet.
$y = 75x + 32,000$; $197,000

22. ◓ The table below shows the projected demand in the United States for high-speed Internet access (*Source:* Cahners In-Stat Group, 2001).

Year	2000	2001	2002	2003	2004
Demand (in millions)	6	13	18	26	32

A linear equation that approximates these data is given by $y = 6.5x + 6$, where $x = 0$ corresponds to the year 2000 and y is in millions of subscribers.

a. Write a sentence that explains the meaning of the slope in the context of this problem.

b. Assuming this model extends beyond 2005, how many high-speed Internet subscribers will there be in the United States in 2006?

a. The demand for high-speed Internet access is increasing by 6.5 million subscribers per year.   b. 45 million subscribers

23. The height $H(t)$, in feet, of the tide at a certain beach in Encinitas, California, can be approximated by

$$H(t) = -0.00013t^5 + 0.00839t^4 - 0.186t^3 + 1.635t^2 - 4.8t + 3.40$$

where t is the number of hours after midnight. Find the height of the tide at 9:00 A.M. Round to the nearest tenth.   4.4 ft

24. A square piece of cardboard is formed into a box by cutting 10-centimeter squares from each of the four corners and then folding up the sides, as shown in the figure at the right. If the volume, V, of the box is to be 49,000 square centimeters, what size square piece of cardboard is needed?   90 cm by 90 cm

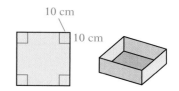

25. Find the vertex and axis of symmetry of the parabola whose equation is $y = \frac{1}{2}x^2 + x - 4$. State the domain and range of the function.

Vertex: $(-1, -\frac{9}{2})$; Axis of symmetry: $x = -1$; Domain: $\{x \mid x \in \text{real numbers}\}$; Range: $\{y \mid y \geq -4.5\}$

10

Exponential and Logarithmic Functions

```
EDIT CALC TESTS
4↑LinReg(ax+b)
 5:QuadReg
 6:CubicReg
 7:QuartReg
 8:LinReg(a+bx)
 9:LnReg
 0↓ExpReg
```

Press **STAT** ◗ to access the LnReg and ExpReg options in the CALC menu.

Do you have a credit card? If so, how long have you had it? Do you pay the balance on your credit card every month? Do you send in your payments on time? If you apply for a loan, the lender will want to know the answers to these and other questions. The lender can get the answers from a credit reporting agency. These agencies provide a credit score based on the information in the report. One of the most widely used credit scores is the FICO® score. The higher the FICO® score, the lower predicted credit risk for lenders. FICO® scores are a topic of **Exploration 2 on page 795**.

Need help? For online student resources, visit this web site:
Math.college.hmco.com

PREP TEST

1. Simplify: 3^{-2} $\dfrac{1}{9}$

2. Simplify: $\left(\dfrac{1}{2}\right)^{-4}$ 16

3. Complete: $\dfrac{1}{8} = 2^?$ -3

4. Evaluate $f(x) = x^4 + x^3$ for $x = -1$ and $x = 3$.
   0; 108

5. Solve: $3x + 7 = x - 5$ -6

6. Solve: $16 = x^2 - 6x$ $-2, 8$

7. Evaluate $A(1 + i)^n$ for $A = 5000$, $i = 0.04$, and $n = 6$. Round to the nearest hundredth.
   6326.60

8. Graph: $f(x) = x^2 - 1$

GO FIGURE

What is the minimum number of times a paper must be folded in half so that its thickness is at least the distance from Earth to the moon? Assume the paper can be folded in half indefinitely and that the paper is 0.01 inch thick. Use a distance from Earth to the moon of 240,000 miles. (*Hint:* 5280 feet = 1 mile)   41 times

---

SECTION 10.1 Algebra of Functions

■ Basic Operations on Functions

■ Composition of Functions

■ Basic Operations on Functions

Just as we can add, subtract, multiply, and divide numbers, we can perform similar operations with functions. For instance, let $f(x) = x^2 - 1$ and $g(x) = -3x + 2$. Then we can create a new function, say h, that is the sum of f and g.

$$h(x) = f(x) + g(x) = (x^2 - 1) + (-3x + 2) = x^2 - 3x + 1$$

Similarly, we can create the difference, product, or quotient of two functions.

> **Operations on Functions**
>
> For all values of x for which both $f(x)$ and $g(x)$ are defined, we define the following functions.
>
> | | |
> |---|---|
> | Sum | $(f + g)(x) = f(x) + g(x)$ |
> | Difference | $(f - g)(x) = f(x) - g(x)$ |
> | Product | $(f \cdot g)(x) = f(x) \cdot g(x)$ |
> | Quotient | $\left(\dfrac{f}{g}\right)(x) = \dfrac{f(x)}{g(x)}$, provided $g(x) \neq 0$ |

INSTRUCTOR NOTE
Exercises 6, 8, 10, and 16 can be used for similar in-class examples.

Suggested Activity

1. Given $f(x) = 4x^{1/2}$ and $g(x) = -2x^{1/2}$, find (a) the sum of the functions, (b) the difference of the functions, and (c) the domains of the sum and difference. [Answer: (a) $2x^{1/2}$ (b) $6x^{1/2}$ (c) The domain of the sum is $\{x \mid x \geq 0\}$; the domain of the difference is $\{x \mid x \geq 0\}$.]
2. Given $f(x) = 5x$ and $g(x) = x^{1/4}$, find (a) the product of the functions, (b) the quotient of the functions, and (c) the domains of the product and quotient. [Answer: (a) $5x^{5/4}$ (b) $5x^{3/4}$ (c) The domain of the product is $\{x \mid x \geq 0\}$; the domain of the quotient is $\{x \mid x > 0\}$.]

EXAMPLE 1

Let $f(x) = x^2 + 4x - 4$ and $g(x) = -x + 15$. Find:

a. $(f + g)(-3)$ **b.** $(f \cdot g)(2)$ **c.** $\left(\dfrac{f}{g}\right)(x)$ **d.** $(f - g)(x)$

Solution

a. $(f + g)(-3) = f(-3) + g(-3) = [(-3)^2 + 4(-3) - 4] + [-(-3) + 15]$
$= -7 + 18 = 11$

b. $(f \cdot g)(2) = f(2) \cdot g(2) = [(2)^2 + 4(2) - 4] \cdot [-2 + 15] = 8 \cdot 13 = 104$

c. $\left(\dfrac{f}{g}\right)(x) = \dfrac{f(x)}{g(x)} = \dfrac{x^2 + 4x - 4}{-x + 15}$

d. $(f - g)(x) = f(x) - g(x) = (x^2 + 4x - 4) - (-x + 15) = x^2 + 5x - 19$

YOU TRY IT 1

Let $f(x) = 3x - 2$ and $g(x) = x + 1$. Find:

a. $(f + g)(3)$ **b.** $\left(\dfrac{f}{g}\right)(4)$ **c.** $(f \cdot g)(x)$ **d.** $(f - g)(x)$

Solution   See page S41.   **a.** 11 **b.** 2 **c.** $3x^2 + x - 2$ **d.** $2x - 3$

The graphs of f and g from Example 1 are shown below. The graph in bold is the graph of the sum of f and g. By using the TRACE feature, we can verify that the values of the function and the sum of those values are as calculated in Example 1a.

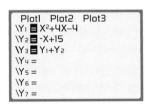

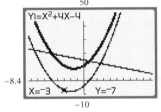

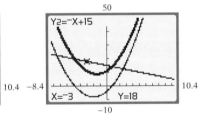

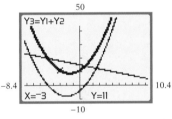

INSTRUCTOR NOTE
Before proceeding with Example 2, give students a simple example of the concept that profit equals revenue minus cost. For instance, if a company earns $25 on the sale of a computer game, and it costs the company $20 to produce the game, what is the profit on the sale of one game? What is the profit on the sale of n games?

INSTRUCTOR NOTE
Exercise 18 can be used for a similar in-class example.

In economics, profit is equal to revenue minus cost. We can represent this symbolically by letting $P(x)$, $R(x)$, and $C(x)$ represent, respectively, the profit, revenue, and cost to produce x products. Then $\boldsymbol{P(x) = R(x) - C(x)}$. Thus to find a profit function, a business analyst must find the difference between the revenue and cost functions.

EXAMPLE 2

Suppose the chief financial officer of a surfboard company has determined that the revenue from the sale of x surfboards is given by $R(x) = 250x - 0.1x^2$ and that the cost to produce x surfboards is given by $C(x) = 200x + 3000$. Find the profit function for the surfboards, and use the profit function to determine the profit for selling 120 surfboards.

State the goal.   We want to find the profit function and the profit for selling 120 surfboards.

Devise a strategy. Profit equals revenue minus cost. Subtract the cost function from the revenue function. Then evaluate the profit function at 120 to find the profit for selling 120 surfboards.

Solve the problem. Profit = Revenue − Cost

$$P(x) = R(x) - C(x)$$
$$= (250x - 0.1x^2) - (200x + 3000)$$
$$= -0.1x^2 + 50x - 3000$$

The profit function is given by $P(x) = -0.1x^2 + 50x - 3000$.

To determine the profit for selling 120 surfboards, evaluate the profit function when $x = 120$.

$$P(x) = -0.1x^2 + 50x - 3000$$
$$P(120) = -0.1(120)^2 + 50(120) - 3000 = 1560$$

The profit for selling 120 surfboards is $1560.

Check your work. One way to check your work is to evaluate the revenue function and the cost function at 120. Then subtract those values. The profit should be the same.

$$R(x) = 250x - 0.1x^2$$
$$R(120) = 250(120) - 0.1(120)^2 = 28,560$$
$$C(x) = 200x + 3000$$
$$C(120) = 200(120) + 3000 = 27,000$$
$$R(120) - C(120) = 28,560 - 27,000 = 1560$$

The profit is $1560, the same number we calculated above. The answer is correct.

Suggested Activity

See Section 10.1 of the *Student Activity Manual* for an investigation into the operations on functions.

YOU TRY IT 2

The revenue from the sale of x computers is given by $R(x) = -0.09x^2 + 300x$. The cost to produce x computers is given by $C(x) = 100x + 300$. Find the profit function for selling x computers. Determine the profit for selling 2000 computers.

Solution See page S41. $P(x) = -0.09x^2 + 200x - 300;$ $39,700

▪ Composition of Functions

Composition of functions is another way in which functions can be combined. This method of combining functions uses the output of one function as an input for a second function.

Suppose a forest fire is started by lightning striking a tree, and the spread of the fire can be approximated by a circle whose radius r, in feet, is given by $r(t) = 24\sqrt{t}$, where t is the number of hours after the tree is struck by the lightning. The area of the fire is the area of a circle and is given by the formula $A(r) = \pi r^2$. Because the area of the fire depends on the radius of the circle and the radius depends on the time since the tree was struck, there is a relationship between the area of the fire and time. This relationship can be found by evaluating the formula for the area of a circle using $r(t) = 24\sqrt{t}$.

$$A(r) = \pi r^2$$
$$A[r(t)] = \pi[r(t)]^2 \qquad \bullet \text{ Replace } r \text{ by } r(t).$$
$$= \pi[24\sqrt{t}]^2 \qquad \bullet \ r(t) = 24\sqrt{t}.$$
$$= 576\pi t \qquad \bullet \text{ Simplify.}$$

The result is the function $A(t) = 576\pi t$, which gives the area of the fire in terms of the time since the lightning struck. For instance, when $t = 3$, we have

$$A(t) = 576\pi t$$
$$A(3) = 576\pi(3)$$
$$\approx 5429$$

Three hours after the lightning strikes, the area of the fire is approximately 5429 square feet.

The function produced above, in which one function was evaluated with another function, is referred to as the *composition* of A with r. The notation $A \circ r$ is used to denote this composition of functions. That is,

$$(A \circ r)(t) = 576\pi t$$

> **Definition of the Composition of Two Functions**
>
> Let f and g be two functions such that $g(x)$ is in the domain of f for all x in the domain of g. Then the **composition** of the two functions, denoted by $f \circ g$, is the function whose value at x is given by $(f \circ g)(x) = f[g(x)]$.

The function defined by $(f \circ g)(x)$ is also called the **composite** of f and g. We read $(f \circ g)(x)$ or $f[g(x)]$ as "f of g of x."

Consider $f(x) = 3x - 1$ and $g(x) = x^2 + 1$. The expression $(f \circ g)(-2)$ or, equivalently, $f[g(-2)]$ means to evaluate the function f at $g(-2)$.

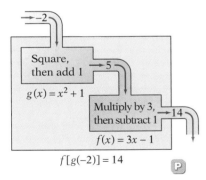

$$g(x) = x^2 + 1$$
$$g(-2) = (-2)^2 + 1 \qquad \bullet \text{ Evaluate } g \text{ at } -2.$$
$$g(-2) = 5$$
$$f(x) = 3x - 1$$
$$f(5) = 3(5) - 1 = 14 \qquad \bullet \text{ Evaluate } f \text{ at } g(-2) = 5.$$

If we apply our function machine analogy, a composition of functions looks something like the figure at the left.

We can find a general expression for $f[g(x)]$ by evaluating f at $g(x)$. For instance, using $f(x) = 3x - 1$ and $g(x) = x^2 + 1$, we have

$$f(x) = 3x - 1$$
$$f[g(x)] = 3[g(x)] - 1 \qquad \bullet \text{ Replace } x \text{ by } g(x).$$
$$= 3[x^2 + 1] - 1 \qquad \bullet \text{ Replace } g(x) \text{ by } x^2 + 1.$$
$$= 3x^2 + 2 \qquad \bullet \text{ Simplify.}$$

In general, the composition of functions is not a commutative operation. That is, $(f \circ g)(x) \neq (g \circ f)(x)$. To verify this, we will compute the composition $(g \circ f)(x) = g[f(x)]$ by again using the functions $f(x) = 3x - 1$ and $g(x) = x^2 + 1$.

$$g(x) = x^2 + 1$$
$$g[f(x)] = [f(x)]^2 + 1 \qquad \bullet \text{ Replace } x \text{ by } f(x).$$
$$= [3x - 1]^2 + 1 \qquad \bullet \text{ Replace } f(x) \text{ by } 3x - 1.$$
$$= 9x^2 - 6x + 2 \qquad \bullet \text{ Simplify.}$$

Thus $f[g(x)] = 3x^2 + 2$, which is not equal to $g[f(x)] = 9x^2 - 6x + 2$. Therefore, in general $(f \circ g)(x) \neq (g \circ f)(x)$ and **composition is not a commutative operation.**

> ? QUESTION   Let $f(x) = 2x - 4$ and $g(x) = \frac{1}{2}x + 2$. Then $f[g(x)] = g[f(x)]$. (You should verify this statement.) Does this contradict the statement we made that composition is not a commutative operation?

The requirement in the definition of the composition of two functions that $g(x)$ be in the domain of f for all x in the domain of g is important. For instance, let

$$f(x) = \frac{x}{x + 1} \qquad \text{and} \qquad g(x) = 2x - 3$$

When $x = 1$,

$$g(1) = 2(1) - 3 = -1$$
$$f[g(1)] = f(-1) = \frac{-1}{-1 + 1} = \frac{-1}{0} \qquad \bullet \text{ Undefined}$$

In this case, $g(1)$ is not in the domain of f. Thus the composition is not defined at $x = 1$.

INSTRUCTOR NOTE
Exercises 28 and 30 can be used for similar in-class examples.

EXAMPLE 3

Given $f(x) = 1 - 2x$ and $g(x) = 2x^2 + x - 1$, evaluate each composite function.   **a.** $f[(g(-2)]$ **b.** $g[f(x)]$

Solution

a. $g(x) = 2x^2 + x - 1$

$$g(-2) = 2(-2)^2 + (-2) - 1 \qquad \bullet \text{ To evaluate } f[g(-2)], \text{ first evaluate } g(-2).$$
$$g(-2) = 5$$

$$f(x) = 1 - 2x$$
$$f[g(-2)] = f(5) = 1 - 2(5) \qquad \bullet \text{ Substitute the value of } g(-2) \text{ for } x \text{ in } f(x).$$
$$f[g(-2)] = -9 \qquad \bullet \text{ Simplify.}$$

> ? ANSWER   No. When we say that composition is not a commutative operation, we mean that given any two functions, generally $(f \circ g)(x) \neq (g \circ f)(x)$. However, there may be particular instances for which $(f \circ g)(x) = (g \circ f)(x)$. It turns out that these particular instances are quite important, as we shall see later.

See Appendix A:
Evaluating Functions

Graphical Check:

Begin by entering the two functions into Y_1 and Y_2. Some typical screens are shown at the right. The evaluation $Y_1(Y_2(-2)) = -9$ and $f[g(-2)] = -9$. The answer checks.

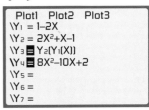

b. $g(x) = 2x^2 + x - 1$

$g[f(x)] = 2[f(x)]^2 + [f(x)] - 1$

• To find $g[f(x)]$, replace x in $g(x)$ by $f(x)$.

$= 2[1 - 2x]^2 + [1 - 2x] - 1$ • $f(x) = 1 - 2x$

$= 2(1 - 4x + 4x^2) + 1 - 2x - 1$

$= 8x^2 - 10x + 2$ • Simplify.

Graphical Check:

Draw the graph of the composition of the two functions and the graph of the answer you found. The graphs of the functions should overlap. You may need to adjust the viewing window to see the graphs.

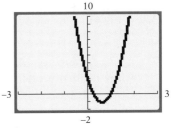

YOU TRY IT 3

Given $g(x) = 3x - 2$ and $h(x) = x^2 - 2x$, evaluate each composite function.
a. $g[h(-1)]$ **b.** $h[g(x)]$

Solution See page S41. **a.** 7 **b.** $9x^2 - 18x + 8$

10.1 EXERCISES Suggested Assignment: 5–45, odds

Topics for Discussion

1. Explain the meaning of the notation $f[g(3)]$.
   $f[g(3)]$ means to evaluate the function f at the value of the function g when g is evaluated at 3.

2. What is the meaning of the notation $(f \circ g)(x)$?
   $(f \circ g)(x)$ means to evaluate the function f after replacing x by the function g.

3. Correct each of the following statements.
   a. We read $(f \circ g)(x)$ as "f times g of x."
   b. The function $A \circ r$ is referred to as the *composition* of r *with A*.
   c. The expression $(f \circ g)(-2)$ means to evaluate the function g at $f(-2)$.
   d. $(f + g)(x) = f(g + x)$
   a. We read $(f \circ g)(x)$ as "f of g of x." b. The function $A \circ r$ is the *composition* of A with r. c. The expression $(f \circ g)(-2)$ means to evaluate the function f at $g(-2)$.
   d. $(f + g)(x) = f(x) + g(x)$

4. Given $f(x) = x - 3$ and $g(x) = x + 5$, find the domain of $\dfrac{f(x)}{g(x)}$.

$(x \mid x \neq -5)$

■ Basic Operations on Functions

In Exercises 5 to 16, let $f(x) = x^2 - 3x + 1$ and $g(x) = 3x - 1$. Evaluate each of the following.

5. $(f + g)(3)$ 9

6. $(f - g)(-1)$ 9

7. $(f - g)\left(\dfrac{1}{2}\right)$ $-\dfrac{3}{4}$

8. $(f + g)\left(-\dfrac{1}{2}\right)$ $\dfrac{1}{4}$

9. $(f \cdot g)(2)$ -5

10. $(f \cdot g)(-3)$ -190

11. $(g \cdot f)\left(\dfrac{3}{2}\right)$ $-\dfrac{35}{8}$

12. $(g \cdot f)\left(\dfrac{1}{2}\right)$ $-\dfrac{1}{8}$

13. $\left(\dfrac{f}{g}\right)(4)$ $\dfrac{5}{11}$

14. $\left(\dfrac{f}{g}\right)(1)$ $-\dfrac{1}{2}$

15. $(f + g)(x)$ x^2

16. $(f - g)(x)$ $x^2 - 6x + 2$

17. *Manufacturing* A company manufactures and sells snowmobiles. The total monthly cost, in dollars to produce n snowmobiles, is given by $C(n) = 150n + 3000$. The company's revenue, in dollars, obtained from selling all n snowmobiles is given by $R(n) = -0.6n^2 + 450n$. Express the company's monthly profit in terms of n, and use the profit function to determine the profit for selling 150 snowmobiles.
   $P(n) = -0.6n^2 + 300n - 3000$; \$28,500

18. *Manufacturing* A company's total monthly cost, in dollars, for manufacturing and selling n portable CD players per month is given by $C(n) = 85n + 9000$. The company's revenue, in dollars, from selling all n portable CD players is given by $R(n) = -0.4n^2 + 700n$. Express the company's monthly profit in terms of n, and use the profit function to determine the profit for selling 400 portable CD players.
   $P(n) = -0.4n^2 + 615n - 9000$; \$173,000

■ Composition of Functions

In Exercises 19 to 24, given $f(x) = 3x - 2$ and $g(x) = -2x + 1$, evaluate the composite function.

19. $(f \circ g)(2)$ -11

20. $g[f(0)]$ 5

21. $f[g(-3)]$ 19

22. $(f \circ g)(-2)$ 13

23. $g[f(x)]$ $-6x + 5$

24. $(f \circ g)(x)$ $-6x + 1$

In Exercises 25 to 30, given $g(x) = x^2 + 3$ and $h(x) = 2x - 5$, evaluate the composite function.

25. $g[h(2)]$ 4

26. $(g \circ h)(-1)$ 52

27. $h[g(-2)]$ 9

28. $g[h(-2)]$ 84

29. $(h \circ g)(x)$ $2x^2 + 1$

30. $g[h(x)]$ $4x^2 - 20x + 28$

In Exercises 31 to 36, given $f(x) = x^2 + 2x$ and $h(x) = 2x - 1$, evaluate the composite function.

31. $(f \circ h)(-1)$ 3

32. $h[f(-1)]$ -3

33. $(h \circ f)(2)$ 15

34. $(f \circ h)(2)$ 15

35. $f[h(x)]$ $4x^2 - 1$

36. $(h \circ f)(x)$ $2x^2 + 4x - 1$

In Exercises 37 to 42, given $f(x) = \dfrac{x}{x + 1}$ and $g(x) = 2x + 5$, evaluate the composite function.

37. $(f \circ g)(3)$ $\dfrac{11}{12}$

38. $g[f(-4)]$ $\dfrac{23}{3}$

39. $(g \circ f)(-1)$ Undefined

40. $f[g(2)]$ $\dfrac{9}{10}$

41. $(f \circ g)(x)$ $\dfrac{2x + 5}{2x + 6}$

42. $(g \circ f)(x)$ $\dfrac{7x + 5}{x + 1}$

43. *Oil Spill*   Suppose the spread of an oil leak from a tanker can be approximated by a circle with the tanker at its center and radius r, in feet. The radius of the spill t hours after the beginning of the leak is given by $r(t) = 45t$.

   a. Find the area of the spill as a function of time.

   b. What is the area of the spill after 3 hours? Round to the nearest whole number.

   a. $A(t) = 2025\pi t^2$ **b.** 57,256 ft^2

44. *Manufacturing*   Suppose the manufacturing cost, in dollars, per digital camera is given by the function $M(x) = \dfrac{50x + 10,000}{x}$. A camera store will sell the cameras by marking up the manufacturing cost per camera, $M(x)$, by 60%.

   a. Express the selling price of a camera as a function of the number of cameras to be manufactured. That is, find $S \circ M$.

   b. Find $(S \circ M)(5000)$.

   c. Explain the meaning of the answer to part b.

   a. $S(M(x)) = 80 + \dfrac{16,000}{x}$ **b.** \$83.20   **c.** When 5000 digital cameras are manufactured, the camera store sells each camera for \$83.20.

45. *Manufacturing*   The number of fax machines m that a factory can produce per day is a function of the number of hours h it operates.

$$m(h) = 250h \text{ for } 0 \le h \le 10$$

The daily cost c to manufacture m fax machines is given by the function

$$c(m) = 0.05m^2 + 60m + 1000$$

a. Find $(c \circ m)(h)$.
b. Evaluate $(c \circ m)(10)$.
c. Write a sentence that explains the meaning of the answer to part b.

a. $c[m(h)] = 3125h^2 + 15{,}000h + 1000$ b. $463{,}500   c. If the plant is run for 10 h, the cost to produce the fax machines is $463,500.

46. *Automobile Rebates*   A car dealership offers a $1500 rebate and a 10% discount off the price of a new car. Let p be the sticker price of a new car on the dealer's lot, r be the price after the rebate, and d be the price after the discount. Then $r(p) = p - 1500$ and $d(p) = 0.90p$.

a. Write a composite function for the dealer taking the rebate first and then the discount.
b. Write a composite function for the dealer taking the discount first and then the rebate.
c. Which composite function would you prefer the dealer to use when you buy a new car?

a. $d[r(p)] = 0.90p - 1350$ b. $r[d(p)] = 0.90p - 1500$ c. $r[d(p)]$ (The cost is less.)

Applying Concepts

47. Let $f(x) = 2x - 3$. Find $(f \circ f)(x)$. $4x - 9$

48. Let $f(x) = x^2 + 1$. Find $(f \circ f)(x)$. $x^4 + 2x^2 + 2$

Given $f(x) = 2x - 1$, $g(x) = x + 3$, and $h(x) = x^2$, find:

49. $f(g[h(2)])$ 13

50. $g(h[f(-2)])$ 28

51. $h(f[g(-3)])$ 1

52. $f(h[g(3)])$ 71

53. $g(f[h(x)])$ $2x^2 + 2$

54. $h(g[f(x)])$ $4x^2 + 8x + 4$

55. Sets R, S, and T are shown below, along with functions f and g.

a. Evaluate $f(4)$.
b. Evaluate $g(9)$.
c. Evaluate $(g \circ f)(2)$.
d. Evaluate $(g \circ f)(1)$.
e. Write a rule for f.
f. Write a rule for g.
g. Write a rule for $g \circ f$.
h. Create your own two functions f and g, diagram them, and find the composition of at least two elements of $g \circ f$.

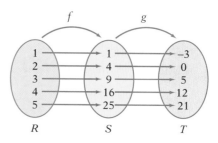

a. 16   b. 5   c. 0   d. -3 e. The function f squares an element in set R.   f. The function g subtracts 4 from an element in set S.   g. The composite of functions f and g first squares an element in set R and then subtracts 4 from the result.   h. Answers will vary.

The graphs of *f* and *g* are shown at the right. Use the graphs to determine the values of the following composite functions.

56. $f[g(-1)]$ 6   **57.** $g[f(1)]$ 0   **58.** $(g \circ f)(2)$ −3

59. $(f \circ g)(3)$ −2   **60.** $g[f(3)]$ −4   **61.** $f[g(0)]$ 7

f(x)

g(x)

EXPLORATION

1.   *Business*   A computer outlet store is offering a 15% discount on the purchase of any computer. The total cost *T* to a customer is the sale price *S* of the computer plus a sales tax that is 5.5% of the sale price. Let *r* be the regular price of a computer.
   a.   Explain the meaning of $S(r) = 0.85r$.
   b.   Explain the meaning of $T(S) = 1.055S$.
   c.   Find $(T \circ S)(r)$.
   d.   What is the meaning of the value of $(T \circ S)(r)$?
   e.   Does the composition of S with T, $S \circ T$, make sense in the context of this problem? Why or why not?

   a. The sale price is a function of the regular price. The sale price is 85% of the regu-lar price.   b. The total cost to a customer is a function of the sale price. The total cost is 105.5% of the sale price.   c. $(T \circ S)(r) = T[S(r)] = T(0.85r) = 1.055(0.85r) = 0.89675r$ d. $(T \circ S)(r)$ represents the total cost as a function of the regular price.
   e. $S \circ T = S[T(S)]$. This does not make sense in the context of this problem because it represents sale price as a function of sale price.

2.   *Temperature*   The function $F(x) = \frac{9}{5}x + 32$ converts *x* degrees Celsius into degrees Fahrenheit. The function $C(k) = k - 273.16$ converts *k* degrees Kelvin into degrees Celsius.
   a.   Explain the meaning of $(F \circ C)(k)$.
   b.   Does $(C \circ F)(x)$ make sense in the context of this exercise?
   c.   Part b refers to an application of a property that composition of functions does not have. What is that property?

   a. $(F \circ C)(k) = F[C(k)]$. This converts *k* degrees Kelvin into degrees Fahrenheit.
   b. $(C \circ F)(x) = C[F(x)]$. This does not make sense in the context of this problem. $F(x)$ converts degrees Celsius into degrees Fahrenheit. Degrees Fahrenheit cannot be used to evaluate the function C.   c. The Commutative Property

SECTION **10.2** ## Inverse Functions

■ One-to-One Functions
■ Inverse Functions

■ One-to-One Functions

Recall that a function is a set of ordered pairs in which no two ordered pairs that have the same first component have different second components. This means that given any input *x* in the domain of the function, there is only one output *y* that can be paired with that *x*. **A one-to-one function** satisfies the additional condition that given any *y*, there is only one *x* that can be paired with the given *y*. One-to-one functions are commonly expressed by writing 1-1.

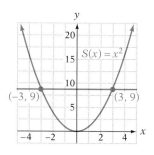

FIGURE 10.1

The function given by $S(x) = x^2$ is not a 1-1 function since, given $y = 9$, there are two possible values of x, -3 and 3, that produce the y value of 9. Two ordered pairs of this function are $(-3, 9)$ and $(3, 9)$. The graph of S in Figure 10.1 illustrates that a horizontal line intersects the graph more than once.

Just as the vertical-line test can be used to determine whether a graph represents a function, the **horizontal-line test** can be used to determine whether the graph of a function represents a 1-1 function.

> **Horizontal-Line Test**
>
> A graph of a function is the graph of a 1-1 function if any horizontal line intersects the graph at no more than one point.

The graph in Figure 10.1 is not the graph of a 1-1 function because a horizontal line intersects the graph at more than one point.

INSTRUCTOR NOTE
Ask students if the graph in Exercise 58 is the graph of a 1-1 function.

EXAMPLE 1

Determine whether or not the graph represents the graph of a 1-1 function.

a.

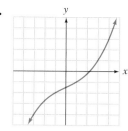

b.

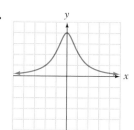

c.

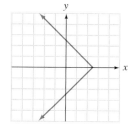

Solution

a.

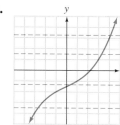

• No horizontal line intersects the graph at more than one point.

The graph is the graph of a 1-1 function.

b.

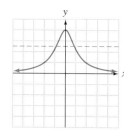

• A horizontal line intersects the graph at more than one point.

The graph is not the graph of a 1-1 function.

c.

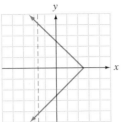

• A vertical line intersects the graph at more than one point.

The graph does not represent a function and is therefore not the graph of a 1-1 function.

YOU TRY IT 1

Determine whether or not the graph represents the graph of a 1-1 function.

a.

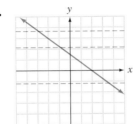

b.

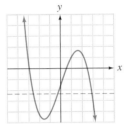

Solution See page S41. **a.** Yes **b.** No

■ Inverse Functions

Consider the "doubling" function $f(x) = 2x$. This function doubles every input. Some of the ordered pairs of this function are

$$\{(-4, -8), (-1, -2), (2, 4), (3.5, 7), (5, 10)\}$$

Now consider the "halving" function $g(x) = \frac{1}{2}x$. This function takes one-half of every input. Some of the ordered pairs of this function are

$$\{(-8, -4), (-2, -1), (4, 2), (7, 3.5), (10, 5)\}$$

Observe that the coordinates of the ordered pairs of g are the reverse of the coordinates of the ordered pairs of f. This is always the case for f and g. Here is one more instance.

$$f(4) = 2(4) = 8 \qquad g(8) = \frac{1}{2}(8) = 4$$

Ordered pair: $(4, 8)$ Ordered pair: $(8, 4)$

For these functions, f and g are called *inverse functions* of one another.

> **TAKE NOTE**
>
> Observe that when the output (8) of f is used as the input for g, the output of g is the original input (4) of f. A characteristic of inverse functions is that this relationship is true for all x in the domain of f.

> **Inverse Function**
>
> If the coordinates of the ordered pairs of a function g are the reverse of the coordinates of the ordered pairs of a function f, then g is said to be the **inverse function** of f.

TAKE NOTE

It is important to re-member the informa-tion in the paragraph at the right. If *f* is a function and *g* is the inverse of *f*, then

Domain of *g* = Range of *f*

and

Range of *g* = Domain of *f*.

Because the coordinates of the ordered pairs of the inverse function *g* are the reverse of the coordinates of the ordered pairs of the function *f*, the domain of *g* is the range of *f* and the range of *g* is the domain of *f*.

Not all functions have an inverse that is a function. Consider, for instance, the "square" function $S(x) = x^2$. Some of the ordered pairs of S are

$$\{(-3, 9), (-1, 1), (0, 0), (1, 1), (3, 9), (5, 25)\}$$

If we reverse the coordinates of the ordered pairs, we have

$$\{(9, -3), (1, -1), (0, 0), (1, 1), (9, 3), (25, 5)\}$$

This set of ordered pairs is not a function because there are ordered pairs, for instance $(9, -3)$ and $(9, 3)$, with the same first coordinate and different second coordinates. In this case, there is an inverse *relation* for S but not an inverse *function*. A graph of S is shown at the left. Note that $x = -3$ and $x = 3$ produce the same value of *y*. Thus the graph of S fails the horizontal-line test, and therefore S is not a 1-1 function. This observation is used in the following.

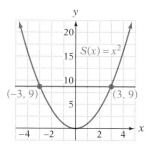

Condition for an Inverse Function

A function *f* has an inverse function if and only if *f* is a one-to-one function.

TAKE NOTE

$f^{-1}(x)$ does not mean $\frac{1}{f(x)}$. For $f(x) = 2x$,
$f^{-1}(x) = \frac{1}{2}x$ and
$\frac{1}{f(x)} = \frac{1}{2x}$.

If a function *g* is the inverse of a function *f*, we usually denote the inverse function as f^{-1} rather than as *g*. For the doubling and halving functions *f* and *g* discussed earlier, we write

$$f(x) = 2x \qquad f^{-1}(x) = \frac{1}{2}x$$

? QUESTION If *f* is a one-to-one function and $f(4) = 5$, find $f^{-1}(5)$.

If a one-to-one function *f* is defined by an equation, then we can use the following method to find the equation for f^{-1}.

TAKE NOTE

If the ordered pairs of *f* are given by (*x*, *y*), then the ordered pairs of f^{-1} are given by (*y*, *x*). That is, *x* and *y* are interchanged. This is the reason for Step 2 at the right.

Steps for Finding the Inverse of a Function

1. Substitute *y* for $f(x)$.
2. Interchange *x* and *y*.
3. Solve, if possible, for *y* in terms of *x*.
4. Substitute $f^{-1}(x)$ for *y*.

? ANSWER Because f^{-1} is the inverse function of *f*, the coordinates of the ordered pairs of f^{-1} are the reverse of the coordinates of the the ordered pairs of *f*. Therefore, $f^{-1}(5) = 4$.

INSTRUCTOR NOTE
Exercise 18 can be used for
a similar in-class example.

Suggested Activity

1. To convert square feet to acres, we can use the equation $A = \frac{s}{43,560}$, where A is the number of acres and s is the number of square feet. **a.** Write an equation for the inverse function. **b.** What does the inverse function represent?
[Answer: **a.** $s = 43,560A$ **b.** It represents the equation to use to convert acres to square feet.]

2. On a given day, the exchange rate for converting the euro to U.S. dollars was given by $d = 0.898E$, where d is the number of U.S. dollars and E is the number of euros. **a.** Write an equation for the inverse function. Write the coefficient as a decimal rounded to the nearest thousandth. **b.** What does the inverse function represent?
[Answer: **a.** $E = 1.114d$ **b.** It represents the exchange rate on that day for converting U.S. dollars to the euro.]

INSTRUCTOR NOTE
One way of helping students with mirror image is to have them think of one graph as drawn with wet ink. Then folding the paper along the line $y = x$ produces a mirror image.

INSTRUCTOR NOTE
It may help some students to show other instances of reversing operations. For example, if you take a number, say 7, add 5 and then subtract 5, you are back to 7.

EXAMPLE 2

Find the inverse function of $f(x) = -2x + 8$.

Solution

$$f(x) = -2x + 8$$
$$y = -2x + 8 \qquad \bullet \text{ Replace } f(x) \text{ by } y.$$
$$x = -2y + 8 \qquad \bullet \text{ Interchange } x \text{ and } y.$$
$$x - 8 = -2y \qquad \bullet \text{ Solve for } y.$$
$$\frac{x - 8}{-2} = y$$
$$-\frac{1}{2}x + 4 = f^{-1}(x) \qquad \bullet \text{ Replace } y \text{ by } f^{-1}(x).$$

The inverse function is given by $f^{-1}(x) = -\dfrac{1}{2}x + 4$.

YOU TRY IT 2

Find the inverse function of $f(x) = 3x + 5$.

Solution   See page S42. $f^{-1}(x) = \dfrac{1}{3}x - \dfrac{5}{3}$

The fact that the coordinates of the ordered pairs of the inverse of a function are the reverse of those of the function has a graphical interpretation. The function graphed in Figure 10.2a includes the points $(-2, 0)$, $(-1, 2)$, $(1, 4)$, and $(5, 6)$. The graph in Figure 10.2b shows the points with the coordinates reversed. The inverse function is graphed by drawing a smooth curve through those points, as shown in Figure 10.2c.

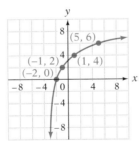

FIGURE 10.2a

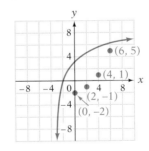

FIGURE 10.2b

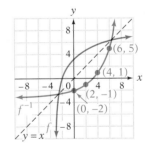

FIGURE 10.2c

Note that the dashed graph of $y = x$ is shown in Figure 10.2c. **The graph of a function and its inverse are mirror images with respect to the graph of $y = x$.**

Note the effect, as shown below, of taking the composition of functions that are inverses of one another.

$$f(x) = 2x \qquad\qquad\qquad g(x) = \frac{1}{2}x$$

$$f[g(x)] = 2\left[\frac{1}{2}x\right] \quad \bullet \text{ Replace } x \text{ by } g(x). \qquad g[f(x)] = \frac{1}{2}[2x] \quad \bullet \text{ Replace } x \text{ by } f(x).$$

$$f[g(x)] = x \qquad\qquad\qquad g[f(x)] = x$$

TAKE NOTE

If we use the idea of a function as a machine, then the Composition of Inverse Function Property can be represented as shown below. Take any input, x, for f. Use the output of f as the input for f^{-1}. The result is the original input, x.

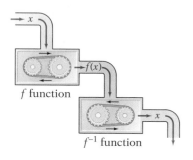

f function

f^{-1} function

INSTRUCTOR NOTE
Exercise 36 can be used for a similar in-class example.

This property of the composition of inverse functions is always true. In other words, **the composition of inverse functions says that an inverse function reverses the effect of the function.** For these two functions, f doubles a number; g halves a number. If you double a number and then take one-half of the result, you are back to the original number.

Composition of Inverse Functions Property

If f is a one-to-one function, then f^{-1} is the inverse function of f if and only if

$$(f \circ f^{-1})(x) = f[f^{-1}(x)] = x \qquad \text{for all } x \text{ in the domain of } f^{-1}.$$

and

$$(f^{-1} \circ f)(x) = f^{-1}[f(x)] = x \qquad \text{for all } x \text{ in the domain of } f.$$

EXAMPLE 3

Show that $f^{-1}(x) = 3x - 6$ is the inverse function for $f(x) = \dfrac{1}{3}x + 2$.

Solution   We must show that $f[f^{-1}(x)] = x$ and $f^{-1}[f(x)] = x$.

$$f(x) = \frac{1}{3}x + 2 \qquad\qquad f^{-1}(x) = 3x - 6$$

$$f[f^{-1}(x)] = \frac{1}{3}[3x - 6] + 2 \qquad f^{-1}[f(x)] = 3\left[\frac{1}{3}x + 2\right] - 6$$

$$f[f^{-1}(x)] = x \qquad\qquad f^{-1}[f(x)] = x$$

YOU TRY IT 3

Show that $f^{-1}(x) = \dfrac{1}{2}x - 2$ is the inverse function for $f(x) = 2x + 4$.

Solution   See page S42.   Show that $f[f^{-1}(x)] = x$ and $f^{-1}[f(x)] = x$.

 See Appendix A:
Inverse Function

A graphing calculator can be used to produce the graph of the inverse of a function. Select the square viewing window. Enter the function whose inverse you want to graph in Y₁. We will use the function from Example 3. Then select DrawInv from the [DRAW] menu to create the graph of the function and its inverse.

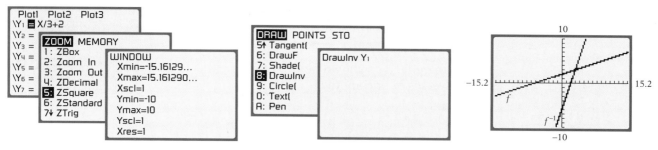

The reason why the square viewing window is used is that in the standard viewing window of a calculator, the distance between two tick marks on the *x*-axis is not equal to the distance between two tick marks on the *y*-axis. As a result, the graph of $y = x$ does not appear to bisect the first and third quadrants. See Figure 10.3. Because the graph of $y = x$ does not quite bisect the first and third quadrants, the graphs of f and f^{-1} will not appear to be mirror images of each other. The graphs of $f(x) = \frac{1}{3}x + 2$ and $f^{-1}(x) = 3x - 6$ from Example 3 are shown in Figure 10.4. Note that the graphs do not appear to be symmetric about the graph of $y = x$.

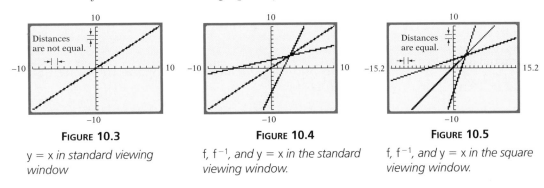

FIGURE 10.3	**FIGURE 10.4**	**FIGURE 10.5**
y = x *in standard viewing window*	f, f⁻¹, *and* y = x *in the standard viewing window.*	f, f⁻¹, *and* y = x *in the square viewing window.*

To get a better view of a function and its inverse, it is necessary to use the square viewing window as in Figure 10.5. In this window, the distance between two tick marks on the *x*-axis is equal to the distance between two tick marks on the *y*-axis.

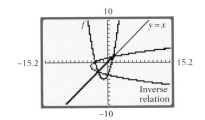

Although we have focused on functions that have an inverse function, the **DrawInv** command will draw the inverse *relation* of any function entered in **Y1**.

In the graph at the left, the graph of $f(x) = x^2 + 2x - 3$ and its inverse relation are drawn. Because a vertical line can intersect the graph of the inverse relation more than once, the inverse relation is not a function.

There are practical applications of finding the inverse of a function. Here is one that is related to converting a shirt size in the United States to a shirt size in Italy. Finding the inverse function gives the function that converts a shirt size in Italy to a shirt size in the United States.

INSTRUCTOR NOTE
Exercise 40 can be used for a similar in-class example.

Suggested Activity

See Section 10.2 of the *Student Activity Manual* for an activity which reinforces the concept of inverse functions.

EXAMPLE 4

The function $IT(x) = 2x + 8$ converts a man's shirt size *x* in the United States to the equivalent shirt size in Italy. Find IT^{-1} and use IT^{-1} to determine the U.S. men's shirt size that is equivalent to an Italian shirt size of 36.

Solution

State the goal.   The goal is to find IT^{-1}, the inverse function of *IT*.

Devise a strategy.   Use the steps for finding an inverse function. Then evaluate the inverse function at the Italian shirt size.

Solve the problem. $IT(x) = 2x + 8$

$$y = 2x + 8 \qquad \bullet \text{ Replace } IT(x) \text{ by } y.$$

$$x = 2y + 8 \qquad \bullet \text{ Interchange } x \text{ and } y.$$

$$x - 8 = 2y \qquad \bullet \text{ Solve for } y.$$

$$\frac{x - 8}{2} = y$$

$$IT^{-1}(x) = \frac{x - 8}{2} \qquad \text{or} \qquad IT^{-1}(x) = \frac{1}{2}x - 4$$

To find the equivalent size shirt in the United States, substitute 36 for x in IT^{-1}.

$$IT^{-1}(x) = \frac{1}{2}x - 4$$

$$IT^{-1}(36) = \frac{1}{2}(36) - 4 = 18 - 4 = 14$$

A size 36 Italian shirt is equivalent to a size 14 U.S. shirt.

Check your work.   One way to check your work is to evaluate the function $IT(x)$ at 14, the output of the inverse function. The result should be 36, the Italian shirt size.

$$IT(x) = 2x + 8$$

$$IT(14) = 2(14) + 8 = 28 + 8 = 36$$

The answer checks.

YOU TRY IT 4

The function $h(x) = 3x - 10$ converts a woman's hat size x in the United States to the equivalent hat size in France. Find h^{-1} and use h^{-1} to determine the hat size in the United States of a French hat size of 41.

Solution   See page S42. $h^{-1}(x) = \frac{1}{3}x + \frac{10}{3};\ 17$

10.2 EXERCISES   Suggested Assignment: 7–45, odds; 57, 59

Topics for Discussion

1. How are the ordered pairs of the inverse of a function related to the function?

    Answers may vary. For example, the coordinates of each ordered pair of the inverse are the reverse of the coordinates of the ordered pairs of the function.

2. If f and f^{-1} are inverse functions of one another, is it possible to determine the value of $f[f^{-1}(4)]$? If so, what is the value?   Yes. 4.

3. Is the inverse of a constant function a function? Explain your answer.

No. Explanations will vary. For example, reversing the coordinates of the ordered pairs of a constant function results in an equation of the form $x = a$, where a is a constant. An equation of this form is not a function.

4. Determine whether the statement is always true, sometimes true, or never true.

   a. A function has an inverse if and only if it is a 1-1 function.

   b. The inverse of the function $\{(2, 3), (4, 5), (6, 3)\}$ is the function $\{(3, 2), (5, 4), (3, 6)\}$.

   c. The inverse of a function is a relation.

   d. The inverse of a function is a function.

   a. Always true   **b.** Never true   **c.** Always true   **d.** Sometimes true

5. What is a 1-1 function? What is the horizontal-line test for a 1-1 function?

For a 1-1 function, for any given y there is exactly one x that can be paired with that y. The horizontal-line test says that if every horizontal line intersects a graph of a function at most once, then the graph is the graph of a 1-1 function.

■ **One-to-One Functions and Inverse Functions**

For Exercises 6 to 9, assume that the given function has an inverse function.

6. Given $f(4) = -5$, find $f^{-1}(-5)$.   4

7. Given $g(-1) = 3$, find $g^{-1}(3)$. -1

8. Given $h^{-1}(0) = 4$, find $h(4)$.   0

9. Given $f^{-1}(-3) = 0$, find $f(0)$. -3

10. If f is a 1-1 function and $f(-3) = 0, f(-1) = -3$, and $f(0) = -4$, find: **a.** $f^{-1}(0)$ **b.** $f^{-1}(-3)$

   a. -3 **b.** -1

11. If f is a 1-1 function and $f(0) = 5, f(5) = 7$, and $f(7) = 9$, find: **a.** $f^{-1}(7)$ **b.** $f^{-1}(5)$

   a. 5   **b.** 0

12. The domain of the inverse function f^{-1} is the _____ of f.   range

13. The range of the inverse function f^{-1} is the _____ of f.   domain

In Exercises 14 to 17, find the inverse of the function. If the function does not have an inverse function, write "No inverse."

14. $\{(-4, 5), (-2, 6), (0, 7), (2, -7)\}$

   $\{(-7, 2), (5, -4), (6, -2), (7, 0)\}$

15. $\{(-1, 2), (0, 0), (1, 0), (2, 4), (3, 8)\}$

   No inverse

16. $\{(1, 2), (2, 2), (3, 4), (4, 5), (5, 5)\}$

   No inverse

17. $\{(0, 0), (2, 1), (4, 2), (8, 3), (16, 4)\}$

   $\{(0, 0), (1, 2), (2, 4), (3, 8), (4, 16)\}$

In Exercises 18 to 31, find $f^{-1}(x)$.

18. $f(x) = 2x + 2$ $f^{-1}(x) = \dfrac{1}{2}x - 1$

19. $f(x) = -3x - 9$ $f^{-1}(x) = -\dfrac{1}{3}x - 3$

20. $f(x) = -4x + 8$ $f^{-1}(x) = -\dfrac{1}{4}x + 2$

21. $f(x) = 3x + 7$ $f^{-1}(x) = \dfrac{1}{3}x - \dfrac{7}{3}$

22. $f(x) = -2x + 5$ $f^{-1}(x) = -\dfrac{1}{2}x + \dfrac{5}{2}$

23. $f(x) = 4x + 2$ $f^{-1}(x) = \dfrac{1}{4}x - \dfrac{1}{2}$

24. $f(x) = 4 - 3x$ $f^{-1}(x) = -\frac{1}{3}x + \frac{4}{3}$

25. $f(x) = 2 + 3x$ $f^{-1}(x) = \frac{1}{3}x - \frac{2}{3}$

26. $f(x) = \frac{2}{3}x + 2$ $f^{-1}(x) = \frac{3}{2}x - 3$

27. $f(x) = -\frac{1}{2}x - 3$ $f^{-1}(x) = -2x - 6$

28. $f(x) = 2x - 2$ $f^{-1}(x) = \frac{1}{2}x + 1$

29. $f(x) = \frac{5}{4}x - 1$ $f^{-1}(x) = \frac{4}{5}x + \frac{4}{5}$

30. $f(x) = x$ $f^{-1}(x) = x$

31. $f(x) = \frac{1}{x}$ $f^{-1}(x) = \frac{1}{x}$

In Exercises 32 to 39, use composition of functions to determine whether f and g are inverses of one another.

32. $f(x) = -2x; g(x) = -\frac{1}{2}x$ Inverse functions

33. $f(x) = 5x; g(x) = \frac{1}{5x}$ Not inverse functions

34. $f(x) = 4x; g(x) = -4x$ Not inverse functions

35. $f(x) = x - 1; g(x) = -x + 1$ Not inverse functions

36. $f(x) = 6x - 3; g(x) = \frac{1}{6}x + \frac{1}{2}$ Inverse functions

37. $f(x) = \frac{3}{4}x - \frac{1}{2}; g(x) = \frac{4}{3}x + \frac{2}{3}$ Inverse functions

38. $f(x) = -\frac{1}{2}x - \frac{1}{2}; g(x) = -2x + 1$
Not inverse functions

39. $f(x) = 3x + 2; g(x) = \frac{1}{3}x - \frac{2}{3}$
Inverse functions

40. *Unit conversions*   The function $f(x) = 12x$ converts feet, x, into inches, $f(x)$. Find f^{-1} and explain what it does.
$f^{-1}(x) = \frac{1}{12}x$. The inverse function converts inches to feet.

41. *Unit conversions*   A conversion function, such as in Exercise 40, converts a measurement in one unit into a measurement in another unit. Is a conversion function always a one-to-one function? Does a conversion function always have an inverse function? Explain your answer.
Yes. Yes. Because a conversion function is 1-1, it has an inverse function.

42. *Grading scale*   Does the grading scale function given below have an inverse function? Explain your answer.

Score	Grade
45–50	A
40–44	B
35–39	C
30–34	D
0–29	F

No. There are two ordered pairs with the same first element and different second elements [for instance, (B, 43) and (B, 44)]. Therefore, the relation is not a function.

43.  *Postage*   Does the first-class postage rate function given below have an inverse function? Explain your answer.

Weight (in ounces)	Cost
$0 < w \le 1$	$.37
$1 < w \le 2$	$.60
$2 < w \le 3$	$.83
$3 < w \le 4$	$1.06

No. There are two ordered pairs with the same first element and different second elements (for instance, (0.37, 0.5) and (0.37, 0.25)). Therefore, the relation is not a function.

44. *Fashion*   The function $s(x) = 2x + 24$ can be used to convert a U.S. woman's shoe size to an Italian woman's shoe size. Determine a function s^{-1} that can be used to convert an Italian woman's shoe size to her equivalent U.S. woman's shoe size. $s^{-1}(x) = \frac{1}{2}x - 12$

45. *Fashion*   The function $K(x) = \frac{13}{10}x - \frac{47}{10}$ converts a man's shoe size in the United States to the equivalent shoe size in the United Kingdom. Determine the function K^{-1} that can be used to convert a man's shoe size in the United Kingdom to his equivalent U.S. shoe size. $K^{-1}(x) = \frac{10}{13}x + \frac{47}{13}$

46. *Compensation*   The monthly earnings $E(s)$, in dollars, of a software sales executive is given by $E(s) = 0.025s + 3000$, where s is the value, in dollars, of the software sold by the executive during the month. Find E^{-1} and explain how the executive could use that function.
$E^{-1}(s) = 40s - 120{,}000$. Given monthly earnings, this function can be used to find the value of the software sold during the month.

Applying Concepts

47. Suppose that f is a linear function, $f(3) = 6$, and $f(9) = 15$. If $f(5) = c$, is c less than 6, between 6 and 15, or greater than 15?   Between 6 and 15

48. Suppose that f is a linear function, $f(-3) = 8$, and $f(-1) = 2$. If $f(0) = c$, is c less than 2, between 2 and 8, or greater than 8?   Less than 2

49. Suppose that f is a linear function, $f(1) = 3$, and $f(3) = 7$. If $f(5) = c$, is c less than 3, between 3 and 7, or greater than 7?   Greater than 7

50. Suppose that f is a linear function, $f^{-1}(2) = 4$, and $f^{-1}(5) = 7$. If $f(8) = c$, express the value of c as an inequality. $c > 5$

51. Suppose that f is a 1-1 function, $f(2) = 5$, and $f(6) = 9$. Between which two numbers is $f^{-1}(7)$?   2 and 6

52. Suppose that f is a linear function, $f(3) = -1$, and $f(7) = 5$. Between which two numbers is $f^{-1}(3)$?   3 and 7

53. Suppose that g is a linear function, $g^{-1}(4) = 5$, and $g^{-1}(8) = 10$. Between which two numbers is $g(7)$?   4 and 8

54. Suppose that g is a linear function, $g^{-1}(-2) = 5$, and $g^{-1}(0) = -3$. Between which two numbers is $g(0)$?   −2 and 0

55. True or false: If f is some function and $f(a) = f(b)$, then $a = b$.
False. For instance, let $f(x) = |x|$. Then $f(-2) = f(2)$ but $2 \neq -2$.

Given the graph of the 1-1 function, draw the graph of the inverse of the function using the technique shown on page 715.

56.

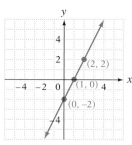 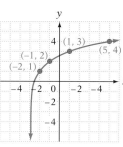

57.

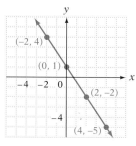

58.

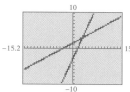

59.

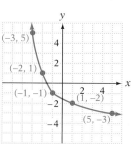

The function and its inverse have the same graph.

For each of the following functions, **a.** use the DRAW INVERSE feature and the square window of a graphing calculator to draw the graph of the function and the inverse relation for that function and **b.** determine whether the inverse relation is a function.

60. $f(x) = 2x - 3$ **a.**

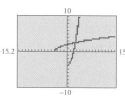

b. Yes

61. $f(x) = -\dfrac{2}{3}x + 2$ **a.**

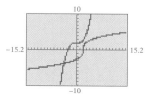

b. Yes

62. $f(x) = \sqrt{x + 4}$ **a.**

b. Yes

63. $f(x) = 0.1x^3 + 2$ **a.**

b. Yes

64. $f(x) = x^2 - 4$ **a.**

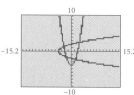

b. No

65. $f(x) = x^3 - 6x - 1$ **a.**

b. No

EXPLORATION

1. *Intersection Point of the Graphs of f and f^{-1}* For each of the following, graph f and its inverse function.

   i. $f(x) = 2x - 4$ **ii.** $f(x) = \dfrac{2}{3}x + 6$

   iii. $f(x) = -x + 2$ **iv.** $f(x) = x^3 + 1$

   a. What do you notice about the intersection point?
   b. If the graph of a 1-1 function and its inverse function intersect, must it always be as suggested in part a? Explain your answer.
   c. Do the graph of a one-to-one function and the graph of its inverse always intersect? Support your answer.

   a. The x- and y-coordinates of the point of intersection are the same number.
   b. Yes. The point lies on the line $y = x$, so the x- and y-coordinates are always the same number. **c.** No. The graph of a line parallel to $f(x) = x$ and its inverse will not intersect.

2. *Cryptography* Functions and their inverses can be used to make secret codes so that secure business transactions can be made over the Internet. Let A = 10, B = 11, . . . , Z = 35. A coding function is one that takes the numerical input for a word and changes it to some other number.

   a. Use the coding function $f(x) = 2x + 3$ to code the word MATH (M − 22, A − 10, T − 29, H − 17) by finding $f(22102917)$.
   b. Find the inverse of f and show that applying f^{-1} to the output of f returns the original word (22102917).

   The function we used is quite simple. When this is used on the Internet, the functions are more complicated, and the inverses of the functions are extremely difficult to find.

   c. A friend is using a letter–number correspondence and the coding function $f(x) = 2x - 1$. Suppose this friend sends you the coded message 5658602667. Decode this message.

   a. $f(22102917) = 2(22102917) + 3 = 44205837$

   b. $f^{-1}(x) = \dfrac{1}{2}x - \dfrac{3}{2}; f^{-1}(44205837) = 22102917$

   c. $f^{-1}(x) = \dfrac{1}{2}x + \dfrac{1}{2}; f^{-1}(5658602667) = 2829301334;$ STUDY

SECTION **10.3** # Exponential Functions

- Evaluate Exponential Functions
- Graphs of Exponential Functions
- Applications of Exponential Functions

■ Evaluate Exponential Functions

The growth of a $500 savings account that earns 5% annual interest compounded daily is shown at the right. In 14 years, the savings account contains approximately $1000, twice the initial amount. The growth of this savings account is modeled by an exponential function.

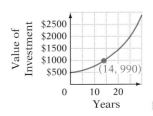

TAKE NOTE

It is important to distinguish between $F(x) = 2^x$ and $P(x) = x^2$. The first is an exponential function; the second is a polynomial function. Exponential functions are characterized by a constant base and a variable exponent. Polynomial functions have a variable base and a constant exponent.

The pressure of the atmosphere at a certain height is shown in the graph at the right. This is another situation that is modeled by an exponential function. From the graph, we can read that the air pressure is approximately 6.5 pounds per square inch at an altitude of 20,000 feet.

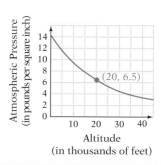

Definition of an Exponential Function

The **exponential function** with base b is defined by

$$f(x) = b^x$$

where $b > 0$, $b \neq 1$, and x is any real number.

In the definition of an exponential function, it states that $b > 0$, which means that the base is required to be positive. If the base were a negative number, the value of the function would be a complex number for some values of x. For instance, consider the value of $f(x) = (-4)^x$ when $x = \frac{1}{2}$. Enter the expression $(-4)^{1/2}$ into your calculator in the complex number mode, and press [ENTER]. The display reads $2i$. This is because $f\left(\frac{1}{2}\right) = (-4)^{1/2} = \sqrt{-4} = 2i$. **To avoid complex number values of a function, the base of the exponential function is a positive number.**

? QUESTION   Which of the following cannot be the base of an exponential function?

a. 7   **b.** $\dfrac{1}{4}$ **c.** -5 **d.** 0.01

➡ Evaluate $f(x) = 2^x$ at $x = 3$ and $x = -2$.

Substitute 3 for x and simplify. $f(3) = 2^3 = 8$

Substitute -2 for x and simplify. $f(-2) = 2^{-2} = 0.25$ ⬅

The definition of the exponential function states that $b \neq 1$. Evaluate $f(x) = 1^x$ at $x = -3, -1, 0, 2,$ and 4. The value of the function is 1 for each value of x. Note that this is the constant function $f(x) = 1$ and not an exponential function. Therefore, **the value of b in the exponential function $f(x) = b^x$ cannot be 1.**

There are situations in which we want to evaluate an exponential expression for an irrational number such as $\sqrt{2}$. We can find an approximation to the

TAKE NOTE

For $f(x) = 2^x$, $f(-2) =$
$2^{-2} = \dfrac{1}{2^2} = \dfrac{1}{4} = 0.25.$

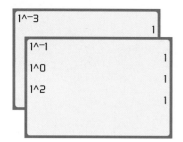

? ANSWER -5 cannot be the base of an exponential function because it is not a positive number.

CALCULATOR NOTE

On a graphing calculator, enter 4^√(2) to evaluate the function at the right.

INSTRUCTOR NOTE
Exercise 8 can be used for a similar in-class example.

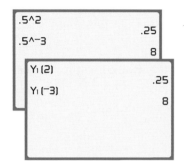

See Appendix A:
Evaluating Functions

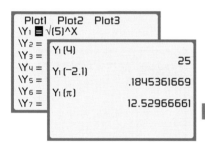

value of the function by using a calculator. For instance, the va when $x = \sqrt{2}$ is

$$f(\sqrt{2}) = 4^{\sqrt{2}} \approx 7.1030$$

EXAMPLE 1

a. Evaluate $f(x) = \left(\dfrac{1}{2}\right)^x$ at $x = 2$ and $x = -3$.

b. Evaluate $f(x) = 2^{3x-1}$ at $x = 1$ and $x = -1$.

c. Evaluate $f(x) = (\sqrt{5})^x$ at $x = 4$, $x = -2.1$, and $x = \pi$. Round to the nearest ten-thousandth.

Solution

a. $f(x) = \left(\dfrac{1}{2}\right)^x$

$f(2) = \left(\dfrac{1}{2}\right)^2 = 0.25$

$f(-3) = \left(\dfrac{1}{2}\right)^{-3} = 2^3 = 8$

• Two methods of evaluating these expressions on the home screen are shown at the far left. The second method requires entering the expression in the equation editor in Y_1. You can also use the TABLE feature with the independent variable set on ASK.

b. $f(x) = 2^{3x-1}$

$f(1) = 2^{3(1)-1} = 2^2 = 4$ $f(-1) = 2^{3(-1)-1} = 2^{-4} = 0.0625$

c. $f(x) = (\sqrt{5})^x$

$f(4) = (\sqrt{5})^4$ $f(-2.1) = (\sqrt{5})^{-2.1}$ $f(\pi) = (\sqrt{5})^{\pi}$

$= 25$ ≈ 0.1845 ≈ 12.5297

YOU TRY IT 1

a. Evaluate $f(x) = \left(\dfrac{2}{3}\right)^x$ at $x = 3$ and $x = -2$.

b. Evaluate $f(x) = 2^{2x+1}$ at $x = 0$ and $x = -2$.

c. Evaluate $f(x) = \pi^x$ at $x = 3$, $x = -2$, and $x = \pi$. Round to the nearest ten-thousandth.

Solution   See page S42.   **a.** $\dfrac{8}{27}$ and $\dfrac{9}{4}$ **b.** 2 and $\dfrac{1}{8}$ **c.** 31.0063, 0.1013, and 36.4622

▼ *Point of Interest*

The natural exponential function is an extremely important function. It is used extensively in applied problems in virtually all disciplines from archaeology to zoology.

Leonhard Euler (1707–1783) was the first to use the letter e as the base of the natural exponential function.

A frequently used base in applications of exponential functions is an irrational number designated by e. The number e is approximately 2.71828183. Because it is an irrational number, it has a nonterminating, nonrepeating decimal representation.

Natural Exponential Function

The function defined by $f(x) = e^x$ is called the **natural exponential function.**

The e^x key on a calculator is used to evaluate the natural exponential function.

INSTRUCTOR NOTE
Exercise 12 can be used for
a similar in-class example.

EXAMPLE 2

Evaluate $f(x) = e^x$ at $x = 2$, $x = -3$, and $x = \pi$. Round to the nearest ten-thousandth.

Solution $f(x) = e^x$

$$f(2) = e^2 \qquad\qquad f(-3) = e^{-3} \qquad\qquad f(\pi) = e^{\pi}$$
$$\approx 7.3891 \qquad\qquad\qquad \approx 0.0498 \qquad\qquad\quad \approx 23.1407$$

YOU TRY IT 2

Evaluate $f(x) = e^x$ at $x = 1.4$, $x = -0.5$, and $x = \sqrt{2}$. Round to the nearest ten-thousandth.

Solution   See page S42.   4.0552, 0.6065, 4.1133

■ Graphs of Exponential Functions

Some of the properties of an exponential function can be seen by considering its graph. We will begin by looking at the graph of $f(x) = 2^x$. Note that this is an exponential function of the form $f(x) = b^x$ in which $b > 1$.

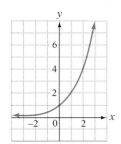

To graph $f(x) = 2^x$, think of the function as the equation $y = 2^x$.

We can enter Y₁ = 2^X in a graphing calcula-tor and use the TABLE feature to find or-dered pairs of the function.

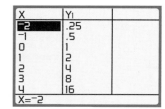

The ordered pairs can be graphed on a rec-tangular coordinate system and the points connected with a smooth curve, as shown at the left.

Graph the equation $y = 2^x$ on a graphing calculator using the standard view-ing window and verify the graph shown above.

INSTRUCTOR NOTE
Graph $F(x) = x^2$ so that students can see the difference
between this graph and
that of $f(x) = 2^x$.
    Also have students look
for and then describe the
pattern in the Y values in
the table to the right. (Each
entry is twice the preceding
Y value.) Make the connection between this and the
base of the exponential
function.

Note that a vertical line would intersect the graph of $f(x) = 2^x$ at only one point. Therefore, by the vertical-line test, $f(x) = 2^x$ is the graph of a function. Also, a horizontal line would intersect the graph at only one point. Therefore, $f(x) = 2^x$ is the graph of a 1-1 function.

Note also that as x increases, the values of the function increase. Change the viewing window on your calculator to see the graph of this function over larger values of y.

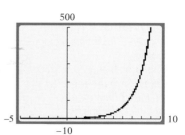

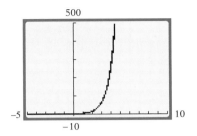

Use the same viewing window to graph $f(x) = 4^x$. Note that this is another example of an exponential function in which $b > 1$. The graph of this function is shown at the left. Note that it, too, is the graph of an increasing function.

Now let's look at the function $f(x) = \left(\frac{1}{2}\right)^x$. This is an exponential function in which the value of b is between 0 and 1; that is, $0 < b < 1$.

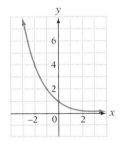

Think of the function as the equation $y = \left(\frac{1}{2}\right)^x$.

Enter Y₁ = (1/2)^X in a graphing calculator and use the TABLE feature to find ordered pairs of the function.

The ordered pairs can be graphed on a rectangular coordinate system and the points connected with a smooth curve, as shown at the left.

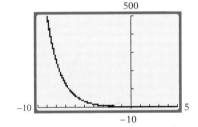

Graph the equation $y = \left(\frac{1}{2}\right)^x$ on a graphing calculator using the standard viewing window and verify the graph shown at the left.

Note that as x increases, the values of the function decrease. Change the viewing window on your calculator to see the graph of this function over larger values of y.

INSTRUCTOR NOTE
Have students look for and then describe the pattern in the Y values in the table. (Each entry is one-half the preceding Y value.) Make the connection between this and the base of the exponential function.

TAKE NOTE

Applying the vertical- and horizontal-line tests reveals that $f(x) = \left(\frac{1}{2}\right)^x$ is also the graph of a 1-1 function.

Use the same viewing window to graph $f(x) = \left(\frac{1}{3}\right)^x$. Note that this is another example of an exponential function in which $0 < b < 1$. The graph of this function is shown at the left below. Note that it, too, is the graph of a decreasing function.

? **QUESTION**   How would you enter the equations $y = 2^{x-1}$ and $y = 2^x - 1$ on your calculator?

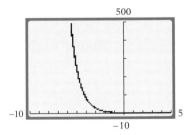

EXAMPLE 3

a. Graph $f(x) = 3^{\frac{1}{2}x-1}$ using the standard viewing window.
b. What is the value of b in this exponential function? Is $b > 1$ or $0 < b < 1$?

INSTRUCTOR NOTE
Exercise 22 can be used for a similar in-class example.

? ANSWER   For $y = 2^{x-1}$, enter 2^(X − 1). For $y = 2^x - 1$, enter 2^X − 1. In the first function, the "−1" is a part of the exponent. In the second function, it is not part of the exponent.

c. Is this an increasing or a decreasing function?
d. For what values in the domain are the corresponding values in the range less than 0?

Solution

a.

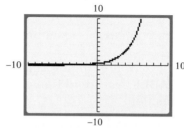

b. The value of b is 3; therefore, $b > 1$.
c. As x increases, the values of y increase. This is an increasing function.
d. There are no values in the domain for which the corresponding values in the range are less than 0.

YOU TRY IT 3

a. Graph $f(x) = 0.4^x - 1$ using the standard viewing window.
b. What is the value of b in this exponential function? Is $b > 1$ or $0 < b < 1$?
c. Is this an increasing or a decreasing function?
d. For what values in the domain are the corresponding values in the range less than 0?

Solution See page S42. **a.** 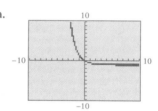 **b.** 0.4; $0 < b < 1$
c. Decreasing **d.** $x > 0$

INSTRUCTOR NOTE
Exercise 24 can be used for a similar in-class example.

EXAMPLE 4

a. Graph $f(x) = e^x - 2$ using the standard viewing window.
b. What is the value of b in this exponential function? Is $b > 1$ or $0 < b < 1$?
c. Is this an increasing or a decreasing function?
d. What is the y-intercept of the graph of the function? Round to the nearest hundredth.

Solution

a.

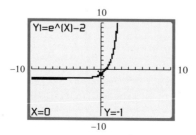

b. The value of b is $e \approx 2.71828$; therefore, $b > 1$.
c. As x increases, the values of y increase. This is an increasing function.
d. The y-intercept is $(0, -1)$.

YOU TRY IT 4

a. Graph $f(x) = e^{0.5x+1}$ using the standard viewing window.
b. What is the value of b in this exponential function? Is $b > 1$ or $0 < b < 1$?
c. Is this an increasing or a decreasing function?
d. What is the y-intercept of the graph of the function? Round to the nearest ten-thousandth.

Solution See page S43. **a.**

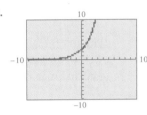

b. $b = e \approx 2.718; b > 1$
c. Increasing
d. (0, 2.7182)

In Example 4 and You Try It 4, we found the y-intercept of an exponential function. The y-intercept is the value of y when $x = 0$. Recall that **a zero of a function is a value of x for which $y = 0$. It is the x-coordinate of an x-intercept of the graph of the function.** Example 5 and You-Try-It 5 illustrate approximating the zero of a function.

INSTRUCTOR NOTE
Exercise 36 can be used for a similar in-class example.

See Appendix A: Zero

EXAMPLE 5

Graph $f(x) = 1.5^x - 2$ and approximate the zero of f to the nearest hundredth.

Solution

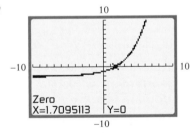

• Use the features of a graphing calculator to determine the x-intercept of the graph, which is the zero of f.

To the nearest hundredth, the zero of f is 1.71.

YOU TRY IT 5

Graph $f(x) = \left(\dfrac{3}{4}\right)^x - 3$ and approximate the zero of f to the nearest hundredth.

Solution See page S43. -3.82

On page 726, we graphed $f(x) = 2^x$. We are now going to look at the graph of $f(x) = a \cdot 2^x$, where $a > 0$.

```
Plot1   Plot2   Plot3
\Y1 ▤ 2^X
\Y2 ▤ 4*2^X
\Y3 ▤ .25*2^X
\Y4 =
\Y5 =
\Y6 =
\Y7 =
```

TAKE NOTE

There are no values of x for which the y-values are less than or equal to 0. The range is $y > 0$.

Suggested Activity

1. Ask students to sketch any graph of $f(x) = b^x$, $b > 0$. Have them label the y-intercept. Describe the graph as exponential growth or exponential decay.
2. Ask students to sketch any graph of $f(x) = b^x$, $0 < b < 1$. Have them label the y-intercept. Describe the graph as exponential growth or exponential decay.

Enter all three of the following functions in the equation editor. Then graph the functions.

$$f(x) = 2^x$$
$$f(x) = 4 \cdot 2^x$$
$$f(x) = \frac{1}{4} \cdot 2^x$$

For each graph,

1. The function is an increasing function.
2. The y-intercept is $(0, a)$.
3. The domain is all real numbers.
4. The range of f is the positive real numbers.

These characteristics are always true for an exponential function of the form $f(x) = ab^x$, $a > 0$, $b > 1$. An exponential function of this form is called an **exponential growth function.**

On page 727, we graphed $f(x) = \left(\frac{1}{2}\right)^x$. We are now going to look at the graph of $f(x) = a \cdot \left(\frac{1}{2}\right)^x$, where $a > 0$.

Enter all three of the following functions in the equation editor. Then graph the functions.

$$f(x) = \left(\frac{1}{2}\right)^x$$
$$f(x) = 3 \cdot \left(\frac{1}{2}\right)^x$$
$$f(x) = \frac{1}{3} \cdot \left(\frac{1}{2}\right)^x$$

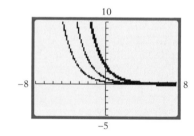

For each graph,

1. The function is an decreasing function.
2. The y-intercept is $(0, a)$.
3. The domain is all real numbers.
4. The range of f is the positive real numbers.

These characteristics are always true for an exponential function of the form $f(x) = ab^x$, $a > 0$, $0 < b < 1$. An exponential function of this form is called an **exponential decay function.**

Exponential Growth and Exponential Decay Functions

The function $f(x) = ab^x$, $a > 0$, is an **exponential growth function** when $b > 1$. It is an **exponential decay function** when $0 < b < 1$.

INSTRUCTOR NOTE
Exercise 40b can be used for a similar in-class example.

EXAMPLE 6

Is the function $y = 200(5^x)$ an exponential growth function or an exponential decay function?

Solution The function is of the form $y = ab^x$, where $a = 200$ and $b = 5$.
$b > 1$

The function is an exponential growth function.

YOU TRY IT 6

Is the function $y = 30(0.5^x)$ an exponential growth function or an exponential decay function?

Solution See page S43.   Exponential decay

■ Applications of Exponential Functions

A biologist places one single-celled bacterium in a culture, and each hour that particular species of bacterium divides into two bacteria. After one hour there will be two bacteria. After two hours, each of the two bacteria will divide and there will be four bacteria. After three hours, each of the four bacteria will divide and there will be eight bacteria.

Time, t	Number of Bacteria, N
0	1
1	2
2	4
3	8
4	16

The table at the left shows the number of bacteria in the culture after various intervals of time t, in hours. Values in this table could also be found by using the exponential equation $N = 2^t$.

We can write the equation $N = 2^t$ as $N = 1 \cdot 2^t$. Therefore, it is of the form $y = ab^x$, where $a = 1$ and $b = 2$. It is an exponential growth function. Exponential growth functions are important not only in population growth studies but also in physics, chemistry, psychology, and economics.

Recall that interest is the amount of money one pays (or receives) when borrowing (or investing) money. **Compound interest** is interest that is computed not only on the original principal but also on the interest already earned. The frequency with which the interest is compounded is called the **compounding period.** The compound interest formula is an exponential growth equation.

The **compound interest formula** is $A = P\left(1 + \dfrac{r}{n}\right)^{nt}$, where

TAKE NOTE

The compound interest formula is an equation of the form $y = ab^x$, where A corresponds to y, P corresponds to a, $\left(1 + \dfrac{r}{n}\right)$ corresponds to b, and nt corresponds to x.

$A =$ the future value of an investment
$P =$ the principal, or the original value of the investment
$r =$ the annual interest rate
$n =$ the number of compounding periods per year
$t =$ the number of years the money is invested

As stated above, n is the number of times interest is calculated each year. Possible values of n are listed in the table below.

Suggested Activity

See Section 10.3 of the *Student Activity Manual*, for an activity that focuses on the characteristics of exponential functions and their graphs.

If interest is	then n =
compounded annually	1
compounded semiannually	2
compounded quarterly	4
compounded monthly	12
compounded daily	365

There are two methods for calculating interest that is compounded daily: the exact method and the ordinary method. By the **exact method,** the number of days in a year is 365. The **ordinary method** uses a 360-day year. The exact method is used in this text.

INSTRUCTOR NOTE
Exercise 46 can be used for a similar in-class example.

EXAMPLE 7

An investment broker deposits $1000 into an account that earns 8% annual interest compounded quarterly. What is the value of the investment after 2 years? Round to the nearest dollar.

State the goal. The goal is to find out how much money will be in the account after 2 years.

Devise a strategy. • Determine the values of P, r, n, and t.
• Use the compound interest formula.

Solve the problem.

$P = 1000$ • The original value of the investment is $1000.

$r = 8\% = 0.08$ • r is the annual interest rate.
Write the interest rate as a decimal.

$n = 4$ • The investment is compounded quarterly.

$t = 2$ • The money is invested for 2 years.

$A = P\left(1 + \dfrac{r}{n}\right)^{nt}$ • Use the compound interest formula.

$A = 1000\left(1 + \dfrac{0.08}{4}\right)^{4(2)}$ • Replace P, r, n and t by their values.

$A \approx 1172$ • Evaluate the expression.

The value of the investment after 2 years is $1172.

Check your work.

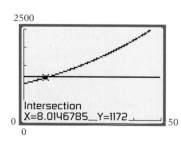

2500

Intersection
X=8.0146785 Y=1172

0   50
0

• Graph the function $y = 1000\left(1 + \dfrac{0.08}{4}\right)^{x}$.
Because the exponent in the equation
$A = 1000\left(1 + \dfrac{0.08}{4}\right)^{4(2)}$ is $4(2) = 8$, we
want to determine the y value when $x = 8$
in the equation $y = 1000\left(1 + \dfrac{0.08}{4}\right)^{x}$.

See Appendix A:
Trace

YOU TRY IT 7

A financial advisor recommends that a client deposit $2500 into a fund that earns 7.5% annual interest compounded monthly. What will be the value of the investment after 3 years? Round to the nearest cent.

Solution   See page S43. $3128.62

One of the most common illustrations of exponential decay is the decay of a radioactive substance.

Time, t	Amount, A
0	10
5	5
10	2.5
15	1.25
20	0.625

A radioactive isotope of cobalt has a half-life of approximately 5 years. This means that one-half of any given amount of the cobalt isotope will disintegrate in 5 years. The table at the left indicates the amount of the initial 10 milligrams of a cobalt isotope that remains after various intervals of time t, in years. The values in this table could also be found by using the exponential equation $A = 10\left(\frac{1}{2}\right)^{t/5}$.

The equation $A = 10\left(\frac{1}{2}\right)^{t/5}$ is an equation of the form $y = ab^x$, where $a > 0$ and $0 < b < 1$. Therefore, it is an exponential decay equation.

INSTRUCTOR NOTE
Exercise 52 can be used for a similar in-class example.

EXAMPLE 8

A sample of iodine-131 contains 100 grams. The number of grams of the sample that remains after t days is given by the equation $A = 100(0.92)^t$. Find the number of grams of the sample that remains after 30 days. Round to the nearest hundredth.

State the goal.   The goal is to determine how many grams of the sample of iodine-131 remain after 30 days.

Devise a strategy.   Replace t by 30 in the given equation. Then evaluate the resulting exponential expression.

Solve the problem. $A = 100(0.92)^t$ • Use the given formula.

$A = 100(0.92)^{30}$ • Replace t by 30.

$A \approx 8.20$ • Evaluate the exponential expression.

After 30 days, 8.20 grams of the original 100 grams in the sample remain.

Check your work.

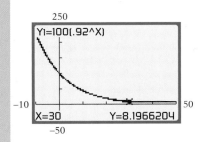

• Use a graphing calculator to graph the function $y = 100(0.92)^x$. Use the TRACE feature to find the value of y when $x = 30$. Note that the value of y will be approximately 8.2. Alternatively, graph both $y = 100(0.92)^x$ and $y = 8.2$. Use the INTERSECT feature to verify that, at the point of intersection, $X \approx 30$.

See Appendix A:
Trace or Intersect

> ### YOU TRY IT 8
>
> You purchase a new car for $21,000. Each year after the purchase, the value of the car decreases by 15%. The value of the car after t years is given by the equation $A = 21,000(1 - 0.15)^t$.
>
> **a.** Find the value of the car after 5 years. Round to the nearest dollar.
> **b.** After 5 years, is the car worth more or less than one-half its original value?
>
> **Solution**   See page S43.   **a.** $9318   **b.** Less than one-half its original value

10.3 EXERCISES   Suggested Assignment: 7–55, odds

Topics for Discussion

1. **a.** What is an exponential function?
    b. How does an exponential function differ from a polynomial function?
    a. Answers may vary. For example, an exponential function with base b is defined by $f(x) = b^x$, $b > 0$, $b \neq 1$, and x is any real number.   **b.** Answers will vary. For example, an exponential function has a constant base and a variable exponent, whereas a polynomial function has a variable base and a constant exponent.

2. **a.** Explain why the condition $b > 0$ is given for $f(x) = b^x$.
    b. Why is the condition $b \neq 1$ given for $f(x) = b^x$?
    a. Answers may vary. For example, $b > 0$ to avoid complex numbers.
    b. Answers may vary. For example, $b \neq 1$ to avoid the constant function $f(x) = 1$.

3. What is the natural exponential function?
    It is the function defined by $f(x) = e^x$, where e is an irrational number approximately equal to 2.71828183.

4. For an exponential function of the form $y = ab^x$, $x > 0$, how can you distinguish between an exponential growth function and an exponential decay function?
    For an exponential growth function, $b > 1$, whereas for an exponential decay function, $0 < b < 1$.

5. Determine whether the statement is always true, sometimes true, or never true.
    a. The domain of an exponential function $f(x) = b^x$, $b > 0$, $b \neq 1$, is the set of positive numbers.
    b. An exponential function $f(x) = b^x$, $b > 0$, $b \neq 1$, is a 1-1 function.
    c. The graph of an exponential function $f(x) = b^x$, $b > 0$, $b \neq 1$, passes through the point $(0, 0)$.
    d. For the function $f(x) = b^x$, $b > 0$, $b \neq 1$, the base b is a positive integer.
    e. An exponential function $f(x) = b^x$, $b > 0$, $b \neq 1$, has two x-intercepts.
    a. Never true   b. Always true   c. Never true   d. Sometimes true   e. Never true

■ Evaluate Exponential Functions

Evaluate the function for the given values. Round to the nearest ten-thousandth.

6. $f(x) = 3^x$

   a. $f(2)$ **b.** $f(0)$ **c.** $f(-2)$

   a. 9 b. 1 c. 0.1111

7. $H(x) = 2^x$

   a. $H(-3)$ **b.** $H(0)$ **c.** $H(2)$

   a. 0.125 b. 1 c. 4

8. $g(x) = 2^{x+1}$

   a. $g(3)$ **b.** $g(1)$ **c.** $g(-3)$

   a. 16 b. 4 c. 0.25

9. $F(x) = 3^{x-2}$

   a. $F(-4)$ **b.** $F(-1)$ **c.** $F(0)$

   a. 0.0014 b. 0.0370 c. 0.1111

10. $G(r) = \left(\dfrac{1}{2}\right)^{2r}$

   a. $G(0)$ **b.** $G\left(\dfrac{3}{2}\right)$ **c.** $G(-2)$

   a. 1 b. 0.125 c. 16

11. $R(t) = \left(\dfrac{1}{3}\right)^{3t}$

   a. $R\left(-\dfrac{1}{3}\right)$ **b.** $R(1)$ **c.** $R(-2)$

   a. 3 b. 0.0370 c. 729

12. $h(x) = e^{x/2}$

   a. $h(4)$ **b.** $h(-2)$ **c.** $h\left(\dfrac{1}{2}\right)$

   a. 7.3891 b. 0.3679 c. 1.2840

13. $f(x) = e^{2x}$

   a. $f(-2)$ **b.** $f\left(-\dfrac{2}{3}\right)$ **c.** $f(2)$

   a. 0.0183 b. 0.2636 c. 54.5982

14. $H(x) = e^{x+3}$

   a. $H(-1)$ **b.** $H(3)$ **c.** $H(5)$

   a. 7.3891 b. 403.4288 c. 2980.9580

15. $g(x) = e^{x-4}$

   a. $g(-3)$ **b.** $g(4)$ **c.** $g(8)$

   a. 0.0009 b. 1 c. 54.5982

16. $F(x) = (\sqrt{7})^x$

   a. $F(2)$ **b.** $F(-1)$ **c.** $F(3.5)$

   a. 7 b. 0.3780 c. 30.1246

17. $Q(x) = (\sqrt{3})^x$

   a. $Q(3)$ **b.** $Q(-2)$ **c.** $Q(4.8)$

   a. 5.1962 b. 0.3333 c. 13.9666

18. Evaluate $\left(1 + \dfrac{1}{n}\right)^n$ for $n = 100$, 1000, $10{,}000$, and $100{,}000$ and compare the results with the value of e, the base of the natural exponential function. On the basis of your evaluation, complete the following sentence:

As n increases, $\left(1 + \dfrac{1}{n}\right)^n$ becomes closer to _____. e

■ Graphs of Exponential Functions

19. Each of the graphs below is an exponential function. Which are graphs of exponential growth functions?

a.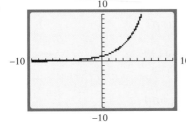

b.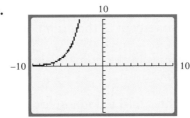

c.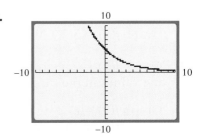

a, b

20. Each of the graphs below is an exponential function. Which are graphs of exponential decay functions?

a. 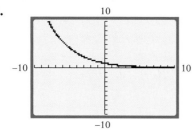 **b.** 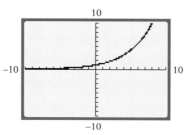 **c.**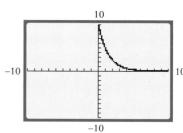

a, c

For Exercises 21 to 24, state (a) the value of b in the exponential function and (b) whether the function is increasing or decreasing.

21. $f(x) = 5^{.5x+1}$ **a.** 5 **b.** Increasing

22. $f(x) = 0.6^x - 2$ **a.** 0.6 **b.** Decreasing

23. $f(x) = e^{2x} - 1$ **a.** e **b.** Increasing

24. $f(x) = e^{\frac{1}{2}x+3}$ **a.** e **b.** Increasing

25. Which of the following are exponential growth functions?
   a. $y = 0.2(5^x)$ **b.** $y = 2(0.5^x)$ **c.** $y = 2 + 5x$ **d.** $y = 2.5^x$ a, d

26. Which of the following are exponential decay functions?
   a. $y = 0.8(4^x)$ **b.** $y = 8(0.4^x)$ **c.** $y = 0.4x + 8$ **d.** $y = 0.84^x$ b, d

27. For which value of x is the value of the function $f(x) = 6(1.05^x)$ larger, 10 or 20? Explain your answer.
   20. $f(x) = 6(1.05^x)$ is an increasing function; therefore, as values of x increase, values of y increase.

28. For which value of x is the value of the function $f(x) = 80(0.25^x)$ larger, 5 or 25? Explain your answer.
   5. $f(x) = 80(0.25^x)$ is a decreasing function; therefore, as values of x increase, values of y decrease.

29. For which value of x is the value of the function $f(x) = 0.7(0.95^x)$ smaller, 6 or 12? Explain your answer.
   12. $f(x) = 0.7(0.95^x)$ is a decreasing function; therefore, as values of x increase, values of y decrease.

30. For which value of x is the value of the function $f(x) = 0.5(2.8^x)$ smaller, 3 or 6? Explain your answer.
   3. $f(x) = 0.5(2.8^x)$ is an increasing function; therefore, as values of x increase, values of y increase.

31. Graph $f(x) = 3\left(\dfrac{1}{3}\right)^x$. What is the y-intercept of the graph of the function?
   (0, 3)

32. Graph $f(x) = 5(2^x)$. What is the y-intercept of the graph of the function?
   (0, 5)

33. Graph $f(x) = 2^x - 1$. For what values in the domain are the corresponding values in the range less than 0? $x < 0$

34. Graph $f(x) = e^x + 1$. For what values in the domain are the corresponding values in the range less than 0?   There are no values in the domain for which the corresponding values in the range are less than 0.

35. Graph $f(x) = 2^x - 3$ and approximate the zero of f to the nearest tenth.
1.6

36. Graph $f(x) = 5 - 3^x$ and approximate the zero of f to the nearest tenth.
1.5

37. Graph $f(x) = e^x - 2$ and approximate, to the nearest tenth, the value of x for which $f(x) = 3$.   1.6

38. Graph $f(x) = e^{2x-3}$ and approximate, to the nearest tenth, the value of x for which $f(x) = 2$.   1.8

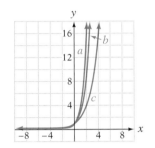

39. Three exponential functions are graphed in the rectangular coordinate system at the right.

   a. Match the letter with the function $y = 2^x$, $y = 3^x$, or $y = 4^x$. Explain why you made the choices you did.

   b. Are these exponential growth or exponential decay functions?

   a. Graph a is $y = 4^x$, graph b is $y = 3^x$, and graph c is $y = 2^x$. Explanations will vary. For example, the greater the base of the exponential function, the more rapidly the value of y increases.   **b.** Exponential growth

40. Three exponential functions are graphed in the rectangular coordinate system at the right.

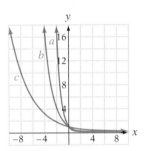

   a. Match the letter with the function $y = 0.25^x$, $y = 0.5^x$, or $y = 0.75^x$. Explain why you made the choices you did.

   b. Are these exponential growth or exponential decay functions?

   a. Graph a is $y = 0.25^x$, graph b is $y = 0.5^x$, and graph c is $y = 0.75^x$. Explanations will vary. For example, the greater the base of the exponential function, the more rapidly the value of y decreases.   **b.** Exponential decay

41. Determine whether the data in each table represent

   (A) an exponential growth function
   (B) an exponential decay function
   (C) a linear function
   (D) a quadratic function

a.

X	Y_1
0	1
1	4
2	16
3	64
4	256

(A)

b.

X	Y_1
0	1
1	.2
2	.04
3	.008
4	.0016

(B)

c.

X	Y_1
0	2
1	−1
2	−4
3	−7
4	−10

(C)

d.

X	Y_1
0	−3
1	−6
2	−7
3	−6
4	−3

(D)

e.

X	Y_1
0	0
1	7
2	14
3	21
4	28

(C)

f.

X	Y_1
0	3
1	6
2	12
3	24
4	48

(A)

■ **Applications of Exponential Functions**

42. *Investments*   The exponential function $F(n) = 500(1.00021918)^{365n}$ gives the value in n years of a $500 investment in a certificate of deposit that earns 8% annual interest compounded daily. Graph F and determine in how many years the investment will be worth $1000.
    9 years

43. ⬤ *Population*   Assuming that the current population of Earth is 6 billion people and that Earth's population is growing at an annual rate of 1.5%, the exponential function $P(t) = 6(1.015)^t$ gives the size, in billions of people, of the population t years from now. Graph P and determine the number of years before Earth's population reaches 7 billion people.   10 years

44. *Light*   The percent of light that reaches m meters below the surface of the ocean is given by the equation $P(m) = 100\left(\dfrac{1}{e}\right)^{1.38m}$. Graph P and determine the depth to which 50% of the light will reach. Round to the nearest tenth.   0.5 m

45. *Radioactivity*   The number of grams of radioactive cesium that remain after t years from an original sample of 30 grams is given by $N(t) = 30\left(\dfrac{1}{2}\right)^{0.0322t}$. Graph N and determine in how many years there will be 20 grams of cesium remaining.   18 years

For Exercises 46 to 49, use the compound interest formula $A = P\left(1 + \dfrac{r}{n}\right)^{nt}$, where A is the value of an original investment of P dollars invested at an annual interest rate r compounded n times per year for t years. Round to the nearest cent.

46. *Investments*   An accountant deposits $15,000 into an account that earns 8.5% annual interest compounded monthly. What is the value of the investment after 10 years? $34,989.71

47. *Investments*   An annuity earns 6% annual interest compounded semi-annually. What is the value of a $12,000 investment in this annuity after 15 years? $29,127.15

48. *Investments*   A computer network specialist deposits $2500 into a retirement account that earns 7.5% annual interest compounded daily. What is the value of the investment after 20 years? $11,202.50

49. *Investments*   A $10,000 certificate of deposit (CD) earns 5% annual interest compounded daily. What is the value of the investment after 20 years? $27,180.96

50. *Investments*   Some banks now use continuous compounding of an amount invested. In this case, the equation that relates the value of an

initial investment of P dollars in t years at an annual interest rate of r is given by $A = Pe^{rt}$. Using this equation, find the value in 5 years of an investment of \$2500 into an account that earns 5% annual interest.
\$3210.06

For Exercises 51 and 52, use the exponential decay equation $A = A_0\left(\dfrac{1}{2}\right)^{t/k}$, where A is the amount of a radioactive material present after time t, k is the half-life, and A_0 is the original amount of radioactive substance. Round to the nearest tenth.

51. *Biology*   An isotope of technetium is used to prepare images of internal body organs. This isotope has a half-life of approximately 6 hours. If a patient is injected with 30 milligrams of this isotope, what will be the technetium level in the patient after 3 hours?   21.2 mg

52. *Biology*   Iodine-131 is an isotope that is used to study the functioning of the thyroid gland. This isotope has a half-life of approximately 8 days. If a patient is given an injection that contains 8 micrograms of iodine-131, what will be the iodine level in the patient after 5 days?
5.2 micrograms

53. *Education*   The percent of correct welds that a student can make will increase with practice and can be approximated by the equation $P = 100[1 - (0.75)^t]$, where P is the percent of correct welds and t is the number of weeks of practice. Find the percent of correct welds that a student will make after 4 weeks of practice. Round to the nearest percent.   68%

54. *Education*   The number of words per minute that a student can type increases with practice and can be approximated by the equation $N = 100[1 - (0.9)^t]$, where N is the number of words typed per minute after t weeks of instruction. How many words per minute can a student type after 14 weeks of instruction? Round to the nearest whole number.
77 words per minute

55. ● *Postage*   In 1962 the cost of a first-class stamp was \$.04. In 2003 the cost was \$.37. The increase in cost can be modeled by the equation $C = 0.04e^{0.057t}$, where C is the cost and t is the number of years after 1962. According to this model, what was the cost of a first-class stamp in 1974? Round to the nearest cent.   \$.08

56. *Earth Science*   Earth's atmospheric pressure changes as you rise above the surface. At an altitude of h kilometers, where $0 < h < 80$, the pressure P in newtons per square centimeter is approximately modeled by the equation $P(h) = 10.13\left(\dfrac{1}{e}\right)^{0.116h}$.

   a.   What is the approximate pressure at 40 kilometers above Earth? Round to the nearest thousandth.

   b.   What is the approximate pressure on Earth's surface?

   c.   Does atmospheric pressure increase or decrease as you rise above Earth's surface?

   a. 0.098 newtons/cm²   **b.** 10.13 newtons/cm²   **c.** decrease

Applying Concepts

57. *Consumer Loans*   You borrow $16,000 to purchase a car. The bank charges an annual interest rate of 9%, compounded monthly, on the 4-year loan. Find the amount of interest you pay on the loan. Use the formula $A = P\left(1 + \dfrac{r}{n}\right)^{nt}$, where A is the total cost of a loan of P dollars when money is borrowed at an annual interest rate r compounded n times per year for t years. Round to the nearest cent. $6902.49

58. *Investments*   At the age of 25, you deposit $5000 in a retirement account that earns 8% annual interest compounded daily. How much more money is in the account when you reach the age of 60 than when you reach the age of 35? Use the compound interest formula $A = P\left(1 + \dfrac{r}{n}\right)^{nt}$, where A is the value of an original investment of P dollars invested at an annual interest rate r compounded n times per year for t years. Round to the nearest cent. $71,071.28

59. Consider the functions $g(x) = 3^x$ and $h(x) = 4^x$.
   a.   Which function has the greater values when $x > 0$?
   b.   Which function has the greater values when $x < 0$?
   a. $h(x) = 4^x$ b. $g(x) = 3^x$

60. **a.**   Between what two consecutive integers does the value of the function $f(x) = 3^x$ first exceed the value of the function $P(x) = x^3$?
   b.   For what values of x does $f(x) = P(x)$? Round to the nearest tenth.
   a. 2 and 3   b. 2.5 and 3

EXPLORATIONS

1.   *Exponential Functions of the Form f(x) = ab^{-x}*

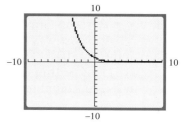

The graph of the function $f(x) = \left(\dfrac{1}{2}\right)^x$ is shown at the right as graphed in the standard viewing window.

   a.   Graph the function $f(x) = 2^{-x}$ in the standard viewing window. How does the graph of $f(x) = \left(\dfrac{1}{2}\right)^x$ compare with the graph of $f(x) = 2^{-x}$? Explain. [*Hint:* $2^{-x} = (2^{-1})^x$. Rewrite 2^{-1} without a negative exponent.]
   b.   Which of the following functions have the same graph?

$$f(x) = 3^x \qquad f(x) = \left(\frac{1}{3}\right)^x \qquad f(x) = x^3 \qquad f(x) = 3^{-x}$$

   c.   Write a function of the form $f(x) = ab^x$, $b > 1$, that has the same graph as the function $f(x) = \left(\dfrac{1}{4}\right)^x$.

   d. Write a function of the form $f(x) = ab^x$, $0 < b < 1$, that has the same graph as the function $f(x) = 5^{-x}$.

   e. Describe the graph of $f(x) = 3^x$ in terms of the graph of $f(x) = 3^{-x}$.

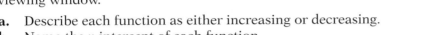

a. They are the same. $2^{-x} = (2^{-1})^x = \left(\frac{1}{2}\right)^x$ **b.** $f(x) = \left(\frac{1}{3}\right)^x$ and $f(x) = 3^{-x}$ **c.** $f(x) = 4^{-x}$ **d.** $f(x) = \left(\frac{1}{5}\right)^x$ **e.** The graph of $f(x) = 3^{-x}$ is a reflection of the graph of $f(x) = 3^x$ over the y-axis.

2. *Negative Values of a in Exponential Functions of the Form $f(x) = ab^x$*

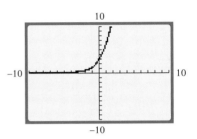

The graph of the function $f(x) = 3 \cdot 2^x$ is shown at the right as graphed in the standard viewing window.

Graph the functions $f(x) = 3 \cdot 2^x$ and $g(x) = -3 \cdot 2^x$ in the standard viewing window.

   a. Describe each function as either increasing or decreasing.

   b. Name the y-intercept of each function.

   c. Describe the graph of $g(x) = -3 \cdot 2^x$ in terms of the graph of $f(x) = 3 \cdot 2^x$.

   d. State the domain of each function.

   e. State the range of each function.

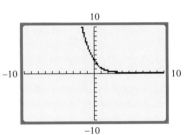

The graph of the function $f(x) = 3(.5^x)$ is shown at the right as graphed in the standard viewing window.

Graph the functions $f(x) = 3(.5^x)$ and $g(x) = -3(.5^x)$ in the standard viewing window.

   f. Describe each function as either increasing or decreasing.

   g. Name the y-intercept of each function.

   h. Describe the graph of $g(x) = -3(.5^x)$ in terms of the graph of $f(x) = 3(.5^x)$.

   i. State the domain of each function.

   j. State the range of each function.

   k. Make at least four general statements regarding the graphs of $f(x) = ab^x$ and $f(x) = -ab^x$, $a > 0$, $b > 0$, $b \neq 1$.

a. $f(x) = 3 \cdot 2^x$ is increasing, $f(x) = -3 \cdot 2^x$ is decreasing. **b.** The y-intercept of the graph of $f(x) = 3 \cdot 2^x$ is (0, 3). The y-intercept of the graph of $f(x) = -3 \cdot 2^x$ is (0, −3). **c.** The graph of $f(x) = -3 \cdot 2^x$ is a reflection of the graph of $f(x) = 3 \cdot 2^x$ over the x-axis.
d. The domain of $f(x) = 3 \cdot 2^x$ is all real numbers. The domain of $f(x) = -3 \cdot 2^x$ is all real numbers. **e.** The range of $f(x) = 3 \cdot 2^x$ is $\{y|y > 0\}$. The range of $f(x) = -3 \cdot 2^x$ is $\{y|y < 0\}$. **f.** $f(x) = 3(.5^x)$ is decreasing. $f(x) = -3(.5^x)$ is increasing. **g.** The y-intercept of the graph of $f(x) = 3(.5^x)$ is (0, 3). The y-intercept of the graph of $f(x) = -3(.5^x)$ is (0, −3). **h.** The graph of $f(x) = -3(.5^x)$ is a reflection of the graph of $f(x) = 3(.5^x)$ over the x-axis. **i.** The domain of $f(x) = 3(.5^x)$ is all real numbers. The domain of $f(x) = -3(.5^x)$ is all real numbers. **j.** The range of $f(x) = 3(.5^x)$ is $\{y|y > 0\}$. The range of $f(x) = -3(.5^x)$ is $\{y|y < 0\}$. **k.** Answers will vary. See examples in the Solutions Manual.

3. *Graphs of Exponential Functions*

   a. Given $h(x) = 5^x$, write a function g that is the reflection of h over the y-axis.

   b. Given $h(x) = 5^x$, write a function g that is the reflection of h over the x-axis.

   c. Given $F(x) = 6^{-x}$, write a function g that is the reflection of F over the y-axis.

   d. Given $F(x) = 6^{-x}$, write a function g that is the reflection of F over the x-axis.

   e. Write a function g that is a reflection of $f(x) = a^x$ over the y-axis.

   f. Write a function g that is a reflection of $f(x) = a^x$ over the x-axis.

a. $g(x) = \left(\frac{1}{5}\right)^x$ **b.** $g(x) = -(5^x)$ **c.** $g(x) = \left(\frac{1}{6}\right)^{-x}$ **d.** $g(x) = -(6^{-x})$ **e.** $g(x) = \left(\frac{1}{a}\right)^x$, $a \neq 0$ **f.** $g(x) = -(a^x)$

SECTION **10.4**

■ Write Exponential Models
■ Exponential Regression

Exponential Models and Exponential Regression

■ Write Exponential Models

INSTRUCTOR NOTE
In this portion of the lesson, students are learning to write exponential models without the use of a graphing calculator. Writing exponential models using regression and a graphing calculator follows this discussion.

In the last section, we discussed compound interest. For example, in Example 7, we found the value of $1000, invested at 8% annual interest, after 2 years.

When a quantity increases or decreases by a fixed percent each year, the amount of that quantity after t years can be modeled by an exponential function. (In some cases, the time period will be not in years but rather in some other time period, such as hours, days, or months.)

The general form of an exponential function is $f(x) = ab^x$ or $y = ab^x$.

Suggested Activity

See Section 10.4 of the *Student Activity Manual* for an investigation into compound interest.

 QUESTION Name the values of a and b in the exponential functions.

   a. $y = 4(0.8)^x$ **b.** $y = \dfrac{1}{2} \cdot 6^x$

Here is an example of writing an exponential function given the initial quantity and the fixed percent by which that quantity *increases* each year.

INSTRUCTOR NOTE
Exercise 6 can be used for a similar in-class example.

EXAMPLE 1

In 1990 the population of the United States was 250 million. Since then the population has increased at a rate of about 0.8% per year.

a. Write an exponential function to model this population growth.
b. Use the function to approximate, to the nearest million, the population of the United States in 1995.

INSTRUCTOR NOTE
You might also ask the students to use the function to predict the year during which the population of the United States will reach 290 million. This requires that they determine the x value given a y value.

[Answer: 2009]

Solution

a. The initial quantity (the value of a) is 250 million, the population in 1990.

   Find the growth factor (the value of b).

     The population at the end of each year is found by multiplying the population at the beginning of the year by

$$100\% + 0.8\% = 1.00 + 0.008 = 1.008$$

$y = ab^x$
$y = 250(1.008)^x$ • Replace a by 250 and b by 1.008.

   The exponential function is $y = 250(1.008)^x$, where y is the population of the United States, in millions, x years after 1990.

 ANSWER   **a.** $a = 4, b = 0.8$ **b.** $a = \dfrac{1}{2}, b = 6$

TAKE NOTE

The *y*-intercept of the graph of the equation should be 250.

 See Appendix A: Trace

b. 1995 is 5 years after 1990 (1995 − 1990 = 5).

$$y = 250(1.008)^x$$
$$y = 250(1.008)^5 \qquad \bullet \text{ Replace } x \text{ by 5 in the exponential function.}$$
$$y \approx 260 \qquad \bullet \text{ Evaluate the exponential expression.}$$

The population of the United States in 1995 was approximately 260 million.

Graphical Check:

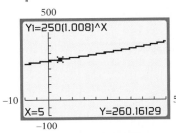

- Use a graphing calculator to graph $y = 250(1.008^x)$. Use the TRACE feature. When $x = 5$, $y \approx 260$.

YOU TRY IT 1

Suppose you invest $25,000 in 2010, and the investment earns an average annual interest rate of 6%, compounded annually.

a. Write an exponential function to model the value of the investment.
b. Use the function to determine the value of the investment in 2020.

Solution See page S43. **a.** $y = 25{,}000(1.06)^x$ **b.** $44,771.19

Here is an example of writing an exponential function given the initial quantity and the fixed percent by which that quantity *decreases* each year.

INSTRUCTOR NOTE
Exercise 8 can be used for a similar in-class example.

Suggested Activity

1. Ten years ago, the cost of tuition at a state university was $5000 per year. Since then the cost of tuition has risen 4% per year. Write an equation to model the exponential growth.
[Answer: $y = 5000(1.04)^x$]

2. An adult takes 500 milligrams of ibuprofen. Each hour the amount of ibuprofen in the adult's system decreases by 30%. Write an equation to model the exponential decay.
[Answer: $y = 500(0.70)^x$]

EXAMPLE 2

In 1980 the average amount of fuel used annually by each car in the United States was about 590 gallons. Since then, the average amount of fuel used has decreased by about 2% each year.

a. Write a function to model this exponential decay.
b. Use the function to estimate the average amount of fuel used by each car in the United States in 1992. Round to the nearest whole number.

Solution

a. The initial quantity (the value of *a*) is 590 gallons, the average annual fuel consumption per car in 1980.

Find the decay factor (the value of *b*).

The consumption at the end of each year is found by multiplying the consumption at the beginning of the year by

$$100\% - 2\% = 1.00 - 0.02 = 0.98$$

$$y = ab^x$$
$$y = 590(0.98)^x \qquad \bullet \text{ Replace } a \text{ with 590 and } b \text{ with 0.98.}$$

INSTRUCTOR NOTE
You might also ask students to use the function to predict the year during which average fuel consumption per car was 500 gallons. This requires that they determine the x value given a y value. [Answer: 1988]

TAKE NOTE

The y-intercept of the graph of the equation should be 590.

See Appendix A: Trace

The exponential function is $y = 590(0.98)^x$, where y is the average annual fuel consumption, in gallons, per car x years after 1980.

b. $y = 590(0.98)^x$

$y = 590(0.98)^{12}$ • 1992 – 1980 = 12. Replace x by 12.

$y \approx 463$ • Evaluate the exponential expression.

The average amount of fuel used per car in the United States in 1992 was 463 gallons.

Graphical Check:

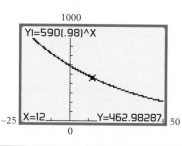

• Use a graphing calculator to graph the function $y = 590(0.98)^x$. Use the TRACE feature. When $x = 12$, $y \approx 463$.

YOU TRY IT 2

The half-life of iodine-123 is approximately 13 hours. Suppose the initial amount of a sample is 40 grams.

a. Write a function to model the exponential decay. (*Hint:* Let x = the number of 13-hour periods.)

b. Find the amount of iodine-123 in the sample after 52 hours.

Solution See page S43. **a.** $y = 40(0.5)^x$ **b.** 2.5 g

■ Exponential Regression

Given data that approximate an exponential function, you can use a graphing calculator to perform a regression and determine an exponential model. An example follows.

INSTRUCTOR NOTE
Exercise 16 can be used for a similar in-class example.

EXAMPLE 3

The growth of online sales of recorded music, including CDs, cassettes, and albums, during the last four years of the 1990s is shown below (*Source:* Jupiter Communications).

Year	1996	1997	1998	1999
Sales (in millions of dollars)	18	47	110	240

a. Use regression to determine an exponential model for these data. Use $x = 6$ for 1996, $x = 7$ for 1997, and so on.

b. What is the correlation coefficient for your model? What does that imply about the fit of the data to the graph of the equation?

c. Use the equation to predict online sales of recorded music in 2005. Round to the nearest million dollars.

d. According to the equation, online sales of recorded music were $750 million in which year?

See Appendix A: Regression

Solution

a. Enter the data into a graphing calculator. Use a calculator to determine the regression line for the data. Some typical screens are shown at the right.

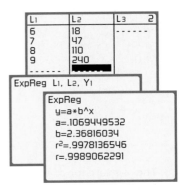

The exponential regression equation is given in the form $y = a * b^x$.
To the nearest millionth, the equation is $y = 0.106945 * 2.368160^x$.

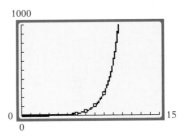

b. The correlation coefficient (the value of r) is approximately 0.998906. This is very close to 1, which means that the data points lie very close to the graph of $y = 0.106945 * 2.368160^x$.

You can verify this by graphing both the data points and the regression line on a graphing calculator. Note how the data points lie either on or very, very close to the graph of the regression equation.

 See Appendix A: Regression

c. The year 2005 corresponds to $x = 15$. Graph the regression equation. (It is already stored in Y₁.) Use the TRACE feature to find y when $x = 15$.

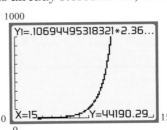

- The coordinates printed at the bottom of the screen are approximately (15, 44190).

The equation predicts that online sales of recorded music in 2005 will be approximately $44,190 million.

d. The regression equation is already stored in Y₁. Enter 750 for Y₂. Find the point of intersection of Y₁ and Y₂ to determine the x value when $y = 750$.

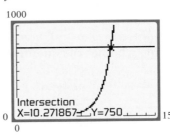

- The coordinates printed at the bottom of the screen are approximately (10.27, 750). An x value of 10 corresponds to the year 2000.

 See Appendix A: Intersect

According to the equation, sales were $750 million in the year 2000.

YOU TRY IT 3

Net sales, in billions, for Wal-Mart are shown below.

Year	1978	1988	1998
Net Sales (in billions of dollars)	0.7	16	118

a. Use regression to determine an exponential model for these data. Use $x = 8$ for 1978, $x = 18$ for 1988, and $x = 28$ for 1998.
b. What is the correlation coefficient for your model? What does that imply about the fit of the data to the graph of the equation?
c. Use the equation to predict net sales for Wal-Mart in 1975. Round to the nearest million.
d. Use the equation to predict during which year net sales were $50 billion.
e. Actual net sales for Wal-Mart in 2002 were $217.8 billion. Find the difference between the model's prediction for net sales in 2002 and actual net sales in 2002. Round to the nearest billion.

Solution See page S44. **a.** $y = 0.1087(1.2922)^x$ **b.** $r \approx 0.9919856$. The data points lie very close to the graph of $y = 0.1087(1.2922)^x$. **c.** $392 million **d.** 1994 **e.** $179 billion

---

10.4 EXERCISES   Suggested Assignment: 7–19, odds; 10

Topics for Discussion

1. Indicate which variable in the general form of an exponential function $y = ab^x$ represents
  a. the growth or decay factor
  b. the number of time periods
  c. the initial amount
  a. b b. x c. a

2. In an exponential model of the form $y = ab^x$, $a > 0$, $b > 0$, $b \neq 1$, $x > 0$, how can the value of b be used to determine whether the quantity y is increasing each year or decreasing each year?
If $b > 1$, the quantity y is increasing each year. If $0 < b < 1$, the quantity y is decreasing each year.

3. Will the graphs of $f(x) = 10\left(\dfrac{1}{5}\right)^x$ and $f(x) = 10(5^x)$ intersect? If so, where do they intersect? If not, explain why not.   They will intersect at (0, 10).

4. In an exponential model of the form $y = ab^x$, $a > 0$, $b > 0$, $b \neq 1$, $x > 0$, what is the y-intercept of the graph of the function?   (0, a)

5. Describe a difference between the graph of an exponential growth function and the graph of an exponential decay function.

Answers may vary. For example, for an exponential growth function, as *x* increases, *y* increases. For an exponential decay function, as *x* increases, *y* decreases.

■ Write Exponential Models

6.  *Population Growth* In 1990 the population of Dallas-Forth Worth, Texas, was 4.037 million. From 1990 to 2000, the population increased at a rate of about 2.7% per year.

   a. Write an exponential function to model this population growth.
   b. Use the model to approximate the population of Dallas-Forth Worth in 1995. Round to the nearest thousand.
   c. Use the model to predict what the population of Dallas will be in 2008 if the growth continues at the same rate. Round to the nearest thousand.

   a. $y = 4.037(1.027)^x$ **b.** 4.612 million or 4,612 thousand **c.** 6.521 million or 6,521 thousand

Texas

7.  *Population Growth* In 1990 the population of Atlanta, Georgia, was 2.960 million. From 1990 to 2000, the population increased at a rate of about 3.35% per year.

   a. Write an exponential function to model this population growth.
   b. Use the model to approximate the population of Atlanta in 1998. Round to the nearest thousand.
   c. Use the model to predict what the population of Atlanta will be in 2007 if the growth continues at the same rate. Round to the nearest thousand.

   a. $y = 2.960(1.0335)^x$ **b.** 3.853 million or 3,853 thousand **c.** 5.183 million or 5,183 thousand

Georgia

8. *Depreciation* A car that cost $24,000 when new depreciates at the rate of 14% each year.

   a. Write an exponential function to model the depreciation.
   b. Use the model to determine the value of the car after 4 years.
   c. After how many years will the car's value be one-quarter of its original value?

   a. $y = 24,000(0.86)^x$ **b.** $13,128.20 **c.** after 9 years

9. *Trust Funds* A retired dental hygienist has a trust fund valued at $100,000. Each year the hygienist is paid 20% of the value of the trust.

   a. Write an exponential function to model the value of the trust at the end of each year.
   b. Use the model to determine the value of the trust after 5 years.
   c. After how many years will the trust's value be one-tenth of its original value? Round to the nearest year.

   a. $y = 100,000(0.80)^x$ **b.** $32,768 **c.** after 10 years

10. Complete the table.

Exponential Function	G for Growth, D for Decay	Initial Amount	Growth or Decay Factor	Rate of Growth or Decay
$y = 100(1.5)^x$	G	100	1.5	(50%)
$y = 5000(2.5)^x$	G	5000	2.5	(150%)
$y = 40(0.90)^x$	D	40	0.9	(10%)
$y = 8(0.25)^x$	D	8	0.25	(75%)
$y = 300(1.05)^x$	G	300	1.05	(5%)
$y = 40,000(1.0675)^x$	G	40,000	1.0675	(6.75%)
$y = 7000(0.75)^x$	D	7000	0.75	(25%)
$y = 20(0.5)^x$	D	20	0.5	(50%)
$y = 6(1.20)^x$	G	6	1.20	20%
$y = 100,000(1.0625)^x$	G	100,000	1.0625	6.25%
$y = 500(0.96)^x$	D	500	0.96	4%
$y = 80(0.995)^x$	D	80	0.995	0.5%

11. *Investments*   Assume you invest $1000 at 5%, compounded annually.
  a. Write an exponential function to model the growth of your investment.
  b. Compare your model to the compound interest formula. What are the values of P, $\frac{r}{n}$, and nt in your model?
  c. Use the model to determine the value of the investment after 10 years.
  d. ✎   If you invest $10,000 instead of $1000, is the value of your investment after 10 years 10 times as great? Explain.
  a. $y = 1000(1.05)^x$ b. The model can be written in the form of the compound interest formula by writing it as $A = 1000(1 + 0.05)^t$. Here $P = 1000$, $r = 0.05$, $n = 1$, $r/n = 0.05/1 = 0.05$, $nt = 1 \cdot t = t$.   c. $1628.89   d. Yes. The model equation for the $10,000 is $y = 10,000(1.05)^x$. The factor 1000 is changed to 10,000, so the exponential expression 1.05^x is multiplied by 10 times 1000.

12. *Internet Access*   Of the 150,000 people in a city that have Internet access in their homes, 125,000 of them do not have cable access. Assume that 15% of those without cable access are expected to switch to cable access service each year.

  a. Write an exponential function to model the number of customers without cable access service at the end of each year.
  b. Use the model to determine the number of customers without cable access service after 3 years.
  c. After how many years will the number of customers without cable access service first fall below 50,000? Round to the nearest year.
  a. $y = 125,000(0.85)^x$ b. 76,766 customers   c. after 6 years

13. Determine whether the data in each table represent

  (A) an exponential growth function
  (B) an exponential decay function
  (C) a linear function
  (D) a quadratic function

If the data represent an exponential growth or exponential decay function, write the exponential function. For your convenience, functions for all tables are provided.

a.

X	Y
0	1
1	3
2	9
3	27
4	81

(A); $y = 3^x$

b.

X	Y
0	1
1	.5
2	.25
3	.125
4	.0625

(B); $y = \left(\dfrac{1}{2}\right)^x$ or $y = 0.5^x$

c.

X	Y
0	5
1	3
2	1
3	−1
4	−3

(C); $y = -2x + 5$

d.

X	Y
0	−3
1	−4
2	−3
3	0
4	5

(D); $y = x^2 - 2x - 3$

e.

X	Y
0	−1
1	−.1
2	−.01
3	−.001
4	−.0001

(B); $y = -\left(\dfrac{1}{10}\right)^x$ or $y = -(0.1)^x$

f.

X	Y
0	0
1	1
2	3
3	7
4	15

(A); $y = 2^x - 1$

g.

X	Y
0	1
1	4
2	5
3	4
4	1

(D); $y = -x^2 + 4x + 1$

h.

X	Y
0	−4
1	−1
2	2
3	5
4	8

(C); $y = 3x - 4$

i.

X	Y
0	2
1	8
2	32
3	128
4	512

(A); $y = 2(4^x)$

14. *Biology* Data from a biology experiment are shown in the table below.

Number of Hours Passed, N	0	1	2	3	4	5
Population, P		150	300	600	1200	2400	4800

a. Write an exponential function to model the population growth of the bacteria.
b. What will the population be after 10 hours?
c. What will the population be after 12 hours and 30 minutes?

a. $y = 150(2)^x$ b. 153,600   c. 868,893

■ Exponential Regression

15. 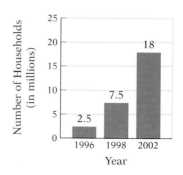 *The Internet* The bar graph shows the actual and projected number of households doing online banking (*Sources: Working Woman*, Jupiter Communications).

a. Use regression to determine an exponential model for these data. Use $x = 6$ for the year 1996, $x = 8$ for 1998, and $x = 12$ for 2002.

b. Use the function to predict the number of households that will be doing online banking in 2005. Round to the nearest million.

c. Use the function to predict during which year the number of households doing online banking will reach 70 million.

d. ✎ Provide an explanation for the increase in online banking.

a. To the nearest ten-thousandth: $y = 0.4609(1.3679)^x$ **b.** 51 million households **c.** 2006 **d.** Answers will vary. For example, the increase in the number of households with computers and the convenience of not having to go to the bank.

16. *The Military* Actual and projected numbers of U.S. strategic nuclear warheads are shown in the bar graph (*Sources:* Arms Control Association, Natural Resources Defense Council).

a. Use regression to determine an exponential model for these data. Use $x = 10$ for the year 1990.

b. What is the correlation coefficient for your model? What does that imply about the fit of the data to the graph of the equation?

c. Use the equation to predict the number of strategic nuclear warheads in the United States in 2005. Round to the nearest hundred.

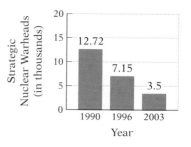

a. To the nearest ten-thousandth: $y = 34.5898(0.9054)^x$ **b.** $r \approx -0.9998$. This is very close to -1, which means that the data points lie very close to the graph of the regression equation. **c.** 2900 warheads

17. *Sociology* The number of interracial marriages in the United States has been increasing. The bar graph shows the number of interracial marriages, in thousands, in 1980, 1990, and 2000 (*Source:* U.S. Census Bureau).

a. Use regression to determine an exponential model for these data. Use $x = 80$ for the year 1980, $x = 90$ for 1990, and $x = 100$ for 2000.

b. What is the correlation coefficient for your model? What does that imply about the fit of the data to the graph of the equation?

c. Use the model to determine the number of interracial marriages in 1995. Round to the nearest thousand.

d. Use the equation to predict during which year the number of interracial marriages was 800,000.

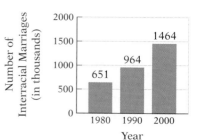

a. To the nearest ten-thousandth: $y = 25.3462(1.0414)^x$ **b.** $r \approx 0.9998$. This is very close to 1, which means that the data points lie very close to the graph of the regression equation. **c.** 1,190,000 interracial marriages **d.** 1985

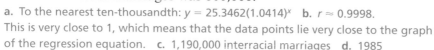

18. *Business* Shown below are estimates given by Paul Kagan Associates for revenue earned, in millions of dollars, by web radio.

Year	1999	2000	2001	2002	2003	2004
Revenue (in millions of dollars)	76	190	384	678	1095	2004

a. Use a graphing calculator to draw a scatter diagram of these data. Describe the curve that would go through the points. Use $x = 9$ for the year 1999.

b. Use regression to determine an exponential model for these data.

c. What is the correlation coefficient for your model? What does that imply about the fit of the data to the graph of the equation?

d. Graph both the scatter plot and the regression line. Describe the fit of the line to the scatter plot.

e. Use the model to predict revenue earned by web radio stations in 2005. Round to the nearest million dollars.

f. Use the model to predict the year in which Internet radio stations will earn revenue of $1095 million. How does this compare to the given data?

a. Answers will vary. For example, the curve is an exponential growth function.
b. To the nearest ten-thousandth: $y = 0.3070(1.8848)^x$ **c.** $r \approx 0.994$. This is very close to 1, which means that the data points lie very close to the graph of the regression equation.   **d.** The data points lie either on or very, very close to the graph of the regression equation.   **e.** $4,132 million   **f.** 2003. It is the same year.

19. *The Federal Government* The graph shows antiterrorism spending by the FBI (*Source:* FBI).

a. Use regression to determine an exponential model for these data. Use $x = 3$ for the year 1993.

b. What is the correlation coefficient for your model? What does that imply about the fit of the data to the graph of the equation?

c. Use the model to predict the antiterrorism spending by the FBI in 1990. Round to the nearest million.

d. The actual antiterrorism spending by the FBI in 2001 was $359 million. Find the difference between the model's prediction for spending in 2001 and the actual spending in 2001. Round to the nearest million.

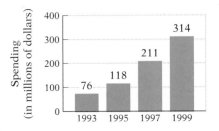

a. To the nearest ten-thousandth: $y = 36.5807(1.2736)^x$ **b.** $r \approx 0.9976$. This is very close to 1, which means that the data points lie very close to the graph of the regression equation.   **c.** $37 million   **d.** $164

20. *The Internet* Computer technology is growing rapidly. The table below shows the increase in Internet access devices over the last couple of decades (*Source:* The President's Commission on Critical Infrastructure Protection).

Year	1982	1996	2002
Internet Access Devices (in millions)	0	32	300

a. Use regression to determine an exponential model for the data. Use $x = 2$ for the year 1982.

Note: you will get an error message if you enter the *y* value of 0 for 1982. Instead, add 1 to each *y* value. Perform the regression. Then subtract 1 from the regression function.

b. Use the equation to estimate the number of Internet access devices in 1990. Round to the nearest million.

c. Use the equation to predict during which year the number of Internet access devices was 200 million.

d. ✎ Provide an explanation for the increase in Internet access devices.

a. To the nearest ten-thousandth: $y = 0.5206(1.3219)^x - 1$ **b.** 7 million Internet access devices **c.** 2001 **d.** Answers will vary. For example, the increase in the number of elementary schools, high schools, and colleges that now provide Internet access for their students.

Applying Concepts

21. ✎ *Sports*   There are 64 teams in the first round of a state high school basketball tournament. In each round, every team in the round plays a game against one other team. Only the winning teams advance to the next round.

a. Why can the number of teams in each round be modeled by an exponential decay equation?

b. Write an exponential equation that can be used to find the number of teams in any round. Describe what each variable represents.

a. After each round, the number of teams is halved. **b.** $t(r) = 64\left(\frac{1}{2}\right)^r$, where *t* is the number of teams in the round and *r* is the number of rounds that have been played

22. ✎ Three exponential functions are graphed in the rectangular coordinate system at the right.

a. Match each letter with the function $y = 2^x$, $y = 4(2^x)$, or $y = \frac{1}{4}(2^x)$. Explain why you made the choices you did.

b. Explain the effect of the value of *a*, *a* > 0, on the graph of the function $y = a \cdot 2^x$.

a. Graph a is $y = 4(2^x)$, graph b is $y = 2^x$, and graph c is $y = \frac{1}{4}(2^x)$. Explanations will vary. For example, the *y*-intercept of $y = \frac{1}{4}(2^x)$ is (0, 0.25), the *y*-intercept of $y = 2^x$ is (0, 1), and the *y*-intercept of $y = 4(2^x)$ is (0, 4). **b.** Answers will vary. For example, greater values of *a* correspond to greater values of *y* for the same *x* values. If *a* > 1, the graph of $y = a \cdot 2^x$ lies above the graph of $y = 2^x$. If 0 < *a* < 1, the graph of $y = a \cdot 2^x$ lies between the graph of $y = 2^x$ and the *x*-axis.

23. *Biology*   A biologist places 10 single-celled bacteria in a culture. Each hour the number of bacteria increase by 100%.

a. Create a table showing the number of bacteria at the beginning of each hour for the first 4 hours.

b. What is the growth factor when the percent increase is 100%?

c. Write an exponential function to model the population growth of the bacteria.

d. What is the growth factor when the percent increase is 200%?

a.

Hour	Number of Bacteria
0	10
1	20
2	40
3	80
4	160

b. 2   **c.** $f(x) = 10 \cdot 2^x$ **d.** 3

24. *Biology*   Suppose a culture of bacteria starts with 30 bacteria, and the population doubles each hour.

   a. Write an equation to model the exponential growth.
   b. What is the rate of growth between $x = 1$ and $x = 2$?
   c. What is the rate of growth between $x = 2$ and $x = 3$?
   d. What is the rate of growth between $x = 3$ and $x = 4$?
   e. Express the rate of growth between $x = n$ and $x = n + 1$.

   a. $y = 30(2^x)$ **b.** 60 bacteria per hour   **c.** 120 bacteria per hour   **d.** 240 bacteria per hour   **e.** $(60)2^{n-1}$

EXPLORATION

1.   *Modeling Exponential Growth and Exponential Decay*   You will need the following materials for each group completing this exploration:

a paper plate
two cups, one labeled A and one labeled B
a small bag, such as a plastic sandwich bag
approximately 150 M&Ms (or other candy pieces with one side marked and the other unmarked)

Model 1

   a. Count and record the number of M&Ms in the bag. This is the "initial population" and is Trial 0. Record this in the table below.
   b. Put the M&Ms into a cup. Then shake them onto a paper plate in a single layer. Remove any pieces with the M facing upward. Put these in Cup A. Count the remaining M&Ms. Record this as Trial 1.
   c. Put the remaining M&Ms into Cup B. Then shake them onto a paper plate. Remove any pieces with the M facing upward. Put these in Cup A. Count the remaining M&Ms. Record this as the next trial.
   d. Repeat Step c until you have one or two M&Ms remaining. (True exponential functions never go to 0.)
   e. Make a scatter diagram. (Place the trial number in L1 and the number of M&Ms remaining in L2.)
   f. Write an exponential function that approximates the data. It will be of the form $y = a \cdot b^x$. Define the variables x and y. Explain the values of a and b.
   g. Describe the function as either exponential growth or exponential decay. What is the growth or decay rate?
   h. What is the correlation coefficient? What does this mean?

Trial Number	0	1	2						
Number of M&Ms Remaining									

Model 2

   a. Take two M&Ms out of the bag and place them in either one of the cups. Shake the M&Ms onto a paper plate. This is the "initial population" and Trial 0. Record this in the table below.

b.   For every letter that appears facing up, add another M&M from the bag. Count the total number of M&Ms on the plate. Record this as Trial 1.

c.   Put the M&Ms that are on the plate into a cup. Then shake them onto a paper plate. For every letter that appears facing up, add another M&M from the bag. Count the total number of M&Ms on the plate. Record this as the next trial.

d.   Repeat Step c until you can no longer add the correct number of candies.

e.   Make a scatter diagram. (Place the trial number in L1 and the total number of M&Ms in L2.)

f.   Write an exponential function that approximates the data. It will be of the form $y = a \cdot b^x$. Define the variables x and y. Explain the values of a and b.

g.   Describe the function as either exponential growth or exponential decay. What is the growth or decay rate?

h.   What is the correlation coefficient? What does this mean?

Trial Number	0	1	2						
Number of M&Ms									

SECTION **10.5** **Logarithmic Functions**

- Logarithmic Functions
- Graphs of Logarithmic Functions
- Applications of Logarithmic Functions

■ Logarithmic Functions

Because the exponential function is a 1-1 function, it has an inverse function that is called a *logarithm*. A logarithm is used to answer a question similar to the following: "If $16 = 2^y$, what is the value of y?" Because $16 = 2^4$, the value of y is 4. Therefore, the logarithm, base 2, of 16 is 4. Note that a logarithm is an exponent that solves a certain equation.

INSTRUCTOR NOTE
Some students see the definition of logarithm as artificial. The following may help. Defining a logarithm as the inverse of the exponential function is similar to defining square roots as the inverse of squaring.
   "If $49 = x^2$, what is x?" The answer is the square root of 49. Now consider "If $8 = 2^x$, what is x?" The answer is the logarithm, base 2, of 8.
   Extend this analogy to explain the need for calculators.
   If $19 = x^2$, what is x?"
   If $21 = 10^x$, what is x?"

> **Definition of Logarithm**
>
> For $b > 0$, $b \neq 1$, $y = \log_b x$ is equivalent to $x = b^y$.

Read $\log_b x$ as "the logarithm of x, base b" or "log base b of x."

The table below shows equivalent statements written in both exponential and logarithmic form.

Exponential Form		Logarithmic Form
$2^4 = 16$	$\longleftrightarrow$	$\log_2 16 = 4$
$\left(\dfrac{2}{3}\right)^2 = \dfrac{4}{9}$	$\longleftrightarrow$	$\log_{\frac{2}{3}}\left(\dfrac{4}{9}\right) = 2$
$10^{-1} = 0.1$	$\longleftrightarrow$	$\log_{10}(0.1) = -1$

INSTRUCTOR NOTE
Exercises 8 and 16 can be used for similar in-class examples.

EXAMPLE 1

a. Write $4^5 = 1024$ in logarithmic form.
b. Write $\log_7 343 = 3$ in exponential form.

Solution

a. $4^5 = 1024$ is equivalent to $\log_4 1024 = 5$.
b. $\log_7 343 = 3$ is equivalent to $7^3 = 343$.

YOU TRY IT 1

a. Write $3^{-4} = \dfrac{1}{81}$ in logarithmic form.

b. Write $\log_{10} 0.0001 = -4$ in exponential form.

Solution   See page S44.   **a.** $\log_3 \dfrac{1}{81} = -4$ **b.** $10^{-4} = 0.0001$

Recalling the equations $y = \log_b x$ and $x = b^y$ from the definition of a logarithm, note that because $b^y > 0$ for all values of y, x is always a positive number. Therefore, **in the equation $y = \log_b x$, x is a positive number. The logarithm of a negative number is not a real number.**

The 1-1 property of exponential functions can be used to evaluate some logarithms.

TAKE NOTE

In the expression 2^y, there is no value of y that results in 2^y being a negative number or 0. 2^y is a positive number for all values of y.

Suggested Activity

Determine the domain of the function. Recall that the logarithm of a negative number is not defined.

1. $f(x) = \log_3 (x - 4)$
   [Answer: $\{x \mid x > 4\}$]
2. $f(x) = \log_2 (x + 2)$
   [Answer: $\{x \mid x > -2\}$]
3. $f(x) = \ln (x^2 + 4)$
   [Answer:
   $\{x \mid x \in \text{real numbers}\}$]
4. $f(x) = \log_2 x + \log_2 (x - 1)$
   [Answer: $\{x \mid x > 1\}$]

Equality of Exponents Property

For $b > 0$, $b \neq 1$, if $b^u = b^v$, then $u = v$.

INSTRUCTOR NOTE
Exercise 22 can be used for a similar in-class example.

EXAMPLE 2

Evaluate: $\log_3 \left(\dfrac{1}{9} \right)$

Solution $\log_3 \left(\dfrac{1}{9} \right) = x$ • Write an equation.

$\dfrac{1}{9} = 3^x$ • Write the equation in its equivalent exponential form.

$3^{-2} = 3^x$ • Write $\dfrac{1}{9}$ in exponential form using 3 as the base.

$-2 = x$ • Use the Equality of Exponents Property.

$\log_3 \left(\dfrac{1}{9} \right) = -2$

YOU TRY IT 2

Evaluate: $\log_4 64$

Solution   See page S44.   3

```
4^-2
                    .0625
Ans►Frac
                     1/16
```

➡ Solve $\log_4 x = -2$ for x.

$$\log_4 x = -2$$
$$4^{-2} = x \quad \bullet \text{ Write the equation in its equivalent exponential form.}$$
$$\frac{1}{16} = x \quad \bullet \text{ Solve for } x.$$

The solution is $\frac{1}{16}$. ⬅

INSTRUCTOR NOTE
Exercise 30 can be used for a similar in-class example.

EXAMPLE 3

Solve $\log_6 x = 2$ for x.

Solution $\log_6 x = 2$
$$6^2 = x \quad \bullet \text{ Write } \log_6 x = 2 \text{ in its equivalent exponential form.}$$
$$36 = x$$

The solution is 36.

YOU TRY IT 3

Solve $\log_2 x = -4$ for x.

Solution See page S44. $\frac{1}{16}$

Suggested Activity

Given $\log_b a = x$ and $1 < a < b$, between what two integers is the value of x?
[Answer: between 0 and 1]

▼ **Point of Interest**

Logarithms were developed independently by Jobst Burgi (1552–1632) and John Napier (1550–1617) as a means of simplifying the calculations of astronomers. The idea was to devise a method by which two numbers could be multiplied by performing additions. Napier is usually given credit for logarithms because he published his results first.

In Napier's original work, the logarithm of 10,000,000 was 0. After this work was published, Napier, in discussions with Henry Briggs (1561–1631), decided that tables of logarithms would be easier to use if the logarithm of 1 was 0. Napier died before new tables could be prepared, and Briggs took on the task. His table consisted of logarithms accurate to 30 decimal places, all accomplished without a calculator! The logarithms Briggs calculated are the common logarithms mentioned on this page.

In Example 3 above, 36 is called the **antilogarithm** base 6 of 2. In general, if $\log_b M = N$, then M is the antilogarithm base b of N. The antilogarithm of a number can be determined by rewriting the logarithmic equation in exponential form. For instance, if $\log_5 x = 3$, then x, which is the antilogarithm, base 5, of 3, is $5^3 = 125$.

Definition of Antilogarithm

If $\log_b M = N$, the **antilogarithm,** base b, of N is M. In exponential form, $M = b^N$.

Logarithms with base 10 are called **common logarithms.** Usually the base, 10, is omitted when the common logarithm of a number is written. Therefore, $\log_{10} x$ is written $\log x$. To find the common logarithm of most numbers, a calculator or table is necessary. Because the logarithms of most numbers are irrational numbers, the value in the display of a calculator is an approximation of the logarithm of the number.

Using a calculator,

Mantissa

$$\log 384 \approx 2.584331224$$

Characteristic

The decimal part of a *common logarithm* is called the **mantissa;** the integer part is called the **characteristic.**

When e (the base of the natural exponential function) is used as a base of a logarithm, the logarithm is referred to as the **natural logarithm** and is abbreviated ln x. This is read "el en x." Using a calculator, we find that

$$\ln 23 \approx 3.135494216$$

The integer and decimal parts of a natural logarithm do not have names associated with them as they do in common logarithms.

? QUESTION   Use a calculator to evaluate **a.** log 78 and **b.** ln 65. What are the answers to the nearest ten-thousandth?

■ Graphs of Logarithmic Functions

The graph of a logarithmic function can be drawn by using the relationship between the exponential and logarithmic functions.

To graph $f(x) = \log_2 x$, think of the function as the equation $y = \log_2 x$.

$$f(x) = \log_2 x$$
$$y = \log_2 x$$

Write the equivalent exponential equation.

$$x = 2^y$$

Because the equation is solved for x in terms of y, it is easier to choose values of y and find the corresponding values of x. Some ordered-pair solutions are recorded in the input/output table below.

x	$\dfrac{1}{4}$	$\dfrac{1}{2}$	1	2	4
y	-2	-1	0	1	2

The ordered pairs can be graphed on a rectangular coordinate system and the points connected with a smooth curve, as shown at the left.

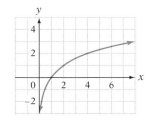

Applying the vertical- and horizontal-line tests reveals that $f(x) = \log_2 x$ is the graph of a 1-1 function.

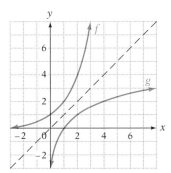

Recall that the graph of the inverse of a function f is the mirror image of f with respect to the line $y = x$. The graph of $f(x) = 2^x$ is shown on page 726. Because $g(x) = \log_2 x$ is the inverse of $f(x) = 2^x$ the graphs of these functions are mirror images of each other with respect to the line $y = x$. This is shown at the left.

INSTRUCTOR NOTE
Exercise 56 can be used for a similar in-class example.

See Appendix A:
Zero

> **EXAMPLE 4**
>
> Graph $f(x) = 2 \ln x + 3$. Find the zero of the function. Round to the nearest hundredth.
>
> **Solution**
>
>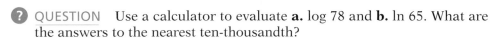
>
> • Use the features of a graphing calculator to determine the x-intercept of the graph, which is the zero of f.
>
> To the nearest hundredth, the zero of f is 0.22.

? ANSWER   **a.** 1.8921   **b.** 4.1744

INSTRUCTOR NOTE
Exercise 44 can be used for a similar in-class example.

YOU TRY IT 4

Graph $f(x) = 10 \log(x - 2)$. Find the zero of the function.

Solution   See page S44.   3

EXAMPLE 5

Solve $\ln x = -1$ for x. Round to the nearest ten-thousandth. Verify your answer using a graphing calculator.

Solution $\ln x = -1$ • $\ln x$ is the abbreviation for $\log_e x$.

$e^{-1} = x$ • Write the equation in its equivalent exponential form.

$0.3679 = x$

Graphical Check:

See Appendix A:
Intersect

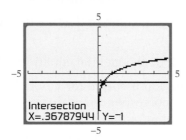

• Graph $Y_1 = \ln x$ and $Y_2 = -1$. The point of intersection is approximately $(0.3679, -1)$. The solution checks.

YOU TRY IT 5

Solve $\log x = 1.5$ for x. Round to the nearest ten-thousandth. Verify your answer using a graphing calculator.

Solution   See page S44.   31.6228

INSTRUCTOR NOTE
Exercise 54 can be used for a similar in-class example.

Suggested Activity

See Section 10.5 of the *Student Activity Manual* for an activity intended to help students understand the concept of logarithms.

EXAMPLE 6

What value in the domain of $f(x) = 2 \log_3 x$ corresponds to the range value of -4?

Solution

$f(x) = 2 \log_3 x$

$y = 2 \log_3 x$ • Substitute y for $f(x)$.

$\dfrac{y}{2} = \log_3 x$ • Solve the equation for $\log_3 x$.

$x = 3^{y/2}$ • Write the equivalent exponential equation.

$x = 3^{-4/2} = 3^{-2} = \dfrac{1}{3^2} = \dfrac{1}{9}$ • Substitute -4 for y and solve for x.

The value $\dfrac{1}{9}$ in the domain corresponds to the range value of -4.

YOU TRY IT 6

What value in the domain of $f(x) = \log_2 (x - 1)$ corresponds to the range value of -2?

Solution   See page S44. $\dfrac{5}{4}$

■ Applications of Logarithmic Functions

The first application of logarithms (and the main reason why they were developed) was to reduce computational drudgery. Today, with the widespread use of calculators and computers, the computational uses of logarithms have diminished. However, a number of other applications of logarithms have emerged.

INSTRUCTOR NOTE
The Exploration at the end of the exercise set for this section deals with the application of logarithms to the measurement of earthquakes.

The percent of light that will pass through a substance is given by the equation **log $P = -kd$,** where P is the percent of light passing through the substance, k is a constant that depends on the substance, and d is the thickness of the substance in centimeters.

➡ Find the percent of light that will pass through glass for which k is 0.4 and d is 0.5 centimeter. Round to the nearest tenth of a percent.

$$\log P = -kd$$
$$\log P = -(0.4)(0.5)$$ • Replace k and d in the equation by their given values, and solve for P.
$$\log P = -0.2$$
$$P = 10^{-0.2}$$ • Use the relationship between the logarithmic and exponential functions.
$$P \approx 0.631$$

Approximately 63.1% of the light will pass through the glass.

See Appendix A:
Intersect

Graphical Check:

Enter $Y_1 = \log(X)$ and $Y_2 = -0.2$. Use the INTERSECT feature. The point of intersection of the two graphs is approximately $(0.631, -0.2)$. The solution checks.

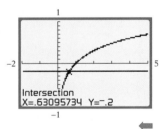

INSTRUCTOR NOTE
Exercise 62 can be used for a similar in-class example.

EXAMPLE 7

Astronomers use the **distance modulus** of a star as a method of determining how far the star is from Earth. The formula is $M = 5 \log r - 5$, where M is the distance modulus and r is the distance the star is from Earth in parsecs. (One parsec is approximately 3.3 light-years, or 1.9×10^{13} miles.) How many parsecs from Earth is a star that has a distance modulus of 4?

State the goal.   The goal is to determine the number of parsecs in a distance modulus of 4.

Devise a strategy.   Replace M in the given formula by 4. Then solve the resulting equation for r.

Solve the problem.

$$M = 5 \log r - 5$$
$$4 = 5 \log r - 5$$ • Replace M by 4.
$$9 = 5 \log r$$ • Add 5 to each side of the equation.
$$\frac{9}{5} = \log r$$ • Divide each side of the equation by 5.
$$r = 10^{9/5}$$ • Write the logarithmic equation in exponential form.
$$r \approx 63.095734$$ • Use a calculator to simplify $10^{9/5}$.

The star is approximately 63 parsecs from Earth.

Check your work.

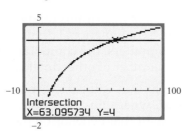

Intersection
X=63.095734 Y=4

 See Appendix A:
Trace or Intersect

- Use a graphing calculator and enter $Y_1 = 5 \log(X) - 5$. Use the TRACE feature to find the value of y when x is 63. Alternatively, graph both $Y_1 = 5 \log(X) - 5$ and $Y_2 = 4$. Use the INTERSECT feature to verify that at the point of intersection, $X \approx 63$.

YOU TRY IT 7

The **expiration time** T of a natural resource is the time remaining before it is completely consumed. A model for the expiration time of the world's oil supply is given by $T = 14.29 \ln (0.00411r + 1)$, where r is the estimated number of billions of barrels of oil remaining in the world's oil supply. According to this model, how many billion barrels of oil are needed to last 25 years?

Solution   See page S44.   1156 billion barrels of oil

10.5 EXERCISES   Suggested Assignment: 7–69, odds

Topics for Discussion

1. **a.** What does the Equality of Exponents Property state?
   b. Provide an example of a situation in which you would use this property.
   a. For $b > 0$, $b \neq 1$, if $b^u = b^v$, then $u = v$.   **b.** Examples will vary.

2. Determine whether the statement is always true, sometimes true, or never true.
   a. For $b > 0$, $b \neq 1$, $y = \log_b x$ is equivalent to $b^y = x$.
   b. The inverse of an exponential function is a logarithmic function.
   c. If x and y are positive real numbers, $x < y$, and $b > 1$, then $\log_b x < \log_b y$.
   a. Always true   **b.** Always true   **c.** Always true

3. **a.** What is a common logarithm?
   b. How is the common logarithm of $4z$ written?
   a. A common logarithm is a logarithm with base 10.   **b.** log $4z$

4. **a.** What is a natural logarithm?
   b. How is the natural logarithm of x written?
   a. A natural logarithm is a logarithm with base e.   **b.** ln x

5. What is the relationship between the graphs of $y = 3^x$ and $y = \log_3 x$?
   They are mirror images of each other with respect to the line $y = x$.

■ Logarithmic Functions

Write the exponential equation in logarithmic form.

6. $7^2 = 49$

$\log_7 49 = 2$

7. $10^3 = 1000$

$\log 1000 = 3$

8. $4^{-2} = \dfrac{1}{16}$

$\log_4 \dfrac{1}{16} = -2$

9. $3^{-3} = \dfrac{1}{27}$

$\log_3 \dfrac{1}{27} = -3$

10. $10^y = x$

$\log x = y$

11. $e^y = x$

$\ln x = y$

12. $a^x = w$

$\log_a w = x$

13. $b^y = c$

$\log_b c = y$

Write the logarithmic equation in exponential form.

14. $\log_3 9 = 2$

$3^2 = 9$

15. $\log_2 32 = 5$

$2^5 = 32$

16. $\log 0.01 = -2$

$10^{-2} = 0.01$

17. $\log_5 \dfrac{1}{5} = -1$

$5^{-1} = \dfrac{1}{5}$

18. $\ln x = y$

$e^y = x$

19. $\log x = y$

$10^y = x$

20. $\log_b u = v$

$b^v = u$

21. $\log_c x = y$

$c^y = x$

Evaluate.

22. $\log_3 81$

4

23. $\log_7 49$

2

24. $\log_2 128$

7

25. $\log_8 1$

0

26. $\log 100$

2

27. $\log 0.001$

−3

28. $\ln e^3$

3

29. $\ln e^2$

2

Solve for x.

30. $\log_3 x = 2$

9

31. $\log_5 x = 1$

5

32. $\log_4 x = 3$

64

33. $\log_2 x = 6$

64

34. $\log_7 x = -1$

$\dfrac{1}{7}$

35. $\log_8 x = -2$

$\dfrac{1}{64}$

36. $\log_6 x = 0$

1

37. $\log_4 x = 0$

1

■ Graphs of Logarithmic Functions

38. Describe two characteristics of the graph of $y = \log_b x$, where $b > 1$.

Descriptions will vary. For example: The domain is the set of positive real numbers; the range is the set of real numbers. The graph never touches the y-axis. The x-intercept is (1, 0).

39. Match each function with its graph.

a. $f(x) = \log_3(2 - x)$ **b.** $f(x) = -\log_2 x$ **c.** $f(x) = \log_2(x - 1)$ **d.** $f(x) = -\log_2(1 - x)$

i.
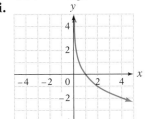

ii.
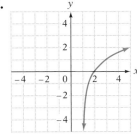

iii.
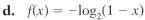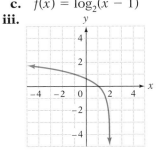

iv.
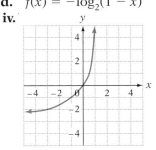

a and iii, b and i, c and ii, d and iv

40. Which of the following graphs is the graph of the function $f(x) = \log_2 (x + 1)$? Explain how you determined the answer.

a.

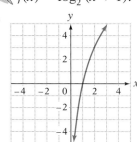

b.

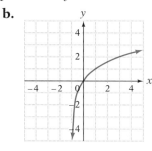

c.

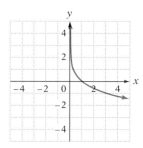

b. Explanations will vary.

Graph the functions on the same rectangular coordinate system.

41. $f(x) = 3^x$; $g(x) = \log_3 x$ **42.** $f(x) = 4^x$; $g(x) = \log_4 x$ **43.** $f(x) = \left(\dfrac{1}{2}\right)^x$; $g(x) = \log_{1/2} x$

                           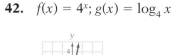

Solve for x. Round to the nearest hundredth. Verify your answer using a graphing calculator.

44. $\log x = 2.5$ 316.23   **45.** $\log x = 3.2$ 1584.89   **46.** $\log x = -1.75$ 0.02   **47.** $\log x = -2.1$ 0.01

48. $\ln x = 2$ 7.39   **49.** $\ln x = 4$ 54.60   **50.** $\ln x = -\dfrac{1}{2}$ 0.61   **51.** $\ln x = -1.7$ 0.18

52. What value in the domain of $f(x) = 3 \log_2 x$ corresponds to the range value of -6? $\frac{1}{4}$

53. What value in the domain of $f(x) = \frac{1}{2} \log_2 x$ corresponds to the range value of -1? $\frac{1}{4}$

54. What value in the domain of $f(x) = \log_3 (2 - x)$ corresponds to the range value of 1? -1

55. What value in the domain of $f(x) = -\log_2 (x - 1)$ corresponds to the range value of -2?   5

56. Graph $f(x) = 3 \ln x + 2$ and find the zero of the function. Round to the nearest hundredth.   0.51

57. Graph $f(x) = 5 \log x - 4$ and find the zero of the function. Round to the nearest hundredth.   6.31

58. Graph $f(x) = 2 \ln (x + 1) - 2$. What are the x- and y-intercepts of the graph of the function? Round to the nearest hundredth.

The x-intercept is approximately (1.72, 0). The y-intercept is (0, −2).

59. Graph $f(x) = 5 \log(x - 1)$. What are the x- and y-intercepts of the graph of the function? Round to the nearest hundredth.

The x-intercept is (2, 0). There is no y-intercept.

60. Graph $f(x) = 8 \log x + 1$. Describe where the graph is increasing and where it is decreasing.

It is increasing on (0, ∞). There is no interval on which the graph is decreasing.

61. Graph $f(x) = -3 \log x - 2$. Describe where the graph is increasing and where it is decreasing.

It is decreasing on (0, ∞). There is no interval on which the graph is increasing.

■ Applications of Logarithmic Functions

The percent of light that will pass through a material is given by the equation $\log P = -kd$, where P is the percent of light passing through the material, k is a constant that depends on the material, and d is the thickness of the material in centimeters.

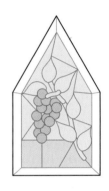

62. *Light* The constant k for a piece of glass that is 0.5 centimeter thick is 0.2. Find the percent of light that will pass through the glass. Round to the nearest percent.   79%

63. *Light* The constant k for a piece of tinted glass is 0.5. How thick is a piece of this glass that allows 60% of the light incident to the glass to pass through it? Round to the nearest hundredth.   0.44 cm

The number of decibels, D, of a sound can be given by the equation $D = 10(\log I + 16)$, where I is the power of a sound measured in watts. Round to the nearest whole number.

64. *Sound* Find the number of decibels of normal conversation. The power of the sound of normal conversation is approximately 3.2×10^{-10} watt.   65 decibels

65. *Sound* The loudest sound made by any animal is made by the blue whale and can be heard over 500 miles away. The power of the sound is 630 watts. Find the number of decibels of sound emitted by the blue whale.   188 decibels

Astronomers use the distance modulus formula $M = 5 \log r - 5$, where M is the distance modulus and r is the distance of a star from Earth in parsecs. (One parsec is approximately 1.92×10^{13} miles, or approximately 20 trillion miles.) Round to the nearest tenth.

66. *Astronomy* The distance modulus of the star Betelgeuse is 5.89. How many parsecs from Earth is this star?   150.7 parsecs

67. *Astronomy* The distance modulus of Alpha Centauri is −1.11. How many parsecs from Earth is this star?   6.0 parsecs

One model for the time it will take for the world's oil supply to be depleted is given by the equation $T = 14.29 \ln (0.00411r + 1)$, where r is the estimated world oil reserves in billions of barrels and T is the time before that amount of oil is depleted. Round to the nearest tenth.

68. *Ecology* How many barrels of oil are necessary to last 20 years?
742.9 billion barrels

69. *Ecology* How many barrels of oil are necessary to last 50 years?
7,805.5 billion barrels

Applying Concepts

70. Solve for x: $\log_2 (\log_2 x) = 3$ 256

71. Solve for x: $\ln (\ln x) = 1$ 15.1543

72. Solve the equation $T = 14.29 \ln (0.00411r + 1)$ for r. $r = \dfrac{1}{0.00411}(e^{\frac{T}{14.29}} - 1)$

When a rock is tossed into the air, the mass of the rock remains constant and a reasonable model for the height of the rock can be given by a quadratic function. However, when a rocket is launched straight up from Earth's surface, the rocket is burning fuel, so the mass of the rocket is always changing. The height of the rocket above Earth can be approximated by the equation

$$y(t) = \left(At - 16t^2 + \frac{A}{k}(M + m - kt) \right)\ln\left(1 - \frac{k}{M + m}t \right)$$

where M is the mass of the rocket, m is the mass of the fuel, A is the rate at which fuel is ejected from the engines, k is the rate at which fuel is burned, t is the time in seconds, and $y(t)$ is the height in feet after t seconds.

73. *Rocketry* During the development of the V-2 rocket program in the United States, approximate values for a V-2 rocket were $M = 8000$ pounds, $m = 16{,}000$ pounds, $A = 8000$ feet per second, and $k = 250$ pounds per second.

   a. Use a graphing calculator to estimate, to the nearest second, the time required for the rocket to reach a height of 1 mile (5280 feet).

   b. Use $v(t) = -32t + A \ln \left(\dfrac{M + m}{M + m - kt} \right)$, and the answer to part a, to determine the velocity of the rocket. Round to the nearest whole number.

   a. 14 s   b. 813 ft/s

74. Given $\log(\log x) = 3$, determine the number of digits in x.   1001

75. ✎ Without using a calculator, determine whether $\log 20 > \ln 20$ or $\ln 20 > \log 20$. Explain how you arrived at your answer.
ln 20 > log 20. Explanations will vary.

76. Evaluate $\log(\log(\log 10))$.
$\log(\log(\log 10)) = \log(\log 1) = \log 0$, which is not a real number.

EXPLORATION

1. *Earthquakes*

    The **Richter scale** measures the magnitude M of an earthquake in terms of the intensity of its shock waves. As measured on the Richter scale, the magnitude M of an earthquake that has a shock wave T times greater than the smallest shock wave that can be measured on a seismograph is given by the formula

    $$M = \log T$$

    When we refer to the size of a shock wave, we are referring to its amplitude. Look at the graph below. It is a **seismogram,** which is used to measure the magnitude of an earthquake. The magnitude is determined by the amplitude A of the wave and the difference in time t between the occurrence of two types of waves, called primary waves and secondary waves. As you can see on the graph, a primary wave is abbreviated p-wave, and a secondary wave is abbreviated s-wave. The **amplitude** A of a wave is one-half the difference between its highest and lowest points. For this graph, A is 23 millimeters. The equation is

    $$M = \log A + 3 \log 8t - 2.92$$

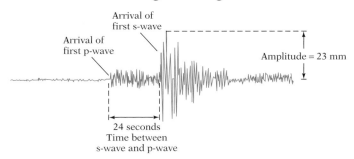

    a. Determine the magnitude of the earthquake for the seismogram shown in the figure above. Round to the nearest tenth.
    b. Find the magnitude of an earthquake that has a seismogram with an amplitude of 30 millimeters and for which t is 21 seconds.
    c. Find the magnitude of an earthquake that has a seismogram with an amplitude of 28 millimeters and for which t is 28 seconds.

    Returning to the equation $M = \log T$, let's look at the magnitude of an earthquake that produces a shock wave that is 1000 times greater than the smallest shock wave that can be measured on a seismograph.

    $$M = \log T$$
    $$M = \log 1000$$
    $$10^M = 1000$$
    $$10^M = 10^3$$
    $$M = 3$$

    The magnitude of the earthquake is 3.

d. An earthquake has magnitude 4 on the Richter scale. What can be said about the size of its shock wave?

e. Find the magnitude M of an earthquake that produces a shock wave for which $T = 100$.

f. Find the magnitude M of an earthquake that produces a shock wave for which $T = 10{,}000$.

g. How many times greater are the shock waves of an earthquake that has magnitude 5 on the Richter scale than those of an earthquake that has magnitude 4?

h. How many times greater are the shock waves of an earthquake that has magnitude 6 on the Richter scale than one that has magnitude 5?

i. An earthquake has magnitude 7 on the Richter scale. How many times greater is its shock wave than the smallest shock wave measurable on a seismogram?

Some scientists use a scale that measures the total amount of energy released by an earthquake. A formula that relates the number on the Richter scale to the energy of an earthquake is

$$r = 0.67 \log E - 7.6$$

where r is the number on the Richter scale and E is the energy in ergs.

j. What is the Richter number of an earthquake that releases 3.9×10^{15} ergs of energy?

k. What is the Richter number of an earthquake that releases 2.5×10^{20} ergs of energy?

a. 5.3   b. 5.2   c. 5.6   d. Its shock wave is 10,000 times greater than the smallest shock wave that can be measured on a seismograph.   e. 2   f. 4   g. 10 times greater   h. 10 times greater   i. 10,000,000 times greater   j. 2.8   k. 6.1

SECTION **10.6** **Properties of Logarithms**

■ Properties of Logarithms
■ The Change-of-Base Formula

■ Properties of Logarithms

Because a logarithm is an exponent, the Properties of Logarithms are similar to the Properties of Exponents.

The table at the right shows some powers of 2 and the equivalent logarithmic forms.

The table can be used to show that $\log_2 4 + \log_2 8$ equals $\log_2 32$.

$2^0 = 1$	$\log_2 1 = 0$
$2^1 = 2$	$\log_2 2 = 1$
$2^2 = 4$	$\log_2 4 = 2$
$2^3 = 8$	$\log_2 8 = 3$
$2^4 = 16$	$\log_2 16 = 4$
$2^5 = 32$	$\log_2 32 = 5$

$$\log_2 4 + \log_2 8 = 2 + 3 = 5$$
$$\log_2 32 = 5$$
$$\log_2 4 + \log_2 8 = \log_2 32$$

Note that $\log_2 32 = \log_2 (4 \times 8) = \log_2 4 + \log_2 8$.

The property of logarithms that states that the logarithm of the product of two numbers equals the sum of the logarithms of the two numbers is similar to the property of exponents that states that to multiply two exponential expressions with the same base, we add the exponents.

TAKE NOTE

Pay close attention to this property. Note, for instance, that it states that

$\log_3(4p)$
$= \log_3 4 + \log_3 p$

It also states that

$\log_5 9 + \log_5 z$
$= \log_5 (9z)$

It does *not* state any relationship that involves $\log_b (x + y)$. **This expression cannot be simplified.**

> **The Product Property of Logarithms**
>
> For any positive real numbers x, y, and b, $b \neq 1$,
>
> $$\log_b (xy) = \log_b x + \log_b y$$

The Product Property of Logarithms is used to rewrite logarithmic expressions.

The $\log_b 6z$ is written in **expanded form** as $\log_b 6 + \log_b z$.

The $\log_b 12 + \log_b r$ is written as a single logarithm as $\log_b 12r$.

The Logarithm Property of Products can be extended to include the logarithm of the product of more than two factors. For example,

$$\log_b xyz = \log_b (xy)z = \log_b xy + \log_b z = \log_b x + \log_b y + \log_b z$$

To write $\log_b 5st$ in expanded form, use the Logarithm Property of Products.

$$\log_b 5st = \log_b 5 + \log_b s + \log_b t$$

A second property of logarithms involves the logarithm of the quotient of two numbers. This property of logarithms is also based on the fact that a logarithm is an exponent and that to divide two exponential expressions with the same base, we subtract the exponents.

> **The Quotient Property of Logarithms**
>
> For any positive real numbers x, y, and b, $b \neq 1$,
>
> $$\log_b \frac{x}{y} = \log_b x - \log_b y$$

The Quotient Property of Logarithms is used to rewrite logarithmic expressions.

The $\log_b \frac{p}{8}$ is written in expanded form as $\log_b p - \log_b 8$.

The $\log_b y - \log_b v$ is written as a single logarithm as $\log_b \frac{y}{v}$.

A third property of logarithms, useful especially in the computation of a power of a number, is based on the fact that a logarithm is an exponent and that the power of an exponential expression is found by multiplying the exponents.

The table of the powers of 2 given at the beginning of this section can be used to show that $\log_2 2^3$ equals $3 \log_2 2$.

$$\log_2 2^3 = \log_2 8 = 3$$
$$3 \log_2 2 = 3 \cdot 1 = 3$$
$$\log_2 2^3 = 3 \log_2 2$$

> **The Power Property of Logarithms**
> For any positive real numbers x and b, $b \neq 1$, and for any real number r,
> $$\log_b x^r = r \log_b x$$

The Power Property of Logarithms is used to rewrite logarithmic expressions.

The $\log_b x^3$ is written in terms of $\log_b x$ as $3 \log_b x$.

$\dfrac{2}{3} \log_4 z$ is written with a coefficient of 1 as $\log_4 z^{2/3}$.

? QUESTION   Does the equation illustrate the Product Property of Logarithms, the Quotient Property of Logarithms, or the Power Property of Logarithms?

a. $\log_2 z^3 = 3 \log_2 z$ **b.** $\log_2 \dfrac{3}{z} = \log_2 3 - \log_2 z$ **c.** $\log_2 3z = \log_2 3 + \log_2 z$

The Properties of Logarithms can be used in combination to simplify expressions that contain logarithms.

INSTRUCTOR NOTE
Exercise 6 can be used for a similar in-class example.

EXAMPLE 1

Write the logarithm in expanded form.

a. $\log_b \dfrac{xy}{z}$ **b.** $\ln \dfrac{x^2}{y^3}$ **c.** $\log_8 \sqrt{x^3 y}$

Solution

INSTRUCTOR NOTE
Exercises such as these give students practice in applying the Properties of Logarithms. Facility with these operations will help students when they solve logarithmic equations later in the text.

a. $\log_b \dfrac{xy}{z}$

$= \log_b (xy) - \log_b z$ • Use the Quotient Property of Logarithms.

$= \log_b x + \log_b y - \log_b z$ • Use the Product Property of Logarithms.

b. $\ln \dfrac{x^2}{y^3} = \ln x^2 - \ln y^3$ • Use the Quotient Property of Logarithms.

$= 2 \ln x - 3 \ln y$ • Use the Power Property of Logarithms.

c. $\log_8 \sqrt{x^3 y} = \log_8 (x^3 y)^{1/2}$ • Write the radical expression as an exponential expression.

$= \dfrac{1}{2} \log_8 x^3 y$ • Use the Power Property of Logarithms.

$= \dfrac{1}{2} (\log_8 x^3 + \log_8 y)$ • Use the Product Property of Logarithms.

? ANSWERS   **a.** The Power Property of Logarithms   **b.** The Quotient Property of Logarithms
c. The Product Property of Logarithms

$$= \frac{1}{2}(3 \log_8 x + \log_8 y)$$ • Use the Power Property of Logarithms.

$$= \frac{3}{2} \log_8 x + \frac{1}{2} \log_8 y$$ • Use the Distributive Property.

YOU TRY IT 1

Write the logarithm in expanded form.

a. $\log_b \dfrac{x^2}{y}$ **b.** $\ln y^{1/3} z^3$ **c.** $\log_8 \sqrt[3]{xy^2}$

Solution   See page S45.   **a.** $2 \log_b x - \log_b y$ **b.** $\frac{1}{3} \ln y + 3 \ln z$ **c.** $\frac{1}{3} \log_8 x + \frac{2}{3} \log_8 y$

INSTRUCTOR NOTE
Exercise 24 can be used for
a similar in-class example.

EXAMPLE 2

Express as a single logarithm with a coefficient of 1.

a. $3 \log_5 x + \log_5 y - 2 \log_5 z$

b. $\dfrac{1}{3}(2 \ln x - 4 \ln y)$

Solution

a. $3 \log_5 x + \log_5 y - 2 \log_5 z$

$$= \log_5 x^3 + \log_5 y - \log_5 z^2$$ • Use the Power Property of Logarithms.

$$= \log_5 x^3 y - \log_5 z^2$$ • Use the Product Property of Logarithms.

$$= \log_5 \frac{x^3 y}{z^2}$$ • Use the Quotient Property of Logarithms.

b. $\dfrac{1}{3}(2 \ln x - 4 \ln y)$

$$= \frac{1}{3}(\ln x^2 - \ln y^4)$$ • Use the Power Property of Logarithms.

$$= \frac{1}{3}\left(\ln \frac{x^2}{y^4}\right)$$ • Use the Quotient Property of Logarithms.

$$= \ln \left(\frac{x^2}{y^4}\right)^{\frac{1}{3}} = \ln \sqrt[3]{\frac{x^2}{y^4}}$$ • Use the Power Property of Logarithms. Write the exponential expression as a radical expression.

Suggested Activity

See Section 10.6 of the *Student Activity Manual* for activities which require use of
the properties of logarithms.

YOU TRY IT 2

Express as a single logarithm with a coefficient of 1.

a. $2 \log_b x - 3 \log_b y - \log_b z$

b. $\dfrac{1}{3}(\log_4 x - 2 \log_4 y + \log_4 z)$

c. $\dfrac{1}{2}(2 \ln x - 5 \ln y)$

Solution   See page S45.   **a.** $\log_b \dfrac{x^2}{y^3 z}$ **b.** $\log_4 \sqrt[3]{\dfrac{xz}{y^2}}$ **c.** $\ln \sqrt{\dfrac{x^2}{y^5}}$

There are three other properties of logarithms that are useful in simplifying logarithmic expressions.

Suggested Activity

Have students explain why each of the following is true:

$$\ln e^x = x$$
$$\ln e = 1$$
$$e^{\ln x} = x$$

Properties of Logarithms

The Logarithmic Property of One

For any positive real number b, $b \neq 1$, $\log_b 1 = 0$.

The Inverse Property of Logarithms

For any positive real numbers x and b, $b \neq 1$, $\log_b b^x = x$.

The 1-1 Property of Logarithms

For any positive real numbers x, y, and b, $b \neq 1$, if $\log_b x = \log_b y$, then $x = y$.

INSTRUCTOR NOTE
Ask students to simplify $6 \log_3 3$ and $\log_5 1$.

EXAMPLE 3

Simplify.　**a.** $8 \log_4 4$　**b.** $\log_8 1$

Solution

a. $8 \log_4 4 = \log_4 4^8$　• Use the Power Property of Logarithms.

$\qquad\qquad\quad = 8$　• Use the Inverse Property of Logarithms.

b. $\log_8 1 = 0$　• Use the Logarithmic Property of One.

YOU TRY IT 3

Simplify.　**a.** $\log_{16} 1$　**b.** $12 \log_3 3$

Solution　See page S45.　**a.** 0　**b.** 12

The Change-of-Base Formula

Although only common logarithms and natural logarithms are programmed into a calculator, the logarithms for other positive bases can be found.

➡ Evaluate $\log_5 22$. Round to the nearest ten-thousandth.

INSTRUCTOR NOTE
Have students perform the same steps shown at the right to rewrite $\log_3 11$ as $\frac{\log 11}{\log 3}$. Doing so will help them better understand the Change-of-Base Formula.

$\log_5 22 = x$　• Write an equation.

$\qquad 5^x = 22$　• Write the equation in its equivalent exponent form.

$\log 5^x = \log 22$　• Apply the common logarithm to each side of the equation.

$x \log 5 = \log 22$　• Use the Power Property of Logarithms.

$x = \dfrac{\log 22}{\log 5}$　• Divide each side by log 5. This is an exact answer.

$x \approx 1.9206$　• This is an approximate answer.

$\log_5 22 \approx 1.9206$ ⬅

TAKE NOTE

Look closely at the example at the right. Note that $\log_5 22 = \dfrac{\log 22}{\log 5}$.

In the third step in the example above, the natural logarithm, instead of the common logarithm, could have been applied to each side of the equation. The same result would have been obtained.

Using a procedure similar to the one used above to evaluate $\log_5 22$, a formula for changing bases can be derived.

TAKE NOTE

Because graphing calculators have only preprogrammed common and natural logarithms, the Change-of-Base Formula is used to graph logarithms with other bases. This is illustrated after Example 5.

Change-of-Base Formula

$$\log_a N = \frac{\log_b N}{\log_b a}$$

INSTRUCTOR NOTE
Exercise 48 can be used for a similar in-class example.

EXAMPLE 4

Evaluate $\log_7 32$ using both common logarithms and natural logarithms. Round to the nearest ten-thousandth.

Solution $\log_7 32 = \dfrac{\log 32}{\log 7} \approx 1.7810$ • Use the Change-of-Base Formula.
  $N = 32, a = 7, b = 10$

$\log_7 32 = \dfrac{\ln 32}{\ln 7} \approx 1.7810$ • Use the Change-of-Base Formula.
  $N = 32, a = 7, b = e$

Note in Example 4 that whether common logarithms or natural logarithms are used, the result is the same.

YOU TRY IT 4

Evaluate $\log_4 2.4$ using both common logarithms and natural logarithms. Round to the nearest ten-thousandth.

Solution   See page S45.   0.6315

INSTRUCTOR NOTE
Exercise 60 can be used for a similar in-class example.

EXAMPLE 5

Rewrite $f(x) = -3 \log_7 (2x - 5)$ in terms of natural logarithms.

Solution $f(x) = -3 \log_7 (2x - 5)$

$= -3 \dfrac{\ln (2x - 5)}{\ln 7}$ • Use the Change-of-Base Formula to
  rewrite $\log_7 (2x - 5)$ as $\frac{\ln (2x - 5)}{\ln 7}$.

$= -\dfrac{3 \ln (2x - 5)}{\ln 7}$

YOU TRY IT 5

Rewrite $f(x) = 4 \log_8 (3x + 4)$ in terms of natural logarithms.

Solution   See page S45. $\dfrac{4 \ln (3x + 4)}{\ln 8}$

In Example 5, it is important to understand that

$$-\frac{3 \ln (2x - 5)}{\ln 7} \quad \text{and} \quad -3 \log_7 (2x - 5)$$

are *exactly* equal. If common logarithms had been used, the result would have been $f(x) = -\frac{3 \log (2x - 5)}{\log 7}$. The expressions

$$-\frac{3 \log (2x - 5)}{\log 7} \quad \text{and} \quad -3 \log_7 (2x - 5)$$

are also *exactly* equal.

If you are working in a base other than base 10 or base e, the Change-of-Base Formula enables you to calculate the value of the logarithm in that base just as though that base were programmed into the calculator.

Also, the graph of logarithmic functions to other than base e or base 10 can be drawn with a graphing calculator by first using the Change-of-Base Formula to rewrite the logarithmic function in terms of base e or base 10.

➡ Use a graphing calculator to graph $f(x) = \log_3 x$.

$$\log_3 x = \frac{\ln x}{\ln 3}$$

- Use the Change-of-Base Formula to rewrite $\log_3 x$ in terms of $\log x$ or $\ln x$. The natural logarithm is used here.

- To graph $f(x) = \log_3 x$ using a graphing calculator, use the equivalent form $f(x) = \frac{\ln x}{\ln 3}$.

The examples that follow are graphed by rewriting the logarithmic function in terms of the natural logarithmic function. The common logarithmic function could also have been used.

TAKE NOTE

The graph of $f(x) = \log_3 x$ in this example can be drawn by rewriting $\log_3 x$ in terms of $\log x$ as $\frac{\log x}{\log 3}$ or in terms of $\ln x$ as $\frac{\ln x}{\ln 3}$. The graph of $f(x) = \log_3 x$ is identical to the graphs of $f(x) = \frac{\log x}{\log 3}$ and $f(x) = \frac{\ln x}{\ln 3}$.

INSTRUCTOR NOTE
Exercise 64 can be used for a similar in-class example.

EXAMPLE 6

Use a graphing calculator to graph $f(x) = -3 \log_2 x$.

Solution $f(x) = -3 \log_2 x$

$$= -3 \frac{\ln x}{\ln 2} = -\frac{3 \ln x}{\ln 2}$$

- Rewrite $\log_2 x$ in terms of $\ln x$. $\log_2 x = \frac{\ln x}{\ln 2}$

- The graph of $f(x) = -3 \log_2 x$ is the same as the graph of $f(x) = -\frac{3 \ln x}{\ln 2}$.

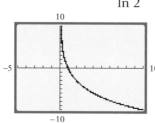

YOU TRY IT 6

Use a graphing calculator to graph $f(x) = 2 \log_4 x$.

Solution   See page S45.

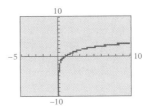

INSTRUCTOR NOTE
Exercise 68 can be used for
a similar in-class example.

EXAMPLE 7

Graph $f(x) = 3 \log_4 (x + 3)$ and estimate, to the nearest tenth, the value of x for which $f(x) = 4$.

Solution

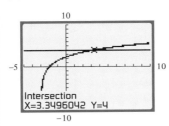

- $f(x) = 3 \log_4 (x + 3) = \dfrac{3 \ln(x + 3)}{\ln 4}$

- Graph $Y_1 = \dfrac{3 \ln(x + 3)}{\ln 4}$ and $Y_2 = 4$.

- Find the point of intersection.
  $f(x) = 4$ when $x \approx 3.3$.

The value of x for which $f(x) = 4$ is approximately 3.3.

YOU TRY IT 7

Graph $f(x) = -2 \log_5 (3x - 4)$ and estimate, to the nearest tenth, the value of x for which $f(x) = 1$.

Solution   See page S46.   1.5

An algebraic solution to Example 7 can be determined by using the relationship between the exponential function and the logarithmic function.

Suggested Activity

Have students show an algebraic solution to You Try It 7.

$$f(x) = 3 \log_4 (x + 3)$$

$$4 = 3 \log_4 (x + 3) \quad \text{• Replace } f(x) \text{ by 4.}$$

$$\frac{4}{3} = \log_4 (x + 3) \quad \text{• Solve for } \log_4 (x + 3).$$

$$4^{4/3} = x + 3 \quad \text{• Rewrite the logarithmic equation in}$$
$$\text{exponential form and then solve for } x.$$

$$x = 4^{4/3} - 3 \approx 3.3$$

The algebraic solution confirms the graphical solution.

10.6 EXERCISES   Suggested Assignment: 5–63, odds; 69–81, odds

Topics for Discussion

1. What is the Product Property of Logarithms?
   Answers may vary. For example, the log of a product is equal to the sum of the logs: $\log_b (xy) = \log_b x + \log_b y$.

2. What is the Quotient Property of Logarithms?
   Answers may vary. For example, the log of a quotient is equal to the difference of the logs: $\log_b \frac{x}{y} = \log_b x - \log_b y$.

3. Explain how to use the Change-of-Base Formula to evaluate $\log_5 12$.
   Explanations will vary. For example, rewrite $\log_5 12$ as $\frac{\log 12}{\log 5}$ and then use a calculator to evaluate this quotient.

4. Is the statement a property of logarithms?
   a. $\log_b \dfrac{x}{y} = \dfrac{\log_b x}{\log_b y}$ **b.** $\dfrac{\log x}{\log y} = \dfrac{x}{y}$ **c.** $\log(x + y) = \log x + \log y$ **d.** $\log_b \sqrt{x} = \dfrac{1}{2}\log_b x$
   a. No   **b.** No   **c.** No   **d.** Yes

▪ Properties of Logarithms

Write the logarithm in expanded form.

5. $\log_3 (x^2 y^6)$ $2\log_3 x + 6\log_3 y$
6. $\log_4 (t^4 u^2)$ $4\log_4 t + 2\log_4 u$
7. $\log_7 \left(\dfrac{u^3}{v^4}\right)$ $3\log_7 u - 4\log_7 v$

8. $\log \left(\dfrac{s^5}{t^2}\right)$ $5\log s - 2\log t$
9. $\log_2 (rs)^2$ $2\log_2 r + 2\log_2 s$
10. $\log_3 (x^2 y)^3$ $6\log_3 x + 3\log_3 y$

11. $\log_9 x^2 yz$

    $2\log_9 x + \log_9 y + \log_9 z$
12. $\log_6 xy^2 z^3$

    $\log_6 x + 2\log_6 y + 3\log_6 z$
13. $\ln \left(\dfrac{xy^2}{z^4}\right)$

    $\ln x + 2\ln y - 4\ln z$

14. $\ln \left(\dfrac{r^2 s}{t^3}\right)$ $2\ln r + \ln s - 3\ln t$
15. $\log_7 \sqrt{xy}$ $\dfrac{1}{2}\log_7 x + \dfrac{1}{2}\log_7 y$
16. $\log_8 \sqrt[3]{xz}$ $\dfrac{1}{3}\log_8 x + \dfrac{1}{3}\log_8 z$

17. $\log_2 \sqrt{\dfrac{x}{y}}$ $\dfrac{1}{2}\log_2 x - \dfrac{1}{2}\log_2 y$
18. $\log_3 \sqrt[3]{\dfrac{r}{s}}$ $\dfrac{1}{3}\log_3 r - \dfrac{1}{3}\log_3 s$
19. $\ln \sqrt{x^3 y}$ $\dfrac{3}{2}\ln x + \dfrac{1}{2}\ln y$

20. $\ln \sqrt{x^5 y^3}$ $\dfrac{5}{2}\ln x + \dfrac{3}{2}\ln y$
21. $\log_7 \sqrt{\dfrac{x^3}{y}}$ $\dfrac{3}{2}\log_7 x - \dfrac{1}{2}\log_7 y$
22. $\log_b \sqrt[3]{\dfrac{r^2}{t}}$ $\dfrac{2}{3}\log_b r - \dfrac{1}{3}\log_b t$

Write as a single logarithm with a coefficient of 1.

23. $3\log_5 x + 4\log_5 y$

    $\log_5 x^3 y^4$
24. $2\log_6 x + 5\log_6 y$

    $\log_6 x^2 y^5$
25. $2\log_3 x - \log_3 y + 2\log_3 z$

    $\log_3 \dfrac{x^2 z^2}{y}$

26. $4\log_5 r - 3\log_5 s + \log_5 t$

    $\log_5 \dfrac{r^4 t}{s^3}$
27. $\log_b x - (2\log_b y + \log_b z)$

    $\log_b \dfrac{x}{y^2 z}$
28. $2\log_2 x - (3\log_2 y + \log_2 z)$

    $\log_2 \dfrac{x^2}{y^3 z}$

29. $2(\ln x + \ln y)$ $\ln x^2 y^2$
30. $3(\ln r + \ln t)$ $\ln r^3 t^3$
31. $\dfrac{1}{2}(\log_6 x - \log_6 y)$ $\log_6 \sqrt{\dfrac{x}{y}}$

32. $\dfrac{1}{3}(\log_8 x - \log_8 y)$

$\log_8 \sqrt[3]{\dfrac{x}{y}}$

33. $2(\log_4 s - 2\log_4 t + \log_4 r)$

$\log_4 \dfrac{s^2 r^2}{t^4}$

34. $3(\log_9 x + 2\log_9 y - 2\log_9 z)$

$\log_9 \dfrac{x^3 y^6}{z^6}$

35. $3\ln t - 2(\ln r - \ln v)$

$\ln \dfrac{t^3 v^2}{r^2}$

36. $2\ln x - 3(\ln y - \ln z)$

$\ln \dfrac{x^2 z^3}{y^3}$

37. $\dfrac{1}{2}(3\log_4 x - 2\log_4 y + \log_4 z)$

$\log_4 \sqrt{\dfrac{x^3 z}{y^2}}$

Evaluate.

38. $\log_8 2 - \log_6 216 + \log_3 81 - \log_5 (625)^{1/3}$ 0

39. $\log_9 9^7 - \log_4 64 + \log_2 \dfrac{1}{2} - \log_3 1$ 3

Use the Properties of Logarithms to solve for x.

40. $\log_8 x = 3\log_8 2$ 8

41. $\log_5 x = 2\log_5 3$ 9

42. $\log_4 x = \log_4 2 + \log_4 3$ 6

43. $\log_3 x = \log_3 4 + \log_3 7$ 28

44. $\log_6 x = 3\log_6 2 - \log_6 4$ 2

45. $\log_9 x = 5\log_9 2 - \log_9 8$ 4

46. $\log x = \dfrac{1}{3}\log 27$ 3

47. $\log_2 x = \dfrac{3}{2}\log_2 4$ 8

■ **The Change-of-Base Formula**

Evaluate. Round to the nearest ten-thousandth.

48. $\log_8 6$ 0.8617

49. $\log_4 8$ 1.5000

50. $\log_5 30$ 2.1133

51. $\log_6 28$ 1.8597

52. $\log_3 (0.5)$ −0.6309

53. $\log_5 (0.6)$ −0.3174

54. $\log_7 (1.7)$ 0.2727

55. $\log_6 (3.2)$ 0.6492

Rewrite each function in terms of common logarithms.

56. $f(x) = \log_3 (3x - 2)$ $\dfrac{\log(3x - 2)}{\log 3}$

57. $f(x) = \log_5 (x^2 + 4)$ $\dfrac{\log(x^2 + 4)}{\log 5}$

58. $f(x) = 5\log_9 (6x + 7)$ $\dfrac{5\log(6x + 7)}{\log 9}$

59. $f(x) = 3\log_2 (2x^2 - x)$ $\dfrac{3\log(2x^2 - x)}{\log 2}$

Rewrite each function in terms of natural logarithms.

60. $f(x) = \log_3 (x^2 + 9)$ $\dfrac{\ln(x^2 + 9)}{\ln 3}$

61. $f(x) = \log_7 (3x + 4)$ $\dfrac{\ln(3x + 4)}{\ln 7}$

62. $f(x) = 7\log_8 (10x - 7)$ $\dfrac{7\ln(10x - 7)}{\ln 8}$

63. $f(x) = 7\log_3 (2x^2 - x)$ $\dfrac{7\ln(2x^2 - x)}{\ln 3}$

Use a graphing calculator to graph the function.

64. $f(x) = \log_2 x - 3$

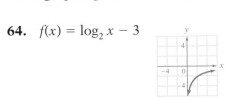

65. $(x) = -\dfrac{1}{2}\log_2 x - 1$

66. $f(x) = x + \log_3 (2 - x)$

67. $f(x) = \dfrac{x}{3} - 3 \log_2 (x + 3)$

68. Given $f(x) = 3 \log_6 (2x - 1)$, determine $f(7)$ to the nearest hundredth.
   4.29

69. Given $S(t) = 8 \log_5 (6t + 2)$, determine $S(2)$ to the nearest hundredth.
   13.12

70. Given $P(v) = -3 \log_6 (4 - 2v)$, determine $P(-4)$ to the nearest hundredth.   −4.16

71. Given $G(x) = -5 \log_7 (2x + 19)$, determine $G(-3)$ to the nearest hundredth.   −6.59

72. Graph $f(x) = \dfrac{\ln x}{x}$ and determine the maximum value of f. Round to the nearest tenth.   0.4

73. Graph $f(x) = x^2 - \ln x$ and determine the minimum value of f. Round to the nearest tenth.   0.8

74. Graph $f(x) = 2 \log_3 (x - 1)$ and estimate, to the nearest hundredth, the value of x for which $f(x) = 3$.   6.20

75. Graph $f(x) = 3 \log_2 (4x)$ and estimate, to the nearest hundredth, the value of x for which $f(x) = 2$.   0.40

76. *Astronomy* Astronomers use the *distance modulus* of a star as a method of determining the star's distance from Earth. The formula is $M = 5 \log s - 5$, where M is the distance modulus and s is the star's distance from Earth in parsecs. (One parsec $\approx 2.1 \times 10^{13}$ miles.)

   a. Graph the equation.
   b. The point whose approximate coordinates are (25.1, 2) is on the graph. Write a sentence that describes the meaning of this ordered pair.

   b. A star that is 25.1 parsecs from Earth has a distance modulus of 2.

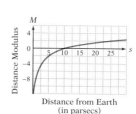

Distance from Earth (in parsecs)

77. *Employment* Without practice, the proficiency of a typist decreases over time. An equation that approximates this decrease is given by $S = 60 - 7 \ln (t + 1)$, where S is the typing speed in words per minute and t is the number of months without typing.

   a. Graph the equation.
   b. The point whose approximate coordinates are (4, 49) is on the graph. Write a sentence that describes the meaning of this ordered pair.

   b. After 4 months, the typist's proficiency has dropped to 49 words per minute.

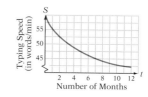

Number of Months

78. *Biology*   To discuss the variety of species that live in a certain environment, a biologist needs a precise definition of *diversity*. Let $p_1, p_2, \ldots, p_n$ be the proportions of n species that live in an environment. The biological diversity, D, of this system is

$$D = -(p_1 \log_2 p_1 + p_2 \log_2 p_2 + \ldots + p_n \log_2 p_n)$$

The larger the value of D, the greater the diversity of the system. Suppose an ecosystem has exactly five different varieties of grass: rye (R), Bermuda (B), blue (L), fescue (F), and St. Augustine (A).

	R	B	L	F	A
Table 1	$\frac{1}{5}$	$\frac{1}{5}$	$\frac{1}{5}$	$\frac{1}{5}$	$\frac{1}{5}$
Table 2	$\frac{1}{8}$	$\frac{3}{8}$	$\frac{1}{16}$	$\frac{1}{8}$	$\frac{5}{16}$
Table 3	0	$\frac{1}{4}$	0	0	$\frac{3}{4}$
Table 4	0	0	0	0	1

 a. Calculate the diversity of the ecosystem if the proportions are as given in Table 1.

 b. Because Bermuda and St. Augustine are virulent grasses, after a time the proportions are as given in Table 2. Does this system have more or less diversity than the one given in Table 1?

 c. After an even longer time period, the Bermuda and St. Augustine completely overrun the environment, and the proportions are as given in Table 3. Calculate the diversity of the system. (*Note:* For purposes of the diversity definition, $0 \log_2 0 = 0$.) Does it have more or less diversity than the system given in Table 2?

 d. Finally, the St. Augustine overruns the Bermuda, and the proportions are as in Table 4. Calculate the diversity of this system. Write a sentence that describes your answer.

 a. 2.3219281 **b.** Less **c.** 0.8112781; Less **d.** 0. This system has only one species, so there is no diversity in the system.

79. *Business*   The table below shows candle sales, in billions (*Source: The Candle Report: The Market, the Industry, the Trends, 2000*, Unity Marketing).

Year	1997	1998	1999	2000	2001
Candle Sales (in billions of dollars)	1.8	2.1	2.2	2.4	2.6

Draw a scatter diagram for the data. Would the equation that best fits the points graphed be the equation of a linear function, an exponential function, or a logarithmic function? A linear function.

80. *Economics*   General interest rate theory suggests that short-term interest rates (less than 2 years) are lower than long-term interest rates (more than 10 years) because short-term securities are less risky than long-term ones. In periods of high inflation, however, the situation is reversed and economists discuss *inverted-yield* curves. During the early 1980s, inflation was very high in the United States. The rates for short-term and long-term U.S. Treasury securities during 1980 are shown in the table at the right. An equation that models these data is $y = 14.33759 - 0.62561 \ln x$, where x is the term of the security in years and y is the interest rate as a percent.

Term (in years)	Interest Rate
0.5	15.0%
1	14.0%
5	13.5%
10	12.8%
20	12.5%

 a. Graph the equation.

 b. According to this model, what is the term, to the nearest tenth of a year, of a security that has a yield of 13%?

 c. Determine the interest rate, to the nearest tenth of a percent, that this model predicts for a security that has a 30-year maturity.

 b. 8.5 years **c.** 12.2%

a.
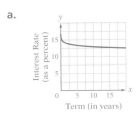

Applying Concepts

Determine the domain of the function. Recall that the logarithm of a negative number is not defined.

81. $f(x) = \log_3 (x - 4)$
$\{x \mid x > 4\}$

82. $f(x) = \log_2 (x + 2)$
$\{x \mid x > -2\}$

83. $f(x) = \ln (x^2 - 4)$
$\{x \mid x < -2 \text{ or } x > 2\}$

84. $f(x) = \ln (x^2 + 4)$
$\{x \mid x \in \text{real numbers}\}$

85. $f(x) = \log_2 x + \log_2 (x - 1)$
$\{x \mid x > 1\}$

86. $f(x) = \log_4 \dfrac{x}{x + 2}$
$\{x \mid x < -2 \text{ or } x > 0\}$

Find $f^{-1}(x)$.

87. $f(x) = e^{2x} - 1$
$f^{-1}(x) = \dfrac{\ln(x + 1)}{2}$

88. $f(x) = e^{-x+2}$
$f^{-1}(x) = -\ln x + 2$

89. $f(x) = \ln (2x + 3)$
$f^{-1}(x) = \dfrac{e^x - 3}{2}$

90. $f(x) = \ln (2x) + 3$
$f^{-1}(x) = \dfrac{e^{x-3}}{2}$

91. Find all values of x such that $\log_2 x = \log_4 x$.   1

92. When expanded, 3^{1999} has d more digits than 2^{1999}. Find d. Use $\log 2 = 0.30103$ and $\log 3 = 0.47712$.   352

EXPLORATION

1. *The Properties of Logarithms*

   a. Use the Properties of Logarithms to show that $\log_a a^x = x$, $a > 0$.

   b. Use the Properties of Logarithms to show that $a^{\log_a x} = x$, $a > 0$, $x > 0$.

   c. Show that $\log_b a = \dfrac{1}{\log_a b}$.

   d. Show that $\log \left(\dfrac{x - \sqrt{x^2 - a^2}}{a^2} \right) = -\log(x + \sqrt{x^2 - a^2})$.

The complete solutions are in the Solutions Manual.

2. *Fractals*[1]

Fractals have a wide variety of applications. They have been used to create special effects for movies, such as the *Star Wars* and *Star Trek* movies, and to explain the behavior of some biological and economic systems. One aspect of fractals that has fascinated mathematicians is that they apparently have fractional dimension.

To understand the idea of fractional dimension, one must first understand the terms "scale factor" and "size." Consider a unit square (a square of length 1). By joining four of these squares, we can create another square, the length of which is 2 and the size of which is 4. (Here, size = number of square units.) Four of these larger squares can in turn be put together to make a third square of length 4 and size 16. This

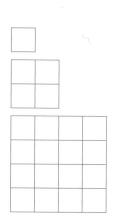

[1] Adapted with permission from "Student Math Notes," by Tami Martin, *New Bulletin*, November 1991.

process of grouping together four squares can in theory be done an infinite number of times; yet at each step, the following quantities will be the same:

$$\text{Scale factor} = \frac{\textbf{new length}}{\textbf{old length}} \qquad \text{Size ratio} = \frac{\textbf{new size}}{\textbf{old size}}$$

Consider the unit square as Step 1, the four unit squares as Step 2, etc.

a. Calculate the scale factor going from Step 1 to Step 2, Step 2 to Step 3, and Step 3 to Step 4.

b. Calculate the size ratio going from Step 1 to Step 2, Step 2 to Step 3, and Step 3 to Step 4.

c. What is the scale factor and the size ratio going from Step n to Step $n + 1$?

Mathematicians have defined dimension using the formula

$$d = \frac{\textbf{log(size ratio)}}{\textbf{log(scale factor)}}$$

For the squares discussed above, $d = \dfrac{\log(\text{size ratio})}{\log(\text{scale factor})} = \dfrac{\log 4}{\log 2} = 2.$

Thus by this definition of dimension, squares are two-dimensional figures.

Now consider a unit cube (Step 1). Group eight unit cubes to form a cube that is 2 units on each side (Step 2). Group eight of the cubes from Step 2 to form a cube that is 4 units on each side (Step 3).

d. Calculate the scale factor and the size ratio for this process.

e. Show that the cubes are three-dimensional figures.

In each of the previous examples, if the process is continued indefinitely, we still have a square or a cube. Consider a process that is more difficult to envision. Let Step 1 be an equilateral triangle whose base has length 1 unit, and let Step 2 be a grouping of three of these equilateral triangles such that the space between them is another equilateral triangle with a base of length 1 unit. Three shapes from Step 2 are arranged with an equilateral triangle in their center, and so on. It is hard to imagine the result if this is done an infinite number of times, but mathematicians have shown that the result is a single figure of fractional dimension. (Similar processes have been used to create fascinating artistic patterns and to explain scientific phenomena.)

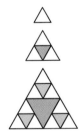

f. Show that for this process the scale factor is 2 and the size ratio is 3.

g. Calculate the dimension of the fractal. (Note that it is a *fractional* dimension!)

a. 2, 2, 2   **b.** 4, 4, 4   **c.** 2, 4   **d.** 2, 8   **e.** $d = \frac{\log 8}{\log 2} = 3$

f. scale factor $= \frac{\text{new length}}{\text{old length}} = \frac{2}{1} = 2$, size ratio $= \frac{\text{new size}}{\text{old size}} = \frac{3}{1} = 3$ **g.** $d = \frac{\log 3}{\log 2} \approx 1.58$

SECTION **10.7**

■ Solve Exponential and Logarithmic Equations

Exponential and Logarithmic Equations

■ Solve Exponential and Logarithmic Equations

In Section 10.3, we used the compound interest formula $A = P\left(1 + \dfrac{r}{n}\right)^{nt}$, where A is the value of an original investment of P dollars invested at an annual interest rate r compounded n times per year for t years, to find the value of $1000 deposited in an account earning 8% annual interest compounded quarterly for 2 years.

Suppose that, instead of being given the time period of the investment, we wanted to know how long it would take for the value of the investment to double. In this section we will develop methods to solve problems such as this one, which requires solving an exponential equation.

An **exponential equation** is one in which a variable occurs in the exponent. The examples at the right are exponential equations.

$$6^{2x+1} = 6^{3x-2}$$
$$4^x = 3$$
$$2^{x+1} = 7$$

An exponential equation in which both sides of the equation can be expressed in terms of the same base can be solved by using the Equality of Exponents Property.

> **The Equality of Exponents Property**
> If $b^u = b^v$, then $u = v$.

INSTRUCTOR NOTE
Exercise 18 can be used for a similar in-class example.

EXAMPLE 1

Solve and check: $9^{x+1} = 27^{x-1}$

Solution

$9^{x+1} = 27^{x-1}$	
$(3^2)^{x+1} = (3^3)^{x-1}$	• Rewrite each side of the equation using the same base.
$3^{2x+2} = 3^{3x-3}$	• Use the property of exponents $(a^m)^n = a^{mn}$.
$2x + 2 = 3x - 3$	• Use the Equality of Exponents Property to equate the exponents.
$2 = x - 3$	• Solve the resulting equation.
$5 = x$	

Check:

$9^{x+1} = 27^{x-1}$	
9^{5+1}	27^{5-1}
9^6	27^4

The solution is 5.

> **TAKE NOTE**
> Although it is *possible* to check this solution using a graphing calculator, it may not be practical. Note that when $x = 5$ (the solution), $y = 531,441$.

Point of Interest

Oncologists specialize in the study and treatment of tumors. In a tumor's early stages of development, doctors use an exponential growth curve to model its growth. They refer to the "doubling time" that is a characteristic of certain types of tumors. They use the growth curve of the cancer cell population in a tumor to determine the type of treatment to recommend for a malignant tumor.

> **YOU TRY IT 1**
>
> Solve and check: $4^{2x+3} = 8^{x+1}$
>
> **Solution**   See page S46. -3

When both sides of an exponential equation cannot easily be expressed in terms of the same base, logarithms are used to solve the exponential equation.

INSTRUCTOR NOTE
Exercise 6 can be used for a similar in-class example.

EXAMPLE 2

Solve for x. Round to the nearest ten-thousandth.

a. $4^x = 7$ **b.** $3^{2x} = 4$

Solution   **a.** $4^x = 7$

$\log 4^x = \log 7$ • Take the common or natural logarithm of each side of the equation. The common logarithm is used here.

$x \log 4 = \log 7$ • Use the Power Property of Logarithms.

$x = \dfrac{\log 7}{\log 4}$ • Solve for x. Divide each side by log 4.

$x \approx 1.4037$

> **TAKE NOTE**
>
> The solution 1.4037 is an approximation. Therefore, an algebraic check of this solution will be an approximation.

ALGEBRAIC CHECK:

$$4^x = 7$$
$$\frac{}{4^{1.4037}\ \Big|\ 7}$$
$$7.0002 \approx 7$$

GRAPHICAL CHECK:

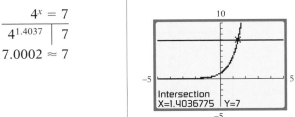

• Graph $Y_1 = 4^x$ and $Y_2 = 7$, and find the x-coordinate of the point of intersection. Alternatively, graph $Y_1 = 4^x - 7$ and find the zero of the function.

The solution is 1.4037.

b. $3^{2x} = 4$

$\log 3^{2x} = \log 4$ • Take the common logarithm of each side of the equation.

$2x \log 3 = \log 4$ • Use the Power Property of Logarithms.

$2x = \dfrac{\log 4}{\log 3}$ • Solve for x. Divide each side by log 3.

$x = \dfrac{\log 4}{2 \log 3}$ • Divide each side by 2.

$x \approx 0.6309$

> **TAKE NOTE**
>
> Alternatively, the equation at the right can be solved by using the following steps:
>
> $2x \log 3 = \log 4$
>
> $(2 \log 3)x = \log 4$
>
> $x = \dfrac{\log 4}{2 \log 3}$

The solution is 0.6309.

See Appendix A:
Intersect or Zero

ALGEBRAIC CHECK: | **GRAPHICAL CHECK:**

$$\frac{3^{2x} = 4}{3^{2(0.6309)} \,\big|\, 4}$$

$$3.9997 \approx 4$$

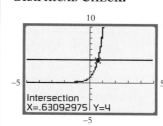

• Graph $Y_1 = 3^{2x}$ and $Y_2 = 4$, and find the x-coordinate of the point of intersection. Alternatively, graph $Y_1 = 3^{2x} - 4$ and find the zero of the function.

YOU TRY IT 2

Solve for x. Round to the nearest ten-thousandth.

a. $4^{3x} = 25$ **b.** $(1.06)^x = 1.5$

Solution   See page S46.   **a.** 0.7740   **b.** 6.9585

In Example 2a, the equation was solved algebraically and checked using a graphing calculator. Example 3 illustrates an equation that is appropriately solved using a graphing calculator and checked algebraically.

INSTRUCTOR NOTE
Exercise 50 can be used for a similar in-class example.

EXAMPLE 3

Solve $e^x = 2x + 1$ for x. Round to the nearest hundredth.

Solution

$$e^x = 2x + 1$$
$$e^x - 2x - 1 = 0$$

• Rewrite the equation by subtracting $2x + 1$ from each side.

See Appendix A:
Zero

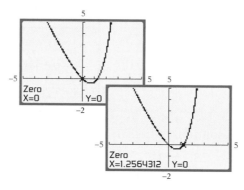

• The zeros of $f(x) = e^x - 2x - 1$ are the solutions of $e^x = 2x + 1$. Graph f and use the ZERO feature of a graphing calculator to estimate the solutions to the nearest hundredth.

The zeros are 0 and 1.26.

Algebraic Check:

$$\frac{e^x = 2x + 1}{e^0 \,\big|\, 2(0) + 1}$$
$$1 \,\big|\, 0 + 1$$
$$1 = 1$$

$$\frac{e^x = 2x + 1}{e^{1.26} \,\big|\, 2(1.26) + 1}$$
$$3.5254 \,\big|\, 2.52 + 1$$
$$3.5254 \approx 3.52$$

INSTRUCTOR NOTE
You might reinforce the concept that if a solution is an approximation, it will not check exactly by asking why the algebraic check of 1.26 is not exact.

The solutions are 0 and approximately 1.26.

YOU TRY IT 3

Solve $e^x = x$ for x. Round to the nearest hundredth.

Solution   See page S46.   No real number solution

A logarithmic equation can be solved by using the Properties of Logarithms.

➡ Solve: $\log_9 x + \log_9 (x - 8) = 1$

$$\log_9 x + \log_9 (x - 8) = 1$$ • Use the Product Property of Logarithms
$$\log_9 x(x - 8) = 1$$ to rewrite the left side of the equation.
$$9^1 = x(x - 8)$$ • Write the equation in exponential form.
$$9 = x^2 - 8x$$ • Simplify and solve for x.
$$0 = x^2 - 8x - 9$$
$$0 = (x - 9)(x + 1)$$
$$x - 9 = 0 \qquad x + 1 = 0$$
$$x = 9 \qquad x = -1$$

When x is replaced by 9 in the original equation, 9 checks as a solution. When x is replaced by -1, the original equation contains the expression $\log_9(-1)$. Because the logarithm of a negative number is not a real number, -1 does not check as a solution.

The solution of the equation is 9.   ⬅

In the solution of the equation $\log_9 x + \log_9 (x - 8) = 1$, the algebraic solution showed that although -1 and 9 may be solutions, only 9 satisfies the equation. The extraneous solution was introduced at the second step. The Product Property of logarithms [$\log_b x + \log_b y = \log_b (xy)$] applies only when both x and y are positive numbers. For $\log_9 x + \log_9 (x - 8) = 1$, this occurs when $x > 8$; therefore, a solution to this equation must be greater than 8.

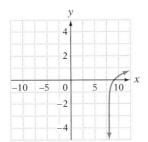

$f(x) = \log_9 x + \log_9(x - 8) - 1$

The graphs of $f(x) = \log_9 x + \log_9 (x - 8) - 1$ and $g(x) = \log_9 x(x - 8) - 1$ are shown at the left. Note that the only zero of f is 9, whereas the zeros of g are -1 and 9.

? **QUESTION** What are the possible values of x in the equation $\log_b (x - 4) = 3$?

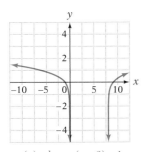

$g(x) = \log_9 x(x - 8) - 1$

Some logarithmic equations can be solved by using the 1-1 Property of Logarithms. The use of this property is illustrated in Example 4b.

INSTRUCTOR NOTE
Exercise 26 can be used for a similar in-class example.

EXAMPLE 4

Solve for x.   **a.** $\log_3 (2x - 1) = 2$ **b.** $\log_2 x - \log_2 (x - 1) = \log_2 2$

Solution   **a.** $\log_3 (2x - 1) = 2$

$$3^2 = 2x - 1$$ • Rewrite in exponential form.
$$9 = 2x - 1$$ • Solve for x.
$$10 = 2x$$
$$5 = x$$

? ANSWER   The logarithm of a negative number is not a real number, so $x - 4$ must be greater than 0. The possible values of x are $x > 4$.

See Appendix A:
Intersect or Zero

> **TAKE NOTE**
>
> Use the Change-of-Base Formula
> $$\log_a N = \frac{\log_b N}{\log_b a}$$
> to rewrite
> $\log_3 (2x - 1)$ as
> $$\frac{\log(2x - 1)}{\log 3}.$$

Graphical Check:

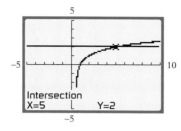

The solution is 5.

• Graph $Y_1 = \frac{\log(2x - 1)}{\log 3}$ and $Y_2 = 2$.
Find the x-coordinate of the point of intersection of the two graphs. Alternatively, graph $Y_1 = \frac{\log(2x - 1)}{\log 3} - 2$ and find the zero of the function.

b. $\log_2 x - \log_2 (x - 1) = \log_2 2$

$$\log_2 \frac{x}{x - 1} = \log_2 2$$ • Use the Quotient Property of Logarithms.

$$\frac{x}{x - 1} = 2$$ • Use the 1-1 Property of Logarithms.

$$(x - 1)\left(\frac{x}{x - 1}\right) = (x - 1)2$$ • Solve for x.

$$x = 2x - 2$$

$$-x = -2$$

$$x = 2$$

ALGEBRAIC CHECK:

$$\begin{array}{c|c} \log_2 x - \log_2 (x - 1) = \log_2 2 \\ \hline \log_2 2 - \log_2 (2 - 1) & \log_2 2 \\ \log_2 2 - \log_2 1 & \log_2 2 \\ 1 - 0 & 1 \\ 1 = 1 \end{array}$$

The solution is 2.

 See Appendix A:
Intersect

> **TAKE NOTE**
>
> By the Change-of-Base Formula,
> $$\log_2 2 = \frac{\log 2}{\log 2}.$$
> Any number divided by itself is 1,
> so $\frac{\log 2}{\log 2} = 1$.

GRAPHICAL CHECK:

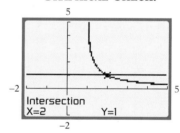

The intersection of the functions $Y_1 = \frac{\log x}{\log 2} - \frac{\log(x - 1)}{\log 2}$ and $Y_2 = \frac{\log 2}{\log 2}$ is $(2, 1)$.

YOU TRY IT 4

Solve for x.   **a.** $\log_4 (x^2 - 3x) = 1$ **b.** $\log_3 x + \log_3 (x + 3) = \log_3 4$

Solution   See page S46.   a. −1, 4   b. 1

Some logarithmic equations cannot be solved algebraically. In these cases, a graphical approach may be appropriate.

INSTRUCTOR NOTE
Exercise 58 can be used for a similar in-class example.

See Appendix A: Intersect or Zero

Suggested Activity

See Section 10.7 of the *Student Activity Manual* for one activity on graphs of exponential and logarithmic functions and another on applications of these functions.

EXAMPLE 5

Solve $\ln(2x + 4) = x^2$ for x. Round to the nearest hundredth.

Solution $\ln(2x + 4) = x^2$ • Rewrite the equation by subtracting x^2 from each side.

$\ln(2x + 4) - x^2 = 0$

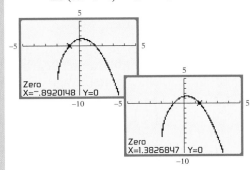

• The zeros of $f(x) = \ln(2x + 4) - x^2$ are the solutions of $\ln(2x + 4) = x^2$. Graph f and use the ZERO feature of a graphing calculator to estimate the solutions to the nearest hundredth. Alternatively, graph $Y_1 = \ln(2x + 4)$ and $Y_2 = x^2$; then find the x-coordinates of the points of intersection of the graphs.

The zeros are approximately -0.89 and 1.38.

Algebraic Check:

$\ln(2x + 4) = x^2$	
$\ln[2(-0.89) + 4]$	$(-0.89)^2$
$\ln 2.22$	0.7921
$0.7975 \approx 0.7921$	

$\ln(2x + 4) = x^2$	
$\ln[2(1.38) + 4]$	$(1.38)^2$
$\ln 6.76$	1.9044
$1.911 \approx 1.9044$	

The solutions are -0.89 and 1.38.

YOU TRY IT 5

Solve $\log(3x - 2) = -2x$ for x. Round to the nearest hundredth.

Solution   See page S46.   0.68

TAKE NOTE

Recall that in the compound interest formula

$$A = P\left(1 + \frac{r}{n}\right)^{nt},$$

A is the future value of an investment, P is the original value of the investment, r is the annual interest rate, n is the number of compounding periods per year, and t is the number of years the money is invested.

At the beginning of this section, we asked how long it would take for a $1000 investment to double in value when invested at 8% annual interest compounded quarterly. To answer this question, we will solve the compound interest formula for t. We are given that P is 1000. When $1000 doubles in value, it is worth $2000, so A is 2000. $r = 8\% = 0.08$, $n = 4$, and $\frac{r}{n} = \frac{0.08}{4} = 0.02$.

$$A = P\left(1 + \frac{r}{n}\right)^{nt}$$

Substitute the known values. $2000 = 1000(1 + 0.02)^{4t}$

Divide each side by 1000. $2 = (1.02)^{4t}$

Take the common logarithm of each side of the equation. $\log 2 = \log(1.02)^{4t}$

Use the Power Property of Logarithms. $\log 2 = 4t \log(1.02)$

Divide each side by 4 and $\log(1.02)$. $$\frac{\log 2}{4 \log(1.02)} = t$$

$$8.75 \approx t$$

The investment will double in value in approximately 9 years.

Suggested Activity

Another exponential function from finance that will be of interest to students is the one that enables them to calculate the amount of an amortized loan payment, such as a car loan.

$$P = B\left[\dfrac{\dfrac{i}{12}}{1 - \left[1 + \dfrac{i}{12}\right]^{-n}}\right]$$

B is the amount borrowed, i is the annual interest rate as a decimal, and n is the number of months to repay the loan. Ask students to use this formula to find the monthly payment on a 30-year mortgage of $100,000 with an annual interest rate of 8%. [Answer: $733.76]

▼ **Point of Interest**

Willard Libby (1908–1980), a professor at the University of California, received the Nobel Prize in chemistry in 1960 for developing the carbon-14 dating technique.

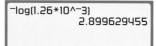

A method by which an archaeologist can measure the age of a bone is called **carbon dating.** Carbon dating is based on a radioactive isotope of carbon called carbon-14, which has a half-life of approximately 5570 years. The exponential decay equation is given by $A = A_0\left(\dfrac{1}{2}\right)^{t/5570}$, where A_0 is the original amount of carbon-14 present in the bone, t is the age of the bone, and A is the amount present after t years.

For example, consider a bone that originally contained 100 milligrams of carbon-14 and now has 70 milligrams of carbon-14. We can use the exponential decay equation to approximate the age of the bone.

$$A = A_0\left(\frac{1}{2}\right)^{t/5570}$$

$$70 = 100\left(\frac{1}{2}\right)^{t/5570}$$ • Replace A and A_0 by their given values.

$$70 = 100(0.5)^{t/5570}$$

$$0.7 = (0.5)^{t/5570}$$ • Divide each side of the equation by 100.

$$\log 0.7 = \log(0.5)^{t/5570}$$ • Take the common logarithm of each side of the equation.

$$\log 0.7 = \frac{t}{5570}\log 0.5$$ • Use the Power Property of Logarithms.

$$\frac{5570 \log 0.7}{\log 0.5} = t$$ • Solve for t by multiplying each side by 5570 and dividing each side by log 0.5.

$$2866 \approx t$$

The age of the bone is approximately 2866 years.

A chemist measures the acidity or alkalinity of a solution by using the formula **pH $= -\log(H^+)$**, where H^+ is the concentration of hydrogen ions in the solution. A neutral solution such as distilled water has a pH of 7, acids have a pH less than 7, and alkaline solutions (also called basic solutions) have a pH greater than 7.

We can use this formula to find the pH of vinegar for which $H^+ = 1.26 \times 10^{-3}$.

$$\text{pH} = -\log(H^+)$$

$$\text{pH} = -\log(1.26 \times 10^{-3})$$ • Replace H^+ with 1.26×10^{-3}.

$$\text{pH} \approx 2.8996$$

The pH of vinegar is approximately 2.9.

The **Richter scale** measures the magnitude, M, of an earthquake in terms of the intensity, I, of its shock waves. This can be expressed as the logarithmic equation $M = \log\dfrac{I}{I_0}$, where I_0 is a constant.

We can use this equation to answer the following question: How many times stronger is an earthquake that has a magnitude 4 on the Richter scale than one that has magnitude 2 on the scale?

Let I_1 represent the intensity of the earthquake that has magnitude 4, and let I_2 represent the intensity of the earthquake that has magnitude 2. The ratio of I_1 to I_2, written $\frac{I_1}{I_2}$, measures how much stronger I_1 is than I_2.

$$4 = \log \frac{I_1}{I_0}$$

$$2 = \log \frac{I_2}{I_0}$$

Use the Richter equation to write a system of equations, one equation for magnitude 4 and one for magnitude 2. Then rewrite the system using the Properties of Logarithms.

$$4 = \log I_1 - \log I_0$$
$$2 = \log I_2 - \log I_0$$

Use the addition method to eliminate $\log I_0$.

$$\begin{aligned} 4 &= \log I_1 - \log I_0 \\ -2 &= -\log I_2 + \log I_0 \\ \hline 2 &= \log I_1 - \log I_2 \end{aligned}$$

Rewrite the equations using the Properties of Logarithms.

$$2 = \log \frac{I_1}{I_2}$$

Solve for the ratio using the relationship between logarithms and exponents.

$$\frac{I_1}{I_2} = 10^2$$

$$\frac{I_1}{I_2} = 100$$

$$I_1 = 100 I_2$$

An earthquake that has magnitude 4 on the Richter scale is 100 times stronger than an earthquake that has magnitude 2.

INSTRUCTOR NOTE
Exercise 60 can be used for a similar in-class example.

EXAMPLE 6

The number of words per minute that a student can type will increase with practice and can be approximated by the equation $N = 100[1 - (0.9)^t]$, where N is the number of words typed per minute after t weeks of instruction. In how many weeks will the student be able to type 50 words per minute?

State the goal. The goal is to find out how many weeks it will be before a student is able to type 50 words per minute.

Devise a strategy. Replace N by 50 in the given formula and solve for t.

Solve the problem. $N = 100[1 - (0.9)^t]$

$50 = 100[1 - (0.9)^t]$ • Replace N by 50.

$0.5 = 1 - (0.9)^t$ • Divide each side of the equation by 100.

$-0.5 = -(0.9)^t$ • Subtract 1 from each side of the equation.

$0.5 = (0.9)^t$ • Multiply each side of the equation by -1.

$$\log 0.5 = \log(0.9)^t$$

- Take the common logarithm of each side of the equation.

$$\log 0.5 = t \log 0.9$$

- Use the Power Property of Logarithms.

$$t = \frac{\log 0.5}{\log 0.9}$$

- Divide each side of the equation by log 0.9.

$$t \approx 6.578813479$$

After approximately 7 weeks the student will type 50 words per minute.

Check your work. We can check the reasonableness of the answer: 7 weeks of instruction is reasonable for attaining that typing speed. For example, if your solution was 0.65788 weeks of instruction, you would know that was inadequate instruction to reach a speed of 50 words per minute.

GRAPHICAL CHECK:

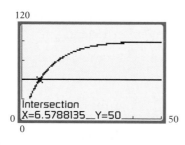

- Use a graphing calculator to graph $Y_1 = 100[1 - (0.9)^X]$ and $Y_2 = 50$. Use the INTERSECT feature to verify that at the point of intersection $X \approx 6.5788135$.

See Appendix A: Intersect

YOU TRY IT 6

In 1958 the cost of a first-class stamp was $.04. In 2003 the cost was $.37. The increase in cost can be modeled by the equation $C = 4.50e^{0.05t}$, where C is the cost, in cents, and t is the number of years after 1958. According to this model, in what year did a first-class stamp cost $.22?

Solution See page S46. 1990

10.7 EXERCISES Suggested Assignment: 7–55, every other odd; 59–81, odds

Topics for Discussion

1. What is an exponential equation? Give an example of an exponential equation.

   An exponential equation is one in which a variable occurs in the exponent. Examples will vary.

2. Explain how to solve the equation $7^{x+1} = 7^5$.

   Explanations will vary. For example, use the 1-1 property of an exponential function to equate the exponents. Solve the resulting linear equation for x.

3. Explain how to solve the equation $2 = \log_3 x$.

Explanations will vary. For example, write the equation in exponential form: $3^2 = x$.
Solve for x: $x = 9$.

4. **a.** Explain why 1 cannot be a value of x in the equation
$\log_b x + \log_b(x - 5) = 2$.

b. What are the possible values of x?

a. If $x = 1$, then $\log_b(x - 5) = \log_b(-4)$, which is not a real number.   **b.** $x > 5$

5. Determine whether the statement is always true, sometimes true, or never true, given that x, y, and b are positive real numbers and $b \neq 1$.

a. If $b^x = b^y$, then $x = y$.

b. For $x > 1$, $\log(2x - 2) - \log x = 4$ is equivalent to $\log \dfrac{2x - 2}{x} = 4$.

c. $\log(2x) + \log 4 = 6$ is equivalent to $\log(8x) = 6$.

a. Always true   **b.** Always true   **c.** Always true

■ Solve Exponential and Logarithmic Equations

Solve for x. Round to the nearest ten-thousandth.

6. $5^x = 6$ 1.1133

7. $7^x = 10$ 1.1833

8. $12^x = 6$ 0.7211

9. $10^x = 5$ 0.6990

10. $\left(\dfrac{1}{2}\right)^x = 3$ -1.5850

11. $\left(\dfrac{1}{3}\right)^x = 2$ -0.6309

12. $(1.5)^x = 2$ 1.7095

13. $(2.7)^x = 3$ 1.1061

14. $2^{-x} = 7$ -2.8074

15. $3^{-x} = 14$ -2.4022

16. $3^{2x-1} = 4$ 1.1309

17. $4^{-x+2} = 12$ 0.2075

18. $9^x = 3^{x+1}$ 1

19. $2^{x-1} = 4^x$ -1

20. $8^{x+2} = 16^x$ 6

21. $9^{3x} = 81^{x-4}$ -8

22. $5^{x^2} = 21$

1.3754, $-$ 1.3754

23. $3^{x^2} = 40$

1.8324, -1.8324

24. $3^{-x+2} = 18$

-0.6309

25. $5^{-x+1} = 15$

-0.6826

Solve for x.

26. $\log_2(2x - 3) = 3$ $\dfrac{11}{2}$

27. $\log_4(3x + 1) = 2$ 5

28. $\log_2(x^2 + 2x) = 3$ 2, -4

29. $\log_3(x^2 + 6x) = 3$ -9, 3

30. $\dfrac{3}{4} \log x = 3$ 10,000

31. $\dfrac{2}{3} \log x = 6$ 1,000,000,000

32. $\log_6\left(\dfrac{3x}{x + 1}\right) = 1$

-2

33. $\log_6\left(\dfrac{2x}{x - 1}\right) = 1$

1.5

34. $\log(x - 2) - \log x = 3$

No solution

35. $\log_7 x = \log_7 (1 - x)$

$\dfrac{1}{2}$

36. $\log_3 (x + 4) = \log_3(2 - x)$

-1

37. $\log_4 2x - \log_4(6 - x) = 0$

2

38. $\log_5(3x - 4) - \log_5(4x) = 0$

No solution

39. $\log_3 x + \log_3(x - 1) = \log_3 6$

3

40. $\log_4 x + \log_4(x - 2) = \log_4 15$

5

41. $\log_9 x + \log_9(2x - 3) = \log_9 2$ 2

42. $\log_6 x + \log_6(3x - 5) = \log_6 2$ 2

43. $\log_8(6x) = \log_8 2 + \log_8(x - 4)$ No solution

44. $\log_7 (5x) = \log_7 3 + \log_7(2x + 1)$ No solution

45. $\log_2(8x) - \log_2(x^2 - 1) = \log_2 3$ 3

46. $\log_5(3x) - \log_5(x^2 - 1) = \log_5 2$ 2

47. $\log_9(7x) = \log_9 2 + \log_9(x^2 - 2)$ 4

48. $\log_3 x = \log_3 2 + \log_3(x^2 - 3)$ 2

Solve for x by graphing. Round to the nearest hundredth.

49. $2^x = 2x + 4$ −1.86, 3.44

50. $3^x = -x - 1$ −1.25

51. $e^x = -2x - 2$ −1.16

52. $e^x = 3x + 4$ −1.24, 2.42

53. $\log x = -x + 2$ 1.76

54. $\log x = -2x$ 0.28

55. $\log(2x - 1) = -x + 3$ 2.42

56. $\log(x + 4) = -2x + 1$ 0.19

57. $\ln(x + 2) = x^2 - 3$ −1.51, 2.10

58. $\ln x = -x^2 + 1$ 1.00

59. If $3^x = 5$, find the value of 3^{2x+3}.   675

60. *Population Growth* According to population studies, the population of India can be approximated by the equation $P(t) = 0.984(1.02)^t$, where $t = 2$ corresponds to 2000 and $P(t)$ is the population, in billions, of India in t years after 1998. Use this equation to predict when the population of India will be 1.25 billion.   2010

New Delhi

61. *Physics* If air resistance is ignored, the speed v, in feet per second, of an object t seconds after it has been dropped is given by $v = 32t$. This is true regardless of the mass of the object. However, if air resistance is considered, then the speed depends on the mass (and on other things). For a certain mass, the speed t seconds after it has been dropped is given by $v = 64(1 - e^{-t/2})$. Use this equation to find the time when the speed of the object reaches 55.5 feet per second. Round to the nearest tenth. (*Hint:* Graph the equation using the domain $[-0.5, 10]$ and the range $[-0.5, 70]$.)   4.0 s

62. *Physics* A model for the distance s, in feet, an object that is experiencing air resistance will fall in t seconds is given by $s = 312.5 \ln\left(\dfrac{e^{0.32t} + e^{-0.32t}}{2}\right)$. Determine, to the nearest hundredth of a second, the time it takes the object to travel 100 feet. (*Hint:* Graph the equation using the domain $[-0.5, 5]$ and the range $[-0.5, 150]$.)   2.64 s

63. *Demography* The U.S. Census Bureau provides information on the various segments of the population in the United States. The following table gives the number of people, in millions, aged 80 and older at the beginning of each decade from 1910 to 2000.

Year	1910	1920	1930	1940	1950	1960	1970	1980	1990	2000
80-year-olds (in millions)	0.3	0.4	0.5	0.8	1.1	1.6	2.3	2.9	3.9	9.3

a. Use regression to determine an exponential model for these data. Use $x = 0$ for the year 1900.

b. What is the correlation coefficient for your model? What does that imply about the fit of the data to the graph of the equation?

   c. According to the model, what is the predicted population of this age group in the year 2020? Round to the nearest tenth of a million.

   d. In what year does the model predict that the population of this age group will be 15 million? Determine the answer algebraically. Then check the answer using a graphing calculator. Round to the nearest year.

   a. To the nearest hundred-thousandths, the equation is $y = 0.18808(1.03652)^x$.

   b. $r \approx 0.9917$. This is very close to 1, which means that the data points lie very close to the graph of the regression equation.   c. 13.9 million people   d. 2022

For Exercises 64 to 67, use the compound interest formula $A = P\left(1 + \dfrac{r}{n}\right)^{nt}$, where A is the value of an original investment of P dollars invested at an annual interest rate r compounded n times per year for t years.

64. *Investments* To save for college tuition, the parents of a preschooler invest $5000 in a bond fund that earns 6% annual interest compounded monthly. In approximately how many years will the investment be worth $15,000?   18 years

65. *Investments* A hospital administrator deposits $10,000 into an account that earns 9% annual interest compounded monthly. In approximately how many years will the investment be worth $15,000?   5 years

66. *Inflation* If the average annual rate of inflation is 5%, in how many years will prices double? Round to the nearest whole number.   14 years

67. *Investments* An investment of $1000 earns $177.23 in interest in 2 years. If the interest is compounded annually, find the annual interest rate. Round to the nearest tenth of a percent.   8.5%

For Exercises 68 to 71, use the exponential decay equation $A = A_0\left(\dfrac{1}{2}\right)^{t/k}$, where A is the amount of a radioactive material present after time t, k is the half-life, and A_0 is the original amount of radioactive substance. Round to the nearest tenth.

68. *Biology* An isotope of technetium is used to prepare images of internal body organs. This isotope has a half-life of approximately 6 hours. If a patient is injected with 30 milligrams of this isotope, how long (in hours) will it take for the technetium level to reach 20 milligrams?   3.5 h

69. *Biology* Iodine-131 is an isotope that is used to study the functioning of the thyroid gland. This isotope has a half-life of approximately 8 days. If a patient is given an injection that contains 8 micrograms of iodine-131, how long (in days) will it take for the iodine level to reach 5 micrograms?   5.4 days

70. *Physics* A sample of promethium-147 (used in some luminous paints) contains 25 milligrams. One year later, the sample contains 18.95 milligrams. What is the half-life of promethium-147, in years?   2.5 years

71. *Physics* Francium-223 is a very rare radioactive isotope discovered in 1939 by Marguerite Percy. A 3-microgram sample of francium-223 decays to 2.54 micrograms in 5 minutes. What is the half-life of francium-223, in minutes?   20.8 min

For Exercises 72 and 73, use the equation pH $= -\log(H^+)$, where H^+ is the hydrogen ion concentration of a solution. Round to the nearest hundredth.

72. *Chemistry* Find the pH of the digestive solution of the stomach, for which the hydrogen ion concentration is 0.045.   1.35

73. *Chemistry* Find the pH of a morphine solution used to relieve pain, for which the hydrogen ion concentration is 3.2×10^{-10}.   9.49

74. *Education* The percent of correct welds that a student can make will increase with practice and can be approximated by the equation $P = 100[1 - (0.75)^t]$, where P is the percent of correct welds and t is the number of weeks of practice. After how many weeks of practice will the student make 75% of the welds correctly? Round to the nearest whole number.

5 weeks

75. *Earth Science* The atmospheric pressure P decreases exponentially with height above sea level. The equation relating the pressure P, in pounds per square inch, and height h, in feet, is $P = 14.7e^{-0.00004h}$. Find the height of Mt. Everest if the atmospheric pressure at the top is 4.6 pounds per square inch. Round to the nearest foot.   29,045 ft

The intensity I of an X-ray after it has passed through a material that is x centimeters thick is given by $I = I_0e^{-kx}$, where I_0 is the initial intensity and k is a number that depends on the material. Use this equation for Exercises 76 and 77.

76. *Radiology* The constant k for aluminum is 3.2. Find the thickness of aluminum that is needed so that the intensity of the X-ray after passing through the aluminum is 25% of the original intensity. Round to the nearest tenth.   0.4 cm

77. *Radiology* Radiologists (physicians who specialize in the use of radioactive substances in diagnosis and treatment of disease) wear lead shields when giving a patient an X-ray. The constant k for lead is 43. Explain, using the given equation, why a piece of lead the same thickness as a piece of copper ($k = 3.2$) makes a better shield than the piece of copper.   Answers will vary.

For Exercises 78 and 79, use the Richter equation $M = \log \frac{I}{I_0}$, where M is the magnitude of an earthquake, I is the intensity of the shock waves, and I_0 is a constant. Round to the nearest tenth.

78. *Geology* On March 2, 1933, the largest earthquake ever recorded struck Japan. The earthquake measured 8.9 on the Richter scale. In

October 1989, an earthquake of magnitude 7.1 on the Richter scale struck San Francisco, California. How many times stronger was the earthquake in Japan than the San Francisco earthquake? Round to the nearest tenth.   63.1 times

China

Japan

79.  *Geology*   An earthquake that occurred in China in 1978 measured 8.2 on the Richter scale. In 1988, an earthquake in America measured 6.9 on the Richter scale. How many times stronger was the earthquake in China? Round to the nearest tenth.   20.0 times

80. *Mathematics*   When all eight positive integral factors of 30 are multiplied together, the product is 30^k. Find k.   4

81. *Biology*   At 9 A.M., a culture of bacteria had a population of 1.5×10^6. At noon, the population was 3.0×10^6. If the population is growing exponentially, at what time will the population be 9×10^6? Round to the nearest hour.   5 P.M.

Applying Concepts

Solve for x. Round to the nearest ten-thousandth.

82. $4^{\frac{x}{3}} = 2$ 1.5
83. $9^{\frac{2x}{3}} = 8$ 1.4196
84. $1.2^{\frac{x}{2}-1} = 1.4$ 5.6910
85. $5.6^{\frac{x}{3}+1} = 7.8$ 0.5770

Solve the system of equations.

86. $\log(x + y) = 3$
$x = y + 4$
(502, 498)

87. $\log(x + y) = 3$
$x - y = 20$
(510, 490)

88. $8^{3x} = 4^{2y}$
$x - y = 5$
(−4, −9)

89. $9^{3x} = 81^{3y}$
$x + y = 3$
(2, 1)

90. Solve $215 = e^{(x + \frac{4.723}{2})}$ for x. Round to the nearest ten-thousandth.   3.0091

91. Solve for x: $5^x = 5^{99} + 5^{99} + 5^{99} + 5^{99} + 5^{99}$ 100

92. Given $2^x = 8^{y+1}$ and $9^y = 3^{x-9}$, find the value of $x + y$.   27

93. Solve $(\ln x)^2 + 5 \ln x - 6 = 0$ for x. Round to the nearest ten-thousandth.   0.0025, 2.7183

94. Solve for x: $3^x - 3^{x-1} = 162$ 5

95. Find the greatest integral value of x for which $3^{x+2} < 3^x + 2$.   −2

96. The following "proof" appears to show that $0.04 < 0.008$. Explain the error.

$$2 < 3$$
$$2 \log 0.2 < 3 \log 0.2$$
$$\log(0.2)^2 < \log(0.2)^3$$
$$(0.2)^2 < (0.2)^3$$
$$0.04 < 0.008$$

The error is in the second step. Because log 0.2 < 0, multiplying each side of an inequality by this quantity changes the direction of the inequality.

97. Exponential equations of the form $y = Ab^{kt}$ are frequently rewritten in the form $y = Ae^{mt}$, where the base e is used rather than the base b. Rewrite $y = 10(2^{0.12t})$ in the form $y = Ae^{mt}$. $y = 10e^{0.08317766t}$

98. Rewrite $y = A2^{kt}$ in the form $y = Ae^{mt}$. (See Exercise 97.) $y = Ae^{(k \ln 2)t}$

99. *Investments*   The value of an investment in an account that earns an annual interest rate of 10% compounded daily grows according to the equation $A = A_0\left(1 + \dfrac{0.10}{365}\right)^{365t}$. Find the time for the investment to double in value. Round to the nearest year.   7 years

100. *Mortgages*   When you purchase a car or home and make monthly payments on the loan, you are amortizing the loan. Part of each monthly payment is interest on the loan, and the remaining part of the payment is a repayment of the loan amount. The amount remaining to be repaid on the loan after x months is given by $y = A(1 + i)^x + B$, where y is the amount of the loan to be repaid. In this equation, $A = \dfrac{Pi - M}{i}$ and $B = \dfrac{M}{i}$, where P is the original loan amount, i is the monthly interest rate $\left(\dfrac{\text{annual interest rate}}{12}\right)$, and M is the monthly payment. For a 30-year home mortgage of \$100,000 with an annual interest rate of 8%, $i = 0.00667$ and $M = 733.76$.

   a. How many months are required to reduce the loan amount to \$90,000? Round to the nearest month.

   b. How many months are required to reduce the loan amount to one-half the original amount? Round to the nearest month.

   c. The total amount of interest paid after x months is given by $I = Mx + A(1 + i)^2 + B - P$. Determine the month in which the total interest paid exceeds \$100,000. Round to the nearest month.

   a. 104 months   **b.** 269 months   **c.** month 136

101. *Investments*   An annuity is a fixed amount of money that is either paid or received over equal intervals of time. A retirement plan in which a certain amount is deposited each month is an example of an annuity; equal deposits are made over equal intervals of time (monthly). The equation that relates the amount of money available for retirement to the monthly deposit is $V = P\left[\dfrac{(1 + i)^x - 1}{i}\right]$, where i is the interest rate per month, x is the number of months deposits are made, P is the payment, and V is the value (called the *future value*) of the retirement fund after x payments. Suppose \$100 is deposited each month into an account that earns interest at the rate of 0.5% per month (6% per year). For how many years must the investor make deposits in order to have a retirement account worth \$20,000?   12 years

EXPLORATION

1. *Earth's Carrying Capacity*

   a. One scientific study suggested that the *carrying capacity* of Earth is around 10 billion people. What is meant by "carrying capacity"?

b. Find the current world population and project when Earth's population will reach 10 billion, assuming population growth rates of 1%, 2%, 3%, 4%, and 5%.

c. Find the current rate of world population growth and use that number to determine when the population will reach 10 billion.

a. The carrying capacity of Earth is the maximum number of humans that can be supported indefinitely on Earth.

b. Answers may vary. For example, using a world population of 6 billion and rounding to the nearest year: 1% growth rate: 51 years; 2% growth rate: 26 years; 3% growth rate: 17 years; 4% growth rate: 13 years; 5% growth rate: 10 years.

c. According to the United Nations, the current rate of world population growth is 77 million people per year. At this rate, the world population will reach 10 billion people in 52 years.

2. *Credit Reports and FICO® Scores*   When a consumer applies for a loan, the lender generally wants to know the consumer's credit history. For this, the lender turns to a credit reporting agency. These agencies maintain files on millions of borrowers. They provide the lender with a credit report, which lists information such as the types of credit the consumer uses, the length of time the consumer's credit accounts have been open, the amounts owed by the consumer, and whether the consumer has paid his or her bills on time. Along with the credit report, the lender can buy a credit score based on the information in the report. The credit score gives the lender a fast, objective measure of the consumer's credit risk, or how likely that consumer is to repay a debt. It answers the lender's question, "If I lend this person money (or give this person a credit card), what is the probability that I will be paid back in a timely manner?"

One of the most widely used credit bureau scores is the FICO® score, so named because scores are produced from software developed by Fair, Isaac and Company (FICO). FICO scores range from 300 to 850. The higher the score, the lower the predicted credit risk for lenders.

The chart below shows the delinquency rate, or credit risk, associated with ranges of FICO scores. For example, the delinquency rate of consumers in the 550–599 range of scores is 51%. This means that within the next two years, for every 100 borrowers in this range, approximately 51 will default on a loan, file for bankruptcy, or fail to pay a credit card bill within 90 days of the due date.

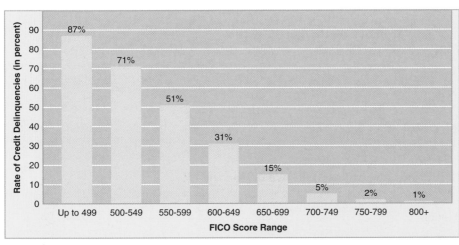

Because the delinquency rate depends on the FICO score, we can let the independent variable x represent the FICO score and the dependent variable y represent the delinquency rate. We will use the middle of each range for the x values. The resulting ordered pairs are recorded in the table at the right.

x	y
400	87
525	71
575	51
625	31
675	15
725	5
775	2
825	1

a. Use your graphing calculator to find a logarithmic regression equation that approximates these data. Use the natural logarithm.

b. Use your equation to predict the delinquency rate of a consumer with a score of 500. Round to the nearest whole number.

c. The equation pairs a delinquency rate of 36% with what score?

d. Explain why your credit rating is important.

e. 🌐 Use the Internet to find the name of at least one of the major credit reporting agencies.

a. To the nearest hundredth, the equation is $y = 918.22 - 137.49 \ln x$. b. 64%

c. 612 d. Answers will vary. e. Three major credit reporting agencies are Equifax, Experian, and Trans Union.

Source: **www.myfico.com**

CHAPTER **10** *SUMMARY*

Key Terms

antilogarithm [p. 756]
carbon dating [p. 786]
characteristic [p. 756]
common logarithm [p. 756]
composite [p. 705]
composition of two functions [p. 705]
compounding period [p. 731]
compound interest [p. 731]
distance modulus [p. 759]
exact method [p. 732]
expanded form [p. 767]
exponential decay function [p. 730]
exponential equation [p. 780]

exponential function [p. 724]
exponential growth function [p. 730]
exponential regression [p. 744]
horizontal-line test [p. 712]
inverse of a function [p. 713]
logarithm [p. 754]
mantissa [p. 756]
natural exponential function [p. 725]
natural logarithm [p. 757]
one-to-one function [p. 711]
ordinary method [p. 732]
pH [p. 786]
Richter scale [pp. 765 and 786]

Essential Concepts

Operations on Functions
For all values of x for which both $f(x)$ and $g(x)$ are defined, we define the following functions. [p. 702]

Sum $(f + g)(x) = f(x) + g(x)$

Difference $(f - g)(x) = f(x) - g(x)$

Product $(f \cdot g)(x) = f(x) \cdot g(x)$

Quotient $\left(\dfrac{f}{g}\right)(x) = \dfrac{f(x)}{g(x)}$, provided $g(x) \neq 0$

Steps for Finding the Inverse of a Function
1. Substitute y for $f(x)$.
2. Interchange x and y.
3. Solve, if possible, for y in terms of x.
4. Substitute $f^{-1}(x)$ for y. **[p. 714]**

Condition for an Inverse Function
A function f has an inverse function if and only if f is a one-to-one function. **[p. 714]**

Composition of Inverse Functions Property
$(f \circ f^{-1})(x) = f[f^{-1}(x)] = x$ for all x in the domain of f^{-1} and $(f^{-1} \circ f)(x) = f^{-1}[f(x)] = x$ for all x in the domain of f. **[p. 716]**

General Form of an Exponential Function [p. 724]
$f(x) = ab^x$ or $y = ab^x$

Definition of Logarithm
For $b > 0$, $b \neq 1$, $y = \log_b x$ is equivalent to $x = b^y$. **[p. 754]**

Equality of Exponents Property
For $b > 0$, $b \neq 1$, if $b^u = b^v$, then $u = v$. **[pp. 755 and 780]**

Properties of Logarithms [pp. 767–770]

The Product Property of Logarithms
For any positive real numbers x, y, and b, $b \neq 1$, $\log_b(xy) = \log_b x + \log_b y$.

The Quotient Property of Logarithms
For any positive real numbers x, y, and b, $b \neq 1$, $\log_b \dfrac{x}{y} = \log_b x - \log_b y$.

The Power Property of Logarithms
For any positive real numbers x and b, $b \neq 1$, and for any real number r, $\log_b x^r = r \log_b x$.

The Logarithmic Property of One
For any positive real number b, $b \neq 1$, $\log_b 1 = 0$.

The Inverse Property of Logarithms
For any positive real numbers x and b, $b \neq 1$, $\log_b b^x = x$.

The 1-1 Property of Logarithms
For any positive real numbers x, y, and b, $b \neq 1$, if $\log_b x = \log_b y$, then $x = y$.

Change-of-Base Formula [p. 771]
$\log_a N = \dfrac{\log_b N}{\log_b a}$

CHAPTER **10** *REVIEW EXERCISES*

1. Graph $f(x) = \left(\frac{1}{2}\right)^x$. What are the x- and y-intercepts of the graph of the function?
There is no x-intercept. The y-intercept is (0, 1).

2. Graph $f(x) = 3 - 2^x$ and approximate the zeros of f to the nearest tenth.
1.6

3. Given $f(x) = x^2 + 4$ and $g(x) = 4x - 1$, find $f[g(0)]$.   5

4. Given $f(x) = 3x^2 - 4$ and $g(x) = 2x + 1$, find $f[g(x)]$. $12x^2 + 12x - 1$

5. Find the inverse of the function $f(x) = \frac{1}{2}x + 8$.

$f^{-1}(x) = 2x - 16$

6. Are the functions $f(x) = -\frac{1}{4}x + \frac{5}{4}$ and $g(x) = -4x + 5$ inverses of each other?
Yes

7. Write $2^5 = 32$ in logarithmic form.

$\log_2 32 = 5$

8. Evaluate the function $f(x) = 3^{x+1}$ at $x = -2$.
$\frac{1}{3}$

9. Solve for x: $\log_5 x = 3$

125

10. Write $\log_6 \sqrt{xy^3}$ in expanded form.
$\frac{1}{2}\log_6 x + \frac{3}{2}\log_6 y$

11. What value in the domain of $f(x) = \log_2 (2x)$ corresponds to the range value of 3?

4

12. Write $\frac{1}{2}(\log_3 x - \log_3 y)$ as a single logarithm with a coefficient of 1.
$\log_3 \sqrt{\dfrac{x}{y}}$

13. Evaluate $\log_2 5$. Round to the nearest ten-thousandth.   2.3219

14. Solve for x: $3^{x^2} = 9^{2x+6}$ $-2, 6$

15. Let $f(x) = 2x + 5$ and $g(x) = x^2 - 1$. Find $(f - g)(3)$.   3

16. Graph $f(x) = \dfrac{\log x}{x}$ and approximate the zeros of f. Round to the nearest tenth.   1.0

17. Solve $3^{x+2} = 5$ for x. Round to the nearest thousandth. -0.5350

18. Solve for x: $\log_6 2x = \log_6 2 + \log_6 (3x - 4)$ 2

19. Solve $\log x = -2x + 3$ for x by graphing. Round to the nearest hundredth.   1.42

20. Given $f(-3) = 6$, find $f^{-1}(6)$. -3

21. Evaluate the function $f(x) = e^{x-2}$ at $x = 2$.   1

22. Evaluate: $\log_4 16$ 2

23. Which of the following are exponential growth functions?
a. $y = 4(2x)$ **b.** $y = 4(0.2^x)$ **c.** $y = 0.4(2^x)$ **d.** $y = 4.2^x$
c, d

24. Rewrite $f(x) = \log_2 (x + 5)$ in terms of natural logarithms.
$f(x) = \dfrac{\ln(x + 5)}{\ln 2}$

25. Find the value of $10,000 invested for 6 years at 7.5% compounded monthly. Use the compound interest formula $A = P\left(1 + \dfrac{r}{n}\right)^{nt}$, where A is the value of an original investment of P dollars invested at an annual interest rate r compounded n times per year for t years. Round to the nearest dollar. $15,661

26. In typography, a point is a unit used to measure type size. To convert points to inches we can use the equation $I = 0.0138337P$, where I is the number of inches and P is the number of points.

    a. Write an equation for the inverse function of the equation $I = 0.0138337P$. Round the coefficient to the nearest hundredth.

    b. What does the inverse function represent?

    a. $P = 72.29I$ b. It represents the equation for converting inches to points.

27. A student wants to achieve a typing speed of 50 words per minute. The length of time t, in days, that it takes to achieve this goal is given by the equation $t = -62.5 \ln (1 - 0.0125N)$, where N is the number of words typed per minute. Determine the amount of time it will take the student to achieve the goal of 50 words per minute. Round to the nearest whole number.   61 days

28. Acid rain has a pH of less than 5.6. If a rain's concentration of hydrogen ions, $[H^+]$, is 10^{-4}, is it acid rain? Use the equation $pH = -\log(H^+)$.

    Yes. (The pH is 4.)

29. The power output P, in watts, of a satellite is given by $P = 50e^{-t/250}$, where t is the time, in days, the satellite will operate. Determine how long the satellite will continue to operate if the equipment on board the satellite requires 20 watts of power. Round to the nearest whole number.

    229 days

30.  In 1990 the population of Orlando, Florida, was 1.2 million. From 1990 to 2000, the population increased at a rate of about 3.2% per year.

    a. Write an exponential function to model this population growth.

    b. Use the model to approximate the population of Orlando in 1997. Round to the nearest hundred thousand.

    c. Use the model to predict what the population of Orlando will be in 2008 if the growth continues at the same rate. Round to the nearest hundred thousand.

    a. $y = 1.2(1.032)^x$ b. 1.5 million or 1,500 thousand   c. 2.1 million or 2,100 thousand

Florida   Orlando

CHAPTER **10** *TEST*

1. Evaluate $f(x) = \left(\dfrac{2}{3}\right)^x$ at $x = 0$.   1

2. Write $\dfrac{1}{2} (\log_3 x - \log_3 y)$ as a single logarithm with a coefficient of 1. $\log_3 \sqrt{\dfrac{x}{y}}$

3. Solve for x: $\log_5 \dfrac{7x + 2}{3x} = 1$ $\dfrac{1}{4}$

4. Evaluate: $\log_2 16$ 4

5. Find the inverse of the function $f(x) = \dfrac{2}{3}x - 12$.

    $f^{-1}(x) = \dfrac{3}{2}x + 18$

6. Write $\log_6 \sqrt[3]{x^2 y^5}$ in expanded form.

    $\dfrac{2}{3} \log_6 x + \dfrac{5}{3} \log_6 y$

7. Solve for x: $8^x = 2^{x-6}$ -3

8. Graph $f(x) = e^{x+1} - 2$ and approximate the zeros of f. Round to the nearest tenth. -0.3

9. Given $f(x) = x^2 + 2x - 3$ and $g(x) = x^2 - 2$, find $(f + g)(2)$ and $(f \cdot g)(-4)$. 7; 70

10. Solve for x: $\log_3 x = -2$ $\dfrac{1}{9}$

11. What value in the domain of $f(x) = \log_4 (x + 2)$ corresponds to the range value of 2? 14

12. Given $S(t) = 8 \ln (2t - 1)$, determine $S(3)$ to the nearest hundredth. 12.88

13. Evaluate $\log_6 22$. Round to the nearest ten-thousandth. 1.7251

14. Solve for x: $\log x + \log(x - 4) = \log 12$ 6

15. Write $5^4 = 625$ in logarithmic form.
$\log_5 625 = 4$

16. Solve for x: $\log x + \log(2x + 3) = \log 2$
$\dfrac{1}{2}$

17. Solve $\log x = -3x + 2$ for x by graphing. Round to the nearest hundredth. 0.72

18. Given $f(10) = -9$, find $f^{-1}(-9)$. 10

19. Given $g(x) = 2x - 3$ and $h(x) = x^2 - 3x$, evaluate $g[h(-2)]$ and $h[g(x)]$.
17; $4x^2 - 18x + 18$

20. Find the inverse of the function
$\{(-2, 1), (2, 3), (5, -4), (7, 9)\}$.
$\{(1, -2), (3, 2), (-4, 5), (9, 7)\}$

21. Which of the following are exponential decay functions?
  a. $y = 6(3x)$ **b.** $y = 0.6(0.3^x)$ **c.** $y = 0.3(0.6^x)$ **d.** $y = 6.3^x$
  b, c

22. Rewrite $f(x) = \log_3 (2x - 1)$ in terms of natural logarithms.
$f(x) = \dfrac{\ln(2x - 1)}{\ln 3}$

23. The percent of light that will pass through a translucent material is given by the equation $\log P = -0.5d$, where P is the percent of light that passes through the material and d is the thickness of the material in centimeters. How thick is a translucent material that only 50% of light will pass through? Round to the nearest thousandth. 0.602 cm

24. Use the equation $A = A_0\left(\dfrac{1}{2}\right)^{t/k}$, where A is the amount of a radioactive material present after time t, k is the half-life, and A_0 is the original amount of radioactive material, to find the half-life of a material that decays from 40 mg to 30 mg in 10 hours. Round to the nearest whole number. 24 h

25. Shown below are estimates given by Jupiter Media Metrix of households in the United States that make bill payments online.

Year	1999	2000	2001	2002	2003
Number of U.S. Households (in millions)	2.6	5.3	7.7	11.4	17

  a. Use regression to determine an exponential model for these data. Use $x = 9$ for the year 1999.
  b. What is the correlation coefficient for your model? What does that imply about the fit of the data to the graph of the equation?
  c. Use the model to predict the number of U.S. households that will make bill payments online in 2006. Round to the nearest million.
  d. Use the model to predict the year in which 30 million U.S. households will be making bill payments online.
    a. To the nearest ten-thousandth: $y = 0.0504(1.5717)^x$ **b.** $r \approx 0.990$. This is very close to 1, which means that the data points lie very close to the graph of the regression equation.   **c.** 70 million   **d.** 2004

Cumulative Review Exercises

1. Simplify: $3(8a - 2) - (4a + 6)$
   $20a - 12$

2. Graph: $\{x \mid x \le -2\} \cup \{x \mid x > 4\}$

3. Solve: $x^2 + 3x - 5 = 0$
   $\dfrac{-3 \pm \sqrt{29}}{2}$

4. Find the range of $f(x) = 2x^2 - 4x$ when the domain is $\{-4, -2, 0, 2, 4\}$.
   $\{0, 16, 48\}$

5. Find the equation of the line that passes through the points $(-2, 5)$ and $(4, -1)$.
   $y = -x + 3$

6. Solve: $3x - 3y = 2$
   $\quad\;\;\, 6x - 4y = 5$
   $\left(\dfrac{7}{6}, \dfrac{1}{2}\right)$

7. Solve: $x + 2y + z = 3$
   $\quad\;\;\, 2x - y + 2z = 6$
   $\quad\;\;\, 3x + y - z = 5$
   $(2, 0, 1)$

8. Simplify: $125^{2/3}$
   25

9. Solve: $\sqrt{3x - 5} - 2 = 3$ 10

10. Solve: $x^4 - 8x^2 - 9 = 0$ $\pm 3, \pm i$

11. Find the maximum value of the function $f(x) = -2x^2 + 4x + 1$.
    3

12. Evaluate the function $f(x) = 2^{-x-1}$ at $x = -3$.
    4

13. Given $f(x) = 6x + 8$ and $g(x) = 4x + 2$, find $g[f(-1)]$. 10

14. Find the zeros of the function $f(x) = 2x^2 - 9x - 5$. $-\dfrac{1}{2}, 5$

15. Solve: $3x - 2(x - 4) = 8 - 5(4 - x)$
    5

16. Find the inverse of the function $f(x) = \dfrac{2}{3}x - 12$.
    $f^{-1}(x) = \dfrac{3}{2}x + 18$

17. Factor: $3x^2y + 10xy - 8y$ $y(3x - 2)(x + 4)$

18. Simplify: $(ab)^{-2}(2ab^{-3})$ $\dfrac{2}{ab^5}$

19. Write 0.0000786 in scientific notation.
    7.86×10^{-5}

20. Simplify: $\dfrac{x^2 + 2x - 8}{x^2 + x - 12}$
    $\dfrac{x - 2}{x - 3}$

21. Solve: $\dfrac{2x + 5}{x - 1} + x = 11$ $2, 8$

22. Multiply: $(5 - 7i)(4 + 2i)$ $34 - 18i$

23. A new printer can print checks three times faster than an older printer. The older printer can print the checks in 30 minutes. How long would it take to print the checks with both printers operating?   7.5 min

24. An alloy containing 25% tin is mixed with an alloy containing 50% tin. How much of each were used to make 2000 pounds of an alloy containing 40% tin?   800 lb of the 25% alloy, 1200 lb of the 50% alloy

25. A plane travels 960 miles in 3 hours. Determine a linear model that will predict the number of miles the plane can travel in a given amount of time. Use this model to predict the distance the plane will travel in 5.5 hours. $y = 320x$; 1760 mi

Additional Topics in Algebra

Need help? For online student resources,
visit this web site:
Math.college.hmco.com

803

SECTION **1**

■ Sequences
■ Arithmetic and Geometric Sequences
■ Series
■ Arithmetic and Geometric Series

INSTRUCTOR NOTE
The grains-of-wheat problem is revisited in the Exploration on page 818.

▼ *Point of Interest*

Leonardo of Pisa, commonly known as Fibonacci, was an Italian mathematician of the 13th century. Fibonacci formulated what is known as the Fibonacci sequence. The first two terms of the sequence are 1. Each successive term is found by adding the two previous terms of the sequence: 1, 1, 2, 3, 5, 8, 13, 21, 34, . . .

Suggested Activity

Write a formula for the *n*th term of the sequence.
1. the sequence of the positive integers
2. the sequence of the positive multiples of 8
3. the sequence of the natural numbers that are divisible by 5
4. the sequence of the odd integers greater than 4
5. the sequence of the negative integers less than −6
6. the sequence of the odd integers less than 0
[Answer: **1.** $a_n = n$
2. $a_n = 8n$ **3.** $a_n = 5n$
4. $a_n = 2n + 3$
5. $a_n = -n - 6$
6. $a_n = -2n + 1$]

Introduction to Sequences and Series

■ Sequences

According to legend, when Sissa Ben Dahir of India invented the game of chess, King Shirham was so impressed with the game that he summoned the game's inventor and offered him the reward of his choosing. The inventor pointed to the chess board and requested that, for his reward, he would like one grain of wheat on the first square, two grains of wheat on the second square, four grains on the third square, eight grains on the fourth square, and so on for all 64 squares on the chessboard. The king considered this a very modest reward and said he would grant the inventor's wish.

The number of grains of wheat on each of the first 6 squares is shown below.

Square	1	2	3	4	5	6
Grains of wheat	1	2	4	8	16	32

The list of numbers 1, 2, 4, 8, 16, 32 is called a sequence. As defined in Section 1.1, a **sequence** is an ordered list of numbers. The list 1, 2, 4, 8, 16, 32 is ordered because the position of a number in this list indicates the square which holds that number of grains of wheat. Each of the numbers in a sequence is called a **term** of the sequence.

? <u>QUESTION</u> What is the seventh term of the sequence listed above?*

Examples of other sequences are shown below. These sequences are separated into two groups. A **finite sequence** contains a finite number of terms. An **infinite sequence** contains an infinite number of terms.

Finite Sequences	**Infinite Sequences**
10, 9, 8, 7, 6, 5, 4, 3, 2, 1	5, 10, 15, 20, 25, . . .
$\frac{1}{5}, \frac{2}{5}, \frac{3}{5}, \frac{4}{5}, 1$	$1, \frac{1}{2}, \frac{1}{4}, \frac{1}{6}, \frac{1}{8}, \ldots$
0, 1, 0, 1, 0, 1	0, 1, 0, 1, 0, 1, . . .

? <u>QUESTION</u> Is the sequence finite or infinite?†
  a. 1, 1, 2, 3, 5, 8   **b.** 1, 1, 2, 3, 5, 8, . . .

For the sequence 3, 6, 9, 12, . . ., the first term is 3, the second term is 6, the third term is 9, and the fourth term is 12. A general sequence is shown below. The first term is a_1, the second term is a_2, the third term is a_3, and the *n*th term, also called the **general term** of the sequence, is a_n.

$$a_1, a_2, a_3, \ldots, a_n, \ldots$$

Frequently, a sequence has a definite pattern that can be expressed by a formula.

? <u>ANSWERS</u> *2(32) = 64. The seventh term of the sequence is 64.   †**a.** finite **b.** infinite

Each term of the sequence shown at the right is paired with a natural number by the formula $a_n = 4n$. The first term, a_1, is 4. The second term, a_2, is 8. The third term, a_3, is 12. The nth term, a_n, is $4n$.

$a_n = 4n$

$$a_1, \quad a_2, \quad a_3, \ldots, \qquad a_n, \ldots$$
$$4(1), \quad 4(2), \quad 4(3), \ldots, \quad 4(n), \ldots$$
$$4, \quad 8, \quad 12, \ldots, \qquad 4n, \ldots$$

INSTRUCTOR NOTE
Exercise 14 can be used for a similar in-class example.

EXAMPLE 1

Write the first three terms of the sequence whose nth term is given by the formula $a_n = 3n + 1$.

Solution $a_n = 3n + 1$

$a_1 = 3(1) + 1 = 3 + 1 = 4$ • Replace n by 1.
$a_2 = 3(2) + 1 = 6 + 1 = 7$ • Replace n by 2.
$a_3 = 3(3) + 1 = 9 + 1 = 10$ • Replace n by 3.

The first three terms of the sequence are 4, 7, 10.

YOU TRY IT 1

Write the first four terms of the sequence whose nth term is given by the formula $a_n = n(n + 2)$.

Solution   See page S47.   3, 8, 15, 24

INSTRUCTOR NOTE
Although subscripting a variable has been shown before, the idea is confusing to students. To help these students, you might try writing the nth term of a sequence with open parentheses, as shown below.

$$a_n = 2n$$
$$a_{(\ )} = 2(\ )$$

Substitute the term number into the parentheses and then simplify.

INSTRUCTOR NOTE
Exercise 24 can be used for a similar in-class example.

EXAMPLE 2

Find the seventh and ninth terms of the sequence whose nth term is given by the formula $a_n = \dfrac{n}{2n + 1}$.

Solution $a_n = \dfrac{n}{2n + 1}$

$a_7 = \dfrac{7}{2(7) + 1} = \dfrac{7}{15}$ • Replace n by 7.

$a_9 = \dfrac{9}{2(9) + 1} = \dfrac{9}{19}$ • Replace n by 9.

The seventh term is $\dfrac{7}{15}$. The ninth term is $\dfrac{9}{19}$.

YOU TRY IT 2

Find the eighth and tenth terms of the sequence whose nth term is given by the formula $a_n = \dfrac{n - 2}{n}$.

Solution   See page S47. $\dfrac{3}{4}, \dfrac{4}{5}$

Suggested Activity

See Additional Topics in Algebra Section 1 of the *Student Activity Manual* for an introduction to sequences and series through geometric and numeric patterns.

See Appendix A:
Sequences and Series

A graphing calculator can be used to find the terms of a sequence. At the right are typical screens for listing the first five terms of the sequence $a_n = n^2 + 6$. Note that you need to enter into the calculator the nth term of the expression, the variable used in the nth term, the number of the term to start with, and the number of the term to end with.

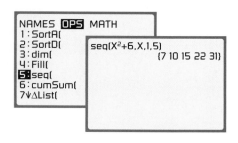

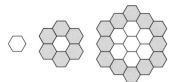

6 cells added   12 cells added

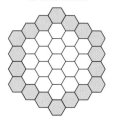

18 cells added

■ Arithmetic and Geometric Sequences

Just as there are different types of functions, such as polynomial functions and absolute value functions, there are different types of sequences.

When bees create a honeycomb, they begin with one cell and then add rings around that cell, as shown in the figure at the left. The first ring requires 6 additional cells, the second ring requires 12 additional cells, the third ring requires 18 additional cells. If we had continued the drawing of the honeycomb, the next ring would have required 24 cells. We can write the terms of this sequence as

$$a_1 = 6, a_2 = 12, a_3 = 18, a_4 = 24$$

The sequence 6, 12, 18, 24 is called an arithmetic sequence. An **arithmetic sequence,** or **arithmetic progression,** is one in which the difference between any two consecutive terms is the same constant. The difference between consecutive terms is called the **common difference** of the sequence. For the sequence of honeycomb cells, we have

$$a_2 - a_1 = 12 - 6 = 6$$
$$a_3 - a_2 = 18 - 12 = 6$$
$$a_4 - a_3 = 24 - 18 = 6$$

Therefore, the common difference is 6.

Each sequence shown below is an arithmetic sequence. To find the common difference of an arithmetic sequence, subtract the first term from the second term.

$2, 7, 12, 17, 22, \ldots$	Common difference: $7 - 2 = 5$
$3, 1, -1, -3, -5, \ldots$	Common difference: $1 - 3 = -2$
$1, \dfrac{3}{2}, 2, \dfrac{5}{2}, 3, \dfrac{7}{2}, \ldots$	Common difference: $\dfrac{3}{2} - 1 = \dfrac{1}{2}$

TAKE NOTE

When used as an adjective, as in *arithmetic sequence* and *arithmetic progression,* the word *arithmetic* is pronounced ăr′ ĭth·mĕt′ ĭk.

Consider an arithmetic sequence in which the first term is a_1 and the common difference is d. Adding the common difference to each successive term of the arithmetic sequence yields a formula for the nth term.

$a_1 = a_1$	• a_1 is the first term.
$a_2 = a_1 + d$	• The second term is the sum of the first term and the common difference d.
$a_3 = a_2 + d$	• The third term is the sum of the second term and the common difference d.
$a_3 = (a_1 + d) + d = a_1 + 2d$	• Replace a_2 by its value, $a_1 + d$.

$$a_4 = a_3 + d$$

• The fourth term is the sum of the third term and the common difference d.

$$a_4 = (a_1 + 2d) + d = a_1 + 3d$$

• Replace a_3 by its value, $a_1 + 2d$.

Note the relationship between the subscript of a term, or the term number, and the coefficient of d.

$$\overbrace{a_2 = a_1 + 1d}^{2 - 1 = 1} \qquad \overbrace{a_3 = a_1 + 2d}^{3 - 1 = 2} \qquad \overbrace{a_4 = a_1 + 3d}^{4 - 1 = 3}$$

Using inductive reasoning, we can conjecture that the coefficient of d for the nth term of an arithmetic sequence is $n - 1$, one less than the subscript.

Formula for the nth Term of an Arithmetic Sequence

The nth term of an arithmetic sequence with common difference d is given by $a_n = a_1 + (n - 1)d$.

EXAMPLE 3

Find the 27th term of the arithmetic sequence $-4, -1, 2, 5, 8, \ldots$.

Solution $d = a_2 - a_1 = -1 - (-4) = 3$ • Find the common difference.

$$a_n = a_1 + (n - 1)d$$

• Use the Formula for the nth Term of an Arithmetic Sequence to find the 27th term. $a_1 = -4, n = 27, d = 3$

$$a_{27} = -4 + (27 - 1)3$$
$$= -4 + 26(3)$$
$$= 74$$

YOU TRY IT 3

Find the 15th term of the arithmetic sequence $9, 3, -3, -9, \ldots$.

Solution   See page S47. -75

EXAMPLE 4

Find the formula for the nth term of the arithmetic sequence $5, 3, 1, -1, \ldots$.

Solution $d = a_2 - a_1 = 3 - 5 = -2$ • Find the common difference.

$$a_n = a_1 + (n - 1)d$$

• Use the Formula for the nth Term of an Arithmetic Sequence. Replace a_1, and d; $a_1 = 5, d = -2$.

$$= 5 + (n - 1)(-2)$$
$$= 5 - 2n + 2$$
$$= -2n + 7$$

YOU TRY IT 4

Find the formula for the nth term of the arithmetic sequence $-3, 1, 5, 9, \ldots$.

Solution   See page S47. $a_n = 4n - 7$

Suggested Activity

The loge seating in a concert hall consists of 27 rows of chairs. There are 73 seats in the first row, 75 seats in the second row, 77 seats in the third row, and so on in an arithmetic sequence. Each seat in rows 1 through 9 costs $24, each seat in rows 10 through 18 costs $20, and each seat in rows 19 through 27 costs $16. Find the revenue if all seats in the loge section of the concert hall are sold.

[Answer: $52,164]

INSTRUCTOR NOTE
Exercise 44 can be used for a similar in-class example.

INSTRUCTOR NOTE
Exercise 54 can be used for a similar in-class example.

INSTRUCTOR NOTE
After introducing the concept of geometric sequence, give students examples of sequences and ask them to identify each sequence as arithmetic, geometric, or neither. Three possible sequences are

$1, 4, 9, 16, \ldots, n^2, \ldots$

$2, 4, 8, 16, \ldots, 2^n, \ldots$

$2, 4, 6, 8, \ldots, 2n, \ldots$

An arithmetic sequence is characterized by a common *difference* between successive terms. A **geometric sequence** is characterized by a common *ratio* between successive terms.

The sequence 6, 18, 54, 162, 486, 1458, . . . is a geometric sequence. Note that the ratio of any two successive terms is 3.

$$\frac{18}{6} = 3 \qquad \frac{54}{18} = 3 \qquad \frac{162}{54} = 3 \qquad \frac{486}{162} = 3 \qquad \frac{1458}{486} = 3$$

Geometric sequences have many different applications. For instance, suppose an ore sample contains 20 milligrams of a radioactive material with a half-life of 1 week. The sequence below represents the amount in the sample at the beginning of each week.

Week	1	2	3	4	5
Amount	20	10	5	2.5	1.25

As shown below, the ratio of succeeding terms of the sequence is $\frac{1}{2}$.

$$\frac{10}{20} = \frac{1}{2} \qquad \frac{5}{10} = \frac{1}{2} \qquad \frac{2.5}{5} = \frac{1}{2} \qquad \frac{1.25}{2.5} = \frac{1}{2}$$

TAKE NOTE
Geometric sequences are different from arithmetic sequences. For a geometric sequence, every two successive terms have the same *ratio*. For an arithmetic sequence, every two successive terms have the same *difference*.

The sequence 20, 10, 5, 2.5, 1.25 is a geometric sequence. A **geometric sequence,** or **geometric progression,** is one in which each successive term of the sequence is the same nonzero constant multiple of the preceding term. The common multiple is called the **common ratio** of the sequence.

Each of the sequences shown below is a geometric sequence. To find the common ratio of a geometric sequence, divide the second term of the sequence by the first term.

$3, 6, 12, 24, 48, \ldots$ Common ratio: $6 \div 3 = 2$

$4, -12, 36, -108, 324, \ldots$ Common ratio: $-12 \div 4 = -3$

$6, 4, \dfrac{8}{3}, \dfrac{16}{9}, \dfrac{32}{27}, \ldots$ Common ratio: $4 \div 6 = \dfrac{2}{3}$

Consider a geometric sequence in which the first term is a_1 and the common ratio is r. Multiplying each successive term of the geometric sequence by the common ratio yields a formula for the nth term.

The first term is a_1. $\qquad a_1 = a_1$

To find the second term, multiply the first term by the common ratio r. $\qquad a_2 = a_1 r$

To find the third term, multiply the second term by the common ratio r. $\qquad a_3 = (a_2)r = (a_1 r)r$
$\qquad a_3 = a_1 r^2$

To find the fourth term, multiply the third term by the common ratio r. $\qquad a_4 = (a_3)r = (a_1 r^2)r$
$\qquad a_4 = a_1 r^3$

Note the relationship between the term number and the number that is the exponent on r. The exponent on r is 1 less than the term number. $\qquad a_n = a_1 r^{n-1}$

> ### The Formula for the nth Term of a Geometric Sequence
> The nth term of a geometric sequence with first term a_1 and common ratio r is given by $a_n = a_1 r^{n-1}$.

? QUESTION   In the Formula for the nth Term of a Geometric Sequence:

   a. What does the variable a_1 represent?
   b. What does the variable r represent?

INSTRUCTOR NOTE
Exercise 72 can be used for a similar in-class example.

Suggested Activity

Have students discover the relationship between the signs of the terms of a geometric sequence and the value of r. [Answer: For $r > 0$, all terms have the same sign. For $r < 0$, the signs of the terms alternate.]

EXAMPLE 5

Find the 6th term of the geometric sequence 3, 6, 12, ...

Solution

$$r = \frac{a_2}{a_1} = \frac{6}{3} = 2 \qquad \bullet \text{ Find the common ratio.}$$

$$a_n = a_1 r^{n-1} \qquad \bullet \text{ Use the Formula for the } n\text{th term of a Geometric}$$
$$a_6 = 3(2)^{6-1} \qquad \quad \text{Sequence. Replace } n, a_1, \text{ and } r\text{: } n = 6, a_1 = 3, r = 2.$$
$$= 3(2)^5$$
$$= 3(32)$$
$$= 96$$

YOU TRY IT 5

Find the 5th term of the geometric sequence $5, 2, \dfrac{4}{5}, \ldots$.

Solution   See page S47. $\dfrac{16}{125}$

INSTRUCTOR NOTE
Exercise 78 can be used for a similar in-class example.

INSTRUCTOR NOTE
Some students may assume that a sequence must be either arithmetic or geometric. The sequence $a_n = \dfrac{1}{n}$ can be used to dispel that belief. As an extra-credit problem, have students prove that this sequence is neither arithmetic nor geometric.

EXAMPLE 6

Find a formula for the nth term of the geometric sequence 8, 12, 18, 27, ...

Solution

$$r = \frac{a_2}{a_1} = \frac{12}{8} = \frac{3}{2} \qquad \bullet \text{ Find the common ratio.}$$

The common ratio is $\dfrac{3}{2}$.

$$a_n = a_1 r^{n-1} \qquad \bullet \text{ Use the Formula for the } n\text{th Term of a Geometric Sequence.}$$
$$\qquad\qquad\qquad \text{Replace } a_1 \text{ by 8 and } r \text{ by } \dfrac{3}{2}.$$

$$a_n = 8\left(\frac{3}{2}\right)^{n-1}$$

? ANSWER   **a.** a_1 represents the first term.   **b.** r represents the common ratio.

YOU TRY IT 6

Find a formula for the nth term of the geometric sequence $2, -1, \dfrac{1}{2}, -\dfrac{1}{4}, \ldots$.

Solution   See page S47. $a_n = 2\left(-\dfrac{1}{2}\right)^{n-1}$

■ Series

At the beginning of this section, the sequence 1, 2, 4, 8, 16, 32 was shown to represent the number of grains of wheat on each of the first six squares of a chessboard. The sum of the terms of this sequence represents the total number of grains of wheat on the first six squares of the chessboard.

$$1 + 2 + 4 + 8 + 16 + 32 = 63$$

The first six squares of the chessboard hold a total of 63 grains of wheat.

The indicated sum of the terms of a sequence is called a **series.** Given the sequence 1, 2, 4, 8, 16, 32, the series $1 + 2 + 4 + 8 + 16 + 32$ can be written.

S_n is used to indicate the sum of the first n terms of a sequence.

For the preceding example, the sums of the series S_1, S_2, S_3, S_4, S_5, and S_6 represent the total number of grains of wheat on the first 1, 2, 3, 4, 5, and 6 squares of the chessboard, respectively.

$$
\begin{aligned}
S_1 &= 1 &&= 1\\
S_2 &= 1 + 2 &&= 3\\
S_3 &= 1 + 2 + 4 &&= 7\\
S_4 &= 1 + 2 + 4 + 8 &&= 15\\
S_5 &= 1 + 2 + 4 + 8 + 16 &&= 31\\
S_6 &= 1 + 2 + 4 + 8 + 16 + 32 &&= 63
\end{aligned}
$$

? QUESTION   What is S_7 for the sequence 1, 2, 4, 8, 16, 32, 64?

For the general sequence $a_1, a_2, a_3, \ldots, a_n$, the series $S_1, S_2, S_3,$ and S_n are shown at the right.

$$
\begin{aligned}
S_1 &= a_1\\
S_2 &= a_1 + a_2\\
S_3 &= a_1 + a_2 + a_3\\
S_n &= a_1 + a_2 + a_3 + \ldots + a_n
\end{aligned}
$$

It is convenient to represent a series in a compact form called **summation notation,** or **sigma notation.** The Greek letter sigma, Σ, is used to indicate a sum.

The first four terms of the sequence whose nth term is given by the formula $a_n = 2n$ are 2, 4, 6, 8. The corresponding series is shown below written in summation notation and is read "the summation from 1 to 4 of $2n$." The letter n is called the **index** of the summation.

$$\sum_{n=1}^{4} 2n$$

? ANSWER $S_7 = 1 + 2 + 4 + 8 + 16 + 32 + 64 = 127$

INSTRUCTOR NOTE
Whether a student takes statistics, finite math, or any other math course after completing this one, summation notation will be a part of that course. This notation is not easy for students, so many examples may be necessary. One important aspect of this notation is that the index is *always* incremented by 1.

INSTRUCTOR NOTE
Exercise 30 can be used for a similar in-class example.

TAKE NOTE

The placement of the parentheses in Example 7b is important. Note that

$$\sum_{n=1}^{3} 2n + 1$$
$$= [2(1) + 2(2)$$
$$+ 2(3)] + 1$$
$$= 2 + 4 + 6 + 1$$
$$= 13 \neq 15$$

To write the terms of the series, replace n by the consecutive integers from 1 to 4. Note that the series is $2 + 4 + 6 + 8$, and the sum of the series is 20.

$$\sum_{n=1}^{4} 2n = 2(1) + 2(2) + 2(3) + 2(4)$$
$$= 2 + 4 + 6 + 8 \qquad \bullet \text{ This is the series.}$$
$$= 20 \qquad \bullet \text{ This is the sum of the series.}$$

EXAMPLE 7

Evaluate the series.

a. $\displaystyle\sum_{i=4}^{8} \frac{i}{2}$ **b.** $\displaystyle\sum_{n=1}^{3} (2n + 1)$

Solution

a. $\displaystyle\sum_{i=4}^{8} \frac{i}{2} = \frac{4}{2} + \frac{5}{2} + \frac{6}{2} + \frac{7}{2} + \frac{8}{2}$ $\bullet$ Replace i by 4, 5, 6, 7, and 8.

$$= \frac{30}{2} = 15 \qquad \bullet \text{ Find the sum of the series.}$$

b. $\displaystyle\sum_{n=1}^{3} (2n + 1)$

$$= [2(1) + 1] + [2(2) + 1] + [2(3) + 1] \quad \bullet \text{ Replace } n \text{ by 1, 2, and 3.}$$
$$= 3 + 5 + 7 \qquad \bullet \text{ Write the series.}$$
$$= 15 \qquad \bullet \text{ Find the sum of the series.}$$

YOU TRY IT 7

Evaluate the series.

a. $\displaystyle\sum_{i=1}^{5} (4 - i)$ **b.** $\displaystyle\sum_{n=3}^{6} (n^2 + 2)$

Solution   See page S47.   **a.** 5   **b.** 94

Note in Example 7 that the index can be any letter. However, the variable used for the index must match the variable used in the expression for the general term of the sequence.

A graphing calculator can be used to evaluate a series. Typical screens for evaluating the series in Example 7b are shown below.

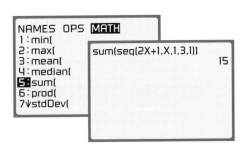

■ **Arithmetic and Geometric Series**

The indicated sum of the terms of an arithmetic sequence is called an **arithmetic series.** The formula for the sum of an arithmetic sequence can be found by pairing terms of the sequence.

Consider the arithmetic sequence 2, 5, 8, 11, 14, 17, 20, and 23. The sum of the terms of the sequence is shown below, along with the sum of the pairs of certain terms.

$$
\begin{array}{c}
2 + 23 = 25 \\
5 + 20 = 25 \\
8 + 17 = 25 \\
11 + 14 = 25
\end{array}
$$

$$2 + 5 + 8 + 11 + 14 + 17 + 20 + 23$$

There are 4 pairs (one-half of 8, the number of terms of the sequence) whose sum is 25. Therefore, the sum of the 8 terms is $4(25) = 100$. This idea can be extended to give the following formula.

▼ *Point of Interest*

This formula was proved in Aryabhatiya, *which was written by Aryabhata around 499. The book is the earliest known Indian mathematical work by an identifiable author. Although the proof of the formula appears in that text, the formula was known before Aryabhata's time.*

The Formula for the Sum of n Terms of an Arithmetic Series

Let a_1 be the first term of a finite arithmetic sequence, let n be the number of terms, and let a_n be the last term of the sequence. Then the sum of the series S_n is given by $S_n = \dfrac{n(a_1 + a_n)}{2}$.

INSTRUCTOR NOTE
Exercise 84 can be used for a similar in-class example.

EXAMPLE 8

Find the sum of the first 15 terms of the arithmetic sequence 2, 4, 6, 8, ...

Solution   To apply the formula given above, we must determine a_n, the nth term of the arithmetic sequence.

$$d = a_2 - a_1 = 4 - 2 = 2$$ • Find the common difference.

$$a_n = a_1 + (n - 1)d$$
$$a_{15} = 2 + (15 - 1)2$$
$$= 2 + 14(2)$$
$$= 30$$

• Use the Formula for the nth Term of an Arithmetic Sequence to find the 15th term. Replace a_1, n_1, and d: $a_1 = 2, n = 15, d = 2$.

$$S_n = \frac{n(a_1 + a_n)}{2}$$

$$S_{15} = \frac{15(2 + 30)}{2}$$

$$= \frac{15(32)}{2} = 240$$

• Use the Formula for the Sum of n Terms of an Arithmetic Series. Replace a_1, n_1, and a_n: $a_1 = 2, n = 15, a_n = 30$.

The sum of the first 15 terms of the sequence is 240.

YOU TRY IT 8

Find the sum of the first 25 terms of the arithmetic sequence $-1, 4, 9, 14, \ldots$.

Solution   See page S48.   1475

▼ *Point of Interest*

Geometric series are used extensively in the mathematics of finance. Finite geometric series are used to calculate loan balances and monthly payments for amortized loans.

The indicated sum of the terms of a geometric sequence is called a **geometric series.**

For the geometric sequence $3, 6, 12, 24$, the corresponding geometric series is $3 + 6 + 12 + 24$.

The sum of a geometric series can be found by a formula.

The Formula for the Sum of n Terms of a Finite Geometric Series

Let a_1 be the first term of a finite geometric sequence, let n be the number of terms, and let r be the common ratio. Then the sum of the series S_n is given by $S_n = \dfrac{a_1(1 - r^n)}{1 - r}$.

INSTRUCTOR NOTE
Exercise 96 can be used for a similar in-class example.

EXAMPLE 9

Find the sum of the geometric sequence 2, 8, 32, 128, 512.

Solution $r = \dfrac{a_2}{a_1} = \dfrac{8}{2} = 4$ • Find the common ratio.

$$S_n = \frac{a_1(1 - r^n)}{1 - r}$$

$$S_5 = \frac{2(1 - 4^5)}{1 - 4}$$

• Use the Formula for the Sum of n Terms of a Finite Geometric Series. Replace n, a_1, and r: $n = 5$, $a_1 = 2$, $r = 4$.

$$= \frac{2(1 - 1024)}{-3}$$

$$= \frac{-2046}{-3}$$

$$= 682$$

YOU TRY IT 9

Find the sum of terms of the geometric sequence $1, -\dfrac{1}{3}, \dfrac{1}{9}, -\dfrac{1}{27}$.

Solution   See page S48. $\dfrac{20}{27}$

---

1 **EXERCISES**   Suggested Assignment: 5–119, odds

Topics for Discussion

1. What is **a.** a sequence? **b.** an arithmetic sequence? **c.** a geometric sequence?   a. A sequence is an ordered list of numbers. b. An arithmetic sequence is one for which successive terms differ by the same constant c. A geometric sequence is one for which the ratio of successive terms is the same constant

2. What is a series?   A series is the indicated sum of the terms of a sequence.

3. Explain the meaning of each of the expressions.
   a. a_1 **b.** a_n **c.** S_4
   a. The first term of a sequence   **b.** The nth term of a sequence   **c.** The sum of the first four terms of a sequence

4. Explain the meaning of $\sum\limits_{n=1}^{6} 5n$.
   This represents the sum of the first six terms of the sequence whose nth term is given by the formula $a_n = 5n$.

▪ Sequences and Series

Write the first four terms of the sequence whose nth term is given by the formula.

5. $a_n = 2n + 1$ 3, 5, 7, 9

6. $a_n = 1 - 2n$ $-1, -3, -5, -7$

7. $a_n = 2^n$ 2, 4, 8, 16

8. $a_n = 3^n$ 3, 9, 27, 81

9. $a_n = n^2 + 1$ 2, 5, 10, 17

10. $a_n = n^2 - 1$ 0, 3, 8 15

11. $a_n = \dfrac{n^2 - 1}{n}$ $0, \dfrac{3}{2}, \dfrac{8}{3}, \dfrac{15}{4}$

12. $a_n = n - \dfrac{1}{n}$ $0, \dfrac{3}{2}, \dfrac{8}{3}, \dfrac{15}{4}$

13. $a_n = (-1)^{n+1}n$ 1, -2, 3, -4

14. $a_n = (-1)^n(n^2 + 2n + 1)$ $-4, 9, -16, 25$

15. $a_n = \dfrac{(-1)^{n+1}}{n^2 + 1}$ $\dfrac{1}{2}, -\dfrac{1}{5}, \dfrac{1}{10}, -\dfrac{1}{17}$

16. $a_n = \dfrac{(-1)^{n+1}}{n + 1}$ $\dfrac{1}{2}, -\dfrac{1}{3}, \dfrac{1}{4}, -\dfrac{1}{5}$

Find the indicated term of the sequence whose nth term is given by the formula.

17. $a_n = 3n + 4; a_{12}$ 40

18. $a_n = 2n - 5; a_{10}$ 15

19. $a_n = (-1)^{n-1}n^2; a_{15}$ 225

20. $a_n = (-1)^{n-1}(n - 1); a_{25}$ 24

21. $a_n = \left(\dfrac{1}{2}\right)^n; a_8$ $\dfrac{1}{256}$

22. $a_n = \left(\dfrac{2}{3}\right)^n; a_5$ $\dfrac{32}{243}$

23. $a_n = (n + 2)(n + 3); a_{17}$ 380

24. $a_n = (n + 4)(n + 1); a_7$ 88

25. $a_n = \dfrac{(-1)^{2n-1}}{n^2}; a_6$ $-\dfrac{1}{36}$

26. $a_n = \dfrac{(-1)^{2n}}{n + 4}; a_{16}$ $\dfrac{1}{20}$

27. $a_n = \dfrac{3}{2}n^2 - 2; a_8$ 94

28. $a_n = \dfrac{1}{3}n + n^2; a_6$ 38

Find the sum of the series.

29. $\sum\limits_{n=1}^{5} (2n + 3)$ 45

30. $\sum\limits_{i=1}^{7} (i + 2)$ 42

31. $\sum\limits_{k=1}^{4} 2k$ 20

32. $\sum\limits_{i=1}^{7} i$ 28

33. $\sum\limits_{n=1}^{6} n^2$ 91

34. $\sum\limits_{n=1}^{5} (n^2 + 1)$ 60

35. $\sum\limits_{k=1}^{6} (-1)^k$ 0

36. $\sum\limits_{n=1}^{4} \dfrac{1}{2n}$ $\dfrac{25}{24}$

37. $\sum\limits_{i=3}^{6} i^3$ 432

38. $\displaystyle\sum_{i=1}^{4} (-1)^{i-1}(i+1)$ -2

39. $\displaystyle\sum_{n=3}^{5} \frac{(-1)^{n-1}}{n-2}$ $\frac{5}{6}$

40. $\displaystyle\sum_{n=4}^{7} \frac{(-1)^{n-1}}{n-3}$ $-\frac{7}{12}$

■ **Arithmetic and Geometric Sequences**

Find the indicated term of the arithmetic sequence.

41. $1, 11, 21, \ldots; a_{15}$ 141

42. $3, 8, 13, \ldots; a_{20}$ 98

43. $-6, -2, 2, \ldots; a_{15}$ 50

44. $-7, -2, 3, \ldots; a_{14}$ 58

45. $3, 7, 11, \ldots; a_{18}$ 71

46. $-13, -6, 1, \ldots; a_{31}$ 197

47. $-\frac{3}{4}, 0, \frac{3}{4}, \ldots; a_{11}$ $\frac{27}{4}$

48. $\frac{3}{8}, 1, \frac{13}{8}, \ldots; a_{17}$ $\frac{83}{8}$

49. $2, \frac{5}{2}, 3, \ldots; a_{31}$ 17

50. $1, \frac{5}{4}, \frac{3}{2}, \ldots; a_{17}$ 5

51. $6, 5.75, 5.50, \ldots; a_{10}$ 3.75

52. $4, 3.7, 3.4, \ldots; a_{12}$ 0.7

Find the formula for the nth term of the arithmetic sequence.

53. $1, 2, 3, \ldots$ $a_n = n$

54. $1, 4, 7, \ldots$ $a_n = 3n - 2$

55. $6, 2, -2, \ldots$ $a_n = -4n + 10$

56. $3, 0, -3, \ldots$ $a_n = -3n + 6$

57. $2, \frac{7}{2}, 5, \ldots$ $a_n = \frac{3n+1}{2}$

58. $7, 4.5, 2, \ldots$ $a_n = -2.5n + 9.5$

59. $-8, -13, -18, \ldots$
   $a_n = -5n - 3$

60. $17, 30, 43, \ldots$
   $a_n = 13n + 4$

61. $26, 16, 6, \ldots$
   $a_n = -10n + 36$

State whether the sequence is arithmetic (A), geometric (G), or neither (N), and write the next term in the sequence.

62. $4, -2, 1, \ldots$ $G, -\frac{1}{2}$

63. $-8, 0, 8, \ldots$ A, 16

64. $5, 6.5, 8, \ldots$ A, 9.5

65. $-7, 14, -28, \ldots$ G, 56

66. $1, 4, 9, 16, \ldots$ N, 25

67. $\sqrt{1}, \sqrt{2}, \sqrt{3}, \sqrt{4}, \ldots$ N, $\sqrt{5}$

68. $x^8, x^6, x^4, \ldots$ G, x^2

69. $5a^2, 3a^2, a^2, \ldots$ A, $-a^2$

70. $\log x, 2 \log x, 3 \log x, \ldots$ A, $4 \log x$

71. $\log x, 3 \log x, 9 \log x, \ldots$ G, $27 \log x$

Find the indicated term of the geometric sequence.

72. $2, 8, 32, \ldots; a_9$ 131,072

73. $4, 3, \frac{9}{4}, \ldots; a_8$ $\frac{2187}{4096}$

74. $6, -4, \frac{8}{3}, \ldots; a_7$ $\frac{128}{243}$

75. $-5, 15, -45, \ldots; a_7$ -3645

76. $1, \sqrt{2}, 2, \ldots; a_9$ 16

77. $3, 3\sqrt{3}, 9, \ldots; a_8$ $81\sqrt{3}$

Find a formula for the nth term of the geometric sequence.

78. $9, 6, 4, \dfrac{8}{3}, \ldots$ $a_n = 9\left(\dfrac{2}{3}\right)^{n-1}$

79. $8, 6, \dfrac{9}{2}, \dfrac{27}{8}, \ldots$ $a_n = 8\left(\dfrac{3}{4}\right)^{n-1}$

80. $3, -2, \dfrac{4}{3}, -\dfrac{8}{9}, \ldots$ $a_n = 3\left(-\dfrac{2}{3}\right)^{n-1}$

81. $6, -12, 24, -48, \ldots$ $a_n = 6(-2)^{n-1}$

82. $-3, 12, -48, 192, \ldots$ $a_n = -3(-4)^{n-1}$

83. $-5, 25, -125, 625, \ldots$ $a_n = -5(-5)^{n-1}$

Find the sum of the indicated number of terms of the arithmetic sequence.

84. $1, 3, 5, \ldots ; n = 50$ 2500

85. $2, 4, 6, \ldots ; n = 25$ 650

86. $20, 18, 16, \ldots ; n = 40$ −760

87. $25, 20, 15, \ldots ; n = 22$ −605

88. $\dfrac{1}{2}, 1, \dfrac{3}{2}, \ldots ; n = 27$ 189

89. $2, \dfrac{11}{4}, \dfrac{7}{2}, \ldots ; n = 10$ $\dfrac{215}{4}$

Find the sum of the arithmetic series.

90. $\displaystyle\sum_{i=1}^{15} (3i - 1)$ 345

91. $\displaystyle\sum_{i=1}^{15} (3i + 4)$ 420

92. $\displaystyle\sum_{n=1}^{17} \left(\dfrac{1}{2}n + 1\right)$ $\dfrac{187}{2}$

93. $\displaystyle\sum_{n=1}^{10} (1 - 4n)$ −210

94. $\displaystyle\sum_{i=1}^{15} (4 - 2i)$ −180

95. $\displaystyle\sum_{n=1}^{10} (5 - n)$ −5

Find the sum of the indicated number of terms of the geometric sequence.

96. $2, 6, 18, \ldots ; n = 7$ 2186

97. $-4, 12, -36, \ldots ; n = 7$ −2188

98. $12, 9, \dfrac{27}{4}, \ldots ; n = 5$ $\dfrac{2343}{64}$

99. $3, 3\sqrt{2}, 6, \ldots ; n = 12$ $189 + 189\sqrt{2}$

Find the sum of the geometric series.

100. $\displaystyle\sum_{i=1}^{5} (2)^i$ 62

101. $\displaystyle\sum_{i=1}^{5} (4)^i$ 1364

102. $\displaystyle\sum_{n=1}^{8} (3)^n$ 9840

103. $\displaystyle\sum_{n=1}^{4} (7)^n$ 2800

104. $\displaystyle\sum_{n=1}^{6} \left(\dfrac{3}{2}\right)^n$ $\dfrac{1995}{64}$

105. $\displaystyle\sum_{n=1}^{5} \left(\dfrac{1}{3}\right)^n$ $\dfrac{121}{243}$

106. $\displaystyle\sum_{i=1}^{3} \left(\dfrac{7}{4}\right)^i$ $\dfrac{651}{64}$

107. $\displaystyle\sum_{i=1}^{6} \left(\dfrac{1}{2}\right)^i$ $\dfrac{63}{64}$

108. *Physics*   The distance that an object dropped from a cliff will fall is 16 feet the first second, 48 feet the next second, 80 feet the third second, and so on in an arithmetic sequence. What is the total distance the object will fall in 6 seconds?   576 ft

109. *Health*   An exercise program calls for walking 12 minutes each day for a week. Each week thereafter, the amount of time spent walking increases by 6 minutes per day. In how many weeks will a person be walking 60 minutes each day?   9 weeks

110. *Business*   A display of cans in a grocery store consists of 20 cans in the bottom row, 18 cans in the next row, and so on in an arithmetic sequence. The top row has 4 cans. Find the total number of cans in the display.   108 cans

111. *The Arts*   A theater in the round has 52 seats in the first row, 58 seats in the second row, 64 seats in the third row, and so on in an arithmetic sequence. Find the total number of seats in the theater if there are 20 rows of seats.   2180 seats

112. *The Arts*   The loge seating section in a concert hall consists of 26 rows of chairs. There are 65 seats in the first row, 71 seats in the second row, 77 seats in the third row, and so on in an arithmetic sequence. How many seats are in the loge seating section?   3640 seats

113. *Salary*   The salary schedule for an engineering assistant is $2200 for the first month and a $150-per-month salary increase for the next 9 months. Find the monthly salary during the tenth month. Find the total salary for the 10-month period. $3550; $28,750

114. *Sports*   To test the bounce of a tennis ball, the ball is dropped from a height of 8 feet. The ball bounces to 80% of its previous height with each bounce. How high does the ball bounce on the fifth bounce? Round to the nearest tenth.   2.6 ft

115. *Recreation*   The temperature of a hot water spa is 75°F. Each hour, the temperature is 10% higher than during the previous hour. Find the temperature of the spa after 3 hours. Round to the nearest tenth.
99.8°F

116. *Real Estate*   Assume the average value of a home increases 5% per year. How much would a house costing $100,000 be worth in 30 years?
$432,194.24

117. *Wages*   Suppose an employee receives a wage of 1¢ the first day of work, 2¢ the second day, 4¢ the third day, and so on in a geometric sequence. Find the total amount of money this employee earns for working 30 days. $10,737,418.23

Day 1   Day 2   Day 3

118. *Mechanics*   A vacuum pump removes one-half the air in a chamber with each stroke. After 11 strokes, what percent of the original amount of air is left in the chamber? Round to the nearest hundredth of a percent.
0.05%

119. *Biology*   A culture of bacteria doubles every 2 hours. If there are 500 bacteria at the beginning, how many bacteria will there be after 24 hours?   2,048,000 bacteria

120. 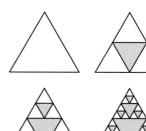 *Art* The fabric designer Jhane Barnes created a fabric pattern based on the *Sierpinski triangle*. This triangle is a fractal, which is a geometric pattern that is repeated at ever smaller scales to produce irregular shapes. The first four stages in the construction of a Sierpinski triangle are shown at the right. The initial triangle is an equilateral triangle with sides 1 unit long. The cut-out triangles are formed by connecting the midpoints of the sides of the unshaded triangles. This pattern is repeated indefinitely. Find a formula for the nth term of the number of unshaded triangles. $a_n = 3^{n-1}$

121. *Art* A Sierpinski carpet is similar to a Sierpinski triangle (see Exercise 120) except that all of the unshaded squares must be divided into nine congruent smaller squares with the one in the center shaded. The first three stages of the pattern are shown at the right. Find a formula for the nth term of the number of unshaded squares. $a_n = 8^{n-1}$

Applying Concepts

122. 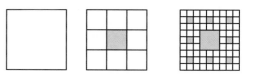 *Epidemiology* A model used by epidemiologists (people who study epidemics) to study the spread of a virus suggests that the number of people in a population who are newly infected on a given day is proportional to the number not yet exposed on the previous day. This can be described by a sequence defined by $a_n - a_{n-1} = k(P - a_{n-1})$, where P is the number of people in the original population exposed to a virus, a_n is the number of people ill with the virus n days after being exposed, a_{n-1} is the number of people ill with the virus on the previous day, and k is a constant that depends on the contagiousness of the disease and is determined from experimental evidence.

   a. Suppose that a population of 5000 people is exposed to a virus and 150 people become ill ($a_0 = 150$). The next day, 344 people are ill ($a_1 = 344$). Determine the value of k.

   b. Substitute the values of k and P into the recursive equation, and solve for a_n.

   c. How many people are infected on the fourth day?

   a. 0.04   **b.** $a_n = 200 + 0.96a_{n-1}$ **c.** 709 people

EXPLORATION

1. *Grains of Wheat on a Chess Board* Reread the paragraph at the beginning of this section. The numbers of grains of wheat on each of the first 6 squares are

Square	1	2	3	4	5	6
Grains of wheat	1	2	4	8	16	32

   a. Describe the pattern in the sequence 1, 2, 4, 8, 16, 32, ...

   b. List the next four terms of the sequence.

The number of grains of wheat on a square can be expressed by a power of 2:

The number of grains of wheat on square $n = 2^{n-1}$.

c. Calculate the number of grains of wheat on square 11.

The total number of grains of wheat on the first few squares is calculated below.

Square 1:   1
Squares 1–2: $1 + 2 = 3$
Squares 1–3: $1 + 2 + 4 = 7$
Squares 1–4: $1 + 2 + 4 + 8 = 15$

d. Calculate the total number of grains of wheat on the first 5 squares, on the first 6 squares, on the first 7 squares, and on the first 8 squares.

e. Compare the total number of grains of wheat on the first n squares with the number of grains of wheat on square $n + 1$. Find a pattern and describe it.

f. How many grains of wheat are on square 12? How many grains of wheat are on the first 12 squares?

g. Approximate the number of grains of wheat on all 64 squares. Write the answer in scientific notation. (*Note:* This is more wheat than has been produced in the world since chess was invented.)

a. Answers will vary.   **b.** 64, 128, 256, 512   **c.** 1024 grains   **d.** 31 grains, 63 grains, 127 grains, and 255 grains
e. The total number of grains of wheat on the first n squares is one less than the number of grains of wheat on square
$n + 1$.   **f.** 2048 grains; 4095 grains   **g.** 1.8×10^{19} grains of wheat

---

SECTION **2**

■ Expand $(a + b)^n$

Binomial Expansions

■ Expand $(a + b)^n$

By carefully observing the expansion of the binomial $(a + b)^n$ shown below, it is possible to identify some patterns.

$$(a + b)^1 = a + b$$
$$(a + b)^2 = a^2 + 2ab + b^2$$
$$(a + b)^3 = a^3 + 3a^2b + 3ab^2 + b^3$$
$$(a + b)^4 = a^4 + 4a^3b + 6a^2b^2 + 4ab^3 + b^4$$
$$(a + b)^5 = a^5 + 5a^4b + 10a^3b^2 + 10a^2b^3 + 5ab^4 + b^5$$

First we will look at the patterns for the variable part of the terms.

TAKE NOTE

Each term in the expansion of a binomial is a monomial. The degree of a monomial is the sum of the exponents of the variables.

Patterns for the Variable Part

1. The first term is a^n. The exponent on a decreases by 1 for each successive term.
2. The exponent on b increases by 1 for each successive term. The last term is b^n.
3. The degree of each term is n.

➡ Write the variable parts of the terms of the expansion of $(a + b)^6$.

The first term is a^6. For each successive term, the exponent on a decreases by 1, and the exponent on b increases by 1. The last term is b^6.

$a^6, a^5b, a^4b^2, a^3b^3, a^2b^4, ab^5, b^6$

⬅

? QUESTION   What are the variable parts of the terms of the expansion of $(a + b)^7$?*

▼ Point of Interest

Blaise Pascal (1623–1662) is given credit for Pascal's Triangle, which he first published in 1654. In that publication, the triangle looked like

```
    1  2  3  4   5...
  1  1  1  1  1  1...
2 1  2  3  4...
3 1  3  6...
4 1  4...
5 1
```

Thus the triangle was rotated 45° from the way it is shown today.

The variable parts of the general expansion of $(a + b)^n$ are

$$a^n, a^{n-1}b, a^{n-2}b^2, \ldots, a^{n-r}b^r, \ldots, ab^{n-1}, b^n$$

A pattern for the coefficients of the terms of the expanded binomial can be found by writing the coefficients in a triangular array known as **Pascal's Triangle.**

Each row begins and ends with the number 1. Any other number in a row is the sum of the two closest numbers above it. For example, $4 + 6 = 10$.

For $(a + b)^1$:   1   1
For $(a + b)^2$:   1   2   1
For $(a + b)^3$:   1   3   3   1
For $(a + b)^4$:   1   4   6   4   1
For $(a + b)^5$:   1   5   10   10   5   1 Ⓟ

▼ Point of Interest

The first European publication of Pascal's Triangle is attributed to Peter Apianus in 1527. However, there are versions of it in a Chinese text by Yang Hui that dates from 1250. In that text, Hui demonstrated how to find the third, fourth, fifth, and sixth roots of a number by using the triangle.

➡ Write the sixth row of Pascal's Triangle.

To write the sixth row, first write the numbers of the fifth row. The first and last numbers of the sixth row are 1. Each of the other numbers of the sixth row can be obtained by finding the sum of the two closest numbers above it in the fifth row.

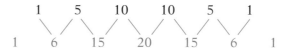

These numbers will be the coefficients of the terms of the expansion of $(a + b)^6$.

? QUESTION   What is the seventh row of Pascal's Triangle?†

Using the numbers of the sixth row of Pascal's Triangle for the coefficients, and using the pattern for the variable part of each term, we can write the expanded form of $(a + b)^6$ as follows:

$$(a + b)^6 = a^6 + 6a^5b + 15a^4b^2 + 20a^3b^3 + 15a^2b^4 + 6ab^5 + b^6$$

Although Pascal's Triangle can be used to find the coefficients for the expanded form of the power of any binomial, this method is inconvenient when the power of the binomial is large. An alternative method for determining those coefficients is based on the concept of *factorial*.

―――――

? ANSWERS   *$a^7, a^6b, a^5b^2, a^4b^3, a^3b^4, a^2b^5, ab^6, b^7$
†This is the sixth row: 1   6   15   20   15   6   1
This is the seventh row: 1   7   21   35   35   21   7   1

▼ *Point of Interest*

The exclamation point was first used as a notation for factorial in 1808 in the book Eléments d'Arithmétique Universelle *for the convenience of the printer of the book. Some English text-books suggested using the phrase "n-admiration" for n!*

> ### *n* Factorial
>
> $n!$ (which is read "*n* factorial") is the product of the first *n* consecutive natural numbers. 0! is defined to be 1.
>
> $$n! = n(n - 1)(n - 2) \ldots 3 \cdot 2 \cdot 1$$

The values of 1!, 5!, and 7! are shown at the right.

$$1! = 1$$
$$5! = 5 \cdot 4 \cdot 3 \cdot 2 \cdot 1 = 120$$
$$7! = 7 \cdot 6 \cdot 5 \cdot 4 \cdot 3 \cdot 2 \cdot 1 = 5040$$

EXAMPLE 1

Evaluate: $\dfrac{7!}{4!3!}$

Solution $\dfrac{7!}{4!3!} = \dfrac{7 \cdot 6 \cdot 5 \cdot 4 \cdot 3 \cdot 2 \cdot 1}{(4 \cdot 3 \cdot 2 \cdot 1)(3 \cdot 2 \cdot 1)}$

• Write each factorial as a product.

$$= 35$$

• Simplify.

INSTRUCTOR NOTE
Exercise 16 can be used for a similar in-class example.

Graphical Check:

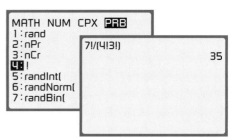

YOU TRY IT 1

Evaluate: $\dfrac{12!}{7!5!}$

Solution   See page S48.   792

INSTRUCTOR NOTE
Show students that factorials do not operate in standard ways. For instance:

$$3! + 4! \neq 7!$$

and

$$\frac{12!}{3!} \neq 4!$$

▼ *Point of Interest*

Leonhard Euler (1707–1783) used the notations $\left(\dfrac{n}{r} \right)$ *and* $\left[\dfrac{n}{r} \right]$ *for binomial coefficients around 1784. The notation* $\dbinom{n}{r}$ *appeared in the late 1820s.*

The coefficients in a binomial expansion can be given in terms of factorials. Note that in the expansion of $(a + b)^5$ shown below, the coefficient of a^2b^3 can be given by $\dfrac{5!}{2!3!}$. The numerator is the factorial of the power of the binomial. The denominator is the product of the factorials of the exponents on *a* and *b*.

$$(a + b)^5 = a^5 + 5a^4b + 10a^3b^2 + \mathbf{10a^2b^3} + 5ab^4 + b^5$$

$$\frac{5!}{2!3!} = \frac{5 \cdot 4 \cdot 3 \cdot 2 \cdot 1}{(2 \cdot 1)(3 \cdot 2 \cdot 1)} = 10$$

In general, the coefficients of $(a + b)^n$ are given as the quotients of factorials. **The coefficient of $a^{n-r}b^r$ is $\dfrac{n!}{(n - r)!r!}$.** The symbol $\dbinom{n}{r}$ is used to express this quotient of factorials.

$$\binom{n}{r} = \frac{n!}{(n - r)!r!}$$

INSTRUCTOR NOTE
Exercise 22 can be used for
a similar in-class example.

❓ QUESTION   How is $\binom{9}{6}$ written as the quotient of factorials?

EXAMPLE 2

Evaluate: $\binom{8}{5}$

Solution $\binom{8}{5} = \dfrac{8!}{(8-5)!5!}$ • Write the quotient of the factorials.

$= \dfrac{8!}{3!5!}$ • Simplify.

$= \dfrac{8 \cdot 7 \cdot 6 \cdot 5 \cdot 4 \cdot 3 \cdot 2 \cdot 1}{(3 \cdot 2 \cdot 1)(5 \cdot 4 \cdot 3 \cdot 2 \cdot 1)} = 56$

Suggested Activity

1. Evaluate $\frac{n!}{(n-1)!}$ for $n = 20$.
   [Answer: 20]
2. Simplify $\frac{n!}{(n-2)!}$.
   [Answer: $n(n-1)$ or $n^2 - n$]

YOU TRY IT 2

Evaluate: $\binom{7}{0}$

Solution   See page S48.   1

Using factorials and the pattern for the variable part of each term, we can write a formula for any natural-number power of a binomial.

INSTRUCTOR NOTE
Students will encounter this
theorem in courses such as
probability and statistics,
genetics, and calculus.

The Binomial Expansion Formula

$(a + b)^n =$

$$\binom{n}{0}a^n + \binom{n}{1}a^{n-1}b + \binom{n}{2}a^{n-2}b^2 + \ldots + \binom{n}{r}a^{n-r}b^r + \ldots + \binom{n}{n}b^n$$

The Binomial Expansion Formula is used below to expand $(a + b)^7$.

$(a + b)^7 =$

$$\binom{7}{0}a^7 + \binom{7}{1}a^6b + \binom{7}{2}a^5b^2 + \binom{7}{3}a^4b^3 + \binom{7}{4}a^3b^4 + \binom{7}{5}a^2b^5 + \binom{7}{6}ab^6 + \binom{7}{7}b^7$$

$$= a^7 + 7a^6b + 21a^5b^2 + 35a^4b^3 + 35a^3b^4 + 21a^2b^5 + 7ab^6 + b^7$$

INSTRUCTOR NOTE
Exercise 34 can be used for
a similar in-class example.

EXAMPLE 3

Write $(3m - n)^4$ in expanded form.

Solution $(3m - n)^4$

$$= \binom{4}{0}(3m)^4 + \binom{4}{1}(3m)^3(-n) + \binom{4}{2}(3m)^2(-n)^2 + \binom{4}{3}(3m)(-n)^3 + \binom{4}{4}(-n)^4$$

TAKE NOTE

Note that there are
$n + 1$ terms in a bino-
mial expansion.

❓ ANSWER $\binom{9}{6} = \dfrac{9!}{(9-6)!6!}$

$$= 1(81m^4) + 4(27m^3)(-n) + 6(9m^2)(n^2)$$
$$+ 4(3m)(-n^3) + 1(n^4)$$
$$= 81m^4 - 108m^3n + 54m^2n^2 - 12mn^3 + n^4$$

YOU TRY IT 3

Write $(4x + 3y)^3$ in expanded form.

Solution   See page S48. $64x^3 + 144x^2y + 108xy^2 + 27y^3$

INSTRUCTOR NOTE
Exercise 46 can be used for a similar in-class example.

Suggested Activity

See Additional Topics in Algebra Section 2 of the *Student Activity Manual* for a lesson that includes the binomial probability theorem.

EXAMPLE 4

Find the first three terms in the expansion of $(x + 3)^{15}$.

Solution $(x + 3)^{15} = \binom{15}{0}x^{15} + \binom{15}{1}x^{14}(3) + \binom{15}{2}x^{13}(3)^2 + \ldots$

$$= 1x^{15} + 15x^{14}(3) + 105x^{13}(9) + \ldots$$
$$= x^{15} + 45x^{14} + 945x^{13} + \ldots$$

YOU TRY IT 4

Find the first three terms in the expansion of $(y - 2)^{10}$.

Solution   See page S48. $y^{10} - 20y^9 + 180y^8 + \ldots$

2 EXERCISES   Suggested Assignment: 9–51, odds

Topics for Discussion

1. What is the factorial of a number n?
   It is the product of the first n consecutive natural numbers.

2. Determine whether the statement is always true, sometimes true, or never true.

   a. $0! \cdot 4! = 0$ **b.** $\dfrac{4!}{0!}$ is undefined.

   a. Never true   b. Never true

3. What does the notation $\binom{n}{r}$ mean? $\dfrac{n!}{(n-r)!r!}$

4. What is the sum of the exponents in each term of the expansion $(a + b)^n$? Write the answer in terms of n. n

5. How many terms are in the expansion of $(a + b)^n$? Write the answer in terms of n. $n + 1$

6. What does it mean to expand $(a + b)^n$?
Explanations may vary. For example: It means to write it as the sum of its terms.

7. What is the purpose of the Binomial Expansion Formula?
It is used to expand $(a + b)^n$.

8. Is the Binomial Expansion Formula used to simplify $(ab)^5$?
No, $(ab)^5$ is a power of a monomial.

■ Factorials

Evaluate.

9. $3!$ 6   **10.** $4!$ 24   **11.** $8!$ 40,320   **12.** $9!$ 362,880   **13.** $0!$ 1

14. $1!$ 1   **15.** $\dfrac{5!}{2!3!}$ 10   **16.** $\dfrac{8!}{5!3!}$ 56   **17.** $\dfrac{6!}{6!0!}$ 1   **18.** $\dfrac{10!}{10!0!}$ 1

19. $\dfrac{9}{6!3!}$ 84   **20.** $\dfrac{10!}{2!8!}$ 45   **21.** $\dbinom{7}{2}$ 21   **22.** $\dbinom{8}{6}$ 28   **23.** $\dbinom{9}{0}$ 1

24. $\dbinom{10}{10}$ 1   **25.** $\dbinom{11}{1}$ 11   **26.** $\dbinom{13}{1}$ 13   **27.** $\dbinom{6}{3}$ 20   **28.** $\dbinom{8}{4}$ 70

29. Evaluate $\dfrac{n!}{(n - 2)!}$ for $n = 50$.   2450   **30.** Simplify $\dfrac{n!}{(n - 1)!}$. n

■ Binomial Expansion

31. Write the 8th row of Pascal's Triangle.   1 8 28 56 70 56 28 8 1

Write in expanded form.

32. $(x + y)^4$ $x^4 + 4x^3y + 6x^2y^2 + 4xy^3 + y^4$ **33.** $(r - s)^3$ $r^3 - 3r^2s + 3rs^2 - s^3$

34. $(x - y)^5$ $x^5 - 5x^4y + 10x^3y^2 - 10x^2y^3 + 5xy^4 - y^5$ **35.** $(y - 3)^4$ $y^4 - 12y^3 + 54y^2 - 108y + 81$

36. $(2m + 1)^4$ $16m^4 + 32m^3 + 24m^2 + 8m + 1$ **37.** $(2x + 3y)^3$ $8x^3 + 36x^2y + 54xy^2 + 27y^3$

38. $(2r - 3)^5$
$32r^5 - 240r^4 + 720r^3 - 1080r^2 + 810r - 243$

39. $(x + 3y)^4$
$x^4 + 12x^3y + 54x^2y^2 + 108xy^3 + 81y^4$

40. *Geometry*   The edge of a cube is x centimeters in length. Each edge is increased by 5 centimeters. Write the volume, in expanded form, of the larger cube. $(x^3 + 15x^2 + 75x + 125)$ cm^3

Find the first three terms in the expansion.

41. $(a + b)^{10}$ $a^{10} + 10a^9b + 45a^8b^2$ **42.** $(a + b)^9$ $a^9 + 9a^8b + 36a^7b^2$

43. $(a - b)^{11}$ $a^{11} - 11a^{10}b + 55a^9b^2$ **44.** $(a - b)^{12}$ $a^{12} - 12a^{11}b + 66a^{10}b^2$

45. $(2x + y)^8$ $256x^8 + 1024x^7y + 1792x^6y^2$ **46.** $(x + 3y)^9$ $x^9 + 27x^8y + 324x^7y^2$

47. $(4x - 3y)^8$ $65,536x^8 - 393,216x^7y + 1,032,192x^6y^2$ **48.** $(2x - 5)^7$ $128x^7 - 2240x^6 + 16,800x^5$

49. $\left(x + \dfrac{1}{x}\right)^7$ $x^7 + 7x^5 + 21x^3$

50. $\left(x - \dfrac{1}{x}\right)^8$ $x^8 - 8x^6 + 28x^4$

51. $(x^2 + 3)^5$ $x^{10} + 15x^8 + 90x^6$

52. $(x^2 - 2)^6$ $x^{12} - 12x^{10} + 60x^8$

Applying Concepts

53. Write the term that contains an x^3 in the expansion of $(x + a)^7$. $35x^3a^4$

54. Find the value of n for which $(3!)(5!)(7!) = n!$ 10

Expand the binomial. *Note:* In Exercise 57, i is the imaginary unit.

55. $(x^{1/2} + 2)^4$

$x^2 + 8x^{3/2} + 24x + 32x^{1/2} + 16$

56. $(x^{-1} + y^{-1})^3$

$\dfrac{1}{x^3} + \dfrac{3}{x^2y} + \dfrac{3}{xy^2} + \dfrac{1}{y^3}$

57. $(1 + i)^6$

$-8i$

58. Note that $6! = (6)(5)(4)(3)(2)(1) = (2^4)(3^2)(5)$. Find the value of n for which $n! = (2^{25})(3^{13})(5^6)(7^4)(11^2)(13^2)(17)(19)(23)$.   28

59. Convert the expression $\displaystyle\sum_{i=0}^{n} \binom{n}{i} x^{n-i}5^i$ to one of the form $(a + b)^n$.

$(x + 5)^n$

EXPLORATION

1. *Approximating Numerical Exponential Expressions*
The Binomial Theorem can be used to quickly approximate numerical exponential expressions. For example,

$$(1.002)^3 = (1 + 0.002)^3$$
$$= 1^3 + 3(1^2)(0.002) + 3(1)(0.002)^2 + (0.002)^3$$
$$= 1 + 0.006 + 0.000012 + 0.000000008$$
$$= 1.006012008$$

The last two terms in the expansion are so small that they can be ignored in an approximation.

$$(1.002)^3 \approx 1.006$$

a. Approximate $(1.004)^3$ to three decimal places. Check your answer using a calculator.

b. Approximate $(1.02)^{10}$ to three decimal places. Check your answer using a calculator.

c. Approximate $(1.001)^{10}$ to three decimal places. Check your answer using a calculator.

d. Suppose you wanted to approximate $(0.98)^{12}$ using the first four terms of a binomial expansion. What binomial would you use?

   a. 1.012   **b.** 1.219   **c.** 1.010   **d.** $1 - 0.02$

2.   *ISBN and UPC Numbers*   Every book that is cataloged for the Library of Congress must have an ISBN (International Standard Book Number). An ISBN is a 10-digit number of the form $a_1\text{-}a_2a_3a_4\text{-}a_5a_6a_7a_8a_9\text{-}c$. For instance, the ISBN for the Windows version of the CD-ROM containing the

American Heritage Children's Dictionary is 0-395-73580-7. The first number, 0, indicates that the book is written in English. The next three numbers, 395, indicate the publisher (Houghton Mifflin Company). The next five numbers, 73580, identify the book (*American Heritage Children's Dictionary*), and the last digit, c, is called a **check digit.** This digit is chosen such that the following sum is divisible by 11.

$$a_1(10) + a_2(9) + a_3(8) + a_4(7) + a_5(6) + a_6(5) + a_7(4) + a_8(3) + a_9(2) + c$$

For the *American Heritage Children's Dictionary.*

$$0(10) + 3(9) + 9(8) + 5(7) + 7(6) + 3(5) + 5(4) + 8(3) + 0(2) + c = 235 + c$$

The last digit of the ISBN is chosen as 7 because $235 + 7 = 242$ and $242 \div 11 = 22$. The value of c could be any number from 0 to 10. The number 10 is coded as X.

One purpose of the ISBN method of coding books is to ensure that orders placed for books are accurately filed. For instance, suppose a clerk sends an order for the *American Heritage Children's Dictionary* and inadvertently enters the number 0-395-75380-7 (the 3 and 5 have been transposed). Now

$$0(10) + 3(9) + 9(8) + 5(7) + 7(6) + 5(5) + 3(4) + 8(3) + 0(2) + 7 = 244$$

and 244 is not divisible by 11. Thus an error in the ISBN was made.

a. Determine the check digit for the book *Reader's Digest Book of Facts.* The first nine digits of the ISBN are 0-895-77692.
b. Is 0-395-12370-4 a possible ISBN?
c. Check the ISBN of this algebra textbook.

Another coding scheme that is closely related to ISBN is the UPC (Universal Product Code). This number is particularly useful in grocery stores. A checkout clerk passes the number by a scanner that reads the number and records the product and its price.

There are two major UPC codes: Type A consists of 12 digits and type E consists of 8 digits. For type A UPC symbols, the first 6 digits identify the manufacturer. The next 5 digits constitute a product identification number given by the manufacturer to identify the product. The final digit is a check digit. The check digit is chosen so that the sum is divisible by 10.

$$a_1(13) + a_2 + a_3(13) + a_4 + a_5(13) + a_6 + a_7(13) + a_8 + a_9(13) + a_{10} + a_{11}(13) + c$$

The UPC for the TI-83 Plus is 0-33317-19865-8. Using the expression above,

$$0(13) + 3 + 3(13) + 3 + 1(13) + 7 + 1(13) + 9 + 8(13) + 6 + 5(13) + c = 262 + c$$

Because $262 + 8$ is a multiple of 10, the check digit is 8.

d. What numbers can the check digit of a type A UPC be?
e. Check the UPC for the *Information Please Almanac* shown at the right.
f. Check the UPC of a product in your home.

a. 8 b. Yes c. Students should use the given equation and note that the ISBN checks. d. 1, 2, 3, 4, 5, 6, 7, 8, 9, or 0 e. $0(13) + 4 + 6(13) + 4 + 4(13) + 2 + 7(13) + 7 + 0(13) + 1 + 8(13) + 7 = 350$, which is a multiple of 10. f. Answers will vary.

ISBN 0–395–75524–7

0 46442 77018 7

65)

7 is the check digit

▼ *Point of Interest*

Menaechmus (375–325 B.C.), a Greek mathematician and a teacher of Alexander the Great, is credited with the discovery of the conic sections. His study of the equations $\frac{c}{x} = \frac{x}{y}$ and $\frac{c}{x} = \frac{y}{c}$, where c is a constant, led him to the equations $x^2 = cy$ and $xy = c^2$. The graphs of these equations produced a parabola and a hyperbola, respectively.

INSTRUCTOR NOTE
Properties of the parabola were covered earlier. Those properties are reviewed here for completeness.

INSTRUCTOR NOTE
Exercise 8 can be used for a similar in-class example.

Conic Sections

■ The Parabola

The **conic sections** are curves that can be constructed from the intersection of a plane and a right circular cone. The parabola, which was introduced earlier, is one of these curves. Here we will review some of that previous discussion and look at equations of parabolas that were not discussed before.

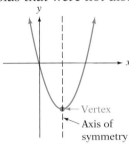

Every parabola has an axis of symmetry and a vertex that is on the axis of symmetry. The graph of the equation $y = ax^2 + bx + c$, $a \neq 0$, is a parabola with the axis of symmetry parallel to the y-axis. The parabola opens up when $a > 0$ and opens down when $a < 0$. When the parabola opens up, the vertex is the lowest point on the parabola. It is the point at which the function has a minimum value. When the parabola opens down, the vertex is the highest point on the parabola. It is the point at which the function has a maximum value.

Recall that the x-coordinate of the vertex of the graph of an equation of the form $y = ax^2 + bx + c$ is $-\frac{b}{2a}$. The y-coordinate of the vertex can be determined by substituting this value of x into $y = ax^2 + bx + c$ and solving for y.

EXAMPLE 1

Find the vertex and the axis of symmetry of the parabola whose equation is $y = x^2 - 4x + 3$. Then sketch its graph.

Solution $x = -\dfrac{b}{2a} = -\dfrac{-4}{2(1)} = 2$
 • Find the x-coordinate of the vertex. $a = 1$ and $b = -4$.

$$y = x^2 - 4x + 3$$
$$y = 2^2 - 4(2) + 3$$
$$y = -1$$
 • Find the y-coordinate of the vertex by replacing x with 2 and solving for y.

The vertex is $(2, -1)$.

The axis of symmetry is the line $x = 2$.
 • The axis of symmetry is the line $x = -\dfrac{b}{2a}$.

 • Because a is positive ($a = 1$), the parabola opens up. Find a few ordered pairs, and use symmetry to sketch the graph.

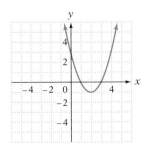

YOU TRY IT 1

Find the vertex and axis of symmetry of the parabola whose equation is $y = x^2 - 2x - 1$. Then sketch its graph.

Solution See page S48. $(1, -2); x = 1$

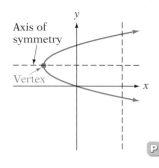

The graph of an equation of the form $x = ay^2 + by + c$, $a \neq 0$, is also a parabola. In this case, the parabola opens to the right when a is positive and opens to the left when a is negative.

For a parabola of this form, the **y-coordinate of the vertex** is $-\frac{b}{2a}$. The **axis of symmetry** is the line $y = -\frac{b}{2a}$.

Using the vertical-line test reveals that the graph of a parabola of this form is not the graph of a function. The graph of $x = ay^2 + by + c$ is a relation. Note that it does not have a minimum or maximum value.

? QUESTION Does the graph of the equation open to the right or to the left?

a. $x = 4y^2 - y - 2$ b. $x = -y^2 + 3y + 5$

INSTRUCTOR NOTE
Exercise 10 can be used for a similar in-class example.

INSTRUCTOR NOTE
Point out that the analysis of $x = ay^2 + by + c$ is essentially the same as that for $y = ax^2 + bx + c$. The difference is that the graph opens right or left instead of up or down.

EXAMPLE 2

Find the vertex and axis of symmetry of the parabola. Then sketch its graph.

a. $x = 2y^2 - 8y + 5$ b. $x = -2y^2 - 4y - 3$

Solution

a. $y = -\dfrac{b}{2a} = -\dfrac{-8}{2(2)} = 2$ • Find the y-coordinate of the vertex. $a = 2, b = -8$

$x = 2y^2 - 8y + 5$ • Find the x-coordinate of the vertex by replacing y with 2 and solving for x.
$x = 2(2)^2 - 8(2) + 5$
$x = -3$

The vertex is $(-3, 2)$.

The axis of symmetry is   • The axis of symmetry is the line
the line $y = 2$. $y = -\dfrac{b}{2a}$.

_____

? ANSWERS   a. $a = 4$. Because $a > 0$, the parabola opens to the right.   **b.** $a = -1$. Because $a < 0$, the parabola opens to the left.

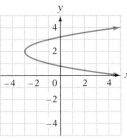

- Because a is positive ($a = 2$), the parabola opens to the right. Find a few ordered pairs, and use symmetry to sketch the graph.

b. $y = -\dfrac{b}{2a} = -\dfrac{-4}{2(-2)} = -1$

- Find the y-coordinate of the vertex. $a = -2, b = -4$

$x = -2y^2 - 4y - 3$

$x = -2(-1)^2 - 4(-1) - 3$

$x = -1$

- Find the x-coordinate of the vertex by replacing y with -1 and solving for x.

The vertex is $(-1, -1)$.

The axis of symmetry is the line $y = -1$.

- The axis of symmetry is the line $y = -\dfrac{b}{2a}$.

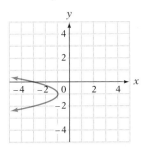

- Because a is negative ($a = -2$), the parabola opens to the left. Find a few ordered pairs, and use symmetry to sketch the graph.

YOU TRY IT 2

Find the vertex and axis of symmetry of the parabola. Then sketch its graph.

a. $x = 2y^2 - 4y + 1$ **b.** $x = -y^2 - 2y + 2$

Solution   See page S48.

a. $(-1, 1)$; $y = 1$

b. $(3, -1)$; $y = -1$

Note that we can determine the domain and range of the relation in Example 2b from the vertex and the fact that the graph of $x = -2y^2 - 4y - 3$ opens to the left. The domain is $\{x \mid x \le -1\}$. The range is $\{y \mid y \in \text{real numbers}\}$. This is verified in the graph of the equation.

? QUESTION   What are the domain and the range of the relation in Example 2a?

? ANSWER   The domain is $\{x \mid x \ge -3\}$. The range is $\{y \mid y \in \text{real numbers}\}$.

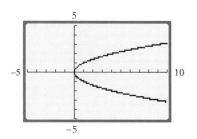

Because the graph of an equation of the form $x = ay^2 + by + c$ is not a function, its equation cannot be entered into a calculator. To examine a graph of this form using a graphing calculator, you must enter two equations.

For example, to graph $x = y^2$, solve the equation for y: $y = \pm\sqrt{x}$. Enter into the calculator the equations $Y_1 = \sqrt{X}$ and $Y_2 = -\sqrt{X}$. The result is shown here, using the window $\{x \mid -5 \leq x \leq 10\}$ and $\{y \mid -5 \leq y \leq 5\}$.

? **QUESTION** What equations would you enter into a graphing calculator to graph $x = -y^2$?

■ The Circle

A **circle** is a conic section formed by the intersection of a cone and a plane parallel to the base of the cone.

> **TAKE NOTE**
>
> As the angle of the plane that intersects the cone changes, different conic sections are formed. For a parabola, the plane is *parallel to the side* of the cone. For a circle, the plane is *parallel to the base* of the cone.

A **circle** can be defined as all the points $P(x, y)$ in the plane that are a fixed distance from a given point $C(h, k)$ called the **center.** The fixed distance is the **radius** of the circle.

The Standard Form of the Equation of a Circle

Let r be the radius of a circle and let $C(h, k)$ be the coordinates of the center of the circle. Then the equation of the circle is given by

$$(x - h)^2 + (y - k)^2 = r^2$$

➡ Sketch a graph of $(x - 1)^2 + (y + 2)^2 = 9$.

> **TAKE NOTE**
>
> Applying the vertical-line test reveals that the graph of a circle is not the graph of a function. The graph of a circle is the graph of a relation.

$(x - 1)^2 + [y - (-2)]^2 = 3^2$ • Rewrite the equation in standard form
Center: $(1, -2)$ to determine the center and radius.
Radius: 3

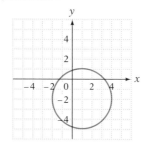

• Graph a circle with center $(1, -2)$ and a radius of 3 units.

? ANSWER   Enter $Y_1 = \sqrt{-X}$ and $Y_2 = -\sqrt{-X}$.

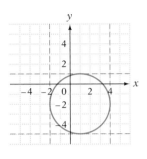

INSTRUCTOR NOTE
Exercise 16 can be used for a similar in-class example.

We can determine the domain and range of the relation

$$(x - 1)^2 + (y + 2)^2 = 9$$

from its graph. The domain is $\{x \mid -2 \leq x \leq 4\}$. The range is $\{y \mid -5 \leq y \leq 1\}$. See the graph at the left.

EXAMPLE 3

Sketch a graph of $(x + 2)^2 + (y - 1)^2 = 4$.

Solution
$$(x - h)^2 + (y - k)^2 = r^2$$
$$[x - (-2)]^2 + (y - 1)^2 = 2^2$$
Center: $(h, k) = (-2, 1)$
Radius: $r = 2$

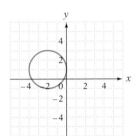

YOU TRY IT 3

Sketch a graph of $(x - 2)^2 + (y + 3)^2 = 9$.
Solution   See page S49.

? **QUESTION**   What are the domain and the range of the relation in Example 3?

Because the graph of a circle is not a function, its equation cannot be entered into a calculator. To examine a graph of this form using a graphing calculator, you must enter two equations.

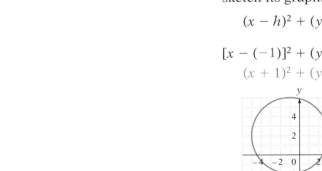

For example, to graph $x^2 + y^2 = 1$, solve the equation for y: $y = \pm \sqrt{1 - x^2}$. Enter into the calculator the equations $Y_1 = \sqrt{1 - X^2}$ and $Y_2 = -\sqrt{1 - X^2}$. The result is shown at the left in the square viewing window. Using the square viewing window ensures that the graph is not distorted.

➡ Find the equation of the circle with radius 4 and center $(-1, 2)$. Then sketch its graph.

$$(x - h)^2 + (y - k)^2 = r^2$$ • Use the standard form of the equation of a circle.

$$[x - (-1)]^2 + (y - 2)^2 = 4^2$$ • Replace h with -1, k with 2, and r with 4.
$$(x + 1)^2 + (y - 2)^2 = 16$$

• Sketch the graph by drawing a circle with center $(-1, 2)$, and radius 4.

? <u>ANSWER</u>   The domain is $\{x \mid -4 \leq x \leq 0\}$. The range is $\{y \mid -1 \leq y \leq 3\}$.

INSTRUCTOR NOTE
Exercise 22 can be used for
a similar in-class example.

EXAMPLE 4

Find the equation of the circle with radius 5 and center $(-1, 3)$. Then sketch its graph.

Solution
$$(x - h)^2 + (y - k)^2 = r^2$$
$$[x - (-1)]^2 + (y - 3)^2 = 5^2$$
$$(x + 1)^2 + (y - 3)^2 = 25$$

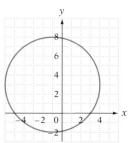

YOU TRY IT 4

Find the equation of the circle with radius 4 and center $(2, -3)$. Then sketch its graph.

Solution   See page S49. $(x - 2)^2 + (y + 3)^2 = 16$

▼ *Point of Interest*

Appollonius, a 3rd-century Greek mathematician, showed in his book The Conics *that conic sections could be produced by slicing a cone in different ways. Edmund Halley (1656–1742), an English astronomer, translated this text. Halley's comet travels along an elliptical orbit, and Halley used this fact to predict the time of the comet's return.*

■ The Ellipse

The orbits of the planets around the sun are "oval" shaped. This oval shape can be described as an **ellipse,** which is another of the conic sections.

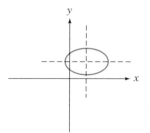

There are two **axes of symmetry** for an ellipse. The intersection of these two axes is the **center** of the ellipse.

An ellipse with center at the origin is shown at the right. Note that there are two x-intercepts and two y-intercepts.

Using the vertical-line test, we find that the graph of an ellipse is not the graph of a function. The graph of an ellipse is the graph of a relation.

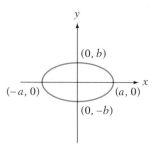

The Standard Form of the Equation of an Ellipse with Center at the Origin

The equation of an ellipse with center at the origin is $\dfrac{x^2}{a^2} + \dfrac{y^2}{b^2} = 1$.

The x-intercepts are $(a, 0)$ and $(-a, 0)$. The y-intercepts are $(0, b)$ and $(0, -b)$.

By finding the x- and y-intercepts for an ellipse and using the fact that the ellipse is "oval" shaped, we can sketch a graph of an ellipse.

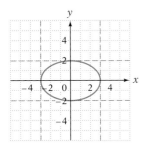

➡ Sketch the graph of the ellipse whose equation is $\frac{x^2}{9} + \frac{y^2}{4} = 1$.

Comparing $\frac{x^2}{9} + \frac{y^2}{4} = 1$ with $\frac{x^2}{a^2} + \frac{y^2}{b^2} = 1$, we have $a^2 = 9$ and $b^2 = 4$.
Therefore, $a = 3$ and $b = 2$.

The x-intercepts are $(3, 0)$ and $(-3, 0)$.
The y-intercepts are $(0, 2)$ and $(0, -2)$.

Use the intercepts to sketch a graph of the ellipse.

We can determine the domain and range of the relation $\frac{x^2}{9} + \frac{y^2}{4} = 1$ from its graph. The domain is $\{x \mid -3 \leq x \leq 3\}$. The range is $\{y \mid -2 \leq y \leq 2\}$. See the graph at the left. ⬅

We can also determine the domain and range from the equation of the ellipse. **In general, the domain of an ellipse is $\{x \mid -|a| \leq x \leq |a|\}$, and the range is $\{y \mid -|b| \leq y \leq |b|\}$.**

INSTRUCTOR NOTE
Exercise 26 can be used for a similar in-class example.

___

Suggested Activity

See Additional Topics in Algebra Section 3 of the *Student Activity Manual* for an investigation involving folding conic sections.

___

EXAMPLE 5

Sketch a graph of the ellipse given by the equation.

a. $\dfrac{x^2}{9} + \dfrac{y^2}{16} = 1$ **b.** $\dfrac{x^2}{16} + \dfrac{y^2}{12} = 1$

Solution

a. x-intercepts:
$(3, 0)$ and $(-3, 0)$
y-intercepts:
$(0, 4)$ and $(0, -4)$

b. x-intercepts:
$(4, 0)$ and $(-4, 0)$
y-intercepts:
$(0, 2\sqrt{3})$ and $(0, -2\sqrt{3})$
$[2\sqrt{3} \approx 3.5]$

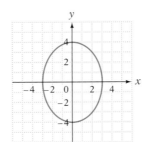

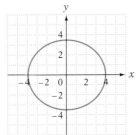

YOU TRY IT 5

Sketch a graph of the ellipse given by the equation.

a. $\dfrac{x^2}{4} + \dfrac{y^2}{25} = 1$ **b.** $\dfrac{x^2}{18} + \dfrac{y^2}{9} = 1$

Solution   See page S49.   a.

b.

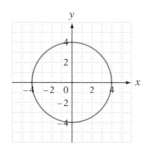

▼ *Point of Interest*

Hyperbolas are used in LORAN (LOng RAnge Navigation) as a method for a ship's navigator to determine the position of the ship.

? QUESTION   What are the domain and the range of the relation in Example 5a?

Shown at the left is the graph of the equation $\frac{x^2}{16} + \frac{y^2}{16} = 1$.

In this equation, $a^2 = 16$ and $b^2 = 16$. Therefore, $a = 4$ and $b = 4$.

The x-intercepts are $(4, 0)$ and $(-4, 0)$. The y-intercepts are $(0, 4)$ and $(0, -4)$. This is the graph of a circle. **A circle is a special case of an ellipse. It occurs when $a^2 = b^2$ in the equation $\frac{x^2}{a^2} + \frac{y^2}{b^2} = 1$.**

■ The Hyperbola

A **hyperbola** is a conic section that is formed by the intersection of a cone and a plane perpendicular to the base of the cone.

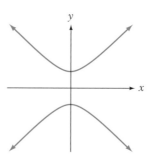

The hyperbola has two **vertices** and an **axis of symmetry** that passes through the vertices. The **center** of a hyperbola is the point halfway between the two vertices.

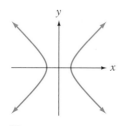
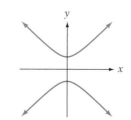

The graphs at the left show two possible graphs of a hyperbola with center at the origin.

In the first graph, the branches open to the left and right, and the vertices are x-intercepts.

In the second graph, the branches open up and down, and the vertices are y-intercepts.

Note that in either case, the graph of a hyperbola is not the graph of a function. The graph of a hyperbola is the graph of a relation.

The Standard Form of the Equation of a Hyperbola with Center at the Origin

The equation of a hyperbola for which the vertices are on the x-axis is $\frac{x^2}{a^2} - \frac{y^2}{b^2} = 1$. The vertices are $(a, 0)$ and $(-a, 0)$.

The equation of a hyperbola for which the vertices are on the y-axis is $\frac{y^2}{b^2} - \frac{x^2}{a^2} = 1$. The vertices are $(0, b)$ and $(0, -b)$.

? ANSWER   The domain is $\{x \mid -3 \leq x \leq 3\}$. The range is $\{y \mid -4 \leq y \leq 4\}$.

To sketch a hyperbola, it is helpful to draw two lines that are "approached" by the hyperbola. These two lines are called **asymptotes.** As the hyperbola gets farther from the origin, the hyperbola "gets closer to" the asymptotes.

Because the asymptotes are straight lines, their equations are linear equations. **The equations of the asymptotes for a hyperbola with center at the origin are $y = \frac{b}{a}x$ and $y = -\frac{b}{a}x$.**

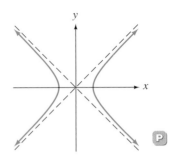

➡ Sketch the graph of the hyperbola whose equation is $\frac{y^2}{9} - \frac{x^2}{4} = 1$.

This is an equation of the form $\frac{y^2}{b^2} - \frac{x^2}{a^2} = 1$ with $b^2 = 9$ and $a^2 = 4$.

The vertices are on the y-axis.
The vertices are $(0, b)$ and $(0, -b)$: $(0, 3)$ and $(0, -3)$.

The asymptotes are $y = \frac{b}{a}x$ and $y = -\frac{b}{a}x$: $y = \frac{3}{2}x$ and $y = -\frac{3}{2}x$.

Sketch the asymptotes. Use symmetry and the fact that the hyperbola approaches the asymptotes to sketch its graph.

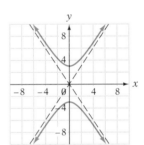

> **TAKE NOTE**
>
> The domain of this relation is $\{x | x \in \text{real numbers}\}$.
> The range is $\{y | y \le -3 \text{ or } y \ge 3\}$.

INSTRUCTOR NOTE
Exercises 32 and 34 can be used for similar in-class examples.

Suggested Activity

Find the equation of the hyperbola with vertices $(0, 4)$ and $(0, -4)$ and asymptotes $y = \frac{1}{2}x$ and $y = -\frac{1}{2}x$.

[Answer: $\frac{y^2}{16} - \frac{x^2}{64} = 1$]

EXAMPLE 6

Sketch a graph of the hyperbola given by the equation.

a. $\dfrac{x^2}{16} - \dfrac{y^2}{4} = 1$ **b.** $\dfrac{y^2}{16} - \dfrac{x^2}{25} = 1$

Solution

a. $a^2 = 16$, $b^2 = 4$
  The vertices are on the x-axis.
  Vertices: $(4, 0)$ and $(-4, 0)$

  Asymptotes: $y = \dfrac{1}{2}x$ and $y = -\dfrac{1}{2}x$

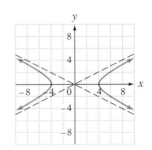

b. $b^2 = 16$, $a^2 = 25$
The vertices are on the y-axis.
Vertices: $(0, 4)$ and $(0, -4)$
Asymptotes: $y = \dfrac{4}{5}x$ and $y = -\dfrac{4}{5}x$

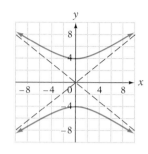

YOU TRY IT 6

Sketch a graph of the hyperbola given by the equation.

a. $\dfrac{x^2}{9} - \dfrac{y^2}{25} = 1$ **b.** $\dfrac{y^2}{9} - \dfrac{x^2}{9} = 1$

Solution   See page S49.   a.   b.

3 EXERCISES   Suggested Assignment: Exercises 7–77, odds

Topics for Discussion

1. How can you determine whether the graph of $x = ay^2 + by + c$, $a \neq 0$, opens to the right or to the left? Provide an example of an equation whose graph opens to the right and an example of an equation whose graph opens to the left.

   If $a > 0$, it opens to the right. If $a < 0$, it opens to the left. Examples will vary.

2. Explain how to determine the vertex of the graph of an equation of the form $x = ay^2 + by + c$, $a \neq 0$.

   Find the y-coordinate of the vertex using the equation $y = -\frac{b}{2a}$. Find the x-coordinate of the vertex by substituting the value of the y-coordinate of the vertex into $x = ay^2 + by + c$ and solving for x.

3. Write the standard form of the equation of a circle. What do the variables h, k, and r in the equation represent?

   $(x - h)^2 + (y - k)^2 = r^2$. h is the x-coordinate of the center of the circle, k is the y-coordinate of the center of the circle, and r is the radius of the circle.

4. How can you distinguish the equation of an ellipse from the equation of a hyperbola?

   Answers may vary. For example, students may note that the terms are added in $\frac{x^2}{a^2} + \frac{y^2}{b^2} = 1$, the equation of an ellipse, and subtracted in $\frac{x^2}{a^2} - \frac{y^2}{b^2} = 1$, the equation of a hyperbola.

5. How can you determine whether the branches in the graph of a hyperbola open to the left and right or open up and down?

    Answers may vary. For example, students may note that the branches open to the left and right when the x^2 term is first and open up and down when the y^2 term is first.

6. How can you determine whether the graph of a hyperbola has x-intercepts or y-intercepts?

    Answers may vary. For example, students may note that the hyperbola has x-intercepts when the x^2 term is first and y-intercepts when the y^2 term is first.

■ Parabolas, Circles, Ellipses, and Hyperbolas

Find the vertex and axis of symmetry of the parabola given by the equation. Then sketch its graph.

7. $y = x^2 - 2x - 4$

Vertex: $(1, -5)$
Axis of symmetry: $x = 1$

8. $y = x^2 + 4x - 4$

Vertex: $(-2, -8)$
Axis of symmetry: $x = -2$

9. $y = -x^2 + 2x - 3$

Vertex: $(1, -2)$
Axis of symmetry: $x = 1$

10. $x = y^2 + 6y + 5$

Vertex: $(-4, -3)$
Axis of symmetry: $y = -3$

11. $x = y^2 - 2y - 5$

Vertex: $(-6, 1)$
Axis of symmetry: $y = 1$

12. $x = -\frac{1}{2}y^2 + 4$

Vertex: $(4, 0)$
Axis of symmetry: $y = 0$

13. $x = -\frac{1}{4}y^2 - 1$

Vertex: $(-1, 0)$
Axis of symmetry: $y = 0$

14. $x = \frac{1}{2}y^2 - y + 1$

Vertex: $\left(\frac{1}{2}, 1\right)$
Axis of symmetry: $y = 1$

15. $x = -\frac{1}{2}y^2 + 2y - 3$

Vertex: $(-1, 2)$
Axis of symmetry: $y = 2$

Sketch a graph of the circle given by the equation.

16. $(x - 2)^2 + (y + 2)^2 = 9$

17. $(x + 2)^2 + (y - 3)^2 = 16$

18. $(x + 3)^2 + (y - 1)^2 = 25$

19. $(x - 2)^2 + (y + 3)^2 = 4$

20. $(x + 5)^2 + (y + 2)^2 = 4$

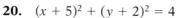

21. $(x + 1)^2 + (y - 1)^2 = 9$

22. Find the equation of the circle with radius 2 and center $(2, -1)$. Then sketch its graph.
$(x - 2)^2 + (y + 1)^2 = 4$

23. Find the equation of the circle with radius 3 and center $(-1, -2)$. Then sketch its graph.
$(x + 1)^2 + (y + 2)^2 = 9$

24. Find the equation of the circle with radius $\sqrt{5}$ and center $(-1, 1)$. Then sketch its graph.
$(x + 1)^2 + (y - 1)^2 = 5$

25. Find the equation of the circle with radius $\sqrt{5}$ and center $(-2, 1)$. Then sketch its graph.
$(x + 2)^2 + (y - 1)^2 = 5$

Sketch a graph of the ellipse given by the equation.

26. $\dfrac{x^2}{4} + \dfrac{y^2}{9} = 1$

27. $\dfrac{x^2}{25} + \dfrac{y^2}{16} = 1$

28. $\dfrac{x^2}{36} + \dfrac{y^2}{16} = 1$

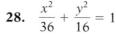

29. $\dfrac{x^2}{49} + \dfrac{y^2}{64} = 1$

30. $\dfrac{x^2}{8} + \dfrac{y^2}{25} = 1$

31. $\dfrac{x^2}{12} + \dfrac{y^2}{4} = 1$

Sketch a graph of the hyperbola given by the equation.

32. $\dfrac{x^2}{9} - \dfrac{y^2}{16} = 1$

33. $\dfrac{x^2}{25} - \dfrac{y^2}{4} = 1$

34. $\dfrac{y^2}{16} - \dfrac{x^2}{9} = 1$

35. $\dfrac{y^2}{4} - \dfrac{x^2}{9} = 1$

36. $\dfrac{x^2}{4} - \dfrac{y^2}{25} = 1$

37. $\dfrac{x^2}{9} - \dfrac{y^2}{49} = 1$

38. $\dfrac{y^2}{25} - \dfrac{x^2}{9} = 1$

39. $\dfrac{y^2}{4} - \dfrac{x^2}{16} = 1$

40. $\dfrac{x^2}{9} - \dfrac{y^2}{9} = 1$

Determine the domain and the range of the relation.

41. $y = x^2 - 4x - 2$
D: $\{x \mid x \in \text{real numbers}\}$
R: $\{y \mid y \geq -6\}$

42. $y = x^2 - 6x + 1$
D: $\{x \mid x \in \text{real numbers}\}$
R: $\{y \mid y \geq -8\}$

43. $y = -x^2 + 2x - 3$
D: $\{x \mid x \in \text{real numbers}\}$
R: $\{y \mid y \leq -2\}$

44. $y = -x^2 - 2x + 4$
D: $\{x \mid x \in \text{real numbers}\}$
R: $\{y \mid y \leq 5\}$

45. $x = y^2 + 6y - 5$
D: $\{x \mid x \geq -14\}$
R: $\{y \mid y \in \text{real numbers}\}$

46. $x = y^2 + 4y - 3$
D: $\{x \mid x \geq -7\}$
R: $\{y \mid y \in \text{real numbers}\}$

47. $x = -y^2 - 2y + 6$
D: $\{x \mid x \leq 7\}$
R: $\{y \mid y \in \text{real numbers}\}$

48. $x = -y^2 - 6y + 2$
D: $\{x \mid x \leq 11\}$
R: $\{y \mid y \in \text{real numbers}\}$

49. $(x + 3)^2 + (y - 6)^2 = 25$
D: $\{x \mid -8 \leq x \leq 2\}$
R: $\{y \mid 1 \leq y \leq 11\}$

50. $(x - 4)^2 + (y + 5)^2 = 36$
D: $\{x \mid -2 \leq x \leq 10\}$
R: $\{y \mid -11 \leq y \leq 1\}$

51. $\dfrac{x^2}{25} + \dfrac{y^2}{9} = 1$
D: $\{x \mid -5 \leq x \leq 5\}$
R: $\{y \mid -3 \leq y \leq 3\}$

52. $\dfrac{x^2}{16} + \dfrac{y^2}{9} = 1$
D: $\{x \mid -4 \leq x \leq 4\}$
R: $\{y \mid -3 \leq y \leq 3\}$

53. $\dfrac{x^2}{25} - \dfrac{y^2}{16} = 1$
D: $\{x \mid x \leq -5 \text{ or } x \geq 5\}$
R: $\{y \mid y \in \text{real numbers}\}$

54. $\dfrac{y^2}{9} - \dfrac{x^2}{36} = 1$
D: $\{x \mid x \in \text{real numbers}\}$
R: $\{y \mid y \leq -3 \text{ or } y \geq 3\}$

55. $\dfrac{y^2}{16} - \dfrac{x^2}{4} = 1$
D: $\{x \mid x \in \text{real numbers}\}$
R: $\{y \mid y \leq -4 \text{ or } y \geq 4\}$

56. Find the equation of a circle that has center $(5, -6)$ and has an area of 49π square units.
$(x - 5)^2 + (y + 6)^2 = 49$

57. Find the equation of a circle that has center $(4, 0)$ and passes through the origin.
$(x - 4)^2 + y^2 = 16$

Sketch a graph of the conic section given by the equation.

58. $y = \dfrac{1}{2}x^2 + 2x - 6$

59. $x = y^2 - y - 6$

60. $(x - 4)^2 + (y + 2)^2 = 1$

61. $(x - 3)^2 + (y - 2)^2 = 16$

62. $\dfrac{x^2}{9} + \dfrac{y^2}{25} = 1$

63. $\dfrac{x^2}{36} + \dfrac{y^2}{9} = 1$

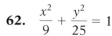

64. $\dfrac{y^2}{9} - \dfrac{x^2}{36} = 1$

65. $\dfrac{x^2}{25} - \dfrac{y^2}{9} = 1$

66. $\dfrac{x^2}{16} - \dfrac{y^2}{25} = 1$

67. A circle has its center at the point (3, 0) and passes through the origin. Find the equation of the circle. $(x - 3)^2 + y^2 = 9$

68. A diameter of a circle has endpoints $P_1(-1, 3)$ and $P_2(5, 5)$. Find the equation of the circle. $(x - 2)^2 + (y - 4)^2 = 10$

69. A diameter of a circle has endpoints $P_1(-2, 4)$ and $P_2(2, -2)$. Find the equation of the circle. $x^2 + (y - 1)^2 = 13$

70. A circle has a radius of 1 unit, is tangent to both the x- and y-axes, and lies in quadrant II. Find the equation of the circle. $(x + 1)^2 + (y - 1)^2 = 1$

71. A circle has a radius of 1 unit, is tangent to both the x- and y-axes, and lies in quadrant IV. Find the equation of the circle. $(x - 1)^2 + (y + 1)^2 = 1$

72. Find the equation of the circle with center at (3, 3) if the circle is tangent to the x-axis. $(x - 3)^2 + (y - 3)^2 = 9$

73. Find the equation of the circle that is a translation of $x^2 + y^2 = 12$ to the right 6 units and down 7 units. $(x - 6)^2 + (y + 7)^2 = 12$

74. 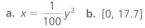 *Communication Satellites*   Many communication satellites orbit our planet at an altitude of approximately 22,500 miles above its surface. Write an equation for the orbit of a communication satellite. Consider Earth's center the origin and the orbit of the satellite circular. (*Hint:* Earth's radius is approximately 4000 miles.) $x^2 + y^2 = (26,500)^2$

75. *Weather Forecasting*   A radar dish used in the Cassegrain radar system has a cross section that is a parabola. The radar dish, used in weather forecasting, has a diameter of 84 feet. It is made of structural steel and has a depth of 17.7 feet. Signals from the radar system are reflected off clouds, collected by the radar system, and then analyzed.

   a. Determine an equation of the radar dish. Round to the nearest whole number.

   b. Over what interval for x is the equation valid?

a. $x = \dfrac{1}{100}y^2$ b. [0, 17.7]

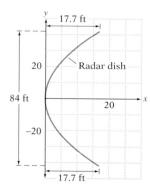

Cassegrain Radar Dish

76. *Comets*   The orbit of Halley's comet is an ellipse with a major axis of approximately 36 AU and a minor axis of approximately 9 AU. (One AU is one astronomical unit and is approximately 92,960,000 miles, the average distance of Earth from the sun.)

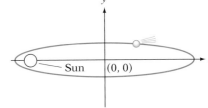

   a. Determine an equation for the orbit of Halley's comet in terms of astronomical units. See the diagram at the right.

   b. The distance of the sun from the center of Halley's comet's elliptical orbit is $\sqrt{a^2 - b^2}$. The aphelion of the orbit (the point at which the comet is farthest from the sun) is a vertex on the major axis. Determine, to the nearest hundred thousand miles, the distance from the sun to the point at the aphelion of Halley's comet.

   c. The perihelion of the orbit (the point at which the comet is closest to the sun) is a vertex on the major axis. Determine, to the nearest hundred thousand miles, the distance from the sun to the point at the perihelion of Halley's comet.

   a. $\dfrac{x^2}{324} + \dfrac{y^2}{20.25} = 1$ **b.** 3,293,400,000 mi   **c.** 53,100,000 mi

77. *Comets*   The orbit of the comet Hale–Bopp is an ellipse, as shown at the right. The units are astronomical units, abbreviated AU. (One AU is one astronomical unit and is approximately 92,960,000 miles, the average distance of Earth from the sun.)

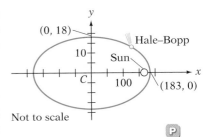

   a. Find the equation of the orbit of the comet.

   b. The distance from the center, C, of the orbit to the sun is approximately 182.085 AU. Find the aphelion (the point at which the comet is farthest from the sun) in miles. Round to the nearest million miles.

   c. Find the perihelion (the point at which the comet is closest to the sun) in miles. Round to the nearest hundred thousand miles.

   a. $\dfrac{x^2}{33,489} + \dfrac{y^2}{324} = 1$ **b.** 33,938,000,000 mi   **c.** 85,100,000 mi

78. *Solar System*   The orbits of the planets in our solar system are elliptical. The length of the major axis of the orbit of Mars is 3.04 AU. (See Exercise 76.) The length of the minor axis is 2.99 AU.

   a. Determine an equation for the orbit of Mars.

   b. Determine the aphelion to the nearest hundred thousand miles.

   c. Determine the perihelion to the nearest hundred thousand miles.

   a. $\dfrac{x^2}{2.310} + \dfrac{y^2}{2.235} = 1$ **b.** 166,800,000 mi   **c.** 115,800,000 mi

Applying Concepts

Sketch a graph of the conic section given by the equation. (*Hint:* Divide each term by the number on the right side of the equation.)

79. $4x^2 + y^2 = 16$

80. $x^2 - y^2 = 9$

81. $y^2 - 4x^2 = 16$

82. $9x^2 + 4y^2 = 144$

83. $9x^2 - 25y^2 = 225$

84. $4y^2 - x^2 = 36$

85. When are the asymptotes of the graph of $\dfrac{x^2}{a^2} - \dfrac{y^2}{b^2} = 1$ perpendicular?

When $a = b$.

86. Find the integer value(s) of $x + y$ if $x^2 + y^2 = 36$ and $xy = -10$. $-4, 4$

87. Find the shortest distance between the graphs of the equations $x^2 + y^2 = 1$ and $(x - 5)^2 + (y - 12)^2 = 1$.   11 units

88. The line $x = 5$ crosses the circle $x^2 + y^2 = 61$ at the points A and B. Determine the length of AB.   12 units

89. Find the area of the smallest region bounded by the graphs of $y = |x|$ and $x^2 + y^2 = 4$. π square units

90. Explain the relationship between the distance formula and the standard form of the equation of a circle.

Answers may vary. For example: The distance formula is used to derive the equation of a circle. If $C(h, k)$ is a fixed point in the plane, and $P(x, y)$ is any other point in the plane, then the distance between C and P is $r = \sqrt{(x - h)^2 + (y - k)^2}$. Squaring each side of this equation gives the equation of a circle in standard form: $r^2 = (x - h)^2 + (y - k)^2$.

91. *The Solar System*   As shown in this section, the graph of the ellipse whose equation is $\dfrac{x^2}{16} + \dfrac{y^2}{16} = 1$ is a circle with a radius of 4 units. For a circle, $a = b$ in the equation $\dfrac{x^2}{a^2} + \dfrac{y^2}{b^2} = 1$. Thus $\dfrac{a}{b} = 1$. Early Greek astronomers thought that each planet had a circular orbit. Today we know that the planets have elliptical orbits. However, in most cases the ellipse is very nearly a circle. For Earth, $\dfrac{a}{b} \approx 1.00014$. The most elliptical orbit is Pluto's. For its orbit, $\dfrac{a}{b} \approx 1.0328$.

   a. Write an equation that approximates Earth's orbit.
   b. Write an equation that approximates Pluto's orbit.

Answers may vary. For example: **a.** $\dfrac{x^2}{(100{,}014)^2} + \dfrac{y^2}{(100{,}000)^2} = 1$ **b.** $\dfrac{x^2}{(10{,}328)^2} + \dfrac{y^2}{(10{,}000)^2} = 1$

92. *Geometry*   The radius of a sphere is 12 inches. What is the radius of the circle that is formed by the intersection of a plane and the sphere at a point 6 inches from the center of the sphere? $6\sqrt{3}$ in.

93. Besides the curves presented in this section, how else might the intersection of a plane and a cone be represented?

Answers may vary. (1) The intersection of a plane perpendicular to the axis of the cone and through the vertex of the cone is a point; (2) the intersection of a plane parallel to the axis of the cone and through the vertex of the cone forms two intersecting straight lines; and (3) the intersection of a plane and the lateral surface of the cone is a line.

EXPLORATION

1. *The Focus of a Parabola*   Parabolas have a unique property that is important in the design of telescopes and antennas. If a parabola has a mirrored surface, then all light rays parallel to the axis of symmetry of the parabola are reflected to a single point called the **focus** of the parabola. The location of this point is p units from the vertex on the axis of symmetry. The value of p is given by $p = \frac{1}{4a}$, where $y = ax^2$ or $x = ay^2$ is the equation of a parabola with vertex at the origin. For the graph of $y = \frac{1}{4}x^2$ shown at the right, $p = \dfrac{1}{4\left(\frac{1}{4}\right)} = 1$. The focus is 1 unit

from the vertex on the axis of symmetry. The coordinates of the focus are $(0, 1)$.

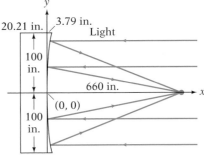

Parallel rays of light are reflected to the focus

$a = \frac{1}{4}$

$p = \frac{1}{4a} = \frac{1}{4(1/4)} = 1$

In parts a through f, find the coordinates of the focus of the parabola.

   a. $y = 2x^2$ **b.** $y = \frac{1}{10}x^2$ **c.** $y = 2x^2 - 4x + 1$

   d. $y = -\frac{1}{4}x^2 + 2$ **e.** $x = \frac{1}{2}y^2 + y - 2$ **f.** $x = -y^2 - 4y + 1$

   g. Find the equation of the parabola with vertex at the origin and focus $(0, -4)$.

   h. Find the equation of the parabola with vertex at the origin and focus $(5, 0)$.

   i.   The 200-inch mirror at the Palomar Observatory in California is made from Pyrex, is 2 feet thick at the ends, and weighs 14.75 tons. The cross section of the mirror has been ground to a true parabola within 0.0000015 inch. No matter where light strikes the parabolic surface, the light is reflected to the focus of the parabola, as shown in the figure at the right. Determine an equation of the mirror.

   a. $(0, \frac{1}{8})$ **b.** $(0, 2.5)$ **c.** $(1, -\frac{7}{8})$ **d.** $(0, 1)$ **e.** $(-2, -1)$ **f.** $(\frac{19}{4}, -2)$

   g. $y = -\frac{1}{16}x^2$ **h.** $x = \frac{1}{20}y^2$ **i.** $x = \frac{1}{2639}y^2$

2. *The Eccentricity and Foci of an Ellipse*   The graph of an ellipse can be long and thin, or it can have a shape that is very close to a circle. The **eccentricity,** *e,* of an ellipse is a measure of its "roundness."

The shapes of ellipses with various eccentricities are shown below.

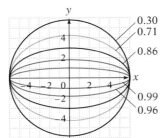

   a. Based on the eccentricities of the ellipses shown above, answer the following question: "As the eccentricity of an ellipse gets closer to 1, do the ellipses get <u>flatter or rounder</u>?"

b. 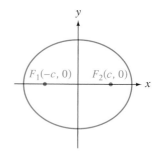 The planets travel around the sun in elliptical orbits. The eccentricities of the orbits of the planets are shown in the table at the right. Which planet has the most nearly circular orbit?

Planet	Eccentricity
Mercury	0.206
Venus	0.007
Earth	0.017
Mars	0.093
Jupiter	0.049
Saturn	0.051
Uranus	0.046
Neptune	0.005
Pluto	0.250

For an ellipse given by the equation $\frac{x^2}{a^2} + \frac{y^2}{b^2} = 1$, $a > b$, a formula for eccentricity is $e = \frac{\sqrt{a^2 - b^2}}{a}$. Use this formula to find the eccentricity of the ellipses in parts c and d. If necessary, round to the nearest hundredth.

c. $\dfrac{x^2}{25} + \dfrac{y^2}{16} = 1$ **d.** $\dfrac{x^2}{9} + \dfrac{y^2}{4} = 1$

Ellipses have a reflective property that has been used in the design of some buildings. The **foci** of an ellipse are two points on the longer axis, called the **major axis,** of the ellipse. The foci are c units from the center of an ellipse, where $c = \sqrt{a^2 - b^2}$ for an ellipse whose equation is $\frac{x^2}{a^2} + \frac{y^2}{b^2} = 1$, $a > b$.

e. Find the foci of an ellipse whose equation is $\dfrac{x^2}{169} + \dfrac{y^2}{144} = 1$.

f. 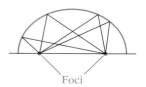 If light or sound emanates from one focus of an ellipse, it is reflected to the other focus. This phenomenon results in what are called "whispering galleries." Statuary Hall in the rotunda of the Capitol Building in Washington, DC, is a whispering gallery; it is an elliptical chamber in which a whisper spoken by a person at one focus can be heard clearly by a person at the other focus. The gallery is approximately 78 feet wide and 95 feet long. Write an equation to describe the ellipse.

a. flatter   **b.** Neptune   **c.** 0.6   **d.** 0.75   **e.** $(-5, 0)$ and $(5, 0)$

f. $\dfrac{x^2}{2256.25} + \dfrac{y^2}{1521} = 1$

Keystroke Guide for the TI-83 and TI-83 Plus

This appendix contains some keystroke suggestions for many graphing calculator operations that are featured in this text. The keystrokes are for the TI-83 and TI-83 Plus calculators. The descriptions in the margin are the same as those used in the text and are arranged alphabetically. Please see your manual for additional information about your calculator.

Basic Operations

Numerical calculations are performed on the **home screen.** You can always return to the home screen by pressing `2nd` QUIT. Pressing `CLEAR` erases the home screen.

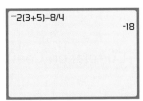

To evaluate the expression $-2(3 + 5) - 8 \div 4$, use the following keystrokes.

`(-)` 2 `(` 3 `+` 5 `)` `−` 8 `÷` 4 `ENTER`

> Note: There is a difference between the key to enter a negative number, `(-)`, and the key for subtraction, `−`. You cannot use these keys interchangeably.

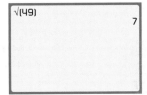

The `2nd` key is used to access the commands in gold writing above a key. For instance, to evaluate the $\sqrt{49}$, press `2nd` $\sqrt{}$ 49 `)` `ENTER`.

The `ALPHA` key is used to place a letter on the screen. One reason to do this is to store a value of a variable. The following keystrokes give A the value of 5.

5 `STO▸` `ALPHA` A `ENTER`

This value is now available in calculations. For instance, we can find the value of $3a^2$ by using the following keystrokes: 3 `ALPHA` A x^2. To display the value of the variable on the screen, press `2nd` RCL `ALPHA` A.

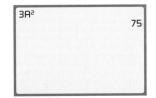

> Note: When you use the `ALPHA` key, only capital letters are available on the TI-83 calculator.

Complex Numbers

To perform operations on complex numbers, first press `MODE` and then use the arrow keys to select a+bi. Then press `ENTER` `2nd` QUIT.

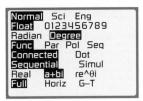

Addition of complex numbers To add $(3 + 4i) + (2 - 7i)$, use the keystrokes

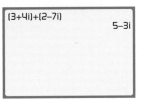

`(` 3 `+` 4 `2nd` i `)` `+`
`(` 2 `−` 7 `2nd` i `)` `ENTER`.

Division of complex numbers. To divide $\dfrac{26 + 2i}{2 + 4i}$, use the keystrokes  26 $+$ 2 2nd i) $\div$ (2 $+$ 4 2nd i) ENTER .

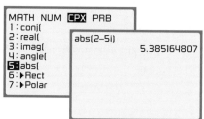

Note: Operations for subtraction and multiplication are similar.

Additional operations on complex numbers can be found by selecting **CPX** under the MATH key.

To find the absolute value of $2 - 5i$, press MATH (scroll to **CPX**) (scroll to **abs**) ENTER (2 $-$ 5 2nd i) ENTER .

Correlation Coefficient

The value of the correlation coefficient for a regression equation calculation is not shown unless the **DiagnosticOn** feature is enabled. To enable this feature, press 2nd CATALOG D (scroll to **DiagnosticOn**) ENTER ENTER .

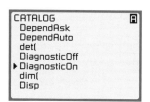 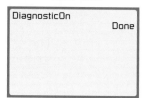

To calculate the correlation coefficient, proceed as if calculating a regression equation.

Evaluating Functions

There are various methods of evaluating a function but all methods require that the expression be entered as one of the ten functions Y₁ to Y₀. To evaluate $f(x) = \dfrac{x^2}{x - 1}$ when $x = -3$, enter the expression into, for instance, Y₁, and then press VARS ▶ 1 1 ((−) 3) ENTER .

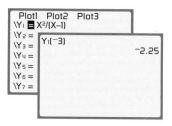

> Note: If you try to evaluate a function at a number that is not in the domain of the function, you will get an error message. For instance, 1 is not in the domain of $f(x) = \dfrac{x^2}{x - 1}$. If we try to evaluate the function at 1, the error screen at the right appears.

TAKE NOTE

Use the down arrow key to scroll past Y₇ to see Y₈, Y₉, and Y₀.

Evaluating Variable Expressions

To evaluate a variable expression, first store the values of each variable. Then enter the variable expression. For instance, to evaluate $s^2 + 2sl$ when $s = 4$ and $l = 5$, use the following keystrokes.

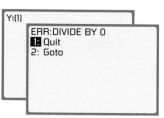

4 STO▸ ALPHA S ENTER 5 STO▸ ALPHA L ENTER ALPHA S

 x^2 $+$ 2 ALPHA S ALPHA L ENTER

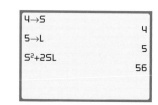

Graph

To graph a function, use the [Y=] key to enter the expression for the function, select a suitable viewing window, and then press [GRAPH]. For instance, to graph $f(x) = 0.1x^3 - 2x - 1$ in the standard viewing window, use the following keystrokes.

[Y=] 0.1 [X,T,θ,n] [^] 3 [−] 2 [X,T,θ,n] [−] 1 [ZOOM] (scroll to 6) [ENTER]

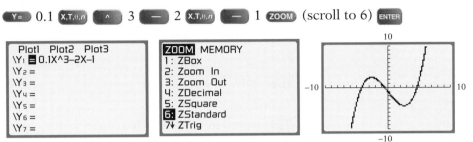

Note: For the keystrokes above, you do not have to scroll to 6. Alternatively, use [ZOOM] 6. This will select the standard viewing window and automatically start the graph. Use the [WINDOW] key to create a custom window for a graph.

Graphing Inequalities

To illustrate this feature, we will graph $y \le 2x - 1$. Enter $2x - 1$ into Y_1. Because $y \le 2x - 1$, we want to shade below the graph. Move the cursor to the left of Y_1 and press [ENTER] three times. Press [GRAPH].

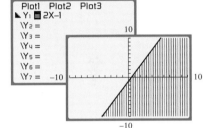

Note: To shade above the graph, move the cursor to the left of Y_1 and press [ENTER] two times. An inequality with the symbol $\le$ or $\ge$ should be graphed with a solid line, and an inequality with the symbol $<$ or $>$ should be graphed with a dashed line. However, the graph of a linear inequality on a graphing calculator does not distinguish between a solid line and a dashed line.

To graph the solution set of a system of inequalities, solve each inequality for y and graph each inequality. The solution set is the intersection of the two inequalities. The solution set of $\begin{aligned} 3x + 2y &> 10 \\ 4x - 3y &\le 5 \end{aligned}$ is shown at the right.

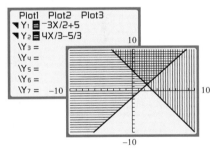

Intersect

The INTERSECT feature is used to solve a system of equations. To illustrate this feature, we will use the system of equations $\begin{aligned} 2x - 3y &= 13 \\ 3x + 4y &= -6 \end{aligned}$.

Note: Some equations can be solved by this method. See the section "Solve an equation" below. Also, this method is used to find a number in the domain of a function for a given number in the range. See the section "Find a domain element."

Solve each of the equations in the system of equations for y. In this case, we have $y = \frac{2}{3}x - \frac{13}{3}$ and $y = -\frac{3}{4}x - \frac{3}{2}$.

Use the Y-editor to enter $\frac{2}{3}x - \frac{13}{3}$ into Y₁ and $-\frac{3}{4}x - \frac{3}{2}$ into Y₂. Graph the two functions in the standard viewing window. (If the window does not show the point of intersection of the two graphs, adjust the window until you can see the point of intersection.)

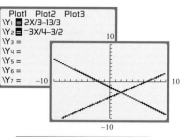

Press <kbd>2nd</kbd> CALC (scroll to 5, intersect) <kbd>ENTER</kbd>.

Alternatively, you can just press <kbd>2nd</kbd> CALC 5.

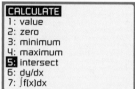

First curve? is shown at the bottom of the screen and identifies one of the two graphs on the screen. Press <kbd>ENTER</kbd>.

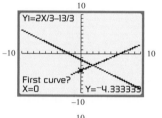

Second curve? is shown at the bottom of the screen and identifies the second of the two graphs on the screen. Press <kbd>ENTER</kbd>.

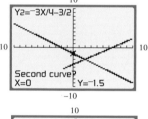

Guess? shown at the bottom of the screen asks you to use the left or right arrow key to move the cursor to the *approximate* location of the point of intersection. (If there are two or more points of intersection, it does not matter which one you choose first.) Press <kbd>ENTER</kbd>.

The solution of the system of equations is $(2, -3)$.

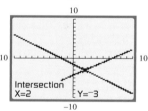

Solve an equation To illustrate the steps involved, we will solve the equation $2x + 4 = -3x - 1$. The idea is to write the equation as the system of equations $\begin{array}{l} y = 2x + 4 \\ y = -3x - 1 \end{array}$ and then use the steps for solving a system of equations.

Use the Y-editor to enter the left and right sides of the equation into Y₁ and Y₂. Graph the two functions and then follow the steps for Intersect.

The solution is -1, the x-coordinate of the point of intersection.

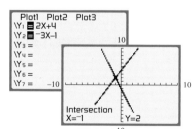

Find a domain element For this example, we will find a number in the domain of $f(x) = -\frac{2}{3}x + 2$ that corresponds to 4 in the range of the function. This is like solving the system of equations $y = -\frac{2}{3}x + 2$ and $y = 4$.

Use the Y= editor to enter the expression for the function in Y₁ and the desired output, 4, in Y₂. Graph the two functions and then follow the steps for Intersect.

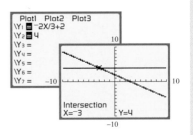

The point of intersection is (−3, 4). The number −3 in the domain of *f* produces an output of 4 in the range of *f*.

Math

Pressing **MATH** gives you access to many built-in functions. The following keystrokes will convert 0.125 to a fraction: .125 **MATH** 1 **ENTER** .

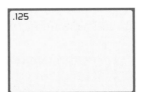 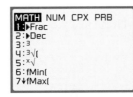 

Additional built-in functions under **MATH** can be found by pressing **MATH** **▶** . For instance, to evaluate −|−25|, press **(−)** **MATH** **▶** 1 **(−)** 25 **)** **ENTER** .

 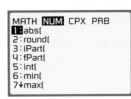 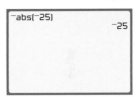

See your owner's manual for assistance with other functions under the **MATH** key.

Matrix

On a TI-83, **matrix operations** are accessed by pressing **MATRIX**. On a TI-83 Plus, press **2nd** MATRX to access the matrix menu.

To enter the elements of a matrix, select the matrix key. Then use the right arrow to select EDIT. Now use the down arrow key to select the name of the matrix. There are 10 matrices with names A through J. By pressing the down arrow key, you can see the additional names. Once you have selected the name of the matrix, press **ENTER** .

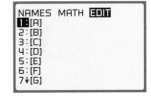

For instance, to enter the matrix $\begin{bmatrix} 2 & -3 & 4 \\ 1 & 5 & 3 \end{bmatrix}$ with 2 rows and 3 columns, access the matrix menu, arrow right to EDIT, and press **ENTER** . Now enter the dimension and the elements of the matrix, pressing **ENTER** after each number. You can change an element by using the arrow keys to select that element. After you have entered all the elements, press **2nd** QUIT to return to the home screen.

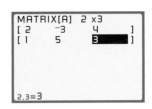

Elementary row operations Elementary row operations are performed by selecting MATH from the matrix menu. Use the down arrow key to scroll to those operations. Your screen should look something like this:

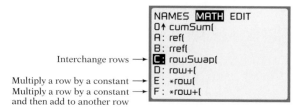

Interchange rows →
Multiply a row by a constant →
Multiply a row by a constant and then add to another row →

The operation row+(shown by D: is to add two rows. This is really the same as F: where the constant is 1.

Here are keystrokes for each elementary row operation. We will use the matrix $\begin{bmatrix} 1 & 3 & -4 & 6 \\ 3 & 2 & 0 & -1 \\ -2 & -5 & 3 & 4 \end{bmatrix}$ for this demonstration and assume it is stored in matrix [B].

Interchange rows: Access the matrix menu and highlight MATH. Scroll down to C:rowSwap(. Press ENTER. Access the matrix menu. Scroll to [B]; then press ENTER. Press ⟨,⟩ 1 ⟨,⟩ 3 ⟨)⟩ ENTER. (This interchanges row 1 and row 3. Change these numbers to interchange other rows.)

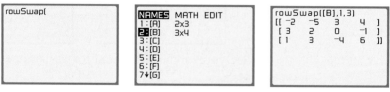

Multiply a row by a constant: Access the matrix menu and highlight MATH. Scroll down to E:*row(. Press ENTER (−) 2. (This is the constant that will multiply a row.) Press ⟨,⟩. Access the matrix menu. Scroll to [B] and then press ENTER. Press ⟨,⟩ 3 ⟨)⟩ ENTER. (Row 3 is being multiplied.)

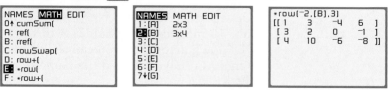

Multiply a row by a constant and then add it to another row: Access the matrix menu and highlight MATH. Scroll down to F:*row+(. Press ENTER 2. (This is the constant that will multiply a row.) Press ⟨,⟩. Access the matrix menu. Scroll to [B]; then press ENTER. Press ⟨,⟩ 1 ⟨,⟩ 3 ⟨)⟩ ENTER. (Row 1 is being multiplied by 2 and then added to row 3.)

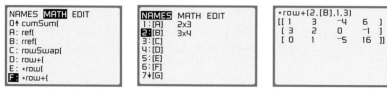

Row echelon form   The ref(function performs all of the elementary row operations on a matrix and directly produces a row echelon form of a matrix. The abbreviation ref stands for <u>r</u>ow <u>e</u>chelon <u>f</u>orm.

To write $\begin{bmatrix} 2 & 1 & 3 & -1 \\ 1 & 3 & 5 & -1 \\ -3 & -1 & 1 & 2 \end{bmatrix}$ in row echelon form, enter the matrix in, for instance, [A].

Press (2nd) QUIT. Then access the matrix menu and highlight MATH, scroll to ref(and press (ENTER), access the matrix menu, select [A], and press (ENTER) ( ) (ENTER). This will produce a matrix in row echelon form. Pressing (MATH) 1 (ENTER) will rewrite the matrix with fractions rather than decimals. (See MATH for assistance with the fraction command.)

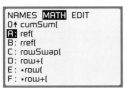
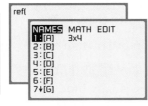

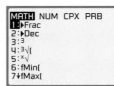
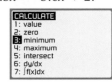

Min and Max

The local minimum and the local maximum values of a function are calculated by accessing the CALC menu. For this demonstration, we will find the minimum value and the maximum value of $f(x) = 0.2x^3 + 0.3x^2 - 3.6x + 2$.

Enter the function into Y_1. Press (2nd) CALC (scroll to 3 for minimum of the function) (ENTER).

Alternatively, you can just press (2nd) CALC 3.

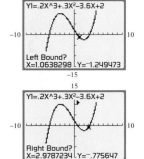

Left Bound? shown at the bottom of the screen asks you to use the left or right arrow key to move the cursor to the *left* of the minimum. Press (ENTER).

Right Bound? shown at the bottom of the screen asks you to use the left or right arrow key to move the cursor to the *right* of the minimum. Press (ENTER).

Guess? shown at the bottom of the screen asks you to use the left or right arrow key to move the cursor to the *approximate* location of the minimum. Press (ENTER).

The minimum value of the function is the *y*-coordinate. For this example, the minimum value of the function is -2.4.

The *x*-coordinate for the minimum is 2. However, because of rounding errors in the calculation, it is shown as a number close to 2.

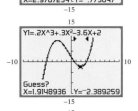
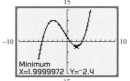

To find the maximum value of the function, follow the same steps as above except select maximum under the CALC menu. The screens for this calculation are shown below.

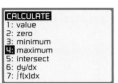
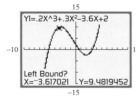
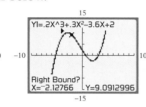
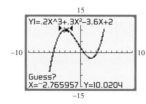
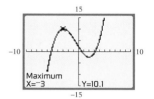

The maximum value of the function is 10.1.

Radical Expressions

To evaluate a square-root expression, press **2nd** √ .

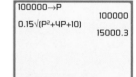

For instance, to evaluate $0.15\sqrt{p^2 + 4p + 10}$ when $p = 100,000$, first store 100,000 in P. Then press 0.15 **2nd** √ **ALPHA** P x^2 **+** 4 **ALPHA** P **+** 10 **)** **ENTER**.

To evaluate a radical expression other than a square root, access $\sqrt[x]{}$ by pressing **MATH**. For instance, to evaluate $\sqrt[4]{67}$, press 4 (the index of the radical) **MATH** (scroll to 5) **ENTER** 67 **ENTER**.

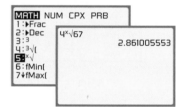

Regression

For the discussion of linear regression, we will use the data in this table.

Temperature, x (in °C)	20	35	50	60	75	90	100
Grams of sugar, y	50	80	120	145	175	205	230

All calculations and graphs involving statistical data begin by entering the data using the Edit option, which is accessed by pressing **STAT**.

For the data above, press **STAT** to access the statistics menu. Press 1 to Edit or enter data. To delete data already in a list, press the up arrow until the cursor is highlighting the list name. For instance, to delete data in L₁, highlight L₁. Then press **CLEAR** and **ENTER**. Now enter each value of the independent variable in L₁, pressing **ENTER** after each entry. Use the up and down arrow keys to change a value. When all values of the independent variable are entered, press **▶**. This will put you in the next column to enter the values of the dependent variable in L₂.

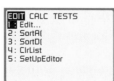

Create a scatter diagram Press **2nd** STATPLOT (use the down arrow key to select Plot1, Plot2, or Plot3) **ENTER**. Use arrow keys to move the cursor to ON and then press **ENTER**. The first graph type is for a scatter diagram. Move the cursor over that symbol and press **ENTER**. Be sure that Xlist and Ylist are the names of the lists into which you stored data. You can change these by pressing 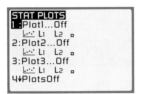 and then selecting the appropriate list, L₁ through L₆. Prepare to graph the data by adjusting the viewing window by pressing WINDOW and entering appropriate values. Now press GRAPH.

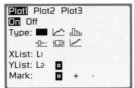
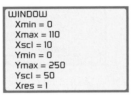
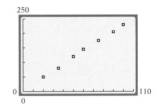
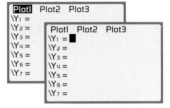

Note: You can tell that **STAT PLOTS** is active by pressing Y= . For one screen at the right, observe that **PLOT1** is highlighted, indicating it is active. To turn **STAT PLOTS** off, use the up arrow key to highlight it, and then press ENTER . Now use the arrow key to move the cursor to the right of the equals sign for Y₁.

Find a linear regression equation   Press STAT ▶ (scroll to 4) ENTER 2nd L₁ , 2nd L₂ , VARS ▶ 1 1 ENTER . The values of the slope and *y*-intercept of the linear regression equation will be displayed on the screen. If **DiagnosticOn** is enabled (see Correlation coefficient), then the coefficient of determination r^2 and the correlation coefficient *r* are also shown.

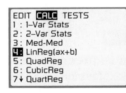
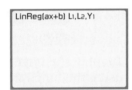
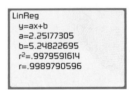

Note: If data are stored in L₁ and L₂, the keystrokes 2nd L₁ , 2nd L₂ are not necessary. The keystrokes VARS ▶ 11 ENTER place the regression equation in Y₁. These keystrokes are not necessary but are helpful if you need to graph the regression equation or evaluate the equation at a given value of the independent variable. See below for more details.

Graph a linear regression equation   Press STAT ▶ (scroll to 4) ENTER 2nd L₁ , 2nd L₂ , VARS ▶ 11 ENTER . This will store the regression equation in Y₁. Now press GRAPH . It may be necessary to adjust the viewing window.

Evaluate a regression equation   Complete the steps to graph a regression equation, but do not graph the equation. To evaluate the equation when $x = 50$, press VARS ▶ 11 (50) ENTER .

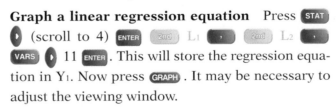
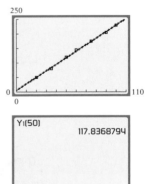

Other regression equations can be calculated. For instance, to find a regression equation of the form $y = ax^b$, called a power regression equation, enter the data and then select PwrReg from the CALC menu under the STAT menu.

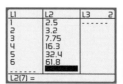 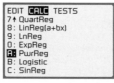 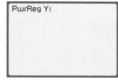 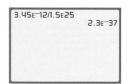

Note: Because the data were entered into L_1 and L_2, it was not necessary to include them in PwrReg. We did include the optional Y_1. This is good practice because it makes evaluating and graphing a regression equation much easier.

Scientific Notation

To enter a number in scientific notation, use 2nd EE. For instance, to find $\frac{3.45 \times 10^{-12}}{1.5 \times 10^{25}}$, press 3.45 2nd EE (−) 12 ÷ 1.5 2nd EE 25 ENTER. The answer is 2.3×10^{-37}.

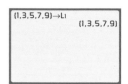

Sequences and Series

The terms of a sequence and the sum of a series can be calculated by using the 2nd LIST feature.

Store a sequence A sequence is stored in one of the lists L_1 through L_6. For instance, to store the sequence 1, 3, 5, 7, 9 in L_1, use the following keystrokes.

2nd { 1 , 3 , 5 , 7 ,
9 2nd } STO→ 2nd L1 ENTER

Display the terms of a sequence The terms of a sequence are displayed by using the function seq(expression, variable, begin, end, increment). For instance, to display the 3rd through 8th terms of the sequence given by $a_n = n^2 + 6$, enter the following keystrokes.

 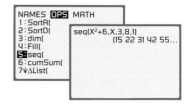

2nd LIST ▶ (scroll to 5)

ENTER X,T,θ,n x^2 + 6
, X,T,θ,n , 3 , 8
, 1 ENTER STO→ 2nd L1 ENTER

The keystrokes STO→ 2nd L1 ENTER store the terms of the sequence in L_1. This is not necessary but is sometimes helpful if additional work will be done with that sequence.

Find a sequence of partial sums To find a sequence of partial sums, use the cumSum(function. For instance, to find the sequence of partial sums for 2, 4, 6, 8, 10, use the following keystrokes.

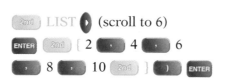 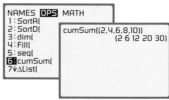

2nd LIST ▶ (scroll to 6)

ENTER 2nd { 2 , 4 , 6
, 8 , 10 2nd }) ENTER

If a sequence is stored as a list in L₁, then the sequence of partial sums can be calculated by pressing [2nd] LIST [▶] (scroll to 6 [or press 6]) [ENTER] [2nd] L1 [)] [ENTER].

Find the sum of a series   The sum of a series is calculated using sum<list, start, end>. For instance, to find $\sum_{n=3}^{6} (n^2 + 2)$, enter the following keystrokes.

[2nd] LIST [▶] [▶] (scroll to 5)

[ENTER] [2nd] LIST [▶] (scroll to 5 [or press 5])

[ENTER] [X,T,θ,n] [x²] [+] 2 [,] [X,T,θ,n] [,] 3

[,] 6 [,] 1 [)] [ENTER]

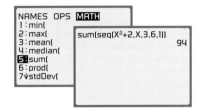

Table

There are three steps in creating an input/output table for a function. First use the [Y=] editor to input the function. The second step is setting up the table, and the third step is displaying the table.

To set up the table, press [2nd] TBLSET. TblStart is the first value of the independent variable in the input/output table. △Tbl is the difference between successive values. Setting this to 1 means that, for this table, the input values are −2, −1, 0, 1, 2. . . . If △Tbl = 0.5, then the input values are −2, −1.5, −1, −0.5, 0, 0.5, . . .

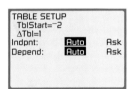

Indpnt is the independent variable. When this is set to Auto, values of the independent variable are automatically entered into the table. Depend is the dependent variable. When this is set to Auto, values of the dependent variable are automatically entered into the table.

To display the table, press [2nd] TABLE. An input/output table for $f(x) = x^2 - 1$ is shown at the right.

Once the table is on the screen, the up and down arrow keys can be used to display more values in the table. For the table at the right, we used the up arrow key to move to $x = -7$.

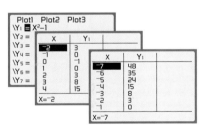

An input/output table for any given input can be created by selecting Ask for the independent variable. The table at the right shows an input/output table for $f(x) = \dfrac{4x}{x - 2}$ for selected values of x. Note the word ERROR when 2 was entered. This occurred because f is not defined when $x = 2$.

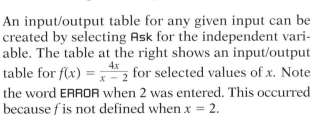
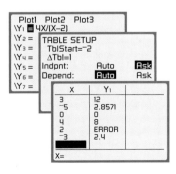

Note: Using the table feature in Ask mode is the same as evaluating a function for given values of the independent variable. For instance, from the table at the right, we have $f(4) = 8$.

Test   The TEST feature has many uses, one of which is to graph the solution set of a linear inequality in one variable. To illustrate this feature, we will graph the solution set of $x - 1 < 4$. Press (Y=) (X,T,θ,n) (—) 1 (2nd) TEST (scroll to 5) (ENTER) 4 (GRAPH).

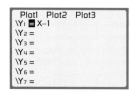
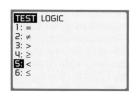
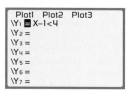
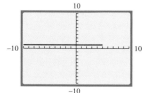

Trace   Once a graph is drawn, pressing (TRACE) will place a cursor on the screen, and the coordinates of the point below the cursor are shown at the bottom of the screen. Use the left and right arrow keys to move the cursor along the graph. For the graph at the right, we have $f(4.8) = 3.4592$, where $f(x) = 0.1x^3 - 2x + 2$ is shown at the top left of the screen.

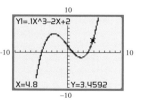

In TRACE mode, you can evaluate a function at any value of the independent variable that is within Xmin and Xmax. To do this, first graph the function. Now press (TRACE) (the value of x) (ENTER). For the graph at the left below, we used $x = -3.5$. If a value of x is chosen outside the window, an error message is displayed.

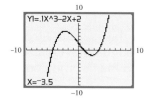
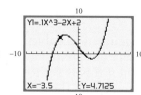
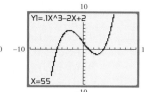

In the example above where we entered -3.5 for x, the value of the function was calculated as 4.7125. This means that $f(-3.5) = 4.7125$. The keystrokes (2nd) QUIT (VARS) (▸) 11 (MATH) 1 (ENTER) will convert the decimal value to a fraction.

When the TRACE feature is used with two or more graphs, the up and down arrow keys are used to move between the graphs. The graphs below are for the functions $f(x) = 0.1x^3 - 2x + 2$ and $g(x) = 2x - 3$. By using the up and down arrows, we can place the cursor on either graph. The right and left arrows are used to move along the graph.

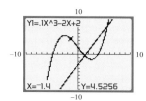
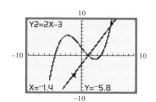

Window   The viewing window for a graph is controlled by pressing (WINDOW). Xmin and Xmax are the minimum value and maximum value, respectively, of the independent variable shown on the graph. Xscl is the distance between tic marks on the x-axis. Ymin and Ymax are the minimum value and maximum value, respectively, of the dependent variable shown on the graph. Yscl is the distance between tic marks on the y-axis. Leave Xres as 1.

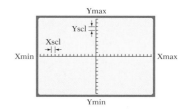

Note: In the standard viewing window, the distance between tic marks on the x-axis is different from the distance between tic marks on the y-axis. This will distort a graph. A more accurate picture of a graph can be created by using a square viewing window. See ZOOM.

Y=

The Y= editor is used to enter the expression for a function. There are ten possible functions, labeled Y₁ to Y₀, that can be active at any one time. For instance, to enter $f(x) = x^2 + 3x - 2$ as Y₁, use the following keystrokes.

Y= X,T,θ,n x² + 3 X,T,θ,n − 2

Note: If an expression is already entered for Y₁, place the cursor anywhere on that expression and press CLEAR.

To enter $s = \dfrac{2v - 1}{v^3 - 3}$ into Y₂, place the cursor to the right of the equals sign for Y₂. Then press (2 X,T,θ,n −

1) ÷ (X,T,θ,n ^ 3 − 3) .

Note: When we enter an equation, the independent variable, v in the expression above, is entered using X,T,θ,n. The dependent variable, s in the expression above, is one of Y₁ to Y₀. Also note the use of parentheses to ensure the correct order of operations.

Observe the black rectangle that covers the equals sign for the two examples we have shown. This rectangle means that the function is "active." If we were to press GRAPH, then the graph of both functions would appear. You can make a function inactive by using the arrow keys to move the cursor over the equals sign of that function and then pressing ENTER. This will remove the black rectangle. We have done that for Y₂, as shown at the right. Now if GRAPH is pressed, only Y₁ will be graphed.

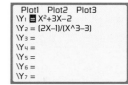

It is also possible to control the appearance of the graph by moving the cursor on the Y= screen to the left of any Y. With the cursor in this position, pressing ENTER will change the appearance of the graph. The options are shown at the right.

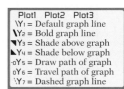

Zero

The ZERO feature of a graphing calculator is used for various calculations: to find the x-intercepts of a function, to solve some equations, and to find the zero of a function.

x-intercepts To illustrate the procedure for finding x-intercepts, we will use $f(x) = x^2 + x - 2$.

First, use the Y-editor to enter the expression for the function and then graph the function in the standard viewing window. (It may be necessary to adjust this window so that the intercepts are visible). Once the graph is displayed, use the keystrokes below to find the x-intercepts of the graph of the function.

Press 2nd CALC (scroll to 2 for **zero** of the function) ENTER .

Alternatively, you can just press 2nd CALC 2.

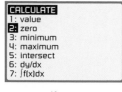

Left Bound? shown at the bottom of the screen asks you to use the left or right arrow key to move the cursor to the *left* of the desired x-intercept. Press ENTER .

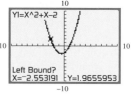

Right Bound? shown at the bottom of the screen asks you to use the left or right arrow key to move the cursor to the *right* of the desired x-intercept. Press ENTER .

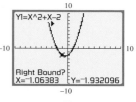

Guess? shown at the bottom of the screen asks you to use the left or right arrow key to move the cursor to the *approximate* location of the desired x-intercept. Press ENTER .

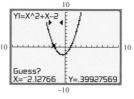

The x-coordinate of an x-intercept is -2. Therefore, an x-intercept is $(-2, 0)$.

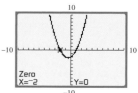

To find the other x-intercept, follow the same steps as above. The screens for this calculation are shown below.

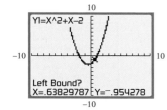
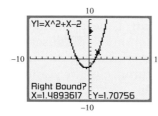
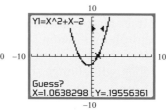
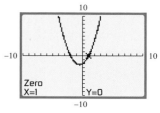

A second x-intercept is $(1, 0)$.

Solve an equation To use the ZERO feature to solve an equation, first rewrite the equation with all terms on one side. For instance, one way to solve $x^3 - x + 1 = -2x + 3$ is first to rewrite the equation as $x^3 + x - 2 = 0$. Enter $x^3 + x - 2$ into Y_1 and then follow the steps for finding x-intercepts.

Find the real zeros of a function   To find the real zeros of a function, follow the steps for finding x-intercepts.

Zoom

Pressing ZOOM allows you to select some preset viewing windows. This key also gives you access to ZBox, Zoom In, and Zoom Out. These functions enable you to redraw a selected portion of a graph in a new window. Some windows used frequently in this text are shown below.

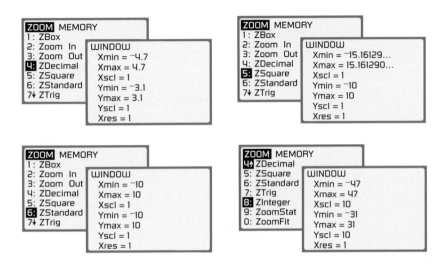

SECTION 1.1

YOU TRY IT 1

Goal We want to find three numbers whose product is 4590 and that are elements of the set {13, 14, 15, 16, 17, 18, 19}. None of the three numbers are the same.

Strategy By dividing 4590 by each element of the set, we can determine which elements of the set are factors of 4590 and which are not. (If a number is not a factor of 4590, it cannot be one of the three numbers whose product equals 4590.)

Solution 4590 is not evenly divisible by 13.
4590 is not evenly divisible by 14.
$4590 \div 15 = 306$
4590 is not evenly divisible by 16.
$4590 \div 17 = 270$
$4590 \div 18 = 255$
4590 is not evenly divisible by 19.

Only 15, 17, and 18 are factors of 4590.
The ages of the teenagers are 15, 17, and 18.
The oldest of the teens is 18 years old.

Check $15(17)(18) = 4590$

The solution checks.

YOU TRY IT 2

The pattern of the black beads is

$$1, 2, 3, 4, 5, 6, 7, \ldots$$

The pattern of the white beads is

$$2, 4, 8, 16, 32, 64, 128, \ldots$$

We can see the group of 4 black beads before the break in the string, and we can see the group of 7 black beads after the break. Therefore, not shown along the break in the string are

5 black beads

6 black beads

We can see 2 of the group of 16 white beads before the break. We can see 5 of the group of 64 white beads after the break (and before the 7 black beads). Therefore, not shown along the break in the string are

14 white beads in the group of 16

32 white beads in the group of 32

59 white beads in the group of 64

$5 + 6 + 14 + 32 + 59 = 116$

Along the dashed portion of the string, 116 beads are not shown.

YOU TRY IT 3

$$\frac{2}{33} = 0.060606\ldots; \frac{10}{33} = 0.303030\ldots; \frac{25}{33} = 0.757575\ldots$$

Note that $2(3) = 6$, $10(3) = 30$, and $25(3) = 75$. The repeating digits of the decimal representation of the fraction equal 3 times the numerator of the fraction.

The decimal representation of a proper fraction with a denominator of 33 is a repeating decimal in which the repeating digits are the product of the numerator and 3.

$$19(3) = 57$$

By this reasoning, $\frac{19}{33} = 0.575757\ldots$.

YOU TRY IT 4

Because ¥¥¥ = △△△△ and △△△△ = ΩΩ, ¥¥¥ = ΩΩ.

Because 3 ¥'s = 2 Ω's, 9 ¥'s = 6 Ω's.

That is, ¥¥¥¥¥¥¥¥¥ = ΩΩΩΩΩΩ.

YOU TRY IT 5

The conclusion is based on a principle. Therefore, it is an example of deductive reasoning.

YOU TRY IT 6

From statement 1, Mike is not the treasurer. In the chart on page S2, write X1 for this condition.

From statement 2, Clarissa is not the secretary or the president. Roger is not the secretary or the president. In the chart, write X2 for these conditions.

From statement 3, Betty is not the president, since we know from statement 2 that the president has lived there the longest. Write X3 for this condition. There are now X's for three of the four people in the president's column; therefore, Mike must be the president. Place a √ in that box. Since Mike is the president, he cannot be either the vice president or the secretary. Write X3 for these conditions. There are now three X's in the secretary's column. Therefore, Betty must be the secretary. Place a √ in that box. Since Betty is the secretary, she cannot be either the vice president or the treasure. Write X3 for these conditions.

From statement 4, together with statement 2, Clarissa is the vice president. Place a √ in that box. Now Clarissa cannot be the treasurer. Write an X4 for that condition. Since there are three X's in the treasurer's column, Roger must be the treasurer. Place a √ in that box.

	President	Vice Pres.	Secretary	Treasurer
Mike	√	X3	X3	X1
Clarissa	X2	√	X2	X4
Roger	X2	X4	X2	√
Betty	X3	X3	√	X3

Therefore, Mike is the president, Clarissa is the vice president, Roger is the treasurer, and Betty is the secretary.

SECTION 1.2

YOU TRY IT 1 {1, 3, 5, 7, 9}

YOU TRY IT 2 {$x | x > 19$, $x \in$ real numbers}

YOU TRY IT 3 The set is the real numbers greater than −3. Draw a left parenthesis at −3, and darken the number line to the right of −3.

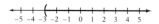

YOU TRY IT 4 $E \cup F = \{-5, -2, -1, 0, 1, 2, 5\}$

YOU TRY IT 5 The set is the numbers greater than or equal to 1 and less than or equal to −3.

YOU TRY IT 6 **a.** $A \cap B = \{0\}$

b. There are no odd integers that are also even integers.
$$C \cap D = \varnothing$$

YOU TRY IT 7 The set is $\{x | -1 \le x \le 2\}$.

YOU TRY IT 8 **a.** The set is the real numbers greater than or equal to −8 or less than −1.
$$[-8, -1)$$
b. The set is the numbers greater than −12.
$$\{x | x > -12\}$$

YOU TRY IT 9 $(-\infty, -2) \cup (-1, \infty)$ is the set of real numbers less than −2 and greater than −1.

SECTION 1.3

YOU TRY IT 1 **a.** $|47| = 47$

b. $|-50| = 50$

c. $-|-89| = -89$

YOU TRY IT 2 **a.** The input variable is t, the number of hours since the plane left Los Angeles. The output variable is d, the distance, in miles, the plane is from Boston.

t	d
0	2650
0.5	2387.5
1	2125
1.5	1862.5
2	1600
2.5	1337.5
3	1075
3.5	812.5
4	550

b. The number 1862.5 is the output when the input is 1.5.

The number 1862.5 means that the plane is 1862.5 miles from Boston 1.5 hours after the plane leaves Los Angeles.

YOU TRY IT 3 **a.** $|-52| = 52$, $|36| = 36$
$52 - 36 = 16$
$|-52| > |36|$
$-52 + 36 = -16$

b. $c + d$
$-18 + 9$
$|-18| = 18$, $|9| = 9$
$|-18| > |9|$
$-18 + 9 = -9$

YOU TRY IT 4 **a.** $-8 - (-26) = -8 + 26$
$= 18$

b. $-15 - 12 - 9 - (-36)$
$= -15 + (-12) + (-9) + 36$
$= -27 + (-9) + 36$
$= -36 + 36$
$= 0$

c. $a - b$
$46 - 72 = 46 + (-72)$
$= -26$

YOU TRY IT 5 **a.** $-5(33) = -165$

b. pr
$-18(-21) = 378$

You Try It 6 **a.** $-121 \div (-11) = 11$

b. $-24 \div 0$ is undefined.

c. $\dfrac{m}{n}$

$\dfrac{96}{-8} = -12$

You Try It 7 **a.** $(-8)^4 = (-8)(-8)(-8)(-8) = 4096$

b. $y^3 z^2$

$3^3(5^2) = (3 \cdot 3 \cdot 3)(5 \cdot 5) = 27 \cdot 25$
$= 675$

You Try It 8 **a.** $a^2 b \div c^3 - bc$

$4^2(3) \div 2^3 - 3(2)$
$16(3) \div 8 - 3(2)$
$48 \div 8 - 3(2)$
$6 - 3(2)$
$6 - 6$
0

b. $a(b - a)^2 - |c \div a|$

$-4[-6 - (-4)]^2 - |8 \div (-4)|$
$-4(-6 + 4)^2 - |-2|$
$-4(-2)^2 - |-2|$
$-4(-2)^2 - 2$
$-4(4) - 2$
$-16 - 2$
$-16 + (-2)$
-18

SECTION 1.4

You Try It 1 $-\dfrac{3}{8} + \left(-\dfrac{1}{3}\right) = -\dfrac{3}{8} \cdot \dfrac{3}{3} + \left(-\dfrac{1}{3} \cdot \dfrac{8}{8}\right)$

$= -\dfrac{9}{24} + \left(-\dfrac{8}{24}\right)$

$= \dfrac{-9 + (-8)}{24} = \dfrac{-17}{24} = -\dfrac{17}{24}$

You Try It 2 $-\dfrac{3}{4} - \dfrac{3}{16} = -\dfrac{12}{16} - \dfrac{3}{16}$

$= \dfrac{-12 - 3}{16} = \dfrac{-15}{16} = -\dfrac{15}{16}$

You Try It 3 $x - y + z$

$-\dfrac{7}{8} - \dfrac{5}{6} + \dfrac{3}{4} = -\dfrac{21}{24} - \dfrac{20}{24} + \dfrac{18}{24}$

$= \dfrac{-21 - 20 + 18}{24} = -\dfrac{23}{24}$

You Try It 4 $a - b$

$-16.127 - 67.91 = -16.127 + (-67.91)$
$= -84.037$

You Try It 5 **a.** st

$-\dfrac{3}{8}\left(-\dfrac{5}{12}\right) = \dfrac{3}{8} \cdot \dfrac{5}{12} = \dfrac{3 \cdot 5}{8 \cdot 12} = \dfrac{5}{32}$

b. $a \div d$

$-\dfrac{5}{8} \div \left(-\dfrac{5}{40}\right) = \dfrac{5}{8} \div \dfrac{5}{40}$

$= \dfrac{5}{8} \cdot \dfrac{40}{5}$

$= \dfrac{5 \cdot 40}{8 \cdot 5}$

$= 5$

You Try It 6 **a.** $-cd$

$-(4.027)(0.49) \approx -1.97$

b. $\dfrac{g}{h}$

$\dfrac{-2.835}{-1.35} = 2.1$

You Try It 7

Goal We want to find the average monthly net income for Friendly Ice Cream for the first quarter of 2001.

Strategy To find the average monthly net income, divide Friendly Ice Cream's net income for the first quarter of 2001 (-3.203) by 3, the number of months in one quarter of a year.

Solution $-3.203 \div 3 \approx -1.068$

Friendly Ice Cream's average monthly net income for the first quarter of 2001 was $-\$1.068$ million.

Check $-1.068(3) = -3.204 \approx -3.203$

You Try It 8 $x(x - y)^2 \div z$

$4.5(4.5 - 6.2)^2 \div (-0.5)$
$= 4.5(-1.7)^2 \div (-0.5)$
$= 4.5(2.89) \div (-0.5)$
$= 13.005 \div (-0.5)$
$= -26.01$

You Try It 9 $P = 2L + 2W$
$P = 2(8.5) + 2(3.5)$
$P = 17 + 7$
$P = 24$

The perimeter is 24 meters.

You Try It 10 **a.** Because the question asks for the amount of garbage generated per person per day (Y_1) in 1990 (X), look in the table for an input value of 1990. The corresponding output value is 4.5. Thus 4.5 pounds of garbage was generated per person per day in 1990.

b. Because the question asks for the year (X) when the amount of garbage

S3

generated per person per day (Y1) will
be 5.75 pounds, look in the table for an
output value of 5.75. You need to scroll
down the table. The corresponding input value is 2015. Thus the amount of
garbage generated per person per day
will be 5.75 pounds in 2015.

SECTION 1.5

You Try It 1 **a.** $4(3x) = (4 \cdot 3)x$
b. $12 + (-12) = 0$

You Try It 2 **a.** $-5(-3a) = [-5(-3)]a = 15a$

b. $\left(-\dfrac{1}{2}c\right)2 = 2\left(-\dfrac{1}{2}c\right) = \left[2\left(-\dfrac{1}{2}\right)\right]c$
$$= -1c = -c$$

You Try It 3 **a.** $3a - 2b + 5a = 3a + 5a - 2b$
$$= (3a + 5a) - 2b$$
$$= 8a - 2b$$

b. $2z^2 - 5z - 3z^2 + 6z$
$$= 2z^2 - 3z^2 + 6z - 5z$$
$$= (2z^2 - 3z^2) + (6z - 5z)$$
$$= -1z^2 + 1z$$
$$= -z^2 + z$$

You Try It 4 **a.** $-3(5y - 2) = -3(5y) - (-3)(2)$
$$= -15y + 6$$

b. $-(6c + 5) = -1(6c + 5)$
$$= -1(6c) + (-1)(5)$$
$$= -6c - 5$$

c. $(3p - 7)(-3) = 3p(-3) - 7(-3)$
$$= -9p + 21$$

d. $-2(4x + 2y - 6z)$
$$= -2(4x) + (-2)(2y) - (-2)(6z)$$
$$= -8x - 4y + 12z$$

You Try It 5 **a.** $7(-3x - 4y) - 3(3x + y)$
$$= -21x - 28y - 9x - 3y$$
$$= -30x - 31y$$

b. $2y - 3[5 - 3(3 + 2y)]$
$$= 2y - 3[5 - 9 - 6y]$$
$$= 2y - 3[-4 - 6y]$$
$$= 2y + 12 + 18y$$
$$= 20y + 12$$

You Try It 6 **a.** Let the unknown number be x.
seven <u>more than</u> the <u>product</u> of a number and twelve

the product of a number and 12: $12x$
$12x + 7$

b. Let the unknown number be x.
the <u>total</u> of eighteen and the <u>quotient</u>
of a number and nine
the quotient of a number and nine: $\dfrac{x}{9}$
$18 + \dfrac{x}{9}$

You Try It 7 Let the unknown number be x.
a number <u>minus</u> the <u>difference</u> between
the number and seventeen
the difference between the number and
seventeen: $x - 17$
$x - (x - 17)$
$$= x - x + 17$$
$$= 0 + 17$$
$$= 17$$

You Try It 8 one number: x
the other number: $10 - x$

You Try It 9 Let h represent the number of hours of
overtime worked.
$640 + 32$ for each hour of overtime
worked
$640 + 32h$

SOLUTIONS to Chapter 2 You Try Its

SECTION 2.1

You Try It 1 Plot the points $A(-2, 4)$, $B(4, 0)$, $C(0, 3)$,
and $D(-3, -4)$.

You Try It 2 The input/output table for $y = x^2 + 2x$ for
$x = -4, -3, -2, -1, 0, 1,$ and 2 is shown
to the right in a vertical format.

Input, x	Output, $x^2 + 2x = y$
-4	$(-4)^2 + 2(-4) = 8$
-3	$(-3)^2 + 2(-3) = 3$
-2	$(-2)^2 + 2(-2) = 0$
-1	$(-1)^2 + 2(-1) = -1$
0	$(0)^2 + 2(0) = 0$
1	$(1)^2 + 2(1) = 3$
2	$(2)^2 + 2(2) = 8$

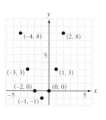

YOU TRY IT 3 The input/output table for $y = -\frac{x}{2} - 2$ for $x = -6, -4, -2, 0, 2,$ and 4 is shown below in a vertical format.

Input, x	Output, $-\dfrac{x}{2} - 2 = y$
-6	$-\dfrac{(-6)}{2} - 2 = 1$
-4	$-\dfrac{(-4)}{2} - 2 = 0$
-2	$-\dfrac{(-2)}{2} - 2 = -1$
0	$-\dfrac{(0)}{2} - 2 = -2$
2	$-\dfrac{(2)}{(2)} - 2 = -3$
4	$-\dfrac{(4)}{2} - 2 = -4$

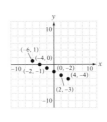

YOU TRY IT 4 The temperature T, in degrees Fahrenheit, h hours after 4:00 P.M. one summer day was given by

$$T = \frac{960}{h + 12}.$$

a. The input/output table for $T = \frac{960}{h + 12}$ is shown below in a horizontal format.

Input, time h	0	0.5	1	1.5	2	2.5	3
Output, temperature T	80	76.8	73.8	71.1	68.6	66.2	64

b. At 6:00 P.M., the temperature was 68.6° F.

YOU TRY IT 5

a. The input/output table for $y = \frac{2}{3}x - 3$ is shown below in a horizontal format.

x	-6	-3	0	3	6	9
y	-7	-5	-3	-1	1	3

b.

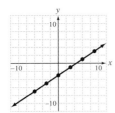

c.

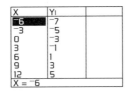

YOU TRY IT 6

a. Input $-\frac{2}{3}x + 4$ into Y₁ and select the integer viewing window.

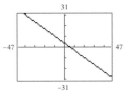

b. Trace along the curve until the x-coordinate is 9.

The value of y is -2 when $x = 9$.

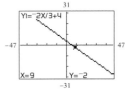

c. Trace along the curve until the y-coordinate is 8.

The value of x is -6 when $y = 8$.

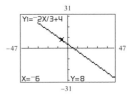

YOU TRY IT 7

Enter $\frac{1}{2}x + 2$ into Y₁ and then graph the equation in the integer viewing window.

a. Use the TRACE feature of the calculator to find the ordered-pair solution corresponding to $x = -5$.

The ordered-pair solutions is $(-5, -0.5)$.

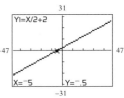

b. To check the results algebraically, evaluate $\frac{1}{2}x + 2$ when $x = -5$.

$$y = \frac{1}{2}x + 2$$

$$y = \frac{1}{2}(-5) + 2 \qquad \bullet \text{ Replace } x \text{ by } -5.$$

$$= -\frac{5}{2} + 2 \qquad \bullet \text{ Simplify.}$$

$$= -\frac{1}{2}$$

$$= -0.5$$

The solution checks.

SECTION 2.2

YOU TRY IT 1

The domain of a relation is the set of the first coordinates of the ordered pairs of the relation. The range of a relation is the set of second coordinates of the relation. For the relation {(1, 1), (2, 1), (3, 1), (4, 1), (5, 1), (6, 1), (7, 1)}, the domain is {1, 2, 3, 4, 5, 6, 7}. The range is {1}.

Because no two ordered pairs have the same first coordinate, the relation is a function.

YOU TRY IT 2

a. $f(z) = 2z^3 - 4z$
$f(-1) = 2(-1)^3 - 4(-1)$
$f(-1) = 2(-1) - 4(-1)$
$f(-1) = -2 + 4$
$f(-1) = 2$

b. To find the value of f when $z = -3$ means to evaluate the function when z is -3.

$f(z) = 2z^3 - 4z$
$f(-3) = 2(-3)^3 - 4(-3)$
$f(-3) = 2(-27) - 4(-3)$
$f(-3) = -54 + 12$
$f(-3) = -42$

YOU TRY IT 3

a. Evaluate $h(x) = 2x - 3$ for the given values of x.

x	-2	-1	0	1	2
$h(x)$	-7	-5	-3	-1	1

b. Graph the ordered pairs $(-2, -7)$, $(-1, -5)$, $(0, -3)$, $(1, -1)$, and $(2, 1)$. Then draw a line through the points.

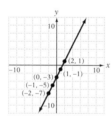

YOU TRY IT 4

The graph of $g(x) = 2$ is a horizontal line through $(0, 2)$.

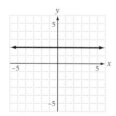

YOU TRY IT 5

To find the range of $f(x) = -x^2 + 2x + 2$ with domain $\{-2, -1, 0, 1, 2, 3\}$, evaluate the function at each element of the domain. The set of outputs is the range of the function for the given domain.

$f(x) = -x^2 + 2x + 2$
$f(-2) = -(-2)^2 + 2(-2) + 2 = -4 - 4 + 2 = -6$
$f(-1) = -(-1)^2 + 2(-1) + 2 = -1 - 2 + 2 = -1$
$f(0) = -(0)^2 + 2(0) + 2 = 0 + 0 + 2 = 2$
$f(1) = -(1)^2 + 2(1) + 2 = -1 + 2 + 2 = 3$
$f(2) = -(2)^2 + 2(2) + 2 = -4 + 4 + 2 = 2$
$f(3) = -(3)^2 + 2(3) + 2 = -9 + 6 + 2 = -1$

The range is $\{-6, -1, 2, 3\}$.

YOU TRY IT 6

Evaluate $P(t) = \dfrac{t}{t^2 + 1}$ for each of the given values 4, -4, 0, 3.

$P(t) = \dfrac{t}{t^2 + 1}$

$P(4) = \dfrac{4}{(4)^2 + 1} = \dfrac{4}{17}$ A real number

$P(-4) = \dfrac{-4}{(-4)^2 + 1} = -\dfrac{4}{17}$ A real number

$P(0) = \dfrac{0}{(0)^2 + 1} = 0$ A real number

$P(3) = \dfrac{3}{(3)^2 + 1} = \dfrac{3}{10}$ A real number

Each of the given numbers is in the domain of P. All are included in the domain.

YOU TRY IT 7

Graph $f(x) = \dfrac{3}{x^2 - x - 6}$ in the decimal viewing window. Then trace along the curve to find the two x-coordinates for which there is no y-coordinate.

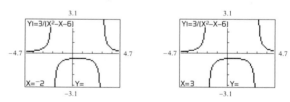

The two numbers that are not in the domain of f are -2 and 3. To verify this algebraically, attempt to evaluate the function for these two numbers.

$f(x) = \dfrac{3}{x^2 - x - 6}$

$f(-2) = \dfrac{3}{(-2)^2 - (-2) - 6}$

$= \dfrac{3}{4 + 2 - 6}$

$= \dfrac{3}{0}$ Not a real number

$f(x) = \dfrac{3}{x^2 - x - 6}$

$f(3) = \dfrac{3}{(3)^2 - (3) - 6}$

$= \dfrac{3}{9 - 3 - 6}$

$= \dfrac{3}{0}$ Not a real number

YOU TRY IT 8

Evaluate the function for $m = 12$.

$$N(m) = \frac{m(m-1)}{2}$$

$$N(12) = \frac{12(12-1)}{2} = \frac{12(11)}{2} = 66$$

66 different line segments can be drawn between 12 different points in the plane.

YOU TRY IT 9

a. Because p is in thousands, a value of 100,000 is given as $p = 100$. Evaluate $C(p) = 0.15\sqrt{p^2 + 4p + 10}$ when $p = 100$.

$$C(p) = 0.15\sqrt{p^2 + 4p + 10}$$
$$C(100) = 0.15\sqrt{100^2 + 4(100) + 10}$$
$$\approx 15.3$$

The carbon monoxide concentration for a city of 100,000 people is 15.3 ppm.

b. Evaluate $C(p)$ when $p = 0$.
$$C(p) = 0.15\sqrt{p^2 + 4p + 10}$$
$$C(0) = 0.15\sqrt{0^2 + 4(0) + 10}$$
$$\approx 0.5$$

If there were no people in an area, the carbon monoxide concentration would be approximately 0.5 ppm.

SECTION 2.3

YOU TRY IT 1

a.

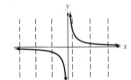

All vertical lines intersect the graph at most once. The graph is the graph of a function.

b.

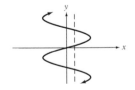

There is at least one vertical line that intersects the graph at more that one point. The graph is not the graph of a function.

YOU TRY IT 2

a. To find the y-intercept, evaluate $g(x) = 2x^2 - 5x + 2$ at $x = 0$.

$$g(x) = 2x^2 - 5x + 2$$
$$g(0) = 2(0)^2 - 5(0) + 2 = 2$$

The y-intercept is $(0, 2)$.

b. To find the x-intercept, graph $g(x) = 2x^2 - 5x + 2$ and then use the ZERO feature of a graphing calculator to find the x-intercepts.

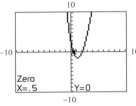
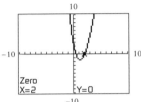

One x-intercept is $(0.5, 0)$.   A second x-intercept is $(2, 0)$.

YOU TRY IT 3

To find the number in the domain of $G(x) = 1 - 2x$ for which the output is -6, use a graphing calculator to graph $G(x) = 1 - 2x$ and $F(x) = -6$ on the same coordinate grid, and then find the point of intersection.

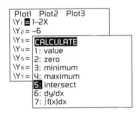
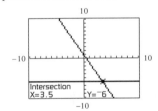

The output of $G(x) = 1 - 2x$ is -6 when $x = 3.5$.

YOU TRY IT 4

To find the element in the domain of $f(x) = 2x - 3$ and $g(x) = \frac{x}{2} + 3$ for which the values of the functions are equal, use a graphing calculator to graph each equation, and then find the point of intersection. The x-coordinate of the point of intersection is the desired value.

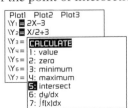
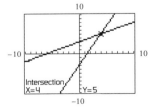

The values of the functions are equal when $x = 4$. To algebraically verify the result, evaluate each function for $x = 4$.

$$f(x) = 2x - 3 \qquad g(x) = \frac{x}{2} + 3$$

$$f(4) = 2(4) - 3 \qquad g(4) = \frac{4}{2} + 3$$

$$= 5 \qquad\qquad = 5$$

The value of both functions is 5 when $x = 4$.

You Try It 5

To solve this problem, we need to determine the value of t for which $s(t) = 10$. This is similar to Example 3. Graph $s(t) = 26 - 8t$ and $g(t) = 10$, and determine the point of intersection. Use a domain of $[0, 5]$ and a range of $[0, 30]$.

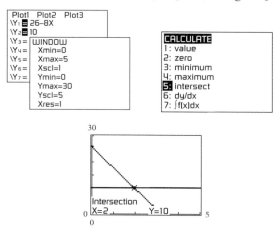

The marathon runner will be 10 miles from the finish line in 2 hours.

We can verify this algebraically as follows:

$$s(t) = 26 - 8t$$
$$s(2) = 26 - 8(2) \qquad \bullet \text{ Replace } t \text{ by 2.}$$
$$= 26 - 16$$
$$= 10$$

The marathon runner will be 10 miles from the finish line in 2 hours.

You Try It 6

To solve this problem, we need to find the value of t for which $h(t) = f(t)$. This is similar to Example 4. Graph each function on the same coordinate grid, and then determine the point of intersection.

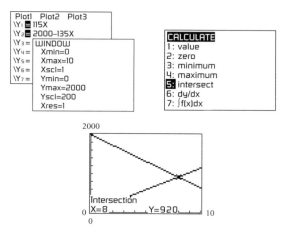

The planes are at the same height after 8 minutes.

SOLUTIONS to Chapter 3 You Try Its

SECTION 3.1

You Try It 1

$$9 + n = 4$$
$$9 - 9 + n = 4 - 9$$
$$n = -5$$

The solution is -5.

You Try It 2

$$-4x = -20$$
$$\frac{-4x}{-4} = \frac{-20}{-4}$$
$$x = 5$$

The solution is 5.

You Try It 3

$$5 - 4z = 15$$
$$5 - 5 - 4z = 15 - 5$$
$$-4z = 10$$
$$\frac{-4z}{-4} = \frac{10}{-4}$$
$$z = -\frac{5}{2}$$

The solution is $-\dfrac{5}{2}$.

$6y - 3 + y = 2y + 7$

$7y - 3 = 2y + 7$

$7y - 2y - 3 = 2y - 2y + 7$

$5y - 3 = 7$

$5y - 3 + 3 = 7 + 3$

$5y = 10$

$\dfrac{5y}{5} = \dfrac{10}{5}$

$y = 2$

The solution is 2.

YOU TRY IT 5

$2(3x + 1) = 4x + 8$

$6x + 2 = 4x + 8$

$6x - 4x + 2 = 4x - 4x + 8$

$2x + 2 = 8$

$2x + 2 - 2 = 8 - 2$

$2x = 6$

$\dfrac{2x}{2} = \dfrac{6}{2}$

$x = 3$

The solution is 3.

SECTION 3.2

YOU TRY IT 1

Goal   The goal is to find how many ounces of gold costing $320 per ounce should be mixed with 100 ounces of an alloy costing $100 per ounce to produce a new alloy that costs $160 per ounce.

Strategy   Let x represent the number of ounces of gold that are needed.

Find the value of each of the metals.

Value of the gold: $V = AC = x(320) = 320x$

Value of the $100-per-ounce alloy:
$V = AC = 100(100) = 10,000$

Value of the $160-per-ounce alloy (the mixture)
= value of the gold + value of the
$100-per-ounce alloy
= $320x + 10,000$

Find the amount of the mixture.

Amount of the mixture = amount of gold +
amount of the
$100-per-ounce
alloy
= $x + 100$

To find the value of x, use the equation $V = AC$ for the mixture. The unit cost of the mixture is $160 per ounce. Solve the equation for x.

Solution $V = AC$

$320x + 10,000 = (x + 100)160$ • V is the value of the mixture, A is the amount of the mixture, C is the unit cost of the mixture.

$320x + 10,000 = 160x + 16,000$

$160x + 10,000 = 16,000$

$160x = 6000$

$x = 37.5$

The jeweler must use 37.5 ounces of gold.

Check   One way to check the solution is to substitute the value of x into the original equation and determine whether the left and right sides of the equation are equal.

A second way to check the solution is to calculate the value of the mixture to ensure that its value is $160 per ounce.

Value of 37.5 ounces of gold:
$V = AC = 37.5(320) = 12,000$

Value of 100 ounces of the $100-per-ounce alloy:
$V = AC = 100(100) = 10,000$

Value of the two ingredients =
$12,000 + 10,000 = 22,000$

The mixture contains
100 ounces + 37.5 ounces = 137.5 ounces.

$V = AC$

$\begin{array}{c|c} \hline 22,000 & 137.5(160) \\ 22,000 & = 22,000 \end{array}$ ✓ The value of the mixture is $22,000. The solution checks.

YOU TRY IT 2

Goal   The goal is to determine the speed of each of the two cyclists.

Strategy   Let r represents the rate of the first cyclist. Then the rate of the second cyclist is $r + 5$.

Use the equation $d = rt$ to represent the distance traveled by each cyclist in 4 hours.

First cyclist: $d = rt$
$d = r(4)$
$d = 4r$

Second cyclist: $d = rt$
$d = (r + 5)4$
$d = 4r + 20$

The total distance traveled by the two cyclists is 140 miles.

$$\underset{4r}{\underset{\text{by first cyclist}}{\text{Distance traveled}}} + \underset{4r + 20}{\underset{\text{by second cyclist}}{\text{Distance traveled}}} = \underset{= 140}{140 \text{ miles}}$$

Solve this equation for r.

Solution
$$4r + 4r + 20 = 140$$
$$8r + 20 = 140$$
$$8r = 120$$
$$r = 15$$

Substitute the value of r into the expression representing the rate of the second cyclist.

$$r + 5 = 15 + 5 = 20$$

The first cyclist is traveling 15 mph. The second cyclist is traveling 20 mph.

Check In 4 hours, the first cyclist travels $4(15) = 60$ miles and the second cyclist travels $4(20) = 80$.

The total distance traveled by the two cyclists in 4 hours is 60 miles + 80 miles = 140 miles.

SECTION 3.3

You Try It 1

Goal The goal is to find the measures of the two complementary angles.

Strategy Let x represent the measure of one angle.

The measure of the complement of x is $90° - x$.

It is given that x is $3°$ less than the measure of its complement.

"x is $3°$ less than the measure of its complement" is translated as

$$x = (90 - x) - 3$$

Solve this equation for x.

Solution

$$x = (90 - x) - 3$$
$$x = 90 - x - 3$$
$$x = 87 - x$$
$$x + x = 87 - x + x$$
$$2x = 87$$
$$x = 43.5$$

Substitute the value of x into the expression for the complement of x.

$$90 - x = 90 - 43.5 = 46.5$$

The angles measure $43.5°$ and $46.5°$.

Check

$$43.5° + 46.5° = 90°;$$
the sum of the angles is $90°$.

$46.5° - 43.5° = 3°$; one angle is $3°$ less than the other angle.

S10

You Try It 2

Goal The goal is to find the measure of the larger of two adjacent angles for a pair of intersecting lines.

Strategy Adjacent angles of intersecting lines are supplementary angles (their sum is $180°$).

$$(2x + 20) + (3x + 50) = 180$$

Solve this equation for x.
Then substitute the value of x into the expressions $2x + 20$ and $3x + 50$ to determine the larger angle.

Solution
$$(2x + 20) + (3x + 50) = 180$$
$$5x + 70 = 180$$
$$5x = 110$$
$$x = 22$$
$$2x + 20 = 2(22) + 20 = 44 + 20 = 64$$
$$3x + 50 = 3(22) + 50 = 66 + 50 = 116$$

The measure of the larger angle is $116°$.

Check $64 + 116 = 180$; the sum of the two angles is $180°$.

You Try It 3

Goal We want to determine the value of x, given two of the angles formed by a transversal and two parallel lines.

Strategy The angles given are alternate interior angles. Alternate interior angles have the same measure.

$$4x - 50 = 2x + 10$$

Solve this equation for x.

Solution
$$4x - 50 = 2x + 10$$
$$4x - 2x - 50 = 2x - 2x + 10$$
$$2x - 50 = 10$$
$$2x - 50 + 50 = 10 + 50$$
$$2x = 60$$
$$x = 30$$

The value of x is $30°$.

Check We can use the value of x to check that the two angles have the same measure.

$$4x - 50 = 4(30) - 50 = 120 - 50 = 70$$
$$2x + 10 = 2(30) + 10 = 60 + 10 = 70$$

You Try It 4

Goal The goal is to find $m\angle b$, the measure of an exterior angle of the triangle pictured.

Strategy We are given the measure of $\angle a$.

We can use the measure of $\angle a$ to find the measure of the adjacent interior angle of the triangle.

Use the fact that the sum of the measures of an interior angle and the adjacent exterior angle of a triangle is 180°.

We will represent the measure of the adjacent interior angle by $m\angle y$.

Then find the measure of the interior angle adjacent to $\angle b$.

Use the $m\angle y$, the fact that the triangle is a right triangle, and the fact that the sum of the measures of the interior angles of a triangle is 180°.

We will represent the measure of the interior angle adjacent to $\angle b$ by $m\angle z$.

Then find the measure of $\angle b$.

Use the fact that the sum of the measures of an interior angle and the adjacent exterior angle of a triangle is 180°.

Solution
$$m\angle a + m\angle y = 180°$$
$$112° + m\angle y = 180°$$
$$m\angle y = 68°$$
$$m\angle y + m\angle z + 90° = 180°$$
$$68° + m\angle z + 90° = 180°$$
$$158° + m\angle z = 180°$$
$$m\angle z = 22°$$
$$m\angle z + m\angle b = 180°$$
$$22° + m\angle b = 180°$$
$$m\angle b = 158°$$

Check Check all the steps of the solution.

You Try It 5

Goal We want to determine the measure of $\angle AEB$.

Strategy $\angle AEB$ is an inscribed angle because its vertex is on the circumference of the circle and its sides are chords.

According to the Inscribed-Angle Theorems:

If $\angle AEB$ is an inscribed angle of a circle, then $m\angle AEB = \frac{1}{2}m\widehat{AB}$.

The measure of an arc is the measure of the central angle that intersects it.

Therefore, the measure of $\widehat{AB}$ is equal to the measure of central angle ACB, or 138°.

Use the Inscribed-Angle Theorems to find the measure of $\angle AEB$

Solution
$$m\angle AEB = \frac{1}{2}m\widehat{AB}$$
$$m\angle AEB = \frac{1}{2}(138°)$$
$$m\angle AEB = 69°$$

The measure of $\angle AEB$ is 69°.

Check Be sure to check the calculations.

You Try It 6

Goal The goal is to find the value of x in the expression $2x + 20$ in the diagram.

Strategy According to the Inscribed-Angle Theorems:

If $\angle BAC$ is an inscribed angle of a circle, then $m\angle BAC = \frac{1}{2}m\widehat{BC}$.

We are given the measure of $\angle BAC$, so we can use the theorem to write an equation.
$$60° = \frac{1}{2}(2x + 20)°$$

Solve this equation for x.

Solution
$$60 = \frac{1}{2}(2x + 20)$$
$$60 = x + 10$$
$$50 = x$$

The value of x is 50°.

Check Use the value of x to find the measure of the arc: $2x + 20 = 2(50) + 20 = 100 + 20 = 120$.

One-half the measure of the arc is equal to the measure of the inscribed angle: $\frac{1}{2}(120) = 60$.

SECTION 3.4

You Try It 1

$$x - 4 \le 1$$
$$x - 4 + 4 \le 1 + 4$$
$$x \le 5$$

In set-builder notation, the solution set is written $\{x \mid x \le 5\}$.

In interval notation, the solution set is written $(-\infty, 5]$.

You Try It 2

$$-3x \ge 6$$
$$\frac{-3x}{-3} \le \frac{6}{-3}$$
$$x \le -2$$

In set-builder notation, the solution set is written $\{x \mid x \le -2\}$.

In interval notation, the solution set is written $(-\infty, -2]$.

$$3x - 1 \le 5x - 7$$
$$3x - 5x - 1 \le 5x - 5x - 7$$
$$-2x - 1 \le -7$$
$$-2x - 1 + 1 \le -7 + 1$$
$$-2x \le -6$$
$$\frac{-2x}{-2} \ge \frac{-6}{-2}$$
$$x \ge 3$$
$$\{x \mid x \ge 3\}$$

YOU TRY IT 4 $\quad 3 - 2(3x + 1) < 7 - 2x$
$$3 - 6x - 2 < 7 - 2x$$
$$-6x + 1 < 7 - 2x$$
$$-4x + 1 < 7$$
$$-4x < 6$$
$$x > -\frac{3}{2}$$
$$\left(-\frac{3}{2}, \infty\right)$$

YOU TRY IT 5

Goal The goal is to find the maximum height of the triangle.

Strategy Substitute the given values in the inequality $\frac{1}{2}bh < A$ and solve for x.

Solution
$$\frac{1}{2}bh < A$$
$$\frac{1}{2}(12)(x + 2) < 50$$
$$6(x + 2) < 50$$
$$6x + 12 < 50$$
$$6x < 38$$
$$x < \frac{19}{3}$$

The largest integer less than $\frac{19}{3}$ is 6.
$$x + 2 = 6 + 2 = 8$$
The maximum height of the triangle is 8 inches.

Check When the height is 8 inches, the area is $\frac{1}{2}bh = \frac{1}{2}(12)(8) = 48$ square inches.

When the height is 9 inches, the area is $\frac{1}{2}bh = \frac{1}{2}(12)(9) = 54$ square inches, which is greater than 50 square inches.

The solution checks.

YOU TRY IT 6

$$5x - 1 \ge -11 \quad \text{and} \quad 4 - 6x > -14$$
$$5x - 1 + 1 \ge -11 + 1 \qquad 4 - 4 - 6x > -14 - 4$$
$$5x \ge -10 \qquad\qquad -6x > -18$$
$$\frac{5x}{5} \ge \frac{-10}{5} \qquad\qquad \frac{-6x}{-6} < \frac{-18}{-6}$$
$$x \ge -2 \qquad\qquad\qquad x < 3$$
$$\{x \mid x \ge -2\} \qquad\qquad (x \mid x < 3)$$

The solution of the compound inequality is the intersection of the solution sets for each inequality.
$$\{x \mid x \ge -2\} \cap \{x \mid x < 3\} = \{x \mid -2 \le x < 3\}$$

YOU TRY IT 7

$$3 - 4x > 7 \qquad \text{or} \qquad 4x + 5 > 9$$
$$3 - 3 - 4x > 7 - 3 \qquad 4x + 5 - 5 > 9 - 5$$
$$-4x > 4 \qquad\qquad 4x > 4$$
$$\frac{-4x}{-4} < \frac{4}{-4} \qquad\qquad \frac{4x}{4} > \frac{4}{4}$$
$$x < -1 \qquad\qquad x > 1$$
$$(-\infty, -1) \qquad\qquad (1, \infty)$$

The solution set is the union of the two intervals.
$$(-\infty, -1) \cup (1, \infty)$$

SECTION 3.5

YOU TRY IT 1

a. $|5 - 6x| = 1$
$$5 - 6x = 1 \qquad\qquad 5 - 6x = -1$$
$$-6x = -4 \qquad\qquad -6x = -6$$
$$x = \frac{2}{3} \qquad\qquad x = 1$$

The solutions are $\frac{2}{3}$ and 1.

b. $|3x - 7| + 4 = 2$
$$|3x - 7| = -2$$
The absolute value of a number is positive or zero.
There is no solution.

$|2x - 5| \leq 7$

$-7 \leq 2x - 5 \leq 7$

$-7 + 5 \leq 2x - 5 + 5 \leq 7 + 5$

$-2 \leq 2x \leq 12$

$\dfrac{-2}{2} \leq \dfrac{2x}{2} \leq \dfrac{12}{2}$

$-1 \leq x \leq 6$

The solution set is $\{x | -1 \leq x \leq 6\}$.

YOU TRY IT 3

$|5x + 4| \geq 16$

$5x + 4 \geq 16$	$5x + 4 \leq -16$
$5x + 4 - 4 \geq 16 - 4$	$5x + 4 - 4 \leq -16 - 4$
$5x \geq 12$	$5x \leq -20$
$\dfrac{5x}{5} \geq \dfrac{12}{5}$	$\dfrac{5x}{5} \leq \dfrac{-20}{5}$
$x \geq \dfrac{12}{5}$	$x \leq -4$
$\left\{ x \middle| x \geq \dfrac{12}{5} \right\}$ or	$\{x | x \leq -4\}$

The solution set is the union of the solution sets of the two inequalities.

$\left\{ x \middle| x \geq \dfrac{12}{5} \right\} \cup \{x | x \leq -4\} = \left\{ x \middle| x \leq -4 \text{ or } x \geq \dfrac{12}{5} \right\}$

YOU TRY IT 4

Goal   The goal is to find the lower and upper limits of the diameter of a bushing that has a tolerance of 0.003 inch.

Strategy   Let b represent the desired diameter of the bushing, T the tolerance, and d the actual diameter of the bushing. Solve the absolute value inequality $|d - b| \leq T$ for d.

Solution $|d - b| \leq T$

$|d - 2.55| \leq 0.003$

$-0.003 \leq d - 2.55 \leq 0.003$

$-0.003 + 2.55 \leq d - 2.55 + 2.55 \leq 0.003 + 2.55$

$2.547 \leq d \leq 2.553$

The lower and upper limits of the diameter of the bushing are 2.547 inches and 2.553 inches.

Check   Be sure to check your work by doing a check of your calculations. As an estimate, the answers appear reasonable in that the diameters are close to 2.55 inches.

YOU TRY IT 5

Goal   The goal is to determine the SAT scores, x, that satisfy the inequality $\left| \dfrac{x - 950}{98} \right| < 1.96$.

Strategy   Solve the inequality $\left| \dfrac{x - 950}{98} \right| < 1.96$ for x.

Solution $\left| \dfrac{x - 950}{98} \right| < 1.96$

$-1.96 < \dfrac{x - 950}{98} < 1.96$

$98(-1.96) < 98 \left(\dfrac{x - 950}{98} \right) < 98(1.96)$

$-192.08 < x - 950 < 192.08$

$-192.08 + 950 < x - 950 + 950 < 192.08 + 950$

$757.92 < x < 1142.08$

The values of x that the registrar expects from a student applicant are $\{x | 757.92 < x < 1142.08\}$.

Check   Be sure to check the calculations.

SOLUTIONS to Chapter 4 You Try Its

SECTION 4.1

YOU TRY IT 1   The function is of the form $f(x) = mx + b$, where m is the slope.

For the function $g(t) = -20t + 8000$, the slope m is -20.

The slope means the plane is descending 20 feet per second.

YOU TRY IT 2   **a.** $(x_1, y_1) = (-6, 5)$, $(x_2, y_2) = (4, -5)$

$m = \dfrac{y_2 - y_1}{x_2 - x_1} = \dfrac{-5 - 5}{4 - (-6)} = \dfrac{-10}{10} = -1$

The slope is -1.

b. $(x_1, y_1) = (-5, 0)$, $(x_2, y_2) = (-5, 7)$

$m = \dfrac{y_2 - y_1}{x_2 - x_1} = \dfrac{7 - 0}{-5 - (-5)} = \dfrac{7}{0}$

The slope is undefined.

You Try It 3 Rewrite the slope -1 as $\dfrac{-1}{1}$.

Draw a dot at $(2, 4)$.

Starting at $(2, 4)$, move 1 unit down (the change in y) and then 1 unit to the right (the change in x). Draw a dot at $(3, 3)$.

Draw a line through the two points.

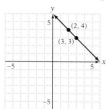

You Try It 4 $y = \dfrac{3}{4}x - 1$

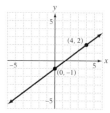

You Try It 5

$$3x + 2y = -6$$
$$3x - 3x + 2y = -3x - 6$$
$$2y = -3x - 6$$
$$\dfrac{2y}{2} = \dfrac{-3x - 6}{2}$$
$$y = -\dfrac{3}{2}x - 3$$

The slope is $-\dfrac{3}{2}$. The y-intercept is $(0, -3)$.

You Try It 6 $3x + y = 6$

To find the x-intercept, let $y = 0$ and solve for x.

$$3x + y = 6$$
$$3x + 0 = 6$$
$$3x = 6$$
$$x = 2$$

The x-intercept is $(2, 0)$.

To find the y-intercept, let $x = 0$ and solve for y.

$$3x + y = 6$$
$$3(0) + y = 6$$
$$y = 6$$

The y-intercept is $(0, 6)$.

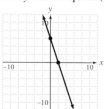

You Try It 7 $y - 5 = 0$
$$y = 5$$

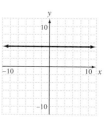

You Try It 8 $x = 1$

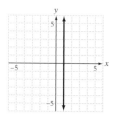

SECTION 4.2

You Try It 1 Let x represent the number of kilometers above sea level and y represent the boiling point of water.

Since the boiling point of water at sea level is 100°C, $x = 0$ when $y = 100$. The y-intercept is $(0, 100)$.

The slope is the decrease in the boiling point per kilometer increase in altitude.

Since the boiling point decreases 3.5°C per 1-kilometer increase in altitude, the slope is negative; $m = -3.5$.

To find the linear function, replace m and b in $f(x) = mx + b$ by their values.

$$f(x) = mx + b$$
$$f(x) = -3.5x + 100$$

The linear function is $f(x) = -3.5x + 100$, where $f(x)$ is the boiling point of water x kilometers above sea level.

You Try It 2 $y - y_1 = m(x - x_1)$

$$y - 2 = -\dfrac{1}{2}[x - (-2)]$$

$$y - 2 = -\dfrac{1}{2}x - 1$$

$$y = -\dfrac{1}{2}x + 1$$

You Try It 3 A line whose slope is undefined is a vertical line that passes through the point $(a, 0)$.

The equation of the line is $x = a$.

The value of x in the given point, $(4, 3)$, is 4.

The equation of the line is $x = 4$.

Goal Find a linear model that predicts the population of adults 65 years old or older in terms of the year.

Then use the model to approximate the population of these adults in 2005.

Strategy Because the function will predict the population, let y represent the population in year x.

Then $y = 13$ million when $x = 1950$.

The population is increasing 0.5 million per year. Therefore, the slope is 0.5.

Use the point-slope formula to find the linear model.

To find the population in 2005, evaluate the function when $x = 2005$.

Solution
$$y - y_1 = m(x - x_1)$$
$$y - 13 = 0.5(x - 1950)$$
$$y - 13 = 0.5x - 975$$
$$y = 0.5x - 962$$

A linear function that models the population is $f(x) = 0.5x - 962$.

$$f(x) = 0.5x - 962$$
$$f(2005) = 0.5(2005) - 962$$
$$= 1002.5 - 962$$
$$= 40.5$$

The predicted population of adults 65 years old or older in 2005 is 40.5 million.

Check The population is increasing 0.5 million per year. $2005 - 1950 = 55$; 2005 is 55 years after 1950.

The expected increase is $(0.5)(55) = 27.5$ million people.

The population in 1950 + the increase in population = 13 million + 27.5 million = 40.5 million.

Our solution checks.

YOU TRY IT 5 $(x_1, y_1) = (-2, 3)$, $(x_2, y_2) = (4, 1)$

$$m = \frac{y_2 - y_1}{x_2 - x_1} = \frac{1 - 3}{4 - (-2)} = \frac{-2}{6} = -\frac{1}{3}$$
$$y - y_1 = m(x - x_1)$$
$$y - 1 = -\frac{1}{3}(x - 4)$$
$$y - 1 = -\frac{1}{3}x + \frac{4}{3}$$
$$y = -\frac{1}{3}x + \frac{7}{3}$$

Goal Find a linear model that gives the number of calories in lean hamburger in terms of the number of ounces in the serving.

Then use the model to find the number of calories in a 5-ounce serving of lean hamburger.

Strategy Because the function will predict the number of calories, let y represent the number of calories.

Then x represents the number of ounces in a serving.

From the given data, two ordered pairs of the function are (2, 126) and (3, 189).

Use the two ordered pairs to find the slope of the line.

Use the point-slope formula to find the linear model.

To find the number of calories in a 5-ounce serving of lean hamburger, evaluate the linear function at $x = 5$.

Solution Let $(x_1, y_1) = (2, 126)$ and $(x_2, y_2) = (3, 189)$.

$$m = \frac{y_2 - y_1}{x_2 - x_1} = \frac{189 - 126}{3 - 2} = \frac{63}{1} = 63$$
$$y - y_1 = m(x - x_1)$$
$$y - 126 = 63(x - 2)$$
$$y - 126 = 63x - 126$$
$$y = 63x$$

The linear function is $f(x) = 63x$.

$$f(x) = 63x$$
$$f(5) = 63(5)$$
$$= 315$$

There are 315 calories in a 5-ounce serving of lean hamburger.

Check A 1-ounce serving contains 63 calories, a 2-ounce serving contains $63 + 63 = 126$ calories, a 3-ounce serving contains $126 + 63 = 189$ calories, a 4-ounce serving contains $189 + 63 = 252$ calories, and a 5-ounce serving contains $252 + 63 = 315$ calories. Our solution checks.

YOU TRY IT 7 The slope of the given line is -3.

The slope of any parallel line is also -3.

$$y - y_1 = m(x - x_1)$$
$$y - (-4) = -3[x - (-5)]$$
$$y + 4 = -3(x + 5)$$
$$y + 4 = -3x - 15$$
$$y = -3x - 19$$

YOU TRY IT 8

$$3x + 5y = 15$$
$$5y = -3x + 15$$
$$\frac{5y}{5} = \frac{-3x + 15}{5}$$
$$y = -\frac{3}{5}x + 3$$

The slope of the given line is $-\frac{3}{5}$.

The slope of any parallel line is also $-\frac{3}{5}$.

$$y - y_1 = m(x - x_1)$$
$$y - 3 = -\frac{3}{5}[x - (-2)]$$
$$y - 3 = -\frac{3}{5}(x + 2)$$
$$y - 3 = -\frac{3}{5}x - \frac{6}{5}$$
$$y = -\frac{3}{5}x + \frac{9}{5}$$

YOU TRY IT 9   The slope of the given line is $-\frac{4}{3}$.

The slope of any perpendicular line is $\frac{3}{4}$.

$$y - y_1 = m(x - x_1)$$
$$y - 3 = \frac{3}{4}[x - (-4)]$$
$$y - 3 = \frac{3}{4}(x + 4)$$
$$y - 3 = \frac{3}{4}x + 3$$
$$y = \frac{3}{4}x + 6$$

YOU TRY IT 10

$$5x - 3y = 15$$
$$-3y = -5x + 15$$
$$\frac{-3y}{-3} = \frac{-5x + 15}{-3}$$
$$y = \frac{5}{3}x - 5$$

The slope of the given line is $\frac{5}{3}$.

The slope of any perpendicular line is $-\frac{3}{5}$.

$$y - y_1 = m(x - x_1)$$
$$y - (-2) = -\frac{3}{5}[x - (-5)]$$
$$y + 2 = -\frac{3}{5}(x + 5)$$

$$y + 2 = -\frac{3}{5}x - 3$$
$$y = -\frac{3}{5}x - 5$$

YOU TRY IT 11

Goal   The goal is to find the equation of the line that is perpendicular to the line containing the points (0, 0) and (2, 8) and goes through (2, 8).

Strategy   The initial path of the ball is perpendicular to the line through *OP*. Therefore, the slope of the initial path of the ball is the negative reciprocal of the slope of the line between *O* and *P*.

We need to find the slope of the line through *OP*.

The slope of the line we are looking for is the negative reciprocal of that slope.

We will then have the slope of the line and a point on the line. We can use the point–slope formula to find the equation of the line.

Solution   Slope of the line through *OP*:

$$m = \frac{y_2 - y_1}{x_2 - x_1} = \frac{8 - 0}{2 - 0} = \frac{8}{2} = 4$$

The slope of the line that is the initial path of the ball is the negative reciprocal of 4.

Therefore, the slope of a perpendicular line is $-\frac{1}{4}$.

$$y - y_1 = m(x - x_1)$$
$$y - 8 = -\frac{1}{4}(x - 2)$$
$$y - 8 = -\frac{1}{4}x + \frac{1}{2}$$
$$y = -\frac{1}{4}x + \frac{17}{2}$$

Check   One way to check the solution is to graph $f(x) = -\frac{1}{4}x + \frac{17}{2}$ and $f(x) = -\frac{1}{4}x$ in the square viewing window of a graphing calculator. The lines should appear to be perpendicular. Use the TRACE feature to check that the ordered pair (2, 8) is on the graph.

SECTION 4.3

YOU TRY IT 1

We chose (1999, 23.15) to be P_1 and (2002, 19.91) to be P_2. (Other points are possible.)

$$m = \frac{y_2 - y_1}{x_2 - x_1} = \frac{19.91 - 23.15}{2002 - 1999} = \frac{-3.24}{3} = -1.08$$

$$y - y_1 = m(x - x_1)$$

$$y - 23.15 = -1.08(x - 1999)$$

$$y - 23.15 = -1.08x + 2158.92$$

$$y = -1.08x + 2182.07$$

The equation for our line is $y = -1.08x + 2182.07$.
Other equations are possible.

YOU TRY IT 2   **a.** Use a calculator to determine the regression line for the data. The regression equation is $y = 5.6\overline{3}x - 252.8\overline{6}$.

b. $y = 5.6\overline{3}x - 252.8\overline{6}$

$y = 5.6\overline{3}(63) - 252.8\overline{6}$

$y \approx 102$

The weight of a woman on a college swim team who is 63 inches tall would be approximately 102 pounds.

c. The slope indicates the increase in weight for every 1-inch increase in height.

d. A woman 0 in. tall is predicted to weight -253 lb.

YOU TRY IT 3   The points $(-2, 3)$ and $(4, 1)$ are represented in the following input/output table.

x	-2	4
y	3	1

Enter the x values in one list of a calculator and the y values in another. Then use the calculator to determine the regression line for the data.
The equation is $y = -0.\overline{3}x + 2.\overline{3}$.

SECTION 4.4

YOU TRY IT 1 $2x - 3y < 12$

$$-3y < -2x + 12$$

$$\frac{-3y}{-3} > \frac{-2x + 12}{-3}$$

$$y > \frac{2}{3}x - 4$$

Graph $y = \frac{2}{3}x - 4$ as a dashed line.

Shade the upper half-plane.

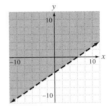

From the graph, we can see that the point $(3, -1)$ is in the solution set of the inequality.

YOU TRY IT 2   **a.** $x \geq 1$

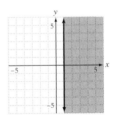

b. $y < -5$

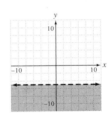

SOLUTIONS to Chapter 5 You Try Its

SECTION 5.1

YOU TRY IT 1

$$y = -\frac{2}{3}x + 1$$

$$2x + y = -3$$

$$y = -\frac{2}{3}x + 1$$

$$y = -2x - 3$$

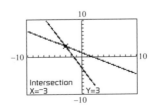

The solution is $(-3, 3)$.

You Try It 2

(1) $\qquad y = 2x + 3$

(2) $\quad 2x + 3y = 17$

Substitute $2x + 3$ for y in Equation (2) and solve for x.

$\quad 2x + 3y = 17$ • This is Equation (2).

$2x + 3(2x + 3) = 17$

$\quad 2x + 6x + 9 = 17$ • Solve for x.

$\qquad 8x + 9 = 17$

$\qquad 8x = 8$

$\qquad x = 1$

Replace x in Equation (1) by 1 and solve for y.

$y = 2x + 3$ • This is Equation (1).

$\ = 2(1) + 3$ • Replace x by 1.

$\ = 5$

The solution is $(1, 5)$.

You Try It 3

(1) $\quad 3x + \ y = 2$

(2) $\quad 9x + 3y = 6$

Solve Equation (1) for y.

$3x + y = 2$

$\qquad y = -3x + 2$

Replace y in Equation (2) by $-3x + 2$ and solve for x.

$\qquad 9x + 3y = 6$

$9x + 3(-3x + 2) = 6$

$\quad 9x - 9x + 6 = 6$

$\qquad 6 = 6$ • This is a true equation.

This means that if x is any real number and $y = -3x + 2$, then the ordered pair (x, y) is a solution of the system of equations. The solutions are the ordered pairs $(x, -3x + 2)$.

You Try It 4

Goal The goal is to find the measure of each angle of an isosceles triangle.

Strategy An isosceles triangle has two angles of equal measure. Let x be the measure of one of the equal angles, and let y be the measure of the third angle.

The sum of the measures of the angles of a triangle is 180°. Therefore,

$$x + x + y = 180 \text{ or } 2x + y = 180.$$

We are also given that the sum of the measures of the two equal angles is equal to the measure of the third angle. Therefore,

$$x + x = y \text{ or } 2x = y.$$

Solve the system of equations $\quad \begin{aligned} 2x + y &= 180 \\ 2x &= y \end{aligned}$ using the substitution method.

Solution $\quad 2x + y = 180$

$\quad 2x + 2x = 180$ • $y = 2x$

$\qquad 4x = 180$

$\qquad x = 45$

Substitute this value of x into $y = 2x$ and solve for y.

$\qquad y = 2x$

$\qquad y = 2(45) = 90$

The measures of the angles are 45°, 45°, and 90°.

Check Be sure to check your work. In this case, the sum of the measures of the angles is 180°, which indicates that the solution is correct.

You Try It 5

Goal The goal is to find the amount of money that should be invested at 4.2% and the amount at 6% so that both accounts earn the same interest.

Strategy Let x represent the amount invested at 4.2%, and let y represent the amount invested at 6%.

The total amount invested is $13,600. Therefore,

$$x + y = 13{,}600$$

Using the equation $I = Pr$, we can determine the interest earned from each account.

Interest earned at 4.2%: $0.042x$

Interest earned at 6%: $0.06y$

Both accounts are to earn the same interest, so $0.042x = 0.06y$

Now solve the system of equations formed by the two equations.

Solution (1) $\quad x + y = 13{,}600$

(2) $0.042x = 0.06y$

Solve Equation (2) for y.

$$0.042x = 0.06y$$

(3) $0.7x = y$

Replace y in Equation (1) by $0.7x$ and solve for x.

$\qquad x + y = 13{,}600$

$\quad x + 0.7x = 13{,}600$

$\qquad 1.7x = 13{,}600$

$\qquad x = 8000$

Replace x in Equation (3) by 8000 and solve for y.

$$0.7x = y$$
$$0.7(8000) = y$$
$$5600 = y$$

$8000 should be invested in the 4.2% account, and $5600 should be invested in the 6% account.

Check Be sure to check your work.

SECTION 5.2

YOU TRY IT 1

$$\begin{array}{l} 3 \diagdown (2x - 5y) = 3(4) \\ -2 \diagup (3x - 7y) = -2(15) \end{array}$$
• Multiply Equation (1) by 3 and multiply Equation (2) by −2.

$$6x - 15y = 12$$ • 3 times Equation (1).
$$\underline{-6x + 14y = -30}$$ • −2 times Equation (2).
$$-y = -18$$ • Add the equations.
$$y = 18$$ • Solve for y.

Substitute the value of y into one of the equations and solve for x. Equation (1) is used here.

$$2x - 5y = 4$$ • This is Equation (1).
$$2x - 5(18) = 4$$ • Replace y by 18.
$$2x - 90 = 4$$ • Solve for x.
$$2x = 94$$
$$x = 47$$

The solution is (47, 18).

YOU TRY IT 2

$$(1) \quad x + 2y = 6$$
$$(2) \quad 3x + 6y = 6$$
$$-3x - 6y = -18$$ • Multiply Equation (1) by −3.
$$\underline{3x + 6y = 6}$$
$$0 = -12$$ • This is not a true equation.

The system of equations has no solution.

YOU TRY IT 3

$$(1) \quad 2x + 5y = 10$$
$$(2) \quad 8x + 20y = 40$$
$$-8x - 20y = -40$$ • Multiply Equation (1) by −4.
$$\underline{8x + 20y = 40}$$
$$0 = 0$$ • This is a true equation.

The system of equations is dependent. To find the ordered-pair solutions, solve one of the equations for y. Equation (1) will be used here.

$$2x + 5y = 10$$
$$5y = -2x + 10$$
$$y = -\frac{2}{5}x + 2$$

The ordered-pair solutions are $\left(x, -\frac{2}{5}x + 2\right)$.

YOU TRY IT 4

You can choose any variable to eliminate first. We will choose x. We first eliminate x from Equation (1) and Equation (2) by multiplying Equation (2) by −3 and then adding it to Equation (1).

$$3x - y - 2z = 11$$ • Equation (1)
$$-3(x - 2y + 3z) = -3(12)$$ • −3 times Equation (2).
$$3x - y - 2z = 11$$
$$\underline{-3x + 6y - 9z = -36}$$
$$5y - 11z = -25$$ • Add the equations. This is Equation (4).

Eliminate x from Equation (2) and Equation (3) by multiplying Equation (3) by −1 and then adding to Equation (2).

$$x - 2y + 3z = 12$$ • Equation (2)
$$-1(x + y - 2z) = -1(5)$$ • −1 times Equation (3).
$$x - 2y + 3z = 12$$
$$\underline{-x - y + 2z = -5}$$
$$-3y + 5z = 7$$ • Add the equations. This is Equation (5).

Now form a system of two equations in two variables using Equation (4) and Equation (5). We will solve this system of equations by multiplying Equation (4) by 3 and Equation (5) by 5 and then adding the equations.

$$5y - 11z = -25$$ • Equation (4)
$$-3y + 5z = 7$$ • Equation (5)

$$3(5y - 11z) = 3(-25)$$ • 3 times Equation (4).
$$5(-3y + 5z) = 5(7)$$ • 5 times Equation (5).

$$15y - 33z = -75$$
$$\underline{-15y + 25z = 35}$$
$$-8z = -40$$ • Add the equations. Then solve for z.
$$z = 5$$

Substitute 5 for z into Equation (4) or (5) and solve for y. We will use Equation (5).

$$-3y + 5z = 7$$ • Equation (5)
$$-3y + 5(5) = 7$$ • Replace z by 5.
$$-3y + 25 = 7$$
$$-3y = -18$$
$$y = 6$$

Now replace y by 6 and z by 5 in one of the original equations of the system. Equation (1) will be used here.

$$3x - y - 2z = 11 \quad \bullet \text{ Equation (1)}$$
$$3x - 6 - 2(5) = 11 \quad \bullet \text{ Replace } y \text{ by 6 and replace } z \text{ by 5.}$$
$$3x - 16 = 11$$
$$3x = 27$$
$$x = 9$$

The solution of the system of equations is (9, 6, 5).

YOU TRY IT 5

Goal The goal is to find the rate of the plane in calm air and the rate of the wind.

Strategy Let x represent the rate of the plane in calm air, and let y represent the rate of the wind. Flying with the wind, the speed of the plane in calm air is increased by the rate of the wind. Flying against the wind, the speed of the plane in calm air is decreased by the rate of the wind. This can be expressed as follows:

Rate of plane with the wind: $x + y$

Rate of plane against the wind: $x - y$

Now use the equation $rt = d$ to express the distance traveled by the plane with the wind and the distance traveled against the wind in terms of the rate of the plane and the time traveled.

Distance traveled with the wind: $5(x + y) = 1000$

Distance traveled against the wind: $5(x - y) = 500$

These two equations form a system of equations.

Solution
$$5(x + y) = 1000 \Rightarrow x + y = 200 \quad \bullet \text{ Divide each side by 5.}$$
$$5(x - y) = 500 \Rightarrow x - y = 100 \quad \bullet \text{ Divide each side by 5.}$$
$$2x = 300 \quad \bullet \text{ Add the equations.}$$
$$x = 150$$

Substitute 150 for x in one of the equations and solve for y. We will use $x + y = 200$.
$$x + y = 200$$
$$150 + y = 200$$
$$y = 50$$

The rate of the plane in calm air is 150 miles per hour; the rate of the wind is 50 miles per hour.

Check Be sure to check your work.

YOU TRY IT 6

Goal The goal is to determine how many liters of each solution must be mixed to produce a 50-liter, 25% disinfectant solution.

Strategy Let x represent the number of liters of the 55% disinfectant solution, and let y represent

the number of liters of the 15% disinfectant solution. After mixing, the new solution is 50 liters. Therefore,
$$x + y = 50 \quad \bullet \text{ Equation 1}$$
The amount of disinfectant in each solution can be found by using $Q = Ar$.

Quantity of disinfectant in a solution: $Ar = rA$

Quantity of disinfectant in the 55% solution: $0.55x$

Quantity of disinfectant in the 15% solution: $0.15y$

Quantity of disinfectant in 25% solution: $0.25(50)$

Write an equation using the fact that the sum of the quantity of disinfectant in the 55% solution and the quantity of disinfectant in the 15% solution equals the quantity of disinfectant in the 25% solution.
$$0.55x + 0.15y = 0.25(50)$$
$$0.55x + 0.15y = 12.5 \quad \bullet \text{ Equation 2}$$
Equations (1) and (2) form a system of equations.

Solve (1) $\qquad x + y = 50$
(2) $\quad 0.55x + 0.15y = 12.5$

Eliminate x by multiplying Equation (1) by -0.55 and then adding it to Equation (2).
$$-0.55x - 0.55y = -27.5$$
$$\underline{0.55x + 0.15y = 12.5}$$
$$-0.4y = -15 \quad \bullet \text{ Add the equations.}$$
$$y = 37.5 \quad \bullet \text{ Solve for } y.$$

Substitute the value of y into Equation (1) and solve for x.
$$x + y = 50$$
$$x + 37.5 = 50$$
$$x = 12.5$$

Because x represents the number of liters of the 55% solution, the hospital staff must use 12.5 liters of the 55% solution. Because y represents the number of liters of the 15% solution, the hospital staff must use 37.5 liters of the 15% solution.

Check Quantity of disinfectant in the 55% solution
$$Q_1 = 0.55(12.5) = 6.875$$
Quantity of disinfectant in the 15% solution
$$Q_2 = 0.15(37.5) = 5.625$$
The quantity of disinfectant in the mixture is $6.875 + 5.625 = 12.5$ liters. The amount of mixture is 50 liters. To find the percent concentration of disinfectant, solve $Q = Ar$ for r given that $Q = 12.5$ and $A = 50$.

$$Q = Ar$$
$$12.5 = 50r$$
$$\frac{12.5}{50} = r$$
$$0.25 = r$$

The percent concentration is 25%. The solution checks.

SECTION 5.3

You Try It 1

$$\begin{bmatrix} 2 & -3 & 1 & | & 4 \\ 1 & 0 & -2 & | & 3 \\ 0 & 1 & 2 & | & -3 \end{bmatrix}$$

$$2x - 3y + z = 4$$
$$x - 2z = 3$$
$$y + 2z = -3$$

You Try It 2

$$B = \begin{bmatrix} 1 & 8 & -2 & 3 \\ 2 & -3 & 4 & 1 \\ 3 & 5 & -7 & 3 \end{bmatrix}$$

a. $\begin{bmatrix} 1 & 8 & -2 & 3 \\ 2 & -3 & 4 & 1 \\ 3 & 5 & -7 & 3 \end{bmatrix} \xrightarrow{R_2 \leftrightarrow R_3} \begin{bmatrix} 1 & 8 & -2 & 3 \\ 3 & 5 & -7 & 3 \\ 2 & -3 & 4 & 1 \end{bmatrix}$

b. $\begin{bmatrix} 1 & 8 & -2 & 3 \\ 2 & -3 & 4 & 1 \\ 3 & 5 & -7 & 3 \end{bmatrix} \xrightarrow{3R_2} \begin{bmatrix} 1 & 8 & -2 & 3 \\ 6 & -9 & 12 & 3 \\ 3 & 5 & -7 & 3 \end{bmatrix}$

c.

$\begin{bmatrix} 1 & 8 & -2 & 3 \\ 2 & -3 & 4 & 1 \\ 3 & 5 & -7 & 3 \end{bmatrix} \xrightarrow{-3R_1 + R_3} \begin{bmatrix} 1 & 8 & -2 & 3 \\ 2 & -3 & 4 & 1 \\ 0 & -19 & -1 & -6 \end{bmatrix}$

You Try It 3

$\begin{bmatrix} 1 & -3 & 2 & 1 \\ -4 & 14 & 0 & -2 \\ 2 & -5 & -3 & 16 \end{bmatrix} \xrightarrow{4R_1 + R_2} \begin{bmatrix} 1 & -3 & 2 & 1 \\ 0 & 2 & 8 & 2 \\ 2 & -5 & -3 & 16 \end{bmatrix}$

$\begin{bmatrix} 1 & -3 & 2 & 1 \\ 0 & 2 & 8 & 2 \\ 2 & -5 & -3 & 16 \end{bmatrix} \xrightarrow{-2R_1 + R_3} \begin{bmatrix} 1 & -3 & 2 & 1 \\ 0 & 2 & 8 & 2 \\ 0 & 1 & -7 & 14 \end{bmatrix}$

$\begin{bmatrix} 1 & -3 & 2 & 1 \\ 0 & 2 & 8 & 2 \\ 0 & 1 & -7 & 14 \end{bmatrix} \xrightarrow{\frac{1}{2}R_2} \begin{bmatrix} 1 & -3 & 2 & 1 \\ 0 & 1 & 4 & 1 \\ 0 & 1 & -7 & 14 \end{bmatrix}$

$\begin{bmatrix} 1 & -3 & 2 & 1 \\ 0 & 1 & 4 & 1 \\ 0 & 1 & -7 & 14 \end{bmatrix} \xrightarrow{-1R_2 + R_3} \begin{bmatrix} 1 & -3 & 2 & 1 \\ 0 & 1 & 4 & 1 \\ 0 & 0 & -11 & 13 \end{bmatrix}$

$\begin{bmatrix} 1 & -3 & 2 & 1 \\ 0 & 1 & 4 & 1 \\ 0 & 0 & -11 & 13 \end{bmatrix} \xrightarrow{-\frac{1}{11}R_3} \begin{bmatrix} 1 & -3 & 2 & 1 \\ 0 & 1 & 4 & 1 \\ 0 & 0 & 1 & -\frac{13}{11} \end{bmatrix}$

A row echelon form of the matrix is $\begin{bmatrix} 1 & -3 & 2 & 1 \\ 0 & 1 & 4 & 1 \\ 0 & 0 & 1 & -\frac{13}{11} \end{bmatrix}$.

You Try It 4

$$4x - 5y = 17$$
$$3x + 2y = 7$$

$\begin{bmatrix} 4 & -5 & | & 17 \\ 3 & 2 & | & 7 \end{bmatrix} \xrightarrow{\frac{1}{4}R_1} \begin{bmatrix} 1 & -\frac{5}{4} & | & \frac{17}{4} \\ 3 & 2 & | & 7 \end{bmatrix}$

$\begin{bmatrix} 1 & -\frac{5}{4} & | & \frac{17}{4} \\ 3 & 2 & | & 7 \end{bmatrix} \xrightarrow{-3R_1 + R_2} \begin{bmatrix} 1 & -\frac{5}{4} & | & \frac{17}{4} \\ 0 & \frac{23}{4} & | & -\frac{23}{4} \end{bmatrix}$

$\begin{bmatrix} 1 & -\frac{5}{4} & | & \frac{17}{4} \\ 0 & \frac{23}{4} & | & -\frac{23}{4} \end{bmatrix} \xrightarrow{\frac{4}{23}R_2} \begin{bmatrix} 1 & -\frac{5}{4} & | & \frac{17}{4} \\ 0 & 1 & | & -1 \end{bmatrix}$

(1) $\qquad x - \dfrac{5}{4}y = \dfrac{17}{4}$

(2) $\qquad y = -1$

$$x - \frac{5}{4}(-1) = \frac{17}{4}$$
$$x + \frac{5}{4} = \frac{17}{4}$$
$$x = \frac{12}{4}$$
$$x = 3$$

The solution is $(3, -1)$.

You Try It 5

$$2x + 3y + 3z = -2$$
$$x + 2y - 3z = 9$$
$$3x - 2y - 4z = 1$$

$$\begin{bmatrix} 2 & 3 & 3 & -2 \\ 1 & 2 & -3 & 9 \\ 3 & -2 & -4 & 1 \end{bmatrix} \quad R_1 \leftrightarrow R_2 \quad \begin{bmatrix} 1 & 2 & -3 & 9 \\ 2 & 3 & 3 & -2 \\ 3 & -2 & -4 & 1 \end{bmatrix}$$

$$\begin{bmatrix} 1 & 2 & -3 & 9 \\ 2 & 3 & 3 & -2 \\ 3 & -2 & -4 & 1 \end{bmatrix} \quad -2R_1 + R_2 \quad \begin{bmatrix} 1 & 2 & -3 & 9 \\ 0 & -1 & 9 & -20 \\ 3 & -2 & -4 & 1 \end{bmatrix}$$

$$\begin{bmatrix} 1 & 2 & -3 & 9 \\ 0 & -1 & 9 & -20 \\ 3 & -2 & -4 & 1 \end{bmatrix} \quad -3R_1 + R_3 \quad \begin{bmatrix} 1 & 2 & -3 & 9 \\ 0 & -1 & 9 & -20 \\ 0 & -8 & 5 & -26 \end{bmatrix}$$

$$\begin{bmatrix} 1 & 2 & -3 & 9 \\ 0 & -1 & 9 & -20 \\ 0 & -8 & 5 & -26 \end{bmatrix} \quad -1R_2 \quad \begin{bmatrix} 1 & 2 & -3 & 9 \\ 0 & 1 & -9 & 20 \\ 0 & -8 & 5 & -26 \end{bmatrix}$$

$$\begin{bmatrix} 1 & 2 & -3 & 9 \\ 0 & 1 & -9 & 20 \\ 0 & -8 & 5 & -26 \end{bmatrix} \quad 8R_2 + R_3 \quad \begin{bmatrix} 1 & 2 & -3 & 9 \\ 0 & 1 & -9 & 20 \\ 0 & 0 & -67 & 134 \end{bmatrix}$$

$$\begin{bmatrix} 1 & 2 & -3 & 9 \\ 0 & 1 & -9 & 20 \\ 0 & 0 & -67 & 134 \end{bmatrix} \quad -\frac{1}{67}R_3 \quad \begin{bmatrix} 1 & 2 & -3 & 9 \\ 0 & 1 & -9 & 20 \\ 0 & 0 & 1 & -2 \end{bmatrix}$$

$$x + 2y - 3z = 9$$
$$y - 9z = 20$$
$$z = -2$$

$$y - 9z = 20$$
$$y - 9(-2) = 20$$
$$y + 18 = 20$$
$$y = 2$$
$$x + 2y - 3z = 9$$
$$x + 2(2) - 3(-2) = 9$$
$$x + 4 + 6 = 9$$
$$x + 10 = 9$$
$$x = -1$$

The solution is $(-1, 2, -2)$.

YOU TRY IT 6

Goal The goal is to find the values of a, b, and c for the equation $y = ax^2 + bx + c$ given that the graph passes through the points $(2, 3)$, $(-1, 0)$, and $(0, -3)$.

Strategy Substitute the coordinates of the three given points into $y = ax^2 + bx + c$. Each point will create one equation of a system of equations.

$$y = ax^2 + bx + c$$
$$3 = a(2)^2 + b(2) + c$$
$$0 = a(-1)^2 + b(-1) + c$$
$$-3 = a(0)^2 + b(0) + c$$

Simplify the three equations.
$$4a + 2b + c = 3$$
$$a - b + c = 0$$
$$c = -3$$

Write the system as an augmented matrix.

$$\begin{bmatrix} 4 & 2 & 1 & 3 \\ 1 & -1 & 1 & 0 \\ 0 & 0 & 1 & -3 \end{bmatrix}$$

Solve the system of equations by using the Gaussian elimination method.

Solution

$$\begin{bmatrix} 4 & 2 & 1 & 3 \\ 1 & -1 & 1 & 0 \\ 0 & 0 & 1 & -3 \end{bmatrix} \quad R_1 \leftrightarrow R_2 \quad \begin{bmatrix} 1 & -1 & 1 & 0 \\ 4 & 2 & 1 & 3 \\ 0 & 0 & 1 & -3 \end{bmatrix}$$

$$\begin{bmatrix} 1 & -1 & 1 & 0 \\ 4 & 2 & 1 & 3 \\ 0 & 0 & 1 & -3 \end{bmatrix} \quad -4R_1 + R_2 \quad \begin{bmatrix} 1 & -1 & 1 & 0 \\ 0 & 6 & -3 & 3 \\ 0 & 0 & 1 & -3 \end{bmatrix}$$

$$\begin{bmatrix} 1 & -1 & 1 & 0 \\ 0 & 6 & -3 & 3 \\ 0 & 0 & 1 & -3 \end{bmatrix} \quad \frac{1}{6}R_2 \quad \begin{bmatrix} 1 & -1 & 1 & 0 \\ 0 & 1 & -\frac{1}{2} & \frac{1}{2} \\ 0 & 0 & 1 & -3 \end{bmatrix}$$

$$a - b + c = 0$$
$$b - \frac{1}{2}c = \frac{1}{2}$$
$$c = -3$$

$$b - \frac{1}{2}(-3) = \frac{1}{2}$$

$$b + \frac{3}{2} = \frac{1}{2}$$

$$b = -1$$
$$a - (-1) + (-3) = 0$$
$$a - 2 = 0$$
$$a = 2$$

The equation is $y = 2x^2 - x - 3$.

Check Verify that each given ordered pair is a solution of the equation by substituting the coordinates into the equation $y = 2x^2 - x - 3$.

YOU TRY IT 7

Goal The goal is to find the number of each type of ticket sold.

Strategy There are three unknowns in this problem. Using the information from the problem, write a system of three equations in three unknowns. Let x be the number of regular admission tickets sold, let y be the number of member discount tickets sold, and let z be the number of student tickets sold.

Because there were 750 tickets sold, we have $x + y + z = 750$.

The receipts for selling these tickets were $5400. Therefore, $10x + 7y + 5z = 5400$.

Because 20 more student tickets than full-price tickets were sold, we have $z = x + 20$, or $-x + z = 20$. Solve the system of equations

$$x + y + z = 750$$
$$10x + 7y + 5z = 5400$$
$$-x + z = 20$$

Write the system as an augmented matrix.

$$\begin{bmatrix} 1 & 1 & 1 & 750 \\ 10 & 7 & 5 & 5400 \\ -1 & 0 & 1 & 20 \end{bmatrix}$$

Solve the system of equations by using the Gaussian elimination method.

Solution

```
[[1   .7    .5        ...
 [0   1    2.142857  ...
 [0   0    1          ...
Ans▶Frac
[[1   7/10  1/2    5...
 [0   1    15/7    8...
 [0   0    1       2...
```

$$x + \frac{7}{10}y + \frac{1}{2}z = 540$$

$$y + \frac{15}{7}z = 800$$

$$z = 210$$

$$y + \frac{15}{7}(210) = 800$$

$$y + 450 = 800$$

$$y = 350$$

$$x + \frac{7}{10}(350) + \frac{1}{2}(210) = 540$$

$$x + 245 + 105 = 540$$

$$x + 350 = 540$$

$$x = 190$$

There were 190 regular admission tickets, 350 member tickets, and 210 student tickets sold.

Check Check the solution by verifying that these numbers satisfy each of the conditions of the problem.

YOU TRY IT 1

Shade above the solid line $y = 2x - 3$. Shade above the dashed line $y = -3x$.

YOU TRY IT 2

Goal Find the number of kiloliters of each solvent that the company should make to maximize profit.

Strategy Let $x = $ the number of kiloliters of S_1 and $y = $ the number of kiloliters of S_2.

The objective function is the profit function $P = 100x + 85y$.

Because x kiloliters of S_1 require $12x$ liters of chemical 1 and y kiloliters of S_2 require $24y$ liters of chemical 1, the total amount of chemical 1 needed is $12x + 24y$. There are 480 liters of chemical 1 in inventory, so $12x + 24y \le 480$. Following similar reasoning, we have the constraints

$$\begin{cases} 12x + 24y \le 480 \\ 9x + 5y \le 180 \\ 30x + 30y \le 720 \\ x \ge 0, y \ge 0 \end{cases}$$

Graph the constraints and determine the vertices of the set of feasible solutions.

Solve Two of the vertices of the set of feasible solutions can be found by solving two systems of equations. These systems are formed by the equations of the lines that intersect to form a vertex of the set of feasible solutions.

$$12x + 24y = 480$$
$$30x + 30y = 720$$

The solution is $(8, 16)$.

$$9x + 5y = 180$$
$$30x + 30y = 720$$

The solution is $(15, 9)$.

The vertices on the x- and y-axes are the x- and y-intercepts, $(20, 0)$ and $(0, 20)$.

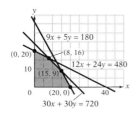

$$30x + 30y = 720$$

Substitute the coordinates of the vertices into the objective function.

(x, y)	$P = 100x + 85y$
$(0, 20)$	$P = 100(0) + 85(20) = 1700$
$(8, 16)$	$P = 100(8) + 85(16) = 2160$
$(15, 9)$	$P = 100(15) + 85(9) = 2265$
$(20, 0)$	$P = 100(20) + 85(0) = 2000$

The maximum value of the objective function is \$2265 when the company produces 15 kiloliters of S_1 and 9 kiloliters of S_2.

Check Be sure to check your work.

SOLUTIONS to Chapter 6 You Try Its

SECTION 6.1

YOU TRY IT 1 $\quad 16a^4b^3 + 10a^4b^3 + 5a^4b^3 = 31a^4b^3$

YOU TRY IT 2 $\quad 37m^3n^2p - 14m^3n^2p = 23m^3n^2p$

YOU TRY IT 3 $\quad t^3 \cdot t^8 = t^{3+8} = t^{11}$

YOU TRY IT 4 $\quad n^6 \cdot n \cdot n^2 = n^{6+1+2} = n^9$

YOU TRY IT 5 $\quad c^9(c^5d^8) = c^{9+5}d^8 = c^{14}d^8$

YOU TRY IT 6 $\quad (5y^4)(3y^2) = (5 \cdot 3)(y^4 \cdot y^2) = 15y^{4+2} = 15y^6$

YOU TRY IT 7 $\quad (12p^4q^3)(-3p^5q^2) = [12(-3)]p^{4+5}q^{3+2}$
$$= -36p^9q^5$$

YOU TRY IT 8 $\quad (t^3)^6 = t^{3 \cdot 6} = t^{18}$

YOU TRY IT 9 $\quad (bc^7)^8 = b^{1 \cdot 8}c^{7 \cdot 8} = b^8c^{56}$

YOU TRY IT 10 $\quad (4y^6)^3 = 4^{1 \cdot 3}y^{6 \cdot 3} = 4^3y^{18} = 64y^{18}$

YOU TRY IT 11 $\quad (2v^6w^9)^5 = 2^{1 \cdot 5}v^{6 \cdot 5}w^{9 \cdot 5} = 2^5v^{30}w^{45}$
$$= 32v^{30}w^{45}$$

YOU TRY IT 12 $\quad (-2x^3y^7)^3 = (-2)^{1 \cdot 3}x^{3 \cdot 3}y^{7 \cdot 3} = (-2)^3x^9y^{21}$
$$= -8x^9y^{21}$$

YOU TRY IT 13 $\quad (-xy^4)(-2x^3y^2)^2 = (-xy^4)[(-2)^{1 \cdot 2}x^{3 \cdot 2}y^{2 \cdot 2}]$
$$= (-xy^4)[(-2)^2x^6y^4]$$
$$= (-xy^4)(4x^6y^4)$$
$$= (-1 \cdot 4)(x^{1+6})(y^{4+4})$$
$$= -4x^7y^8$$

YOU TRY IT 14 $\quad \dfrac{t^{10}}{t^4} = t^{10-4} = t^6$

YOU TRY IT 15 $\quad \dfrac{a^7b^6}{ab^3} = a^{7-1}b^{6-3} = a^6b^3$

YOU TRY IT 16 $\quad (-8x^2y^7)^0 = 1$

YOU TRY IT 17 $\quad -(9c^7d^4)^0 = -1$

YOU TRY IT 18 $\quad \dfrac{2}{c^{-4}} = 2 \cdot \dfrac{1}{c^{-4}} = 2 \cdot c^4 = 2c^4$

YOU TRY IT 19 $\quad \dfrac{12x^{-8}y}{-16xy^{-3}} = -\dfrac{3x^{-8}y}{4xy^{-3}} = -\dfrac{3}{4}x^{-8-1}y^{1-(-3)}$
$$= -\dfrac{3}{4}x^{-9}y^4 = -\dfrac{3y^4}{4x^9}$$

YOU TRY IT 20 $\quad \left(\dfrac{m^{-6}}{n^{-8}}\right)^3 = \dfrac{m^{-6 \cdot 3}}{n^{-8 \cdot 3}} = \dfrac{m^{-18}}{n^{-24}} = \dfrac{n^{24}}{m^{18}}$

YOU TRY IT 21 **a.** $(-2ab)(2a^3b^{-2})^{-3} = (-2ab)(2^{-3}a^{-9}b^6)$
$$= (-2 \cdot 2^{-3})(a \cdot a^{-9})(b \cdot b^6)$$
$$= \left(-2 \cdot \dfrac{1}{2^3}\right)a^{-8}b^7$$
$$= \left(-2 \cdot \dfrac{1}{8}\right)a^{-8}b^7 = -\dfrac{b^7}{4a^8}$$

b. $\left(\dfrac{2x^2y^{-4}}{4x^{-2}y^{-5}}\right)^{-3} = \left(\dfrac{x^2y^{-4}}{2x^{-2}y^{-5}}\right)^{-3} = \dfrac{x^{-6}y^{12}}{2^{-3}x^6y^{15}}$
$$= 2^3x^{-6-6}y^{12-15} = 8x^{-12}y^{-3} = \dfrac{8}{x^{12}y^3}$$

YOU TRY IT 22 **a.** $57,000,000,000 = 5.7 \times 10^{10}$

b. $0.000000017 = 1.7 \times 10^{-8}$

YOU TRY IT 23 **a.** $5 \times 10^{12} = 5,000,000,000,000$

b. $4.0162 \times 10^{-9} = 0.0000000040162$

YOU TRY IT 24 **a.** $(2.4 \times 10^{-9})(1.6 \times 10^3) = 3.84 \times 10^{-6}$

b. $\dfrac{5.4 \times 10^{-2}}{1.8 \times 10^{-4}} = 3 \times 10^2$

SECTION 6.2

YOU TRY IT 1

$$V(r) = 2000r^3 + 6000r^2 + 6000r + 2000$$
$$V(0.07) = 2000(0.07)^3 + 6000(0.07)^2 + 6000(0.07) + 2000$$
$$V(0.07) = 0.686 + 29.4 + 420 + 2000$$
$$V(0.07) = 2450.086$$

After 3 years, the value of $2000 deposited in an IRA that earns 7% interest is $2450.09.

YOU TRY IT 2 $(-4d^2 - 3d + 2) + (3d^2 - 4d)$
$$= (-4d^2 + 3d^2) + (-3d - 4d) + 2$$
$$= -d^2 - 7d + 2$$

YOU TRY IT 3 $(5x^2 - 3x + 4) - (-6x^3 - 2x + 8)$
$$= (5x^2 - 3x + 4) + (6x^3 + 2x - 8)$$
$$= 6x^3 + 5x^2 + (-3x + 2x) + (4 - 8)$$
$$= 6x^3 + 5x^2 - x - 4$$

YOU TRY IT 4

Goal   The goal is to write a variable expression for the company's monthly profit.

Strategy   Use the formula $P = R - C$. Substitute the given polnomials for R and C. Then subtract the polynomials.

Solution $P = R - C$
$$P = (-0.2n^2 + 175n) - (35n + 2000)$$
$$P = (-0.2n^2 + 175n) + (-35n - 2000)$$
$$P = -0.2n^2 + (175n - 35n) - 2000$$
$$P = -0.2n^2 + 140n - 2000$$

The company's monthly profit is $(-0.2n^2 + 140n - 2000)$ dollars.

Check $\checkmark$

SECTION 6.3

YOU TRY IT 1   a. $(-2d + 3)(-4d) = -2d(-4d) + 3(-4d)$
$$= 8d^2 - 12d$$

b. $-a^3(3a^2 + 2a - 7)$
$$= -a^3(3a^2) + (-a^3)(2a) - (-a^3)(7)$$
$$= -3a^5 - 2a^4 + 7a^3$$

YOU TRY IT 2 $(3c^3 - 2c^2 + c - 3)(2c + 5)$

$$
\begin{array}{r}
3c^3 - 2c^2 + c - 3 \\
2c + 5 \\
\hline
15c^3 - 10c^2 + 5c - 15 \\
6c^4 - 4c^3 + 2c^2 - 6c \\
\hline
6c^4 + 11c^3 - 8c^2 - c - 15
\end{array}
$$

YOU TRY IT 3   a. $(4y - 5)(3y - 3)$
$$= 4y(3y) + 4y(-3) + (-5)(3y) + (-5)(-3)$$
$$= 12y^2 - 12y - 15y + 15$$
$$= 12y^2 - 27y + 15$$

b. $(3a + 2b)(3a - 5b)$
$$= 3a(3a) + 3a(-5b) + 2b(3a) + 2b(-5b)$$
$$= 9a^2 - 15ab + 6ab - 10b^2$$
$$= 9a^2 - 9ab - 10b^2$$

YOU TRY IT 4 $(3x + 2y)^2 = (3x + 2y)(3x + 2y)$
$$= 9x^2 + 6xy + 6xy + 4y^2$$
$$= 9x^2 + 12xy + 4y^2$$

YOU TRY IT 5

Goal   The goal is to determine the area of the circle in terms of x.

Strategy   To determine the area of the circle in terms of x, use the formula for the area of a circle, $A = \pi r^2$. Substitute the variable expression $(x - 4)$ for r. Then simplify.

Solution $A = \pi r^2$
$$A = \pi(x - 4)^2$$
$$A = \pi(x - 4)(x - 4)$$
$$A = \pi(x^2 - 4x - 4x + 16)$$
$$A = \pi(x^2 - 8x + 16)$$
$$A = \pi x^2 - 8\pi x + 16\pi$$

The area of the circle is $(\pi x^2 - 8\pi x + 16\pi)$ square feet.

Check   Check each step of the solution.

YOU TRY IT 6 $\dfrac{4x^3y + 8x^2y^2 - 4xy^3}{2xy} = \dfrac{4x^3y}{2xy} + \dfrac{8x^2y^2}{2xy} - \dfrac{4xy^3}{2xy}$
$$= 2x^2 + 4xy - 2y^2$$

YOU TRY IT 7 $(x^3 - 7 - 2x) \div (x - 2)$
$$= (x^3 + 0x^2 - 2x - 7) \div (x - 2)$$

$$
\begin{array}{r}
x^2 + 2x + 2 \\
x - 2 \overline{)x^3 + 0x^2 - 2x - 7} \\
\underline{x^3 - 2x^2} \\
2x^2 - 2x \\
\underline{2x^2 - 4x} \\
2x - 7 \\
\underline{2x - 4} \\
-3
\end{array}
$$

$(x^3 - 7 - 2x) \div (x - 2) = x^2 + 2x + 2 - \dfrac{3}{x - 2}$

YOU TRY IT 8

$$
\begin{array}{r}
x^3 + 2 \\
x + 5 \overline{)x^4 + 5x^3 + 2x + 10} \\
\underline{x^4 + 5x^3} \\
0 2x + 10 \\
\underline{2x + 10} \\
0
\end{array}
$$

Another factor of $x^4 + 5x^3 + 2x + 10$ is $x^3 + 2$.

YOU TRY IT 9

Goal   The goal is to write a variable expression for the height of a parallelogram that has an area of $(3x^2 + 2x - 8)$ square feet and a base of length $(x + 2)$ feet.

S25

Strategy Using the formula $h = \frac{A}{b}$, substitute the given polynomials for A and b. Then divide the polynomials.

Solution $h = \dfrac{A}{b}$

$h = \dfrac{3x^2 + 2x - 8}{x + 2}$

$$\begin{array}{r} 3x - 4 \\ x+2\overline{)3x^2 + 2x - 8} \\ \underline{3x^2 + 6x} \\ -4x - 8 \\ \underline{-4x - 8} \\ 0 \end{array}$$

The height is $(3x - 4)$ ft.

Check $(3x - 4)(x + 2) = 3x^2 + 6x - 4x - 8$
$$= 3x^2 + 2x - 8$$

YOU TRY IT 10

a. $\begin{array}{r|rrr} -2 & 6 & 8 & -5 \\ & & -12 & 8 \\ \hline & 6 & -4 & 3 \end{array}$

$(6x^2 + 8x - 5) \div (x + 2) = 6x - 4 + \dfrac{3}{x + 2}$

b. $\begin{array}{r|rrrrr} 3 & 2 & -3 & -8 & 0 & -2 \\ & & 6 & 9 & 3 & 9 \\ \hline & 2 & 3 & 1 & 3 & 7 \end{array}$

$(2x^4 - 3x^3 - 8x^2 - 2) \div (x - 3)$
$$= 2x^3 + 3x^2 + x + 3 + \dfrac{7}{x - 3}$$

YOU TRY IT 11 $\begin{array}{r|rrrr} 3 & 2 & 0 & -4 & -5 \\ & & 6 & 18 & 42 \\ \hline & 2 & 6 & 14 & 37 \end{array}$

By the Remainder Theorem, $P(3) = 37$.

YOU TRY IT 12 $\begin{array}{r|rrrr} 1 & -1 & 4 & -5 & 2 \\ & & -1 & 3 & -2 \\ \hline & -1 & 3 & -2 & 0 \end{array}$

The remainder is 0.
Yes, $x - 1$ is a factor of $P(x) = -x^3 + 4x^2 - 5x + 2$.

SECTION 6.4

YOU TRY IT 1 The GCF is $3x^2y^2$.

$\dfrac{6x^4y^2 - 9x^3y^2 + 12x^2y^4}{3x^2y^2} = 2x^2 - 3x + 4y^2$

$6x^4y^2 - 9x^3y^2 + 12x^2y^4 = 3x^2y^2(2x^2 - 3x + 4y^2)$

YOU TRY IT 2 $a(b - 7) + b(b - 7) = (b - 7)(a + b)$

YOU TRY IT 3
$$\begin{aligned} y^5 - 5y^3 + 4y^2 - 20 &= (y^5 - 5y^3) + (4y^2 - 20) \\ &= y^3(y^2 - 5) + 4(y^2 - 5) \\ &= (y^2 - 5)(y^3 + 4) \end{aligned}$$

YOU TRY IT 4
$$\begin{aligned} 3xy - 6y - 8 + 4x &= (3xy - 6y) - (8 - 4x) \\ &= 3y(x - 2) - 4(2 - x) \\ &= 3y(x - 2) + 4(x - 2) \\ &= (x - 2)(3y + 4) \end{aligned}$$

YOU TRY IT 5 **a.** $x^2 + 9x + 20 = (x + 4)(x + 5)$
b. $x^2 + 7x - 18 = (x + 9)(x - 2)$

YOU TRY IT 6 $2x^2 - x - 3$

Positive Factors of 2	Factors of -3
1, 2	1, -3
	-1, 3

Trial Factors	Middle Term
$(x + 1)(2x - 3)$	$-3x + 2x = -x$
$(x - 3)(2x + 1)$	$x - 6x = -5x$
$(x - 1)(2x + 3)$	$3x - 2x = x$
$(x + 3)(2x - 1)$	$-x + 6x = 5x$

$2x^2 - x - 3 = (x + 1)(2x - 3)$

YOU TRY IT 7 $2x^2 + 13x - 7$
$a \cdot c = -14$

Factors of -14	Sum
1, -14	-13
-1, 14	13
2, -7	-5
-2, 7	5

$$\begin{aligned} 2x^2 + 13x - 7 &= 2x^2 - x + 14x - 7 \\ &= (2x^2 - x) + (14x - 7) \\ &= x(2x - 1) + 7(2x - 1) \\ &= (2x - 1)(x + 7) \end{aligned}$$

SECTION 6.5

YOU TRY IT 1 $x^2 - 36y^4 = (x)^2 - (6y^2)^2$
$$= (x + 6y^2)(x - 6y^2)$$

YOU TRY IT 2 $4x^2 - 20x + 25 = (2x - 5)^2$

YOU TRY IT 3 $x^3y^3 - 1 = (xy - 1)(x^2y^2 + xy + 1)$

YOU TRY IT 4 $3x^4 + 4x^2 - 4$
$$\begin{aligned} &= 3u^2 + 4u - 4 \qquad \bullet \text{ Let } u = x^2. \\ &= (3u - 2)(u + 2) \\ &= (3x^2 - 2)(x^2 + 2) \end{aligned}$$

You Try It 5 $4a^3 - 4a^2b - 24ab^2$
$$= 4a(a^2 - ab - 6b^2)$$
$$= 4a(a - 3b)(a + 2b)$$

You Try It 6 $4x - 4y - x^3 + x^2y$
$$= (4x - 4y) - (x^3 - x^2y)$$
$$= 4(x - y) - x^2(x - y)$$
$$= (x - y)(4 - x^2)$$
$$= (x - y)(2 + x)(2 - x)$$

You Try It 7

Goal The goal is to find the length, width, and height of a box (a rectangular solid) that has a volume of $(3x^2y + 21xy + 36y)$ cubic inches. We are looking for three expressions that when multiplied equal $3x^2y + 21xy + 36y$.

Strategy Factor the polynomial $3x^2y + 21xy + 36y$.

Solution $3x^2y + 21xy + 36y = 3y(x^2 + 7x + 12)$
$$= 3y(x + 3)(x + 4)$$

The dimensions of the box are $3y$ inches by $(x + 3)$ inches by $(x + 4)$ inches.

Check $3y(x + 3)(x + 4) = (3xy + 9y)(x + 4)$
$$= 3x^2y + 12xy + 9xy + 36y$$
$$= 3x^2y + 21xy + 36y$$

SECTION 6.6

You Try It 1 $3x^2 + 5x = 2$
$$3x^2 + 5x - 2 = 0$$
$$(3x - 1)(x + 2) = 0$$

$3x - 1 = 0$ $x + 2 = 0$
$3x = 1$ $x = -2$
$$x = \frac{1}{3}$$

The solutions are -2 and $\frac{1}{3}$.

You Try It 2 $(x - 2)(x + 5) = 8$
$$x^2 + 3x - 10 = 8$$
$$x^2 + 3x - 18 = 0$$
$$(x + 6)(x - 3) = 0$$

$x + 6 = 0$ $x - 3 = 0$
$x = -6$ $x = 3$

The solutions are -6 and 3.

You Try It 3 $x^3 + 4x^2 - 9x - 36 = 0$
$$(x^3 + 4x^2) - (9x + 36) = 0$$
$$x^2(x + 4) - 9(x + 4) = 0$$
$$(x + 4)(x^2 - 9) = 0$$
$$(x + 4)(x + 3)(x - 3) = 0$$

$x + 4 = 0$ $x + 3 = 0$ $x - 3 = 0$
$x = -4$ $x = -3$ $x = 3$

The solutions are -4, -3, and 3.

You Try It 4

Goal The goal is to find how many consecutive natural numbers (the numbers 1, 2, 3, 4, . . .) beginning with 1 have a sum of 78.

Strategy Substitute 78 for S in the given formula and solve for n.

Solution $2S = n^2 + n$
$$2(78) = n^2 + n$$
$$156 = n^2 + n$$
$$0 = n^2 + n - 156$$
$$0 = (n + 13)(n - 12)$$

$n + 13 = 0$ $n - 12 = 0$
$n = -13$ $n = 12$

-13 is not a natural number.

The first 12 natural numbers have a sum of 78.

Check One method of checking the solution is to add the first 12 natural numbers.
Use one of the problem-solving techniques presented in Chapter 1.
$$1 + 2 + 3 + 4 + 5 + 6 + 7 + 8 + 9 + 10 + 11 + 12 =$$
$$6(13) = 78$$

SOLUTIONS to Chapter 7 You Try Its

SECTION 7.1

You Try It 1

$$g(x) = \frac{5 - x}{x^2 - 4}$$
$$x^2 - 4 = 0$$
$$(x + 2)(x - 2) = 0$$

$x + 2 = 0$ $x - 2 = 0$
$x = -2$ $x = 2$

The domain is $\{x \mid x \neq -2, 2\}$.

You Try It 2

$$p(x) = \frac{6x}{x^2 + 4}$$

The domain must exclude values of x for which $x^2 + 4 = 0$. It is not possible for $x^2 + 4 = 0$, because $x^2 \geq 0$, and a positive number added to a number equal to or greater than zero cannot equal zero. Therefore, there are no real numbers that must be excluded from the domain of p.

The domain is $\{x \mid x \in \text{real numbers}\}$.

You Try It 3

a. $g(x) = \dfrac{3x^2 + 5}{x^2 - 25}$

$x^2 - 25 = 0$

$(x + 5)(x - 5) = 0$

$x + 5 = 0 \qquad x - 5 = 0$

$ x = -5 \qquad\quad x = 5$

The lines $x = -5$ and $x = 5$ are vertical asymptotes of the graph of g.

b. $h(x) = \dfrac{4}{x^2 + 9}$

There are no zeros of the denominator.
The graph of h has no vertical asymptotes.

You Try It 4

a. $\dfrac{6x^4 - 24x^3}{12x^3 - 48x^2} = \dfrac{6x^3(x - 4)}{12x^2(x - 4)} = \dfrac{\overset{1}{\cancel{6x^3}}\cancel{(x-4)}}{\underset{1}{\cancel{12x^2}}\cancel{(x-4)}} = \dfrac{x}{2}$

b. $\dfrac{20x - 15x^2}{15x^3 - 5x^2 - 20x} = \dfrac{5x(4 - 3x)}{5x(3x^2 - x - 4)}$

$\phantom{\dfrac{20x - 15x^2}{15x^3}} = \dfrac{5x(4 - 3x)}{5x(3x - 4)(x + 1)}$

$\phantom{\dfrac{20x - 15x^2}{15x^3}} = \dfrac{\overset{-1}{\cancel{5x(4 - 3x)}}}{\underset{1}{\cancel{5x(3x - 4)}}(x + 1)} = -\dfrac{1}{x + 1}$

c. $\dfrac{x^{2n} + x^n - 12}{x^{2n} - 3x^n} = \dfrac{(x^n + 4)(x^n - 3)}{x^n(x^n - 3)}$

$\phantom{\dfrac{x^{2n} + x^n - 12}{x^{2n}}} = \dfrac{(x^n + 4)\overset{1}{\cancel{(x^n - 3)}}}{x^n\underset{1}{\cancel{(x^n - 3)}}} = \dfrac{x^n + 4}{x^n}$

SECTION 7.2

You Try It 1

a. $\dfrac{12 + 5x - 3x^2}{x^2 + 2x - 15} \cdot \dfrac{2x^2 + x - 45}{3x^2 + 4x}$

$= \dfrac{(4 + 3x)(3 - x)}{(x + 5)(x - 3)} \cdot \dfrac{(2x - 9)(x + 5)}{x(3x + 4)}$

$= \dfrac{(4 + 3x)(3 - x)(2x - 9)(x + 5)}{(x + 5)(x - 3) \cdot x(3x + 4)}$

$= -\dfrac{2x - 9}{x}$

b. $\dfrac{2x^2 - 13x + 20}{x^2 - 16} \cdot \dfrac{2x^2 + 9x + 4}{6x^2 - 7x - 5}$

$= \dfrac{(2x - 5)(x - 4)}{(x - 4)(x + 4)} \cdot \dfrac{(2x + 1)(x + 4)}{(3x - 5)(2x + 1)}$

$= \dfrac{(2x - 5)(x - 4)\ (2x + 1)(x + 4)}{(x - 4)(x + 4)\ (3x - 5)(2x + 1)}$

$= \dfrac{2x - 5}{3x - 5}$

You Try It 2

a. $\dfrac{6x^2 - 3xy}{10ab^4} \div \dfrac{16x^2y^2 - 8xy^3}{15a^2b^2}$

$= \dfrac{6x^2 - 3xy}{10ab^4} \cdot \dfrac{15a^2b^2}{16x^2y^2 - 8xy^3}$

$= \dfrac{3x(2x - y)}{10ab^4} \cdot \dfrac{15a^2b^2}{8xy^2(2x - y)}$

$= \dfrac{45a^2b^2x(2x - y)}{80ab^4xy^2(2x - y)} = \dfrac{9a}{16b^2y^2}$

b. $\dfrac{6x^2 - 7x + 2}{3x^2 + x - 2} \div \dfrac{4x^2 - 8x + 3}{5x^2 + x - 4}$

$= \dfrac{6x^2 - 7x + 2}{3x^2 + x - 2} \cdot \dfrac{5x^2 + x - 4}{4x^2 - 8x + 3}$

$= \dfrac{(2x - 1)(3x - 2)}{(x + 1)(3x - 2)} \cdot \dfrac{(x + 1)(5x - 4)}{(2x - 1)(2x - 3)}$

$= \dfrac{(2x - 1)(3x - 2)(x + 1)(5x - 4)}{(x + 1)(3x - 2)(2x - 1)(2x - 3)}$

$= \dfrac{5x - 4}{2x - 3}$

You Try It 3

$\dfrac{5}{y - 3} - \dfrac{2}{y + 1} = \dfrac{5}{y - 3} \cdot \dfrac{y + 1}{y + 1} - \dfrac{2}{y + 1} \cdot \dfrac{y - 3}{y - 3}$

$\phantom{\dfrac{5}{y - 3} - \dfrac{2}{y + 1}} = \dfrac{5y + 5}{(y - 3)(y + 1)} - \dfrac{2y - 6}{(y - 3)(y + 1)}$

$\phantom{\dfrac{5}{y - 3} - \dfrac{2}{y + 1}} = \dfrac{(5y + 5) - (2y - 6)}{(y - 3)(y + 1)}$

$\phantom{\dfrac{5}{y - 3} - \dfrac{2}{y + 1}} = \dfrac{3y + 11}{(y - 3)(y + 1)}$

You Try It 4

$x - \dfrac{5}{6x} = \dfrac{x}{1} - \dfrac{5}{6x} = \dfrac{x}{1} \cdot \dfrac{6x}{6x} - \dfrac{5}{6x} = \dfrac{6x^2}{6x} - \dfrac{5}{6x} = \dfrac{6x^2 - 5}{6x}$

You Try It 5

a. $\dfrac{3 + \dfrac{16}{x} + \dfrac{16}{x^2}}{6 + \dfrac{5}{x} - \dfrac{4}{x^2}} = \dfrac{3 + \dfrac{16}{x} + \dfrac{16}{x^2}}{6 + \dfrac{5}{x} - \dfrac{4}{x^2}} \cdot \dfrac{x^2}{x^2}$

$= \dfrac{3 \cdot x^2 + \dfrac{16}{x} \cdot x^2 + \dfrac{16}{x^2} \cdot x^2}{6 \cdot x^2 + \dfrac{5}{x} \cdot x^2 - \dfrac{4}{x^2} \cdot x^2}$

$$= \frac{3x^2 + 16x + 16}{6x^2 + 5x - 4}$$

$$= \frac{(3x + 4)(x + 4)}{(2x - 1)(3x + 4)} = \frac{x + 4}{2x - 1}$$

b. $\dfrac{2x + 5 + \dfrac{14}{x - 3}}{4x + 16 + \dfrac{49}{x - 3}} = \dfrac{2x + 5 + \dfrac{14}{x - 3}}{4x + 16 + \dfrac{49}{x - 3}} \cdot \dfrac{x - 3}{x - 3}$

$$= \frac{2x(x - 3) + 5(x - 3) + \dfrac{14}{x - 3}(x - 3)}{4x(x - 3) + 16(x - 3) + \dfrac{49}{x - 3}(x - 3)}$$

$$= \frac{2x^2 - 6x + 5x - 15 + 14}{4x^2 - 12x + 16x - 48 + 49}$$

$$= \frac{2x^2 - x - 1}{4x^2 + 4x + 1}$$

$$= \frac{(2x + 1)(x - 1)}{(2x + 1)(2x + 1)} = \frac{x - 1}{2x + 1}$$

You Try It 6

$\dfrac{\dfrac{1}{x + 2} + \dfrac{4}{x - 3}}{\dfrac{2}{x - 3} - \dfrac{7}{x + 2}} = \dfrac{\dfrac{1}{x + 2} \cdot \dfrac{x - 3}{x - 3} + \dfrac{4}{x - 3} \cdot \dfrac{x + 2}{x + 2}}{\dfrac{2}{x - 3} \cdot \dfrac{x + 2}{x + 2} - \dfrac{7}{x + 2} \cdot \dfrac{x - 3}{x - 3}}$

$$= \frac{\dfrac{x - 3}{x^2 - x - 6} + \dfrac{4x + 8}{x^2 - x - 6}}{\dfrac{2x + 4}{x^2 - x - 6} - \dfrac{7x - 21}{x^2 - x - 6}}$$

$$= \frac{\dfrac{5x + 5}{x^2 - x - 6}}{\dfrac{-5x + 25}{x^2 - x - 6}}$$

$$= \frac{5x + 5}{x^2 - x - 6} \div \frac{-5x + 25}{x^2 - x - 6}$$

$$= \frac{5x + 5}{x^2 - x - 6} \cdot \frac{x^2 - x - 6}{-5x + 25}$$

$$= \frac{5(x + 1)}{(x + 2)(x - 3)} \cdot \frac{(x + 2)(x - 3)}{-5(x - 5)}$$

$$= -\frac{x + 1}{x - 5}$$

SECTION 7.3

You Try It 1

a.

$$\frac{5}{2x - 3} = \frac{-2}{x + 1}$$

$$(x + 1)(2x - 3)\left(\frac{5}{2x - 3}\right) = (x + 1)(2x - 3)\left(\frac{-2}{x + 1}\right)$$

$$5(x + 1) = -2(2x - 3)$$

$$5x + 5 = -4x + 6$$

$$9x + 5 = 6$$

$$9x = 1$$

$$x = \frac{1}{9}$$

The solution is $\dfrac{1}{9}$.

b.

$$3y + \frac{25}{3y - 2} = -8$$

$$(3y - 2)\left(3y + \frac{25}{3y - 2}\right) = (3y - 2)(-8)$$

$$(3y - 2)(3y) + (3y - 2)\left(\frac{25}{3y - 2}\right) = (3y - 2)(-8)$$

$$9y^2 - 6y + 25 = -24y + 16$$

$$9y^2 + 18y + 9 = 0$$

$$9(y^2 + 2y + 1) = 0$$

$$y^2 + 2y + 1 = 0$$

$$(y + 1)(y + 1) = 0$$

$$y + 1 = 0 \qquad y + 1 = 0$$

$$y = -1 \qquad y = -1$$

The solution is -1.

You Try It 2

Goal The goal is to determine the amount of time it would take the smaller pipe, working alone, to fill the tank.

Strategy • Let x represent the amount of time it takes the smaller pipe, working alone, to fill the tank.

• Write an equation using the fact that the sum of the part of the task completed by the large pipe and the part of the task completed by the small pipe equals 1, the complete task. Solve this equation for x.

Solution

Part of task completed by large pipe

$= \text{rate of work} \cdot \text{time worked} = \dfrac{1}{9} \cdot 6 = \dfrac{6}{9} = \dfrac{2}{3}$

Part of task completed by small pipe

$= \text{rate of work} \cdot \text{time worked} = \dfrac{1}{x} \cdot 6 = \dfrac{6}{x}$

The sum of the part of the task completed by the large pipe and the part of the task completed by the small pipe is 1.

$$\frac{2}{3} + \frac{6}{x} = 1$$

$$3x\left(\frac{2}{3} + \frac{6}{x}\right) = 3x(1)$$

$$2x + 18 = 3x$$

$$18 = x$$

The small pipe working alone will fill the tank in 18 hours.

Check √

Goal The goal is to find the rate of the wind.

Strategy • Let r represent the rate of the wind. Then the plane flies at a rate of $(150 + r)$ mph when traveling with the wind and at a rate of $(150 - r)$ mph when traveling against the wind.

• Write an equation using the fact that the time spent traveling with the wind equals the time spent traveling against the wind. Solve this equation for r.

Solution

Time spent traveling with the wind: $\dfrac{\text{Distance}}{\text{Rate}} = \dfrac{700}{150 + r}$

Time spent traveling against the wind: $\dfrac{\text{Distance}}{\text{Rate}} = \dfrac{500}{150 - r}$

The time spent traveling with the wind equals the time spent traveling against the wind.

$$\frac{700}{150 + r} = \frac{500}{150 - r}$$

$$(150 + r)(150 - r)\left(\frac{700}{150 + r}\right) = (150 + r)(150 - r)\left(\frac{500}{150 - r}\right)$$

$$(150 - r)(700) = (150 + r)(500)$$

$$105,000 - 700r = 75,000 + 500r$$

$$105,000 = 75,000 + 1200r$$

$$30,000 = 1200r$$

$$25 = r$$

The rate of the wind is 25 mph.

Check √

SECTION 7.4

$$\frac{n}{5} = \frac{12}{25}$$

$$5(12) = n(25)$$

$$60 = 25n$$

$$2.4 = n$$

S30

Check $\dfrac{2.4}{5} = \dfrac{12}{25}$

$$5(12) = 2.4(25)$$

$$60 = 60$$

The solution is 2.4.

Goal **a.** The goal is to find the number of workers in 1960.

b. The goal is to find the expected number of beneficiaries in 2030.

Strategy **a.** Write and solve a proportion to find the number of workers in 1960.

b. Write and solve a proportion to find the number of beneficiaries in 2030.

Solution **a.** $\dfrac{5 \text{ workers}}{1 \text{ beneficiary}} = \dfrac{W \text{ workers}}{14,000,000 \text{ beneficiaries}}$

$$1 \cdot W = 5(14,000,000)$$

$$W = 70,000,000$$

There were 70,000,000 workers in 1960.

b. $\dfrac{2 \text{ workers}}{1 \text{ beneficiary}} = \dfrac{167,000,000 \text{ workers}}{B \text{ beneficiaries}}$

$$1 \cdot 167,000,000 = 2 \cdot B$$

$$167,000,000 = 2B$$

$$83,500,000 = B$$

There will be 83,500,000 beneficiaries in 2030.

Check √

Goal The goal is to find the area of triangle AOB.

Strategy • Triangle AOB is similar to triangle DOC. Write and solve a proportion to find AO, the height of triangle AOB.

• Use the equation for the area of a triangle, $A = \frac{1}{2}bh$.

Solution $\dfrac{AB}{CD} = \dfrac{AO}{DO}$

$$\frac{10}{4} = \frac{AO}{3}$$

$$4(AO) = 10(3)$$

$$4(AO) = 30$$

$$AO = 7.5$$

$$A = \frac{1}{2}bh$$

$$A = \frac{1}{2}(10)(7.5) = 5(7.5) = 37.5$$

The area of the triangle is 37.5 square centimeters.

Check √

YOU TRY IT 4

Goal   The goal is to find w, the width of the river.

Strategy   Write and solve a proportion to find w.

Solution
$$\frac{120}{150} = \frac{36}{w}$$
$$150(36) = 120(w)$$
$$5400 = 120w$$
$$45 = w$$

The width of the river is 45 feet.

Check √

YOU TRY IT 5

For feet and yards, the conversion factors are $\frac{1 \text{ yard}}{3 \text{ feet}}$ and $\frac{3 \text{ feet}}{1 \text{ yard}}$. Because we are converting from feet to yards, we use the conversion factor that contains yards in the numerator. We use the abbreviations ft for feet and yd for yards.

$$29 \text{ ft} = 29 \text{ ft} \cdot 1 = \frac{29 \text{ ft}}{1} \cdot \frac{1 \text{ yd}}{3 \text{ ft}} = \frac{29}{3} \text{ yd} = 9\frac{2}{3} \text{ yards}$$

YOU TRY IT 6

Goal   The goal is to determine the number of housing lots, each 10,000 square feet in area, that are possible on $8\frac{1}{2}$ acres of land.

Strategy   To determine the number of housing lots, we need to find how many 10,000-square-foot lots there are in $8\frac{1}{2}$ acres of land. This requires that we divide $8\frac{1}{2}$ acres by 10,000 square feet. Because the units, acres and square feet, are not the same, part of the solution must be to convert acres to square feet. It may require some research to find that there are 43,560 square feet in 1 acre.

Solution
$$8\frac{1}{2} \text{ acres} = 8\frac{1}{2} \text{ acres} \cdot \frac{43,560 \text{ square feet}}{1 \text{ acre}}$$
$$= 8.5 \text{ acres} \cdot \frac{43,560 \text{ square feet}}{1 \text{ acre}}$$
$$= 370,260 \text{ square feet}$$

There are 370,260 square feet in $8\frac{1}{2}$ acres.

Find the number of 10,000-square-foot housing lots that are possible on 370,260 square feet of land.

$$370,260 \div 10,000 = 37.026$$

37 housing lots are possible on the $8\frac{1}{2}$ acres of land.

Check √

SECTION 7.5

YOU TRY IT 1

$$y = kx$$
$$120 = k \cdot 8$$
$$\frac{120}{8} = \frac{k \cdot 8}{8}$$
$$15 = k$$

The constant of variation is 15.

YOU TRY IT 2

Goal   We want to find the nurse's wage for working 18 hours.

Strategy   • This is a direct variation. To find the value of k, write the basic variation equation, replace the variables by the given values, and solve for k.
• Write the direct variation equation, replacing k by its value. Substitute 18 for h, and solve for w.

Solution
$$w = kh$$
$$264 = k \cdot 12$$
$$\frac{264}{12} = \frac{k \cdot 12}{12}$$
$$22 = k$$
$$w = 22h$$
$$w = 22(18)$$
$$w = 396$$

The nurse's wage for working 18 hours is $396.

Check √

YOU TRY IT 3

Goal   We want to find the distance a body will fall in 5 seconds.

Strategy   • This is a direct variation. To find the value of k, write the basic variation equation, replace the variables by the given values, and solve for k.

- Write the direct variation equation, replacing k by its value. Substitute 5 for t, and solve for s.

Solution
$$s = kt^2$$
$$64 = k \cdot 2^2$$
$$64 = k \cdot 4$$
$$\frac{64}{4} = \frac{k \cdot 4}{4}$$
$$16 = k$$
$$s = 16t^2$$
$$s = 16(5^2)$$
$$s = 16(25)$$
$$s = 400$$

A body will fall 400 feet in 5 seconds.

Check √

You Try It 4

Goal We want to determine how many hours it takes four assembly machines to complete the daily quota.

Strategy
- This is an inverse variation. To find the value of k, write the basic inverse variation equation, replace the variables by the given values, and solve for k.
- Write the inverse variation equation, replacing k by its value. Substitute 4 for m, and solve for h.

Solution
$$h = \frac{k}{m}$$
$$9 = \frac{k}{5}$$
$$5 \cdot 9 = 5 \cdot \frac{k}{5}$$
$$45 = k$$
$$h = \frac{45}{m}$$

$$h = \frac{45}{4}$$
$$h = 11.25$$

Four assembly machines will take 11.25 hours to complete the daily quota.

Check √

You Try It 5

Goal We want to find the intensity of a light that is 5 feet from the source.

Strategy
- This is an inverse variation. To find the value of k, write the basic inverse variation equation, replace the variables by the given values, and solve for k.
- Write the inverse variation equation, replacing k by its value. Substitute 5 for d, and solve for I.

Solution
$$I = \frac{k}{d^2}$$
$$20 = \frac{k}{8^2}$$
$$20 = \frac{k}{64}$$
$$64 \cdot 20 = 64 \cdot \frac{k}{64}$$
$$1280 = k$$

$$I = \frac{1280}{d^2}$$
$$I = \frac{1280}{5^2}$$
$$I = \frac{1280}{25}$$
$$I = 51.2$$

The intensity of the light source at 5 feet is 51.2 footcandles.

Check √

SOLUTIONS to Chapter 8 You Try Its

SECTION 8.1

You Try It 1

a. $16^{3/4} = (2^4)^{3/4} = 2^3 = 8$

b. $64^{-2/3} = (2^6)^{-2/3} = 2^{-4} = \frac{1}{2^4} = \frac{1}{16}$

c. $(-100)^{3/4}$

The base of the exponential expression, -100, is a negative number, and the denominator of the exponent is a positive even number.

$(-100)^{3/4}$ is not a real number.

a. $p^{3/4}(p^{-1/8})(p^{1/2}) = p^{3/4-1/8+1/2} = p^{6/8-1/8+4/8} = p^{9/8}$

b. $(a^{5/3}b^{1/6})^6 = a^{(5/3)6}b^{(1/6)6} = a^{10}b$

c. $\left(\dfrac{2a^{-2}b}{50a^6b^{-3}}\right)^{1/2} = \left(\dfrac{a^{-8}b^4}{25}\right)^{1/2} = \left(\dfrac{b^4}{5^2a^8}\right)^{1/2} = \dfrac{b^2}{5a^4}$

YOU TRY IT 3

a. $f(x) = 3605x^{-39/40}$

$f(12) = 3605(12)^{-39/40} \approx 320$

The monthly savings for a child who will be entering college in 12 years is $320.

b. $f(x) = 3605x^{-39/40}$

$f(15) = 3605(15)^{-39/40} \approx 257$

$f(5) = 3605(5)^{-39/40} \approx 751$

$751 - 257 = 494$

The difference in the monthly savings for a child who is 15 years from entering college and a child who will be going to college in 5 years is $494.

c. $f(x) = 3605x^{-39/40}$

$f(1) = 3605(1)^{-39/40} = 3605$

$f(20) = 3605(20)^{-39/40} \approx 194$

The range of the function is

$\{y \mid 194 \le y \le 3605, y \in \text{integers}\}$.

YOU TRY IT 4

a. $b^{3/7} = (b^3)^{1/7} = \sqrt[7]{b^3}$

b. $(3y)^{2/5} = \sqrt[5]{(3y)^2} = \sqrt[5]{9y^2}$

c. $-9d^{5/8} = -9(d^5)^{1/8} = -9\sqrt[8]{d^5}$

YOU TRY IT 5

a. $\sqrt[5]{p^9} = (p^9)^{1/5} = p^{9/5}$

b. $\sqrt[3]{26} = (26)^{1/3} = 26^{1/3}$

c. $\sqrt[3]{c^3 + d^3} = (c^3 + d^3)^{1/3}$

YOU TRY IT 6

$\sqrt[5]{243^4} = 81$

$243^{4/5} = 81$

SECTION 8.2

YOU TRY IT 1

a. $\sqrt{121x^{10}y^4} = \sqrt{11^2x^{10}y^4} = (11^2x^{10}y^4)^{1/2} = 11x^5y^2$

b. $\sqrt[3]{-8x^{12}y^3} = \sqrt[3]{(-2)^3x^{12}y^3} = [(-2)^3x^{12}y^3]^{1/3} = -2x^4y$

c. $-\sqrt[4]{81a^{12}b^8} = -\sqrt[4]{3^4a^{12}b^8} = -(3^4a^{12}b^8)^{1/4} = -3a^3b^2$

YOU TRY IT 2

$h = 0.9\sqrt[5]{p^3}$

$h = 0.9\sqrt[5]{12^3}$

$h = 0.9\sqrt[5]{1728}$

$h \approx 4.0$

The time required to cook a 12-pound pot roast is 4.0 hours.

YOU TRY IT 3

a. $\sqrt[5]{x^7} = \sqrt[5]{x^5 \cdot x^2} = \sqrt[5]{x^5} \cdot \sqrt[5]{x^2} = x\sqrt[5]{x^2}$

b. $\sqrt[4]{32x^{10}} = \sqrt[4]{16x^8 \cdot 2x^2}$

$\qquad = \sqrt[4]{16x^8} \cdot \sqrt[4]{2x^2} = 2x^2\sqrt[4]{2x^2}$

c. $\sqrt[3]{-64c^8d^{18}} = \sqrt[3]{(-4)^3c^8d^{18}}$

$\qquad = \sqrt[3]{(-4)^3c^6d^{18} \cdot c^2}$

$\qquad = \sqrt[3]{(-4)^3c^6d^{18}} \cdot \sqrt[3]{c^2} = -4c^2d^6\sqrt[3]{c^2}$

SECTION 8.3

YOU TRY IT 1

$c^2 = a^2 + b^2$

$c^2 = 7^2 + 14^2$

$c^2 = 49 + 196$

$c^2 = 245$

$c = \sqrt{245}$

$c \approx 15.65$

The length of the hypotenuse is 15.65 meters.

YOU TRY IT 2

$a^2 + b^2 = c^2$

$5^2 + b^2 = 16^2$

$25 + b^2 = 256$

$\qquad b^2 = 231$

$\qquad b = \sqrt{231}$

$\qquad b \approx 15.20$

The measure of the other leg is approximately 15.20 feet.

YOU TRY IT 3

Goal The goal is to determine whether the 30-foot ladder is long enough to reach the gutters when the gutters are 28 feet above the ground and the bottom of the ladder is 7 feet from the base of the side of the house.

Strategy
- Use the Pythagorean Theorem to find the hypotenuse of a right triangle with legs that measure 7 feet and 28 feet.
- Compare the length of the hypotenuse with 30 feet. If the hypotenuse is shorter than 30 feet, the ladder is long enough to reach the gutters. If the hypotenuse is longer than 30 feet, the ladder will not reach the gutters.

Solution

$$c^2 = a^2 + b^2$$
$$c^2 = 7^2 + 28^2$$
$$c^2 = 49 + 784$$
$$c^2 = 833$$
$$c = \sqrt{833}$$
$$c \approx 28.9$$

$28.9 < 30$ The hypotenuse is shorter than 30 feet.

The ladder will reach the gutters.

Check $\sqrt{}$

YOU TRY IT 4

Let $(x_1, y_1) = (5, -2)$ and $(x_2, y_2) = (-4, 3)$.

$$d = \sqrt{(x_1 - x_2)^2 + (y_1 - y_2)^2}$$
$$d = \sqrt{[5 - (-4)]^2 + (-2 - 3)^2}$$
$$d = \sqrt{(9)^2 + (-5)^2}$$
$$d = \sqrt{81 + 25}$$
$$d = \sqrt{106}$$
$$d \approx 10.3$$

The distance between the points is approximately 10.3 units.

SECTION 8.4

YOU TRY IT 1

$3xy \sqrt[3]{81x^5y} - \sqrt[3]{192x^8y^4}$

$= 3xy \sqrt[3]{27x^3 \cdot 3x^2y} - \sqrt[3]{64x^6y^3 \cdot 3x^2y}$

$= 3xy \sqrt[3]{27x^3} \cdot \sqrt[3]{3x^2y} - \sqrt[3]{64x^6y^3} \cdot \sqrt[3]{3x^2y}$

$= 3xy \cdot 3x \sqrt[3]{3x^2y} - 4x^2y \sqrt[3]{3x^2y}$

$= 9x^2y \sqrt[3]{3x^2y} - 4x^2y \sqrt[3]{3x^2y}$

$= 5x^2y \sqrt[3]{3x^2y}$

YOU TRY IT 2

$\sqrt{5b}(\sqrt{3b} - \sqrt{10}) = \sqrt{15b^2} - \sqrt{50b}$

$= \sqrt{b^2 \cdot 15} - \sqrt{25 \cdot 2b}$

$= \sqrt{b^2} \cdot \sqrt{15} - \sqrt{25} \cdot \sqrt{2b}$

$= b\sqrt{15} - 5\sqrt{2b}$

YOU TRY IT 3

$(2\sqrt[3]{2x} - 3)(\sqrt[3]{2x} - 5) = 2\sqrt[3]{4x^2} - 10\sqrt[3]{2x} - 3\sqrt[3]{2x} + 15$

$= 2\sqrt[3]{4x^2} - 13\sqrt[3]{2x} + 15$

YOU TRY IT 4

$(\sqrt{a} - 3\sqrt{y})(\sqrt{a} + 3\sqrt{y}) = (\sqrt{a})^2 - (3\sqrt{y})^2$

$= a - 9y$

YOU TRY IT 5

a. $\sqrt{\dfrac{48p^7}{q^4}} = \dfrac{\sqrt{48p^7}}{\sqrt{q^4}} = \dfrac{\sqrt{16p^6 \cdot 3p}}{\sqrt{q^4}}$

$= \dfrac{\sqrt{16p^6} \cdot \sqrt{3p}}{\sqrt{q^4}} = \dfrac{4p^3\sqrt{3p}}{q^2}$

b. $\dfrac{\sqrt[3]{54y^8z^4}}{\sqrt[3]{2y^5z}} = \sqrt[3]{\dfrac{54y^8z^4}{2y^5z}} = \sqrt[3]{27y^3z^3} = 3yz$

YOU TRY IT 6

a. $\dfrac{b}{\sqrt{3b}} = \dfrac{b}{\sqrt{3b}} \cdot \dfrac{\sqrt{3b}}{\sqrt{3b}} = \dfrac{b\sqrt{3b}}{(\sqrt{3b})^2} = \dfrac{b\sqrt{3b}}{3b} = \dfrac{\sqrt{3b}}{3}$

b. $\dfrac{3}{\sqrt[3]{3y^2}} = \dfrac{3}{\sqrt[3]{3y^2}} \cdot \dfrac{\sqrt[3]{9y}}{\sqrt[3]{9y}} = \dfrac{3\sqrt[3]{9y}}{\sqrt[3]{27y^3}} = \dfrac{3\sqrt[3]{9y}}{3y} = \dfrac{\sqrt[3]{9y}}{y}$

YOU TRY IT 7

a. $\dfrac{6}{5 - \sqrt{7}} = \dfrac{6}{5 - \sqrt{7}} \cdot \dfrac{5 + \sqrt{7}}{5 + \sqrt{7}} = \dfrac{30 + 6\sqrt{7}}{5^2 - (\sqrt{7})^2}$

$= \dfrac{30 + 6\sqrt{7}}{25 - 7} = \dfrac{30 + 6\sqrt{7}}{18} = \dfrac{6(5 + \sqrt{7})}{6 \cdot 3}$

$= \dfrac{5 + \sqrt{7}}{3}$

b. $\dfrac{3 + \sqrt{6}}{2 - \sqrt{6}} = \dfrac{3 + \sqrt{6}}{2 - \sqrt{6}} \cdot \dfrac{2 + \sqrt{6}}{2 + \sqrt{6}}$

$= \dfrac{6 + 3\sqrt{6} + 2\sqrt{6} + (\sqrt{6})^2}{2^2 - (\sqrt{6})^2} = \dfrac{6 + 5\sqrt{6} + 6}{4 - 6}$

$= \dfrac{12 + 5\sqrt{6}}{-2} = -\dfrac{12 + 5\sqrt{6}}{2}$

SECTION 8.5

YOU TRY IT 1

a. $f(x) = 2\sqrt[3]{6x}$

f contains an odd root.

The radicand can be positive or negative.

x can be any real number.

The domain is $(-\infty, \infty)$.

The graph of the function confirms this answer.

b. $F(x) = (5x - 10)^{1/2}$

$F(x) = \sqrt{5x - 10}$

F contains an even root.

The radicand must be greater than or equal to zero.

$$5x - 10 \geq 0$$
$$5x \geq 10$$
$$x \geq 2$$

The domain is $[2, \infty)$.

The graph of the function confirms this answer.

YOU TRY IT 2

a.

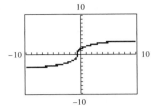

b. The expression $\sqrt[3]{5x + 2}$ is a real number for all values of x.

The domain of g is $\{x \mid x \in$ real numbers$\}$.

The range is $\{y \mid y \in$ real numbers$\}$.

c. Use the TRACE feature on the calculator to find the y value of the function when $x = 0$.

$$f(0) \approx 1.3$$

YOU TRY IT 3

a. Graph $v = \sqrt{12r}$ on a graphing calculator.

Use the TRACE feature to find Y when X = 6.

The speed of a rider who is sitting 6 feet from the center of the merry-go-round is 8.5 feet per second.

b. Use the TRACE feature to find X when Y = 10.

The distance between the rider and the center of the merry-go-round is 8.3 feet.

c. Answers will vary.

SECTION 8.6

YOU TRY IT 1

a.
$$\sqrt[4]{2x - 9} = 3$$
$$(\sqrt[4]{2x - 9})^4 = 3^4$$
$$2x - 9 = 81$$
$$2x = 90$$
$$x = 45$$
The solution checks.
The solution is 45.

b.
$$\sqrt{4x + 5} - 12 = -5$$
$$\sqrt{4x + 5} = 7$$
$$(\sqrt{4x + 5})^2 = 7^2$$
$$4x + 5 = 49$$
$$4x = 44$$
$$x = 11$$
The solution checks.
The solution is 11.

YOU TRY IT 2

$$8 + 3\sqrt{x + 2} = 5$$
$$3\sqrt{x + 2} = -3$$

$$\sqrt{x + 2} = -1$$
$$(\sqrt{x + 2})^2 = (-1)^2$$
$$x + 2 = 1$$
$$x = -1$$

-1 does not check as a solution.

There is no solution.

YOU TRY IT 3

$$\sqrt{x + 5} + \sqrt{x} = 5$$
$$\sqrt{x + 5} = 5 - \sqrt{x}$$
$$(\sqrt{x + 5})^2 = (5 - \sqrt{x})^2$$
$$x + 5 = 25 - 10\sqrt{x} + x$$
$$5 = 25 - 10\sqrt{x}$$
$$-20 = -10\sqrt{x}$$
$$2 = \sqrt{x}$$
$$2^2 = (\sqrt{x})^2$$
$$4 = x$$

The solution checks.

The solution is 4.

YOU TRY IT 4

Goal The goal is to determine the value of x in the expression $\sqrt{x + 8}$.

Strategy We are given the perimeter of the triangle. Therefore, we need to use the formula for the perimeter of a triangle to write an equation. We can do this by substituting, in the formula, 15 for P (the perimeter) and $\sqrt{x + 8}$ for each of the three sides. We can then solve the equation for x.

Solution
$$P = a + b + c$$
$$15 = \sqrt{x + 8} + \sqrt{x + 8} + \sqrt{x + 8}$$
$$15 = 3(\sqrt{x + 8})$$
$$5 = \sqrt{x + 8}$$
$$5^2 = (\sqrt{x + 8})^2$$
$$25 = x + 8$$
$$17 = x$$
The solution 17 checks.
The value of x is 17.

Check When $x = 17$, $\sqrt{x + 8} = \sqrt{17 + 8} = \sqrt{25} = 5$.

If each side measures 5 centimeters, then the perimeter of the equilateral triangle is $5 + 5 + 5 = 15$. This is the perimeter we are given in the problem statement. The solution checks.

SECTION 8.7

YOU TRY IT 1

a. $\sqrt{-60} = i\sqrt{60} = i\sqrt{4 \cdot 15} = 2i\sqrt{15}$

b. $\sqrt{40} - \sqrt{-80} = \sqrt{40} - i\sqrt{80}$
$$= \sqrt{4 \cdot 10} - i\sqrt{16 \cdot 5}$$
$$= 2\sqrt{10} - 4i\sqrt{5}$$

YOU TRY IT 2

a. $(-10 + 6i) + (9 - 4i)$
$$= (-10 + 9) + [6 + (-4)]i$$
$$= -1 + 2i$$

b. $(3 + i) - (8i) = (3 - 0) + (1 - 8)i = 3 - 7i$

c. $(4 - 2i) + (-4 + 2i) = 0 + 0i = 0$

YOU TRY IT 3

a. $-7i \cdot 2i = -14i^2 = -14(-1) = 14$

b. $5i(2 - 8i) = 10i - 40i^2 = 10i - 40(-1) = 40 + 10i$

YOU TRY IT 4

a. $(3 - 4i)(2 + 5i) = 6 + 15i - 8i - 20i^2$
$$= 6 + 7i - 20i^2$$
$$= 6 + 7i - 20(-1)$$
$$= 6 + 7i + 20$$
$$= 26 + 7i$$

b. $\left(\dfrac{9}{10} + \dfrac{3}{10}i\right)\left(1 - \dfrac{1}{3}i\right) = \dfrac{9}{10} - \dfrac{3}{10}i + \dfrac{3}{10}i - \dfrac{1}{10}i^2$
$$= \dfrac{9}{10} - \dfrac{1}{10}i^2$$
$$= \dfrac{9}{10} - \dfrac{1}{10}(-1)$$
$$= \dfrac{9}{10} + \dfrac{1}{10} = 1$$

YOU TRY IT 5

$(6 + 5i)(6 - 5i) = 6^2 + 5^2 = 36 + 25 = 61$

YOU TRY IT 6

$\dfrac{4 + 5i}{3i} = \dfrac{4 + 5i}{3i} \cdot \dfrac{i}{i}$
$$= \dfrac{4i + 5i^2}{3i^2}$$
$$= \dfrac{4i + 5(-1)}{3(-1)}$$
$$= \dfrac{-5 + 4i}{-3} = \dfrac{5}{3} - \dfrac{4}{3}i$$

YOU TRY IT 7

$\dfrac{5 - 3i}{4 + 2i} = \dfrac{5 - 3i}{4 + 2i} \cdot \dfrac{4 - 2i}{4 - 2i}$
$$= \dfrac{20 - 10i - 12i + 6i^2}{4^2 + 2^2}$$
$$= \dfrac{20 - 22i + 6(-1)}{16 + 4}$$
$$= \dfrac{14 - 22i}{20}$$
$$= \dfrac{14}{20} - \dfrac{22}{20}i = \dfrac{7}{10} - \dfrac{11}{10}i$$

SOLUTIONS to Chapter 9 You Try Its

SECTION 9.1

YOU TRY IT 1

$(x + 3)(x - 7) = 11$
$x^2 - 4x - 21 = 11$
$x^2 - 4x - 32 = 0$
$(x + 4)(x - 8) = 0$
$x + 4 = 0 \qquad x - 8 = 0$
$x = -4 \qquad x = 8$

The solutions are -4 and 8.

YOU TRY IT 2

$s(c) = 4$
$c^2 - c - 2 = 4$
$c^2 - c - 6 = 0$
$(c + 2)(c - 3) = 0$
$c + 2 = 0 \qquad c - 3 = 0$
$c = -2 \qquad c = 3$

The values of c are -2 and 3.

YOU TRY IT 3

Goal The goal is to find how long it takes for the projectile to return to Earth.

Strategy When the projectile returns to Earth, its height is 0 feet. Replace s in the given formula by 0. The initial velocity is 200 feet per second. Replace v_0 by 200. Then solve the equation for t.

Solution

$$s = v_0 t - 16t^2$$
$$0 = 200t - 16t^2$$
$$16t^2 - 200t = 0$$
$$8t(2t - 25) = 0$$

$$8t = 0 \qquad 2t - 25 = 0$$
$$t = 0 \qquad 2t = 25$$
$$t = \frac{25}{2} = 12.5$$

At $t = 0$, the projective is just being fired.

The projectile returns to Earth after 12.5 seconds.

Check √

YOU TRY IT 4

$$(x - r_1)(x - r_2) = 0$$
$$\left[x - \left(-\frac{1}{6}\right)\right]\left(x - \frac{2}{3}\right) = 0$$
$$\left(x + \frac{1}{6}\right)\left(x - \frac{2}{3}\right) = 0$$
$$x^2 - \frac{1}{2}x - \frac{1}{9} = 0$$
$$18\left(x^2 - \frac{1}{2}x - \frac{1}{9}\right) = 0$$
$$18x^2 - 9x - 2 = 0$$

SECTION 9.2

YOU TRY IT 1

$$2(x + 1)^2 + 24 = 0$$
$$2(x + 1)^2 = -24$$
$$(x + 1)^2 = -12$$
$$\sqrt{(x + 1)^2} = \sqrt{-12}$$
$$x + 1 = \pm\sqrt{-12}$$
$$x + 1 = \pm 2i\sqrt{3}$$

$$x + 1 = 2i\sqrt{3} \qquad x + 1 = -2i\sqrt{3}$$
$$x = -1 + 2i\sqrt{3} \qquad x = -1 - 2i\sqrt{3}$$

The solutions are $-1 + 2i\sqrt{3}$ and $-1 - 2i\sqrt{3}$.

YOU TRY IT 2

Goal The goal is to find the diameter of the circle.

Strategy
- Use the formula for the area of a circle. Substitute 64 for A. Solve the equation for r.
- Convert the radius to a diameter by multiplying the radius by 2.

Solution

$$A = \pi r^2$$
$$64 = \pi r^2$$
$$\frac{64}{\pi} = r^2$$
$$\sqrt{\frac{64}{\pi}} = \sqrt{r^2}$$
$$\frac{8}{\sqrt{\pi}} = r$$
$$d = 2r = 2\left(\frac{8}{\sqrt{\pi}}\right) = \frac{16}{\sqrt{\pi}} \approx 9.0$$

The diameter of the circle is approximately 9.0 inches.

Check √

YOU TRY IT 3

$$4x^2 - 4x - 1 = 0$$
$$4x^2 - 4x = 1$$
$$\frac{4x^2 - 4x}{4} = \frac{1}{4}$$
$$x^2 - x = \frac{1}{4}$$
$$x^2 - x + \frac{1}{4} = \frac{1}{4} + \frac{1}{4}$$
$$\left(x - \frac{1}{2}\right)^2 = \frac{1}{2}$$
$$\sqrt{\left(x - \frac{1}{2}\right)^2} = \sqrt{\frac{1}{2}}$$
$$x - \frac{1}{2} = \pm\sqrt{\frac{1}{2}}$$
$$x - \frac{1}{2} = \pm\frac{\sqrt{2}}{2}$$

$$x - \frac{1}{2} = \frac{\sqrt{2}}{2} \qquad x - \frac{1}{2} = -\frac{\sqrt{2}}{2}$$
$$x = \frac{1}{2} + \frac{\sqrt{2}}{2} \qquad x = \frac{1}{2} - \frac{\sqrt{2}}{2}$$
$$x = \frac{1 + \sqrt{2}}{2} \qquad x = \frac{1 - \sqrt{2}}{2}$$

The solutions are $\dfrac{1 + \sqrt{2}}{2}$ and $\dfrac{1 - \sqrt{2}}{2}$.

Goal The goal is to determine the length and width of the rectangle.

Strategy Let $L = W + 12$.

Use the formula for the area of a rectangle. Substitute 620 for A. Solve the equation for W. Find L by substituting the value of W in the equation $L = W + 12$.

Solution
$$A = LW$$
$$A = (W + 12)W$$
$$A = W^2 + 12W$$
$$620 = W^2 + 12W$$
$$620 + 36 = W^2 + 12W + 36$$
$$656 = (W + 6)^2$$
$$\sqrt{656} = \sqrt{(W + 6)^2}$$
$$\pm\sqrt{656} = W + 6$$
$$-6 \pm \sqrt{656} = W$$
$$W \approx 19.61 \qquad W \approx -31.6$$

W cannot be negative.
$$L = W + 12$$
$$L \approx 19.61 + 12 = 31.61$$

The length of the rectangle is 31.61 feet.
The width of the rectangle is 19.61 feet.

Check √

SECTION 9.3

YOU TRY IT 1

$$4x^2 = 4x - 1$$
$$4x^2 - 4x + 1 = 0 \qquad a = 4, b = -4, c = 1$$
$$x = \frac{-b \pm \sqrt{b^2 - 4ac}}{2a}$$
$$= \frac{-(-4) \pm \sqrt{(-4^2) - 4(4)(1)}}{2(4)}$$
$$= \frac{4 \pm \sqrt{16 - 16}}{8}$$
$$= \frac{4 \pm \sqrt{0}}{8} = \frac{4}{8} = \frac{1}{2}$$

The solution is $\frac{1}{2}$.

YOU TRY IT 2

$$3x^2 - x - 1 = 0$$
$$b^2 - 4ac = (-1)^2 - 4(3)(-1)$$
$$= 1 - (-12)$$
$$= 13$$

$13 > 0$

The equation has two real number solutions.

S38

$$y = x^2 - x - 6$$
$$b^2 - 4ac = (-1)^2 - 4(1)(-6)$$
$$= 1 - (-24)$$
$$= 25 \qquad \bullet \text{ The discriminant is greater than 0.}$$

The parabola has two x-intercepts.

Goal The goal is to find the base and the height of the triangle to be cut from the fabric.

Strategy Let $h = b - 10$.

Use the formula for the area of a triangle. Substitute 150 for A. Solve the equation for b. Find h by using the equation $h = b - 10$.

Solution
$$A = \frac{1}{2}bh$$
$$150 = \frac{1}{2}b(b - 10)$$
$$150 = \frac{1}{2}b^2 - 5b$$
$$0 = 0.5b^2 - 5b - 150$$
$$b = \frac{-b \pm \sqrt{b^2 - 4ac}}{2a}$$
$$= \frac{-(-5) \pm \sqrt{(-5)^2 - 4(0.5)(-150)}}{2(0.5)}$$
$$= \frac{5 \pm \sqrt{25 + 300}}{1}$$
$$= 5 \pm \sqrt{325}$$
$$5 + \sqrt{325} \approx 23.028$$
$$5 - \sqrt{325} \approx -13.028 \qquad \bullet \text{ A negative base is not possible.}$$
$$h = b - 10$$
$$h = 23.028 - 10 = 13.028$$

The base of the triangle is 23.028 inches.
The height of the triangle is 13.028 inches.

Check √

YOU TRY IT 5

Goal The goal is to find the year in which consumer-to-consumer transactions will reach $50 billion.

Strategy Use a graphing calculator to graph
$Y_1 = 0.6144x^2 - 7.7658x + 23.9343$
and $Y_2 = 50$. Find the point of intersection.

Solution We want the rightmost point of intersection, because we are asking a question about the years after 2001.

The x-coordinate of the intersection of $Y_1 = 0.6144x^2 - 7.7658x + 23.9343$ and $Y_2 = 50$ is approximately 15.40. Because $x = 10$ corresponds to 2000, $x = 15$ corresponds to 2005.

The model predicts that transactions will reach \$50 billion in 2005.

Check $\sqrt{}$

SECTION 9.4

You Try It 1

a.
$$x - 5x^{1/2} + 6 = 0$$
$$(x^{1/2})^2 - 5(x^{1/2}) + 6 = 0$$
$$u^2 - 5u + 6 = 0$$
$$(u - 2)(u - 3) = 0$$

$$u - 2 = 0 \qquad u - 3 = 0$$
$$u = 2 \qquad u = 3$$
$$x^{1/2} = 2 \qquad x^{1/2} = 3$$
$$(x^{1/2})^2 = 2^2 \qquad (x^{1/2})^2 = 3^2$$
$$x = 4 \qquad x = 9$$

The solutions check.
The solutions are 4 and 9.

b.
$$4x^4 + 35x^2 - 9 = 0$$
$$4(x^2)^2 + 35(x^2) - 9 = 0$$
$$4u^2 + 35u - 9 = 0$$
$$(u + 9)(4u - 1) = 0$$

$$u + 9 = 0 \qquad 4u - 1 = 0$$
$$u = -9 \qquad 4u = 1$$
$$u = \frac{1}{4}$$

$$x^2 = -9 \qquad x^2 = \frac{1}{4}$$

$$\sqrt{x^2} = \sqrt{-9} \qquad \sqrt{x^2} = \sqrt{\frac{1}{4}}$$

$$x = \pm 3i \qquad x = \pm\frac{1}{2}$$

The solutions check.
The solutions are $3i$, $-3i$, $\frac{1}{2}$, and $-\frac{1}{2}$.

You Try It 2
$$\sqrt{2x + 1} + x = 7$$
$$x - 7 = -\sqrt{2x + 1}$$
$$(x - 7)^2 = (-\sqrt{2x + 1})^2$$
$$x^2 - 14x + 49 = 2x + 1$$
$$x^2 - 16x + 48 = 0$$

$$(x - 12)(x - 4) = 0$$
$$x - 12 = 0 \qquad x - 4 = 0$$
$$x = 12 \qquad x = 4$$

The solution 12 does not check.
The solution 4 checks.
The solution is 4.

You Try It 3

Goal The goal is to find the unknown number.

Strategy Translate the sentence into an equation and solve.

Solution
$$x^3 + 12x = 7x^2$$
$$x^3 - 7x^2 + 12x = 0$$
$$x(x^2 - 7x + 12) = 0$$
$$x(x - 3)(x - 4) = 0$$
$$x = 0 \qquad x - 3 = 0 \qquad x - 4 = 0$$
$$x = 3 \qquad x = 4$$

The solutions 0, 3, and 4 check.
The number is 0, 3, or 4.

Check $\sqrt{}$

SECTION 9.5

You Try It 1
$$-\frac{b}{2a} = -\frac{0}{2(1)} = 0$$
$$y = x^2 - 2$$
$$y = (0)^2 - 2$$
$$y = -2$$

The vertex is $(0, -2)$.
The axis of symmetry is the line $x = 0$.

You Try It 2

$$g(x) = x^2 + 4x - 2$$

Because a is positive ($a = 1$), the graph of g will open up.
The x-coordinate of the vertex is

$$x = -\frac{b}{2a} = -\frac{4}{2(1)} = -2.$$

The y-coordinate of the vertex is

$$g(-2) = (-2)^2 + 4(-2) - 2 = -6.$$

The vertex is $(-2, -6)$.

Because $g(x) = x^2 + 4x - 2$ is a real number for all values of x, the domain of the function is $\{x | x \in \text{real numbers}\}$.

The vertex of the parabola is the lowest point on the graph. Because the y-coordinate at that point is -6, the range is $\{y | y \geq -6\}$.

You Try It 3

a. $y = 2x^2 - 5x + 2$

$0 = 2x^2 - 5x + 2$

$0 = (2x - 1)(x - 2)$

$2x - 1 = 0 \qquad x - 2 = 0$

$x = \dfrac{1}{2} \qquad\quad x = 2$

The x-intercepts are $\left(\dfrac{1}{2}, 0\right)$ and $(2, 0)$.

b. $y = x^2 + 4x + 4$

$0 = x^2 + 4x + 4$

$0 = (x + 2)(x + 2)$

$x + 2 = 0 \qquad x + 2 = 0$

$x = -2 \qquad\quad x = -2$

The x-intercept is $(-2, 0)$.

You Try It 4

$f(x) = 2x^2 - 3x + 1$

$x = -\dfrac{b}{2a} = -\dfrac{-3}{2(2)} = \dfrac{3}{4}$

$f(x) = 2x^2 - 3x + 1$

$f\left(\dfrac{3}{4}\right) = 2\left(\dfrac{3}{4}\right)^2 - 3\left(\dfrac{3}{4}\right) + 1$

$f\left(\dfrac{3}{4}\right) = \dfrac{9}{8} - \dfrac{9}{4} + 1 = -\dfrac{1}{8}$

The minimum value of the function is $-\dfrac{1}{8}$.

You Try It 5

Goal   The goal is first to find the time it takes the ball to reach its maximum height and then to find the maximum height the ball reaches.

Strategy   • Find the t-coordinate of the vertex.

• Evaluate the function at the t-coordinate of the vertex.

Solution $s(t) = -16t^2 + 64t$

$t = -\dfrac{b}{2a} = -\dfrac{64}{2(-16)} = 2$

It takes the ball 2 seconds to reach its maximum height.

$s(t) = -16t^2 + 64t$

$s(2) = -16(2)^2 + 64(2) = -64 + 128 = 64$

The maximum height the ball reaches is 64 feet.

Check   √

You Try It 6

Goal   The goal is to find the dimensions that will yield the maximum area for the floor.

Strategy   The perimeter is 44 feet.

Use the equation for the perimeter of a rectangle.

Substitute 44 for P and solve for W.

$$P = 2L + 2W$$
$$44 = 2L + 2W$$
$$22 = L + W$$
$$22 - L = W$$

The area is $LW = L(22 - L) = 22L - L^2$.

• To find the length, find the L-coordinate of the vertex of the function $f(x) = -L^2 + 22L$.

• To find the width, replace L in $22 - L$ by the L-coordinate of the vertex and evaluate.

Solution

$$L = -\dfrac{b}{2a} = -\dfrac{22}{2(-1)} = 11$$

The length of the rectangle is 11 feet.

$$22 - L = 22 - 11 = 11$$

The width of the rectangle is 11 feet.

Check   √

You Try It 7

a. Use a graphing calculator to perform the regression. The quadratic regression equation is given in the form $y = ax^2 + bx + c$.
To the nearest hundredth, the equation is $y = -0.08x^2 + 6.58x + 2.46$.

b. Graph the regression equation. Use the TRACE feature to find the value of y when $x = 40$.
The coordinates printed at the bottom of the screen are approximately $(40, 142.33)$.
The distance the ball travels when it is hit at an angle of 40° is approximately 142.33 meters.

c. The regression equation is already stored in Y_1. Enter 120 for Y_2. Find the points of intersection of Y_1 and Y_2 to determine the x values when $y = 120$.
The coordinates printed at the bottom of the screen are approximately $(25.53, 120)$.
At the second point of intersection, the coordinates are approximately $(59.91, 120)$.
The equation predicts that the angle at which the ball must be hit to travel a distance of 120 meters is 25.53° or 59.91°.

d. Because the parabola opens down, the vertex is the highest point on the parabola; the value of the function at this point is a maximum. To find the vertex, use a graphing calculator to find the maximum.

The coordinates printed at the bottom of the screen are approximately (42.68, 142.88).

The vertex is (42.68, 142.88).

The ball will travel a maximum distance of 142.88 meters when it is hit at an angle of 42.68°.

You Try It 1

$$2x^2 - x - 10 \leq 0$$
$$(2x - 5)(x + 2) \leq 0$$

2x – 5 – – – | – – – – – – – – | +++
x + 2 – – – | +++++++ | +++

  –3 –2 –1 0 1 2 3

$$\left\{ x \,\middle|\, -2 \leq x \leq \frac{5}{2} \right\}$$

–5 –4 –3 –2 –1 0 1 2 3 4 5

SOLUTIONS to Chapter 10 You Try Its

SECTION 10.1

You Try It 1

a. $(f + g)(3) = f(3) + g(3) = [3(3) - 2] + [3 + 1]$
$$= 7 + 4 = 11$$

b. $\left(\dfrac{f}{g}\right)(4) = \dfrac{f(4)}{g(4)} = \dfrac{3(4) - 2}{4 + 1} = \dfrac{10}{5} = 2$

c. $(f \cdot g)(x) = f(x) \cdot g(x) = (3x - 2)(x + 1)$
$$= 3x^2 + x - 2$$

d. $(f - g)(x) = f(x) - g(x) = (3x - 2) - (x + 1)$
$$= 2x - 3$$

You Try It 2

Goal   We want to find the profit function and to find the profit for selling 2000 computers.

Strategy   Profit equals revenue minus cost. Subtract the cost function from the revenue function. Then evaluate the profit function at 2000 to find the profit for selling 2000 computers.

Solution   Profit = Revenue – Cost

$$P(x) = R(x) - C(x)$$
$$= (-0.09x^2 + 300x) - (100x + 300)$$
$$= -0.09x^2 + 200x - 300$$

The profit function is given by $P(x) = -0.09x^2 + 200x - 300$. To determine the profit for selling 2000 computers, evaluate the profit function when $x = 2000$.

$$P(x) = -0.09x^2 + 200x - 300$$

$$P(2000) = -0.09(2000)^2 + 200(2000) - 300$$
$$= 39,700$$

The profit was $39,700.

Check   Be sure to check your work.

You Try It 3

a. $h(x) = x^2 - 2x$
$$h(-1) = (-1)^2 - 2(-1) = 1 + 2$$
$$h(-1) = 3$$
$$g(x) = 3x - 2$$
$$g[h(-1)] = g(3) = 3(3) - 2$$
$$= 9 - 2$$
$$g[h(-1)] = 7$$

b. $h[g(x)] = [g(x)]^2 - 2[g(x)]$
$$= [3x - 2]^2 - 2[3x - 2]$$
$$= [9x^2 - 12x + 4] - [6x - 4]$$
$$h[g(x)] = 9x^2 - 18x + 8$$

SECTION 10.2

You Try It 1

a.

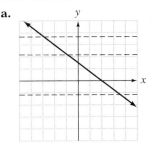

No horizontal line intersects the graph at more than one point. The graph is the graph of a 1-1 function.

b.

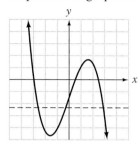

A horizontal line intersects the graph at more than one point. The graph is not the graph of a 1-1 function.

YOU TRY IT 2

$$f(x) = 3x + 5$$

$$y = 3x + 5 \qquad \bullet \text{ Replace } f(x) \text{ by } y.$$

$$x = 3y + 5 \qquad \bullet \text{ Interchange } x \text{ and } y.$$

$$x - 5 = 3y \qquad \bullet \text{ Solve for } y.$$

$$\frac{x-5}{3} = \frac{3y}{3}$$

$$\frac{1}{3}x - \frac{5}{3} = y$$

$$\frac{1}{3}x - \frac{5}{3} = f^{-1}(x) \qquad \bullet \text{ Replace } y \text{ by } f^{-1}(x).$$

YOU TRY IT 3

$$f(x) = 2x + 4 \qquad\qquad f^{-1}(x) = \frac{1}{2}x - 2$$

$$f[f^{-1}(x)] = 2\left[\frac{1}{2}x - 2\right] + 4 \qquad f^{-1}[f(x)] = \frac{1}{2}[2x + 4] - 2$$

$$f[f^{-1}(x)] = x - 4 + 4 \qquad\qquad f^{-1}[f(x)] = x + 2 - 2$$

$$f[f^{-1}(x)] = x \qquad\qquad\qquad f^{-1}[f(x)] = x$$

YOU TRY IT 4

Goal The goal is to find h^{-1}, the inverse function of h.

Strategy Use the steps for finding an inverse function. Then evaluate the inverse function at the French hat size.

Solution
$$h(x) = 3x - 10$$
$$y = 3x - 10$$
$$x = 3y - 10$$
$$x + 10 = 3y$$
$$\frac{x + 10}{3} = y$$

$$h^{-1}(x) = \frac{x + 10}{3} = \frac{1}{3}x + \frac{10}{3}$$

To find the equivalent U.S. hat size, substitute 41 for x in h^{-1}.

$$h^{-1}(x) = \frac{1}{3}x + \frac{10}{3}$$

$$h^{-1}(41) = \frac{1}{3}(41) + \frac{10}{3} = \frac{41}{3} + \frac{10}{3} = 17$$

A size 41 French hat is equivalent to a size 17 U.S. hat.

Check Be sure to check your work.

SECTION 10.3

YOU TRY IT 1

a. $f(x) = \left(\frac{2}{3}\right)^x$

$$f(3) = \left(\frac{2}{3}\right)^3 = \frac{8}{27} \qquad f(-2) = \left(\frac{2}{3}\right)^{-2} = \left(\frac{3}{2}\right)^2 = \frac{9}{4}$$

b. $f(x) = 2^{2x+1}$

$$f(0) = 2^{2(0)+1} = 2^1 = 2$$

$$f(-2) = 2^{2(-2)+1} = 2^{-3} = \frac{1}{2^3} = \frac{1}{8}$$

c. $f(x) = \pi^x$

$$f(3) = \pi^3 \approx 31.0063$$
$$f(-2) = \pi^{-2} \approx 0.1013$$
$$f(\pi) = \pi^\pi \approx 36.4622$$

YOU TRY IT 2

$$f(x) = e^x$$
$$f(1.4) = e^{1.4} \approx 4.0552$$
$$f(-0.5) = e^{-0.5} \approx 0.6065$$
$$f(\sqrt{2}) = e^{\sqrt{2}} \approx 4.1133$$

YOU TRY IT 3

a.

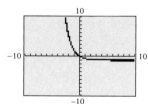

b. $b = 0.4, 0 < b < 1$

c. As the values of x increase, the values of y decrease. This is a decreasing function.

d. The graph is below the x-axis when $x > 0$. When $x > 0$, the corresponding values in the range are less than 0.

YOU TRY IT 4

a.

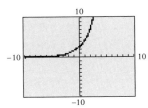

b. $b = e \approx 2.718, b > 1$

c. As the values of x increase, the values of y increase. This is an increasing function.

d. Use the TRACE feature to find the value of y when $x = 0$. The y-intercept is approximately $(0, 2.7182)$.

YOU TRY IT 5

$f(x) = \left(\dfrac{3}{4}\right)^x - 3$

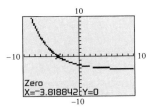

Use the ZERO function to approximate the zero of f. The zero of f is approximately -3.82.

YOU TRY IT 6

The base of the exponential function $y = 30(0.5^x)$ is 0.5.

$$0 < 0.5 < 1$$

The function is an exponential decay function.

YOU TRY IT 7

Goal The goal is to find the value of the investment after 3 years.

Strategy • Determine the values of P, r, n, and t.
• Use the compound interest formula.

Solution

$$P = 2500$$
$$r = 7.5\% = 0.075$$
$$n = 12$$
$$t = 3$$
$$A = P\left(1 + \frac{r}{n}\right)^{nt}$$
$$A = 2500\left(1 + \frac{0.075}{12}\right)^{12(3)}$$
$$A \approx 3128.62$$

Check √

YOU TRY IT 8

Goal The goal is to **a.** find the value of the car after 5 years and **b.** determine whether the car is worth more or less than one-half its original value after the 5 years.

Strategy **a.** Replace t in the given equation by 5. Then evaluate the resulting expression.

b. Compare the value of the car after 5 years with half the original value of the car.

Solution **a.** $A = 21{,}000(1 - 0.15)^t$

$A = 21{,}000(1 - 0.15)^5$

$A \approx 9318$

The value of the car after 5 years is \$9318.

b. $\dfrac{1}{2}(21{,}000) = 10{,}500$

$9318 < 10{,}500$

After 5 years, the car is worth less than one-half its original value.

Check √

SECTION 10.4

YOU TRY IT 1

a. The initial quantity (the value of a) is 25,000, the amount invested in 2010.

Find the growth factor (the value of b).

The value of the investment at the end of each year is found by multiplying the value of the investment at the beginning of the year by

$$100\% + 6\% = 1.00 + 0.06 = 1.06$$
$$y = ab^x$$
$$y = 25{,}000(1.06)^x$$

The exponential function is $y = 25{,}000(1.06)^x$, where y is the value of the investment x years after 2010.

b. 2020 is 10 years after 2010.

$$y = 25{,}000(1.06)^x$$
$$y = 25{,}000(1.06)^{10}$$
$$y \approx 44{,}771.19$$

The value of the investment in 2020 is approximately \$44,771.19.

YOU TRY IT 2

a. The initial quantity (the value of a) is 40 grams.

Find the decay factor (the value of b).

The problem refers to the half-life of iodine-123. The decay factor is 50%.

S43

$100\% - 50\% = 1.00 - 0.5 = 0.5$

$y = ab^x$

$y = 40(0.5)^x$

The exponential function is $y = 40(0.5)^x$, where y is the amount of iodine-123 in the sample, in grams, and x is the number of 13-hour periods.

b. $52 \div 13 = 4$. There are four 13-hour periods in 52 hours.

$y = 40(0.5)^x$

$y = 40(0.5)^4$

$y = 2.5$

The amount of iodine-123 in the sample after 52 hours is 2.5 grams.

YOU TRY IT 3

a. Use a graphing calculator to determine the regression line for the data.

The exponential regression equation is $y = 0.1087(1.2922)^x$.

b. The correlation coefficient (the value of r) is approximately 0.9919856. This is very close to 1, which means that the data points lie very close to the graph of $y = 0.1087(1.2922)^x$.

c. The year 1975 corresponds to $x = 5$. Graph the regression equation. (It is already stored in Y_1.) Use the TRACE feature to find y when $x = 5$.

The equation predicts that net sales for Wal-Mart in 1975 were $392 million.

d. The regression equation is already stored in Y_1. Enter 50 for Y_2. Find the point of intersection of Y_1 and Y_2 to determine the x value when $y = 50$.

According to the equation, net sales were $50 billion in the year 1994.

e. The year 2002 corresponds to $x = 32$. Graph the regression equation. (It is already stored in Y_1.) Use the TRACE feature to find y when $x = 32$.

The equation predicts that net sales for Wal-Mart in 2002 were $397.29715 billion.

$$397.29715 - 217.8 = 179.49715 \approx 179$$

The difference between the model's prediction for net sales in 2002 and actual net sales in 2002 was $179 billion.

SECTION 10.5

YOU TRY IT 1

a. $3^{-4} = \dfrac{1}{81}$ is equivalent to $\log_3 \dfrac{1}{81} = -4$.

b. $\log_{10} 0.0001 = -4$ is equivalent to $10^{-4} = 0.0001$.

YOU TRY IT 2

$\log_4 64 = x$

$64 = 4^x$

$4^3 = 4^x$

$3 = x$

$\log_4 64 = 3$

YOU TRY IT 3

$\log_2 x = -4$

$2^{-4} = x$

$\dfrac{1}{2^4} = x$

$\dfrac{1}{16} = x$

The solution is $\dfrac{1}{16}$.

YOU TRY IT 4

$f(x) = 10 \log(x - 2)$

Use the features of a graphing calculator to determine the x-intercept of the graph, which is the zero of f.

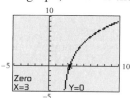

The zero of the function is 3.

YOU TRY IT 5

$\log x = 1.5$

$10^{1.5} = x$

$31.6228 \approx x$

YOU TRY IT 6

$f(x) = \log_2 (x - 1)$

$y = \log_2 (x - 1)$

$y = \log_2 (x - 1)$ is equivalent to $2^y = x - 1$.

$2^y + 1 = x$

$x = 2^{-2} + 1 = \dfrac{1}{4} + 1 = \dfrac{5}{4}$

The value $\dfrac{5}{4}$ in the domain corresponds to the range value of -2.

YOU TRY IT 7

Goal The goal is to determine how many billion barrels of oil are needed to last 25 years.

Strategy   Replace T in the given formula by 25. Then solve the resulting equation for r.

Solution
$$T = 14.29 \ln (0.00411r + 1)$$
$$25 = 14.29 \ln (0.00411r + 1)$$
$$\frac{25}{14.29} = \ln (0.00411r + 1)$$
$$e^{\frac{25}{14.29}} = 0.00411r + 1$$
$$e^{\frac{25}{14.29}} - 1 = 0.00411r$$
$$\frac{e^{\frac{25}{14.29}} - 1}{0.00411} = r$$
$$1156 \approx r$$

1156 billion barrels of oil are needed to last 25 years.

Check   √

SECTION 10.6

You Try It 1

a. $\log_b \dfrac{x^2}{y} = \log_b x^2 - \log_b y$
$$= 2 \log_b x - \log_b y$$

b. $\ln y^{1/3} z^3 = \ln y^{1/3} + \ln z^3$
$$= \frac{1}{3} \ln y + 3 \ln z$$

c. $\log_8 \sqrt[3]{xy^2} = \log_8 (xy^2)^{1/3}$
$$= \frac{1}{3} \log_8 (xy^2)$$
$$= \frac{1}{3} (\log_8 x + \log_8 y^2)$$
$$= \frac{1}{3} (\log_8 x + 2 \log_8 y)$$
$$= \frac{1}{3} \log_8 x + \frac{2}{3} \log_8 y$$

You Try It 2

a. $2 \log_b x - 3 \log_b y - \log_b z$
$$= \log_b x^2 - \log_b y^3 - \log_b z$$
$$= \log_b \frac{x^2}{y^3} - \log_b z$$
$$= \log_b \frac{x^2}{y^3 z}$$

b. $\dfrac{1}{3} (\log_4 x - 2 \log_4 y + \log_4 z)$
$$= \frac{1}{3} (\log_4 x - \log_4 y^2 + \log_4 z)$$
$$= \frac{1}{3} \left(\log_4 \frac{x}{y^2} + \log_4 z \right)$$
$$= \frac{1}{3} \left(\log_4 \frac{xz}{y^2} \right)$$
$$= \log_4 \left(\frac{xz}{y^2} \right)^{\frac{1}{3}} = \log_4 \sqrt[3]{\frac{xz}{y^2}}$$

c. $\dfrac{1}{2}(2 \ln x - 5 \ln y)$
$$= \frac{1}{2} (\ln x^2 - \ln y^5)$$
$$= \frac{1}{2} \left(\ln \frac{x^2}{y^5} \right)$$
$$= \ln \left(\frac{x^2}{y^5} \right)^{\frac{1}{2}} = \ln \sqrt{\frac{x^2}{y^5}}$$

You Try It 3

a. $\log_{16} 1 = 0$

b. $12 \log_3 3 = \log_3 3^{12} = 12$

You Try It 4

$$\log_4 2.4 = \frac{\log 2.4}{\log 4} \approx 0.6315$$

$$\log_4 2.4 = \frac{\ln 2.4}{\ln 4} \approx 0.6315$$

You Try It 5

$$f(x) = 4 \log_8 (3x + 4)$$
$$= 4 \frac{\ln(3x + 4)}{\ln 8}$$
$$= \frac{4 \ln(3x + 4)}{\ln 8}$$

You Try It 6

$$f(x) = 2 \log_4 x = 2 \frac{\ln x}{\ln 4} = \frac{2 \ln x}{\ln 4}$$

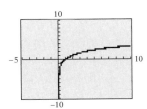

YOU TRY IT 7

$$f(x) = -2 \log_5 (3x - 4) = -\frac{2 \ln(3x - 4)}{\ln 5}$$

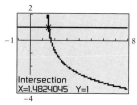

The value of x for which $f(x) = 1$ is 1.5.

SECTION 10.7

YOU TRY IT 1

$$4^{2x+3} = 8^{x+1}$$
$$(2^2)^{2x+3} = (2^3)^{x+1}$$
$$2^{4x+6} = 2^{3x+3}$$
$$4x + 6 = 3x + 3$$
$$x + 6 = 3$$
$$x = -3$$

The solution −3 check.
The solution is −3.

YOU TRY IT 2

a.
$$4^{3x} = 25$$
$$\log 4^{3x} = \log 25$$
$$3x \log 4 = \log 25$$
$$3x = \frac{\log 25}{\log 4}$$
$$x = \frac{\log 25}{3 \log 4}$$
$$x \approx 0.7740$$

The solution is 0.7740.

b.
$$(1.06)^x = 1.5$$
$$\log(1.06)^x = \log 1.5$$
$$x \log 1.06 = \log 1.5$$
$$x = \frac{\log 1.5}{\log 1.06}$$
$$x \approx 6.9585$$

The solution is 6.9585.

YOU TRY IT 3

$$e^x = x$$
$$e^x - x = 0$$

Graph $f(x) = e^x - x$.

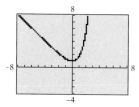

The equation has no real number solutions.

YOU TRY IT 4

a. $\log_4 (x^2 - 3x) = 1$

$$4^1 = x^2 - 3x$$
$$4 = x^2 - 3x$$
$$0 = x^2 - 3x - 4$$
$$0 = (x + 1)(x - 4)$$
$$x + 1 = 0 \qquad x - 4 = 0$$
$$x = -1 \qquad x = 4$$

The solutions −1 and 4 check.
The solutions are −1 and 4.

b. $\log_3 x + \log_3 (x + 3) = \log_3 4$

$$\log_3 [x(x + 3)] = \log_3 4$$
$$x(x + 3) = 4$$
$$x^2 + 3x = 4$$
$$x^2 + 3x - 4 = 0$$
$$(x + 4)(x - 1) = 0$$
$$x + 4 = 0 \qquad x - 1 = 0$$
$$x = -4 \qquad x = 1$$

−4 does not check as a solution.
The solution is 1.

YOU TRY IT 5

$$\log(3x - 2) = -2x$$
$$\log(3x - 2) + 2x = 0$$

Graph $f(x) = \log(3x - 2) + 2x$.

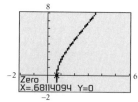

The solution is 0.68.

YOU TRY IT 6

Goal The goal is to determine the year in which a first-class stamp cost $.22.

S46

Strategy Replace C by 22 and solve for t. Then add 1958 to the value of t.

Solution

$$C = 4.50e^{0.05t}$$
$$22 = 4.50e^{0.05t}$$
$$\frac{22}{4.50} = e^{0.05t}$$
$$\ln \frac{22}{4.50} = \ln e^{0.05t}$$
$$\ln \frac{22}{4.50} = 0.05t \ln e$$
$$\ln \frac{22}{4.50} = 0.05t(1)$$

$$\ln \frac{22}{4.50} = 0.05t$$
$$\frac{\ln \frac{22}{4.50}}{0.05} = t$$
$$32 \approx t$$
$$1958 + 32 = 1990$$

According to the model, a first-class stamp cost $.22 in 1990.

Check Use a graphing calculator to graph $Y_1 = 4.50e^{0.05x}$ and $Y_2 = 22$. Use the intersect feature to verify that at the point of intersection $X \approx 32$.

SOLUTIONS to Additional Topics in Algebra You Try Its

SECTION 1

YOU TRY IT 1

$a_n = n(n + 2)$
$a_1 = 1(1 + 2) = 1(3) = 3$
$a_2 = 2(2 + 2) = 2(4) = 8$
$a_3 = 3(3 + 2) = 3(5) = 15$
$a_4 = 4(4 + 2) = 4(6) = 24$

The first four terms of the sequence are 3, 8, 15, 24.

YOU TRY IT 2

$$a_n = \frac{n - 2}{n}$$
$$a_8 = \frac{8 - 2}{8} = \frac{6}{8} = \frac{3}{4}$$
$$a_{10} = \frac{10 - 2}{10} = \frac{8}{10} = \frac{4}{5}$$

The eighth term is $\frac{3}{4}$. The tenth term is $\frac{4}{5}$.

YOU TRY IT 3

Find the common difference.

$d = a_2 - a_1 = 3 - 9 = -6$

Use the Formula for the nth Term of an Arithmetic Sequence to find the 15th term.

$a_n = a_1 + (n - 1)d$
$a_{15} = 9 + (15 - 1)(-6)$ • $a_1 = 9, n = 15, d = -6$.
  $= 9 + 14(-6)$
  $= -75$

YOU TRY IT 4

Find the common difference.

$d = a_2 - a_1 = 1 - (-3) = 4$

Use the Formula for the nth Term of an Arithmetic Sequence.

$a_n = a_1 + (n - 1)d$
$a_n = -3 + (n - 1)(4)$ • $a_1 = -3, d = 4$.
$a_n = -3 + 4n - 4$
$a_n = 4n - 7$

YOU TRY IT 5

$5, 2, \dfrac{4}{5}, \ldots$

$$r = \frac{a_2}{a_1} = \frac{2}{5}$$
$$a_n = a_1 r^{n-1}$$
$$a_5 = 5\left(\frac{2}{5}\right)^{5-1} = 5\left(\frac{2}{5}\right)^4 = 5\left(\frac{16}{625}\right) = \frac{16}{125}$$

YOU TRY IT 6

$$r = \frac{a_2}{a_1} = \frac{-1}{2} = -\frac{1}{2}$$
$$a_n = 2\left(-\frac{1}{2}\right)^{n-1}$$

YOU TRY IT 7

a. $\displaystyle\sum_{i=1}^{5} (4 - i) = (4 - 1) + (4 - 2) + (4 - 3)$
$\qquad + (4 - 4) + (4 - 5)$
$\qquad = 3 + 2 + 1 + 0 + -1 = 5$

S47

b. $\displaystyle\sum_{n=3}^{6} (n^2 + 2) = (3^2 + 2) + (4^2 + 2)$
$$+ (5^2 + 2) + (6^2 + 2)$$
$$= 11 + 18 + 27 + 38 = 94$$

You Try It 8

Determine a_n, the nth term of the arithmetic sequence. We begin by finding the common difference d.

$d = a_2 - a_1 = 4 - (-1) = 5$

Use the Formula for the nth Term of an Arithmetic Sequence to find the 25th term.

$a_n = a_1 + (n - 1)d$
$a_{25} = -1 + (25 - 1)5$ $\quad\bullet\ a_1 = -1, n = 25, d = 5.$
$\quad\ = -1 + (24)5 = 119$

Use the Formula for the Sum of n Terms of an Arithmetic Series to find the sum.

$S_n = \dfrac{n(a_1 + a_n)}{2}$

$S_{25} = \dfrac{25(-1 + 119)}{2} = \dfrac{25(118)}{2} = 1475$

The sum of the first 25 terms of the sequence is 1475.

You Try It 9

$1, -\dfrac{1}{3}, \dfrac{1}{9}, -\dfrac{1}{27}$

$r = \dfrac{a_2}{a_1} = \dfrac{-\dfrac{1}{3}}{1} = -\dfrac{1}{3}$

$S_n = \dfrac{a_1(1 - r^n)}{1 - r}$

$S_4 = \dfrac{1\left[1 - \left(-\dfrac{1}{3}\right)^4\right]}{1 - \left(-\dfrac{1}{3}\right)} = \dfrac{1 - \dfrac{1}{81}}{\dfrac{4}{3}} = \dfrac{\dfrac{80}{81}}{\dfrac{4}{3}} = \dfrac{80}{81} \cdot \dfrac{3}{4} = \dfrac{20}{27}$

SECTION 2

You Try It 1

$\dfrac{12!}{7!5!} = \dfrac{12 \cdot 11 \cdot 10 \cdot 9 \cdot 8 \cdot 7 \cdot 6 \cdot 5 \cdot 4 \cdot 3 \cdot 2 \cdot 1}{(7 \cdot 6 \cdot 5 \cdot 4 \cdot 3 \cdot 2 \cdot 1)(5 \cdot 4 \cdot 3 \cdot 2 \cdot 1)} = 792$

You Try It 2

$\dbinom{7}{0} = \dfrac{7!}{(7 - 0!)0!} = \dfrac{7!}{7!0!} = \dfrac{7 \cdot 6 \cdot 5 \cdot 4 \cdot 3 \cdot 2 \cdot 1}{(7 \cdot 6 \cdot 5 \cdot 4 \cdot 3 \cdot 2 \cdot 1)(1)} = 1$

You Try It 3

$(4x + 3y)^3$

$= \dbinom{3}{0}(4x)^3 + \dbinom{3}{1}(4x)^2(3y) + \dbinom{3}{2}(4x)(3y)^2 + \dbinom{3}{3}(3y)^3$

$= 1(64x^3) + 3(16x^2)(3y) + 3(4x)(9y^2) + 1(27y^3)$

$= 64x^3 + 144x^2y + 108xy^2 + 27y^3$

You Try It 4

$(y - 2)^{10}$

$= \dbinom{10}{0}y^{10} + \dbinom{10}{1}y^9(-2) + \dbinom{10}{2}y^8(-2)^2 + \ldots$

$= 1(y^{10}) + 10y^9(-2) + 45y^8(4) + \ldots$

$= y^{10} - 20y^9 + 180y^8 + \ldots$

SECTION 3

You Try It 1

$y = x^2 - 2x - 1$

$x = -\dfrac{b}{2a} = -\dfrac{-2}{2(1)} = 1$

$y = x^2 - 2x - 1$
$y = (1)^2 - 2(1) - 1$
$y = -2$

The vertex is $(1, -2)$.
The axis of symmetry is the line $x = 1$.

You Try It 2

a. $x = 2y^2 - 4y + 1$

$y = -\dfrac{b}{2a} = -\dfrac{-4}{2(2)} = 1$

$x = 2y^2 - 4y + 1$
$x = 2(1)^2 - 4(1) + 1$
$x = -1$

The vertex is $(-1, 1)$.
The axis of symmetry is the line $y = 1$.

b. $x = -y^2 - 2y + 2$

$$y = -\frac{b}{2a} = -\frac{-2}{2(-1)} = -1$$

$x = -y^2 - 2y + 2$

$x = -(-1)^2 - 2(-1) + 2$

$x = 3$

The vertex is $(3, -1)$.

The axis of symmetry is the line $y = -1$.

You Try It 3

$(x - h)^2 + (y - k)^2 = r^2$

$(x - 2)^2 + (y + 3)^2 = 9$

$(x - 2)^2 + [y - (-3)]^2 = 3^2$

Center: $(h, k) = (2, -3)$

Radius: $r = 3$

You Try It 4

Radius $= 4$

Center $= (2, -3)$

$(x - h)^2 + (y - k)^2 = r^2$

$(x - 2)^2 + [y - (-3)]^2 = 4^2$

$(x - 2)^2 + (y + 3)^2 = 16$

You Try It 5

a. $\dfrac{x^2}{4} + \dfrac{y^2}{25} = 1$

x-intercepts: $(2, 0)$ and $(-2, 0)$

y-intercepts: $(0, 5)$ and $(0, -5)$

b. $\dfrac{x^2}{18} + \dfrac{y^2}{9} = 1$

x-intercepts: $(3\sqrt{2}, 0)$ and $(-3\sqrt{2}, 0)$

y-intercepts: $(0, 3)$ and $(0, -3)$

$$\left[3\sqrt{2} \approx 4\frac{1}{4}\right]$$

You Try It 6

a. $\dfrac{x^2}{9} - \dfrac{y^2}{25} = 1$

$a^2 = 9, b^2 = 25$

Vertices are on the x-axis.

Vertices: $(3, 0)$ and $(-3, 0)$

Asymptotes:

$$y = \frac{5}{3}x \text{ and } y = -\frac{5}{3}x$$

b. $\dfrac{y^2}{9} - \dfrac{x^2}{9} = 1$

$b^2 = 9, a^2 = 9$

Vertices are on the y-axis.

Vertices: $(0, 3)$ and $(0, -3)$

Asymptotes: $y = x$ and $y = -x$

PREP TEST

1. 924   **2.** 1244   **3.** 15,873   **4.** 24   **5.** 127.16   **6.** a, c, d   **7.** a and C; b and D; c and A; d and B
8. 24   **9.** 4   **10.** $3 \cdot 7$

1.1 Exercises

1. Understand the problem and state the goal, devise a strategy to solve the problem, solve the problem, and review the solution and check your work.   **3.** Answers may vary.   **5.** Deductive reasoning involves drawing a conclusion that is based on given facts. Examples will vary.   **7.** 2601 tiles   **9.** 1   **11.** M, N   **13.** 6 students   **15.** 8
17. 55 mph   **19.** $\frac{101}{99}$ **21.** 7 children   **23.** $\frac{1}{8}$ and $\frac{1}{10}$ **25.** 28 minutes   **27.** 41   **29.** 216   **31.** 93
33. u   **35.** 111,111,111; 222,222,222; 333,333,333; 444,444,444; 555,555,555. Explanations will vary.
12,345,679 · 54 = 666,666,666; 12,345,679 · 63 = 777,777,777   **37.**    **39.** 12   **41.** 3

43. The difference is always 3087.   **45.** February and March   **47.** deductive reasoning
49. inductive reasoning   **51.** Maria owns the utility stock, Jose the automotive stock, Anita the technology stock, and Tony the oil stock.   **53.** Atlanta held the stamp convention, Chicago the baseball card convention, Philadelphia the coin convention, and Seattle the comic book convention.   **55.** No   **57.** 6   **59.** 3

1.2 Exercises

1. Explanations will vary.   **3.** $\{x \mid x < 5\}$ does not include the element 5, whereas $\{x \mid x \leq 5\}$ does include the element 5.
5a. No. Explanations will vary.   **b.** Yes. Explanations will vary.   **7a.** 31, 8600   **b.** 31, 8600
c. 31, −45, −2, 8600   **d.** 31, 8600   **e.** −45, −2   **f.** 31   **9a.** −17   **b.** −17, 0.3412, $\frac{27}{91}$, 6.1$\overline{2}$
c. $\frac{3}{\pi}$, −1.010010001 . . .   **d.** all   **11.** $\{-3, -2, -1\}$ **13.** $\{1, 3, 5, 7, 9, 11, 13\}$ **15.** $\{a, b, n\}$ **17.** $\varnothing$
19. $\{x \mid x < -5, x \in \text{integers}\}$ **21.** $\{x \mid x \geq -4\}$ **23.** $\{x \mid -2 < x < 5\}$ **25.** False   **27.** False   **29.** False
31. ![number line](−5 to 5)   **33.** ![number line](−5 to 5)   **35.** $\{2, 3, 5, 8, 9, 10\}$ **37.** $\{x \mid x \in \text{real numbers}\}$
39. $\{4, 6\}$ **41.** $\varnothing$ **43.** $M \cup C = \{1, 2, 3, 4, 5, 6, 7, 8, 9, 10\}$; $M \cap C = \varnothing$ **45.** ![number line](−5 to 5)
47. ![number line](−5 to 5)   **49.** ![number line](−5 to 5)   **51.** ![number line](−5 to 5)
53. ![number line](−5 to 5)   **55.** $\{x \mid -5 \leq x \leq 7\}$ **57.** $\{x \mid -9 < x \leq 5\}$ **59.** $\{x \mid x \geq -2\}$ **61.** $[0, 3]$
63. $[-2, 7)$ **65.** $(-\infty, -5]$ **67.** $(23, \infty)$ **69.** ![number line](−5 to 5)   **71.** ![number line](−5 to 5)
73. ![number line](−5 to 5)   **75.** ![number line](−5 to 5)   **77.** ![number line](−5 to 5)
79. Explanations may vary.   **81.** A set is well defined if it is possible to determine whether any given item is an element of the set. Examples will vary.

1.3 Exercises

1. Sometimes true   **3.** Never true   **5.** Sometimes true   **7.** It means to replace the variables in a variable expression with numbers and then simplify the resulting numerical expression.
9. The distance from −24 to 0 on the number line is greater than the distance from 7 to 0.   **11.** −25   **13.** 34
15. 0   **17.** −12   **19.** 16   **21.** −49   **23.** 16   **25.** 32   **27.** −86   **29.** −54   **31a.** 11   **b.** 1
33a. 19   **b.** 4   **35a.** 1550 ft   **b.** 5 s   **37.** 3   **39.** −20   **41.** −10   **43.** −7   **45.** 59   **47.** 18
49. 13   **51.** −8   **53.** 5   **55.** 2   **57.** 5   **59.** −2   **61.** −11   **63.** −17   **65.** 6   **67.** 4   **69.** 7
71. 7   **73.** Calcavecchia: 278; Duval: 274; Els: 279; Furyk: 279; Izawa: 278; Langer: 279; Michelson: 275; Triplett: 279; Woods: 272   **75.** 252   **77.** −162   **79.** 90   **81.** −7   **83.** undefined   **85.** 24   **87.** −20

89. 192   **91.** 90   **93.** 800   **95.** 9   **97.** -8 **99.** 6   **101.** -8 **103.** 108, -324, 972   **105.** -375, -1875, -9375 **107.** $-\$180,688,000$ **109.** $\$4,231,000$ **111.** -49 **113.** -8 **115.** -12 **117.** 32   **119.** -864 **121.** 1008   **123.** -16 **125.** -36 **127.** -20 **129.** 31   **131.** 4   **133.** 6   **135.** 16   **137.** -9 **139.** -15 **141.** 1   **143.** 5   **145.** 10   **147.** -2 **149.** 20   **151.** 24   **153.** $-1°C$ **155.** 214 points   **157.** -7 **159.** 11°F   **161a.** 2 **b.** -2 **163.** -1 **165.** 26; 3   **167.** Answers will vary.

1.4 Exercises

1. Never true   **3.** Never true   **5.** Always true   **7.** Rewrite the fraction as a decimal. Then compare the two decimals.   **9a.** No **b.** No   **11.** $-\frac{1}{8}$ **13.** $-\frac{1}{12}$ **15.** $-\frac{5}{24}$ **17.** $-\frac{19}{24}$ **19.** $-\frac{25}{18}$ **21.** $\frac{11}{8}$ **23.** $-\frac{7}{16}$ **25.** $\frac{11}{24}$ **27.** -13.309 **29.** -10.03 **31.** -60.03 **33.** $\frac{5}{8} - \left(-\frac{5}{6}\right)$ **35.** $-\frac{4}{9}$ **37.** $-\frac{3}{2}$ **39.** $\frac{1}{21}$ **41.** $-\frac{7}{30}$ **43.** $-\frac{10}{9}$ **45.** $-\frac{3}{10}$ **47.** 1.136   **49.** -10.759 **51.** -1.104 **53.** -3.5 **55.** $\left(-\frac{8}{9}\right)\left(-\frac{3}{4}\right)$ **57.** $-\$103.408$ million   **59.** $-\$43.584$ million   **61.** $\$30.481$ million   **63.** 2000; 1975   **65.** $\$288.6$ billion   **67.** 1999–2000   **69.** $-\$92.425$ billion   **71.** 0.51   **73.** -1 **75.** $-\frac{15}{2}$ **77.** 4.96   **79.** -5.68 **81.** -4 **83.** $\$138.499$ billion   **85.** 4 times greater   **87.** $-\$710.5$ million   **89.** $\$27.2848$ billion   **91.** 220 cm   **93.** 3 m^2 **95.** $\$48,000$ **97.** 136 ft   **99a.** 4   **b.** 3.5   **101a.** 7.125   **b.** 2.75   **103.** The architect charges a fee of $\$4740$ to design a 1600-square-foot house.   **105a.** 91°F   **b.** 2450 ft   **107a.** 75.4 min   **b.** 2.5 min
109a.

s	2	4	6	8	10	12	14
P	8	16	24	32	40	48	56

b. When the length of a side of a square is 12 in., the perimeter is 48 in.
111a.

s	1	2	3	4	5	6	7
A	1	8	27	64	125	216	343

b. When the length of a side of a cube is 3 m, the volume of the cube is 27 m^3.   **113.** greater than
115. Row 1: $-\frac{1}{6}$, 0; row 2: $-\frac{1}{2}$; row 3: $\frac{1}{3}$, $\frac{1}{2}$ **117.** As the denominator increases, the value of the fraction decreases.

1.5 Exercises

1. Always true   **3.** Always true   **5.** Sometimes true   **7.** Always true   **9.** Sometimes true   **11.** Never true
13. The Commutative Property says that two numbers can be added in either order. The Associative Property states that when three numbers are added together, the numbers can be grouped in any order.   **15.** 2   **17.** 17
19. 4   **21.** 4   **23.** 0   **25.** The Multiplication Property of One   **27.** The Addition Property of Zero
29. The Inverse Property of Multiplication   **31.** The Associative Property of Addition
33. The Commutative Property of Multiplication   **35.** $2x^2$, $5x$, $\underline{-8}$ **37.** $-n^4$, $\underline{6}$ **39.** $\underline{7x^2y}$, $\underline{6xy^2}$ **41.** 1, -9
43. 1, -4, -1 **45.** $21y$ **47.** $-6y$ **49.** The expression is in simplest form.   **51.** $5xy$ **53.** $-14x^2$
55. $-3x - 8y$ **57.** $-2x$ **59.** $22y^2$ **61.** 0   **63.** $-\frac{7}{20}y$ **65.** $12x$ **67.** $-6a$ **69.** x **71.** $2x$ **73.** $3y$
75. $-2a - 14$ **77.** $15x^2 + 6x$ **79.** $-6y^2 + 21$ **81.** $-12a^2 - 20a + 28$ **83.** $a - 7$ **85.** $18y - 51$
87. $4x - 4$ **89.** $\frac{4}{p - 6}$ **91.** $\frac{3}{8}(t + 15)$ **93.** $13 - x$ **95.** $\frac{3}{7}x$ **97.** $5x - 8$ **99.** $7x + 14$ **101.** $(n + n^3) - 6$
103. $11 + \frac{1}{2}x$ **105.** $80 - 13x$ **107.** $7x^2 - 4$ **109.** $x + (x + 10)$; $2x + 10$ **111.** $x - (9 - x)$; $2x - 9$
113. $\frac{1}{5}x - \frac{3}{8}x$; $-\frac{7}{40}x$ **115.** $(x + 9) + 4$; $x + 13$ **117.** $2(3x + 40)$; $6x + 80$ **119.** $16\left(\frac{1}{4}x\right)$; $4x$
121. $9x - 2x$; $7x$ **123.** $(x - 5) + 19$; $x + 14$ **125.** Let p be the cruising speed of a propeller-driven plane; $2p$
127. Let c be the amount of cashews; $4c$ **129.** Let a be the age of the 8¢ stamp; $a + 25$ **131.** Let L be the measure of the largest angle; $\frac{1}{2}L - 3$ **133.** x and $35 - x$ **135.** $640 + 24h$ **137.** $0.3C$ **139.** $-\frac{1}{2}x + \frac{5}{2}y$
141. For example, four less than five times a number.   **143.** For example, the product of five and four less than a number.   **145.** $5n + 10d$ **147.** $2x$

CHAPTER 1 REVIEW EXERCISES

1. 13 first cousins [1.1]　**2.** 22 [1.1]　**3.** 4 [1.1]　**4.** $\{-8, -7, -6, -5, -4, -3\}$ [1.2]　**5.** $\{x \mid x \leq -10\}$ [1.2]
6. $\{1, 2, 3, 4, 5, 6, 7, 8\}$ [1.2]　**7.** $\{2, 3\}$ [1.2]　**8.** $\{x \mid -2 \leq x \leq 3\}$ [1.2]　**9.** $(-\infty, -44)$ [1.2]
10. ⊢+++(+++++)++ [1.2]　**11.** ⊢+++++++++ +++ [1.2]　**12.** ⊢+++(+++++)++ [1.2]
    −5 −4 −3 −2 −1 0 1 2 3 4 5　　　　−5 −4 −3 −2 −1 0 1 2 3 4 5　　　　−5 −4 −3 −2 −1 0 1 2 3 4 5
13. $-\frac{1}{7}$ [1.4]　**14.** 0.339 [1.4]　**15.** 5.8 [1.4]　**16.** −441.2 [1.4]　**17.** $\frac{41}{24}$ [1.4]　**18.** $-\frac{5}{6}$ [1.4]　**19.** 18 [1.3]
20. $\frac{2}{7}$ [1.4]　**21.** −3°C [1.3]　**22.** 395.45°C [1.4]　**23.** 37.5 cm [1.4]
24a.

D	2	4	6	8	10	12	14
P	16	17	18	19	20	21	22

    b. At a depth of 6 ft, the pressure is 18 lb/in². [1.4]
25. The Commutative Property of Multiplication [1.5]　**26.** 24d [1.5]　**27.** $3a^2 + 10a$ [1.5]
28. $19a - 13$ [1.5]　**29.** $8\left(\frac{2x}{16}\right)$; x [1.5]　**30.** Let d be the distance from Earth to the Sun; 30d [1.5]

CHAPTER 1 TEST

1. 211 [1.1]　**2.** deductive reasoning [1.1]　**3.** $\{-6, -5, -4, -3, -2, -1, 0\}$ [1.2]　**4.** $\{x \mid x \geq -2\}$ [1.2]
5. $\{-1, 0, 1\}$ [1.2]　**6.** $\{x \mid -4 \leq x \leq 6\}$ [1.2]　**7.** $[-20, \infty)$ [1.2]　**8.** ⊢+++++]+++++++ [1.2]
                                                         −5 −4 −3 −2 −1 0 1 2 3 4 5
9. ⊢+++]+++(+++++ [1.2]　**10.** $\frac{1}{10}$ [1.4]　**11.** $\frac{1}{12}$ [1.4]　**12.** $-\frac{3}{10}$ [1.4]　**13.** 33 [1.3]
    −5 −4 −3 −2 −1 0 1 2 3 4 5
14. −5° [1.3]　**15.** $38,669 [1.4]　**16a.** 49 ft　**b.** 1.25 s [1.4]　**17.** $-8y^2 + 9y$ [1.5]　**18.** $5w - 17$ [1.5]
19. $(x - 3) + (x + 2)$; $2x - 1$ [1.5]　**20.** Let s be the speed of the second car; $s + 15$ [1.5]

ANSWERS to Chapter 2 Exercises

PREP TEST

1. 2 [1.3]　**2.** 11 [1.3]　**3.** 2.5 [1.4]　**4.** 5 [1.3]　**5.** 0 [1.3]　**6.** −7 [1.4]

2.1 Exercises

1. Answers will vary.
3. A solution of an equation in two variables is an ordered pair that makes the equation a true statement.
5. The input variable is t. The output variable is s.
7.

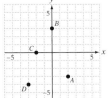

9. −2, 0; 0, −3

11. Yes　**13.** No

15.

x	−3	−2	−1	0	1	2	3
y	7	5	3	1	−1	−3	−5

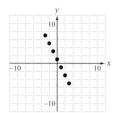

17.

x	−8	−4	0	4	8
y	−5	−2	1	4	7

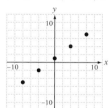

19.

x	−3	−2	−1	0	1	2	3
y	10	5	2	1	2	5	10

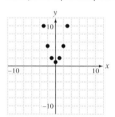

21.

t	−5	−4	−3	−2	−1	0	1
s	2	−3	−6	−7	−6	−3	2

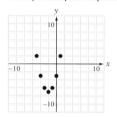

23a.

Input, time t (in seconds)	0	5	10	15	20	25	30
Output, distance d (in feet)	0	55	110	165	220	275	330

b. In 20 s, the jogger runs 220 ft.

25a.

Input, time t (in seconds)	0	0.5	1	1.5	2	2.5	3
Output, distance d (in feet)	0	4	16	36	64	100	144

b. In 1.5 s, the object will fall 36 ft.

27a.

Input, weight of jewelry w (in grams)	0	5	10	15	20	25	30
Output, quantity of gold Q (in grams)	0	3.75	7.5	11.25	15	18.75	22.5

b. In a 15-gram piece of 18-carat gold jewelry, there are 11.25 g of gold.

29a.

Input, time t (in seconds)	0	0.5	1	1.5	2	2.5	3
Output, height h (in feet)	5	36	59	74	81	80	71

b. The ball is 80 ft above the ground 2.5 s after it is released.

31. $y = 2x - 4$

x	−2	−1	0	1	2
y	−8	−6	−4	−2	0

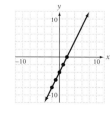

33. $y = \frac{x}{2} + 1$

x	−4	−2	0	2	4
y	−1	0	1	2	3

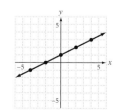

35. $y = \dfrac{-5x}{4}$

x	-8	-4	0	4	8
y	10	5	0	-5	-10

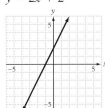

37. $y = \dfrac{3}{4}x - 4$

x	-8	-4	0	4	8
y	-10	-7	-4	-1	2

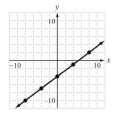

39. $y = 2x + 2$

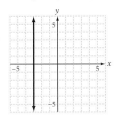

41. $y = \dfrac{3}{2}x - 3$

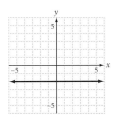

43. $y = -\dfrac{3}{4}x + 1$

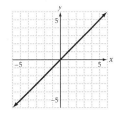

45a. 10 **b.** 5 **47a.** 0 **b.** 22 **49a.** 19 **b.** 8 **51a.** -11 **b.** 28 **53a.** 10 **b.** -6 **55.** $(7.5, 20)$

57. $(-6.3, 20.9)$ **59.** $\left(-\dfrac{61}{3}, -1\right)$ **61.** $\left(-\dfrac{57}{10}, -\dfrac{9}{2}\right)$ **63.** $(12.1, -23.03)$ **65.** 0 **67.** Answers will vary.

For example, $(-3, 2)$ and $(5, 2)$. **69.** $(5, 4)$ and $(-3, -1)$ **71a.** 5 **b.** 3 **73a.** 4 **b.** 7 **75a.** 0 **b.** 2

77.

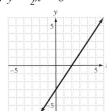

79.

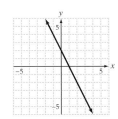

81.

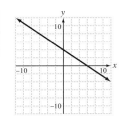

2.2 Exercises

1. A relation is a set of ordered pairs. A function is a relation in which no two ordered pairs have the same first coordinate and different second coordinates. **3.** The domain is the set of first coordinates of the function; the range is the set of second coordinates of the function. **5.** To evaluate a function means to replace the independent variable by a given number and simplify the resulting expression. **7.** No **9.** Domain: $\{-3, -2, -1, 0, 1\}$, Range: $\{-13, -11, -9, -7, -5\}$, Yes **11.** Domain: $\{-4, -2, 0, 2\}$, Range: $\{6, 8, 10, 12\}$, No **13.** Domain: $\{2, 3, 4, 5, 6\}$, Range: $\{-6, -3, 6\}$, Yes

15. Domain: $\{-4, -2, 0, 3, 5\}$, Range: $\{0\}$, Yes **17.** 13 **19.** 14 **21.** 15 **23.** $\dfrac{4}{3}$ **25.** 13

27. $f(x) = 2 - 2x$

x	-2	-1	0	1	2
y	6	4	2	0	-2

29. $f(x) = -\dfrac{2x}{3} + 4$

x	-6	-3	0	3	6
y	8	6	4	2	0

31. $f(x) = x^2 - 2$

x	-3	-2	-1	0	1	2	3
y	7	2	-1	-2	-1	2	7

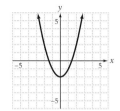

33. $f(x) = -x^2 + 2x - 1$

x	-2	-1	0	1	2	3	4
y	-9	-4	-1	0	-1	-4	-9

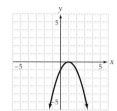

35. $f(x) = -2$

x	-2	-1	0	1	2	3	4
y	-2	-2	-2	-2	-2	-2	-2

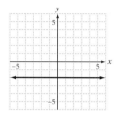

37. The graph is a horizontal line through $(0, 1)$. **39.** 10 **41.** -27 **43.** 8 **45.** 10 **47.** 7 **49.** $\frac{71}{8}$
51. $\{-6, 1, 6, 9, 10\}$ **53.** $\{1, 3, 6, 10, 15, 21, 28\}$ **55.** $\left\{-\frac{3}{10}, -\frac{2}{5}, -\frac{1}{2}, 0, \frac{1}{2}, \frac{2}{5}, \frac{3}{10}\right\}$ **57.** None **59.** -5

61. $-1, 3$ **63.** None **65.** -4 **67.** -2 **69.** None **71.** 1 **73.** -3 **75.** 1.5 **77.** $-1, 1$
79. $-1, 3$ **81.** $-0.5, 1$ **83a.** 16 m **b.** 20 ft **85a.** 100 ft **b.** 68 ft **87a.** 1136 ft/s **b.** increases
89a. 30 games **b.** 45 games **91a.** 1.92 s **b.** 0.96 s **93.** Answers will vary. Possibilities are: **a.** $\{(1, 2), (4, 7),$
$(6, 9)\}$ **b.** $\{(1, 4), (2, 4), (3, 4)\}$ **95.** Answers will vary. One possibility is $f(x) = \frac{1}{x - 5}$.

2.3 Exercises

1. If every vertical line intersects a graph at most once, then the graph is the graph of a function. **3a.** 0 **b.** 0
5a. The domain of a function is the set of all values of the independent variable. The range of a function is the set of all
values of the dependent variable. **b.** The domain of a function is the set of all input values. The range of a function is
the set of all output values. **7.** Yes **9.** No **11.** Yes **13.** $(-2, 0); (0, 6)$ **15.** $(-2, 0), (3, 0); (0, -6)$
17. $(-3, 0), (0.5, 0); (0, -3)$ **19.** $(-1, 0), (5, 0); (0, 5)$ **21.** $(-2, 0), (1, 0), (5, 0); (0, 10)$ **23.** $(-3, 0), (0, 0),$
$(4, 0); (0, 0)$ **25.** 3 **27.** 2 **29.** 1.6 **31.** 3 **33.** 0 **35.** 1.71 **37.** 3 **39.** 8 **41.** -2 **43.** 2
45. -2 **47.** -2 **49.** 2.5 **51.** -2.35 **53.** 2.1 **55a.** -9 **b.** 1 **57a.** 4 **b.** 9 **59.** 44 g
61a. 3.3 s **b.** The rock hits the bottom of the ravine in 3.3 s. **63.** 5 years **65.** \$22 **67.** 2 h **69.** 2011
71. 2 **73.** -8 **75.** 4 **77.** $-3, 1$ **79.** $-4, 2$

CHAPTER 2 REVIEW EXERCISES

1.

x	-6	-4	-2	0	2	4
y	-6	-5	-4	-3	-2	-1

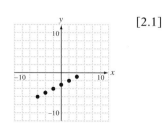

[2.1]

2.

x	−3	−2	−1	0	1	2	3
y	3	−1	−3	−3	−1	3	9

[2.1]

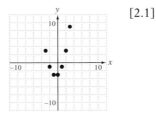

3.

x	−9	−6	−3	0	3
y	5	4	3	2	1

[2.1]

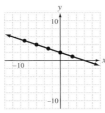

4a. −7 **b.** −5 [2.1]   **5a.** 4 **b.** −3 [2.1]   **6.** Domain: {−1, 0, 1, 2, 3, 4}, Range: {−1, 1, 3, 5}; Yes [2.2]
7. [−27, −13, −3, 3, 5] [2.2]   **8.** −2 [2.2]   **9.** −3, 3 [2.2]
10. $f(x) = -2x + 3$ **11.** $f(x) = -x^2 + 4x - 1$

x	−3	−2	−1	0	1
y	9	7	5	3	1

x	−1	0	1	2	3	4	5
y	−6	−1	2	3	2	−1	−6

[2.2]

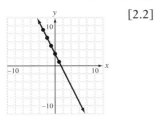

[2.2]

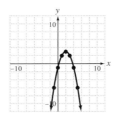

12. $f(x) = -4.$

x	−2	−1	0	1	2
y	−4	−4	−4	−4	−4

[2.2]

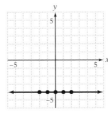

13. 16 [2.2]   **14.** −10 [2.2]   **15.** −7 [2.2]   **16.** −10.625 [2.2]   **17.** −1 [2.2]   **18.** 2 [2.2]
19. Yes; it is a function [2.3]   **20.** (6, 0); (0, −3) [2.3]   **21.** (−4, 0), (2, 0); (0, −8) [2.3]   **22.** 3.5 [2.3]
23. −2 [2.3]   **24a.** 6 min **b.** 300 ft [2.3]   **25.** 2005 [2.3]

CHAPTER 2 TEST

1.

x	−2	−1	0	1	2	3	4	5	6
y	−7	0	5	8	9	8	5	0	−7

[2.1]

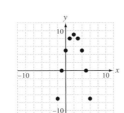

2.

x	−6	−3	0	3	6	9
y	−6	−4	−2	0	2	4

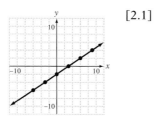

[2.1]

3. 4 [2.1] **4.** Domain: {−4, −2, 0, 2, 4}, Range: {−2, −1, 0}; Yes [2.2] **5.** {−3, −1, 3, 9} [2.2] **6.** 2 [2.2]

7. $f(x) = -\frac{3}{2}x + 4$

x	−2	0	2	4	6
y	7	4	1	−2	−5

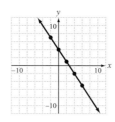

[2.2]

8. $f(x) = -x^2 - 2x + 3$

x	−4	−3	−2	−1	0	1	2
y	−5	0	3	4	3	0	−5

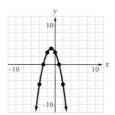

[2.2]

9. −6 [2.2]   **10.** −12 [2.2]   **11.** −3, 3 [2.2]   **12.** Answers will vary. [2.3]   **13.** (−4, 0); (0, 3) [2.3]
14. (−1, 0), (3, 0); (0, −3) [2.3]   **15.** −4 [2.3]   **16.** 4 [2.3]   **17a.** 50% **b.** 20% [2.2]
18. 35 s [2.3]   **19.** 3.1 s [2.3]   **20.** 2007 [2.3]

CUMULATIVE REVIEW EXERCISES

1. {x | −2 ≤ x ≤ 3} [1.2]   **2.** True [1.2]   **3.** 34 [1.3]   **4.** 160 [1.3]   **5.** $-\frac{13}{12}$ [1.4]
6. −10a + 38 [1.5]   **7.** Commutative Property of Multiplication [1.5]
8.

x	−4	−2	0	2	4	6
y	−9	−6	−3	0	3	6

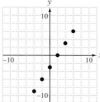

[2.1]

9. a.

Input, distance driven m (in miles)	0	100	150	200	250	300	350
Output, NO$_x$ g (in grams)	0	40	60	80	100	120	140

b. When this car is driven 150 mi, it emits 60 g of NO$_x$. [2.1]

10. 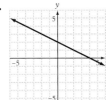 [2.1] **11.** 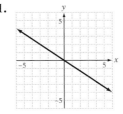 [2.1]

12.

x	-1	0	1	2	3	4	5
y	8	1	-4	-7	-8	-7	-4

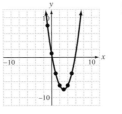 [2.2]

13.

x	-4	-3	-2	-1	0	1	2
y	-7	-2	1	2	1	-2	-7

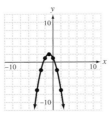 [2.2]

14. -9 [2.1]   **15.** 4 [2.3]   **16.** 2 [2.2]   **17.** $(-3, 0), (1, 0); (0, -3)$ [2.3]   **18.** 2 [2.3]
19. 8 min [2.3]   **20.** 2004 [2.3]

ANSWERS to Chapter 3 Exercises

PREP TEST

1. -4 [1.3]   **2.** -6 [1.3]   **3.** 3 [1.3]   **4.** 1 [1.4]   **5.** $10x - 5$ [1.5]   **6.** -9 [1.5]   **7.** $9x - 18$ [1.5]
8. 8 [1.3]   **9.** $20 - n$ [1.5]

3.1 Exercises

1. An equation has an equals sign; an expression does not have an equals sign.   **3.** The goal of solving an equation is to find its solutions. The goal of simplifying an expression is to combine like terms and write the expression in simplest form.   **5.** The same number can be added to each side of an equation without changing the solution of the equation. This property is used to remove a term from one side of an equation.   **7.** 9   **9.** 4   **11.** $\frac{15}{2}$ **13.** -2
15. 4.8   **17.** $\frac{3}{2}$ **19.** 3   **21.** 5   **23.** -3 **25.** -3 **27.** $\frac{4}{3}$ **29.** $\frac{5}{3}$ **31.** 7   **33.** -3 **35.** $\frac{1}{2}$
37. $\frac{5}{6}$ **39.** $\frac{2}{3}$ **41.** 6   **43.** -12 **45.** The equation has no solution.   **47.** 9   **49.** $-\frac{15}{2}$
51a. $40,000; $25,000   **b.** $60,000; $75,000   **c.** $50,000

3.2 Exercises

1. $V = x(0.85) = 0.85x$ **3.** Yes. Sometimes; it depends on the ratio of the juices used. No.   **5.** The second jogger
7. The total distance between the two objects = the distance traveled by the first object + the distance traveled by the second object.   **9.** 800 mph   **11.** $.12   **13.** 17.5 lb of $6 coffee; 7.5 lb of $3.50 coffee
15. 72 adult tickets   **17.** 250 bushels of soybeans; 750 bushels of wheat   **19.** 100 oz   **21.** $6.85 per ounce

23. 15.6 kg of walnuts; 34.4 kg of cashews   **25.** $d = 12t$

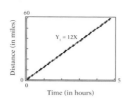

27a. Imogene   **b.** No. Her graph does not intersect Imogene's graph.   **29.** First car: 54 mph; second car: 64 mph
31. 336 mi   **33.** one hiker: 3 mph; other hiker: 3.5 mph   **35.** Freight train: 30 mph; passenger train: 48 mph
37. 44 mi   **39.** Walnuts: 10 lb; cashews: 20 lb   **41.** 1600 years   **43.** 10:15 A.M.

3.3 Exercises

1. A right angle is an angle whose measure is 90°. An acute angle is an angle whose measure is between 0° and 90°. An obtuse angle is an angle whose measure is between 90° and 180°. A straight angle is an angle whose measure is 180°.
3. Vertical angles are nonadjacent angles formed by intersecting lines.   **5.** They are not parallel.
7. A central angle is formed by two radii of the circle. The measure of the central angle is equal to the measure of the intercepted arc.   **9.** 47°   **11.** 82°   **13.** 28° and 62°   **15.** 48° and 132°   **17.** 33°   **19.** 5°   **21.** 25°
23. 42°   **25.** 78°   **27.** 4   **29.** $m\angle a = 44°$, $m\angle b = 136°$ **31.** $m\angle a = 122°$, $m\angle b = 58°$ **33.** 40
35. 20   **37.** $m\angle x = 125°$, $m\angle y = 135°$ **39.** $m\angle x = 65°$, $m\angle y = 155°$ **41.** 22° and 68°   **43.** 35   **45.** 33
47. 10   **49.** 28   **51.** 64   **53.** The three angles form a straight angle. The sum of the measures of the interior angles of a triangle is 180°.   **55.** 140°

3.4 Exercises

1.) and (indicate that the endpoint of an interval is not included in the solution set;] and [indicate that the endpoint point of an interval is included in the solution set.   **3.** The Multiplication Property of Inequalities states that when each side of an inequality is multiplied by a positive number, the inequality symbol remains the same; when each side of an inequality is multiplied by a negative number, the inequality symbol must be reversed.   **5.** A number cannot be less than -3 *and* greater than 4.   **7.** $\{x \mid x > 3\}$

9. $\{n \mid n \geq 2\}$ **11.** $\{x \mid x > -2\}$

13. $\{n \mid n \geq 4\}$ **15.** $\{x \mid x \leq 2\}$ **17.** $\{x \mid x > 3\}$ **19.** $\{x \mid x \leq 2\}$ **21.** $\{x \mid x \geq 2\}$

23. $\{x \mid x \leq 3\}$ **25.** $\{x \mid x < -3\}$ **27.** $\left[-\frac{1}{2}, \infty\right)$ **29.** $\left(-\infty, \frac{8}{3}\right)$ **31.** $(-\infty, 8.125)$ **33.** $(-\infty, 1]$

35. $\left(-\infty, \frac{7}{4}\right]$ **37.** $(-\infty, 2]$ **39.** 3150 or more   **41.** More than $5714   **43.** More than 60 min
45. More than 38 mi   **47.** The TopPage plan is less expensive for more than 460 pages per month.
49. 32°F to 86°F   **51.** more than 200 checks   **53.** 58 to 100   **55.** $\{x \mid x < 3 \text{ or } x > 5\}$

57. $\{x \mid x < -3\}$ **59.** $\{x \mid x < -2 \text{ or } x > 2\}$ **61.** $\left\{x \mid x > 5 \text{ or } x < -\frac{5}{3}\right\}$ **63.** $\{x \mid x \in \text{real numbers}\}$

65. $\{x \mid x \in \text{real numbers}\}$ **67.** $\{1, 2\}$ **69.** $\{1, 2, 3\}$ **71.** $\{3, 4, 5\}$ **73.** $\{10, 11, 12, 13\}$
75a. Always true   **b.** Sometimes true   **c.** Sometimes true   **d.** Sometimes true   **e.** Always true

3.5 Exercises

1a. Always true   **b.** Never true   **c.** Never true   **d.** Sometimes true   **e.** Never true   **3.** Parts b, d, and f have no solution. For parts a, c, and e, the solution set is all real numbers.   **5.** The solution set of $|ax + b| \leq c$ contains the endpoints of the interval. The solution set of $|ax + b| < c$ does not contain the endpoints.   **7.** $-2, 2$ **9.** $-9, 9$
11. The equation has no solution.   **13.** $-7, -3$ **15.** 4, 12   **17.** -7 **19.** The equation has no solution.
21. $0, \frac{8}{3}$ **23.** $\frac{3}{2}$ **25.** The equation has no solution.   **27.** $-3, 7$ **29.** $-\frac{10}{3}, 2$ **31.** $\frac{3}{2}$ **33.** The equation has no solution.   **35.** $\frac{11}{6}, -\frac{1}{6}$ **37.** $-\frac{1}{3}, -1$ **39.** The equation has no solution.   **41.** The equation has no solution.
43. $\frac{7}{3}, \frac{1}{3}$ **45.** $-\frac{1}{2}$ **47.** $-\frac{8}{3}, \frac{10}{3}$ **49.** $\{x \mid -5 < x < 5\}$ **51.** $\{x \mid x < 1 \text{ or } x > 3\}$ **53.** $\{x \mid 1 \leq x \leq 7\}$

55. $\{x | x \le 1 \text{ or } x \ge 5\}$ **57.** $\left\{x | -\frac{2}{3} < x < 2\right\}$ **59.** $\left\{x | x < -\frac{12}{7} \text{ or } x > 2\right\}$ **61.** $\varnothing$ **63.** $\{x | x \in \text{real numbers}\}$

65. $\{x | x < -1 \text{ or } x > 8\}$ **67.** $\left\{x | -2 < x < \frac{20}{7}\right\}$ **69.** $\{x | x \in \text{real numbers}\}$ **71.** $\left\{x | -3 < x < \frac{17}{5}\right\}$

73. 3.476 in.; 3.484 in. **75.** 93.5 volts; 126.5 volts **77.** $10\frac{11}{32}$ in.; $10\frac{13}{32}$ in. **79.** 28,420 ohms; 29,580 ohms

81. 23,750 ohms; 26,250 ohms **83.** $228 < h < 272$ **85.** $x < 7.3 \text{ or } x > 17.5$ **87.** $-0.25, 1.5$ **89.** $-\frac{1}{3}, 3$

91. $\frac{4}{3}$ **93.** $\{x | -0.5 < x < 1.25\}$ **95a.** $-b - a \le x \le b - a$ **b.** $x < a - b \text{ or } x > a + b$ **c.** $x < -2a \text{ or } x > 0$
d. $0 \le x \le 2a$ **97.** $|x + 2| = 5$ **99.** For $c > 0$, the solution set is $-c \le x - 4 \le c$. For $c = 0$, the solution is 4. For
$c < 0$, the solution set is the empty set.

CHAPTER 3 REVIEW EXERCISES

1. $\frac{7}{20}$ [3.1] **2.** $\frac{8}{3}$ [3.1] **3.** $\frac{1}{9}$ [3.1] **4.** 0 [3.1] **5.** $\{x | x < 3\}$ [3.4] **6.** $[-2, 1]$ [3.4]

7. $(-\infty, -2) \cup (2, \infty)$ [3.4] **8.** $1, -\frac{7}{3}$ [3.5] **9.** $\left\{x | x < \frac{1}{2} \text{ or } x > 2\right\}$ [3.5] **10.** $\left\{x | -\frac{19}{3} < x < 7\right\}$ [3.5]

11. 28 [3.3] **12.** $m\angle x = 140°; m\angle y = 77°$ [3.3] **13.** $m\angle a = 138°; m\angle b = 42°$ [3.3] **14.** 58 [3.3]
15. 148° [3.3] **16.** $50.4 < x < 89.6$ [3.5] **17.** 2:20 P.M. [3.2] **18.** 2.747 in.; 2.753 in. [3.5]
19. $82 \le N \le 100$, where N is the score on the fifth exam [3.4] **20.** 52 gal [3.2]

CHAPTER 3 TEST

1. $-\frac{1}{12}$ [3.1] **2.** $\frac{2}{3}$ [3.1] **3.** $\frac{8}{3}$ [3.1] **4.** -3 [3.1] **5.** $\{x | x \le 2\}$ [3.4] **6.** $\{x | x \le 1\}$ [3.4]

7. $(-1, 2)$ [3.4] **8.** $(-\infty, \infty)$ [3.4] **9.** $\{x | 1 < x < 5\}$ [3.4] **10.** $-\frac{5}{2}, \frac{11}{2}$ [3.5] **11.** $\{x | 1 \le x \le 4\}$ [3.5]

12. $\left\{x | x \le \frac{1}{2} \text{ or } x \ge 3\right\}$ [3.5] **13.** $\varnothing$ [3.5] **14.** 6 [3.3] **15.** $m\angle a = 141°, m\angle b = 39°$ [3.3]

16. 82 [3.3] **17.** 30°, 60°, 90° [3.3] **18.** 440 mph, 520 mph [3.2] **19.** $4.79 per pound [3.2]
20. 4.8 mg; 5.2 mg [3.5]

CUMULATIVE REVIEW EXERCISES

1. 6 [1.1] **2.** B, C, A [1.1] **3.** $\{1, 3, 5, 7\}$ [1.2] **4.** $E \cup F = \{-6, -4, -2, 0, 2, 3, 4, 6, 9, 12\}$;
$E \cap F = \{0, 6\}$ [1.2] **5.** 0 [1.3] **6.** 4°C [1.3] **7.** 1.229 [1.4] **8.** 40 m² [1.4] **9.** $23y - 44$ [1.5]
10. $x - (x + 20); -20$ [1.5] **11a.** 0, 56, 112, 168, 224, 280, 336 **b.** The car can travel 280 mi on 10 gal of gasoline. [2.1]
12. -2 [2.2] **13.** $\{-128, -10, 4, 10, 104\}$ [2.2] **14.** $(-3, 0), (2, 0); (0, -6)$ [2.3] **15.** 1 [2.3]

16. 1 [3.1] **17.** $(-\infty, -3]$ [3.4] **18.** $-4, 7$ [3.5] **19.** $\left\{x | \frac{1}{3} \le x \le 3\right\}$ [3.5] **20.** 340 mph [3.2]

ANSWERS to Chapter 4 Exercises

PREP TEST

1. $-4x + 12$ [1.5] **2.** $y + 5$ [1.5] **3.** $\frac{3}{4}x - 4$ [1.5] **4.** -2 [1.3] **5.** 240 [1.3] **6.** -1 [1.3]

7. 4 [3.1] **8.** -2 [3.1] **9.** -8 [3.1] **10.** a, b, c [3.4]

4.1 Exercises

1. Answers will vary. For example, **a.** $y = 3x - 4$, **b.** $2x - 5y = 10$. **3.** The graph of a line with zero slope is horizontal. The graph of a line with no slope is vertical. **5.** No. For instance, the graph of $x = 3$ is a line but not the graph of a function.   **7.** The graph of $x = a$ is a vertical line passing through $(a, 0)$. The graph of $y = b$ is a horizontal line passing through $(0, b)$.   **9.** Yes. $-\frac{3x}{4} = -\frac{3}{4}x$. It is a function of the form $f(x) = mx + b$.   **11.** No. The exponent on the variable is 2, not 1.   **13.** -1 **15.** $\frac{1}{3}$ **17.** $-\frac{2}{3}$ **19.** $-\frac{3}{4}$ **21.** Undefined   **23.** $\frac{7}{5}$ **25.** 0   **27.** $-\frac{1}{2}$ **29.** Undefined   **31.** 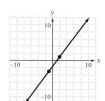   **33.** 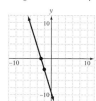   **35.**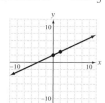

37. 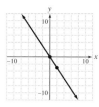   **39.** 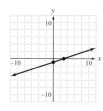   **41.** $m = 0.40$. The cellular call costs \$.40 per minute.

43. $m = 0.04$. Each second, 0.04 megabyte is downloaded.   **45.** $m = -5$. The temperature of the oven decreases 5° per minute.   **47.** $m = 40$. The average speed of the motorist is 40 mph.   **49.** $m = 0.28$. The tax rate is 28%.
51. $m = -0.05$. For each mile the car is driven, 0.05 gallon of fuel is used.

53a. The x-intercept is $\left(\frac{30}{7}, 0\right)$. This means that when the temperature is $\frac{30}{7}$°C, the number of chirps per minute is 0. In other words, the cricket no longer chirps.   **b.** The slope of 7 means that the number of chirps per minute increases by 7 chirps for every 1°C increase in the temperature.   **55a.** The intercept on the vertical axis is $(0, -100)$. This means that when the object was taken from the freezer, its temperature was -100°F. The intercept on the horizontal axis is $(5, 0)$. This means that 5 hours after the object was removed from the freezer, its temperature was 0°F.   **b.** The slope of 20 means that the temperature of the object increases 20° per hour.   **57.** $m = -6.5$. The slope of -6.5 means that the temperature is decreasing 6.5°C for each 1-kilometer increase in height above sea level.   **59.** $m = 50$. The slope of 50 means that the pigeon flies 50 mph.   **61.** 168 in.   **63.** -4 **65.** -7 **67.** Line A represents the depth of water for Can 1. Line B represents the depth of water for Can 2.   **69.** $y = -2x + 5$ **71.** $y = 5x - 7$ **73.** $y = -\frac{3}{2}x + 3$ **75.** $y = \frac{2}{5}x - 2$ **77.** $y = -\frac{1}{3}x + 2$

79. $y = \frac{6}{5}x - 2$ **81.**   **83.** 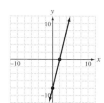   **85.**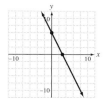

87. 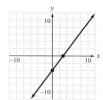   **89.** 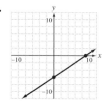   **91.** 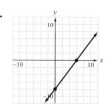   **93.**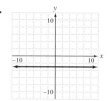

95. -6 **97.** From the three points, A, B, and C, create three pairs of points, A and B, A and C, and B and C. Determine whether the lines containing each pair of points have the same slope.   **a.** The points lie on the same line.   **b.** The points do not lie on the same line.   **99.** No; for example, $x = 2$.   **101.** 3   **103.** Answers will vary.

4.2 Exercises

1. The slope, m, and the y-intercept, $(0, b)$, can be read directly from the equation. **3.** m would be positive because as the age of the tree increases, the height of the tree increases. **5a.** Parallel lines have equal slopes.
b. The product of the slopes of perpendicular lines is -1; that is, their slopes are negative reciprocals.

7. $y = -2x - 1$ **9.** $y = -\frac{1}{4}x + 2$ **11.** $y = \frac{1}{6}x$ **13.** $y = 2x + 5$ **15.** $y = -\frac{5}{4}x + 5$ **17.** $y = 3x - 9$

19. $y = -3$ **21.** $x = 3$ **23.** $y = x + 2$ **25.** $y = \frac{3}{4}x$ **27.** $y = x - 1$ **29.** $y = -\frac{3}{2}x + 3$ **31.** $y = \frac{1}{3}x + \frac{10}{3}$

33. $y = -4$ **35.** $x = -2$ **37.** $y = 1200x$; 13,200 ft **39.** $y = 1000x + 5200$; 13,200 ft

41. $y = -3.5x + 100$; 69°C **43.** $y = 2.4x - 194.4$; 55.2 million **45.** $y = 0.017x + 1463$; 1506 m/s

47. $y = -20x + 230,000$; 60,000 trucks **49.** $y = -\frac{3}{5}x + 545$; 485 rooms

51a. $y = -0.0186x$ **b.** For every 1 g of sugar added, the freezing point decreases 0.0186°C. **c.** -0.93°C **53.** Yes

55. No **57.** Yes **59.** No **61.** $y = -3x + 7$ **63.** $y = \frac{2}{3}x - \frac{8}{3}$ **65.** No **67.** Yes **69.** $y = \frac{1}{3}x - \frac{1}{3}$

71. $y = -\frac{5}{3}x - \frac{14}{3}$ **73.** $y = -\frac{1}{9}x + \frac{82}{9}$ **75.** Yes; $y = -x + 6$ **77.** No **79.** 7 **81.** -1 **83.** $\frac{A_1}{B_1} = \frac{A_2}{B_2}$

85. Possible answers are $(0, 3)$, $(1, 2)$, and $(3, 0)$. **87.** Any equation of the form $y = 2x + b$, where $b \neq -13$, or of the form $y = -\frac{3}{2}x + c$, where $c \neq 8$. **89.** $x = 3$; 3° up

4.3 Exercises

1. Answers will vary. **3.** Answers will vary. For example, we can use the equation to project possible future outcomes. **5.** r would be positive because as the number of months increases, the weight increases.
7. r would be close to 0 because there is no correlation between height and history exam scores.
9a. Answers will vary. The equation of the line through the points $(95, 14)$ and $(99, 141)$ is $y = 31.75x - 3002.25$.
b. The slope of 31.75 means that the wolf population increased by approximately 32 wolves per year from 1995 to 2000. **11a.** Answers will vary. The regression equation is $y = 66.4857x - 129,626.\overline{6}$. **b.** The slope of 66.4857 means that electricity sales increased 66.4857 billion kilowatt hours per year from 1990 to 2000.
13a. $y = -0.1376741486 + 31.23316563$ **b.** 22 miles per gallon **c.** The slope is negative; as x increases, y decreases.
15a. $y = 0.3213184476x + 0.400318979$ **b.** 3.6 m/s **c.** With an increase of 1 cm in body length, an animal's running speed increases approximately 0.32 m/s. **d.** The y-intercept of approximately 0.4 represents the running speed of an animal of length 0 cm. **17a.** $y = 0.3325054x + 37.2985961$ **b.** $r \approx 0.99999$; The fit of the data to the regression line is very good. **c.** 73.9 million children **d.** The slope of approximately 0.333 means that the number of children in the United States is increasing at a rate of about 0.333 million per year. **e.** The y-intercept of approximately 37.299 means that in 1900 there were 37.299 million children in the United States. **19a.** $y = 1.551862378x + 1.986390721$
b. The slope of approximately 1.55 means that a state receives 1.55 electoral college votes per 1 million residents.
c. The y-intercept of approximately 1.986 means that a state with a population of 0 people would have 1.986 electoral college votes. **d.** The r value is not exactly 1 because the data are not completely linear. States cannot have a fractional part of a vote. The number of votes a state receives is rounded to a whole number. **21a.** ii **b.** iii **c.** iv **d.** i
23. $y = -2.5x + 9$ **25a.** $y = 34.142857x - 3234.4286$ **b.** The r^2 value would decrease because the data values would not as closely fit a straight line. **27.** The value of r is between -1 and 1; it cannot be greater than 1.
29a. $y = 0.36720042x + 315.62414$; $r \approx 0.6491$ **b.** $y = 1.1474246x - 66.269762$; $r \approx 0.6491$ **c.** No. The r values indicate that there is not a strong relationship between the two variables.

4.4 Exercises

1. No **3.** No. There are ordered pairs with the same first component and different second components.
5.   Yes **7.** 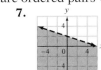 No **9.**   Yes **11.** 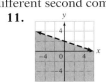 No

13. Yes **15.** No **17.** Yes **19.** Yes

21. No **23.** Yes **25.** 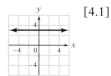 No **27.**   **29.**

31. $y < -\frac{2}{3}x + 2$

33. No. The solution sets do not intersect, which means that there are no ordered pairs that satisfy both inequalities.

CHAPTER 4 REVIEW EXERCISES

1. -1 [4.1]   **2.** Undefined [4.1]

3. [4.1] **4.** [4.1] **5.** [4.1] **6.** [4.1]

7. $y = -\frac{4}{3}x - 5$ [4.2]   **8.** $y = 4x - 2$ [4.2]   **9.** $y = -\frac{1}{2}x + 7$ [4.2]   **10.** $y = -4$ [4.2]   **11.** No [4.2]

12. $y = -3x + 7$ [4.2]   **13.** No [4.2]   **14.** $y = \frac{3}{2}x + 2$ [4.2]   **15.**  [4.4]

16. $y = 90x + 60,000$; $285,000 [4.2]   **17.** The slope is -0.72. The maximum recommended exercise heart rate decreases -0.72 beat per minute for every year older. [4.1]   **18a.** $y = 25x + 1000$ **b.** Water is being added to the pond at a rate of 25 gal/min. **c.** 10,000 gal   [4.2]   **19a.** It costs $.25 per minute to use the phone. **b.** The y-intercept is 19.95. When the phone is used for 0 minutes during the month, the phone bill is $19.95. [4.1]
20a. $y = 0.28928571x + 74.07142857$ **b.** 95.8°F [4.3]

CHAPTER 4 TEST

1. -2 [4.1]   **2.** 0 [4.1]

3. [4.1] **4.** [4.1] **5.** 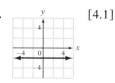 [4.1] **6.** [4.1]

7. $y = \frac{3}{4}x - 2$ [4.1]   **8.** [4.4]   **9.** $y = -\frac{5}{3}x - 4$ [4.2]   **10.** $y = \frac{1}{2}x - \frac{7}{2}$ [4.2]

11. $y = x + 1$ [4.2]   **12.** $y = -\frac{5}{4}x + 3$ [4.2]   **13.** Yes [4.2]   **14.** $y = 4x + 13$ [4.2]   **15.** Yes [4.2]

16. $y = -\frac{1}{3}x + 5$ [4.2]   **17a.** $y = 1500x + 18{,}500$ **b.** The slope of 1500 means that the profit is increasing by $1500 per month.   **c.** The profit in December will be $36,500. [4.2]   **18.** $m = 7$. The prices are increasing 7 cents per month. [4.1]   **19a.** $y = -20x + 2800$ **b.** The slope of -20 means the metal is cooling 20°F per minute.
c. 400°F [4.2]   **20a.** $y = 0.098x - 80.079$ **b.** The equation predicts that a car weighs 2700 lb has an engine that delivers approximately 186 hp. [4.3]

CUMULATIVE REVIEW EXERCISES

1. 8 [1.1]   **2.** 40 [1.1]   **3.** $E \cup F = \{-10, -5, 0, 5, 10, 15\}$; $E \cap F = \{0, 5, 10\}$ [1.2]   **4.** 17 [1.3]

5. 15 ft^3 [1.4]   **6.** $-6x - 3$ [1.5]   **7.** $20\left(\frac{1}{5}x\right)$; $4x$ [1.5]

8.

x	-2	-1	0	1	2
y	2	-1	-2	-1	2

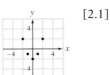 [2.1]

9. Domain: $\{-1, 0, 1\}$, Range: $\{-1, 0, 1, 2\}$, No [2.2]   **10.** -11 [2.2]   **11.** $(-1, 0)$, $(3, 0)$; $(0, -3)$ [2.3]
12. -3 [2.3]   **13.** 0 [3.1]   **14.** 2 [3.1]   **15.** 15 lb of the $6 coffee; 45 lb of the $4 coffee [3.2]
16. $m\angle a = 46°$, $m\angle b = 134°$ [3.3]   **17.** $\left(-\infty, \frac{1}{2}\right]$ [3.4]   **18.** $-2, \frac{4}{3}$ [3.5]

19. [4.1]   **20.** $y = \frac{3}{2}x + 2$ [4.2]

ANSWERS to Chapter 5 Exercises

PREP TEST

1. $6x + 5y$ [1.5]   **2.** 7 [1.3]   **3.** 0 [3.1]   **4.** -3 [3.1]   **5.** 1000 [3.1]
6. [4.1]   **7.** [4.1]   **8.** [4.4]

5.1 Exercises

1. Explanations will vary. For example, the solution is represented by an ordered pair (x, y).
3. Explanations will vary. For example, for a dependent system, the resulting equation is true; for an inconsistent system, the resulting equation is false.   **5.** $(2, 3)$ **7.** $(-1, 2)$ **9.** $(-1, 4)$ **11.** $\left(x, \frac{2}{5}x - 2\right)$

13. The system of equations has no solution.   **15.** $(-2.5, 3)$ **17.** $(2, 1)$ **19.** $(2, 1)$ **21.** $(3, -4)$

23. $\left(\frac{1}{2}, 3\right)$ **25.** $(-1, 2)$ **27.** $(0, 0)$ **29.** $(1, 5)$ **31.** $(x, -2x + 1)$ **33.** $.16   **35.** 30° and 60°

37. First powder: 200 mg; second powder: 450 mg   **39.** $6000 at 9.5%; $4000 at 7.5%   **41.** $\frac{3}{2}$ **43.** 4

45. $\left(6, \frac{3}{2}\right)$ **47.** $(-1, 2)$ **49.** 6

5.2 Exercises

1. The result of one of the steps will be an equation that is never true. **3.** A three-dimensional coordinate system is one formed by three mutually perpendicular axes. **5.** The planes intersect at one point. **7.** Between 10% and 20% apple juice **9.** 12 mph **11.** (6, 1) **13.** (2, 1) **15.** $\left(-\frac{1}{2}, 2\right)$ **17.** (−1, −2) **19.** $\left(\frac{1}{2}, \frac{3}{4}\right)$

21. $\left(\frac{2}{3}, -\frac{2}{3}\right)$ **23.** $\left(x, \frac{4}{3}x - 6\right)$ **25.** $\left(\frac{1}{3}, -1\right)$ **27.** No solution **29.** (2, 1, 3) **31.** (1, −1, 2)

33. (0, 2, 0) **35.** The system of equations has no solution. **37.** (6, −2, 2) **39.** (−2, 1, 1) **41.** (1, 4, 1)
43. (1, 1, 3) **45.** Plane: 250 mph; wind: 50 mph
47. Boat: 19 kilometers per hour; current: 3 kilometers per hour **49.** Boat: 16.5 mph; current: 1.5 mph

51. $33\frac{1}{3}\%$ **53.** 20 ml of the 3% hydrogen peroxide solution; 30 ml of the 12% hydrogen peroxide solution

55. 40 g of the 25% alloy; 80 g of pure gold **57.** $4000 **59.** $8000 at 9%, $6000 at 7%, $4000 at 5%
61. $14,000 at 12%; $10,000 at 8%; $9000 at 9% **63.** For the first diet, 100 g of food type I, 200 g of food type II, and 200 g of food type III. For the second diet, 200 g of food type I, 200 g of food type II, and 100 g of food type III.
65. $A = 3, B = -1$ **67.** 2 **69.** 6
71. The system of equations has a unique solution when $2A + 13 \neq 0$ or $A \neq -\frac{13}{2}$.

5.3 Exercises

1. A matrix is a rectangular array of numbers. **3.** 1. Interchange two rows. 2. Multiply a row by a constant.
3. Replace a row by the sum of that row and a nonzero multiple of another row.

5. b **7.** $\begin{bmatrix} 1 & 4 & -1 \\ 0 & 1 & -4 \end{bmatrix}$ **9.** $\begin{bmatrix} 1 & \frac{1}{2} & -\frac{1}{2} \\ 0 & 1 & 5 \end{bmatrix}$ **11.** $\begin{bmatrix} 1 & \frac{5}{2} & -2 \\ 0 & 1 & -\frac{16}{13} \end{bmatrix}$ **13.** $\begin{bmatrix} 1 & 2 & 2 & -1 \\ 0 & 1 & -\frac{7}{2} & \frac{1}{2} \\ 0 & 0 & 1 & -\frac{2}{11} \end{bmatrix}$

15. $\begin{bmatrix} 1 & -3 & \frac{1}{2} & -\frac{3}{2} \\ 0 & 1 & \frac{3}{2} & \frac{5}{2} \\ 0 & 0 & 1 & 3 \end{bmatrix}$ **17.** $\begin{bmatrix} 1 & -\frac{3}{2} & \frac{9}{4} & 1 \\ 0 & 1 & -\frac{47}{30} & -\frac{4}{15} \\ 0 & 0 & 1 & -\frac{17}{13} \end{bmatrix}$ **19.** (9, 1, −1) **21.** (2, −1) **23.** (2, 4)

25. (−2, 5) **27.** The system of equations has no solution. **29.** (−1, 4) **31.** (1, 3, −2) **33.** (2, 4, 1)

35. The system of equations has no solution. **37.** $\left(\frac{1}{2}, 0, \frac{2}{3}\right)$ **39.** $\left(\frac{2}{3}, -1, \frac{1}{2}\right)$ **41.** (2, −1, 3)

43. (0, 1, −2) **45.** $z = \frac{3}{2}x - \frac{5}{2}y + 1$ **47.** $y = -x^2 + 3x - 4$ **49.** $d_1 = 8$ in., $d_2 = 6$ in., $d_3 = 12$ in.
51a. Rabbits do not prey on hawks. **b.** Snakes prey on rabbits. **c.** A coyote is not prey for hawks, rabbits, snakes, or coyotes. **d.** A rabbit does not prey on hawks, rabbits, snakes, or coyotes. **53.** $A = 2, B = 3, C = -3$ **55.** (1, 1)
57a. A plane parallel to the yz-plane passing through $x = 3$. **b.** A plane parallel to the xy-plane passing through $y = 4$. **c.** A plane parallel to the xy-plane passing through $x = 2$. **d.** A plane perpendicular to the xy-plane along the line $y = x$ in the xy-plane.

5.4 Exercises

1. Graph each inequality. Then determine the intersection of the solution sets of the individual inequalities.
3. The objective function is the function that is to be maximized or minimized. **5.** b

7. **9.** **11.** **13.** **15.** **17.**

19. **21.** **23.** **25.** **27.** **29.**

31. The minimum is 16 at (0, 8).   **33.** The maximum is 71 at (6, 5).   **35.** The maximum is 70 at (0, 10).
37. The minimum is 18 at (2, 6).   **39.** 20 acres of wheat; 40 acres of barley   **41.** 0 starter sets; 18 pro sets
43. $y \geq -2$ **45.** $y > x$ **47.** 24 oz of food group B, 0 oz of food group A; \$2.40
     $x \geq 1$ $y < -x + 2$
49. Two 4-cylinder engines, seven 6-cylinder engines; \$2050

CHAPTER 5 REVIEW EXERCISES

1. $\left(6, -\frac{1}{2}\right)$ [5.1]   **2.** $(-4, 7)$ [5.2]   **3.** $(0, 3)$ [5.1]   **4.** $(x, 2x - 4)$ [5.1]   **5.** $\left(x, \frac{1}{3}x - 2\right)$ [5.2]

6. $(3, -1, -2)$ [5.2]   **7.** $\begin{bmatrix} 2 & -3 & -1 & | & 1 \\ 3 & 0 & -4 & | & -2 \\ 0 & 4 & -5 & | & 0 \end{bmatrix}$ [5.3]   **8.** $\begin{aligned} x + 3y &= -2 \\ 2x - y + z &= 0 \\ 3x + 2y - 5z &= 4 \end{aligned}$ [5.3]   **9.** $\begin{bmatrix} 1 & 2 & -1 & 3 \\ 0 & 1 & -2 & 13 \\ 0 & 0 & 1 & \frac{9}{5} \end{bmatrix}$ [5.3]

10. $(2, -1)$ [5.3]   **11.** $(3, -2)$ [5.3]   **12.** $(2, 3, -5)$ [5.3]   **13.** $(-1, -3, 4)$ [5.3]
14. The minimum is 8 at (0, 8). [5.4]   **15.** The maximum is 18 at (4, 5). [5.4]
16. [5.4]   **17.** [5.4]   **18.** Cabin cruiser: 16 mph; current: 4 mph [5.2]

19. 100 children [5.2]   **20.** $\frac{680}{7}$ acres [5.4]

CHAPTER 5 TEST

1. $\left(\frac{3}{4}, \frac{7}{8}\right)$ [5.1]   **2.** The system of equations has no solution. [5.2]   **3.** $(2, 1)$ [5.1]   **4.** $(-3, -4)$ [5.1]

5. $\left(x, \frac{1}{3}x - 1\right)$ [5.2]   **6.** $(0, -2, 3)$ [5.2]   **7.** $\begin{bmatrix} 3 & -1 & 2 & | & 4 \\ 1 & 4 & 0 & | & -1 \\ 0 & 5 & -1 & | & 3 \end{bmatrix}$ [5.3]   **8.** $\begin{aligned} 2x - y + 3z &= 4 \\ x + 5y - 2z &= 6 \\ -3x - 4z &= 1 \end{aligned}$ [5.3]

9. $\begin{bmatrix} 1 & \frac{3}{2} & -1 \\ 0 & 1 & -1 \end{bmatrix}$ [5.3]   **10.** $(-2, 1)$ [5.3]   **11.** $\left(-\frac{1}{3}, -\frac{10}{3}\right)$ [5.3]   **12.** $(2, -1, -2)$ [5.3]

13. $(1, -6, 3)$ [5.3]   **14.** The maximum is 44 at (6, 4). [5.4]   **15.** The minimum is 12 at (2, 3). [5.4]
16. [5.4]   **17.** [5.4]   **18.** Plane: 150 mph; wind: 25 mph [5.2]

19. Cotton: \$9; wool: \$14 [5.2]   **20.** 4 standard gloves, 0 professional gloves [5.4]

1. 0 [1.1] **2.** 17 votes [1.1] **3.** $4 = 2^2$, $9 = 3^2$, $16 = 4^2$, $25 = 5^2$. Answers may vary; for example, each number is the square of the number of odd consecutive integers added. $1 + 3 + 5 + 7 + 9 + 11 = 36 = 6^2$. [1.1]
4. Deductive reasoning [1.1] **5.** $\{-4, -3, 0, 5, 12, 21\}$ [2.2] **6.** 0, 2 [2.3] **7.** $(-4, 0), (-2, 0), (1, 0), (0, -8)$ [2.3]
8.

s	1500	1600	1700	1800	1900	2000	2100
F	6000	6350	6700	7050	7400	7750	8100

[2.1]

9. -13 [3.1] **10.** $[-6, \infty)$ [3.4] **11.** $\{x \mid 3 < x < 6\}$ [3.4] **12.** 1, 3 [3.5] **13.** $y = -x + 1$ [4.2]

14. $(2, -3)$ [5.1] **15.** $(-2, -1)$ [5.2] **16.** $(2, 0)$ [5.1] **17.** $\begin{bmatrix} 1 & -1 & 1 & | & 1 \\ 0 & 1 & -3 & | & 5 \\ 0 & 0 & 1 & | & -2 \end{bmatrix}$ [5.3]

18. $(1, -1, 2)$ [5.3] **19a.** $y = -0.357142857x + 766.71429$ **b.** 45% **c.** The slope is negative; as x increases, y decreases. [4.3] **20.** Ship, 25 mph; current, 5 mph [5.2]

ANSWERS to Chapter 6 Exercises

PREP TEST

1. $-12y$ [1.5] **2.** -8 [1.3] **3.** $3a - 8b$ [1.5] **4.** $11x - 2y - 2$ [1.5] **5.** $-x + y$ [1.5] **6.** $2 \cdot 2 \cdot 2 \cdot 5$ [1.4]
7. 4 [1.4] **8.** -13 [1.3] **9.** $-\frac{1}{3}$ [3.1]

6.1 Exercises

1. a. This is a monomial because it is the product of a number, 32, and variables, a and b.

b. This is a monomial because it is the product of a number, $\frac{5}{7}$, and a variable, n.

c. This is not a monomial because there is a variable in the denominator.

3. a. The exponent on 6 is positive; it should not be moved to the denominator. $6x^{-3} = \dfrac{6}{x^3}$

b. The exponent on x is positive; it should not be moved to the denominator. $xy^{-2} = \dfrac{x}{y^2}$

c. The exponent on 8 is positive; it should not be moved to the numerator. $\dfrac{1}{8a^{-4}} = \dfrac{a^4}{8}$

d. The exponent on c is positive; it should not be moved to the numerator. $\dfrac{1}{b^{-5}c} = \dfrac{b^5}{c}$

5. a^9 **7.** z^8 **9.** x^{15} **11.** $x^{12}y^{18}$ **13.** $17s^4t^3$ **15.** 1 **17.** a^6 **19.** $-m^9n^3$ **21.** $24x^7$ **23.** $-8a^6$
25. $4p^4q^5$ **27.** $-24r^3$ **29.** $8a^9b^3c^6$ **31.** mn^2 **33.** 1 **35.** $-\frac{2a}{3}$ **37.** $-27m^2n^4p^3$ **39.** $-15x^3y^8$
41. $\frac{1}{x^5}$ **43.** $54n^{14}$ **45.** $\frac{7xz}{8y^3}$ **47.** $-8x^{13}y^{14}$ **49.** $\frac{1}{w^8}$ **51.** a^5 **53.** $\frac{1}{64}$ **55.** 243 **57.** $\frac{4}{x^7}$ **59.** $\frac{2}{x^2y^4}$
61. 1 **63.** $\frac{1}{x^5}$ **65.** $\frac{9}{x^4}$ **67.** $\frac{x^2}{3}$ **69.** $\frac{x^6}{y^{12}}$ **71.** $\frac{9}{x^2y^4}$ **73.** $\frac{2}{x^4}$ **75.** $\frac{1}{2x^3}$ **77.** $\frac{1}{2x^2y^6}$ **79.** $\frac{1}{x^6}$ **81.** $\frac{y^8}{x^4}$
83. $\frac{b^{10}}{4a^{10}}$ **85.** $11xy$ **87.** $15a^2b$ **89.** $64x^4y^2$ m² **91.** $50c^3d^4$ mi **93.** $54m^2n^4$ km² **95.** $6a^2b^3$ yd
97. $3a^2$ **99.** 2.37×10^6 **101.** 4.5×10^{-4} **103.** 3.09×10^5 **105.** 6.01×10^{-7} **107.** 5.7×10^{10}
109. 1.7×10^{-8} **111.** 710,000 **113.** 0.000043 **115.** 671,000,000 **117.** 0.00000713
119. 5,000,000,000,000 **121.** 0.00801 **123.** 1.6×10^{10} **125.** $\$1.317 \times 10^8$ **127.** 1.6×10^{-19}
129. 6.023×10^{23} **131.** 3.086×10^{18} **133.** 6.65×10^{19} **135.** 3.22×10^{-14} **137.** 3.6×10^5
139. 1.8×10^{-18} **141a.** $\frac{3}{64}$ **b.** $\frac{4}{81}$ **143.** $\frac{1}{4}, \frac{1}{2}, 1, 2, 4; 4, 2, 1, \frac{1}{2}, \frac{1}{4}$ **145.** a

6.2 Exercises

1a. This is a binomial. It contains two terms, $8x^4$ and $-6x^2$ **b.** This is a trinomial. It contains three terms, $4a^2b^2$, $9ab$, and 10. **c.** This is a monomial. It is one term, $7x^3y^4$. (*Note:* It is a product of a number and variables. There is no addition or subtraction operation in the expression.) **3a.** Yes. Both $\frac{1}{5}x^3$ and $\frac{1}{2}x$ are monomials. (*Note:* The coefficients of variables can be fractions.) **b.** No. A polynomial does not have a variable in the denominator of a fraction. **c.** Yes. Both x and $\sqrt{5}$ are monomials. (*Note:* The variable is not under a radical sign.)
5. 67.02 ft^3 **7.** 576.99 m **9.** 50 ft **11.** 191.7 lb **13.** $18{,}819 \text{ foot-pounds}$ **15.** $-2x^2 + 3x$ **17.** $4x$
19. $7b^2 + b - 4$ **21.** $3y^2 - 4y - 2$ **23.** $3a^2 - 3a + 17$ **25.** $-7x - 7$ **27.** $-2x^3 + x^2 + 2$
29. $x^3 + 2x^2 - 6x - 6$ **31.** $5x^3 + 10x^2 - x - 4$ **33.** $y^3 - y^2 + 6y - 6$ **35.** $4y^3 - 2y^2 + 2y - 4$
37. $11x^2 + 2x + 4$ **39.** $5a^2 + 3a - 9$ **41.** $(8d^2 + 12d + 4) \text{ km}$ **43.** $(-2n^2 + 160n - 1200) \text{ dollars}$
45. $-4x^2 + 6x + 3$ **47.** $4x^2 - 6x + 3$ **49a.** Sometimes true **b.** Always true **c.** Sometimes true
51a. $k = 8$ **b.** $k = -4$

6.3 Exercises

1. The FOIL method is used to multiply two binomials. **3.** $x + 1$ and $x^2 - x + 1$
5a. Sometimes true **b.** Never true **c.** Always true **7.** The degree of the first term of the quotient is one degree less than the degree of the first term of the dividend. **9.** $-5b^3 - 7b^2 + 35b$ **11.** $-4y^6 - 6y^4 + 7y^3$
13. $-2b^3 + 7b^2 + 19b - 20$ **15.** $x^4 - 4x^3 - 3x^2 + 14x - 8$ **17.** $y^4 + 4y^3 + y^2 - 5y + 2$ **19.** $x^2 + 4x + 3$
21. $a^2 + a - 12$ **23.** $y^2 - 10y + 21$ **25.** $2x^2 + 15x + 7$ **27.** $3x^2 + 11x - 4$ **29.** $4x^2 - 31x + 21$
31. $3y^2 - 2y - 16$ **33.** $21a^2 - 83a + 80$ **35.** $15b^2 + 47b - 78$ **37.** $2a^2 + 7ab + 3b^2$ **39.** $6a^2 + ab - 2b^2$
41. $d^2 - 36$ **43.** $4x^2 - 9$ **45.** $x^2 + 2x + 1$ **47.** $9a^2 - 30a + 25$ **49.** $x + 1$ **51.** $2a - 5$ **53.** $3a + 2$
55. $4b^2 - 3$ **57.** $x - 2$ **59.** $-x + 2$ **61.** $x^2 + 3x - 5$ **63.** $x^4 - 3x^2 - 1$ **65.** $xy + 2$ **67.** $b - 7$
69. $2x + 1$ **71.** $x + 1 + \dfrac{2}{x - 1}$ **73.** $2x - 1 - \dfrac{2}{3x - 2}$ **75.** $a + 3 + \dfrac{4}{a + 2}$ **77.** $y - 6 + \dfrac{26}{2y + 3}$
79. $2x + 5 + \dfrac{8}{2x - 1}$ **81.** $x^2 + 2x + 3$ **83.** $x^2 - 3$ **85.** $(4x^2 + 4x + 1)\text{m}^2$ **87.** $(10x^2 - 35x) \text{ mi}^2$
89. $(\pi x^2 + 8\pi x + 16\pi) \text{ in}^2$ **91.** $(64x^3 + 48x^2 + 12x + 1) \text{ in}^3$ **93.** $(4x^2 + 10x) \text{ m}^2$ **95.** $(18x^2 + 12x + 2) \text{ in}^2$
97. $(3x + 2) \text{ ft}$ **99.** $(2x^2 + x - 1) \text{ in.}$ **101.** $(90x + 2025) \text{ ft}^2$ **103.** $(4x^3 - 160x^2 + 1600x) \text{ in}^3$; 3468 in^3
105. $x^3 + 5x^2 + 5x - 3$ **107.** $x^2 + 7x + 1$ **109.** $x^3 - 7x^2 - 7$ **111.** $4x^2 - 6x + 3$
113. The quotient is $x - 4$, and the graph is $y = x - 4$, except when $x = 3$. **115.** $x^2 - 5x + 6$ **117.** $3x^2 + 4x - 2$
119. $x^2 + 7x + 16 + \dfrac{53}{x - 3}$ **121.** $x^2 + 2x + 6 + \dfrac{17}{x - 2}$ **123.** $4x + 8 + \dfrac{8}{x - 2}$ **125.** $-2x^3 + 3x^2 + 4x - 1$
127. $4x^3 - x + 2 - \dfrac{4}{x + 3}$ **129.** $x^3 - 5x^2 + 10x - 20 + \dfrac{10}{x + 2}$ **131.** $\pi(x^2 + 6x + 9) \text{ cm}^2$ **133.** $x - 2$
135. -58 **137.** 4 **139.** 22 **141.** 525 **143.** 157 **145.** 11 **147.** No **149.** Yes **151.** No
153. No **155a.** Always true **b.** Always true **157.** $(4x^3 - 80x^2 + 400x) \text{ in}^3$; No; Explanations will vary.
159. -2 **161.** 0 **163.** -9 **165.** $(3, -7)$ **167.** $x + 2$

6.4 Exercises

1. A factor is a number or expression in a multiplication. To factor means to write a polynomial as a product of other polynomials. **3a.** The common factor is 3. **b.** $3x^2 + 15$ is a sum; it is not a product. It can be factored as $3(x^2 + 5)$.
c. In $x(2x - 1)$, x is multiplied times $2x - 1$. Both x and $2x - 1$ are factors. **d.** It cannot be expressed as the product of two binomials. **e.** $(2x + 3)(x - 5)$ is a product, both $2x + 3$ and $x - 5$ are nonfactorable, and when the two binomials are multiplied, the result is the trinomial. **5.** When the constant term of the trinomial is positive, the constant terms of the binomials have the same sign. When the constant term of the trinomial is negative, the constant terms of the binomials have different signs.
7. $y(12y - 5)$ **9.** $5xyz(2xz + 3y^2)$ **11.** $5(x^2 - 3x + 7)$ **13.** $3y^2(y^2 - 3y - 2)$ **15.** $x^2y^2(x^2y^2 - 3xy + 6)$
17. $8x^2y(2 - xy^3 - 6y)$ **19.** $(x - 4)(x^2 - 3)$ **21.** $(y - 4)(3y^2 + 1)$ **23.** $(y + 6)(z - 3)$ **25.** $(y + 4)(x^2 + 3)$
27. $(t + 4)(t - s)$ **29.** $(2a - 3)(5ab - 2)$ **31.** $(x - 1)(x + 2)$ **33.** $(a + 4)(a - 3)$ **35.** $(a - 1)(a - 2)$
37. $(b + 8)(b - 1)$ **39.** $(z + 5)(z - 9)$ **41.** $(z - 5)(z - 9)$ **43.** $(b + 4)(b + 5)$
45. Nonfactorable over the integers **47.** $(x - 7)(x - 8)$ **49.** $(y + 3)(2y + 1)$ **51.** $(a - 1)(3a - 1)$
53. $(x - 3)(2x + 1)$ **55.** $(2t + 1)(5t + 3)$ **57.** $(2z - 1)(5z + 4)$ **59.** Nonfactorable over the integers
61. $(t + 2)(2t - 5)$ **63.** $(3y + 1)(4y + 5)$ **65.** $(a - 5)(11a + 1)$ **67.** $(2b - 3)(3b - 2)$ **69.** $2r^2(4 - \pi)$
71. $r^2(\pi - 2)$ **73.** $(3x + 2) \text{ mi by } (x + 5) \text{ mi}$ **75.** $(4x + 3) \text{ ft by } (x + 5) \text{ ft}$ **77.** P doubles

79a. $-7, 7, -5, 5$ **b.** $-5, 5, -1, 1$ **c.** $-7, 7, -5, 5$ **d.** $-5, 5, -1, 1$ **e.** $-11, 11, -7, 7$ **f.** $-9, 9, -3, 3$
81. An infinite number of different values of k are possible. Explanations will vary.
83. No. The leading term of the product of these four factors will be x^4, not x^3.

6.5 Exercises

1a. Never true **b.** Always true **3.** Examples will vary. **5.** Examples will vary. **7.** $(a + 7)(a - 7)$
9. $(3x + 1)(3x - 1)$ **11.** $(1 + 8x)(1 - 8x)$ **13.** Nonfactorable over the integers **15.** $(8 + xy)(8 - xy)$
17. $(b^n + 4)(b^n - 4)$ **19.** $(y - 3)^2$ **21.** $(7x + 2)^2$ **23.** Nonfactorable over the integers **25.** $(2xy + 3)^2$
27. $(2a - 9b)^2$ **29.** $(y^n - 8)^2$ **31.** $(x - 3)(x^2 + 3x + 9)$ **33.** $(4a + 3)(16a^2 - 12a + 9)$
35. $(x - 2y)(x^2 + 2xy + 4y^2)$ **37.** $(xy + 4)(x^2y^2 - 4xy + 16)$ **39.** Nonfactorable over the integers
41. $(a^n + 4)(a^{2n} - 4a^n + 16)$ **43.** $(xy + 3)(xy - 11)$ **45.** $(y^2 + 2)(y^2 - 8)$ **47.** $(a^2b^2 + 13)(a^2b^2 - 2)$
49. $(a^n + 3)(a^n - 4)$ **51.** $(5xy - 4)(xy - 11)$ **53.** $(3x^2 + 8)(x^2 + 4)$ **55.** $(x^n + 3)^2$ **57.** $3(x + 2)(x + 3)$
59. $a(b + 8)(b - 1)$ **61.** $2x^2(2x - 5)^2$ **63.** $2y^2(y + 3)(y - 16)$ **65.** $(x + 1)(x - 1)(x - 2)$
67. Nonfactorable over the integers **69.** $x(x - 5)(2x - 1)$ **71.** $5(t + 2)(2t - 5)$ **73.** $p(2p + 1)(3p + 1)$
75. $(2x + 3)(2x - 3)(y + 1)(y - 1)$ **77.** $2(5 + x)(5 - x)$ **79.** $b^2(b + a)(b - a)$ **81.** $(2x + 3)$ cm
83. $(2x - 3)$ by $(2x + 3)$ **85.** $(3a + 5)$ by $(3a - 5)$ **87.** $4y$ in. by $(x + 5)$ in. by $(x + 3)$ in.
89. $(x - 1)(x^2 + x + 1)(a + b)$ **91.** $(4x + 3)$ ft by $(4x + 3)$ ft; yes; $x > -\frac{3}{4}$

6.6 Exercises

1. A quadratic equation is an equation of the form $ax^2 + bx + c = 0, a \neq 0$. A linear equation is an equation of the form $ax + b = 0, a \neq 0$. A quadratic equation has a term of degree 2. In a linear equation, the highest exponent on a variable is 1. Examples will vary. $3x^2 - 4x + 5 = 0$ is an example of a quadratic equation. $6x + 1 = 0$ is an example of a linear equation. **3.** The Principle of Zero Products is based on the Multiplication Property of Zero, which states that the product of a number and zero is zero. The Principle of Zero Products begins with the conclusion of the Multiplication Property of Zero, stating that if the product of two or more factors is zero, then at least one of the factors must be equal to zero. **5.** Yes, because it is the product of two factors, $x + 5$ and $x - 6$, and their product is zero. It can be solved by setting each factor equal to zero and solving the resulting equations for x. **7.** No **9.** $x^2 - 5x + 6 = 0$
11. $2x^2 + 3x - 5 = 0$ **13.** No **15.** Yes **17.** Yes **19.** $2, 5$ **21.** $-8, 0$ **23.** $0, \frac{2}{3}$ **25.** $-9, \frac{1}{4}$
27. $-\frac{4}{3}, \frac{4}{3}$ **29.** $-5, 1$ **31.** $0, 9$ **33.** $0, 4$ **35.** $-4, 7$ **37.** $-5, \frac{2}{3}$ **39.** $\frac{2}{5}, 3$ **41.** $-\frac{1}{4}, \frac{3}{2}$ **43.** $-5, 7$
45. $3, 9$ **47.** $-\frac{4}{3}, 2$ **49.** $-5, \frac{4}{3}$ **51.** $-4, 8$ **53.** $2, 8$ **55.** $-4, 1, 4$ **57.** $-2, -\frac{1}{2}, 2$ **59.** $-\frac{1}{2}, \frac{1}{2}, \frac{2}{3}$
61. 10 s **63.** 10 teams **65.** 8 sides **67.** 6 s **69.** 12.5 s **71.** 175 or 15
73. 2, 4, and 6 or $-2, -4$, and -6

CHAPTER 6 REVIEW EXERCISES

1. $24a^2b^3c$ [6.1] **2.** $3x^3 + 6x^2 - 8x + 3$ [6.2] **3.** $-20x^3y^5$ [6.1] **4.** $-4x^6$ [6.1] **5.** $(2a - 3)^2$ [6.5]
6. $5(x^2 - 9x - 3)$ [6.4] **7.** $\frac{6b}{a}$ [6.1] **8.** $4x^4 - 2x^2 + 5$ [6.3] **9.** $(a - 3)(a - 16)$ [6.4] **10.** $x(x + 5)(x - 3)$ [6.5]
11. $6y^4 - 9y^3 + 18y^2$ [6.3] **12.** $4x^2 - 20x + 25$ [6.3] **13.** $(2x + 1)(3x + 8)$ [6.4] **14.** $2(b + 4)(b - 4)$ [6.5]
15. $\frac{9a^4}{b^6}$ [6.1] **16.** 2.9×10^{-6} [6.1] **17.** $(b + 6)(a - 3)$ [6.4] **18.** $16y^2 - 9$ [6.3] **19.** $(4x + 5)(4x - 5)$ [6.5]
20. $10a^3 - 39a^2 + 20a - 21$ [6.3] **21.** Nonfactorable over the integers [6.4] **22.** $-7, 2$ [6.6]
23. $-\frac{1}{4a^2b^5}$ [6.1] **24.** $4x + 8 + \frac{21}{2x - 3}$ [6.3] **25.** 0.000000035 [6.1] **26.** $(3p - 2)(9p^2 + 6p + 4)$ [6.5]
27. $2y^2(y + 1)(y - 8)$ [6.5] **28.** $-4, -\frac{1}{2}$ [6.6] **29.** $5y^2 - 10y$ [6.2] **30.** $(4x^2 + 12x + 9)$ m^2 [6.3]
31. $81,030,000,000,000,000,000$ [6.1] **32.** 56 ft [6.2] **33.** $(4x + 1)$ in. by $(x + 3)$ in. [6.4]

CHAPTER 6 TEST

1. $21y^2 + 4y - 1$ [6.2] **2.** $-18a^8b^6$ [6.1] **3.** $12x^5 + 8x^3 - 28x^2$ [6.3] **4.** $-\frac{2y^3}{x^3}$ [6.1] **5.** $(3a - 5)^2$ [6.5]
6. $2x(3x^2 - 4x + 5)$ [6.4] **7.** $-\frac{10a}{b}$ [6.1] **8.** $4b^4 + 12b^2 - 1$ [6.3] **9.** $(a + 3)(a - 12)$ [6.4]
10. $3x(2x + 5)(2x - 3)$ [6.5] **11.** $-\frac{8a^{12}}{b^{15}}$ [6.1] **12.** 7.8×10^{10} [6.1] **13.** $(a + 2b)(2x - 3y)$ [6.4]
14. $36y^2 - 25$ [6.3] **15.** $(9x + 1)(9x - 1)$ [6.5] **16.** $6a^3 + 17a^2 - 2a - 21$ [6.3] **17.** $-\frac{1}{3}, \frac{1}{2}$ [6.6]

A20

18. $(4a - b)(16a^2 + 4ab + b^2)$ [6.5]   **19.** $2x^3 - 4x^2 + 6x - 14$ [6.2]   **20.** $x^2 - 2x - 1 + \dfrac{2}{x - 3}$ [6.3]
21. $(6y^4 - 5)(6y^4 - 1)$ [6.5]   **22.** 3, 5 [6.6]   **23.** $(10x^2 - 29x - 21)$ cm² [6.3]
24. 240,000,000,000,000 [6.1]   **25.** 12 sides [6.6]

CUMULATIVE REVIEW EXERCISES

1. $\dfrac{1}{2}$ [1.1]   **2.** -7 [1.3]   **3.** The Associative Property of Addition [1.5]   **4.** $18x^2$ [1.5]
5. Domain: $\{-5, -3, -1, 1, 3\}$; Range: $\{-4, -2, 0, 2, 4\}$; Yes [2.2]   **6.** $\{-11, -7, -3, 1, 5\}$ [2.2]   **7.** 5 [3.1]
8. $(-\infty, -3]$ [3.4]   **9.** [4.1]   **10.** [4.1]   **11.** 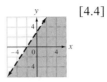 [4.4]
12. $y = \dfrac{2}{5}x + 4$ [4.2]   **13.** $(1, 2)$ [5.2]   **14.** $-\dfrac{8y^6}{x^{12}}$ [6.1]   **15.** $3y^3 - 7y^2 + 8y - 7$ [6.2]   **16.** $4x + 8 + \dfrac{21}{2x - 3}$ [6.3]
17. $(5a + 6b)(5a - 6b)$ [6.5]   **18.** $-5, \dfrac{4}{3}$ [6.6]   **19.** 40 oz [3.2]
20. $m = 50$. A slope of 50 means the average speed was 50 mph. [4.1]

ANSWERS to Chapter 7 Exercises

PREP TEST

1. 50 [1.4]   **2.** $-\dfrac{1}{6}$ [1.4]   **3.** $-\dfrac{3}{2}$ [1.4]   **4.** $\dfrac{1}{24}$ [1.4]   **5.** $\dfrac{5}{24}$ [1.4]   **6.** $\dfrac{1}{3}$ [1.4]   **7.** -2 [3.1]
8. $\dfrac{10}{7}$ [3.1]   **9.** 110 mph, 130 mph [3.2]

7.1 Exercises

1. A rational expression is one in which the numerator and denominator are polynomials. Examples will vary.
3. Explanations will vary. For example: Because $x^2 \geq 0$, $x^2 + 5$ is positive, and there are no values of x for which the
denominator is equal to zero.   **5.** Explanations will vary. For example: Factor the numerator and denominator,
divide by the common factors, and then write the answer in simplest form.   **7.** -2 **9.** $\dfrac{1}{9}$ **11.** $\dfrac{1}{35}$ **13.** $\dfrac{3}{4}$
15. $\{x \mid x \neq -4\}$ **17.** $\{x \mid x \neq -3\}$ **19.** $\{x \mid x \neq 0\}$ **21.** $\{x \mid x \in \text{real numbers}\}$ **23.** $\{x \mid x \neq -5, 1\}$
25. $\{x \mid x \neq -5, 0, \dfrac{1}{2}\}$ **27a.** First table: $-5, -11, -23, -119, -5999$; second table: 7, 13, 25, 121, 6001
b. It decreases. **c.** It increases.   **29.** decreases   **31.** Discontinuous   **33.** Continuous   **35.** $x = 3$
37. $x = -\dfrac{5}{2}, x = 2$ **39.** $x = -2, x = 4$ **41.** $x = -3, x = 2$ **43.** No vertical asymptotes   **45.** $x = \dfrac{2}{3}, x = \dfrac{3}{2}$
47a. \$1333.33 **b.** \$8000 **c.** The cost increases. **d.** The cost to remove 50% of the salt in a tank of sea water is
\$2000. **e.** When $p = 100$, the denominator of the rational function is 0, and division by zero is undefined.
49a. 1136 cars per hour **b.** No **c.** The flow of traffic is 1125 cars per hour when the speed of the cars is 45 mph.
51. $3x - 1$ **53.** $-\dfrac{2}{x}$ **55.** $\dfrac{x}{2}$ **57.** $\dfrac{x^n - 3}{4}$ **59.** $\dfrac{x - 3}{x - 5}$ **61.** $\dfrac{x + y}{x - y}$ **63.** $-\dfrac{x + 4}{4 - x}$ **65.** Simplest form
67. $\dfrac{x^2 + 2}{(x + 1)(x - 1)}$ **69.** $\dfrac{xy + 7}{xy - 7}$ **71.** $\dfrac{a^n - 2}{a^n + 2}$ **73.** Answers will vary. For example, $\dfrac{x}{x^2 - 4x - 12}$ **75.** 5
77. No. Explanations may vary. For example: Because $g(a) = h(a) = 0$, $x - a$ is a factor of both $g(x)$ and $h(x)$; therefore,
$F(x)$ is not in simplest form.

7.2 Exercises

1. Explanations may vary. For example: Multiply the numerators and denominators, divide by the common factors, and write the answer in simplest form. **3.** Answers will vary. **5.** The goal is to rewrite the expression with no fractions in the numerator or denominator. **7.** $\frac{abx}{2}$ **9.** $\frac{x}{2}$ **11.** $\frac{y(x-1)}{x^2(x+1)}$ **13.** $-\frac{x+5}{x-2}$ **15.** 1

17. $\frac{x^n+4}{x^n-1}$ **19.** $\frac{4by}{3ax}$ **21.** $\frac{x+2}{x+5}$ **23.** $(x^n+1)^2$ **25.** $\frac{1}{x-1}$ **27.** $\frac{15-16xy-9x}{10x^2y}$ **29.** $-\frac{6x+19}{36x}$

31. $-\frac{x^2+x}{(x-3)(x-5)}$ **33.** $\frac{5xy+8}{5y}$ **35.** $\frac{8(x-5)}{x(x-4)}$ **37.** $\frac{x+2}{x+3}$ **39.** $\frac{8}{15}\neq\frac{1}{8}$ **41.** $\frac{2x+1}{x(x+1)}$ **43.** $\frac{5}{b}+\frac{8}{a}$

45. $\frac{4x-12}{18x^2-45x}$ **47.** $\frac{x}{x-1}$ **49.** $-\frac{a-1}{a+1}$ **51.** $\frac{2}{5}$ **53.** $\frac{x^2-x-1}{x^2+x+1}$ **55.** $\frac{x+2}{x-1}$ **57.** $\frac{x-2}{x+2}$

59. $\frac{3n^2+12n+8}{n(n+2)(n+4)}$ **61a.** $\frac{2ab}{a+b}$ **b.** 12 **63a.** $P(x)=\frac{Cx(x+1)^{60}}{(x+1)^{60}-1}$ **b.**

Monthly payment (in dollars) / *Monthly interest rate (as a decimal)*

c. 0% to 22.8%

d. The ordered pair (0.006, 198.96) means that when the monthly interest rate on a car loan is 0.6%, the monthly payment on the loan is $198.96. **e.** $203 **65.** $\frac{8m}{3(m+3)}$ **67.** $\frac{4}{3(a-2)}$ **69.** $\frac{y+x}{y-x}$

71. Answers will vary. **73a.** $f(x)=\frac{2x^3+528}{x}$ **b.**

Surface area (in square inches) / *Height of box (in inches)*

c. When the height of the box is 4 in., 164 in² of cardboard will be needed. **d.** 5.1 in. **e.** 155.5 in²

7.3 Exercises

1. A rational equation is an equation that contains fractions. Examples will vary. **3.** $\frac{x}{20}$ of the lawn
5. $\frac{t}{10}+\frac{t}{15}=1$ **7.** -1 **9.** $\frac{5}{2}$ **11.** $-3, 3$ **13.** 8 **15.** 5 **17.** No solution **19.** 4, 10 **21.** $-3, 4$
23. $-3, 1$ **25.** $-\frac{2}{3}, 6$ **27.** 2 **29.** $-2, 2$ **31.** $\frac{1}{4}, 8$ **33.** 5 : 4 **35.** 24 lb **37.** 12 h **39.** 1.8 h
41. 10 h **43.** 20 h **45.** 45 h **47.** 5 h **49.** 60 min **51.** 4.8 h **53.** Smaller unit: 12 h; larger unit: 4 h
55. Newer sorter: 14 min; older sorter: 35 min **57.** Commercial: 540 mph; corporate: 420 mph **59.** 8 mph
61. 288 mph **63.** 360 mph **65.** 12 mph **67.** 2 mph **69.** 105 mph **71.** 2 mph **73.** 12 mph
75. 16 mph **77.** $5\frac{1}{3}$ min **79.** 50 mph **81.** $t=\frac{AB}{A+B}$ **83.** $A=3, B=-1$

7.4 Exercises

1. Answers may vary. For example: **a.** A rate is the quotient of two quantities that have different units. **b.** A ratio is the quotient of two quantities that have the same unit. **c.** A proportion is the equality of two rates or ratios.
3. Answers will vary. For example, the second and third terms are the means. The first and fourth terms are the extremes. **5a.** Always true **b.** Sometimes true **c.** Always true **d.** Always true **e.** Sometimes true
7. Answers will vary. For example, the conversion factor is equal to 1, and multiplying an expression by 1 does not change its value. **9.** 9 **11.** 7.5 **13.** 8 **15.** 5 **17.** 7 **19.** -1 **21.** $-6, 2$ **23.** $-\frac{2}{3}, 5$
25. 17,500 homes **27.** $1202.50 **29.** 800 fish **31a.** 398,200 divorces **b.** 1980 **c.** 437,500 divorces
d. 65% **e.** 11% **33a.** 13,253 crashes **b.** 35- to 44-year-olds; 2811 more **c.** Greater than **d.** Answers will vary.
35. 12.8 in. **37.** 4.7 ft **39.** 18 m **41.** 48 cm² **43.** 15 m **45.** 10 ft **47.** 12 m **49.** 36 m
51. $2\frac{1}{3}$ ft **53.** 12 qt **55.** 54 in. **57.** $1\frac{3}{4}$ lb **59.** $\frac{1}{2}$ pt **61.** 30 mph **63.** $\frac{1}{20}$ gal/s

65. 1,103,760,000 s   **67.** $10\frac{5}{12}$ lb   **69.** 750 mph   **71.** 18 pleats   **73.** $.46   **75.** $234.38
77. Answers will vary. For example, the population of the United States was greater in 1960.   **79.** $36.80
81a. 52,173,913 skier days   **b.** Answers will vary.   **c.** Answers will vary.

7.5 Exercises

1a. When an increase in one quantity leads to a proportional increase in the other quantity.   **b.** When an increase in one quantity leads to a proportional decrease in the other quantity.   **3.** Answers will vary.   **5a.** True   **b.** False
c. False   **7.** 70   **9.** 16   **11.** 1.25   **13.** 10   **15.** 154   **17.** 4   **19.** 20 lb   **21.** 975 words   **23.** 2.7 h
25. 800 ft   **27.** 975 items   **29.** 52 mph   **31.** $66.\overline{6}$ psi   **33.** 2160 computers   **35.** 112.5 lb
37. 18.75 rpm   **39.** 2.25 s   **41a.** Never true   **b.** Always true   **c.** Always true   **d.** Never true   **43a.** 8   **b.** $\frac{1}{8}$

CHAPTER 7 REVIEW EXERCISES

1. $\frac{2}{21}$ [7.1]   **2.** $\{x \mid x \neq -3, 2\}$ [7.1]   **3.** $\frac{x+4}{x(x+2)}$ [7.1]   **4.** $\frac{x^n+6}{x^n-3}$ [7.2]   **5.** $\frac{4x-1}{x+2}$ [7.2]   **6.** $\frac{21a+40b}{24a^2b^4}$ [7.2]

7. $\frac{x-1}{4}$ [7.2]   **8.** $\frac{1}{a-1}$ [7.2]   **9.** 2, 3 [7.3]   **10.** -1 [7.3]   **11.** $6.\overline{6}$ tanks [7.4]   **12.** Continuous [7.1]
13. -1 [7.3]   **14.** $x = 3$ [7.1]   **15.** 28,800 [7.5]   **16.** $\frac{17}{21}$ [7.3]   **17.** 20 h [7.3]
18. First car: 40 mph; second car: 50 mph [7.3]   **19.** 48 mi [7.4]   **20.** $5.\overline{45}$ min [7.3]   **21.** 2 mph [7.3]
22. 10,035 ft [7.5]   **23.** 300 footcandles [7.5]   **24.** 4.27 ft [7.4]   **25.** 127.6 ft/s [7.4]

CHAPTER 7 TEST

1. 2 [7.1]   **2.** $\{x \mid x \neq -3, 3\}$ [7.1]   **3.** $\frac{v(v+2)}{2v-1}$ [7.1]   **4.** $\frac{x+1}{3x-4}$ [7.2]   **5.** $\frac{22x-1}{(3x-4)(2x+3)}$ [7.2]   **6.** $\frac{2x}{x+2}$ [7.2]
7. $\frac{x+9}{4x}$ [7.2]   **8.** $\frac{x+3}{x-4}$ [7.2]   **9.** No solution [7.3]   **10.** 10 [7.3]   **11.** $12,000 [7.4]   **12.** Continuous [7.1]
13. $x = -2, x = 3$ [7.1]   **14.** 6 h [7.3]   **15.** 20 mph [7.3]   **16.** 20 oz [7.4]   **17.** 45 mph [7.3]
18. 11.2 in. [7.5]   **19.** 24 in. [7.4]   **20.** 76.27 ft/s [7.4]

CUMULATIVE REVIEW EXERCISES

1. $(0, -8)$; $(-4, 0)$ and $(2, 0)$ [2.3]   **2.** $(-\infty, 1]$ [3.4]   **3.** $\{x \mid x < -8 \text{ or } x > 15\}$ [3.4]   **4.** $\{x \mid x \leq \frac{1}{3} \text{ or } x \geq 3\}$ [3.5]
5. $(3, 4)$ [5.2]   **6.** $(2, 0, -1)$ [5.2]   **7.** $\frac{q^{10}}{p^{20}}$ [6.1]   **8.** -2 [6.6]   **9.** $\frac{(3x-2)(x+4)}{(x+2)(x-2)}$ [7.2]
10. $4x + 7 + \frac{10}{3x-2}$ [6.3]   **11.** $x^2 - 6x + 18 - \frac{42}{x+3}$ [6.3]   **12.** $-\frac{3}{2}, -1$ [7.3]   **13.** 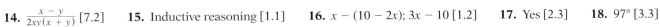 [4.1]

14. $\frac{x-y}{2xy(x+y)}$ [7.2]   **15.** Inductive reasoning [1.1]   **16.** $x - (10 - 2x); 3x - 10$ [1.2]   **17.** Yes [2.3]   **18.** $97°$ [3.3]
19. 50 lb [3.2]   **20.** $y = \frac{4}{3}x - 10$ [4.2]   **21.** $y = -\frac{4}{5}x + 272$; 176 rooms [4.2]   **22.** 1 [7.1]   **23a.** $\{x \mid x \neq -1, 4\}$
b. $x = -1, x = 4$ [7.1]   **24.** First car: 30 mph; second car: 40 mph [7.3]   **25.** 3 h [7.3]

ANSWERS to Chapter 8 Exercises

PREP TEST

1. 16 [1.3]   **2.** 32 [1.3]   **3.** 9 [1.4]   **4.** $\frac{1}{12}$ [1.4]   **5.** $-5x - 1$ [1.5]   **6.** $\frac{xy^5}{4}$ [6.1]   **7.** $9x^2 - 12x + 4$ [6.3]
8. $-12x^2 + 14x + 10$ [6.3]   **9.** $36x^2 - 1$ [6.3]   **10.** $-1, 15$ [6.6]

8.1 Exercises

1. There is no real number that, when squared, equals a negative number. **3.** Answers will vary.
5. Because $(x + y)^3 \neq x^3 + y^3$. **7.** 125 **9.** $\frac{1}{16}$ **11.** 32 **13.** Not a real number **15.** t **17.** a^2b^6
19. $a^6b^3c^9$ **21.** $\frac{b^3}{a}$ **23.** $\frac{a^4}{b^6}$ **25.** $\frac{ab}{6}$ **27a.** 14.4 births per 1000 people per year
b. 2.7 births per 1000 people per year **c.** Decreasing **29a.** 136.4 crashes per 1000 licensed drivers
b. 132.2 crashes per 1000 licensed drivers **c.** Answers will vary. For example, older drivers have more driving
experience. **31a.** \$71 **b.** \$4 **c.** $\{85, 71, 64, 60, 57\}$ **33.** $\sqrt[3]{b^4}$ **35.** $\sqrt[3]{9x^2}$ **37.** $-3\sqrt[5]{a^2}$ **39.** $\sqrt[3]{3x - 2}$
41. $\frac{1}{\sqrt[4]{b^3}}$ **43.** $y^{1/4}$ **45.** $d^{3/4}$ **47.** $(4y^7)^{1/5}$ **49.** $2pr^{1/5}$ **51.** $(a^2 + 5)^{1/6}$ **53.** 8 **55.** 49 **57.** 25
59. 27 **61.** No. Explanations will vary.

8.2 Exercises

1. Parts a and b are irrational numbers. Explanations will vary. **3.** Explanations will vary. **5.** $4a^2b^6$
7. Not a real number **9.** $3x^3y^4$ **11.** $-4c^3d^4$ **13.** $-x^2y^3$ **15.** $3p^4q$ **17.** $2wx^2$ **19.** $7\sqrt{2}$
21. $2\sqrt[3]{9}$ **23.** $2cd^4\sqrt{2c}$ **25.** $3xyz^2\sqrt{5yz}$ **27.** $-5d\sqrt[3]{c^2d}$ **29.** $2p^2q^3r^5\sqrt[3]{2p^2q^2}$ **31.** $2x^2y\sqrt[4]{2xy}$
33. Answers may vary. For example, 216, 1728, or 5832. **35.** 63.1% **37a.** 9.0 million property crimes
b. 12.0 million property crimes **c.** 12.8 million property crimes **d.** Answers will vary.
e. 2.4 million property crimes **39a.** Answers will vary. **b.** Answers will vary. **41.** 16 **43.** 32

8.3 Exercises

1. The right angle symbol must be at the 90° angle. The right angle must be labeled C and the hypotenuse labeled c.
One acute angle should be labeled A with the side opposite it labeled a. The other acute angle should be labeled B with
the side opposite it labeled b. **3.** No. The triangle is not a right triangle. **5.** If (x_1, y_1) and (x_2, y_2) are two points
in the plane, then the distance between the two points is given by $d = \sqrt{(x_1 - x_2)^2 + (y_1 - y_2)^2}$. **7.** No
9. 10.3 cm **11.** 9.75 ft **13.** 13.9 mi **15.** More than halfway **17.** 1000 mi **19.** 19.8 in. **21.** Yes
23. $2\sqrt{13}$ **25.** 9.1 **27.** 6.4 **29.** 6.7 **31a.** 17.2 units **b.** 12 square units **33a.** 20.9 units
b. 24 square units **35a.** 27.5 units **b.** 42 square units **37.** $r^2(\pi - 2)$ **39.** Answers will vary.
41. Yes. Explanations will vary. **43a.** Sometimes true **b.** Sometimes true **c.** Always true **d.** Always true
e. Always true

8.4 Exercises

1. Answers will vary. **3.** No. Examples will vary. For example, $\frac{\sqrt{x}}{\sqrt[3]{x}} = \frac{x^{1/2}}{x^{1/3}} = x^{1/6} = \sqrt[6]{x}$. **5.** Explanations will vary.
7. $11\sqrt{3x}$ **9.** $20y\sqrt{2y}$ **11.** $-2xy\sqrt{2y}$ **13.** $6ab^2\sqrt{3ab} + 3ab\sqrt{3ab}$ **15.** $-\sqrt[3]{2}$ **17.** $8b\sqrt[3]{2b^2}$
19. $3a\sqrt[4]{2a}$ **21.** $6\sqrt{3} - 6\sqrt{2}$ **23.** $2y\sqrt[3]{2x}$ **25.** $5x^3y^2\sqrt{2y}$ **27.** $2ab^2\sqrt[3]{4b^2}$ **29.** $2ab\sqrt[4]{27a^3b^3}$
31. $2ab^2\sqrt[3]{36ab^2}$ **33.** $10 - 5\sqrt{2}$ **35.** $9a\sqrt{a} - a\sqrt{3}$ **37.** $-8\sqrt{5}$ **39.** $12x - y$ **41.** $\sqrt[3]{x^2} + \sqrt[3]{x} - 20$
43. $y\sqrt{5y}$ **45.** $\frac{m\sqrt[3]{49m^2}}{n^4}$ **47.** $\frac{\sqrt{5x}}{x}$ **49.** $\frac{\sqrt{5x}}{5}$ **51.** $\frac{3\sqrt[3]{2x}}{2x}$ **53.** $\frac{\sqrt{2xy}}{2y}$ **55.** $\frac{3\sqrt[4]{2x}}{2x}$ **57.** $2\sqrt{5} - 4$
59. $-5 + 2\sqrt{6}$ **61.** $\frac{9 - 6\sqrt{x} + x}{9 - x}$ **63a.** $8\sqrt{10}$ units **b.** 40 square units **65a.** $8\sqrt{5} + 4\sqrt{10}$ units
b. 40 square units **67a.** $4\sqrt{2}$ units **b.** $(-1, 4)$ and $(4, -1)$ **69.** 12 **71.** $38 + 17\sqrt{5}$
73. $\frac{3\sqrt{y + 1} - 3}{y}$ **75.** $\frac{b + 18 - 6\sqrt{b + 9}}{b}$

8.5 Exercises

1. Parts a and b are radical functions because the radicand contains a variable expression. No variable appears within
a radical in part c or d. **3.** Write the inequality $4x + 16 \geq 0$. Then solve for x. **5.** $\{x \mid x \in \text{real numbers}\}$
7. $\{x \mid x \geq 0\}$ **9.** $\{x \mid x \in \text{real numbers}\}$ **11.** $\{x \mid x \geq 0\}$ **13.** $(-\infty, \infty)$ **15.** $(-\infty, 3]$ **17.** $[2, \infty)$
19. $\left(-\infty, \frac{2}{3}\right]$

21a. 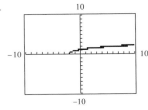   **b.** Domain: {$x \mid x \in$ real numbers}; Range: {$y \mid y \in$ real numbers}

c. $f(-4) \approx -1.4$

23a. 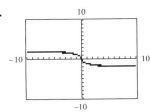   **b.** Domain: {$x \mid x \geq -2$}; Range: {$y \mid y \geq 0$}   **c.** $f(6) \approx 1.7$

25a. 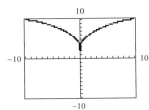   **b.** Domain: $(-\infty, \infty)$; Range: $(-\infty, \infty)$ **c.** $f(-9) \approx 2.1$

27a. 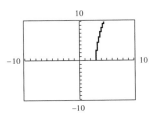   **b.** Domain: $(-\infty, \infty)$; Range: $[2, \infty)$ **c.** $f(-5) \approx 7.7$

29. 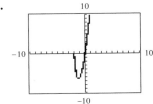   **31.**

33a. 15.5 mph   **b.** 52.1 ft   **c.** Answers will vary.   **35.** I = E, II = D, III = A, IV = F, V = G,
VI = H, VII = C, VIII = B   **37a.** 39.7 s   **b.** {$h \mid 0 \leq h \leq 16$}   **39.** 1, 4   **41.** There are no zeros.   **43.** −64, 8
45a.   **b.** 400   **c.** Approximately (233, 233)   **d.** 330 ft

8.6 Exercises

1. When the periscope is 20 feet above the surface of the water, the lookout can see 6.26 miles.   **3.** The resulting
equation may have a solution that is not a solution of the original equation.   **5.** Answers will vary. For example:
Algebraically: Substitute 4 for x in the equation; evaluate the left side of the equation; if the result is 6, the solution
checks. Graphically: On a graphing calculator, graph the equations $Y_1 = \sqrt{2x + 1} + 3$ and $Y_2 = 6$; find the point of

intersection; if the point of intersection is (4, 6), the solution checks.   **7.** $-\frac{7}{4}$ **9.** 14   **11.** 45   **13.** 7
15. 1   **17.** -4 **19.** 3   **21.** 21   **23.** $-5, 1$ **25.** 2   **27.** 11   **29.** No solution   **31.** No solution
33. $\frac{9}{16}$ **35.** $\frac{3}{4}$ **37.** 44.4 ft   **39.** 180 m   **41.** 121.5 lb   **43.** 3   **45.** 68 million miles   **47a.** 1973
b. 2020   **c.** Answers will vary.   **49.** 3700 lb   **51.** 27   **53.** 4 and 16   **55.** 1

8.7 Exercises

1. The letter i represents the number whose square is -1.   **3.** Explanations may vary. For example: A real number is a complex number in which $b = 0$.   **5.** The radical expressions must first be rewritten in terms of i before the radicands can be multiplied.   **7.** $2i$ **9.** $7i\sqrt{2}$ **11.** $3i\sqrt{3}$ **13.** $3ai$ **15.** $7x^6i$ **17.** $3a^5b^4i\sqrt{2b}$ **19.** $5 + 3i$
21. $2\sqrt{15} - 4i\sqrt{3}$ **23.** $10 - 7i$ **25.** $-5 - 3i$ **27.** $6 - 4i$ **29.** 0   **31.** $11 + 6i$ **33.** $16 + 7i$
35. -24 **37.** -15 **39.** $-70 - 40i$ **41.** $-15 - 12i$ **43.** $-10i$ **45.** $29 - 2i$ **47.** 1   **49.** 1

51. 89   **53.** 50   **55.** $-5\sqrt{2}$ **57.** $-2\sqrt{6}$ **59.** $-\frac{4}{5}i$ **61.** $\frac{3}{4} + \frac{1}{2}i$ **63.** $\frac{2}{3} - \frac{5}{3}i$ **65.** $-1 + 4i$

67. $\frac{30}{29} - \frac{12}{29}i$ **69.** $\frac{20}{17} + \frac{5}{17}i$ **71.** $\frac{11}{13} + \frac{29}{13}i$ **73.** $\frac{4}{5} + \frac{1}{10}i$ **75.** $-9 + 12i, -36 - 27i, 81 - 108i$

77a. Yes   **b.** Yes   **79.** $\frac{5\sqrt{6}}{6}i$ **81.** Two integers (0 and 1)   **83a.** $\frac{a}{a^2 + b^2} - \frac{b}{a^2 + b^2}i$ **b.** $-a - bi$ **c.** $\frac{a}{a^2 + b^2} - \frac{b}{a^2 + b^2}i$
85. $(x + 5i)(x - 5i)$ **87.** $(4x + yi)(4x - yi)$

CHAPTER 8 REVIEW EXERCISES

1. 27 [8.1]   **2.** b [8.1]   **3.** $\frac{x^6}{y^4}$ [8.1]   **4.** $\frac{a^6}{b^{16}}$ [8.1]   **5.** $3x^2y^3$ [8.2]   **6.** $-2ab^2\sqrt[5]{2a^3b^2}$ [8.2]   **7.** $\frac{8\sqrt{3y}}{3y}$ [8.4]
8. $\frac{x\sqrt{x} - x\sqrt{2} + 2\sqrt{x} - 2\sqrt{2}}{x - 2}$ [8.4]   **9.** $5i\sqrt{2}$ [8.7]   **10.** $\frac{14}{5} + \frac{7}{5}i$ [8.7]   **11.** $2a^2b\sqrt{2b}$ [8.4]   **12.** $3x + 3\sqrt{3x}$ [8.4]
13. $-13 + 6\sqrt{3}$ [8.4]   **14.** $5 + i$ [8.7]   **15.** $-12 + 10i$ [8.7]   **16.** 90 [8.7]   **17.** $39 - 2i$ [8.7]
18. $3\sqrt[8]{x^3}$ [8.1]   **19.** 23 [8.6]   **20.** -3 [8.6]   **21.** 4 [8.6]   **22a.** 4.92 reports per 1000 passengers
b. Decreasing [8.1]   **23a.** $12\sqrt{2} + 4\sqrt{13}$ units   **b.** 20 square units [8.3]   **24.** 3 ft [8.6]   **25.** $\sqrt{97}$ [8.3]
26. $(-\infty, \infty)$ [8.5]   **27a.**   **b.** Domain: $\{x \mid x \geq 1\}$; Range: $\{y \mid y \leq 0\}$

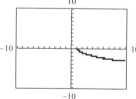

c. $f(4) \approx -1.7$ [8.5]   **28a.** 214 mi   **b.** 35,267 ft [8.6]

CHAPTER 8 TEST

1. $\frac{1}{32}$ [8.1]   **2.** c^6d^9 [8.1]   **3.** $\frac{x^3}{2y^3}$ [8.1]   **4.** $-2a^2b^4$ [8.2]   **5.** $xy^2z^2\sqrt[4]{x^2z^2}$ [8.2]   **6.** $5x\sqrt{x}$ [8.4]
7. $\frac{3\sqrt{y} - 3}{y - 1}$ [8.4]   **8.** $7 - 4i$ [8.7]   **9.** $2 - 3i$ [8.7]   **10.** $-2 + 7i$ [8.7]   **11.** $5\sqrt{6}$ [8.4]   **12.** $4x\sqrt[3]{x^2}$ [8.4]
13. $-17 + \sqrt{3}$ [8.4]   **14.** $9 - i$ [8.7]   **15.** $3 - 9i$ [8.7]   **16.** 48 [8.7]   **17.** $7 + 3i$ [8.7]   **18.** $-6\sqrt{2}$ [8.7]
19. $15 - 10i$ [8.7]   **20.** $7yz^{2/5}$ [8.1]   **21.** 4 [8.6]   **22.** $\frac{9}{16}$ [8.6]   **23.** 4 [8.6]   **24.** 14.4 in. [8.3]
25. 9.5 units [8.3]   **26a.** 222.4 million people   **b.** 272.2 million people   **c.** 306.8 million people
d. Answers will vary. [8.2]   **27a.** 1978   **b.** 1989 [8.6]

28a. 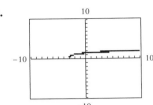   **b.** Domain: $[-3, \infty)$; Range: $[0, \infty)$ **c.** $f(13) = 2$ [8.5]

CUMULATIVE REVIEW EXERCISES

1. u [1.1]   **2.** 9 [1.1]   **3.** 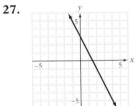 [1.2]   **4.** 47 [1.3]   **5.** $2b^3 - 4b^2 + 6b - 15$ [6.2]
6. 304,000,000,000 [6.1]   **7.** $\{-4, -1, 8\}$ [2.2]   **8.** -2 [4.1]   **9.** -2 [7.1]   **10.** $(0, -4)$; $(-4, 0)$ and $(1, 0)$ [2.3]
11. $\frac{3}{2}$ [3.1]   **12.** $\frac{2}{3}$ [3.1]   **13.** $\{x \mid 4 < x < 5\}$ [3.4]   **14.** $\{x \mid x < 2 \text{ or } x > \frac{8}{3}\}$ [3.5]
15. $y = \frac{3}{2}x + 3$; $m = \frac{3}{2}$; y-intercept: $(0, 3)$ [4.1]   **16.** $\frac{1}{64}$ [8.1]   **17.** $14 + 10\sqrt{3}$ [8.4]   **18.** $14 + 2i$ [8.7]   **19.** $-\frac{3}{5}$ [1.4]
20. $10y - 19$ [1.5]   **21.** $\frac{4b^{12}}{a^7}$ [6.1]   **22.** $(3x + 1)(x - 4)$ [6.4]   **23.** $(x^2\pi + 4\pi x + 4\pi)$ m² [6.3]   **24.** $\sqrt{26}$ [8.3]
25. 86 mi [8.3]   **26.** 40 oz [3.2]
27.   [4.1]   **28.** $y = \frac{3}{2}x + 2$ [4.2]   **29.** $(1, -1)$ [5.1]   **30.** $\left(-\frac{9}{7}, \frac{2}{7}, \frac{11}{7}\right)$ [5.2]

ANSWERS to Chapter 9 Exercises

PREP TEST

1. $3\sqrt{2}$ [8.2]   **2.** $3i$ [8.7]   **3.** $\frac{2x - 1}{x - 1}$ [7.2]   **4.** 8 [1.3]   **5.** Yes [6.5]   **6.** $(2x - 1)^2$ [6.5]
7. $(3x + 2)(3x - 2)$ [6.5]   **8.**   [1.2]   **9.** $-3, 5$ [6.6]   **10.** 4 [7.4]

9.1 Exercises

1a. A quadratic equation is an equation of the form $ax^2 + bx + c = 0$, $a \neq 0$.   **b.** Examples will vary.
3a. The polynomial $ax^2 + bx + c$ is in descending order and equal to zero.   **b.** Explanations may vary. For example, use the Commutative Property to rearrange the terms: $4x^2 + 6x - 3 = 0$.   **c.** Examples will vary.
5a. If the product of two factors is zero, then at least one of the factors must be zero.   **b.** It is used to solve some quadratic equations.   **7.** $a = -1, b = 1, c = -8$ **9.** $x = 1, b = -7, c = 0$
11. Yes; $5p^2 + 3p - 4 = 0$ or $-5p^2 - 3p + 4 = 0$ **13.** Yes; $z^2 + z - 5 = 0$ **15.** $-1, 16$ **17.** $\frac{3}{2}, 3$ **19.** $-2, \frac{3}{2}$
21. $\frac{1}{2}, \frac{2}{3}$ **23.** $-5, 0$ **25.** $-\frac{1}{3}, 6$ **27.** $-4, -\frac{3}{2}$ **29.** $-5, 1$ **31.** $-\frac{2}{3}, \frac{3}{2}$ **33.** 3
35. Length: 18 m; width: 10 m   **37.** 4 m   **39.** 2, 4, and 6 or $-6, -4,$ and -2 **41.** 8 and 10 or -10 and -8
43. 10 s   **45.** 12   **47.** 7 teams   **49.** 1 s and 3 s   **51.** Length: 20 in.; width: 10 in.   **53.** $x^2 - 5x - 6 = 0$
55. $x^2 - 9 = 0$ **57.** $x^2 - 5x = 0$ **59.** $3x^2 - 8x + 4 = 0$ **61.** $10x^2 - 7x - 6 = 0$ **63.** $8x^2 + 6x + 1 = 0$
65. 8   **67.** $-12y, 3y$ **69.** $-\frac{b}{2}, -b$ **71.** $x^2 + 4 = 0$ **73.** $x^2 - 12 = 0$ **75.** $x^2 + 12 = 0$ **77.** The equation should first be written in standard form: $x^2 - x = 0$. Solve by factoring: $x(x - 1) = 0$. $x = 0$ or $x = 1$.   **79.** 1, 2, 3, 4, 5

9.2 Exercises

1. Answers will vary.   **3.** Explanations will vary. For example, add the square of one-half of b to the expression.
5. $\pm\frac{9}{2}$ **7.** $\pm 7i$ **9.** $\pm 4\sqrt{3}$ **11.** $\pm 3i\sqrt{2}$ **13.** $-5, 7$ **15.** $-6, 0$ **17.** $-2 \pm 5i$ **19.** $-\frac{1}{5}, 1$

A27

21. $\frac{7}{3}, 1$ **23.** $-5 \pm \sqrt{6}$ **25.** $-1 \pm 2\sqrt{3}$ **27.** $\frac{1 \pm 4\sqrt{5}}{2}$ **29.** No ($v \approx 27.0$ mph)   **31.** 4.47 m/s   **33.** 4 h

35. $\sqrt{2}$ or $-\sqrt{2}$ **37.** $-2 \pm \sqrt{11}$ **39.** $3 \pm \sqrt{2}$ **41.** $1 \pm i$ **43.** $\frac{1 \pm \sqrt{5}}{2}$ **45.** $-3 \pm 2i$ **47.** $1 \pm 3\sqrt{2}$

49. $-1, -\frac{1}{2}$ **51.** $\frac{1}{2} \pm i$ **53.** $\frac{2 \pm \sqrt{14}}{2}$ **55.** $-3.236, 1.236$ **57.** $0.293, 1.707$ **59.** $-0.809, 0.309$

61. $\frac{1 + \sqrt{5}}{2}$ or $\frac{1 - \sqrt{5}}{2}$ **63.** \$444.80   **65a.** 4.431 s   **b.** No. The ball will have gone only 197.2 ft when it hits the

ground.   **67.** $\pm \frac{4b}{a}$ **69.** $-a \pm 2$ **71.** $-\frac{1}{2}$ **73.** The complete solution is in the Solutions Manual.
75. 0.0086 mole per liter   **77.** No (When $18x + 144$ is substituted for y^2 in $x^2 + y^2 = 40$, the solutions of the resulting equation are complex numbers.)

9.3 Exercises

1a. $x = \frac{-b \pm \sqrt{b^2 - 4ac}}{2a}$ **b.** It is used to solve quadratic equations.   **c.** a is the coefficient of x^2, b is the coefficient of x, and c is the constant in the quadratic equation $ax^2 + bx + c = 0$.   **3.** When a quadratic equation has two solutions that are the same number, it is called a double root of the equation.   **5.** Answers will vary.   **7.** $4 \pm \sqrt{17}$

9. $-9, 4$ **11.** $4 \pm 2\sqrt{22}$ **13.** $1 \pm 2\sqrt{2}$ **15.** $\frac{1 \pm \sqrt{2}}{2}$ **17.** $2 \pm i$ **19.** $-\frac{3}{2}, -\frac{1}{2}$ **21.** $-3 \pm 2i$

23. $-\frac{1}{2} \pm \frac{5}{2}i$ **25.** $0.394, 7.606$ **27.** $-4.236, 0.236$ **29.** $-1.351, 1.851$ **31.** Two complex number solutions

33. One real number solution   **35.** Two real number solutions   **37.** No. (The discriminant is less than zero.)
39. $\{p \,|\, p < 25, p \in \text{real numbers}\}$ **41.** $\{p \,|\, p > 4, p \in \text{real numbers}\}$ **43.** Two   **45.** One   **47.** None
49. Height: 9.348 m, base: 5.348 m   **51.** Between 2.76 s and 7.24 s   **53a.** 5, 6, or 7 watches   **b.** 3 to 9 watches
c. 0; -1200. No less than 0 watches can be manufactured. When $x = 0$, there are 0 watches manufactured and the manufacturer experiences a loss of \$1200.   **55a.** 24 cars per minute   **b.** 9:00 A.M.   **57.** 2004   **59.** 1991 and 2000

61a. \$22 trillion   **b.** 2017   **63.** 4.3 m to spare   **65.** -7 and 9   **67a.** Answers may vary. For example, from
the quadratic formula, $r_1 = \frac{-b + \sqrt{b^2 - 4ac}}{2a}$ and $r_2 = \frac{-b - \sqrt{b^2 - 4ac}}{2a}$. Use these expressions to show that $r_1 + r_2 = -\frac{b}{a}$
and $r_1 r_2 = \frac{c}{a}$.   **b.** Answers will vary. For example, use the solutions $r_1 = 1 + \sqrt{3}$ and $r_2 = 1 - \sqrt{3}$ of the equation
$x^2 - 2x - 2 = 0$ to show that $r_1 + r_2 = 2 = -\frac{b}{a}$ and $r_1 r_2 = -2 = \frac{c}{a}$.   **69.** $b^2 - 4ac = b^2 + 4 > 0$ for any real number b.

9.4 Exercises

1a. Yes. Explanations will vary.   **b.** No. Explanations will vary.   **3a.** Always true   **b.** Sometimes true
c. Always true   **5.** $\pm 2, \pm 3$ **7.** $\pm 2, \pm \sqrt{2}$ **9.** 1, 4   **11.** 16   **13.** $\pm 1, \pm 2i$ **15.** $\pm 2, \pm 4i$ **17.** 16

19. 1, 512   **21.** $1, 2, -1 \pm i\sqrt{3}, -\frac{1}{2} \pm \frac{\sqrt{3}}{2}i$ **23.** $\pm 1, \pm 2, \pm i, \pm 2i$ **25.** $-64, 8$ **27.** $\frac{1}{4}, 1$ **29.** 32, 243

31. 5   **33.** 5   **35.** 4   **37.** 4   **39.** 1   **41.** 9   **43.** -3 **45.** $\sqrt{2}, i$ **47.** $\sqrt{5}$ or $-\sqrt{5}$ **49.** 10.2 in.

51. 0, 1, or 7   **53.** 419.2 ft   **55.** $(5 - \sqrt{33}, 6)$ and $(5 + \sqrt{33}, 6)$, or $(-0.74, 6)$ and $(10.74, 6)$ **57.** $-1, 0, 1$

59. 2, 3   **61.** Explanations will vary.   **63.** -400

9.5 Exercises

1. A quadratic function is a function of the form $f(x) = ax^2 + bx + c$, $a \neq 0$.   **3.** When $a > 0$, it is the point with the smallest y-coordinate. When $a < 0$, it is the point with the largest y-coordinate.   **5.** Descriptions may vary. For example: Use the equation $x = -\frac{b}{2a}$ to find the x-coordinate of the vertex. Then find the y-coordinate of the vertex by replacing x in the equation of the parabola by the value found for x.   **7.** Answers may vary. For example, it is the value of the function at the vertex of the graph of the function.   **9.** -5 **11.** $x = 7$ **13.** Vertex: $(0, -2)$, Axis of symmetry: $x = 0$ **15.** Vertex: $(0, 3)$; Axis of symmetry: $x = 0$ **17.** Vertex: $(0, 0)$; Axis of symmetry: $x = 0$

19. Vertex: $(0, -1)$; Axis of symmetry: $x = 0$ **21.** Vertex: $(-1, -1)$; Axis of symmetry: $x = -1$ **23.** Vertex: $\left(-1, \frac{1}{2}\right)$;

Axis of symmetry: $x = -1$ **25.** Vertex: $\left(\frac{3}{2}, -\frac{1}{4}\right)$; Axis of symmetry: $x = \frac{3}{2}$ **27.** Vertex: $\left(\frac{1}{4}, -\frac{25}{8}\right)$; Axis of symmetry:

$x = \frac{1}{4}$ **29.** The graph becomes wider. **31.** The graph is lower on the rectangular coordinate system. **33.** 16

35. Sketches for parts a through h will vary. Examples are provided in the Solutions Manual. Parts i and j are not possible. **37.** Domain: $\{x \mid x \in \text{real numbers}\}$; Range: $\{y \mid y \geq -5\}$ **39.** Domain: $\{x \mid x \in \text{real numbers}\}$; Range: $\{y \mid y \leq 0\}$ **41.** Domain: $\{x \mid x \in \text{real numbers}\}$; Range: $\{y \mid y \geq -7\}$ **43.** Domain: $\{x \mid x \in \text{real numbers}\}$; Range: $\{y \mid y \leq 5\}$ **45.** $(0, 0), (2, 0)$ **47.** $(3, 0), \left(-\frac{1}{2}, 0\right)$ **49.** $(\sqrt{2}, 0), (-\sqrt{2}, 0)$ **51.** $(-1 + \sqrt{2}, 0)$,

$(-1 - \sqrt{2}, 0)$ **53.** No x-intercepts **55.** $(-1 + \sqrt{2}, 0), (-1 - \sqrt{2}, 0)$ **57.** $-3.3, 0.3$ **59.** $-1.3, 2.8$

61. No real zeros **63.** Minimum: 2 **65.** Maximum: -1 **67.** Minimum: $-\frac{11}{4}$ **69.** Maximum: $\frac{9}{4}$

71. Maximum: $\frac{9}{8}$ **73.** c **75.** 150 ft **77.** 100 lenses **79.** 24.36 ft **81.** Yes **83.** 141.6 ft

85. 41 mph; 33.658 mpg **87a.** **b.** 21.3 mph; 70 calories per hour

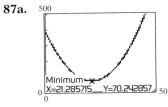

c. The function is decreasing on $0 < x < 21$. This means that the number of calories burned is decreasing as the speed increases. The function is increasing on $21 < x < 50$. This means that the number of calories burned is increasing as the speed increases. **89.** 7 and -7 **91a.** To the nearest ten-thousandth: $y = 0.0475x^2 + 2.2107x - 0.0701$

b. 182.9 ft **c.** 73.6 mph **93a.** $y = 0.006x^2 - 1.402x + 93.76$ **b.** 12.1% **c.** 1998 **d.** 2016 **95a.** To the nearest hundredth-thousandth: $y = -0.06998x^2 + 6.22845x - 0.26229$ **b.** $r \approx 0.999$. This is very close to 1, which means that the data points lie very close to the graph of the regression equation. **c.** 108.9 m **d.** 24.4° and 64.6° **e.** The vertex is $(44.5, 138.3)$. The vertex represents that angle at which the ball must be hit in order to travel the maximum distance. When hit at an angle of 44.5°, the ball travels 138.3 m. **97.** $(2, -3)$ **99.** $(0, -4)$ **101.** 7 **103.** -8

105a. Above **b.** Yes **c.** Maximum **107.** -2 **109.** $-\frac{3}{2}$ **111.** 7 **113.** 15 units

9.6 Exercises

1. Both $x - 3$ and $x - 5$ are positive or both $x - 3$ and $x - 5$ are negative. **3.** The trinomial must be written as a product so that we can determine what the signs of the factors must be in order for their product to be greater than zero. **5.** $\{x \mid x < -2 \text{ or } x > 4\}$ **7.** $\{x \mid x \leq 1 \text{ or } x \geq 2\}$

9. $\{x \mid -3 < x < 4\}$ **11.** $\{x \mid x < -2 \text{ or } 1 < x < 3\}$

13. $\{x \mid x < -4 \text{ or } x > 4\}$ **15.** $\{x \mid x \neq 2\}$ **17.** $\{x \mid -3 \leq x \leq 12\}$ **19.** $\left\{x \mid x \leq \frac{1}{2} \text{ or } x \geq 2\right\}$

21. $\left\{x \mid \frac{1}{2} < x < \frac{3}{2}\right\}$ **23.** $\{x \mid x \leq -3 \text{ or } 2 \leq x \leq 6\}$ **25.** $\left\{x \mid -\frac{3}{2} < x < \frac{1}{2} \text{ or } x > 4\right\}$ **27.** $\{x \mid x \leq -3 \text{ or } -1 \leq x \leq 1\}$

29. $\{x \mid -2 \leq x \leq 1 \text{ or } x \geq 2\}$ **31.** **33.**

35. **37.** **39.** 7 **41.** All real numbers

CHAPTER 9 REVIEW EXERCISES

1. $-2, 9$ [9.1] **2.** $\pm 5\sqrt{3}$ [9.2] **3.** $-7, 3$ [9.2] **4.** $\frac{3 \pm \sqrt{5}}{2}$ [9.2] **5.** $1 \pm 2i\sqrt{3}$ [9.2] **6.** $\frac{1 \pm 2\sqrt{2}}{2}$ [9.3]

7. $3 \pm i$ [9.3] **8.** $\pm\sqrt{5}, \pm i$ [9.4] **9.** $-8, \frac{1}{8}$ [9.4] **10.** 2 [9.4] **11.** 9 [9.5] **12.** $\frac{-3 \pm \sqrt{41}}{2}$ [9.5]

13. $\left\{x \mid -3 < x < \frac{5}{2}\right\}$ [9.6]   **14.** $(-4, 0), \left(\frac{3}{2}, 0\right)$ [9.5]   **15.** Vertex: $(1, 2)$; Axis of symmetry: $x = 1$ [9.5]

16. Domain: $\{x \mid x \in \text{real numbers}\}$; Range: $\{y \mid y \geq -5\}$ [9.5]   **17.** 2 and -7 [9.1]   **18.** $3x^2 + 8x - 3 = 0$ [9.1]
19. Two real number solutions [9.3]   **20.** None [9.3]   **21.** 33 mph [9.3]   **22a.** 0.46 s and 2.04 s
b. 30 ft   **c.** after 1.25 s [9.3]   **23.** 324.7 ft [9.4]   **24.** $\frac{1 + \sqrt{19}}{3}$ [9.3]   **25a.** $y = 0.1x^2 - 18.04x + 809.44$
b. 35.04 million U.S. households   **c.** 2007 [9.5]

CHAPTER 9 TEST

1. $-5, \frac{1}{2}$ [9.1]   **2.** $\pm 4\sqrt{3}$ [9.2]   **3.** $-\frac{1}{2} \pm 2i$ [9.2]   **4.** $3 \pm \sqrt{11}$ [9.2]   **5.** $-3 \pm i$ [9.2]   **6.** $\frac{1 \pm \sqrt{3}}{2}$ [9.3]
7. $\frac{2}{3} \pm \frac{\sqrt{2}}{3}i$ [9.3]   **8.** $\pm 2, \pm \sqrt{2}$ [9.4]   **9.** $\frac{1}{4}$ [9.4]   **10.** 1 [9.4]   **11.** 3 [9.5]   **12.** $-\frac{1}{3} \pm \frac{\sqrt{5}}{3}i$ [9.5]

13. $\{x \mid x < -3 \text{ or } 0 < x < 2\}$ [9.6]   **14.** $(-2, 0), \left(\frac{1}{2}, 0\right)$ [9.5]

15. Vertex: $\left(\frac{3}{2}, \frac{1}{4}\right)$; Axis of symmetry: $x = \frac{3}{2}$ [9.5]   **16.** Domain: $\{x \mid x \in \text{real numbers}\}$; Range: $\{y \mid y \geq -4.5\}$ [9.5]

17. -5 and 6 [9.1]   **18.** $2x^2 + 3x - 5 = 0$ [9.1]   **19.** Two complex number solutions [9.3]   **20.** Two [9.3]
21. Length: 12 cm; width: 5 cm [9.1]   **22.** 272 mi by 383 mi [9.3]   **23a.** 0.39 s and 2.43 s
b. 36.64 ft   **c.** after 1.41 s [9.3]   **24.** $x^2 - 30x + 200 = 0$ [9.1]   **25a.** To the nearest ten-thousandth,
$y = 0.1964x^2 - 35.2964x + 1603.4929$ **b.** $82 billion   **c.** 2010 [9.5]

CUMULATIVE REVIEW EXERCISES

1. -28 [3.1]   **2.** $8a\sqrt{2a}$ [8.4]   **3.** $(7, -4)$ [5.2]   **4.** $18 + 16i$ [8.7]   **5.** $2x(2x - 3)(x + 4)$ [6.5]
6. $x^2 - 3x - 4 - \frac{6}{3x - 4}$ [6.3]   **7.** $\{x \mid x < -2 \text{ or } x > 5\}$ [3.5]   **8.** $-8, 3$ [6.6]   **9.** 0, 6 [9.2]   **10.** $2 \pm 3i$ [9.2/9.3]
11. $-\frac{3}{2}, -1$ [7.3]   **12.** 4 [9.4]   **13.** $\frac{3}{x^2 y}$ [8.1]   **14.** $\{-1, 15, 63\}$ [2.2]   **15.** $y = x + 7$ [4.2]   **16.** 1, 3, 7 [1.1]
17. $2\sqrt{5}$ [8.3]   **18.** 350 mi [3.2]   **19.** 600 ml of the 12% solution; 300 ml of the 6% solution [5.2]   **20.** 75 oz [3.2]
21. $y = 75x + 32,000$; $197,000 [4.2]   **22a.** The demand for high-speed Internet access is increasing by 6.5 million
subscribers per year.   **b.** 45 million subscribers [4.1]   **23.** 4.4 ft [2.2]   **24.** 90 cm by 90 cm [9.1]
25. Vertex: $\left(-1, -\frac{9}{2}\right)$; Axis of symmetry: $x = -1$; Domain: $\{x \mid x \in \text{real numbers}\}$, Range: $\{y \mid y \geq -4.5\}$ [9.5]

ANSWERS to Chapter 10 Exercises

PREP TEST

1. $\frac{1}{9}$ [6.1]   **2.** 16 [6.1]   **3.** -3 [6.1]   **4.** 0; 108 [2.2]   **5.** -6 [3.1]   **6.** $-2, 8$ [6.6/9.1]
7. 6326.60 [1.4]   **8.**   [9.5]

10.1 Exercises

1. $f[g(3)]$ means to evaluate the function f at the value of the function g when g is evaluated at 3.
3a. We read $(f \circ g)(x)$ as "f of g of x."   **b.** The function $A \circ r$ is the *composition* of A with r.   **c.** The expression
$(f \circ g)(-2)$ means to evaluate the function f at $g(-2)$.   **d.** $(f + g)(x) = f(x) + g(x)$ **5.** 9   **7.** $-\frac{3}{4}$ **9.** -5

11. $-\frac{35}{8}$ **13.** $\frac{5}{11}$ **15.** x^2 **17.** $P(n) = -0.6n^2 + 300\,n - 3000$; \$28,500 **19.** -11 **21.** 19

23. $-6x + 5$ **25.** 4 **27.** 9 **29.** $2x^2 + 1$ **31.** 3 **33.** 15 **35.** $4x^2 - 1$ **37.** $\frac{11}{12}$

39. Undefined **41.** $\frac{2x + 5}{2x + 6}$ **43a.** $A(t) = 2025\pi t^2$ **b.** 57,256 ft² **45a.** $c[m(h)] = 3125h^2 + 15{,}000h + 1000$
b. \$463,500 **c.** If the plant is run for 10 h, the cost to produce the fax machines is \$463,500. **47.** $4x - 9$
49. 13 **51.** 1 **53.** $2x^2 + 2$ **55a.** 16 **b.** 5 **c.** 0 **d.** -3 **e.** The function f squares an element in set R.
f. The function g subtracts 4 from an element in set S. **g.** The composite of functions f and g first squares an element in set R and then subtracts 4 from the result. **h.** Answers will vary. **57.** 0 **59.** -2 **61.** 7

10.2 Exercises

1. Answers will vary. **3.** No. Explanations will vary. For example, reversing the coordinates of the ordered pairs of a constant function results in an equation of the form $x = a$, where a is a constant. An equation of this form is not a function. **5.** For a 1-1 function, for any given y there is exactly one x that can be paired with that y. The horizontal-line test says that if every horizontal line intersects a graph of a function at most once, then the graph is the graph of a 1-1 function. **7.** -1 **9.** -3 **11a.** 5 **b.** 0 **13.** domain **15.** No inverse **17.** $\{(0, 0), (1, 2), (2, 4), (3, 8), (4, 16)\}$

19. $f^{-1}(x) = -\frac{1}{3}x - 3$ **21.** $f^{-1}(x) = \frac{1}{3}x - \frac{7}{3}$ **23.** $f^{-1}(x) = \frac{1}{4}x - \frac{1}{2}$ **25.** $f^{-1}(x) = \frac{1}{3}x - \frac{2}{3}$ **27.** $f^{-1}(x) = -2x - 6$
29. $f^{-1}(x) = \frac{4}{5}x + \frac{4}{5}$ **31.** $f^{-1}(x) = \frac{1}{x}$ **33.** Not inverse functions **35.** Not inverse functions

37. Inverse functions **39.** Inverse functions **41.** Yes. Yes. Because a conversion function is 1-1, it has an inverse function. **43.** No. There are two ordered pairs with the same first element and different second elements (for instance, (0.37, 0.5) and (0.37, 0.25)). Therefore, the relation is not a function.

45. $K^{-1}(x) = \frac{10}{13}x + \frac{47}{13}$ **47.** Between 6 and 15 **49.** Greater than 7 **51.** 2 and 6 **53.** 4 and 8

55. False. For instance, let $f(x) = |x|$. Then $f(-2) = f(2)$ but $2 \neq -2$.

57. 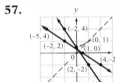 **59.** **61a.** 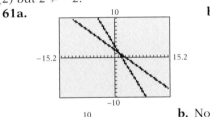 **b.** Yes

63a. 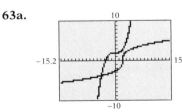 **b.** Yes **65a.** 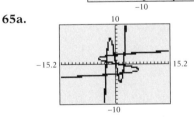 **b.** No

10.3 Exercises

1a. Answers may vary. For example, an exponential function with base b is defined by $f(x) = b^x$, $b > 0$, $b \neq 1$, and x is any real number. **b.** Answers will vary. For example, an exponential function has a constant base and a variable exponent, whereas a polynomial function has a variable base and a constant exponent. **3.** It is the function defined by $f(x) = e^x$, where e is an irrational number approximately equal to 2.71828183. **5a.** Never true **b.** Always true
c. Never true **d.** Sometimes true **e.** Never true **7a.** 0.125 **b.** 1 **c.** 4 **9a.** 0.0014 **b.** 0.0370
c. 0.1111 **11a.** 3 **b.** 0.0370 **c.** 729 **13a.** 0.0183 **b.** 0.2636 **c.** 54.5982 **15a.** 0.0009 **b.** 1
c. 54.5982 **17a.** 5.1962 **b.** 0.3333 **c.** 13.9666 **19.** a, b **21a.** 5 **b.** Increasing
23a. e **b.** Increasing **25.** a, d **27.** 20. $f(x) = 6(1.05^x)$ is an increasing function; therefore, as values of x increase, values of y increase **29.** 12. $f(x) = 0.7(0.95^x)$ is a decreasing function; therefore, as values of x increase, values of y decrease. **31.** (0, 3) **33.** $x < 0$ **35.** 1.6 **37.** 1.6 **39a.** Graph a is $y = 4^x$, graph b is $y = 3^x$, and graph c is $y = 2^x$. Explanations will vary. For example, the greater the base of the exponential function, the more rapidly the value of y increases. **b.** Exponential growth **41a.** (A) **b.** (B) **c.** (C) **d.** (D) **e.** (C) **f.** (A) **43.** 10 years
45. 18 years **47.** \$29,127.15 **49.** \$27,180.96 **51.** 21.2 mg **53.** 68% **55.** \$.08 **57.** \$6902.49
59a. $h(x) = 4^x$ **b.** $g(x) = 3^x$

10.4 Exercises

1a. b **b.** x **c.** a **3.** They will intersect at (0, 10).   **5.** Answers may vary. For example, for an exponential growth function, as x increases, y increases. For an exponential decay function, as x increases, y decreases.
7a. $y = 2.960(1.0335)^x$ **b.** 3.853 million or 3,853 thousand **c.** 5.183 million or 5,183 thousand
9a. $y = 100,000(0.80)^x$ **b.** \$32,768 **c.** after 10 years   **11a.** $y = 1000(1.05)^x$ **b.** The model can be written in the form of the compound interest formula by writing it as $A = 1000(1 + 0.05)^t$. Here $P = 1000$, $r = 0.05$, $n = 1$, $r/n = 0.05/1 = 0.05$, $nt = 1 \cdot t = t$. **c.** \$1628.89 **d.** Yes. The model equation for the \$10,000 is $y = 10,000(1.05)^x$. The factor 1000 is changed to 10,000, so the exponential expression 1.05^x is multiplied by 10 times 1000.

13a. (A); $y = 3^x$ **b.** (B); $y = \left(\dfrac{1}{2}\right)^x$ or $y = 0.5^x$ **c.** (C) **d.** (D) **e.** (B); $y = -\left(\dfrac{1}{10}\right)^x$ or $y = -(0.1)^x$ **f.** (A); $y = 2^x - 1$

g. (D) **h.** (C) **i.** (A); $y = 2(4^x)$ **15a.** To the nearest ten-thousandth: $y = 0.4609(1.3679)^x$ **b.** 51 million households **c.** 2006 **d.** Answers will vary. For example, the increase in the number of households with computers and the convenience of not having to go to the bank.   **17a.** To the nearest ten-thousandth: $y = 25.3462(1.0414)^x$
b. $r \approx 0.9998$. This is very close to 1, which means that the data points lie very close to the graph of the regression equation. **c.** 1,190,000 interracial marriages **d.** 1985   **19a.** To the nearest ten-thousandth: $y = 36.5807(1.2736)^x$
b. $r \approx 0.9976$. This is very close to 1, which means that the data points lie very close to the graph of the regression

equation. **c.** \$37 million **d.** \$164   **21a.** After each round, the number of teams is halved. **b.** $t(r) = 64\left(\dfrac{1}{2}\right)^r$,
where t is the number of teams in the round and r is the number of rounds that have been played.
23a.   **b.** 2 **c.** $f(x) = 10 \cdot 2^x$ **d.** 3

Hour	Number of Bacteria
0	10
1	20
2	40
3	80
4	160

10.5 Exercises

1a. For $b > 0$, $b \neq 1$, if $b^u = b^v$, then $u = v$. **b.** Examples will vary.   **3a.** A common logarithm is a logarithm with base 10. **b.** $\log 4z$ **5.** They are mirror images of each other with respect to the line $y = x$.   **7.** $\log 1000 = 3$
9. $\log_3 \dfrac{1}{27} = -3$ **11.** $\ln x = y$ **13.** $\log_b c = y$ **15.** $2^5 = 32$ **17.** $5^{-1} = \dfrac{1}{5}$ **19.** $10^y = x$

21. $c^y = x$ **23.** 2   **25.** 0   **27.** -3 **29.** 2   **31.** 5   **33.** 64   **35.** $\dfrac{1}{64}$ **37.** 1

39. a and iii, b and i, c and ii, d and iv   **41.**   **43.**   **45.** 1584.89

47. 0.01   **49.** 54.60   **51.** 0.18   **53.** $\dfrac{1}{4}$ **55.** 5   **57.** 6.31
59. The x-intercept is (2, 0). There is no y-intercept.   **61.** It is decreasing on $(0, \infty)$. There is no interval on which the graph is increasing.   **63.** 0.44 cm   **65.** 188 decibels   **67.** 6.0 parsecs   **69.** 7,805.5 billion barrels
71. 15.1543   **73a.** 14 s **b.** 813 ft/s   **75.** $\ln 20 > \log 20$. Explanations will vary.

10.6 Exercises

1. Answers may vary. For example, the log of a product is equal to the sum of the logs: $\log_b (xy) = \log_b x + \log_b y$.
3. Explanations will vary. For example, rewrite $\log_5 12$ as $\dfrac{\log 12}{\log 5}$ and then use a calculator to evaluate this quotient.
5. $2 \log_3 x + 6 \log_3 y$ **7.** $3 \log_7 u - 4 \log_7 v$ **9.** $2 \log_2 r + 2 \log_2 s$ **11.** $2 \log_9 x + \log_9 y + \log_9 z$

13. $\ln x + 2 \ln y - 4 \ln z$ **15.** $\frac{1}{2} \log_7 x + \frac{1}{2} \log_7 y$ **17.** $\frac{1}{2} \log_2 x - \frac{1}{2} \log_2 y$ **19.** $\frac{3}{2} \ln x + \frac{1}{2} \ln y$

21. $\frac{3}{2} \log_7 x - \frac{1}{2} \log_7 y$ **23.** $\log_5 x^3 y^4$ **25.** $\log_3 \frac{x^2 z^2}{y}$ **27.** $\log_b \frac{x}{y^2 z}$ **29.** $\ln x^2 y^2$ **31.** $\log_6 \sqrt{\frac{x}{y}}$ **33.** $\log_4 \frac{s^2 r^2}{t^4}$

35. $\ln \frac{t^3 v^2}{r^2}$ **37.** $\log_4 \sqrt{\frac{x^3 z}{y^2}}$ **39.** 3   **41.** 9   **43.** 28   **45.** 4   **47.** 8   **49.** 1.5000   **51.** 1.8597

53. −0.3174   **55.** 0.6492   **57.** $\frac{\log (x^2 + 4)}{\log 5}$ **59.** $\frac{3 \log (2x^2 - x)}{\log 2}$ **61.** $\frac{\ln (3x + 4)}{\ln 7}$ **63.** $\frac{7 \ln (2x^2 - x)}{\ln 3}$

65. 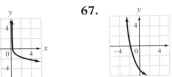   **67.**   **69.** 13.12   **71.** −6.59   **73.** 0.8   **75.** 0.40

77a.   **b.** After 4 months, the typist's proficiency has dropped to 49 words per minute.

79. A linear function   **81.** $\{x \mid x > 4\}$ **83.** $\{x \mid x < -2 \text{ or } x > 2\}$ **85.** $\{x \mid x > 1\}$

87. $f^{-1}(x) = \frac{\ln (x + 1)}{2}$ **89.** $f^{-1}(x) = \frac{e^x - 3}{2}$ **91.** 1

10.7 Exercises

1. An exponential equation is one in which a variable occurs in the exponent. Examples will vary.

3. Explanations will vary. For example, write the equation in exponential form: $3^2 = x$. Solve for x: $x = 9$.

5a. Always true   **b.** Always true   **c.** Always true   **7.** 1.1833   **9.** 0.6990   **11.** −0.6309   **13.** 1.1061

15. −2.4022   **17.** 0.2075   **19.** −1   **21.** −8   **23.** 1.8324, −1.8324   **25.** −0.6826   **27.** 5   **29.** −9, 3

31. 1,000,000,000   **33.** 1.5   **35.** $\frac{1}{2}$ **37.** 2   **39.** 3   **41.** 2   **43.** No solution   **45.** 3   **47.** 4

49. −1.86, 3.44   **51.** −1.16   **53.** 1.76   **55.** 2.42   **57.** −1.51, 2.10   **59.** 675   **61.** 4.0 s

63a. To the nearest hundred-thousandths, the equation is $y = 0.18808(1.03652)^x$.   **b.** $r \approx 0.9917$. This is very close to 1, which means that the data points lie very close to the graph of the regression equation.   **c.** 13.9 million people

d. 2022   **65.** 5 years   **67.** 8.5%   **69.** 5.4 days   **71.** 20.8 min   **73.** 9.49   **75.** 29,045 ft

77. Answers will vary.   **79.** 20.0 times   **81.** 5 P.M.   **83.** 1.4196   **85.** 0.5770   **87.** (510, 490)   **89.** (2, 1)

91. 100   **93.** 0.0025, 2.7183   **95.** −2   **97.** $y = 10e^{0.08317766t}$ **99.** 7 years   **101.** 12 years

CHAPTER 10 REVIEW EXERCISES

1. There is no x-intercept. The y-intercept is (0, 1). [10.3]   **2.** 1.6 [10.3]   **3.** 5 [10.1]   **4.** $12x^2 + 12x - 1$ [10.1]

5. $f^{-1}(x) = 2x - 16$ [10.2]   **6.** Yes [10.2]   **7.** $\log_2 32 = 5$ [10.5]   **8.** $\frac{1}{3}$ [10.3]   **9.** 125 [10.5]

10. $\frac{1}{2} \log_6 x + \frac{3}{2} \log_6 y$ [10.6]   **11.** 4 [10.5]   **12.** $\log_3 \sqrt{\frac{x}{y}}$ [10.6]   **13.** 2.3219 [10.6]   **14.** −2, 6 [10.7]

15. 3 [10.1]   **16.** 1.0 [10.5]   **17.** −0.5350 [10.7]   **18.** 2 [10.7]   **19.** 1.42 [10.7]   **20.** −3 [10.2]

21. 1 [10.3]   **22.** 2 [10.5]   **23.** c, d [10.3]   **24.** $f(x) = \frac{\ln (x + 5)}{\ln 2}$ [10.6]   **25.** $15,661 [10.3]

26a. $P = 72.29I$ **b.** It represents the equation for converting inches to points. [10.2]

27. 61 days [10.5]   **28.** Yes. (The pH is 4.) [10.7]   **29.** 229 days [10.7]

30a. $y = 1.2(1.032)^x$ **b.** 1.5 million or 1,500 thousand   **c.** 2.1 million or 2,100 thousand   [10.4]

CHAPTER 10 TEST

1. 1 [10.3]   **2.** $\log_3 \sqrt{\frac{x}{y}}$ [10.6]   **3.** $\frac{1}{4}$ [10.7]   **4.** 4 [10.5]   **5.** $f^{-1}(x) = \frac{3}{2}x + 18$ [10.2]

6. $\frac{2}{3} \log_6 x + \frac{5}{3} \log_6 y$ [10.6]   **7.** −3 [10.7]   **8.** −0.3 [10.3]   **9.** 7; 70 [10.1]   **10.** $\frac{1}{9}$ [10.5]   **11.** 14 [10.5]

12. 12.88 [10.5]   **13.** 1.7251 [10.6]   **14.** 6 [10.7]   **15.** $\log_5 625 = 4$ [10.5]   **16.** $\frac{1}{2}$ [10.7]   **17.** 0.72 [10.7]

18. 10 [10.2]   **19.** 17; $4x^2 - 18x + 18$ [10.1]   **20.** $[(1, -2), (3, 2), (-4, 5), (9, 7)]$ [10.2]   **21.** b, c [10.3]
22. $f(x) = \frac{\ln(2x - 1)}{\ln 3}$ [10.6]   **23.** 0.602 cm   [10.7]   **24.** 24 h   [10.7]
25a. To the nearest ten-thousandth: $y = 0.0504(1.5717)^x$ **b.** $r \approx 0.990$. This is very close to 1, which means that the data points lie very close to the graph of the regression equation.   **c.** 70 million   **d.** 2004   [10.4]

CUMULATIVE REVIEW EXERCISES

1. $20a - 12$ [1.5]   **2.** [1.2]   **3.** $\frac{-3 \pm \sqrt{29}}{2}$ [9.2/9.3]   **4.** $\{0, 16, 48\}$ [2.2]

5. $y = -x + 3$ [4.2]   **6.** $\left(\frac{7}{6}, \frac{1}{2}\right)$ [5.2]   **7.** $(2, 0, 1)$ [5.2]   **8.** 25 [8.1]   **9.** 10 [8.6]   **10.** $\pm 3, \pm i$ [9.4]

11. 3 [9.5]   **12.** 4 [10.3]   **13.** 10 [10.1]   **14.** $-\frac{1}{2}, 5$ [2.3]   **15.** 5 [3.1]   **16.** $f^{-1}(x) = \frac{3}{2}x + 18$ [10.2]

17. $y(3x - 2)(x + 4)$ [6.5]   **18.** $\frac{2}{ab^5}$ [6.1]   **19.** 7.86×10^{-5} [6.1]   **20.** $\frac{x - 2}{x - 3}$ [7.1]   **21.** 2, 8 [9.4]

22. $34 - 18i$ [8.7]   **23.** 7.5 min [7.3]   **24.** 800 lb of the 25% alloy, 1200 lb of the 50% alloy [5.2]
25. $y = 320x$; 1760 mi [4.2]

ANSWERS to Additional Topics in Algebra Exercises

Section 1 Exercises

1a. A sequence is an ordered list of numbers.   **b.** An arithmetic sequence is one for which successive terms differ by the same constant.   **c.** A geometric sequence is one for which the ratio of successive terms is the same constant.
3a. The first term of a sequence   **b.** The nth term of a sequence   **c.** The sum of the first four terms of a sequence
5. 3, 5, 7, 9   **7.** 2, 4, 8, 16   **9.** 2, 5, 10, 17   **11.** $0, \frac{3}{2}, \frac{8}{3}, \frac{15}{4}$ **13.** 1, −2, 3, −4   **15.** $\frac{1}{2}, -\frac{1}{5}, \frac{1}{10}, -\frac{1}{17}$ **17.** 40
19. 225   **21.** $\frac{1}{256}$ **23.** 380   **25.** $-\frac{1}{36}$ **27.** 94   **29.** 45   **31.** 20   **33.** 91   **35.** 0   **37.** 432   **39.** $\frac{5}{6}$
41. 141   **43.** 50   **45.** 71   **47.** $\frac{27}{4}$ **49.** 17   **51.** 3.75   **53.** $a_n = n$ **55.** $a_n = -4n + 10$
57. $a_n = \frac{3n + 1}{2}$ **59.** $a_n = -5n - 3$ **61.** $a_n = -10n + 36$ **63.** A, 16   **65.** G, 56   **67.** N, $\sqrt{5}$ **69.** A, $-a^2$
71. G, $27 \log x$ **73.** $\frac{2187}{4096}$ **75.** −3645   **77.** $81\sqrt{3}$ **79.** $a_n = 8\left(\frac{3}{4}\right)^{n-1}$ **81.** $a_n = 6(-2)^{n-1}$ **83.** $a_n = -5(-5)^{n-1}$
85. 650   **87.** −605   **89.** $\frac{215}{4}$ **91.** 420   **93.** −210   **95.** −5   **97.** −2188   **99.** $189 + 189\sqrt{2}$
101. 1364   **103.** 2800   **105.** $\frac{121}{243}$ **107.** $\frac{63}{64}$ **109.** 9 weeks   **111.** 2180 seats   **113.** \$3550; \$28,750
115. 99.8°F   **117.** \$10,737,418.23   **119.** 2,048,000 bacteria   **121.** $a_n = 8^{n-1}$

Section 2 Exercises

1. It is the product of the first n consecutive natural numbers.   **3.** $\frac{n!}{(n - r)!r!}$ **5.** $n + 1$
7. It is used to expand $(a + b)^n$.   **9.** 6   **11.** 40,320   **13.** 1   **15.** 10   **17.** 1   **19.** 84   **21.** 21
23. 1   **25.** 11   **27.** 20   **29.** 2450   **31.** 1 8 28 56 70 56 28 8 1   **33.** $r^3 - 3r^2s + 3rs^2 - s^3$
35. $y^4 - 12y^3 + 54y^2 - 108y + 81$ **37.** $8x^3 + 36x^2y + 54xy^2 + 27y^3$ **39.** $x^4 + 12x^3y + 54x^2y^2 + 108xy^3 + 81y^4$
41. $a^{10} + 10a^9b + 45a^8b^2$ **43.** $a^{11} - 11a^{10}b + 55a^9b^2$ **45.** $256x^8 + 1024x^7y + 1792x^6y^2$
47. $65,536x^8 - 393,216x^7y + 1,032,192x^6y^2$ **49.** $x^7 + 7x^5 + 21x^3$ **51.** $x^{10} + 15x^8 + 90x^6$ **53.** $35x^3a^4$
55. $x^2 + 8x^{3/2} + 24x + 32x^{1/2} + 16$ **57.** $-8i$ **59.** $(x + 5)^n$

A34

Section 3 Exercises

1. If $a > 0$, it opens to the right. If $a < 0$, it opens to the left. Examples will vary.   **3.** $(x - h)^2 + (y - k)^2 = r^2$. h is the x-coordinate of the center of the circle, k is the y-coordinate of the center of the circle, and r is the radius of the circle.

5. Answers may vary.   **7.**

Vertex: $(1, -5)$;
Axis of symmetry: $x = 1$

9.

Vertex: $(1, -2)$;
Axis of symmetry: $x = 1$

11.

Vertex: $(-6, 1)$;
Axis of symmetry: $y = 1$

13.

Vertex: $(-1, 0)$;
Axis of symmetry: $y = 0$

15.

Vertex: $(-1, 2)$;
Axis of symmetry: $y = 2$

17.

19.

21.

23. $(x + 1)^2 + (y + 2)^2 = 9$ **25.** $(x + 2)^2 + (y - 1)^2 = 5$ **27.**

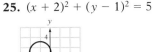

29.

31.

33.

35.

37.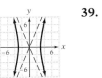

39.

41. Domain: $\{x \mid x \in \text{real numbers}\}$; range: $\{y \mid y \geq -6\}$ **43.** Domain: $\{x \mid x \in \text{real numbers}\}$; range: $\{y \mid y \leq -2\}$
45. Domain: $\{x \mid x \geq -14\}$; range: $\{y \mid y \in \text{real numbers}\}$ **47.** Domain: $\{x \mid x \leq 7\}$; range: $\{y \mid y \in \text{real numbers}\}$
49. Domain: $\{x \mid -8 \leq x \leq 2\}$; range: $\{y \mid 1 \leq y \leq 11\}$ **51.** Domain: $\{x \mid -5 \leq x \leq 5\}$; range: $\{y \mid -3 \leq y \leq 3\}$
53. Domain: $\{x \mid x \leq -5 \text{ or } x \geq 5\}$; range: $\{y \mid y \in \text{real numbers}\}$ **55.** Domain: $\{x \mid x \in \text{real numbers}\}$;
range: $\{y \mid y \leq -4 \text{ or } y \geq 4\}$ **57.** $(x - 4)^2 + y^2 = 16$

59.

61.

63.

65.

67. $(x - 3)^2 + y^2 = 9$ **69.** $x^2 + (y - 1)^2 = 13$ **71.** $(x - 1)^2 + (y + 1)^2 = 1$ **73.** $(x - 6)^2 + (y + 7)^2 = 12$

75a. $x = \frac{1}{100}y^2$ **b.** $[0, 17.7]$ **77a.** $\frac{x^2}{33,489} + \frac{y^2}{324} = 1$ **b.** 33,938,000,000 miles   **c.** 85,100,000 miles

79.

81.

83.

85. When $a = b$.   **87.** 11 units   **89.** π square units

91. Answers may vary.   **93.** Answers may vary.

INDEX

INDEX OF APPLICATIONS